Autodata

Timing Belts

for petrol and diesel engines 1991-03

2003

GW00494693

Autodata Limited, Priors Way, Maidenhead, Berkshire SL6 2HP, England
Tel: 01628 634321   Fax: 01628 770385   email: sales@autodata.ltd.uk   technical@autodata.ltd.uk

Argentina
Condistelec S.A, Carlos Pellegrini 1785 (1602), Florida, Pcia de Buenos Aires
Tel: 0054-11-4730 3533   Fax: 0054-11-4760 0596
email: condistelec@condistelec.com.ar

Australia
Bookworks Pty. Ltd., 56 Bonds Road, Punchbowl, NSW 2196
Tel: (02) 9740 6766   Fax: (02) 9740 6591   email: sales@bookworks.com.au

Bookworks Pty. Ltd., 46 Isabella Street, Moorabbin, Vic 3189
Tel: (03) 9555 6555   Fax: (03) 9553 3897   email: sales@bookworks.com.au

Austria
Buchhandlung Helmut Godai, 1150 Wien, Mariahilfer Str. 169
Tel: 01 893 41 05   Fax: 01 893 41 63   email: info@godai.at

Belgium
Autodata, Thillostraat 3, 2920 Kalmthout
Tel: 03 666 45 36   Fax: 03 666 58 99   email: autodata@pi.be

Canada
Autodata Publications Inc., 19 Bonazzoli Avenue, Hudson, MA 01749 , U.S.A.
Tel: 978 562-9511   Fax: 978 562-9533   email: sales@autodatapubs.com

Croatia
Autoelektrika Novak d.o.o., T. Ujevića 29, HR - 40323 Prelog
Tel: (00385-40) 645 861   Fax: (00385-40) 645 861

Cyprus
Pergamon Bookhouse, 12-12A King Paul Street, PO Box 25062, Nicosia
Tel: (02) 676 343   Fax: (02) 676 773

Czech Republic
AUTOservis akademie s.r.o., Přepeřská 1809, 511 01 Turnov
Tel: (+420) 481 323 931   Fax: (+420) 481 323 712
email: info@autoservisakademie.cz

Denmark
Robert Bosch A/S, Telegrafvej 1, 2750 Ballerup
Tlf: 44 89 83 80   Fax: 44 89 86 87

Autodata Skandinavien ApS, Box 96, Trehøjevej 2, 7200 Grindsted
Tlf: 75 32 55 57   Fax: 75 31 02 41

Estonia
AS Megastar, Kanali tee 1, 10112 Tallinn
Tel: +372 601 6026   Fax: +372 601 6027   email: megastar@megastar.ee

Finland
Autodata Oy, PL 29, Onkkaalantie 71, 36601 PÄLKÄNE
Tel: (03) 53 43 980   Fax: (03) 53 43 983   email: autodata@autodata.fi

France
S.F.T.A. Autodata, 13 rue Paul Sabatier, Z.A. de Faveyrolles n°1,
26700 Pierrelatte
tel: 04.75.96.96.96   fax: 04.75.96.96.95   email: contact@autodata-sfta.fr

Germany
Fust, Wever & Co. GmbH, Maxstr. 9, 45127 Essen
Tel: 0201 82774-0 / 22 79 12   Fax: 0201 82774-39 / 23 25 56
email: info@fust-wever.de

Greece
B.D. Papathanassiou SA, 30 Venizelou, Kalithea, 176 76 Athena
Tel: (01) 9566.503   Fax: (01) 9582.500

Hong Kong
Kenterton Limited, 2nd Floor, 210 Wong Chuk Wan, Sai Kung, New Territories
Tel: (852) 9022 5665   Fax: (852) 2550 2043   email: mjackson@netvigator.com

Hungary
Maróti Könyvkereskedés Kft. 1205 Budapest, Nagykőrösi út 91.
Tel.: 285-6608   Fax: 285-0116   e-mail: maroti.konyvker@axelero.hu

Iceland
Bilgreinasambandid, Hus Verslunarinnar, 103 Reykjavik
Tel: (01) 68 15 50   Fax: (01) 68 98 82   email: bgs@centrum.is

Ireland
Hella Ireland Ltd., Unit 6.1, Woodford Business Park, Santry, Dublin 17
Tel: (01) 862 0000   Fax: (01) 862 1133   email: sales@hella.ie

Israel
Esco Engineering Supplies Ltd., 22 Harakevet Street, PO Box 45, Tel Aviv 61000
Tel: (03) 560 3472   Fax: (03) 560 2153

Italy
Tecnodata, Via le Petrene, 53017 Radda in Chianti (Si)
Tel: (0577) 738 239   Fax: (0577) 738 790   email: tecnodata@albaclick.com

Latvia
Autodati, 6 - 216 Ezermalas Street, Riga, LV-1006
Tel: 371 7089 741   Fax: 371 7089 744   email: autodati@rtu.lv

Robert Bosch SIA, A. Deglava Str. 60, Riga, LV-1035,
Tel: 371 7802 080   Fax: 371 7548 441   email: rigaoffice@bosch.lv

Netherlands
Autodata, Postbus 581, 4645 ZX Putte
Tel: 00 32 3 666 45 36   Fax: 00 32 3 666 58 99   email: autodata@pi.be

New Zealand
Autodata (NZ) Ltd., 59 Roberts Road, Whangaparaoa 1463
Tel: (09) 424 8990   Fax: (09) 424 8990   email: marilyn@ihug.co.nz

Norway
Autodata Skandinavien A/S, Kongensgate 6, Postboks 2047, 3202 Sandefjord
Tlf: 33 46 73 70   Faks: 33 46 45 40

Poland
Precyzja-Service Sp. zo.o., ul. Gdańska 99, 85 022 Bydgoszcz
Tel: (052) 325 1026   Fax: (052) 321 0571   email: sales@precyzja-service.pl

Portugal
Autodata, Thillostraat 3, B-2920 Kalmthout, Belgium
Tel: 0032 3 666 45 36   Fax: 0032 3 666 58 99   email: autodata@pi.be

Russia
Legion-Autodata, Shosse Entusiastov 15/16, 111024, Moscow
Tel: (095) 273-42-61   Fax: (095) 362-18-19   email: legion@autodata.ru

Barclay Auto, Bros Group, Barclay Str. 13/1, Office 500, Moscow
Tel: (095) 1451698   Fax: (095) 1451698/(095) 1451048   email: abarclay@dol.ru

South Africa
The Phoenix Exchange, PO Box 273, Edenvale 1610,
Tel: (011) 452 0875   Fax: (011) 452 8372   email: tyhogan@icon.co.za

Spain
Autodata, Thillostraat 3, B-2920 Kalmthout, Belgium
Tel: 0032 3 666 45 36   Fax: 0032 3 666 58 99   email: autodata@pi.be

Sweden
Autometric AB, Tillverkarvägen 16, 187 66 Täby
Tel: (08) 630 00 77   Fax: (08) 756 11 51   email: sales@autometric.se

Switzerland
Autodata GmbH, Bahnhofstrasse 28, 8153 Rümlang
Tel: (01) 880 7400   Fax: (01) 880 7434   email: autodata@swissonline.ch

U.S.A.
Autodata Publications Inc., 19 Bonazzoli Avenue, Hudson, MA 01749
Tel: 978 562-9511   Fax: 978 562-9533   email: sales@autodatapubs.com

Product No.: 03-1800	ISBN: 1-904473-20-2	010503

This manual brings together into a single source comprehensive information on timing belt replacement for the majority of European and Asian cars fitted with petrol or diesel engines.

Many problems associated with timing belt failure can be attributed to incorrect fitting and tensioning or mishandling of the belt during replacement. It is important to follow the general instructions and the specific instructions for each engine.

Contents

Important notice	1
Introduction	3
How to use the illustration	4
Timing belt replacement intervals	5
Index	6
Safety precautions	31
Abbreviations	31
General instructions	32
Belt tension measurement	36
Tension gauge suppliers	37
Troubleshooter	38
Replacement and adjustment	40

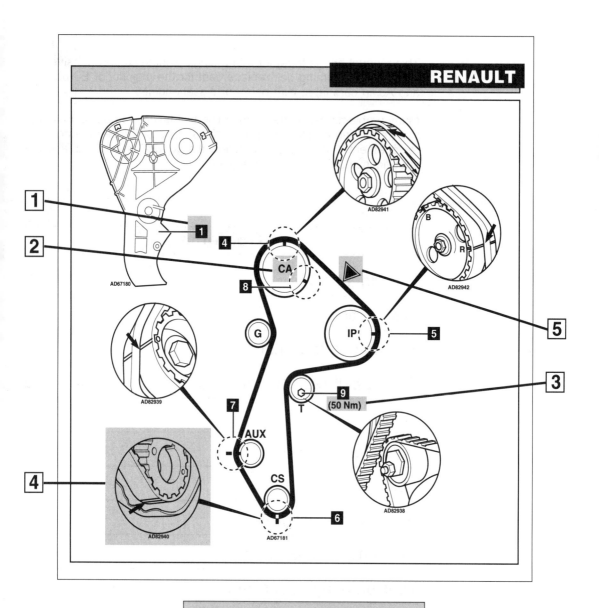

RENAULT

1	Illustration identification number
2	Abbreviated sprocket identification
3	Tightening torque - Newton metres
4	Detail of timing marks
5	Timing belt tension checking point

The information relating to timing belt replacement intervals is additional to the main purpose of this manual but is included to provide guidance to garages and for customer advice.

Where possible the recommended intervals have been compiled from vehicle manufacturers' information. In a few instances no recommendation has been made by the manufacturer and the decision to replace the belt must be made from the evidence of a thorough examination of the condition of the existing belt.

Apart from the visible condition of the belt, which is explained fully later in this section, there are several other factors which must be considered when checking a timing belt:

1. Is the belt an original or a replacement.

2. When was the belt last replaced and was it at the correct mileage.

3. Is the service history of the vehicle known.

4. Has the vehicle been operated under arduous conditions which might warrant a shorter replacement interval.

5. Is the general condition of other components in the camshaft drive, such as the tensioner, pulleys, and other ancillary components driven by the timing belt, typically the water pump, sound enough to ensure that the life of the replacement belt will not be affected.

6. If the condition of the existing belt appears good, can you be satisfied that the belt will not fail before the next check or service is due.

7. If the belt does fail, have you considered the consequences. If the engine is an INTERFERENCE type then considerable expensive damage may well be the result.

8. The cost of replacing a belt as part of a routine service could be as little as 5 to 10% of the repair cost following a belt failure. Make sure your customer is aware of the consequences.

9. If in doubt about the condition of the belt - RENEW it.

Note: Models without camshaft timing belts are identified as having CHAIN *or* GEARS *.*
For more information, refer to Autodata Timing Chains & Gears manual.

ALFA ROMEO

■ Petrol models

6 2,5	1980-83	CHAIN
75 1,8/2,0	1985-90	CHAIN
145 1,3/1,6 8V	1994-97	40
145 1,4/1,6/1,8 Twin Spark	1996-01	42
145 1,7 16V	1994-97	46
145 2,0 Twin Spark	1996-01	48
146 1,3/1,6 8V	1994-97	40
146 1,4/1,6/1,8 Twin Spark	1996-01	42
146 1,7 16V	1994-97	46
146 2,0 Twin Spark	1996-01	48
147 1,6 Twin Spark	2001-03	42
147 2,0 Twin Spark	2001-03	48
155 1,6/1,8 Twin Spark	1996-98	42
155 1,8/2,0 Twin Spark 671.02/672.02 engines	1992-97	CHAIN
155 2,0 Twin Spark	1995-98	48
155 2,5 V6	1992-97	52
156 1,6/1,8 Twin Spark	1997-02	42
156 2,0 Twin Spark	1997-02	48
156 2,5 V6	1997-02	54
164 2,0 Turbo	1987-91	CHAIN
164 2,0 Twin Spark	1987-93	CHAIN
166 2,5 V6	1998-01	54
166 3,0 V6 24V	1998-01	54
Alfetta/GTV 1,6/1,8/2,0	1980-85	CHAIN
GTV 1,8 Twin Spark	1998-01	42
GTV 2,0 16V	1996-02	48
GTV 2,0 Twin Spark	1998-02	48
GTV 3,0 V6 24V	1998-00	54
Giulia Super/Spider 1,6	1980-81	CHAIN
Giulietta 1,3/1,6/1,8/2,0	1980-81	CHAIN
Spider 1,8 Twin Spark	1998-01	42
Spider 2,0 Twin Spark	1998-02	48
Spider 2,0 16V	1996-02	48
ZR 3,0	1993-01	CHAIN

■ Diesel models

6 2,5 Turbo D	1983-87	GEARS
33 1,8/Turbo D	1986-94	GEARS
75 2,0/2,4 Turbo D	1985-92	GEARS
90 2,4 Turbo D	1984-87	GEARS
155 2,4 Turbo D	1993-97	GEARS
164 2,5 Turbo D	1987-98	GEARS
Alfetta 2,0/2,4 Turbo D	1980-87	GEARS
Giulietta 2,0 Turbo D	1980-85	GEARS

AUDI

■ Petrol models

80 1,6/2,0	1991-95	64
80 2,3	1991-94	72
80 2,0 16V manual tensioner	1990-94	68
80 2,0 16V automatic tensioner	1994-95	70
80 V6 2,6/2,8	1991-96	74
100 2,0	1991-94	64
100 2,0 16V manual tensioner	1992-94	68
100 2,0 16V automatic tensioner	1994	70
100 2,3	1991-94	72
100 2,6/2,8	1992-94	74
A2 1,4	2000-03	56
A3 1,6	1996-03	60
A3 1,8	1996-03	62
A3 1,8T	1996-03	62
A4 1,6	1994-97	76
A4 1,6 AHL/ALZ engines	1996-03	78
A4 1,8	1994-07/97	80
A4 1,8	08/97-01	82
A4 1,8T AEB/AJL engines	1994-01	80
A4 1,8T AVJ engine	2001-03	84
A4 2,0	2001-03	84
A4 2,4	1997-01	86
A4 2,6/2,8	1994-01	74

A4 2,8 *ALG engine*	1997-01	**86**
A4 2,8 *ACK engine*	1995-01	**90**
A4 3,0	2001-03	**92**
A6 1,8 *ADR engine*	1995-97	**66**
A6 1,8 *AJP engine*	1997-03	**96**
A6 1,8T	1997-03	**80**
A6 2,0	1994-97	**64**
A6 2,0 16V	1994-97	**70**
A6 2,4	1997-02	**86**
A6 2,6/2,8	1994-97	**74**
A6 2,8 *ALG engine*	1997-03	**86**
A6 2,8 *ACK engine*	1995-03	**90**
A6 3,0	2001-03	**92**
A6 4,2	1999-03	**98**
A8 2,8	1991-02	**78**
A8 2,8 *ACK engine*	1995-02	**90**
A8 2,8 *ALG/APR/AQD engines*	1998-02	**86**
A8 4,2	1998-03	**102**
Cabriolet 1,8	1997-00	**68**
Cabriolet 2,3	1991-94	**72**
Cabriolet 2,6	1994-00	**74**
Coupe 2,3 *automatic tensioner*	1991-94	**72**
S3 1,8 Turbo	1999-03	**62**
S4 2,7 quattro	1997-01	**92**
TT 1,8 Turbo	1998-03	**106**

■ Diesel models

80 1,9 TDI	1991-95	**116**
80 1,9 TD	1991-94	**118**
100 2,5 TDI	1991-95	**120**
A2 1,2 TDI PD	2001-03	**108**
A2 1,4 TDI PD	2000-03	**108**
A3 1,9 TDI	1996-03	**112**
A3 1,9 TDI PD	2001-03	**102**
A4 1,9 TDI	1994-01	**122**
A4 1,9 TDI *AHH engine*	1997-01	**126**

A4 1,9 TDI PD	1999-03	**130**
A4 2,5 V6 TDI	1997-03	**134**
A6 1,9 TDI	1994-03	**122**
A6 1,9 TDI PD	2000-03	**130**
A6 2,5 TDI	1994-97	**138**
A6 2,5 V6 TDI	1997-03	**134**
A8 2,5 V6 TDI	1997-02	**134**
Allroad 2,5 V6 TDI	2000-03	**134**

BEDFORD see VAUXHALL-OPEL

BMW

■ Petrol models

316/315/318 (E21)	1980-87	CHAIN
316/318 (E30)	1982-89	CHAIN
318i 16V (E30)	1989-91	CHAIN
316/318 (E30)	1988-93	**140**
316/318 (E36)	1991-93	**140**
3 Series (E36)	1993-98	CHAIN
3 Series (E46)	1998-03	CHAIN
320/323 (E21)	1977-83	**142**
320/323/325 (E30)	1986-93	**142**
518/520/525/528/535 (E12)	1980-81	CHAIN
518/525/528 (E28)	1981-84	CHAIN
518 (E34)	1989-93	**140**
520/525 (E28/E34)	1983-90	**142**
5 Series (E34)	1993-96	CHAIN
5 Series (E39)	1996-03	CHAIN
6 Series (E24)	1980-89	CHAIN
7 Series (E23)	1980-86	CHAIN
7 Series (E32)	1986-94	CHAIN
7 Series (E38)	1994-03	CHAIN
8 Series (E31)	1990-99	CHAIN
M3 (E30)	1987-91	CHAIN
X5	2000-03	CHAIN
Z1	1989-93	**142**
Z3/Z8	1997-03	CHAIN

Index

■ Diesel models

318/320/325/525/725 d/td/tds	1991-03	CHAIN
X5	2000-03	CHAIN

CHRYSLER

■ Petrol models

Jeep Cherokee/Wrangler 2,5	1992-03	CHAIN
Jeep Cherokee/Wrangler 4,0	1992-03	CHAIN
Neon 1,8	1997-99	**144**
Neon 2,0 16V	1995-03	**144**
PT Cruiser 2,0	2000-03	**148**
Voyager 2,0	1997-01	**144**
Voyager/Grand Voyager 3,3	1996-03	CHAIN
Jeep Grand Cherokee 4,0/4,7/5,2/5,9	1995-03	CHAIN
Viper GTS V10	1997-01	CHAIN

■ Diesel models

Jeep Cherokee 2,5 D/TD	1996-03	GEARS
Voyager 2,5 D/TD	1996-03	GEARS
Jeep Grand Cherokee 2,5 D	1997-99	GEARS
Jeep Grand Cherokee 3,1 TD	1999-03	GEARS

CITROEN

■ Petrol models

2CV	1980-90	GEARS
AX 1,0/1,1/1,4	1986-93	**150**
AX GTi	1991-93	**150**
AX 1,0/1,1/1,4	1994-97	**152**
AX GTi	1991-97	**152**
BX 14	1989-93	**150**
BX 16/19	1992-94	**160**
C5 1,8/2,0	2000-03	**174**
C5 2,0 HPi	2000-03	**174**
C15 1,0/1,1/1,4	1986-94	**150**
C25	1987-93	CHAIN
CX 2400	1980-84	CHAIN
CX 25	1984-89	CHAIN
Berlingo 1,1/1,4	1997-03	**152**
Berlingo 1,8	1997-02	**160**
LNA 1124 cc	1982-84	CHAIN

Relay 2,0	1994-02	**186**
Saxo 1,0/1,1/1,4/1,6	1996-03	**152**
Saxo 1,6 16V	1997-03	**156**
Synergie 1,8	1997-00	**160**
Synergie 2,0/Turbo	1995-00	**186**
Visa 1124 cc	1980-84	CHAIN
Visa 1219 cc	1980-81	CHAIN
Xantia 1,6/1,8/2,0	1993-98	**160**
Xantia 1,8 16V	1995-01	**164**
Xantia 2,0 16V	1993-01	**168**
Xantia 3,0 V6	1997-01	**162**
XM 2,0	1989-94	**160**
XM 2,0 16V	1993-00	**168**
XM 2,0 Turbo	1993-00	**160**
XM V6	1989-98	CHAIN
XM 3,0 V6	1997-00	**162**
Xsara 1,4/1,6	1997-03	**152**
Xsara 1,6 16V	2000-03	**156**
Xsara 1,8	1997-00	**144**
Xsara 1,8 16V	1997-00	**164**
Xsara 2,0 16V *XU10 engine*	1997-00	**170**
Xsara 2,0 16V *XU10J4RS engine*	2000-03	**178**
Xsara 2,0 16V *EW10 engine*	2000-03	**174**
Xsara Picasso 1,8 16V	1999-03	**174**
ZX 1,1/1,4	1991-94	**150**
ZX 1,1/1,4	1994-98	**152**
ZX 1,6/1,8/1,9/2,0 *eccentric tensioner*	1991-98	**160**
ZX 1,8 16V	1997-98	**164**
ZX 2,0 16V	1992-97	**168**
ZX 2,0 16V *RFS engine*	1996-98	**170**

■ Diesel models

AX 1,4 D	1989-94	**188**
AX 1,5 D	1994-97	**190**
Berlingo 1,8/1,9 D *XUD engine*	1996-99	**192**
Berlingo 1,9 D *DW8 engine*	1998-03	**196**
Berlingo 2,0 HDi *with adjustable camshaft sprocket*	1999-01	**200**
Berlingo 2,0 HDi *with adjustable crankshaft sprocket*	2001-03	**204**

Autodata

BX 17D/19 D/Turbo	1984-94	192
C5 2,0 HDi *with adjustable camshaft sprocket*	2000-01	200
C5 2,0 HDi *with adjustable crankshaft sprocket*	2001-03	204
C5 2,2 HDi	2000-02	212
C15 D	1984-97	192
C25 1,9 D	1988-94	192
Dispatch 1,9 D/TD *XUD engine*	1995-01	192
Dispatch 1,9 D *DW8 engine*	1998-03	196
Dispatch 2,0 HDi	1999-03	200
Dispatch 2,1 TD	1996-99	208
Relay 1,9D/TD	1994-02	192
Relay 2,0 HDi	2001-03	200
Relay 2,5 D	1994-02	216
Relay 2,5 TD	1994-01	216
Relay 2,5 TD *THX engine*	1997-02	218
Saxo 1,5 D	1996-03	190
Synergie 1,9 TD	1995-00	192
Synergie 2,0 HDi	1999-02	200
Synergie 2,1 TD	1996-00	208
Visa 17 D	1984-88	192
Xantia 1,9 D/TD	1993-01	192
Xantia 2,0 HDi	1998-01	200
Xantia 2,1 TD	1996-98	208
XM 2,1 TD	1989-01	208
XM 2,2 D	1989-96	208
Xsara 1,5 D	1997-00	190
Xsara 1,9 D/TD *XUD engine*	1997-00	192
Xsara 1,9 D *DW8 engine*	1998-01	196
Xsara 2,0 HDi *with adjustable camshaft sprocket*	1998-01	200
Xsara 2,0 HDi *with adjustable crankshaft sprocket*	2001-03	204
Xsara Picasso 2,0 HDi *with adjustable camshaft sprocket*	1999-01	200
Xsara Picasso 2,0 HDi *with adjustable crankshaft sprocket*	2001-03	204
ZX 1,8/1,9 D/TD	1991-98	192

DAEWOO

■ Petrol models

Espero 1,5	1995-97	224
Espero 1,8/2,0	1995-97	230
Korando	1999-03	CHAIN
Lanos 1,4	1997-03	226
Lanos 1,5	1997-03	226
Lanos 1,6	1997-03	228
Leganza 2,0	1997-03	234
Matiz	1998-03	220
Musso	1999-03	CHAIN
Nexia 1,5 SOHC	1995-97	222
Nexia 1,5 DOHC	1995-97	224
Nubira 1,6	1997-03	232
Nubira 2,0	1997-03	234
Tacuma 1,6	2001-03	232
Tacuma 1,8	2001-03	236
Tacuma 2,0	2001-03	234

■ Diesel models

Korando	1999-03	CHAIN
Musso	1999-03	CHAIN

DAIHATSU

■ Petrol models

Applause	1988-96	242
Charade 1,3 16V	1988-02	240
Charade 1,5	1994-00	240
Charade 16V	1993-00	240
Charade GSXi	1993-00	240
Charmant 1,3	1982-89	CHAIN
Charmant 1,6	1982-83	CHAIN
Cuore (L501)	1996-99	238
Grand Move	1997-02	240
Hi-Jet	1986-98	244
Mira	1993-97	238
Move	1997-02	226
Sirion 1,3	2000-03	CHAIN
Sportrak	1988-98	242
Terios	1997-00	240

| Terios | 2000-03 | CHAIN |
| YRV 1,3 | 2000-03 | CHAIN |

■ Diesel models

| Fourtrak 2,8 D/TD | 1988-03 | **246** |

FIAT

■ Petrol models

126	1980-95	CHAIN
127 903 cc	1980-81	CHAIN
127 965 cc	1984-85	CHAIN
900T 903 cc	1980-82	CHAIN
Brava/Bravo 1,2 16V	1998-02	**252**
Brava/Bravo 1,4 12V	1995-98	**260**
Brava/Bravo 1,6 16V	1995-02	**262**
Brava/Bravo 1,8 16V	1995-02	**264**
Bravo 2,0 20V	1995-01	**266**
Cinquecento Sporting 1,1	1994-98	**248**
Cinquecento 704/903 cc	1991-98	CHAIN
Coupe 2,0 16V	1994-97	**258**
Coupe 2,0 Turbo	1994-01	**258**
Coupe 2,0 20V/Turbo	1996-01	**266**
Croma 2,0 i.e./Turbo 16V with balancer shaft	1993-97	**256**
Ducato 1800/2000	1982-85	CHAIN
Ducato 2,0	1994-02	**270**
Marea/Weekend 1,6 16V	1996-02	**262**
Marea/Weekend 1,8 16V	1996-02	**268**
Marea/Weekend 2,0 20V	1996-01	**266**
Multipla 1,6 16V	1999-03	**262**
Panda 30	1980-85	CHAIN
Panda 903 cc	1980-85	CHAIN
Panda FIRE	1986-93	**236**
Punto 55 1,1	1993-99	**236**
Punto 60 1,2	1993-99	**248**
Punto 1,2	1999-02	**248**
Punto 1,2 16V	1997-03	**252**
Punto 75 1,2	1994-99	**248**
Punto 1,4 GT	1993-99	**250**
Punto 90 1,6	1989-97	**250**
Seicento 899 cc	1997-00	CHAIN
Seicento Sporting	1997-00	**248**

Stilo 1,2 16V	2001-03	**252**
Stilo 1,6 16V	2001-03	**262**
Tempra 1,4	1990-95	**250**
Tempra 1,6	1988-97	**250**
Tempra 2,0 i.e.	1993-97	**256**
Tipo 1100 FIRE	1986-94	**248**
Tipo 1,4	1988-95	**250**
Tipo 1,6	1988-97	**250**
Tipo 1,8 i.e	1993-95	**256**
Tipo 2,0 i.e	1993-95	**256**
Tipo 2,0 16V	1993-95	**258**
Ulysse 1,8	1996-00	**270**
Ulysse 2,0/Turbo	1994-00	**270**
Ulysse 2,0 16V	1998-01	**272**
Uno 1,4/Turbo	1993-95	**250**
Uno 45 FIRE	1985-89	**248**
Uno 903 cc	1982-93	CHAIN

■ Diesel models

Brava/Bravo 1,9 D Turbo	1995-00	**282**
Brava/Bravo 1,9 D Turbo	1995-02	**292**
Brava/Bravo 1,9 JTD	1998-02	**282**
Croma 1,9 TD	1988-90	**286**
Ducato 1,9 D/TD to engine No.1723290	1987-98	**286**
Ducato 1,9 D/TD from engine No.1723291	1987-98	**288**
Ducato 1,9 D/TD	1998-02	**294**
Ducato 2,8 D	1998-02	**302**
Ducato 2,8 TD	1997-02	**302**
Duna 1,7 D	1986-90	**276**
Fiorino 1,7 D	1986-98	**276**
Fiorino 1,7 TD	1997-01	**300**
Marea 1,9 D Turbo	1996-99	**282**
Marea 1,9 D Turbo	1996-03	**292**
Marea 1,9 JTD	1999-03	**282**
Marea Weekend 1,9 D Turbo	1996-99	**282**
Marea Weekend 1,9 D Turbo	1996-03	**292**
Marea Weekend 1,9 JTD	1999-03	**282**
Marea 2,4 JTD	1999-03	**282**
Marea Weekend 2,4 JTD	1999-03	**282**
Multipla 1,9 JTD	1999-03	**282**
Punto 1,7 D/TD	1993-95	**278**
Punto 1,7 D/TD	1996-97	**280**

Punto 1,9 D	1999-02	**282**
Punto 1,9 JTD	1999-02	**282**
Regata 1,7 D	1980-90	**276**
Ritmo/Strada 1,7 D	1980-90	**276**
Scudo 1,9 D/TD	1996-01	**280**
Stilo 1,9 JTD	2001-02	**282**
Tempra 1,9 D/TD *to engine No.1723290*	1990-95	**286**
Tempra 1,9 D/TD *from engine No.1723291*	1990-95	**288**
Tipo 1,7 D	1990-95	**276**
Tipo 1,9 D/TD *to engine No.1723290*	1987-95	**286**
Tipo 1,9 D/TD *from engine No.1723291*	1987-95	**288**
Ulysse 1,9 TD	1994-00	**294**
Ulysse 2,0 JTD	1999-02	**296**
Uno 1,7 D	1989-96	**276**

FORD

■ Petrol models

Capri 1,3 OHV	1980-81	CHAIN
Capri 2,3/2,8 V6	1980-87	GEARS
Cortina 1,3 OHV	1980-82	CHAIN
Cortina 2,3 V6	1980-82	GEARS
Cougar 2,0	1998-01	**318**
Cougar 2,5	1998-01	CHAIN
Courier 1,3	1991-02	CHAIN
Escort MkII 1,1/1,3/1,6 OHV	1980-81	CHAIN
Escort/Orion 1,1/1,3 OHV	1980-97	CHAIN
Escort/Orion 1,1/1,3/1,4/1,6	1981-98	**314**
Escort/Orion 1,6i/1,8 16V	1992-01	**312**
Escort RS 2000	1991-96	CHAIN
Escort RS Cosworth	1990-96	**316**
Escort RS Turbo 1,6	1980-90	**314**
Escort Van 1,1/1,3 OHV	1974-97	CHAIN
Escort XR3/XR3i	1980-90	**314**
Explorer 4,0 V6	1997-02	CHAIN
Fiesta/Van 950/1,0/1,1/1,3 OHV	1977-02	CHAIN
Fiesta 1,25/1,4 *automatic tensioner*	1995-97	**306**
Fiesta 1,25/1,4/1,6 *semi-automatic tensioner*	1997-02	**308**
Fiesta 1,3/1,4/1,6/RS	1983-95	**304**
Fiesta XR2	1982-84	CHAIN
Fiesta XR2/XR2i	1983-93	**304**
Fiesta 1,6i/XR2i/RS1800 16V	1991-95	**312**
Focus 1,4/1,6	1998-03	**308**
Focus 1,8/2,0	1998-03	**318**
Galaxy 2,0	1995-00	CHAIN
Galaxy 2,3	1997-03	CHAIN
Galaxy 2,8	1995-03	CHAIN
Granada 2,0 DOHC	1989-91	CHAIN
Granada 2,3/2,8 V6	1980-85	GEARS
Granada 2,4/2,9 V6	1986-91	CHAIN
Granada/Scorpio 1,8/2,0	1981-89	**320**
Ka 1,3	1996-03	CHAIN
Maverick 2,4	1993-96	CHAIN
Mondeo 1,6/1,8/2,0 16V	1993-4/98	**326**
Mondeo 1,6/1,8/2,0	5/98-00	**318**
Mondeo 1,8/2,0	2000-03	CHAIN
Mondeo 2,5 V6	1994-03	CHAIN
P100 1,6/2,0	1982-94	**320**
Probe 2,0 16V	1994-98	**332**
Probe 2,5 24V	1994-98	**334**
Puma 1,4	1997-00	**308**
Puma 1,6	2000-01	**328**
Puma 1,7	1997-02	**328**
Scorpio 2,0 8V	1994-98	CHAIN
Scorpio 2,0 16V	1994-96	CHAIN
Scorpio 2,3 16V	1996-98	CHAIN
Scorpio 2,9 V6	1994-96	CHAIN
Scorpio 2,9 V6 24V	1994-98	CHAIN
Sierra 2,0 DOHC	1989-93	CHAIN
Sierra 2,3/2,8 V6	1985-89	GEARS
Sierra 2,9 V6	1989-93	CHAIN
Sierra/Sapphire 1,3 (OHC)	1982-87	**320**
Sierra/Sapphire 1,6 (OHC)	1982-90	**320**
Sierra/Sapphire 1,6/1,8 (CVH)	1988-93	**322**
Sierra/Sapphire 2,0 (OHC)	1982-90	**320**
Sierra/Sapphire RS Cosworth	1986-94	**324**
Transit 1,6 OHV	1978-83	CHAIN
Transit 1,6/2,0	1982-95	**320**
Transit 2,0 DOHC	1994-00	CHAIN
Transit 2,3	2000-03	CHAIN
Transit 2,9 V6	1989-95	CHAIN

■ Diesel models

Courier 1,8 D *with guide pulley*	1991-99	**336**
Courier 1,8 D *with additional tensioner*	1996-02	**340**
Escort/Orion 1,8 D/TD	1989-96	**336**
Escort 1,8 D/TD	1996-00	**324**
Escort Van 1,8 D/TD	1989-99	**336**
Fiesta/Van 1,8 D	1988-96	**336**
Fiesta/Van 1,8 D	1996-00	**340**
Fiesta 1,8 TD	1999-02	**344**
Flareside 2,5 D	1995-00	**362**
Focus 1,8 TD	1998-03	**344**
Galaxy 1,9 TD	1995-99	**350**
Galaxy 1,9 TDI	1997-00	**352**
Galaxy 1,9 TDI *ANU/AUY engines*	1999-03	**354**
Granada/Scorpio 2,5 TD	1985-91	CHAIN
Granada/Scorpio 2,5 TD	1992-94	GEARS
Maverick 2,7 TD	1993-98	GEARS
Mondeo 1,8 TD	1993-96	**336**
Mondeo 1,8 TD	1996-00	**340**
Mondeo 2,0 TD	2000-01	CHAIN
P100 1,8 TD	1988-94	**346**
Ranger 2,5 D/TD	1999-03	**358**
Scorpio 2,5 TCI	1994-98	GEARS
Sierra 1,8 TD	1990-93	**346**
Sierra 2,3 D	1987-89	CHAIN
Transit 2,0 TD	2000-03	CHAIN
Transit 2,4 D Turbo 16V	1999-03	CHAIN
Transit 2,5 DI	1983-86	**360**
Transit 2,5 DI/Turbo	1986-94	**360**
Transit 2,5 DI/Turbo	1995-99	**362**
Transit 2,5 TCi	1997-00	**362**

FSO

■ Petrol models

Caro 1,5/1,6	1994-98	CHAIN

HONDA

■ Petrol models

Accord 1,6	1998-03	**366**
Accord 1,8	1990-91	**378**
Accord 1,8	1995-03	**380**
Accord 2,0	1990-93	**378**
Accord 2,0	1993-03	**380**
Accord 2,0 16V	1987-90	**382**
Accord 2,2	1990-93	**378**
Accord 2,2 V-TEC	1993-98	**380**
Accord 2,3 DOHC	1993-95	**386**
Accord Type R	1998-03	**388**
Accord Aerodeck 2,0/2,2	1993-97	**380**
Accord Coupe 2,0/2,2	1993-97	**380**
Accord Coupe 2,0	1998-03	**384**
Civic 1,2/1,3/1,4/1,5/1,6	1988-95	**364**
Civic 1,4	1995-01	**366**
Civic 1,4	2001-03	**368**
Civic 1,5 V-TEC	1995-98	**370**
Civic 1,5 V-TEC	1997-00	**366**
Civic 1,6 SOHC	1997-00	**366**
Civic 1,6	2001-03	**368**
Civic 1,6 VTi DOHC	1990-95	**372**
Civic 1,6 SOHC V-TEC	1995-98	**370**
Civic 1,6 SOHC V-TEC	1997-03	**366**
Civic 1,6 DOHC V-TEC	1995-96	**372**
Civic 1,6 DOHC V-TEC	1996-00	**374**
Civic 1,6 Coupe SOHC	1996-00	**366**
Civic 1,7	2001-03	**368**
Civic 1,8 V-TEC	1997-00	**374**
Civic CRX 1,5/1,6	1991-94	**364**
Civic Shuttle 1,4/1,5	1988-91	**364**
Concerto 1,4/1,5/1,6 SOHC	1990-95	**364**
Concerto 1,6 DOHC	1990-95	**376**
CR-V	1997-02	**394**
CRX 1,6	1996-98	**366**
HR-V	1999-02	**366**
Insight	2000-02	CHAIN
Integra 1,6 DOHC	1988-89	**376**
Integra Type R	1998-01	**372**
Legend 3,5 V6	1996-02	**390**
Logo	1999-02	**366**
Prelude 2,0 SOHC	1988-92	**364**

Autodata

Prelude 2,0 SOHC	1990-96	**378**
Prelude 2,0 SOHC	1997-01	**384**
Prelude 2,0 16V	1987-92	**382**
Prelude 2,2 V-TEC	1993-01	**388**
Prelude 2,3 DOHC	1992-96	**386**
Shuttle 2,2	1995-98	**392**
Shuttle 2,3	1998-01	**392**
Stream 2,0	2000-02	CHAIN
S2000	1999-03	CHAIN

■ Diesel models		
Civic 2,0 TD	1997-00	**396**
Accord 2,0 TD	1996-03	**396**

HYUNDAI

■ Petrol models		
Accent 1,3	1994-02	**402**
Accent 1,5	1994-02	**402**
Accent 1,5 16V	1996-02	**404**
Atoz	1998-03	**400**
Coupe 1,6/2,0 16V	1996-02	**410**
Elantra 1,6 16V	2000-03	**404**
Elantra 1,8/2,0 16V	2001-03	**410**
Lantra 1,5	1991-94	**402**
Lantra 1,6 16V	1991-92	**406**
Lantra 1,6 16V G4-R engine	1993-96	**408**
Lantra 1,6 16V G4DR engine	1995-01	**410**
Lantra 1,8 16V G4-N engine	1992-96	**408**
Lantra 1,8 16V G4DM engine	1995-01	**410**
Lantra 2,0 16V	1996-01	**410**
Matrix 1,6 16V	2001-03	**404**
Matrix 1,8 16V	2001-03	**410**
Pony 1,2/1,4	1980-85	CHAIN
Santa Fe 2,4	2000-03	**414**
S-Coupe 1,5	1990-92	**402**
Sonata 2,0 16V	1992-98	**408**
Sonata 1,8/2,0/2,4	1988-96	**412**
Stellar 1,6	1984-91	CHAIN
Trajet 2,0	2000-03	**414**

■ Diesel models		
H100 2,5 D/TD	1993-03	**416**

ISUZU

■ Petrol models		
Trooper 3,2 V6	1992-98	**418**
Trooper 3,5 V6	1998-03	**418**

■ Diesel models		
Trooper 2,8 Turbo D	1988-92	**420**
Trooper 3,1 Turbo D	1992-98	**420**
N Series 4,3/4,5/4,7 D/TD	1996-03	GEARS

IVECO

■ Diesel models		
Daily 35 2,5 D	1981-96	**422**
Daily 2,5 TD	1986-96	**422**
Daily 2,5 D	1996-00	**424**
Turbo Daily 2,5 TD	1996-00	**424**
Turbo Daily 2,8 TD	1996-00	**424**

JAGUAR (DAIMLER)

All models	1980-03	CHAIN

KIA

■ Petrol models		
Mentor 1,5/1,6 SOHC	1994-99	**426**
Mentor 1,5 DOHC	1996-02	**428**
Sephia 1,5/1,6 SOHC	1994-99	**426**
Sephia 1,5 DOHC	1996-98	**428**
Sportage 2,0 SOHC	1995-01	**430**
Sportage 2,0 16V	1995-02	**432**

LAND ROVER

■ Petrol models

90/110/Defender V8	1980-03	CHAIN
2,25/2,6	1980-83	CHAIN
Discovery V8	1990-02	CHAIN
Freelander 1,8 *manual tensioner*	1997-00	**434**
Freelander 1,8 *automatic tensioner*	1997-00	**436**
Freelander 2,5 V6	2000-03	**438**
Range Rover V8	1980-03	CHAIN

■ Diesel models

88/109 2,25 D	1980-83	CHAIN
90/110/Defender 200 Tdi	1991-94	**446**
Defender Tdi	1994-99	**448**
Defender TD5	1998-03	CHAIN
Discovery 200/300 Tdi	1990-98	**448**
Discovery TD5	1998-03	CHAIN
Freelander 2,0D Turbo	1997-00	**442**
Freelander 2,0 TD4	2000-02	CHAIN
Range Rover 2,4 Turbo D	1986-92	GEARS
Range Rover 200 Tdi	1993-94	**448**
Range Rover Classic 300 Tdi	1994-95	**448**
Range Rover 2,5 DT/SE (6 cyl)	1994-02	CHAIN
Range Rover 3,0 TD6	2002-03	CHAIN

LDV

■ Petrol models

300/400 Series - V8	1987-94	CHAIN
Cub	1998-02	**452**
Sherpa 200 1,7/2,0	1987-94	**712**
Sherpa 300/400 2,0	1987-94	**712**

■ Diesel models

200 1,9 D	1994-98	**456**
400/Convoy 2,5 D/TD	1989-96	CHAIN
Convoy 2,4 TD	2002-03	CHAIN
Convoy 2,5 D/TD	1997-02	**458**
Pilot 1,9 D	1994-98	**456**
Sherpa 200 Di	1986-94	**454**

LEXUS

■ Petrol models

IS 200	1999-03	**460**
IS 300	2001-03	**462**
GS 300	1993-97	**464**
GS 300	1997-03	**462**
LS 400	1990-97	**466**
LS 400	1997-00	**468**

MAZDA

■ Petrol models

121 1,1/1,3	1988-96	**470**
121 1,3	1995-01	CHAIN
323 (BF) 1,1	1985-89	CHAIN
323 (BF) 1,3	1987-89	CHAIN
323 (BG) 1,3	1988-94	**470**
323 (BJ) 1,3	1998-03	**472**
323 (BF) 1,5 *E5 engine*	1987-89	CHAIN
323 (BF) 1,5 *B5 engine*	1987-89	**470**
323 (BW) 1,5 Estate	1990-93	**470**
323 (BA) 1,5	1994-98	**474**
323 (BJ) 1,5	1998-03	**476**
323 (BG) 1,6	1989-94	**470**
323 (BG) 1,8 16V	1989-94	**478**
323 (BA) 1,8 16V	1995-98	**482**
323 (BJ) 1,8	1998-03	**480**
323 (BA) 2,0 V6	1994-98	**484**
626 1,6/1,8 *without guide pulley*	1983-89	**486**
626 1,8/2,0 *with guide pulley*	1987-89	**488**
626 1,8/2,0 DOHC	1992-02	**480**
626 2,0	1983-87	**486**
626 2,0 8V	1986-92	**488**
626 2,0/2,2 12V	1987-93	**488**
626 2,5 V6	1992-98	**490**
B2000	1983-93	**486**
Demio 1,3	1998-03	**472**
E1400 1,4	1982-94	CHAIN
E1800/E2000	1984-93	**486**
E2000 (FE EGi)	1994-99	**488**

Autodata

MX-3 1,6 SOHC	1991-94	**472**
MX-3 1,6 16V DOHC	1994-98	**482**
MX-3 1,8 V6	1991-98	**492**
MX-5 1,6/1,8 DOHC	1989-03	**494**
MX-6 2,5 V6	1992-98	**490**
Premacy 1,8	1999-03	**480**
Tribute 3,0	2001-03	CHAIN
Xedos 6 2,0	1992-00	**490**
Xedos 9 2,0/2,5	1994-00	**490**

■ Diesel models		
323 (BJ) 2,0 TD	1998-03	**496**
626 2,0 TD	1998-02	**496**
B2200 D	1985-97	**500**
B Series 2,5D/2,5TD	1998-03	**498**
E2200 D	1985-03	**500**
MPV	1998-01	**498**
Premacy 2,0 TD	1999-03	**496**

MCC

Smart	1998-03	CHAIN

MERCEDES-BENZ

All models	1980-03	CHAIN

MINI

One 1,6	2001-03	CHAIN
Cooper 1,6	2001-03	CHAIN
Cooper S 1,6	2002-03	CHAIN

MITSUBISHI

■ Petrol models		
3000 GT 24V DOHC	1991-00	**516**
Carisma 1,6 16V	1996-03	**508**
Carisma 1,8 16V SOHC	1996-00	**508**
Carisma 1,8 16V DOHC	1996	**510**
Carisma 1,8 GDI	1997-03	**510**

Colt 1200/1300/1500	1984-96	**502**
Colt 1,3 12V	1996-03	**504**
Colt 1,6	1996-02	**508**
Colt 1,6 16V SOHC	1991-96	**506**
Colt 1,8 DOHC	1992-95	**510**
Galant 1,8 16V	1991-95	**506**
Galant 2,0 16V	1992-96	**514**
Galant 2,0 V6	1993-97	**520**
Galant 2000/Turbo	1984-94	**518**
Galant 2,4 GDI	1999-00	**522**
Galant 2,5 24V DOHC	1993-97	**526**
Galant 2,5 V6 SOHC	1997-03	**528**
L200/L300 2,0 SOHC	1985-93	**518**
Lancer 1200/1300/1500	1984-96	**502**
Lancer 1,3 12V	1996-03	**504**
Lancer 1,6	1996-01	**508**
Lancer 1,6 16V SOHC	1991-96	**506**
Lancer 1,8 DOHC	1992-95	**510**
Sapporo 2,4	1987-90	**518**
Sapporo Turbo 2,0 SOHC	1984-87	**518**
Shogun Pinin 1,8 GDI	1999-03	**510**
Shogun 2,4	1991-94	**518**
Shogun 2,6	1988-91	CHAIN
Shogun 3,0 V6 12V	1989-94	**530**
Shogun 3,0 V6 24V	1994-03	**532**
Shogun 3,5 V6 24V DOHC	1994-03	**534**
Sigma 3,0 V6 24V DOHC	1991-96	**526**
Space Gear 2,4 GDI	1997-01	**522**
Space Runner 1,8 16V	1991-99	**506**
Space Runner 2,4 GDI	1997-03	**522**
Space Star 1,8 GDI	1999-03	**510**
Space Wagon 1,8 16V	1991-99	**506**
Space Wagon 2,0 SOHC	1985-91	**518**
Space Wagon 2,0 16V	1992-99	**514**
Space Wagon 2,4 GDI	1997-03	**522**
Starion Turbo 2,0 SOHC	1982-89	**518**
Starion Turbo 2,6	1989-90	CHAIN

■ Diesel models		
Canter 2,3/2,5 D	1983-98	**540**
Carisma 1,9 TD	1997-02	**538**
Galant 2,0 Turbo D	1993-03	**536**
Galant 2,3 Turbo D	1980-84	**540**
L200 2,5 D	1986-03	**540**

L300 2,3/2,5 D	1986-99	**540**
Shogun 2,3/2,5 Turbo D	1982-03	**540**
Shogun 2,8/3,2 TD	1994-03	CHAIN
Space Wagon 2,0 Turbo D	1992-00	**536**

NISSAN

■ Petrol models

200 SX 2,0 Turbo	1994-02	CHAIN
240K-GT	1980-83	CHAIN
280C	1980-87	CHAIN
280 ZX 2,8	1980-87	CHAIN
Almera/Tino 1,4/1,5	1995-03	CHAIN
Almera/Tino 1,6/1,8	1995-03	CHAIN
Almera/Tino/GTI 2,0	1996-03	CHAIN
Bluebird 1,6/1,8	1980-84	CHAIN
Cabstar 2,0 (F22)	1982-94	CHAIN
Cabstar 2,2 (H40)	1982-91	CHAIN
Cabstar/Urvan (F20/E20)	1980-82	CHAIN
Laurel 2,0/2,4	1980-89	CHAIN
Maxima QX 2,0/3,0 V6	2000-03	CHAIN
Micra (K11) 1,0/1,3/1,4	1992-02	CHAIN
Micra	2002-03	CHAIN
Patrol 2,8/3,0	1982-93	CHAIN
Patrol GR 4,2	1992-99	CHAIN
Pick-up 1,6/1,8 (720)	1980-86	CHAIN
Pick-up 2,0/2,4 (D21)	1986-93	CHAIN
Prairie 2,4	1992-95	CHAIN
Primera 1,6/1,8/2,0	1990-03	CHAIN
QX 2,0/3,0 V6	1994-00	CHAIN
Serena 1,6	1992-02	CHAIN
Serena 2,0	1992-02	CHAIN
Silvia 2,0	1984-88	CHAIN
Skyline GT-R	1998-01	CHAIN
Sunny 1,4/1,6	1988-95	CHAIN
Sunny 2,0	1990-93	CHAIN
Terrano II 2,4	1993-96	CHAIN
Urvan (E23)	1980-82	CHAIN
Urvan 2,0 (E23)	1982-87	CHAIN
Urvan 2,0 (E24)	1987-94	CHAIN
Vanette 1,5 (C120)	1983-86	CHAIN
Vanette 1,5 (C22)	1986-94	CHAIN
X-Trail 2,0	2001-03	CHAIN

■ Diesel models

Almera 2,0 D (CD20)	1995-96	**546**
Almera 2,0 D (CD20E)	1996-00	**548**
Almera 2,2 D	2000-03	CHAIN
Bluebird 2,0 D (T12/T72)	1986-91	**552**
Bluebird 2,0 D (U11)	1984-86	**552**
Bluebird 2,0 D/TD	1984-90	CHAIN
Bluebird Turbo D (U11)	1984-86	**552**
Cabstar 2,5 D	1984-00	GEARS
Cabstar 3,3 D	1982-87	GEARS
Laurel/Cedric 2,8 D	1985-88	CHAIN
Micra 1,5 D	1998-02	**544**
Patrol (260) 2,8 D/TD	1989-94	**556**
Patrol (Y60) 2,8 TD	1989-00	**556**
Patrol GR (Y61) 2,8 TD	1998-00	**556**
Patrol 4,2 TD	1992-98	GEARS
Pick-up 2,3/2,5 D (D21 with SD23/25 engines)	1986-88	GEARS
Pick-up 2,5 D (D21 with TD25 engine)	1990-98	GEARS
Primera 2,0 D	1992-96	**546**
Primera 2,0 TD	1996-02	**548**
Serena 2,0 D	1992-94	**554**
Serena 2,3 D	1994-02	**554**
Sunny 2,0 D (N14)	1991-96	**546**
Sunny Wagon 2,0 D (Y10)	1991-96	**546**
Terrano 2,4 (D21)	1987-91	CHAIN
Terrano 2,7 TD	1990-91	GEARS
Terrano II 2,7 TD	1993-01	GEARS
Urvan 2,3/2,5 D	1984-96	GEARS
Vanette 2,0 D (C220)	1984-87	CHAIN
Vanette 2,0 D (C220)	1987-96	**552**
Vanette Cargo 2,3 D	1995-02	**554**
X-Trail 2,2 TD	2001-03	CHAIN

PEUGEOT

■ Petrol models

104 1,0/1,1/1,3/1,4	1980-83	CHAIN
106 1,0/1,1/1,4	1991-94	**558**
106 1,0/1,1/1,4/1,6	1993-03	**560**
106 1,6 GTi	1997-03	**564**
205 1,1/1,3 (E1/G1 engines)	1987-91	CHAIN

Autodata

205 1,0/1,1/1,4	1987-94	**558**
205 1,1/1,4	1993-96	**560**
205 1,6/1,9 *eccentric tensioner*	1990-96	**568**
205 GT 1,4	1982-83	CHAIN
206 1,1/1,4/1,6	1998-03	**572**
206 1,6 16V	2000-03	**574**
206 2,0 16V *adjustable camshaft sprockets*	1998-00	**576**
206 2,0 16V *non-adjustable camshaft sprockets*	2000-03	**580**
305 1,3/1,5	1980-85	CHAIN
306 1,1/1,4/1,6	1993-01	**560**
306 1,8/2,0	1990-01	**568**
306 1,8 16V	1997-98	**584**
306 1,8 16V	1999-01	**588**
306 2,0 16V	1993-97	**592**
306 2,0 16V	1997-98	**594**
306 2,0 16V	1998-01	**576**
307 1,4	2001-03	**560**
307 1,6 16V	2001-03	**574**
307 2,0 16V	2001-03	**580**
309 1,1/1,4	1989-93	**558**
309 1,3 (G2 engines)	1986-89	CHAIN
309 1,6/1,9 *eccentric tensioner*	1990-93	**568**
405 1,4	1987-94	**558**
405 1,4	1993-94	**560**
405 1,6/1,8/1,9/2,0 *eccentric tensioner*	1989-95	**568**
405 Mi16 2,0	1993-97	**592**
406 1,6	1995-97	**598**
406 1,8	1997-03	**568**
406 1,8 16V	1995-98	**584**
406 1,8 16V	1999-03	**588**
406 2,0 16V	1995-98	**594**
406/Coupé 2,0 16V *adjustable camshaft sprockets*	1999-00	**576**
406/Coupé 2,0 16V *non-adjustable camshaft sprockets*	2000-03	**580**
406 2,0 Turbo	1997-00	**598**
406/Coupé 2,2 16V	1999-03	**602**
406/Coupé 3,0 V6	1997-03	**606**
504 1,8	1980-90	CHAIN
504 2,0	1980-85	CHAIN
504 2,6 V6	1980-83	CHAIN
505 2,0	1980-86	CHAIN
505 2,9 V6	1987-90	CHAIN
604 2,6	1980-83	CHAIN
605 2,0	1989-98	**568**
605 2,0 Turbo	1992-99	**568**
605 2,0 16V	1995-98	**592**
605 3,0 V6	1989-98	CHAIN
605 3,0 V6	1997-99	**606**
607 2,0 16V	2001-03	**580**
607 2,2 16V	2000-03	**602**
806 1,8	1997-01	**568**
806 2,0	1994-01	**598**
806 2,0 16V	2000-02	**580**
806 2,0 Turbo	1994-01	**598**
Boxer 2,0	1995-02	**568**
J5 1,8/2,0	1983-91	CHAIN
Partner 1,1/1,4	1996-03	**560**
Partner 1,8	1996-02	**568**

■ **Diesel models**

106 1,4 D	1993-95	**610**
106 1,5 D	1994-03	**612**
106 Van 1,5 D	1994-03	**612**
205 1,8 D/TD/1,9 D	1987-96	**614**
206 1,9 D	1998-03	**618**
206 2,0 HDi	1999-03	**622**
305 1,5 D	1980-82	CHAIN
305 1,8/1,9 D	1987-90	**614**
306 1,8 D/1,9 D/TD *XUD engine*	1993-01	**614**
306 1,9 D *DW8 engine*	1998-01	**618**
306 2,0 HDi	1999-01	**622**
307 2,0 HDi	2001-03	**622**
309 1,8 D/TD/1,9 D	1987-93	**614**
405 1,8/1,9 D/TD	1988-96	**614**
406 1,9 TD	1995-00	**626**
406 2,0 HDi	1997-03	**622**
406 2,1 TD	1995-98	**628**
406 2,2 HDi	2000-03	**630**
504 2,3 D	1987-91	CHAIN
505 2,5 D/TD	1982-93	CHAIN
605 2,1 D/TD	1989-98	**634**
605 2,1 TD *P8C engine*	1994-99	**628**

607 2,0 HDi	2000-03	**622**
607 2,2 HDi	2000-03	**630**
806 1,9 TD	1995-00	**626**
806 2,0 HDi	1999-02	**622**
806 2,1 TD	1995-00	**628**
Boxer 1,9 D/TD	1994-00	**626**
Boxer 2,5 D	1994-02	**636**
Boxer 2,5 TD	1994-97	**636**
Boxer 2,5 TD THX engine	1997-02	**638**
Boxer 2,8 HDi	2000-02	**640**
Expert 1,9 D/TD XUD engine	1995-02	**626**
Expert 1,9 D DW8 engine	1999-03	**618**
J5 1,9 D	1986-94	**614**
J7/J9 2,1 D	1981-87	CHAIN
Partner 1,8/1,9 D XUD engine	1996-99	**614**
Partner 1,9 D DW8 engine	1998-03	**618**

PORSCHE

911	1980-03	CHAIN
Boxster	1996-03	CHAIN

PROTON

■ Petrol models

1,3/1,5	1989-97	**642**
313/315/413/415	1993-03	**642**
416/418	1993-03	**644**
418 DOHC	1996-03	**646**
Impian 1,6	2001-03	**648**
Persona 1,3/1,5	1993-03	**642**
Persona Compact 1,3	1995-03	**642**
Persona 1,6/1,8	1993-02	**644**
Persona 1,8 DOHC	1996-03	**646**

■ Diesel models

420 D/TD	1996-00	**650**
Persona 2,0 D/TD	1996-00	**650**

RENAULT

■ Petrol models

R4 (782/845/1108 cc)	1980-90	CHAIN
R5 (except 1,7)	1980-85	CHAIN
R5 1,7	1986-93	**654**
R9/11 1,1/1,4	1982-89	CHAIN
R11 1,7	1986-89	**654**
R14	1980-83	CHAIN
R18/Fuego 1,4/1,6/1,7	1980-85	CHAIN
R18/Fuego 2,0	1980-86	**668**
R19 1,4	1988-91	CHAIN
R19 1,4	1988-96	**656**
R19 1,4 ECO (X535)	1995-96	**656**
R19 1,7	1988-96	**654**
R19 1,8 16V	1989-96	**660**
R19 1,8	1992-94	**662**
R20 2,0	1981-84	**668**
R20/30	1980-84	CHAIN
R21 1,4	1989-91	CHAIN
R21 1,7	1986-93	**654**
R21 2,0 8V/12V	1986-94	**668**
R25 2,0 8V/12V	1986-93	**668**
R25 V6/Turbo	1984-92	CHAIN
Alpine	1980-85	CHAIN
Alpine 2,5/2,9	1986-91	CHAIN
Clio 1,1	1990-93	CHAIN
Clio 1,2 E5F/E7F engines	1990-98	**656**
Clio 1,2 D7F engine	1995-01	**652**
Clio 1,4	1990-98	**656**
Clio 1,4 16V	1998-03	**658**
Clio 1,6 16V	1998-03	**658**
Clio 1,7	1990-93	**654**
Clio 1,8	1992-98	**662**
Clio 1,8 16V	1991-97	**660**
Clio Sport 2,0 16V	1999-03	**664**
Clio Williams 2,0 16V	1993-95	**660**
Espace 2,0/2,2	1984-97	**668**
Espace 2,0	1996-99	**662**
Espace 2,0 16V	1999-02	**664**
Espace V6	1991-97	CHAIN
Espace V6	1997-02	**682**
Extra 1,4 (C3J engine)	1992-97	CHAIN

Extra/Express 1,1/1,2 (C1E/C1G/C3G engines)	1992-99	CHAIN
Extra/Express 1,4	1990-02	656
Kangoo 1,2	1998-03	652
Kangoo 1,4	1998-03	656
Laguna 1,6 16V	1998-01	658
Laguna 1,8/2,0	1994-01	662
Laguna 1,8 16V	1998-01	664
Laguna 2,0 16V	1996-99	678
Laguna 2,0 16V	1999-01	664
Laguna 3,0	1994-98	CHAIN
Laguna 3,0	1997-03	682
Master 2,0/2,2	1986-98	668
Mégane/Scénic 1,4	1996-99	672
Mégane/Scénic 1,4 16V	1998-02	658
Mégane/Scénic 1,6	1996-99	672
Mégane/Scénic 1,6 16V	1998-00	658
Mégane/Scénic 2,0	1996-99	674
Mégane 2,0 16V	1996-99	676
Mégane/Scénic 2,0 16V	1999-00	664
Safrane 2,0 16V	1996-01	678
Safrane 2,5 20V	1996-01	678
Safrane 3,0	1992-97	CHAIN
Safrane 3,0	1998-01	682
Trafic 1,3	1980-91	CHAIN
Trafic 1,7	1986-91	CHAIN
Trafic 1,7	1988-94	654
Trafic 2,0/2,2	1986-98	668
Twingo 1,2 D7F engine	1996-03	652
Twingo 1,2 C3G engine	1993-97	CHAIN

■ Diesel models

R5/9/11/Extra 1,6 D	1988-96	686
R5 1,9 D	1991-95	686
R5 Van 1,9 D	1991-95	686
R18 D	1981-96	692
R19/Turbo/Chamade 1,9 D with non-adjustable or three bolt adjustable injection pump sprocket	1988-96	686
R19 1,9 D/TD Lucas pump with centrally adjustable sprocket	1992-96	688
R20/30 D/Turbo	1980-83	692
R21 1,9 D	1988-95	686
R21/25 D/Turbo	1986-95	692
Clio 1,9 D with non-adjustable or three bolt adjustable injection pump sprocket	1990-01	686
Clio 1,9 D Lucas pump with centrally adjustable sprocket	1990-01	688
Clio Van 1,9D	1994-98	686
Clio 1,9 TD	1999-02	690
Espace 2,1D Turbo	1984-97	692
Espace 2,2D Turbo	1996-00	694
Extra 1,9 D with non-adjustable or three bolt adjustable injection pump sprocket	1991-02	686
Extra 1,9 D Lucas pump with centrally adjustable sprocket	1991-02	688
Fuego D Turbo	1981-86	692
Kangoo 1,9 D	1998-00	688
Kangoo 1,9 TD	1999-03	690
Laguna 1,9 TD/TDi with manual tensioner	1998-01	696
Laguna 1,9 TD with automatic tensioner	1999-01	690
Laguna 1,9 TD Common Rail injection system	1999-02	698
Laguna 2,2 D	1994-96	700
Laguna 2,2 D/TD	1996-00	702
Master 2,1 D	1979-00	692
Master 2,5 D/TD	1980-98	704
Master 2,5 D	1998-02	706
Master 2,8 D Turbo	1998-02	706
Mégane/Scénic 1,9 D/TD non-adjustable injection pump sprocket	1995-99	694
Mégane/Scénic 1,9 TD Lucas pump with centrally adjustable sprocket	1996-00	688
Mégane/Scénic 1,9 TD F9Q 730/734 engines	1997-00	696
Mégane/Scénic 1,9 TD F9Q 732 engine	1999-01	698
Mégane 1,9 TD with automatic tensioner	1999-02	690
Safrane 2,2 TD	1996-00	694
Trafic 1,9 D	1998-00	688
Trafic 2,1 D	1979-00	692
Trafic 2,5 D	1980-00	704

ROVER

■ Petrol models

Model	Years	Page
25 1,1	1999-02	**710**
25 1,1 16V	2000-03	**720**
25 1,4/1,6/1,8	1999-03	**720**
25 1,8 VVC	1999-03	**728**
45 1,4/1,6/1,8	1999-03	**720**
45 2,0 V6	1999-03	**736**
75 1,8	1999-03	**720**
75 2,0 V6	1999-02	**736**
75 2,5 V6	1999-03	**736**
111	1990-95	**708**
111	1995-98	**710**
114	1990-95	**708**
114 16V	1990-95	**714**
114	1995-98	**710**
200vi	1995-99	**728**
211	1997-99	**710**
214	1995-99	**710**
214/216/218 16V manual tensioner	1995-99	**716**
214/216/218 16V automatic tensioner	1998-99	**720**
214S SOHC	1990-92	**708**
214 16V DOHC	1989-95	**714**
216	1985-89	**704**
216 16V SOHC	1989-95	**724**
216 Cabrio/Coupé manual tensioner	1996-99	**716**
216 Cabrio/Coupé automatic tensioner	1998-99	**720**
216 GTi 16V DOHC	1990-95	**726**
218 16V manual tensioner	1997-99	**716**
218 16V automatic tensioner	1998-99	**720**
218 Coupe	1996-99	**728**
220 16V	10/92-96	**732**
220 16V Turbo	10/92-96	**732**
414/416 manual tensioner	1995-99	**716**
414/416 automatic tensioner	1998-99	**720**
414 16V DOHC	1989-95	**714**
416 SLi Automatic	1995-99	**734**
416 16V SOHC	1990-96	**724**
416 GTi 16V DOHC	1990-92	**726**
416 Tourer SOHC	1989-96	**724**
416 Tourer DOHC manual tensioner	1996-99	**716**
416 Tourer DOHC automatic tensioner	1998-99	**720**
420i	1995-99	**732**
420 16V	10/92-96	**732**
618	1995-99	**740**
620 SOHC	1993-99	**740**
620 16V Turbo	1994-99	**742**
623 DOHC	1993-99	**744**
820 16V	11/91-99	**732**
825 2,5 V6	1986-91	**746**
825 2,5 V6	1995-99	**748**
827 2,7 V6	1991-96	**746**
3500 (SD1)	1980-84	CHAIN
Allegro	1980-82	CHAIN
BRM 1,8 VVC	1998-99	**728**
Maestro 1,6	1983-84	CHAIN
Maestro 2,0	1984-93	**712**
Maestro/Montego 1,3	1983-93	CHAIN
Marina 1,3/1,8	1980-85	CHAIN
Maxi	1980-82	CHAIN
Metro	1980-91	CHAIN
Metro 1,1/1,4 SOHC	1990-95	**708**
Metro 1,4 16V DOHC	1990-95	**714**
MG Maestro 2,0	1984-93	**712**
MG Montego 2,0	1984-93	**712**
MG RV8	1993-95	CHAIN
MGF 1,6/1,8/Trophy	1995-02	**752**
MGF 1,8 VVC	1995-02	**728**
MG ZR 1,4/1,8	2001-03	**720**
MG ZS 1,6/1,8	2001-03	**720**
MG ZR 1,8 VVC	2001-03	**728**
MG ZS 180 2,5 V6	2001-03	**736**
MG ZT/ZT-T 160 2,5 V6	2001-02	**736**
MG ZT/ZT-T 180 2,5 V6	2002-03	**736**
MG ZT/ZT-T 190 2,5 V6	2001-03	**736**
Mini	1980-00	CHAIN
Montego 2,0	1984-93	**712**
Princess 2200	1980-82	CHAIN
Triumph Dolomite	1980-81	CHAIN
Vitesse 2,0 16V Turbo	12/91-99	**732**

■ Diesel models

25 2,0 TD	1999-01	**764**
25 2,0 TD	2001-03	**768**
45 2,0 TD	1999-01	**764**
45 2,0 TD	2001-03	**768**
75 2,0 TD	1999-03	CHAIN
115	1995-98	**756**
218 SD	1991-95	**762**
218 D Turbo	1991-95	**762**
220 D Turbo	1995-99	**764**
418 SD	1991-98	**762**
418 D Turbo	1991-98	**762**
418 Tourer D Turbo	1991-98	**762**
420 D Turbo	1995-99	**772**
620 D Turbo	1995-99	**772**
825 TD	1990-99	GEARS
2400 Turbo D (SD1)	1982-86	GEARS
Maestro/Van 2,0 D	1986-95	**760**
Maestro 2,0 D Turbo	1992-95	**760**
MG ZR 2,0 TD	2001-03	**768**
MG ZS 2,0 TD	2001-03	**768**
Montego 2,0 D Turbo	1989-94	**760**

SAAB

■ Petrol models

90 2,0	1984-91	CHAIN
99 2,0	1980-84	CHAIN
900 2,0/Turbo	1980-93	CHAIN
900 2,0	1995-98	CHAIN
900 2,1/2,3	1991-98	CHAIN
900 2,5 V6 without raised outer edge on tensioner pulley	1993-96	**776**
900 2,5 V6 with raised outer edge on tensioner pulley	1995-98	**780**
9000 2,0/2,3/Turbo	1984-98	CHAIN
9000 3,0 V6 without raised outer edge on tensioner pulley	1995-96	**776**
9000 3,0 V6 with raised outer edge on tensioner pulley	1995-97	**780**
9-3 2,0/2,3	1998-00	CHAIN
9-5 2,0/2,3	1997-03	CHAIN
9-5 3,0 V6	1998-03	**784**

■ Diesel models

9-3 2,2 TiD	1998-02	CHAIN

SEAT

■ Petrol models

Alhambra 1,8 Turbo	1997-03	**812**
Alhambra 2,0	1996-99	**798**
Alhambra 2,0	2000-03	**814**
Arosa 1,4 16V	1999-03	**790**
Arosa 1,0/1,4	1997-03	**788**
Arosa 1,0 AHT engine	1998-03	CHAIN
Cordoba 1,0/1,4/1,6 automatic tensioner	1996-02	**788**
Cordoba 1,4/1,6 water pump tensioner	1995-99	**794**
Cordoba 1,4 16V AFH engine	1998-99	**796**
Cordoba 1,4 16V APE/AQQ/AUA/AUB engines	1999-02	**790**
Cordoba 1,6/1,8/2,0 timing mark - front of camshaft sprocket	1995-99	**798**
Cordoba 1,6 AFT engine	1996-99	**800**
Cordoba 1,6	1999-02	**802**
Cordoba 1,8 16V	1994-97	**804**
Cordoba 1,8 Turbo	1999-02	**806**
Cordoba 2,0 16V	1996-99	**808**
Ibiza 1,0	1996-02	**788**
Ibiza 1,05/1,3/1,4/1,6 water pump tensioner	1993-99	**794**
Ibiza 1,4/1,6 automatic tensioner	1996-02	**788**
Ibiza 1,4 16V AFH engine	1998-99	**796**
Ibiza 1,4 16V APE/AQQ/AUA/AUB engines	1999-03	**790**
Ibiza 1,6/1,8/2,0 timing mark - front of camshaft sprocket	1995-99	**798**
Ibiza 1,6 AFT engine	1996-99	**800**
Ibiza 1,6	1999-02	**802**
Ibiza 1,8 16V	1994-97	**804**
Ibiza 1,8 Turbo	1999-02	**806**
Ibiza 2,0 16V	1996-99	**808**
Inca 1,4/1,6 automatic tensioner	1995-02	**788**

Index

Inca 1,6	1995-02	**798**
Leon 1,4 16V	1999-03	**790**
Leon 1,6	1999-03	**802**
Leon 1,8/Turbo	1999-03	**806**
Marbella 850/903	1980-88	CHAIN
Toledo 1,4 16V	1999-03	**790**
Toledo 1,6	1991-97	**810**
Toledo 1,6 *AFT/AKS engines*	1996-99	**800**
Toledo 1,6	1999-03	**802**
Toledo 1,8 *timing mark - rear of camshaft sprocket*	1991-94	**810**
Toledo 1,8/2,0 *timing mark - front of camshaft sprocket*	1991-99	**798**
Toledo 1,8	1999-03	**806**
Toledo 1,8 16V	1991-94	**804**
Toledo 2,0 16V	1991-99	**804**
Toledo 2,3 V5	1999-03	CHAIN

■ Diesel models

Arosa 1,4 TDI PD	1999-03	**816**
Arosa 1,7 SDI	1998-03	**820**
Alhambra 1,9 TDI	1996-00	**822**
Alhambra 1,9 TDI PD	1999-03	**832**
Cordoba 1,7 SDI *automatic tensioner*	1998-99	**822**
Cordoba 1,9 D/Turbo D *manual tensioner*	1993-95	**824**
Cordoba 1,9 D/TD *automatic tensioner*	1994-99	**822**
Cordoba 1,9 D/TD *two piece injection pump sprocket*	1994-99	**826**
Cordoba 1,9 SDI/TDI	1996-99	**822**
Cordoba 1,9 SDI/TDI	1999-02	**828**
Ibiza 1,7 SDI *automatic tensioner*	1998-99	**822**
Ibiza 1,9 D/TD *manual tensioner*	1993-95	**824**
Ibiza 1,9 D/TD *automatic tensioner*	1994-99	**822**
Ibiza 1,9 D/TD *two piece injection pump sprocket*	1994-99	**826**
Ibiza 1,9 SDI/TDI	1996-99	**822**
Ibiza 1,9 SDI/TDI	1999-02	**828**
Inca 1,9 D	1996-00	**826**
Inca 1,9 SDI	1996-02	**822**
Leon 1,9 SDI/TDI	1999-03	**828**

Toledo 1,9 D/TD *manual tensioner*	1991-95	**824**
Toledo 1,9 D/TD *automatic tensioner*	1994-99	**822**
Toledo 1,9 D/TD *two piece injection pump sprocket*	1994-99	**826**
Toledo 1,9 TDI	1995-99	**822**
Toledo 1,9 TDI	1999-03	**828**

SKODA

■ Petrol models

135/136	1988-91	CHAIN
Cube Van 1,3	1993-99	CHAIN
Estelle	1981-91	CHAIN
Fabia 1,4	1999-03	**836**
Forman Pickup 1,3	1989-96	CHAIN
Favorit 1,1/1,3	1989-95	CHAIN
Felicia 1,3	1995-01	CHAIN
Felicia Pickup 1,3	1995-01	CHAIN
Felicia 1,6	1996-01	**840**
Octavia 1,4	1999-03	CHAIN
Octavia 1,6 *AEE engine*	1996-03	**840**
Octavia 1,6 *AEH/AKL engines*	1997-03	**842**
Octavia 1,8	1996-00	**844**
Octavia 1,8 Turbo	1998-03	**844**
Octavia 2,0	1999-03	**846**
Rapid	1982-85	CHAIN

■ Diesel models

Felicia 1,9 D	1996-01	**848**
Felicia Pick-up 1,9D	1997-01	**848**
Felicia Cube Van 1,9D	1997-01	**848**
Octavia 1,9 SDI	1997-02	**852**
Octavia 1,9 TDI	1996-03	**856**

SSANGYONG

All models	1980-99	CHAIN

SUBARU

■ Petrol models

1400/1600 (E63/71 engines)	1977-79	GEARS
1800	1981-93	GEARS
Forester 2,0/Turbo	1997-03	862
Impreza 1,6	1993-96	864
Impreza 1,6	1997-00	862
Impreza 1,8	1993-96	864
Impreza 2,0	1995-96	866
Impreza 2,0	1997-00	862
Impreza 2,0 Turbo DOHC	1994-96	868
Impreza 2,0 Turbo DOHC	1997-00	870
Justy 1,0/1,2	1987-96	860
Legacy 1,8	1989-93	866
Legacy 2,0	1989-96	866
Legacy 2,0	1997-03	862
Legacy 2,0 Turbo DOHC	1991-94	868
Legacy 2,2	1989-96	866
Legacy 2,2	1997-98	862
Legacy 2,5	1996-99	868
Legacy 2,5	2000-03	862
Legacy Outback 2,5	1997-99	868
Legacy Outback 2,5	2000-03	862
Legacy 3,0	2001-03	CHAIN
MV Pick-up 1600/1800	1986-93	GEARS
Sumo (E10/E12)	1989-97	860

SUZUKI

■ Petrol models

Alto	1994-01	872
Baleno 1,3/1,6 16V	1995-02	876
Baleno 1,8 16V	1996-02	CHAIN
Grand Vitara 2,0	1999-03	CHAIN
Grand Vitara 2,5/2,7 V6	1998-03	CHAIN
Ignis	2000-03	CHAIN
Jimny 1,3 16V	1998-03	880
Liana 1,3/1,6	2001-03	CHAIN
SJ413/Samurai	1986-93	872
Swift (SA310) 1,0	1985-89	872
Swift (SA413) 1,3	1985-89	872
Swift (SF413) 1,3	1989-03	872

Swift 1,6 16V	1990-92	876
Swift GTi (SA413)	1986-89	874
Swift GTi (SF413)	1989-97	874
Vitara 1,6 8V	1988-03	878
Vitara 1,6 16V	1991-03	880
Vitara 2,0 V6	1995-03	CHAIN
Wagon R+ 1,0/1,2	1997-00	CHAIN
X-90	1996-98	880

■ Diesel models

Vitara 2,0 TD	1995-00	882
Grand Vitara 2,0 TD	1998-03	882

TALBOT

■ Diesel models

Express 1,9 D	1988-93	614
Horizon 1,9 D	1982-86	614
Solara 1,9 D	1982-88	614

TOYOTA

■ Petrol models

1000/Starlet	1974-84	CHAIN
Avensis 1,6/1,8	1998-00	890
Avensis 1,6/1,8	2000-03	CHAIN
Avensis 2,0	1998-00	902
Avensis 2,0	2000-03	CHAIN
Camry 1,8/2,0	1983-91	894
Camry 2,2 16V	1992-01	896
Camry 2,4	2001-03	CHAIN
Camry 3,0 24V	1991-96	904
Camry 3,0 24V	1996-01	906
Carina 1400/1600/1800	1974-77	CHAIN
Carina 1600 (TA40)	1982-84	CHAIN
Carina 1800 (TA62)	1982-84	CHAIN
Carina II (AT171) 1,6	1984-88	888
Carina II (AT151) 1,6	1988-92	888
Carina II (ST171) 1,8/2,0	1984-92	894
Carina E 1,6i/1,8	1992-98	890
Carina E 2,0	1992-98	896
Carina E (ST191) 2,0 GTi	1992-97	898

Celica (AT160)	1988-90	**892**
Celica 1,8	1999-03	CHAIN
Celica 2,0	1982-85	CHAIN
Celica 2000	1980-82	CHAIN
Celica GT (ST182) 2,0	1990-99	**898**
Celica GT 1,8 16V	1994-99	**890**
Celica GT-4 (ST185) 2,0	1990-97	**898**
Celica GT/ST	1980-85	CHAIN
Corolla 1,3	1983-85	**888**
Corolla 1,3	1985-92	**884**
Corolla 1,3 16V	1992-00	**886**
Corolla 1,4	2000-03	CHAIN
Corolla (AE82) 1,6	1984-86	**888**
Corolla (AE92/95) 1,6	1987-92	**888**
Corolla (AE101/111) 1,6 16V	1992-99	**890**
Corolla 1,6/1,8 VVT-i	1999-03	CHAIN
Corolla 1,8 16V	1993-95	**890**
Corolla 1300	1980-87	CHAIN
Corolla 1600	1980-82	CHAIN
Corolla GT Coupe (AE86)	1984-87	**892**
Corolla GTi (AE92)	1987-92	**892**
Corona 1800-2000	1980-83	CHAIN
Cressida	1981-83	CHAIN
Crown 2600/Super	1980-83	CHAIN
Hi-Ace 2,0	1983-94	CHAIN
Hi-Ace 2,4	1989-03	CHAIN
Hi-Ace PowerVan 2,4	1996-03	CHAIN
Hi-Lux 1,6	1977-82	GEARS
Hi-Lux 1,8/2,0/2,2	1977-88	CHAIN
Hi-Lux 1,8/2,2	1989-96	CHAIN
Hi-Lux 2,4	1988-02	CHAIN
Landcruiser (2F engine)	1981-84	GEARS
Landcruiser (RJ70)	1987-94	CHAIN
Landcruiser Colorado 3,4 V6	1996-02	**912**
Landcruiser 4,5	1995-98	CHAIN
Lite-Ace	1980-91	CHAIN
Lite-Ace 1,5	1985-96	CHAIN
MR2 (AW11) 1,6	1985-90	**892**
MR2 1,8	2000-03	CHAIN
MR2 (SW20) 2,0	1990-94	**894**
MR2 (SW20) 2,0 GT	1990-00	**898**
Picnic	1996-01	**902**

Previa 2,4	1990-03	CHAIN
Prius 1,5	2000-03	CHAIN
RAV-4	1994-00	**902**
RAV-4 1,8/2,0	2000-03	CHAIN
Space Cruiser/Model-F	1983-90	CHAIN
Starlet 1,0	1985-93	**884**
Starlet 1,3	1990-99	**882**
Starlet 1,3 16V	1996-99	**886**
Starlet 1200/1300	1980-85	CHAIN
Supra 3,0 (JZM80) Turbo	1993-97	**908**
Tercel (AL20/25) 1,3/1,5	1982-88	**888**
Yaris/Verso 1,0/1,3/1,5	1998-03	CHAIN
4-Runner 3,0 V6	1989-96	**910**

■ Diesel models

Avensis 2,0 TD	1998-99	**914**
Avensis 2,0 TD *1CD-FTV engine*	1999-03	**916**
Avensis Verso 2,0 TD *1CD-FTV engine*	2001-03	**916**
Camry Turbo D	1984-91	**914**
Carina II D	1984-92	**914**
Carina E D/TD	1992-98	**914**
Corolla D	1987-99	**914**
Corolla 2,0 TD	2000-03	**916**
Corolla Verso 2,0 TD	2001-03	**916**
Dyna 100/150	1988-98	**922**
Hi-Ace (LH61) 2,4 D	1989-95	**922**
Hi-Ace PowerVan (LXH12) 2,4 D/TD	1996-00	**922**
Hi-Lux D/TD	1988-02	**922**
Landcruiser (BJ) 3,5 D/TD	1985-89	GEARS
Landcruiser 2,4 TD	1988-93	**922**
Landcruiser 3,0 TD	1993-96	**920**
Landcruiser 4,0 D	1981-90	GEARS
Landcruiser 4,2 TD *1HD-FT*	1995-98	**920**
Landcruiser Amazon 4,2 TD	1998-03	**920**
Landcruiser Colorado 3,0 TD	1997-00	**920**
Lite-Ace D	1985-95	**914**
Picnic 2,2 TD	1998-01	**918**
4-Runner 3,0 TD	1993-96	**920**

VAUXHALL-OPEL

■ Petrol models

Agila 1,0/1,2	2000-03	CHAIN
Ascona 1,3/1,4/1,6/1,8/,2,0 water pump tensioner	1982-93	**938**
Astra-E 1,3/1,4/1,6/1,8/2,0	1984-91	**938**
Astra-E 2,0 16V	1986-91	**940**
Astra-F 1,4/1,6	1991-96	**926**
Astra-F 1,4 X14NZ engine	1993-99	**928**
Astra-F 1,6 X16SZ/X16SZR engines	1993-99	**928**
Astra-F 1,8/2,0 water pump tensioner	1991-92	**938**
Astra-F 1,6/1,8/2,0 automatic tensioner	1992-95	**942**
Astra-F 1,4/1,6 16V	1993-98	**930**
Astra-F 2,0 16V manual tensioner	1991-92	**940**
Astra-F 1,8/2,0 16V automatic tensioner	1992-99	**944**
Astra-F Van 1,4/1,6	1991-94	**926**
Astra-F Van 1,6	1994-99	**928**
Astra-G 1,2	1998-03	CHAIN
Astra-G 1,4/1,6/1,8 16V	1998-03	**932**
Astra-G 1,6 X16SZR/Z16SE engines	1998-03	**936**
Astra-G 2,0 16V	1998-00	**946**
Astra-G 2,0 Turbo	2000-03	**950**
Astra-G 2,2	2000-03	CHAIN
Astra/Kadett 1,2	1979-91	CHAIN
Astramax 1,3/1,4/1,6	1986-94	**938**
Astravan 1,3/1,4/1,6	1986-91	**938**
Belmont 1,4/1,6/1,8	1984-91	**938**
Belmont 1,4/1,6	1991	**926**
Calibra 2,0	1990-93	**938**
Calibra 2,0 8V	1993-97	**942**
Calibra 2,0 16V manual tensioner	1990-92	**940**
Calibra 2,0 16V/Turbo automatic tensioner	1992-98	**944**
Calibra 2,5 V6 (C25XE engine)	1993-96	**954**
Calibra 2,5 V6 (X25XE engine) without raised outer edge on tensioner pulley	1996-97	**958**
Calibra 2,5 V6 (X25XE engine) with raised outer edge on tensioner pulley	1997-98	**962**
Carlton/Omega-A 1,8/2,0	1986-94	**938**
Carlton/Omega-A 2,6	1990-94	CHAIN
Carlton/Omega-A 3,0	1987-94	CHAIN
Carlton/Omega-A 3,0 24V	1989-94	CHAIN
Carlton/Rekord 1,9/2,0	1980-86	CHAIN
Cavalier 1,3/1,4/1,6/1,8/,2,0 water pump tensioner	1982-93	**938**
Cavalier 1,4/1,6	1991-95	**926**
Cavalier 1,6 C16NZ2 engine	1993-95	**942**
Cavalier 1,6/1,8/2,0 automatic tensioner	1990-95	**942**
Cavalier 2,0 16V manual tensioner	1988-92	**940**
Cavalier 2,0 16V automatic tensioner	1992-95	**944**
Cavalier 2,5 V6	1993-95	**954**
Cavalier/Ascona 1,6/1,9/2,0	1975-81	CHAIN
Chevette/Chevanne 1,3	1980-85	CHAIN
Combo 1,4	1993-99	**928**
Corsa-A 1,2/1,4/1,6 additional automatic tensioner	1991-93	**926**
Corsa-A GSi additional automatic tensioner	1991-93	**926**
Corsa-B 1,0	1997-00	CHAIN
Corsa-B 1,2 X12XE/Z12XE engines	1997-00	CHAIN
Corsa-B 1,2/1,4	1993-00	**928**
Corsa-B 1,4/1,6 8V	1993-00	**926**
Corsa-B 1,4/1,6 16V	1993-00	**930**
Corsa-B Van 1,4	1994-00	**928**
Corsa-C 1,0	2000-03	CHAIN
Corsa-C 1,2	2000-03	CHAIN
Corsa-C 1,4 16V	2000-03	**932**
Corsa-C 1,6	2000-03	**936**
Corsa-C 1,8 16V	2000-03	**932**
Frontera Sport 2,0 water pump tensioner	1991-93	**938**
Frontera Sport 2,0 automatic tensioner	1992-98	**942**
Frontera 2,2 16V	1995-00	**966**
Frontera 2,2 16V	2000-03	CHAIN
Frontera 2,4	1991-95	CHAIN
Frontera 3,2 V6	1998-03	**968**
Kadett-E 1,3/1,4/1,6/1,8/2,0	1984-91	**938**
Kadett-E 1,4/1,6	1991	**926**
Kadett-E 2,0 16V	1986-91	**940**
Manta-B 1,8	1983-88	**938**
Manta-B 2,0	1981-88	CHAIN

Monterey 3,2 V6 24V	1994-98	**968**
Nova 1,0	1982-93	CHAIN
Nova 1,2/1,4/1,6 *additional automatic tensioner*	1991-93	**926**
Nova GTE *additional automatic tensioner*	1991-93	**926**
Omega-B 2,0	1994-96	**942**
Omega-B 2,0 16V	1994-00	**944**
Omega-B 2,2 16V	1999-03	**966**
Omega-B 2,5 V6 *without raised outer edge on tensioner pulley*	1994-96	**958**
Omega-B 2,5 V6 *with raised outer edge on tensioner pulley*	1996-00	**962**
Omega-B 2,6 V6	2000-03	**962**
Omega-B 3,0 V6 *without raised outer edge on tensioner pulley*	1994-96	**958**
Omega-B 3,0 V6 *with raised outer edge on tensioner pulley*	1996-00	**962**
Omega-B 3,2 V6	2000-03	**962**
Rekord 1,8/2,0	1986-94	**938**
Senator-B 2,5/2,6/3,0	1987-93	CHAIN
Senator-B 3,0 24V	1989-93	CHAIN
Senator/Monza 2,5	1980-84	CHAIN
Senator/Monza 3,0	1980-84	CHAIN
Senator/Monza/Royale 2,8	1982-83	CHAIN
Sintra 2,2 16V	1996-99	**966**
Sintra 3,0 V6 *without raised outer edge on tensioner pulley*	1996	**958**
Sintra 3,0 V6 *with raised outer edge on tensioner pulley*	1996-99	**962**
Tigra 1,4/1,6	1998-00	**926**
Tigra 1,4/1,6 16V	1995-00	**930**
Vectra-A 1,3/1,4/1,6/1,8/,2,0 *water pump tensioner*	1982-93	**938**
Vectra-A 1,4/1,6 *additional automatic tensioner*	1991-95	**926**
Vectra-A 1,6/1,8/2,0 *automatic tensioner*	1990-95	**942**
Vectra-A 2,0 16V *manual tensioner*	1988-92	**940**
Vectra-A 2,0 16V *automatic tensioner*	1992-95	**944**
Vectra-A 2,5 V6	1993-95	**954**
Vectra-B 1,6 *16LZ2 engine*	1995-01	**942**
Vectra-B 1,6 *X16SZR engine*	1995-00	**936**
Vectra-B 1,6 16V	1995-02	**932**
Vectra-B 1,8 16V *X18XE1 engine*	1998-00	**932**
Vectra-B 1,8/2,0 16V	1995-02	**946**
Vectra-B 2,0	1995-00	**942**
Vectra-B 2,2	2000-02	CHAIN
Vectra-B 2,5 V6 *without raised outer edge on tensioner pulley*	1995-96	**958**
Vectra-B 2,5 V6 *with raised outer edge on tensioner pulley*	1996-00	**962**
Vectra-B 2,6 V6	2000-02	**962**
VX220	2000-03	CHAIN
Zafira 1,6/1,8 16V	1998-03	**932**
Zafira 2,0 Turbo	2000-03	**950**
Zafira 2,2 16V	2000-03	CHAIN

■ **Diesel models**

Arena 1,9 D	1997-01	**996**
Arena 2,5 D	1997-01	**998**
Ascona-C 1,6 D *water pump tensioner*	1984-88	**976**
Astra-E 1,6/1,7 D	1986-91	**976**
Astra-F 1,7 D/TD *17DR/X17DTL engines*	1992-98	**978**
Astra-F 1,7 TD *17DT/X17DT engine*	1994-98	**972**
Astra-G 1,7 TD *X17DTL engine*	1998-00	**982**
Astra-G 1,7 TD *Y17DT/Y17DTI engine*	2000-03	**974**
Astra-G 2,0 TD	1998-03	CHAIN
Astravan/Astramax 1,6 D	1986-89	**976**
Astravan/Astramax 1,7 D *17D engine*	1989-96	**976**
Astravan 1,7 D *17DR engine*	1992-96	**978**
Astravan 1,7 TD	1994-98	**978**
Belmont 1,6/1,7 D	1986-91	**976**
Brava 2,5 D/TD	1990-00	**994**
Carlton/Omega-A 2,3 D/TD	1990-94	CHAIN
Cavalier 1,6/1,7 D *water pump tensioner*	1984-92	**976**
Cavalier 1,7 D *automatic tensioner*	1992-95	**978**
Cavalier 1,7 TD	1992-95	**972**
Corsa-A 1,5 D/TD	1988-93	**970**
Corsa-B 1,5 D/TD	1993-00	**970**
Corsa-B 1,7 D	1993-00	**972**
Corsa-C 1,7 TD	2000-03	**974**

Corsavan/Combo 1,7 D	1993-00	972
Kadett-E 1,6/1,7 D	1986-91	976
Frontera 2,2 TD	1998-03	CHAIN
Frontera 2,3 TD	1991-94	CHAIN
Frontera 2,5 TD	1996-98	GEARS
Frontera 2,8 TD	1995-96	986
Midi 2,0 TD	1988-95	992
Midi 2,2 D	1988-95	992
Midi 2,4 TD	1995-97	992
Monterey 3,0 TD	1998-01	988
Monterey 3,1 Turbo D	1992-98	990
Movano 2,2 DTi	2000-03	1000
Movano 2,5 D	1998-00	998
Movano 2,8 D/TD	1998-00	998
Nova 1,5 TD	1988-93	970
Omega-B 2,0/2,5 TD	1994-03	CHAIN
Senator 2,3 TD	1984-86	CHAIN
Sintra 2,2 TD	1998-00	CHAIN
Vectra-A 1,7 D *water pump tensioner*	1988-92	976
Vectra-A 1,7 D *automatic tensioner*	1992-95	978
Vectra-B 1,7 TD	1992-95	972
Vectra-B 1,7 TD	1995-96	974
Vectra-B 2,0/2,2 TD	1997-02	CHAIN
Zafira 1,7 TD	2000-03	974
Zafira 2,0 TD	1998-03	CHAIN

VOLKSWAGEN

■ **Petrol models**

Beetle 1,6	1999-03	1016
Beetle 1,8 Turbo	1999-03	1022
Beetle 2,0	1998-03	1024
Bora 1,4/1,6 16V	1998-03	1004
Bora 1,6	1998-03	1016
Bora 1,8/Turbo	1998-03	1022
Bora 2,0	1998-03	1024
Bora 2,3 V5	1998-03	CHAIN
Bora 2,8 VR6	1998-03	CHAIN
Caddy 1,4	1995-03	1002
Caddy 1,4 16V	1999-03	1004
Caddy 1,6/1,8 *1F/ADZ engine*	1996-03	1012

Caddy 1,6 *AEE engine*	1997-03	1002
Caddy 1,6/1,8 *timing mark - front of camshaft sprocket*	1988-92	1012
Caravelle 2,8 VR6	1996-02	CHAIN
Corrado 1,8/2,0 16V *manual tensioner*	1989-94	1020
Corrado 2,0	1993-95	1012
Corrado 2,0 16V *automatic tensioner*	1994-95	1028
Corrado 2,8/2,9 VR6	1992-95	CHAIN
Golf 1,1/1,3/1,4/1,6 *water pump tensioner*	1981-95	1008
Golf 1,4/1,6 *automatic tensioner*	1994-97	1002
Golf 1,4/1,6 16V	1997-03	1004
Golf 1,6/1,8/2,0 *timing mark - front of camshaft sprocket*	1988-98	1012
Golf 1,6 *AEK/AFT/AKS engines*	1994-98	1014
Golf 1,6	1997-03	1020
Golf 1,8/Turbo	1997-03	1022
Golf 2,0 *APK/AQY engines*	1997-03	1024
Golf Cabriolet 2,0	1999-03	1030
Golf 1,8/2,0 16V *manual tensioner*	1985-94	1020
Golf 2,0 16V *automatic tensioner*	1994-98	1028
Golf 2,3 V5	1997-03	CHAIN
Golf 2,8/2,9 VR6	1991-97	CHAIN
Golf 2,8 VR6	1997-02	CHAIN
Jetta 1,1/1,3	1980-91	1008
Jetta 1,6/1,8/2,0 *timing mark - front of camshaft sprocket*	1988-92	1012
Jetta 1,8/2,0 16V	1985-92	1020
LT 2,3	1996-03	CHAIN
LT 2,4	1983-96	1048
Lupo 1,0 *AHT engine*	1998-03	CHAIN
Lupo 1,0	1998-03	1002
Lupo 1,4 16V	1998-03	1004
Passat 1,3	1981-87	1008
Passat 1,6 *AEK/AFT engines*	1994-96	1014
Passat 1,6 *ADP engine*	1996-00	1030
Passat 1,6 *AHL/ALZ/ANA/ARM engines*	1996-03	1032

Passat 1,8/2,0 *timing mark - front of camshaft sprocket*	1988-96	**1012**
Passat 1,8	1996-00	**1034**
Passat 1,8 Turbo	1996-02	**1036**
Passat 1,8/2,0 16V *manual tensioner*	1988-94	**1020**
Passat 2,0 16V *automatic tensioner*	1994-96	**1028**
Passat 2,3 V5	1996-03	CHAIN
Passat 2,8/2,9 VR6	1991-96	CHAIN
Passat 2,8 30V	1996-02	**1038**
Passat W8 *camshaft drive*	2002-03	CHAIN
Polo 1,0/1,1/1,3/1,4/1,6 *automatic tensioner*	1994-02	**1002**
Polo 1,1/1,3 *water pump tensioner*	1981-94	**1008**
Polo 1,4 16V *AFH engine*	1995-00	**1010**
Polo 1,4 16V	1999-03	**1004**
Polo Classic 1,4	1995-02	**1002**
Polo Classic 1,4 16V	1999-02	**1004**
Polo Classic 1,6/1,8 *1F/ADZ engines*	1995-00	**1012**
Polo Classic 1,6 *AFT engine*	1996-00	**1014**
Polo Classic 1,6 *AEE engine*	1997-00	**1002**
Polo Classic 1,6 *AEH/AKL/APF/AUR engines*	2000-02	**1016**
Polo G40	1987-94	**1008**
Santana 1,3	1981-87	**1008**
Scirocco 1,3	1981-84	**1008**
Scirocco 1,6/1,8 *timing mark - front of camshaft sprocket*	1986-93	**1012**
Scirocco 1,8 16V	1986-93	**1020**
Sharan 1,8 Turbo	1997-00	**1040**
Sharan 2,0	1995-00	**1014**
Sharan 2,8 VR6	1995-03	CHAIN
Taro 1,8/2,2	1989-91	CHAIN
Transporter 1,6/1,9/2,1	1980-92	GEARS
Transporter 1,8	1974-76	GEARS
Transporter 2,0	1975-83	GEARS
Transporter 2,0	1990-03	**1042**
Transporter 2,5	10/91-7/95	**1044**
Transporter 2,5	08/95-03	**1046**
Transporter 2,8 VR6	1996-03	CHAIN

Vento 1,4/1,6 *water pump tensioner*	1991-95	**1008**
Vento 1,4/1,6 *automatic tensioner*	1994-98	**1002**
Vento 1,6 *AEK/AFT/AKS engines*	1994-98	**1014**
Vento 1,6/1,8/2,0	1991-98	**1012**
Vento 2,0 16V *manual tensioner*	1992-94	**1020**
Vento 2,0 16V *automatic tensioner*	1994-98	**1028**
Vento 2,8 VR6	1991-98	CHAIN

■ **Diesel models**

Beetle 1,9 TDI	1998-03	**1064**
Beetle 1,9 TDI PD	2000-03	**1050**
Bora 1,9 SDI/TDI	1998-03	**1064**
Bora 1,9 TDI PD	1999-03	**1050**
Caddy 1,6 D	1984-92	**1068**
Caddy 1,7/1,9 SDI	1996-03	**1058**
Caddy 1,9 SD	1996-03	**1062**
Caddy 1,9 TDI	1999-03	**1064**
Caddy Pick-up 1,9D	1996-03	**1060**
Golf 1,5 D	1976-80	**1068**
Golf 1,6 D	1980-92	**1068**
Golf 1,6 Turbo D	1985-92	**1068**
Golf 1,9 SDI	1995-97	**1058**
Golf 1,9 D/TD *two piece injection pump sprocket*	1994-97	**1072**
Golf 1,9 D/TD/TDI *manual tensioner*	1991-95	**1068**
Golf 1,9 TD/TDI *automatic tensioner*	1994-97	**1070**
Golf 1,9 SDI/TDI	1997-03	**1064**
Golf 1,9 TDI PD	1999-03	**1050**
Golf Cabrio 1,9 TDI	1993-01	**1070**
Jetta 1,6 D	1980-92	**1068**
Jetta 1,6 Turbo D	1985-92	**1068**
LT 2,4 D/TD *auxiliary tensioner*	1992-96	**1098**
LT 2,5 SDI	1996-03	**1100**
LT 2,5 TDI	1996-03	**1100**
LT 2,8 TD/TDI	1997-03	GEARS
Lupo 3L 1,2 TDI PD	1999-03	**1050**
Lupo 1,4 TDI PD	1999-03	**1050**
Lupo 1,7 SDI	1998-03	**1054**

Passat D/TD *manual tensioner*	1976-80	**1068**
Passat 1,9 TD/TDI *automatic tensioner*	1991-96	**1070**
Passat 1,9 TD *two piece injection pump sprocket*	1994-96	**1072**
Passat 1,9 TDI *single piece injection pump sprocket*	1996-00	**1074**
Passat 1,9 TDI *two piece injection pump sprocket*	1997-00	**1076**
Passat 1,9 TDI PD	1999-03	**1080**
Passat 2,5 V6 TDI	1998-03	**1084**
Polo 1,3 D/1,4 D	1986-94	**1056**
Polo 1,4 TDI PD	1999-03	**1050**
Polo 1,7 SDI	1995-00	**1054**
Polo 1,9 SDI	1995-02	**1054**
Polo 1,9 D	1995-02	**1060**
Polo 1,9 TDI	1999-03	**1064**
Polo Classic 1,7/1,9 SDI	1996-02	**1058**
Polo Classic 1,9 SD	1996-00	**1062**
Polo Classic 1,9 TDI	1997-02	**1058**
Polo Classic 1,9 SDI/TDI	1999-03	**1064**
Santana D/TD	1982-88	**1068**
Sharan 1,9 TDI	1995-00	**1074**
Sharan 1,9 TDI PD	1999-03	**1050**
Transporter D/TD *manual tensioner*	1981-94	**1068**
Transporter 1,9 TD *two piece injection pump sprocket*	1994-03	**1072**
Transporter 2,4 D	1990-01/95	**1088**
Transporter 2,4 D	02/95-03	**1090**
Transporter 2,5 TDI	1995-03	**1094**
Vento 1,9 SDI	1995-98	**1058**
Vento 1,9 D/TD *two piece injection pump sprocket*	1994-98	**1072**
Vento 1,9 D/TD/TDI *manual tensioner*	1991-95	**1068**
Vento 1,9 TD/TDI *automatic tensioner*	1994-98	**1070**

VOLVO

■ **Petrol models**

240 2,0/2,3	1985-92	**1106**
262/264/265 2,8 V6	1981-85	CHAIN
340 1,6/1,7/1,8/2,0	1984-91	**1104**

340/343/345 1,4	1980-82	CHAIN
360	1985-89	**1106**
440 1,6/1,7/1,8/2,0	1988-97	**1104**
460 1,6/1,7/1,8/2,0	1989-97	**1104**
480 1,6/1,7/1,8/2,0	1986-95	**1104**
740/Turbo 2,0/2,3	1985-91	**1106**
740 2,3 16V	1990-96	**1120**
760 2,3 Turbo	1983-91	**1106**
760 2,8 V6	1982-91	CHAIN
850 2,0 10V/20V	1991-97	**1124**
850 2,0 Turbo	1994-97	**1124**
850 2,3 Turbo	1993-97	**1124**
850 2,5 10V/20V	1991-97	**1124**
850 2,5 Turbo	1996-97	**1124**
850 R	1996-97	**1124**
940 16V	1991-95	**1120**
940/Turbo 2,0/2,3	1990-97	**1106**
960 2,3 Turbo	1990-97	**1106**
960 2,5 24V	1995-97	**1124**
960 3,0 24V	1990-97	**1124**
S40/V40 1,6/1,8/2,0 *hydraulic tensioner*	1996-99	**1108**
S40/V40 1,6/1,8/2,0 *mechanical tensioner*	1999-03	**1110**
S40/V40 1,8 GDI	1998-02	**1112**
S40/V40 1,9/2,0 Turbo *hydraulic tensioner*	1998-99	**1108**
S40/V40 1,9/2,0 Turbo *mechanical tensioner*	1999-03	**1110**
S60 2,0/2,3 Turbo	2000-03	**1116**
S60 2,4/Turbo	2000-03	**1116**
S70/V70 2,0 10V *hydraulic tensioner*	1997-98	**1124**
S70/V70 2,0 10V *mechanical tensioner*	1998-99	**1126**
S70/V70/C70 2,0 Turbo *hydraulic tensioner*	1997-98	**1124**
S70/V70/C70 2,0 Turbo *mechanical tensioner*	1998	**1126**
S70/V70/C70 2,0 Turbo *variable valve timing*	1999-03	**1116**
S70/V70 2,3 20V *hydraulic tensioner*	1997-98	**1124**
S70/V70 2,3 20V *mechanical tensioner*	1998	**1126**
S70/V70 2,3 20V	1999	**1116**
S70/V70/C70 2,3 Turbo *hydraulic tensioner*	1997-98	**1124**

S70/V70/C70 2,3 Turbo *mechanical tensioner*	1998	**1126**
S70/V70/C70 2,3 Turbo	1999-03	**1116**
S70/V70/C70 2,4/Turbo	2000-03	**1116**
S70/V70/C70 2,5 10V/20V *hydraulic tensioner*	1997-98	**1124**
S70/V70/C70 2,5 10V *mechanical tensioner*	1998-99	**1126**
S70/V70/C70 2,5 20V *mechanical tensioner*	1998	**1126**
S70/V70/C70 2,5 20V	1999	**1116**
S70/V70/C70 2,5 Turbo *hydraulic tensioner*	1997-98	**1124**
S70/V70/C70 2,5 Turbo *mechanical tensioner*	1998	**1126**
S70/V70/C70 2,5 Turbo	1999	**1116**
S80 2,0/2,3/2,4 Turbo	1998-03	**1116**
S80 2,4	1998-03	**1116**
S80 2,8 Turbo	1998-99	**1128**
S80 2,8 Turbo	2000-01	**1116**
S80 2,9	1998-99	**1128**
S80 2,9	2000-03	**1116**
S80 2,9 Turbo	2002-03	**1116**
S90/V90 3,0 24V	1997-99	**1124**
V70 XC (Cross Country) 2,0 Turbo	1997-99	**1116**
V70 XC (Cross Country) 2,3/2,4 Turbo	2000-03	**1116**
V70 XC (Cross Country) 2,5 Turbo *hydraulic tensioner*	1997-98	**1124**
V70 XC (Cross Country) 2,5 Turbo *mechanical tensioner*	1998-99	**1126**

■ **Diesel models**

240 D	1979-89	**1142**
440 1,9 TD	1994-97	**1130**
460 1,9 TD	1994-97	**1130**
740 D	1984-90	**1142**
740 D Turbo	1985-90	**1142**
760 D Turbo	1983-90	**1142**
850 2,5 TDI	1996-97	**1144**
940 D/Turbo *without EGR*	1990-94	**1142**
940 D Turbo *with EGR*	1993-96	**1148**
960 D Turbo	1990-96	**1142**
S40/V40 1,9 TD *D4192T engine*	1996-99	**1132**
S40/V40 1,9 TD *D4192T2 engine*	1999-00	**1134**
S40/V40 1,9 TD *D4192T3/4 engine*	2001-03	**1138**
S70/V70 2,5 TDI	1996-01	**1144**
S80 2,5 TDI	1998-01	**1144**

Autodata

■ Engine damage - where it is stated that engine damage will result from belt failure it is still possible that by chance no damage will have occurred, therefore before removing the cylinder head check the compressions.

■ Before disconnecting the battery earth lead ascertain if the vehicle is fitted with a coded radio. If it is, ensure that the owner has a record of the code or that an auxiliary power supply is connected to the set.

■ Always disconnect the battery earth lead before starting work.

■ Remove spark plugs (petrol engines) or glow plugs (diesel engines), to ease turning the engine.

■ Always turn the engine in normal direction of rotation (clockwise - unless otherwise stated).

■ Do NOT turn the camshaft, crankshaft or diesel injection pump, once the toothed belt has been removed (unless otherwise stated).

■ Do NOT use timing pins to lock the engine when slackening or tightening crankshaft pulley bolt(s).

■ Do NOT turn the crankshaft from the camshaft or other drive sprockets via the drive belt.

■ Do NOT use cleaning fluids on belts, sprockets or rollers.

■ Ensure that the replacement belt has the correct tooth profile. Trapezoidal, curvilinear and modified curvilinear types are NOT interchangeable.

■ Do NOT forcibly twist the belt, turn inside out, or bend through a radius of less than 25 mm.

■ Check the pulley alignment.

■ Check the free running of auxiliary drives, such as water pump, oil pump and balancer shaft.

■ Check the free running of tensioner and guide rollers.

■ Always mark the belt with the direction of running before removal.

■ Always refit a used belt so that its original direction of running is maintained.

■ Do NOT lever or force the belt onto its sprockets.

■ Always check the diesel injection pump timing, after replacing the drive belt - refer to AUTODATA Diesel Injection Manual.

■ Observe all tightening torques.

■ Check the ignition timing after belt replacement - see AUTODATA Technical Data Manual.

Abbreviations

AC	Air conditioning compressor	**GEN**	Alternator	**Nm**	Newton metres
AP	Air pump	**HP**	Hydraulic pump	**OP**	Oil pump
AT	Automatic transmission	**INT**	Intermediate shaft	**PAS**	Power assisted steering pump
AUX	Auxiliary drive shaft	**IP**	Injection pump (diesel)	**RH**	Right-hand (as seen from driver's seat facing forward)
BS	Balancer shaft	**Kg**	Kilogrammes		
CA	Camshaft	**LH**	Left-hand (as seen from driver's seat facing forward)		
CS	Crankshaft			**RWD**	Rear wheel drive
F	Cooling fan			**SC**	Supercharger
FP	Fuel pump (diesel)	**MT**	Manual transmission	**T**	Tensioner
FWD	Front wheel drive	**MY**	Model year	**VP**	Vacuum pump
G	Guide pulley	**mm**	Millimetres	**WP**	Water pump

Toothed timing belts

Since its introduction the toothed (or synchronous) belt has been increasingly used for driving camshafts, balancer shafts and diesel injection pumps, instead of the traditional roller chain or gears.

1. Construction

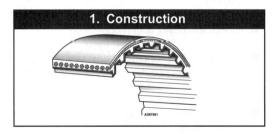

These belts are of complex construction (1), being manufactured from a fibre-glass, Kevlar or steel braided laminated inner core, coated with synthetic rubber, neoprene or highly saturated nitrile (HSN) which is wear and heat resistant.

The teeth, which may be curvilinear (2), modified curvilinear (with a more rounded form between teeth) or trapezoidal (3), are moulded integrally, to close tolerances and have a durable fabric facing for long life.

2. Rounded teeth

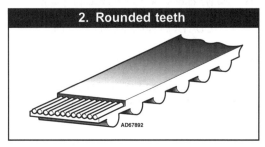

3. Trapezoidal teeth

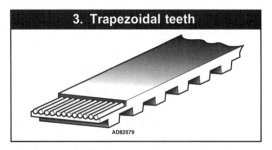

This combination of design and construction results in a belt that will stretch little in use, requires no lubrication, is relatively inexpensive to manufacture, is almost silent in use and has a very high working efficiency.

Service replacement

Important - see Timing Belt Replacement intervals on page 5

When a replacement interval is recommended by the vehicle manufacturer, this appears as a mileage or time interval in the Replacement Interval Guide box of each model-related page. These intervals should be strictly observed to avoid the possibility of belt failure and consequential expensive engine damage.

NOTE: Reference to a special, lower, recommended replacement interval for vehicles used under arduous or adverse conditions relates to the following types of use:

Taxi operations
Permanent door-to-door use.
Frequent short trips with a cold engine in low temperatures.
Use in hot countries with temperatures often over +30 degrees C.
Use in cold countries with temperatures often below -15 degrees C.
Use in countries with a dusty atmosphere.
Towing a trailer or caravan.
Sustained high-speed driving.
Using poor quality fuel or oil.

If there is no manufacturer's recommended replacement interval, this does not mean that the belt can be ignored, or that it will last forever.

Belts should be inspected at regular intervals and always replaced if their condition is in any way suspect.

Contamination

In use the belt is protected from oil or water contamination by a cover, but should a seal or hose fail it is possible that damage to the belt could result, and in such cases the belt should be replaced.

The belt should not be allowed to come into contact with petrol, water, or oil and under no circumstances should any solvents be used to clean it.

If any doubt exists about the condition of the belt it should be replaced as the cost is low compared to the cost for repairing engine damage resulting from belt failure.

Inspection

Important - see Timing Belt Replacement intervals on page 5

At the recommended service intervals, and whenever the timing belt is removed, it should be carefully inspected for wear or damage, however minor, which could lead to failure - with possibly expensive results.

WARNING: In the majority of cases, failure of the drive belt will result in piston and valve contact with resulting engine damage.

Timing belt damage may be visible as cracking or scuffing on the outer surface (4), possibly caused by deposits on the tensioner roller, which often runs on the back of the belt, or by the tensioner binding at some time. Any such damage should be investigated further, to ascertain possible causes before fitting a new belt.

4. Cracking and scuffing

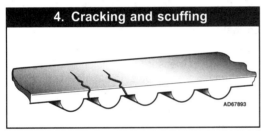

AD67893

5. Damaged teeth

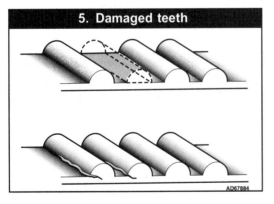

AD67884

The teeth should be checked for signs of cracking or other damage (5), and the sides of the belt also inspected for wear or damage (6) which could indicate that the sprockets on which it runs, may be out of alignment.

Cracking or damage to teeth may indicate that the camshaft or one of the ancillaries, such as the water pump, which may be driven by the belt, has locked up, if only briefly. Therefore these should be checked before replacing the belt.

6. Side wear and damage

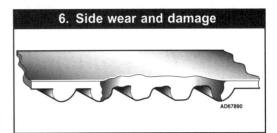

AD67890

The teeth on the sprockets should also be checked for damage, and cleaned only with a soft brush.

A wire brush must not be used, nor any form of metal scraper. If there is a build up of dirt or dust in the corners of the teeth, this may be gently removed with a soft wooden scraper.

Cleaning

Solvents should never be used to clean oil deposits from the surface of the belt, and if in any doubt, the belt should be replaced.

Any cleaning of the belt must be undertaken carefully using a dry soft bristle brush, such as a toothbrush. The belt should be laid on a flat surface and care taken not to twist or crush the belt.

Under no circumstances should the belt be turned inside out for cleaning or inspection. Any maltreatment of the belt could result in premature failure.

Fitting

When fitting a belt, the tensioner must be released, and the belt slid into place.

It may be necessary to stretch the belt slightly over the first sprocket, ensuring that the timing marks remain aligned.

Under no circumstances should any form of lever be applied to the belt to force it into place. Once fitted, the engine should always be turned in the direction of normal rotation (except in special cases, where instructed in the text), never backwards as this could cause the belt to slip and the timing to 'jump'.

At each stage of belt fitting, carefully check the timing marks are aligned correctly.

Some belts have timing marks identified which will relate to marks on the sprockets (7). These may be used in conjunction with other timing marks on the engine castings and sprockets, or may be used solely as the timing reference marks. Again, the specific fitting instructions must be followed.

7. Belt timing marks

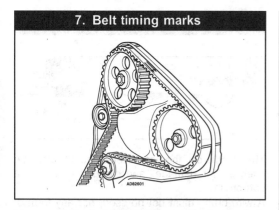

AD82601

Never use the belt to lock the camshaft sprockets when undoing the sprocket retaining bolts, as this will damage the belt teeth.

Use a sprocket holding tool, or the hexagons or flats which are provided on some camshafts for this purpose.

8. Direction arrows

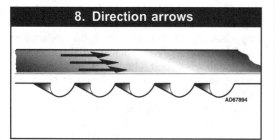

AD67894

Certain manufacturers specify the direction of use for the belt, this being identified by arrows on the top surface (8), and should be strictly observed. If removing a belt for any reason which may be re-used, the direction of rotation should be marked on it, in chalk, for reference when re-fitting.

Any recommended belt checking or replacement intervals should be followed.

Tension

The tension of the belt is important and can be set in a number of different ways.

Engine manufacturers issue specific instructions for each application, which are described in each section. These should always be followed, as incorrect tension can lead to premature failure.

In many cases the tensioner operates automatically, but the installation procedure must be carefully followed to ensure that the correct tension is achieved.

Manually tensioned belts normally require the use of a gauge to measure the tension and these are listed under special tools, in each section.

In some cases an alternative, such as the Burroughs tension gauge can be used and a comparison table, between this and some other gauges, is included in this manual.

Auxiliary drive belts

V-Belts

V-belts used for driving auxiliary components, such as alternator, water pump, air conditioning compressor and cooling fan are normally of 'raw-edged' construction.

In this type of belt, sub-surface fibres are laid perpendicular to the direction of belt rotation (9),

9. V-belt construction

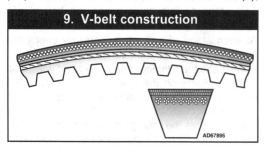

AD67895

10. V-belt & pulley contact

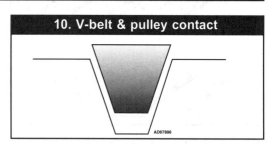

AD67896

providing a high degree of flexibility, but having extreme transverse stiffness and high wear resistance.

Drive is transmitted from the crankshaft pulley through the side wall contact between the belt and the pulley 'V' groove (10).

Several belts may be used, where the loads are high and there are several auxiliary components to be driven.

Poly V-Belts

Introduced in 1979, Poly V (or Micro-V) belts are now fitted to an increasing number of engines.

These belts are considerably wider and thinner than their V-belt counterpart and usually have between three and six ribs (11).

Additionally the back of the belt can be used to drive components or for tensioning purposes.

Due to its reduced thickness, compared to the traditional V-belt, it can drive smaller diameter pulleys, resulting in the so-called 'serpentine drive' layout (12), in which all the auxiliaries are driven by one long belt.

11. Poly V-belt construction

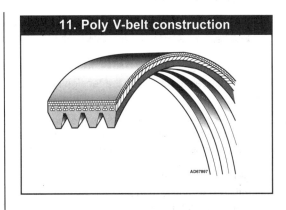

12. Typical serpentine drive belt layout

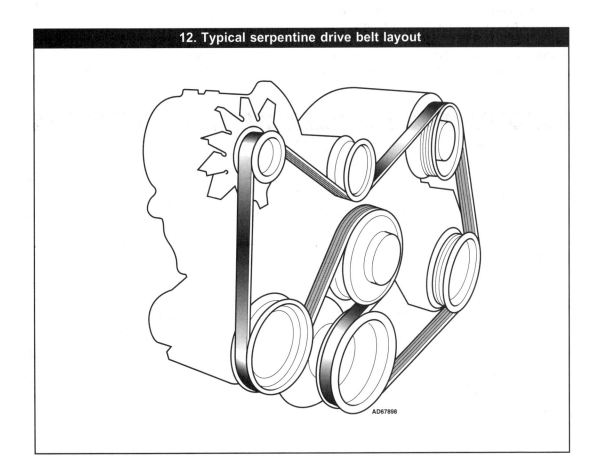

Belt tension measurement

Vehicle manufacturers specify a wide range of different gauges for checking the tension in timing belts.

It is not possible to directly measure the tension in an installed belt so most tension gauges measure belt deflection for a known load or, in some cases, the load for a known deflection. A few manufacturers specify special tools which are essentially a torque arm to load the belt tensioner by a predetermined amount.

To assist workshops offering all-makes service Autodata has prepared a comparison chart for a range of the gauges specified by different car manufacturers. Using this chart it is possible to convert readings from one gauge to another,

enabling one gauge to be used for a variety of makes.

The tests were carried out by comparing known belt tensions with a Burroughs gauge, calibrated in Newtons and other gauges calibrated in arbitrary units.

The restricted ranges on some gauges has reduced the readings available for comparison.

With any tension gauge it is important to take care when using the gauge, following the manufacturer's instructions.

It is often difficult to obtain consistent readings with these gauges and it is good practice to take several readings to establish the correct setting.

Tension gauge comparison

Burroughs (BT33-73F/ BT33-86J)	Sykes-Pickavant 316690	SEEM C.Tronic G2 105.5	SEEM C.Tronic 87 (Pin S)	Lowener (Ford 21-113 & Vauxhall 510-2)	Peiseler (VAG 210, Volvo 5197 & SEAT U.10.028)	BMW (11.2.080) & (Volvo 9988500*)	Burroughs/ Rover - (KM 4088AR)
Newtons	Kg	Units	Units	Units	Units	Units	Units
50			13				
100			22,5				
150		33	31,5			28	
200	16	40	38,5		11,7	35	1,7
250	20	47	46	4	12,5	41,5	2,7
300	25	53	52	7,4	13,1	46,5	4
350	29	61	58	9,6	13,4	51	5
400	35	67	63	11	13,6	53	6
450	40	72	68,5	12			7
500	46	78	73,5				7,5
550	51						8
600	58						8,5
650	63						9
670	65						

*NOTE: This gauge is sensitive to belt thickness and should only be used as recommended by the vehicle manufacturer.

Tension gauge şuppliers

BMW   (11.2.080)

BMW (GB) Ltd,
Ellesfield Avenue,
Bracknell,
Berks.,
RG12 4TA
Tel: 01344 426565

Burroughs   (BT-33-73F/86J)

Snap-On Tools Ltd,
Telford Way,
Kettering,
Northants,
NN16 85N
Tel: 01536 413800

Clavis   (Tension meter type 4)

Integrated Display Systems Ltd.,
Maurice Road,
Wallsend,
Tyne and Wear,
NE28 6BY
Tel: 0191 2620091

Lowener   (Ford 21-113/Vauxhall 510-2)

V.L. Churchill Ltd,
SPX Corporation,
PO Box 38,
London Road,
Daventry,
Northants,
NN11 4YT
Tel: 01327 704461

SEEM   (C. Tronic 87/G2 105.5)

Sud Est Electro Mécanique,
Lot n°1 - ZAC Saint Estève,
06640 Saint Jeannet,
France
Tel: 0033 4 92 12 04 66

SEAT   (U.10.028)

SEAT (UK) Ltd,
Yeomans Drive,
Blakelands,
Milton Keynes,
MK14 5AN
Tel: 01908 679121

Sykes-Pickavant   (316690)

Sykes-Pickavant Ltd,
Kilnhouse Lane,
Lytham St. Annes,
Lancashire,
FY8 3DU
Tel: 01253 783400

Volkswagen   (VAG 210)

VAG (UK) Ltd,
Yeomans Drive,
Blakelands,
Milton Keynes,
MK14 5AN
Tel: 01908 679121

Volvo   (5197/9988500)

Volvo Cars (UK) Ltd,
Globe Business Park,
Marlow,
Bucks
SL7 1YQ
Tel: 01628 6477977

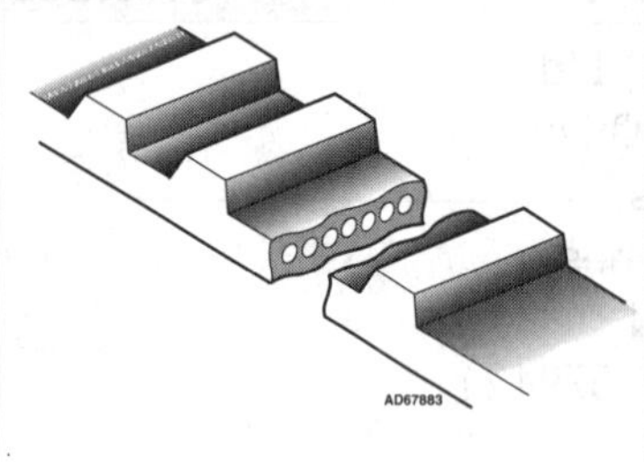

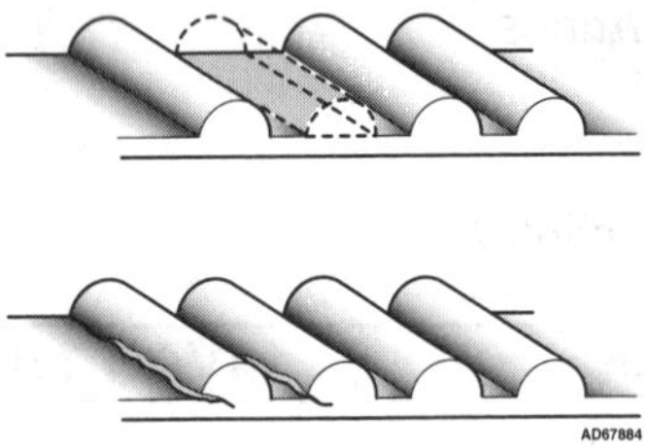

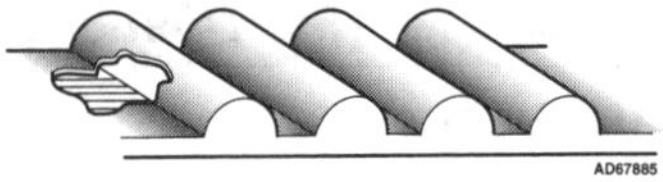

Symptom	Probable Cause	Remedy

Broken belt

AD67883

☐ Foreign body in drive	■ Ensure cover is fitted correctly
☐ Excessive tension	■ Set tension correctly
☐ Belt crimped during installation	■ Avoid mishandling belt

Sheared teeth

AD67884

☐ Insufficient tension	■ Set tension correctly
☐ Seized drive sprocket	■ Remedy seizure
☐ Sprockets misaligned	■ Align correctly

Worn teeth

AD67885

☐ Incorrect tension	■ Set tension correctly
☐ Worn sprockets	■ Replace sprockets

Hollow teeth

AD67886

☐ Pre-set tension very low	■ Set tension correctly
☐ Running tension too low	■ Ensure tensioner is operating correctly

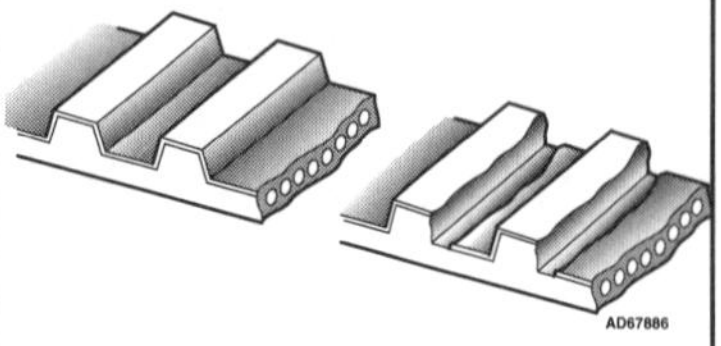

Symptom	Probable Cause	Remedy
Top surface cracks	☐ Excessive heat	■ Check cause
	☐ Tensioner pulley/guide pulley binding	■ Free tensioner pulley/guide pulley
Land wear	☐ Excessive tension	■ Set tension correctly
	☐ Rough sprocket surface	■ Replace sprockets
Oil contamination	☐ Engine oil leak	■ Remedy oil leak
Edge wear	☐ Sprocket flange damaged	■ Replace sprocket
	☐ Sprockets misaligned	■ Align correctly
Noisy operation	☐ Excessive tension	■ Set tension correctly
	☐ Insufficient tension	■ Set tension correctly
	☐ Sprockets misaligned	■ Align correctly
	☐ Sprocket flange damaged	■ Replace sprocket

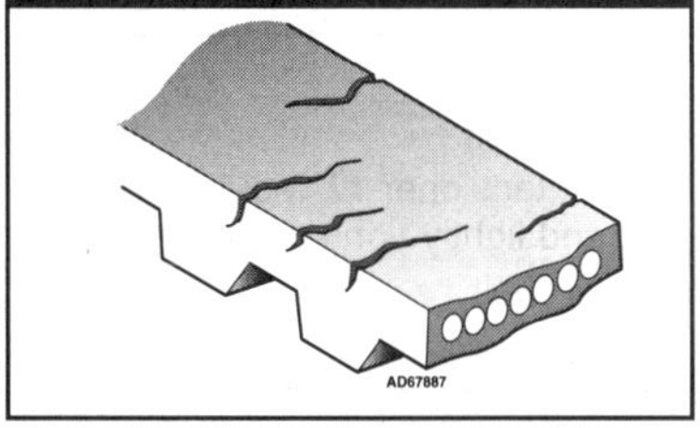

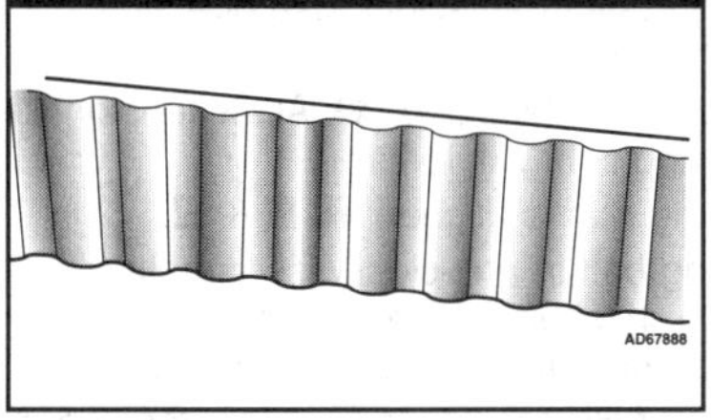

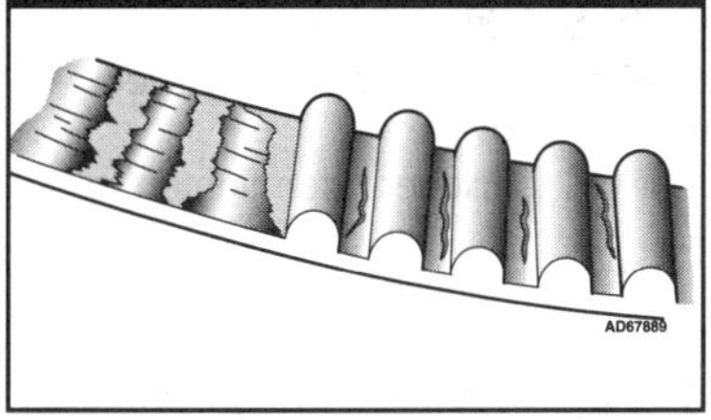

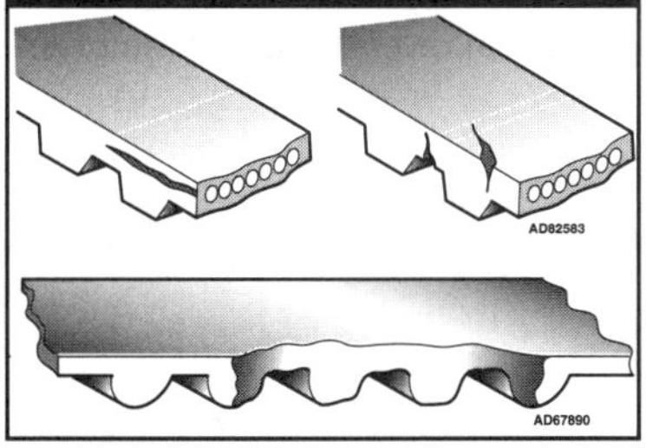

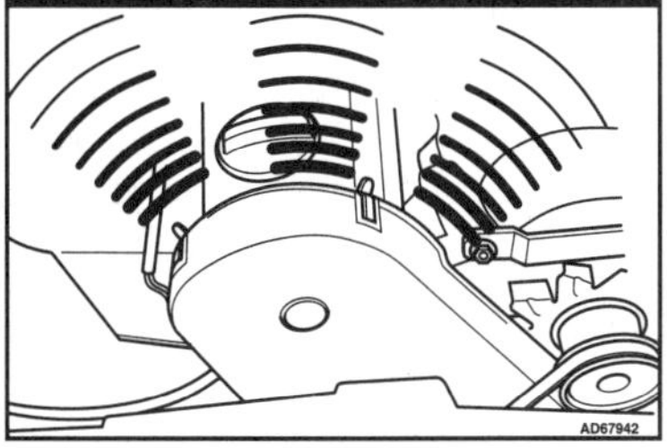

AD67887
AD67888
AD67889
AD82583
AD67890
AD67942

ALFA ROMEO

Model:	145 1,3/1,6 8V • 146 1,3/1,6 8V
Year:	1994-97
Engine Code:	AR 332.01, 335.01

Replacement Interval Guide

Alfa Romeo recommend check & adjust every 48,000 miles and replacement every 72,000 miles.
The previous use and service history of the vehicle must always be taken into account.
Refer to Timing Belt Replacement Intervals at the front of this manual.

Check For Engine Damage

CAUTION: This engine has been identified as an INTERFERENCE engine in which the possibility of valve-to-piston damage in the event of a timing belt failure is MOST LIKELY to occur.
A compression check of all cylinders should be performed before removing the cylinder head.

Repair Times – hrs

Check & adjust	0,90
Remove & install	1,35

Special Tools

■ None required.

Special Precautions

■ Disconnect battery earth lead.
■ DO NOT turn crankshaft or camshaft when timing belt removed.
■ Remove spark plugs to ease turning engine.
■ Turn engine in normal direction of rotation (unless otherwise stated).
■ DO NOT turn engine via camshaft or other sprockets.
■ Observe all tightening torques.

Removal

1. Remove:
   ❏ Cooling fan.
   ❏ Air filter intake pipe.
   ❏ Spark plugs.
2. Turn crankshaft clockwise until flywheel timing marks and timing marks on camshafts aligned **1**, **2** & **3**.
   *NOTE: Timing marks of camshafts aligned when sprocket tooth with half-circle marks either side visible through hole in timing belt rear cover **2** & **3**.*
3. Remove:
   ❏ Auxiliary drive belts.
   ❏ Water pump pulley **4**.
   ❏ Timing belt covers **5** & **6**.
4. Slacken nut of RH tensioner **7**. Move tensioner away from belt and lightly tighten nut.

5. Remove RH timing belt.
6. Slacken nut of LH tensioner **8**. Move tensioner away from belt and lightly tighten nut.
7. Remove LH timing belt.

Installation

1. Ensure timing marks aligned **1**, **2** & **3**.
2. Fit LH timing belt in anti-clockwise direction, starting at crankshaft sprocket.
3. Slacken nut of LH tensioner **8**. Allow tensioner to operate and lightly tighten nut.
4. Ensure timing marks aligned **1**, **2** & **3**.
5. Fit RH timing belt in anti-clockwise direction, starting at crankshaft sprocket. Ensure sprockets do not turn during this operation.
6. Slacken nut of RH tensioner **7**. Allow tensioner to operate and lightly tighten nut.
7. Turn crankshaft clockwise several times. Align timing marks **1**, **2** & **3**.
8. Turn crankshaft 90° clockwise until flywheel mark aligned with arrow **9**.
9. Slacken nut of RH tensioner, then tighten to 37-46 Nm **7**.
10. Turn crankshaft 360° clockwise until flywheel mark aligned **9**.
11. Slacken nut of LH tensioner, then tighten to 37-46 Nm **8**.
12. Turn crankshaft clockwise. Ensure timing marks aligned **1**, **2** & **3**.
13. Install components in reverse order of removal.

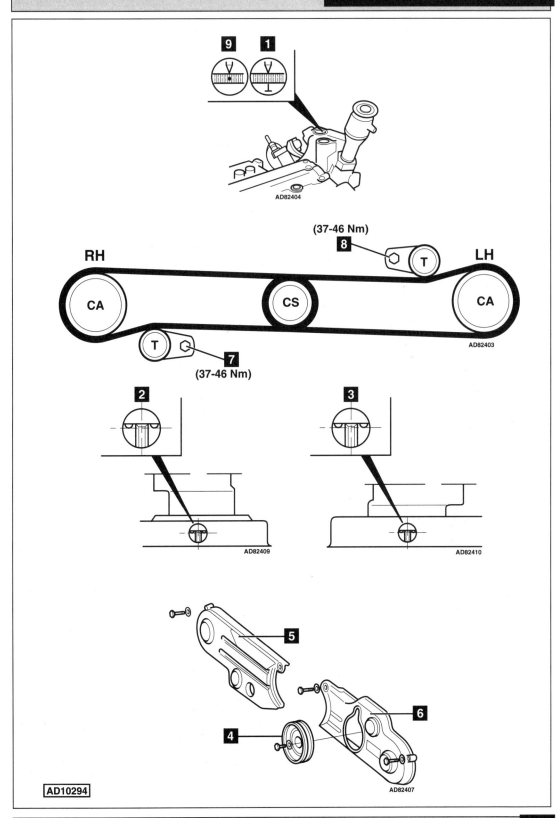

RH

LH

(37-46 Nm)

8

T

CA

CS

CA

AD82403

T

7

(37-46 Nm)

AD82404

9

1

2

AD82409

3

AD82410

5

6

4

AD82407

AD10294

ALFA ROMEO

Model:	145/146 1,4/1,6/1,8 Twin Spark • 147 1,6 Twin Spark
	155 1,6/1,8 Twin Spark • 156 1,6/1,8 Twin Spark
	Spider/GTV 1,8 Twin Spark
Year:	1996-03
Engine Code:	AR 322.01, 335.03, 671.06, 676.01, 372.03, 321.04

Replacement Interval Guide

Alfa Romeo recommend:

→12/00:
Except 156: Check every 48,000 miles or 4 years.
Replacement every 72,000 miles or 6 years.

01/01→:
Check every 36,000 miles or 3 years.
Replacement every 72,000 miles or 5 years.
The previous use and service history of the vehicle must always be taken into account.
Refer to Timing Belt Replacement Intervals at the front of this manual.

Check For Engine Damage

CAUTION: This engine has been identified as an INTERFERENCE engine in which the possibility of valve-to-piston damage in the event of a timing belt failure is MOST LIKELY to occur.
A compression check of all cylinders should be performed before removing the cylinder head(s).

Repair Times – hrs

Remove & install:

145/146	2,75
155	2,75
156	2,25
Spider/GTV	2,75

Special Tools

- 1,4/1,6: Camshaft locking tools –
  Alfa Romeo No.1.825.042.000.
- 1,8: Camshaft locking tools –
  Alfa Romeo No.1.825.041.000.
- Except 147: Inlet camshaft holding tool –
  Alfa Romeo No.1.822.155.000.
- 147: Inlet camshaft holding tool –
  Alfa Romeo No.1.822.156.000.
- Exhaust camshaft holding tool –
  Alfa Romeo No.1.822.146.000.
- Timing belt tensioning tool –
  Alfa Romeo No.1.822.149.000.

Special Precautions

- Disconnect battery earth lead.
- DO NOT turn crankshaft or camshaft when timing belt removed.
- Remove spark plugs to ease turning engine.
- Turn engine in normal direction of rotation (unless otherwise stated).
- DO NOT turn engine via camshaft or other sprockets.
- Observe all tightening torques.

Removal

1. Remove:
   - ❑ RH front wheel.
   - ❑ Engine cover.
   - ❑ Auxiliary drive belt.
   - ❑ Auxiliary drive belt guide pulley **1**.
   - ❑ Crankshaft pulley **2**.
   - ❑ Timing belt upper cover **3**.
   - ❑ Ignition coils.
   - ❑ Cylinder head cover **4**.
   - ❑ Centre spark plug – cylinder No.1.
2. Insert dial gauge in No.1 cylinder centre plug hole **5**. Turn crankshaft slowly to TDC on No.1 cylinder.
4. Ensure marks on belt aligned with marks on sprockets **6** & **7**.
5. Slacken timing belt tensioner nut **8**.
6. Remove timing belt.

Installation

1. Ensure crankshaft at TDC on No.1 cylinder. Use dial gauge **5**.
2. Hold camshaft sprockets. Except 147: Use tool Nos.1.822.155.000 & 1.822.146.000. 147: Use tool Nos.1.822.156.000 & 1.822.146.000 **9** & **10**. Slacken bolt(s) of each camshaft sprocket.
3. Except 147: Remove third bearing cap from each camshaft **11** & **12**.
4. 147: Remove second inlet camshaft bearing cap and third exhaust camshaft bearing cap **16** & **17**.
   NOTE: Mark bearing caps before removal for identification.
5. Fit locking tools in place of bearing caps **11** & **12** or **16** & **17**. 1,4/1,6: Tool No. 1.825.042.000. 1,8: Tool No. 1.825.041.000.
   NOTE: Ensure locking tools aligned with respective cam profiles to prevent damage. Before fitting belt ensure camshaft sprockets turned fully clockwise.

➡

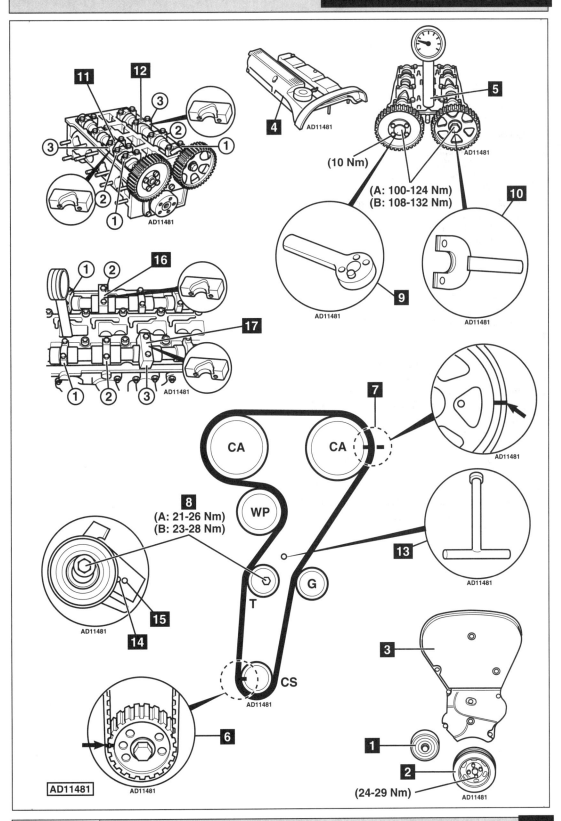

(10 Nm)

(A: 100-124 Nm)
(B: 108-132 Nm)

8
(A: 21-26 Nm)
(B: 23-28 Nm)

(24-29 Nm)

CA   CA

WP

T   G

CS

←

6. Fit timing belt in anti-clockwise direction, starting at crankshaft sprocket. Ensure directional arrows point in direction of rotation.

7. Ensure marks on belt aligned with marks on sprockets **6** & **7**.

8. Tension timing belt to maximum. Use tool No.1.822.149.000 **13**.

9. Tighten tensioner nut **8**.

10. Hold camshaft sprockets. Except 147: Use tool Nos.1.822.155.000 & 1.822.146.000. 147: Use tool Nos.1.822.156.000 & 1.822.146.000 **9** & **10**.

11. Tighten bolt(s) of each camshaft sprocket.
    ❏ (A) Except 147: 100-124 Nm.
    ❏ (B) 147: 108-132 Nm.
    ❏ M6 bolts: 10 Nm.

12. Remove:
    ❏ Dial gauge **5**.
    ❏ Locking tools **11** & **12** or **16** & **17**.

13. Fit bearing caps in correct locations.

14. Lubricate camshaft bearing cap bolts. Tighten bolts. Except 147: 13-16 Nm. 147: 14-17 Nm.

15. Turn crankshaft two turns clockwise to TDC on No.1 cylinder **6**.

16. Ensure timing marks aligned **6** & **7**.

17. Fit tensioning tool **13**. Tool No.1.822.149.000.

18. Slacken tensioner nut **8**. Turn tensioner anti-clockwise until pointer **14** aligned with hole **15**.

19. Tighten tensioner nut **8**.
    ❏ (A) Except 147: 21-26 Nm.
    ❏ (B) 147: 23-28 Nm.

20. Turn crankshaft two turns clockwise to TDC on No.1 cylinder **6**.

21. Ensure timing marks aligned **6** & **7**.

22. Install components in reverse order of removal.

23. Tighten crankshaft pulley bolts.
    Tightening torque: 24-29 Nm.

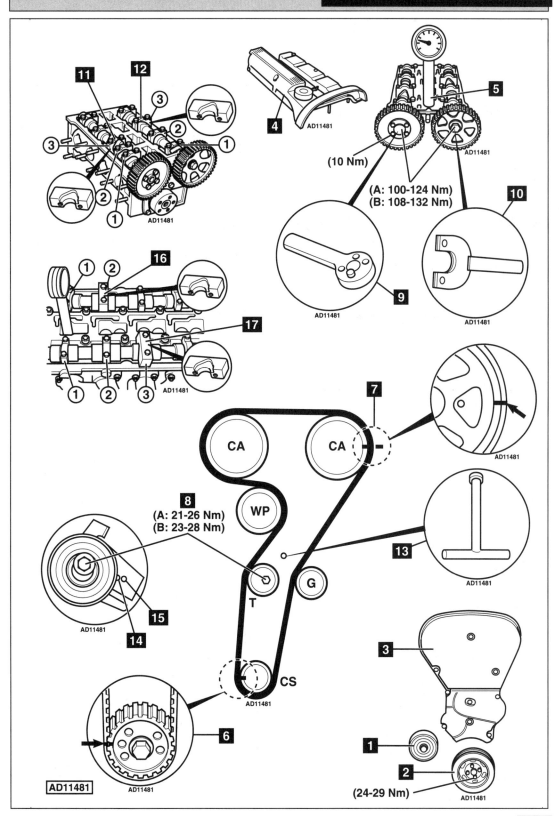

11 **12**

③ ② ① ③ ② ② ① AD11481

4 AD11481

5

(10 Nm) AD11481

(A: 100-124 Nm)
(B: 108-132 Nm)

10

9 AD11481

AD11481

① ② **16**

17

① ② ③ AD11481

7 AD11481

CA CA ─

8
(A: 21-26 Nm)
(B: 23-28 Nm)

WP

13 AD11481

T

G

15
AD11481

14

CS AD11481

3

6

AD11481

AD11481

1

2

(24-29 Nm) AD11481

ALFA ROMEO

Model:	**145 1,7 16V • 146 1,7 16V**
Year:	**1994-97**
Engine Code:	**334.01**

Replacement Interval Guide

Alfa Romeo recommend check & adjust every 48,000 miles and replacement every 72,000 miles.
The previous use and service history of the vehicle must always be taken into account.
Refer to Timing Belt Replacement Intervals at the front of this manual.

Check For Engine Damage

CAUTION: This engine has been identified as an INTERFERENCE engine in which the possibility of valve-to-piston damage in the event of a timing belt failure is MOST LIKELY to occur.
A compression check of all cylinders should be performed before removing the cylinder head.

Repair Times – hrs

Check & adjust (including AC)	2,40
Remove & install	1,50
AC	+ 0,95

Special Tools

- None required.

Special Precautions

- Disconnect battery earth lead.
- DO NOT turn crankshaft or camshaft when timing belt removed.
- Remove spark plugs to ease turning engine.
- Turn engine in normal direction of rotation (unless otherwise stated).
- DO NOT turn engine via camshaft or other sprockets.
- Observe all tightening torques.

Removal

NOTE: On models fitted with air conditioning, specialist equipment will be required to drain/recharge AC system.

1. Drain AC system (if fitted).
2. Remove:
   - Radiator grille and bumper.
   - Bonnet release cable from lock.
   - Radiator upper crossmember.
   - Coolant hose from expansion tank.
3. Disconnect radiator hose from thermostat housing.
4. Disconnect electrical connections from cooling fans.
5. Disconnect AC condenser inlet/outlet hoses (if fitted).

6. Remove:
   - Radiator complete with fans (and condenser, if fitted).
   - Auxiliary drive belts.
   - Water pump pulley **1**.
   - Timing belt covers **2**.
   - Spark plugs.
7. Turn crankshaft clockwise until flywheel timing marks and timing marks on camshaft sprockets aligned **3**, **4** & **5**.
8. Slacken nuts of RH tensioner **6**. Move tensioner away from belt and lightly tighten nuts.
9. Remove RH timing belt.
10. Slacken nuts of LH tensioner **7**. Move tensioner away from belt and lightly tighten nuts.
11. Remove LH timing belt.

Installation

1. Ensure timing marks aligned **3**, **4** & **5**.
2. Fit LH timing belt in anti-clockwise direction, starting at crankshaft sprocket. Ensure belt is taut between crankshaft sprocket, guide pulley and camshaft sprockets.
3. Slacken nuts of LH tensioner **7**. Allow tensioner to operate and lightly tighten nuts.
4. Ensure timing marks aligned **3**, **4** & **5**.
5. Fit RH timing belt in anti-clockwise direction, starting at crankshaft sprocket. Ensure belt is taut between crankshaft sprocket, guide pulley and camshaft sprockets.
6. Slacken nuts of RH tensioner **6**. Allow tensioner to operate and lightly tighten nuts.
7. Turn crankshaft several times clockwise to allow belts to settle. Ensure timing marks aligned **3**, **4** & **5**.
8. Turn crankshaft 90° clockwise until flywheel mark and tensioning marks on RH camshaft sprockets aligned **8** & **9**.
9. Slacken nuts of RH tensioner **6**. Allow tensioner to operate and tighten nuts.
   Tightening torque: 29-36 Nm.
10. Turn crankshaft 360° clockwise until flywheel mark and tensioning marks on LH camshaft sprockets aligned **8** & **10**.
11. Slacken nuts of LH tensioner **7**. Allow tensioner to operate and tighten nuts.
   Tightening torque: 29-36 Nm.
12. Turn crankshaft several times clockwise. Ensure timing marks aligned **3**, **4** & **5**.
13. Install components in reverse order of removal.
14. Recharge AC system (if fitted).

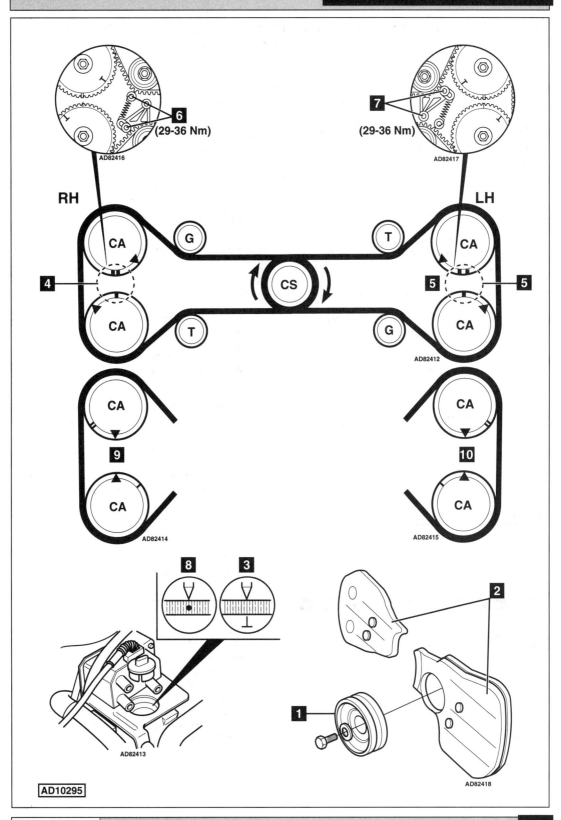

RH

LH

6 (29-36 Nm)

AD82416

7 (29-36 Nm)

AD82417

CA   G   T   CA

4   CS   5   5

CA   T   G   CA

AD82412

CA   CA

9   10

CA   CA

AD82414   AD82415

8   3

AD82413

2

1

AD82418

ALFA ROMEO

Model:	**145/146 2,0 Twin Spark • 147 2,0 Twin Spark • 155 2,0 Twin Spark 156 2,0 Twin Spark • Spider/GTV 2,0 Twin Spark**
Year:	**1995-03**
Engine Code:	**AR 323.01, 672.04, 162.01, 323.10**

Replacement Interval Guide

Alfa Romeo recommends:

Balancer shaft belt
→12/00:
Except 156: Check every 24,000 miles or
2 years.
Replacement every 72,000 miles or 6 years.
01/01→:
Check every 24,000 miles or 2 years.
Replacement every 72,000 miles or 5 years.

Timing belt
→12/00:
Except 156: Check every 48,000 miles or
4 years.
Replacement every 72,000 miles or 6 years.
01/01→:
Check every 36,000 miles or 3 years.
Replacement every 72,000 miles or 5 years.
*The previous use and service history of the vehicle
must always be taken into account.*
*Refer to Timing Belt Replacement Intervals at the front
of this manual.*

Check For Engine Damage

*CAUTION: This engine has been identified as an
INTERFERENCE engine in which the possibility of
valve-to-piston damage in the event of a timing belt
failure is MOST LIKELY to occur.*
*A compression check of all cylinders should be
performed before removing the cylinder head(s).*

Repair Times – hrs

Remove & install:

145/146/155	2,90
156	2,95
Spider/GTV	2,90

Special Tools

- Camshaft locking tools –
  Alfa Romeo No.1.825.041.000.
- Inlet camshaft holding tool –
  Alfa Romeo No.1.822.155.000.
- Exhaust camshaft holding tool –
  Alfa Romeo No.1.822.146.000.
- Timing belt tensioning tool –
  Alfa Romeo No.1.822.149.000.
- Balancer shaft belt tensioning tool –
  Alfa Romeo No.1.822.154.000.

Special Precautions

- Disconnect battery earth lead.
- DO NOT turn crankshaft or camshaft when timing belt removed.
- Remove spark plugs to ease turning engine.
- Turn engine in normal direction of rotation (unless otherwise stated).
- DO NOT turn engine via camshaft or other sprockets.
- Observe all tightening torques.

Removal

1. Remove:
   - RH wheel.
   - Engine undershield.
   - RH wheel arch liner.
   - Auxiliary drive belt and tensioner.
   - Crankshaft pulley **1**.
   - Timing belt upper cover **2**.
   - Timing belt lower cover **3**.
   - Ignition coils.
   - Cylinder head cover **5**.
   - Centre spark plug – cylinder No.1.
2. Insert dial gauge in No.1 cylinder centre plug hole **6**.
3. Turn crankshaft slowly to TDC on No.1 cylinder. Use dial gauge **6**.
4. Ensure balancer shaft marks aligned **7** & **8**.
5. Ensure marks on belt aligned with marks on sprockets **9** & **10**.
6. Slacken balancer shaft belt tensioner nut **11**.
7. Remove balancer shaft belt and sprocket from crankshaft **4**.
8. Slacken timing belt tensioner nut **12**.
9. Remove timing belt.

Installation

1. Ensure crankshaft at TDC on No.1 cylinder. Use dial gauge **6**.
2. Hold camshaft sprockets. Use tool Nos.1.822.155.000 & 1.822.146.000 **13** & **14**. Slacken bolt(s) of each camshaft sprocket.
3. Remove third bearing cap from each camshaft **15** & **16**.
   NOTE: Mark bearing caps before removal for identification.

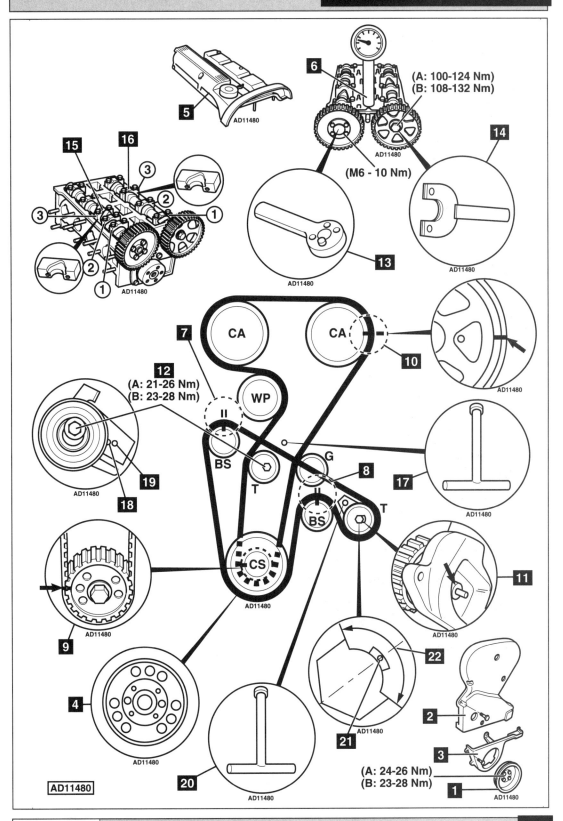

5 AD11480

6 (A: 100-124 Nm)
(B: 108-132 Nm)

AD11480

(M6 - 10 Nm)

14

15 **16**
3
2
1
3
2
1
AD11480

13 AD11480

AD11480

7 CA   CA   **10**

12
(A: 21-26 Nm)
(B: 23-28 Nm)

WP

II

19 BS
18   T
AD11480

G
8   **17**
BS   T
11
AD11480

CS
AD11480

AD11480

9

4 AD11480

22
AD11480

2

21

3
(A: 24-26 Nm)
(B: 23-28 Nm)
1   AD11480

20 AD11480

AD11480

4. Fit locking tools in place of bearing caps **15** & **16**. Tool No.1.825.041.000.

   NOTE: Ensure locking tools aligned with respective cam profiles to prevent damage. Before fitting belt ensure camshaft sprockets turned fully clockwise.

5. Fit timing belt in anti-clockwise direction, starting at crankshaft sprocket. Ensure directional arrows point in direction of rotation.

6. Ensure marks on belt aligned with marks on sprockets **9** & **10**.

7. Tension timing belt to maximum. Use tool No.1.822.149.000 **17**.

8. Tighten tensioner nut **12**.

9. Hold camshaft sprockets. Use tool Nos.1.822.155.000 & 1.822.146.000 **13** & **14**.

10. Tighten bolt(s) of each camshaft sprocket.
    - ❏ (A) Except 147: 100-124 Nm.
    - ❏ (B) 147: 108-132 Nm.
    - ❏ M6 bolts: 10 Nm.

11. Remove dial gauge and locking tools **6**, **15** & **16**.

12. Fit bearing caps in correct locations.

13. Lubricate camshaft bearing cap bolts. Tighten bolts. Except 147: 13-16 Nm. 147: 14-17 Nm.

14. Turn crankshaft two turns clockwise to TDC on No.1 cylinder **6**.

15. Fit tensioning tool **17**. Tool No.1.822.149.000.

16. Slacken tensioner nut **12**.

17. Turn tensioner anti-clockwise until pointer **18** aligned with hole **19**.

18. Tighten tensioner nut **12**.
    - ❏ (A) Except 147: 21-26 Nm.
    - ❏ (B) 147: 23-28 Nm.

19. Ensure crankshaft at TDC on No.1 cylinder **6**.

20. Align balancer shaft timing marks **7** & **8**.

21. Fit balancer shaft belt and tensioning tool No.1.822.154.000 **20**.

22. Turn tensioner until hole **21** aligned with centre of tensioner **22**. Tighten tensioner nut **11**.

23. Turn crankshaft two turns clockwise to TDC on No.1 cylinder **6**.

24. Ensure timing marks aligned **7**, **8**, **9** & **10**.

25. Install components in reverse order of removal.

26. Tighten crankshaft pulley bolts.
    - ❏ (A) Except 147: 24-26 Nm.
    - ❏ (B) 147: 23-28 Nm.

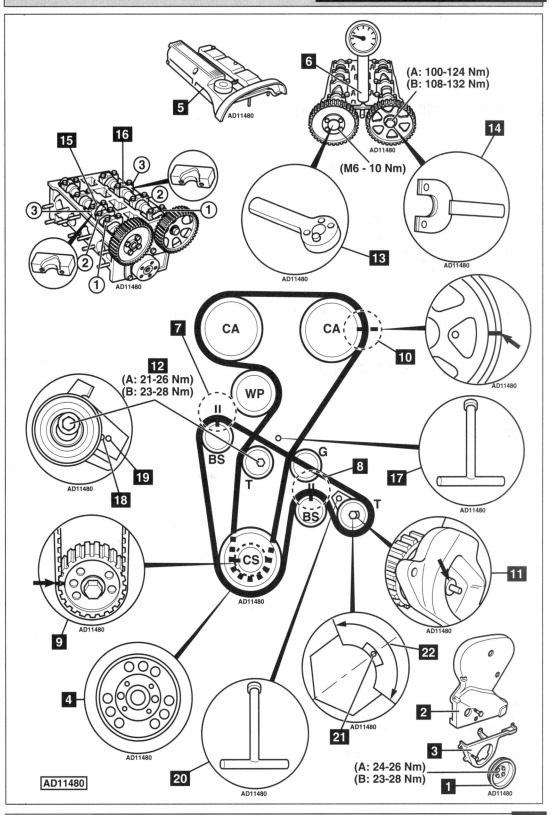

5 AD11480

6 (A: 100-124 Nm) (B: 108-132 Nm)

AD11480

(M6 - 10 Nm)

15 **16** **3** **2** **1** **3** **2** **1**

AD11480

13 AD11480

14 AD11480

7 CA CA

10 AD11480

12 (A: 21-26 Nm) (B: 23-28 Nm)

WP

II

BS

T

G

8

II

BS

T

17 AD11480

19 **18** AD11480

CS AD11480

11 AD11480

9 AD11480

4 AD11480

22 AD11480

21

2

3

20 AD11480

1 (A: 24-26 Nm) (B: 23-28 Nm)

AD11480

AD11480

ALFA ROMEO

Model:	**155 2,5i**
Year:	**1992-97**
Engine Code:	**673.01**

Replacement Interval Guide

Alfa Romeo recommend check and replace if necessary every 48,000 miles or 4 years and replacement every 72,000 miles or 6 years.
The previous use and service history of the vehicle must always be taken into account.
Refer to Timing Belt Replacement Intervals at the front of this manual.

Check For Engine Damage

CAUTION: This engine has been identified as an INTERFERENCE engine in which the possibility of valve-to-piston damage in the event of a timing belt failure is MOST LIKELY to occur.
A compression check of all cylinders should be performed before removing the cylinder head.

Repair Times – hrs

Check	0,25
Remove & install	4,00

Special Tools

■ Tensioner locking pin –
Alfa Romeo No.1.820.053.000 (A.2.0363).

Special Precautions

■ Disconnect battery earth lead.
■ DO NOT turn crankshaft or camshaft when timing belt removed.
■ Remove spark plugs to ease turning engine.
■ Turn engine in normal direction of rotation (unless otherwise stated).
■ DO NOT turn engine via camshaft or other sprockets.
■ Observe all tightening torques.

Removal

1. Remove:
   ❏ Air intake trunking.
   ❏ Idle speed control (ISC) actuator.
   ❏ Crankcase ventilation pipe.
2. Disconnect:
   ❏ Spark plug leads.
   ❏ Idle speed control (ISC) actuator multi-plug.
   ❏ Throttle position (TP) sensor multi-plug.
   ❏ Accelerator cable.
   ❏ Vacuum pipe.
   ❏ Exhaust gas recirculation (EGR) pipes.
   ❏ Coolant hose from throttle body.
   ❏ Brake servo vacuum pipe.
   ❏ Timing belt cover ventilation pipe.
3. Raise and support vehicle.

4. Remove:
   ❏ Air intake pipes.
   ❏ Timing belt upper cover **1**.
   ❏ RH front wheel.
   ❏ Inner wing panel.
   ❏ Cover from automatic tensioner unit (auxiliary drive belts) **2**.
   ❏ Auxiliary drive belts.
   ❏ Automatic tensioner unit.
   ❏ Water pump pulley.
   ❏ Timing belt lower cover **3**.
   ❏ TDC sensor with its bracket.
5. Lower vehicle.
6. Remove cylinder head covers.
7. Turn crankshaft clockwise until timing marks on camshafts aligned **4**.
8. Ensure timing mark on trigger wheel aligned with TDC pointer **5**.
9. Raise and support vehicle.
10. Lift tensioner pulley arm with screwdriver and retain in position with locking pin **6**. Tool No.1.820.053.000.
11. Slacken two tensioner pulley nuts. Turn tensioner pulley upwards and lightly tighten nuts.
12. Lower vehicle.
13. Remove timing belt.

Installation

1. Ensure timing marks aligned **4**.
2. Ensure timing mark on trigger wheel aligned with TDC pointer **5**.
3. Raise and support vehicle.
4. Fit timing belt in anti-clockwise direction, starting at crankshaft sprocket. Ensure belt is taut between sprockets.
5. Slacken two tensioner pulley nuts.
6. Turn crankshaft one turn clockwise. Tighten tensioner pulley nuts.
7. Turn crankshaft clockwise to TDC on No.1 cylinder.
8. Ensure timing marks aligned **4**.
9. Ensure crankshaft sprocket timing marks aligned **5**.
10. Lift tensioner pulley arm and remove locking pin **6**.
11. Install components in reverse order of removal.

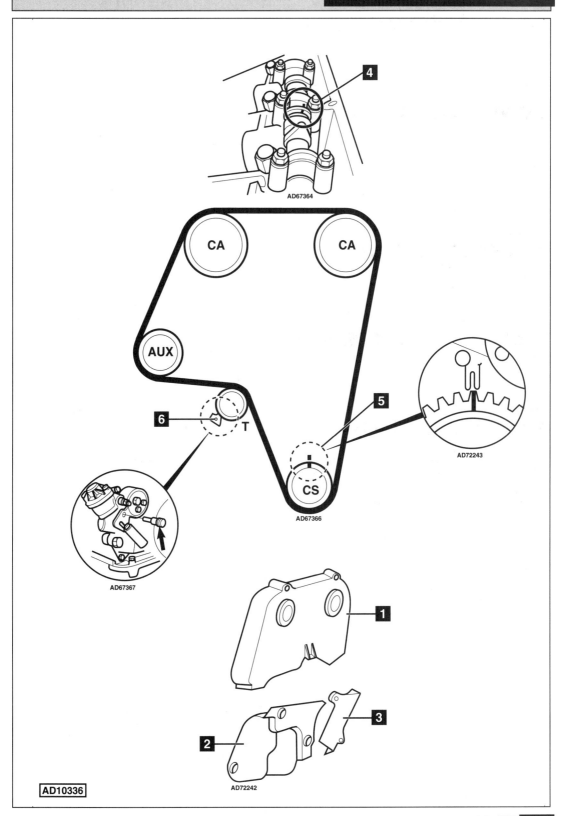

AD67364

CA

CA

AUX

6

T

5

AD72243

AD67367

CS

AD67366

1

3

2

AD72242

AD10336

ALFA ROMEO

Model:	156 2,5 V6 24V • 166 2,5 V6 24V • 166 3,0 V6 24V • GTV 3,0 V6 24V
Year:	1997-02
Engine Code:	161.02, 324.01, 342.01, 343.01

Replacement Interval Guide

Alfa Romeo recommends:
→12/00:
Replacement every 72,000 miles or 6 years.
01/01→:
Check every 36,000 miles or 3 years.
Replacement every 72,000 miles or 5 years.
The previous use and service history of the vehicle must always be taken into account.
Refer to Timing Belt Replacement Intervals at the front of this manual.

Check For Engine Damage

CAUTION: This engine has been identified as an INTERFERENCE engine in which the possibility of valve-to-piston damage in the event of a timing belt failure is MOST LIKELY to occur.
A compression check of all cylinders should be performed before removing the cylinder head(s).

Repair Times – hrs

Remove & install:	
156	4,40
166	4,40
GTV	4,50
GTV (with AC)	+0,90

Special Tools

- Camshaft locking tools – No.1825040000.
- Camshaft sprocket puller – No.1860954001.
- Crankshaft pulley locking tool – No.1870646000.
- Dial gauge and adaptor – No.1825013000.
- Flywheel locking tool (GTV 1997) – No.1820088000.
- Sprocket holding tool – No.1822146000.
- Timing belt tensioning tool – No.1860950000.

Removal

1. Raise and support front of vehicle.
2. Remove:
   - ❏ 166/GTV: Upper engine cover(s).
   - ❏ RH front wheel & inner wing panel.
3. Turn auxiliary drive belt tensioner anti-clockwise **1**.
4. Lock auxiliary drive belt tensioner. Use suitable pin.
5. Remove:
   - ❏ Auxiliary drive belt & tensioner.
   - ❏ GTV: Auxiliary drive belt guide pulley.
   - ❏ 166: PAS reservoir and bracket.
   - ❏ Inlet manifold **2**.
   - ❏ Ignition coils & cover **3** & **4**.
   - ❏ Cylinder head covers **5**.
   - ❏ Timing belt upper cover **6**.
   - ❏ Timing belt lower covers **7**.
   - ❏ Spark plug – cylinder No.1.
   - ❏ GTV 1997: Exhaust downpipe.
   - ❏ GTV 1997: Flywheel cover.
6. Except GTV 1997: Fit crankshaft pulley locking tool. Tool No.1870646000.
7. GTV 1997: Fit flywheel locking tool. Tool No.1820088000.
8. Slacken crankshaft pulley nut **20**.
9. Remove:
   - ❏ Locking tool. Tool No.1870646000 or 1820088000.
   - ❏ Crankshaft pulley **8**.
10. Fit dial gauge and adaptor to No.1 cylinder **9**. Tool No.1825013000.
11. Turn crankshaft clockwise until No.1 cylinder at TDC of compression stroke. Use dial gauge.
12. Slacken timing belt tensioner bolts **10**.
13. Remove timing belt.

Installation

1. Ensure No.1 cylinder at TDC of compression stroke. Use dial gauge **9**.
2. Remove four camshaft bearing caps **11**.
   NOTE: Mark bearing caps before removal.
3. Fit locking tools in place of bearing caps **11**. Tool No.1825040000.
   NOTE: Ensure locking tools aligned with respective cam profiles to prevent damage.
4. GTV: Remove engine steady bar and bracket.
5. Hold camshaft sprockets. Use tool No.1822146000 **12**.
6. Slacken bolt of each camshaft sprocket **13** & **14**.
7. Ensure camshaft sprockets can turn freely.
8. If not: Use puller. Tool No.1860954001.
9. Fit timing belt in anti-clockwise direction, starting at crankshaft sprocket.
10. GTV 1997: Support engine and remove:
    - ❏ Bolts securing rear gearbox mounting to crossmember.
    - ❏ Lower rear of engine slightly.
11. Remove lower alternator bolt and water pump bolt.
12. Fit tensioning tool **15**. Tool No.1860950000.
    NOTE: Ensure tensioning tool correctly located on tensioner lever **16.**
13. Turn tensioner nut **17** until notch on lever positioned below fixed mark **18**.
14. Hold camshaft sprockets. Use tool No.1822146000 **12**.
15. Tighten bolt of each camshaft sprocket **13** & **14**. Tightening torque: 68-84 Nm.
16. Remove camshaft locking tools **11** and dial gauge and adaptor **9**.
17. Lubricate and fit bearing caps in correct locations.
18. Tighten bearing cap bolts to 18-20 Nm.
19. Turn crankshaft two turns clockwise.
20. Check tensioner marks aligned **19**.
21. If not: Turn tensioner nut **17** until marks aligned.
22. Tighten tensioner bolts to 17-21 Nm **10**.
23. Remove tensioning tool **15**.
24. Install components in reverse order of removal.
25. Tighten crankshaft pulley nut **20**. Tightening torque: 200-247 Nm.

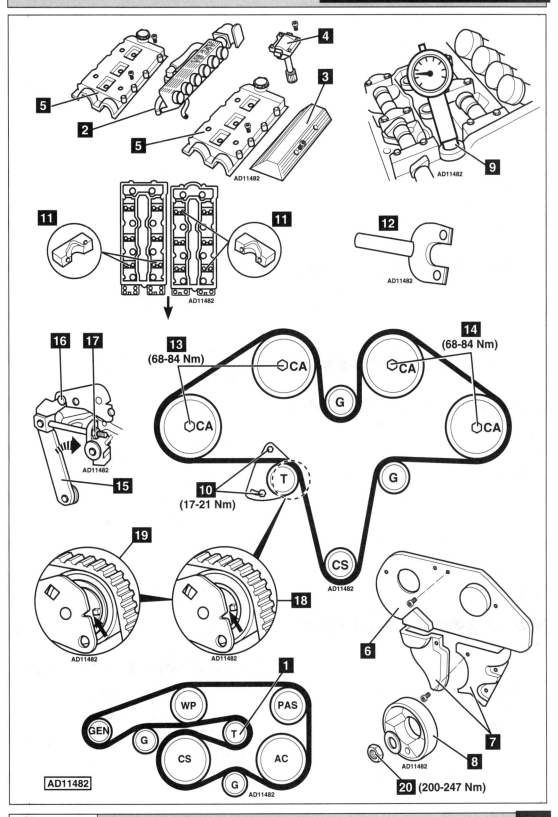

AUDI

Model:	**A2 1,4**
Year:	**2000-03**
Engine Code:	**AUA**

Replacement Interval Guide

Audi recommend check at the first 90,000 kilometres (55,923 miles) and then every 30,000 kilometres (18,641 miles) (replace if necessary).

NOTE: The vehicle manufacturer publishes this information only in kilometres. The conversion to miles is included for reference purposes only. *The previous use and service history of the vehicle must always be taken into account.*
Refer to Timing Belt Replacement Intervals at the front of this manual.

Check For Engine Damage

CAUTION: This engine has been identified as an INTERFERENCE engine in which the possibility of valve-to-piston damage in the event of a timing belt failure is MOST LIKELY to occur.
A compression check of all cylinders should be performed before removing the cylinder head(s).

Repair Times – hrs

Remove & install	2,50

Special Tools

- Camshaft locking tool – No.T10016.
- Crankshaft pulley holding tool – No.3415.

Special Precautions

- Disconnect battery earth lead.
- DO NOT turn crankshaft or camshaft when timing belt removed.
- Remove spark plugs to ease turning engine.
- Turn engine in normal direction of rotation (unless otherwise stated).
- DO NOT turn engine via camshaft or other sprockets.
- Observe all tightening torques.

Removal

Timing Belt

1. Remove bonnet:
   - ❏ Release bonnet catch in driver's compartment.
   - ❏ Release catches behind front service grille.
   - ❏ Remove bonnet. Fit protective cover to bonnet.
   - ❏ Store in horizontal position.
2. Remove:
   - ❏ Upper engine cover.
   - ❏ Air filter assembly.
   - ❏ Timing belt upper cover **1**.
3. Raise and support front of vehicle.

4. Turn crankshaft clockwise to TDC on No.1 cylinder. Ensure timing marks on crankshaft pulley aligned **2**.
5. Ensure locating holes aligned **3**.
6. If locating holes are not aligned: Turn crankshaft one turn clockwise.
7. Fit locking tool to camshaft sprockets **4**. Tool No.T10016 **5**.
   NOTE: Ensure locking tool located correctly in cylinder head.
8. Support engine.
9. Remove:
   - ❏ PAS reservoir. DO NOT disconnect hoses.
   - ❏ RH engine mounting.
   - ❏ RH engine mounting bracket.
   - ❏ RH engine undershield.
10. Lower engine until crankshaft pulley bolt accessible.
11. Remove auxiliary drive belt.
12. Fit crankshaft pulley holding tool. Tool No.3415.
13. Slacken crankshaft pulley bolt **6**.
14. Remove:
    - ❏ Holding tool. Tool No.3415.
    - ❏ Crankshaft pulley bolt **6**.
    - ❏ Crankshaft pulley **7**.
15. Fit two washers to crankshaft pulley bolt **6**.
16. Fit crankshaft pulley bolt **6**. Lightly tighten bolt.
17. Remove:
    - ❏ Auxiliary drive belt guide pulley (models with AC).
    - ❏ Auxiliary drive belt tensioner.
    - ❏ Timing belt lower cover **8**.
18. Slacken tensioner pulley bolt **9**.
19. Turn tensioner pulley anti-clockwise to release tension on belt.
20. Remove timing belt.
    NOTE: Mark direction of rotation on belt with chalk if belt is to be reused.

Installation

Timing Belt

1. Ensure locking tool fitted to camshaft sprockets **4**. Tool No.T10016 **5**.
2. Ensure timing mark on crankshaft sprocket aligned **10**.
   NOTE: Align ground tooth on crankshaft sprocket.
3. Fit timing belt in anti-clockwise direction, starting at water pump sprocket.

➡

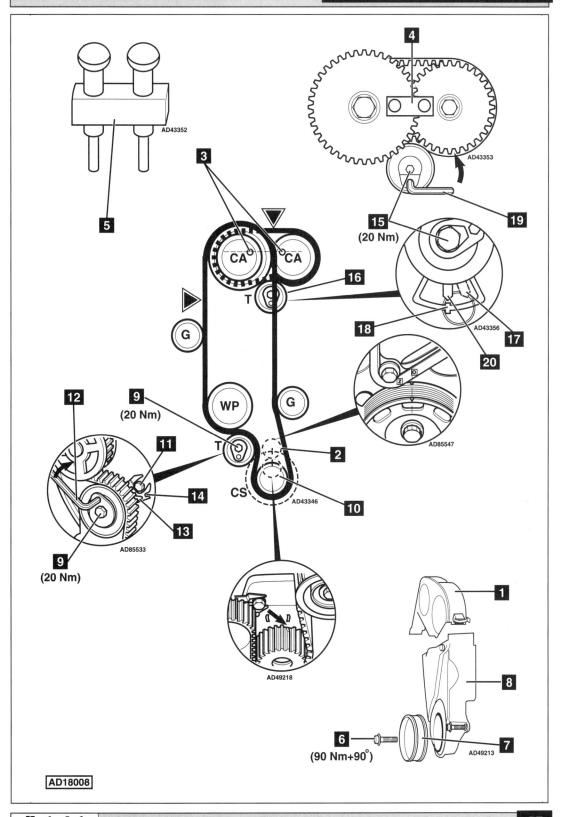

AD43352

AD43353

AD43356

AD85547

AD85533

AD49218

AD49213

AD43346

5

4

3

15
(20 Nm)

19

16

18

17

20

12

9
(20 Nm)

11

14

13

9
(20 Nm)

2

10

1

8

6
(90 Nm+90°)

7

CA   CA   T

G

WP   G

T

CS

AD18008

←

4. Tighten tensioner pulley bolt finger tight **9**. Ensure baseplate is supported by bolt **11**.
5. Turn tensioner pulley clockwise **12** until pointer **13** aligned with notch in baseplate **14**.
6. Remove locking tool from camshaft sprockets **4**. Tool No.T10016 **5**.
7. Tighten tensioner pulley bolt to 20 Nm **9**.
8. Turn crankshaft two turns clockwise to TDC on No.1 cylinder. Ensure timing mark on crankshaft sprocket aligned **10**.
9. Ensure locking tool can be inserted into camshaft sprockets **4**. Tool No.T10016 **5**.
10. Ensure pointer **13** aligned with notch in baseplate **14**.
11. If not: Repeat tensioning procedure.
12. Apply firm thumb pressure to belt at ▽. Pointer **13** and notch in baseplate **14** must move apart.
13. Release thumb pressure from belt at ▽.
14. Turn crankshaft two turns clockwise to TDC on No.1 cylinder.
15. Ensure pointer **13** aligned with notch in baseplate **14**.
16. Remove crankshaft pulley bolt **6**.
17. Install:
    - ❑ Timing belt lower cover **8**.
    - ❑ Crankshaft pulley **7**.
    - ❑ New oiled crankshaft pulley bolt **6**.
18. Fit crankshaft pulley holding tool. Tool No.3415.
19. Tighten crankshaft pulley bolt **6**. Tightening torque: 90 Nm + 90°.
20. Remove holding tool. Tool No.3415.
21. Install components in reverse order of removal.

Removal

Exhaust Camshaft Drive Belt

1. Remove timing belt as described previously.
2. Slacken tensioner pulley bolt **15**.
3. Turn tensioner pulley clockwise to release tension on belt.
4. Remove:
    - ❑ Tensioner pulley bolt **15**.
    - ❑ Tensioner pulley **16**.
    - ❑ Drive belt.

    NOTE: Mark direction of rotation on belt with chalk if belt is to be reused.

Installation

Exhaust Camshaft Drive Belt

1. Ensure locking tool fitted to camshaft sprockets **4**. Tool No.T10016 **5**.
2. Fit drive belt in clockwise direction, starting at top of inlet camshaft sprocket.
3. Ensure belt is taut between sprockets on non-tensioned side.
4. Turn tensioner pulley clockwise until pointer in position as shown **17**.
5. Install:
    - ❑ Tensioner pulley **16**.
    - ❑ Tensioner pulley bolt **15**.
6. Tighten tensioner pulley bolt finger tight **15**.
    *NOTE: Ensure lug in baseplate **18** is located in cylinder head hole.*
7. Turn tensioner anti-clockwise **19** until pointer **20** aligned with lug in baseplate **18**.
8. Tighten tensioner pulley bolt to 20 Nm **15**.
9. Fit timing belt as described previously.
10. Remove locking tool from camshaft sprockets **4**. Tool No.T10016 **5**.
11. Turn crankshaft two turns clockwise to TDC on No.1 cylinder. Ensure timing marks on crankshaft sprocket aligned **10**.
12. Ensure locking tool can be inserted into camshaft sprockets **4**. Tool No.T10016 **5**.
13. Ensure pointer **20** aligned with lug in baseplate **18**.
14. If not: Repeat tensioning procedure.
15. Apply firm thumb pressure to belt at ▽. Pointer **20** and lug in baseplate **18** must move apart.
16. Release thumb pressure from belt at ▽.
17. Turn crankshaft two turns clockwise to TDC on No.1 cylinder.
18. Ensure pointer **20** aligned with lug in baseplate **18**.
19. Install components in reverse order of removal.

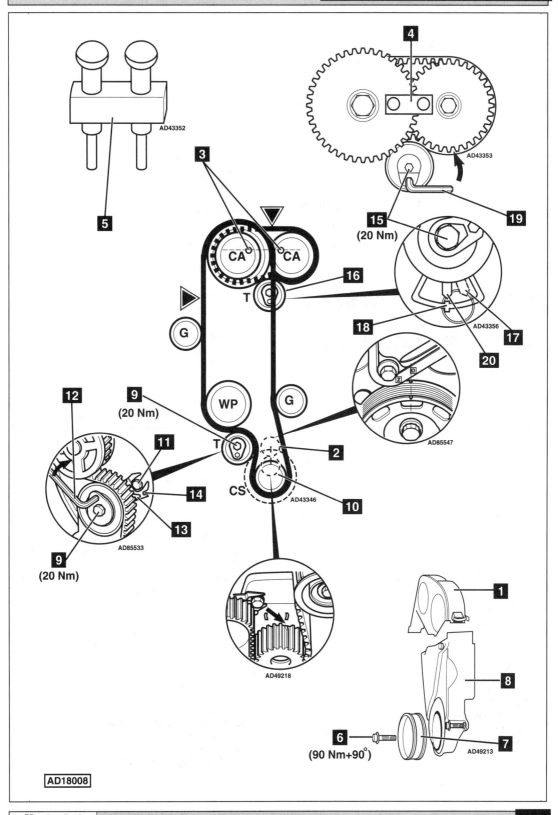

AD43352

AD43353

AD43356

AD85547

AD85533

AD49218

AD49213

AD18008

5

4

3

15
(20 Nm)

16

19

18

17

20

12

9
(20 Nm)

11

14

13

9
(20 Nm)

2

10

1

8

6
(90 Nm+90°)

7

CA CA

T

G

WP

G

T

CS

Model:	A3 1,6
Year:	1996-03
Engine Code:	AEH, AKL, APF

Replacement Interval Guide

The vehicle manufacturer has not recommended a timing belt replacement interval for this engine.
The previous use and service history of the vehicle must always be taken into account.
Refer to Timing Belt Replacement Intervals at the front of this manual.

Check For Engine Damage

CAUTION: This engine has been identified as an INTERFERENCE engine in which the possibility of valve-to-piston damage in the event of a timing belt failure is MOST LIKELY to occur.
A compression check of all cylinders should be performed before removing the cylinder head.

Repair Times – hrs

Remove & install	2,50

Special Tools

- Two-pin wrench – No.T10020.

Special Precautions

- Disconnect battery earth lead.
- DO NOT turn crankshaft or camshaft when timing belt removed.
- Remove spark plugs to ease turning engine.
- Turn engine in normal direction of rotation (unless otherwise stated).
- DO NOT turn engine via camshaft or other sprockets.
- Observe all tightening torques.

Removal

1. Remove:
   - ❏ Engine top cover(s).
   - ❏ Auxiliary drive belt.
   - ❏ Auxiliary drive belt tensioner.
2. Turn crankshaft clockwise to TDC on No.1 cylinder. Ensure flywheel timing marks aligned **1**.
3. Support engine.
4. Remove:
   - ❏ Crankshaft pulley bolts **2**.
   - ❏ Crankshaft pulley **3**.
   - ❏ Timing belt covers **4**, **5** & **6**.
   - ❏ PAS reservoir. DO NOT disconnect hoses.
   - ❏ RH engine mounting **7**.

NOTE: It may be necessary to raise engine slightly to remove engine mounting bolt.

5. Ensure camshaft sprocket timing marks aligned **8**.
6. Slacken tensioner nut **9**. Move tensioner away from belt.
7. Remove timing belt.

NOTE: Mark direction of rotation on belt with chalk if belt is to be reused.

Installation

1. Ensure timing marks aligned **1** & **8**.
2. Fit timing belt in following order:
   - ❏ Crankshaft sprocket.
   - ❏ Water pump sprocket.
   - ❏ Tensioner pulley.
   - ❏ Camshaft sprocket.
3. Ensure belt is taut between sprockets on non-tensioned side.
4. Ensure retaining lug located in slot in cylinder head **10**.
5. Turn tensioner 5 times fully anti-clockwise and clockwise from stop to stop. Use tool No.T10020 **11**.
6. Turn tensioner fully anti-clockwise then slowly clockwise until pointer **12** aligned with notch **13** in baseplate. Use tool No.T10020 **11**.

   NOTE: Engine must be COLD.
7. Tighten tensioner nut **9**.
   Tightening torque: 20 Nm.
8. Turn crankshaft two turns clockwise. Ensure timing marks aligned **1** & **8**.
9. Apply firm thumb pressure to belt at ▽. Ensure pointer **12** moves away from notch **13**.
10. Release thumb pressure. Pointer must realign with notch.
11. Install components in reverse order of removal.
12. Tighten crankshaft pulley bolts **2**.
    Tightening torque: 25 Nm.
13. Fit and align RH engine mounting:
    - ❏ Engine mounting clearance: 13 mm **17**.
    - ❏ Ensure engine mounting bolts **16** aligned with edge of mounting **18**.
14. Tighten:
    - ❏ Engine mounting bolts **14**.
      Tightening torque: 40 Nm + 90°. Use new bolts.
    - ❏ Engine mounting bolts **15**.
      Tightening torque: 25 Nm.
    - ❏ Engine mounting bolts **16**.
      Tightening torque: 60 Nm + 90°.

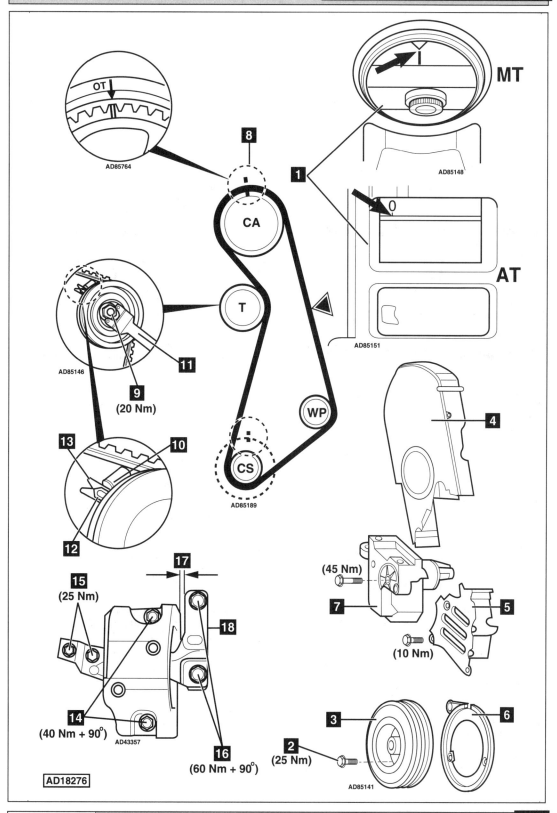

OT

AD85764

MT

AD85148

8

CA

1

0

AT

AD85151

T

11

AD85146

9
(20 Nm)

4

13

10

CS

AD85189

WP

12

17

15
(25 Nm)

18

(45 Nm)

7

5

14
(40 Nm + 90°)
AD43357

16
(60 Nm + 90°)

3

2
(25 Nm)

6

AD85141

AD18276

Model:	**A3 1,8/Turbo • S3 1,8 Turbo**
Year:	**1996-03**
Engine Code:	**AGN, AGU, APY**

Replacement Interval Guide

Audi recommend replacement every
180,000 kilometres (111,846 miles).
**NOTE: The vehicle manufacturer publishes this
information only in kilometres. The conversion
to miles is included for reference purposes only.**
The previous use and service history of the vehicle
must always be taken into account.
Refer to Timing Belt Replacement Intervals at the
front of this manual.

Check For Engine Damage

CAUTION: This engine has been identified as an
INTERFERENCE engine in which the possibility of
valve-to-piston damage in the event of a timing belt
failure is MOST LIKELY to occur.
A compression check of all cylinders should be
performed before removing the cylinder head.

Repair Times – hrs

Remove & install	2,20

Special Tools

■ 1 x M5 x 55 mm stud, nut and washer.

Special Precautions

■ Disconnect battery earth lead.
■ DO NOT turn crankshaft or camshaft when timing belt
  removed.
■ Remove spark plugs to ease turning engine.
■ Turn engine in normal direction of rotation (unless
  otherwise stated).
■ DO NOT turn engine via camshaft or other sprockets.
■ Observe all tightening torques.

Removal

1. Remove engine top cover.
2. Disconnect:
   ❏ Coolant expansion tank multi-plug.
   ❏ Turbo: Evaporative emission (EVAP) canister
     purge valve multi-plug.
3. Raise and support front of vehicle.
4. Remove:
   ❏ Engine undershield.
   ❏ Turbo: Fuel vapour hose from charcoal canister
     to throttle body.
   ❏ Auxiliary drive belt.
   ❏ Auxiliary drive belt tensioner.
   ❏ Coolant expansion tank. DO NOT disconnect
     hoses.
   ❏ PAS reservoir. DO NOT disconnect hoses.
   ❏ Timing belt upper cover **1**.
5. Support engine.
6. Remove:
   ❏ RH engine mounting **2**.
   ❏ Engine mounting bracket **3**.
   ❏ Timing belt centre cover **4**.
7. Turn crankshaft clockwise to TDC on No.1 cylinder.

8. Ensure crankshaft timing marks aligned **5** or **6**.
9. Ensure camshaft sprocket timing marks aligned **7**.
10. Remove:
    ❏ Crankshaft pulley bolts (4 bolts) **8**.
    ❏ Crankshaft pulley **9**.
    ❏ Timing belt lower cover **10**.
11. Insert M5 stud into tensioner **11**.
12. Fit nut and washer **12** to stud. Tighten nut sufficiently
    to allow a suitable locking pin to be inserted **13**.
    NOTE: DO NOT overtighten nut.
13. Remove timing belt.
    **NOTE: Mark direction of rotation on belt with chalk
    if belt is to be reused.**

Installation

1. Ensure timing marks aligned **6** & **7**.
2. Fit timing belt in following order:
   ❏ Crankshaft sprocket.
   ❏ Water pump sprocket.
   ❏ Tensioner pulley.
   ❏ Camshaft sprocket.
   **NOTE: Ensure belt is taut between sprockets on
   non-tensioned side.**
3. Remove locking pin **13**.
4. Remove nut, washer and stud **11** & **12**.
5. Turn crankshaft two turns clockwise.
6. Ensure timing marks aligned **6** & **7**.
7. Install:
   ❏ Timing belt lower cover **10**.
   ❏ Crankshaft pulley **9**.
   ❏ Crankshaft pulley bolts **8**.
   ❏ Timing belt centre cover **4**.
   ❏ Engine mounting bracket **3**.
8. Tighten crankshaft pulley bolts **8**. Tightening torque:
   ❏ 8.8 = 10 Nm + 90°. Use new bolts.
   ❏ 10.9 = 40 Nm.
9. Tighten engine mounting bracket bolts **14**.
   Tightening torque: 45 Nm.
10. Fit and align RH engine mounting **2**.
    ❏ Engine mounting clearance: 13 mm **18**.
    ❏ Ensure engine mounting bolts **17** aligned with
      edge of mounting **19**.
11. Tighten:
    ❏ Engine mounting bolts **15**.
      Tightening torque: 40 Nm + 90°. Use new bolts.
    ❏ Engine mounting bolts **16**.
      Tightening torque: 25 Nm.
    ❏ Engine mounting bolts **17**.
      Tightening torque: 60 Nm + 90°.
12. Install remainder of components in reverse order of
    removal.

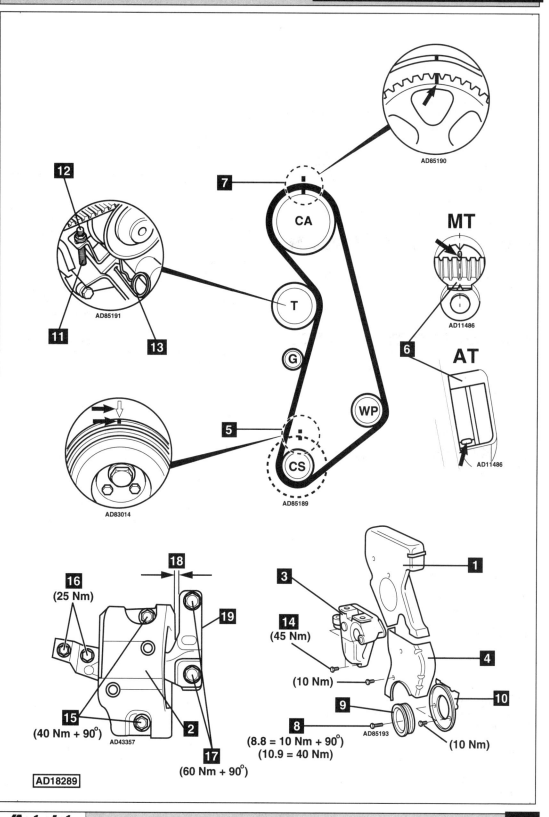

AD85190

12

7

CA

MT

AD11486

T

11

AD85191

13

G

6

AT

AD11486

WP

5

CS

AD85189

AD83014

18

16
(25 Nm)

3

1

19

14
(45 Nm)

4

(10 Nm)

9

10

15

2

8

AD85193

(10 Nm)

(40 Nm + 90°)

AD43357

(8.8 = 10 Nm + 90°)
(10.9 = 40 Nm)

17

(60 Nm + 90°)

AD18289

AUDI

Model:	80 1,6/2,0 • 100 2,0 • A6 2,0
Year:	1991-97
Engine Code:	AAD, AAE, ABB, ABK, ABM, ABT, ADA

Replacement Interval Guide

Audi recommend replacement every 120,000 kilometres (74,564 miles).

NOTE: The vehicle manufacturer publishes this information only in kilometres. The conversion to miles is included for reference purposes only. The previous use and service history of the vehicle must always be taken into account.
Refer to Timing Belt Replacement Intervals at the front of this manual.

Check For Engine Damage

CAUTION: This engine has been identified as an INTERFERENCE engine in which the possibility of valve-to-piston damage in the event of a timing belt failure is MOST LIKELY to occur.
A compression check of all cylinders should be performed before removing the cylinder head.

Repair Times – hrs

80:
Check & adjust	0,50
Remove & install	1,00

100/A6:
Check & adjust	0,50
Remove & install	1,10

Special Tools

■ Two-pin wrench – Matra V.159.

Special Precautions

■ Disconnect battery earth lead.
■ DO NOT turn crankshaft or camshaft when timing belt removed.
■ Remove spark plugs to ease turning engine.
■ Turn engine in normal direction of rotation (unless otherwise stated).
■ DO NOT turn engine via camshaft or other sprockets.
■ Observe all tightening torques.

Removal

1. Remove auxiliary drive belts.
2. Remove:
   ❏ Crankshaft damper.
   ❏ Timing belt upper cover **2**.
   ❏ Timing belt lower cover **3**.
3. Temporarily fit crankshaft damper.
4. Turn crankshaft until timing mark on crankshaft damper aligned with mark on auxiliary shaft sprocket **4**.
5. Ensure 'OT' mark on camshaft sprocket aligned with mark on timing belt rear cover **5**.
6. Remove crankshaft damper.
7. Slacken tensioner bolt **6**. Turn tensioner away from belt. Lightly tighten bolt.
8. Remove timing belt.

Installation

1. Ensure 'OT' mark on camshaft sprocket aligned with mark on timing belt rear cover **5**.
2. Fit timing belt to crankshaft sprocket and auxiliary shaft sprocket.
3. Temporarily fit crankshaft damper with one bolt **1**. Lightly tighten bolt.
4. Ensure timing mark on crankshaft pulley aligned with mark on auxiliary shaft sprocket **4**.
5. Fit timing belt to camshaft sprocket and tensioner pulley.
   *NOTE: Belt tension must only be adjusted when engine is COLD, as tensioner is fitted with temperature compensation unit **7**.*
6. Fit wrench Matra V.159 to tensioner pulley **8**.
7. Turn tensioner anti-clockwise until belt can just be twisted with finger and thumb through 90° at ▽.
8. Tighten tensioner bolt to 20 Nm **6**.
9. Turn crankshaft two turns clockwise.
10. Ensure timing marks aligned **4** & **5**.
11. Recheck belt tension.
12. Remove crankshaft damper.
13. Install components in reverse order of removal.
14. Fit crankshaft damper. Tighten bolts to 20 Nm **1**.

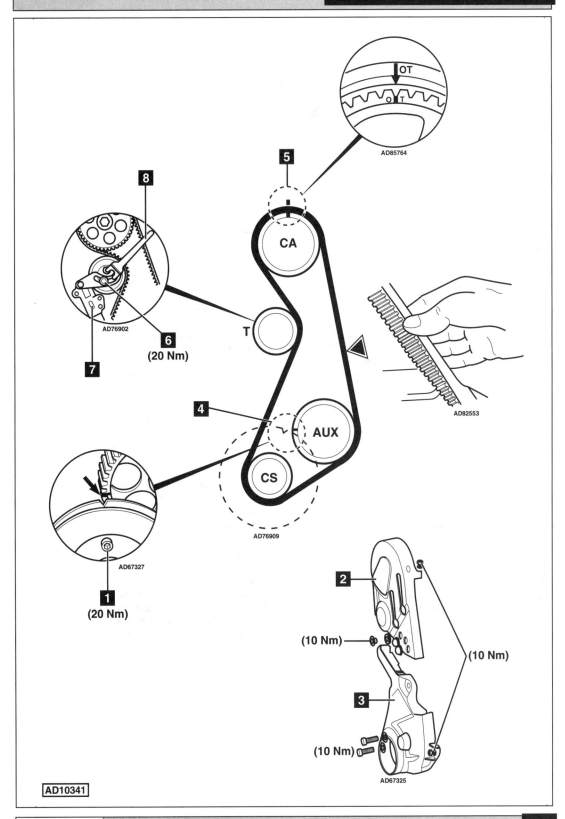

OT

AD85764

5

8

CA

AD76902

6
(20 Nm)

7

T

AD82553

4

AUX

CS

AD76909

AD67327

1
(20 Nm)

2

(10 Nm)

(10 Nm)

3

(10 Nm)

AD67325

AD10341

AUDI

Model:	Cabriolet 1,8 • A6 1,8
Year:	1995-00
Engine Code:	ADR

Replacement Interval Guide

Audi recommend:

Cabriolet
→1999MY: Replacement every
120,000 kilometres (74,564 miles).
2000MY→: Replacement every
180,000 kilometres (111,846 miles).

A6
Replacement every 120,000 kilometres
(74,564 miles).

NOTE: The vehicle manufacturer publishes this information only in kilometres. The conversion to miles is included for reference purposes only.
The previous use and service history of the vehicle must always be taken into account.
Refer to Timing Belt Replacement Intervals at the front of this manual.

Check For Engine Damage

CAUTION: This engine has been identified as an INTERFERENCE engine in which the possibility of valve-to-piston damage in the event of a timing belt failure is MOST LIKELY to occur.
A compression check of all cylinders should be performed before removing the cylinder head.

Repair Times – hrs

Remove & install:

A6	1,30
AC	+0,30
Cabriolet	1,60

Special Tools

■ None required.

Special Precautions

■ Disconnect battery earth lead.
■ DO NOT turn crankshaft or camshaft when timing belt removed.
■ Remove spark plugs to ease turning engine.
■ Turn engine in normal direction of rotation (unless otherwise stated).
■ DO NOT turn engine via camshaft or other sprockets.
■ Observe all tightening torques.

Removal

1. Remove:
   ❏ Auxiliary drive belts.
   ❏ Timing belt upper cover **1**.
2. Turn crankshaft clockwise to TDC on No.1 cylinder. Ensure timing marks aligned **2** & **3**.
3. Remove:
   ❏ Crankshaft pulley bolts **4**.
   ❏ Crankshaft pulley **5**.
   ❏ Timing belt lower cover **6**.
4. Turn tensioner pulley anti-clockwise until holes in pushrod and tensioner body aligned **8**. Use 8 mm Allen key **7**. Retain pushrod with a suitable pin through hole in tensioner body.
   *NOTE: DO NOT slacken bolts **9**.*
5. Remove timing belt.
   NOTE: Mark direction of rotation on belt with chalk if belt is to be reused.

Installation

1. Ensure timing marks aligned **3**.
2. Fit timing belt to crankshaft sprocket.
3. Install:
   ❏ Timing belt lower cover **6**.
   ❏ Crankshaft pulley **5**.
   ❏ Crankshaft pulley bolts **4**.
   Tightening torque: 20 Nm.
4. Ensure timing marks aligned **2**.
5. Fit timing belt to remaining sprockets and pulleys.
6. Ensure belt is taut between sprockets on non-tensioned side.
7. Turn tensioner pulley slightly anti-clockwise. Use 8 mm Allen key **7**. Remove pin from tensioner body to release pushrod **8**.
8. Turn crankshaft two turns clockwise. Ensure timing marks aligned **2** & **3**.
9. Install components in reverse order of removal.

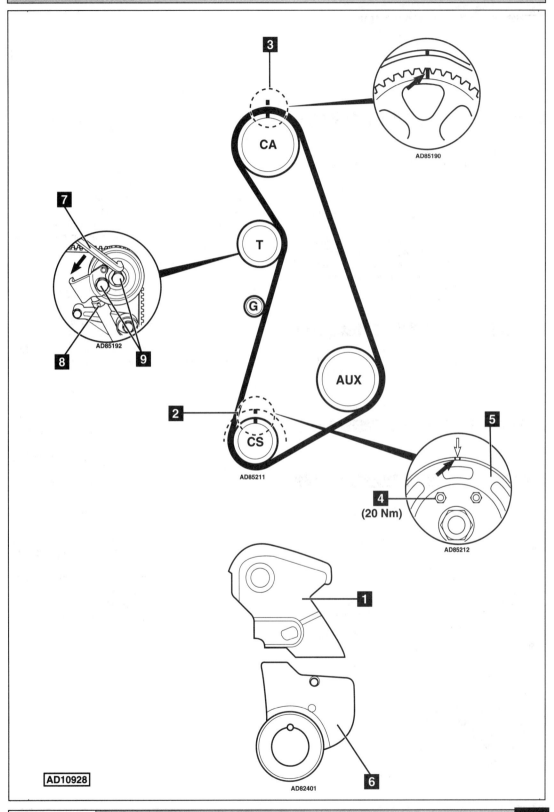

3

CA

AD85190

7

T

G

AD85192

8

9

2

CS

AD85211

AUX

5

4
(20 Nm)

AD85212

1

6

AD10928

AD82401

AUDI

Model:	80/100 2,0 16V
Year:	1990-94
Engine Code:	Manual tensioner 6A, ACE

Replacement Interval Guide

Audi recommend check and adjust or replacement if necessary every 30,000 kilometres (18,641 miles). Replacement as follows:

→1991: No manufacturer's recommended replacement interval.

1991→: Replacement every 120,000 kilometres (74,564 miles).

NOTE: The vehicle manufacturer publishes this information only in kilometres. The conversion to miles is included for reference purposes only. The previous use and service history of the vehicle must always be taken into account. Refer to Timing Belt Replacement Intervals at the front of this manual.

Check For Engine Damage

CAUTION: This engine has been identified as an INTERFERENCE engine in which the possibility of valve-to-piston damage in the event of a timing belt failure is MOST LIKELY to occur.
A compression check of all cylinders should be performed before removing the cylinder head.

Repair Times – hrs

Check & adjust	0,50
Remove & install	1,10

Special Tools

- Tension gauge – VAG No.210.
- Two-pin wrench – Matra V.159.

Special Precautions

- Disconnect battery earth lead.
- DO NOT turn crankshaft or camshaft when timing belt removed.
- Remove spark plugs to ease turning engine.
- Turn engine in normal direction of rotation (unless otherwise stated).
- DO NOT turn engine via camshaft or other sprockets.
- Observe all tightening torques.

Removal

1. Remove:
   - ❏ Engine undershield (if fitted).
   - ❏ Auxiliary drive belts.
   - ❏ Timing belt upper cover **1**.
   NOTE: 80 – Remove cooling fan.
2. Turn crankshaft to TDC on No.1 cylinder. Ensure timing marks aligned **2**, **3** & **4**.

3. If cylinder head cover removed: Use camshaft sprocket rear timing mark **9**.
   NOTE: Align notch with upper cylinder head face.
4. Remove:
   - ❏ Crankshaft pulley bolts **5**.
   - ❏ Crankshaft pulley **6**.
   - ❏ Water pump pulley (if required).
   - ❏ Timing belt lower cover **7**.
5. Slacken tensioner bolt **8**. Turn tensioner anti-clockwise to release tension on belt. Lightly tighten bolt.
6. Remove timing belt.

Installation

1. Ensure timing marks aligned **2** & **4**.
2. If cylinder head cover removed: Ensure camshaft sprocket rear timing mark aligned **9**.
   NOTE: Notch aligned with upper cylinder head face.
3. Fit timing belt in anti-clockwise direction, starting at crankshaft sprocket.
4. Attach tension gauge to belt at ▽. Tool No.210.
5. Turn tensioner clockwise until tension gauge indicates 13-14 units. Use wrench Matra V.159.
6. Tighten tensioner bolt. Tightening torque: 25 Nm (M8)/45 Nm (M10) **8**.
7. Remove tension gauge. Turn crankshaft two turns clockwise.
8. Ensure timing marks aligned **2** & **4**.
9. Recheck belt tension.
   NOTE: Belt is correctly tensioned when it can just be twisted with finger and thumb through 90° at ▽.
10. Install components in reverse order of removal.
11. Tighten crankshaft pulley bolts. Tightening torque: 20 Nm **5**.
    NOTE: Crankshaft pulley bolt holes are offset.

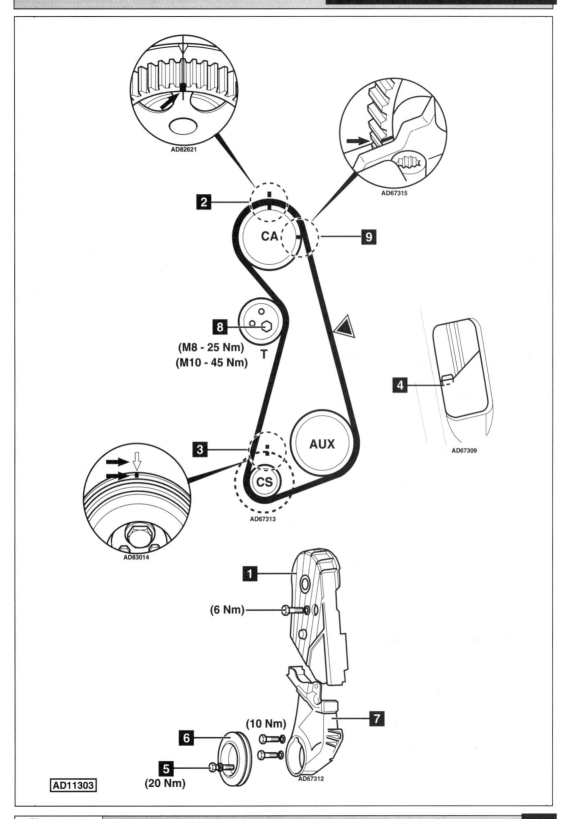

AD82621

AD67315

2

CA

9

8

(M8 - 25 Nm)
(M10 - 45 Nm)
T

4

AD67309

3

AUX

CS

AD67313

AD83014

1

(6 Nm)

(10 Nm)

7

6

5

(20 Nm)

AD67312

AD11303

AUDI

Model:	80/100/A6 2,0 16V
Year:	1994-97
Engine Code:	**Automatic tensioner pulley** **ACE**

Replacement Interval Guide

Audi recommend replacement every
120,000 kilometres (74,564 miles).

**NOTE: The vehicle manufacturer publishes this
information only in kilometres. The conversion
to miles is included for reference purposes only.**
*The previous use and service history of the vehicle
must always be taken into account.*
*Refer to Timing Belt Replacement Intervals at the front
of this manual.*

Check For Engine Damage

*CAUTION: This engine has been identified as an
INTERFERENCE engine in which the possibility of
valve-to-piston damage in the event of a timing belt
failure is MOST LIKELY to occur.*
*A compression check of all cylinders should be
performed before removing the cylinder head.*

Repair Times – hrs

Remove & install	1,10

Special Tools

■ Two-pin wrench – Matra V.159.

Special Precautions

■ Disconnect battery earth lead.
■ DO NOT turn crankshaft or camshaft when timing belt removed.
■ Remove spark plugs to ease turning engine.
■ Turn engine in normal direction of rotation (unless otherwise stated).
■ DO NOT turn engine via camshaft or other sprockets.
■ Observe all tightening torques.

Removal

1. Remove:
   ❏ Engine undershield.
   ❏ Auxiliary drive belts.
   ❏ Timing belt upper cover **1**.
   NOTE: 80 – Remove cooling fan.
2. Turn crankshaft to TDC on No.1 cylinder. Ensure timing marks aligned **2**, **3** & **4**.
3. If cylinder head cover removed: Use camshaft sprocket rear timing mark **11**.
   NOTE: Align notch with upper cylinder head face.
4. Remove:
   ❏ Crankshaft pulley bolts **5**.
   ❏ Crankshaft pulley **6**.
   ❏ Timing belt lower cover **7**.

5. Slacken tensioner nut **8**. Turn tensioner clockwise to release tension on belt. Lightly tighten nut.
6. Remove timing belt.

Installation

1. Ensure timing marks aligned **2** & **4**.
2. If cylinder head cover removed: Ensure camshaft sprocket rear timing mark aligned **11**.
   NOTE: Notch aligned with upper cylinder head face.
3. Fit timing belt in anti-clockwise direction, starting at crankshaft sprocket.
4. Turn tensioner anti-clockwise until it touches stop. Use wrench Matra V.159 **9**.
5. Release tension until marks align exactly **10**.
   NOTE: Engine must be cold.
6. Tighten tensioner nut.
   Tightening torque: 25 Nm **8**.
7. Turn crankshaft two turns clockwise. Ensure timing marks aligned **2** & **4**.
8. Check tensioner marks aligned **10**.
9. If not: Repeat tensioning procedure.
10. Apply firm thumb pressure to belt at ▽. Tensioner marks must move apart **10**.
11. Release thumb pressure from belt at ▽.
12. Turn crankshaft two turns clockwise. Ensure timing marks aligned **2** & **4**.
13. Tensioner marks should realign **10**.
14. Install components in reverse order of removal.
15. Tighten crankshaft pulley bolts.
    Tightening torque: 20 Nm **5**.
    NOTE: Crankshaft pulley bolt holes are offset.

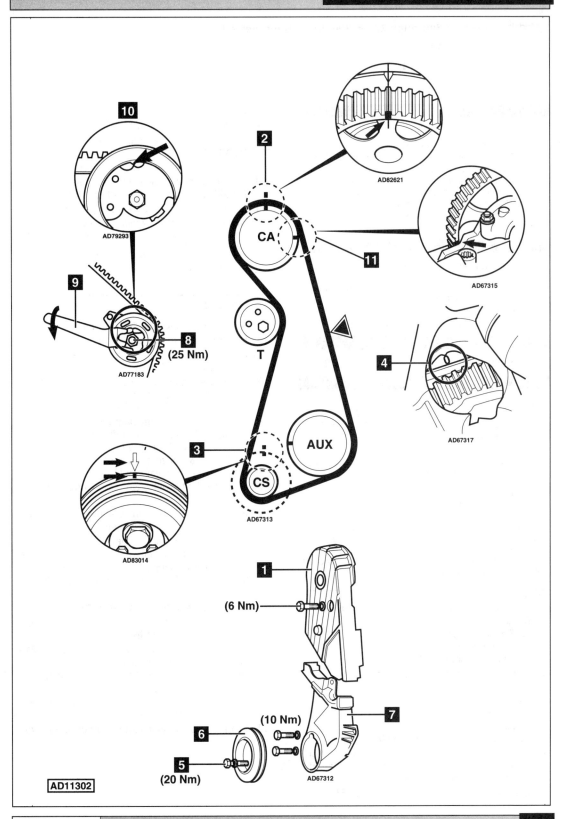

AD82621

AD67315

10

AD79293

2

CA

11

9

8
(25 Nm)

AD77183

T

4

AD67317

3

AUX

CS

AD67313

AD83014

1

(6 Nm)

7

AD67312

6

(10 Nm)

5
(20 Nm)

AD11302

AUDI

Replacement Interval Guide

Audi recommend replacement every 120,000 kilometres (74,564 miles).

NOTE: The vehicle manufacturer publishes this information only in kilometres. The conversion to miles is included for reference purposes only. The previous use and service history of the vehicle must always be taken into account.
Refer to Timing Belt Replacement Intervals at the front of this manual.

Check For Engine Damage

CAUTION: This engine has been identified as an INTERFERENCE engine in which the possibility of valve-to-piston damage in the event of a timing belt failure is MOST LIKELY to occur.
A compression check of all cylinders should be performed before removing the cylinder head.

Repair Times – hrs

Remove & install	2,20

Special Tools

- Extension spanner – VAG No.2079.
- Radiator support bracket – VAG No.3251.
- Crankshaft damper locking tool – VAG No.3256.
- Tension gauge – VAG No.210.

Special Precautions

- Disconnect battery earth lead.
- DO NOT turn crankshaft or camshaft when timing belt removed.
- Remove spark plugs to ease turning engine.
- Turn engine in normal direction of rotation (unless otherwise stated).
- DO NOT turn engine via camshaft or other sprockets.
- Observe all tightening torques.

Removal

1. Remove:
   - ❏ Engine undershield.
   - ❏ Front bumper assembly and trim.
   - ❏ Radiator mounting bolts.
2. Move radiator forward and support with tool. Tool No.3251.
3. Remove:
   - ❏ Viscous fan.
   - ❏ Auxiliary drive belt.
   - ❏ Auxiliary drive belt tensioner.
4. Turn crankshaft clockwise until timing mark on camshaft sprocket aligned with upper edge of cylinder head cover gasket **1**.
5. Ensure flywheel timing marks aligned **2**.

6. Remove timing belt upper cover **3**.
7. Slacken bolts one turn **4** & **5**.
8. Slacken bolt **6**. Move tensioner pulley away from belt.
9. Remove timing belt lower cover **7**.
10. Lock crankshaft pulley/damper. Use tool No.3256. Remove crankshaft bolt **8**.
11. Remove crankshaft pulley/damper **9**.

    NOTE: Pull crankshaft pulley/damper with crankshaft sprocket and timing belt from engine as an assembly 10.
    Mark direction of rotation on belt with chalk if belt is to be reused.

Installation

NOTE: Engine must be COLD when installing belt.
1. Ensure timing marks aligned **1** & **2**.
2. Fit timing belt to crankshaft sprocket. Fit assembly to crankshaft **10**.
3. Fit crankshaft bolt **8**. Coat threads and contact face with sealing compound (AMV 188 001 02 or similar).
4. Lock crankshaft pulley/damper. Use tool No.3256. Tighten crankshaft bolt to 350 Nm **8**. Use tool No.2079 and torque wrench.
   NOTE: If tool No.2079 not available: Tighten crankshaft bolt to 450 Nm 8.
5. Fit timing belt lower cover **7**. Tighten bolts to 10 Nm.
6. Fit timing belt to remaining sprockets. Ensure belt is taut between sprockets.
7. Ensure timing marks aligned **1** & **2**.
8. Apply clockwise torque of 25 Nm (NG engine) or 35 Nm (AAR engine) to hexagon of automatic tensioner unit **11**. Ensure torque wrench held in a vertical position.
9. Tighten bolts to 20 Nm **4**, **5** & **6**.
10. Turn crankshaft two turns clockwise. Ensure timing marks aligned **1** & **2**.
11. Attach tension gauge to belt at ▽. Tool No.210. Check tension gauge reading.
    AAR: 13,0-13,5 units. NG: 14,0-14,5 units.
12. If tension not as specified: Slacken bolts **4**, **5** & **6**. Adjust position of automatic tensioner unit. Tighten bolts to 20 Nm.
13. Install components in reverse order of removal.

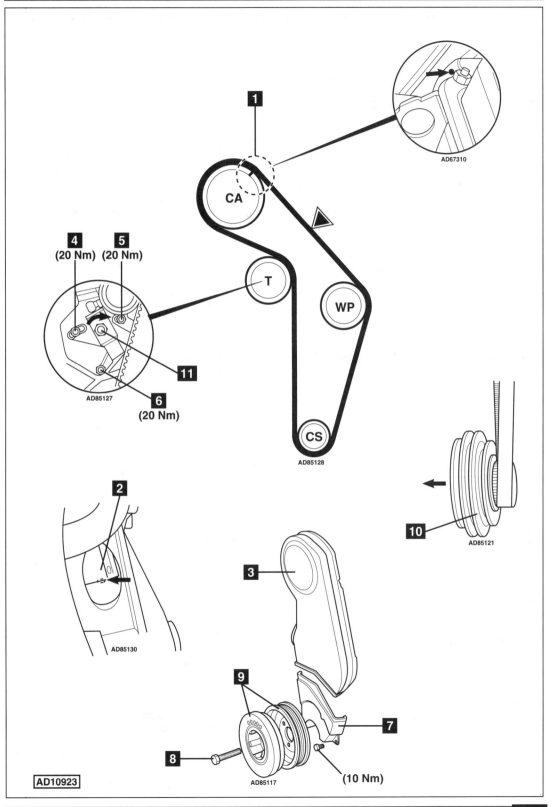

Model:	80 2,6/2,8 • A4 2,6/2,8 • 100 2,6/2,8 • A6 2,6/2,8 • A8 2,8 • Cabrio 2,6
Year:	1991-02
Engine Code:	AAH, ABC

Replacement Interval Guide

Audi recommend check every 60,000 kilometres (37,282 miles).

NOTE: The vehicle manufacturer publishes this information only in kilometres. The conversion to miles is included for reference purposes only.

The vehicle manufacturer has not recommended a timing belt replacement interval for this engine.
The previous use and service history of the vehicle must always be taken into account.
Refer to Timing Belt Replacement Intervals at the front of this manual.

Check For Engine Damage

CAUTION: This engine has been identified as an INTERFERENCE engine in which the possibility of valve-to-piston damage is MOST LIKELY to occur.
A compression check of all cylinders should be performed before removing the cylinder head.

Repair Times – hrs

Remove & install	2,20

Special Tools

- ■ Crankshaft locking tool – VAG No.3242.
- ■ Camshaft locking tool – VAG No.3243.
- ■ Support guides – No.3369.

Special Precautions

- ■ Disconnect battery earth lead.
- ■ DO NOT turn crankshaft or camshaft when timing belt removed.
- ■ Remove spark plugs to ease turning engine.
- ■ Turn engine in normal direction of rotation (unless otherwise stated).
- ■ DO NOT turn engine via camshaft or other sprockets.
- ■ Observe all tightening torques.

Removal

1. Raise and support front of vehicle.
2. A4: Move radiator support panel into service position:
   - ❏ Remove front bumper.
   - ❏ Remove air intake pipe between front panel and air filter.
   - ❏ Remove front panel bolts.
   - ❏ Install support guides No.3369 in front panel.
   - ❏ Slide front panel forward.
   - ❏ Refit upper rear bolts in front holes to steady panel.
3. Remove auxiliary drive belt.
4. Unclip timing belt LH and RH covers **7** & **8**.

5. Turn crankshaft to TDC on No.3 cylinder.
6. Ensure crankshaft pulley timing marks aligned **1**.
7. Ensure large holes in locking plates of camshaft sprockets face in towards each other.
8. If not: Turn crankshaft one turn clockwise.
9. Remove TDC sensor from crankcase. Ensure TDC hole in crankshaft web aligned.
10. Screw in crankshaft locking tool **2**. Tool No.3242.
11. Remove:
    - ❏ Auxiliary drive belt tensioner.
    - ❏ Timing belt LH and RH covers.
    - ❏ Crankshaft pulley.
    - **NOTE: Centre bolt does NOT have to be removed to remove crankshaft pulley.**
    - ❏ Timing belt lower cover **9**.
12. Slacken bolt of each camshaft sprocket. DO NOT remove.
13. Slacken tensioner bolt **4**.
14. Remove timing belt.

Installation

1. Ensure camshaft sprockets can turn on taper but not tilt.
2. Ensure crankshaft locking tool fitted **2**.
3. Fit timing belt to camshaft sprockets.
4. Fit locking tool to camshafts **3**. Tool No.3243.
5. Fit timing belt in following order:
   - ❏ Guide pulley.
   - ❏ Crankshaft sprocket.
   - ❏ Tensioner pulley.
6. Lightly tighten tensioner bolt **4**. Ensure tensioner pulley can still be turned by hand.
7. Slide 23 mm diameter spacer between timing belt cover and belt **6**.
8. Install torque wrench to hexagon of tensioner **5**. Tension timing belt to 4 Nm. Tighten tensioner bolt simultaneously **4**.
9. Remove torque wrench and spacer **6**.
10. Tighten tensioner bolt to 45 Nm **4**.
11. Tighten bolt of each camshaft sprocket to 70 Nm.
12. Remove:
    - ❏ Camshaft locking tool **3**.
    - ❏ Crankshaft locking tool.
13. Fit TDC sensor.
14. Fit crankshaft pulley. Ensure notch aligned with tab on crankshaft sprocket **10**.
15. Install components in reverse order of removal.

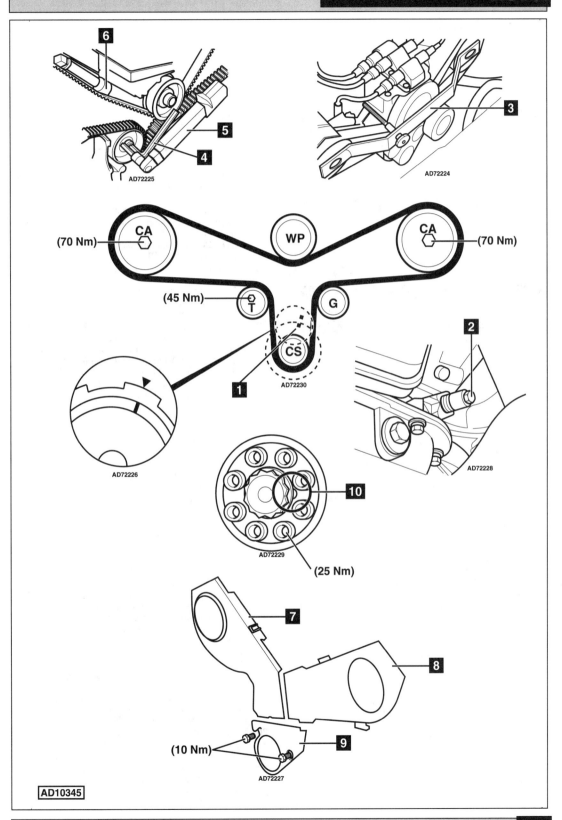

(70 Nm) CA

WP

CA (70 Nm)

(45 Nm)

G

CS

AD72230

AD72226

AD72225

AD72224

AD72228

AD72229

(25 Nm)

(10 Nm)

AD72227

AD10345

Model:	**A4 1,6**
Year:	**1994-97**
Engine Code:	**ADP**

Replacement Interval Guide

The vehicle manufacturer has not recommended a timing belt replacement interval for this engine.
The previous use and service history of the vehicle must always be taken into account.
Refer to Timing Belt Replacement Intervals at the front of this manual.

Check For Engine Damage

CAUTION: This engine has been identified as an INTERFERENCE engine in which the possibility of valve-to-piston damage in the event of a timing belt failure is MOST LIKELY to occur.
A compression check of all cylinders should be performed before removing the cylinder head.

Repair Times – hrs

Remove & install	2,20

Special Tools

- Support guides – No.3369.
- Two-pin wrench – Hazet No.2587.

Special Precautions

- Disconnect battery earth lead.
- DO NOT turn crankshaft or camshaft when timing belt removed.
- Remove spark plugs to ease turning engine.
- Turn engine in normal direction of rotation (unless otherwise stated).
- DO NOT turn engine via camshaft or other sprockets.
- Observe all tightening torques.

Removal

1. Raise and support front of vehicle.
2. Remove:
   - ❏ Engine undershield.
   - ❏ Front bumper.
   - ❏ Air filter intake duct.
   - ❏ Front panel bolts **1** & **2**.
3. Install support guides No.3369 in front panel **3**.
4. Slide front panel forward.
5. Refit upper rear bolts in front holes to steady panel.
6. Remove:
   - ❏ Auxiliary drive belt.
   - ❏ Auxiliary drive belt tensioner.
7. Lock cooling fan viscous coupling with 5 x 60 mm bolt **11**.
8. Unscrew viscous coupling with Allen key **12**.
9. Remove timing belt upper cover **4**.
10. Turn crankshaft to TDC on No.1 cylinder.
11. Ensure timing marks aligned **5** & **6**.
12. Remove:
    - ❏ Crankshaft pulley bolts **7**.
    - ❏ Crankshaft pulley **8**.
    - ❏ Timing belt lower cover **9**.
13. Slacken tensioner bolt **10**.
14. Move tensioner away from belt. Lightly tighten bolt **10**.

15. Remove timing belt.
    NOTE: Mark direction of rotation on belt with chalk if belt is to be reused.

Installation

1. Ensure camshaft sprocket timing marks aligned **5**.
2. Fit timing belt to crankshaft sprocket and auxiliary shaft sprocket.
3. Temporarily fit crankshaft pulley (with one bolt).
4. Ensure timing mark on crankshaft pulley aligned with mark on auxiliary shaft sprocket **13**.
   *NOTE: DO NOT use auxiliary shaft sprocket timing mark labelled 'OT' **14**.*
5. Fit timing belt to tensioner pulley and camshaft sprocket.
   NOTE: Ensure belt is taut between sprockets on non-tensioned side.
6. Slacken tensioner bolt **10**.
7. Fit two-pin wrench **16**. Use wrench Hazet No.2587.
   *NOTE: Two-pin wrench locates in hole **19** and against lug **20**. When turning two-pin wrench clockwise, the tensioner pulley will turn anti-clockwise and apply tension to belt.*
8. Turn two-pin wrench clockwise **17** until piston **15 A** is fully extended and piston **15 B** has risen approximately 1 mm. Use tool No.2587 **16**.
9. Lightly tighten tensioner bolt **10**.
10. Remove:
    - ❏ Crankshaft pulley bolt **7**.
    - ❏ Crankshaft pulley **8**.
11. Install:
    - ❏ Timing belt lower cover **9**.
    - ❏ Crankshaft pulley **8**.
    - ❏ Crankshaft pulley bolts **7**.
12. Lightly tighten crankshaft pulley bolts **7**.
13. Ensure timing mark on crankshaft pulley aligned **6**.
14. Turn crankshaft two turns clockwise. Ensure timing marks aligned **5** & **6**.
15. Remove distributor cap.
16. Ensure distributor rotor arm aligned with mark for cylinder No.1 on distributor body **18**.
17. Check alignment of tensioner lever **15 C** in relation to upper edge of piston **15 B** as follows:
    - ❏ **15 D** – adjustment correct.
    - ❏ **15 E** – wear area, re-adjust.
    - ❏ **15 F** – check tensioner and re-adjust.
18. When adjustment correct, dimension **15 G** will be 25-29 mm.
19. If alignment incorrect or dimension **15 G** not as specified: Repeat tensioning procedure.
    *NOTE: After installation, correct adjustment is maintained providing upper edge of piston **15 B** is positioned within areas **15 D** and **15 E**.*
20. Tighten tensioner bolt to 25 Nm **10**.
21. Tighten crankshaft pulley bolts to 25 Nm **7**.
22. Install components in reverse order of removal.
23. Push front panel back into place. Remove support guides No.3369 **3**. Fit bolts.
24. Tighten 8 mm bolts to 45 Nm **2**. Tighten 6 mm bolts to 10 Nm **1**.

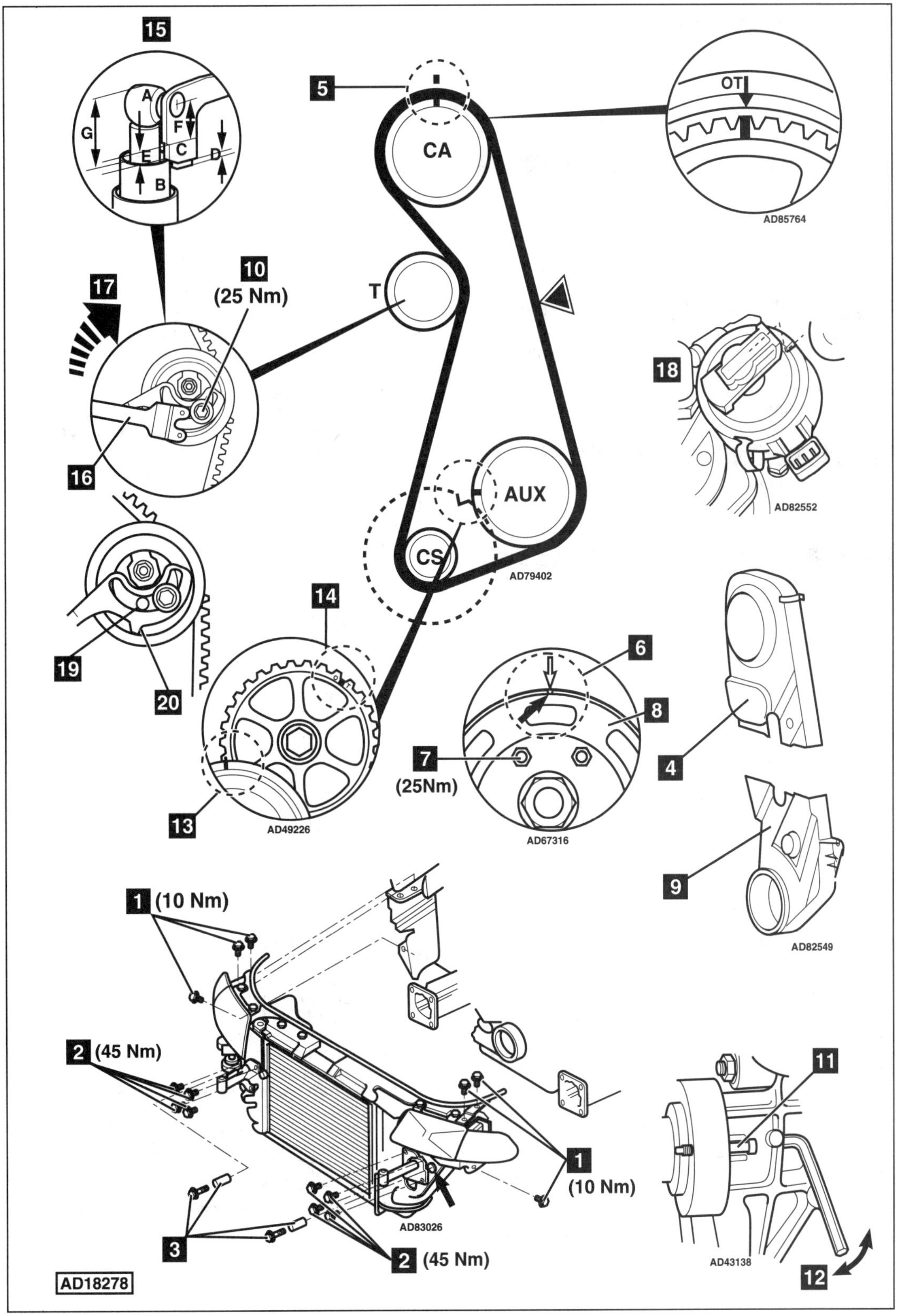
15
A
G
F
E
C
D
B
5
CA
OT
AD85764
17
10
(25 Nm)
T
16
18
AD82552
AUX
CS
AD79402
19
20
14
6
8
7
(25Nm)
13
AD49226
AD67316
4
9
AD82549
1 (10 Nm)
2 (45 Nm)
1
(10 Nm)
3
AD83026
2 (45 Nm)
11
AD43138
12
AD18278

AUDI

Model:	**A4 1,6**
Year:	**1996-03**
Engine Code:	**AHL, ALZ**

Replacement Interval Guide

The vehicle manufacturer has not recommended a timing belt replacement interval for this engine.
The previous use and service history of the vehicle must always be taken into account.
Refer to Timing Belt Replacement Intervals at the front of this manual.

Check For Engine Damage

CAUTION: This engine has been identified as an INTERFERENCE engine in which the possibility of valve-to-piston damage in the event of a timing belt failure is MOST LIKELY to occur.
A compression check of all cylinders should be performed before removing the cylinder head(s).

Repair Times – hrs

Remove & install:	
AHL	2,50
ALZ	2,20

Special Tools

- ■ Two-pin wrench – Matra V.159 or T10020.
- ■ Support guides – No.3369.

Removal

1. Raise and support front of vehicle.
2. Remove:
   - ❏ ALZ: Engine cover.
   - ❏ Engine undershield.
   - ❏ Front bumper.
   - ❏ Air filter intake duct.
   - ❏ Front panel bolt **1**.
3. Install support guides No.3369 in front panel **2**.
4. Remove front panel bolts **3** & **4**.
5. Slide front panel forward.
6. Remove auxiliary drive belt.
7. AHL: Lock cooling fan viscous coupling with 5 mm pin **17**.
8. AHL: Unscrew viscous coupling with Allen key.
9. Remove:
   - ❏ Auxiliary drive belt tensioner **5**.
   - ❏ Timing belt upper cover **6**.
10. Turn crankshaft to TDC on No.1 cylinder.
11. Ensure timing marks aligned **7** or **18**.
12. Ensure camshaft sprocket timing marks aligned **8**.
13. Remove:
   - ❏ Crankshaft pulley bolts **9**.
   - ❏ Crankshaft pulley **10**.
   - ❏ Timing belt centre cover **11**.
   - ❏ Timing belt lower cover **12**.
14. Slacken tensioner nut **13**. Turn tensioner clockwise away from belt. Lightly tighten nut.
15. Remove timing belt.
    NOTE: Mark direction of rotation on belt with chalk if belt is to be reused.

Installation

1. Ensure camshaft sprocket timing marks aligned **8**.
2. Fit timing belt to crankshaft sprocket.
3. Fit:
   - ❏ Timing belt lower cover **12**.
   - ❏ Crankshaft pulley **10**.
   - ❏ Crankshaft pulley bolts **9**.
4. Lightly tighten crankshaft pulley bolts **9**.
5. Ensure timing marks aligned **7** or **18**.
6. Fit timing belt in following order:
   - ❏ Water pump sprocket.
   - ❏ Tensioner pulley.
   - ❏ Camshaft sprocket.
   NOTE: If reusing old belt: Observe direction of rotation marks on belt.
7. Ensure tensioner retaining lug is properly engaged **14**.
8. Slacken tensioner nut **13**.
9. AHL: Adjust tensioner:
   NOTE: Engine must be cold.
   - ❏ Turn tensioner anti-clockwise to maximum position. Use wrench Matra V.159.
   - ❏ Slacken tensioner until pointer **15** is approximately 10 mm below notch **16**.
   - ❏ Tighten tensioner until pointer aligned with notch **15** & **16**.
   - ❏ Tighten tensioner nut to 15 Nm **13**.
10. ALZ: Adjust tensioner:
    NOTE: Engine must be cold.
    - ❏ Turn tensioner fully in both directions 5 times. Use tool No.T10020.
    - ❏ Turn tensioner anti-clockwise to maximum position. Use tool No.T10020.
    - ❏ Slacken tensioner until pointer aligned with notch **15** & **16**.
    - ❏ Tighten tensioner nut to 20 Nm **13**.
11. Turn crankshaft two turns clockwise to TDC on No.1 cylinder.
    NOTE: Turn crankshaft last 45° smoothly without stopping.
12. Ensure pointer aligned with notch **15** & **16**.
13. Ensure timing marks aligned **7** or **18** & **8**.
14. Check tensioner operation:
    - ❏ Apply firm thumb pressure to belt at ▽.
    - ❏ Pointer **15** and notch **16** must move apart.
    - ❏ Release thumb pressure from belt.
    - ❏ Turn crankshaft two turns clockwise to TDC on No.1 cylinder.
    NOTE: Turn crankshaft last 45° smoothly without stopping.
    - ❏ Ensure pointer aligned with notch **15** & **16**.
15. Tighten crankshaft pulley bolts **9**:
    - ❏ ALZ – 25 Nm.
    - ❏ AHL – 40 Nm.
16. Install components in reverse order of removal.

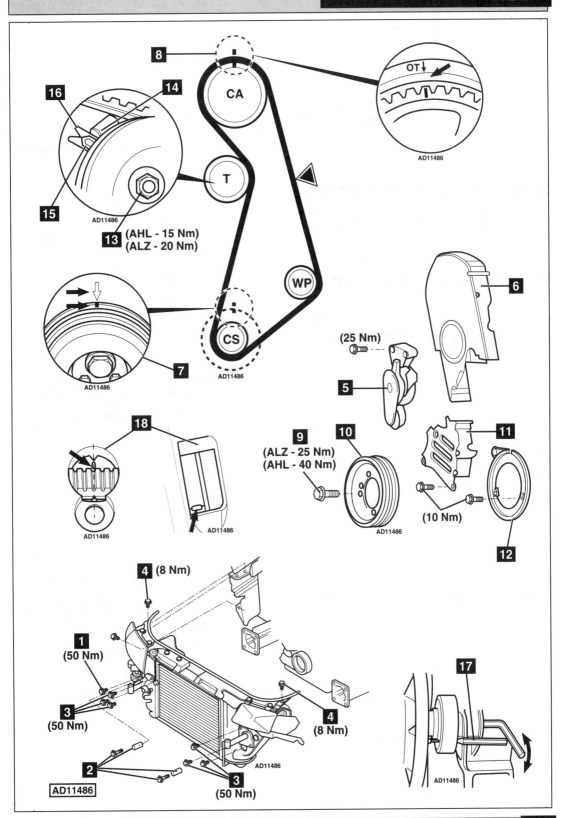

8

16 14

CA

OT↓

AD11486

13 (AHL - 15 Nm)
   (ALZ - 20 Nm)

15

AD11486

T

WP

6

7

CS

AD11486

(25 Nm)

5

18

9 (ALZ - 25 Nm)
  (AHL - 40 Nm)

10

11

AD11486

AD11486

(10 Nm)

12

4 (8 Nm)

1 (50 Nm)

3 (50 Nm)

4 (8 Nm)

17

2

3 (50 Nm)

AD11486

AD11486

Model:	A4 1,8/Turbo • A6 1,8 Turbo
Year:	1994-03
Engine Code:	ADR (→07/97), AEB, AJL

Replacement Interval Guide

Audi recommend replacement every 75,000 miles.
The previous use and service history of the vehicle must always be taken into account.
Refer to Timing Belt Replacement Intervals at the front of this manual.

Check For Engine Damage

CAUTION: This engine has been identified as an INTERFERENCE engine in which the possibility of valve-to-piston damage in the event of a timing belt failure is MOST LIKELY to occur.
A compression check of all cylinders should be performed before removing the cylinder head(s).

Repair Times – hrs

Remove & install:

A4	2,20
A6	2,70

Special Tools

- Two-pin wrench – Matra V.159.
- Support guides – No.3369.

Special Precautions

- Disconnect battery earth lead.
- DO NOT turn crankshaft or camshaft when timing belt removed.
- Remove spark plugs to ease turning engine.
- Turn engine in normal direction of rotation (unless otherwise stated).
- DO NOT turn engine via camshaft or other sprockets.
- Observe all tightening torques.

Removal

1. Raise and support front of vehicle.
2. Remove:
   - ❏ Engine undershield.
   - ❏ Front bumper.
   - ❏ Air filter intake duct.
   - ❏ Front panel bolts **1** & **2**.
3. Install support guides No.3369 in front panel **3**.
4. Slide front panel forward.
5. A4: Refit upper rear bolts in front holes to steady panel.
6. Remove:
   - ❏ Auxiliary drive belt.
   - ❏ Auxiliary drive belt tensioner.
7. Lock cooling fan viscous coupling with 5 x 60 mm bolt **11**.
8. Unscrew viscous coupling with Allen key **12**.
9. Remove timing belt upper cover **4**.
10. Turn crankshaft to TDC on No.1 cylinder.
11. Ensure timing marks aligned **5** & **6**.
12. Remove:
    - ❏ Crankshaft pulley bolts **7**.
    - ❏ Crankshaft pulley **8**.
    - ❏ Timing belt lower cover **9**.
13. Slacken tensioner bolt **10**.

14. Move tensioner away from belt. Lightly tighten bolt **10**.
15. Remove timing belt.
    NOTE: Mark direction of rotation on belt with chalk if belt is to be reused.

Installation

1. Ensure timing mark aligned **5**.
2. Fit timing belt to crankshaft sprocket and auxiliary shaft sprocket.
   NOTE: If reusing old belt: Observe direction of rotation marks on belt.
3. Install:
   - ❏ Timing belt lower cover **9**.
   - ❏ Crankshaft pulley **8**.
   - ❏ Crankshaft pulley bolts **7**.
4. Lightly tighten crankshaft pulley bolts **7**.
5. Ensure timing mark on crankshaft pulley aligned **6**.
6. Fit timing belt to tensioner pulley and camshaft sprocket.
7. Slacken tensioner bolt **10**.
8. Fit two-pin wrench **13**. Use wrench Matra V.159.
   *NOTE: Two-pin wrench locates in hole **16** and against lug **17**. When turning two-pin wrench clockwise, the tensioner pulley will turn anti-clockwise and apply tension to belt.*
9. Turn two-pin wrench clockwise **15** until piston **14** **A** is fully extended and piston **14** **B** has risen approximately 1 mm. Use wrench Matra V.159 **13**.
10. Lightly tighten tensioner bolt **10**.
11. Turn crankshaft two turns clockwise.
12. Ensure timing marks aligned **5** & **6**.
13. Check alignment of tensioner lever **14** **C** in relation to upper edge of piston **14** **B** as follows:
    - ❏ **14** **D** – adjustment correct.
    - ❏ **14** **E** – wear area, re-adjust.
    - ❏ **14** **F** – check tensioner and re-adjust.
14. When adjustment correct, dimension **14** **G** will be 25-29 mm.
15. If alignment incorrect or dimension **14** **G** not as specified: Repeat tensioning procedure.
    *NOTE: After installation, correct adjustment is maintained providing upper edge of piston **14** **B** is positioned within areas **14** **D** and **14** **E**.*
16. Tighten tensioner bolt to 25 Nm **10**.
17. Tighten crankshaft pulley bolts to 25 Nm **7**.
18. Install components in reverse order of removal.
19. Push front panel back into place. Remove support guides No.3369 **3**. Fit bolts.
20. Tighten bolts. A4: 45 Nm, A6: 50 Nm **2**.
    Tighten bolts to 10 Nm **1**.

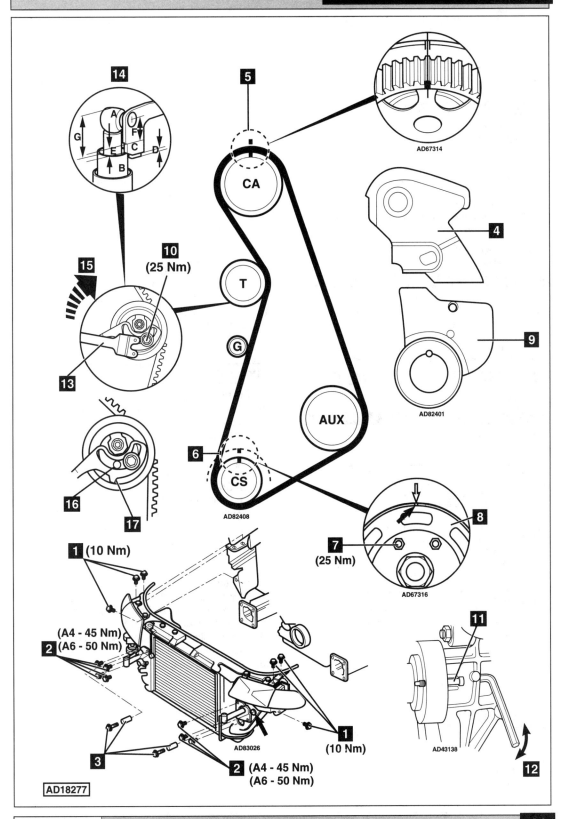

14

A

G
F

E C D

B

5

AD67314

4

10
(25 Nm)

CA

15

T

9

G

AD82401

13

AUX

16

6

CS

17

AD82408

8

7
(25 Nm)

AD67316

1 (10 Nm)

(A4 - 45 Nm)
2 (A6 - 50 Nm)

11

1 (10 Nm)

3

AD83026

12

2 (A4 - 45 Nm)
(A6 - 50 Nm)

AD43138

AD18277

Model:	A4 1,8
Year:	08/97-01
Engine Code:	ADR

Replacement Interval Guide

Audi recommend:
→1999MY: Replacement every
120,000 kilometres (74,564 miles).
2000MY→: Replacement every
180,000 kilometres (111,846 miles).

NOTE: The vehicle manufacturer publishes this information only in kilometres. The conversion to miles is included for reference purposes only.
The previous use and service history of the vehicle must always be taken into account.
Refer to Timing Belt Replacement Intervals at the front of this manual.

Check For Engine Damage

CAUTION: This engine has been identified as an INTERFERENCE engine in which the possibility of valve-to-piston damage in the event of a timing belt failure is MOST LIKELY to occur.
A compression check of all cylinders should be performed before removing the cylinder head(s).

Repair Times – hrs

Remove & install	2,20

Special Tools

■ Support guides – No.3369.

Special Precautions

■ Disconnect battery earth lead.
■ DO NOT turn crankshaft or camshaft when timing belt removed.
■ Remove spark plugs to ease turning engine.
■ Turn engine in normal direction of rotation (unless otherwise stated).
■ DO NOT turn engine via camshaft or other sprockets.
■ Observe all tightening torques.

Removal

1. Raise and support front of vehicle.
2. Remove:
   ❑ Engine undershield.
   ❑ Front bumper.
   ❑ Air filter intake duct.
   ❑ Front panel bolts **1** & **2**.
3. Install support guides No.3369 in front panel **3**.
4. Slide front panel forward.
5. Refit upper rear bolts in front holes to steady panel.
6. Remove:
   ❑ Auxiliary drive belt.
   ❑ Auxiliary drive belt tensioner.

7. Lock cooling fan viscous coupling with 5 x 60 mm bolt **4**.
8. Unscrew viscous coupling with Allen key **5**.
9. Remove timing belt upper cover **6**.
10. Turn crankshaft to TDC on No.1 cylinder.
11. Ensure timing marks aligned **7** & **8**.
12. Remove:
    ❑ Crankshaft pulley bolts **9**.
    ❑ Crankshaft pulley **10**.
    ❑ Timing belt lower cover **11**.
13. Turn tensioner pulley anti-clockwise until holes in pushrod and tensioner body aligned **12**. Use 8 mm Allen key **13**. Retain pushrod with a suitable pin through hole in tensioner body.
    *NOTE: DO NOT slacken bolts **14**.*
14. Remove timing belt.
    NOTE: Mark direction of rotation on belt with chalk if belt is to be reused.

Installation

1. Ensure timing mark aligned **8**.
2. Fit timing belt to crankshaft sprocket and auxiliary shaft sprocket.
   NOTE: If reusing old belt: Observe direction of rotation marks on belt.
3. Install:
   ❑ Timing belt lower cover **11**.
   ❑ Crankshaft pulley **10**.
   ❑ Crankshaft pulley bolts **9**.
4. Lightly tighten crankshaft pulley bolts **9**.
5. Ensure timing marks aligned **7**.
6. Fit timing belt to tensioner pulley and camshaft sprocket.
7. Ensure belt is taut between sprockets on non-tensioned side.
8. Turn tensioner pulley slightly anti-clockwise. Use 8 mm Allen key **13**. Remove pin from tensioner body to release pushrod **12**.
9. Turn crankshaft two turns clockwise. Ensure timing marks aligned **7** & **8**.
10. Install components in reverse order of removal.

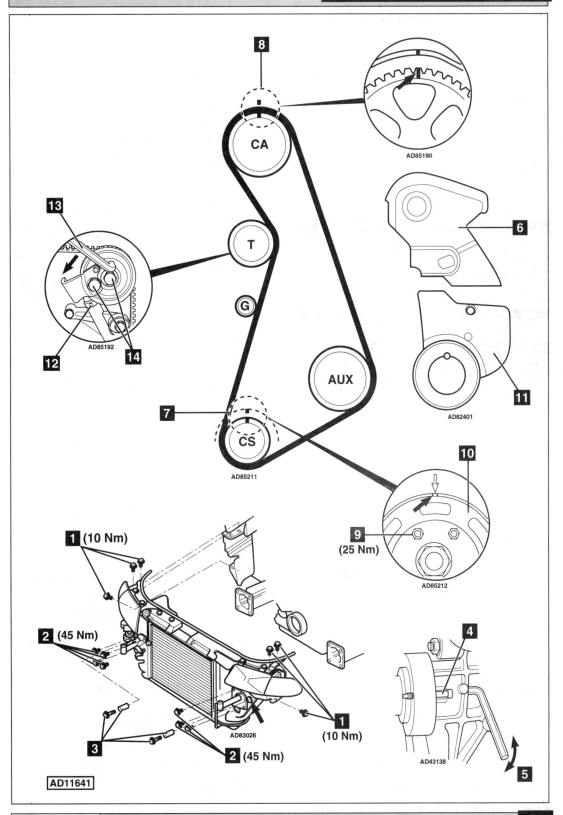

AUDI

Model:	**A4 1,8 Turbo • A4 2,0**
Year:	**2001-03**
Engine Code:	**ALT, AVJ**

Replacement Interval Guide

Audi recommend replacement every 180,000 kilometres (111,846 miles).

NOTE: The vehicle manufacturer publishes this information only in kilometres. The conversion to miles is included for reference purposes only. The previous use and service history of the vehicle must always be taken into account.
Refer to Timing Belt Replacement Intervals at the front of this manual.

Check For Engine Damage

CAUTION: This engine has been identified as an INTERFERENCE engine in which the possibility of valve-to-piston damage in the event of a timing belt failure is MOST LIKELY to occur. A compression check of all cylinders should be performed before removing the cylinder head(s).

Repair Times – hrs

Remove & install	2,40

Special Tools

- Auxiliary drive belt tensioner locking pin – No.T10060.
- Support guides – No.3369.
- Tensioner locking tool – No.T10008.
- Two-pin wrench – No.3387.
- 8 mm drill bit.

Special Precautions

- Disconnect battery earth lead.
- DO NOT turn crankshaft or camshaft when timing belt removed.
- Remove spark plugs to ease turning engine.
- Turn engine in normal direction of rotation (unless otherwise stated).
- DO NOT turn engine via camshaft or other sprockets.
- Observe all tightening torques.

Removal

1. Raise and support front of vehicle.
2. Disconnect exhaust pipe for auxiliary heater from engine undershield (if fitted).
3. Remove:
   - ❑ Engine undershield.
   - ❑ Air filter intake duct.
   - ❑ Front bumper.
   - ❑ Front panel bolts **1**.
4. Install support guides No.3369 in front panel **2**.
5. Remove front panel bolts **3**.
6. Slide front panel forward into service position.
7. Remove:
   - ❑ Engine cover.
   - ❑ Auxiliary drive belt. Use tool No.T10060.
   - ❑ Auxiliary drive belt tensioner **4**.
   - ❑ Timing belt upper cover **5**.
8. Turn crankshaft clockwise to TDC on No.1 cylinder. Ensure timing marks aligned **6** & **7**.
9. Remove:
   - ❑ Crankshaft pulley bolts **8**.
   - ❑ Crankshaft pulley **9**.
   - ❑ Timing belt centre cover **10**.
   - ❑ Timing belt lower cover **11**.
10. Fully insert Allen key into tensioner pulley **12**.
11. Turn tensioner pulley slowly anti-clockwise until locking tool can be inserted **13**. Tool No.T10008.
12. Slacken tensioner nut **14**.
13. Turn tensioner pulley slowly clockwise **15**. Use tool No.3387 **16**.
    NOTE: DO NOT damage tensioner lug **17.
14. Remove timing belt.
    NOTE: Mark direction of rotation on belt with chalk if belt is to be reused.

Installation

1. Ensure camshaft sprocket timing marks aligned **7**.
2. Ensure automatic tensioner unit locked with tool **13**. Tool No.T10008.
3. Fit timing belt to crankshaft sprocket.
4. Install:
   - ❑ Timing belt lower cover **11**.
   - ❑ Crankshaft pulley **9**.
   - ❑ Crankshaft pulley bolts **8**.
5. Lightly tighten crankshaft pulley bolts **8**.
6. Ensure timing marks aligned **6**.
7. Fit timing belt in following order:
   - ❑ Water pump sprocket.
   - ❑ Tensioner pulley.
   - ❑ Camshaft sprocket.
   NOTE: If reusing old belt: Observe direction of rotation marks on belt.
8. Turn tensioner pulley slowly anti-clockwise **18** until locking tool **13** can be removed. Use tool No.3387 **16**.
9. Turn tensioner pulley slowly clockwise **15** until dimension **19** is 8 mm. Use drill bit **20**.
   NOTE: Engine must be COLD.
10. Tighten tensioner nut **14**. Tightening torque: 27 Nm.
11. Turn crankshaft slowly two turns clockwise to TDC on No.1 cylinder.
12. Ensure timing marks aligned **6** & **7**.
13. Ensure dimension **19** is 6-10 mm. Use drill bit **20**.
14. If not:
    - ❑ Fully insert Allen key into tensioner pulley **12**.
    - ❑ Turn tensioner pulley slowly anti-clockwise until locking tool can be inserted **13**. Tool No.T10008.
    - ❑ Slacken tensioner nut **14**.
    - ❑ Turn tensioner pulley slowly anti-clockwise **18** until locking tool **13** can be removed. Use tool No.3387 **16**.
    - ❑ Turn tensioner pulley slowly clockwise **15** until dimension **19** is 8 mm. Use drill bit **20**.
    - ❑ Tighten tensioner nut **14**. Tightening torque: 27 Nm.
    - ❑ Turn crankshaft slowly two turns clockwise to TDC on No.1 cylinder.
    - ❑ Ensure dimension **19** is 6-10 mm. Use drill bit **20**.
15. Install components in reverse order of removal.
16. Use new crankshaft pulley bolts **8**. Tightening torque: 10 Nm + 90°.

/Autodata

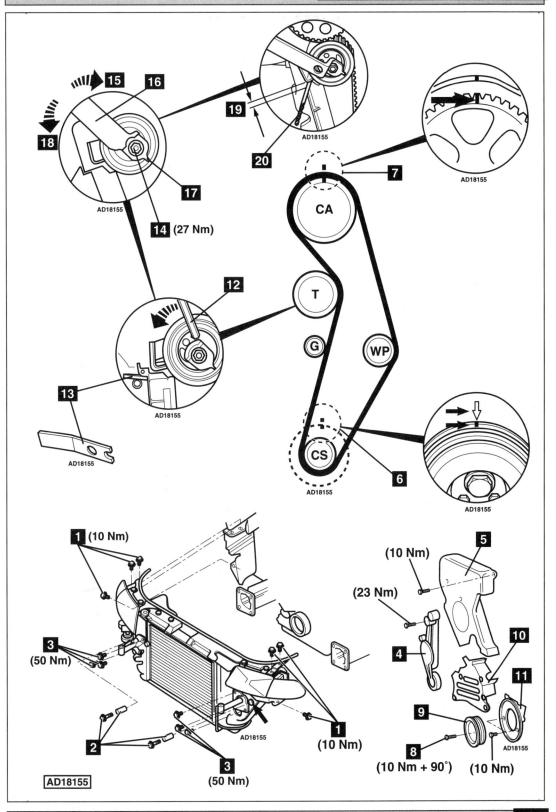

15 **16**

18

19

20

7

17

14 (27 Nm)

CA

12

T

G   WP

13

CS

6

1 (10 Nm)

5

(10 Nm)

(23 Nm)

3
(50 Nm)

10

4

11

9

2

1
(10 Nm)

8

3
(50 Nm)

(10 Nm + 90°)   (10 Nm)

AD18155

Model:	A4 2,4 • A4 2,8 • A6 2,4 • A6 2,8 • A8 2,8 • S4 2,7 Turbo
Year:	1997-03
Engine Code:	AGA, AJG, ALF, AGB, ALG, ALW, APR, AQD

Replacement Interval Guide

Audi recommend:
Replacement every 120,000 kilometres (74,564 miles).
2,7: Replacement of tensioner pulley, guide pulley and tensioner lever every 120,000 kilometres (74,564 miles).
2,8: Replacement of tensioner pulley every 120,000 kilometres (74,564 miles).

NOTE: The vehicle manufacturer publishes this information only in kilometres. The conversion to miles is included for reference purposes only.
The previous use and service history of the vehicle must always be taken into account.
Refer to Timing Belt Replacement Intervals at the front of this manual.

Check For Engine Damage

CAUTION: This engine has been identified as an INTERFERENCE engine in which the possibility of valve-to-piston damage in the event of a timing belt failure is MOST LIKELY to occur.
A compression check of all cylinders should be performed before removing the cylinder head(s).

Repair Times – hrs

Remove & install:

A4	2,90
A6	3,50
A8	1,90
S4	3,70

Special Tools

- Crankshaft locking pin – No.3242.
- Camshaft sprocket puller – No.T40001.
- Camshaft locking tool – No.3391.
- Viscous fan holder – No.3212.
- Viscous fan spanner – No.3312.
- Support guides – No.3369.

Special Precautions

- Disconnect battery earth lead.
- DO NOT turn crankshaft or camshaft when timing belt removed.
- Remove spark plugs to ease turning engine.
- Turn engine in normal direction of rotation (unless otherwise stated).
- DO NOT turn engine via camshaft or other sprockets.
- Observe all tightening torques.

Removal

1. Remove:
   - ❏ Engine cover.
   - ❏ Engine undershield.
   - ❏ Intercooler hoses (Turbo).

A4/A6

2. Remove:
   - ❏ Front bumper.
   - ❏ Air intake pipe between radiator support panel and air filter.
   - ❏ Radiator support panel bolt **1**.
3. Install support guides No.3369 through holes in panel **2**.
4. Remove radiator support panel bolts **3** & **4**.
5. Move radiator support panel into service position.

All models

6. Disconnect secondary air injection hose (if fitted).
7. Remove:
   - ❏ Viscous fan (LH thread). Use tool Nos.3212 & 3312.
   - ❏ Auxiliary drive belt.
   - ❏ Auxiliary drive belt tensioner.
   - ❏ Timing belt covers **5**, **6** & **7**.
8. Turn crankshaft clockwise to TDC on No.3 cylinder. Ensure crankshaft pulley timing marks aligned **8**.
9. Ensure large holes in locking plates of camshaft sprockets face in towards each other **9**.
10. If large holes face outwards: Turn crankshaft one turn clockwise.
11. Remove blanking plug from cylinder block. Fit locking pin No.3242 **10**.
    NOTE: TDC hole in crankshaft web must be aligned with blanking plug hole.
12. Turn tensioner pulley slowly clockwise until holes in pushrod and tensioner body aligned. Use Allen key **11**.
13. Lock tensioner pushrod in position with a 2 mm diameter pin **12**.

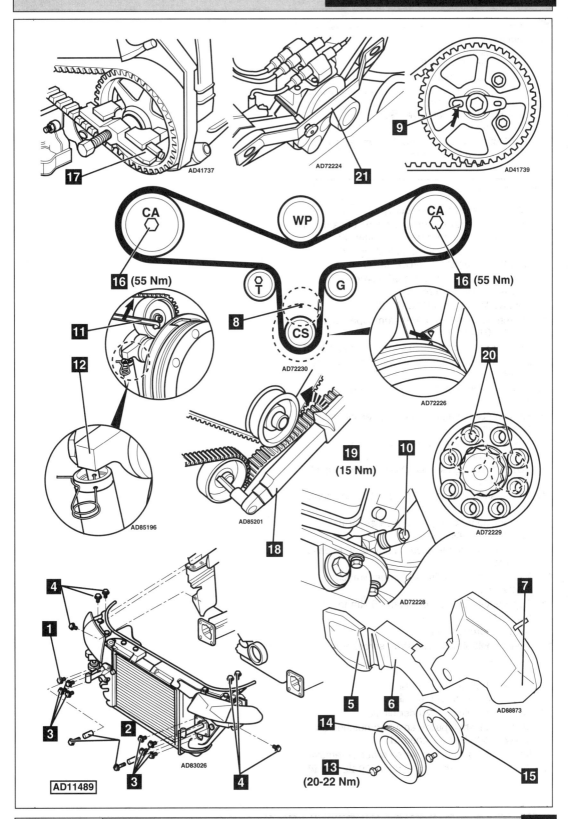

AD41737

17

AD72224

21

9

AD41739

CA WP CA

16 (55 Nm)   16 (55 Nm)

T   G

8

11

12

CS

AD72230

AD72226

20

19
(15 Nm)

10

AD85196

AD85201

18

AD72229

AD72228

4

1

5   6   7

AD88873

3

2

14

3   AD83026

13
(20-22 Nm)

4

15

AD11489

←

14. Remove:
   ❑ Crankshaft pulley bolts **13**.
   ❑ Crankshaft pulley **14**.
   ❑ Viscous fan mounting bracket.
   NOTE: Two bolts of mounting bracket accessed through hole in pulley.
   ❑ Timing belt lower cover **15**.
   ❑ Timing belt.
   NOTE: Mark direction of rotation on belt with chalk if belt is to be reused.

Installation

NOTE: Replace tensioner pulley (2,7/2,8).

1. Ensure large holes in locking plates of camshaft sprockets face in towards each other **9**.
2. Temporarily install camshaft locking tool No.3391 to camshaft sprockets **21**.
3. Slacken bolt of each camshaft sprocket approximately five turns **16**.
4. Remove locking tool No.3391 **21**.
5. Loosen camshaft sprockets from taper. Use tool No.T40001 **17**.
6. Lightly tighten bolt of each camshaft sprocket **16**.
7. Ensure camshaft sprockets can turn freely but not tilt.
8. Ensure crankshaft locking pin located correctly **10**.
9. Fit timing belt in anti-clockwise direction, starting at crankshaft sprocket.
10. Ensure belt is taut between sprockets on non-tensioned side.
    NOTE: If reusing old belt: Observe direction of rotation marks.
11. Fit camshaft locking tool No.3391 to camshaft sprockets **21**.
12. Turn tensioner pulley clockwise as far as possible. Use Allen key **11**.
13. Remove pin from tensioner body **12** to release pushrod.
14. Remove Allen key. Install torque wrench to tensioner **18**.
15. Apply anti-clockwise torque of 15 Nm to tensioner **19**.
16. Remove torque wrench.
17. Tighten bolt of each camshaft sprocket to 55 Nm **16**.
18. Remove:
    ❑ Camshaft locking tool **21**.
    ❑ Crankshaft locking pin **10**.
19. Fit blanking plug to cylinder block.
20. Install crankshaft pulley ensuring notches on pulley and hub are aligned **20**.
21. Install components in reverse order of removal.
22. Tighten crankshaft pulley bolts to 20-22 Nm **13**.

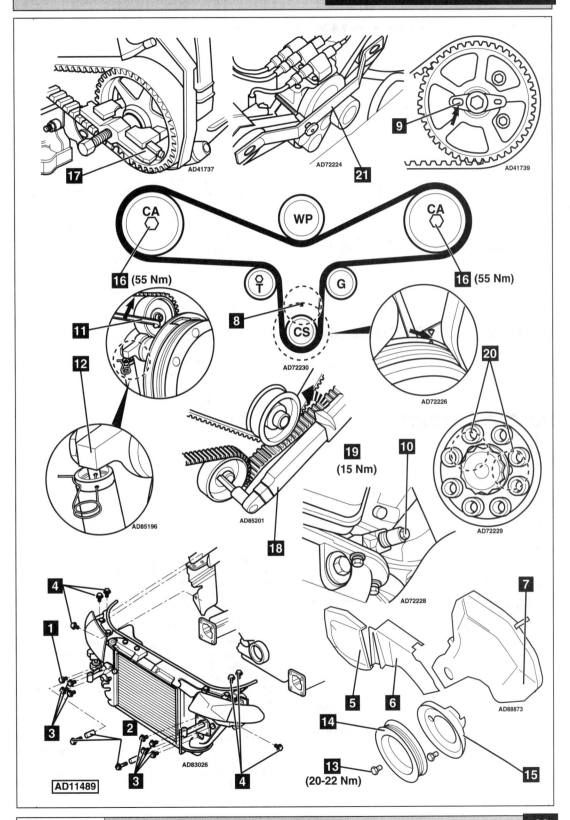

17 AD41737

21 AD72224

9 AD41739

CA WP CA

16 (55 Nm) **16** (55 Nm)

11

12

8

CS

AD72230

AD72226

20 AD72229

AD85196

19 (15 Nm)

10

AD85201

18

AD72228

7

5 **6** AD88873

14

13 (20-22 Nm)

15

4

1

3

2

3

4

AD83026

AD11489

Model:	A4 2,8 30V • A6 2,8 30V • A8 2,8 30V
Year:	1995-03
Engine Code:	ACK

Replacement Interval Guide

Audi recommend replacement every 120,000 kilometres (74,564 miles) (tensioner pulley must also be replaced).

NOTE: The vehicle manufacturer publishes this information only in kilometres. The conversion to miles is included for reference purposes only.
The previous use and service history of the vehicle must always be taken into account.
Refer to Timing Belt Replacement Intervals at the front of this manual.

Check For Engine Damage

CAUTION: This engine has been identified as an INTERFERENCE engine in which the possibility of valve-to-piston damage in the event of a timing belt failure is MOST LIKELY to occur.
A compression check of all cylinders should be performed before removing the cylinder head.

Repair Times – hrs

Remove & install:

A4	2,90
A6 →1997	2,50
A6 1997→	3,50
A8	1,90

Special Tools

- Crankshaft locking tool – No.3242.
- Camshaft locking tool – No.3391.
- Camshaft sprocket puller – No.3032.
- Support guides – No.3369.

Special Precautions

- Disconnect battery earth lead.
- DO NOT turn crankshaft or camshaft when timing belt removed.
- Remove spark plugs to ease turning engine.
- Turn engine in normal direction of rotation (unless otherwise stated).
- DO NOT turn engine via camshaft or other sprockets.
- Observe all tightening torques.

Removal

1. Raise and support front of vehicle.
2. A6 1997→/A4: Move radiator support panel into service position.
   - Remove front bumper.
   - Remove air intake pipe between front panel and air filter.
   - Remove front panel bolts.
   - Install support guides No.3369 in front panel.
   - Slide front panel forward.
   - Refit upper rear bolts in front holes to steady panel.
3. Remove:
   - Viscous fan (LH thread).
   - Auxiliary drive belt.
   - Timing belt LH and RH covers **1** & **2**.

4. Turn crankshaft clockwise to TDC on No.3 cylinder. Ensure timing marks aligned **3**.
5. Ensure large holes in locking plates of camshaft sprockets face in towards each other **4**.
6. If not: Turn crankshaft one turn clockwise.
7. Remove blanking plug from crankcase. Screw in crankshaft locking tool **5**. Tool No.3242.
   NOTE: TDC hole in crankshaft web must be aligned with blanking plug hole.
8. Remove auxiliary drive belt tensioner.
9. Turn tensioner pulley clockwise until holes in pushrod and tensioner body aligned. Use 8 mm Allen key **6**. Retain pushrod with 2 mm diameter pin through hole in tensioner body **7**.
10. Remove:
    - Crankshaft pulley bolts **8**.
    - Crankshaft pulley.
    - Viscous fan mounting bracket.
    NOTE: Two bolts of mounting bracket accessed through hole in pulley.
    - Timing belt lower cover **9**.
    - Timing belt.
    NOTE: Mark direction of rotation on belt with chalk if belt is to be reused.

Installation

1. Remove bolt of each camshaft sprocket **10**.
2. Screw M10 bolt into camshaft to act as support for puller.
3. Remove both camshaft sprockets. Use puller No.3032 **11**.
4. Install:
   - Camshaft sprockets.
   - Locking plates **4**.
5. Lightly tighten bolt of each camshaft sprocket **10**.
6. Ensure camshaft sprockets can turn but not tilt.
7. Fit timing belt to camshaft sprockets and water pump pulley.
8. Fit locking tool to camshafts. Tool No.3391.
9. Ensure crankshaft locking tool fitted **5**.
10. Fit timing belt to guide pulley, crankshaft sprocket and tensioner pulley.
11. Turn tensioner pulley slightly clockwise. Use 8 mm Allen key **6**. Remove pin from tensioner body to release pushrod **7**.
12. Install torque wrench to hexagon of tensioner.
13. Tension timing belt in anti-clockwise direction to 15 Nm **12**.
14. Remove torque wrench.
15. Tighten bolt of each camshaft sprocket to 55 Nm **10**.
16. Remove:
    - Camshaft locking tool.
    - Crankshaft locking tool **5**.
17. Fit blanking plug.
18. Fit crankshaft pulley. Ensure notches aligned with tab on crankshaft sprocket **13**.
19. Tighten crankshaft pulley bolts to 20 Nm **8**.
20. Install components in reverse order of removal.

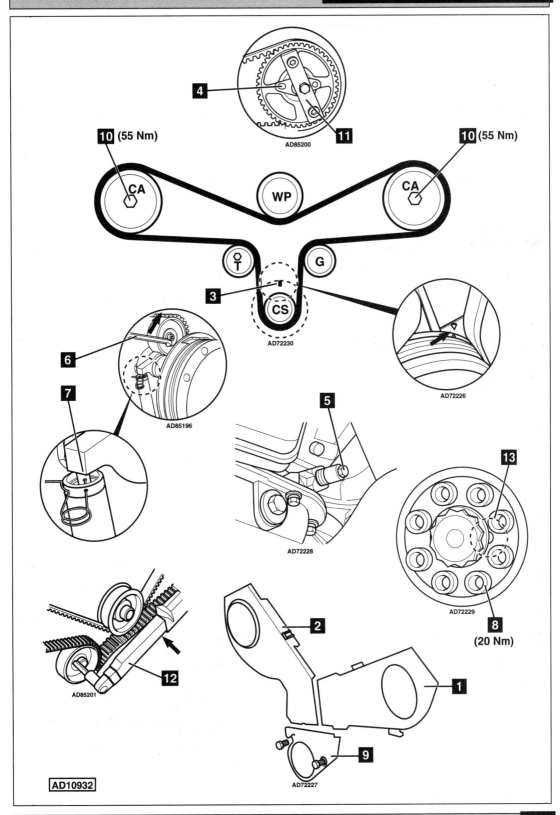

AD85200

10 (55 Nm)

10 (55 Nm)

CA

WP

CA

O T

G

CS

AD72230

AD72226

6

AD85196

7

5

AD72228

13

AD72229

2

8
(20 Nm)

12

AD85201

1

9

AD72227

AD10932

AUDI

Model:	**A4 3,0 V6 • A6 3,0 V6**
Year:	**2001-03**
Engine Code:	**ASN, BBJ**

Replacement Interval Guide

Audi recommend replacement every 120,000 kilometres (74,564 miles).

NOTE: The vehicle manufacturer publishes this information only in kilometres. The conversion to miles is included for reference purposes only. The previous use and service history of the vehicle must always be taken into account.
Refer to Timing Belt Replacement Intervals at the front of this manual.

Check For Engine Damage

CAUTION: This engine has been identified as an INTERFERENCE engine in which the possibility of valve-to-piston damage in the event of a timing belt failure is MOST LIKELY to occur.
A compression check of all cylinders should be performed before removing the cylinder head.

Repair Times – hrs

Remove & install	4,50

Special Tools

- Crankshaft locking tool – No.T40026.
- Camshaft locking tools – No.T40030.
- Camshaft sprocket adjusting tool – No.T40028.
- Tensioner locking pin – No.T40011.
- PAS pump pulley locking tool – No.3212.
- Two-pin wrench – No.3387.
- Support guides – No.3369.
- Auxiliary drive belt tensioner tool – No.3299.

Special Precautions

- Disconnect battery earth lead.
- DO NOT turn crankshaft or camshaft when timing belt removed.
- Remove spark plugs to ease turning engine.
- Turn engine in normal direction of rotation (unless otherwise stated).
- DO NOT turn engine via camshaft or other sprockets.
- Observe all tightening torques.

Removal

1. Raise and support front of vehicle.
2. Disconnect exhaust pipe for auxiliary heater from engine undershield (if fitted).
3. Remove:
   - ❏ Engine undershield.
   - ❏ Air filter cover.
   - ❏ Air intake pipe between front panel and air filter.
   - ❏ LH engine cover.
   - ❏ Front bumper.
   - ❏ Front panel bolts **1**.

4. Install support guides No.3369 in front panel **2**.
5. Remove front panel bolts **3**.
6. Slide front panel forward.
7. Refit upper rear bolts in rear holes to steady front panel.
8. Remove:
   - ❏ Front engine cover.
9. Turn auxiliary drive belt tensioner clockwise. Use tool No.3299.
10. Remove auxiliary drive belt.
    NOTE: Mark direction of rotation on auxiliary drive belt with chalk.
11. Remove:
    - ❏ Auxiliary drive belt tensioner.
    - ❏ PAS pump pulley. Use tool No.3212.
    - ❏ Crankshaft pulley bolts **4**.
    - ❏ Crankshaft pulley **5**.
    *NOTE: Thrust washer **6** only fitted to crankshaft sprocket with part No. 06C 105 063 A.*
12. Remove:
    - ❏ Timing belt covers **7**.
    - ❏ Rear engine cover.
13. Move coolant expansion tank to one side. DO NOT disconnect hoses.
14. Remove:
    - ❏ Air filter housing and hoses.
    - ❏ Dipstick and tube.
    - ❏ Air hose from RH cylinder head.
    - ❏ Ignition coils.
    - ❏ Both rear crankcase breather hoses.
    - ❏ Both cylinder head covers.
15. Turn crankshaft clockwise until camshaft lobes for No.3 cylinder (CA1 & CA2) angled upwards **8**.
16. Fit locking tool to camshafts. No.3 cylinder **9**. Tool No.T40030.
17. Fit locking tool to camshafts. No.4 cylinder **10**. Tool No.T40030.
    NOTE: Rock crankshaft slightly to ensure locking tools located correctly.
    Tighten locking tools no more than 10 Nm.
18. Disconnect wiring from engine coolant blower run-on motor (if fitted).
19. Remove blanking plug from cylinder block **11**.
20. Screw in crankshaft locking tool. Tool No.T40026.
21. Remove circlip and centre blanking cap from each camshaft sprocket.
22. Slacken bolt of each camshaft sprocket slightly **12**.

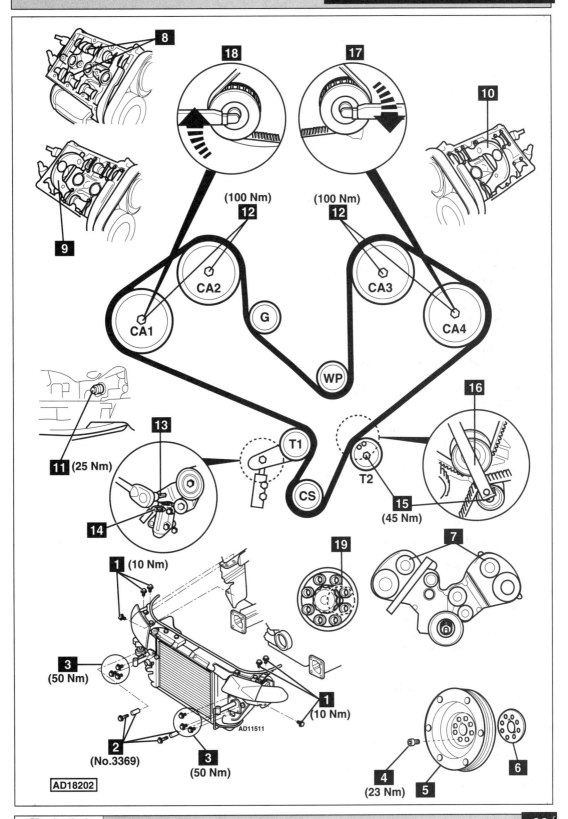

←

23. Turn tensioner pulley (T1) anti-clockwise until holes in pushrod and tensioner body aligned. Use 8 mm Allen key **13**.

24. Retain pushrod with locking pin through hole in tensioner body **14**. Tool No.T40011.

25. Slacken tensioner pulley bolt (T2) **15**.

26. Remove timing belt.

NOTE: Mark direction of rotation on belt with chalk if belt is to be reused.

Installation

1. Lightly tighten bolt of each camshaft sprocket **12**. Ensure camshaft sprockets can turn freely but not tilt.

2. Fit timing belt in clockwise direction, starting at crankshaft sprocket.

NOTE: Check edge of timing belt aligned with edge of sprockets.

3. Turn tensioner pulley (T2) clockwise until tool aligned with centre of water pump pulley **16**. Tool No.3387.

4. Hold tool and tighten tensioner pulley bolt **15**. Tightening torque: 45 Nm.

5. Install torque wrench to hexagon of tensioner pulley (T1) **13**.

6. Apply clockwise torque of 45 Nm to tensioner pulley (T1) **13**.

7. Turn tensioner pulley (T1) slowly anti-clockwise until locking pin can be removed **14**.

8. Remove locking pin from tensioner body to release pushrod **14**.

9. Tension timing belt in clockwise direction to 25 Nm **13**.

10. Remove torque wrench.

11. Fit adjusting tool to camshaft sprocket CA4. Tool No.T40028.

12. Turn camshaft sprocket clockwise and apply a torque of 10 Nm **17**.

13. Fit adjusting tool to camshaft sprocket CA1. Tool No.T40028.

14. Turn camshaft sprocket clockwise and apply a torque of 10 Nm **18**.

15. Tighten bolt of each camshaft sprocket **12**. Tightening torque: 100 Nm.

16. Fit centre blanking cap and circlip to each camshaft sprocket.

17. Remove:
   ❑ Camshaft locking tools **9** & **10**.
   ❑ Crankshaft locking tool **11**.

18. Fit blanking plug. Tightening torque: 25 Nm.

19. Fit crankshaft pulley.

*NOTE: Ensure notches aligned with tab on crankshaft sprocket **19**.*

20. Tighten crankshaft pulley bolts to 23 Nm **4**.

21. Install components in reverse order of removal.

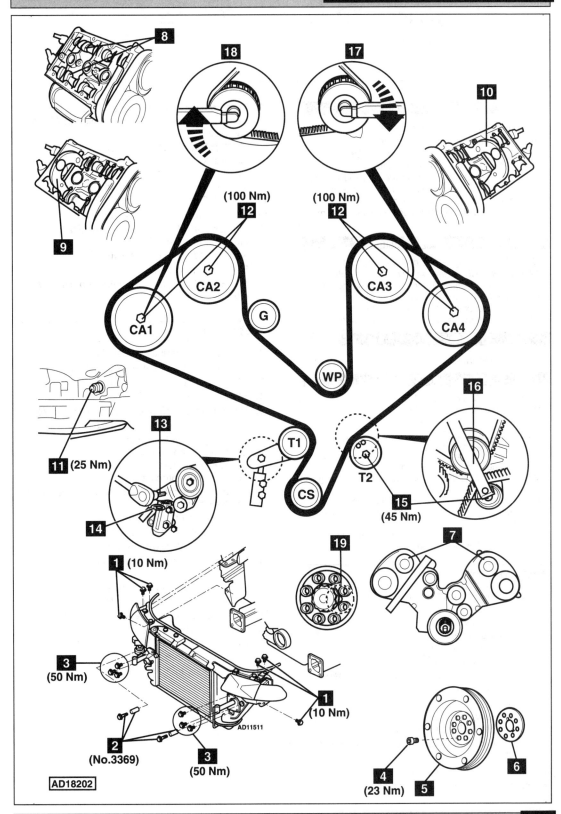

8

18

17

10

9

(100 Nm)
12

(100 Nm)
12

CA2

CA3

CA1

G

CA4

WP

11 (25 Nm)

13

T1

16

14

T2

15
(45 Nm)

CS

1 (10 Nm)

19

7

3
(50 Nm)

1
(10 Nm)

AD11511

2
(No.3369)

3
(50 Nm)

4
(23 Nm)

5

6

AD18202

AUDI

Model:	A6 1,8
Year:	1997-03
Engine Code:	AJP

Check For Engine Damage

*CAUTION: This engine has been identified as an
INTERFERENCE engine in which the possibility of
valve-to-piston damage in the event of a timing belt
failure is MOST LIKELY to occur.*
*A compression check of all cylinders should be
performed before removing the cylinder head(s).*

Repair Times – hrs

Remove and install	2,90

Special Tools

- 1 x M5 x 55 mm stud and nut.
- Support guides – No.3369.

Special Precautions

- Disconnect battery earth lead.
- DO NOT turn crankshaft or camshaft when timing belt removed.
- Remove spark plugs to ease turning engine.
- Turn engine in normal direction of rotation (unless otherwise stated).
- DO NOT turn engine via camshaft or other sprockets.
- Observe all tightening torques.

Removal

1. Raise and support front of vehicle.
2. Remove:
   - Engine undershield.
   - Front bumper.
   - Air filter intake duct.
   - Front panel bolts **1** & **2**.
3. Install support guides No.3369 in front panel **3**.
4. Slide front panel forward.
5. Remove:
   - Auxiliary drive belt.
   - Auxiliary drive belt tensioner.
6. Lock cooling fan viscous coupling with 5 mm pin **4**.
7. Unscrew viscous coupling with Allen key.
8. Remove timing belt upper cover **5**.

9. Turn crankshaft clockwise to TDC on No.1 cylinder. Ensure timing marks aligned **6** & **7**.
10. Remove:
    - Crankshaft pulley bolts **8**.
    - Crankshaft pulley **9**.
    - Timing belt centre cover **10**.
    - Timing belt lower cover **11**.
11. Insert M5 stud into tensioner **12**.
12. Fit nut and washer to stud **13**.
13. Tighten nut sufficiently to allow a suitable locking pin to be inserted **14**.
    NOTE: DO NOT overtighten nut.
14. Remove timing belt.
    NOTE: Mark direction of rotation on belt with chalk if belt is to be reused.

Installation

1. Ensure timing marks aligned **6**.
2. Fit timing belt to crankshaft sprocket.
3. Fit timing belt lower cover **11**.
4. Fit crankshaft pulley **9**.
5. Tighten crankshaft pulley bolts to 25 Nm **8**.
6. Ensure timing marks aligned **7**.
7. Fit timing belt in following order:
   - Water pump sprocket.
   - Tensioner pulley.
   - Camshaft sprocket.
   NOTE: Ensure belt is taut between sprockets on non-tensioned side.
8. Remove locking pin **14**.
9. Remove nut and stud **12** & **13**.
10. Turn crankshaft two turns clockwise.
11. Ensure timing marks aligned **6** & **7**.
12. Install:
    - Timing belt centre cover **10**.
    - Timing belt upper cover **5**.
13. Install components in reverse order of removal.
14. Push front panel back into place. Remove support guides No.3369 **3**. Fit bolts.
15. Tighten bolts to 50 Nm **2**.
16. Tighten bolts to 10 Nm **1**.

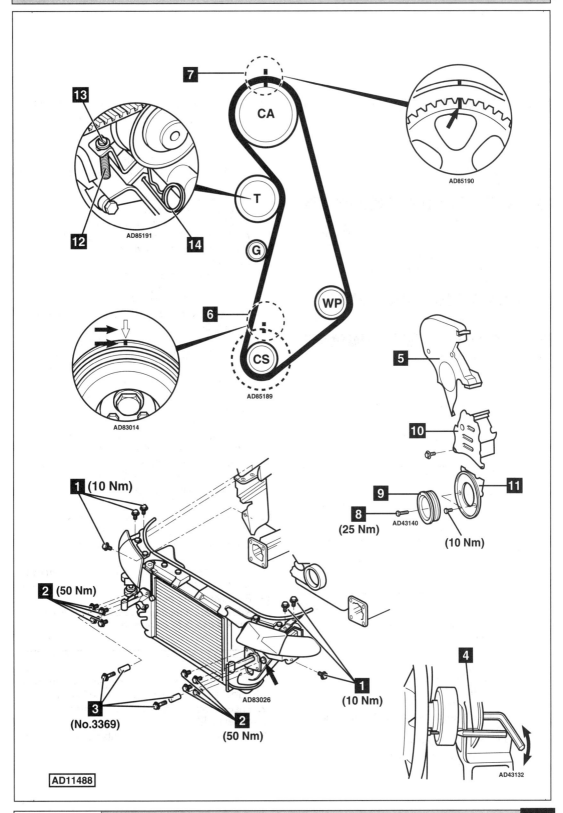

7

CA

13

T

G

12 AD85191 14

WP

6

CS

AD85189

AD85190

AD83014

5

10

9

8
AD43140
(25 Nm)

11

(10 Nm)

1 (10 Nm)

2 (50 Nm)

3

(No.3369)

AD83026

1
(10 Nm)

2
(50 Nm)

4

AD43132

AD11488

AUDI

Model:	**A6 4,2 V8**
Year:	**1999-03**
Engine Code:	**ARS**

Replacement Interval Guide

Audi recommend replacement every 80,000 miles.
The previous use and service history of the vehicle must always be taken into account.
Refer to Timing Belt Replacement Intervals at the front of this manual.

Check For Engine Damage

CAUTION: This engine has been identified as an INTERFERENCE engine in which the possibility of valve-to-piston damage in the event of a timing belt failure is MOST LIKELY to occur.
A compression check of all cylinders should be performed before removing the cylinder head.

Repair Times – hrs

Remove & install	3,30

Special Tools

- Support guides – No.3369.
- Viscous fan holding tool – No.3212.
- Viscous fan spanner – No.3312.
- Crankshaft locking pin – No.3242.
- Camshaft sprocket locking tool – No.T40005.
- Camshaft sprocket puller – No.T4000.
- Tensioner pushrod locking pin – No.2024A.
- Two-pin wrench – No.40009.

Special Precautions

- Disconnect battery earth lead.
- DO NOT turn crankshaft or camshaft when timing belt removed.
- Remove spark plugs to ease turning engine.
- Turn engine in normal direction of rotation (unless otherwise stated).
- DO NOT turn engine via camshaft or other sprockets.
- Observe all tightening torques.

Removal

1. Raise and support front of vehicle.
2. Drain coolant.
3. Remove:
   - ❏ Engine top cover.
   - ❏ Engine undershield.
   - ❏ Front bumper.
   - ❏ Top and bottom hoses.
   - ❏ Air intake hose between air filter and throttle body.
   - ❏ Radiator support panel bolt **1**.
4. Install support guides No.3369 through holes in support panel **2**.

5. Remove radiator support panel bolts **3** & **4**.
6. Move radiator support panel into service position.
7. Remove:
   - ❏ Upper air cowl for fans.
   - ❏ Viscous fan (LH thread). Use tool Nos.3212 & 3312.
   - ❏ Viscous fan casing.
   - ❏ Auxiliary drive belt.
   - ❏ Auxiliary drive belt tensioner.

   NOTE: Mark direction of rotation on belt with chalk if belt is to be reused.
8. Remove:
   - ❏ Crankshaft pulley bolts **5**.
   - ❏ Crankshaft pulley **6**.
   - ❏ Timing belt covers **7**.
9. Turn crankshaft clockwise until timing marks aligned **8**.
10. Ensure large holes in locking plates of camshaft sprockets face in towards each other **9**.
11. Remove blanking plug from LH cylinder block. Fit locking pin No.3242 **10**.
12. Remove front RH torque reaction link support bracket.
13. Turn tensioner pulley (T1) slowly anti-clockwise until holes in pushrod and tensioner body aligned. Use 8 mm Allen key **11**.
14. Lock tensioner pushrod in position with a 2 mm diameter pin **12**. Tool No.2024A.
15. Lock camshaft sprockets. Use tool No.T40005 **13**.
16. Slacken bolt of each camshaft sprocket 5 turns **14**. Remove locking tool.
17. Loosen camshaft sprockets from taper **15**. Use tool No.T4000.
18. Slacken tensioner pulley bolt (T2) **16**. Remove tensioner pulley.
19. Remove timing belt.

   NOTE: Mark direction of rotation on belt with chalk if belt is to be reused.

Installation

1. Ensure large holes in locking plates of camshaft sprockets face in towards each other **9**.
2. Ensure crankshaft locking pin located correctly **10**.
3. Fit tensioner pulley (T2). Tighten bolt finger tight.

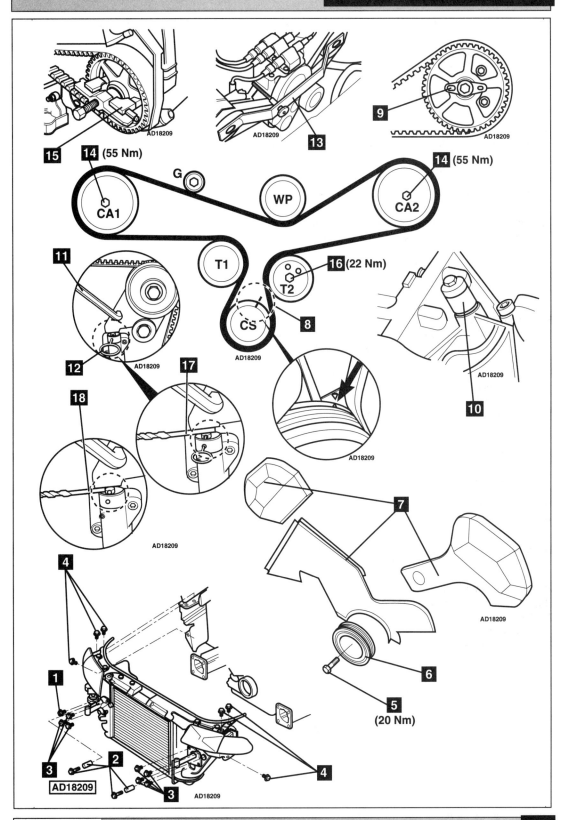

15 **14** (55 Nm) **14** (55 Nm)

G

CA1 **WP** **CA2**

11 **T1** **16** (22 Nm)

T2

12 **CS** **8**

17 **10**

18

7

4

1

6

5
(20 Nm)

3 **2** **3**

AD18209

4. Fit timing belt in following order:
   - ❏ Crankshaft sprocket.
   - ❏ Tensioner pulley (T1).
   - ❏ Tensioner pulley (T2).
   - ❏ Camshaft sprocket (CA1).
   - ❏ Guide pulley.
   - ❏ Water pump pulley.

5. Remove camshaft sprocket (CA2). Fit camshaft sprocket to belt, then install camshaft sprocket with belt onto end of camshaft.

6. Tighten bolt of each camshaft sprocket finger tight.

7. Ensure camshaft sprockets can turn freely but not tilt.

8. Lock camshaft sprockets. Tool No.T40005 **13**.

9. Insert a 5 mm drill bit between tensioner arm and pushrod **17**.

10. Apply anti-clockwise torque of 4 Nm to tensioner pulley (T2). Use tool No.T40009.

11. Tighten tensioner pulley bolt (T2) **16**.
    Tightening torque: 22 Nm.

12. Remove 5 mm drill bit **17**.

13. Turn tensioner pulley (T1) slowly anti-clockwise until locking pin can be removed **12**. Use 8 mm Allen key **11**.

14. Turn tensioner pulley (T1) slowly clockwise until a 7 mm drill bit can be inserted between tensioner arm and pushrod **18**.

15. Tighten bolt of each camshaft sprocket **14**.
    Tightening torque: 55 Nm.

16. Remove:
    - ❏ Camshaft sprocket locking tool **13**.
    - ❏ 7 mm drill bit.
    - ❏ Crankshaft locking pin **10**.

17. Turn crankshaft two turns clockwise.

18. Insert crankshaft locking pin **10**.

19. Ensure large holes in locking plates of camshaft sprockets face in towards each other **9**.

20. Ensure 5 mm drill bit can be inserted between tensioner arm and pushrod **18**. If not: Repeat tensioning procedure.

21. Remove crankshaft locking pin **10**.

22. Install components in reverse order of removal.

23. Tighten crankshaft pulley bolts **5**.
    Tightening torque: 20 Nm.

24. Refill cooling system.

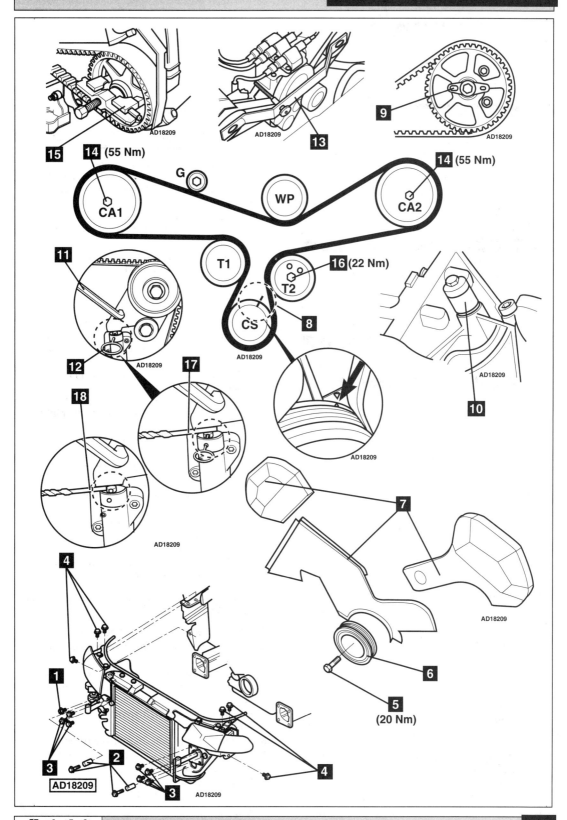

AUDI

Model:	A8 4,2 V8
Year:	1998-03
Engine Code:	AKG, AKH

Replacement Interval Guide

Audi recommend replacement every 120,000 kilometres (74,564 miles).

NOTE: The vehicle manufacturer publishes this information only in kilometres. The conversion to miles is included for reference purposes only. *The previous use and service history of the vehicle must always be taken into account.*
Refer to Timing Belt Replacement Intervals at the front of this manual.

Check For Engine Damage

CAUTION: This engine has been identified as an INTERFERENCE engine in which the possibility of valve-to-piston damage in the event of a timing belt failure is MOST LIKELY to occur.
A compression check of all cylinders should be performed before removing the cylinder head.

Repair Times – hrs

Remove & install	4,20

Special Tools

- Crankshaft pulley holding tool – No.3197.
- Crankshaft pulley bolt installer – No.2079.
- Camshaft locking tool – 2 x No.3341.
- Sprocket holding tool – No.3036.
- Two-pin wrench – Matra V.159.
- Viscous fan holding tool – No.3212.

Special Precautions

- Disconnect battery earth lead.
- DO NOT turn crankshaft or camshaft when timing belt removed.
- Remove spark plugs to ease turning engine.
- Turn engine in normal direction of rotation (unless otherwise stated).
- DO NOT turn engine via camshaft or other sprockets.
- Observe all tightening torques.

Removal

1. Drain coolant.
2. Remove:
   - Engine undershield.
   - Engine top cover.
   - Top and bottom hoses.
   - Air intake hose between air filter and throttle body.
   - Upper air cowl for fans.
   - Cooling fan.
   - Viscous fan (LH thread). Use tool No.3212.
   - Viscous fan casing.
   - Auxiliary drive belt.
   - Front engine mounting support bar.

3. Hold crankshaft pulley. Use tool No.3197. Remove crankshaft bolt **1**.
4. Turn crankshaft to TDC on No.1 cylinder. Ensure timing marks aligned **2**.
5. Remove Hall effect sensor cover on rear of LH camshaft. Check Hall effect sensor gap aligned. If not: Turn crankshaft one turn clockwise.
6. Remove:
   - Timing belt LH cover **3**.
   - Timing belt RH cover and auxiliary drive belt tensioner **4**.
7. Hold camshaft sprockets. Use tool No.3036. Slacken bolts two turns **5**.
8. Remove:
   - Hall effect sensor from LH camshaft **6**.
   - Cover plate for rear of RH camshaft.
9. Fit locking tool to each camshaft **7**. Tool No.3341. If necessary: Remove cylinder head covers and turn end of camshafts with spanner.
   NOTE: DO NOT use camshaft locking tools to counterhold camshafts.
10. Slacken tensioner pulley nut **8**. Turn tensioner pulley away from belt **9**. Use wrench Matra V.159 **10**.
11. Compress tensioner damper **13**. Remove tensioner pulley.
12. Remove timing belt from camshaft sprockets.
13. Tap each camshaft sprocket gently with plastic hammer to loosen sprocket from taper.
14. Remove:
    - Crankshaft pulley bolts **11**.
    - Crankshaft pulley **12**.
15. Remove timing belt.
    NOTE: Mark direction of rotation on belt with chalk if belt is to be reused.

Installation

1. Fit timing belt to crankshaft sprocket.
2. Fit crankshaft pulley **12**. Tighten bolts finger tight **11**.
   NOTE: Fit new crankshaft pulley centre bolt.
3. Fit crankshaft bolt **1**. Coat threads with locking compound. Tighten bolt finger tight.
4. Ensure each camshaft sprocket can turn freely on taper.
5. Fit timing belt in anti-clockwise direction.
6. Fit tensioner pulley with timing belt **9**. Tighten tensioner pulley nut finger tight **8**.
7. Tighten crankshaft pulley bolts **11**. Tightening torque: 25 Nm.

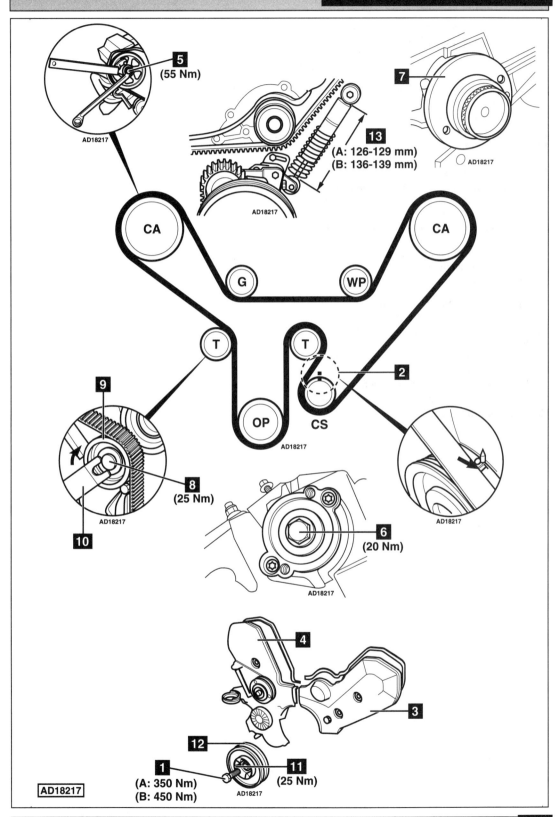

5 (55 Nm)

13 (A: 126-129 mm) (B: 136-139 mm)

7

CA   CA

G   WP

T   T

2

9   OP   CS

8 (25 Nm)

10

6 (20 Nm)

4

3

12   **11** (25 Nm)

1 (A: 350 Nm) (B: 450 Nm)

AD18217

←

8. Turn tensioner pulley ⑨ clockwise until tensioner damper length ⑬ is as follows:
   ❏ Warm engine: (A) 126-129 mm. Use wrench Matra V.159 ⑩.
   ❏ Cold engine: (B) 136-139 mm. Use wrench Matra V.159 ⑩.

9. Tighten tensioner pulley nut ⑧.
   Tightening torque: 25 Nm.

10. Hold camshaft sprockets. Use tool No.3036. Tighten bolts to 55 Nm ⑤.

11. Remove locking tools ⑦. Tool No.3341.

12. Fit Hall effect sensor to LH camshaft. Ensure drive dog locates in camshaft ⑥. Tighten bolt to 20 Nm.

13. Turn crankshaft two turns clockwise. Ensure timing marks aligned ②.

14. Hold crankshaft pulley. Use tool No.3197. Tighten crankshaft bolt ①:
    ❏ With tool No.2079 – (A):
       Tightening torque: 350 Nm.
    ❏ Without tool No.2079 – (B):
       Tightening torque: 450 Nm.

15. Check length of tensioner damper ⑬ is as follows:
    ❏ Warm engine: (A) 126-129 mm.
    ❏ Cold engine: (B) 136-139 mm.

16. If not: Repeat tensioning procedure.

17. Install components in reverse order of removal.

18. Refill cooling system.

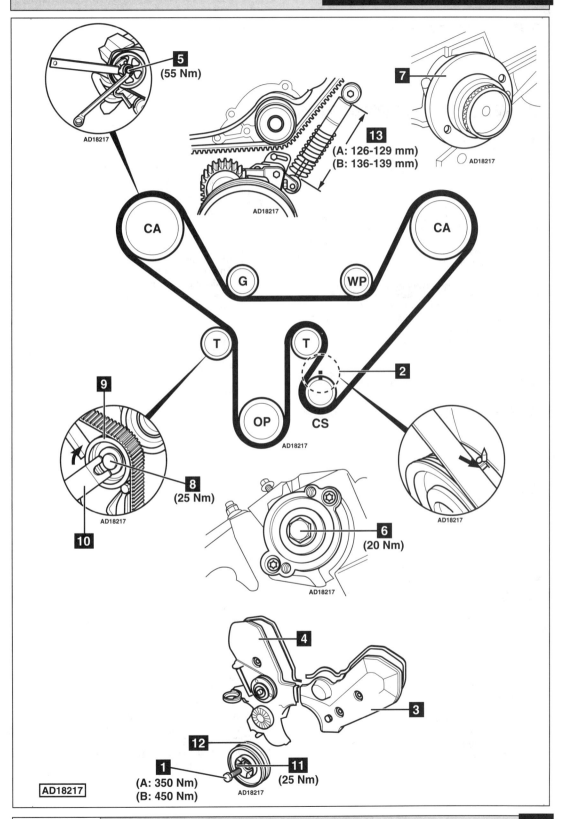

5
(55 Nm)

AD18217

7

AD18217

13
(A: 126-129 mm)
(B: 136-139 mm)

AD18217

CA

CA

G

WP

T

T

2

9

OP

CS

AD18217

8
(25 Nm)

AD18217

10

6
(20 Nm)

AD18217

AD18217

4

3

12

1
(A: 350 Nm)
(B: 450 Nm)

11
(25 Nm)

AD18217

AD18217

AUDI

Model:	**TT 1,8 Turbo**
Year:	**1998-03**
Engine Code:	**AJQ, APX**

Replacement Interval Guide

Audi recommend:
AJQ: Replacement every 180,000 kilometres (111,846 miles).
APX (→2000MY): Replacement every 180,000 kilometres (111,846 miles).
APX (2001MY→): Replacement every 120,000 kilometres (74,564 miles).

NOTE: The vehicle manufacturer publishes this information only in kilometres. The conversion to miles is included for reference purposes only. The previous use and service history of the vehicle must always be taken into account.
Refer to Timing Belt Replacement Intervals at the front of this manual.

Check For Engine Damage

CAUTION: This engine has been identified as an INTERFERENCE engine in which the possibility of valve-to-piston damage in the event of a timing belt failure is MOST LIKELY to occur. A compression check of all cylinders should be performed before removing the cylinder head(s).

Repair Times – hrs

Remove & install	2,50

Special Tools

- Engine support – Audi No.10-222A.
- Engine support legs – Audi No.10-222A/1.
- Engine support bracket – Audi No.3180.
- 1 x M5 x 55 mm stud, nut and washer.

Special Precautions

- Disconnect battery earth lead.
- DO NOT turn crankshaft or camshaft when timing belt removed.
- Remove spark plugs to ease turning engine.
- Turn engine in normal direction of rotation (unless otherwise stated).
- DO NOT turn engine via camshaft or other sprockets.
- Observe all tightening torques.

Removal

1. Raise and support front of vehicle.
2. Remove:
   - Lower splash guard.
   - RH splash guard.
   - Upper engine cover.
   - Auxiliary drive belt.
   - Auxiliary drive belt tensioner.
3. Disconnect:
   - Coolant expansion tank multi-plug.
   - Evaporative emission (EVAP) canister purge valve multi-plug.
   - Fuel vapour hose from charcoal canister to throttle body.
4. Remove:
   - Coolant expansion tank. DO NOT disconnect hoses.
   - PAS reservoir. DO NOT disconnect hoses.
   - Timing belt upper cover **1**.
5. Support engine. Use tool Nos.10-222A/A1 & 3180.
6. Turn crankshaft clockwise to TDC on No.1 cylinder. Ensure timing marks aligned **2** & **3**.
7. Remove:
   - Crankshaft pulley bolts **4**.
   - Crankshaft pulley **5**.
   - Timing belt centre cover **6**.
   - Timing belt lower cover **7**.
   - RH engine mounting **8**.
   - Engine mounting bracket **9**.
8. Insert M5 stud into tensioner **10**.
9. Fit nut and washer **11** to stud. Tighten sufficiently to compress automatic tensioner unit.
   NOTE: DO NOT overtighten nut.
10. Insert locking pin to retain in position **12**.
11. Remove timing belt.
   NOTE: Mark direction of rotation on belt with chalk if belt is to be reused.

Installation

1. Ensure timing marks aligned **3**.
2. Fit timing belt to crankshaft sprocket.
3. Fit timing belt lower cover **7**.
4. Fit crankshaft pulley **5**. Tighten bolts to 25 Nm **4**.
5. Ensure timing marks aligned **2**.
6. Fit timing belt in following order:
   - Water pump sprocket.
   - Guide pulley.
   - Tensioner pulley.
   - Camshaft sprocket.
   NOTE: Ensure belt is taut between sprockets on non-tensioned side.
7. Ensure timing marks aligned **2** & **3**.
8. Remove locking pin **12**.
9. Remove nut and stud **10** & **11**. Allow tensioner to operate.
10. Turn crankshaft slowly two turns clockwise to TDC on No.1 cylinder.
11. Ensure timing marks aligned **2** & **3**.
12. Install:
   - Timing belt centre cover **6**.
   - Engine mounting bracket **9**.
13. Tighten engine mounting to bracket bolts **13**. Tightening torque: 45 Nm.
14. Renew RH engine mounting bolts **14** & **15**.
15. Fit engine mounting **8**.
16. Check engine mounting clearance: 13 mm **16**.
17. Tighten engine mounting to bracket bolts **14**. Tightening torque: 60 Nm + 90°.
18. Tighten engine mounting to body bolts **15**. Tightening torque: 40 Nm + 90°.
19. Install components in reverse order of removal.

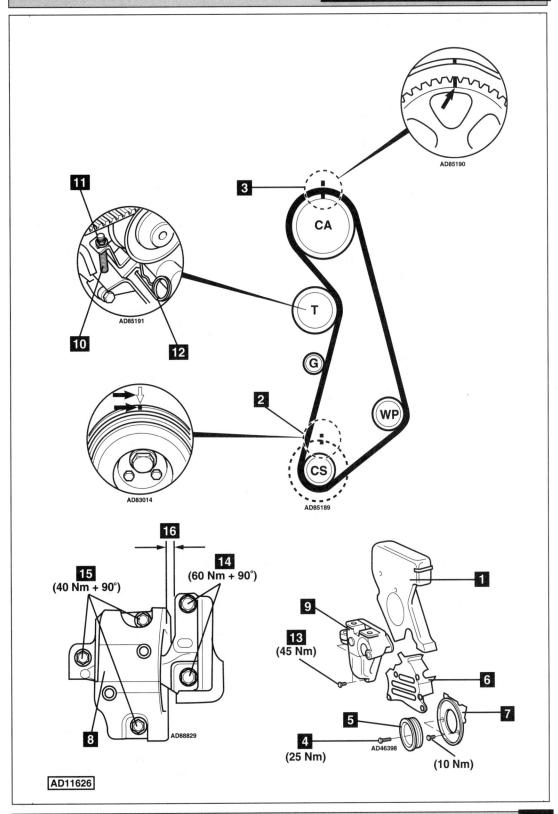

AD85190

11

3

CA

AD85191

10

12

T

G

2

WP

CS

AD83014

AD85189

16

15
(40 Nm + 90°)

14
(60 Nm + 90°)

1

9

13
(45 Nm)

6

5

7

8

4
(25 Nm)

AD46398

(10 Nm)

AD88829

AD11626

AUDI

Model:	**A2 1,2 TDI PD • A2 1,4 TDI PD • A3 1,9 TDI PD**
Year:	**2001-03**
Engine Code:	**AMF, ANY, ASZ, ATD**

Replacement Interval Guide

Audi recommend replacement every
90,000 kilometres (55,923 miles) (tensioner
pulley must also be replaced).

**NOTE: The vehicle manufacturer publishes this
information only in kilometres. The conversion
to miles is included for reference purposes only.**
The previous use and service history of the vehicle
must always be taken into account.
Refer to Timing Belt Replacement Intervals at the front
of this manual.

Check For Engine Damage

CAUTION: This engine has been identified as an
INTERFERENCE engine in which the possibility of
valve-to-piston damage in the event of a timing belt
failure is MOST LIKELY to occur. A compression check
of all cylinders should be performed before removing
the cylinder head(s).

Repair Times – hrs

Remove & install	2,90

Special Tools

- Camshaft locking tool – No.3359.
- Crankshaft sprocket locking tool – No.T10050.
- Tensioner locking tool – No.T10008.
- Two-pin wrench – No.3387.
- AMF/ASZ/ATD – 4 mm drill bit.
- ANY – 7 mm drill bit.

Special Precautions

- Disconnect battery earth lead.
- DO NOT turn crankshaft or camshaft when timing belt removed.
- Remove glow plugs to ease turning engine.
- Turn engine in normal direction of rotation (unless otherwise stated).
- DO NOT turn engine via camshaft or other sprockets.
- Observe all tightening torques.

Removal

1. A2: Remove bonnet:
   - Release bonnet catch in driver's compartment.
   - Release catches behind front service grille.
   - Remove bonnet. Fit protective cover to bonnet.
   - Store in horizontal position.
2. Remove:
   - Engine top cover.
   - Turbocharger air hoses.
   - A2: Mass air flow (MAF) sensor.

3. Raise and support front of vehicle.
4. Remove:
   - Engine undershield.
   - A2: Lower engine steady bar.
   - Auxiliary drive belt.
   - Auxiliary drive belt tensioner.
   - Timing belt upper cover **1**.
   - A3: Coolant expansion tank. DO NOT disconnect hoses.
   - A3: PAS reservoir. DO NOT disconnect hoses.
5. A2: Detach fuel pipes from cylinder head cover.
6. A3: Detach fuel pipes from fuel filter.
7. Support engine.
8. Remove:
   - RH engine mounting.
   - RH engine mounting bracket **2**.
   - Timing belt centre cover **3**.
9. Lower engine slightly.
10. Remove:
    - Crankshaft pulley bolts **4**.
    - Crankshaft pulley **5**.
    - Timing belt lower cover **6**.
11. Turn crankshaft clockwise to TDC on No.1 cylinder. Ensure timing mark aligned with notch on camshaft sprocket hub:
    - AMF/ANY: 3Z **7**.
    - ASZ/ATD: 4Z **8**.
    NOTE: Notch located behind camshaft sprocket teeth.
12. Lock crankshaft sprocket **9**. Use tool No.T10050.
13. Ensure timing marks aligned **10**.
14. Lock camshaft **11**. Use tool No.3359.
15. Fully insert Allen key into tensioner pulley **12**.
16. Turn tensioner pulley slowly anti-clockwise until locking tool can be inserted **13**. Tool No.T10008.
17. Slacken tensioner nut **14**.
18. Remove:
    - Automatic tensioner unit **15**.
    - Timing belt.

Installation

1. Ensure camshaft locked with tool **11**. Tool No.3359.
2. Ensure crankshaft sprocket locking tool located correctly **9**. Tool No.T10050.
3. Ensure timing marks aligned **10**.
4. Ensure automatic tensioner unit locked with tool **13**. Tool No.T10008.

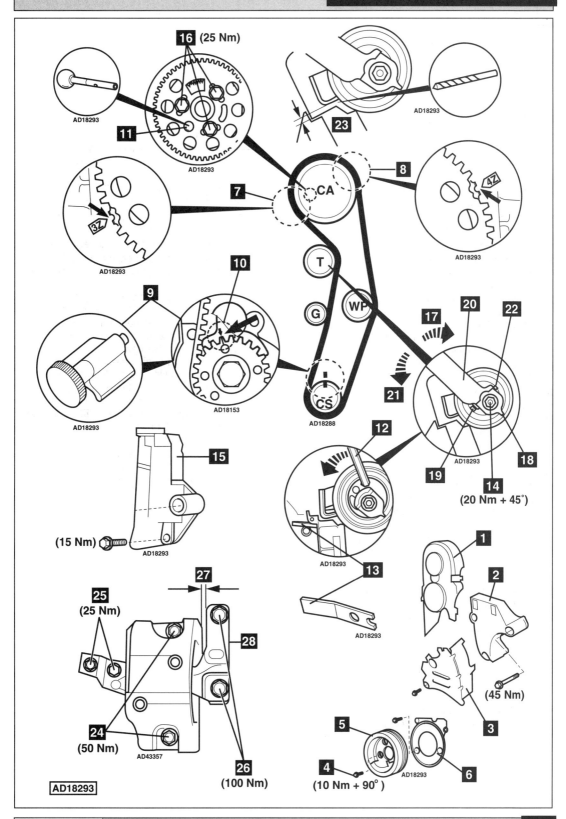

16 (25 Nm)

11

AD18293

AD18293

AD18293

23

AD18293

7

CA

8

3Z

4Z

AD18293

10

9

T

9

AD18153

G

WP

AD18293

20

22

17

21

12

CS

AD18288

AD18293

19

18

AD18293

14
(20 Nm + 45°)

15

13

AD18293

1

(15 Nm)

AD18293

AD18293

2

27

25
(25 Nm)

28

24

50 Nm)

(50 Nm)

AD43357

26
(100 Nm)

5

4

6

AD18293

(10 Nm + 90°)

(45 Nm)

3

AD18293

5. Slacken camshaft sprocket bolts 🔢.

6. Turn camshaft sprocket fully clockwise in slotted holes. Tighten bolts finger tight 🔢.

7. Turn tensioner pulley slowly clockwise 🔢 until lug 🔢 just reaches stop 🔢. Use tool No.3387 🔢.

8. Fit timing belt in following order:
   - ❏ Camshaft sprocket.
   - ❏ Tensioner pulley.
   - ❏ Crankshaft sprocket.
   - ❏ Water pump sprocket.

9. Install automatic tensioner unit 🔢.
   NOTE: Ensure belt is taut between sprockets on non-tensioned side.

10. Turn tensioner pulley slowly anti-clockwise 🔢 (lug 🔢 moves towards stop 🔢). Use tool No.3387 🔢.

11. Remove locking tool without force 🔢.
    NOTE: DO NOT release tensioner pulley.

12. Turn tensioner pulley slowly clockwise 🔢 (lug 🔢 moves towards stop 🔢) until dimension 🔢 as specified:
    - ❏ AMF/ASZ/ATD – 4±1 mm. Use drill bit.
    - ❏ ANY – 7±1 mm. Use drill bit.
    NOTE: Engine must be COLD.

13. Tighten tensioner nut 🔢.
    Tightening torque: 20 Nm + 45°.

14. Tighten camshaft sprocket bolts 🔢.
    Tightening torque: 25 Nm.

15. Remove:
    - ❏ Camshaft locking tool 🔢.
    - ❏ Crankshaft sprocket locking tool 🔢.
    - ❏ Drill bit.

16. Turn crankshaft slowly two turns clockwise to TDC on No.1 cylinder.

17. Ensure dimension 🔢 as specified:
    - ❏ AMF/ASZ/ATD – 4±1 mm. Use drill bit.
    - ❏ ANY – 7±1 mm. Use drill bit.

18. If not: Slacken tensioner nut 🔢. Turn tensioner pulley until dimension as specified 🔢. Tighten tensioner nut 🔢.
    Tightening torque: 20 Nm + 45°.

19. Lock crankshaft sprocket 🔢. Use tool No.T10050.

20. Ensure timing marks aligned 🔢.

21. Ensure camshaft locking tool can be inserted easily 🔢. Tool No.3359.

22. If not:
    - ❏ Slacken camshaft sprocket bolts 🔢.
    - ❏ Turn sprocket hub until locking tool can be inserted 🔢.
    - ❏ Tighten bolts 🔢. Tightening torque: 25 Nm.
    - ❏ Remove locking tools 🔢 & 🔢.
    - ❏ Turn crankshaft slowly two turns clockwise to TDC on No.1 cylinder.
    - ❏ Ensure locking tools can be fitted correctly 🔢 & 🔢.

23. Remove:
    - ❏ Camshaft locking tool 🔢.
    - ❏ Crankshaft locking tool 🔢.
    - ❏ Drill bit.

24. Install:
    - ❏ Timing belt lower cover 🔢.
    - ❏ Crankshaft pulley 🔢.
    - ❏ Crankshaft pulley bolts 🔢.

25. Tighten crankshaft pulley bolts 🔢.
    Tightening torque: 10 Nm + 90°. Use new bolts.

26. Install:
    - ❏ Timing belt centre cover 🔢.
    - ❏ RH engine mounting bracket 🔢.

27. Tighten RH engine mounting bracket bolts.
    Tightening torque: 45 Nm.

28. Install components in reverse order of removal.

29. A2 – AMF: Tighten RH engine mounting:
    - ❏ Bolts securing engine mounting to body – 23 Nm + 90°. Use new bolts.
    - ❏ Bolts securing engine mounting to engine bracket – 50 Nm + 90°. Use new bolts.

30. A2 – ANY: Tighten RH engine mounting:
    - ❏ Bolts securing engine mounting to body – 20 Nm + 45°. Use new bolts.
    - ❏ Bolts securing engine mounting to engine bracket – 40 Nm + 90°. Use new bolts.

31. A3: Fit and align RH engine mounting:
    - ❏ Engine mounting clearance: 13 mm 🔢.
    - ❏ Ensure engine mounting bolts 🔢 aligned with edge of mounting 🔢.

32. A3: Tighten RH engine mounting:
    - ❏ Engine mounting bolts 🔢.
      Tightening torque: 50 Nm.
    - ❏ Engine mounting bolts 🔢.
      Tightening torque: 25 Nm.
    - ❏ Engine mounting bolts 🔢.
      Tightening torque: 100 Nm. Use new bolts.

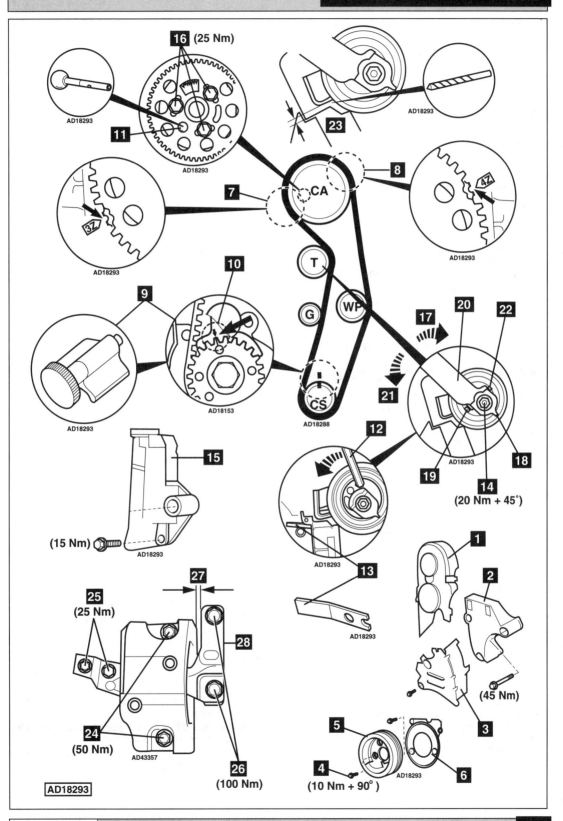

16 (25 Nm)

AD18293

11

AD18293

23

AD18293

8

4Z

AD18293

7

CA

3Z

AD18293

T

10

AD18153

9

G   WP

AD18293

17

20   **22**

21

AD18293   **18**

19

14
(20 Nm + 45°)

15

12

AD18293   **13**

(15 Nm)

AD18293

CS

AD18288

1

2

AD18293

(45 Nm)

27

25
(25 Nm)

28

24
(50 Nm)

AD43357

26
(100 Nm)

3

5

4
(10 Nm + 90°)   AD18293   **6**

AD18293

Model:	**A3 1,9 TDI**
Year:	**1996-03**
Engine Code:	**AGR, AHF, ALH**

Replacement Interval Guide

Audi recommend:
→1999MY: Check condition and width every 15,000 kilometres (9,320 miles) (replacement width less than 22 mm).
→1999MY: Replacement every 60,000 kilometres (37,282 miles) (tensioner pulley must also be replaced).
2000-01MY: Replacement every 90,000 kilometres (55,923 miles).
2002MY→: Replacement every 120,000 kilometres (74,564 miles) (guide pulley below water pump must also be replaced).
NOTE: The vehicle manufacturer publishes this information only in kilometres. The conversion to miles is included for reference purposes only.
The previous use and service history of the vehicle must always be taken into account.
Refer to Timing Belt Replacement Intervals at the front of this manual.

Check For Engine Damage

CAUTION: This engine has been identified as an INTERFERENCE engine in which the possibility of valve-to-piston damage in the event of a timing belt failure is MOST LIKELY to occur.
A compression check of all cylinders should be performed before removing the cylinder head.
Refer to Timing Belt Replacement Intervals at the front of this manual.

Repair Times – hrs

Remove & install	3,30

Special Tools

- Camshaft setting bar – No.3418.
- Injection pump locking pin – No.3359.
- Sprocket holding tool – No.3036.
- Two-pin wrench – Matra V.159.

Special Precautions

- Disconnect battery earth lead.
- DO NOT turn crankshaft or camshaft when timing belt removed.
- Remove glow plugs to ease turning engine.
- Turn engine in normal direction of rotation (unless otherwise stated).
- DO NOT turn engine via camshaft or other sprockets.
- Observe all tightening torques.
- Check diesel injection pump timing after belt replacement.

Removal

1. Remove:
   - ❑ Auxiliary drive belt.
   - ❑ Auxiliary drive belt tensioner.
   - ❑ Intercooler intake hose.
   - ❑ Cylinder head cover.
   - ❑ Timing belt upper cover **1**.
   - ❑ Vacuum pump.
2. Turn crankshaft clockwise to TDC on No.1 cylinder. Ensure flywheel timing marks aligned **2**.
3. Fit setting bar No.3418 to rear of camshaft **3**. Centralise camshaft using feeler gauges.
4. If setting bar cannot be fitted: Turn crankshaft one turn clockwise.
5. Insert locking pin in injection pump. Tool No.3359 **4**.
6. Slacken injection pump sprocket bolts **5**.
   NOTE: DO NOT slacken injection pump centre hub nut.
7. Slacken tensioner nut **6**.
8. Support engine.
9. Remove:
   - ❑ RH engine mounting bolt **7**.
   - ❑ RH engine mounting **8**.
   - ❑ Crankshaft pulley bolts **9**.
   - ❑ Crankshaft pulley **10**.
   - ❑ Timing belt lower cover **11**.
   - ❑ Timing belt centre cover **12**.
   - ❑ Timing belt.
   NOTE: Mark direction of rotation on belt with chalk if belt is to be reused.

Installation

1. Ensure crankshaft at TDC on No.1 cylinder. Ensure flywheel timing marks aligned **2**.
2. Ensure camshaft setting bar fitted correctly **3**.
3. Ensure locking pin located correctly in injection pump **4**.
4. Hold camshaft sprocket. Use tool No.3036. Slacken bolt 1/2 turn **13**.
5. Loosen sprocket from taper using a drift through hole in timing belt rear cover. Ensure sprocket can turn on taper.
   NOTE: DO NOT use setting bar to hold camshaft when slackening sprocket bolt.
6. Turn injection pump sprocket fully clockwise until bolts at end of slotted holes.

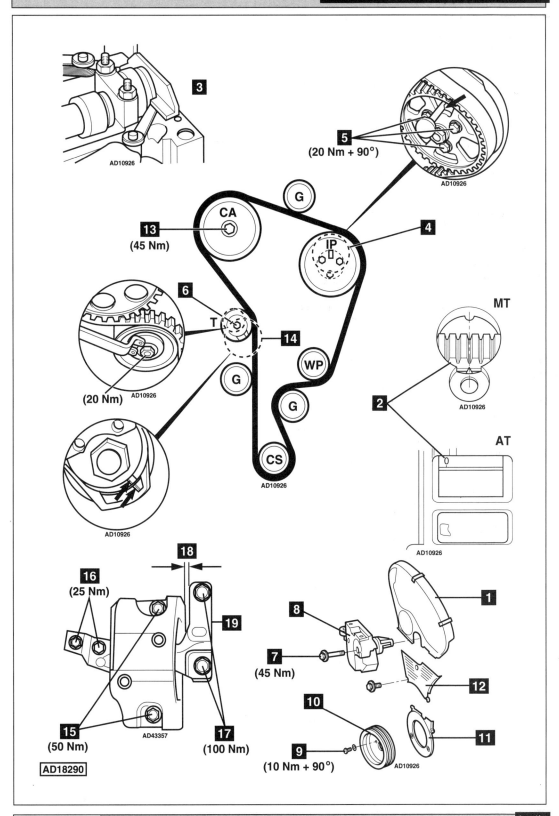

AD10926

3

5
(20 Nm + 90°)

AD10926

G

CA

13
(45 Nm)

4

IP

6

T

14

(20 Nm) AD10926

G

WP

G

MT

2

AD10926

AT

CS

AD10926

AD10926

18

16
(25 Nm)

19

8

1

7
(45 Nm)

12

15
(50 Nm) AD43357

17
(100 Nm)

10

9
(10 Nm + 90°) AD10926

11

AD18290

←

7. Fit timing belt in anti-clockwise direction, starting at crankshaft sprocket. Ensure belt is taut between sprockets.

8. Turn automatic tensioner pulley clockwise until notch and raised mark on tensioner aligned **14**. Use wrench Matra V.159.

9. Tighten tensioner nut to 20 Nm **6**.

10. Ensure flywheel timing marks aligned **2**.

11. Hold camshaft sprocket. Use tool No.3036. Tighten bolt to 45 Nm **13**.

12. Fit new injection pump sprocket bolts **5**. Tightening torque: 20 Nm + 90°.

13. Remove:
    ❏ Injection pump locking pin **4**.
    ❏ Camshaft setting bar **3**.

14. Turn crankshaft two turns clockwise to TDC on No.1 cylinder.

15. Ensure flywheel timing marks aligned **2**.

16. Ensure locking pin can be inserted in injection pump **4**.

17. If not: Slacken injection pump sprocket bolts **5**. Turn sprocket hub until locking pin can be inserted. Tighten bolts **5**. Tightening torque: 20 Nm + 90°.

18. Install components in reverse order of removal.

19. Tighten crankshaft pulley bolts **9**. Tightening torque: 10 Nm + 90°.

20. Fit and align RH engine mounting:
    ❏ Engine mounting clearance: 13 mm **18**.
    ❏ Ensure engine mounting bolts **17** aligned with edge of mounting **19**.

21. Tighten:
    ❏ Engine mounting bracket bolts **7**. Tightening torque: 45 Nm.
    ❏ Engine mounting bolts **15**. Tightening torque: 50 Nm.
    ❏ Engine mounting bolts **16**. Tightening torque: 25 Nm.
    ❏ Engine mounting bolts **17**. Tightening torque: 100 Nm. Use new bolts.

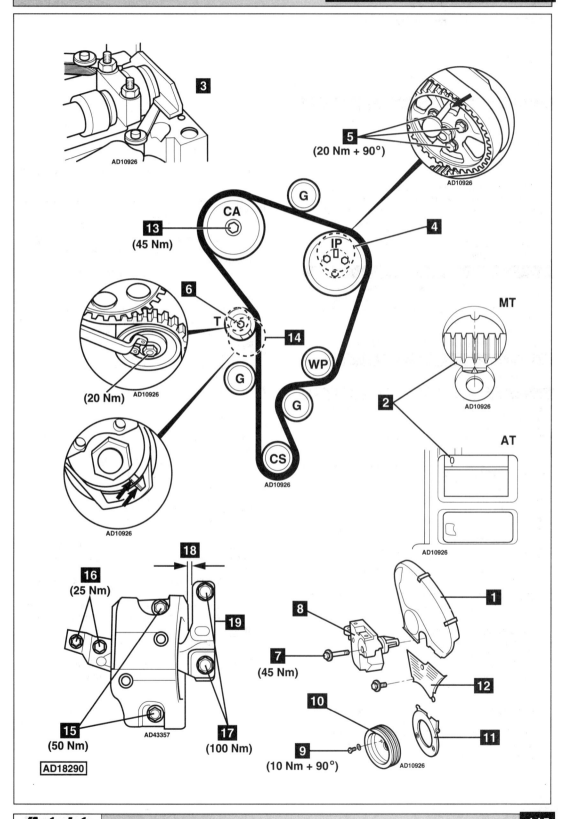

3

AD10926

5
(20 Nm + 90°)

AD10926

G

CA

13
(45 Nm)

4

IP

MT

6

T

14

(20 Nm)   AD10926

WP

G

2

AD10926

G

G

AT

CS

AD10926

AD10926

18

16
(25 Nm)

19

8

1

7
(45 Nm)

12

10

15
(50 Nm)

AD43357

17
(100 Nm)

9
(10 Nm + 90°)

11

AD10926

AD18290

AUDI

Model:	80 1,9 TDI
Year:	1991-95
Engine Code:	1Z

Replacement Interval Guide

Audi recommend check condition and width every 15,000 kilometres (9,320 miles) (replacement width less than 22 mm).
Replacement every 90,000 kilometres (55,923 miles) (tensioner pulley must also be replaced).

NOTE: The vehicle manufacturer publishes this information only in kilometres. The conversion to miles is included for reference purposes only. The previous use and service history of the vehicle must always be taken into account.
Refer to Timing Belt Replacement Intervals at the front of this manual.

Check For Engine Damage

CAUTION: This engine has been identified as an INTERFERENCE engine in which the possibility of valve-to-piston damage in the event of a timing belt failure is MOST LIKELY to occur.
A compression check of all cylinders should be performed before removing the cylinder head.

Repair Times – hrs

Remove & install	1,90

Special Tools

- Injection pump sprocket locking pin – VAG No.2064.
- Camshaft setting bar – VAG No.2065A.
- Two-pin wrench – Matra V.159.
- Sprocket holding tool – VAG No.3036.

Special Precautions

- Disconnect battery earth lead.
- DO NOT turn crankshaft or camshaft when timing belt removed.
- Remove glow plugs to ease turning engine.
- Turn engine in normal direction of rotation (unless otherwise stated).
- DO NOT turn engine via camshaft or other sprockets.
- Observe all tightening torques.
- Check diesel injection pump timing after belt replacement.

Removal

1. Remove:
   - ❑ Cooling fan cowling.
   - ❑ Auxiliary drive belts.
   - ❑ Timing belt upper cover **1**.
   - ❑ Cylinder head cover.
   - ❑ Crankshaft pulley bolts **5**.
   - ❑ Crankshaft pulley **6**.
   - ❑ Timing belt lower cover **7**.
2. Turn crankshaft to TDC on No.1 cylinder. Ensure flywheel timing marks aligned **2**.
   NOTE: Engine is at TDC when centre of timing mark 0 on flywheel aligns with lower edge of gearbox housing.
3. Fit setting bar No.2065A to rear of camshaft **3**. Centralise camshaft using feeler gauges.

4. Lock injection pump sprocket. Use tool No.2064 **4**.
5. Slacken tensioner nut **8**. Turn tensioner anti-clockwise away from belt. Lightly tighten nut.
6. Remove:
   - ❑ Guide pulley **9**.
   - ❑ Timing belt.
   NOTE: Mark direction of rotation on belt with chalk if belt is to be reused.

Installation

1. Ensure flywheel TDC marks aligned **2**.
   NOTE: Engine is at TDC when centre of timing mark 0 on flywheel aligns with lower edge of gearbox housing.
2. Ensure camshaft setting bar fitted correctly **3**.
3. Ensure locking pin located correctly in injection pump sprocket **4**.
4. Slacken camshaft sprocket bolt 1/2 turn **10**.
5. Loosen sprocket from taper using a drift through hole in timing belt rear cover. Ensure sprocket can turn on taper.
6. Fit timing belt in anti-clockwise direction, starting at crankshaft sprocket. Ensure belt is taut between sprockets.
7. Fit guide pulley **9**. Tightening torque: 25 Nm.
8. Remove locking pin from injection pump sprocket.
9. Slacken tensioner nut **8**.
10. Turn automatic tensioner pulley clockwise until notch and raised mark on tensioner aligned **11**. Use wrench Matra V.159 **12**.
11. Tighten tensioner nut to 20 Nm **8**.
12. Ensure flywheel timing marks aligned **2**.
13. Hold camshaft sprocket. Use tool No.3036. Tighten bolt to 45 Nm **10**.
14. Remove camshaft setting bar **3**.
15. Turn crankshaft two turns clockwise. Ensure timing marks aligned **2**.
16. Ensure camshaft setting bar can be fitted **3**.
17. Ensure locking pin can be inserted in injection pump **4**.
18. Apply firm thumb pressure to belt. Tensioner marks should move out of alignment **11**.
19. Release thumb pressure from belt. Tensioner marks should realign. If not: Repeat tensioning procedure.
20. Install:
    - ❑ Timing belt lower cover **7**.
    - ❑ Crankshaft pulley **6**.
21. Tighten crankshaft pulley bolts to 25 Nm **5**.
22. Install components in reverse order of removal.

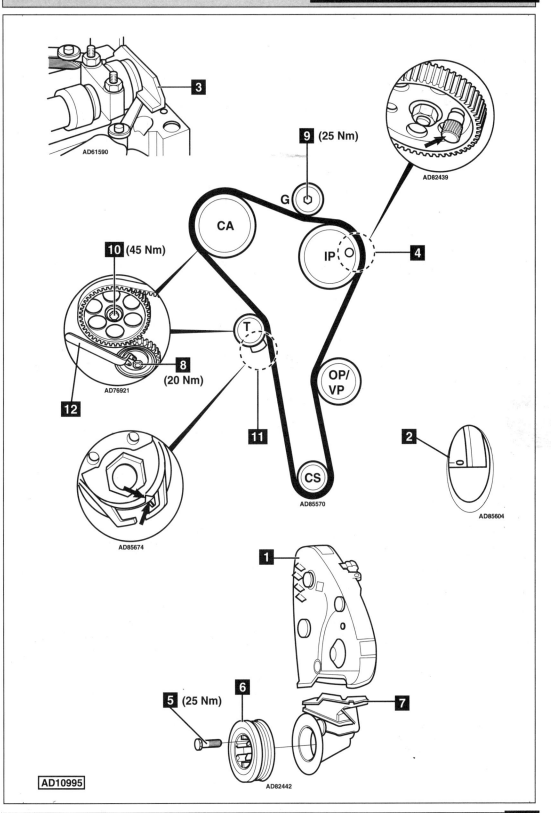

AD61590

3

9 (25 Nm)

AD82439

4

G

CA

IP

10 (45 Nm)

12

T

8
(20 Nm)

AD76921

11

OP/
VP

2

AD85604

CS

AD85570

AD85674

1

5 (25 Nm)

6

7

AD82442

AD10995

AUDI

Model:	**80 1,9 Turbo D**
Year:	**1991-94**
Engine Code:	**AAZ**

Replacement Interval Guide

Audi recommend check & adjust every 30,000 kilometres (18,641 miles) and replacement every 120,000 kilometres (74,564 miles).

NOTE: The vehicle manufacturer publishes this information only in kilometres. The conversion to miles is included for reference purposes only. The previous use and service history of the vehicle must always be taken into account.
Refer to Timing Belt Replacement Intervals at the front of this manual.

Check For Engine Damage

CAUTION: This engine has been identified as an INTERFERENCE engine in which the possibility of valve-to-piston damage in the event of a timing belt failure is MOST LIKELY to occur.
A compression check of all cylinders should be performed before removing the cylinder head.

Repair Times – hrs

Remove & install	1,90

Special Tools

- Tension gauge – VAG No.210.
- Injection pump sprocket locking pin – VAG No.2064.
- Camshaft setting bar – VAG No.2065A.
- Two-pin wrench – Matra V.159.

Special Precautions

- Disconnect battery earth lead.
- DO NOT turn crankshaft or camshaft when timing belt removed.
- Remove glow plugs to ease turning engine.
- Turn engine in normal direction of rotation (unless otherwise stated).
- DO NOT turn engine via camshaft or other sprockets.
- Observe all tightening torques.
- Check diesel injection pump timing after belt replacement.

Removal

1. Remove:
   - ❏ Cooling fan cowling.
   - ❏ Auxiliary drive belts.
   - ❏ Cylinder head cover.
   - ❏ Timing belt upper cover **1**.
   - ❏ Crankshaft pulley bolts **2**.
   - ❏ Crankshaft pulley **3**.
   - ❏ Timing belt lower cover **4**.
2. Turn crankshaft to TDC on No.1 cylinder. Ensure flywheel timing marks aligned **6**.
3. Fit setting bar No.2065A to rear of camshaft **5**. Centralise camshaft using feeler gauges.
4. Lock injection pump sprocket. Use tool No.2064 **8**.
5. Slacken tensioner nut **9**. Turn tensioner anti-clockwise away from belt. Lightly tighten nut.
6. Remove timing belt.

Installation

1. Ensure flywheel TDC marks aligned **6**.
2. Ensure camshaft setting bar fitted correctly **5**.
3. Ensure locking pin located correctly in injection pump sprocket **8**.
4. Slacken camshaft sprocket bolt **7**.
5. Loosen sprocket from taper using a drift through hole in timing belt rear cover. Ensure sprocket can turn on taper.
6. Fit timing belt, starting at crankshaft sprocket. Ensure belt is taut between sprockets on non-tensioned side.
7. Attach tension gauge to belt at ▽. Tool No.210.
8. Slacken tensioner nut **9**.
9. Turn tensioner clockwise until tension gauge indicates 12-13 units. Use wrench Matra V.159. Tighten nut to 45 Nm **9**.
10. Tighten camshaft sprocket bolt to 45 Nm **7**.
11. Remove:
    - ❏ Locking pin **8**.
    - ❏ Camshaft setting bar **5**.
12. Turn crankshaft two turns clockwise. Strike belt once with a rubber faced mallet at ▽. Recheck belt tension.
13. Install:
    - ❏ Timing belt lower cover **4**.
    - ❏ Crankshaft pulley **3**.
14. Tighten crankshaft pulley bolts to 20 Nm **2**.
15. Install components in reverse order of removal.

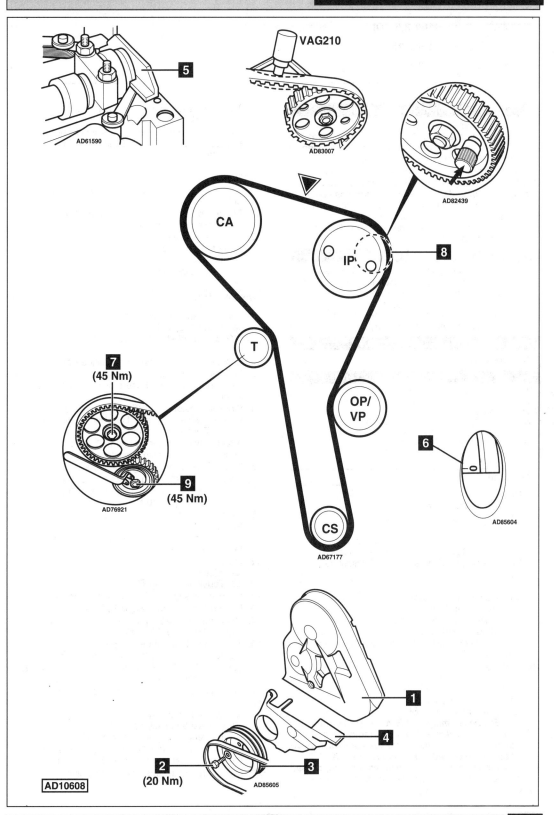

VAG210

AD61590

AD83007

AD82439

5

CA

IP **8**

7
(45 Nm)

T

OP/
VP

6

9
(45 Nm)

AD76921

CS

AD67177

AD85604

1

4

AD10608

2
(20 Nm)

3

AD85605

Model:	**100 2,5 TDI**
Year:	**1991-95**
Engine Code:	**ABP, AAT**

Replacement Interval Guide

Audi recommend check & adjust every 30,000 kilometres (18,641 miles), replacement at the first 60,000 kilometres (37,282 miles) and then every 120,000 kilometres (74,564 miles).

NOTE: The vehicle manufacturer publishes this information only in kilometres. The conversion to miles is included for reference purposes only.
The previous use and service history of the vehicle must always be taken into account.
Refer to Timing Belt Replacement Intervals at the front of this manual.

Check For Engine Damage

CAUTION: This engine has been identified as an INTERFERENCE engine in which the possibility of valve-to-piston damage in the event of a timing belt failure is MOST LIKELY to occur.
A compression check of all cylinders should be performed before removing the cylinder head.

Repair Times – hrs

Remove & install	3,70

Special Tools

- Tension gauge – VAG No.210.
- Camshaft setting bar – VAG No.2065A.
- Support bar – VAG No.3251.
- Sprocket holding tool – VAG No.3036.
- Crankshaft pulley locking tool – VAG No.3256.
- Two-pin wrench – Matra Nr. V.159

Removal

1. Remove:
   - ❏ Engine undershield.
   - ❏ Front bumper and brackets.
   - ❏ Radiator grille and front trim.
2. Fit support bar to RH bumper bracket position. Tool No.3251. Undo radiator lower mountings. Swivel radiator forwards. DO NOT disconnect hoses.
3. Remove:
   - ❏ Viscous fan and pulley.
   - ❏ Auxiliary drive belt.
   - ❏ Auxiliary drive belt tensioner.
   - ❏ Cylinder head cover.
4. Lock crankshaft pulley. Use tool No.3256. Remove centre bolt **13**.
5. Remove:
   - ❏ Crankshaft pulley locking tool.
   - ❏ Crankshaft pulley bolts **14**.
   - ❏ Crankshaft pulley **1**.
   - ❏ Timing belt covers **2**.
   NOTE: Check timing belt is running true on camshaft sprocket – if not this indicates a distorted or badly fitted crankshaft sprocket which should be replaced. If this is the case check for swarf in the bores of the oil pump housing, caused by the sprocket rubbing on the pump housing.
6. Remove injection pump belt cover.

7. Turn crankshaft to TDC on No.1 cylinder. Ensure timing marks aligned **3** & **4**.
8. Slacken injection pump belt tensioner pulley nut **5**. Remove tensioner spring **6**. Move tensioner pulley away from belt. Lightly tighten nut.
9. Hold camshaft rear sprocket. Use tool No.3036 **7**. Remove bolt and washer **8**.
10. Remove camshaft rear sprocket and injection pump belt.
11. Fit setting bar No.2065A to rear of camshaft **9**. Centralise camshaft using feeler gauges **12**.
12. Slacken tensioner nut **10**. Turn tensioner clockwise away from belt. Use wrench Matra V.159. Lightly tighten nut.
13. Remove timing belt.

Installation

1. Slacken camshaft sprocket bolt 1/2 turn **11**. Tap sprocket gently to loosen it from taper.
   NOTE: Sprocket should turn freely on taper but not tilt.
2. Ensure flywheel timing marks aligned **3**.
3. Fit timing belt in anti-clockwise direction, starting at crankshaft sprocket. Ensure belt is taut between sprockets.
4. Slacken tensioner nut **10**. Turn tensioner anti-clockwise to tension belt. Use wrench Matra V.159. Tighten nut to 45 Nm **10**.
5. Attach tension gauge to belt at ▽. Tool No.210. Tension gauge should indicate 12-13 units.
6. If not: Repeat tensioning procedure.
7. Ensure flywheel timing marks aligned **3**.
8. Hold camshaft sprocket. Tighten bolt **11**. Tightening torque: 30 Nm + 90°.
9. Remove camshaft setting bar **9**.
10. Fit camshaft rear sprocket. Fit bolt and washer **8**. Lightly tighten bolt. Ensure sprocket can turn on taper but not tilt.
11. Install:
    - ❏ Injection pump belt.
    - ❏ Tensioner spring **6**.
12. Slacken tensioner pulley nut to tension belt **5**.
13. Ensure timing marks aligned **3** & **4**.
14. Hold camshaft rear sprocket. Use tool No.3036 **7**. Tighten bolt to 160 Nm **8**.
15. Turn crankshaft 1/2 turn clockwise. Tighten tensioner pulley nut to 45 Nm **5**.
16. Turn crankshaft to TDC on No.1 cylinder. Ensure timing marks aligned **3** & **4**.
17. Install components in reverse order of removal.
18. Install:
    - ❏ Crankshaft pulley.
    - ❏ Crankshaft pulley bolts **14**.
      Tightening torque: 20 Nm + 90°.
19. Fit new crankshaft pulley centre bolt **13**. Lightly oil threads. Tightening torque: 160 Nm + 180°.
20. Check and adjust injection pump timing.

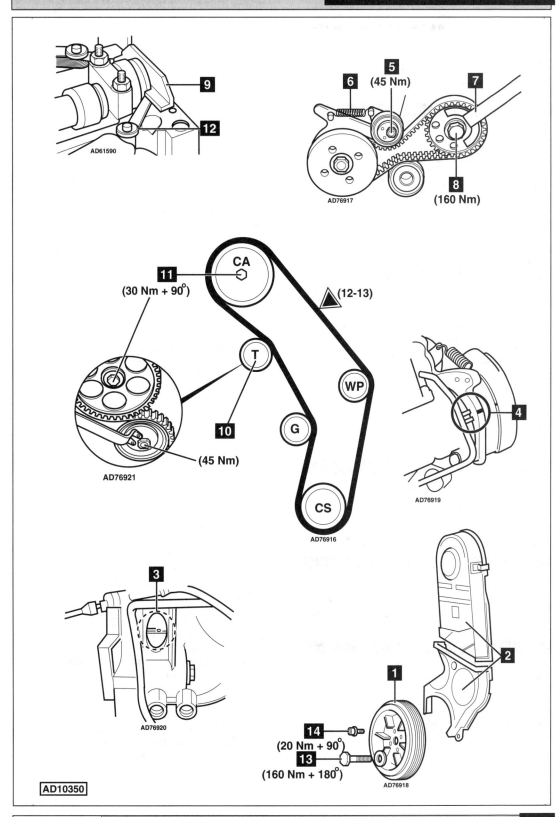

9

6 **5** (45 Nm) **7**

AD61590

12

AD76917

8 (160 Nm)

11 CA
(30 Nm + 90°)

▲(12-13)

T

10

WP

(45 Nm)

AD76921

G

CS

AD76916

4

AD76919

3

AD76920

14
(20 Nm + 90°)
13
(160 Nm + 180°)

1

2

AD76918

AD10350

Model:	**A4 1,9 TDI • A6 1,9 TDI**
Year:	**1994-03**
Engine Code:	**1Z, AHU, AFN, AFF, AVG**

Replacement Interval Guide

Audi recommend:

A4:

→1999MY: Check condition and width every 15,000 kilometres (9,320 miles) (replacement width less than 22 mm).

→1999MY: Replacement every 90,000 kilometres (55,923 miles) (tensioner pulley must also be replaced).

2000MY→: Replacement every 90,000 kilometres (55,923 miles).

A6:

Chassis No.→4BYN002887: Check condition and width every 15,000 kilometres (9,320 miles) (replacement width less than 22 mm).

→1999MY: Replacement every 90,000 kilometres (55,923 miles) (tensioner pulley must also be replaced).

2000MY→: Replacement every 90,000 kilometres (55,923 miles).

NOTE: The vehicle manufacturer publishes this information only in kilometres. The conversion to miles is included for reference purposes only. The previous use and service history of the vehicle must always be taken into account. Refer to Timing Belt Replacement Intervals at the front of this manual.

Check For Engine Damage

CAUTION: This engine has been identified as an INTERFERENCE engine in which the possibility of valve-to-piston damage in the event of a timing belt failure is MOST LIKELY to occur.

A compression check of all cylinders should be performed before removing the cylinder head.

Repair Times – hrs

Remove & install:

A4	2,90
A6 →1997	1,90
A6 1997→	3,50

Special Tools

- Camshaft setting bar – No.2065A.
- Injection pump sprocket locking pin – No.2064.
- Sprocket holding tool – No.3036.
- Support guides – No.3369.
- Two-hole wrench – No.3212.
- Two-pin wrench – Matra V.159.

Special Precautions

- Disconnect battery earth lead.
- DO NOT turn crankshaft or camshaft when timing belt removed.
- Remove glow plugs to ease turning engine.
- Turn engine in normal direction of rotation (unless otherwise stated).
- DO NOT turn engine via camshaft or other sprockets.
- Observe all tightening torques.
- Check diesel injection pump timing after belt replacement.

Removal

1. Except A6 →1997: Move radiator support panel into service position:
   - Remove front bumper.
   - Remove air intake pipe between front panel and air filter.
   - Remove front panel bolts **1** & **2**.
   - Install support guides No.3369 in front panel **3**.
   - Slide front panel forward.
   - Refit upper rear bolts in front holes to steady panel.

2. Remove:
   - Auxiliary drive belt(s).
   - Viscous fan. Use tool No.3212 and 8 mm Allen key (except A6 →1997).
   - Auxiliary drive belt tensioner (except A6 →1997).
   - Timing belt upper cover **4**.
   - Cylinder head cover.

3. Turn crankshaft to TDC on No.1 cylinder. Ensure flywheel timing marks aligned **5**.

4. Fit setting bar No.2065A to rear of camshaft **6**. Centralise camshaft using feeler gauges.

5. Lock injection pump sprocket. Use tool No.2064 **7**.

6. Remove:
   - Crankshaft pulley bolts **8**.
   - Crankshaft pulley **9**.
   - Timing belt lower cover **10**.

7. Slacken tensioner nut **11**. Turn tensioner anti-clockwise away from belt. Lightly tighten nut.

8. Remove:
   - Guide pulley **14**.
   - Timing belt.

   NOTE: Mark direction of rotation on belt with chalk if belt is to be reused.

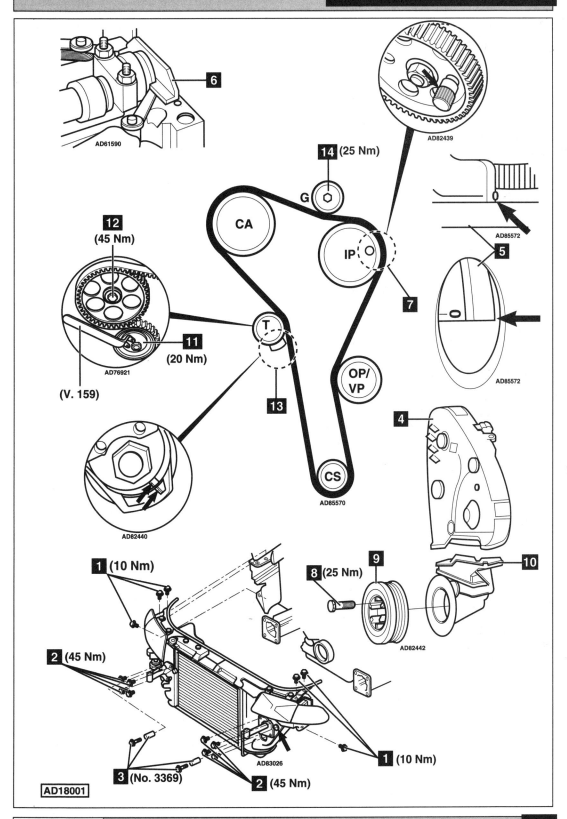

AD61590

6

AD82439

14 (25 Nm)

G

CA

IP

AD85572

5

7

12
(45 Nm)

T

11
(20 Nm)

AD76921

(V. 159)

13

AD82440

**OP/
VP**

AD85572

4

CS

AD85570

9

8 (25 Nm)

10

AD82442

1 (10 Nm)

2 (45 Nm)

1 (10 Nm)

3 (No. 3369)

AD83026

2 (45 Nm)

AD18001

Installation

1. Ensure flywheel timing marks aligned **5**.
2. Ensure camshaft setting bar fitted correctly **6**.
3. Ensure locking pin located correctly in injection pump sprocket **7**.
4. Hold camshaft sprocket. Use tool No.3036.
5. Slacken camshaft sprocket bolt 1/2 turn **12**.
6. Loosen sprocket from taper using a drift through hole in timing belt rear cover. Ensure sprocket can turn on taper.
7. Fit timing belt in anti-clockwise direction, starting at crankshaft sprocket. Ensure belt is taut between sprockets.
8. Fit guide pulley **14**. Tighten bolt to 25 Nm.
9. Remove locking pin from injection pump sprocket.
10. Slacken tensioner nut **11**.
11. Turn automatic tensioner pulley clockwise until notch and raised mark on tensioner aligned **13**. Use wrench Matra V.159.
12. Tighten tensioner nut to 20 Nm **11**.
13. Ensure flywheel timing marks aligned **5**.
14. Hold camshaft sprocket. Use tool No.3036.
15. Tighten camshaft sprocket bolt to 45 Nm **12**.
16. Remove camshaft setting bar **6**.
17. Turn crankshaft two turns clockwise. Ensure timing marks aligned **5**.
18. Ensure camshaft setting bar can be fitted **6**.
19. Ensure locking pin can be inserted in injection pump sprocket **7**.
20. Apply firm thumb pressure to belt. Tensioner marks should move out of alignment **13**.
21. Release thumb pressure from belt. Tensioner marks should realign. If not: Repeat tensioning procedure.
22. Install:
    - ❑ Timing belt lower cover **10**.
    - ❑ Crankshaft pulley **9**.
23. Tighten crankshaft pulley bolts to 25 Nm **8**.
24. Install components in reverse order of removal.
25. Push radiator support panel back into place. Remove support guides No.3369. Screw in bolts **1** & **2**.
26. Tighten 8 mm bolts to 45 Nm **2**. Tighten 6 mm bolts to 10 Nm **1**.

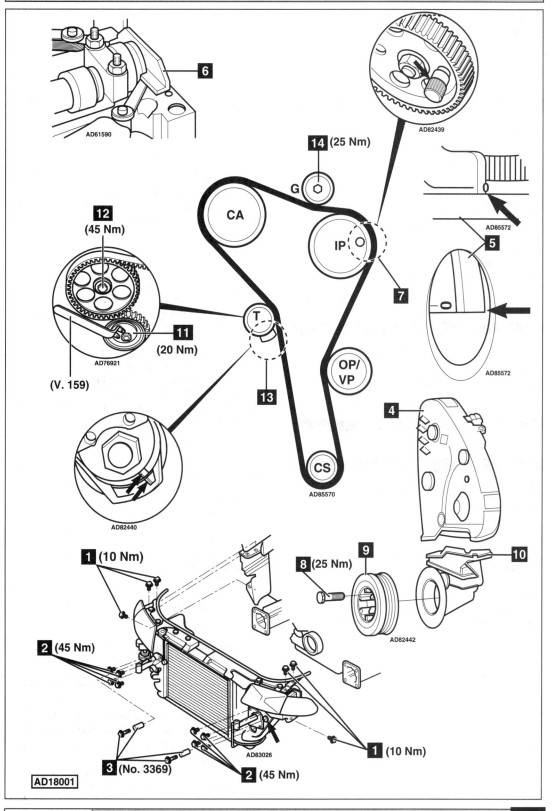

AD61590

6

AD82439

14 (25 Nm)

G

CA

12
(45 Nm)

IP

5

AD85572

7

11
(20 Nm)

AD76921

T

13

(V. 159)

OP/
VP

AD85572

4

CS

AD85570

AD82440

1 (10 Nm)

8 (25 Nm)

9

10

AD82442

2 (45 Nm)

1 (10 Nm)

3 (No. 3369)

2 (45 Nm)

AD83026

AD18001

AUDI

Model:	**A4 1,9 TDI**
Year:	**1997-01**
Engine Code:	**AHH**

Replacement Interval Guide

Audi recommend:
→1999MY: Check condition and width every 15,000 kilometres (9,320 miles) (replacement width less than 22 mm).
→1999MY: Replacement every 60,000 kilometres (37,282 miles) (tensioner pulley must also be replaced).
2000MY→: Replacement every 90,000 kilometres (55,923 miles).

NOTE: The vehicle manufacturer publishes this information only in kilometres. The conversion to miles is included for reference purposes only.
The previous use and service history of the vehicle must always be taken into account.
Refer to Timing Belt Replacement Intervals at the front of this manual.

Check For Engine Damage

CAUTION: This engine has been identified as an INTERFERENCE engine in which the possibility of valve-to-piston damage in the event of a timing belt failure is MOST LIKELY to occur.
A compression check of all cylinders should be performed before removing the cylinder head.

Repair Times – hrs

Remove & install	2,90

Special Tools

- Support guides – No.3369.
- Camshaft setting bar – No.2065A.
- Injection pump locking pin – No.3359.
- Sprocket holding tool – No.3036.
- Two-pin wrench – Matra V.159.

Special Precautions

- Disconnect battery earth lead.
- DO NOT turn crankshaft or camshaft when timing belt removed.
- Remove glow plugs to ease turning engine.
- Turn engine in normal direction of rotation (unless otherwise stated).
- DO NOT turn engine via camshaft or other sprockets.
- Observe all tightening torques.
- Check diesel injection pump timing after belt replacement.

Removal

1. Remove:
   - ❑ Front bumper.
   - ❑ Engine undershield.
   - ❑ Air intake pipe between front panel and air filter.
   - ❑ Front panel bolts **1** & **2**.
2. Install support guides No.3369 in front panel **3**.
3. Slide front panel forward.
4. Remove:
   - ❑ Engine top cover.
   - ❑ Auxiliary drive belt.
   - ❑ Viscous fan.
   - ❑ Viscous fan pulley.
   - ❑ Auxiliary drive belt tensioner.
5. Move turbocharger control solenoid aside. DO NOT disconnect hoses.
6. Remove:
   - ❑ Crankshaft pulley bolts **4**.
   - ❑ Crankshaft pulley **5**.
   - ❑ Timing belt upper cover **6**.
   - ❑ Timing belt lower cover **7**.
7. Turn crankshaft to TDC on No.1 cylinder. Ensure flywheel timing marks aligned **8**.
8. Remove:
   - ❑ Breather hose.
   - ❑ Cylinder head cover.
9. Fit setting bar No.2065A to rear of camshaft **9**. Centralise camshaft using feeler gauges.
10. Insert locking pin in injection pump **10**. Tool No.3359.
11. Slacken injection pump sprocket locking bolts **11**.
12. Slacken tensioner nut **12**. Turn tensioner anti-clockwise away from belt. Lightly tighten nut **12**.
13. Remove:
    - ❑ Guide pulley bolt **13**.
    - ❑ Guide pulley **14**.
    - ❑ Timing belt.
    NOTE: Mark direction of rotation on belt with chalk if belt is to be reused.

Installation

1. Ensure flywheel timing marks aligned **8**.
2. Ensure camshaft setting bar fitted correctly **9**.
3. Ensure locking pin located correctly in injection pump **10**.
4. Hold camshaft sprocket. Use tool No.3036.
5. Slacken camshaft sprocket bolt 1/2 turn **15**.

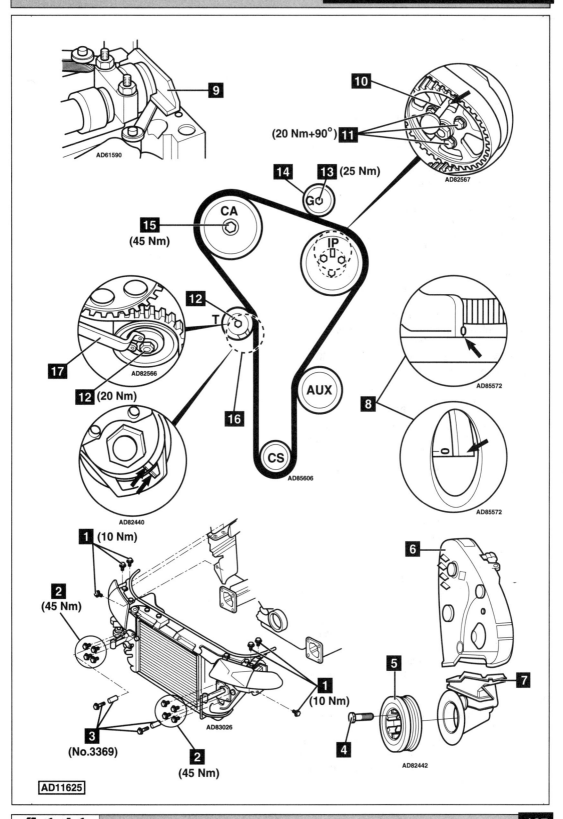

9

10

(20 Nm+90°) 11

AD61590

AD82567

14 13 (25 Nm)

G

15 CA
(45 Nm)

IP

12

T

17

12 (20 Nm)

AD82566

AD82440

16

AUX

CS

AD85606

AD85572

8

0

AD85572

1 (10 Nm)

2
(45 Nm)

3

(No.3369)

2
(45 Nm)

AD83026

6

5

4

7

AD82442

AD11625

←

6. Loosen sprocket from taper using a drift through hole in timing belt rear cover.

7. Remove:
   ❏ Camshaft sprocket bolt **15**.
   ❏ Camshaft sprocket.

8. Remove locking bolts from injection pump sprocket **11**. Fit new bolts. Tighten bolts finger tight.

9. Align injection pump sprocket with bolts in centre of slotted holes.

10. Fit timing belt in following order:
    ❏ Crankshaft sprocket.
    ❏ Auxiliary shaft sprocket.
    ❏ Injection pump sprocket.
    ❏ Tensioner pulley.
    NOTE: Turn injection pump sprocket slightly to engage timing belt teeth.

11. Ensure belt is taut between sprockets.

12. Fit camshaft sprocket to belt, then install camshaft sprocket with belt onto end of camshaft.

13. Fit camshaft sprocket bolt **15**.

14. Lightly tighten camshaft sprocket bolt **15**. Sprocket should turn freely on taper but not tilt.

15. Fit guide pulley **14**. Tighten bolt to 25 Nm **13**.

16. Slacken tensioner nut **12**.

17. Turn automatic tensioner pulley clockwise until notch and raised mark on tensioner aligned **16**. Use wrench Matra V.159 **17**.

18. If tensioner pulley turned too far: Turn fully anti-clockwise and repeat tensioning procedure.

19. Tighten tensioner nut to 20 Nm **12**.

20. Ensure flywheel timing marks aligned **8**.

21. Hold camshaft sprocket. Use tool No.3036.

22. Tighten camshaft sprocket bolt to 45 Nm **15**.

23. Temporarily tighten injection pump sprocket locking bolts to 20 Nm **11**.
    NOTE: DO NOT tighten further 90°.

24. Remove:
    ❏ Camshaft setting bar **9**.
    ❏ Injection pump locking pin **10**.

25. Turn crankshaft two turns clockwise until timing marks aligned **8**.

26. Ensure camshaft setting bar can be fitted **9**.

27. Ensure locking pin can be inserted in injection pump **10**.

28. If not: Slacken injection pump sprocket locking bolts **11**. Turn sprocket hub until locking pin can be inserted **10**.

29. Tighten bolts to 20 Nm + 90° **11**.

30. Ensure notch and raised mark on tensioner aligned **16**.

31. If not: Slacken tensioner nut **12**. Turn automatic tensioner pulley clockwise until notch and raised mark on tensioner aligned **16**. Tighten tensioner nut to 20 Nm **12**.

32. Install components in reverse order of removal.

33. Tighten crankshaft pulley bolts **4**:
    Class 8.8 – 25 Nm. Class 10.9 – 35 Nm.
    NOTE: Class number identified on bolt head.

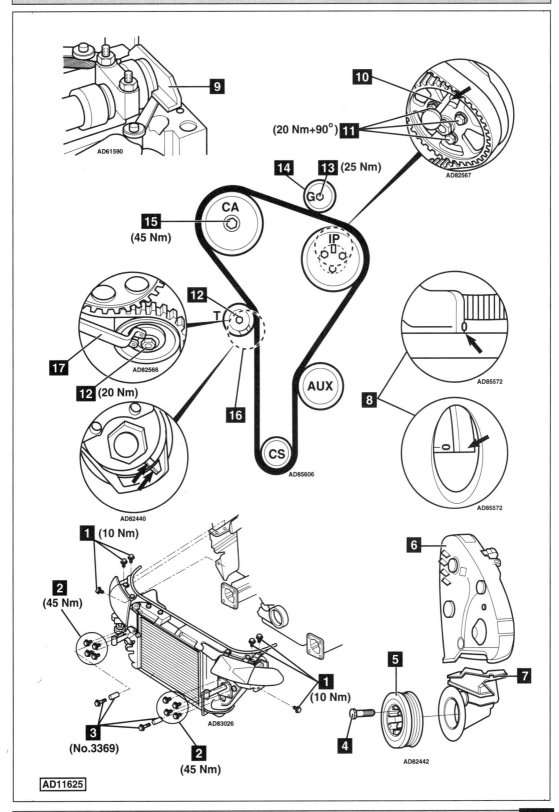

9

AD61590

10

(20 Nm+90°) **11**

AD82567

14 **13** (25 Nm)

G

15
(45 Nm)

CA

IP

12

T

AD85572

17

AD82566

12 (20 Nm)

16

AUX

8

CS

AD85606

AD82440

AD85572

1 (10 Nm)

6

2
(45 Nm)

1
(10 Nm)

5

7

3

4

AD83026

2
(45 Nm)

(No.3369)

AD82442

AD11625

AUDI

Model:	A4 1,9 TDI PD • A6 1,9 TDI PD
Year:	1999-03
Engine Code:	AJM, ATJ, AVB, AVF, AWX

Replacement Interval Guide

Audi recommend replacement every 90,000 kilometres (55,923 miles) (tensioner pulley must also be replaced).

NOTE: The vehicle manufacturer publishes this information only in kilometres. The conversion to miles is included for reference purposes only. The previous use and service history of the vehicle must always be taken into account.
Refer to Timing Belt Replacement Intervals at the front of this manual.

Check For Engine Damage

CAUTION: This engine has been identified as an INTERFERENCE engine in which the possibility of valve-to-piston damage in the event of a timing belt failure is MOST LIKELY to occur. A compression check of all cylinders should be performed before removing the cylinder head(s).

Repair Times – hrs

Remove & install:
A4 →2001	2,50
A4 2001→	2,90
A6	3,20

Special Tools

- Camshaft locking tool – No.3359.
- Crankshaft sprocket locking tool – No.T10050.
- Support guides – No.3369.
- Tensioner locking tool – No.T10008.
- Two-pin wrench – No.3387.
- Viscous fan pulley holding tool – No.3212.
- 4 mm drill bit.

Special Precautions

- Disconnect battery earth lead.
- DO NOT turn crankshaft or camshaft when timing belt removed.
- Remove glow plugs to ease turning engine.
- Turn engine in normal direction of rotation (unless otherwise stated).
- DO NOT turn engine via camshaft or other sprockets.
- Observe all tightening torques.

Removal

1. Move radiator support panel into service position:
   - ❑ Remove front bumper.
   - ❑ Remove air intake pipe between front panel and air filter.
   - ❑ Remove front panel bolts **1** & **2**.
   - ❑ Install support guides No.3369 in front panel **3**.
   - ❑ Slide front panel forward.
2. Raise and support front of vehicle.
3. Disconnect exhaust pipe for auxiliary heater from engine undershield (if fitted).
4. Remove:
   - ❑ Engine top cover.
   - ❑ Engine undershield.
   - ❑ Auxiliary drive belt.
   - ❑ Cooling fan assembly. Use tool No.3212.
   - ❑ Auxiliary drive belt tensioner.
   - ❑ Bolts for coolant pipe brackets (if required).
   - **NOTE: DO NOT disconnect hoses.**
   - ❑ Timing belt upper cover **4**.
5. Move turbocharger (TC) wastegate regulating valve.
6. Remove:
   - ❑ Crankshaft pulley bolts **5**.
   - ❑ Crankshaft pulley **6**.
   - ❑ Timing belt centre cover **7**.
   - ❑ Timing belt lower cover **8**.
7. Turn crankshaft clockwise to TDC on No.1 cylinder. Ensure timing mark aligned with notch on camshaft sprocket hub **9**.
   NOTE: Notch located behind camshaft sprocket teeth.
8. Lock crankshaft sprocket **10**. Use tool No.T10050.
9. Ensure timing marks aligned **11**.
10. Lock camshaft **12**. Use tool No.3359.
11. Fully insert Allen key into tensioner pulley **13**.
12. Turn tensioner pulley slowly anti-clockwise until locking tool can be inserted **14**. Tool No.T10008.
13. Slacken tensioner nut **15**.
14. Remove:
    - ❑ Automatic tensioner unit **16**.
    - ❑ Timing belt.

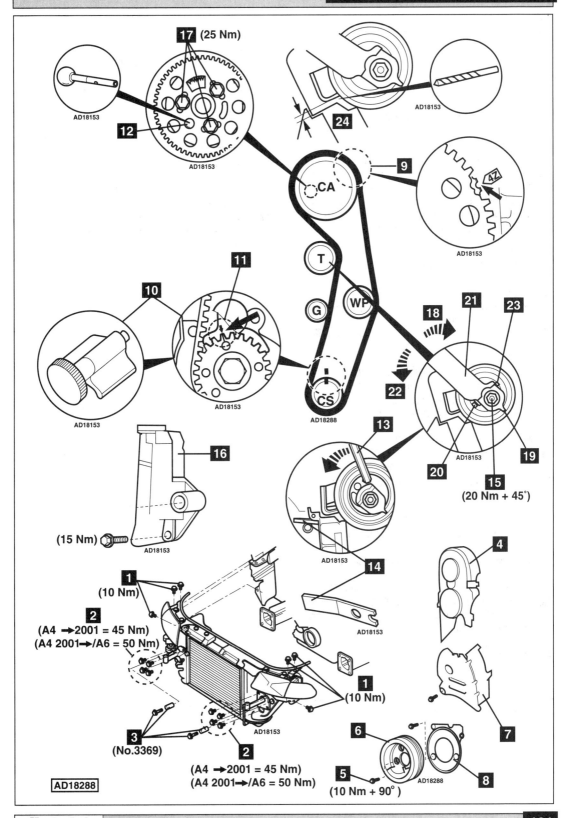

17 (25 Nm)

AD18153

12

AD18153

24

AD18153

CA

9

AZ

AD18153

11

T

10

G

WP

18

21

23

AD18153

CS

22

20

AD18153

19

13

15
(20 Nm + 45°)

16

AD18153

14

AD18153

4

(15 Nm)

AD18153

1
(10 Nm)

2
(A4 ➞2001 = 45 Nm)
(A4 2001➞/A6 = 50 Nm)

AD18153

1
(10 Nm)

7

3
(No.3369)

2
(A4 ➞2001 = 45 Nm)
(A4 2001➞/A6 = 50 Nm)

6

5
(10 Nm + 90°)

AD18288

8

AD18288

Installation

1. Ensure camshaft locked with tool **12**. Tool No.3359.
2. Ensure crankshaft sprocket locking tool located correctly **10**. Tool No.T10050.
3. Ensure timing marks aligned **11**.
4. Ensure automatic tensioner unit locked with tool **14**. Tool No.T10008.
5. Slacken camshaft sprocket bolts **17**.
6. Turn camshaft sprocket fully clockwise in slotted holes. Tighten bolts finger tight **17**.
7. Turn tensioner pulley slowly clockwise **18** until lug **19** just reaches stop **20**. Use tool No.3387 **21**.
8. Fit timing belt in following order:
   - ❏ Camshaft sprocket.
   - ❏ Tensioner pulley.
   - ❏ Crankshaft sprocket.
   - ❏ Water pump sprocket.
9. Install automatic tensioner unit **16**.

   NOTE: Ensure belt is taut between sprockets on non-tensioned side.
10. Turn tensioner pulley slowly anti-clockwise **22** (lug **19** moves towards stop **23**). Use tool No.3387 **21**.
11. Remove locking tool without force **14**.

    NOTE: DO NOT release tensioner pulley.
12. Turn tensioner pulley slowly clockwise **21** (lug **19** moves towards stop **20**) until dimension **24** is 4±1 mm. Use drill bit.

    NOTE: Engine must be COLD.
13. Tighten tensioner nut **15**. Tightening torque: 20 Nm + 45°.
14. Tighten camshaft sprocket bolts **17**. Tightening torque: 25 Nm.
15. Remove:
    - ❏ Camshaft locking tool **12**.
    - ❏ Crankshaft sprocket locking tool **10**.
    - ❏ Drill bit.
16. Turn crankshaft slowly two turns clockwise to TDC on No.1 cylinder.
17. Ensure dimension **24** is 4±1 mm. Use drill bit.
18. If not: Slacken tensioner nut **15**. Turn tensioner pulley until dimension as specified **24**. Tighten tensioner nut **15**. Tightening torque: 20 Nm + 45°.
19. Lock crankshaft sprocket **10**. Use tool No.T10050.
20. Ensure timing marks aligned **11**.
21. Ensure camshaft locking tool can be inserted easily **12**. Tool No.3359.
22. If not:
    - ❏ Slacken camshaft sprocket bolts **17**.
    - ❏ Turn sprocket hub until locking tool can be inserted **12**.
    - ❏ Tighten bolts **17**. Tightening torque: 25 Nm.
    - ❏ Remove locking tools **10** & **12**.
    - ❏ Turn crankshaft slowly two turns clockwise to TDC on No.1 cylinder.
    - ❏ Ensure locking tools can be fitted correctly **10** & **12**.
23. Remove:
    - ❏ Camshaft locking tool **12**.
    - ❏ Crankshaft locking tool **10**.
    - ❏ Drill bit.
24. Install components in reverse order of removal.
25. Tighten crankshaft pulley bolts **5**. Tightening torque: 10 Nm + 90°. Use new bolts.

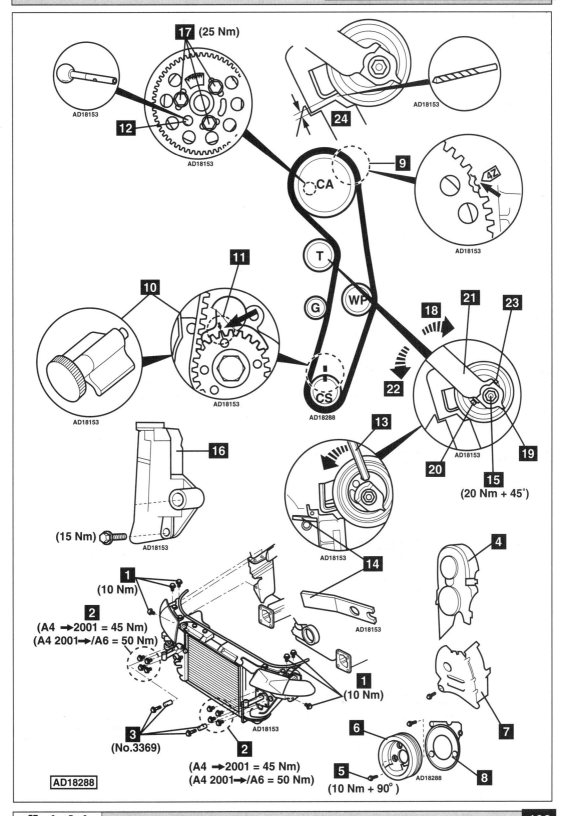

17 (25 Nm)

AD18153

12

AD18153

24

AD18153

9

AD18153

CA

11

10

T

G

WP

18

21

23

22

AD18153

20

19

AD18153

CS

AD18288

13

AD18153

15
(20 Nm + 45°)

16

(15 Nm)

AD18153

14

AD18153

4

1
(10 Nm)

2
(A4 ➡2001 = 45 Nm)
(A4 2001➡/A6 = 50 Nm)

1
(10 Nm)

AD18153

7

3
(No.3369)

2
(A4 ➡2001 = 45 Nm)
(A4 2001➡/A6 = 50 Nm)

AD18153

6

5
(10 Nm + 90°)

AD18288

8

AD18288

AUDI

Model:	A4 2,5 V6 TDI • A6 2,5 V6 TDI • A8 2,5 V6 TDI • Allroad 2,5 V6 TDI
Year:	1997-03
Engine Code:	AFB, AKN, AKE, AYM

Replacement Interval Guide

Audi recommend:

Fixed servicing:
Check timing belt every 30,000 kilometres (18,641 miles).
Check and adjust injection pump belt every 30,000 kilometres (18,641 miles).
Replace timing belt and injection pump belt every 120,000 kilometres (74,564 miles).

Longlife servicing:
Check timing belt at every service.
Check and adjust injection pump belt at every service.
Replace timing belt and injection pump belt every 120,000 kilometres (74,564 miles).

NOTE: The vehicle manufacturer publishes this information only in kilometres. The conversion to miles is included for reference purposes only. The previous use and service history of the vehicle must always be taken into account.
Refer to Timing Belt Replacement Intervals at the front of this manual.

Check For Engine Damage

CAUTION: This engine has been identified as an INTERFERENCE engine in which the possibility of valve-to-piston damage in the event of a timing belt failure is MOST LIKELY to occur.
A compression check of all cylinders should be performed before removing the cylinder head(s).

Repair Times – hrs

Remove & install:

A4 →2001	3,70
A4 2001→	4,10
A6/Allroad	4,50
A8	3,30

Special Tools

- Auxiliary drive belt tensioner locking pin – No.T10060.
- Crankshaft locking tool – No.3242.
- Camshaft aligning tool – 2 x No.3458.
- Holding tool – No.3036.
- Injection pump locking pin – No.3359.
- Puller – No.T40001.
- Support guides – No.3369.
- Tensioner socket – No.3078.
- 2 mm diameter locking pin – No.T40011.

Removal

Injection Pump Belt

1. Raise and support front of vehicle.

A4/A6/Allroad:

2. Disconnect exhaust pipe for auxiliary heater from engine undershield (A4 – if fitted).
3. Remove:
   - ❏ Front bumper.
   - ❏ Engine undershield.

- ❏ Intercooler hoses.
- ❏ Air filter intake duct.
- ❏ Front panel bolts **1**.
- ❏ Front panel bolts **2** (except A4 2001→).
4. Install support guides No.3369 in front panel **3**.
5. Remove front panel bolts **27**.
6. Slide front panel forward into service position.

A8:

7. Remove:
   - ❏ Engine undershield.

A4 2001→:

8. Remove:
   - ❏ Turbocharger heat shield.
   - ❏ LH front wheel.
   - ❏ LH splash guard.
   - ❏ LH driveshaft heat shield.
   - ❏ Pre-catalytic converter.

All models:

9. Remove:
   - ❏ Viscous fan.
   - ❏ Viscous fan air ducting.
   - ❏ Upper engine cover.
   - ❏ Timing belt upper covers.
   - ❏ Auxiliary drive belt cover.
   - ❏ Auxiliary drive belt(s). Use tool No.T10060.
   NOTE: Mark direction of rotation on belt(s).
   - ❏ Oil filler cap.
10. Turn crankshaft clockwise to TDC on No.3 cylinder.
11. Ensure camshaft timing mark aligned with centre of oil filler cap hole **4**.
12. Remove blanking plug from cylinder block.
13. Screw in crankshaft locking tool. Tool No.3242 **5**.
   NOTE: TDC hole in crankshaft web must be aligned with blanking plug hole.
14. Remove:
   - ❏ Coolant expansion tank. DO NOT disconnect hoses.
   - ❏ Vacuum pump. DO NOT disconnect hoses.
   - ❏ A4: Air filter housing.
   - ❏ A4 2001→: Fuel hose support brackets.
   - ❏ A4 2001→: Fuel filter housing. DO NOT disconnect hoses.
   - ❏ A4: Turbocharger air hoses.
   - ❏ Cover plate from rear of RH camshaft **6**.
   NOTE: Cover plate is destroyed when removed. DO NOT damage sealing edge.
15. Install camshaft aligning tools to rear of camshafts. Tool No.3458 **7**.
   NOTE: Retain in position with chains provided, to prevent tools falling out.
16. Remove:
   - ❏ Injection pump vibration damper bolts **8**.
   - ❏ Injection pump vibration damper **9**.
   *NOTE: DO NOT slacken injection pump centre hub nut **10**.*
17. Insert locking pin in injection pump sprocket. Tool No.3359 **11**.
18. Slacken injection pump belt tensioner nut **12**. Tool No.3078.

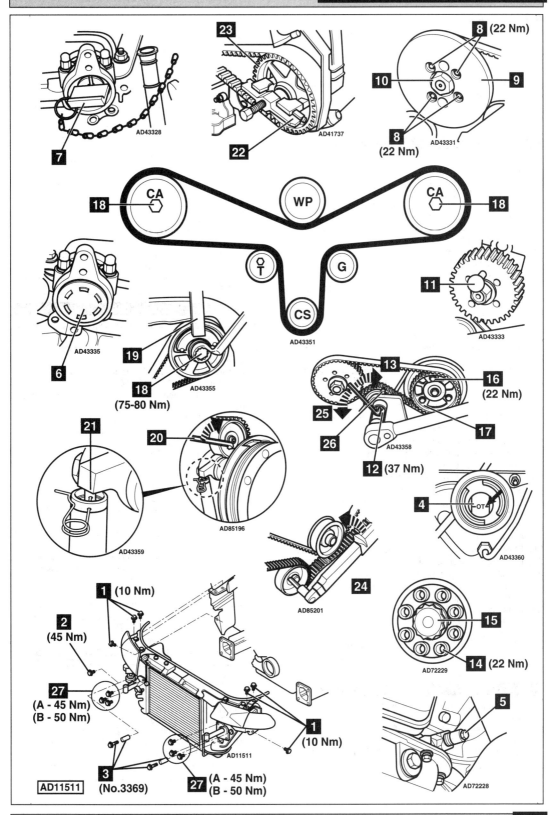

AD43328

7

23

AD41737

22

8 (22 Nm)

10

9

8 (22 Nm)

AD43331

18 CA

WP

CA **18**

AD43333

11

AD43335

6

O/T

G

CS

AD43351

19

18 (75-80 Nm)

AD43355

13

25

26

16 (22 Nm)

17

AD43358

12 (37 Nm)

21

20

AD85196

AD43359

4 OT

AD43360

24

AD85201

15

14 (22 Nm)

AD72229

1 (10 Nm)

2 (45 Nm)

27 (A - 45 Nm) (B - 50 Nm)

3 (No.3369)

AD11511

1 (10 Nm)

27 (A - 45 Nm) (B - 50 Nm)

5

AD72228

←

19. Turn tensioner pulley clockwise away from belt 🔞. Use Allen key.
20. Lightly tighten injection pump belt tensioner nut 🔢.
21. Remove injection pump belt.
    NOTE: Mark direction of rotation on belt with chalk if belt is to be reused.

Installation

Injection Pump Belt

1. Ensure camshaft timing mark aligned with centre of oil filler cap hole 🔢.
2. Ensure crankshaft locking tool located correctly. Tool No.3242 🔢.
3. Ensure camshaft aligning tools fitted to rear of camshafts. Tool No.3458 🔢.
4. Ensure locking pin located correctly in injection pump. Tool No.3359 🔢.
5. Slacken camshaft sprocket bolts 🔢.
    NOTE: DO NOT use camshaft aligning tool to prevent sprocket from turning.
6. Align camshaft outer sprocket with bolts in centre of slotted holes.
7. Lightly tighten bolts 🔢. Ensure sprocket can turn.
8. Fit injection pump belt.
    NOTE: Observe direction of rotation marks.
9. Slacken injection pump belt tensioner nut 🔢. Use tool No.3078.
10. Turn tensioner pulley anti-clockwise 🔢. Use Allen key. Check tensioner pointers align 🔢.
11. Tighten tensioner nut to 37 Nm 🔢. Use tool No.3078.
12. Tighten camshaft sprocket bolts to 22 Nm 🔢.
    NOTE: DO NOT use camshaft aligning tool to prevent sprocket from turning.
13. Remove injection pump locking pin 🔢.
14. Remove camshaft aligning tools 🔢.
15. Remove crankshaft locking tool 🔢. Fit blanking plug.
16. Turn crankshaft slowly two turns clockwise.
17. Screw in crankshaft locking tool. Tool No.3242 🔢.
18. Check camshaft aligning tools can be fitted correctly 🔢.
19. Check locking pin can be fitted correctly in injection pump sprocket 🔢.
20. Check tensioner pointers align 🔢. Adjust if necessary.
21. Remove tools 🔢, 🔢 & 🔢.
22. Install injection pump vibration damper 🔢.
23. Tighten injection pump vibration damper bolts 🔢. Tightening torque: 22 Nm.
24. Install components in reverse order of removal.
25. Tighten front panel bolts 🔢:
    ❑ (A) A4 →2001 – 45 Nm.
    ❑ (B) A4 2001→/A6/Allroad – 50 Nm.

Removal

Timing Belt

1. Remove injection pump belt as described previously.
2. Remove:
    ❑ Crankshaft pulley bolts 🔢.
    ❑ Crankshaft pulley.
    NOTE: DO NOT remove crankshaft centre bolt 🔢.
    ❑ Timing belt lower cover.
    ❑ Viscous fan pulley.
    ❑ Viscous fan bracket.

3. Remove:
    ❑ Camshaft sprocket bolts 🔢.
    ❑ Camshaft outer sprocket (for injection pump belt) 🔢.
    NOTE: DO NOT use camshaft aligning tool to prevent sprocket from turning.
4. Hold sprockets. Use tool No.3036 🔢. Slacken bolt of each camshaft sprocket 🔢.
5. Turn tensioner pulley slowly clockwise until holes in pushrod and tensioner body aligned. Use 8 mm Allen key 🔢.
6. Retain pushrod with 2 mm diameter locking pin through hole in tensioner body 🔢. Tool No.T40011.
7. Loosen camshaft sprockets from camshafts. Use tool No.T40001 🔢.
8. Remove:
    ❑ LH camshaft sprocket bolt 🔢.
    ❑ LH camshaft sprocket 🔢.
    ❑ Timing belt.
    NOTE: Mark direction of rotation on belt with chalk if belt is to be reused.

Installation

Timing Belt

1. Ensure camshaft timing mark aligned with centre of oil filler cap hole 🔢.
2. Ensure crankshaft locking tool located correctly. Tool No.3242 🔢.
3. Ensure camshaft aligning tools fitted to rear of camshafts. Tool No.3458 🔢.
4. Ensure RH camshaft sprocket can turn on taper but not tilt.
5. Fit timing belt in following order:
    NOTE: Observe direction of rotation marks.
    ❑ Crankshaft sprocket.
    ❑ RH camshaft sprocket.
    ❑ Tensioner pulley.
    ❑ Guide pulley.
    ❑ Water pump pulley.
6. Fit LH camshaft sprocket to belt, then install camshaft sprocket with belt onto end of camshaft.
7. Fit camshaft sprocket bolt 🔢.
8. Ensure LH camshaft sprocket can turn on taper but not tilt.
9. Turn tensioner pulley slightly clockwise. Use 8 mm Allen key 🔢. Remove locking pin from tensioner body to release pushrod 🔢.
10. Remove Allen key from tensioner pulley 🔢.
11. Install torque wrench to hexagon of tensioner.
12. Apply anti-clockwise torque (in direction of arrow) of 15 Nm to tensioner pulley 🔢.
13. Remove torque wrench.
14. Hold sprockets. Use tool No.3036 🔢. Tighten bolt of each camshaft sprocket to 75-80 Nm 🔢.
15. Install:
    ❑ Timing belt lower cover.
    ❑ Viscous fan bracket.
    ❑ Viscous fan pulley.
    ❑ Crankshaft pulley.
    ❑ Crankshaft pulley bolts 🔢.
    NOTE: Ensure notches aligned with tabs on crankshaft sprocket.
16. Tighten crankshaft pulley bolts to 22 Nm 🔢.
17. Fit injection pump belt as described previously.
18. Install components in reverse order of removal.

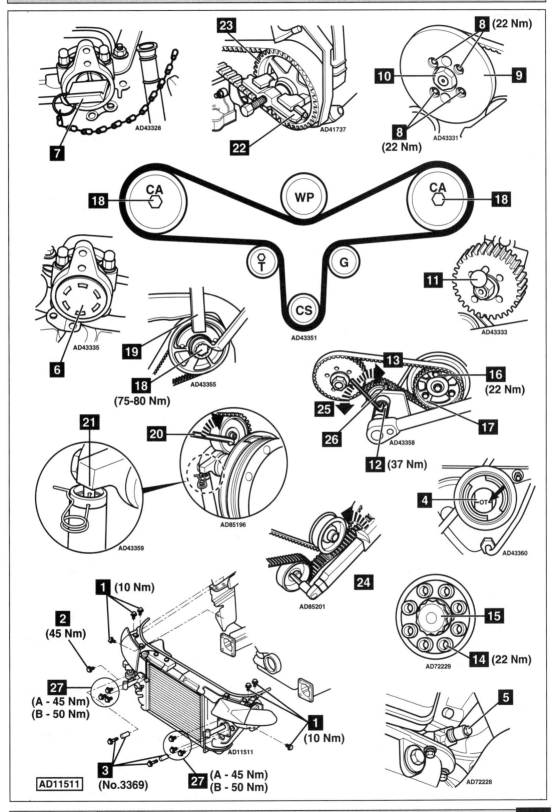

23

8 (22 Nm)

10

9

8
(22 Nm)

AD43331

AD43328

7

22

AD41737

18 CA

WP

CA **18**

T

G

CS

AD43351

11

AD43333

6

AD43335

19

18
(75-80 Nm)

AD43355

13

16
(22 Nm)

25

17

26

AD43358

12 (37 Nm)

21

20

AD85196

4

OT

AD43360

AD43359

24

AD85201

1 (10 Nm)

2
(45 Nm)

15

27
(A - 45 Nm)
(B - 50 Nm)

1
(10 Nm)

14 (22 Nm)

AD72229

5

3
(No.3369)

AD11511

27 **(A - 45 Nm)**
(B - 50 Nm)

AD11511

AD72228

AUDI

Model:	A6 2,5 TDI
Year:	1994-97
Engine Code:	AEL, AAT

Replacement Interval Guide

Audi recommend check and adjust timing belt and injection pump belt every 30,000 kilometres (18,641 miles) and replacement every 120,000 kilometres (74,564 miles).

NOTE: The vehicle manufacturer publishes this information only in kilometres. The conversion to miles is included for reference purposes only.
The previous use and service history of the vehicle must always be taken into account.
Refer to Timing Belt Replacement Intervals at the front of this manual.

Check For Engine Damage

CAUTION: This engine has been identified as an INTERFERENCE engine in which the possibility of valve-to-piston damage in the event of a timing belt failure is MOST LIKELY to occur.
A compression check of all cylinders should be performed before removing the cylinder head.

Repair Times – hrs

Remove & install	3,70

Special Tools

- Crankshaft pulley holding tool – No.3256.
- Tensioner pulley spanner – No.3355.
- Sprocket holding wrench – No.3036.
- Camshaft setting bar – No.2065A.
- Viscous fan holding tool – No.3212.

Removal

1. Remove:
   - ❏ Upper engine cover.
   - ❏ Bonnet shut panel.
   - ❏ PAS pump pulley guard.
   - ❏ Auxiliary drive belt tensioner.
   - ❏ Viscous fan. Use tool No.3212. (LH thread).
   - ❏ Auxiliary drive belt.
   - ❏ Air pipe to air filter.
   - ❏ Viscous fan air ducting.
   - ❏ Cooling fan and cowling.
   - ❏ Timing belt upper cover **1**.
2. Remove two bolts from oil pipe bracket. Fit crankshaft pulley holding tool. Tool No.3256.
3. Slacken crankshaft pulley centre bolt **2**.
4. Turn crankshaft to TDC on No.1 cylinder. Ensure timing marks aligned **3**.
   NOTE: Use a mirror to check that the smaller of two notches on the crankshaft pulley is aligned.
5. Ensure injection pump timing marks aligned **4**.
6. If not: Turn crankshaft one turn clockwise.
7. Remove:
   - ❏ Crankshaft pulley bolts **5**.
   - ❏ Crankshaft pulley centre bolt **2**.
   - ❏ Crankshaft pulley **6**.
   - ❏ Timing belt lower cover **7**.
   - ❏ Injection pump belt cover.
   - ❏ Cylinder head cover.

8. Slacken injection pump belt tensioner nut **10**. Remove tensioner spring **11**. Move tensioner away from belt. Lightly tighten nut.
9. Hold camshaft rear sprocket. Use tool No.3036 **12**. Remove bolt and washer **13**.
10. Remove camshaft rear sprocket and injection pump belt.
11. Fit setting bar No.2065A to rear of camshaft **14**. Centralise camshaft using feeler gauges **15**.
12. Slacken tensioner bolt **8**. Turn tensioner pulley away from belt. Use tool No.3355 **9**.
13. Lightly tighten tensioner bolt **8**.
14. Remove timing belt.
    NOTE: Ensure crankshaft does not move.

Installation

1. Slacken camshaft sprocket bolt 1/2 turn **16**. Tap sprocket gently to loosen it from taper.
   NOTE: Sprocket should turn freely on taper but not tilt.
2. Ensure crankshaft has not moved.
3. Fit timing belt in anti-clockwise direction, starting at crankshaft sprocket. Ensure belt is taut between sprockets.
4. Slacken tensioner bolt **8**.
5. Turn tensioner pulley until pointer has moved approximately 3 mm to the right **17**. Use tool No.3355 **9**.
6. Slacken tensioner bolt until pointers aligned **18**.
7. Tighten tensioner bolt to 25 Nm **8**.
8. Install:
   - ❏ Timing belt lower cover **7**.
   - ❏ Crankshaft pulley **6**. Lightly tighten centre bolt **2**.
9. Ensure timing marks aligned **3**.
10. Tighten camshaft sprocket bolt **16**. Tightening torque: 30 Nm + 90°.
11. Remove camshaft setting bar **14**.
12. Fit camshaft rear sprocket. Fit bolt and washer **13**. Tighten bolt until sprocket can just be turned by hand.
13. Install:
    - ❏ Injection pump belt.
    - ❏ Tensioner spring **11**.
14. Slacken tensioner nut to tension belt **10**.
15. Ensure timing marks aligned **3** & **4**.
16. Hold camshaft rear sprocket. Use tool No.3036 **12**. Tighten bolt to 160 Nm **13**.
17. Turn crankshaft 1/2 turn clockwise. Tighten tensioner nut to 45 Nm **10**.
18. Turn crankshaft 1 1/2 turns to TDC on No.1 cylinder. Ensure timing marks aligned **3** & **4**.
19. Recheck belt tension. Ensure pointers no more than 3 mm apart **17**.
20. Fit new crankshaft pulley centre bolt **2**. Lightly oil threads. Tightening torque: 220 Nm + 270° ($^3/_4$ of a turn).
21. Fit crankshaft pulley bolts **5**. Tightening torque: 20 Nm + 90°.
22. Install components in reverse order of removal.

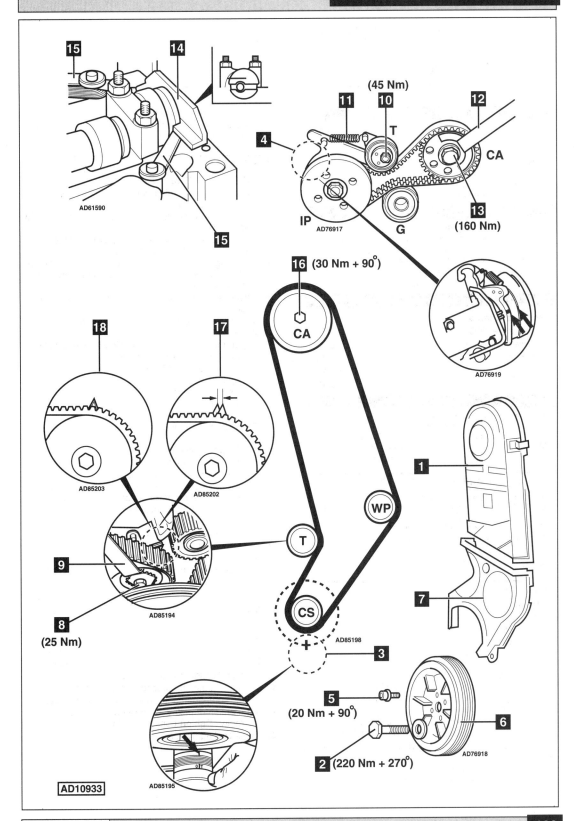

15 **14** **15** AD61590

(45 Nm) **11** **10** T **12** **4** CA IP AD76917 G **13** **(160 Nm)**

16 (30 Nm + 90°) CA

AD76919

18 **17** AD85203 AD85202

9 AD85194 T **8** **(25 Nm)**

1

7

WP

CS AD85198 **+** **3**

5 (20 Nm + 90°)

2 (220 Nm + 270°) **6** AD76918

AD10933 AD85195

Model:	**316i • 318i • 518i**
Year:	**1988-93**
Engine Code:	**M40**

Replacement Interval Guide

BMW recommend replacement every Inspection II (approx. every 28,000 miles) or 3 years, whichever occurs first.
The previous use and service history of the vehicle must always be taken into account.
Refer to Timing Belt Replacement Intervals at the front of this manual.

Check For Engine Damage

CAUTION: This engine has been identified as an INTERFERENCE engine in which the possibility of valve-to-piston damage in the event of a timing belt failure is MOST LIKELY to occur.
A compression check of all cylinders should be performed before removing the cylinder head.

Repair Times – hrs

Remove & install:

3 Series (→1990)	1,70
3 Series (1991→)	2,00
5 Series	1,75
AC	+0,10
PAS	+0,10

Special Tools

- ■ Flywheel timing pin – BMW No.11.2.300.
- ■ Tension gauge – BMW No.11.2.080.
- ■ Camshaft setting tool – BMW No.11.3.190.
- ■ Temperature sensor – BMW No.11.5.060.

Special Precautions

- ■ Disconnect battery earth lead.
- ■ DO NOT turn crankshaft or camshaft when timing belt removed.
- ■ Remove spark plugs to ease turning engine.
- ■ Turn engine in normal direction of rotation (unless otherwise stated).
- ■ DO NOT turn engine via camshaft or other sprockets.
- ■ Observe all tightening torques.

Removal

1. Drain coolant.
2. Remove:
   - ❑ Cooling fan and cowling.
   - ❑ Radiator.
   - ❑ Auxiliary drive belt(s).
   - ❑ Water pump pulley.
   - ❑ Distributor.
   - ❑ Top hose and thermostat housing.
   - ❑ Engine speed (RPM) sensor.
   - ❑ Cylinder head cover.
3. Turn crankshaft to TDC on No.1 cylinder.
4. Insert flywheel timing pin **1**. Tool No.11.2.300.
5. Fit setting tool to camshaft **4**. Tool No.11.3.190.
6. Remove:
   - ❑ Crankshaft pulley.
   - ❑ Timing belt upper cover **2**.
   - ❑ Timing belt lower cover **3**.

7. Slacken camshaft sprocket bolt.
8. Slacken tensioner bolt to release tension on belt.
9. Remove timing belt.

Installation

NOTE: DO NOT refit used belt.
1. Ensure flywheel timing pin located correctly **1**. Tool No.11.2.300.
2. Ensure setting tool located correctly on camshaft **4**. Tool No.11.3.190.
3. Ensure camshaft sprocket can just be turned by hand and turn clockwise against stop.
4. Fit timing belt in anti-clockwise direction, starting at crankshaft sprocket. Ensure belt is taut between sprockets on non-tensioned side.
5. Attach tension gauge to belt at ▽. Zero tension gauge.
6. Turn tensioner pulley anti-clockwise until tension gauge indicates 45 units.
7. Tighten tensioner bolt. Tightening torque: 22 Nm.
8. Tighten camshaft sprocket bolt. Tightening torque: 63 Nm.
9. Remove:
   - ❑ Tension gauge.
   - ❑ Camshaft setting tool.
   - ❑ Flywheel timing pin.
10. Turn crankshaft two turns clockwise.
11. Install:
    - ❑ Flywheel timing pin **1**.
    - ❑ Camshaft setting tool **4**.
12. Attach tension gauge to belt at ▽.
13. Check cylinder head temperature above coolant outlet. Use temperature sensor No.11.5.060.
14. Slacken camshaft sprocket bolt.
15. Slacken tensioner bolt. Turn tensioner pulley anti-clockwise until tension gauge indicates 45 units.
16. Turn tensioner pulley slowly clockwise until tension gauge indicates belt tension as follows:

Cylinder head temperature	Tension
20°C	32 units
25°C	33 units
30°C	34 units
35°C	35 units
40°C	36 units
45°C	38 units
50°C	39 units

17. Tighten tensioner bolt to 22 Nm.
18. Tighten camshaft sprocket bolt to 63 Nm.
19. Install components in reverse order of removal.
20. Fit crankshaft pulley. Ensure dowel pin located correctly. Tighten bolts to 23 Nm.
21. Refill cooling system.

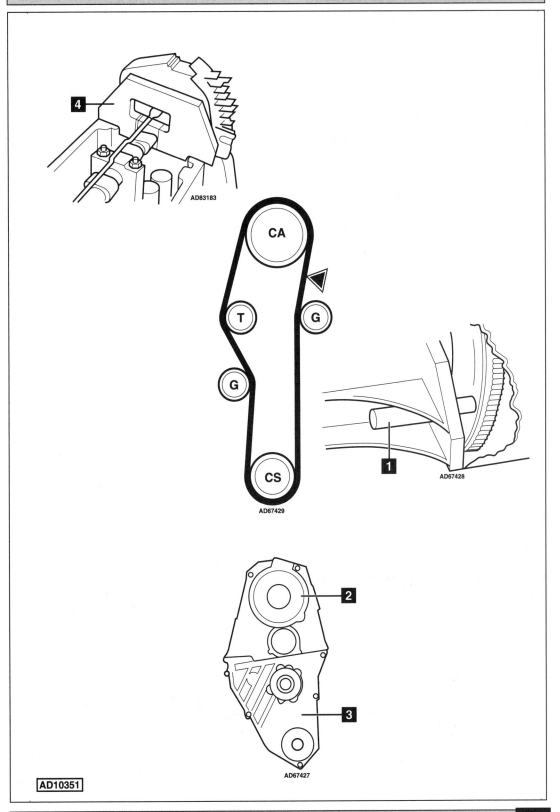

AD83183

CA

T

G

G

CS

AD67429

1

AD67428

2

3

AD67427

AD10351

BMW

Model:	320i/323i (E21) • 320i/323i/325i (E30) • 520i/525i/e (E28) • 520i/525i (E34) Z1
Year:	1977-93
Engine Code:	M20

Replacement Interval Guide

BMW recommend replacement every 2nd Inspection II (approx. every 56,000 miles) or 4 years, whichever occurs first.
The previous use and service history of the vehicle must always be taken into account.
Refer to Timing Belt Replacement Intervals at the front of this manual.

Check For Engine Damage

CAUTION: This engine has been identified as an INTERFERENCE engine in which the possibility of valve-to-piston damage in the event of a timing belt failure is MOST LIKELY to occur.
A compression check of all cylinders should be performed before removing the cylinder head.

Repair Times – hrs

Remove & install:

320i (E21)	1,85
320i (E30 →08/87)	1,60
320i (E30 09/87→)	1,85
320i Cabrio	2,20
323i (E21)	1,60
323i (E30)	1,60
325i (→08/87)	1,75
325i (09/87→)	1,85
325i Cabrio (→08/87)	1,75
325i Cabrio (09/87→)	2,10
520i/525i/e (E28 1983-88)	1,85
520/525i (E34 1988-90)	1,95
AC	+0,10
PAS	+0,10
Z1	1,85

Special Tools

- Water pump pulley holding tool – BMW No.11.5.030.
- Fan wrench – BMW No.11.5.040.

Special Precautions

- Disconnect battery earth lead.
- DO NOT turn crankshaft or camshaft when timing belt removed.
- Remove spark plugs to ease turning engine.
- Turn engine in normal direction of rotation (unless otherwise stated).
- DO NOT turn engine via camshaft or other sprockets.
- Observe all tightening torques.

Removal

1. Turn crankshaft to TDC on No.1 cylinder. Ensure timing marks aligned **1**.
2. E28/E30: Remove distributor cap. Ensure distributor rotor arm aligned with mark on distributor body. E34: Remove distributor cap, rotor arm and backplate.
3. Hold water pump pulley. Use tool No.11.5.030. Using wrench No.11.5.040, undo fan coupling and remove viscous fan.
4. Remove water pump pulley.
   NOTE: Fan coupling nut has LH thread.
5. Remove:
   - Auxiliary drive belts.
   - Engine speed (RPM) sensor (if applicable).
   - Crankshaft pulley/damper bolts **2**.
   - Crankshaft pulley/damper **3**.
   - Timing belt upper cover **4**.
   - Timing belt lower cover **5**.
6. Slacken tensioner bolts. Move tensioner away from belt. Lightly tighten bolts.
7. Remove timing belt.

Installation

NOTE: DO NOT refit used belt.

1. Ensure timing marks aligned **6** & **7**. E28/E30: Ensure distributor rotor arm aligned with mark on distributor body.
2. Fit timing belt in anti-clockwise direction, starting at crankshaft sprocket. Ensure belt is taut between sprockets.
3. Ensure timing marks aligned **6** & **7**.
4. Slacken tensioner bolts.
5. Turn crankshaft two turns clockwise.
6. Ensure timing marks aligned **6** & **7**. Tighten tensioner bolts to 23 Nm.
7. Install:
   - Timing belt upper cover **4**.
   - Timing belt lower cover **5**.
8. Fit crankshaft pulley/damper **3**. Ensure locating pin located correctly.
9. Tighten crankshaft pulley bolts to 23 Nm **2**.
10. Install:
    - Water pump pulley.
    - Auxiliary drive belts.
    - Viscous fan and coupling. Tighten nut to 40 Nm. Use tool No.11.5.040.
    - Engine speed (RPM) sensor and distributor (if applicable).

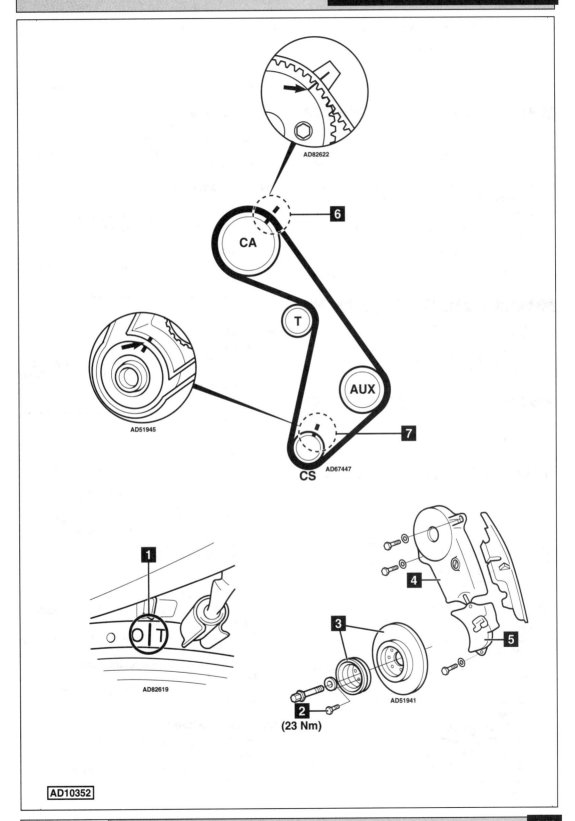

AD82622

6

CA

T

AUX

AD51945

CS

AD67447

7

1

O|T

AD82619

4

3

2
(23 Nm)

AD51941

5

AD10352

CHRYSLER

Model:	Neon 1,8/2,0 16V • Voyager 2,0
Year:	1995-03
Engine Code:	ECB

Special Precautions

- Disconnect battery earth lead.
- DO NOT turn crankshaft or camshaft when timing belt removed.
- Remove spark plugs to ease turning engine.
- Turn engine in normal direction of rotation (unless otherwise stated).
- DO NOT turn engine via camshaft or other sprockets.
- Observe all tightening torques.

Removal

1. Raise and support front of vehicle.
2. Remove:
   - ❏ RH front wheel.
   - ❏ RH splash guard.
   - ❏ Auxiliary drive belt(s).
   - ❏ Crankshaft pulley bolt **1**.
   - ❏ Crankshaft pulley **2**.
     Use tool Nos.1026 & 6827.

3. Support engine – remove:
   - ❏ RH engine mounting and bracket.
   - ❏ Timing belt cover **3**.
4. Turn crankshaft clockwise until timing marks aligned **4** & **5**.

Hydraulic tensioner

5. Remove:
   - ❏ Automatic tensioner unit bolts **6**.
   - ❏ Automatic tensioner unit **7**.

Mechanical tensioner

6. Insert 3 mm Allen key in tensioner pulley **8**.
7. Turn tensioner pulley anti-clockwise until it locks in position **9**. Use 6 mm Allen key.
8. Remove 6 mm Allen key **9**.

All models

9. Remove timing belt.
   NOTE: DO NOT slacken tensioner pulley bolt 10.
   Mark direction of rotation on belt with chalk if belt is to be reused.

Installation

Hydraulic tensioner

1. Check and reset automatic tensioner unit as follows:
   - ❏ Check tensioner body for leakage or damage. Replace if necessary.
   - ❏ Slowly compress pushrod into tensioner body until holes aligned. Use vice.
   - ❏ Retain pushrod with Allen key through hole in tensioner body **11**.

All models

2. Ensure timing marks aligned **5**.
3. Set crankshaft sprocket timing mark 1/2 tooth before timing mark **4**.
4. Fit timing belt in anti-clockwise direction, starting at crankshaft sprocket.
5. Turn crankshaft clockwise until timing marks aligned **4**.
6. Ensure belt is taut between sprockets on non-tensioned side.

Hydraulic tensioner

7. Install automatic tensioner unit to cylinder block **7**.
8. Fit automatic tensioner unit bolts **6**. Finger tighten bolts.

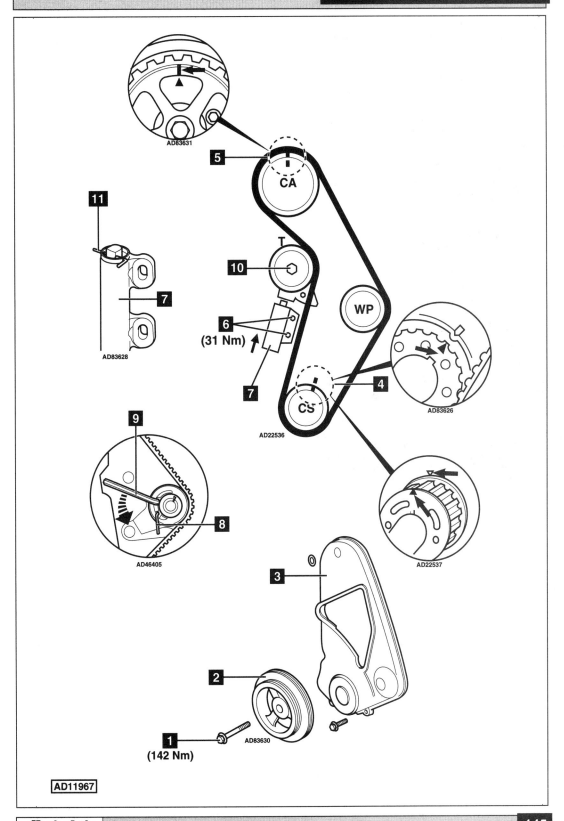

AD83631

5

CA

11

10

T

7

6

(31 Nm)

7

WP

AD83628

4

CS

AD83626

AD22536

9

8

AD46405

AD22537

3

2

AD83630

1

(142 Nm)

AD11967

9. Apply clockwise torque of 28 Nm to tensioner pulley bolt **10**.
10. Maintain torque. Push automatic tensioner unit up towards tensioner pulley.
11. Tighten automatic tensioner unit bolts **6**. Tightening torque: 31 Nm.
12. Remove Allen key from tensioner body to release pushrod **11**. Ensure Allen key can be withdrawn easily.

Mechanical tensioner

13. Remove Allen key from tensioner pulley **8**. Allow tensioner to operate.

All models

14. Turn crankshaft two turns clockwise until timing marks aligned **4** & **5**.

Hydraulic tensioner

15. Wait 5 minutes.
16. Ensure Allen key can be inserted and withdrawn easily **11**.
17. If not: Repeat tensioning procedure.

All models

18. Install components in reverse order of removal.
19. Fit crankshaft pulley **2**. Use tool No.6792.
20. Tighten crankshaft pulley bolt **1**. Tightening torque: 142 Nm.

    NOTE: If belt was replaced because of breakage, trouble codes will have been recorded and engine management system will need to relearn basic values. Connect DRB scan tool to data link connector (DLC) and follow manufacturer's instructions or return vehicle to dealer.

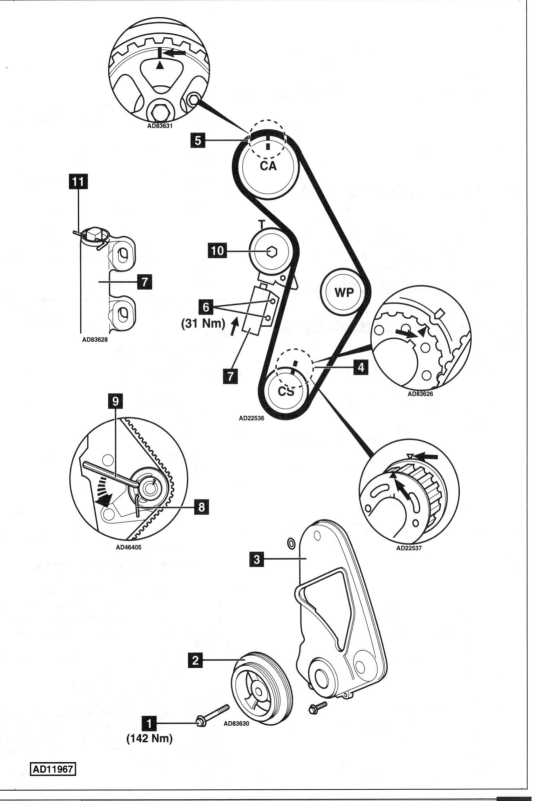

AD83631

5

CA

11

10 T

7

6
(31 Nm)

AD83628

7

WP

AD83626

9

4

CS

AD22536

8

AD46405

AD22537

3

2

1
(142 Nm)

AD83630

AD11967

CHRYSLER

Model:	**PT Cruiser 2,0**
Year:	**2000-03**
Engine Code:	**VIN code position 8 = 9**

Replacement Interval Guide

Chrysler recommend replacement every 120,000 miles or 8 years, whichever occurs first.
The previous use and service history of the vehicle must always be taken into account.
Refer to Timing Belt Replacement Intervals at the front of this manual.

Check For Engine Damage

CAUTION: This engine has been identified as an INTERFERENCE engine in which the possibility of valve-to-piston damage in the event of a timing belt failure is MOST LIKELY to occur.
A compression check of all cylinders should be performed before removing the cylinder head.

Repair Times – hrs

Remove & install   2,60

Special Tools

- Crankshaft pulley puller – Chrysler No.1026.
- Insert for puller – Chrysler No.6827.
- Crankshaft pulley installer – Chrysler No.6792.

Special Precautions

- Disconnect battery earth lead.
- DO NOT turn crankshaft or camshaft when timing belt removed.
- Remove spark plugs to ease turning engine.
- Turn engine in normal direction of rotation (unless otherwise stated).
- DO NOT turn engine via camshaft or other sprockets.
- Observe all tightening torques.

Removal

1. Raise and support front of vehicle.
2. Remove:
   - RH front wheel.
   - RH splash guard.
   - Auxiliary drive belt(s).
   - Crankshaft pulley bolt **1**.
   - Crankshaft pulley **2**.
     Use tool Nos.1026 & 6827.
3. Support engine – remove:
   - RH engine mounting and bracket.
   - PAS pump and bracket. DO NOT disconnect hoses.
   - Timing belt cover **3**.
4. Turn crankshaft clockwise until timing marks aligned **4** & **5**.
5. Remove:
   - Automatic tensioner unit bolts **6**.
   - Automatic tensioner unit **7**.

6. Remove timing belt.
   NOTE: DO NOT slacken tensioner pulley bolts 8.
   Mark direction of rotation on belt with chalk if belt is to be reused.

Installation

1. Check and reset automatic tensioner unit as follows:
   - Check tensioner body for leakage or damage. Replace if necessary.
   - Slowly compress pushrod into tensioner body until holes aligned. Use vice.
   - Retain pushrod with suitable pin through hole in tensioner body **9**.
2. Ensure timing marks aligned **5**.
   NOTE: If camshaft sprockets require aligning: Set crankshaft sprocket timing mark 3 teeth before timing mark 10, then reposition crankshaft sprocket to TDC.
3. Set crankshaft sprocket timing mark 1/2 tooth before timing mark **11**.
4. Fit timing belt in anti-clockwise direction, starting at crankshaft sprocket.
5. Turn crankshaft clockwise until timing marks aligned **4**.
6. Ensure belt is taut between sprockets on non-tensioned side.
7. Install automatic tensioner unit to cylinder block **7**.
8. Fit automatic tensioner unit bolts **6**. Finger tighten bolts.
9. Apply clockwise torque of 28 Nm to tensioner pulley bolt **8**.
10. Maintain torque. Push automatic tensioner unit up towards tensioner pulley.
11. Tighten automatic tensioner unit bolts **6**. Tightening torque: 31 Nm.
12. Remove pin from tensioner body to release pushrod **9**. Ensure pin can be withdrawn easily.
13. Turn crankshaft two turns clockwise until timing marks aligned **4** & **5**. If not: Repeat tensioning procedure.
14. Install components in reverse order of removal.
15. Fit crankshaft pulley **2**. Use tool No.6792.
16. Tighten crankshaft pulley bolt **1**. Tightening torque: 142 Nm.
    NOTE: If the timing belt was replaced because of breakage, trouble codes may have been recorded and the engine management system may need to relearn basic values. Connect a suitable scan tool to the data link connector and follow manufacturer's instructions or return vehicle to dealer.

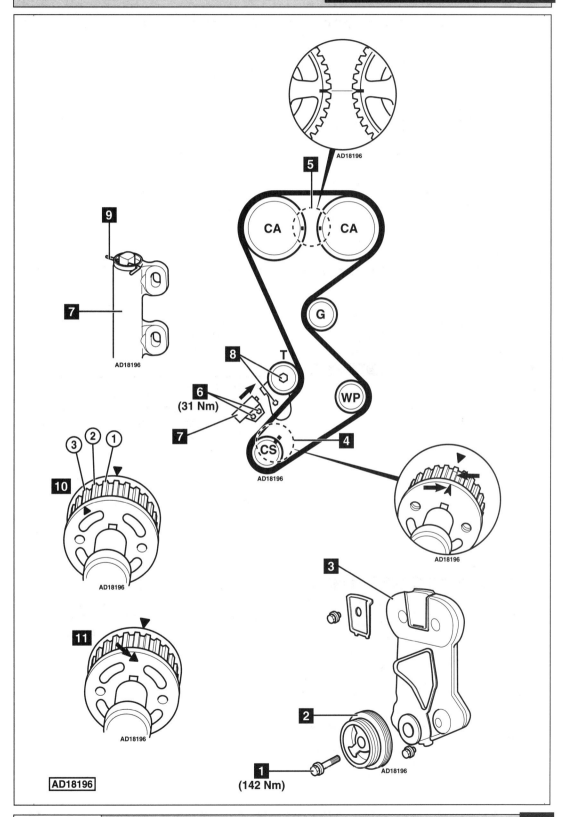

AD18196

5

9

CA — CA

7

G

AD18196

8 T

6
(31 Nm)

7

WP

4

AD18196

CS

3 ② ①

10

AD18196

AD18196

3

11

2

AD18196

AD18196

1

(142 Nm)

AD18196

Model:	AX 1,0/1,1/1,4 • AX GTi • ZX 1,1/1,4 • BX 1,4 • C15 1,0/1,1/1,4
Year:	1986-94
Engine Code:	CDY, CDZ, C1A, HDY, HDZ, H1A, KDX, KDY, KDZ, KFY
	KFZ, K1A, K1F, K1G, K2A, K2B, K2D, K3A, K6B, M4A

Replacement Interval Guide

Citroen recommend check & replacement if necessary every 50,000 miles only under adverse conditions (pre '92 model year). Replacement every 72,000 miles under normal conditions or 36,000 miles under adverse conditions (1992 model year→).
The previous use and service history of the vehicle must always be taken into account.
Refer to Timing Belt Replacement Intervals at the front of this manual.

Check For Engine Damage

CAUTION: This engine has been identified as an INTERFERENCE engine in which the possibility of valve-to-piston damage in the event of a timing belt failure is MOST LIKELY to occur.
A compression check of all cylinders should be performed before removing the cylinder head.

Repair Times – hrs

AX:

Check & adjust:	
Carburettor/Monopoint	1,00
GTi	1,30
Remove & install:	
Carburettor	1,60
Monopoint/GTi	1,90
BX:	
Check & adjust	1,00
Remove & install	2,00
ZX/C15:	
Check & adjust	1,00
Remove & install	1,60

Special Tools

- Flywheel timing pin – Citroen No.4507-T.A.
- Camshaft timing pin – Citroen No.4507-T.B.
- Tensioning tool – Citroen No.4507-T.J.

Special Precautions

- Disconnect battery earth lead.
- DO NOT turn crankshaft or camshaft when timing belt removed.
- Remove spark plugs to ease turning engine.
- Turn engine in normal direction of rotation (unless otherwise stated).
- DO NOT turn engine via camshaft or other sprockets.
- Observe all tightening torques.

Removal

1. Raise and support front of vehicle.
2. Remove:
   - ❑ RH front wheel.
   - ❑ RH wheel arch liner.
   - ❑ Auxiliary drive belt.
   - ❑ Crankshaft pulley (3 bolts) **1**.
   - ❑ Crankshaft pulley housing.
   - ❑ Timing belt upper cover **2**.
   - ❑ Timing belt lower cover **3**.
3. Turn crankshaft clockwise to setting position.
4. Insert timing pin in camshaft sprocket **4**. Tool No.4507-T.B.
5. Insert timing pin in flywheel **5**. Tool No.4507-T.A.
6. Slacken tensioner nut **6**.
7. Remove timing belt.

Installation

1. Ensure timing pins located correctly **4** & **5**.
2. Fit timing belt in clockwise direction, starting at camshaft sprocket. Observe direction of rotation marks on belt.
3. Fit tensioning tool to tensioner pulley **7**. Tool No.4507-T.J.
4. Allow weight to react and tighten tensioner nut.
5. Remove timing pins **4** & **5**.
6. Turn crankshaft four turns clockwise.
7. Slacken tensioner nut slowly and allow weight to react.
8. Tighten tensioner nut **6**. Tightening torque: 23 Nm.
9. Remove tensioning tool.
10. Ensure timing pins can be inserted **4** & **5**.
11. Install components in reverse order of removal.
12. Tighten crankshaft pulley bolts **8**. Tightening torque: 8 Nm.

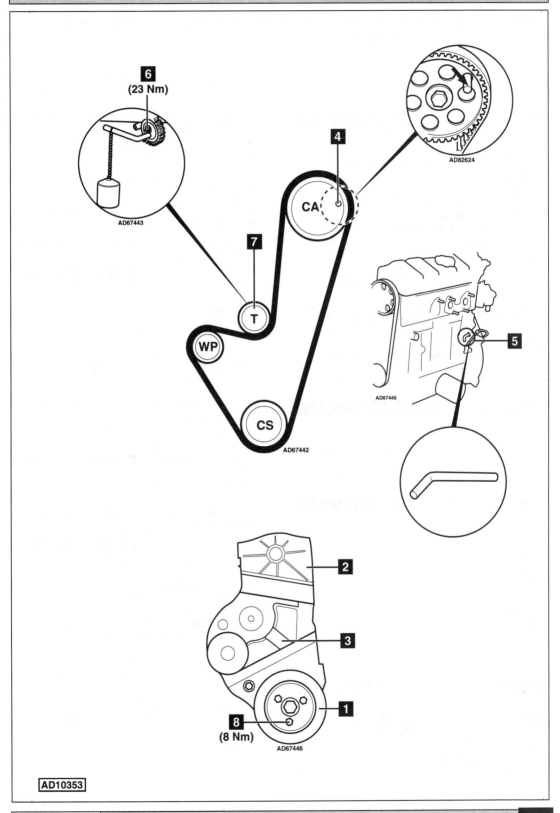

6
(23 Nm)

AD67443

4

AD82624

7

CA

T

WP

AD67445

5

CS

AD67442

2

3

1

8
(8 Nm)

AD67446

AD10353

CITROEN

Model:	AX 1,0/1,1/1,4 • AX GTi • Saxo 1,0/1,1/1,4/1,6 • ZX 1,1/1,4 • Xsara 1,4/1,6 Berlingo 1,1/1,4
Year:	1994-03
Engine Code:	CDY, CDZ, HDY, HDZ, KDX, KDY, KFW, KFX, KFY, KFZ, NFZ

Replacement Interval Guide

Citroen recommend:
→1996MY: Replacement every 72,000 miles under normal conditions or every 48,000 miles under adverse conditions.
1997MY: Replacement every 72,000 miles under normal conditions or every 54,000 miles under adverse conditions.
1998MY: Replacement every 80,000 miles under normal conditions or every 48,000 miles under adverse conditions.
1999MY: Replacement every 75,000 miles under normal conditions or every 60,000 miles under adverse conditions.
2000MY→: Replacement every 75,000 miles or every 10 years under normal conditions, or every 60,000 miles or every 10 years under adverse conditions.
The previous use and service history of the vehicle must always be taken into account.
Refer to Timing Belt Replacement Intervals at the front of this manual.

Check For Engine Damage

CAUTION: This engine has been identified as an INTERFERENCE engine in which the possibility of valve-to-piston damage in the event of a timing belt failure is MOST LIKELY to occur.
A compression check of all cylinders should be performed before removing the cylinder head.

Repair Times – hrs

Remove & install:

AX	1,90
Saxo 1,0/1,1/1,4	1,60
Saxo 1,6	1,80
AC	+0,10
ZX	1,60
Xsara →2001	1,90
Xsara 2001→	2,20
AC/PAS	+0,30
Berlingo	1,40
AC	+0,30

Special Tools

- Flywheel timing pin – Citroen No.4507-T.A.
- Camshaft timing pin – Citroen No.4507-T.B.
- Valve spring compressor (Xsara/Berlingo) – Citroen No.4533-T.Z.
- Tension gauge – Citroen No.4122-T (SEEM C.Tronic 105.5).

Special Precautions

- Disconnect battery earth lead.
- DO NOT turn crankshaft or camshaft when timing belt removed.
- Remove spark plugs to ease turning engine.
- Turn engine in normal direction of rotation (unless otherwise stated).
- DO NOT turn engine via camshaft or other sprockets.
- Observe all tightening torques.

Removal

NOTE: AX/Saxo/ZX: Valve clearances must be correct before fitting new timing belt.
1. Raise and support front of vehicle.
2. Remove:
   - ❏ RH front wheel.
   - ❏ RH inner wing panel (ZX/Xsara).
   - ❏ Auxiliary drive belt.
   - ❏ Crankshaft pulley (3 bolts) **1**.
   - ❏ Timing belt covers **2**, **3** & **4**.
3. Turn crankshaft clockwise to setting position.
4. Insert timing pin in camshaft sprocket **5**. Tool No.4507-T.B.
5. Insert timing pin in flywheel **6**. Tool No.4507-T.A.
6. Slacken tensioner nut **7**. Release tension on belt. Lightly tighten nut.
7. Remove timing belt.

Installation

1. Ensure timing pins located correctly **5** & **6**.
2. Fit timing belt in following order:
   - ❏ Crankshaft sprocket.
   - ❏ Camshaft sprocket.
   - ❏ Water pump sprocket.
   - ❏ Tensioner pulley.
   NOTE: Observe direction of rotation marks on belt. Ensure belt is taut between sprockets on non-tensioned side.
3. Slacken tensioner nut **7**.
4. Turn tensioner anti-clockwise against belt.
5. Tighten tensioner nut **7**.
6. Remove timing pins **5** & **6**.
7. Attach tension gauge to belt at ▽ **8**. Tool No.4122-T.
8. Slacken tensioner nut **7**.

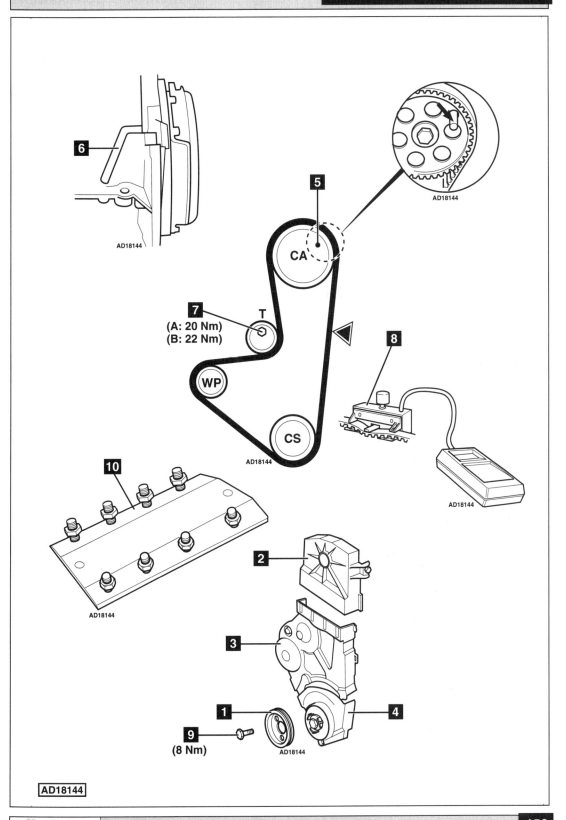

AD18144

6

AD18144

5

CA

AD18144

7

T

(A: 20 Nm)
(B: 22 Nm)

WP

8

CS

AD18144

10

AD18144

2

3

AD18144

1

9

(8 Nm)

AD18144

4

AD18144

9. Turn tensioner anti-clockwise until tension
   gauge indicates the following:
   ❏ 45±3 SEEM units – AX/Saxo/ZX/Xsara
     (→2001).
   ❏ 31 SEEM units – Berlingo.
   ❏ 44 SEEM units – Xsara (2001→).
10. Tighten tensioner nut **7**.
11. Remove tension gauge **8**.
12. Turn crankshaft four turns clockwise to setting
    position.
    *NOTE: DO NOT allow crankshaft to turn
    anti-clockwise.*

AX/Saxo/ZX – proceed as follows:

13. Ensure timing pins can be inserted **5** & **6**.
    NOTE: Ensure timing pins are removed.
14. Slacken tensioner nut **7**.
15. Release belt tension slightly.
16. Attach tension gauge to belt at ▽ **8**.
17. Turn tensioner anti-clockwise and tension belt.
    New belt: 41±3 SEEM units.
    Used belt: 35±3 SEEM units.
18. Tighten tensioner nut **7**.
    Tightening torque: (B) 22 Nm.
19. Remove tension gauge **8**.
20. Turn crankshaft two turns clockwise to setting
    position.
21. Ensure timing pin can be inserted in flywheel **6**.
    NOTE: Ensure timing pin removed.
22. Attach tension gauge to belt at ▽ **8**.
23. Check tension gauge reading.
    New belt: 51±3 SEEM units.
    Used belt: 45±3 SEEM units.
24. If tension not as specified: Repeat operations
    13-23.

Xsara/Berlingo – proceed as follows:

25. Insert timing pins **5** & **6**.
26. Remove cylinder head cover.
27. Install valve spring compressor **10**.
    Tool No.4533-T.Z.
28. Attach tension gauge to belt at ▽ **8**.
    Tool No.4122-T.
29. Slacken tensioner nut **7**.
30. Release belt tension slightly.
31. Turn tensioner anti-clockwise until tension
    gauge indicates 29-33 SEEM units.
32. Tighten tensioner nut **7**. Tightening torque:
    ❏ (A) 20 Nm – Xsara 2001→.
    ❏ (B) 22 Nm – Berlingo/Xsara (→2001).
33. Remove:
    ❏ Tension gauge **8**.
    ❏ Timing pins **5** & **6**.
    ❏ Valve spring compressor **10**.
34. Turn crankshaft two turns clockwise to setting
    position.
35. Ensure timing pins can be inserted **5** & **6**.

All models – proceed as follows:

36. Install components in reverse order of removal.
37. Tighten crankshaft pulley bolts **9**.
    Tightening torque: 8 Nm.

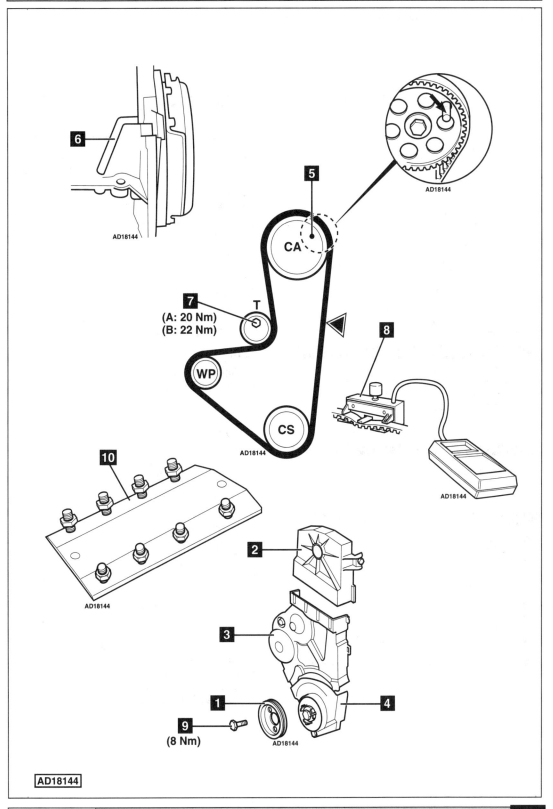

6

AD18144

5

CA

AD18144

7
(A: 20 Nm)
(B: 22 Nm)

T

8

WP

CS

AD18144

AD18144

10

AD18144

2

3

1

9
(8 Nm)

4

AD18144

AD18144

CITROEN

Model:	Saxo 1,6 16V • Xsara 1,6 16V
Year:	1997-03
Engine Code:	NFX (TU5J4/L3), NFU

Replacement Interval Guide

Citroen recommend:

1997MY: Replacement every 72,000 miles under normal conditions or every 54,000 miles under adverse conditions.

1998MY: Replacement every 80,000 miles under normal conditions or every 48,000 miles under adverse conditions.

1999MY: Replacement every 75,000 miles under normal conditions or every 60,000 miles under adverse conditions.

2000MY→: Replacement every 75,000 miles or every 10 years under normal conditions, or every 60,000 miles or every 10 years under adverse conditions.

The previous use and service history of the vehicle must always be taken into account.
Refer to Timing Belt Replacement Intervals at the front of this manual.

Check For Engine Damage

CAUTION: This engine has been identified as an INTERFERENCE engine in which the possibility of valve-to-piston damage in the event of a timing belt failure is MOST LIKELY to occur.
A compression check of all cylinders should be performed before removing the cylinder head(s).

Repair Times – hrs

Remove & install:
Saxo	2,40
Xsara	2,50

Special Tools

- Tension gauge – Citroen No.4122-T (SEEM C.Tronic 105.5).
- Camshaft timing pin (CA1) – Citroen No.4533-TAC1.
- Camshaft timing pin (CA2) – Citroen No.4533-TAC2.
- Flywheel timing pin – Citroen No.4507-T.A.
- Tensioning tool (manual tensioner) – Citroen No.4507-T.J.
- Tensioner pulley locking pin (automatic tensioner) – Citroen No.4200-T.H.
- Timing belt retaining clip – Citroen No.4533-TAD.

Special Precautions

- Disconnect battery earth lead.
- DO NOT turn crankshaft or camshaft when timing belt removed.
- Remove spark plugs to ease turning engine.
- Turn engine in normal direction of rotation (unless otherwise stated).
- DO NOT turn engine via camshaft or other sprockets.
- Observe all tightening torques.

Removal

1. Raise and support front of vehicle.

Saxo:

2. Drain coolant.
3. Remove:
   - ❑ Engine control module (ECM) and tray.
   - ❑ Top radiator hose.

Xsara:

4. Support engine.
5. Remove:
   - ❑ PAS pipe.
   - ❑ RH engine mounting.

All models:

6. Remove:
   - ❑ Air filter housing.
   - ❑ Air filter intake duct.
   - ❑ Auxiliary drive belt.
   - ❑ Crankshaft pulley bolts **1**.
   - ❑ Crankshaft pulley **2**.
   - ❑ Alternator bracket.
   - ❑ Timing belt upper cover **3**.
   - ❑ Timing belt lower cover **4**.
   - ❑ Exhaust manifold heat shield.
7. Unclip AC pipe (if fitted). Push AC pipe to one side.
8. Turn crankshaft clockwise to setting position.
9. Insert timing pins in camshaft sprockets **6**. Tool Nos.4533-TAC1/2.
10. Insert timing pin in flywheel **5**. Tool No.4507-TA.
11. Manual tensioner:
    - ❑ Slacken tensioner bolt **7**. Move tensioner away from belt. Lightly tighten bolt.

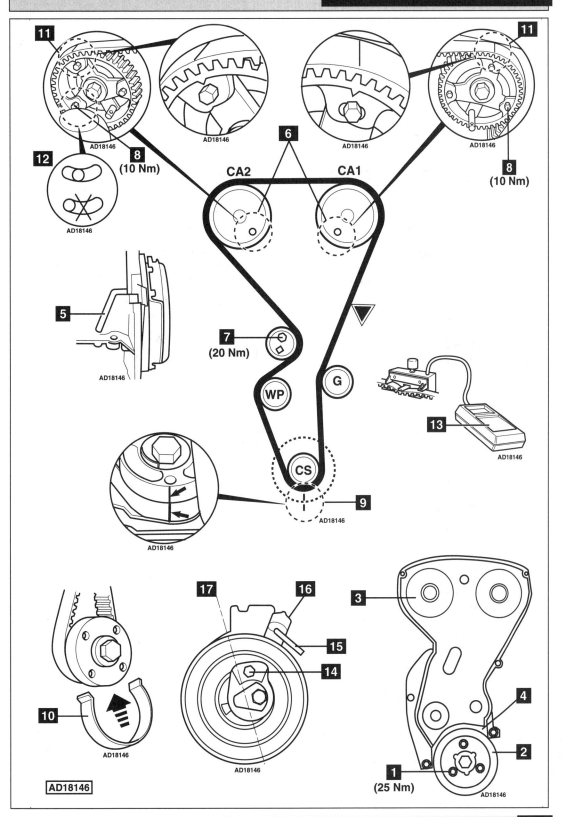

12. Automatic tensioner:
   - ❏ Insert Allen key into tensioner pulley at position **14**.
   - ❏ Turn tensioner until locking pin can be inserted **15**. Tool No.4200-T.H.
   - ❏ Turn tensioner clockwise until pointer **16** aligns with position **17**.
   - ❏ Lock tensioner in this position.

13. Remove timing belt.

Installation

1. Ensure timing pins located correctly **5** & **6**.
2. Two-piece camshaft sprockets:
   - ❏ Slacken bolts of each camshaft sprocket slightly **8**.
   - ❏ Turn camshaft sprockets fully clockwise in slotted holes.

   NOTE: Sprockets should turn with slight resistance.

3. Fit timing belt. Mark on belt should be aligned with mark on crankshaft **9**.
4. Secure belt to crankshaft sprocket with retaining clip **10**. Tool No.4533-TAD.
5. Fit timing belt to guide pulley.
6. Fit timing belt to camshaft sprocket CA1 then CA2. Align marks on belt with marks on sprockets **11**.

   NOTE: Ensure belt is taut between sprockets on non-tensioned side.

7. Fit timing belt to water pump sprocket and tensioner pulley.
8. Manual tensioner:
   - ❏ Attach tension gauge to belt at ▽. Tool No.4122-T **13**.
   - ❏ Insert tensioning tool in square hole of tensioner pulley. Tool No.4507-T.J.
   - ❏ Turn tensioner anti-clockwise until tension gauge indicates 63 units.
   - ❏ Tighten tensioner bolt **7**. Tightening torque: 20 Nm.
   - ❏ Ensure bolts of each camshaft sprocket not at end of slotted holes **12**.
   - ❏ If necessary: Repeat installation procedure.
   - ❏ Tighten bolts of each camshaft sprocket **8**. Tightening torque: 10 Nm.
   - ❏ Ensure marks on belt aligned with marks on sprockets **9** & **11**.
   - ❏ Remove tension gauge **13**.
9. Remove:
   - ❏ Timing belt retaining clip **10**.
   - ❏ Timing pins **5** & **6**.
10. Automatic tensioner:
   - ❏ Insert Allen key into tensioner pulley at position **14**.
   - ❏ Turn tensioner pulley. Remove locking pin **15**.

11. Turn crankshaft four turns clockwise to setting position.

   NOTE: DO NOT allow crankshaft to turn anti-clockwise.

12. Insert timing pin in flywheel **5**. Tool No.4507-TA.
13. Two-piece camshaft sprockets:
   - ❏ Slacken bolts of each camshaft sprocket slightly **8**.
14. Insert timing pins in camshaft sprockets **6**. Tool Nos.4533-TAC1/2.

   NOTE: If timing pins cannot be inserted: Turn camshaft sprocket flanges as required.

15. Manual tensioner:
   - ❏ Attach tension gauge to belt at ▽. Tool No.4122-T **13**.
   - ❏ Slacken tensioner bolt **7**.
   - ❏ Insert tensioning tool in square hole of tensioner pulley. Tool No.4507-T.J.
   - ❏ Turn tensioner anti-clockwise until tension gauge indicates 37 units.
   - ❏ Tighten tensioner bolt **7**. Tightening torque: 20 Nm.
   - ❏ Remove tool **13**.
16. Two-piece camshaft sprockets:
   - ❏ Tighten bolts of each camshaft sprocket **8**. Tightening torque: 10 Nm.
17. Remove tools **5**, **6** & **13**.
18. Install components in reverse order of removal.

Saxo:

19. Refill and bleed cooling system.

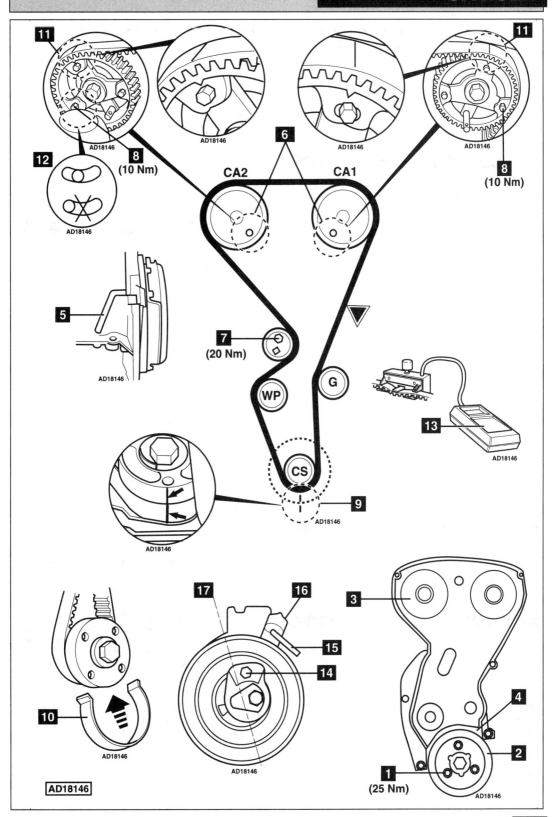

CITROEN

Model:	ZX 1,6/1,8/1,9/2,0 • Xsara 1,8 • BX 16/19 (92-) • Xantia 1,6/1,8/2,0
	XM 2,0/Turbo • Synergie/Evasion 1,8 • Berlingo 1,8
Year:	1989-02
Engine Code:	B1F, B2C, B6E, B4A, BDY, BFZ, D2A, D2F
	D6D, D6E, DDZ, DKZ, LFW, LFX, LFZ, L6A
	RDZ, RFX, RFZ, RGX, RGY, R2A, R6A, R6D

Replacement Interval Guide

Citroen recommend:
→1992MY: The vehicle manufacturer has not recommended a timing belt replacement interval for this engine.
1993-94MY: Replacement every 72,000 miles under normal conditions or every 36,000 miles under adverse conditions.
1995-96MY: Replacement every 72,000 miles under normal conditions or every 48,000 miles under adverse conditions.
1997MY: Replacement every 72,000 miles under normal conditions or every 54,000 miles under adverse conditions.
1998MY: Replacement every 80,000 miles under normal conditions or every 48,000 miles under adverse conditions.
1999MY: Replacement every 75,000 miles under normal conditions or every 60,000 miles under adverse conditions.
2000MY→: Replacement every 75,000 miles or every 10 years under normal conditions, or every 60,000 miles or every 10 years under adverse conditions.
The previous use and service history of the vehicle must always be taken into account.
Refer to Timing Belt Replacement Intervals at the front of this manual.

Check For Engine Damage

CAUTION: This engine has been identified as an INTERFERENCE engine in which the possibility of valve-to-piston damage in the event of a timing belt failure is MOST LIKELY to occur.
A compression check of all cylinders should be performed before removing the cylinder head.

Repair Times – hrs

Check & adjust:	
All models	1,40
Remove & install:	
ZX 1,6/1,9	3,00
ZX 1,8/2,0	2,40
Xsara – MT	2,30
Xsara – AT	2,40
AC	+0,10
BX 16/19	2,40
Xantia	2,20
XM	2,40
Synergie/Evasion	2,40
AC	+0,10
Berlingo	2,30

Special Tools

- Camshaft timing pin – Citroen No.7004-T.G.
- Crankshaft timing pin – Citroen No.7014-T.N.
- Tension gauge – Citroen No.4099/4122-T.

Special Precautions

- Disconnect battery earth lead.
- DO NOT turn crankshaft or camshaft when timing belt removed.
- Remove spark plugs to ease turning engine.
- Turn engine in normal direction of rotation (unless otherwise stated).
- DO NOT turn engine via camshaft or other sprockets.
- Observe all tightening torques.

Removal

1. Raise and support front of vehicle.
2. Remove:
   - ❏ RH front wheel.
   - ❏ RH inner wing panel.
   - ❏ Auxiliary drive belts.
   - ❏ Timing belt upper cover **1**.
3. Turn crankshaft clockwise to setting position.
4. Remove crankshaft pulley bolt **4**.
5. Insert timing pin in camshaft sprocket **3**. Tool No.7004-T.G.
6. Except XM: Insert timing pin in crankshaft pulley **2**. Tool No.7014-T.N.
   NOTE: Timing pin No.7014-T.N is reduced to 8 mm diameter at end. Some models have crankshaft pulley with multiple holes. Ensure correct hole used (8 mm diameter) when inserting timing pin.
7. Prevent crankshaft from turning. Remove crankshaft pulley **5**.
8. Remove timing belt lower covers **6**.
9. XM: Ensure crankshaft correctly aligned **7**. Use tool No.7014-T.N.
10. Slacken tensioner bolt **8**. Move tensioner away from belt. Lightly tighten bolt.
11. Remove timing belt.

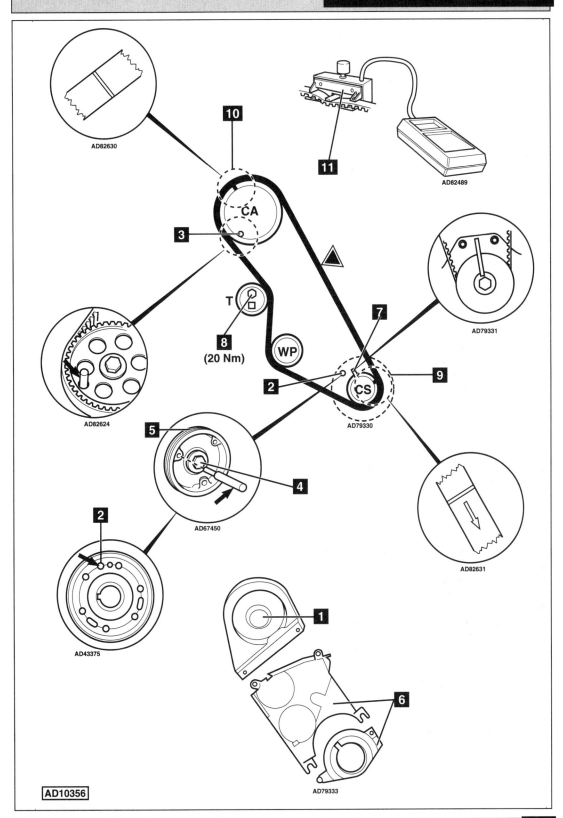

AD82630

10

11

AD82489

CA

3

T

8
(20 Nm)

WP

AD79331

7

2

9

CS

AD79330

5

AD82624

4

AD67450

AD82631

2

AD43375

1

6

AD79333

AD10356

←

Installation

NOTE: Timing belts have yellow/orange marks and 114 teeth. Observe direction of rotation marks on belt. Check tensioner pulley for smooth operation.

1. Ensure timing pin located correctly in camshaft sprocket **3**. Tool No.7004-T.G.

2. Temporarily fit crankshaft pulley. Ensure timing pin can be inserted **2**. Tool No.7014-T.N.
   *NOTE: XM: Ensure crankshaft correctly aligned **7**. Use tool No.7014-T.N.*

3. Remove crankshaft pulley.

4. Fit timing belt in clockwise direction, starting at camshaft sprocket. Ensure belt is taut between sprockets.

5. Ensure marks on belt aligned with marks on sprockets **9** & **10**.

6. Slacken tensioner bolt **8**. Turn tensioner anti-clockwise against belt. Lightly tighten bolt.

7. Remove crankshaft timing pin **2**.
   *NOTE: XM – remove CAMSHAFT timing pin **3**.*

8. Attach tension gauge to belt at ▽ **11**. Tool No.4099/4122-T.

9. Turn tensioner anti-clockwise. Use 8 mm spanner. Tension belt to 30 SEEM units.
   NOTE: 2,0 litre engines: 16 SEEM units.

10. Tighten tensioner bolt to 20 Nm **8**.

11. Remove:
    ❑ Tension gauge.
    ❑ Timing pin **3**.
    NOTE: XM – remove CRANKSHAFT timing pin.

12. Temporarily fit crankshaft pulley **5**.

13. Turn crankshaft two turns clockwise. Ensure timing pins can be inserted easily.

14. Remove timing pins.

15. Turn crankshaft two turns clockwise.

16. Insert camshaft timing pin **3**.
    NOTE: XM – insert CRANKSHAFT timing pin.

17. Attach tension gauge to belt at ▽.

18. Tension gauge should indicate 42-46 SEEM units.

19. If not: Repeat operations 7-17.

20. If tension as specified: Remove timing pins and tension gauge.

21. Remove crankshaft pulley. Fit timing belt covers **6**.

22. Fit crankshaft pulley **5**.

23. Coat crankshaft pulley bolt thread with suitable thread locking compound.

24. Tighten crankshaft pulley bolt **4**.
    ❑ Except Xsara/Berlingo: Tighten bolt to 110 Nm.
    ❑ Xsara/Berlingo: Tighten bolt to 130 Nm.

25. Install components in reverse order of removal.

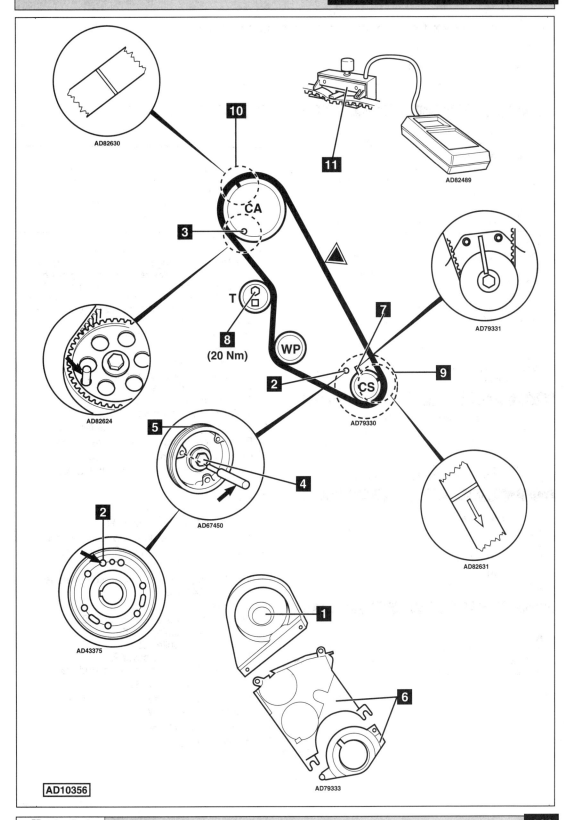

AD82630

10

11

AD82489

CA

3

T

8
(20 Nm)

WP

7

AD79331

2

CS

9

AD79330

AD82624

5

4

AD67450

2

AD82631

AD43375

1

6

AD79333

AD10356

Model:	ZX 1,8 16V • Xsara 1,8 16V • Xantia 1,8 16V
Year:	1995-01
Engine Code:	LFY (XU7JP4)

Replacement Interval Guide

Citroen recommend:
→1996MY: Replacement every 72,000 miles under normal conditions or every 48,000 miles under adverse conditions.
1997MY: Replacement every 72,000 miles under normal conditions or every 54,000 miles under adverse conditions.
1998MY: Replacement every 80,000 miles under normal conditions or every 48,000 miles under adverse conditions.
1999MY: Replacement every 75,000 miles under normal conditions or every 60,000 miles under adverse conditions.
2000MY→: Replacement every 75,000 miles or every 10 years under normal conditions, or every 60,000 miles or every 10 years under adverse conditions.
The previous use and service history of the vehicle must always be taken into account.
Refer to Timing Belt Replacement Intervals at the front of this manual.

Check For Engine Damage

CAUTION: This engine has been identified as an INTERFERENCE engine in which the possibility of valve-to-piston damage in the event of a timing belt failure is MOST LIKELY to occur.
A compression check of all cylinders should be performed before removing the cylinder head.

Repair Times – hrs

Remove & install:
Xantia	2,30
ZX	2,50
Xsara (→11/97)	2,30
Xsara (12/97→)	2,80
AC	+0,10

Special Tools

- Camshaft timing pins – Citroen No.9041-T.Z.
- Crankshaft timing pin – Citroen No.7014-T.N.
- Spanner for tensioner – Citroen No.7017-T.W.
- Flywheel locking tool – Citroen No.9044-T.
- Tension gauge – Citroen No.4122-T (SEEM C.Tronic G2 105.5).

Special Precautions

- Disconnect battery earth lead.
- DO NOT turn crankshaft or camshaft when timing belt removed.
- Remove spark plugs to ease turning engine.
- Turn engine in normal direction of rotation (unless otherwise stated).
- DO NOT turn engine via camshaft or other sprockets.
- Observe all tightening torques.

Removal

WARNING: This engine may suffer from failure of the crankshaft pulley resulting in the possible incorrect alignment of the timing pin hole. The timing belt should be removed and installed with the engine at 90° BTDC. If necessary: Fit new crankshaft pulley.

1. Raise and support front of vehicle.
2. Remove:
   - ❏ RH front wheel.
   - ❏ RH front wheel arch liner.
   - ❏ Auxiliary drive belt.
   - ❏ Auxiliary drive belt tensioner assembly.
   - ❏ Lower flywheel cover.
3. Fit flywheel locking tool **1**. Tool No.9044-T.
4. Remove crankshaft pulley centre bolt **2**.
5. Clean crankshaft pulley centre bolt.
6. Fit crankshaft pulley centre bolt. DO NOT tighten.
7. Remove:
   - ❏ Flywheel locking tool **1**.
   - ❏ Timing belt upper cover **3**.
8. Turn crankshaft clockwise to setting position.
9. Insert timing pin in crankshaft pulley **4**. Tool No.7014-T.N.
10. Insert timing pins in camshaft sprockets **5**. Tool No.9041-T.Z.
11. Remove:
    - ❏ Crankshaft pulley centre bolt **2**.
    - ❏ Crankshaft timing pin **4**.
    - ❏ Crankshaft pulley **6**.
    - ❏ Timing belt lower cover **7**.
    - ❏ Timing belt cover mounting studs **8**.
12. Slacken tensioner bolt **9**.
13. Move tensioner away from timing belt. Use tool No.7017-T.W.
14. Lightly tighten bolt **9**.
15. Remove timing belt.

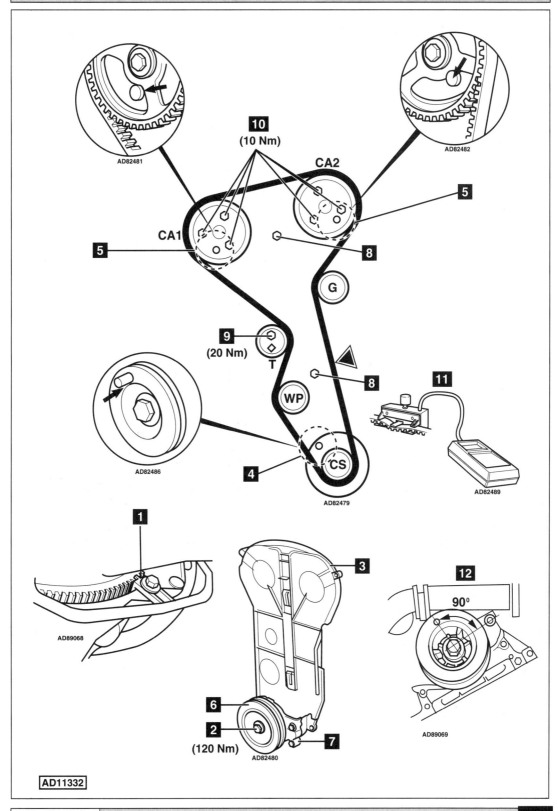

AD82481

10
(10 Nm)

CA2

5

CA1

5

8

G

9
(20 Nm)
T

8

WP

AD82486

4

CS

AD82479

AD82482

11

AD82489

1

AD89068

3

6

2

(120 Nm)

AD82480

7

12

90°

AD89069

AD11332

←

Installation

1. Temporarily fit crankshaft pulley .
2. Insert timing pin in crankshaft pulley ■.
   Tool No.7014-T.N.
3. Ensure timing pins located correctly in
   sprockets ■. Tool No.9041-T.Z.
4. Remove:
   ❏ Crankshaft timing pin ■.
   ❏ Crankshaft pulley ■.
5. Slacken bolts of each camshaft sprocket
   slightly ■.
6. Ensure camshaft sprockets turn freely.
7. Fit timing belt to camshaft sprocket CA1.
8. Retain in position with suitable clip.
9. Turn camshaft sprockets fully clockwise in
   slotted holes.
10. Fit timing belt in following order:
    ❏ Camshaft sprocket CA2.
    ❏ Guide pulley.
    ❏ Crankshaft sprocket.
    ❏ Water pump sprocket.
    ❏ Tensioner pulley.
    *NOTE: Ensure belt is taut between sprockets on
    non-tensioned side.*
11. Attach tension gauge to belt at ▽.
    Tool No.4122-T ■.
12. Slacken tensioner bolt ■.
13. Turn tensioner anti-clockwise until tension
    gauge indicates 45-51 units.
    Use tool No.7017-T.W.
14. Tighten tensioner bolt ■.
    Tightening torque: 20 Nm.
15. Tighten bolts of each camshaft sprocket ■.
    Tightening torque: 10 Nm.
16. Remove:
    ❏ Tension gauge ■.
    ❏ Timing pins ■.
    ❏ Timing belt retaining clip.
17. Install:
    ❏ Timing belt lower cover ■.
    ❏ Crankshaft pulley ■.
    ❏ Flywheel locking tool ■.
    ❏ Crankshaft pulley bolt ■. Use Loctite E6 on
      threads.
18. Tighten crankshaft pulley bolt ■.
    Tightening torque: 120 Nm.
19. Remove flywheel locking tool ■.
20. Turn crankshaft two turns clockwise to setting
    position.
21. Insert timing pin in crankshaft pulley ■.
    Tool No.7014-T.N.
22. Slacken bolts of each camshaft sprocket
    slightly ■.

23. Insert timing pins in camshaft sprockets ■.
    Tool No.9041-T.Z.
    *NOTE: If timing pins cannot be inserted: Turn
    camshafts slightly.*
24. Slacken tensioner bolt ■.
25. Attach tension gauge to belt at ▽.
    Tool No.4122-T ■.
26. Release tension on belt.
27. Turn tensioner anti-clockwise until tension
    gauge indicates 26 units. Use tool No.7017-T.W.
28. Tighten tensioner bolt ■.
    Tightening torque: 20 Nm.
29. Tighten bolts of each camshaft sprocket ■.
    Tightening torque: 10 Nm.
30. Remove:
    ❏ Tension gauge ■.
    ❏ Timing pins ■ & ■.
31. Turn crankshaft two turns clockwise to setting
    position.
32. Insert timing pin in crankshaft pulley ■.
    Tool No.7014-T.N.
33. Slacken bolts of each camshaft sprocket
    slightly ■.
34. Insert timing pins in camshaft sprockets ■.
    Tool No.9041-T.Z.
    *NOTE: If timing pins cannot be inserted: Turn
    camshafts slightly.*
35. Tighten bolts of each camshaft sprocket ■.
    Tightening torque: 10 Nm.
36. Remove:
    ❏ Timing pins ■ & ■.
37. Turn crankshaft 90° clockwise.
    *NOTE: Crankshaft pulley timing pin hole must
    align with timing belt lower cover bolt ■. DO NOT
    allow crankshaft to turn anti-clockwise.*
38. Attach tension gauge to belt at ▽.
    Tool No.4122-T ■.
39. Tension gauge should indicate 32-40 units.
40. Install components in reverse order of removal.

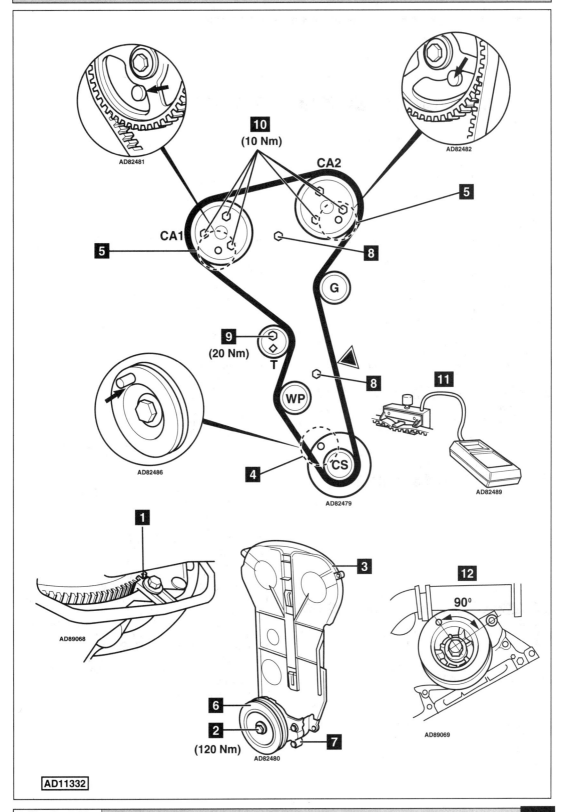

CA2

CA1

10
(10 Nm)

5

5

8

G

9
(20 Nm)

T

8

11

WP

4

CS

AD82479

AD82489

AD82481

AD82482

AD82486

1

AD89068

3

12
90°

AD89069

6

2
(120 Nm)

AD82480

7

AD11332

CITROEN

Model:	ZX 2,0 16V • Xantia 2,0 16V • XM 2,0 16V
Year:	1992-01
Engine Code:	RFY, RFT, RFV

Replacement Interval Guide

Citroen recommend:
→1994MY: Replacement every 72,000 miles under normal conditions or every 36,000 miles under adverse conditions.
1995-96MY: Replacement every 72,000 miles under normal conditions or every 48,000 miles under adverse conditions.
1997MY: Replacement every 72,000 miles under normal conditions or every 54,000 miles under adverse conditions.
1998MY: Replacement every 80,000 miles under normal conditions or every 48,000 miles under adverse conditions.
1999MY: Replacement every 75,000 miles under normal conditions or every 60,000 miles under adverse conditions.
2000MY→: Replacement every 75,000 miles or every 10 years under normal conditions, or every 60,000 miles or every 10 years under adverse conditions.
The previous use and service history of the vehicle must always be taken into account.
Refer to Timing Belt Replacement Intervals at the front of this manual.

Check For Engine Damage

CAUTION: This engine has been identified as an INTERFERENCE engine in which the possibility of valve-to-piston damage in the event of a timing belt failure is MOST LIKELY to occur.
A compression check of all cylinders should be performed before removing the cylinder head.

Repair Times – hrs

Check & adjust:	
ZX/Xantia	1,30
XM	1,60
Remove & install:	
ZX	2,50
Xantia	2,30
XM	3,30
XM with AC	3,00

Special Tools

- Crankshaft timing pin – Citroen No.7014-T.N.
- Camshaft timing pin – Citroen No.7014-T.M.
- Tension gauge – Citroen No.4099-T or 4122-T (SEEM).

Special Precautions

- Disconnect battery earth lead.
- DO NOT turn crankshaft or camshaft when timing belt removed.
- Remove spark plugs to ease turning engine.
- Turn engine in normal direction of rotation (unless otherwise stated).
- DO NOT turn engine via camshaft or other sprockets.
- Observe all tightening torques.

Removal

1. Support engine.
2. Remove:
   - ❏ RH wheel arch liner.
   - ❏ Engine mounting.
   - ❏ Auxiliary drive belt.
   - ❏ Auxiliary drive belt tensioner.
   - ❏ Crankshaft pulley bolts **1**.
   - ❏ Crankshaft pulley **2**.
   - ❏ Timing belt upper cover **3**.
   - ❏ Timing belt lower cover **4**.
3. Turn crankshaft clockwise to setting position.
4. Insert timing pins in camshaft sprockets **5** & **6**. Tool No.7014-T.M.
5. Insert timing pin in crankshaft sprocket **7**. Tool No.7014-T.N.
6. Slacken tensioner bolts **8** & **9**.
7. Remove timing belt.

Installation

1. Ensure timing pins located correctly **5**, **6** & **7**.
2. Fit timing belt in following order:
   - ❏ Crankshaft sprocket.
   - ❏ Water pump sprocket.
   - ❏ Tensioner pulley (T1).
   - ❏ Tensioner pulley (T2).
   - ❏ Camshaft sprocket (CA2) (with no slack in belt).
   - ❏ Camshaft sprocket (CA1).
3. Apply thumb pressure to belt at each tensioner to eliminate any play in timing pins.
4. Attach tension gauge to belt at ▽ **10**. Tool No.4099-T or 4122-T.
5. Push tensioner (T1) against belt **11**. Tension belt to 45 SEEM units.
6. Release tension on belt. Tension belt to 22±2 SEEM units.
7. Tighten tensioner bolts to 20 Nm **9**.
8. Turn tensioner (T2) anti-clockwise until tension gauge indicates 32±2 SEEM units.
9. Tighten tensioner bolt to 20 Nm **8**.
10. Remove all timing pins **5**, **6** & **7**.
11. Turn crankshaft two turns clockwise. Insert timing pin in crankshaft sprocket **7**.
12. If timing pins cannot be inserted easily in camshaft sprockets: Repeat operations 3-11.
13. Remove all timing pins **5**, **6** & **7**.
14. Attach tension gauge to belt at ▽ **10**. Tool No.4099-T or 4122-T. Tension gauge should indicate 53±2 SEEM units.
15. Install components in reverse order of removal.

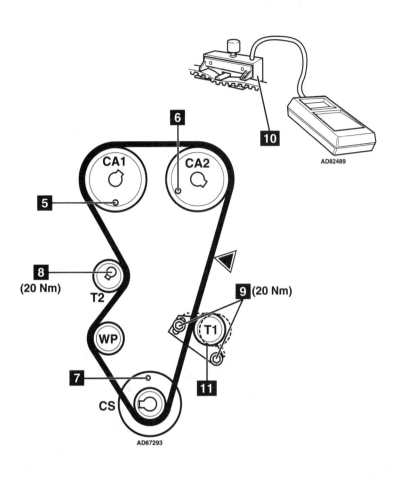

AD82489

CA1

CA2

T2

WP

T1

CS

AD67293

8 (20 Nm)

9 (20 Nm)

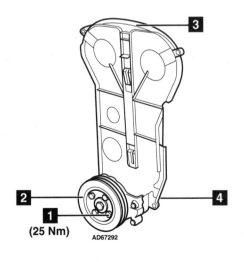

AD67292

1 (25 Nm)

AD10358

CITROEN

Model:	ZX 2,0 16V • Xsara 2,0 16V
Year:	1996-01
Engine Code:	RFS

Replacement Interval Guide

Citroen recommend:
1996MY: Replacement every 72,000 miles under normal conditions or every 48,000 miles under adverse conditions.
1997MY: Replacement every 72,000 miles under normal conditions or every 54,000 miles under adverse conditions.
1998MY: Replacement every 80,000 miles under normal conditions or every 48,000 miles under adverse conditions.
1999MY: Replacement every 75,000 miles under normal conditions or every 60,000 miles under adverse conditions.
2000MY→: Replacement every 50,000 miles or every 10 years under normal conditions, or every 37,500 miles or every 10 years under adverse conditions.
The previous use and service history of the vehicle must always be taken into account.
Refer to Timing Belt Replacement Intervals at the front of this manual.

Check For Engine Damage

CAUTION: This engine has been identified as an INTERFERENCE engine in which the possibility of valve-to-piston damage in the event of a timing belt failure is MOST LIKELY to occur.
A compression check of all cylinders should be performed before removing the cylinder head(s).

Repair Times – hrs

Remove & install:	
ZX	2,50
Xsara	2,90
AC	+0,10

Special Tools

- Camshaft timing pin – Citroen No.9041-T.Z.
- Crankshaft timing pin – Citroen No.7014-T.N.
- Tension gauge – Citroen No.4122-T (SEEM).
- Crankshaft locking tool – Citroen No.9044-T.

Special Precautions

- Disconnect battery earth lead.
- DO NOT turn crankshaft or camshaft when timing belt removed.
- Remove spark plugs to ease turning engine.
- Turn engine in normal direction of rotation (unless otherwise stated).
- DO NOT turn engine via camshaft or other sprockets.
- Observe all tightening torques.

Removal

WARNING: This engine may suffer from failure of the crankshaft pulley resulting in the possible incorrect alignment of the timing pin hole. The timing belt should be removed and installed with the engine at 90° BTDC. If necessary: Fit new crankshaft pulley.

1. Support engine.
2. Remove:
   - ❏ RH wheel arch liner.
   - ❏ Auxiliary drive belt.
3. Disconnect fuel pipes.
4. Raise timing belt cover tab **1**.
5. Remove:
   - ❏ Timing belt upper cover bolts **2**.
   - ❏ Timing belt upper cover **3**.
6. Turn crankshaft clockwise to setting position.
7. Insert timing pin in crankshaft pulley **4**. Tool No.7014-T.N.
8. Insert timing pins in camshaft sprockets **5**. Tool No.9041-T.Z.
9. Lock crankshaft. Use tool No.9044-T **6**.
10. Remove:
    - ❏ Crankshaft timing pin **4**.
    - ❏ Crankshaft pulley bolt **7**.
    - ❏ Crankshaft pulley **8**.
    - ❏ Timing belt lower cover **9**.
11. Slacken tensioner bolt **10**.
12. Move tensioner away from belt and lightly tighten bolt.
13. Remove timing belt.

Installation

*NOTE: If camshafts have been removed, check position of exhaust and inlet camshaft reference holes to ensure correct location **12** & **13**.*

1. Slacken bolts of each camshaft sprocket **11**.
2. Ensure sprockets turn freely on camshafts.
3. Fit timing belt in following order:
   - ❏ Exhaust camshaft sprocket (CA1).
   - ❏ Inlet camshaft sprocket (CA2).
   - ❏ Guide pulley.
   - ❏ Crankshaft sprocket.
   - ❏ Water pump sprocket.
   - ❏ Tensioner pulley.
4. Ensure bolts of each camshaft sprocket not at end of slotted holes **14**.
5. Attach tension gauge to belt at ▽. Tool No.4122-T **15**.
6. Slacken tensioner bolt **10**.

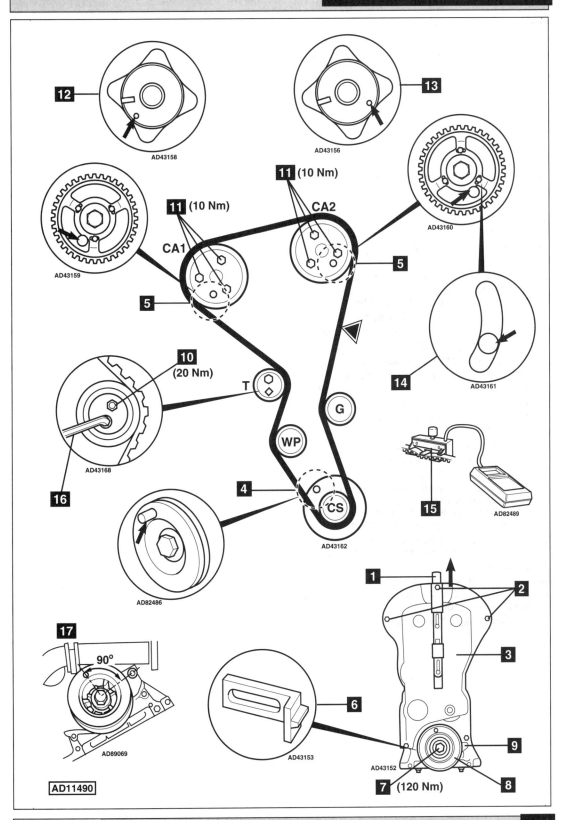

←

7. Turn tensioner anti-clockwise until tension gauge indicates the following : 65 SEEM units (new belt). 35 SEEM units (used belt).

8. Tighten tensioner bolt to 20 Nm ▣.

9. Tighten bolts of each camshaft sprocket to 10 Nm ▣.

10. Remove:
    - ❏ Timing pins ▣.
    - ❏ Tension gauge.
    - ❏ Crankshaft locking tool ▣.

11. Temporarily fit crankshaft pulley.

12. Turn crankshaft six turns clockwise.

13. Insert timing pin in crankshaft pulley ▣.

14. Slacken bolts of each camshaft sprocket ▣.

15. Insert timing pins in camshaft sprockets ▣.

16. Slacken tensioner nut ▣.

17. Attach tension gauge to belt at ▽ ▣.

18. Turn tensioner anti-clockwise until tension gauge indicates the following ▣: 55 SEEM units (new belt). 45 SEEM units (used belt).

19. Tighten tensioner bolt to 20 Nm ▣.

20. Tighten bolts of each camshaft sprocket ▣.

21. Remove timing pins ▣ & ▣.

22. Turn crankshaft two turns clockwise.

23. Insert timing pin in crankshaft pulley ▣.

24. Slacken bolts of each camshaft sprocket ▣.

25. Insert timing pins in camshaft sprockets ▣.

26. Tighten bolts of each camshaft sprocket to 10 Nm ▣.

27. Remove timing pins ▣ & ▣.

28. Check belt tension as follows:

29. Turn crankshaft 90° clockwise.

    NOTE: Crankshaft pulley timing pin hole must align with timing belt lower cover bolt ▣. DO NOT allow crankshaft to turn anti-clockwise.

30. Attach tension gauge to belt at ▽ ▣.

31. Tension gauge should indicate the following: 45±2 SEEM units (new belt). 29±2 SEEM units (used belt).

32. Install components in reverse order of removal.

33. Coat crankshaft pulley bolt thread with suitable thread locking compound.

34. Tighten crankshaft pulley bolt to 120 Nm ▣.

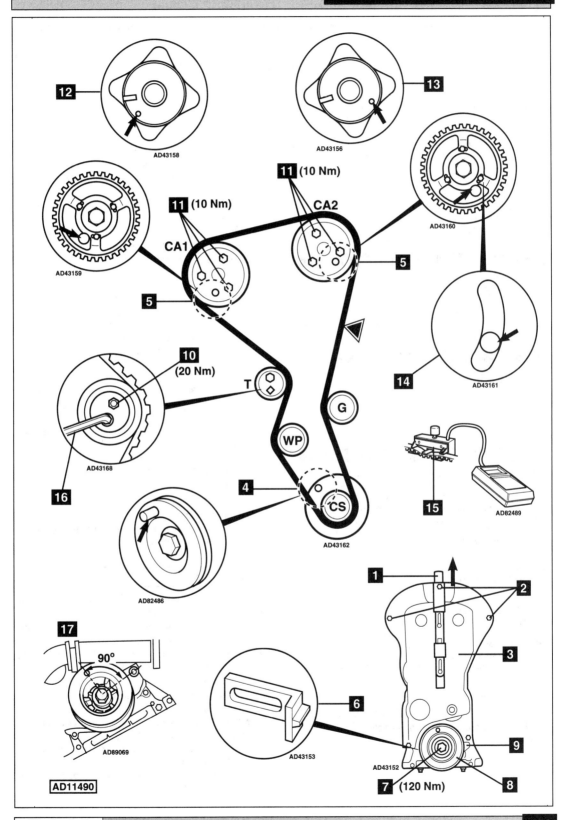

12 AD43158

13 AD43156

11 (10 Nm)

CA2

11 (10 Nm)

CA1

AD43159

AD43160

5

5

14 AD43161

10 (20 Nm)

T

G

WP

AD43168

16

4 **CS**

15 AD82489

AD43162

AD82486

1 **2**

3

17 90°

6

9

AD89069

AD43153

AD43152

7 (120 Nm) **8**

AD11490

CITROEN

Model:	Xsara Picasso 1,8 16V • Xsara 2,0 16V • C5 1,8 • C5 2,0 • C5 2,0 HPI
Year:	2000-03
Engine Code:	6FZ (EW7J4), RFR (EW10J4), RLZ (EW10J4D), RFN

Replacement Interval Guide

Citroen recommend:
Xsara: Replacement every 75,000 miles or
10 years or every 60,000 under severe
conditions.
C5: Replacement every 80,000 miles or
10 years.
*The previous use and service history of the vehicle
must always be taken into account.*
*Refer to Timing Belt Replacement Intervals at the front
of this manual.*

Check For Engine Damage

*CAUTION: This engine has been identified as an
INTERFERENCE engine in which the possibility of
valve-to-piston damage in the event of a timing belt
failure is MOST LIKELY to occur.*
*A compression check of all cylinders should be
performed before removing the cylinder head(s).*

Repair Times – hrs

Remove & install:

Xsara Picasso	2,20
Xsara	3,00
C5 1,8/2,0	2,40
C5 2,0 HPI	2,60

Special Tools

- Crankshaft timing pin – Citroen No.(-).0189-B.
- Camshaft timing pin – Citroen No.(-).0189-A.
- Camshaft timing pin – Citroen No.(-).0189-L.
- Timing belt retaining clip – Citroen No.(-).0189-K.
- Set of blanking plugs – Citroen No.(-).0189-Q.
- Fuel pressure venting hose – Citroen No.4192-T.
- Crankshaft pulley locking tool – Citroen No.6310-T.

Special Precautions

- Disconnect battery earth lead.
- DO NOT turn crankshaft or camshaft when timing belt removed.
- Remove spark plugs to ease turning engine.
- Turn engine in normal direction of rotation (unless otherwise stated).
- DO NOT turn engine via camshaft or other sprockets.
- Observe all tightening torques.

Removal

1. Raise and support front of vehicle.
2. Remove:
   - ❏ RH front wheel.
   - ❏ RH splash guard.
   - ❏ Auxiliary drive belt.
3. Unclip:
   - ❏ C5: Wiring harness (on timing belt cover).
   - ❏ Fuel pipe.
4. C5: Release fuel pressure.
   Use tool No.4192-T **1**.
   ***WARNING: Caution required when working on
   high pressure pipes.***
5. HPi: Disconnect fuel pipe and fit blanking plug.
   Tool No.0189-Q.
6. HPi: Remove coolant expansion tank fixing nut.
7. HPi: Unclip coolant hose from crossmember.
8. Xsara: Support engine.
9. Remove:
   - ❏ HPi: Coolant expansion tank.
   - ❏ HPi/Xsara: Torque reaction link.
   - ❏ Xsara: Exhaust downpipe.
   - ❏ Xsara: RH engine mounting.
   - ❏ Crankshaft pulley bolts **2**.
   - ❏ Crankshaft pulley **3**.
   - ❏ Timing belt upper cover **4**.
   - ❏ Timing belt lower cover **5**.
10. Turn crankshaft clockwise to setting position.
11. Insert timing pin in crankshaft pulley **6**.
    Tool No.(-).0189-B.
12. C5: Insert timing pins in camshaft sprockets **7**
    & **8**. Tool Nos.(-).0189-A & (-).0189-L.
13. Xsara: Insert timing pins in camshaft
    sprockets **7** & **8**. Tool No.(-).0189-A.
14. Slacken tensioner bolt **9**.
15. Turn tensioner clockwise **12**.
16. Remove timing belt.
    ***NOTE: Timing belt must always be renewed once
    it has been removed.***

Installation

1. Ensure timing pins located correctly **6**, **7** & **8**.
2. Fit timing belt to crankshaft sprocket.
3. Secure belt to crankshaft sprocket with retaining
   clip. Tool No.(-).0189-K **11**.
4. Fit timing belt in anti-clockwise direction. Ensure
   belt is taut between sprockets.
5. Remove tools **6**, **7**, **8** & **11**.
6. Turn tensioner pulley anti-clockwise until
   pointer **13** at position **14**. Use Allen key **12**.
   ***NOTE: The arrow should pass notch **17** by at least
   10°. If not, replace tensioner.***
7. Turn tensioner pulley clockwise until pointer and
   notch **17** aligned.
   ***NOTE: If pointer passes notch **17**, repeat
   operations 2-7.***

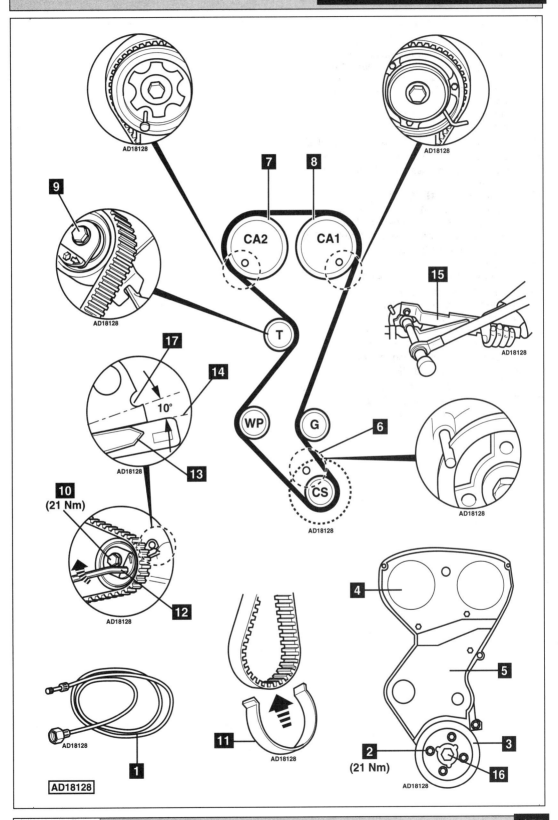

←

8. Tighten tensioner bolt to 21 Nm **10**.

   WARNING: Hold tensioner pulley during tightening to prevent it from turning. Allen key socket 12 must be approximately 15° below cylinder head gasket level. If not, replace tensioner pulley.

9. Turn crankshaft 10 turns clockwise to setting position.

10. Insert timing pin in inlet camshaft sprocket **8**.

11. Ensure tensioner pointer and notch **17** aligned.

    NOTE: If pointer passes notch 17, repeat operations 2-9.

12. Insert timing pins **7**, **8** & **6**.

    NOTE: If timing pin 6 cannot be inserted, reposition crankshaft end plate as follows:

13. Lock crankshaft pulley. Use tool No.6310-T **15**.

14. Slacken crankshaft bolt **16**.

15. Turn end plate until timing pin can be inserted **6**. Use tool No.6310-T **15**.

16. Tighten crankshaft bolt **16**.
    - ❏ Tightening torque: 40 Nm.
    - ❏ Angular tightening torque:
      Gold-coloured steel washer – 53±4°.
      Grey sintered washer – 40±4°.

17. Remove crankshaft locking tool **15**.

18. Remove timing pins **6**, **7** & **8**.

19. Install components in reverse order of removal.

20. Tighten crankshaft pulley bolts **2**.
    Tightening torque: 21 Nm.

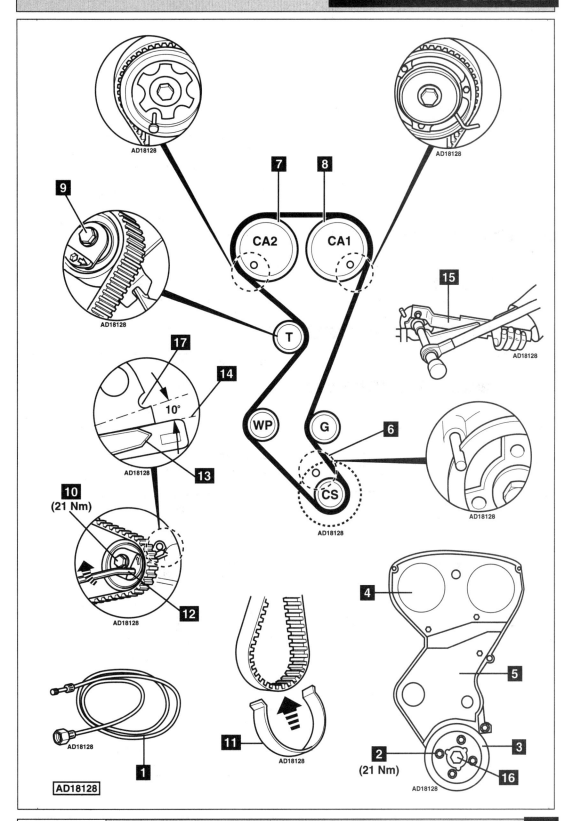

CA2

CA1

T

WP   G

CS

17

14   10°

13

10
(21 Nm)

12

9

15

6

4

5

1

11

2
(21 Nm)   3

16

AD18128

CITROEN

Model:	Xsara 2,0 16V
Year:	2000-03
Engine Code:	XU10J4RS (RFS)

Replacement Interval Guide

Citroen recommend replacement every 50,000 miles or every 10 years under normal conditions: every 40,000 miles or every 10 years under adverse conditions.
The previous use and service history of the vehicle must always be taken into account.
Refer to Timing Belt Replacement Intervals at the front of this manual.

Check For Engine Damage

CAUTION: This engine has been identified as an INTERFERENCE engine in which the possibility of valve-to-piston damage in the event of a timing belt failure is MOST LIKELY to occur.
A compression check of all cylinders should be performed before removing the cylinder head(s).

Repair Times – hrs

Remove & install	2,90
AC	+0,10

Special Tools

- Camshaft timing pin – Citroen No.9041-T.Z.
- Crankshaft timing pin – Citroen No.7014-T.N.
- Tension gauge – Citroen No.4122-T (SEEM).
- Crankshaft locking tool – Citroen No.9044-T.
- Tensioning tool – Citroen No.7017-T.W.
- Camshaft locking tool – Citroen No.4200-T.G.

Special Precautions

- Disconnect battery earth lead.
- DO NOT turn crankshaft or camshaft when timing belt removed.
- Remove spark plugs to ease turning engine.
- Turn engine in normal direction of rotation (unless otherwise stated).
- DO NOT turn engine via camshaft or other sprockets.
- Observe all tightening torques.

Removal

WARNING: This engine may suffer from failure of the crankshaft pulley resulting in the possible incorrect alignment of the timing pin hole. The timing belt should be removed and installed with the engine at 90° BTDC. If necessary, fit new crankshaft pulley.

1. Support engine.
2. Remove:
   - ❑ RH wheel arch liner.
   - ❑ Auxiliary drive belt.
3. Disconnect fuel pipes.
4. Raise timing belt cover tab **1**.

5. Remove:
   - ❑ Timing belt upper cover bolts **2**.
   - ❑ Timing belt upper cover **3**.
6. Turn crankshaft clockwise to setting position.
7. Insert timing pin in crankshaft pulley **4**.
   Tool No.7014-T.N.
8. Insert timing pins in camshaft sprockets **5**.
   Tool No.9041-T.Z.
9. Lock crankshaft. Use tool No.9044-T **6**.
10. Remove:
    - ❑ Crankshaft timing pin **4**.
    - ❑ Crankshaft pulley bolt **7**.
    - ❑ Crankshaft pulley **8**.
    - ❑ Timing belt lower cover **9**.
11. Slacken tensioner bolt **10**. Move tensioner away from belt and lightly tighten bolt.
12. Remove timing belt.

Installation

1. Remove bolt **11**.
2. Lock camshaft sprockets.
   Use tool No.4200-T.G **15**.
3. Slacken bolt of each camshaft sprocket **12**.
4. Remove locking tool **15**.
5. Ensure sprockets turn freely on camshafts.
6. Fit timing belt in following order:
   - ❑ Exhaust camshaft sprocket (CA1).
   - ❑ Inlet camshaft sprocket (CA2).
   - ❑ Guide pulley.
   - ❑ Crankshaft sprocket.
   - ❑ Water pump sprocket.
   - ❑ Tensioner pulley.
7. Attach tension gauge to belt at ▽.
   Tool No.4122-T **13**.
8. Slacken tensioner bolt **10**.
9. Turn tensioner anti-clockwise until tension gauge indicates 55 SEEM units. Use tool No.7017-T.W **14**.
10. Tighten tensioner bolt to 21 Nm **10**.
11. Lock camshaft sprockets.
    Use tool No.4200-T.G **15**.
12. Tighten bolt of each camshaft sprocket **12**.
    Tightening torque: 40 Nm.
13. Remove:
    - ❑ Timing pins **5**.
    - ❑ Tension gauge.
    - ❑ Crankshaft locking tool **6**.
    - ❑ Camshaft locking tool **15**.
14. Temporarily fit crankshaft pulley.
15. Turn crankshaft six turns clockwise.

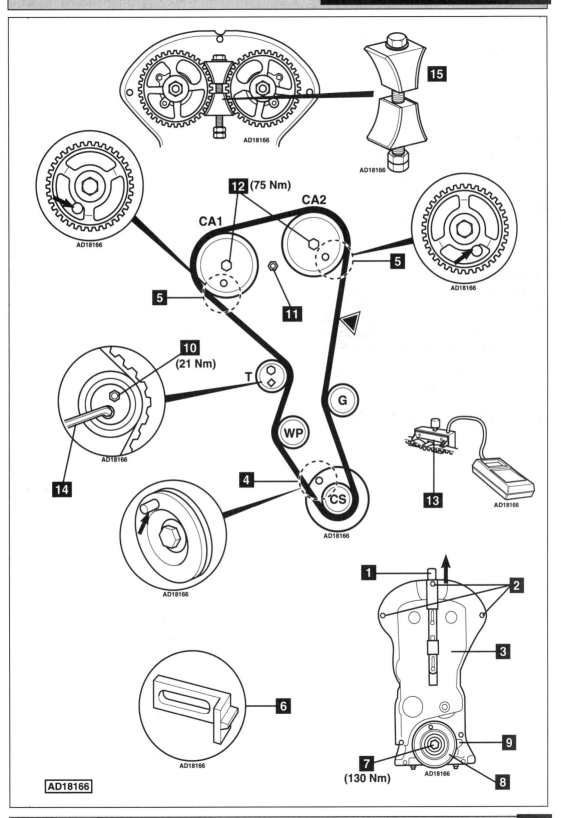

AD18166

12 (75 Nm)

CA1 CA2

5 **5**

11

10
(21 Nm)

T G

WP

4

CS

14

15

13

1 **2**

3

6

9

7
(130 Nm)

8

AD18166

←

16. Insert timing pin in crankshaft pulley ▨.
17. Lock camshaft sprockets.
    Use tool No.4200-T.G ▨.
18. Slacken bolt of each camshaft sprocket ▨.
19. Remove locking tool ▨.
20. Insert timing pins in camshaft sprockets ▨.
21. Slacken tensioner nut ▨.
22. Attach tension gauge to belt at ▽ ▨.
23. Turn tensioner anti-clockwise until tension
    gauge indicates 35 SEEM units.
    Use tool No.7017-T.W ▨.
24. Tighten tensioner bolt to 21 Nm ▨.
25. Lock camshaft sprockets.
    Use tool No.4200-T.G ▨.
26. Tighten bolt of each camshaft sprocket to
    75 Nm ▨.
27. Remove locking tool ▨.
28. Remove timing pins ▨ & ▨.
29. Turn crankshaft two turns clockwise.
30. Insert timing pin in crankshaft pulley ▨.
31. Insert timing pins in camshaft sprockets ▨.
    NOTE: If timing pins cannot be inserted easily,
    repeat installation and tensioning procedures.
32. Remove timing pins ▨ & ▨.
33. Install components in reverse order of removal.
34. Coat crankshaft pulley bolt thread with suitable
    thread locking compound.
35. Tighten crankshaft pulley bolt to 130 Nm ▨.

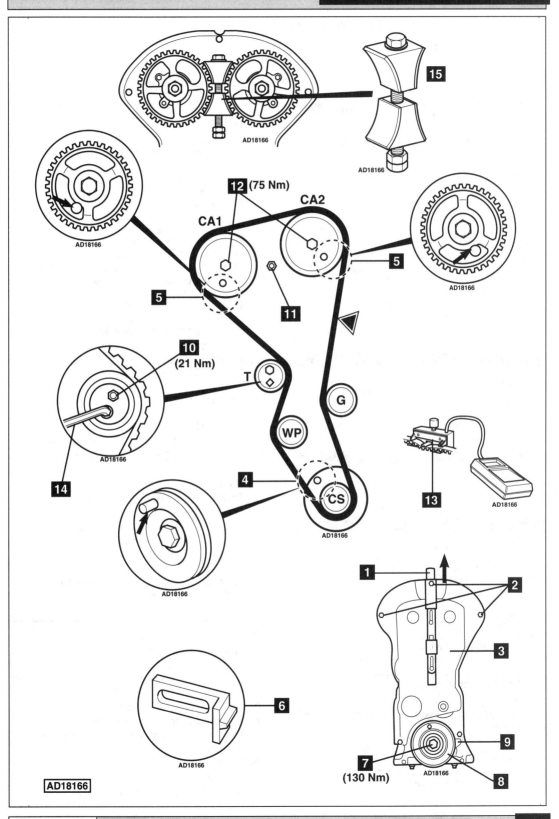

CITROEN

Model:	**Xantia 3,0 V6 • XM 3,0 V6**
Year:	**1997-01**
Engine Code:	**XFZ (ES9J4)**

Replacement Interval Guide

Citroen recommend:
1997MY: Replacement every 72,000 miles under normal conditions or every 54,000 miles under adverse conditions.
1998MY: Replacement every 80,000 miles under normal conditions or every 48,000 miles under adverse conditions.
1999MY: Replacement every 75,000 miles under normal conditions or every 60,000 miles under adverse conditions.
2000MY→: Replacement every 75,000 miles or every 10 years under normal conditions, or every 60,000 miles or every 10 years under adverse conditions.
The previous use and service history of the vehicle must always be taken into account.
Refer to Timing Belt Replacement Intervals at the front of this manual.

Check For Engine Damage

CAUTION: This engine has been identified as an INTERFERENCE engine in which the possibility of valve-to-piston damage in the event of a timing belt failure is MOST LIKELY to occur.
A compression check of all cylinders should be performed before removing the cylinder head(s).

Repair Times – hrs

Remove & install:

Xantia	4,30
XM	No information available

Special Tools

- Crankshaft timing pin – Citroen No.C.0187A.
- Camshaft timing pins – 4 x Citroen No.C.0187B.
- Timing belt retaining clip – Citroen No.C.0187J.
- Camshaft sprocket holding tool – Citroen No.C.0187F.
- Tensioning tool – Citroen No.C.0187E.
- Tension gauge – Citroen No.4122-T (SEEM C.Tronic G2 105.5).
- M8 x 1,25 x 75 mm bolt.
- M8 x 1,25 x 40 mm bolt.

Special Precautions

- Disconnect battery earth lead.
- DO NOT turn crankshaft or camshaft when timing belt removed.
- Remove spark plugs to ease turning engine.
- Turn engine in normal direction of rotation (unless otherwise stated).
- DO NOT turn engine via camshaft or other sprockets.
- Observe all tightening torques.

Removal

1. Remove:
   - ❏ Auxiliary drive belt.
   - ❏ Oil filler cap.
   - ❏ Upper engine cover.
2. Disconnect fuel pipes from fuel rail – XM.
3. Support engine.
4. Remove:
   - ❏ Engine control module (ECM) box.
   - ❏ RH engine mounting.
   - ❏ RH splash guard – XM.
   - ❏ Auxiliary drive belt tensioner assembly.
   - ❏ Timing belt upper covers **1**.
   - ❏ Bracket **2**.
   - ❏ Crankshaft pulley bolts **3**.
   - ❏ Crankshaft pulley **4**.
   - ❏ Timing belt lower cover **5**.
5. Turn crankshaft clockwise to setting position.
6. Insert crankshaft timing pin **6**.
   Tool No.C.0187A.
7. Insert timing pins in camshaft sprockets **7**.
   Tool No.C.0187B.
8. Insert M8 x 1,25 x 75 mm bolt into tensioner bracket **8**.
9. Tighten bolt **8** until it touches bracket **9**.
10. Insert M8 x 1,25 x 40 mm bolt into tensioner bracket **10**.
11. Install special tool No.C.0187E **11**.
12. Tighten bolt **10** until it touches bracket **12**.
13. Fully tighten bolt **10**.
14. Slacken tensioner bolts **13**, **14** & **15**.
    NOTE: DO NOT slacken bolt 16.
15. Slacken bolt **8**.
16. Remove timing belt.
    NOTE: Mark direction of rotation on belt with chalk if belt is to be reused.

Installation

1. Ensure crankshaft timing pin located correctly **6**. Tool No.C.0187A.
2. Ensure timing pins located correctly in camshaft sprockets **7**. Tool No.C.0187B.
3. Slacken camshaft sprocket bolts **17**.
4. Ensure camshaft sprockets turn freely.
5. Turn camshaft sprockets fully clockwise in slotted holes.
6. Tighten bolts to 5 Nm **17**.
7. Slacken bolts 45° **17**.
8. Tighten bolts to 10 Nm **13**, **14** & **15**.

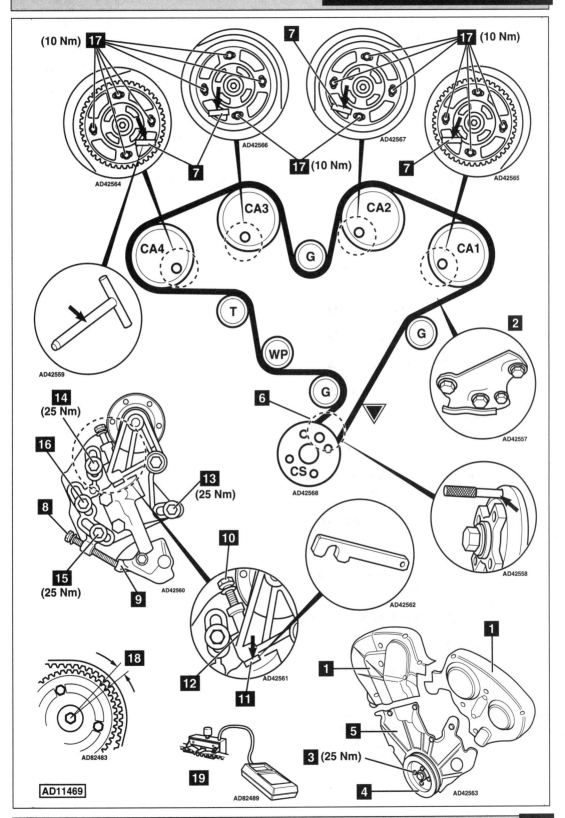

←

9. Slacken bolts 45° **13**, **14** & **15**.
10. Fit timing belt to crankshaft sprocket.
11. Retain in position with special tool.
    Tool No.C.0187J.
12. Fit timing belt in anti-clockwise direction.
    NOTE: If reusing old belt: Observe direction of rotation marks.
13. Turn each camshaft sprocket to engage in nearest belt tooth.
    NOTE: Ensure bolts not at end of slotted holes in sprockets 17. Angular movement of sprockets must not be more than one tooth space 18.
14. Ensure belt is taut between sprockets.
15. Remove retaining clip from timing belt.
    Tool No.C.0187J.
16. Attach tension gauge to belt at ▽.
    Tool No.4122-T **19**.
17. Tighten bolt **8** until tension gauge indicates 83±2 SEEM units.
18. Hold camshaft sprockets. Use tool No.C.0187F.
19. Tighten camshaft sprocket bolts (CA4) **17**.
    Tightening torque: 10 Nm.
20. Tighten camshaft sprocket bolts (CA3) **17**.
    Tightening torque: 10 Nm.
21. Tighten camshaft sprocket bolts (CA2) **17**.
    Tightening torque: 10 Nm.
22. Tighten camshaft sprocket bolts (CA1) **17**.
    Tightening torque: 10 Nm.
23. Tighten bolt **13**. Tightening torque: 25 Nm.
24. Tighten bolt **14**. Tightening torque: 25 Nm.
25. Tighten bolt **15**. Tightening torque: 25 Nm.
26. Remove tension gauge **19**.
27. Remove timing pins from camshaft sprockets **7**.
28. Remove crankshaft timing pin **6**.
29. Turn crankshaft slowly 10 turns clockwise.
30. Insert timing pins in camshaft sprockets **7**.
    Tool No.C.0187B.
31. Insert crankshaft timing pin **6**.
    Tool No.C.0187A.
32. Slacken bolts of each camshaft sprocket **17**.
33. Slacken bolts **13**, **14** & **15**.
34. Remove bolt **10**.
35. Adjust position of bolt **8** until tool **11** slides freely without free play. Tool No.C.0187E.
36. Wait 2 minutes to allow automatic tensioner unit and belt to settle.
37. Check tool **11** slides freely without free play.
    Tool No.C.0187E.
38. Adjust if necessary.
39. Remove special tool **11**.
40. Hold camshaft sprockets. Use tool No.C.0187F.
41. Tighten bolt **13**. Tightening torque: 25 Nm.
42. Tighten bolt **14**. Tightening torque: 25 Nm.
43. Tighten bolt **15**. Tightening torque: 25 Nm.
44. Tighten camshaft sprocket bolts (CA4) **17**.
    Tightening torque: 10 Nm.

45. Tighten camshaft sprocket bolts (CA3) **17**.
    Tightening torque: 10 Nm.
46. Tighten camshaft sprocket bolts (CA2) **17**.
    Tightening torque: 10 Nm.
47. Tighten camshaft sprocket bolts (CA1) **17**.
    Tightening torque: 10 Nm.
48. Remove bolt **8**.
49. Remove crankshaft timing pin **6**.
50. Remove timing pins **7**.
51. Install components in reverse order of removal.
52. Tighten crankshaft pulley bolts **3**.
    Tightening torque: 25 Nm.

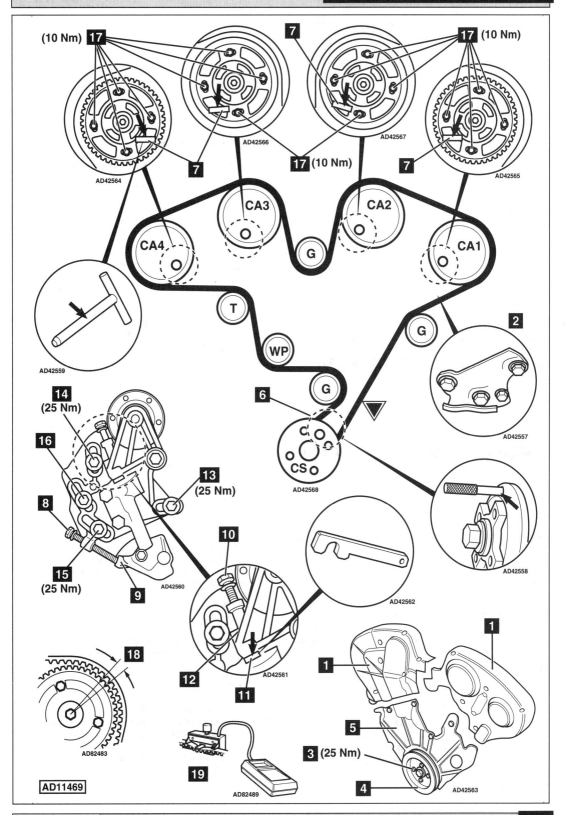

(10 Nm) **17**

7

17 (10 Nm)

AD42564

AD42566

AD42567

AD42565

7

17 (10 Nm)

7

AD42559

CA3

CA2

CA4

G

CA1

T

WP

G

G

6

2

AD42557

C

CS

AD42568

AD42558

14
(25 Nm)

16

13
(25 Nm)

8

AD42560

15
(25 Nm)

9

10

AD42562

18

12

AD42561

11

1

1

AD82483

5

3 (25 Nm)

19

AD82489

4

AD42563

Model:	Synergie/Evasion 2,0/Turbo • Relay/Jumper 2,0
Year:	1994-02
Engine Code:	RFU, RFW, RGX

Replacement Interval Guide

Citroen recommend:
1995-96MY: Replacement every 72,000 miles under normal conditions or every 48,000 miles under adverse conditions.
1997MY: Replacement every 72,000 miles under normal conditions or every 54,000 miles under adverse conditions.
1998MY: Replacement every 80,000 miles under normal conditions or every 48,000 miles under adverse conditions.
1999MY: Replacement every 75,000 miles under normal conditions or every 60,000 miles under adverse conditions.
2000MY→: Replacement every 75,000 miles or every 10 years under normal conditions, or every 60,000 miles or every 10 years under adverse conditions.
The previous use and service history of the vehicle must always be taken into account.
Refer to Timing Belt Replacement Intervals at the front of this manual.

Check For Engine Damage

CAUTION: This engine has been identified as an INTERFERENCE engine in which the possibility of valve-to-piston damage in the event of a timing belt failure is MOST LIKELY to occur.
A compression check of all cylinders should be performed before removing the cylinder head.

Repair Times – hrs

Remove & install:

Evasion/Synergie	2,10
AC	+0,10
Jumper/Relay	2,20
AC	+0,30

Special Tools

- Crankshaft timing pin – Citroen No.(-).0153G.
- Camshaft timing pin (Evasion/Synergie) – Citroen No.(-).0153.AA.
- Camshaft timing pin (Jumper/Relay) – Citroen No.(-).0132R.
- Tension gauge – SEEM C.Tronic 105.5.
- Flywheel locking tool – Facom No.D86.

Removal

1. Raise and support front of vehicle.
2. Remove:
   - RH front wheel.
   - RH wheel arch liner.
   - Auxiliary drive belt.
   - Timing belt upper cover **1**.
3. Turn crankshaft clockwise to setting position.
4. Insert crankshaft timing pin **2**. Tool No.(-).0153G.
5. Insert timing pin in camshaft sprocket **3**.
   Tool No.(-).0153.AA or (-).0132R.

6. Remove clutch housing plate. Lock flywheel. Use tool No.D86 **4**.
7. Remove:
   - Crankshaft pulley bolt **5**.
   - Crankshaft pulley **6**.
   NOTE: An extractor may be required to remove crankshaft pulley.
   - Timing belt centre cover **7**.
   - Timing belt lower cover **8**.
   - Thrust washer **9**.
8. Slacken tensioner bolt **10**. Move tensioner away from belt. Lightly tighten bolt.
9. Remove timing belt.

Installation

1. Temporarily fit crankshaft pulley and timing pin **6** & **2**.
2. Remove:
   - Crankshaft pulley **6**.
   - Crankshaft timing pin **2**.
3. Fit timing belt in clockwise direction, starting at camshaft sprocket.
4. Turn tensioner anti-clockwise. Lightly tension belt. Tighten tensioner bolt **10**.
5. Install:
   - Thrust washer **9**.
   - Crankshaft pulley **6**.
6. Temporarily tighten crankshaft pulley bolt **5**.
7. Attach tension gauge to belt at ▽.
   Tool No.SEEM C.Tronic 105.5.
8. Turn tensioner anti-clockwise until tension gauge indicates 14-18 SEEM units.
9. Tighten tensioner bolt to 20 Nm **10**.
10. Remove:
    - Tension gauge.
    - Timing pin **3**.
    - Flywheel locking tool **4**.
11. Turn crankshaft two turns clockwise.
12. Ensure timing pins can be inserted **2** & **3**.
13. Remove timing pins **2** & **3**. Turn crankshaft two turns clockwise.
14. Insert timing pin in camshaft sprocket **3**.
15. Attach tension gauge to belt at ▽. Check tension gauge reading.
    Evasion/Synergie: 42-46 SEEM units.
    Jumper/Relay: 34-44 SEEM units.
16. If tension not as specified: Repeat installation procedure.
17. Remove:
    - Timing pin **3**.
    - Tension gauge.
    - Crankshaft pulley **6**.
19. Install components in reverse order of removal.
20. Coat crankshaft pulley bolt with suitable thread locking compound.
21. Tighten crankshaft pulley bolt **5**.
    Evasion/Synergie: 110 Nm. Jumper/Relay: 120 Nm.

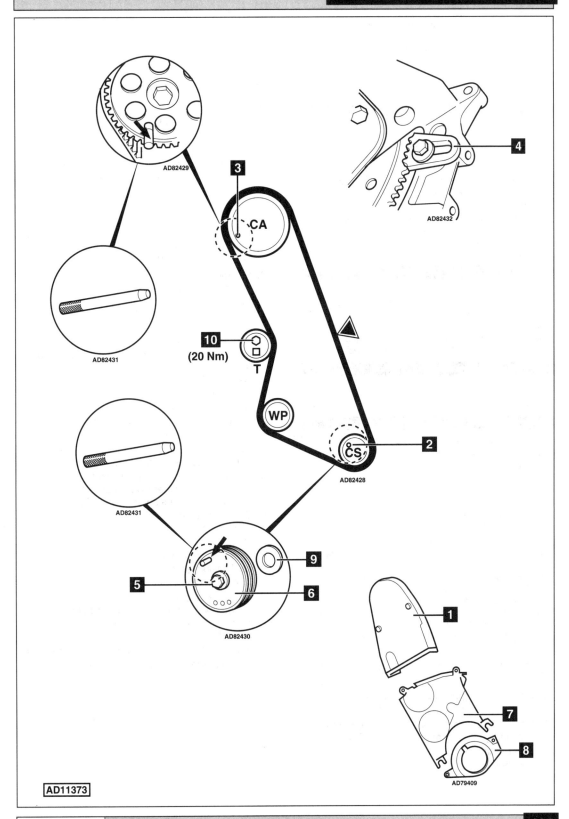

AD82429

CA

3

4

AD82432

AD82431

10
(20 Nm)
T

WP

CS

2

AD82428

AD82431

9

5

6

AD82430

1

7

8

AD79409

AD11373

CITROEN

Model:	**AX 1,4D**
Year:	**1989-94**
Engine Code:	**K9A (TUD3)**

Replacement Interval Guide

Citroen recommend:
→92 MY: Replacement every 48,000 miles under normal conditions or every 40,000 miles under adverse conditions.
93 MY→: Replacement every 72,000 miles under normal conditions or every 36,000 miles under adverse conditions.
The previous use and service history of the vehicle must always be taken into account.
Refer to Timing Belt Replacement Intervals at the front of this manual.

Check For Engine Damage

CAUTION: This engine has been identified as an INTERFERENCE engine in which the possibility of valve-to-piston damage in the event of a timing belt failure is MOST LIKELY to occur.
A compression check of all cylinders should be performed before removing the cylinder head.

Repair Times – hrs

Check & adjust	1,30
Remove & install	1,90

Special Tools

- Timing pin – Citroen No.4507-TA.
- Tension gauge – Citroen No.4099-T or 4122-T (SEEM C.Tronic 105.5).
- Tensioning tool – Citroen No.4507-T.J.
- 3 bolts – M8 x 125 mm.

Special Precautions

- Disconnect battery earth lead.
- DO NOT turn crankshaft or camshaft when timing belt removed.
- Remove glow plugs to ease turning engine.
- Turn engine in normal direction of rotation (unless otherwise stated).
- DO NOT turn engine via camshaft or other sprockets.
- Observe all tightening torques.
- Check diesel injection pump timing after belt replacement.

Removal

1. Remove timing belt upper cover **1**.
2. Turn crankshaft clockwise to setting position.
3. Insert one bolt in camshaft sprocket and two bolts in injection pump sprocket **3** & **4**.
4. Insert timing pin in flywheel **5**.
   Tool No.4507-TA.
5. Remove:
   - ❑ Auxiliary drive belt.
   - ❑ Crankshaft pulley.
   - ❑ Timing belt lower cover **2**.
6. Slacken tensioner nut **6**. Move tensioner away from belt and lightly tighten nut.
7. Remove timing belt.

Installation

1. Ensure timing pin and locking bolts located correctly **5**, **3** & **4**.
2. Fit timing belt in anti-clockwise direction, starting at crankshaft sprocket. Ensure belt is taut between sprockets.
3. Slacken tensioner nut **6**. Insert tensioning tool in square hole of tensioner pulley **7**.
   Tool No.4507-T.J. Turn tensioner pulley anti-clockwise to pre-tension belt. Lightly tighten nut.
4. Remove:
   - ❑ Timing pin **5**.
   - ❑ Locking bolts **3** & **4**.
5. Turn crankshaft four turns clockwise.
6. Insert timing pin in flywheel **5**.
7. Attach tension gauge to belt at ▽. Apply firm thumb pressure several times to belt at ▽.
8. Slacken tensioner nut **6**. Insert tensioning tool in square hole of tensioner pulley **7**. Tool No.4507-T.J. Tension belt to 25 SEEM units.
9. Tighten tensioner nut **6**.
   Tightening torque: 23 Nm.
10. Remove:
    - ❑ Timing pin **5**.
    - ❑ Tension gauge.
11. Turn crankshaft four turns clockwise.
12. Attach tension gauge to belt at ▽. Tension gauge should indicate approximately 38 SEEM units. If not: Repeat tensioning procedure.
13. Install components in reverse order of removal.
14. Tighten crankshaft pulley bolts **8**.
    Tightening torque: 16 Nm.

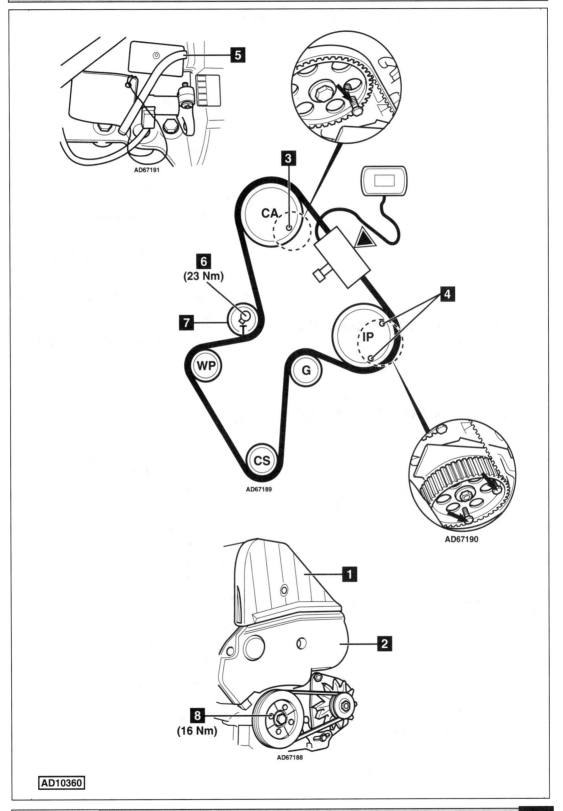

AD67191

CA

6
(23 Nm)

7

WP

G

IP

4

CS

AD67189

AD67190

1

2

8
(16 Nm)

AD67188

AD10360

Model:	Saxo 1,5D • AX 1,5D • Xsara 1,5D
Year:	1994-03
Engine Code:	VJY, VJZ (TUD5)

Replacement Interval Guide

Citroen recommend:
→1995MY: Replacement every 72,000 miles under normal conditions or every 55,000 miles under adverse conditions.
1996-98MY: Replacement every 72,000 miles under normal conditions or every 54,000 miles under adverse conditions.
1999MY: Replacement every 80,000 miles under normal conditions or every 54,000 miles under adverse conditions.
2000MY→: Replacement every 80,000 miles or every 10 years under normal conditions, or every 48,000 miles or every 10 years under adverse conditions.
The previous use and service history of the vehicle must always be taken into account.
Refer to Timing Belt Replacement Intervals at the front of this manual.

Check For Engine Damage

CAUTION: This engine has been identified as an INTERFERENCE engine in which the possibility of valve-to-piston damage in the event of a timing belt failure is MOST LIKELY to occur.
A compression check of all cylinders should be performed before removing the cylinder head.

Repair Times – hrs

Remove & install:	
AX	1,90
AC	+0,20
Saxo	2,20
Xsara	2,00

Special Tools

- Flywheel timing pin – Citroen No.4507-T.A.
- Injection pump timing pin – Citroen No.4527-T.S1.
- Camshaft timing pin – Citroen No.4527-T.S2.
- Camshaft sprocket holding tool – Citroen No.6016-T.
- Tensioning tool – Citroen No.4507-T.J.
- Tension gauge – Citroen No.4099/4122-T (SEEM C.Tronic 105/105.5).

Removal

1. Raise and support front of vehicle.
2. Remove:
   - ❏ RH front wheel.
   - ❏ Inner wing panel.
   - ❏ Glow plug relay. DO NOT disconnect wires.
   - ❏ Timing belt upper cover **1**.
   - ❏ Auxiliary drive belt.
   - ❏ Crankshaft pulley bolts **2**.
   - ❏ Crankshaft pulley **3**.
   - ❏ Timing belt lower cover **4**.
3. Turn crankshaft clockwise to setting position. Insert timing pin in flywheel **5**. Tool No.4507-T.A.
4. Insert timing pin in injection pump sprocket **6**. Tool No.4527-T.S1.

5. Insert timing pin in camshaft sprocket **7**. Tool No.4527-T.S2.
   NOTE: If locating holes for timing pins not aligned: Turn crankshaft one turn clockwise.
6. Slacken tensioner nut **8**. Move tensioner away from belt. Lightly tighten nut.
7. Remove timing belt.

Installation

NOTE: Check tensioner pulley and guide pulley for smooth operation.
1. Ensure timing pins located correctly **5**, **6** & **7**.
2. Hold sprockets. Use tool No.6016-T. Slacken bolts **9** & **10**. Tighten bolts finger tight. Ensure sprockets can still turn.
3. Turn sprockets fully clockwise in slotted holes.
4. Fit timing belt in anti-clockwise direction, starting at crankshaft sprocket. Ensure belt is taut between sprockets.
   NOTE: Turn injection pump and camshaft sprockets slightly anti-clockwise to engage belt teeth.
5. Attach tension gauge to belt at ▽. Tool No.4099/4122-T.
6. Slacken tensioner nut **8**. Insert tensioning tool in square hole of tensioner pulley **11**. Tool No.4507-T.J. Tension belt to 100 SEEM units. Tighten tensioner nut to 23 Nm **8**.
7. Remove tension gauge.
8. Tighten camshaft and injection pump sprocket bolts **9** & **10**. Tightening torque: 23 Nm.
9. Remove timing pins **5**, **6** & **7**.
10. Turn crankshaft ten turns clockwise. Insert timing pins **5**, **6** & **7**.
11. Slacken camshaft sprocket bolts **9**.
12. Slacken injection pump sprocket bolts **10**.
13. Attach tension gauge to belt at ▽.
14. Slacken tensioner nut **8**. Insert tensioning tool in square hole of tensioner pulley **11**. Tool No.4507-T.J. Tension belt.
    New belt: 55±5 SEEM units.
    Used belt: 44 SEEM units.
15. Tighten tensioner nut to 23 Nm **8**.
16. Remove tension gauge.
17. Tighten camshaft and injection pump sprocket bolts **9** & **10**. Tightening torque: 23 Nm.
18. Remove timing pins **5**, **6** & **7**.
19. Turn crankshaft two turns clockwise. Insert flywheel timing pin **5**.
20. Ensure timing pins can be inserted easily in sprockets **6** & **7**. If not: Repeat installation procedure.
21. Install components in reverse order of removal.
22. Tighten crankshaft pulley bolts **2**. Tightening torque: 20 Nm.

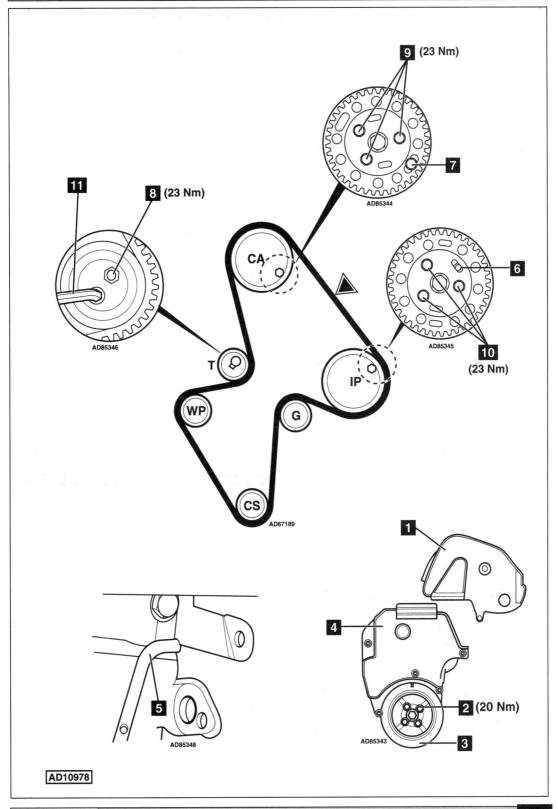

9 (23 Nm)

AD85344

7

8 (23 Nm)

11

AD85346

CA

T

WP

G

IP

CS

AD67189

6

AD85345

10

(23 Nm)

1

4

5

AD85348

2 (20 Nm)

AD85343

3

AD10978

Model:	Visa 17D • BX 17D/TD • BX 19D • ZX 1,8/1,9D/TD • Xantia 1,8/1,9D/TD Xsara 1,9D/TD • Synergie/Evasion 1,9TD • C15D • Berlingo 1,8/1,9D Dispatch/Jumpy 1,9D/TD • C25D 1,9D • Relay/Jumper 1,9D/TD
Year:	1984-02
Engine Code:	161A, 162, D8A, D8B, D8C, D9B, A8A, A9A, AJZ, DJY, DJZ, DHV, DHX, DHW, DHY

Replacement Interval Guide

Citroen recommend:

Except C25D:

→1992MY: Replacement every 48,000 miles under normal conditions or every 40,000 miles under adverse conditions.

1993-94MY: Replacement every 72,000 miles under normal conditions or every 36,000 miles under adverse conditions.

1995MY: Replacement every 72,000 miles under normal conditions or every 55,000 miles under adverse conditions.

1996-98MY: Replacement every 72,000 miles under normal conditions or every 54,000 miles under adverse conditions.

1999MY: Replacement every 80,000 miles under normal conditions or every 54,000 miles under adverse conditions.

2000MY→: Replacement every 80,000 miles or every 10 years under normal conditions, or every 48,000 miles or every 10 years under adverse conditions.

C25D:

→1990MY: No recommended replacement interval.

→1990MY: Check and replacement if necessary every 50,000 miles under adverse conditions.

1991MY→: Replacement every 45,000 miles under normal conditions or every 36,000 miles under adverse conditions.

The previous use and service history of the vehicle must always be taken into account.
Refer to Timing Belt Replacement Intervals at the front of this manual.

Check For Engine Damage

CAUTION: This engine has been identified as an INTERFERENCE engine in which the possibility of valve-to-piston damage in the event of a timing belt failure is MOST LIKELY to occur.
A compression check of all cylinders should be performed before removing the cylinder head.

Repair Times – hrs

Check & adjust:

Visa/C15	1,40
BX 17/19D/Turbo D	1,40
ZX	1,40
Xantia 1,9D	0,90
Xantia 1,9TD	1,10

Remove & install:

Visa/C15	2,90[1]
C25	3,60[1]
BX 17/19D	2,80[1]
BX Turbo D	3,30[1]
ZX	2,80[1]
Xantia 1,9D/TD (→12/93)	2,50
AC	+0,10
Xantia 1,9D (01/94→)	3,20
Xantia 1,9TD (01/94→)	3,70
AC	+0,30
Xsara 1,9D	2,90
Xsara 1,9TD	2,80
AC	+0,10
Berlingo	2,60
AC	+0,10
Dispatch/Jumpy 1,9D	3,30
Dispatch/Jumpy 1,9TD	3,50
AC	+0,10
Evasion/Synergie	3,50
AC	+0,10
Jumper/Relay	3,30

[1]*Includes 0,70 for dynamic pump timing.*

Special Tools

- Except Jumper/Relay/Xsara: Flywheel timing pin – Citroen No.7099-TM.
- Jumper/Relay: Flywheel timing pin – Citroen No.7014-TJ.
- Xsara: Flywheel timing pin – Citroen No.7017-TR.
- 3 bolts – M8 x 125 mm.

Special Precautions

- Disconnect battery earth lead.
- DO NOT turn crankshaft or camshaft when timing belt removed.
- Remove glow plugs to ease turning engine.
- Turn engine in normal direction of rotation (unless otherwise stated).
- DO NOT turn engine via camshaft or other sprockets.
- Observe all tightening torques.
- Check diesel injection pump timing after belt replacement.

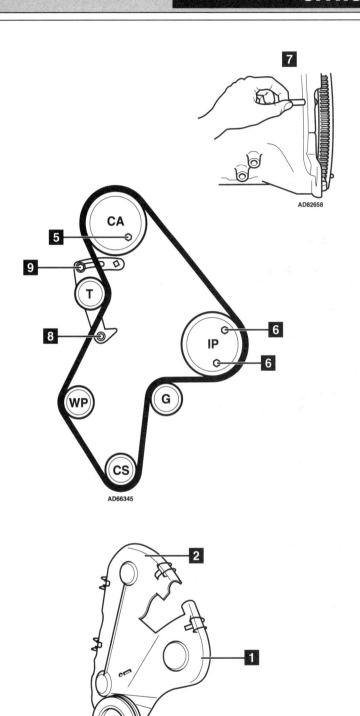

AD82658

AD66345

AD66344

AD11339

Removal

1. Raise and support front of vehicle.
2. Remove:
   - ❏ Some models: RH front wheel.
   - ❏ Some models: Engine undershield.
   - ❏ Auxiliary drive belt.
   - ❏ Some models: Engine mounting.
   - ❏ Timing belt covers **1** & **2**.
   - ❏ Crankshaft pulley bolt **3**.
   - ❏ Crankshaft pulley **4**.
   - ❏ Timing belt lower cover.
3. Fit crankshaft pulley.
4. Turn crankshaft clockwise to setting position.
5. Insert one bolt in camshaft sprocket and two bolts in injection pump sprocket **5** & **6**.
6. Insert timing pin in flywheel **7**.
7. Remove crankshaft pulley.
8. Slacken tensioner bolts **8** & **9**. Move tensioner away from belt. Tighten bolt **9**.
9. Remove timing belt.

Installation

1. Ensure timing pin and locking bolts located correctly **5**, **6** & **7**.
2. Fit timing belt in anti-clockwise direction, starting at crankshaft sprocket. Ensure belt is taut between sprockets.
3. Slacken tensioner bolt **9**. Push tensioner several times against belt.
4. Tighten tensioner bolts **8** & **9**.
5. Remove timing pin and locking bolts. Turn crankshaft two turns clockwise.
6. Insert timing pin **7**.
7. Ensure locking bolts can be inserted in sprockets **5** & **6**.
8. Slacken and tighten tensioner bolts **8** & **9**.
9. Install components in reverse order of removal.
10. Coat crankshaft pulley bolt thread with suitable thread locking compound.
11. Tighten crankshaft pulley bolt **3**.
    - ❏ Visa, BX, ZX, Xantia, C15: 150 Nm.
    - ❏ Except Visa, BX, ZX, Xantia, C15: 40 Nm + 60°.

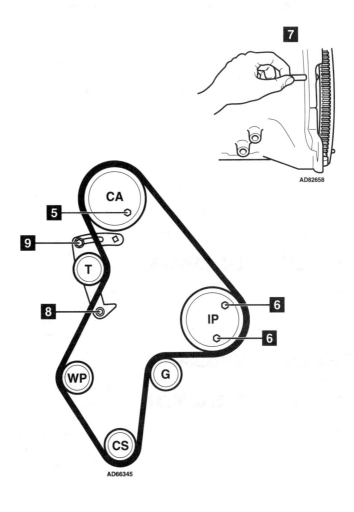

AD82658

AD66345

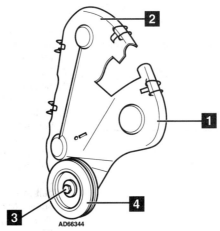

AD66344

AD11339

CITROEN

Model:	Xsara 1,9D • Berlingo 1,9D • Dispatch 1,9D
Year:	1998-03
Engine Code:	WJZ (DW8), WJY (DW8B)

Replacement Interval Guide

Citroen recommend:

→1999MY: Replacement every 80,000 miles under normal conditions or 60,000 miles under adverse conditions.

2000MY→: Replacement every 80,000 miles or every 10 years under normal conditions, or every 48,000 miles or every 10 years under adverse conditions.

The previous use and service history of the vehicle must always be taken into account.

Refer to Timing Belt Replacement Intervals at the front of this manual.

Check For Engine Damage

CAUTION: This engine has been identified as an INTERFERENCE engine in which the possibility of valve-to-piston damage in the event of a timing belt failure is MOST LIKELY to occur.

A compression check of all cylinders should be performed before removing the cylinder head(s).

Repair Times – hrs

Remove & install	3,00

Special Tools

- ■ Flywheel timing pin – Citroen No.(-).0188-D/0153-N.
- ■ Injection pump locking pin – Citroen No.(-).0188-H.
- ■ 1 bolt – M8 x 1,25 mm x 80 mm.
- ■ Timing belt retaining clip – Citroen No.(-).0188-K.
- ■ Tensioning tool – Citroen No.(-).0188-J/J1.
- ■ Tension gauge – SEEM C.Tronic 105.5.

Special Precautions

- ■ Disconnect battery earth lead.
- ■ DO NOT turn crankshaft or camshaft when timing belt removed.
- ■ Remove glow plugs to ease turning engine.
- ■ Turn engine in normal direction of rotation (unless otherwise stated).
- ■ DO NOT turn engine via camshaft or other sprockets.
- ■ Observe all tightening torques.

Removal

1. Raise and support front of vehicle.
2. Disconnect and seal off fuel pipes.
3. Remove:
   - ❏ Engine upper cover.
   - ❏ RH front wheel.
   - ❏ RH splash guard.
   - ❏ Auxiliary drive belt.
   - ❏ Auxiliary drive belt tensioner pulley.
   - ❏ Flywheel housing lower cover.

4. Remove:
   - ❏ Crankshaft pulley bolts **1**.
   - ❏ Crankshaft pulley **2**.
5. Reposition coolant hose from expansion tank.
6. Support engine.
7. Remove:
   - ❏ RH engine mounting and bracket.
   - ❏ Timing belt covers **3**, **4** & **5**.
8. Turn crankshaft clockwise to setting position.
9. Insert timing pin in flywheel **6**.
   Tool No.(-).0188-D/0153N.
10. Insert M8 x 1,25 mm x 80 mm bolt in camshaft sprocket **7**.
11. Insert locking pin in injection pump **8**.
    Tool No.(-).0188-H.
12. Slacken tensioner bolt **9**.
13. Turn tensioner pulley clockwise away from belt. Use tool No.(-).0188-J/J1.
14. Lightly tighten tensioner bolt **9**.
15. Remove timing belt.

Installation

1. Ensure timing pin and locking pin located correctly **6** & **8**.
2. Ensure camshaft sprocket locked with bolt **7**.
3. Slacken camshaft sprocket bolts **10**. Tighten bolts finger tight, then slacken 1/6 turn.
4. Turn camshaft sprocket fully clockwise in slotted holes.
5. Slacken injection pump sprocket bolts **11**. Tighten bolts finger tight, then slacken 1/6 turn.
6. Turn injection pump sprocket fully clockwise in slotted holes.
   NOTE: Sprockets should turn with slight resistance.
7. Fit timing belt to crankshaft sprocket.
8. Secure belt to crankshaft sprocket with retaining clip. Tool No.(-).0188-K **12**.
9. Fit timing belt in anti-clockwise direction. Ensure belt is taut between sprockets.
10. Lay belt on injection pump sprocket teeth. Engage belt teeth by turning sprocket slightly anti-clockwise.
11. Lay belt on camshaft sprocket teeth. Engage belt teeth by turning sprocket slightly anti-clockwise.
    *NOTE: Angular movement of sprockets must not be more than one tooth space **13**.*
12. Fit timing belt to water pump sprocket and tensioner pulley.

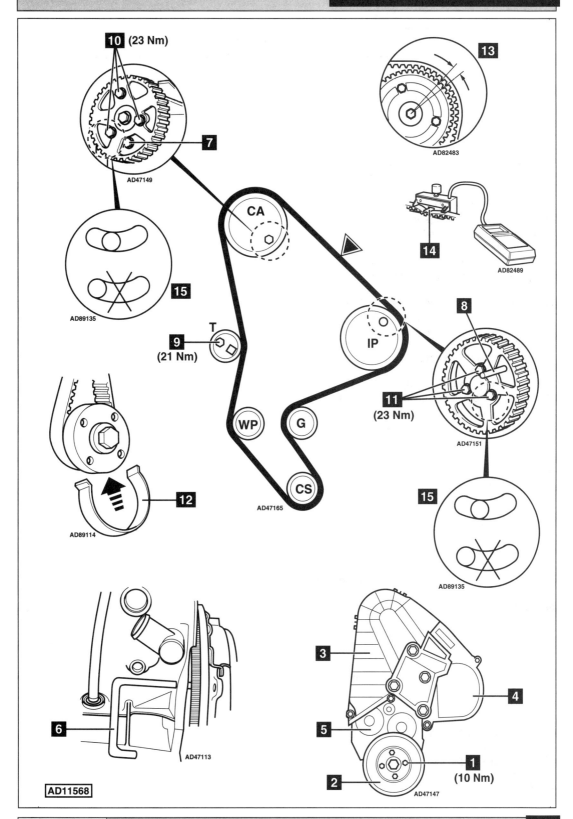

10 (23 Nm)

13

AD82483

7

AD47149

CA

14

AD82489

15

AD89135

8

T

9
(21 Nm)

IP

WP

G

11

(23 Nm)

AD47151

12

AD89114

CS

AD47165

15

AD89135

3

4

5

6

AD47113

1
(10 Nm)

2

AD47147

AD11568

←

13. Slacken tensioner bolt **9**.

14. Turn tensioner pulley anti-clockwise to temporarily tension belt.
    Use tool No.(-).0188-J/J1.

15. Lightly tighten tensioner bolt **9**.
    Tightening torque: 10 Nm.

16. Remove retaining clip **12**.

17. Attach tension gauge to belt at ▽ **14**.
    Tool No.SEEM C-Tronic 105.5.

18. Turn tensioner pulley anti-clockwise until tension gauge indicates 106±2 SEEM units.

19. Tighten tensioner bolt **9**.
    Tightening torque: 21 Nm.

20. Remove tension gauge **14**.

21. Ensure bolts not at end of slotted holes in sprockets **15**.

22. If necessary: Repeat installation procedure.

23. Tighten camshaft and injection pump sprocket bolts **10** & **11**. Tightening torque: 23 Nm.

24. Remove timing pin and locking pin **6** & **8**.

25. Remove locking bolt from camshaft sprocket **7**.

26. Turn crankshaft eight turns clockwise to setting position.
    NOTE: DO NOT allow crankshaft to turn anti-clockwise.

27. Insert timing pin and locking pin **6** & **8**.

28. Insert locking bolt in camshaft sprocket **7**.

29. Slacken camshaft sprocket bolts **10**.

30. Slacken injection pump sprocket bolts **11**.

31. Slacken tensioner bolt to release tension on belt **9**.

32. Attach tension gauge to belt at ▽ **14**.

33. Turn tensioner pulley anti-clockwise until tension gauge indicates 42±2 SEEM units.

34. Hold tensioner pulley in position. Tighten tensioner bolt **9**. Tightening torque: 21 Nm.

35. Tighten camshaft and injection pump sprocket bolts **10** & **11**. Tightening torque: 23 Nm.

36. Remove tension gauge **14**.

37. Check belt tension. Attach tension gauge to belt at ▽ **14**.

38. Tension gauge should indicate 38-46 SEEM units.

39. If not: Repeat tensioning procedure.

40. Remove tension gauge **14**.

41. Remove timing pin and locking pin **6** & **8**.

42. Remove locking bolt from camshaft sprocket **7**.

43. Turn crankshaft two turns clockwise to setting position.

44. Insert timing pin in flywheel **6**.

45. Ensure locking bolt can be inserted easily in camshaft sprocket **7**.

46. Ensure locking pin can be inserted easily in injection pump sprocket **8**.

47. Remove flywheel timing pin **6**.

48. Install components in reverse order of removal.

49. Tighten crankshaft pulley bolts **1**.
    Tightening torque: 10 Nm.

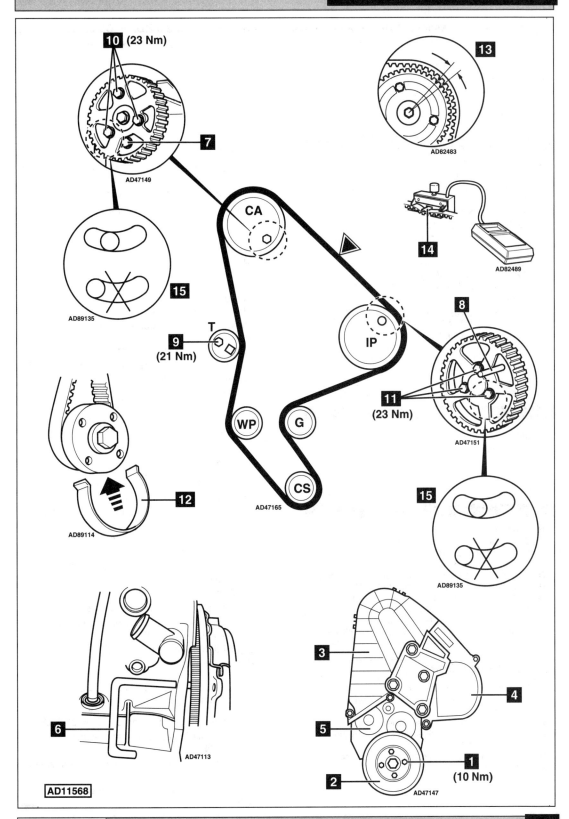

10 (23 Nm)

7

AD47149

13

AD82483

14

AD82489

15

AD89135

9 T
(21 Nm)

CA

IP

8

11
(23 Nm)

AD47151

WP   G

CS

AD47165

12

AD89114

15

AD89135

3

4

5

6

AD47113

1
(10 Nm)

2

AD47147

AD11568

CITROEN

Model:	**Xsara 2,0 HDi • Xsara Picasso 2,0 HDi • Xantia 2,0 HDi • C5 2,0 HDi** **Synergie 2,0 HDi • Berlingo 2,0 HDi • Dispatch 2,0 HDi • Relay 2,0 HDi**
Year:	**1998-02**
Engine Code:	**With adjustable camshaft sprocket** **RHY (DW10ATD), RHZ (DW10ATED/L)**

Replacement Interval Guide

Citroen recommend:
→99MY: Replacement every 100,000 miles under normal conditions or 80,000 miles under adverse conditions.
00MY→: Replacement every 100,000 miles or every 10 years under normal conditions, or every 80,000 miles or every 10 years under adverse conditions.
The previous use and service history of the vehicle must always be taken into account.
Refer to Timing Belt Replacement Intervals at the front of this manual.

Check For Engine Damage

CAUTION: This engine has been identified as an INTERFERENCE engine in which the possibility of valve-to-piston damage in the event of a timing belt failure is MOST LIKELY to occur.
A compression check of all cylinders should be performed before removing the cylinder head(s).

Repair Times – hrs

Remove & install:

Xsara/Picasso	3,80
Xantia	4,20
C5	3,50
Berlingo	3,30

Special Tools

- Flywheel locking tool – Citroen No.(-).0188.F.
- Crankshaft pulley puller – Citroen No.(-).0188.P.
- Flywheel timing pin (except C5) – Citroen No.(-).0288.D.
- Flywheel timing pin (C5) – Citroen No.(-).0118.X.
- Camshaft timing pin – Citroen No.(-).0188.M.
- Timing belt retaining clip – Citroen No.(-).0188.K.
- Tensioning tool – Citroen No.(-).0188.J2.
- Tension gauge – SEEM C.Tronic 105.5.
- Set of blanking plugs – Citroen No.(-).0188.T

Special Precautions

- Disconnect battery earth lead.
- DO NOT turn crankshaft or camshaft when timing belt removed.
- Remove glow plugs to ease turning engine.
- Turn engine in normal direction of rotation (unless otherwise stated).
- DO NOT turn engine via camshaft or other sprockets.
- Observe all tightening torques.

Removal

NOTE: The high-pressure fuel pump fitted to this engine does not require timing.

1. Raise and support front of vehicle.
2. Disconnect exhaust front pipe from manifold.
3. Remove:
   - ❑ RH front wheel.
   - ❑ RH splash guard.
   - ❑ Engine lower cover.
   - ❑ Engine upper cover.
   - ❑ Auxiliary drive belt.
   - ❑ Turbocharger air hoses.
   - ❑ Flywheel housing lower cover.
4. Lock flywheel. Use tool No.(-).0188.F.
5. Remove:
   - ❑ Crankshaft pulley bolt **1**.
   - ❑ Crankshaft pulley **2**. Use tool No.(-).0188.P.
   - ❑ Flywheel locking tool.
   - ❑ Lower torque reaction link.
   - ❑ Engine control module (ECM) and tray.
6. Disconnect and seal off fuel pipes. Use tool No.(-).0188.T.
7. Support engine.
8. Remove:
   - ❑ PAS reservoir (if required).
   - ❑ RH engine mounting and bracket.
   - ❑ Timing belt covers **4**, **5** & **6**.
9. Turn crankshaft clockwise to setting position.
10. Insert timing pin in flywheel **3**.
    - ❑ Except C5: Tool No.(-).0288.D.
    - ❑ C5: Tool No.(-).0188.X.
11. Insert timing pin in camshaft sprocket **7**. Tool No.(-).0188.M.
12. Slacken tensioner bolt **8**.
13. Slacken camshaft sprocket bolts **9**.
14. Turn tensioner pulley clockwise away from belt. Use tool No.(-).0188.J2.
15. Lightly tighten tensioner bolt **8**.
16. Remove timing belt.

Installation

1. Ensure timing pins located correctly **3** & **7**.
2. Tighten bolts finger tight, then slacken 1/6 turn **9**.
3. Turn camshaft sprocket fully clockwise in slotted holes.
   NOTE: Sprocket should turn with slight resistance.

➔

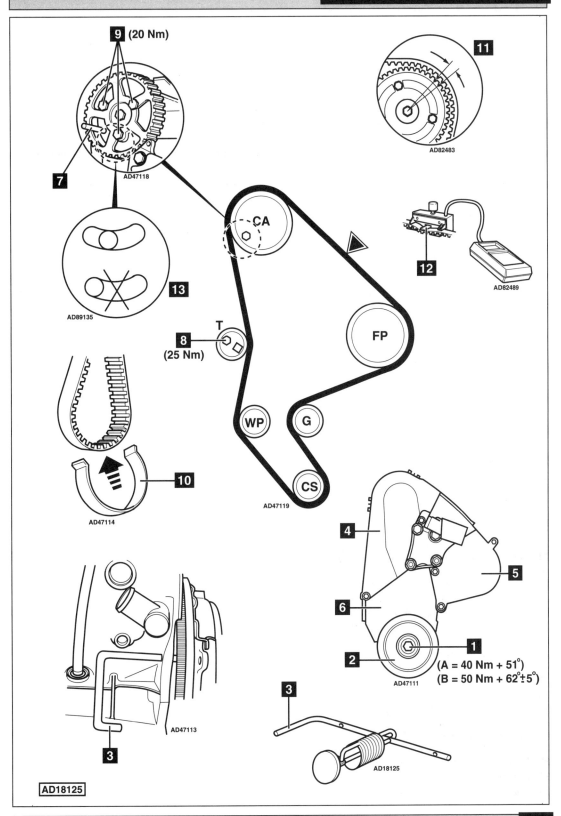

9 (20 Nm)

AD47118

7

11

AD82483

CA

13

AD89135

12

AD82489

T

8
(25 Nm)

FP

10

AD47114

WP

G

CS

AD47119

4

5

6

1

2

(A = 40 Nm + 51°)
(B = 50 Nm + 62° ± 5°)

AD47111

AD47113

3

3

3

AD18125

AD18125

←

4. Fit timing belt to crankshaft sprocket.
5. Secure belt to crankshaft sprocket with retaining clip. Tool No.(-).0188.K 🔟.
6. Fit timing belt in anti-clockwise direction. Ensure belt is taut between sprockets.
7. Lay belt on camshaft sprocket teeth. Engage belt teeth by turning sprocket slightly anti-clockwise.
   NOTE: Angular movement of sprocket must not be more than one tooth space 🔢.
8. Fit timing belt to water pump sprocket and tensioner pulley.
9. Slacken tensioner bolt 🔢.
10. Turn tensioner pulley anti-clockwise to temporarily tension belt. Use tool No.(-).0188.J2.
11. Lightly tighten tensioner bolt 🔢. Tightening torque: 10 Nm.
12. Remove retaining clip 🔟.
13. Attach tension gauge to belt at ▽ 🔢. Tool No.SEEM C.Tronic 105.5.
14. Turn tensioner pulley anti-clockwise until tension gauge indicates 98±2 SEEM units.
15. Tighten tensioner bolt 🔢. Tightening torque: 25 Nm.
16. Remove tension gauge.
17. Ensure sprocket bolts not at end of slotted holes 🔢.
18. If necessary: Repeat installation procedure.
19. Tighten camshaft sprocket bolts 🔢. Tightening torque: 20 Nm.
20. Remove timing pins 🔢 & 🔢.
21. Turn crankshaft eight turns clockwise to setting position.
    NOTE: DO NOT allow crankshaft to turn anti-clockwise.
22. Insert timing pins 🔢 & 🔢.
23. Slacken camshaft sprocket bolts 🔢.
24. Slacken tensioner bolt to release tension on belt 🔢.
25. Attach tension gauge to belt at ▽ 🔢.
26. Turn tensioner pulley anti-clockwise until tension gauge indicates 54±2 SEEM units.
27. Hold tensioner pulley in position. Tighten tensioner bolt 🔢. Tightening torque: 25 Nm.
28. Tighten camshaft sprocket bolts 🔢. Tightening torque: 20 Nm.
29. Remove tension gauge.
30. Attach tension gauge to belt at ▽ 🔢.
31. Check belt tension. Tension gauge should indicate the following:
    ❏ Except C5: 51-57 SEEM units.
    ❏ C5: 54±3 SEEM units.
32. If not: Repeat tensioning procedure.
33. Remove tension gauge.
34. Remove timing pins 🔢 & 🔢.

35. Turn crankshaft two turns clockwise to setting position.
36. Insert timing pin in flywheel 🔢.
37. Ensure timing pin can be inserted easily 🔢.
38. Remove timing pin 🔢.
39. Install components in reverse order of removal.
40. Clean crankshaft pulley bolt and crankshaft threads.
41. Coat crankshaft pulley bolt with suitable thread locking compound.
42. Tighten crankshaft pulley bolt 🔢:
    ❏ A – Dispatch/Synergie/Xsara →02/99: Tightening torque: 40 Nm + 51°.
    ❏ B – Xsara 03/99→/all other models: Tightening torque: 50 Nm + 62°±5°.

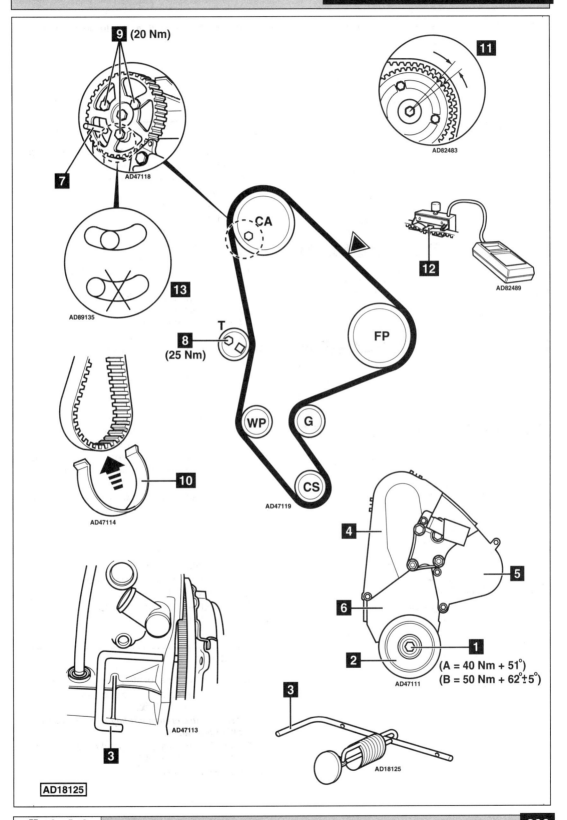

9 (20 Nm)

11

AD82483

7

AD47118

CA

12

AD82489

13

AD89135

T

8
(25 Nm)

FP

10

AD47114

WP

G

CS

AD47119

4

5

6

1

2

(A = 40 Nm + 51°)
(B = 50 Nm + 62°±5°)

AD47111

AD47113

3

3

AD18125

AD18125

CITROEN

Model:	**Xsara 2,0 HDi • Xsara Picasso 2,0 HDi • C5 2,0 HDi • Berlingo 2,0 HDi**
Year:	**2001-03**
Engine Code:	**With adjustable crankshaft sprocket** **RHY (DW10TD), RHZ (DW10ATED)**

Replacement Interval Guide

Citroen recommend replacement every 100,000 miles or every 10 years under normal conditions, or every 80,000 miles or every 10 years under adverse conditions.
The previous use and service history of the vehicle must always be taken into account.
Refer to Timing Belt Replacement Intervals at the front of this manual.

Check For Engine Damage

CAUTION: This engine has been identified as an INTERFERENCE engine in which the possibility of valve-to-piston damage in the event of a timing belt failure is MOST LIKELY to occur.
A compression check of all cylinders should be performed before removing the cylinder head(s).

Repair Times – hrs

Remove & install:

Berlingo →2001	3,30
Berlingo 2002→/C5/Picasso	3,50
Xsara	3,80

Special Tools

- Flywheel locking tool – Citroen No.(-).0188.F.
- Crankshaft pulley puller – Citroen No.(-).0188.P.
- Crankshaft sprocket aligning tool – Citroen No.(-).0188.Q2.
- Flywheel timing pin – Citroen No.(-).0118.Y.
- Camshaft timing pin – Citroen No.(-).0188.M.
- Timing belt clamp – Citroen No.(-).0188.AD.
- Tensioning tool – Citroen No.(-).0188.J2.
- Tension gauge – SEEM C.Tronic 105.5M.
- Set of blanking plugs – Citroen No.(-).0188.T

Special Precautions

- Disconnect battery earth lead.
- DO NOT turn crankshaft or camshaft when timing belt removed.
- Remove glow plugs to ease turning engine.
- Turn engine in normal direction of rotation (unless otherwise stated).
- DO NOT turn engine via camshaft or other sprockets.
- Observe all tightening torques.

Removal

NOTE: The high-pressure fuel pump fitted to this engine does not require timing.

1. Raise and support front of vehicle.
2. Remove:
   - ❏ RH front wheel.
   - ❏ RH splash guard.
   - ❏ Engine lower cover.
   - ❏ Auxiliary drive belt.
   - ❏ Flywheel housing lower cover.
3. Lock flywheel **1**. Use tool No.(-).0188.F.
4. Remove:
   - ❏ Crankshaft pulley bolt **2**.
   - ❏ Crankshaft pulley **3**. Use tool No.(-).0188.P.
   - ❏ Flywheel locking tool.
   - ❏ Lower torque reaction link.
5. Disconnect and seal off fuel pipes. Use tool No.(-).0188.T.
6. Support engine.
7. Remove:
   - ❏ EGR valve (if required).
   - ❏ PAS reservoir (if required).
   - ❏ RH engine mounting and bracket.
   - ❏ Timing belt covers **4**, **5** & **6**.
8. Refit upper timing belt cover bolt & spacer **7**.
   *NOTE: The timing belt cover bolt & spacer **7** is also a water pump mounting bolt.*
9. Turn crankshaft clockwise to setting position.
10. Insert timing pin in flywheel **8**.
    Tool No.(-).0188.Y.
11. Insert timing pin in camshaft sprocket **9**.
    Tool No.(-).0188.M.
12. Slacken tensioner bolt **10**.
13. Turn tensioner pulley clockwise away from belt. Use tool No.(-).0188.J2.
14. Lightly tighten tensioner bolt **10**.
15. Remove timing belt.

Installation

1. Ensure timing pins located correctly **8** & **9**.
2. Fit crankshaft sprocket locking tool.
   Tool No.(-).0188.Q2 **11**.
   *NOTE: Ensure crankshaft key remains against locking tool **11**.*
3. Secure belt to camshaft sprocket with clamp.
   Tool No.(-).0188.AD **12**.
4. Fit timing belt in clockwise direction. Ensure belt is taut between camshaft sprocket and fuel pump sprocket.

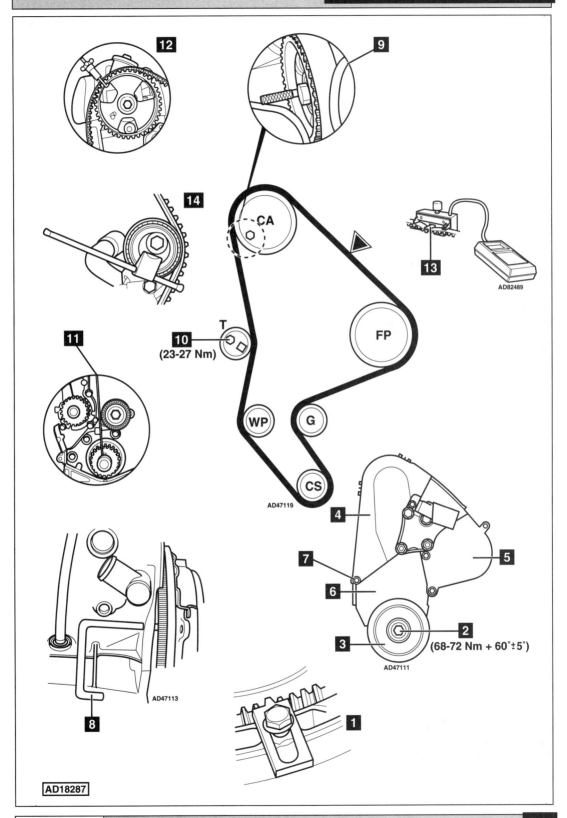

12

9

14

CA

13

AD82489

T

10
(23-27 Nm)

FP

11

WP

G

CS

AD47119

4

7

5

6

2

3

(68-72 Nm + 60°±5°)

AD47111

AD47113

8

1

AD18287

←

5. Attach tension gauge to belt at ▽ **13**.
Tool No.SEEM C.Tronic 105.5M.
6. Remove:
❑ Crankshaft sprocket locking tool **11**.
❑ Timing belt clamp **12**.
7. Turn tensioner pulley anti-clockwise, using tool
No.(-).0188.J2 **14**, until tension gauge indicates
96-100 SEEM units.
8. Tighten tensioner pulley bolt **10**.
Tightening torque: 23-27 Nm.
9. Lock flywheel **1**.
10. Install crankshaft pulley **3**.
11. Tighten crankshaft pulley bolt **2**.
Tightening torque: 63-77 Nm.
12. Remove:
❑ Tension gauge **13**.
❑ Flywheel timing pin **8**.
❑ Camshaft timing pin **9**.
❑ Flywheel locking tool **1**.
13. Turn crankshaft eight turns clockwise to setting
position.
14. Insert timing pin in flywheel **8**.
15. Insert timing pin in camshaft sprocket **9**.
16. Lock flywheel **1**.
17. Slacken:
❑ Crankshaft pulley bolt **2**.
❑ Tensioner pulley bolt **10**.
18. Attach tension gauge to belt at ▽ **13**.
19. Turn tensioner pulley clockwise, using tool
No.(-).0188.J2 **14**, until tension gauge indicates
52-56 SEEM units.
20. Tighten tensioner pulley bolt **10**.
Tightening torque: 23-27 Nm.
21. Remove tension gauge **13**.
22. Attach tension gauge to belt at ▽ **13**.
23. Check belt tension. Tension gauge should
indicate 51-57 SEEM units.
24. If not: Repeat tensioning procedure.
25. Remove:
❑ Tension gauge **13**.
❑ Flywheel timing pin **8**.
❑ Camshaft timing pin **9**.
❑ Flywheel locking tool **1**.
26. Turn crankshaft two turns clockwise to setting
position.
27. Insert timing pin in flywheel **8**.
28. Insert camshaft timing pin **9**.
NOTE: *If timing pins cannot be inserted: Repeat
installation procedure.*
29. Remove:
❑ Flywheel timing pin **8**.
❑ Camshaft timing pin **9**.
30. Install components in reverse order of removal.
31. Clean crankshaft pulley bolt and crankshaft
threads.

32. Coat crankshaft pulley bolt with suitable thread
locking compound.
33. Tighten crankshaft pulley bolt **2**.
Tightening torque: 68-72 Nm + 60°±5°.

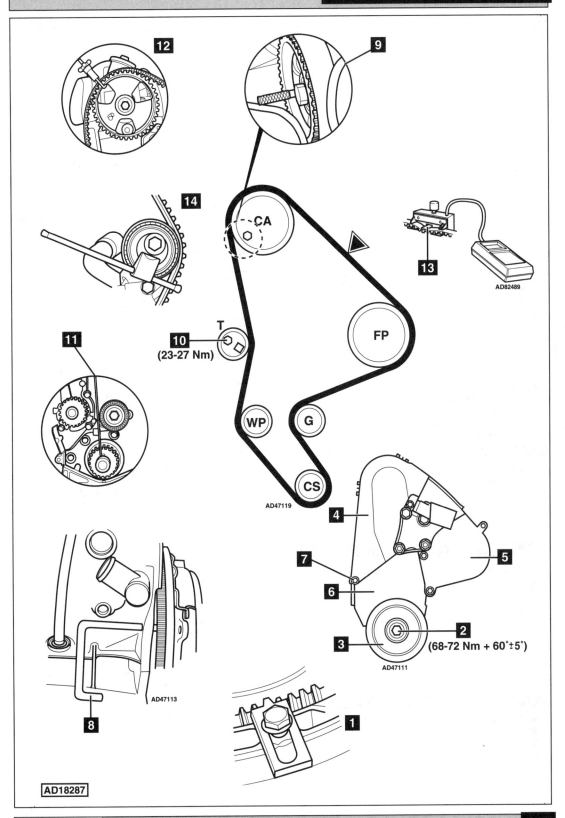

12

9

14

CA

13

AD82489

T

10
(23-27 Nm)

11

FP

WP

G

CS

AD47119

4

7

5

6

3

2
(68-72 Nm + 60˚±5˚)

AD47111

8

AD47113

1

AD18287

CITROEN

Model:	XM 2,1 TD • XM 2,2D • Xantia 2,1D Turbo • Dispatch/Jumpy 2,1D Turbo Synergie/Evasion 2,1D Turbo
Year:	1989-01
Engine Code:	P8A, P8B, P8C, P9A, PHZ (XUD11)

Replacement Interval Guide

Citroen recommend:
→1992MY: Replacement every 48,000 miles under normal conditions or every 40,000 miles under adverse conditions.
1993-94MY: Replacement every 72,000 miles under normal conditions or every 36,000 miles under adverse conditions.
1995MY: Replacement every 72,000 miles under normal conditions or every 55,000 miles under adverse conditions.
1996-98MY: Replacement every 72,000 miles under normal conditions or every 54,000 miles under adverse conditions.
1999MY: Replacement every 80,000 miles under normal conditions or every 54,000 miles under adverse conditions.
2000MY→: Replacement every 80,000 miles or every 10 years under normal conditions, or every 48,000 miles or every 10 years under adverse conditions.
The previous use and service history of the vehicle must always be taken into account.
Refer to Timing Belt Replacement Intervals at the front of this manual.

Check For Engine Damage

CAUTION: This engine has been identified as an INTERFERENCE engine in which the possibility of valve-to-piston damage in the event of a timing belt failure is MOST LIKELY to occur.
A compression check of all cylinders should be performed before removing the cylinder head.

Repair Times – hrs

Remove & install:

XM	4,50[1]

[1]Includes 0,70 for dynamic pump timing.

Xantia	3,80
AC	+0,40
Synergie/Evasion	2,70
AC	+0,10
Dispatch/Jumpy	2,70
AC	+0,10

Special Tools

- Flywheel timing pin – Citroen No.7014-TJ.
- Flywheel locking tool – Citroen No.9044-T.
- 1 bolt – M8 x 1,25 x 40 mm.
- 2 bolts – M8 x 1,25 x 35 mm.

Special Precautions

- Disconnect battery earth lead.
- DO NOT turn crankshaft or camshaft when timing belt removed.
- Remove glow plugs to ease turning engine.
- Turn engine in normal direction of rotation (unless otherwise stated).
- DO NOT turn engine via camshaft or other sprockets.
- Observe all tightening torques.
- Check diesel injection pump timing after belt replacement.

Removal

1. Raise and support front of vehicle.
2. Remove:
   ❏ Engine undershield (if fitted).
   ❏ RH front wheel.
   ❏ RH wheel arch liner.
   ❏ Coolant overflow hose from expansion tank.
   ❏ Auxiliary drive belts.
3. Remove clutch housing plate. Fit flywheel locking tool. Tool No.9044-T.
4. Remove crankshaft pulley **9**.
5. Support engine.
6. Remove:
   ❏ Torque reaction link.
   ❏ Upper engine mounting.
   ❏ Auxiliary drive belt guide pulley (if fitted).
   ❏ Timing belt upper and lower covers **7** & **8**.
   ❏ Timing belt centre cover (except Synergie/Evasion) **6**.
   ❏ Upper engine mounting bracket (if no access to bolt **5**).
   ❏ Flywheel locking tool.
7. Temporarily fit crankshaft pulley bolt without washer.
8. Turn crankshaft clockwise. Insert timing pin in flywheel **1**. Tool No.7014-T.J.
9. Insert locking bolts in camshaft and injection pump sprockets **2** & **3**. Tighten bolts finger tight.
   *NOTE: If locating holes are not aligned: Remove timing pin **1**. Turn crankshaft one turn clockwise.*
10. Slacken tensioner nut and eccentric bolt **4** & **5**.
    NOTE: Accessible through hole in engine mounting bracket (if not removed).
11. Turn eccentric clockwise to release tensioner.
12. Lightly tighten tensioner nut **4**.
13. Remove timing belt.

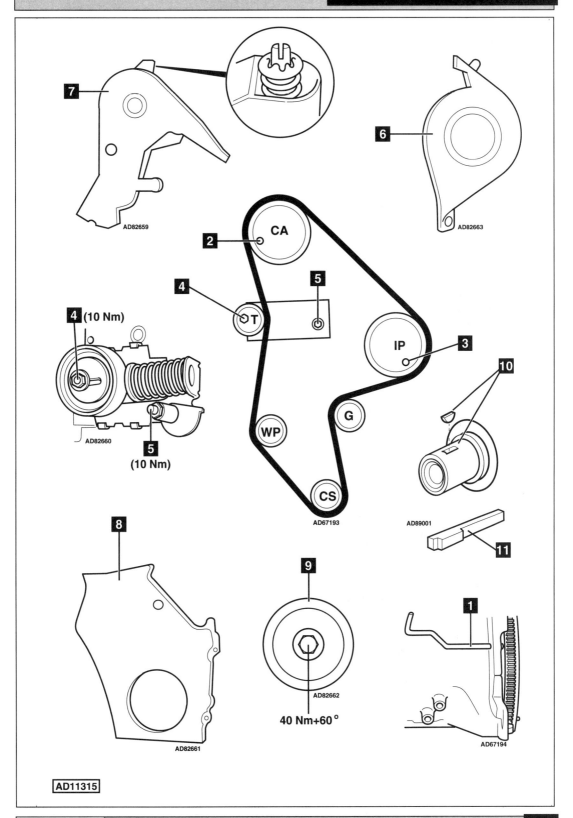

7 AD82659

6 AD82663

2 CA

5

4 **T**

4 (10 Nm)

AD82660

5 (10 Nm)

3 IP

10

WP

G

CS

AD67193

AD89001

11

8

AD82661

9

AD82662

40 Nm+60°

1

AD67194

AD11315

Installation

1. XM 1995→:
   - ❏ Remove crankshaft sprocket.
   - ❏ If crankshaft keyway and Woodruff key as shown **10**:
   - ❏ Replace Woodruff key with modified key **11**.

   NOTE: Align keyways in crankshaft and oil pump flange to allow fitment of modified key.

2. Ensure timing pin and locking bolts located correctly.

3. Check tensioner pulley is held in released position. Turn eccentric anti-clockwise **5**.

4. Fit timing belt in following order:
   - ❏ Injection pump sprocket.
   - ❏ Guide pulley.
   - ❏ Crankshaft sprocket.
   - ❏ Water pump sprocket.
   - ❏ Camshaft sprocket.
   - ❏ Tensioner pulley.

   NOTE: Ensure belt is taut between sprockets. Take care not to twist belt.

5. Slacken tensioner nut **4**. Remove timing pin and locking bolts.

6. Turn crankshaft two turns clockwise.

7. Ensure timing pin can be inserted easily **1**.

8. Lightly tighten tensioner nut **4**.

9. Turn crankshaft a further two turns clockwise.

10. Ensure timing pin can be inserted easily **1**.

11. Slacken tensioner nut one turn **4**.

12. Allow tensioner to operate.

13. Tighten tensioner nut to 10 Nm **4**.

14. Tighten eccentric bolt to 10 Nm **5**.

15. Ensure timing pin and locking bolts can be inserted easily.

16. Install:
    - ❏ Upper engine mounting bracket (if removed).
    - ❏ Timing belt covers **6**, **7** & **8**.
    - ❏ Flywheel locking tool. Tool No.9044-T.
    - ❏ Crankshaft pulley.

17. Except XM 1995→:
    - ❏ Clean crankshaft pulley bolt.
    - ❏ Coat threads with 'Loctite Threadlok'.
    - ❏ Fit bolt and washer.

18. XM 1995→:
    - ❏ Renew crankshaft pulley bolt and washer.
    - ❏ Coat threads with 'Loctite Threadlok'.
    - ❏ Fit bolt and washer.

19. Tighten crankshaft pulley bolt.
    Tightening torque: 40 Nm + 60°.

20. Remove flywheel locking tool. Fit clutch housing plate.

21. Install components in reverse order of removal.

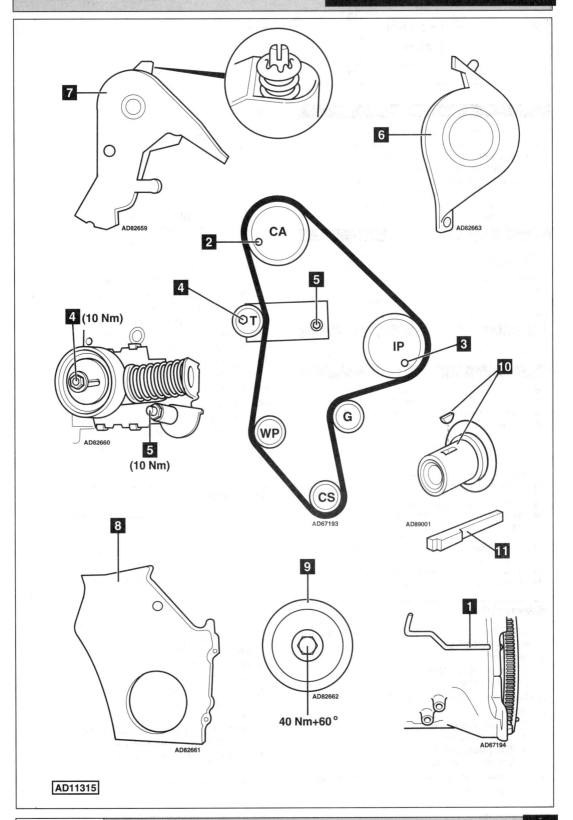

7 AD82659

6 AD82663

2 CA

4 **5**

4 (10 Nm)

AD82660

5 (10 Nm)

T **4**

IP **3**

G

WP

CS
AD67193

10

AD89001

11

8
AD82661

9
AD82662

40 Nm+60°

1
AD67194

AD11315

CITROEN

Model:	C5 2,2 HDi
Year:	2000-02
Engine Code:	4HX (DW12ATED)

Replacement Interval Guide

Citroen recommend replacement every 100,000 miles or every 10 years under normal conditions or 75,000 miles or every 10 years under adverse conditions.

The previous use and service history of the vehicle must always be taken into account.
Refer to Timing Belt Replacement Intervals at the front of this manual.

Check For Engine Damage

CAUTION: This engine has been identified as an INTERFERENCE engine in which the possibility of valve-to-piston damage in the event of a timing belt failure is MOST LIKELY to occur.
A compression check of all cylinders should be performed before removing the cylinder head(s).

Repair Times – hrs

Remove & install	3,50

Special Tools

- Flywheel locking tool – Citroen No.(-).188.F.
- Flywheel timing pin – Citroen No.(-).0188.X.
- Camshaft timing pin – Citroen No.(-).0188.M.
- Tensioning tool – Citroen No.(-).0188.J2.
- Timing belt retaining clip – Citroen No.(-).0188.K.
- Tension gauge – SEEM CTG 105.M.

Special Precautions

- Disconnect battery earth lead.
- DO NOT turn crankshaft or camshaft when timing belt removed.
- Remove glow plugs to ease turning engine.
- Turn engine in normal direction of rotation (unless otherwise stated).
- DO NOT turn engine via camshaft or other sprockets.
- Observe all tightening torques.

Removal

NOTE: The fuel pump fitted to this engine does not require timing.

1. Remove:
   - ❏ Engine upper cover.
   - ❏ Battery cover.
2. Raise and support front of vehicle.
3. Remove:
   - ❏ Engine lower cover.
   - ❏ RH front wheel.
   - ❏ RH splash guard.
   - ❏ Auxiliary drive belt.
4. Disconnect exhaust front pipe from manifold.
5. Support engine.

6. Remove:
   - ❏ Coolant expansion tank (leave hoses connected).
   - ❏ Torque reaction link.
   - ❏ RH engine mounting and bracket.
   - ❏ Flywheel housing lower cover.
7. Lock flywheel. Use tool No.(-).188.F.
8. Remove:
   - ❏ Crankshaft pulley bolt **1**.
   - ❏ Crankshaft pulley **2**.
   - ❏ Flywheel locking tool.
   - ❏ Bolt **3**.
   - ❏ Timing belt upper covers **4**.
   - ❏ Timing belt lower cover **5**.
9. Refit bolt fitted with a 17 mm thick spacer **3**. Tighten bolt to 15 Nm.
10. Turn crankshaft clockwise to setting position.
11. Insert timing pin in flywheel **6**. Tool No.(-).0188.X.
12. Insert timing pin in camshaft sprocket **7**. Tool No.(-).0188.M.
13. Slacken tensioner bolt **8**.
14. Slacken camshaft sprocket bolts **9**.
15. Turn tensioner pulley clockwise away from belt. Use tool No.(-).0188.J2.
16. Lightly tighten tensioner bolt **8**.
17. Remove timing belt.

Installation

SPECIAL NOTE: Supplementary information for engine/cylinder head overhaul.
*Align timing chain links (black) **10** & **11** with marked camshaft sprocket teeth **12** & **13**.*

1. Ensure timing pins located correctly **6** & **7**.
2. Tighten bolts finger tight **9**.
3. Turn camshaft sprocket fully clockwise in slotted holes.
4. Fit timing belt to crankshaft sprocket.
5. Secure belt to crankshaft sprocket with retaining clip. Tool No.(-).0188.K **14**.
6. Fit timing belt in anti-clockwise direction. Ensure belt is taut between sprockets.
7. Lay belt on camshaft sprocket teeth. Engage belt teeth by turning sprocket slightly anti-clockwise.
   *NOTE: Angular movement of sprocket must not be more than one tooth space **15**.*
8. Fit timing belt to water pump sprocket and tensioner pulley.
9. Slacken tensioner bolt **8**.

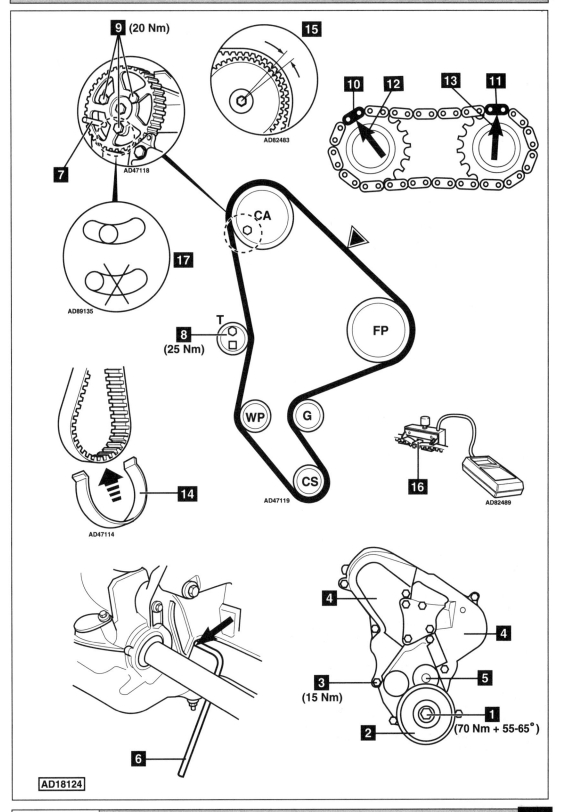

9 (20 Nm)

15

AD82483

10 **12** **13** **11**

7

AD47118

17

AD89135

CA

FP

T

8
(25 Nm)

WP **G**

16

AD82489

14

AD47114

CS

AD47119

4

4

3
(15 Nm)

5

1

2 (70 Nm + 55-65°)

6

AD18124

←

10. Turn tensioner pulley anti-clockwise to temporarily tension belt. Use tool No.(-).0188.J2.

11. Lightly tighten tensioner bolt .

12. Remove retaining clip .

13. Attach tension gauge to belt at ▽ . Tool No.SEEM CTG 105.M.

14. Turn tensioner pulley anti-clockwise until tension gauge indicates 106±2 SEEM units.

15. Tighten tensioner bolt . Tightening torque: 25 Nm.

16. Remove tension gauge .

17. Ensure sprocket bolts not at end of slotted holes .

18. If necessary: Repeat installation procedure.

19. Tighten camshaft sprocket bolts . Tightening torque: 20 Nm.

20. Remove timing pins & .

21. Turn crankshaft eight turns clockwise to setting position.

    NOTE: DO NOT allow crankshaft to turn anti-clockwise.

22. Insert timing pin .

23. Slacken camshaft sprocket bolts .

24. Insert timing pin .

25. Slacken tensioner bolt to release tension on belt .

26. Attach tension gauge to belt at ▽ .

27. Turn tensioner pulley anti-clockwise until tension gauge indicates 51±3 SEEM units.

28. Hold tensioner pulley in position. Tighten tensioner bolt . Tightening torque: 25 Nm.

29. Tighten camshaft sprocket bolts . Tightening torque: 20 Nm.

30. Remove tension gauge .

31. Check belt tension: Attach tension gauge to belt at ▽ . Tension gauge should indicate 51±3 SEEM units.

32. If not: Repeat tensioning procedure.

33. Remove tension gauge .

34. Remove timing pins & .

35. Turn crankshaft two turns clockwise to setting position.

36. Insert timing pin in flywheel .

37. Ensure timing pin can be inserted easily .

38. Remove timing pin .

39. Install components in reverse order of removal.

40. Remove bolt and 17 mm spacer. Refit bolt . Tighten bolt to 15 Nm .

41. Clean crankshaft pulley bolt and crankshaft threads.

42. Coat crankshaft pulley bolt with suitable thread locking compound.

43. Tighten crankshaft pulley bolt . Tightening torque: 70 Nm + 55-65°.

44. Check torque setting of crankshaft pulley bolt . Tightening torque: 260 Nm.

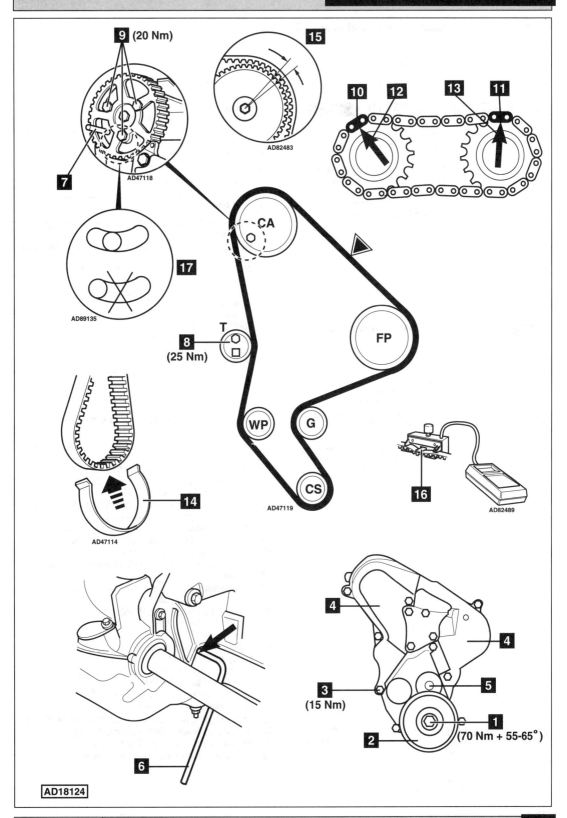

9 (20 Nm)

15

AD82483

10 **12** **13** **11**

7

AD47118

17

AD89135

CA

FP

T

8
(25 Nm)

WP   G

CS

AD47119

16

AD82489

14

AD47114

4

4

3
(15 Nm)

5

1
(70 Nm + 55-65°)

2

6

AD18124

Model:	Relay/Jumper 2,5D/Turbo D
Year:	1994-02
Engine Code:	T9A (DJ5), T8A/THZ (DJ5T)

Replacement Interval Guide

Citroen recommend:
→1995MY: Replacement every 72,000 miles under normal conditions or every 55,000 miles under adverse conditions.
1996-98MY: Replacement every 72,000 miles under normal conditions or every 54,000 miles under adverse conditions.
1999MY: Replacement every 80,000 miles under normal conditions or every 54,000 miles under adverse conditions.
2000MY→: Replacement every 80,000 miles or every 10 years under normal conditions, or every 48,000 miles or every 10 years under adverse conditions.
The previous use and service history of the vehicle must always be taken into account.
Refer to Timing Belt Replacement Intervals at the front of this manual.

Check For Engine Damage

CAUTION: This engine has been identified as an INTERFERENCE engine in which the possibility of valve-to-piston damage in the event of a timing belt failure is MOST LIKELY to occur.
A compression check of all cylinders should be performed before removing the cylinder head.

Repair Times – hrs

Remove & install	2,20

Special Tools

- Flywheel timing pin – Citroen No.7014-T.J.
- Camshaft timing pin – Citroen No.5711-T.A.
- Injection pump timing pin – Citroen No.5711-T.B or T.C.
- Tensioning lever – Citroen No.5711-T.E.
- Tension gauge – Citroen No.4099-T or 4122-T (SEEM C.Tronic).

Removal

1. Raise and support front of vehicle.
2. Remove:
   - ❏ RH front wheel.
   - ❏ Engine undershield.
   - ❏ Auxiliary drive belt.
   - ❏ Timing belt upper cover **1**.
   - ❏ Crankshaft pulley **2**.
   - ❏ Timing belt lower cover **3**.
3. Turn crankshaft clockwise until timing pin can be inserted in flywheel. Tool No.7014-T.J.
   - ❏ Type 1 – **4**.
   - ❏ Type 2 – **12**.
4. Insert timing pin in camshaft sprocket **5**. Tool No.5711-T.A.
5. Insert timing pin in injection pump sprocket **6**.
   NOTE: Bosch pump – use timing pin No. 5711-T.B (9,5 mm). Lucas pump – use timing pin No.5711-T.C (6 mm).
6. Slacken camshaft sprocket bolts **7**. Tighten bolts finger tight. Then slacken 60°.

7. Slacken injection pump sprocket bolts **8**. Tighten bolts finger tight. Then slacken 60°.
8. Slacken tensioner nut **9**. Turn tensioner clockwise. Lightly tighten nut.
9. Remove timing belt.
10. Turn camshaft and injection pump sprockets fully clockwise in slotted holes.

Installation

1. Ensure tensioner pulley and guide pulley in good condition.
2. Fit timing belt in anti-clockwise direction, starting at crankshaft sprocket.
   NOTE: If necessary: Turn crankshaft sprocket to engage in nearest belt tooth.
3. Attach tension gauge to belt at ▽ **10**. Tool No.4099-T/4122-T.
4. Insert tensioning lever in square hole in tensioner. Tool No.5711-T.E **11**.
5. Tension belt to 107 SEEM units. Tighten tensioner nut **9**. Tightening torque: 45 Nm.
6. Remove tension gauge.
7. Tighten camshaft sprocket bolts to 10 Nm and then finally to 25 Nm **7**.
8. Tighten injection pump sprocket bolts to 10 Nm and then finally to 25 Nm **8**.
9. Remove timing pins **4** or **12**, **5** & **6**.
10. Turn crankshaft 10 turns clockwise.
11. Insert flywheel timing pin. Turn crankshaft clockwise until timing pin engages.
    NOTE: DO NOT turn crankshaft anti-clockwise.
12. Slacken camshaft sprocket bolts **7**. Tighten bolts finger tight. Then slacken 60°.
13. Slacken injection pump sprocket bolts **8**. Tighten bolts finger tight. Then slacken 60°.
14. Slacken tensioner nut to release tension on belt.
15. Insert timing pin in camshaft sprocket **5**.
16. Insert timing pin in injection pump sprocket **6**.
    NOTE: Turn sprockets with a spanner if necessary.
17. Attach tension gauge to belt at ▽ **10**.
18. Insert tensioning lever in square hole in tensioner. Tool No.5711-T.E **11**.
19. Tension belt to 58 SEEM units. Tighten tensioner nut **9**. Tightening torque: 45 Nm.
20. Remove tension gauge.
21. Tighten camshaft sprocket bolts to 10 Nm and then finally to 25 Nm **7**.
22. Tighten injection pump sprocket bolts to 10 Nm and then finally to 25 Nm **8**.
23. Remove timing pins **4** or **12**, **5** & **6**.
24. Turn crankshaft two turns clockwise.
25. Ensure timing pins can be inserted easily **4**, **5** & **6**.
26. Check and adjust injection pump timing.
27. Fit crankshaft pulley **2**. Tighten bolts to 20 Nm.
28. Install components in reverse order of removal.

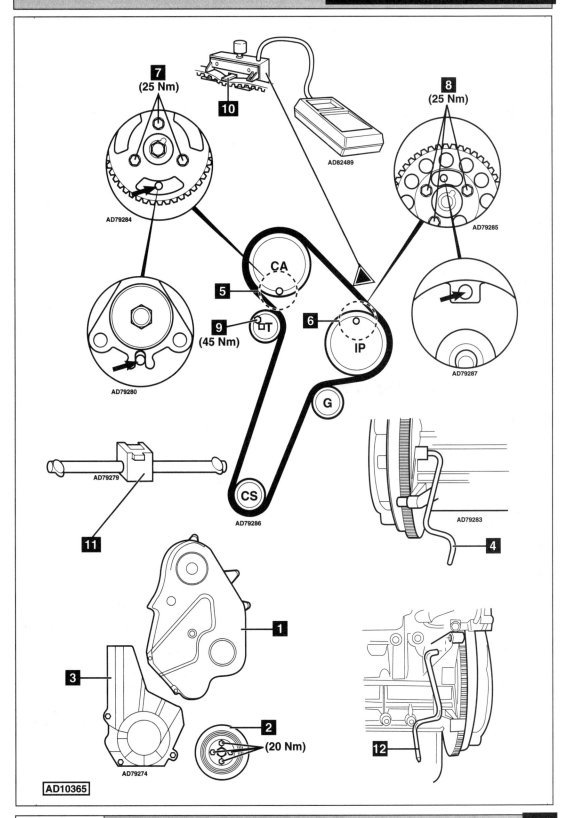

7 (25 Nm)

AD79284

10

AD82489

8 (25 Nm)

AD79285

5

CA

9 (45 Nm)

T

6

IP

AD79287

G

AD79280

CS

AD79286

AD79279

11

AD79283

4

3

1

2 (20 Nm)

AD79274

12

AD10365

CITROEN

Model:	**Relay/Jumper 2,5D Turbo**
Year:	**1997-02**
Engine Code:	**THX (DJ5TED)**

Replacement Interval Guide

Citroen recommend:
→1998MY: Replacement every 80,000 miles under normal conditions or 60,000 miles under adverse conditions.
1999MY: Replacement every 75,000 miles under normal conditions or 60,000 miles under adverse conditions.
2000MY→: Replacement every 75,000 miles or every 10 years under normal conditions, or every 60,000 miles or every 10 years under adverse conditions.
The previous use and service history of the vehicle must always be taken into account.
Refer to Timing Belt Replacement Intervals at the front of this manual.

Check For Engine Damage

CAUTION: This engine has been identified as an INTERFERENCE engine in which the possibility of valve-to-piston damage in the event of a timing belt failure is MOST LIKELY to occur.
A compression check of all cylinders should be performed before removing the cylinder head(s).

Repair Times – hrs

Remove & install	3,00

Special Tools

- Flywheel timing pin – Citroen No.7014-T.J.
- Camshaft timing pin – Citroen No.5711-T.A or T.C.
- Injection pump timing pin – Citroen No.5711-T.B.
- Tensioning lever – Citroen No.5711-T.E.
- Tension gauge – Citroen No.4122-T
  (SEEM C.Tronic 105.5).

Special Precautions

- Disconnect battery earth lead.
- DO NOT turn crankshaft or camshaft when timing belt removed.
- Remove glow plugs to ease turning engine.
- Turn engine in normal direction of rotation (unless otherwise stated).
- DO NOT turn engine via camshaft or other sprockets.
- Observe all tightening torques.
- Check diesel injection pump timing after belt replacement.

Removal

1. Raise and support front of vehicle.
2. Remove:
   - ❏ RH front wheel.
   - ❏ Engine undershield.
   - ❏ Auxiliary drive belt.
   - ❏ Timing belt upper cover **1**.
   - ❏ Crankshaft pulley bolts **2**.
   - ❏ Crankshaft pulley **3**.
   - ❏ Timing belt lower cover **4**.
3. Turn crankshaft clockwise until timing pin can be inserted in flywheel **5**. Tool No.7014-T.J.
   NOTE: Position of timing pin hole varies. Type 1: Behind starter motor. Type 2: Under turbocharger.
4. Insert timing pin in camshaft sprocket **8**. Tool No.5711-T.A/T.C.
5. Insert timing pin in injection pump sprocket **9**. Tool No.5711-T.B.
6. Turn automatic tensioner unit anti-clockwise **6**.
7. Insert suitable pin in automatic tensioner unit to retain pushrod **7**.
8. Slacken camshaft sprocket bolts **10**.
9. Slacken injection pump sprocket bolts **11**.
10. Slacken tensioner nut **12**.
11. Turn tensioner anti-clockwise. Lightly tighten nut.
12. Remove timing belt.

Installation

1. Turn camshaft and injection pump sprockets fully anti-clockwise in slotted holes.
2. Tighten bolts finger tight **10** & **11**.
3. Ensure tensioner pulleys in good condition.
4. Fit timing belt in clockwise direction, starting at crankshaft sprocket.
   NOTE: If necessary: Turn camshaft and injection pump sprockets slightly clockwise to engage in nearest belt tooth. DO NOT exceed one tooth.
5. Turn tensioner pulley **12** to ensure camshaft and injection pump sprockets turn freely.
6. Attach tension gauge to belt at ▽ **13**. Use tool No.4122-T.
7. Insert tensioning lever in square hole in tensioner **14**. Tool No.5711-T.E
8. Tension belt to 41±2 SEEM units.
9. Tighten tensioner nut to 45 Nm **12**.
10. Remove tension gauge.
11. Remove locking pin from automatic tensioner unit **7**.
12. Wait 1 minute.
13. Ensure camshaft and injection pump sprocket bolts are not at end of slotted holes.
14. Tighten camshaft sprocket bolts to 5 Nm and then finally to 23 Nm **10**.
15. Tighten injection pump sprocket bolts to 5 Nm and then finally to 23 Nm **11**.
16. Remove timing pins **5**, **8** & **9**.
17. Turn crankshaft 10 turns clockwise.
18. Insert flywheel timing pin **5**.
19. Insert timing pins in camshaft and injection pump sprockets **8** & **9**.
    NOTE: DO NOT turn crankshaft anti-clockwise.
20. If timing pins cannot be inserted: Repeat installation procedure.
21. Install components in reverse order of removal.

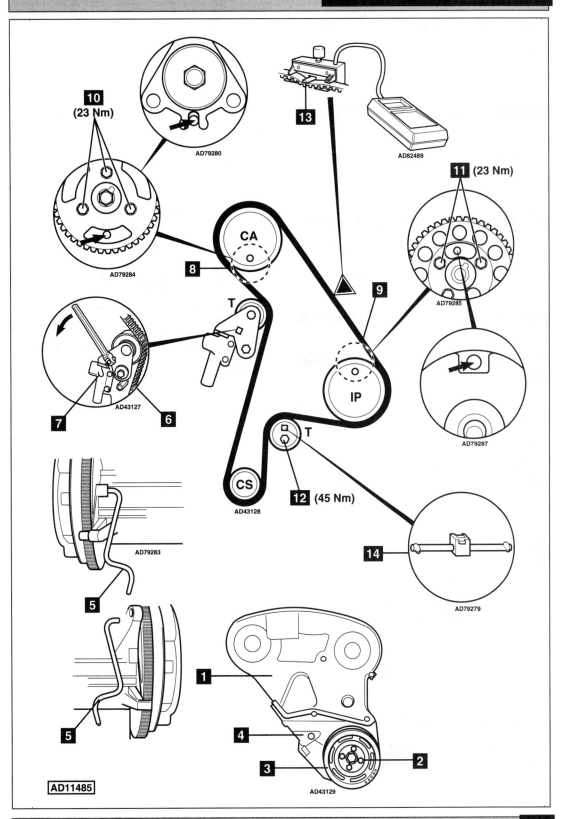

10 (23 Nm)

AD79280

AD79284

13

AD82489

11 (23 Nm)

AD79285

CA

8

9

T

7

AD43127

6

IP

AD79287

T

12 (45 Nm)

CS

AD43128

AD79283

5

14

AD79279

5

1

4

3

2

AD43129

AD11485

DAEWOO

Model:	**Matiz**
Year:	**1998-03**
Engine Code:	**F8CV**

Replacement Interval Guide

Daewoo recommend check and replacement if necessary every 20,000 miles or 2 years. Replacement every 60,000 miles or 6 years.
The previous use and service history of the vehicle must always be taken into account.
Refer to Timing Belt Replacement Intervals at the front of this manual.

Check For Engine Damage

CAUTION: This engine has been identified as an INTERFERENCE engine in which the possibility of valve-to-piston damage in the event of a timing belt failure is MOST LIKELY to occur. A compression check of all cylinders should be performed before removing the cylinder head(s).

Repair Times – hrs

Remove & install	1,10

Special Tools

- None required.

Special Precautions

- Disconnect battery earth lead.
- DO NOT turn crankshaft or camshaft when timing belt removed.
- Remove spark plugs to ease turning engine.
- Turn engine in normal direction of rotation (unless otherwise stated).
- DO NOT turn engine via camshaft or other sprockets.
- Observe all tightening torques.

Removal

1. Remove:
   - RH headlamp.
   - Auxiliary drive belt(s).
   - Timing belt upper cover **1**.
2. Raise and support front of vehicle.
3. Remove:
   - RH front wheel.
4. Slacken crankshaft pulley bolt **2**.
5. Turn crankshaft clockwise to TDC on No.1 cylinder. Ensure timing marks aligned **3** & **4**.
6. Remove:
   - Crankshaft pulley bolt **2**.
   - Crankshaft pulley **5**.
   - Dipstick and tube.
   - Timing belt lower cover **6**.
7. Ensure timing marks aligned **7**.

8. Slacken tensioner bolt **8**. Push tensioner away from belt. Lightly tighten bolt.
9. Remove timing belt.

Installation

1. Ensure timing marks aligned **4** & **7**.
2. Fit timing belt in anti-clockwise direction, starting at crankshaft sprocket. Ensure belt is taut on non-tensioned side.
3. Slacken tensioner bolt **8**. Allow tensioner to operate.
4. Turn crankshaft slowly two turns clockwise to TDC on No.1 cylinder.
5. Ensure timing marks aligned **4** & **7**.
6. Tighten tensioner bolt **8**.
   Tightening torque: 15-23 Nm.
7. Install components in reverse order of removal.
8. Tighten crankshaft pulley bolt **2**.
   Tightening torque: 65-75 Nm.

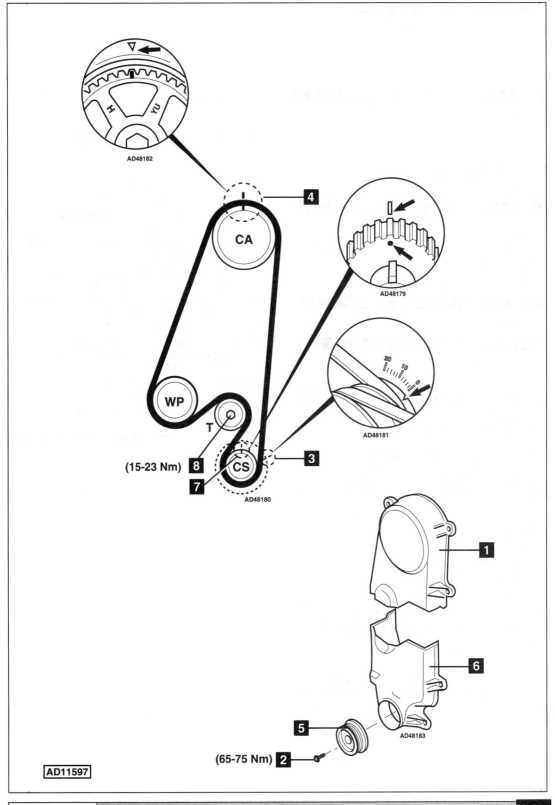

AD48182

4

CA

AD48179

WP

T

AD48181

(15-23 Nm) 8

CS

3

7

AD48180

1

6

5

AD48183

(65-75 Nm) 2

AD11597

Model:	**Nexia 1,5 SOHC**
Year:	**1995-97**
Engine Code:	**L4 SOHC**

Replacement Interval Guide

Daewoo recommend check & adjust every 20,000 miles or 2 years and replacement every 40,000 miles or 4 years.
The previous use and service history of the vehicle must always be taken into account.
Refer to Timing Belt Replacement Intervals at the front of this manual.

Check For Engine Damage

CAUTION: This engine has been identified as a FREEWHEELING engine in which the possibility of valve-to-piston damage in the event of a timing belt failure may be minimal or very unlikely. However, a precautionary compression check of all cylinders should be performed.

Repair Times – hrs

Remove & install	0,90

Special Tools

■ Water pump wrench – No.KM-421-A.

Special Precautions

■ Disconnect battery earth lead.
■ DO NOT turn crankshaft or camshaft when timing belt removed.
■ Remove spark plugs to ease turning engine.
■ Turn engine in normal direction of rotation (unless otherwise stated).
■ DO NOT turn engine via camshaft or other sprockets.
■ Observe all tightening torques.

Removal

1. Remove:
   ❑ Air filter and air intake hose.
   ❑ Auxiliary drive belt(s).
   ❑ PAS pump.
   ❑ Timing belt cover **1**.
2. Turn crankshaft to 10° BTDC on No.1 cylinder **3**. Ensure timing marks aligned **4**.
3. Remove:
   ❑ Crankshaft pulley bolts **5**.
   ❑ Crankshaft pulley **6**.
4. Turn movable part of tensioner until holes aligned **7**. Insert suitable pin to hold tensioner.
5. Remove timing belt.

Installation

1. Ensure timing marks aligned **3** & **4**.
2. Fit timing belt in anti-clockwise direction, starting at crankshaft sprocket. Ensure belt is taut between sprockets.
3. Install:
   ❑ Crankshaft pulley **6**.
   ❑ Crankshaft pulley bolts **5**.
4. Remove locking pin from tensioner **7**.
5. Slacken water pump bolts **8**.
6. Turn water pump clockwise to tension belt **9**. Use tool KM-421-A.
7. Tensioner pointer must be against stop **10**.
8. Lightly tighten water pump bolts **8**.
9. Turn crankshaft two turns clockwise. Ensure timing marks aligned **3** & **4**.
10. Slacken water pump bolts **8**.
11. Turn water pump anti-clockwise **11** until tensioner pointer aligned with notch **12**. Use tool KM-421-A.
12. Tighten water pump bolts **8**.
    Tightening torque: 8 Nm.
13. If tensioner pointer not aligned: Repeat tensioning procedure.
14. Install components in reverse order of removal.

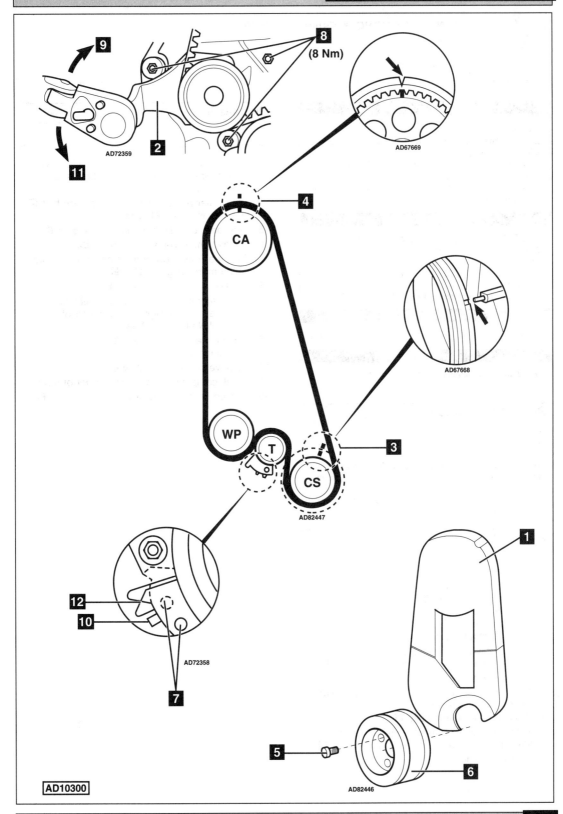

DAEWOO

Model:	**Nexia 1,5 DOHC • Espero 1,5 DOHC**
Year:	**1995-97**
Engine Code:	**L4 DOHC**

Replacement Interval Guide

Daewoo recommend check & adjust every 20,000 miles or 2 years and replacement every 40,000 miles or 4 years.
The previous use and service history of the vehicle must always be taken into account.
Refer to Timing Belt Replacement Intervals at the front of this manual.

Check For Engine Damage

CAUTION: This engine has been identified as an INTERFERENCE engine in which the possibility of valve-to-piston damage in the event of a timing belt failure is MOST LIKELY to occur.
A compression check of all cylinders should be performed before removing the cylinder head(s).

Repair Times – hrs

Remove & install	2,00

Special Tools

■ Water pump wrench – No.KM-421-A.

Special Precautions

■ Disconnect battery earth lead.
■ DO NOT turn crankshaft or camshaft when timing belt removed.
■ Remove spark plugs to ease turning engine.
■ Turn engine in normal direction of rotation (unless otherwise stated).
■ DO NOT turn engine via camshaft or other sprockets.
■ Observe all tightening torques.

Removal

1. Remove:
   ❏ Auxiliary drive belt(s).
   ❏ PAS pump pulley.
   ❏ Timing belt cover **1**.
   ❏ PAS pump.
   ❏ Crankshaft pulley bolt and washer **2**.
   ❏ Crankshaft pulley **3**.
2. Temporarily fit crankshaft pulley bolt. Turn crankshaft to 10° BTDC on No.1 cylinder **4**.
3. Ensure timing marks aligned **5**.
   NOTE: Some engines may only have one timing mark on each sprocket.
4. Slacken water pump bolts **6**.
5. Turn water pump anti-clockwise to release tension on belt **8**. Use tool No.KM-421-A **7**.
6. Remove timing belt.

Installation

1. Ensure timing marks aligned **5**.
2. Ensure crankshaft timing mark at 10° BTDC **4**.
3. Fit timing belt in anti-clockwise direction, starting at crankshaft sprocket. Ensure belt is taut between sprockets.
4. Turn water pump clockwise to tension belt **9**. Use tool No.KM-421-A **7**.
5. Tensioner pointer must be against stop **10**.
6. Lightly tighten water pump bolts **6**.
7. Turn crankshaft two turns clockwise. Ensure timing marks aligned **4** & **5**.
8. Slacken water pump bolts **6**.
9. Turn water pump anti-clockwise **8** until tensioner pointer aligned with notch **11**. Use tool No.KM-421-A **7**.
10. Tighten water pump bolts **6**. Tightening torque: 8 Nm.
11. Remove crankshaft pulley bolt.
12. Install components in reverse order of removal.
13. Tighten crankshaft pulley bolt to 155 Nm **2**.

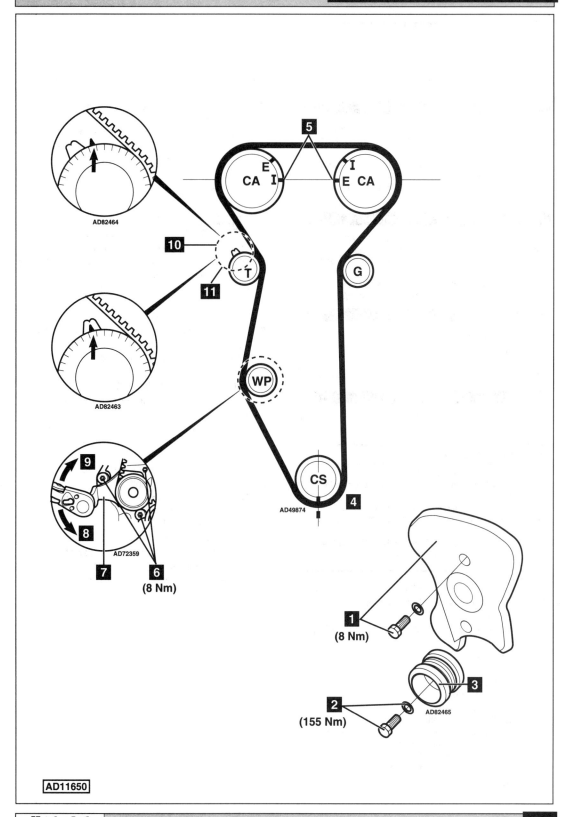

Model:	**Lanos 1,4 • Lanos 1,5**
Year:	**1997-03**
Engine Code:	**A13DM, A15DM**

Replacement Interval Guide

Daewoo recommend check and adjust every 30,000 miles or 3 years and replacement every 40,000 miles or 4 years.
The previous use and service history of the vehicle must always be taken into account.
Refer to Timing Belt Replacement Intervals at the front of this manual.

Check For Engine Damage

CAUTION: This engine has been identified as an INTERFERENCE engine in which the possibility of valve-to-piston damage in the event of a timing belt failure is MOST LIKELY to occur.
A compression check of all cylinders should be performed before removing the cylinder head(s).

Repair Times – hrs

Check & adjust	1,40
Remove & install	1,10
AC	+0,10
PAS	+0,20

Special Tools

- Water pump wrench – No.J-42492.

Special Precautions

- Disconnect battery earth lead.
- DO NOT turn crankshaft or camshaft when timing belt removed.
- Remove spark plugs to ease turning engine.
- Turn engine in normal direction of rotation (unless otherwise stated).
- DO NOT turn engine via camshaft or other sprockets.
- Observe all tightening torques.

Removal

1. Raise and support front of vehicle.
2. Disconnect:
   - ❏ Intake air temperature (IAT) sensor wiring.
   - ❏ Breather hose.
   - ❏ Intake air duct.
3. Remove:
   - ❏ Air filter housing.
   - ❏ RH front wheel.
   - ❏ RH splash guard.
   - ❏ Auxiliary drive belts.
   - ❏ PAS pump. DO NOT disconnect hoses.
   - ❏ Crankshaft pulley bolt **1**.
   - ❏ Crankshaft pulley **2**.
   - ❏ Timing belt covers **3** & **4**.
4. Fit crankshaft pulley bolt.

5. Turn crankshaft clockwise until timing marks aligned **5** & **6**.
6. Slacken water pump bolts slightly **7**.
7. Turn water pump anti-clockwise to release tension on belt **8**. Use tool No.J-42492.
8. Lightly tighten water pump bolts **7**.
9. Remove timing belt.

Installation

1. Ensure timing marks aligned **5** & **6**.
2. Fit timing belt in anti-clockwise direction, starting at crankshaft sprocket.
3. Ensure belt is taut between sprockets on non-tensioned side.
4. Slacken water pump bolts **7**.
5. Turn water pump clockwise to tension belt **8**. Use tool No.J-42492.
6. Movable part of tensioner must be against stop **9**.
7. Tighten water pump bolts **7**.
8. Turn movable part of tensioner until holes aligned **10**.
9. Insert 4,5 mm pin through tensioner holes **10**.
10. Turn crankshaft two turns clockwise until timing marks aligned **5** & **6**.
11. Remove 4,5 mm pin from tensioner **10**.
12. Slacken water pump bolts **7**.
13. Turn water pump anti-clockwise until tensioner pointer aligned with notch in support plate **11**. Use tool No.J-42492.
14. Tighten water pump bolts to 10 Nm **7**.
15. Install components in reverse order of removal.
16. Tighten crankshaft pulley bolt **1**.
    Tightening torque: 95 Nm + 30° + 15°.

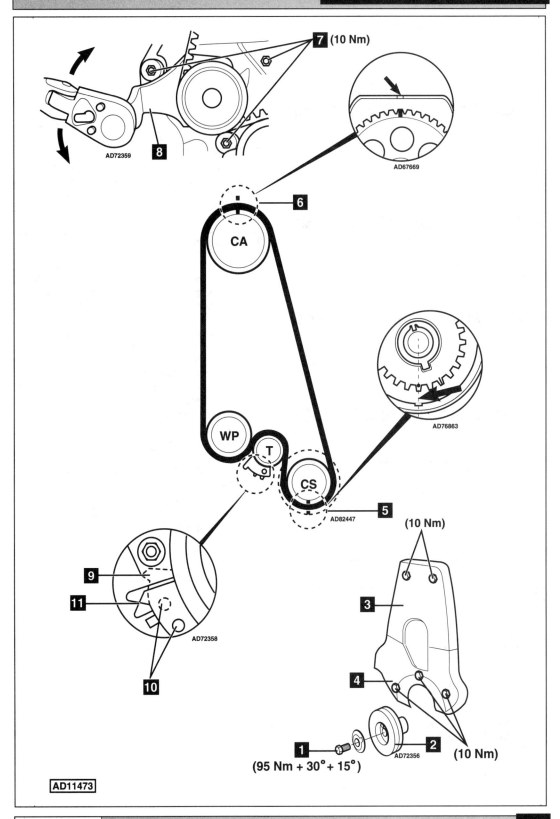

7 (10 Nm)

AD72359

8

AD67669

6

CA

AD76863

WP

T

CS

AD82447

5

(10 Nm)

9

11

10

AD72358

3

4

(10 Nm)

1

2

AD72356

(10 Nm)

(95 Nm + 30° + 15°)

AD11473

DAEWOO

Model:	**Lanos 1,6**
Year:	**1997-03**
Engine Code:	**A16DM**

Replacement Interval Guide

Daewoo recommend check and adjust every 30,000 miles or 3 years and replacement every 40,000 miles or 4 years.
The previous use and service history of the vehicle must always be taken into account.
Refer to Timing Belt Replacement Intervals at the front of this manual.

Check For Engine Damage

CAUTION: This engine has been identified as an INTERFERENCE engine in which the possibility of valve-to-piston damage in the event of a timing belt failure is MOST LIKELY to occur.
A compression check of all cylinders should be performed before removing the cylinder head(s).

Repair Times – hrs

Check & adjust	1,70
AC	+0,10
Remove & install	1,10
AC	+0,10
PAS	+0,60

Special Tools

- Water pump wrench – No.J-42492.

Special Precautions

- Disconnect battery earth lead.
- DO NOT turn crankshaft or camshaft when timing belt removed.
- Remove spark plugs to ease turning engine.
- Turn engine in normal direction of rotation (unless otherwise stated).
- DO NOT turn engine via camshaft or other sprockets.
- Observe all tightening torques.

Removal

1. Remove:
   - ❏ Air filter assembly.
   - ❏ Air intake hose.
2. Raise and support front of vehicle.
3. Remove:
   - ❏ RH front wheel.
   - ❏ RH splash guard.
   - ❏ Auxiliary drive belts.
   - ❏ PAS pump pulley.
   - ❏ Crankshaft pulley bolt **1**.
   - ❏ Crankshaft pulley **2**.
   - ❏ Timing belt upper cover **3**.
   - ❏ Timing belt lower cover **4**.
   - ❏ PAS pump bolts. DO NOT disconnect hoses.

4. Fit crankshaft pulley bolt.
5. Turn crankshaft clockwise until timing marks aligned **5**, **6** & **7**.
6. Slacken water pump bolts **8**.
7. Turn water pump anti-clockwise to release tension on belt **9**. Use tool No.J-42492.
8. Remove timing belt from behind PAS pump (if fitted).

Installation

1. Ensure timing marks aligned **5**, **6** & **7**.
2. Slide timing belt behind PAS pump (if fitted).
3. Fit timing belt in anti-clockwise direction, starting at crankshaft sprocket. Ensure belt is taut between sprockets.
4. Turn water pump clockwise **9** until tensioner pointer aligned with notch in tensioner bracket **10**. Use tool No.J-42492.
5. Tighten water pump bolts **8**.
6. Turn crankshaft two turns clockwise until timing marks aligned **5**, **6** & **7**.
7. Slacken water pump bolts **8**.
8. Turn water pump anti-clockwise **9** until tensioner pointer aligned with pointer on tensioner bracket **11**. Use tool No.J-42492.
9. Tighten water pump bolts to 10 Nm **8**.
10. Install components in reverse order of removal.
11. Tighten PAS pump bolts to 25 Nm.
12. Tighten crankshaft pulley bolt **1**.
    Tightening torque: 95 Nm + 30° + 15°.

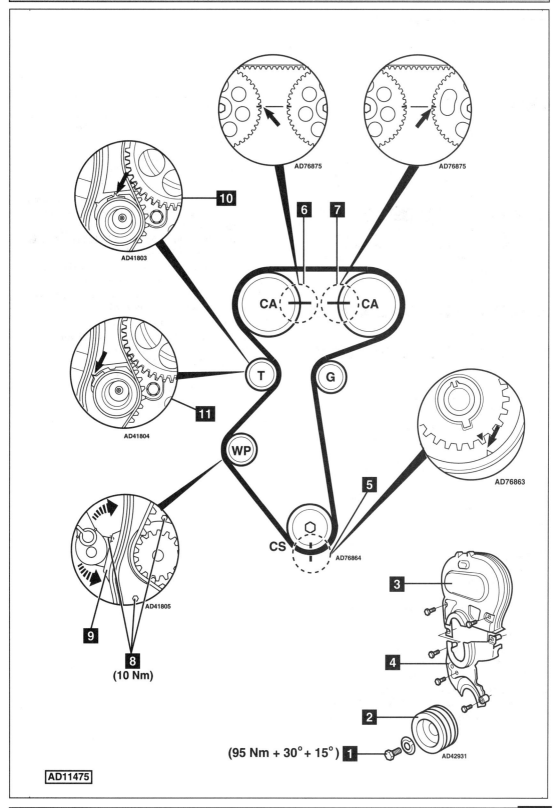

AD76875

AD76875

10

6 **7**

AD41803

CA — — CA

T **G**

11

AD41804

WP

AD76863

5

AD41805

9

CS — AD76864

8
(10 Nm)

3

4

2

(95 Nm + 30° + 15°) **1**

AD42931

AD11475

DAEWOO

Model:	**Espero 1,8/2,0**
Year:	**1995-97**
Engine Code:	**OHC L-4**

Replacement Interval Guide

Daewoo recommend check & adjust every 20,000 miles or 2 years and replacement every 40,000 miles or 4 years.
The previous use and service history of the vehicle must always be taken into account.
Refer to Timing Belt Replacement Intervals at the front of this manual.

Check For Engine Damage

CAUTION: This engine has been identified as a FREEWHEELING engine in which the possibility of valve-to-piston damage in the event of a timing belt failure may be minimal or very unlikely. However, a precautionary compression check of all cylinders should be performed.

Repair Times – hrs

Remove & install   0,90

Special Tools

■ Water pump wrench – MKM-472.

Special Precautions

■ Disconnect battery earth lead.
■ DO NOT turn crankshaft or camshaft when timing belt removed.
■ Remove spark plugs to ease turning engine.
■ Turn engine in normal direction of rotation (unless otherwise stated).
■ DO NOT turn engine via camshaft or other sprockets.
■ Observe all tightening torques.

Removal

1. Remove:
   ❏ Auxiliary drive belt(s).
   ❏ PAS pump/tensioner assembly.
   ❏ Timing belt cover **1**.
2. Turn crankshaft clockwise until timing marks aligned **2** & **3**.
3. Remove:
   ❏ Crankshaft pulley bolts **4**.
   ❏ Crankshaft pulley **5**.
4. Slacken water pump bolts **6**.
5. Turn water pump anti-clockwise to release tension on belt **8**. Use tool No.MKM-472 **7**.
6. Remove timing belt.

Installation

1. Fit timing belt to crankshaft sprocket.
2. Fit crankshaft pulley **5**.
3. Fit crankshaft pulley bolts.
   Tightening torque: 17 Nm **4**.
4. Ensure timing marks aligned **2** & **3**.
5. Fit timing belt in anti-clockwise direction, starting at crankshaft sprocket. Ensure belt is taut between sprockets.
6. Turn water pump clockwise to tension belt **9**. Use tool No.MKM-472 **7**.
7. Tensioner pointer must be against stop **11**.
8. Lightly tighten water pump bolts **6**.
9. Turn crankshaft two turns clockwise. Ensure timing marks aligned **2** & **3**.
10. Slacken water pump bolts **6**.
11. Turn water pump anti-clockwise **8** until tensioner holes aligned **10**.
    Use tool No.MKM-472 **7**.
12. Tighten water pump bolts **6**.
    Tightening torque: 25 Nm.
13. Install components in reverse order of removal.
14. Tighten crankshaft pulley bolts to 17 Nm **4**.

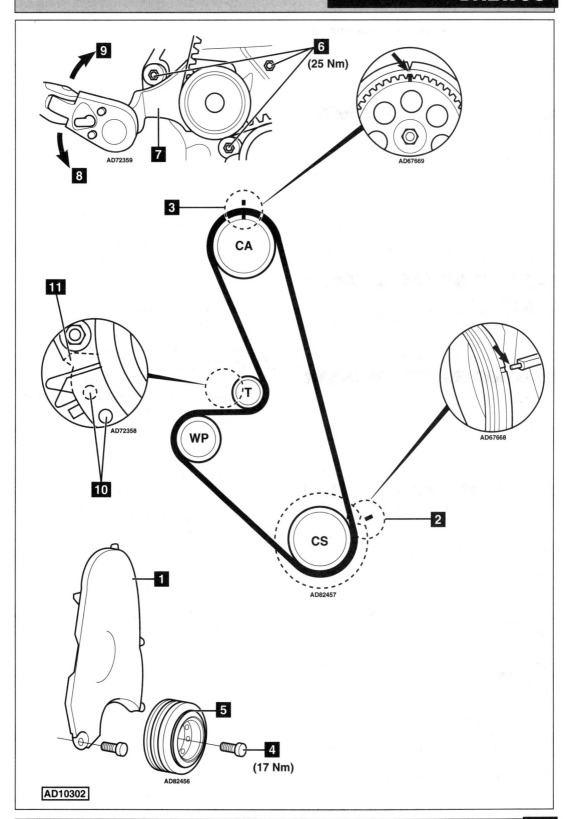

9

6
(25 Nm)

AD72359

7

8

AD67669

3

CA

11

T

WP

AD72358

10

AD67668

2

CS

AD82457

1

5

4
(17 Nm)

AD82456

AD10302

Model:	Nubira 1,6 • Tacuma 1,6
Year:	1997-03
Engine Code:	A16DM, F16S

Replacement Interval Guide

Daewoo recommend:

Nubira:
Check and adjust every 30,000 miles or 3 years and replacement every 40,000 miles or 4 years.

Tacuma:
Check and adjust every 20,000 miles or 2 years and replacement every 40,000 miles or 4 years.
The previous use and service history of the vehicle must always be taken into account.
Refer to Timing Belt Replacement Intervals at the front of this manual.

Check For Engine Damage

CAUTION: This engine has been identified as an INTERFERENCE engine in which the possibility of valve-to-piston damage in the event of a timing belt failure is MOST LIKELY to occur.
A compression check of all cylinders should be performed before removing the cylinder head(s).

Repair Times – hrs

Check & adjust:	
Nubira	1,10
Tacuma	1,30
Remove & install:	
Nubira	1,10
Tacuma	1,30

Special Tools

- Water pump wrench – No.J-42492.

Special Precautions

- Disconnect battery earth lead.
- DO NOT turn crankshaft or camshaft when timing belt removed.
- Remove spark plugs to ease turning engine.
- Turn engine in normal direction of rotation (unless otherwise stated).
- DO NOT turn engine via camshaft or other sprockets.
- Observe all tightening torques.

Removal

1. Remove:
   - Air filter assembly.
   - Air intake hose.
2. Disconnect:
   - Intake air temperature (IAT) sensor wiring.
   - Breather hose.
3. Raise and support front of vehicle.

4. Remove:
   - RH front wheel.
   - RH splash guard.
   - Auxiliary drive belt(s).
   - Crankshaft pulley bolts **1**.
   - Crankshaft pulley **2**.
   - Timing belt upper cover **3**.
   - Timing belt lower cover **4**.
5. Turn crankshaft clockwise until timing marks aligned **5**, **6** & **7**.
6. Slacken water pump bolts **8**.
7. Turn water pump anti-clockwise to release tension on belt **9**. Use tool No.J-42492.
8. Support engine.
9. Remove:
   - RH engine mounting.
   - Timing belt.

Installation

1. Ensure timing marks aligned **5**, **6** & **7**.
2. Fit timing belt in anti-clockwise direction, starting at crankshaft sprocket. Ensure belt is taut between sprockets.
3. Fit RH engine mounting.
4. Turn water pump clockwise **9** until tensioner pointer aligned with notch in tensioner bracket **10**. Use tool No.J-42492.
5. Tighten water pump bolts **8**.
6. Turn crankshaft two turns clockwise until timing marks aligned **5**, **6** & **7**.
7. Slacken water pump bolts **8**.
8. Turn water pump anti-clockwise **9** until tensioner pointer aligned with pointer on tensioner bracket **11**. Use tool No.J-42492.
9. Tighten water pump bolts to 10 Nm **8**.
10. Install components in reverse order of removal.
11. Tighten crankshaft pulley bolts **1**.
    Tightening torque: 20 Nm.

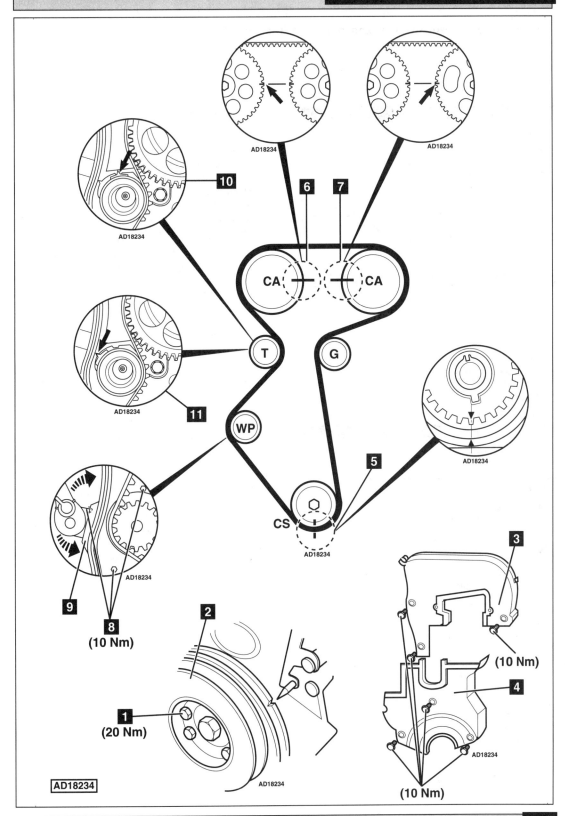

Model:	Nubira 2,0 • Leganza 2,0 • Tacuma 2,0
Year:	1997-03
Engine Code:	C20SE, C20SED, X20SED

Replacement Interval Guide

Daewoo recommend:

Nubira/Leganza:
Check and adjust every 30,000 miles or 3 years and replacement every 40,000 miles or 4 years.

Tacuma/Rezzo:
Check and adjust every 20,000 miles or 2 years and replacement every 40,000 miles or 4 years.
The previous use and service history of the vehicle must always be taken into account.
Refer to Timing Belt Replacement Intervals at the front of this manual.

Check For Engine Damage

CAUTION: This engine has been identified as an INTERFERENCE engine in which the possibility of valve-to-piston damage in the event of a timing belt failure is MOST LIKELY to occur.
A compression check of all cylinders should be performed before removing the cylinder head(s).

Repair Times – hrs

Check & adjust	1,30
Remove & install	1,30

Special Tools

■ None required.

Special Precautions

- ■ Disconnect battery earth lead.
- ■ DO NOT turn crankshaft or camshaft when timing belt removed.
- ■ Remove spark plugs to ease turning engine.
- ■ Turn engine in normal direction of rotation (unless otherwise stated).
- ■ DO NOT turn engine via camshaft or other sprockets.
- ■ Observe all tightening torques.

Removal

1. Raise and support front of vehicle.
2. Disconnect:
   ❏ Intake air temperature (IAT) sensor wiring.
   ❏ Breather hose.
   ❏ Intake air duct.
3. Remove:
   ❏ Air filter housing.
   ❏ RH front wheel.
   ❏ RH front splash guard.
   ❏ Auxiliary drive belt.
   ❏ Crankshaft pulley bolts **1**.
   ❏ Crankshaft pulley **2**.

4. Support engine.
5. Remove:
   ❏ RH engine mounting bracket.
   ❏ Leganza: PAS hose bracket and bolt.
   ❏ Timing belt cover **3**.
6. Turn crankshaft clockwise until timing marks aligned **4**, **5** & **6**.
   *NOTE: The camshaft sprockets are interchangeable and therefore have similar timing marks. Ensure correct dowel location of sprockets on camshafts and timing marks aligned with 'INTAKE' **5** and 'EXHAUST' **6** marks respectively.*
7. Slacken tensioner pulley bolt **7**.
8. Turn tensioner clockwise away from belt **8**. Lightly tighten bolt **7**.
9. Remove timing belt.

Installation

1. Ensure timing marks aligned **4**, **5** & **6**.
2. Fit timing belt in anti-clockwise direction, starting at crankshaft sprocket.
3. Ensure belt is taut between sprockets on non-tensioned side.
4. Slacken tensioner pulley bolt **7**.
5. Turn tensioner pulley anti-clockwise **8** until pointer aligned with notch **9**.
6. Tighten tensioner pulley bolt to 25 Nm **7**.
7. Turn crankshaft slowly two turns clockwise until timing marks aligned **4**, **5** & **6**.
8. Ensure tensioner pointer aligned with notch **9**. If not: Repeat tensioning procedure.
9. Install components in reverse order of removal.
10. Tighten crankshaft pulley bolts to 20 Nm **1**.

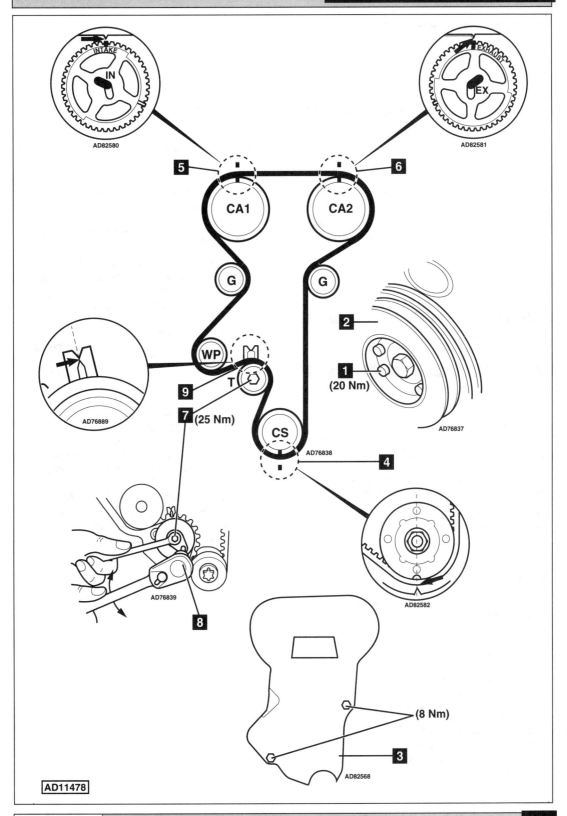

AD82580

AD82581

5

6

CA1

CA2

G

G

2

WP

M

9

T

1
(20 Nm)

AD76889

7 (25 Nm)

CS

AD76838

4

AD76837

AD76839

8

AD82582

(8 Nm)

3

AD82568

AD11478

DAEWOO

Model:	**Tacuma 1,8**
Year:	**2001-03**
Engine Code:	**F18N**

Replacement Interval Guide

Daewoo recommend check and adjust every 20,000 miles or 2 years and replacement every 40,000 miles or 4 years.
The previous use and service history of the vehicle must always be taken into account.
Refer to Timing Belt Replacement Intervals at the front of this manual.

Check For Engine Damage

CAUTION: This engine has been identified as an INTERFERENCE engine in which the possibility of valve-to-piston damage in the event of a timing belt failure is MOST LIKELY to occur.
A compression check of all cylinders should be performed before removing the cylinder head(s).

Repair Times – hrs

Check & adjust	1,10
Remove & install	1,30

Special Tools

- Water pump wrench – No.J-42492.

Special Precautions

- Disconnect battery earth lead.
- DO NOT turn crankshaft or camshaft when timing belt removed.
- Remove spark plugs to ease turning engine.
- Turn engine in normal direction of rotation (unless otherwise stated).
- DO NOT turn engine via camshaft or other sprockets.
- Observe all tightening torques.

Removal

1. Raise and support front of vehicle.
2. Disconnect:
   - ❏ Intake air temperature (IAT) sensor wiring.
   - ❏ Breather hose.
   - ❏ Intake air duct.
3. Remove:
   - ❏ Air filter housing.
   - ❏ RH front wheel.
   - ❏ RH splash guard.
   - ❏ Auxiliary drive belts.
   - ❏ Crankshaft pulley bolt **1**.
   - ❏ Crankshaft pulley **2**.
   - ❏ Timing belt covers **3** & **4**.
4. Fit crankshaft pulley bolt.
5. Turn crankshaft clockwise until timing marks aligned **5** & **6**.

6. Slacken water pump bolts slightly **7**.
7. Turn water pump anti-clockwise to release tension on belt **8**. Use tool No.J-42492.
8. Lightly tighten water pump bolts **7**.
9. Support engine.
10. Remove:
    - ❏ RH engine mounting and bracket.
    - ❏ Timing belt.

Installation

1. Ensure timing marks aligned **5** & **6**.
2. Fit timing belt in anti-clockwise direction, starting at crankshaft sprocket.
3. Ensure belt is taut between sprockets on non-tensioned side.
4. Slacken water pump bolts **7**.
5. Turn water pump clockwise to tension belt **8**. Use tool No.J-42492.
6. Movable part of tensioner must be against stop **9**.
7. Tighten water pump bolts **7**.
8. Turn crankshaft two turns clockwise until timing marks aligned **5** & **6**.
9. Slacken water pump bolts **7**.
10. Turn water pump anti-clockwise until tensioner pointer aligned with notch in support plate **10**. Use tool No.J-42492.
11. Tighten water pump bolts to 10 Nm **7**.
12. Install components in reverse order of removal.
13. Tighten crankshaft pulley bolt **1**.
    Tightening torque: 95 Nm + 30° + 15°.

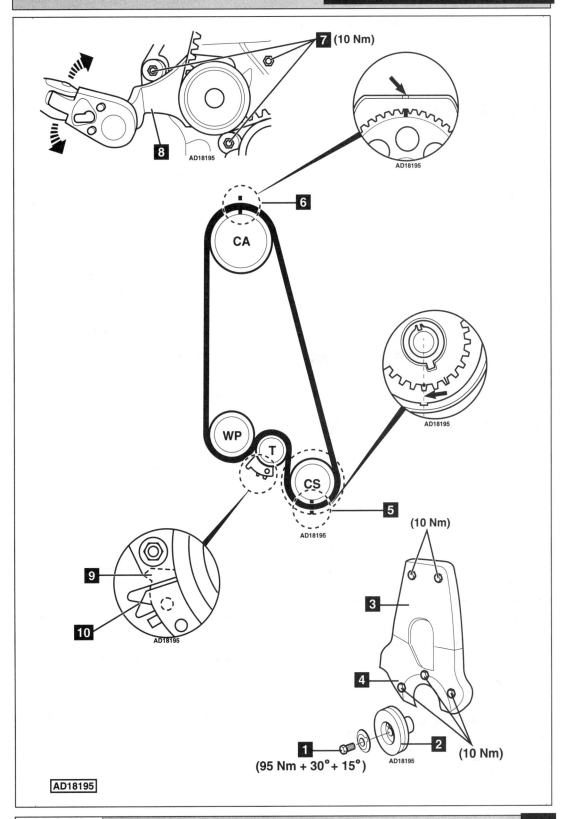

7 (10 Nm)

AD18195

8

AD18195

6

CA

AD18195

WP

T

CS

5

AD18195

(10 Nm)

9

10

AD18195

3

4

1

(95 Nm + 30° + 15°)

2

AD18195

(10 Nm)

AD18195

Model:	Mira • Cuore (L501) • Move
Year:	1994-02
Engine Code:	ED-10, ED-20

Replacement Interval Guide

Daihatsu recommend:
→02/99: Check every 6,000 miles and replacement every 60,000 miles.
03/99→: Check every 12,000 miles and replacement every 60,000 miles or 5 years, whichever occurs first.

The previous use and service history of the vehicle must always be taken into account.
Refer to Timing Belt Replacement Intervals at the front of this manual.

Check For Engine Damage

CAUTION: This engine has been identified as an INTERFERENCE engine in which the possibility of valve-to-piston damage in the event of a timing belt failure is MOST LIKELY to occur.
A compression check of all cylinders should be performed before removing the cylinder head.

Repair Times – hrs

Remove & install:

Mira/Cuore	1,20
Move	1,50

Special Tools

■ None required.

Special Precautions

■ Disconnect battery earth lead.
■ DO NOT turn crankshaft or camshaft when timing belt removed.
■ Remove spark plugs to ease turning engine.
■ Turn engine in normal direction of rotation (unless otherwise stated).
■ DO NOT turn engine via camshaft or other sprockets.
■ Observe all tightening torques.

Removal

1. Remove:
   ❑ Move: Air filter and air duct.
   ❑ Mira: Dipstick.
   ❑ Auxiliary drive belt.
   ❑ Water pump pulley **1**.
   ❑ Crankshaft pulley bolt **2**.
   ❑ Crankshaft pulley **3**.
   ❑ Timing belt cover **4**.
2. Temporarily fit crankshaft pulley bolt **2**.
3. Turn crankshaft clockwise until camshaft sprocket timing hole aligned with notch in timing belt rear cover **5**. Use crankshaft pulley bolt.
4. Remove crankshaft sprocket guide washer **6**.

5. Ensure crankshaft sprocket timing mark aligned **7**.
6. Remove:
   ❑ Tensioner bolt **8**.
   ❑ Tensioner pulley.
   ❑ Timing belt.

Installation

1. Ensure timing marks aligned **5** & **7**.
2. Fit timing belt. Ensure marks on belt **9** aligned with marks on sprockets.
3. Fit tensioner pulley. Lightly tighten bolt **8**.
4. Ensure tensioner free to move.
5. Remove blanking plug from timing belt rear cover. Screw M6 x 25 mm bolt in by hand **10**.
   NOTE: Insert a spacer between end of M6 bolt and tensioner pulley to prevent damage to tensioner guide surface 11.
6. Screw in M6 bolt until distance between rear cover and tensioner pulley is 11-14 mm **12**.
7. Tighten tensioner bolt to 39 Nm **8**.
8. Unscrew M6 bolt **10** two or three turns to clear tensioner pulley.
9. Turn crankshaft two turns clockwise. Ensure timing marks aligned **5** & **7**.
10. If belt not aligned centrally on sprockets: Repeat operations 3-9.
11. Apply a load of 2,7-4,0 kg to belt at ▽. Belt should deflect 5 mm.
12. If not: Slacken tensioner bolt **8**.
13. Insert a spacer between end of M6 bolt and tensioner pulley **11**. Screw bolt in to increase belt tension.
14. Tighten tensioner bolt to 39 Nm **8**.
15. Unscrew M6 bolt **10** two or three turns to clear tensioner pulley.
16. Turn crankshaft two turns clockwise. Ensure timing marks aligned **5** & **7**.
17. Recheck belt tension.
18. Remove M6 bolt **10**. Fit blanking plug.
19. Remove crankshaft pulley bolt **2**. Fit crankshaft sprocket guide washer **6**.
20. Install components in reverse order of removal.
21. Tighten crankshaft pulley bolt to 98 Nm **2**.

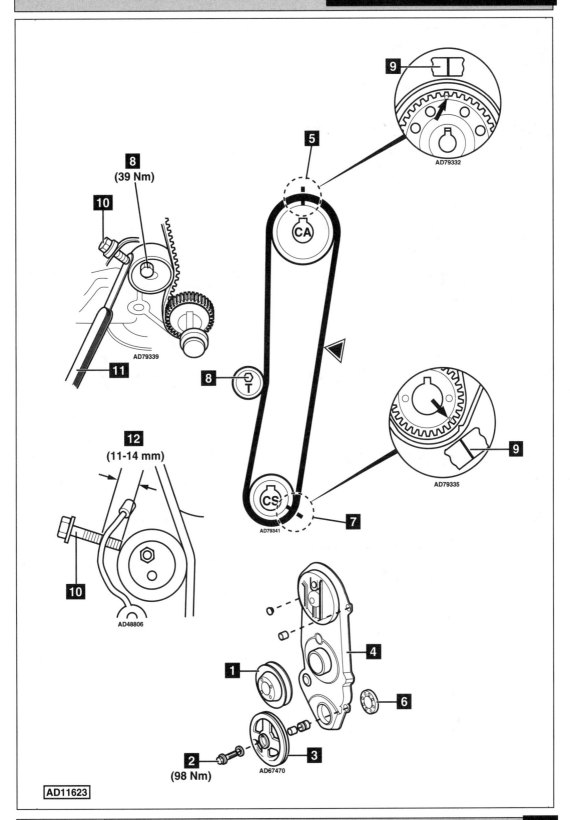

Model:	Charade 1,3 16V • Charade 1,5 • Charade 16V • Charade GSXi Grand Move • Terios
Year:	1988-02
Engine Code:	HC-C, HC-E, HC-EJ, HD-E, HE-EG

Replacement Interval Guide

Daihatsu recommend replacement as follows:
→02/99: Check every 6,000 miles and replacement every 60,000 miles.
03/99→: Check every 12,000 miles and replacement every 60,000 miles or 5 years, whichever occurs first.
The previous use and service history of the vehicle must always be taken into account.
Refer to Timing Belt Replacement Intervals at the front of this manual.

Check For Engine Damage

CAUTION: This engine has been identified as an INTERFERENCE engine in which the possibility of valve-to-piston damage in the event of a timing belt failure is MOST LIKELY to occur.
A compression check of all cylinders should be performed before removing the cylinder head.

Repair Times – hrs

Remove & install:

1,3 16V	0,90
1,5/16V/GSXi	1,70
Grand Move/Terios	1,90

Special Tools

- None required.

Special Precautions

- Disconnect battery earth lead.
- DO NOT turn crankshaft or camshaft when timing belt removed.
- Remove spark plugs to ease turning engine.
- Turn engine in normal direction of rotation (unless otherwise stated).
- DO NOT turn engine via camshaft or other sprockets.
- Observe all tightening torques.

Removal

1. Remove air filter.
2. Disconnect oil pressure switch wire and feed back through engine mounting.
3. Remove:
   - ❏ Auxiliary drive belt(s).
   - ❏ Water pump pulley.
   - ❏ PAS pump pulley (if fitted).
4. Support engine with a jack.

5. Remove:
   - ❏ RH engine mounting(s).
   - ❏ RH engine mounting bracket.
   - ❏ Service hole cover in RH inner wing panel.
   - ❏ Crankshaft pulley bolts **1**.
   - ❏ Crankshaft pulley **2**.
   - ❏ Timing belt upper cover **3**.
   - ❏ Timing belt lower cover **4**.

   NOTE: Differing bolt lengths.
6. Turn crankshaft clockwise until timing marks aligned **5** & **6**.
7. Slacken tensioner bolt **7**. Move tensioner away from belt. Lightly tighten bolt.
8. Remove timing belt.

Installation

1. Ensure timing marks aligned **5** & **6**.
2. Fit timing belt. Ensure belt is taut between sprockets. Observe direction of rotation marks on belt **8**.
3. Slacken tensioner bolt to partially tension belt **7**. Lightly tighten bolt.
4. Turn crankshaft nearly two turns clockwise until 'F' mark **10** on camshaft sprocket is three teeth before indicator on cover.
5. Slacken tensioner bolt **7**. Turn crankshaft until 'F' mark **10** aligns with indicator.
   Tighten tensioner bolt to 40 Nm **7**.
6. Ensure timing marks aligned **5** & **6**.
7. Fit timing belt covers and seals (starting with bolts **9**).
8. Fit crankshaft pulley. Tighten bolts to 25 Nm **1**.
9. Fit water pump pulley. Tighten bolts to 8 Nm.
10. Install components in reverse order of removal.

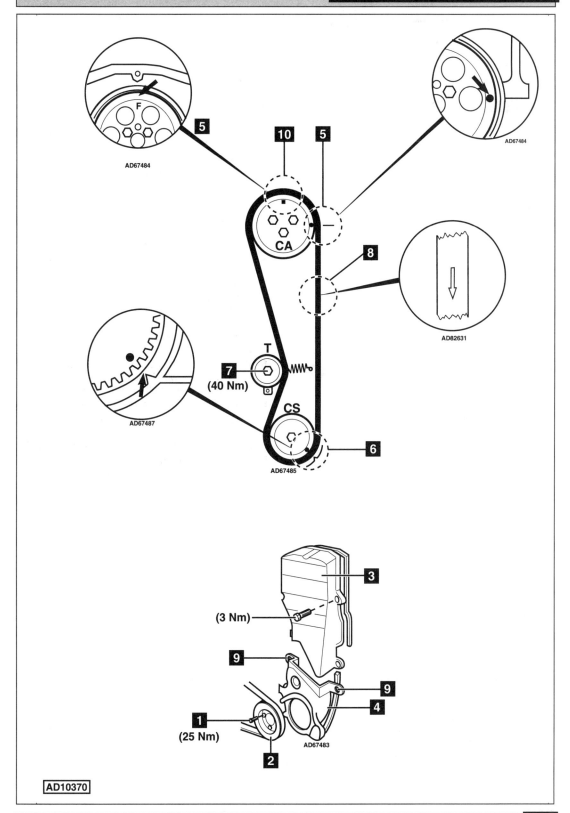

AD67484

5

10

5

AD67484

8

AD82631

7 (40 Nm)

T

CA

CS

AD67487

AD67485

6

3

(3 Nm)

9

9

4

1 (25 Nm)

2

AD67483

AD10370

Model:	**Sportrak 1,6 • Applause 1,6**
Year:	**1988-98**
Engine Code:	**HD-C, HD-E**

Replacement Interval Guide

Daihatsu recommend check every 6,000 miles and replacement every 60,000 miles.
The previous use and service history of the vehicle must always be taken into account.
Refer to Timing Belt Replacement Intervals at the front of this manual.

Check For Engine Damage

CAUTION: This engine has been identified as an INTERFERENCE engine in which the possibility of valve-to-piston damage in the event of a timing belt failure is MOST LIKELY to occur.
A compression check of all cylinders should be performed before removing the cylinder head.

Repair Times – hrs

Remove & install:

Sportrak	1,50
Applause	1,30

Special Tools

■ None required.

Special Precautions

■ Disconnect battery earth lead.
■ DO NOT turn crankshaft or camshaft when timing belt removed.
■ Remove spark plugs to ease turning engine.
■ Turn engine in normal direction of rotation (unless otherwise stated).
■ DO NOT turn engine via camshaft or other sprockets.
■ Observe all tightening torques.

Removal

1. Partially drain coolant.
2. Remove:
   ❏ Coolant overflow hose.
   ❏ Expansion tank and top hose.
   ❏ Air intake duct and hose.
   ❏ Alternator drive belt.
   ❏ PAS drive belt (if fitted).
   ❏ Bolts retaining cooling fan cowling.
   ❏ Bolts retaining cooling fan coupling to water pump.
   ❏ Cooling fan cowling and coupling.
   ❏ Water pump pulley.
   ❏ Crankshaft pulley bolts **1**.
   ❏ Crankshaft pulley **2**.
   ❏ Timing belt upper cover **3**.
   ❏ Timing belt lower cover **4**.

3. Temporarily fit crankshaft pulley. Turn crankshaft clockwise to TDC on No.1 cylinder. Ensure timing marks aligned **5** & **6**. Remove crankshaft pulley.
4. Slacken tensioner bolt **7**. Move tensioner away from belt. Lightly tighten bolt.
5. Remove timing belt.

Installation

1. Ensure timing marks aligned **5** & **6**.
2. Fit timing belt. Ensure belt is taut between sprockets. Observe direction of rotation marks on belt **8**.
3. Slacken tensioner bolt to partially tension belt **7**. Lightly tighten bolt.
4. Turn crankshaft nearly two turns clockwise until 'F' mark **10** on camshaft sprocket is three teeth before indicator on cover.
5. Slacken tensioner bolt. Turn crankshaft until 'F' mark **10** aligns with indicator.
   Tighten tensioner bolt to 40 Nm **7**.
6. Ensure timing marks aligned **5** & **6**.
7. Fit timing belt covers and seals (starting with bolts **9**).
8. Fit crankshaft pulley. Tighten bolts to 25 Nm **1**.
9. Install components in reverse order of removal.
10. Tighten bolts retaining water pump to cooling fan coupling to 15 Nm.
11. Refill cooling system.

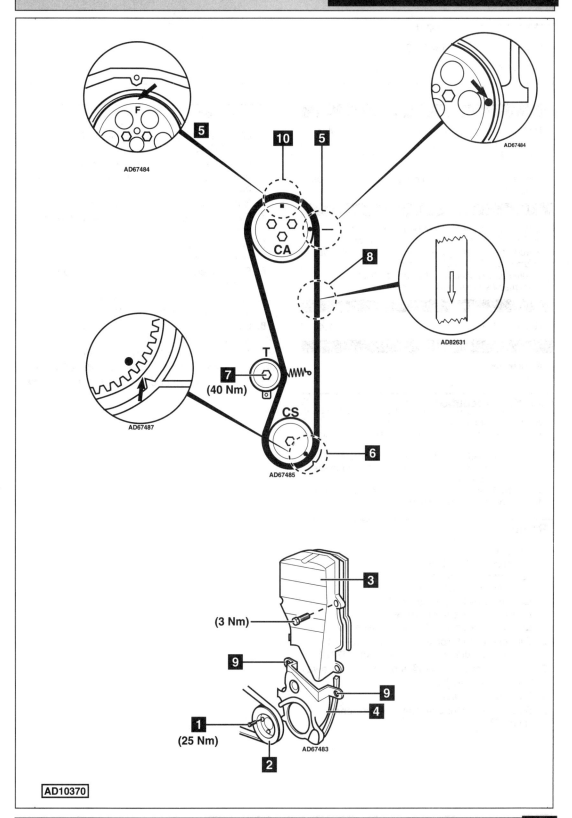

AD67484

5

10

5

AD67484

8

AD82631

T

7 (40 Nm)

CA

AD67487

CS

6

AD67485

3

(3 Nm)

9

9

4

1 (25 Nm)

2

AD67483

AD10370

DAIHATSU

Model:	**Hi-Jet**
Year:	**1986-98**
Engine Code:	**CB-41, CD-20**

Replacement Interval Guide

Daihatsu recommend check every 6,000 miles and replacement every 60,000 miles.
The previous use and service history of the vehicle must always be taken into account.
Refer to Timing Belt Replacement Intervals at the front of this manual.

Check For Engine Damage

CAUTION: This engine has been identified as an INTERFERENCE engine in which the possibility of valve-to-piston damage in the event of a timing belt failure is MOST LIKELY to occur.
A compression check of all cylinders should be performed before removing the cylinder head.

Repair Times – hrs

Remove & install	1,40

Special Tools

■ None required.

Special Precautions

■ Disconnect battery earth lead.
■ DO NOT turn crankshaft or camshaft when timing belt removed.
■ Remove spark plugs to ease turning engine.
■ Turn engine in normal direction of rotation (unless otherwise stated).
■ DO NOT turn engine via camshaft or other sprockets.
■ Observe all tightening torques.

Removal

1. Remove:
   ❏ Auxiliary drive belt.
   ❏ Water pump pulley.
   ❏ Crankshaft pulley **1**.
   ❏ Timing belt upper cover **2**.
   ❏ Timing belt lower cover **3**.
   ❏ Crankshaft sprocket guide washer **4**.
2. Turn crankshaft clockwise to TDC on No.1 cylinder. Ensure timing marks aligned **5** & **6**.
3. Slacken tensioner bolt **7**. Move tensioner away from belt. Lightly tighten bolt.
4. Remove timing belt.
5. Slacken tensioner bolt. Remove tensioner spring **8**.

Installation

1. Check free length of tensioner spring is 54,0 mm **8**.
2. Fit tensioner spring. Push tensioner pulley against spring tension. Lightly tighten tensioner bolt.
3. Ensure timing marks aligned **5** & **6**.
4. Fit timing belt. Ensure belt is taut between sprockets.
5. Fit crankshaft sprocket guide washer (convex side towards belt) **4**.
6. Slacken tensioner bolt **7**. Allow spring to pull tensioner against belt. Tighten bolt.
7. Turn crankshaft two turns clockwise. Ensure timing marks aligned **5** & **6**.
8. Slacken tensioner bolt. Allow spring to pull tensioner against belt.
9. Tighten tensioner bolt to 33-44 Nm **7**.
10. Install components in reverse order of removal.
11. Tighten crankshaft pulley bolt to 88-98 Nm **9**.

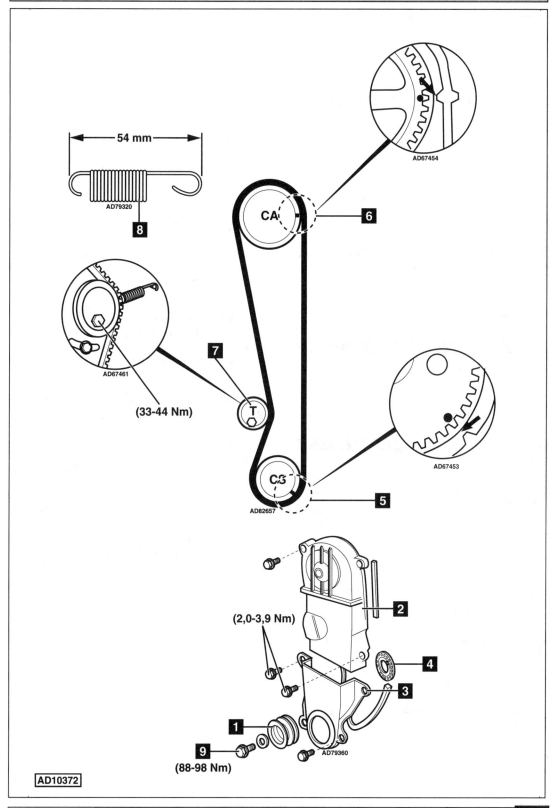

54 mm

AD79320

8

AD67454

6

CA

7

AD67461

(33-44 Nm)

T

AD67453

C3

AD82657

5

2

(2,0-3,9 Nm)

4

3

1

9

(88-98 Nm)

AD79360

AD10372

Model:	Fourtrak/Rocky/Wildcat 2,8 D/TD
Year:	1988-02
Engine Code:	DL

Replacement Interval Guide

Daihatsu recommend replacement as follows:
→02/99: Check every 6,000 miles and replacement every 60,000 miles.
03/99→: Check every 12,000 miles and replacement every 60,000 miles or 5 years, whichever occurs first.
The previous use and service history of the vehicle must always be taken into account.
Refer to Timing Belt Replacement Intervals at the front of this manual.

Check For Engine Damage

CAUTION: This engine has been identified as an INTERFERENCE engine in which the possibility of valve-to-piston damage in the event of a timing belt failure is MOST LIKELY to occur.
A compression check of all cylinders should be performed before removing the cylinder head.

Repair Times – hrs

Remove & install:

→1993	1,30
AC	+0,30
PAS	+0,20
1993→	1,80

Special Tools

■ None required.

Special Precautions

■ Disconnect battery earth lead.
■ DO NOT turn crankshaft or camshaft when timing belt removed.
■ Remove glow plugs to ease turning engine.
■ Turn engine in normal direction of rotation (unless otherwise stated).
■ DO NOT turn engine via camshaft or other sprockets.
■ Observe all tightening torques.
■ Check diesel injection pump timing after belt replacement.

Removal

1. Remove:
   ❏ Auxiliary drive belts.
   ❏ Cooling fan and pulley.
2. Turn crankshaft clockwise to TDC on No.1 cylinder **1**.
3. Remove:
   ❏ Crankshaft pulley bolt.
   ❏ Crankshaft pulley **10**.
   ❏ Timing belt cover **2**.

4. Turn crankshaft clockwise until crankshaft sprocket, camshaft sprocket and injection pump sprocket timing marks aligned **3**, **4** & **5**. No.1 cylinder approximately 30° BTDC.
5. Mark direction of rotation on belt.
6. Slacken tensioner bolts **6** & **7**.
7. Remove crankshaft sprocket guide washer **8**.
8. Remove timing belt.

Installation

NOTE: Observe direction of rotation marks on belt.

1. Ensure timing marks aligned **3**, **4** & **5**. No.1 cylinder approximately 30° BTDC.
2. Fit timing belt with yellow lines on belt aligned with sprocket timing marks.
3. Ensure timing belt at least 2 mm from front edge of sprockets.
4. Insert torque wrench into hexagonal hole in tensioner **9**.
5. Apply anti-clockwise torque of 10 Nm to tensioner pulley. Tighten tensioner bolt **7**.
6. Turn crankshaft slowly clockwise until camshaft sprocket mark has moved three teeth.
7. Slacken tensioner bolt **7**. Repeat operations 4-5.
8. Apply a load of 2-2,6 kg to belt at ▽. Belt should deflect 1,8 mm.
9. Fit crankshaft pulley. Turn crankshaft two turns clockwise. Ensure timing marks aligned **3**, **4** & **5**.
10. Check belt tension. Tighten tensioner bolt **6**.
11. If tension as specified: Remove crankshaft pulley. Fit crankshaft sprocket guide washer (convex side towards belt) **8**.
12. Install components in reverse order of removal.
13. Tighten crankshaft pulley bolt to 217-235 Nm **11**.

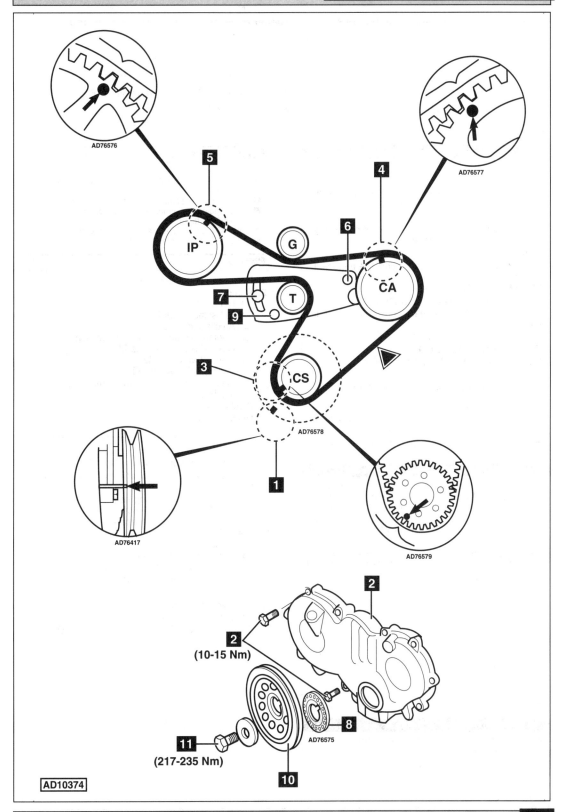

AD76576

5

4

AD76577

IP

G

6

CA

7 T

9

3 CS

AD76578

1

AD76417

AD76579

2

2
(10-15 Nm)

8

AD76575

11
(217-235 Nm)

10

AD10374

FIAT

Model:	Panda 750/1000 FIRE • Cinquecento Sporting • Seicento Sporting Uno 45 FIRE • Punto 1,1/1,2 • Tipo 1100 FIRE
Year:	1986-02
Engine Code:	154 A4.000, 156 A2.000, 156 A2.100, 156 A2.246, 156 A2.048, 156 A3.000 156 A4.000, 156 A4.048, 160 A3.000, 176 A6.000, 176 A.7000, 176 B1.000 176 B2.000, 176 B4.000, 176 A8.000, 188 A4.000

Replacement Interval Guide

Fiat recommend:

1986-12/90:
Every 12,000 miles - check & replace if necessary.
Every 36,000 miles - replace.

01/91-04/91:
Every 24,000 miles - check & replace if necessary.
Every 60,000 miles - replace.

05/91-98:
Every 36,000 miles or 4 years - check & replace if necessary.
Every 63,000 miles - replace.

1999-00 – except New Punto (99-):
Every 36,000 miles or 4 years - check & replace if necessary.
Every 63,000 miles - replace.

1999-00 – New Punto (99-):
Replacement every 72,000 miles or 6 years.

2001→:
Every 36,000 miles or 3 years - check & replace if necessary.
Every 72,000 miles or 5 years - replace.
The previous use and service history of the vehicle must always be taken into account.
Refer to Timing Belt Replacement Intervals at the front of this manual.

Check For Engine Damage

CAUTION: This engine has been identified as a FREEWHEELING engine in which the possibility of valve-to-piston damage in the event of a timing belt failure may be minimal or very unlikely. However, a precautionary compression check of all cylinders should be performed.

Repair Times – hrs

Remove & install:

Cinquecento/Seicento	1,10
AC (Cinquecento)	+0,20
AC (Seicento)	+0,25
New Punto (188A4.000)	1,50
AC	+0,05
Panda	0,70
Uno (→1989)	1,70
Uno (1989→)	1,00
Punto	0,95
Tipo	0,95

Special Tools

- Tensioning tool – Fiat No.1860745300/100.

Special Precautions

- Disconnect battery earth lead.
- DO NOT turn crankshaft or camshaft when timing belt removed.
- Remove spark plugs to ease turning engine.
- Turn engine in normal direction of rotation (unless otherwise stated).
- DO NOT turn engine via camshaft or other sprockets.
- Observe all tightening torques.

Removal

1. Raise and support front of vehicle.
2. Remove:
   - ❑ RH front wheel.
   - ❑ RH splash guard (if necessary).
   - ❑ Auxiliary drive belt.
   - ❑ Coolant expansion tank (if necessary).
   - ❑ Crankshaft pulley bolts **1**.
   - ❑ Crankshaft pulley **2**.
   - ❑ Timing belt cover(s) **3**.
3. Turn crankshaft clockwise until timing marks aligned **4** & **5**.
4. Slacken tensioner nut to release tension on belt **6**.
5. Remove timing belt.

Installation

1. Ensure timing marks aligned **4** & **5**.
2. Fit timing belt. Observe direction of rotation marks on belt.
3. Fit tensioning tool locating pegs to tensioner pulley holes **7** & **8**. Tool No.1860745300/100. Ensure tensioning tool arm is horizontal.
4. Set weight on scale bar as follows:
   - ❑ Seicento: 80 mm.
   - ❑ Except Seicento: 65 mm.
5. Turn crankshaft two turns clockwise.
6. Ensure timing marks aligned **4** & **5**.
7. Tighten tensioner nut to 28 Nm **6**.
8. Remove tensioning tool.
9. Install components in reverse order of removal.
10. Tighten crankshaft pulley bolts to 25 Nm **1**.

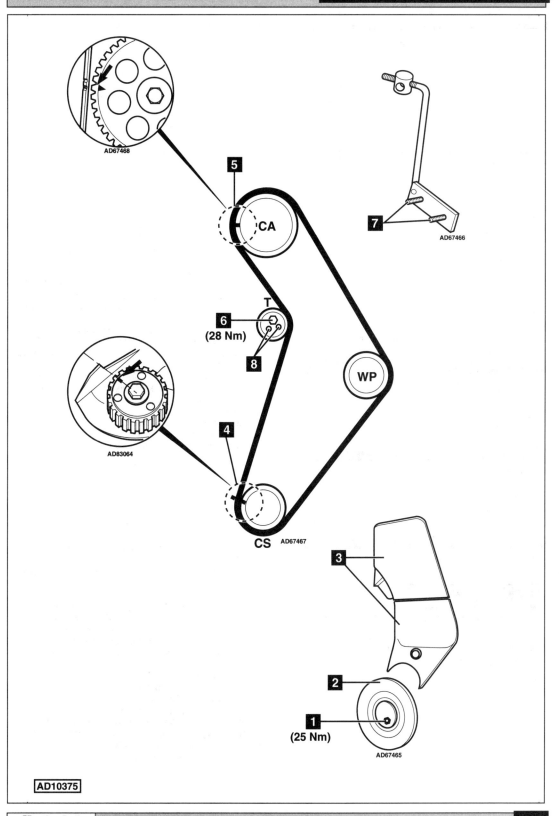

AD67468

5

CA

7

AD67466

T

6
(28 Nm)

8

AD83064

4

WP

CS   AD67467

3

2

1
(25 Nm)

AD67465

AD10375

FIAT

Model:	Uno 1,4/Turbo • Punto 1,4 GT • Punto 1,6 • Tipo 1,4/1,6 Tempra 1,4/1,6
Year:	1989-99
Engine Code:	146 C1.000, 146 A8.000, 146 A8.046, 159 A2.000, 159 A3.000, 159 A3.046 159 A3.048, 160 A1.000, 160 A1.046, 160 A1.048, 160 A2.000, 176 A9.000 176 A4.000, 176 B6.000, 835 C1.000, 836 A4.000

Replacement Interval Guide

Fiat recommend check and/or replacement at the following intervals:

→12/90:
Every 12,000 miles - check & replace if necessary.
Every 36,000 miles - replace.

01/91-04/91:
Every 24,000 miles - check & replace if necessary.
Every 60,000 miles - replace.

05/91→:
Every 36,000 miles or 4 years - check & replace if necessary.
Every 63,000 miles - replace.
The previous use and service history of the vehicle must always be taken into account.
Refer to Timing Belt Replacement Intervals at the front of this manual.

Check For Engine Damage

CAUTION: This engine has been identified as an INTERFERENCE engine in which the possibility of valve-to-piston damage in the event of a timing belt failure is MOST LIKELY to occur.
A compression check of all cylinders should be performed before removing the cylinder head.

Repair Times – hrs

Remove & install:

Uno	1,10
Punto	1,40
Tipo/Tempra	1,00
AC	+0,20

Special Tools

■ Tensioning tool – Fiat No.1860745100/200.

Special Precautions

■ Disconnect battery earth lead.
■ DO NOT turn crankshaft or camshaft when timing belt removed.
■ Remove spark plugs to ease turning engine.
■ Turn engine in normal direction of rotation (unless otherwise stated).
■ DO NOT turn engine via camshaft or other sprockets.
■ Observe all tightening torques.

Removal

1. Raise and support front of vehicle.
2. Remove:
   ❏ Air filter intake pipe.
   ❏ RH front wheel.
   ❏ RH inner wing panel.
   ❏ Timing belt cover upper bolts **1**.
3. Slacken alternator drive belt tensioner (Uno).
4. Slacken alternator.
5. Remove alternator drive belt.
6. Turn crankshaft until timing marks aligned **2** & **3**.
   NOTE: Engines without crankshaft sprocket timing mark: Align crankshaft keyway with raised cast section.
7. Tipo/Tempra/Punto: Ensure timing marks aligned **2** & **3**. Slide peg upwards into locking position **8**.
8. Camshaft timing mark should be aligned with notch on peg **2**.
9. 1,4 engines with distributor: Auxiliary shaft sprocket dowel pin at 2 o'clock **10**. Distributor rotor arm aligned with No.4 HT lead.
10. Remove:
    ❏ Crankshaft pulley nut **7**.
    ❏ Crankshaft pulley **4**.
    ❏ Timing belt cover lower bolts.
    ❏ Timing belt cover.
11. Slacken tensioner bolt **5**.
12. Remove timing belt.

Installation

1. Ensure timing marks aligned **2** & **6**.
   NOTE: Engines without crankshaft sprocket timing mark: Align crankshaft keyway with raised cast section.
2. 1,4 engines with distributor: Auxiliary shaft sprocket dowel pin at 2 o'clock **10**. Distributor rotor arm aligned with No.4 HT lead.
3. If fitted: Ensure camshaft position (CMP) sensor set at timing mark **11**. If not: Adjust auxiliary sprocket.
4. Fit timing belt. Ensure belt is taut between sprockets.
5. Fit tensioning tool to tensioner pulley **5**.
   Tool No. 1860745100/200 **9**. Use tool without weight fitted.
6. Fit crankshaft pulley **4**. Tighten nut.
   A – Punto/Tipo/Tempra: 137 Nm.
   B – Uno: 190 Nm.
7. Turn crankshaft two turns clockwise to TDC.
   NOTE: If scale bar moves from the horizontal: Reset scale bar and repeat tensioning procedure.
8. Tighten tensioner nut to 44 Nm **5**.
9. Install components in reverse order of removal.

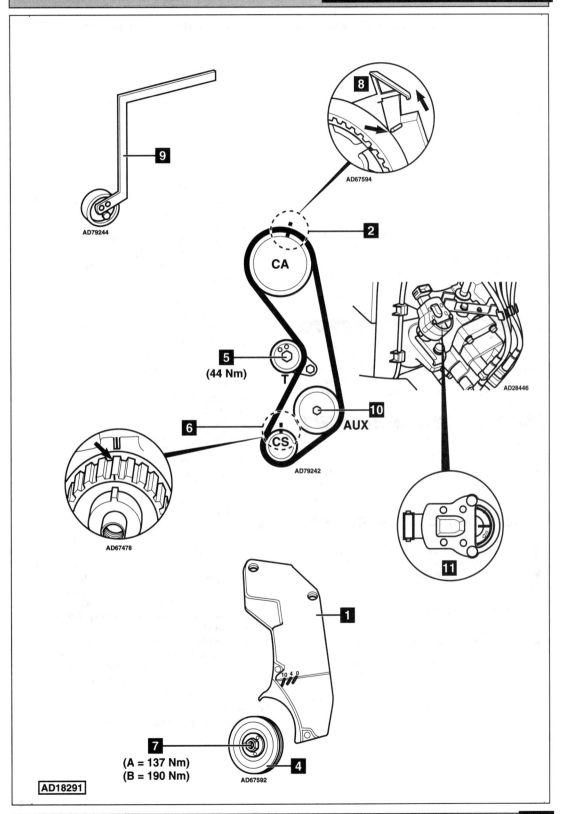

9 — AD79244

8 — AD67594

2

CA

5 (44 Nm) T

6 — AD67478

10 — AUX

CS

AD79242

AD28446

11

1

7
(A = 137 Nm)
(B = 190 Nm)

4

AD67592

AD18291

FIAT

Model:	**Punto 1,2 16V • Bravo 1,2 16V • Brava 1,2 16V • Stilo 1,2 16V**
Year:	**1997-03**
Engine Code:	**176B9.000, 182B2.000, 188A5.000**

Replacement Interval Guide

Fiat recommend:

176B9.000:
Check and replacement if necessary every 36,000 miles or 4 years and replacement every 72,000 miles or 8 years, whichever occurs first.

182B2.000/188A5.000:
→2000: Replacement every 72,000 miles or 6 years.
2001→: Check and replacement if necessary every 36,000 miles and replacement every 72,000 miles or 5 years.
The previous use and service history of the vehicle must always be taken into account.
Refer to Timing Belt Replacement Intervals at the front of this manual.

Check For Engine Damage

CAUTION: This engine has been identified as an INTERFERENCE engine in which the possibility of valve-to-piston damage in the event of a timing belt failure is MOST LIKELY to occur. A compression check of all cylinders should be performed before removing the cylinder head(s).

Repair Times – hrs

Remove & install:	
176B9.000/182B2.000	2,50
AC	+0,20
188A5.000	2,25
AC	+0,05

Special Tools

- ■ Piston position tools – Fiat No.1860992000.
- ■ Camshaft locking tools – Fiat No.1860985000.
- ■ Camshaft holding tool – Fiat No.1860831000.
- ■ Tensioner tool – Fiat No.1860987000.

Special Precautions

- ■ Disconnect battery earth lead.
- ■ DO NOT turn crankshaft or camshaft when timing belt removed.
- ■ Remove spark plugs to ease turning engine.
- ■ Turn engine in normal direction of rotation (unless otherwise stated).
- ■ DO NOT turn engine via camshaft or other sprockets.
- ■ Observe all tightening torques.

Removal

1. Raise and support front of vehicle.
2. Remove:
   - ❏ RH front wheel.
   - ❏ RH wheel arch liner.
   - ❏ Auxiliary drive belts.
3. Turn crankshaft clockwise until dowel pin **1** opposite crankshaft position (CKP) sensor **2**.
4. Remove:
   - ❏ Crankshaft pulley bolts **3**.
   - ❏ Crankshaft pulley **4**.
   - ❏ Timing belt lower cover bolt **5**.
   - ❏ Air filter assembly.
   - ❏ Air resonator box.
   - ❏ Air intake trunking.
   - ❏ Breather hose.
   - ❏ Stilo: RH engine mounting and bracket.
5. Disconnect throttle cable.
6. Disconnect vacuum pipes and wiring connections from upper inlet manifold.
7. Remove:
   - ❏ Upper inlet manifold.
   - ❏ Fuel rail.
   - ❏ Injectors.
   - ❏ Engine control module (ECM). DO NOT disconnect harness multi-plug.
   - ❏ Crankshaft position (CKP) sensor **2**.
   - ❏ Timing belt covers **6** & **7**.
   - ❏ HT leads.
   - ❏ Spark plugs.
   - ❏ Blanking plugs **8**.
8. Screw piston position tools into cylinders No.1 & No.2 **9**. Tool Nos. 1860992000. Tightening torque: 5 Nm.
9. Turn crankshaft slightly until notches of both tools aligned with upper surface **10**.
   NOTE: Dowel pin must remain opposite crankshaft position (CKP) sensor 1. If necessary: Temporarily refit crankshaft pulley. Check alignment.
10. Ensure slots in camshafts are aligned with blanking plug holes **11**. If necessary: Remove tools. Turn crankshaft one turn clockwise.
11. Fit locking tool to each camshaft **12**. Tool No. 1860985000.
12. Slacken tensioner nut **13**.
13. Move tensioner away from belt.
14. Remove timing belt.

/Autodata

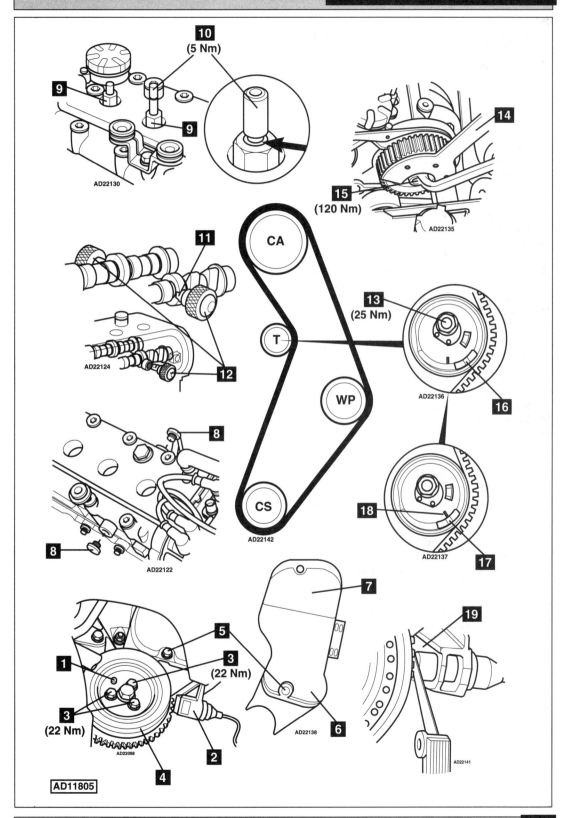

10 (5 Nm)

9

9

AD22130

14

15 (120 Nm)

AD22135

11

AD22124

12

CA

13 (25 Nm)

T

AD22136

16

WP

18

8

AD22137

17

8

AD22122

CS

AD22142

7

5

1

3 (22 Nm)

3 (22 Nm)

2

AD22098

4

6

AD22138

19

AD22141

AD11805

Installation

1. Hold camshaft sprocket. Use tool No.1860831000 **14**.
2. Slacken camshaft sprocket bolt **15**.
3. Ensure piston position tools fitted correctly into cylinders No.1 & No.2 **9**.
4. Ensure locking tools fitted correctly to each camshaft **12**.
5. Fit timing belt in anti-clockwise direction, starting at crankshaft sprocket. Ensure belt is taut between sprockets on non-tensioned side.
6. Turn tensioner pulley until mark at maximum tension position **16**. Use tool No.1860987000.
7. Tighten tensioner nut to 25 Nm **13**.
8. Hold camshaft sprocket. Use tool No.1860831000 **14**.
9. Tighten camshaft sprocket bolt to 120 Nm **15**.
10. Remove locking tools from camshafts **12**. Remove piston position tools **9**.
11. Turn crankshaft two turns clockwise to setting position.
12. Temporarily refit crankshaft pulley. Check alignment of dowel pin opposite crankshaft position (CKP) sensor **1**.
13. Slacken tensioner nut **13**.
14. Turn tensioner pulley until marks aligned **17** & **18**. Use tool No.1860987000.
15. Tighten tensioner nut to 25 Nm **13**.
16. Turn crankshaft two turns clockwise to setting position.
17. Screw piston position tools into cylinders No.1 & No.2 **9**. Tool Nos.1860992000. Tightening torque: 5 Nm.
18. Turn crankshaft slightly until notches of both tools aligned with upper surface **10**.
19. Ensure slots in camshafts are aligned with blanking plug holes **11**.
20. Ensure locking tools fit easily to each camshaft **12**. If not: Repeat installation and tensioning procedures.
21. Install components in reverse order of removal.
22. Tighten crankshaft pulley bolts **3**. Tightening torque: 22 Nm.
23. Ensure crankshaft position (CKP) sensor air gap is 0,5-1,5 mm **19**.

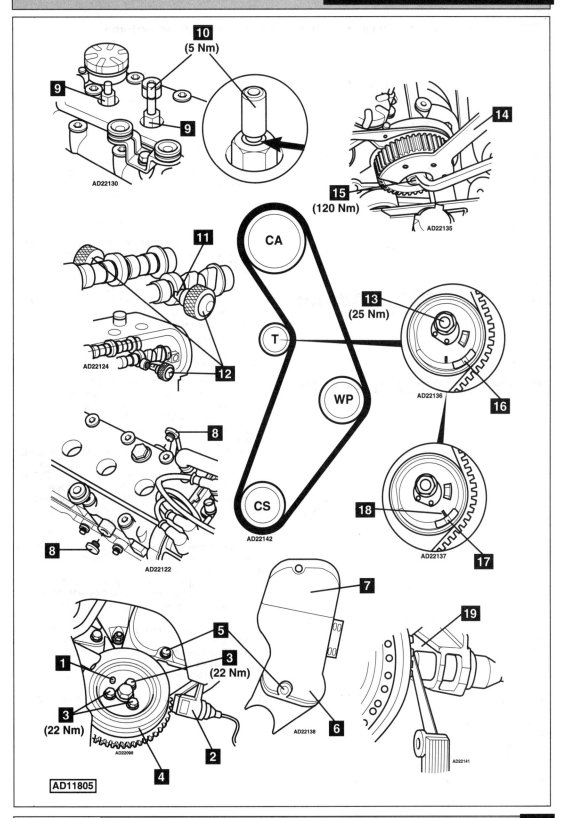

Model:	Tipo 1,8/2,0 i.e. • Tempra 2,0 i.e. • Croma 2,0 i.e/16V/Turbo
Year:	1993-97
Engine Code:	834 B.146, 835 C2.000, 154 C3.000
	154 C4.046, 159 A5.046, 159 A6.046, 154 E1.000

Replacement Interval Guide

Fiat recommend:

Balancer shaft belt
Check & replacement if necessary every 18,000 miles or 2 years and replacement every 63,000 miles.

Timing belt
Check & replacement if necessary every 36,000 miles or 4 years and replacement every 63,000 miles.
The previous use and service history of the vehicle must always be taken into account.
Refer to Timing Belt Replacement Intervals at the front of this manual.

Check For Engine Damage

CAUTION: This engine has been identified as an INTERFERENCE engine in which the possibility of valve-to-piston damage in the event of a timing belt failure is MOST LIKELY to occur.
A compression check of all cylinders should be performed before removing the cylinder head.

Repair Times – hrs

Remove & install:	
Tipo	2,10
Tempra 1,8	1,20
Tempra 2,0	2,10
Croma i.e.	2,40
Croma Turbo (incl. AC)	2,40
Croma i.e. 16V	1,80
AC	+0,30

Special Tools

- Tensioning tool (timing belt) – Fiat No.1860745100/200.
- Tensioning tool (balancer shaft belt) – Fiat No.1860745100/400.

Special Precautions

- Disconnect battery earth lead.
- DO NOT turn crankshaft or camshaft when timing belt removed.
- Remove spark plugs to ease turning engine.
- Turn engine in normal direction of rotation (unless otherwise stated).
- DO NOT turn engine via camshaft or other sprockets.
- Observe all tightening torques.

Removal

1. Support engine.
2. Remove:
   - Auxiliary drive belts.
   - Engine stabiliser bar and mounting **5**.
   - Timing belt cover **1**.
3. Turn crankshaft clockwise to TDC on No.1 cylinder.

4. Remove:
   - Crankshaft pulley bolts **11**.
   - Crankshaft pulley **3**.
5. Ensure balancer shaft timing marks aligned **6** & **7**.
6. Check timing marks on rear of camshaft sprockets aligned with cut-outs in timing belt rear cover and lugs on camshaft bearings **2**.
7. Remove water pump pulley **4**.
8. Slacken balancer shaft belt tensioner sprocket nut **9**.
   *NOTE: If balancer shaft belt is being reused: Mark position of belt on sprockets before removal **6**, **7** & **8**.*
9. Remove balancer shaft belt.
10. Remove crankshaft sprocket **8**.
    NOTE: Crankshaft sprocket bolt has LH thread.
11. Ensure crankshaft mark aligned **12**.
12. Slacken tensioner nut **10**.
13. Remove timing belt.

Installation

1. Ensure timing marks aligned **2** & **12**.
2. Fit timing belt. Ensure belt is taut between sprockets.
3. Fit tensioning tool to tensioner pulley. Tool No.1860745100/200 **13**.
4. With scale bar horizontal set weight at 140 mm mark **15**.
   NOTE: 16V engine – set weight at 100 mm mark.
5. Turn crankshaft two turns clockwise to TDC on No.1 cylinder. Ensure timing marks aligned **2** & **12**.
   NOTE: If scale bar moves from the horizontal: Reset scale bar and repeat tensioning procedure.
6. Tighten timing belt tensioner nut to 44 Nm **10**.
7. Ensure timing marks aligned **2** & **12**.
8. Fit crankshaft sprocket **8**. Tighten bolt to 190 Nm.
9. Ensure crankshaft sprocket TDC mark and balancer shaft timing marks aligned **12**, **6** & **7**.
10. Fit balancer shaft belt.
11. Fit tensioning tool to tensioner sprocket **9** (with weight removed). Tool No.1860745100/400 **14**.
12. With scale bar horizontal set weight at 205 mm mark.
13. Turn crankshaft two turns clockwise to TDC on No.1 cylinder. Ensure timing marks aligned **2** & **12**.
    NOTE: If scale bar moves from the horizontal: Reset scale bar and repeat tensioning procedure.
14. Tighten balancer shaft belt tensioner sprocket nut to 23 Nm **9**.
15. Install components in reverse order of removal.

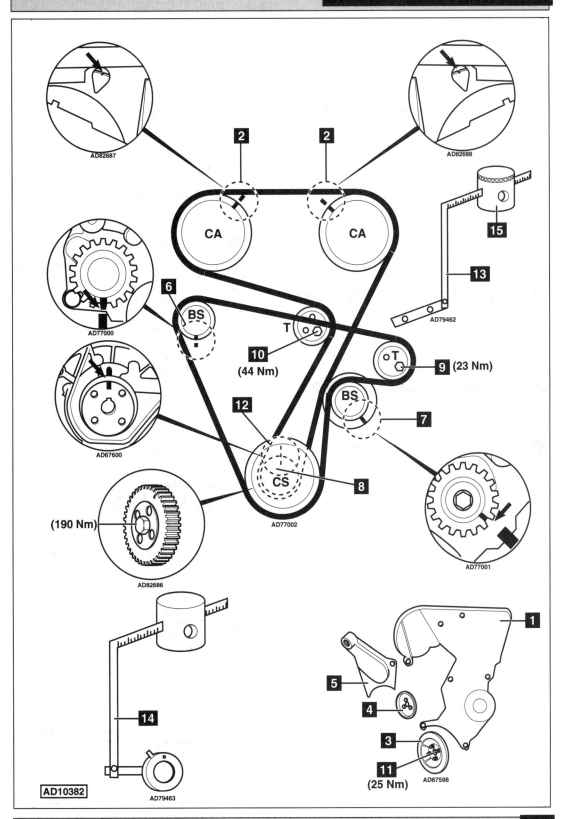

AD82687

AD82688

2

2

CA

CA

15

13

AD79462

6

BS

T

10

(44 Nm)

T

9 (23 Nm)

AD77000

12

BS

7

AD67600

CS

8

AD77002

(190 Nm)

AD77001

AD82686

1

5

4

14

3

11

(25 Nm)

AD67598

AD10382

AD79463

Model:	Coupé 2,0 16V • Coupé 2,0 16V Turbo • Tipo 2,0 16V
Year:	1993-01
Engine Code:	836 A3.000, 175 A1.000

Replacement Interval Guide

Fiat recommend:

→1996

Balancer shaft belt: Check & replacement if necessary every 18,000 miles or 2 years and replacement every 63,000 miles.
Timing belt: Check & replacement if necessary every 36,000 miles or 4 years and replacement every 63,000 miles.

1997→

Replacement every 72,000 miles or 6 years.
The previous use and service history of the vehicle must always be taken into account.
Refer to Timing Belt Replacement Intervals at the front of this manual.

Check For Engine Damage

CAUTION: This engine has been identified as an INTERFERENCE engine in which the possibility of valve-to-piston damage in the event of a timing belt failure is MOST LIKELY to occur.
A compression check of all cylinders should be performed before removing the cylinder head.

Repair Times – hrs

Remove & install:

Coupé	3,15
Tipo	2,75

Special Tools

- Flywheel locking tool – Fiat No.1860771000.
- Tensioning tool (timing belt) – Fiat No.1860745100/200.
- Tensioning tool (balancer shaft belt) – Fiat No.1860745100/400.
- Crankshaft sprocket adaptor – Fiat No.1860768000.

Special Precautions

- Disconnect battery earth lead.
- DO NOT turn crankshaft or camshaft when timing belt removed.
- Remove spark plugs to ease turning engine.
- Turn engine in normal direction of rotation (unless otherwise stated).
- DO NOT turn engine via camshaft or other sprockets.
- Observe all tightening torques.

Removal

1. Raise and support front of vehicle.
2. Remove:
   - ❏ RH front wheel.
   - ❏ RH inner wing panel.
   - ❏ Coolant expansion tank.
   - ❏ Auxiliary drive belt(s).
   - ❏ Crankshaft pulley bolts **1**.
   - ❏ Crankshaft pulley **2**.
   - ❏ Water pump pulley.
   - ❏ Spark plug lead cover.
   - ❏ Coolant pipe support.
3. Disconnect engine steady bar.
4. Remove timing belt covers **3** & **4**.

5. Turn crankshaft clockwise to TDC on No.1 cylinder. Ensure crankshaft sprocket timing marks aligned **5**.
6. Check timing marks on rear of camshaft sprockets aligned **6**.
7. Ensure timing marks on balancer shaft sprockets aligned **7** & **8**.
8. Remove bell housing cover plate. Fit flywheel locking tool. Tool No.1860771000.
9. Slacken balancer shaft belt tensioner sprocket nut **9**. Remove balancer shaft belt.
10. Remove crankshaft sprocket bolt **10**.
    NOTE: Crankshaft sprocket bolt has LH thread.
11. Remove balancer shaft belt sprocket **11**.
12. Ensure crankshaft sprocket timing mark aligned **12**.
13. Slacken timing belt tensioner nut **13**. Remove timing belt.
    NOTE: Mark direction of rotation on belt with chalk if belt is to be reused.

Installation

1. Ensure timing marks aligned **12** & **6**.
2. Fit timing belt in anti-clockwise direction, starting at crankshaft sprocket. Ensure belt is taut between sprockets.
3. Fit tensioning tool to tensioner pulley. Tool No.1860745100/200 **14** & **16**.
4. With scale bar horizontal set weight at 100 mm mark **15**.
5. Remove flywheel locking tool. Fit adaptor to crankshaft sprocket. Tool No.1860768000.
6. Turn crankshaft two turns clockwise to TDC on No.1 cylinder. Use tool No.1860768000. Ensure timing marks aligned **12** & **6**.
   NOTE: If scale bar moves from the horizontal: Reset scale bar and repeat tensioning procedure.
7. Tighten tensioner bolt to 44 Nm **13**. Remove tensioning tool.
8. Fit flywheel locking tool. Remove adaptor from crankshaft sprocket.
9. Fit balancer shaft belt sprocket **11**.
10. Fit bolt **10**. Tightening torque: 190 Nm.
11. Ensure crankshaft at TDC on No.1 cylinder **5**. Ensure timing marks on balancer shaft sprockets aligned **7** & **8**.
12. Fit balancer shaft belt.
13. Fit tensioning tool to tensioner sprocket. Tool No.1860745100/400 **14** & **17**.
14. With scale bar horizontal set weight at 205 mm mark **15**.
15. Remove flywheel locking tool. Turn crankshaft two turns clockwise to TDC on No.1 cylinder.
16. Ensure all timing marks aligned **5**, **6**, **7** & **8**.
    NOTE: If scale bar moves from the horizontal: Reset scale bar and repeat tensioning procedure.
17. Tighten balancer shaft belt tensioner sprocket nut to 23 Nm **9**.
18. Remove tensioning tool.
19. Install components in reverse order of removal.
20. Tighten crankshaft pulley bolts to 25 Nm **1**.

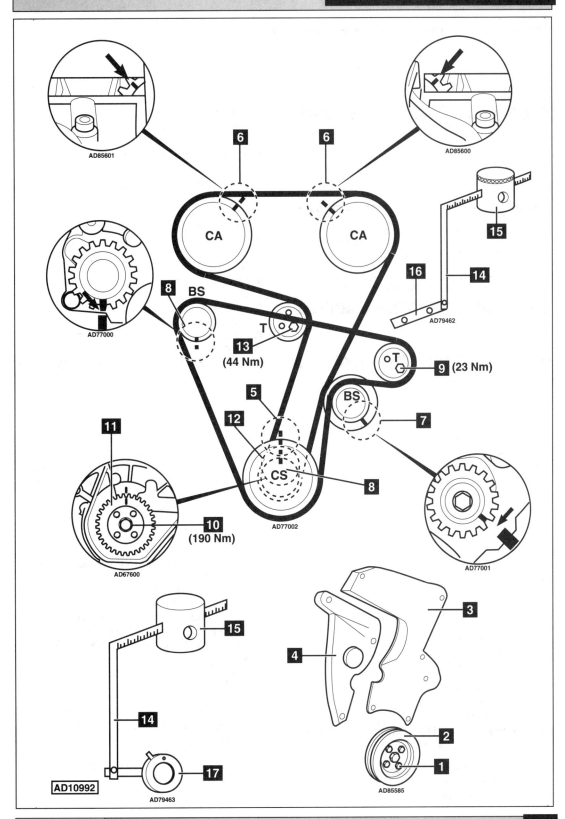

AD85601

AD85600

6

6

CA

CA

15

16

14

AD79462

AD77000

8

BS

T

13

(44 Nm)

T

9 (23 Nm)

5

BS

11

12

7

10

CS

8

(190 Nm)

AD67600

AD77002

AD77001

3

4

15

14

2

17

1

AD10992

AD79463

AD85585

FIAT

Model:	Brava 1,4 12V • Bravo 1,4 12V
Year:	1995-98
Engine Code:	182 A3.000, 182 A5.000

Replacement Interval Guide

Fiat recommend replacement every
72,000 miles or 6 years.
*The previous use and service history of the vehicle
must always be taken into account.*
*Refer to Timing Belt Replacement Intervals at the front
of this manual.*

Check For Engine Damage

*CAUTION: This engine has been identified as an
INTERFERENCE engine in which the possibility of
valve-to-piston damage in the event of a timing belt
failure is MOST LIKELY to occur.*
*A compression check of all cylinders should be
performed before removing the cylinder head.*

Repair Times – hrs

Remove & install	1,40
AC	+0,20

Special Tools

- Camshaft sprocket holding tool –
  Fiat No.1860831000.
- Crankshaft timing tool – Fiat No.1860901000.
- Camshaft timing tool – Fiat No.1860899000.
- Tensioning tool – Fiat No.1860443000.

Special Precautions

- Disconnect battery earth lead.
- DO NOT turn crankshaft or camshaft when timing belt
  removed.
- Remove spark plugs to ease turning engine.
- Turn engine in normal direction of rotation (unless
  otherwise stated).
- DO NOT turn engine via camshaft or other sprockets.
- Observe all tightening torques.

Removal

1. Raise and support front of vehicle.
2. Remove:
   - ❏ RH front wheel.
   - ❏ Wheel arch liner.
   - ❏ Auxiliary drive belt.
   - ❏ Alternator.
   - ❏ Timing belt upper cover.
   - ❏ Cylinder head cover.
3. Slacken crankshaft pulley bolts **1**.
4. Turn crankshaft to TDC on No.1 cylinder. Ensure
   timing marks aligned **2**.
5. Remove:
   - ❏ Crankshaft pulley bolts **1**.
   - ❏ Crankshaft pulley.
   - ❏ Timing belt lower cover.
6. Slacken tensioner sprocket nut **3**.
7. Remove timing belt.

Installation

1. Hold camshaft sprocket. Use tool
   No.1860831000. Slacken bolt **12**.
2. Remove oil pump bolts **4**.
3. Fit timing belt around crankshaft sprocket.
4. Fit timing tool **5** to crankshaft sprocket with
   timing pin **6** located in hole in tool (No.1 cylinder
   at TDC). Tool No.1860901000.
5. Remove bearing cap bolts on exhaust side of
   camshaft.
6. Slacken bearing cap bolts on inlet side of
   camshaft.
7. Carefully raise oil pipe. Remove camshaft
   bearing cap No.2.
8. Install timing tool in place of bearing cap No.2 **7**.
   Tool No.1860899000.
9. Tighten all bearing cap bolts to 10 Nm.
10. Fit timing belt in following order:
    - ❏ Camshaft sprocket.
    - ❏ Water pump pulley.
    - ❏ Tensioner sprocket.
11. Lever tensioner sprocket bracket at position **8**
    until pointer **9** at maximum setting. Use tool
    No.1860443000. Tighten tensioner sprocket
    nut **3**.
12. Hold camshaft sprocket. Use tool
    No.1860831000. Tighten bolt to 113 Nm **12**.
13. Remove timing tool from camshaft bearing No.2.
    Take care not to damage oil pipe.
14. Fit camshaft bearing cap No.2. Tighten bearing
    cap bolts to 15 Nm.
15. Remove timing tool from crankshaft sprocket **5**.
    Fit oil pump bolts **4**.
16. Turn crankshaft two turns clockwise.
17. Slacken tensioner sprocket nut **3**. Lever
    tensioner sprocket bracket at position **8** until
    pointer **9** and mark **10** aligned. Use tool
    No.1860443000.
18. Tighten tensioner sprocket nut **3**.
    Tightening torque: 25 Nm.
19. Install components in reverse order of removal.
20. Ensure crankshaft pulley located correctly on
    pin **11**.
21. Ensure timing marks aligned **2**.
22. Tighten crankshaft pulley bolts **1**.
    Tightening torque: 28 Nm.

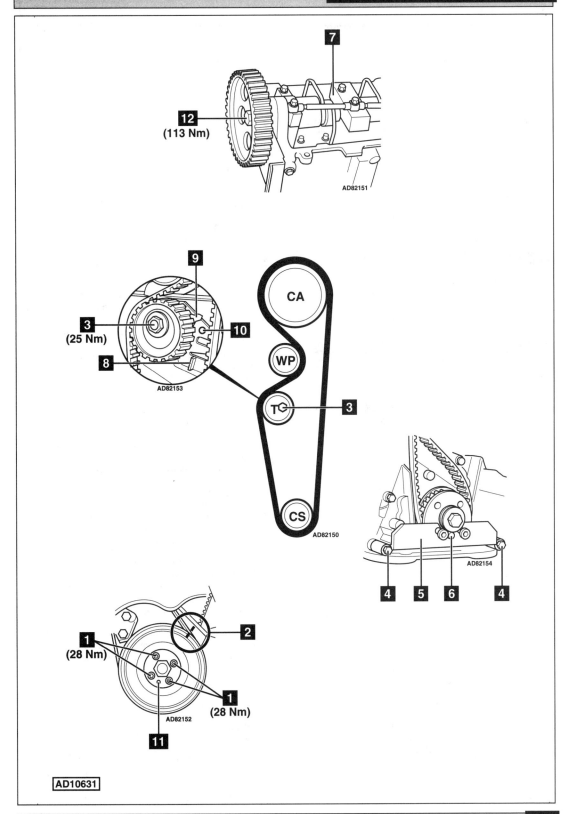

Model:	Brava 1,6 16V • Bravo 1,6 16V • Marea 1,6 16V • Marea Weekend 1,6 16V Multipla 1,6 16V • Stilo 1,6 16V
Year:	1995-03
Engine Code:	182 A4.000, 182 A6.000, 182 B6.000

Replacement Interval Guide

Fiat recommend:
→2000: Replacement every 72,000 miles or 6 years.
2001→: Check and replacement if necessary every 36,000 miles and replacement every 72,000 miles or 5 years.
The previous use and service history of the vehicle must always be taken into account.
Refer to Timing Belt Replacement Intervals at the front of this manual.

Check For Engine Damage

CAUTION: This engine has been identified as an INTERFERENCE engine in which the possibility of valve-to-piston damage in the event of a timing belt failure is MOST LIKELY to occur.
A compression check of all cylinders should be performed before removing the cylinder head.

Repair Times – hrs

Remove & install	1,85
AC	+0,25
Stilo	2,50

Special Tools

- Dial gauge and adaptor – Fiat No.1860895000.
- Flywheel locking tool (except Stilo) – Fiat No.1860771000.
- Flywheel locking tool (Stilo) – Fiat No.1867030000.
- Wrench adaptor – Fiat No.1860885000.
- Wrench – Fiat No.1860831001.
- Camshaft locking tools – Fiat No.1860874000.
- Tensioning tool – Fiat No.1860876000.

Removal

1. Raise and support front of vehicle.
2. Remove:
   - RH front wheel.
   - RH inner wing panel.
   - Air intake hose.
   - Auxiliary drive belts and tensioner.
   - Spark plugs.
   - Ignition coil and HT leads.
   - Camshaft rear covers **1**.
   - Bell housing lower cover.
   - Stilo: Engine control module (ECM). DO NOT disconnect harness multi-plug.
3. Stilo: Support engine. Remove RH engine mounting.
4. Fit dial gauge and adaptor to No.1 cylinder. Tool No.1860895000 **2**.
5. Turn crankshaft to TDC on No.1 cylinder. Set dial gauge to zero. Ensure crankshaft pulley and flywheel timing marks aligned **3** & **4**.
   NOTE: Stilo: Ensure notch on crankshaft pulley is aligned with RPM sensor.
6. Fit flywheel locking tool **5**:
   - Except Stilo: Tool No.18607710000.
   - Stilo: Tool No.1867030000.

7. Remove:
   - Crankshaft pulley nut **6**.
   - Crankshaft pulley.
   - Timing belt cover.
8. Slacken tensioner nut **7**. Release tension on belt.
9. Remove timing belt.

Installation

1. Ensure flywheel locking tool located correctly **5**.
2. Hold camshaft sprockets. Use tool Nos.1860885000 & 1860831001 **8**. Slacken bolts **9**.
3. Fit locking tool to rear of each camshaft with slot in camshaft aligned with lug on tool **10**. Tool No.1860874000.
   *NOTE: As inlet and exhaust camshaft tools differ, ensure notch on tool aligns with blanking plug in cylinder head **11**.*
4. Fit timing belt to crankshaft sprocket. Fit crankshaft pulley and nut **6**.
5. Ensure flywheel locking tool located correctly **5**. Tighten crankshaft pulley nut **6**:
   - A – Brava/Bravo: 220 Nm.
   - B – Marea/Weekend/Multipla: 190 Nm.
   - C – Stilo: 180-200 Nm.
6. Remove flywheel locking tool **5**.
7. Ensure No.1 cylinder at TDC by turning crankshaft slightly clockwise and anti-clockwise. Use dial gauge.
   NOTE: Stilo: Ensure notch on crankshaft pulley is aligned with RPM sensor.
8. Turn camshaft sprockets clockwise.
9. Fit timing belt (already looped round crankshaft sprocket) to remainder of sprockets in anti-clockwise direction. Ensure belt is taut between sprockets.
10. Remove bolt from front casing **12**. Fit tensioning tool **13**. Tool No.1860876000.
11. Use spanner on tensioning tool bolt **16**. Turn anti-clockwise to set automatic tensioner pulley at maximum setting **14**. Tighten tensioner nut **7**.
12. Remove dial gauge and adaptor from No.1 cylinder **2**.
13. Hold camshaft sprockets. Use tool Nos.1860885000 & 1860831001 **8**.
14. Tighten camshaft sprocket bolts **9**:
    - A – Brava/Bravo: 115 Nm.
    - B – Marea/Weekend/Multipla/Stilo: 120 Nm.
15. Remove locking tool from each camshaft **10**.
16. Turn crankshaft two turns clockwise to TDC on No.1 cylinder. Ensure flywheel timing marks aligned **4**.
17. Slacken tensioner nut **7**. Align pointer and mark on automatic tensioner pulley **15**. Use tool No.1860876000.
18. Tighten tensioner nut **7**:
    - A – Brava/Bravo: 23 Nm.
    - B – Stilo: 21-26 Nm.
    - C – Marea/Weekend/Multipla: 25 Nm.
19. Remove tensioning tool. Fit bolt in front casing.
20. Fit timing belt cover.
21. Install components in reverse order of removal.

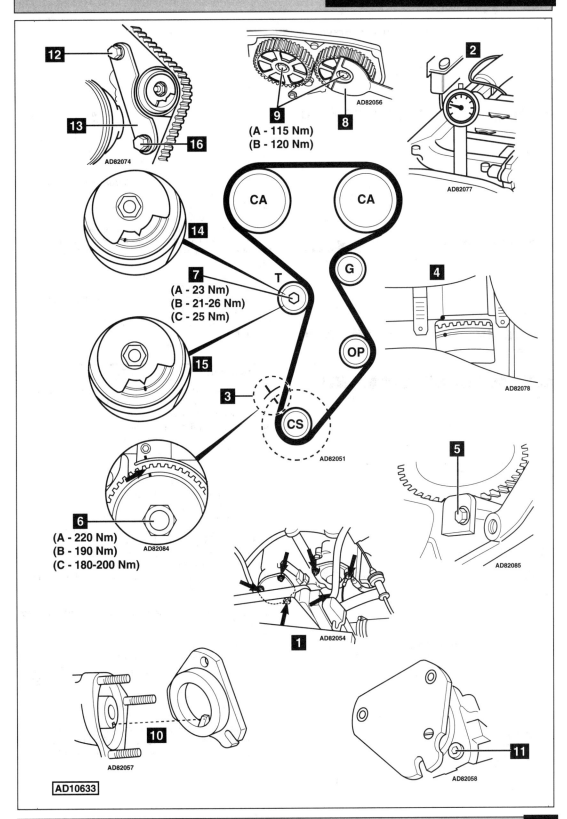

12

13

16

AD82074

9

8

AD82056

(A - 115 Nm)
(B - 120 Nm)

2

AD82077

CA

CA

14

G

4

AD82078

7

T

(A - 23 Nm)
(B - 21-26 Nm)
(C - 25 Nm)

15

OP

3

CS

AD82051

5

AD82085

6

AD82084

(A - 220 Nm)
(B - 190 Nm)
(C - 180-200 Nm)

1 AD82054

10

AD82057

11

AD82058

AD10633

FIAT

Model:	Brava 1,8 16V • Bravo 1,8 16V
Year:	1995-02
Engine Code:	182 A2.000

Replacement Interval Guide

Fiat recommend:
→2000: Replacement every 72,000 miles or
6 years.
2001→: Check and replacement if necessary every
36,000 miles and replacement every 72,000 miles
or 5 years.
*The previous use and service history of the vehicle
must always be taken into account.*
*Refer to Timing Belt Replacement Intervals at the
front of this manual.*

Check For Engine Damage

*CAUTION: This engine has been identified as an
INTERFERENCE engine in which the possibility of
valve-to-piston damage in the event of a timing belt
failure is MOST LIKELY to occur.*
*A compression check of all cylinders should be
performed before removing the cylinder head.*

Repair Times – hrs

Remove & install	1,75
AC	+0,20

Special Tools

- Dial gauge and adaptor – Fiat No.1895879000.
- Camshaft locking tools – Fiat No.1860875000.
- Flywheel locking tool – Fiat No.1860898000.
- Sprocket holding tool – Fiat No.1860831000.
- Tensioning tool – Fiat No.1860845000.

Special Precautions

- Disconnect battery earth lead.
- DO NOT turn crankshaft or camshaft when timing belt removed.
- Remove spark plugs to ease turning engine.
- Turn engine in normal direction of rotation (unless otherwise stated).
- DO NOT turn engine via camshaft or other sprockets.
- Observe all tightening torques.

Removal

1. Raise and support front of vehicle.
2. Remove:
   - ❑ RH front wheel.
   - ❑ RH inner wing panel.
3. Turn auxiliary drive belt tensioner anti-clockwise to release tension on belt. Remove auxiliary drive belt.
4. Remove:
   - ❑ Auxiliary drive belt guide pulley.
   - ❑ Timing belt cover.
   - ❑ Engine top cover.
   - ❑ Disconnect ignition coil multi-plugs and breather hose.
   - ❑ Cylinder head cover.
   - ❑ Ignition coils.
   - ❑ Spark plugs.
   - ❑ Bell housing lower cover.
5. Fit dial gauge and adaptor to No.1 cylinder **1**. Tool No.1895879000.

6. Turn crankshaft to TDC on No.1 cylinder. Use dial gauge and flywheel timing marks **2**.
7. Ensure both camshafts at TDC on No.1 cylinder. If not: Turn crankshaft one turn clockwise.
8. Remove third bearing cap from each camshaft **3** & **4**.
   NOTE: Mark bearing caps before removal for identification.
9. Fit locking tools in place of bearing caps. Tool No.1860875000.
   NOTE: Ensure locking tools aligned with respective cam profiles to prevent damage.
10. Fit flywheel locking tool. Tool No.1860898000.
11. Remove:
    - ❑ Crankshaft pulley bolts **7**.
    - ❑ Crankshaft pulley **6**.
12. Ensure pin in crankshaft aligned centrally with crankcase at 6 o'clock **8**.
13. Slacken tensioner sprocket nut **9**. Release tension on belt.
14. Remove timing belt.

Installation

1. Ensure crankshaft at TDC on No.1 cylinder **1** & **2**.
2. Ensure flywheel locking tool located correctly.
3. Ensure locking tools located correctly in camshafts.
4. Hold camshaft sprockets **10**. Use tool No.1860831000. Slacken bolts.
5. Fit timing belt in following order:
   - ❑ Crankshaft sprocket.
   - ❑ Guide pulley.
   - ❑ Exhaust camshaft sprocket.
   - ❑ Inlet camshaft sprocket.
   - ❑ Water pump pulley.
   - ❑ Tensioner sprocket.
6. Fit tensioning tool into hole adjacent to tensioner sprocket **11**. Tool No.1860845000.
7. Turn tensioning tool to tension belt to maximum. Tighten tensioner sprocket nut **9**.
8. Hold camshaft sprockets. Use tool No.1860831000. Tighten each bolt to 118 Nm.
9. Remove locking tools from camshafts.
10. Fit bearing caps in correct locations. Tighten bolts to 15 Nm **3** & **4**.
11. Remove:
    - ❑ Flywheel locking tool.
    - ❑ Dial gauge and adaptor **1**.
12. Turn crankshaft two turns clockwise to TDC on No.1 cylinder. Ensure flywheel timing marks aligned **2**.
13. Fit tensioning tool **11**. Tool No.1860845000. Slacken tensioner sprocket nut **9**. Align pointer with mark on casing **5**.
14. Tighten tensioner sprocket nut to 25 Nm **9**.
15. Install components in reverse order of removal.
16. Tighten crankshaft pulley bolts to 28 Nm **7**.

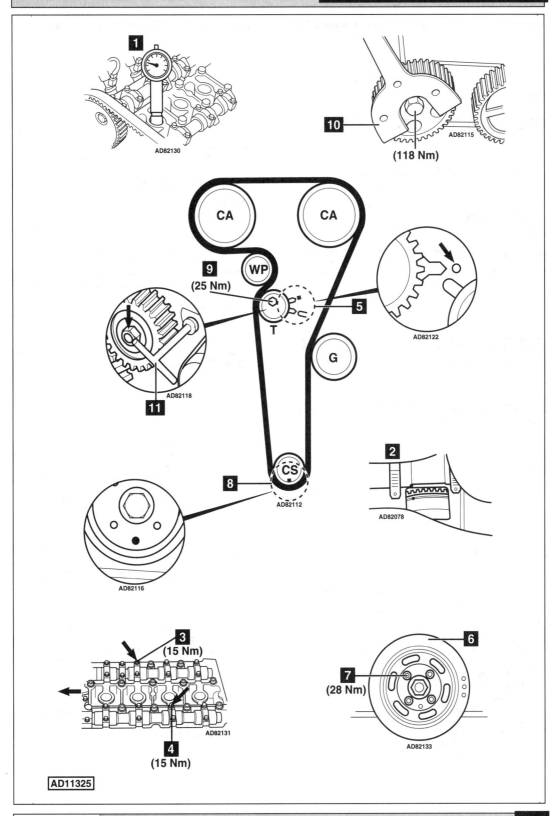

1   AD82130

10   AD82115   (118 Nm)

CA   CA

WP

9   (25 Nm)

5   AD82122

T

11   AD82118

G

CS   AD82112

8

AD82116

2   AD82078

3   (15 Nm)   AD82131

4   (15 Nm)

6

7   (28 Nm)   AD82133

AD11325

Model:	Bravo 2,0 20V • Marea 2,0 20V • Marea Weekend 2,0 20V • Coupé 2,0 20V Coupé 2,0 20V Turbo
Year:	1994-01
Engine Code:	182 A1.000, 175 A3.000

Replacement Interval Guide

Fiat recommend:
→2000: Replacement every 72,000 miles or 6 years.
2001→: Check and replacement if necessary every 36,000 miles and replacement every 72,000 miles or 5 years.
The previous use and service history of the vehicle must always be taken into account.
Refer to Timing Belt Replacement Intervals at the front of this manual.

Check For Engine Damage

CAUTION: This engine has been identified as an INTERFERENCE engine in which the possibility of valve-to-piston damage in the event of a timing belt failure is MOST LIKELY to occur.
A compression check of all cylinders should be performed before removing the cylinder head.

Repair Times – hrs

Remove & install:
Bravo/Marea	10,00
Coupé	13,80

Special Tools

■ Dial gauge and adaptor – Fiat No.1895879000.
■ Camshaft locking tools – Fiat No.1860892000.

Special Precautions

■ Disconnect battery earth lead.
■ DO NOT turn crankshaft or camshaft when timing belt removed.
■ Remove spark plugs to ease turning engine.
■ Turn engine in normal direction of rotation (unless otherwise stated).
■ DO NOT turn engine via camshaft or other sprockets.
■ Observe all tightening torques.

Removal

NOTE: Timing belt cannot be replaced with engine installed in vehicle.
1. Raise and support front of vehicle.
2. Remove engine undershield.
3. Drain coolant.
4. Disconnect hoses, wiring and ancillaries.
5. Remove engine from chassis.
   NOTE: Engine is removed from below vehicle.
6. Place engine on suitable stand or cradle.
7. Remove:
   ❑ Auxiliary drive belt(s).
   ❑ Crankshaft pulley bolts **1**.
   ❑ Crankshaft pulley **2**.
   ❑ Timing belt cover **3**.
   ❑ Spark plugs.
8. Fit dial gauge and adaptor to No.1 cylinder **4**. Use tool No.1895879000.

9. Turn crankshaft clockwise to TDC on No.1 cylinder. Ensure crankshaft sprocket timing marks aligned **5**.
10. Ensure both camshafts at TDC on No.1 cylinder. If not: Turn crankshaft one turn clockwise.
11. Remove fourth bearing cap from inlet camshaft **6**.
12. Remove second bearing cap from exhaust camshaft **7**.
    NOTE: Mark bearing caps before removal for identification.
13. Fit locking tools in place of bearing caps **6** & **7**. Tool No.1860892000.
    NOTE: Locking tools are marked 'A' (inlet) and 'S' (exhaust). Ensure locking tools aligned with respective cam profiles to prevent damage.
14. Slacken tensioner sprocket nut **8**. Release tension on belt.
15. Remove timing belt.

Installation

1. Ensure crankshaft at TDC on No.1 cylinder. Use dial gauge **4**. Ensure timing marks aligned **5**.
2. Ensure locking tools located correctly in camshafts **6** & **7**.
3. Fit timing belt in following order:
   ❑ Crankshaft sprocket.
   ❑ Guide pulley.
   ❑ Exhaust camshaft sprocket.
   ❑ Inlet camshaft sprocket.
   ❑ Tensioner sprocket.
   ❑ Water pump pulley.
   *NOTE: Ensure belt is taut between sprockets and pulleys. Ensure marks on belt aligned with marks on crankshaft sprocket and exhaust camshaft sprocket **9**. Observe direction of rotation mark on belt.*
4. Push against lug on tensioner sprocket bracket until pointer at maximum setting **10**. Use suitable lever.
5. Tighten tensioner sprocket nut to 50 Nm **8**.
6. Remove locking tools from camshafts **6** & **7**. Fit bearing caps in correct locations. Tighten bolts to 15 Nm.
7. Turn crankshaft two turns clockwise to TDC on No.1 cylinder. Ensure timing marks aligned **5**.
8. Hold tensioner sprocket at lug. Use suitable lever. Slacken tensioner sprocket nut **8**.
9. Align pointer with mark **11**. Use suitable lever. Tighten tensioner sprocket nut to 50 Nm **8**.
10. Fit timing belt cover. Tighten bolts to 9 Nm.
11. Fit crankshaft pulley. Tighten bolts to 25 Nm **1**.
12. Install components in reverse order of removal.
13. Install engine to chassis. Reconnect hoses, cables and ancillaries.
14. Refill and bleed cooling system.

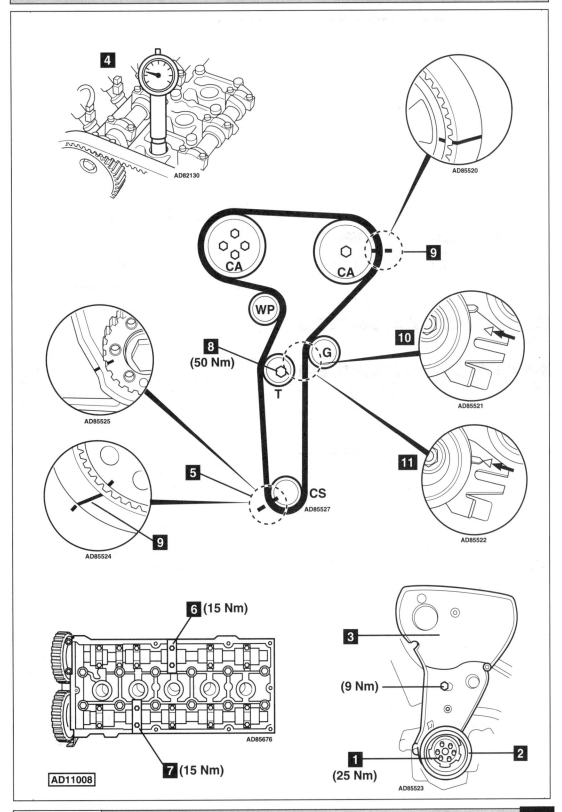

4 AD82130

AD85520

9

CA

CA

WP

8
(50 Nm)

G

10

T

AD85521

AD85525

5

CS

AD85527

11

AD85524

9

AD85522

6 (15 Nm)

3

(9 Nm)

AD85676

1

2

7 (15 Nm)

(25 Nm)

AD85523

AD11008

Model:	Marea 1,8 16V • Marea Weekend 1,8 16V
Year:	1996-02
Engine Code:	182 A2.000

Replacement Interval Guide

Fiat recommend:
→2000: Replacement every 72,000 miles or
6 years.
2001→: Check and replacement if necessary every
36,000 miles and replacement every 72,000 miles
or 5 years.
*The previous use and service history of the vehicle
must always be taken into account.*
*Refer to Timing Belt Replacement Intervals at the
front of this manual.*

Check For Engine Damage

*CAUTION: This engine has been identified as an
INTERFERENCE engine in which the possibility of
valve-to-piston damage in the event of a timing belt
failure is MOST LIKELY to occur.*
*A compression check of all cylinders should be
performed before removing the cylinder head.*

Repair Times – hrs

Remove & install	1,75
AC	+0,20

Special Tools

- Dial gauge and adaptor – Fiat No.1895879000.
- Camshaft locking tools – Fiat No.1860875000.
- Flywheel locking tool – Fiat No.1860898000.
- Sprocket holding tool – Fiat No.1860831000.
- Tensioning tool – Fiat No.1860845000.

Special Precautions

- Disconnect battery earth lead.
- DO NOT turn crankshaft or camshaft when timing belt removed.
- Remove spark plugs to ease turning engine.
- Turn engine in normal direction of rotation (unless otherwise stated).
- DO NOT turn engine via camshaft or other sprockets.
- Observe all tightening torques.

Removal

1. Raise and support front of vehicle.
2. Remove:
   - ❏ RH front wheel.
   - ❏ RH inner wing panel.
3. Turn auxiliary drive belt tensioner anti-clockwise to release tension on belt. Remove auxiliary drive belt.
4. Remove:
   - ❏ Auxiliary drive belt guide pulley.
   - ❏ Timing belt cover.
   - ❏ Engine top cover.
   - ❏ Disconnect ignition coil multi-plugs and breather hose.
   - ❏ Cylinder head cover.
   - ❏ Ignition coils.
   - ❏ Spark plugs.
   - ❏ Bell housing lower cover.
5. Fit dial gauge and adaptor to No.1 cylinder **1**. Tool No.1895879000.
6. Turn crankshaft to TDC on No.1 cylinder. Use dial gauge and flywheel timing marks **2**.
7. Ensure both camshafts at TDC on No.1 cylinder. If not: Turn crankshaft one turn clockwise.
8. Remove third bearing cap from each camshaft **3** & **4**.
   NOTE: Mark bearing caps before removal for identification.
9. Fit locking tools in place of bearing caps. Tool No.1860875000.
   NOTE: Ensure locking tools aligned with respective cam profiles to prevent damage.
10. Fit flywheel locking tool. Tool No.1860898000.
11. Remove:
    - ❏ Crankshaft pulley bolts **7**.
    - ❏ Crankshaft pulley **6**.
12. Ensure pin in crankshaft aligned centrally with crankcase at 6 o'clock **8**.
13. Slacken tensioner sprocket nut **9**. Release tension on belt.
14. Remove timing belt.

Installation

1. Ensure crankshaft at TDC on No.1 cylinder **1** & **2**.
2. Ensure flywheel locking tool located correctly.
3. Ensure locking tools located correctly in camshafts.
4. Hold camshaft sprockets **10**. Use tool No.1860831000. Slacken bolts.
5. Fit timing belt in following order:
   - ❏ Crankshaft sprocket.
   - ❏ Guide pulley.
   - ❏ Exhaust camshaft sprocket.
   - ❏ Inlet camshaft sprocket.
   - ❏ Water pump pulley.
   - ❏ Tensioner sprocket.
6. Fit tensioning tool into hole adjacent to tensioner sprocket **11**. Tool No.1860845000.
7. Turn tensioning tool to tension belt to maximum. Tighten tensioner sprocket nut to 25 Nm **9**.
8. Hold camshaft sprockets. Use tool No.1860831000. Tighten bolts to 120 Nm.
9. Remove locking tools from camshafts.
10. Fit bearing caps in correct locations. Tighten bolts to 15 Nm **3** & **4**.
11. Remove:
    - ❏ Flywheel locking tool.
    - ❏ Dial gauge and adaptor **1**.
12. Turn crankshaft two turns clockwise to TDC on No.1 cylinder. Ensure flywheel timing marks aligned **2**.
13. Fit tensioning tool **11**. Tool No.1860845000. Slacken tensioner sprocket nut **9**. Align pointer with mark on casing **5**.
14. Tighten tensioner sprocket nut to 25 Nm **9**.
15. Install components in reverse order of removal.
16. Tighten crankshaft pulley bolts to 32 Nm **7**.

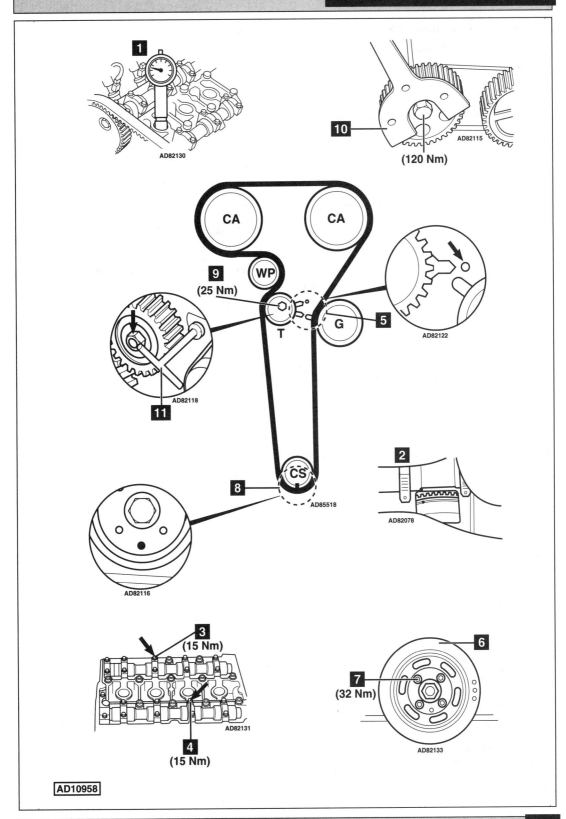

1 AD82130

10 (120 Nm) AD82115

CA   CA

WP

9 (25 Nm)

T   **5** G   AD82122

11 AD82118

8 CS AD85518

AD82116

2 AD82078

3 (15 Nm)

4 (15 Nm)   AD82131

6

7 (32 Nm)   AD82133

AD10958

Model:	Ulysse 1,8 • Ulysse 2,0/Turbo • Ducato 2,0
Year:	1994-02
Engine Code:	LFW, RFU, RGX, RFW

Replacement Interval Guide

Fiat recommend:
→1998: Check & replacement if necessary every 36,000 miles or 4 years and replacement every 63,000 miles.
1999-00: Check & replacement if necessary every 36,000 miles or 3 years and replacement every 72,000 miles or 6 years.
2001→: Check & replacement if necessary every 36,000 miles and replacement every 72,000 miles or 5 years.
The previous use and service history of the vehicle must always be taken into account.
Refer to Timing Belt Replacement Intervals at the front of this manual.

Check For Engine Damage

CAUTION: This engine has been identified as an INTERFERENCE engine in which the possibility of valve-to-piston damage in the event of a timing belt failure is MOST LIKELY to occur.
A compression check of all cylinders should be performed before removing the cylinder head.

Repair Times – hrs

Remove & install:

Ulysse 1,8/2,0	1,90
Ulysse 2,0 Turbo	2,10
Ducato 2,0	1,50

Special Tools

- Crankshaft timing pin – 8 mm diameter.
- Camshaft timing pin – 10 mm diameter.
- Tensioning tool – Fiat No.1860755000
- Flywheel locking tool – Fiat No.1860161000.

Special Precautions

- Disconnect battery earth lead.
- DO NOT turn crankshaft or camshaft when timing belt removed.
- Remove spark plugs to ease turning engine.
- Turn engine in normal direction of rotation (unless otherwise stated).
- DO NOT turn engine via camshaft or other sprockets.
- Observe all tightening torques.

Removal

1. Raise and support front of vehicle.
2. Remove:
   - RH front wheel.
   - RH wheel arch liner.
   - Auxiliary drive belt(s).
   - Ulysse: Auxiliary drive belt tensioner.
   - Auxiliary drive belt cover.
   - Timing belt upper cover **1**.
3. Turn crankshaft clockwise until 10 mm timing pin can be inserted through camshaft sprocket and 8 mm timing pin through crankshaft pulley **3** & **2**.
4. Remove clutch housing plate. Lock flywheel **4**. Use tool No.1860161000.

5. Remove:
   - Crankshaft pulley bolt **5**.
   - Crankshaft pulley **6**.
   - Timing belt centre cover **7**.
   - Timing belt lower cover **8**.
   - Thrust washer **9**.
6. Slacken tensioner bolt **10**. Move tensioner away from belt. Lightly tighten bolt.
7. Remove timing belt.

Installation

1. Temporarily fit crankshaft pulley and timing pin **6** & **2**. Ensure timing pin located correctly.
2. Ensure timing pin located in camshaft sprocket **3**.
3. Remove crankshaft pulley and timing pin **6** & **2**.
4. Fit timing belt in following order:
   - Camshaft sprocket.
   - Crankshaft sprocket.
   - Water pump sprocket.
   - Tensioner pulley.
   NOTE: Ensure belt is taut between sprockets on non-tensioned side.
5. Slacken tensioner bolt **10**. Turn tensioner firmly anti-clockwise to lightly tension belt. Lightly tighten bolt.
6. Install:
   - Thrust washer **9**.
   - Crankshaft pulley **6**. Temporarily tighten bolt **5**.
7. Remove:
   - Camshaft timing pin **3**.
   - Flywheel locking tool **4**.
8. Turn crankshaft two turns clockwise to settle belt.
9. Ensure timing pins can be inserted **2** & **3**.
10. Fit tensioning tool to tensioner pulley **11**. Tool No.1860755000. Set weight position **12**. Ulysse: 18 mm. Ducato: 30 mm.
11. Slacken tensioner bolt **10**.
12. Remove timing pins **2** & **3**.
13. Turn crankshaft slowly two turns clockwise. Ensure timing pins can be inserted **2** & **3**.
14. Tighten tensioner bolt to 21 Nm **10**.
15. Remove:
   - Tensioning tool **11**.
   - Timing pins **2** & **3**.
16. Remove:
   - Crankshaft pulley bolt **5**.
   - Crankshaft pulley **6**.
17. Install components in reverse order of removal.
18. Tighten crankshaft pulley bolt to 120 Nm **5**.

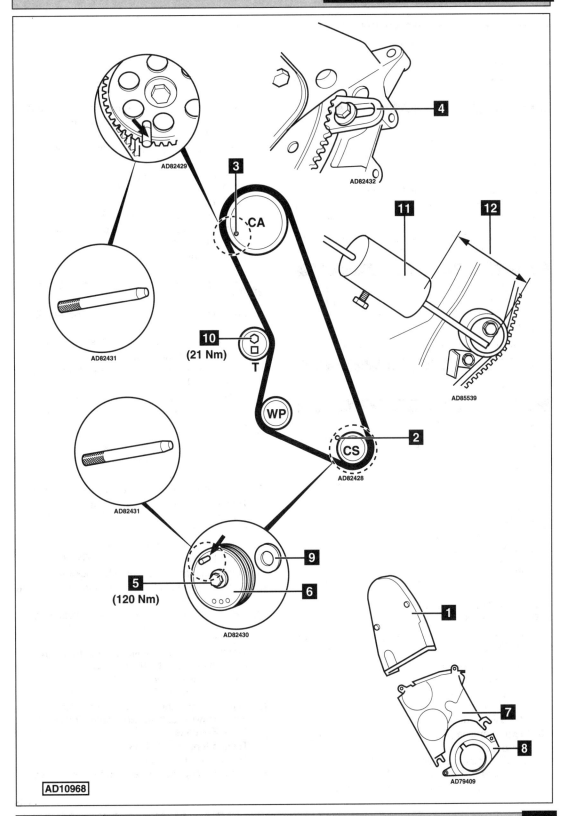

AD82429

AD82432

4

3

CA

11

12

AD85539

10
(21 Nm)
T

WP

CS
AD82428

2

AD82431

AD82431

9

5
(120 Nm)

6

AD82430

1

7

8

AD79409

AD10968

FIAT

Model:	**Ulysse 2,0 16V**
Year:	**1998-02**
Engine Code:	**RFV**

Replacement Interval Guide

Fiat recommend:
→2000: Check and replacement if necessary every 36,000 miles or 3 years and replacement every 72,000 miles or 6 years.
2001→: Check and replacement if necessary every 36,000 miles and replacement every 72,000 miles or 5 years.
The previous use and service history of the vehicle must always be taken into account.
Refer to Timing Belt Replacement Intervals at the front of this manual.

Check For Engine Damage

CAUTION: This engine has been identified as an INTERFERENCE engine in which the possibility of valve-to-piston damage in the event of a timing belt failure is MOST LIKELY to occur.
A compression check of all cylinders should be performed before removing the cylinder head.

Repair Times – hrs

Remove & install	2,35

Special Tools

- 8 mm timing pin – for crankshaft pulley.
- 2 x 6 mm timing pins – for camshaft sprockets.
- Flywheel locking tool – Fiat No.1860766000.
- Camshaft sprocket locking tool – Fiat No.1870723000.
- Tension gauge – SEEM C.Tronic G2 105.5.

Special Precautions

- Disconnect battery earth lead.
- DO NOT turn crankshaft or camshaft when timing belt removed.
- Remove spark plugs to ease turning engine.
- Turn engine in normal direction of rotation (unless otherwise stated).
- DO NOT turn engine via camshaft or other sprockets.
- Observe all tightening torques.

Removal

1. Raise and support front of vehicle.
2. Remove:
   - ❑ RH front wheel.
   - ❑ RH wheel arch liner.
   - ❑ Auxiliary drive belt.
3. Support engine.
4. Remove:
   - ❑ RH engine mounting.
   - ❑ Timing belt upper cover **1**.
5. Turn crankshaft clockwise to setting position.

6. Insert 8 mm timing pin in crankshaft pulley **2**.
7. Insert 6 mm timing pins in camshaft sprockets **3**.
8. Lock flywheel. Use tool No.1860766000.
9. Remove:
   - ❑ 8 mm timing pin **2**.
   - ❑ Crankshaft pulley bolt **4**.
   - ❑ Crankshaft pulley **5**.
   - ❑ Timing belt lower cover **6**.
10. Slacken tensioner bolt **7**. Move tensioner away from belt. Lightly tighten bolt.
11. Remove timing belt.

Installation

1. Install locking tool to camshaft sprockets **8**. Tool No.1870723000.
2. Slacken bolt of each camshaft sprocket **9**.
3. Remove camshaft locking tool **8**.
4. Ensure timing pins located correctly in camshaft sprockets **3**.
5. Fit timing belt to crankshaft sprocket.
   NOTE: Observe direction of rotation marks on belt.
6. Install:
   - ❑ Timing belt lower cover **6**.
   - ❑ Crankshaft pulley **5**.
   - ❑ Crankshaft pulley bolt **4**.
7. Ensure flywheel locking tool located correctly.
8. Tighten crankshaft pulley bolt **4**. Tightening torque: 130 Nm. Use Loctite or similar compound.
9. Remove flywheel locking tool.
10. Insert timing pin in crankshaft pulley **2**.
11. Turn camshaft sprockets fully clockwise.
12. Fit timing belt in anti-clockwise direction. Ensure belt is taut between sprockets.
13. Lay belt on teeth of camshaft sprockets. Engage belt teeth by turning sprockets slightly anti-clockwise.
    *NOTE: Angular movement of sprockets must not be more than one tooth space **10**.*
14. Attach tension gauge to belt at ▽. Tool No.SEEM C.Tronic G2 105.5 **11**.
15. Slacken tensioner bolt **7**. Turn tensioner anti-clockwise until tension gauge indicates 45 SEEM units.
16. Tighten tensioner bolt **7**. Tightening torque: 20 Nm.
17. Install locking tool to camshaft sprockets **8**.

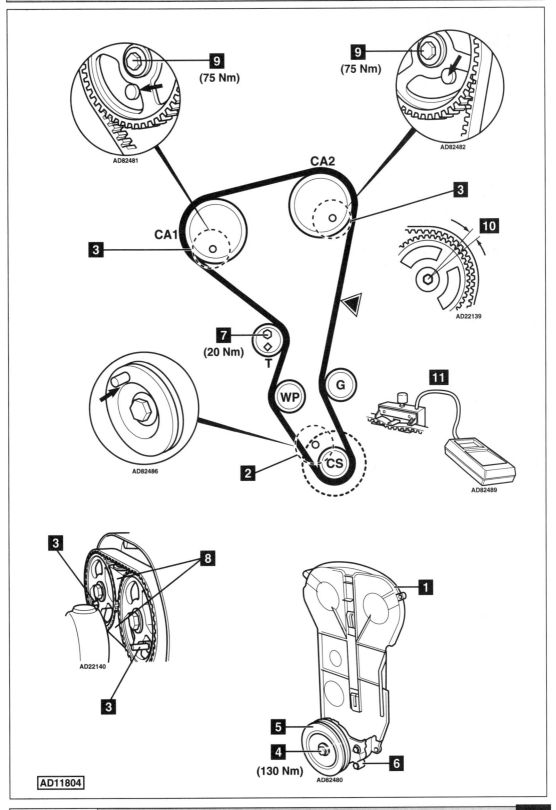

9 (75 Nm)

AD82481

9 (75 Nm)

AD82482

CA2

CA1

3

3

10

AD22139

7 (20 Nm)

T

AD82486

WP

G

11

2

CS

AD82489

3

8

1

AD22140

3

3

5

4 (130 Nm)

AD82480

6

AD11804

←

18. Tighten bolt of each camshaft sprocket **9**.
    Tightening torque: 75 Nm.
19. Remove:
    ❑ Locking tool **8**.
    ❑ Tension gauge **11**.
    ❑ Timing pins **2** & **3**.
20. Turn crankshaft two turns clockwise to setting
    position.
21. Install locking tool to camshaft sprockets **8**.
22. Slacken bolt of each camshaft sprocket **9**.
23. Insert timing pins in camshaft sprockets **3**.
24. Remove camshaft locking tool **8**.
25. Ensure crankshaft pulley timing pin can be
    inserted easily **2**. If not: Repeat installation and
    tensioning procedures.
26. Slacken tensioner bolt **7**.
27. Attach tension gauge to belt at ▽ **11**.
28. Turn tensioner anti-clockwise until tension
    gauge indicates 26 SEEM units.
29. Tighten tensioner bolt **7**.
    Tightening torque: 20 Nm.
30. Install locking tool to camshaft sprockets **8**.
31. Tighten bolt of each camshaft sprocket **9**.
    Tightening torque: 75 Nm.
32. Remove:
    ❑ Tension gauge **11**.
    ❑ Timing pins **2**.
    ❑ Locking tool **8**.
33. Turn crankshaft two turns clockwise to setting
    position.
34. Ensure timing pins can be inserted easily **2**
    & **3**.
35. Install components in reverse order of removal.

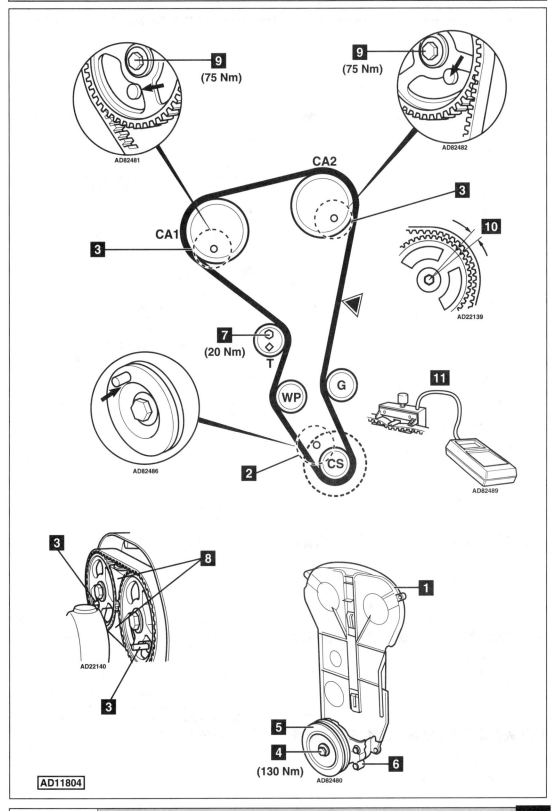

9 (75 Nm)

AD82481

9 (75 Nm)

AD82482

CA2

3

CA1

3

10

AD22139

7 (20 Nm)

T

WP

G

11

AD82486

2

CS

AD82489

3

8

AD22140

3

1

5

4 (130 Nm)

6

AD82480

AD11804

FIAT

Model:	Uno 1,7 D • Duna 1,7 D • Ritmo/Strada 1,7 D • Regata 1,7 D • Tipo 1,7 D Fiorino 1,7 D
Year:	1986-98
Engine Code:	146 B2.000, 149 B3.000, 149 B4.000, 149 B4.046

Replacement Interval Guide

Fiat recommend check and/or replacement at the following intervals:

Tipo/Fiorino:

To 12/88:
Every 54,000 miles - replace.

01/89-04/91:
Every 60,000 miles - replace.

05/91 on:
Every 36,000 miles - check & replace if necessary.
Every 63,000 miles - replace.

Uno/Duna:

To 12/88:
Every 18,000 miles - check & replace if necessary.

01/89-04/91:
Every 60,000 miles - replace.

05/91 on:
Every 36,000 miles - check & replace if necessary.
Every 63,000 miles - replace.

Ritmo/Strada/Regata:

To 12/86:
Every 36,000 miles - replace.

01/87 on:
Every 54,000 miles - replace.
The previous use and service history of the vehicle must always be taken into account.
Refer to Timing Belt Replacement Intervals at the front of this manual.

Check For Engine Damage

CAUTION: This engine has been identified as an INTERFERENCE engine in which the possibility of valve-to-piston damage in the event of a timing belt failure is MOST LIKELY to occur.
A compression check of all cylinders should be performed before removing the cylinder head.

Repair Times – hrs

Remove & install:

Uno/Duna/Fiorino	1,40
Ritmo/Strada	1,60
Regata	1,60
Tipo	1,40

Special Tools

- Injection pump sprocket locking tool – Fiat No.1842128000.
- Tensioning tool – Fiat No.1860722000.

Special Precautions

- Disconnect battery earth lead.
- DO NOT turn crankshaft or camshaft when timing belt removed.
- Remove glow plugs to ease turning engine.
- Turn engine in normal direction of rotation (unless otherwise stated).
- DO NOT turn engine via camshaft or other sprockets.
- Observe all tightening torques.
- Check diesel injection pump timing after belt replacement.

Removal

1. Remove:
   - ❏ Alternator drive belt.
   - ❏ Timing belt upper cover **1**.
   - ❏ Crankshaft pulley **3**.
   NOTE: Crankshaft pulley bolt has LH thread.
2. Remove timing belt lower cover **2**.
3. Turn crankshaft clockwise to TDC. Ensure timing marks aligned **4**, **5** & **6**.
   *NOTE: Camshaft sprocket **4** should be aligned with hole in timing belt cover backplate.*
4. Fit tool No.1842128000 to injection pump sprocket **7**. Lock sprocket with bolt **8**.
5. Slacken tensioner nut **9**. Release tension on belt.
6. Remove timing belt.

Installation

NOTE: DO NOT refit used belt.
1. Ensure timing marks aligned **4**, **5** & **6**.
2. Ensure locking tool located correctly in injection pump sprocket.
3. Fit timing belt.
4. Turn tensioner pulley clockwise. Fit tensioning tool to tensioner pulley **10**. Tool No.1860722000.
5. Remove locking tool from injection pump sprocket **7**.
6. Turn crankshaft two turns clockwise.
7. Tighten tensioner nut to 44 Nm **9**.
8. Turn crankshaft clockwise to TDC. Ensure timing marks aligned **4**, **5** & **6**.
9. Check and adjust injection pump timing.
10. Install components in reverse order of removal.
11. Tighten crankshaft pulley bolt to 180 Nm **11**.

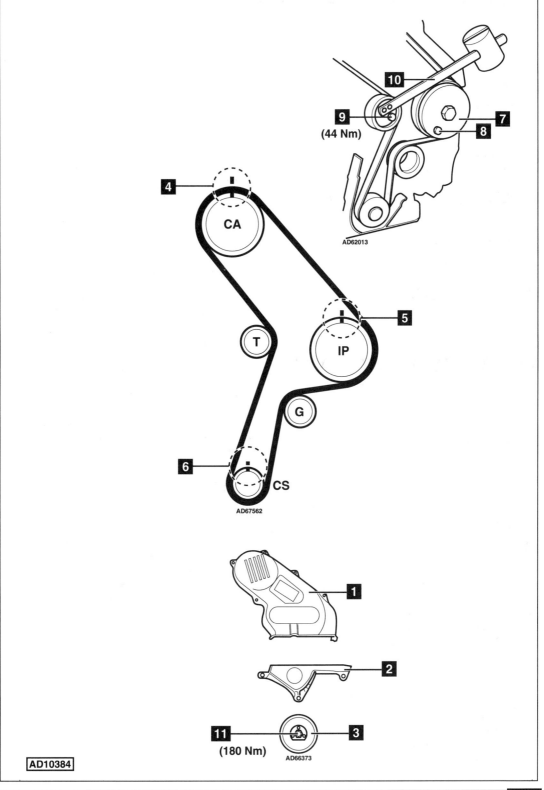

CA

T

IP

G

CS

10

9
(44 Nm)

7

8

4

5

6

AD62013

AD67562

1

2

11
(180 Nm)

3

AD66373

AD10384

FIAT

Model:	Punto 1,7D • Punto 1,7 TD/Cat & EGR
Year:	1993-95
Engine Code:	176 B3.000, 176 A3.000, 176 A5.000

Replacement Interval Guide

Fiat recommend check & replacement if
necessary every 36,000 miles and replacement
every 63,000 miles.
*The previous use and service history of the vehicle
must always be taken into account.*
*Refer to Timing Belt Replacement Intervals at the front
of this manual.*

Check For Engine Damage

*CAUTION: This engine has been identified as an
INTERFERENCE engine in which the possibility of
valve-to-piston damage in the event of a timing belt
failure is MOST LIKELY to occur.*
*A compression check of all cylinders should be
performed before removing the cylinder head.*

Repair Times – hrs

Remove & install:

1,7D	1,35
1,7TD	1,55

Special Tools

- ■ Injection pump sprocket locking tool –
  Fiat No.1842128000.
- ■ Tensioning tool – Fiat No.1860745100/200.

Special Precautions

- ■ Disconnect battery earth lead.
- ■ DO NOT turn crankshaft or camshaft when timing belt removed.
- ■ Remove glow plugs to ease turning engine.
- ■ Turn engine in normal direction of rotation (unless otherwise stated).
- ■ DO NOT turn engine via camshaft or other sprockets.
- ■ Observe all tightening torques.
- ■ Check diesel injection pump timing after belt replacement.

Removal – Engines up to No.1762798

1. Raise and support front of vehicle.
2. Remove:
   - ❏ Coolant expansion tank bolts. Move coolant expansion tank to one side.
   - ❏ RH front wheel.
   - ❏ RH inner wing panel.
   - ❏ Timing belt upper cover **1**.
   - ❏ Timing belt lower cover **2**.
   - ❏ Auxiliary drive belt.
   - ❏ Dipstick and tube.
3. Turn crankshaft slowly clockwise until timing marks aligned **3**, **4** & **5**.

4. Ensure mark on rear of camshaft sprocket aligned with hole in timing belt rear cover **6**.
5. Remove:
   - ❏ Timing belt cover **7**.
   - ❏ Crankshaft pulley bolts **15**.
   - ❏ Crankshaft pulley **8**.
6. Ensure crankshaft sprocket timing marks aligned **9**.
7. Fit tool No.1842128000 to injection pump sprocket **10**. Lock sprocket with bolt **11**.
8. Slacken tensioner nut **12**. Release tension on belt.
9. Remove timing belt.
   NOTE: DO NOT refit a damaged or oil contaminated belt.

Installation

**NOTE: DO NOT reuse a belt which has covered
more than 20,000 miles.**

1. Ensure timing marks aligned **4**, **5** & **9**.
2. Ensure locking tool located correctly in injection pump sprocket.
3. Fit timing belt in anti-clockwise direction, starting at crankshaft sprocket. Ensure belt is taut between sprockets on non-tensioned side.
4. Turn tensioner pulley clockwise. Fit tensioning tool to tensioner pulley **13**.
   Tool No.1860745100/200. With scale bar horizontal set weight at 60 mm mark **14**.
   NOTE: Ensure scale bar remains horizontal during tensioning procedure.
5. Remove locking tool from injection pump sprocket **10**.
6. Turn crankshaft slowly two turns clockwise. Ensure timing marks aligned **4**, **5** & **9**.
7. Tighten tensioner nut to 44 Nm **12**.
8. Install components in reverse order of removal.
9. Tighten crankshaft pulley bolts to 28 Nm **15**.

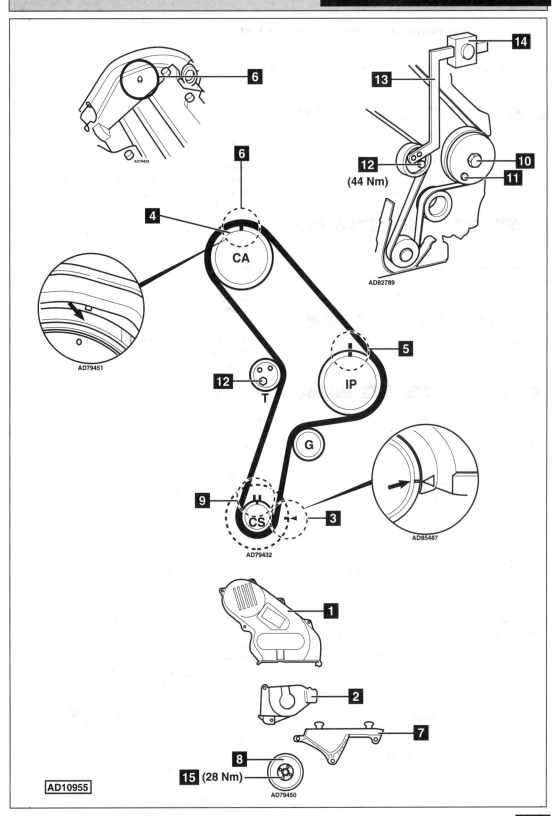

AD79433

6

6

14

13

12
(44 Nm)

10

11

AD82789

4

CA

5

AD79451

12
T

IP

G

9

CS

3

AD85487

AD79432

1

2

7

8

15 (28 Nm)

AD79450

AD10955

Model:	Punto 1,7D • Punto 1,7 TD/Cat & EGR
Year:	1996-97
Engine Code:	176 B3.000, 176 A3.000, 176 A5.000

Replacement Interval Guide

Fiat recommend check & replacement if necessary every 36,000 miles or 4 years and replacement every 63,000 miles.
The previous use and service history of the vehicle must always be taken into account.
Refer to Timing Belt Replacement Intervals at the front of this manual.

Check For Engine Damage

CAUTION: This engine has been identified as an INTERFERENCE engine in which the possibility of valve-to-piston damage in the event of a timing belt failure is MOST LIKELY to occur.
A compression check of all cylinders should be performed before removing the cylinder head.

Repair Times – hrs

Remove & install:

1,7D	1,35
1,7TD	1,55

Special Tools

- Tensioning tool – Fiat No.1860745100/300.
- Crankshaft locking tool – Fiat No.186093300.
- Camshaft aligning tool – Fiat No.1860932000.
- Sprocket holding tool – Fiat No.1860831000.

Special Precautions

- Disconnect battery earth lead.
- DO NOT turn crankshaft or camshaft when timing belt removed.
- Remove glow plugs to ease turning engine.
- Turn engine in normal direction of rotation (unless otherwise stated).
- DO NOT turn engine via camshaft or other sprockets.
- Observe all tightening torques.
- Check diesel injection pump timing after belt replacement.

Removal – Engines from No.1762799

1. Raise and support front of vehicle.
2. Remove:
   - ❏ Coolant expansion tank bolts. Move expansion tank to one side.
   - ❏ RH front wheel.
   - ❏ RH inner wing panel.
   - ❏ Auxiliary drive belts.
   - ❏ Dipstick and tube.
   - ❏ Crankshaft pulley bolts **1**.
   - ❏ Crankshaft pulley **2**.
   - ❏ Timing belt upper cover **3**.
   - ❏ Timing belt lower cover **4**.
   - ❏ Timing belt lower cover **5**.
3. Turn crankshaft slowly clockwise until timing marks aligned **6** & **7**.
   NOTE: Camshaft timing marks may not align exactly, as sprocket has keyhole slot 8 & 17.
4. Slacken tensioner nut to release tension on belt **9**.
5. Remove timing belt.
   NOTE: DO NOT refit a damaged or oil contaminated belt.

Installation

NOTE: DO NOT reuse a belt which has covered more than 20,000 miles.
1. Remove vacuum pump from rear of cylinder head.
2. Install camshaft aligning tool to rear of cylinder head **10**. Tool No.1860932000.
3. If necessary: Centralise dowel **11** by turning bolt **12** slightly.
4. Ensure timing marks aligned **6** & **7**.
5. Fit timing belt to crankshaft sprocket.
6. Remove front cover bolt **13**. Fit crankshaft locking tool **14**. Tool No.186093300.
7. Hold tool in place with bolt **13**.
8. Hold camshaft sprocket **15**.
   Use tool No.1860831000. Slacken bolt **16**.
   NOTE: Camshaft sprocket should be able to turn slightly 17.
9. Fit timing belt in anti-clockwise direction. Ensure belt is taut between sprockets on non-tensioned side.
10. Turn tensioner pulley clockwise. Fit tensioning tool to tensioner pulley **18**.
    Tool No.1860745100/300. With scale bar horizontal set weight at 60 mm mark **19**.
    NOTE: Ensure scale bar remains horizontal during tensioning procedure.
11. Hold camshaft sprocket **15**. Use tool No.1860831000. Tighten bolt to 118 Nm **16**.
12. Remove:
    - ❏ Crankshaft locking tool **14**.
    - ❏ Camshaft aligning tool **10**. Tighten bolt **13**.
13. Turn crankshaft slowly two turns clockwise. Ensure timing marks aligned **6** & **7**.
14. Tighten tensioner nut to 44 Nm **9**.
15. Remove tensioning tool.
16. Check and adjust injection pump timing.
17. Install components in reverse order of removal.
18. Tighten crankshaft pulley bolts **1**.
    Tightening torque: 28 Nm.

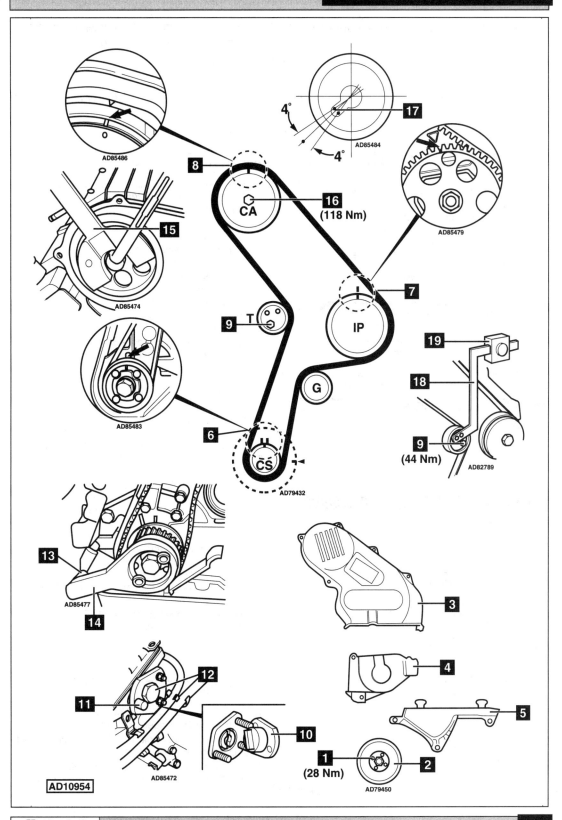

AD85486

17

4°

4°

AD85484

8

CA

16 (118 Nm)

AD85479

15

AD85474

7

9 T

IP

19

18

AD85483

9 (44 Nm)

AD82789

6

CS

AD79432

G

13

AD85477

14

3

12

11

10

4

5

AD85472

1 (28 Nm)

2

AD79450

AD10954

FIAT

Model:	Punto 1,9 D • Punto 1,9 JTD • Brava/Bravo 1,9 TD • Brava/Bravo 1,9 JTD Stilo 1,9 JTD • Marea/Weekend 1,9 TD • Marea/Weekend 1,9 JTD Marea/Weekend 2,4 JTD • Multipla 1,9 JTD
Year:	1995-03
Engine Code:	182 A7.000 (TD 100), 182 B4.000, 188 A2.000, 188 A3.000, 192 A1.000, 192 A3.000

Replacement Interval Guide

Fiat recommend:
→2000: Replacement every 72,000 miles or 6 years.
2001→: Check and replacement if necessary every 36,000 miles and replacement every 72,000 miles or 5 years.
The previous use and service history of the vehicle must always be taken into account.
Refer to Timing Belt Replacement Intervals at the front of this manual.

Check For Engine Damage

CAUTION: This engine has been identified as an INTERFERENCE engine in which the possibility of valve-to-piston damage in the event of a timing belt failure is MOST LIKELY to occur.
A compression check of all cylinders should be performed before removing the cylinder head.

Repair Times – hrs

Remove & install:

JTD	0,80
Except JTD	1,60
AC	+0,10
Stilo	No information available.

Special Tools

- Flywheel locking tool – Fiat No.1860898000.
- Injection pump timing pin – Fiat No.1860965000.
- Crankshaft timing tool – Fiat No.1860905000.

Special Precautions

- Disconnect battery earth lead.
- DO NOT turn crankshaft or camshaft when timing belt removed.
- Remove glow plugs to ease turning engine.
- Turn engine in normal direction of rotation (unless otherwise stated).
- DO NOT turn engine via camshaft or other sprockets.
- Observe all tightening torques.
- Except JTD: Check diesel injection pump timing after belt replacement.

Removal

NOTE: The high-pressure fuel pump fitted to the JTD engine does not require timing.

1. Raise and support front of vehicle.
2. Remove:
   - ❑ RH front wheel.
   - ❑ RH inner wing panel.
   - ❑ Engine undershield.
   - ❑ Auxiliary drive belt.
   - ❑ Flywheel housing lower cover.
3. Fit flywheel locking tool **1**. Tool No.1860898000.
4. Remove:
   - ❑ Crankshaft pulley bolts **2**.
   - ❑ Crankshaft pulley **3**.
   - ❑ Flywheel locking tool **1**.
   - ❑ Timing belt lower cover.
   - ❑ Engine steady bar and bracket.
   - ❑ Timing belt upper cover.
5. Turn crankshaft clockwise to TDC on No.1 cylinder. Ensure pin on crankshaft sprocket aligned centrally with cylinder block **4**.
6. Ensure camshaft sprocket timing marks aligned **5**.
   - ❑ To engine No.416 449: Camshaft sprocket timing mark seven teeth behind cylinder head cover timing mark. Three teeth in front of cylinder head cover lower edge **15**.
   NOTE: Camshaft sprocket timing mark position can be offset by approximately 3° or 1/2 tooth.
   - ❑ From engine No.416 450: Ensure timing marks aligned **16**.
7. If timing marks not aligned: Turn crankshaft one turn clockwise.
8. Slacken tensioner sprocket nut **6**. Move tensioner sprocket away from belt. Lightly tighten nut.
9. Remove timing belt.

Installation

1. Ensure pin on crankshaft sprocket aligned centrally with cylinder block **4**.
2. Remove oil pump bolt. Insert stud of crankshaft timing tool **7**. Tool No.1860905000.
3. Fit timing belt to crankshaft sprocket.
   NOTE: Observe direction of rotation marks on belt.

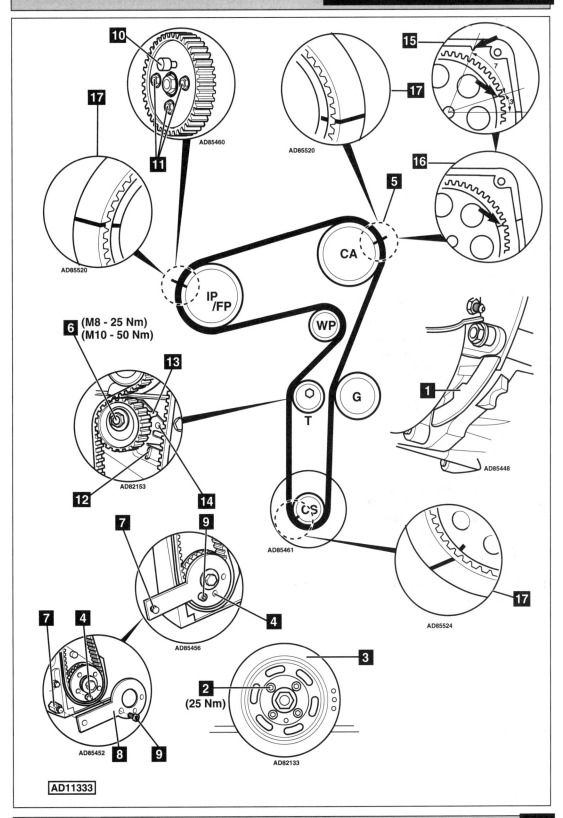

AD11333

Autodata

←

4. Fit timing tool to crankshaft sprocket **8**.
   Tool No.1860905000. Ensure pin on crankshaft
   sprocket located correctly in tool **4**. Secure with
   bolt **9**.

5. Ensure camshaft sprocket timing marks
   aligned **5**.

6. Except JTD: Insert timing pin in injection pump
   sprocket **10**. Tool No.1860965000. Slacken
   bolts **11**.

7. Fit timing belt in following order:
   - ❑ Guide pulley.
   - ❑ Camshaft sprocket.
   - ❑ High-pressure fuel pump/injection pump
     sprocket.
   - ❑ Tensioner sprocket.
   - ❑ Water pump pulley.

8. Ensure timing belt marks aligned:
   - ❑ JTD: Camshaft sprocket **17**.
   - ❑ Except JTD: Camshaft and injection pump
     sprockets **17**.

   ***NOTE: Ensure belt is taut between sprockets on
   non-tensioned side.***

9. Slacken tensioner sprocket nut **6**.

10. Lever tensioner sprocket bracket at position **12**
    until pointer **13** at maximum setting.

11. Tighten tensioner sprocket nut **6**:
    - ❑ M8 nut: 25 Nm.
    - ❑ M10 nut: 50 Nm.

12. Except JTD: Tighten injection pump sprocket
    bolts **11**.

13. Remove:
    - ❑ Except JTD: Timing pin **10**.
    - ❑ Timing tool **8**.

14. Fit oil pump bolt.

15. Turn crankshaft two turns clockwise.

16. Slacken tensioner sprocket nut **6**.

17. Lever tensioner sprocket bracket at position **12**
    until pointer **13** and mark **14** aligned.

18. Tighten tensioner sprocket nut **6**:
    - ❑ M8 nut: 25 Nm.
    - ❑ M10 nut: 50 Nm.

19. Install components in reverse order of removal.

20. Tighten crankshaft pulley bolts **2**.
    Tightening torque: 25 Nm.

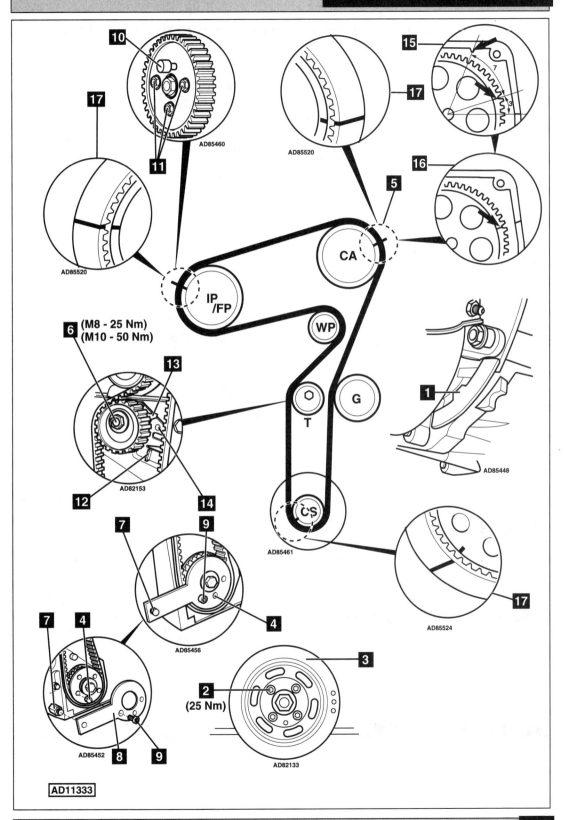

AD85460

AD85520

AD85520

AD82153

AD85456

AD85452

AD82133

AD85461

AD85524

AD85448

CA

IP/FP

WP

T

G

CS

10

17

11

15

17

16

5

6 (M8 - 25 Nm) (M10 - 50 Nm)

13

12

14

7

9

4

7

4

7

4

9

8 9

2 (25 Nm)

3

17

AD11333

FIAT

Model:	Tipo 1,9 D/TD • Tempra 1,9 D/TD • Croma 1,9 TD • Ducato 1,9D/TD
Year:	1987-98
Engine Code:	149 B1.000, 160 A6.000, 160 A7.000, 160 D1.000,154 B.000 230 A2.000, 230 A3.000, 230 A4.000, 280 A1.000 (to engine No.1723290 - non-adjustable camshaft sprocket)

Replacement Interval Guide

Fiat recommend check and/or replacement at the following intervals:

Tipo:

→1988:
Every 54,000 miles - replace.

01/89-04/89:
Every 60,000 miles - replace.

05/89→:
Every 36,000 miles - check & replace if necessary.
Every 63,000 miles - replace.

Tempra:

→04/91:
Every 18,000 miles - check & replace if necessary.
Every 60,000 miles - replace.

05/91→:
Every 36,000 miles - check & replace if necessary.
Every 63,000 miles - replace.

Croma:
Every 60,000 miles - replace.

Ducato:

→05/91:
Every 60,000 miles - replace.

06/91→:
Every 36,000 miles or 4 years - check & replace if necessary.
Every 63,000 miles - replace.
The previous use and service history of the vehicle must always be taken into account.
Refer to Timing Belt Replacement Intervals at the front of this manual.

Check For Engine Damage

CAUTION: This engine has been identified as an INTERFERENCE engine in which the possibility of valve-to-piston damage in the event of a timing belt failure is MOST LIKELY to occur.
A compression check of all cylinders should be performed before removing the cylinder head.

Repair Times – hrs

Remove & install:

Tipo/Tempra 1,9 D	1,35
AC	+0,15
Tipo/Tempra 1,9 TD	1,90
AC	+0,20
Croma 1,9 TD	1,30
Ducato 1,9 D/TD	1,05

Special Tools

- Injection pump sprocket locking tool – Fiat No.1842128000.
- Tensioning tool – Fiat No.1860745100/200.

Special Precautions

- Disconnect battery earth lead.
- DO NOT turn crankshaft or camshaft when timing belt removed.
- Remove glow plugs to ease turning engine.
- Turn engine in normal direction of rotation (unless otherwise stated).
- DO NOT turn engine via camshaft or other sprockets.
- Observe all tightening torques.
- Check diesel injection pump timing after belt replacement.

Removal

1. Remove:
   - Auxiliary drive belt.
   - Timing belt upper cover **1**.
   - Timing belt lower cover **2**.
   - Crankshaft pulley bolts.
   - Crankshaft pulley **3**.
2. Turn crankshaft clockwise to TDC. Ensure timing marks aligned. Non-turbo: **4**, **5** & **6**. Turbo: **13**, **5** & **6**.
3. Fit tool No.1842128000 to injection pump sprocket **7**. Lock sprocket with bolt **8**.
4. Slacken tensioner nut **9**. Release tension on belt.
5. Remove timing belt.

Installation

NOTE: DO NOT refit used belt.

1. Remove camshaft sprocket bolt **11**.
2. Ensure Woodruff key fits tightly into keyway slot **12**.
3. Fit camshaft sprocket bolt **11**.
   Tightening torque: 118 Nm.
4. Ensure timing marks aligned. Non-turbo: **4**, **5** & **6**. Turbo: **13**, **5** & **6**.
5. Ensure locking tool located correctly in injection pump sprocket.
6. Fit timing belt.
7. Turn tensioner pulley clockwise. Fit tensioning tool to tensioner pulley **10**. Tool No.1860745100/200. Set weight at 120 mm mark on scale bar.
   NOTE: Ensure scale bar remains horizontal during tensioning procedure.
8. Remove locking tool from injection pump sprocket **7**.
9. Turn crankshaft two turns clockwise. Ensure timing marks aligned. Non-turbo: **4**, **5** & **6**. Turbo: **13**, **5** & **6**.
10. Tighten tensioner nut to 44 Nm **9**.
11. Turn crankshaft two turns clockwise to TDC.
12. Ensure timing marks aligned. Non-turbo: **4**, **5** & **6**. Turbo: **13**, **5** & **6**.
13. Install components in reverse order of removal.

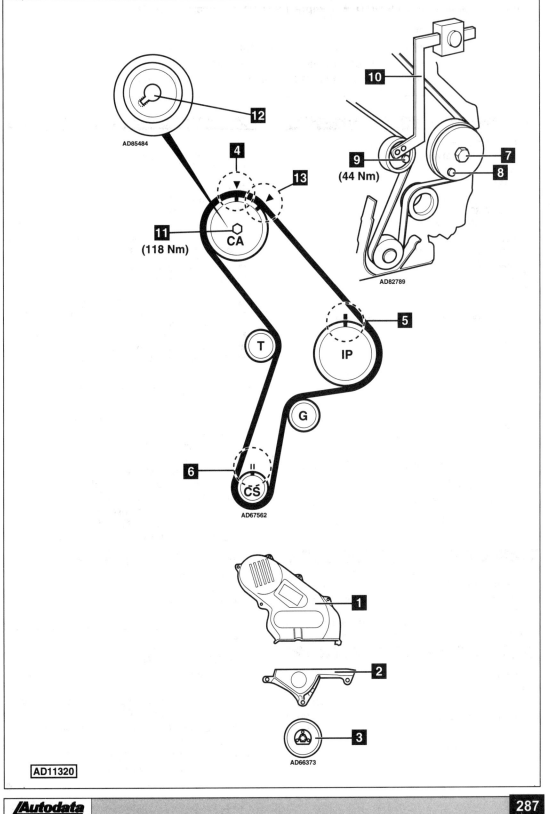

AD85484

12

4

13

10

9
(44 Nm)

7

8

CA

11
(118 Nm)

T

IP

5

G

6

CS

AD67562

AD82789

1

2

3

AD66373

AD11320

FIAT

Model:	Tipo 1,9 D/TD • Tempra 1,9 D/TD • Ducato 1,9D/TD
Year:	1987-98
Engine Code:	149 B1.000, 160 A6.000, 160 A7.000, 160 D1.000 230 A2.000, 230 A3.000, 230 A4.000, 280 A1.000 (from engine No.1723291 - adjustable camshaft sprocket)

Replacement Interval Guide

Fiat recommend check and/or replacement at the following intervals:

Tipo:

→1988:
Every 54,000 miles - replace.

01/89-04/89:
Every 60,000 miles - replace.

05/89→:
Every 36,000 miles - check & replace if necessary.
Every 63,000 miles - replace.

Tempra:

→04/91:
Every 18,000 miles - check & replace if necessary.
Every 60,000 miles - replace.

05/91→:
Every 36,000 miles - check & replace if necessary.
Every 63,000 miles - replace.

Ducato:

→05/91:
Every 60,000 miles - replace.

06/91→:
Every 36,000 miles or 4 years - check & replace if necessary.
Every 63,000 miles - replace.
The previous use and service history of the vehicle must always be taken into account.
Refer to Timing Belt Replacement Intervals at the front of this manual.

Check For Engine Damage

CAUTION: This engine has been identified as an INTERFERENCE engine in which the possibility of valve-to-piston damage in the event of a timing belt failure is MOST LIKELY to occur.
A compression check of all cylinders should be performed before removing the cylinder head.

Repair Times – hrs

Remove & install:	
Tipo/Tempra 1,9 D	1,35
AC	+0,15
Tipo/Tempra 1,9 TD	1,90
AC	+0,20
Ducato 1,9 D/TD	1,05

Special Tools

- Tensioning tool – Fiat No.1860745100/300.
- Crankshaft locking tool – Fiat No.1860933000.
- Camshaft aligning tool (Non-turbo) – Fiat No.1860934000.
- Camshaft aligning tool (Turbo) – Fiat No.1860931000.
- Sprocket holding tool – Fiat No.1860831000.

Special Precautions

- Disconnect battery earth lead.
- DO NOT turn crankshaft or camshaft when timing belt removed.
- Remove glow plugs to ease turning engine.
- Turn engine in normal direction of rotation (unless otherwise stated).
- DO NOT turn engine via camshaft or other sprockets.
- Observe all tightening torques.
- Check diesel injection pump timing after belt replacement.

Removal

1. Raise and support front of vehicle.
2. Remove:
   - ❑ Auxiliary drive belt.
   - ❑ Crankshaft pulley bolts.
   - ❑ Crankshaft pulley **1**.
   - ❑ Timing belt upper cover **3**.
   - ❑ Timing belt lower cover **2**.
3. Turn crankshaft clockwise to TDC. Ensure timing marks aligned **4** & **5**.
   NOTE: Camshaft timing marks may not align exactly, as sprocket has keyhole slot.
   *Non-turbo: **6** & **7**. Turbo: **18** & **7**.*
4. Slacken tensioner nut to release tension on belt **8**.
5. Remove timing belt.

Installation

NOTE: DO NOT refit used belt.

1. Remove vacuum pump from rear of cylinder head.
2. Install camshaft aligning tool to rear of cylinder head **9**. Tool No.1860931000/4000.
3. If necessary: Centralise dowel **10** by turning bolt **11** slightly.
4. Ensure timing marks aligned **4** & **5**.
5. Fit timing belt to crankshaft sprocket.

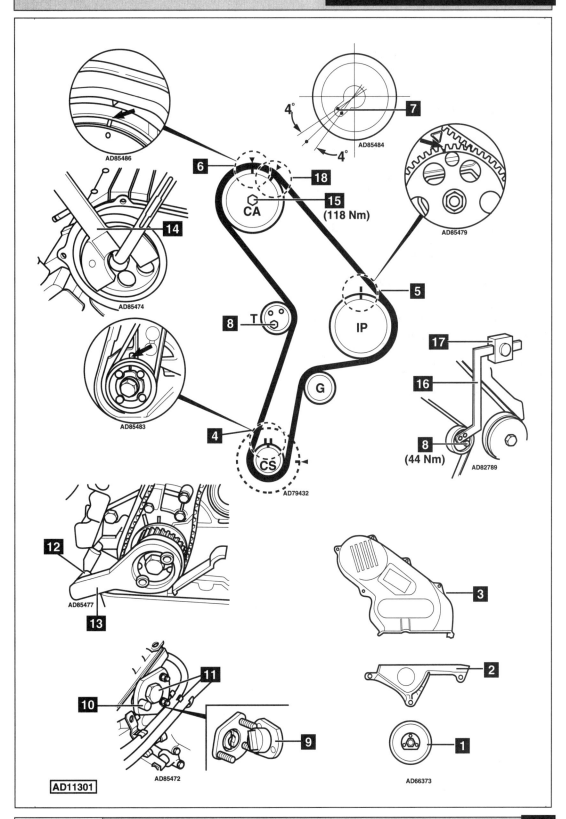

AD85486

4°

4°

AD85484

7

AD85479

6

18

15
(118 Nm)

CA

14

AD85474

8 T

5

IP

17

16

8
(44 Nm)

AD82789

G

AD85483

4

CS

AD79432

12

AD85477

13

3

10

11

AD85472

9

2

1

AD66373

AD11301

6. Remove front cover bolt 🔢. Fit crankshaft locking tool 🔢. Tool No.1860933000.
7. Hold tool in place with bolt 🔢.
8. Hold camshaft sprocket 🔢. Use tool No.1860831000. Slacken bolt 🔢.

    NOTE: Camshaft sprocket should be able to turn slightly 🔢.

9. Fit timing belt in anti-clockwise direction. Ensure belt is taut between sprockets on non-tensioned side.
10. Ensure timing mark aligned 🔢.
11. Turn tensioner pulley clockwise. Fit tensioning tool to tensioner pulley 🔢.
    Tool No.1860745100/300. With scale bar horizontal set weight at 120 mm mark 🔢.

    NOTE: Ensure scale bar remains horizontal during tensioning procedure.

12. Hold camshaft sprocket 🔢. Use tool No.1860831000. Tighten bolt to 118 Nm 🔢.
13. Remove:
    ❏ Crankshaft locking tool 🔢.
    ❏ Camshaft aligning tool 🔢. Tighten bolt 🔢.
14. Turn crankshaft slowly two turns clockwise. Ensure timing marks aligned 🔢 & 🔢.
15. Tighten tensioner nut to 44 Nm 🔢.
16. Remove tensioning tool 🔢.
17. Install components in reverse order of removal.

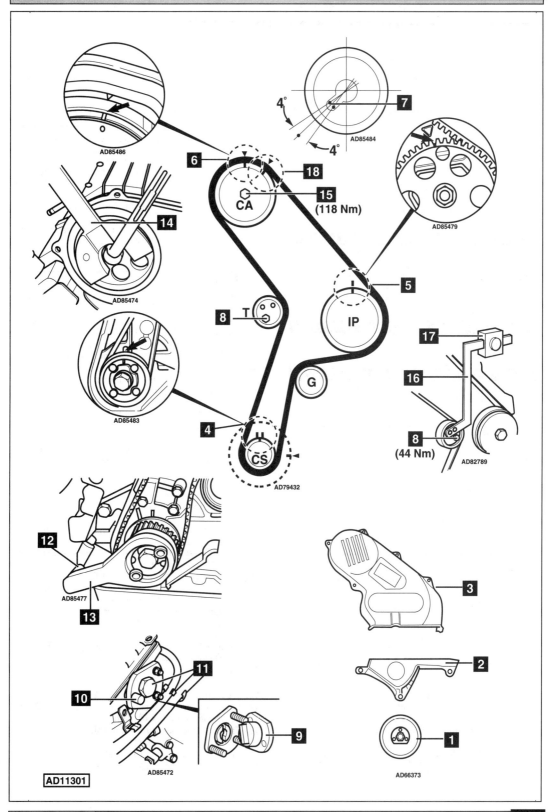

AD85486

4° 7 4°

AD85484

6 18

15
(118 Nm)

CA

AD85479

14

AD85474

5

8 T

IP

17

16

AD85483

8
(44 Nm)

AD82789

4

CS

AD79432

G

12

13

AD85477

3

10 11

2

9

1

AD11301

AD85472

AD66373

FIAT

Model:	**Brava/Bravo 1,9 TD • Marea/Weekend 1,9 TD**
Year:	**1995-03**
Engine Code:	**182 A8.000 (TD75)**

Replacement Interval Guide

Fiat recommend:
→2000: Replacement every 72,000 miles or
6 years.
2001→: Check and replacement if necessary every 36,000 miles and replacement every 72,000 miles or 5 years.
The previous use and service history of the vehicle must always be taken into account.
Refer to Timing Belt Replacement Intervals at the front of this manual.

Check For Engine Damage

CAUTION: This engine has been identified as an INTERFERENCE engine in which the possibility of valve-to-piston damage in the event of a timing belt failure is MOST LIKELY to occur.
A compression check of all cylinders should be performed before removing the cylinder head.

Repair Times – hrs

Remove & install	1,60
AC	+0,10

Special Tools

- Crankshaft timing tool – Fiat No.1860905000.
- Flywheel locking tool – Fiat No.1860898000.

Special Precautions

- Disconnect battery earth lead.
- DO NOT turn crankshaft or camshaft when timing belt removed.
- Remove glow plugs to ease turning engine.
- Turn engine in normal direction of rotation (unless otherwise stated).
- DO NOT turn engine via camshaft or other sprockets.
- Observe all tightening torques.
- Check diesel injection pump timing after belt replacement.

Removal

1. Raise and support front of vehicle.
2. Remove:
   - ❏ RH front wheel.
   - ❏ RH inner wing panel.
   - ❏ Engine undershield.
   - ❏ Auxiliary drive belt.
   - ❏ Flywheel housing lower cover.
3. Fit flywheel locking tool **1**. Tool No.1860898000.
4. Remove:
   - ❏ Crankshaft pulley bolts **2**.
   - ❏ Crankshaft pulley **3**.
   - ❏ Flywheel locking tool **1**.
   - ❏ Timing belt lower cover.
   - ❏ Engine steady bar and bracket.
   - ❏ Timing belt upper cover.
5. Turn crankshaft clockwise to TDC on No.1 cylinder. Ensure pin on crankshaft sprocket aligned centrally with cylinder block **4**.

6. Ensure camshaft sprocket timing marks aligned **5**.
   - ❏ To engine No.416 449: Camshaft sprocket timing mark seven teeth behind cylinder head cover timing mark. Three teeth in front of cylinder head cover lower edge **13**.
   NOTE: Camshaft sprocket timing mark position can be offset by approximately 3° or 1/2 tooth.
   - ❏ From engine No.416 450: Ensure timing marks aligned **14**.
7. If timing marks not aligned: Turn crankshaft one turn clockwise.
8. Slacken tensioner sprocket nut **6**. Move tensioner sprocket away from belt. Lightly tighten nut.
9. Remove timing belt.

Installation

1. Ensure pin on crankshaft sprocket aligned centrally with cylinder block **4**.
2. Remove oil pump bolt. Insert stud of crankshaft timing tool **7**. Tool No.1860905000.
3. Fit timing belt to crankshaft sprocket.
   NOTE: Observe direction of rotation marks on belt.
4. Fit timing tool to crankshaft sprocket **8**. Tool No.1860905000. Ensure pin on crankshaft sprocket located correctly in tool **4**. Secure with bolt **9**.
5. Ensure camshaft sprocket timing marks aligned **5**.
6. Ensure injection pump sprocket timing mark aligned with pointer **16**.
7. Insert two bolts in injection pump sprocket **17**.
8. Fit timing belt in following order:
   - ❏ Guide pulley.
   - ❏ Camshaft sprocket.
   - ❏ Injection pump sprocket.
   - ❏ Tensioner sprocket.
   - ❏ Water pump pulley.
9. Ensure marks on belt aligned with marks on sprockets **15**.
   NOTE: Ensure belt is taut between sprockets on non-tensioned side.
10. Slacken tensioner sprocket nut **6**.
11. Lever tensioner sprocket bracket at position **10** until pointer **11** at maximum setting.
12. Tighten tensioner sprocket nut **6**.
    Tightening torque: 50 Nm.
13. Remove:
    - ❏ Injection pump sprocket locking bolts **17**.
    - ❏ Timing tool **8**.
14. Fit oil pump bolt.
15. Turn crankshaft two turns clockwise.
16. Slacken tensioner sprocket nut **6**.
17. Lever tensioner sprocket bracket at position **10** until pointer **11** and mark **12** aligned.
18. Tighten tensioner sprocket nut **6**.
    Tightening torque: 50 Nm.
19. Install components in reverse order of removal.
20. Tighten crankshaft pulley bolts **2**.
    Tightening torque: 25 Nm.

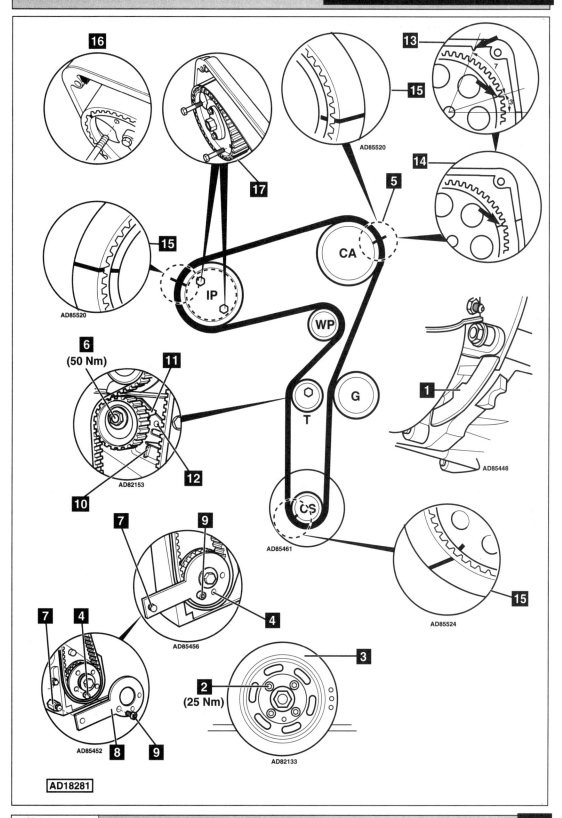

16

13

15

AD85520

14

5

CA

IP

15

AD85520

17

WP

6
(50 Nm)

11

T

G

10

12

AD82153

1

AD85448

CS

AD85461

7

9

7

4

4

3

AD85456

2
(25 Nm)

15

AD85524

AD85452

8

9

AD82133

AD18281

Model:	Ulysse 1,9TD • Scudo 1,9D/TD • Ducato 1,9D/TD
Year:	1994-02
Engine Code:	D8B, D9B, DHX, DJY

Replacement Interval Guide

Fiat recommend:
→1999: Check every 36,000 miles or 4 years and replacement every 63,000 miles or 7 years.
Ulysse 2000→: Check every 36,000 miles or 4 years and replacement every 72,000 miles or 8 years.
Scudo/Ducato 2000: Replacement every 72,000 miles or 8 years.
Scudo/Ducato 2001→: Check every 36,000 miles or 4 years and replacement every 72,000 miles or 5 years.
The previous use and service history of the vehicle must always be taken into account.
Refer to Timing Belt Replacement Intervals at the front of this manual.

Check For Engine Damage

CAUTION: This engine has been identified as an INTERFERENCE engine in which the possibility of valve-to-piston damage in the event of a timing belt failure is MOST LIKELY to occur.
A compression check of all cylinders should be performed before removing the cylinder head(s).

Repair Times – hrs

Remove & install:

Scudo	2,30
Ulysse/Ducato	2,20

Special Tools

- Camshaft sprocket timing pin – 1 x 8 mm bolt.
- Injection pump sprocket timing pins – 2 x 8 mm bolts.
- Flywheel timing pin – Fiat No.1860863000.

Special Precautions

- Disconnect battery earth lead.
- DO NOT turn crankshaft or camshaft when timing belt removed.
- Remove glow plugs to ease turning engine.
- Turn engine in normal direction of rotation (unless otherwise stated).
- DO NOT turn engine via camshaft or other sprockets.
- Observe all tightening torques.
- Check diesel injection pump timing after belt replacement.

Removal

1. Raise and support front of vehicle.
2. Remove:
   ❏ RH front wheel.
   ❏ RH wheel arch liner.
   ❏ Auxiliary drive belt.
   ❏ Engine mounting.
   ❏ Timing belt covers **1** & **2**.
   ❏ Crankshaft pulley bolt **3**.
   ❏ Crankshaft pulley **4**.
   ❏ Timing belt lower cover.
3. Temporarily fit crankshaft pulley.
4. Turn crankshaft clockwise to setting position.
5. Insert one bolt in camshaft sprocket **5** and two bolts in injection pump sprocket **6**.
6. Insert timing pin in flywheel **7**.
   Tool No.1860863000.
7. Remove crankshaft pulley.
8. Slacken tensioner nut and bolt **8** & **9**.
9. Move tensioner away from belt and tighten bolt **9**.
10. Remove timing belt.

Installation

1. Ensure timing pin and locking bolts located correctly **5**, **6** & **7**.
2. Fit timing belt in anti-clockwise direction, starting at crankshaft sprocket.
3. Ensure belt is taut between sprockets.
4. Slacken tensioner bolt **9**. Push tensioner several times against belt.
5. Tighten tensioner nut and bolt **8** & **9**.
6. Remove timing pin and locking bolts.
7. Turn crankshaft two turns clockwise.
8. Insert timing pin **7**.
9. Ensure locking bolts can be inserted in sprockets **5** & **6**.
10. Slacken and tighten tensioner nut and bolt **8** & **9**.
11. Install components in reverse order of removal.
12. Coat crankshaft pulley bolt thread with suitable thread locking compound.
13. Tighten crankshaft pulley bolt **3**.
    Tightening torque: 40 Nm + 60°.

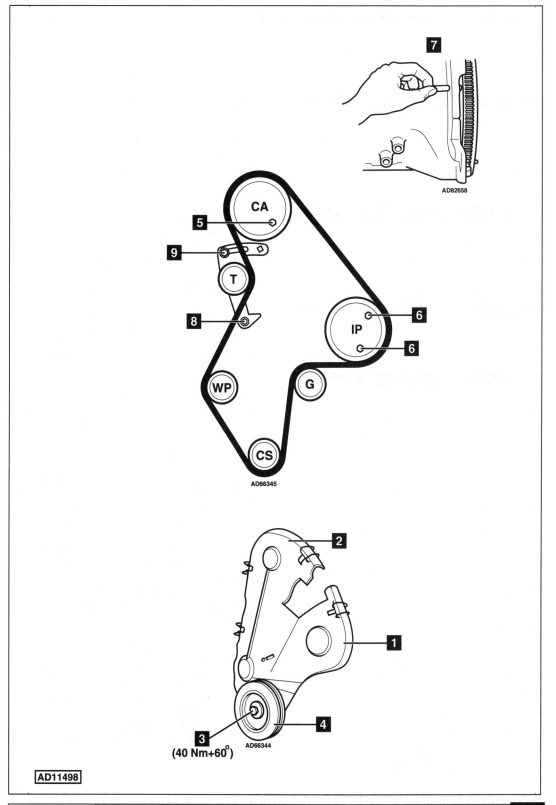

AD82658

CA

5

9

T

8

IP

6

6

WP

G

CS

AD66345

7

2

1

3
(40 Nm+60°)

4

AD66344

AD11498

Model:	Ulysse 2,0 JTD • Scudo 2,0 JTD
Year:	1999-02
Engine Code:	RHZ, RHX

Replacement Interval Guide

Fiat recommend check & replacement if necessary every 36,000 miles and replacement every 72,000 miles or 5 years.
The previous use and service history of the vehicle must always be taken into account.
Refer to Timing Belt Replacement Intervals at the front of this manual.

Check For Engine Damage

CAUTION: This engine has been identified as an INTERFERENCE engine in which the possibility of valve-to-piston damage in the event of a timing belt failure is MOST LIKELY to occur.
A compression check of all cylinders should be performed before removing the cylinder head(s).

Repair Times – hrs

Remove & install:

Ulysse	3,25
Scudo	3,35

Special Tools

- Flywheel locking tool – Fiat No.1867030000.
- Engine support – Fiat No.1860872000/1860872002.
- Flywheel timing pin – Fiat No.1860863000.
- 1/4" square drive wrench.
- Camshaft timing pin.
- Tension gauge – SEEM C.Tronic 105.5.

Special Precautions

- Disconnect battery earth lead.
- DO NOT turn crankshaft or camshaft when timing belt removed.
- Remove glow plugs to ease turning engine.
- Turn engine in normal direction of rotation (unless otherwise stated).
- DO NOT turn engine via camshaft or other sprockets.
- Observe all tightening torques.

Removal

NOTE: The high-pressure fuel pump fitted to this engine does not require timing.

1. Raise and support front of vehicle.
2. Remove:
   - ❑ RH front wheel.
   - ❑ RH splash guards.
   - ❑ Engine lower cover (if fitted).
   - ❑ Engine upper cover.
   - ❑ Auxiliary drive belt.
   - ❑ Turbocharger air hoses.
   - ❑ Flywheel housing lower cover.

3. Lock flywheel. Use tool No.1867030000.
4. Remove:
   - ❑ Crankshaft pulley bolt **1**.
   - ❑ Crankshaft pulley **2**.
   - ❑ Flywheel locking tool.
5. Support engine. Use tool Nos.1860872000/1860872002.
6. Disconnect:
   - ❑ Injectors.
   - ❑ Camshaft position (CMP) sensor.
   - ❑ Fuel temperature sensor.
   - ❑ Fuel pressure sensor.
   - ❑ Fuel pump solenoid wiring.
7. Disconnect and seal off fuel pipes.
8. Move wiring harness to one side.
9. Remove:
   - ❑ RH engine mounting and bracket.
   - ❑ Timing belt covers **3**, **4** & **5**.
10. Turn crankshaft clockwise to setting position.
11. Insert timing pin in camshaft sprocket **6**.
12. Insert timing pin in flywheel **7**.
    Tool No.1860863000.
13. Slacken tensioner bolt **8**.
14. Turn tensioner pulley clockwise away from belt. Use 1/4" square drive wrench **9**.
15. Lightly tighten tensioner bolt **8**.
16. Remove timing belt.

Installation

1. Slacken camshaft sprocket bolts **10**.
   NOTE: Sprocket should turn with slight resistance.
2. Ensure timing pins located correctly **6** & **7**.
3. Turn camshaft sprocket fully clockwise in slotted holes.
4. Fit timing belt to crankshaft sprocket.
5. Fit timing belt in anti-clockwise direction. Ensure belt is taut between sprockets.
6. Lay belt on camshaft sprocket teeth. Engage belt teeth by turning sprocket slightly anti-clockwise.
   *NOTE: Angular movement of sprocket must not be more than one tooth space **11**.*
7. Fit timing belt to water pump sprocket and tensioner pulley.
8. Slacken tensioner bolt **8**.
9. Attach tension gauge to belt at ▽ **12**.
   Tool No.SEEM C.Tronic 105.5.

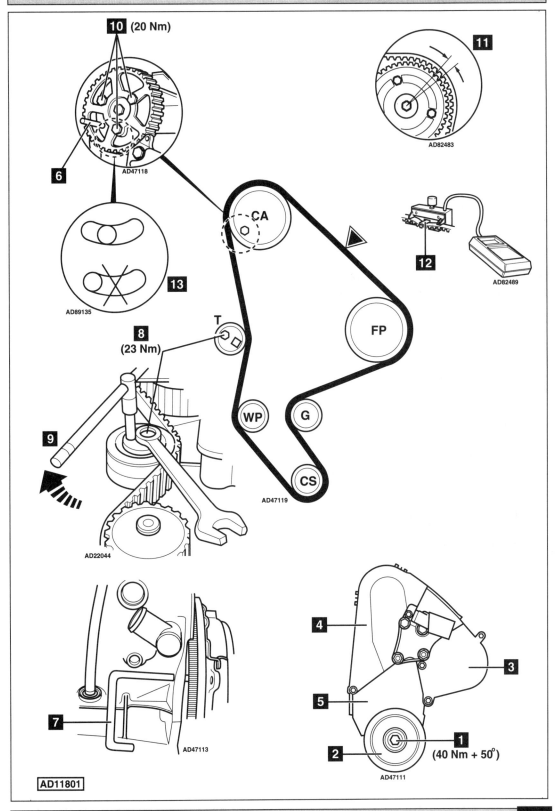

10 (20 Nm)

11

AD82483

6

AD47118

CA

12

AD82489

13

AD89135

FP

8
(23 Nm)

T

9

WP

G

CS

AD47119

AD22044

4

3

5

7

AD47113

1
(40 Nm + 50°)

2

AD47111

AD11801

←

10. Turn tensioner pulley anti-clockwise until tension gauge indicates 108 SEEM units.
11. Tighten tensioner bolt .
    Tightening torque: 23 Nm.
12. Remove tension gauge .
13. Ensure sprocket bolts not at end of slotted holes .
14. Tighten camshaft sprocket bolts .
    Tightening torque: 20 Nm.
15. Remove timing pins & .
16. Turn crankshaft eight turns clockwise to setting position.
    NOTE: DO NOT allow crankshaft to turn anti-clockwise.
17. Insert timing pins & .
18. Slacken camshaft sprocket bolts .
19. Slacken tensioner bolt to release tension on belt .
20. Attach tension gauge to belt at ▽ .
21. Turn tensioner pulley anti-clockwise until tension gauge indicates 54 SEEM units.
22. Tighten tensioner bolt .
    Tightening torque: 23 Nm.
23. Tighten camshaft sprocket bolts .
    Tightening torque: 20 Nm.
24. Remove tension gauge .
25. Check belt tension. Attach tension gauge to belt at ▽ . Tension gauge should indicate 51-57 SEEM units.
26. If not: Repeat tensioning procedure.
27. Remove tension gauge .
28. Remove timing pins & .
29. Turn crankshaft two turns clockwise to setting position.
30. Insert timing pin in flywheel .
31. Ensure timing pin can be inserted easily .
32. Remove timing pin .
33. Install components in reverse order of removal.
34. Tighten crankshaft pulley bolt .
    Tightening torque: 40 Nm + 50°.

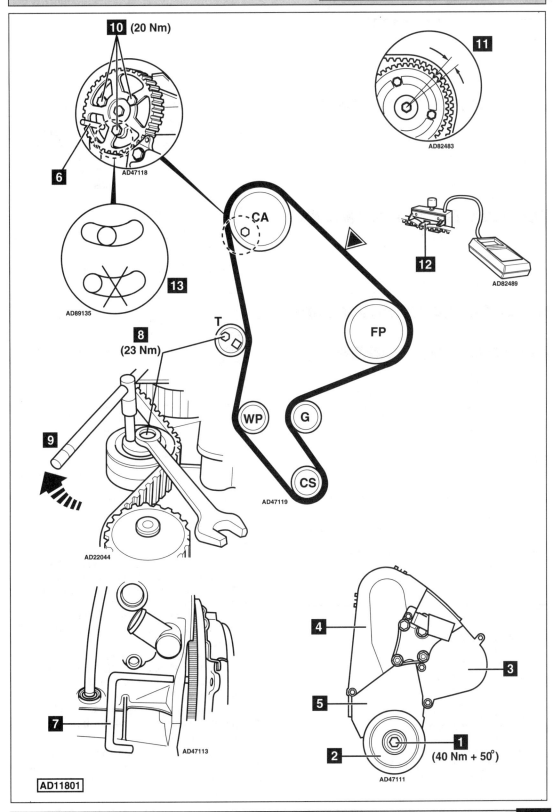

10 (20 Nm)

6

AD47118

11

AD82483

13

AD89135

CA

12

AD82489

8
(23 Nm)

T

FP

9

WP

G

CS

AD47119

AD22044

4

3

5

7

AD47113

1
(40 Nm + 50°)

2

AD47111

AD11801

Model:	**Fiorino 1,7 TD**
Year:	**1997-01**
Engine Code:	**146D7.000**

Replacement Interval Guide

Fiat recommend check & replacement if necessary every 36,000 miles or 4 years and replacement every 63,000 miles.
The previous use and service history of the vehicle must always be taken into account.
Refer to Timing Belt Replacement Intervals at the front of this manual.

Check For Engine Damage

CAUTION: This engine has been identified as an INTERFERENCE engine in which the possibility of valve-to-piston damage in the event of a timing belt failure is MOST LIKELY to occur.
A compression check of all cylinders should be performed before removing the cylinder head.

Repair Times – hrs

Remove & install	1,55

Special Tools

- Tensioning tool – Fiat No.1860745100/200.
- Crankshaft locking tool – Fiat No.1860933000.
- Camshaft aligning tool – Fiat No.1860932000.
- Sprocket holding tool – Fiat No.1860831001/002.
- Injection pump sprocket locking tool – Fiat No.1842128000.

Special Precautions

- Disconnect battery earth lead.
- DO NOT turn crankshaft or camshaft when timing belt removed.
- Remove glow plugs to ease turning engine.
- Turn engine in normal direction of rotation (unless otherwise stated).
- DO NOT turn engine via camshaft or other sprockets.
- Observe all tightening torques.
- Check diesel injection pump timing after belt replacement.

Removal

1. Raise and support front of vehicle.
2. Remove:
   - ❑ RH front wheel.
   - ❑ RH inner wing panel.
   - ❑ Air filter housing.
   - ❑ Air intake pipe.
   - ❑ Auxiliary drive belts.
   - ❑ Dipstick and tube.
   - ❑ Crankshaft pulley bolts **1**.
   - ❑ Crankshaft pulley **2**.
   - ❑ Timing belt upper cover **3**.
   - ❑ Timing belt lower cover **4**.

3. Turn crankshaft slowly clockwise until timing marks aligned **5** & **6**.
   *NOTE: Camshaft timing marks may not align exactly, as sprocket has keyhole slot **7** & **8**.*
4. Fit tool No.1842128000 to injection pump sprocket **9**. Lock sprocket with bolt **10**.
5. Slacken tensioner nut to release tension on belt **11**.
6. Remove timing belt.

Installation

NOTE: DO NOT reuse a belt which has covered more than 20,000 miles.

1. Remove vacuum pump from rear of cylinder head.
2. Install camshaft aligning tool to rear of cylinder head **12**. Tool No.1860932000.
3. If necessary: Centralise dowel **13** by turning bolt **14** slightly.
4. Ensure timing marks aligned **5** & **6**.
5. Fit timing belt to crankshaft sprocket.
6. Remove front cover bolt **15**. Fit crankshaft locking tool **16**. Tool No.1860933000.
7. Hold tool in place with bolt **15**.
8. Hold camshaft sprocket **17**. Use tool No.1860831001/002. Slacken bolt **20**.
   *NOTE: Camshaft sprocket should be able to turn slightly **8**.*
9. Fit timing belt in anti-clockwise direction. Ensure belt is taut between sprockets on non-tensioned side.
10. Turn tensioner pulley clockwise. Fit tensioning tool to tensioner pulley **18**. Tool No.1860745100/200. With scale bar horizontal set weight at 60 mm mark **19**.
    NOTE: Ensure scale bar remains horizontal during tensioning procedure.
11. Hold camshaft sprocket **17**. Use tool No.1860831001/002. Tighten bolt to 118 Nm **20**.
12. Remove:
    - ❑ Crankshaft locking tool **16**.
    - ❑ Camshaft aligning tool **12**. Tighten bolt **15**.
    - ❑ Injection pump locking tool **9**.
13. Turn crankshaft slowly two turns clockwise. Ensure timing marks aligned **5**, **6** & **7**.
14. Tighten tensioner nut to 44 Nm **11**.
15. Remove tensioning tool.
16. Check and adjust injection pump timing.
17. Install components in reverse order of removal.
18. Tighten crankshaft pulley bolts **1**.
    Tightening torque: 28 Nm.

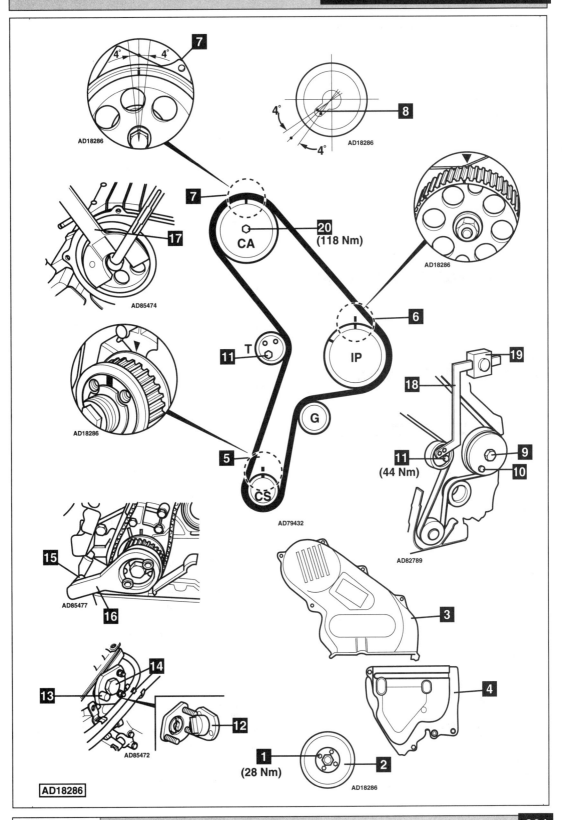

AD18286

Model:	Ducato 2,8 D • Ducato 2,8 TD
Year:	1997-02
Engine Code:	8140.63, 8140.43

Replacement Interval Guide

Fiat recommend:

→2000:
Replacement every 72,000 miles or 6 years.

2001→:
Check and replacement if necessary every 36,000 miles and replacement every 72,000 miles or 5 years.
The previous use and service history of the vehicle must always be taken into account.
Refer to Timing Belt Replacement Intervals at the front of this manual.

Check For Engine Damage

CAUTION: This engine has been identified as an INTERFERENCE engine in which the possibility of valve-to-piston damage in the event of a timing belt failure is MOST LIKELY to occur.
A compression check of all cylinders should be performed before removing the cylinder head.

Repair Times – hrs

Remove & install	1,05
AC	+0,05

Special Tools

- Injection pump locking pin – Fiat No.1860617000.
- Tensioner locking tool – Fiat No.1860638000.
- Flywheel timing pin (8 mm diameter).

Special Precautions

- Disconnect battery earth lead.
- DO NOT turn crankshaft or camshaft when timing belt removed.
- Remove glow plugs to ease turning engine.
- Turn engine in normal direction of rotation (unless otherwise stated).
- DO NOT turn engine via camshaft or other sprockets.
- Observe all tightening torques.
- Check diesel injection pump timing after belt replacement.

Removal

1. Remove:
   - ❏ RH front wheel.
   - ❏ RH front inner wing panel.
   - ❏ Engine undershield.
   - ❏ Auxiliary drive belt.
   - ❏ Water pump pulley.
   - ❏ Oil filler cap.
   - ❏ Engine top cover **1**.
   - ❏ Timing belt upper cover **2**.

2. Slacken crankshaft pulley bolt **3**.
3. Turn crankshaft to TDC on No.1 cylinder. Ensure timing marks aligned **4**.
4. Ensure camshaft sprocket timing mark aligned with mark on cylinder head cover **5**.
5. Insert locking pin in injection pump sprocket **6**. Tool No.1860617000.
6. Insert 8 mm timing pin in flywheel **7**.
7. Remove:
   - ❏ Crankshaft pulley bolt **3**.
   - ❏ Crankshaft pulley **8**.
   - ❏ Bolts **9**.
   - ❏ Timing belt lower cover nut **10**.
8. Ensure crankshaft sprocket timing mark aligned at 6 o'clock position **11**.
9. Slacken tensioner nuts **12** & **13**.
10. Lever tensioner pulley away from belt. Fit tensioner locking tool **14**. Tool No.1860638000.
11. Remove timing belt.

Installation

1. Ensure timing marks aligned **5** & **11**.
2. Ensure flywheel timing pin located correctly **7**.
3. Ensure injection pump locking pin located correctly **6**.
4. Ensure tensioner locking tool fitted **14**.
5. Fit timing belt, starting at crankshaft sprocket. Ensure belt is taut on non-tensioned side.
6. Remove:
   - ❏ Tensioner locking tool **14**.
   - ❏ Flywheel timing pin **7**.
   - ❏ Injection pump locking pin **6**.
7. Turn crankshaft two turns clockwise.
8. Insert 8 mm timing pin in flywheel **7**.
9. Ensure timing marks aligned **5** & **11**.
10. Ensure injection pump locking pin can be inserted easily **6**.
11. Tighten tensioner nut **13**.
    Tightening torque: 40 Nm.
12. Tighten tensioner nut **12**.
    Tightening torque: 20 Nm.
13. Remove injection pump locking pin **6**.
14. Apply a load of 10 kg to belt at ▽. Belt should deflect 7-8 mm.
15. Install components in reverse order of removal.
16. Tighten crankshaft pulley bolt **3**.
    Tightening torque: 200 Nm.

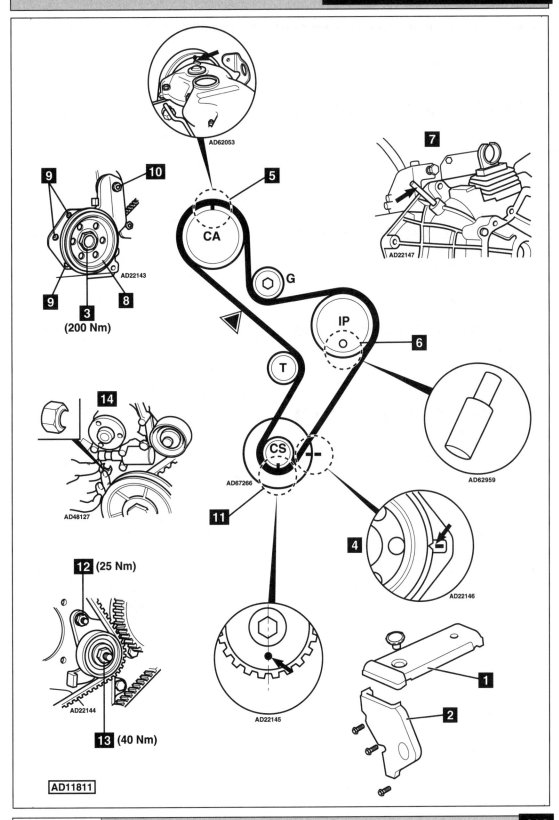

AD62053

7

AD22147

9   10

AD22143

9   8

3
(200 Nm)

5

CA

G

IP

6

T

AD62959

14

AD48127

CS

AD67266

11

4

AD22146

12 (25 Nm)

AD22144

13 (40 Nm)

AD22145

1

2

AD11811

FORD

Model:	Fiesta 1,3/1,4/1,6 • Fiesta XR2i • Fiesta RS Turbo • Fiesta Van 1,4
Year:	1983-96
Engine Code:	F4, F6, FU, JP, LJ, LH, LP, LU

Replacement Interval Guide

Ford recommend replacement every
36,000 miles or 3 years (on 6,000 mile service
intervals) or every 40,000 miles or 5 years
(on 10,000 mile service intervals).
*The previous use and service history of the vehicle
must always be taken into account.*
*Refer to Timing Belt Replacement Intervals at the front
of this manual.*

Check For Engine Damage

*CAUTION: This engine has been identified as an
INTERFERENCE engine in which the possibility of
valve-to-piston damage in the event of a timing belt
failure is MOST LIKELY to occur.*
*A compression check of all cylinders should be
performed before removing the cylinder head.*

Repair Times – hrs

Check & adjust	0,40
Remove & install:	
→1989	0,80
1989→	0,90

Special Tools

- Tension gauge – Ford No.21-113.

Special Precautions

- Disconnect battery earth lead.
- DO NOT turn crankshaft or camshaft when timing belt removed.
- Remove spark plugs to ease turning engine.
- Turn engine in normal direction of rotation (unless otherwise stated).
- DO NOT turn engine via camshaft or other sprockets.
- Observe all tightening torques.

Removal

1. Remove:
   - ❑ Auxiliary drive belt.
   - ❑ Idle air control (IAC) valve and vacuum hose.
   - ❑ Air filter housing.
   - ❑ Air intake pipe.
2. Raise and support RH front of vehicle.
3. 1989MY→: Remove:
   - ❑ RH front wheel.
   - ❑ RH wheel arch liner.
4. Remove:
   - ❑ Crankshaft pulley bolt **2**.
   - ❑ Crankshaft pulley **6**.
   - ❑ Timing belt upper cover **1**.
   - ❑ Later models: Timing belt lower cover **8**.

5. Turn crankshaft to TDC on No.1 cylinder. Ensure timing marks aligned **3** & **4**.
6. Slacken tensioner bolts **5**. Release tension on belt.
7. Remove timing belt.

Installation

*NOTE: Pre 04/88 1,4 and 1,6 engines should be
fitted with a modified (larger diameter) tensioner
pulley.*

1. Ensure timing marks aligned **3** & **4**.
2. Check water pump and tensioner for signs of seizing.
3. Fit timing belt in anti-clockwise direction, starting at crankshaft sprocket.
4. Pretension belt. Tighten tensioner bolts **5**.
5. Temporarily fit crankshaft pulley **6**.
6. Turn crankshaft two turns clockwise to TDC on No.1 cylinder.
7. Turn crankshaft approximately 60° anti-clockwise (3 teeth on camshaft sprocket).
8. Attach tension gauge to belt at ▽ **7**. Tool No.21-113.
9. Check tension gauge reading.
   Used belt: 4-6 units. New belt: 10-11 units.
10. If tension not as specified: Turn crankshaft clockwise to TDC. Repeat tensioning procedure.
11. Turn crankshaft 90° clockwise, then anti-clockwise to 60° BTDC.
12. Check tension gauge reading.
13. Remove crankshaft pulley **6**.
14. Install:
    - ❑ Timing belt upper cover **1**.
    - ❑ Later models: Timing belt lower cover **8**.
15. Install components in reverse order of removal.
16. Tighten crankshaft pulley bolt **2**.
    Tightening torque: 100-115 Nm.

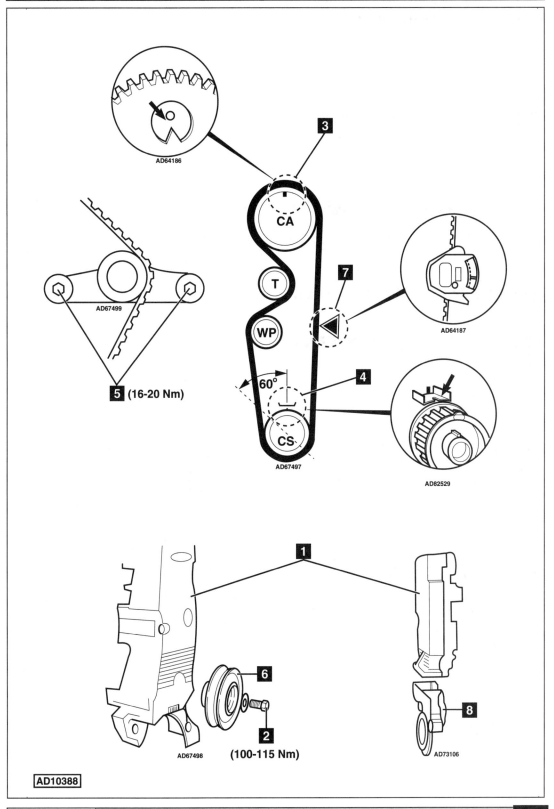

AD64186

3

CA

T

7

AD64187

WP

AD67499

5 (16-20 Nm)

60°

4

CS

AD67497

AD82529

1

6

8

2

AD67498

(100-115 Nm)

AD73106

AD10388

Model:	**Fiesta 1,25/1,4**
Year:	**1995-97**
Engine Code:	**Automatic tensioner** **DHA, FHA**

Replacement Interval Guide

Ford recommend replacement every 100,000 miles or 10 years.
The previous use and service history of the vehicle must always be taken into account.
Refer to Timing Belt Replacement Intervals at the front of this manual.

Check For Engine Damage

CAUTION: This engine has been identified as an INTERFERENCE engine in which the possibility of valve-to-piston damage in the event of a timing belt failure is MOST LIKELY to occur.
A compression check of all cylinders should be performed before removing the cylinder head.

Repair Times – hrs

Remove & install	2,50
PAS & AC	+0,30

Special Tools

- ■ Camshaft setting bar – Ford No.21-162.
- ■ Sprocket holding tool – Ford No.15-030-A.
- ■ Crankshaft pulley puller – Ford No.13-019.
- ■ Crankshaft pulley installer – Ford No.21-214.
- ■ TDC timing pin – Ford No.21-210.
- ■ 1,5 mm drill bit.
- ■ Universal 'G' clamp.

Removal

1. Raise and support front of vehicle.
2. Support engine.
3. Remove:
   - ❏ RH front wheel.
   - ❏ Auxiliary drive belt.
   - ❏ Auxiliary drive belt guide pulley **1**.
   - ❏ Cylinder head cover.
4. Remove blanking plug from cylinder block. Screw in TDC timing pin as far as stop **10**. Tool No.21-210.
5. Turn crankshaft slowly clockwise to TDC on No.1 cylinder until crankshaft web against timing pin.
6. Fit setting bar No.21-162 to rear of camshafts **11**.
   NOTE: If setting bar cannot be fitted: Remove timing pin. Turn crankshaft one turn clockwise. Insert timing pin. Ensure crankshaft web against timing pin.
7. Remove:
   - ❏ Crankshaft pulley bolt **2**.
   - ❏ Crankshaft pulley **3**. Use tool No.13-019.
   NOTE: Crankshaft sprocket NOT keyed to crankshaft. Ensure crankshaft does not turn.
   - ❏ Water pump pulley **4**.
   - ❏ Timing belt upper cover **5**.
   - ❏ Timing belt lower cover **6**.
   - ❏ Thrust washer **7**.
   - ❏ Belt guide device **8**.
   - ❏ RH engine mounting **9**.

8. Slowly compress pushrod into tensioner body until holes aligned. Use universal 'G' clamp **12**.
9. Retain pushrod with 1,5 mm drill bit in hole in tensioner body **13**.
10. Remove:
    - ❏ Universal 'G' clamp.
    - ❏ Timing belt.

Installation

1. Ensure setting bar fitted correctly **11**.
2. Ensure timing pin located correctly **10**.
3. Ensure crankshaft at TDC on No.1 cylinder. Ensure crankshaft web against timing pin.
4. Hold camshaft sprockets. Use tool No.15-030-A. Slacken bolts **14**.
5. Tap each camshaft sprocket gently to loosen it from taper. Ensure sprocket turns freely on camshaft but without free play.
6. Fit timing belt in anti-clockwise direction, starting at crankshaft sprocket. Ensure belt is taut between sprockets.
7. Remove 1,5 mm drill bit from tensioner body to release pushrod.
8. Install:
   - ❏ Belt guide device **8**.
   - ❏ Thrust washer **7**.
   - ❏ Timing belt lower cover **6**.
   - ❏ Crankshaft pulley **3**. Use tool No.21-214.
   NOTE: Ensure crankshaft pulley fully seated against thrust washer.
9. Hold crankshaft pulley **3**. Use tool No.15-030-A. Fit new crankshaft bolt **2**.
   Tightening torque: 40 Nm + 90°.
   NOTE: New crankshaft bolt must be used.
10. Remove:
    - ❏ TDC timing pin **10**.
    - ❏ Setting bar **11**.
11. Hold camshaft sprockets. Use tool No.15-030-A. Tighten bolts **14**. Tightening torque: 60 Nm. Ensure crankshaft and camshafts do not turn.
12. Turn crankshaft slowly two turns clockwise.
13. Screw in timing pin as far as stop **10**. Tool No.21-210.
14. Ensure crankshaft web against timing pin.
15. Ensure setting bar can be fitted **11**.
16. Remove:
    - ❏ TDC timing pin **10**.
    - ❏ Setting bar **11**.
17. Fit blanking plug to cylinder block and tighten to 25 Nm.
18. Install components in reverse order of removal.

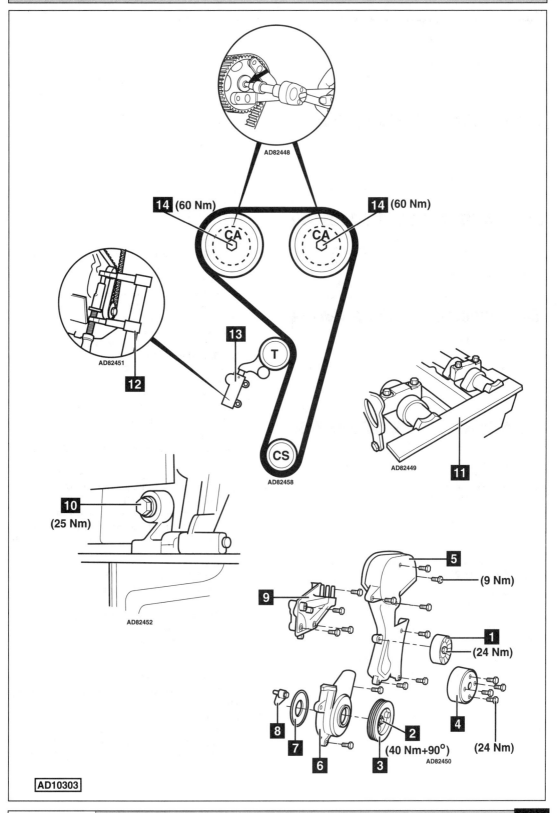

AD82448

14 (60 Nm)

14 (60 Nm)

CA

CA

AD82451

13

T

12

CS

AD82458

AD82449

11

10

(25 Nm)

AD82452

5

(9 Nm)

9

1

(24 Nm)

8

7

4

6

3

2

(40 Nm+90°)

AD82450

(24 Nm)

AD10303

FORD

Model:	**Fiesta 1,25/1,4/1,6 • Focus 1,4/1,6 • Puma 1,4**
Year:	**1997-03**
Engine Code:	**Semi-automatic tensioner** **DHA, DHB, DHC, DHD, FHA, FHD, FHF, FXDA/C, FYDA/C, L1T, L1V**

Replacement Interval Guide

Ford recommend replacement every 100,000 miles or 10 years.
The previous use and service history of the vehicle must always be taken into account.
Refer to Timing Belt Replacement Intervals at the front of this manual.

Check For Engine Damage

CAUTION: This engine has been identified as an INTERFERENCE engine in which the possibility of valve-to-piston damage in the event of a timing belt failure is MOST LIKELY to occur.
A compression check of all cylinders should be performed before removing the cylinder head(s).

Repair Times – hrs

Remove & install:

Fiesta	2,50
PAS & AC	+0,30
Focus	2,40
PAS & AC	+0,30
Puma	2,10

Special Tools

Fiesta/Puma
- Camshaft setting bar – Ford No.21-162B.
- Crankshaft timing pin – Ford No.21-210.
- Crankshaft pulley puller – Ford No.21-215.
- Crankshaft pulley installer – Ford No.21-214.

Focus
- Camshaft setting bar – Ford No.303-376.
- Crankshaft timing pin – Ford No.303-507.

Special Precautions

- Disconnect battery earth lead.
- DO NOT turn crankshaft or camshaft when timing belt removed.
- Remove spark plugs to ease turning engine.
- Turn engine in normal direction of rotation (unless otherwise stated).
- DO NOT turn engine via camshaft or other sprockets.
- Observe all tightening torques.

Removal

1. Raise and support front of vehicle.
2. Support engine.
3. Remove:
   - ❑ RH front wheel.
   - ❑ Cylinder head cover.
   - ❑ Coolant expansion tank. DO NOT disconnect hoses.
4. Disconnect:
   - ❑ Camshaft position (CMP) sensor multi-plug.
   - ❑ Injector multi-plugs.
5. Remove:
   - ❑ Engine mounting bracket.
   - ❑ Engine mounting **9**.
   - ❑ Auxiliary drive belt cover.
   - ❑ Auxiliary drive belt.
   - ❑ Auxiliary drive belt guide pulley **1**.
   - ❑ Alternator (08/98→).
   - ❑ Water pump pulley **4**.
   - ❑ Blanking plug from cylinder block **10**.
6. Insert crankshaft timing pin:
   - ❑ Fiesta/Puma – Tool No.21-210.
   - ❑ Focus – Tool No.303-507.
7. Turn crankshaft slowly clockwise until it stops against timing pin. Ensure crankshaft at TDC on No.1 cylinder.
8. Fit setting bar to rear of camshafts **11**:
   - ❑ Fiesta/Puma – Tool No.21-162B.
   - ❑ Focus – Tool No.303-376.

   NOTE: If setting bar cannot be fitted: Remove timing pin. Turn crankshaft one turn clockwise. Insert timing pin. Ensure crankshaft web against timing pin.
9. Remove:
   - ❑ Crankshaft pulley bolt **2**.
   - ❑ Crankshaft pulley **3**. Use tool No.21-215 (Fiesta/Puma).

   NOTE: Crankshaft sprocket NOT keyed to crankshaft. Ensure crankshaft does not turn.
   - ❑ PAS pump (if AC fitted).
   - ❑ Timing belt upper cover **5**.
   - ❑ Timing belt lower cover **6**.
   - ❑ Thrust washer **7**.

   NOTE: There are two types of tensioner for timing belt.
10. Type A:
    - ❑ Slacken tensioner bolts **13**.
    - ❑ Turn tensioner anti-clockwise to release tension on belt **8**. Use Allen key (8 mm).

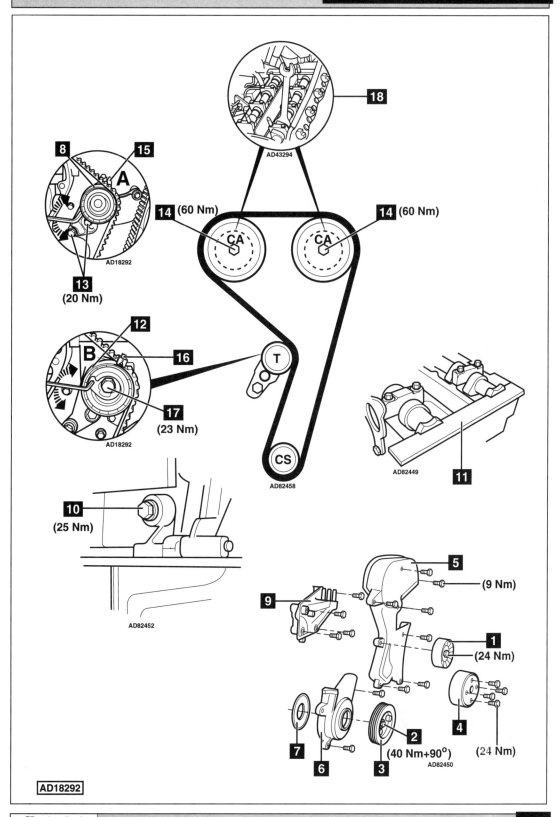

AD43294

18

8 **15** **A**

AD18292

13
(20 Nm)

14 (60 Nm)

14 (60 Nm)

CA

CA

12 **B** **16**

17
(23 Nm)

AD18292

T

CS

AD82458

AD82449

11

10
(25 Nm)

AD82452

5

(9 Nm)

9

1
(24 Nm)

7

6

3

2
(40 Nm+90°)

AD82450

4

(24 Nm)

AD18292

11. Type B:
- ❏ Slacken tensioner bolt **17**.
- ❏ Turn tensioner cam clockwise to release tension on belt **12**. Use Allen key (6 mm).

12. Remove timing belt.

Installation

1. Ensure setting bar fitted correctly **11**.

2. Ensure timing pin located correctly **10**.

3. Ensure crankshaft at TDC on No.1 cylinder. Ensure crankshaft web against timing pin.

4. Remove camshaft setting bar **11**.

5. Hold camshafts with spanner on hexagon **18**.

6. Slacken bolt of each camshaft sprocket **14**.

NOTE: DO NOT use setting bar to hold camshafts when slackening bolts.

7. Tap each camshaft sprocket gently to loosen it from taper.

8. Ensure camshaft sprockets can turn on taper but not tilt.

9. Refit camshaft setting bar **11**.

10. Fit timing belt in anti-clockwise direction, starting at crankshaft sprocket.

NOTE: Ensure directional arrows point in direction of rotation.

11. Ensure belt is taut between sprockets.

NOTE: There are two types of tensioner for timing belt.

12. Type A:
- ❏ Turn tensioner clockwise until indicator central with rectangular recess **15**. Use Allen key (8 mm).
- ❏ Tighten tensioner bolts to 20 Nm **13**.

13. Type B:
- ❏ Turn tensioner cam anti-clockwise until arrow in centre of window **16**. Use Allen key (6 mm).
- ❏ Tighten tensioner bolt to 23 Nm **17**.

14. Install:
- ❏ Thrust washer **7**.
- ❏ Timing belt lower cover **6**.
- ❏ Crankshaft pulley **3**. Use tool No.21-214.

15. Tighten crankshaft pulley bolt **2**. Tightening torque: 40 Nm + 90°. Use new bolt.

16. Ensure crankshaft web against timing pin.

17. Hold camshafts with spanner on hexagon **18**. Tighten bolt of each camshaft sprocket **14**. Tightening torque: 60 Nm.

18. Remove:
- ❏ Crankshaft timing pin **10**.
- ❏ Camshaft setting bar **11**.

19. Turn crankshaft slowly two turns clockwise.

20. Insert crankshaft timing pin **10**.

21. Turn crankshaft until web against timing pin.

22. Ensure camshaft setting bar can be fitted **11**.

23. Remove:
- ❏ Crankshaft timing pin **10**.
- ❏ Camshaft setting bar **11**.

24. Fit blanking plug and tighten to 25 Nm **10**.

25. Install components in reverse order of removal.

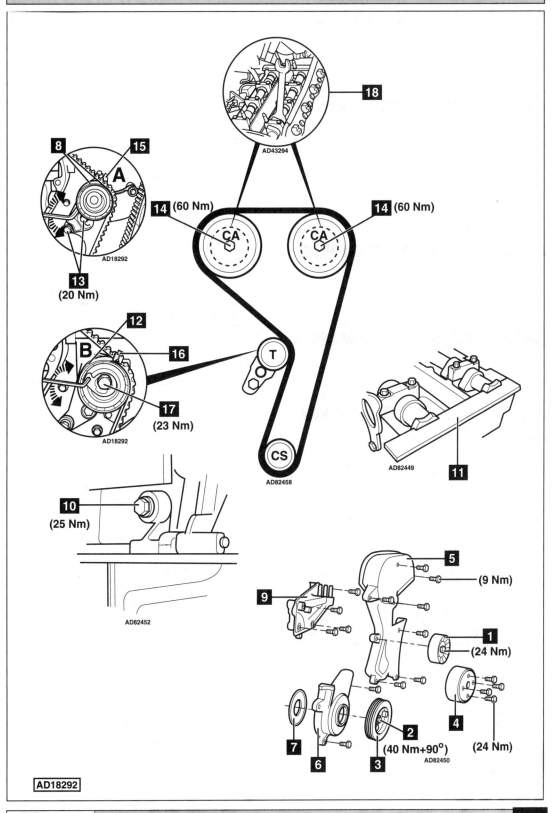

18

AD43294

8   15

A

AD18292

13
(20 Nm)

14 **(60 Nm)**

14 **(60 Nm)**

CA

CA

12

B

16

17
(23 Nm)

AD18292

T

CS

AD82458

11

AD82449

10
(25 Nm)

AD82452

5

(9 Nm)

9

1
(24 Nm)

4
(24 Nm)

7

6

3

2
(40 Nm+90°)

AD82450

AD18292

FORD

Model:	**Fiesta 1,6i/XR2i/RS 1800 1,8i 16V • Escort/Orion 1,6/1,8i 16V**
Year:	**1991-01**
Engine Code:	**L1E, L1G, L1H, L1K, RDA, RDB, RKC, RQB, RQC**

Replacement Interval Guide

Ford recommend
→02/94: Replacement every 60,000 miles or
5 years, whichever occurs first (tensioner and guide
pulleys must also be replaced).
03/94→: Every 80,000 miles or 5 years, whichever
occurs first (tensioner and guides must also be
replaced).
*The previous use and service history of the vehicle
must always be taken into account.*
*Refer to Timing Belt Replacement Intervals at the
front of this manual.*

Check For Engine Damage

*CAUTION: This engine has been identified as an
INTERFERENCE engine in which the possibility of
valve-to-piston damage in the event of a timing belt
failure is MOST LIKELY to occur.*
*A compression check of all cylinders should be
performed before removing the cylinder head.*

Repair Times – hrs

Remove & install:	
Fiesta	2,00
Escort/Orion	1,90
EGR	+0,40

Special Tools

- ■ Camshaft setting bar – Ford No.21-162.
- ■ Sprocket holding tool – Ford No.15-030-A.

Special Precautions

- ■ Disconnect battery earth lead.
- ■ DO NOT turn crankshaft or camshaft when timing belt removed.
- ■ Remove spark plugs to ease turning engine.
- ■ Turn engine in normal direction of rotation (unless otherwise stated).
- ■ DO NOT turn engine via camshaft or other sprockets.
- ■ Observe all tightening torques.

Removal

*NOTE: If timing belt is excessively noisy when
engine is hot and particularly at 1100 rpm, a
modified crankshaft sprocket, tensioner spring
and bolt ᎻᎾ᎒ should be fitted – refer to Ford dealer.
These modified parts were fitted during production
from 23/09/92 on.*

1. Escort/Orion: Drain coolant.
2. Remove:
   - ❏ Air intake pipe from volume air flow (VAF) sensor.
   - ❏ Escort/Orion: Coolant expansion tank.
3. Disconnect:
   - ❏ PAS sensor multi-plug.
   - ❏ Escort/Orion: PAS pump hoses (collect fluid).
   - ❏ Throttle cable.
   - ❏ Crankshaft pulley cover.
4. Fiesta: Support engine. Remove RH engine mounting(s).
5. Remove:
   - ❏ Auxiliary drive belt.
   - ❏ Auxiliary drive belt tensioner (if required).
   - ❏ Water pump pulley.
6. Fiesta: Lower engine slightly.
7. Slacken crankshaft pulley bolt **8**.
8. Turn crankshaft to TDC **1**.
9. Remove:
   - ❏ Crankshaft pulley bolt **8**.
   - ❏ Crankshaft pulley.
   - ❏ Timing belt upper cover **2**.
   - ❏ Timing belt centre cover **3**.
   - ❏ Timing belt lower cover **4**.
   - ❏ Cylinder head cover.
10. Fit setting bar No.21-162 to rear of camshafts **5**.
    NOTE: If necessary: Turn crankshaft one turn.
11. Slacken tensioner bolt **6**. Turn tensioner clockwise away from belt. Lightly tighten bolt.
12. Remove timing belt.

Installation

*NOTE: If not already fitted, a tensioner spring **9**
and retaining pin should be fitted – refer to Ford
dealer.*

1. Fit timing belt in anti-clockwise direction, starting at crankshaft sprocket.
2. Ensure belt is taut between sprockets.
3. Release tensioner to tension belt automatically.
4. Remove setting bar **5**.
5. Temporarily fit crankshaft pulley.
6. Turn crankshaft two turns to TDC **1**.
7. Ensure setting bar can be fitted.
8. If setting bar cannot be fitted: Hold camshaft sprockets. Use tool No.15-030-A. Slacken bolts **7**.
9. Tap each camshaft sprocket gently to loosen it from taper. Ensure each sprocket turns freely on camshaft.
10. Carefully turn camshafts until setting bar can be fitted.
11. Tighten bolt **7**. Tightening torque: 67-72 Nm.
12. Tighten tensioner bolt **6**.
    Tightening torque: 35-40 Nm.
13. Install components in reverse order of removal.
14. Tighten crankshaft pulley bolt **8**.
    Tightening torque: 100-115 Nm.
15. Escort/Orion: Refill cooling system.

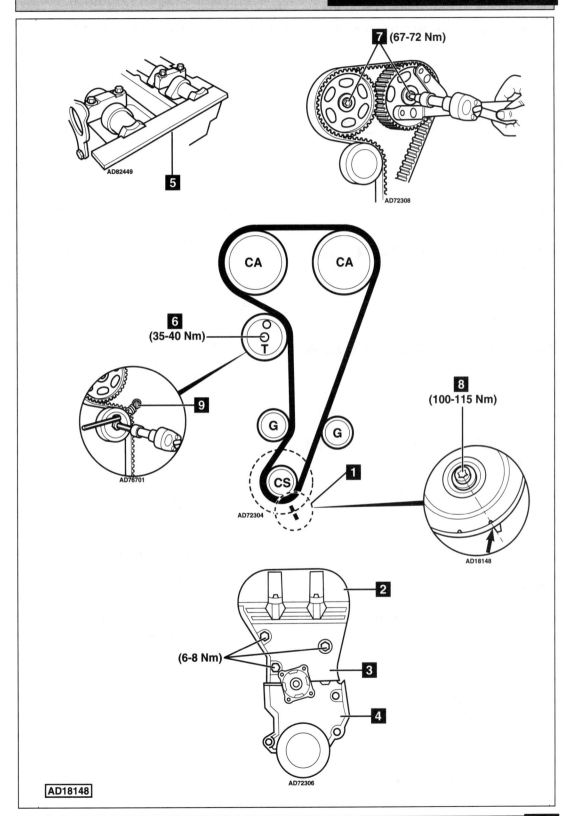

7 (67-72 Nm)

AD72308

5

AD82449

CA   CA

6
(35-40 Nm)   T

8
(100-115 Nm)

9

AD76701

G   G

CS

1

AD72304

AD18148

2

3

(6-8 Nm)

4

AD72306

AD18148

FORD

Model:	Escort 1,1/1,3/1,4/1,6 • Escort XR3/XR3i 1,6 • Escort RS Turbo 1,6 Orion 1,3/1,4/1,6 • Escort Van 1,4/1,6
Year:	1980-98
Engine Code:	F4, F6, FU, JP, LJ, LU, LN, LP, LR, L4, GM, GP, GU

Replacement Interval Guide

Ford recommend replacement every 36,000 miles or 3 years (on 6,000 mile service intervals) or every 40,000 miles or 5 years (on 10,000 mile service intervals).
The previous use and service history of the vehicle must always be taken into account.
Refer to Timing Belt Replacement Intervals at the front of this manual.

Check For Engine Damage

CAUTION: This engine has been identified as an INTERFERENCE engine in which the possibility of valve-to-piston damage in the event of a timing belt failure is MOST LIKELY to occur.
A compression check of all cylinders should be performed before removing the cylinder head.

Repair Times – hrs

Check & adjust	0,40
Fuel injection models	+0,10
Remove & install	0,80
Fuel injection models	+0,10
PAS	+0,10

Special Tools

- ■ Tension gauge – Ford No.21-113.

Special Precautions

- ■ Disconnect battery earth lead.
- ■ DO NOT turn crankshaft or camshaft when timing belt removed.
- ■ Remove spark plugs to ease turning engine.
- ■ Turn engine in normal direction of rotation (unless otherwise stated).
- ■ DO NOT turn engine via camshaft or other sprockets.
- ■ Observe all tightening torques.

Removal

1. Remove:
   - ❑ Alternator drive belt.
   - ❑ Idle air control (IAC) valve and vacuum hose.
   - ❑ Air filter housing.
   - ❑ Air intake pipe.
2. Raise and support RH front of vehicle.
3. Remove:
   - ❑ Lower splash guard.
   - ❑ Crankshaft pulley bolt **2**.
   - ❑ Crankshaft pulley **6**.
   - ❑ Timing belt upper cover **1**.
   - ❑ Later models: Timing belt lower cover **8**.

4. Turn crankshaft to TDC on No.1 cylinder. Ensure timing marks aligned **3** & **4**.
5. Slacken tensioner bolts **5** to release tension on belt.
6. Remove timing belt.

Installation

NOTE: Pre 04/88 1,4 and 1,6 engines should be fitted with a modified (larger diameter) tensioner pulley.

1. Ensure timing marks aligned **3** & **4**.
2. Check water pump and tensioner for signs of seizing.
3. Fit timing belt in anti-clockwise direction, starting at crankshaft sprocket.
4. Pretension belt. Tighten tensioner bolts **5**.
5. Temporarily fit crankshaft pulley **6**.
6. Turn crankshaft two turns clockwise to TDC on No.1 cylinder.
7. Turn crankshaft approximately 60° anti-clockwise (3 teeth on camshaft sprocket).
8. Attach tension gauge to belt at ▽ **7**. Tool No.21-113.
9. Check tension gauge reading. Used belt: 4-6 units. New belt: 10-11 units.
10. If tension not as specified: Turn crankshaft clockwise to TDC. Repeat tensioning procedure.
11. Turn crankshaft 90° clockwise, then anti-clockwise to 60° BTDC.
12. Check tension gauge reading.
13. Remove crankshaft pulley **6**.
14. Install:
    - ❑ Timing belt upper cover **1**.
    - ❑ Later models: Timing belt lower cover **8**.
15. Install components in reverse order of removal.
16. Tighten crankshaft pulley bolt **2**. Tightening torque: 100-115 Nm.

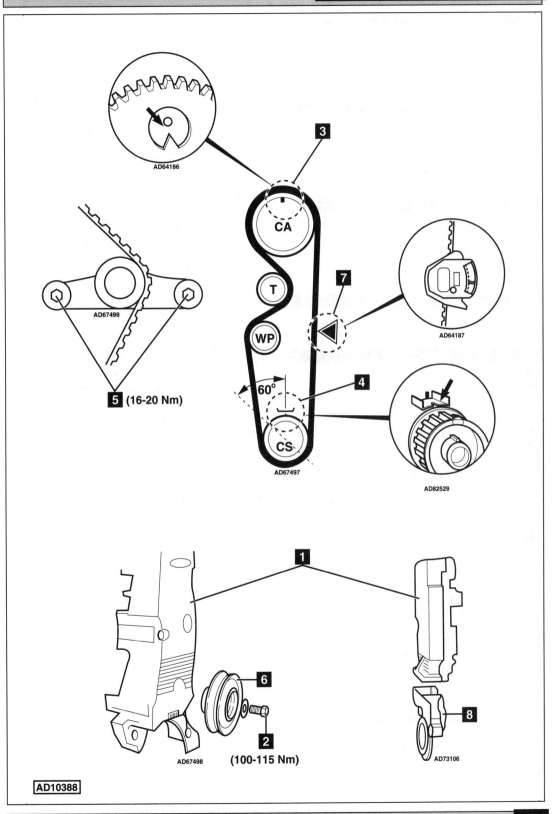

AD64186

3

CA

T

WP

7

AD64187

AD67499

5 (16-20 Nm)

60°

4

CS

AD67497

AD82529

1

6

2

(100-115 Nm)

AD67498

8

AD73106

AD10388

Model:	**Escort RS Cosworth**
Year:	**1992-96**
Engine Code:	**N5E, N5F**

Replacement Interval Guide

Ford recommend replacement every
60,000 miles or 5 years.
*The previous use and service history of the vehicle
must always be taken into account.*
*Refer to Timing Belt Replacement Intervals at the front
of this manual.*

Check For Engine Damage

*CAUTION: This engine has been identified as an
INTERFERENCE engine in which the possibility of
valve-to-piston damage in the event of a timing belt
failure is MOST LIKELY to occur.*
*A compression check of all cylinders should be
performed before removing the cylinder head.*

Repair Times – hrs

Check & adjust	0,50
Remove & install	1,30

Special Tools

■ Tension gauge – Ford No.21-113.

Special Precautions

■ Disconnect battery earth lead.
■ DO NOT turn crankshaft or camshaft when timing belt removed.
■ Remove spark plugs to ease turning engine.
■ Turn engine in normal direction of rotation (unless otherwise stated).
■ DO NOT turn engine via camshaft or other sprockets.
■ Observe all tightening torques.

Removal

1. Drain coolant. Detach coolant hose from thermostat housing.
2. Remove:
   ❏ Engine undershield.
   ❏ Auxiliary drive belt.
   ❏ Timing belt cover **1**.
3. Detach lower radiator mounts, move away from engine and retain with wire.
4. Remove:
   ❏ Crankshaft pulley bolt **2**.
   ❏ Crankshaft pulley **3**.
   ❏ Thrust washer **4**.
5. Temporarily fit crankshaft pulley bolt **2**.
6. Turn crankshaft to TDC on No.1 cylinder. Ensure timing marks aligned **5**, **6** & **7**.
7. Slacken tensioner nut **8**.
8. Remove timing belt.

Installation

1. Ensure timing marks aligned **5**, **6** & **7**.
2. Ensure Hall sender set at timing mark **9**.
3. Fit timing belt in anti-clockwise direction, starting at crankshaft sprocket. Ensure belt is taut between sprockets.
4. Turn tensioner clockwise against belt. Tighten nut to 44 Nm **8**.
   NOTE: DO NOT turn tensioner anti-clockwise.
5. Remove crankshaft pulley bolt **2**.
6. Install:
   ❏ Thrust washer **4**.
   ❏ Crankshaft pulley **3**.
   ❏ Crankshaft pulley bolt **2**.
      Tightening torque: 195 Nm.
7. Check RPM sensor gap is 0,80 mm **10**. Adjust if necessary.
8. Turn crankshaft slowly three turns clockwise.
9. Turn crankshaft one turn anti-clockwise to TDC. DO NOT turn crankshaft past TDC.
10. Attach tension gauge to belt at ▽. Tool No.21-113.
11. Tension gauge should indicate 9,5-10,5 units.
12. If not: Slacken tensioner nut **8**. Turn tensioner clockwise until reading correct.
13. Tighten tensioner nut to 44 Nm **8**.
14. Recheck belt tension. If tension not as specified: Repeat operations 8-11 until tension correct.
15. Fit coolant hose. Refill cooling system.
16. Fit auxiliary drive belt.
17. Start engine and run at idle speed for 10 minutes. Allow engine to cool for 2 hours.
18. Recheck belt tension. If necessary: Repeat tensioning procedure.
19. Install components in reverse order of removal.

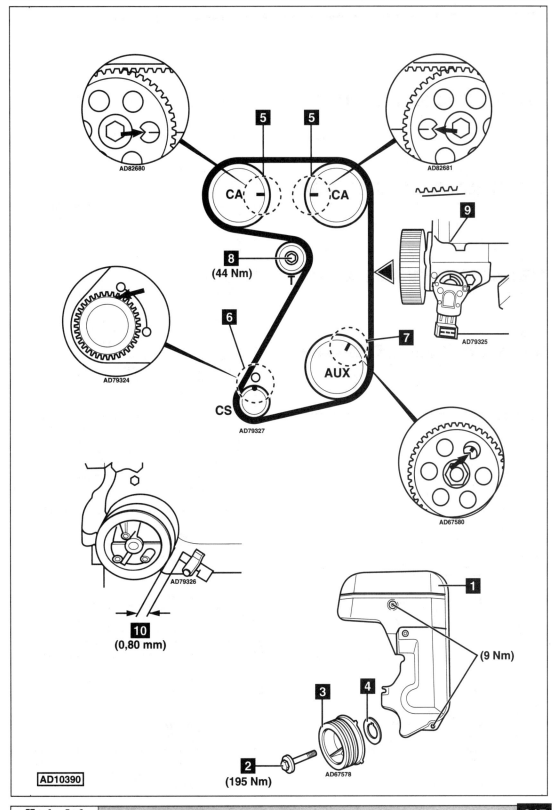

AD82680

AD82681

5 5

CA — — CA

8 (44 Nm)

T

9

AD79325

AD79324

6

7

AUX

CS

AD79327

AD67580

AD79326

10 (0,80 mm)

1

(9 Nm)

3 4

2 (195 Nm)

AD67578

AD10390

Model:	Focus 1,8/2,0 • Mondeo 1,6/1,8/2,0 (05/98→) • Cougar 2,0
Year:	1998-03
Engine Code:	L1N, NGB, NGC, RKF, RKH, EBDA, EDDC, EYDC

Replacement Interval Guide

Ford recommend replacement every 100,000 miles or 10 years.
The previous use and service history of the vehicle must always be taken into account.
Refer to Timing Belt Replacement Intervals at the front of this manual.

Check For Engine Damage

CAUTION: This engine has been identified as an INTERFERENCE engine in which the possibility of valve-to-piston damage in the event of a timing belt failure is MOST LIKELY to occur. A compression check of all cylinders should be performed before removing the cylinder head(s).

Repair Times – hrs

Remove & install:

Mondeo	2,40
Focus	2,20
Cougar	2,40

Special Tools

- ■ Camshaft setting bar – Ford No.303-367 (21-162B).
- ■ Sprocket holding tool – Ford No.205-072 (15-030A).
- ■ Crankshaft timing pin – Ford No.303-620 (21-163).

Removal

1. Raise and support front of vehicle.
2. Remove:
   - ❏ RH front wheel.
   - ❏ Engine undershield.
   - ❏ RH splash guard.
   - ❏ Auxiliary drive belt.
   NOTE: If belt is to be reused: Mark direction of rotation on belt.
   - ❏ Water pump pulley.
   - ❏ Auxiliary drive belt tensioner.
3. Slacken crankshaft pulley bolt **1**.
4. Turn crankshaft clockwise to TDC on No.1 cylinder. Ensure timing marks aligned **2**.
5. Remove:
   - ❏ Crankshaft pulley bolt **1**.
   - ❏ Crankshaft pulley **3**.
   - ❏ Timing belt lower cover **4**.
6. Reposition:
   - ❏ Accelerator cable.
   - ❏ Coolant expansion tank. DO NOT disconnect hoses.
   - ❏ Cruise control cable.
7. Support engine.
8. Remove:
   - ❏ RH engine mounting.
   - ❏ Timing belt upper cover **5**.
   - ❏ PAS hose bracket and clip.
   - ❏ RH engine mounting bracket **6**.
   - ❏ Spark plug cover.
   - ❏ Spark plug leads.
   - ❏ Cylinder head cover.

9. Fit setting bar to rear of camshafts **7**. Tool No.303-367 (21-162B).
10. Slacken tensioner bolt slightly **8**.
11. Turn tensioner pulley clockwise away from belt. Use Allen key **9**.
12. Slacken tensioner bolt four turns **8**.
13. Unhook tensioner from metal clip **10**.
14. Hold camshaft sprockets **11**. Use tool No.205-072 (15-030A).
15. Slacken bolt of each camshaft sprocket **12**.
16. Tap each camshaft sprocket gently to loosen it from taper. Ensure each sprocket turns freely on camshaft.
17. Remove timing belt.
    NOTE: DO NOT refit used belt.

Installation

NOTE: Lower guide pulley (G2) not fitted to some engines approx 01/99 on.
1. Ensure setting bar fitted correctly **7**.
2. Temporarily fit crankshaft pulley **3**.
3. Ensure timing marks aligned **2**.
4. Remove blanking plug from cylinder block **13**.
5. Insert timing pin in cylinder block **14**. Tool No.303-620 (21-163).
6. Ensure crankshaft against timing pin.
7. Remove crankshaft pulley.
8. Ensure tensioner unhooked from metal clip **10**.
9. Fit timing belt in anti-clockwise direction, starting at crankshaft sprocket. Ensure belt is taut between sprockets on non-tensioned side.
10. Hook tensioner into metal clip **10**.
11. Turn tensioner anti-clockwise until pointer **15** aligned with notch **16** in baseplate. Hold tensioner pulley in position.
12. Ensure crankshaft web against timing pin **14**.
13. Tighten tensioner bolt **8**. Tightening torque: 25 Nm.
14. Hold camshaft sprockets **11**. Use tool No.205-072 (15-030A).
15. Tighten bolts **12**. Tightening torque: 68 Nm.
16. Remove:
    - ❏ Setting bar **7**.
    - ❏ Timing pin **14**.
17. Temporarily fit crankshaft pulley **3**.
18. Turn crankshaft slowly two turns clockwise.
19. Ensure timing marks aligned **2**.
20. Insert timing pin in cylinder block **14**.
21. Turn crankshaft clockwise until it stops against timing pin.
22. Ensure setting bar can be fitted **7**.
23. If not: Repeat installation and tensioning procedures.
24. Remove:
    - ❏ Setting bar **7**.
    - ❏ Timing pin **14**.
25. Fit blanking plug **13**. Tightening torque: 24 Nm.
26. Install components in reverse order of removal.
27. Tighten crankshaft pulley bolt **1**. Tightening torque: 115 Nm.

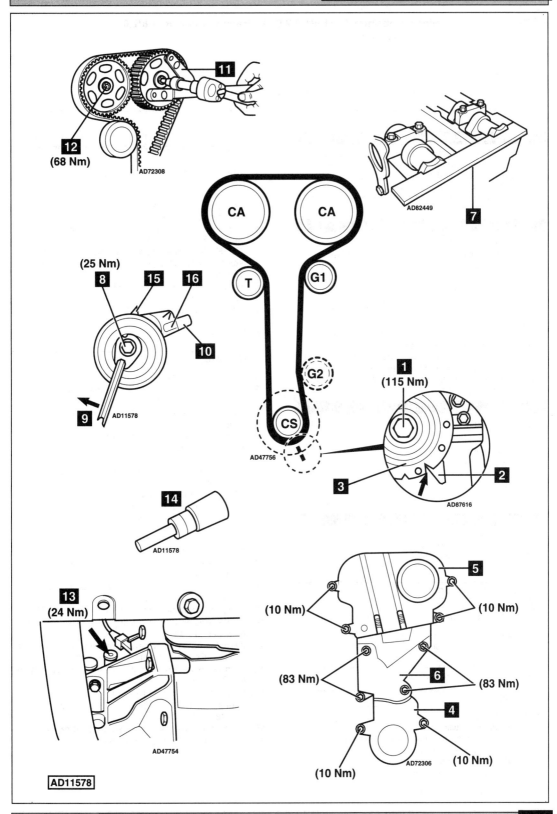

11

12
(68 Nm)

AD72308

7

AD82449

CA   **CA**

T   **G1**

(25 Nm)

8   **15**   **16**

10

9   AD11578

G2

CS

AD47756

1
(115 Nm)

3   **2**

AD87616

14

AD11578

5

(10 Nm)   **(10 Nm)**

(83 Nm)   **6**   **(83 Nm)**

4

13
(24 Nm)

AD47754

(10 Nm)   AD72306   **(10 Nm)**

AD11578

FORD

Model:	Sierra/Sapphire 1,3/1,6/1,8/2,0 • Granada/Scorpio 1,8/2,0
	P100/Transit 1,6/2,0
Year:	1981-95
Engine Code:	JC, LA, LC, LS, RE
	NE, NR, N4, N6T, NA, NBA, NCA, NU

Replacement Interval Guide

The vehicle manufacturer has not recommended a timing belt replacement interval for this engine.
The previous use and service history of the vehicle must always be taken into account.
Refer to Timing Belt Replacement Intervals at the front of this manual.

Check For Engine Damage

1,3/1,6 (except LS), 1,8:
CAUTION: This engine has been identified as an INTERFERENCE engine in which the possibility of valve-to-piston damage in the event of a timing belt failure is MOST LIKELY to occur.
A compression check of all cylinders should be performed before removing the cylinder head.

1,6 (LS)/2,0:
CAUTION: This engine has been identified as a FREEWHEELING engine in which the possibility of valve-to-piston damage in the event of a timing belt failure may be minimal or very unlikely. However, a precautionary compression check of all cylinders should be performed.

Repair Times – hrs

Check & adjust	0,40
Remove & install:	
Sierra/Sapphire	0,70
Granada/Scorpio	0,70
P100	0,70
Transit	1,20
With crankshaft damper	+0,20

Special Tools

- Tension gauge – Ford No.21-113.
- Tensioner spring bolt socket – Ford No.21-012.
- Crankshaft damper puller – Ford No.21-075-A.

Special Precautions

- Disconnect battery earth lead.
- DO NOT turn crankshaft or camshaft when timing belt removed.
- Remove spark plugs to ease turning engine.
- Turn engine in normal direction of rotation (unless otherwise stated).
- DO NOT turn engine via camshaft or other sprockets.
- Observe all tightening torques.

Removal

1. Remove:
   - ❏ Radiator cowlings.
   - ❏ Fuel injection models: Remove radiator.
   - ❏ Auxiliary drive belts.
   - ❏ Cooling fan and pulley.
   - ❏ Radiator hose and thermostat housing (if required).
   - ❏ Timing belt cover **1**.

2. Turn crankshaft to TDC on No.1 cylinder. Ensure timing marks aligned **2** & **3**. Ensure distributor rotor arm aligned with mark on distributor body **4**.
3. Prevent crankshaft from turning.
4. Remove:
   - ❏ Crankshaft pulley bolt **5**.
   - ❏ Crankshaft pulley **6**.
   - ❏ Thrust washer.
5. Fuel injection models: Remove crankshaft damper **7**. Use tool No.21-075-A.
   NOTE: Crankshaft pulley bolt will need to be screwed back into pulley. Allow bolt to protrude 16 mm.
6. Slacken tensioner spring bolt **8**. Use tool No.21-012.
7. Slacken bolt **9**. Move tensioner pulley away from belt and tighten bolt to hold in position.
8. Remove timing belt.

Installation

1. Ensure timing marks aligned **2**, **3** & **4**.
2. Fit timing belt in anti-clockwise direction, starting at crankshaft sprocket.
3. Slacken tensioner bolt **9**. Allow tensioner to operate.
4. Tighten tensioner bolts **8** & **9**.
   Tightening torque: 20-25 Nm.
5. Install:
   - ❏ Thrust washer (thrust side facing crankshaft pulley).
   - ❏ Crankshaft pulley **6**.
   - ❏ Crankshaft pulley bolt **5**.
     Tightening torque: 8.8 = 60 Nm.
     10.9 = 115 Nm.
6. Fuel injection models: Tighten crankshaft damper bolt to 130 Nm.
7. Turn crankshaft two turns clockwise to TDC on No.1 cylinder.
8. Turn crankshaft approximately 60° anti-clockwise (3 teeth on camshaft sprocket).
9. Attach tension gauge to belt at ▽. Tool No.21-113.
10. Check tension gauge reading. Used belt: 4-5 units. New belt: 10-11 units.
11. If tension not as specified: Turn crankshaft clockwise to TDC. Repeat tensioning procedure.
12. Turn crankshaft 90° clockwise, then anti-clockwise to 60° BTDC.
13. Check tension gauge reading.
14. If tension not as specified: Repeat operations 10-12 until tension correct.
15. Install components in reverse order of removal.
16. If necessary: Refill cooling system.

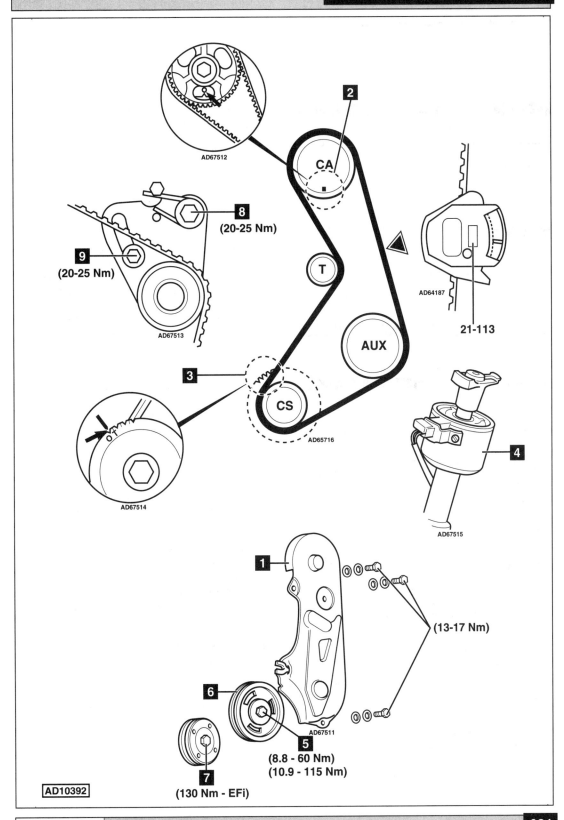

AD67512

8 (20-25 Nm)

9 (20-25 Nm)

AD67513

CA

2

T

AUX

AD64187

21-113

3

CS

AD65716

AD67514

4

AD67515

1

(13-17 Nm)

6

AD67511

5
(8.8 - 60 Nm)
(10.9 - 115 Nm)

7

(130 Nm - EFi)

AD10392

FORD

Model:	**Sierra/Sapphire 1,6/1,8**
Year:	**1988-93**
Engine Code:	**L6, L6B, R2, R6A**

Replacement Interval Guide

Ford recommend replacement every
36,000 miles or 3 years (on 6,000 mile service
intervals) or every 40,000 miles or 5 years
(on 10,000 mile service intervals).
*The previous use and service history of the vehicle
must always be taken into account.*
*Refer to Timing Belt Replacement Intervals at the front
of this manual.*

Check For Engine Damage

*CAUTION: This engine has been identified as an
INTERFERENCE engine in which the possibility of
valve-to-piston damage in the event of a timing belt
failure is MOST LIKELY to occur.*
*A compression check of all cylinders should be
performed before removing the cylinder head.*

Repair Times – hrs

Check & adjust	0,40
Remove & install	0,80

Special Tools

■ Tension gauge – Ford No.21-113.

Special Precautions

■ Disconnect battery earth lead.
■ DO NOT turn crankshaft or camshaft when timing belt
  removed.
■ Remove spark plugs to ease turning engine.
■ Turn engine in normal direction of rotation (unless
  otherwise stated).
■ DO NOT turn engine via camshaft or other sprockets.
■ Observe all tightening torques.

Removal

1. L6 and R2 engines: Remove –
   ❏ Distributor cap **1**.
   ❏ Distributor rotor housing **2**.
2. All models: Remove –
   ❏ Auxiliary drive belts.
   ❏ Crankshaft pulley bolt **3**.
   ❏ Crankshaft pulley **4**.
3. L6B and R6A engines: Remove HT leads.
4. Disconnect crankshaft position (CKP) sensor
   multi-plug.
5. Remove:
   ❏ L6 and R2 engines: Timing belt cover **5**.
   ❏ L6B and R6A engines: Timing belt covers **5**
     & **6**.
6. Temporarily fit crankshaft pulley **4**.
7. Turn crankshaft to TDC on No.1 cylinder.

8. Ensure timing marks aligned **7** & **8**.
9. Slacken tensioner bolts **9**.
10. Move tensioner away from belt. Lightly tighten
    bolts **9**.
11. Remove:
    ❏ Crankshaft pulley **4**.
    ❏ Timing belt.

Installation

1. Ensure timing marks aligned **7** & **8**.
2. Ensure water pump sprocket and tensioner
   pulley are free to turn, with no sign of seizure.
3. Fit timing belt in anti-clockwise direction, starting
   at crankshaft sprocket.
4. Slacken tensioner bolts **9**. Push tensioner
   against belt to pre-tension belt.
5. Tighten tensioner bolts **9**.
   Tightening torque: 20-30 Nm.
6. Temporarily fit crankshaft pulley **4**.
7. Turn crankshaft two turns clockwise to TDC on
   No.1 cylinder.
8. Ensure timing marks aligned **7** & **8**.
9. Turn crankshaft approximately 60° anti-
   clockwise (3 teeth on camshaft sprocket).
10. Attach tension gauge to belt at ▽.
    Tool No.21-113.
11. Check tension gauge reading.
    Used belt: 4-6 units. New belt: 10-11 units.
12. If tension not as specified: Turn crankshaft two
    turns clockwise to TDC on No.1 cylinder.
13. Ensure timing marks aligned **7** & **8**. Repeat
    tensioning procedure.
14. Turn crankshaft 90° clockwise, then
    anti-clockwise to 60° BTDC.
15. Recheck belt tension.
16. If tension not as specified: Repeat operations
    4-15 until tension correct.
17. Remove crankshaft pulley **4**.
18. Fit timing belt cover(s).
19. Fit crankshaft pulley **4**.
20. Fit crankshaft pulley bolt **3**.
    Tightening torque: 110-130 Nm.
21. Install components in reverse order of removal.

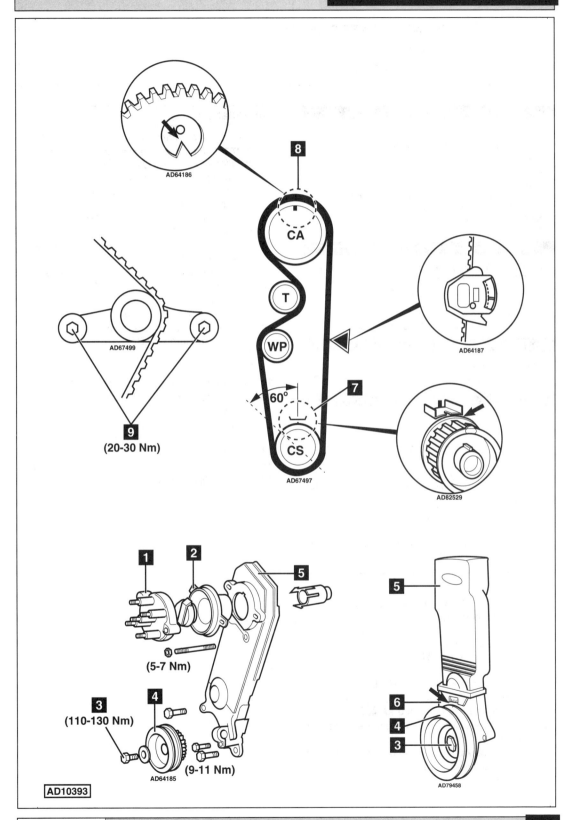

AD64186

8

CA

T

WP

AD64187

60°

7

CS

AD67497

AD82529

AD67499

9

(20-30 Nm)

1 2 5

(5-7 Nm)

3

(110-130 Nm)

4

(9-11 Nm)

AD64185

5

6

4

3

AD79458

AD10393

FORD

Model:	Sierra/Sapphire RS Cosworth
Year:	1986-94
Engine Code:	N5A, N5B, N5C, N5D

Replacement Interval Guide

Ford recommend:
→08/89: Replacement every 48,000 miles or 4 years.
09/89→: Replacement every 60,000 miles or 5 years.
The previous use and service history of the vehicle must always be taken into account.
Refer to Timing Belt Replacement Intervals at the front of this manual.

Check For Engine Damage

CAUTION: This engine has been identified as an INTERFERENCE engine in which the possibility of valve-to-piston damage in the event of a timing belt failure is MOST LIKELY to occur.
A compression check of all cylinders should be performed before removing the cylinder head.

Repair Times – hrs

Check & adjust	0,50
Remove & install	1,00

Special Tools

- Tension gauge – Ford No.21-113.

Special Precautions

- Disconnect battery earth lead.
- DO NOT turn crankshaft or camshaft when timing belt removed.
- Remove spark plugs to ease turning engine.
- Turn engine in normal direction of rotation (unless otherwise stated).
- DO NOT turn engine via camshaft or other sprockets.
- Observe all tightening torques.

Removal

1. Drain coolant. Detach coolant hose from thermostat housing.
2. Remove:
   - ❏ Auxiliary drive belts.
   - ❏ Timing belt cover **1**.
   - ❏ Distributor cap.
3. Remove crankshaft pulley bolt **2**.
4. Turn crankshaft to TDC on No.1 cylinder. Ensure timing marks aligned **3**, **4** & **5**.
5. Ensure distributor rotor arm aligned with mark on distributor body **6**.
6. Remove crankshaft pulley and thrust washer **7**.
7. Slacken tensioner nut **8**.
8. Remove timing belt.

Installation

1. Ensure timing marks aligned **3**, **4**, **5** & **6**.
   *NOTE: Point of fixed crankshaft timing mark must be aligned with LH edge of reluctor tooth **4**.*
2. Fit timing belt to crankshaft sprocket.
3. Fit crankshaft pulley and thrust washer (convex side towards belt).
4. Fit timing belt in anti-clockwise direction, starting at auxiliary shaft sprocket.
5. Turn tensioner clockwise against belt. Tighten tensioner nut **8**.
   NOTE: DO NOT turn tensioner anti-clockwise.
6. Tighten crankshaft pulley bolt **2**.
   Tightening torque: 122-135 Nm.
7. Turn crankshaft several turns.
8. Turn crankshaft back one turn to TDC. DO NOT turn crankshaft past TDC.
9. Attach tension gauge to belt at ▽. Tool No.21-113.
10. Tension gauge should indicate 9,5-10,5 units.
11. Slacken tensioner nut **8**. Turn tensioner until correct tension indicated.
12. Tighten tensioner nut **8**.
    Tightening torque: 40-48 Nm.
13. Repeat operations 7-10.
14. Install components in reverse order of removal.
15. Refill cooling system.

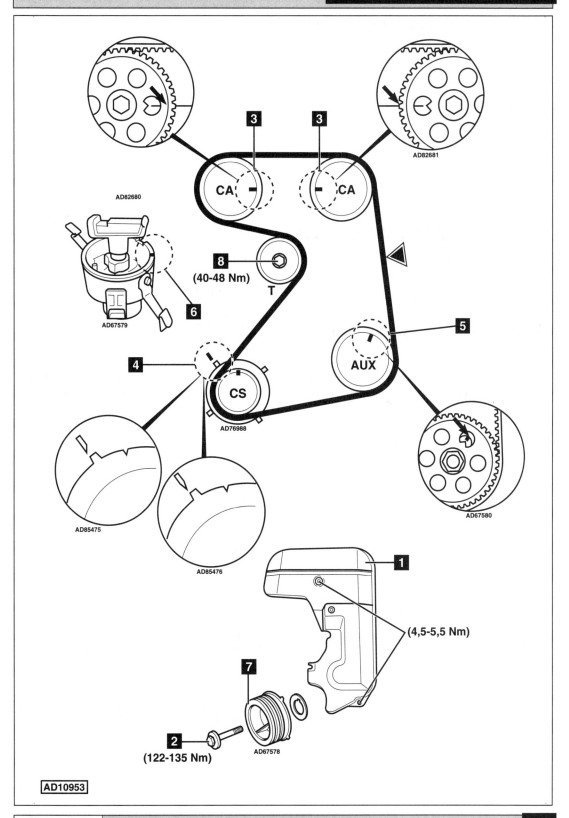

AD82680

AD82681

AD67579

3

3

CA

CA

8
(40-48 Nm)
T

6

4

CS

AD76988

5

AUX

AD85475

AD85476

AD67580

1

(4,5-5,5 Nm)

7

2
(122-135 Nm)

AD67578

AD10953

Model:	**Mondeo 1,6/1,8/2,0 16V**
Year:	**1993-04/98**
Engine Code:	**L1F, L1J, L1L, NGA, RKA, RKB, RKJ**

Replacement Interval Guide

Ford recommend replacement every 80,000 miles or 5 years whichever occurs first (tensioner and guide pulleys must also be replaced).
The previous use and service history of the vehicle must always be taken into account.
Refer to Timing Belt Replacement Intervals at the front of this manual.

Check For Engine Damage

CAUTION: This engine has been identified as an INTERFERENCE engine in which the possibility of valve-to-piston damage in the event of a timing belt failure is MOST LIKELY to occur.
A compression check of all cylinders should be performed before removing the cylinder head.

Repair Times – hrs

Remove & install	2,20
AT & AC	1,90
4x4	2,80

Special Tools

- Camshaft setting bar – Ford No.21-162.
- Sprocket holding tool – Ford No.15-030-A.

Special Precautions

- Disconnect battery earth lead.
- DO NOT turn crankshaft or camshaft when timing belt removed.
- Remove spark plugs to ease turning engine.
- Turn engine in normal direction of rotation (unless otherwise stated).
- DO NOT turn engine via camshaft or other sprockets.
- Observe all tightening torques.

Removal

1. Raise and support front of vehicle.
2. Support engine.
3. Remove:
   - ❏ Air intake pipe.
   - ❏ RH front wheel.
   - ❏ Crankshaft damper cover.
   - ❏ Auxiliary drive belt.
   - ❏ Crankshaft damper **1**.
   - ❏ Water pump pulley.
   - ❏ RH engine mounting.
   - ❏ Cylinder head cover.
   - ❏ Auxiliary drive belt guide pulley.
   - ❏ Timing belt upper cover **2**.
   - ❏ Timing belt centre cover **3**.
   - ❏ Timing belt lower cover **4**.

4. Temporarily fit crankshaft damper **1**.
5. Turn crankshaft to TDC on No.1 cylinder. Ensure crankshaft timing marks aligned **5**.
6. Fit setting bar No.21-162 to rear of camshafts **6**.
   NOTE: If necessary: Turn crankshaft one turn clockwise.
7. Slacken tensioner bolt **7**. Turn tensioner clockwise. Use Allen key **8**. Lightly tighten bolt.
8. Remove:
   - ❏ Crankshaft damper.
   - ❏ Timing belt.

Installation

NOTE: If not already fitted, a tensioner spring and retaining pin should be fitted – refer to Ford dealer.

1. Hold camshaft sprockets **9**.
   Use tool No.15-030-A. Slacken bolts **10**.
2. Tap each camshaft sprocket gently to loosen it from taper. Ensure each sprocket turns freely on camshaft.
3. Check condition and security of tensioner spring **11**.
4. Ensure crankshaft timing marks aligned **5**.
5. Fit timing belt in anti-clockwise direction, starting at crankshaft sprocket. Ensure belt is taut between sprockets.
6. Slacken tensioner bolt **7**. Allow tensioner to operate.
7. Hold camshaft sprockets **9**.
   Use tool No.15-030-A. Tighten bolts **10**.
   Tightening torque: 67-72 Nm.
8. Remove setting bar **6**.
9. Temporarily fit crankshaft damper **1**.
10. Turn crankshaft slowly two turns clockwise. Ensure crankshaft timing marks aligned **5**.
11. Tighten tensioner bolt **7**.
    Tightening torque: 38 Nm.
12. Ensure crankshaft timing marks aligned **5**.
13. Ensure setting bar can be fitted **6**.
14. If not: Repeat tensioning procedure.
15. Install components in reverse order of removal.
16. Tighten crankshaft damper bolt **12**.
    Tightening torque: 115 Nm.

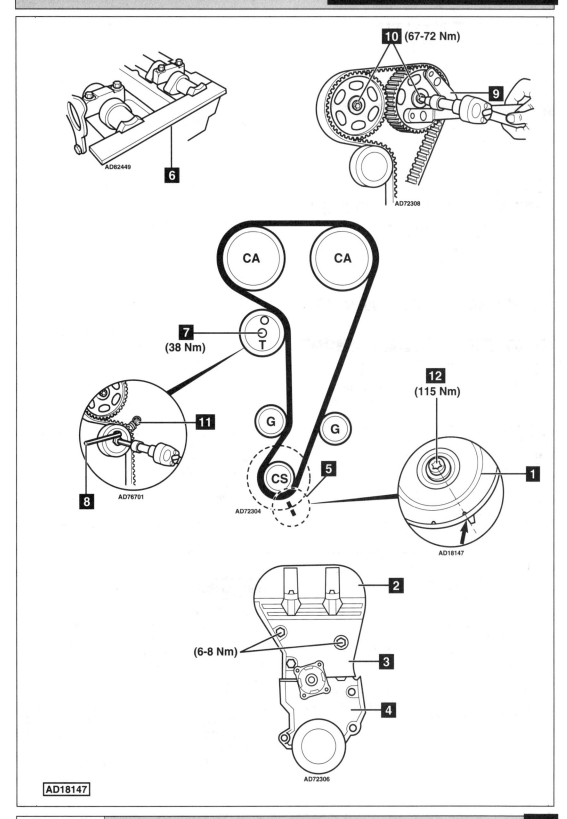

10 (67-72 Nm)

9

AD72308

6

AD82449

CA CA

7
(38 Nm)

T

11

8

AD76701

G G

12
(115 Nm)

CS **5**

AD72304

1

AD18147

2

(6-8 Nm)

3

4

AD72306

AD18147

FORD

Model:	**Puma 1,6/1,7**
Year:	**1997-02**
Engine Code:	**L1W, MHA**

Replacement Interval Guide

Ford recommend replacement every
100,000 miles or 10 years.
*The previous use and service history of the vehicle
must always be taken into account.*
*Refer to Timing Belt Replacement Intervals at the front
of this manual.*

Check For Engine Damage

*CAUTION: This engine has been identified as an
INTERFERENCE engine in which the possibility of
valve-to-piston damage in the event of a timing belt
failure is MOST LIKELY to occur.*
*A compression check of all cylinders should be
performed before removing the cylinder head(s).*

Repair Times – hrs

Remove & install	2,60

Special Tools

- Camshaft setting bar – Ford No.21-162B.
- Crankshaft timing pin – Ford No.21-210.
- Crankshaft pulley puller – Ford No.21-215.
- Crankshaft pulley installer – Ford No.21-214.

Special Precautions

- Disconnect battery earth lead.
- DO NOT turn crankshaft or camshaft when timing belt removed.
- Remove spark plugs to ease turning engine.
- Turn engine in normal direction of rotation (unless otherwise stated).
- DO NOT turn engine via camshaft or other sprockets.
- Observe all tightening torques.

Removal

1. Raise and support front of vehicle.
2. Support engine.
3. Remove:
   - ❏ RH front wheel.
   - ❏ Cylinder head cover.
   - ❏ PAS pump bracket upper bolt.
4. Disconnect:
   - ❏ Breather hose.
   - ❏ Camshaft position (CMP) sensor multi-plug.
   - ❏ Camshaft position (CMP) actuator multi-plug.
   - ❏ Injector multi-plugs.

5. Remove:
   - ❏ Auxiliary drive belt cover.
   - ❏ Auxiliary drive belt.
   - ❏ Auxiliary drive belt guide pulley **1**.
   - ❏ Water pump pulley **4**.
   - ❏ PAS pump bracket.
   - ❏ Timing belt upper cover **5**.
   - ❏ Camshaft position (CMP) sensor.
   - ❏ Ignition coil and HT leads.
   - ❏ Engine lifting eye.
6. Remove blanking plug from cylinder block **10**.
7. Insert crankshaft timing pin. Tool No.21-210.
8. Turn crankshaft slowly clockwise until it stops against timing pin. Ensure crankshaft at TDC on No.1 cylinder.
9. Fit setting bar No.21-162B to rear of camshafts **11**.
   NOTE: If setting bar cannot be fitted: Remove timing pin. Turn crankshaft one turn clockwise. Insert timing pin. Ensure crankshaft web against timing pin.
10. Remove:
    - ❏ Crankshaft pulley bolt **2**.
    - ❏ Crankshaft pulley **3**. Use tool No.21-215.
    NOTE: Crankshaft sprocket NOT keyed to crankshaft. Ensure crankshaft does not turn.
    - ❏ Timing belt lower cover **6**.
    - ❏ RH engine mounting **9**.
    - ❏ Thrust washer **7**.
    - ❏ Timing belt retainer **8**.
11. Turn tensioner until 5 mm drill bit can be inserted in the stop **14**.
    NOTE: Protect flutes of drill bit with tape.
12. Remove tensioner bolt **15**.
13. Remove timing belt.

Installation

1. Ensure setting bar fitted correctly **11**.
2. Ensure timing pin located correctly **10**.
3. Ensure crankshaft at TDC on No.1 cylinder. Ensure crankshaft web against timing pin.
4. Remove camshaft setting bar **11**.
5. Remove blanking plug from inlet camshaft sprocket **16**.
6. Hold camshafts with spanner on hexagon **17**. Slacken bolt of each camshaft sprocket **12** & **13**.
   NOTE: DO NOT use setting bar to hold camshafts when slackening bolts.
7. Tap each camshaft sprocket gently to loosen it from taper.

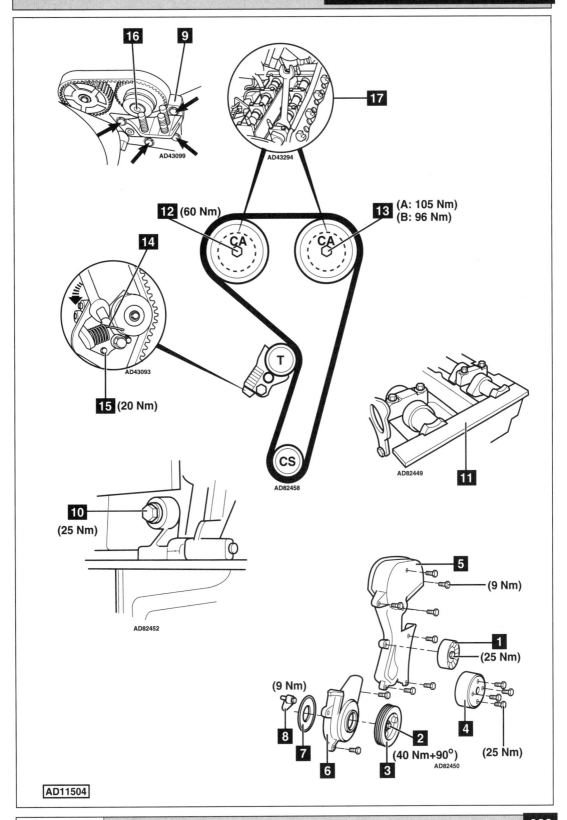

16 **9**

AD43099

17

AD43294

12 (60 Nm)

13 (A: 105 Nm)
(B: 96 Nm)

CA

CA

14

15 (20 Nm)

AD43093

T

CS

AD82458

11

AD82449

10
(25 Nm)

AD82452

5

(9 Nm)

1
(25 Nm)

(9 Nm)

4

8

7

6

2

3
(40 Nm+90°)

(25 Nm)

AD82450

AD11504

←

8. Ensure camshaft sprockets can turn on taper but not tilt.

9. Refit camshaft setting bar .

10. Fit timing belt in anti-clockwise direction, starting at crankshaft sprocket.
    NOTE: Ensure directional arrows point in direction of rotation.

11. Ensure belt is taut between sprockets.

12. Fit timing belt retainer .
    Tightening torque: 9 Nm.

13. Withdraw 5 mm drill bit 2-3 mm from tensioner to tension belt .

14. Tighten tensioner bolt to 20 Nm .

15. Remove 5 mm drill bit completely .

16. Install:
    - ❏ Thrust washer .
    - ❏ Timing belt cover .
    - ❏ Crankshaft pulley . Use tool No.21-214.

17. Tighten crankshaft pulley bolt .
    Tightening torque: 40 Nm + 90°.

18. Ensure crankshaft web against timing pin.

19. Hold camshafts with spanner on hexagon .

20. Tighten bolt of each camshaft sprocket:
    - ❏ Inlet: 1,6 Ⓐ 105 Nm .
    - ❏ Inlet: 1,7 Ⓑ 96 Nm .
    - ❏ Exhaust: 60 Nm .

21. Remove:
    - ❏ Crankshaft timing pin .
    - ❏ Camshaft setting bar .

22. Turn crankshaft slowly two turns clockwise.

23. Insert crankshaft timing pin .

24. Turn crankshaft until web against timing pin.

25. Ensure camshaft setting bar can be fitted .

26. Remove:
    - ❏ Crankshaft timing pin .
    - ❏ Camshaft setting bar .

27. Fit blanking plug and tighten to 25 Nm .

28. Install components in reverse order of removal.

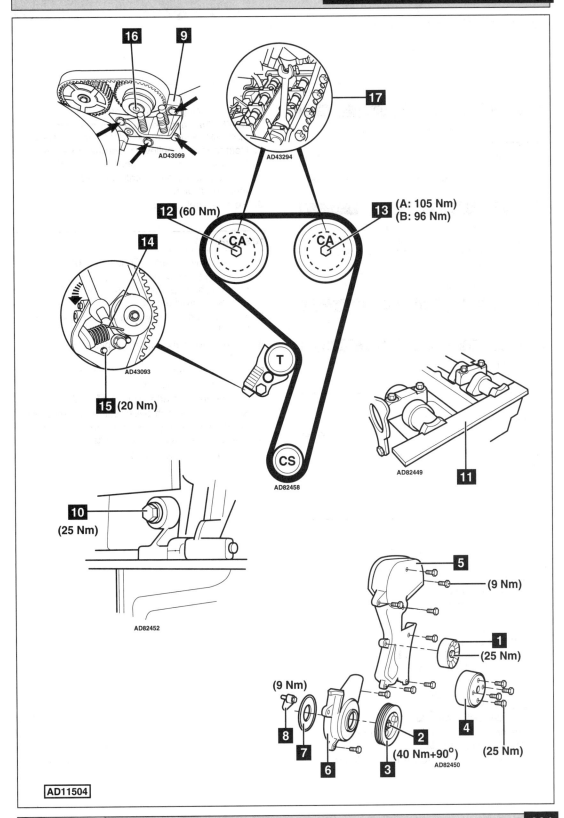

16 **9**

AD43099

17

AD43294

12 (60 Nm)

13 (A: 105 Nm)
(B: 96 Nm)

CA

CA

14

AD43093

15 (20 Nm)

T

CS

AD82458

AD82449

11

10
(25 Nm)

AD82452

5

(9 Nm)

1
(25 Nm)

(9 Nm)

8

7

6

4

2

3

(40 Nm+90°)

(25 Nm)

AD82450

AD11504

FORD

Model:	**Probe 2,0 16V**
Year:	**1994-98**
Engine Code:	**FS**

Replacement Interval Guide

Ford recommend replacement every
60,000 miles or 5 years.
*The previous use and service history of the vehicle
must always be taken into account.
Refer to Timing Belt Replacement Intervals at the front
of this manual.*

Check For Engine Damage

*CAUTION: This engine has been identified as an
INTERFERENCE engine in which the possibility of
valve-to-piston damage in the event of a timing belt
failure is MOST LIKELY to occur.
A compression check of all cylinders should be
performed before removing the cylinder head.*

Repair Times – hrs

Remove & install	2,30

Special Tools

■ None required.

Special Precautions

■ Disconnect battery earth lead.
■ DO NOT turn crankshaft or camshaft when timing belt
   removed.
■ Remove spark plugs to ease turning engine.
■ Turn engine in normal direction of rotation (unless
   otherwise stated).
■ DO NOT turn engine via camshaft or other sprockets.
■ Observe all tightening torques.

Removal

1. Raise and support front of vehicle.
2. Support engine.
3. Remove:
   ❏ RH front wheel.
   ❏ RH inner wing panel.
   ❏ RH engine mounting.
   ❏ Auxiliary drive belts.
   ❏ Water pump pulley.
   ❏ Crankshaft pulley bolt **1**.
   ❏ Crankshaft pulley **2**.
   ❏ HT leads.
   ❏ Cylinder head cover.
   ❏ Dipstick and tube **3**.
   ❏ Timing belt covers **4**.
4. Temporarily fit bolt and washer to retain
   crankshaft sprocket **1**.
5. Turn crankshaft to TDC on No.1 cylinder.

6. Ensure timing marks aligned **5** & **6**.
   ***NOTE: Timing marks on rear of camshaft
   sprockets align with cylinder head upper face.***
7. Hold tensioner pulley **7**. Use Allen key **8**.
   Disconnect tensioner spring from pin **9**.
8. Turn tensioner pulley anti-clockwise to release
   tension on belt.
9. Remove timing belt.

Installation

1. Ensure timing marks aligned **5** & **6**.
2. Fit timing belt in anti-clockwise direction, starting
   at crankshaft sprocket. Ensure belt is taut
   between sprockets.
3. Turn tensioner pulley clockwise **7**. Use Allen
   key **8**.
4. Connect tensioner spring to pin **9**. Ensure belt
   is tensioned.
5. Turn crankshaft slowly two turns clockwise to
   TDC on No.1 cylinder.
6. Ensure timing marks aligned **5** & **6**.
7. Remove crankshaft pulley bolt and washer **1**.
8. Install components in reverse order of removal.
9. Tighten crankshaft pulley bolt **1**.
   Tightening torque: 157-166 Nm.

Autodata

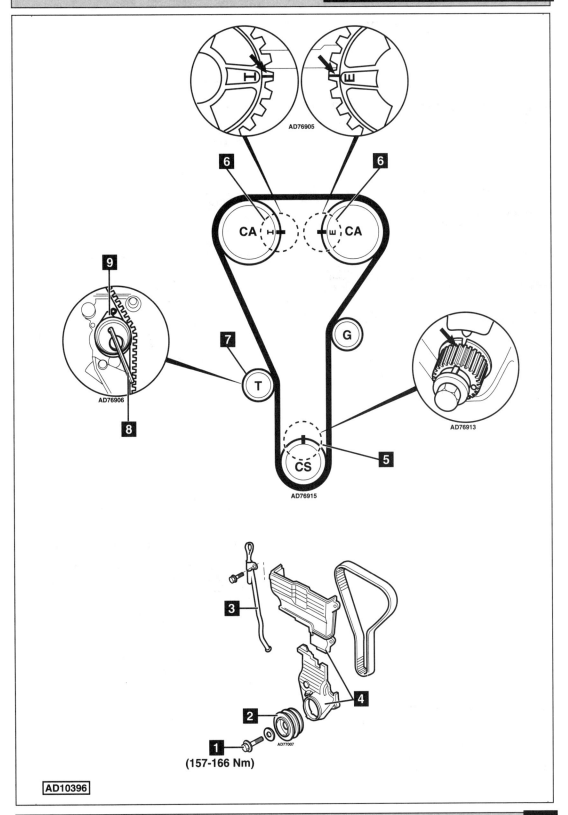

AD76905

AD76906

AD76913

AD76915

AD77007

1 (157-166 Nm)

AD10396

FORD

Model:	**Probe 2,5 24V**
Year:	**1994-98**
Engine Code:	**KL**

Replacement Interval Guide

Ford recommend replacement every 60,000 miles or 5 years.
The previous use and service history of the vehicle must always be taken into account.
Refer to Timing Belt Replacement Intervals at the front of this manual.

Check For Engine Damage

CAUTION: This engine has been identified as an INTERFERENCE engine in which the possibility of valve-to-piston damage in the event of a timing belt failure is MOST LIKELY to occur.
A compression check of all cylinders should be performed before removing the cylinder head.

Repair Times – hrs

Remove & install	3,60

Special Tools

- None required.

Special Precautions

- Disconnect battery earth lead.
- DO NOT turn crankshaft or camshaft when timing belt removed.
- Remove spark plugs to ease turning engine.
- Turn engine in normal direction of rotation (unless otherwise stated).
- DO NOT turn engine via camshaft or other sprockets.
- Observe all tightening torques.

Removal

1. Drain coolant.
2. Raise and support front of vehicle.
3. Support engine.
4. Remove:
   - ❏ RH front wheel.
   - ❏ RH inner wing panel.
   - ❏ Engine undershield.
   - ❏ Radiator hoses.
   - ❏ RH engine mounting.
   - ❏ Auxiliary drive belts.
   - ❏ Water pump pulley **1**.
   - ❏ Auxiliary drive belt tensioner and bracket **2**.
   - ❏ PAS pump and bracket **3**. DO NOT disconnect hoses.
   - ❏ Crankshaft pulley **4**.
   - ❏ Dipstick and tube.
   - ❏ Crankshaft position (CKP) sensor bracket.
   - ❏ Engine wiring harness bracket **5**.
   - ❏ Timing belt LH cover **7**.
   - ❏ Timing belt RH cover **6**.
5. Temporarily fit crankshaft pulley bolt **17**.
6. Turn crankshaft to TDC on No.1 cylinder. Ensure timing marks aligned **8** & **9**.
7. Slacken automatic tensioner unit bolts **10** & **11**.
8. Remove bolt **10**. Allow automatic tensioner unit to swivel inwards towards belt.

9. Remove bolt **11** together with automatic tensioner unit **12**.
   NOTE: Hold automatic tensioner unit firmly to prevent damage to bolt threads.
10. Remove:
    - ❏ Guide pulley (G1).
    - ❏ Timing belt.

Installation

1. Ensure timing marks on camshaft sprockets aligned **9**.
2. Turn crankshaft anti-clockwise until crankshaft sprocket timing mark is one tooth before timing mark on casing.
3. Check tensioner body for leakage or damage **12**.
4. Slowly compress pushrod **13** into tensioner body **12** until holes aligned. Use suitable press.
5. Retain pushrod with 1,6 mm diameter pin through lower hole in tensioner body **14**.
   *NOTE: Place flat washer under tensioner body to avoid damage to body end plug. DO NOT exceed 1000 kg force **15**.*
6. Fit timing belt in following order:
   - ❏ Crankshaft sprocket.
   - ❏ Guide pulley (G2).
   - ❏ Camshaft sprocket (CA1).
   - ❏ Tensioner pulley **16**.
   - ❏ Camshaft sprocket (CA2).
7. Apply firm thumb pressure to belt. Fit guide pulley (G1). Tighten bolt to 38-51 Nm.
8. Turn crankshaft slowly clockwise one tooth. Ensure timing mark aligned **8**.
9. Install automatic tensioner unit to cylinder block. Tighten bolts to 19-25 Nm **10** & **11**.
10. Remove pin from lower hole **14** in tensioner body to release pushrod.
11. Turn crankshaft slowly two turns clockwise to TDC on No.1 cylinder. Ensure timing marks aligned **8** & **9**.
    *NOTE: Crankshaft sprocket timing mark should now be aligned with timing mark on casing **8**. If not: Repeat operations 2-11.*
12. Apply a load of 10 kg to belt at ▽. Belt should deflect 6-8 mm. If not: Replace automatic tensioner unit. Repeat operations 4-10.
13. Install components in reverse order of removal.
14. Tighten crankshaft pulley bolt **17**.
    Tightening torque: 157-166 Nm.
15. Refill cooling system.

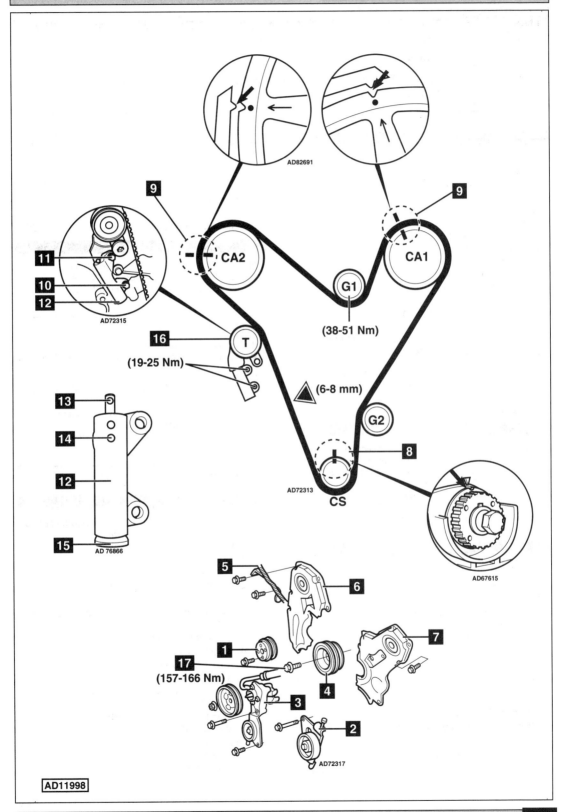

AD82691

9

9

CA2

CA1

G1

(38-51 Nm)

11

10

12

AD72315

16   T

(19-25 Nm)

△ (6-8 mm)

G2

8

13

14

12

15   AD 76866

CS

AD72313

AD67615

5

6

7

1

17
(157-166 Nm)

4

3

2

AD72317

AD11998

FORD

Model:	Fiesta/Van 1,8 D (→1996) • Escort/Van 1,8 D/TD (→1996) • Orion 1,8 D/TD Mondeo 1,8 TD (→1996) • Courier/Kombi 1,8 D
Year:	1988-99
Engine Code:	J4C, RFD, RFK, RFM, RFN, RTA, RTB, RTC, RTD, RTE, RTF, RTG, RTH, RTJ, RTK, RVA

Replacement Interval Guide

Ford recommend replacement as follows:

Timing belt

→8/89: Replacement every 36,000 miles or 3 years.

9/89-8/97: Replacement every 40,000 miles or 5 years.

9/97→: Replacement every 70,000 miles or 5 years (40,000 miles under adverse conditions).

Injection pump belt

→8/89: Replacement every 36,000 miles or 3 years.

9/89-8/97: Replacement every 40,000 miles or 5 years.

9/97→: Replacement every 70,000 miles or 5 years (40,000 miles under adverse conditions).

The previous use and service history of the vehicle must always be taken into account.
Refer to Timing Belt Replacement Intervals at the front of this manual.

Check For Engine Damage

CAUTION: This engine has been identified as an INTERFERENCE engine in which the possibility of valve-to-piston damage in the event of a timing belt failure is MOST LIKELY to occur.
A compression check of all cylinders should be performed before removing the cylinder head.
Refer to Timing Belt Replacement Intervals at the front of this manual.

Repair Times – hrs

Check & adjust:

Fiesta/Courier/Kombi →1995	0,80
PAS	+0,30
Fiesta/Courier/Kombi 1995→	1,10
Escort/Orion →1990	0,80
Escort/Orion 1990→	1,20
Escort TD	1,50
PAS	+0,30
Mondeo	2,90
AC	+0,20

Remove & install:

Fiesta/Courier/Kombi →1995	2,10
PAS	+0,30
Fiesta/Courier/Kombi 1996→	1,90
PAS	+0,30
AC	+0,30

Escort/Orion →1990	1,90
Escort/Orion 1990→	1,70
PAS	+0,30
AC	+0,30
Escort TD	1,90
Mondeo	3,40
Injection pump belt	+0,20

Special Tools

- ■ Crankshaft timing pin – Ford No.21-104.
- ■ Camshaft timing pin – Ford No.23-019.
- ■ Injection pump timing pin – Ford No.23-019.

Special Precautions

- ■ Disconnect battery earth lead.
- ■ DO NOT turn crankshaft or camshaft when timing belt removed.
- ■ Remove glow plugs to ease turning engine.
- ■ Turn engine in normal direction of rotation (unless otherwise stated).
- ■ DO NOT turn engine via camshaft or other sprockets.
- ■ Observe all tightening torques.
- ■ Check diesel injection pump timing after belt replacement.

Removal

Timing Belt

NOTE: Timing belt and injection pump belt should not be removed/installed until engine is cold (stopped for at least 4 hours).

1. Mondeo – Remove:
   - ❑ Engine undershield.
   - ❑ RH lower inner wing panel.
2. Escort – Remove:
   - ❑ PAS reservoir. Disconnect return hose only (if fitted).
   - ❑ Coolant expansion tank.
3. All models – Remove:
   - ❑ Auxiliary drive belt(s).
   - ❑ Alternator cover (if fitted).
   - ❑ Alternator/bracket (if required).
   - ❑ PAS pump pulley (if fitted to injection pump).
   - ❑ PAS pump – leave hoses connected (if required).
4. Fiesta 1996→/Mondeo:
   - ❑ Support front of engine.
   - ❑ Remove RH engine mounting.

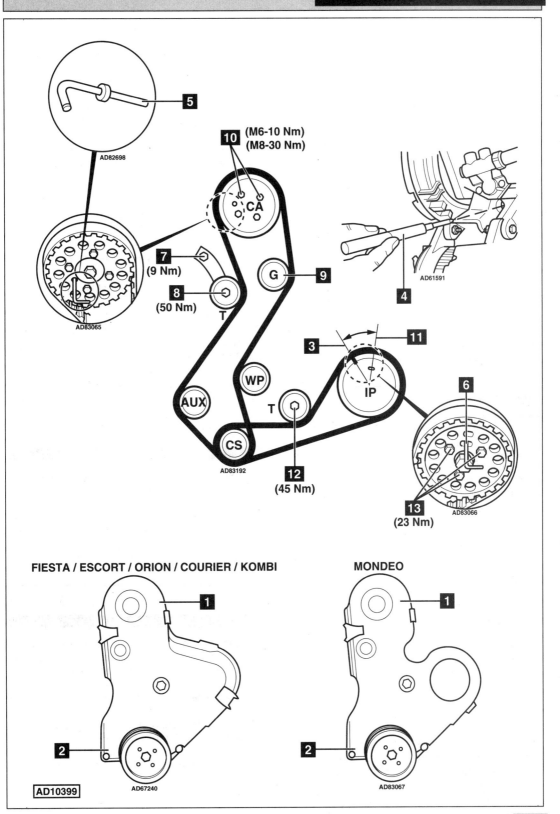

5 AD82698

10 (M6-10 Nm) (M8-30 Nm)

CA

AD83065

7 (9 Nm)

8 (50 Nm) T

G **9**

4 AD61591

3 **11**

WP

AUX **T** **IP**

CS

AD83192

12 (45 Nm)

6

13 (23 Nm) AD83066

FIESTA / ESCORT / ORION / COURIER / KOMBI

1

2

AD10399 AD67240

MONDEO

1

2

AD83067

5. Fiesta 1996→ – Remove:
   - ❏ Crankshaft pulley bolts (4 bolts).
   - ❏ Crankshaft pulley.
6. All models – Remove:
   - ❏ Timing belt covers **1** & **2**.
   - ❏ Injection pump belt cover (Fiesta 1996→).
7. Turn crankshaft clockwise until slot in injection pump sprocket at 11 o'clock **3**.
8. Remove blanking plug from cylinder block. Insert timing pin in cylinder block **4**. Tool No.21-104.
9. Turn crankshaft clockwise until it stops against timing pin.
10. Insert timing pin in camshaft sprocket **5**. Tool No.23-019.
11. Insert timing pin in injection pump sprocket **6**. Tool No.23-019.

   *NOTE: Engines fitted with CAV injection pump: Insert 6 mm drill bit in injection pump sprocket **6**.*
12. Slacken tensioner hexagon bolt and Torx bolt **7** & **8**.
13. Move tensioner away from belt. Lightly tighten bolt **8**.
14. Remove timing belt.

   NOTE: DO NOT refit used belt.

Installation

Timing Belt

NOTE: Refer to Ford dealer for revised timing belt lower cover, to prevent debris entering timing belt drive area.

1. Check tensioner pulley and guide sprocket for cracks, chips or other damage. Replace if necessary.

   *NOTE: If guide sprocket **9** is plastic, it should be replaced with a steel sprocket No.7053802 and bolt No.6701537.*
2. Spin tensioner pulley and guide sprocket. Check for bearing roughness and noise. Replace if necessary.

   *NOTE: If tensioner bolt **8** has 'mushroom' head, check for evidence of removal with a chisel or self-grip wrench. If so, fit a new bolt and tensioner.*
3. Ensure timing pin located correctly in camshaft sprocket **5**.
4. Ensure timing pin or 6 mm drill bit (CAV) located correctly in injection pump sprocket **6**.
5. Ensure crankshaft web against timing pin **4**.
6. Fit timing belt. Slack side of belt must be towards tensioner. Observe direction of rotation marks on belt.
7. Slacken camshaft sprocket bolts **10**.
8. Slacken injection pump sprocket bolts **13**.
9. Slacken tensioner bolts 1/2 turn **7** & **8**. Allow tensioner to operate.
10. Tighten camshaft sprocket bolts **10**.

11. Tighten injection pump sprocket bolts **13**.

   NOTE: Ensure bolts not at end of slotted holes in sprockets.
12. Tighten tensioner bolts **7** & **8**.
13. Remove timing pins **4**, **5** & **6**.
14. Turn crankshaft two turns clockwise. Ensure injection pump sprocket slot at 12 o'clock **11**.
15. Turn crankshaft anti-clockwise until injection pump sprocket slot at 11 o'clock **3**.
16. Insert timing pin in cylinder block **4**. Turn crankshaft clockwise until it stops against timing pin.
17. Insert timing pin in camshaft sprocket **5**.
18. Insert timing pin or 6 mm drill bit (CAV) in injection pump sprocket **6**.
19. Slacken camshaft sprocket bolts **10**.
20. Slacken injection pump sprocket bolts **13**.
21. Slacken tensioner bolts **7** & **8**.
22. Push tensioner firmly against belt and release.
23. Tighten tensioner bolts to specified torque **7** & **8**.
24. Tighten camshaft sprocket bolts to specified torque **10**.
25. Tighten injection pump sprocket bolts to specified torque **13**.
26. Remove timing pins **4**, **5** & **6**.
27. Install components in reverse order of removal.

Removal

Injection Pump Belt

1. Remove timing belt as described previously.
2. Ensure timing pin or 6 mm drill bit (CAV) located correctly in injection pump sprocket **6**.
3. Slacken tensioner bolt **12**. Lift tensioner against spring pressure. Tighten bolt.
4. Remove injection pump belt.
5. Slacken injection pump sprocket bolts **13**.

Installation

Injection Pump Belt

1. Ensure crankshaft web against timing pin **4**.
2. Ensure timing pin or 6 mm drill bit (CAV) located correctly in injection pump sprocket **6**.
3. Fit injection pump belt. Slack side of belt must be towards tensioner. Observe direction of rotation marks.
4. Slacken tensioner bolt **12**. Allow tensioner to operate. Tighten bolt to 45 Nm.
5. Tighten injection pump sprocket bolts **13**. Ensure bolts not at end of slotted holes in sprocket.
6. Fit timing belt as described previously.

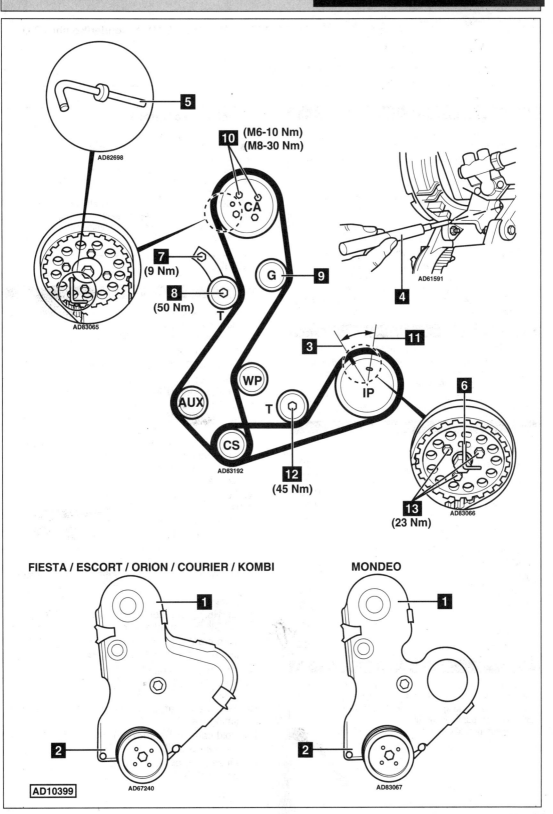

5 AD82698

10 (M6-10 Nm)
(M8-30 Nm)

CA

AD83065

7 (9 Nm)

8 (50 Nm)
T

G **9**

AD61591

4

WP

AUX

T

CS

AD83192

3 **11**

IP

6

12
(45 Nm)

13
(23 Nm)

AD83066

FIESTA / ESCORT / ORION / COURIER / KOMBI

1

MONDEO

1

2

AD10399

AD67240

2

AD83067

FORD

Model:	Fiesta/Van 1,8 D • Escort 1,8 D/TD • Mondeo 1,8 TD • Courier/Kombi 1,8 D
Year:	1996-02
Engine Code:	RFN, RTJ, RTK, RVA, RFK, RKD

Replacement Interval Guide

Ford recommend replacement as follows:

Timing belt
→97 MY: Replacement every 40,000 miles or 5 years.
98 MY→: Replacement every 70,000 miles or 5 years (40,000 miles under adverse conditions).

Injection pump belt
→97 MY: Replacement every 40,000 miles.
98 MY→: Replacement every 70,000 miles (40,000 miles under adverse conditions).
The previous use and service history of the vehicle must always be taken into account.
Refer to Timing Belt Replacement Intervals at the front of this manual.

Check For Engine Damage

CAUTION: This engine has been identified as an INTERFERENCE engine in which the possibility of valve-to-piston damage in the event of a timing belt failure is MOST LIKELY to occur.
A compression check of all cylinders should be performed before removing the cylinder head.

Repair Times – hrs

Check & adjust:	
Fiesta/Courier/Kombi	1,10
Escort	1,50
AC	+ 0,30
Mondeo	2,90
AC	+ 0,20
Remove & install:	
Fiesta/Courier/Kombi	1,70
PAS	+ 0,30
AC	+ 0,30
Escort	1,90
PAS	+ 0,30
AC	+ 0,30
Mondeo	3,40
Injection pump belt	+0,20

Special Tools

- Sprocket holding tool – Ford No.15-030-A.
- Crankshaft timing pin – Ford No.21-104.
- Camshaft setting bar – Ford No.21-162B.
- Injection pump timing pin
  (Bosch pump/Non-turbo) – Ford No.23-029.

Special Precautions

- Disconnect battery earth lead.
- DO NOT turn crankshaft or camshaft when timing belt removed.
- Remove glow plugs to ease turning engine.
- Turn engine in normal direction of rotation (unless otherwise stated).
- DO NOT turn engine via camshaft or other sprockets.
- Observe all tightening torques.
- Check diesel injection pump timing after belt replacement.

Removal

Timing Belt

NOTE: Timing belt and injection pump belt should not be removed/installed until engine is cold (stopped for at least 4 hours).

1. Raise and support front of vehicle.
2. Mondeo: Drain cooling system.
3. Remove:
   - RH front wheel.
   - RH lower inner wing panel.
   - Engine undershield.
   - Auxiliary drive belt(s).
   - Auxiliary drive belt tensioner (if necessary).
   - Mondeo: Alternator cover.
   - Fiesta: Alternator.
   - PAS pump pulley.
4. Support engine.
5. Remove:
   - RH engine mounting.
   - RH engine mounting bracket.
   - Engine steady bar.
   - Mondeo: Coolant expansion tank.
6. Mondeo: Disconnect coolant hose from water pump.
7. Remove:
   - Timing belt covers **1** & **2**.
   - Cylinder head cover.
8. Turn crankshaft clockwise until slot in injection pump sprocket at 11 o'clock **3**.
9. Remove blanking plug from cylinder block. Insert timing pin in cylinder block **4**. Tool No.21-104.
10. Turn crankshaft clockwise until it stops against timing pin.
11. Hold camshaft sprocket. Use tool No.15-030-A. Slacken bolt **10**.
12. Fit setting bar No.21-162B to rear of camshaft **5**.

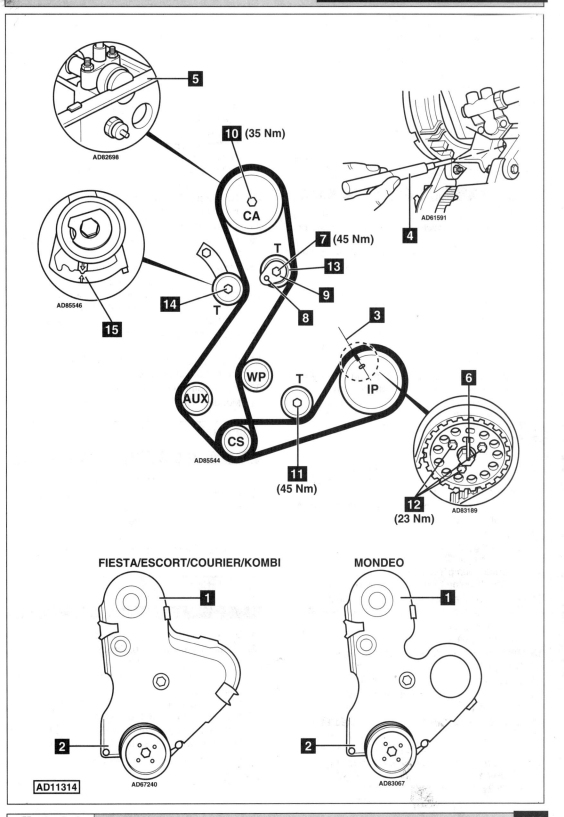

10 (35 Nm)

CA

AD82698

AD85546

15

14 T

7 (45 Nm)

T

13

9

8

AD61591

4

3

IP

WP

T

AUX

CS

AD85544

11
(45 Nm)

6

12
(23 Nm)

AD83189

FIESTA/ESCORT/COURIER/KOMBI

1

2

AD67240

MONDEO

1

2

AD83067

AD11314

←

13. Lock injection pump. Lucas pump/Turbo: Insert 6 mm drill bit in injection pump sprocket **6**. Bosch pump/Non-turbo: Insert timing pin in injection pump sprocket **6**. Tool No.23-029.
14. Slacken tensioner bolt **7**.
15. Turn tensioner anti-clockwise.
16. Remove timing belt.
    NOTE: DO NOT refit used belt.

Installation

Timing Belt

NOTE: Refer to Ford dealer for revised timing belt lower cover, to prevent debris entering timing belt drive area.

1. Check tensioner sprocket and tensioner pulley for cracks, chips or other damage **13** & **14**. Replace if necessary.
2. Spin tensioner sprocket **13** and tensioner pulley **14**. Check for bearing roughness and noise. Replace if necessary.
3. Ensure setting bar fitted correctly **5**.
4. Ensure injection pump sprocket locked **6**.
5. Ensure crankshaft web against timing pin **4**.
6. Loosen camshaft sprocket from taper using a drift through hole in timing belt rear cover.
    NOTE: If no hole in rear cover: Insert strong screwdriver between rear cover and sprocket and tap rear cover gently with hammer.
7. Fit timing belt. Observe direction of rotation marks on belt.
8. Turn tensioner adjusting cam **9** clockwise to 9 o'clock position. Use Allen key at point **8**. Tighten tensioner bolt.
9. Hold camshaft sprocket. Use tool No.15-030-A. Tighten bolt **10**.
10. Remove:
    ❑ Setting bar **5**.
    ❑ Timing pin **4**.
    ❑ Lucas pump/Turbo: 6 mm drill bit **6**.
    ❑ Bosch pump/Non-turbo: Timing pin **6**.
11. Turn crankshaft six turns clockwise until injection pump sprocket slot at 11 o'clock **3**.
12. Insert timing pin in cylinder block **4**. Tool No.21-104.
13. Turn crankshaft clockwise until it stops against timing pin.
14. Slacken camshaft sprocket bolt **10**.
15. Fit setting bar No.21-162B to rear of camshaft **5**.
16. Ensure injection pump sprocket locked **6**.
17. Loosen camshaft sprocket from taper using a drift through hole in timing belt rear cover.
    NOTE: If no hole in rear cover: Insert strong screwdriver between rear cover and sprocket and tap rear cover gently with hammer.

18. Slacken tensioner bolt **7**.
19. Turn tensioner adjusting cam **9** clockwise, using Allen key at point **8**, until arrows **15** on automatic tensioner pulley **14** align.
20. Tighten tensioner bolt to 45 Nm **7**.
21. Hold camshaft sprocket. Use tool No.15-030-A. Tighten bolt to 35 Nm **10**.
22. Remove:
    ❑ Timing pin **4**.
    ❑ Setting bar **5**.
    ❑ Lucas pump/Turbo: 6 mm drill bit **6**.
    ❑ Bosch pump/Non-turbo: Timing pin **6**.
23. Turn crankshaft four turns clockwise until injection pump sprocket slot at 11 o'clock **3**.
24. Insert timing pin in cylinder block **4**.
25. Turn crankshaft clockwise until it stops against timing pin.
26. Ensure setting bar can be fitted **5**.
27. Lucas pump/Turbo: Ensure 6 mm drill bit can be inserted in injection pump sprocket **6**. Bosch pump/Non-turbo: Ensure timing pin can be inserted in injection pump sprocket **6**.
28. If tools cannot be inserted: Repeat operations 10-27.
29. Check arrows on automatic tensioner pulley **15** are not offset by more than 3 mm.
30. Install components in reverse order of removal.
31. Mondeo: Refill cooling system.

Removal

Injection Pump Belt

1. Remove timing belt as described previously.
2. Ensure injection pump sprocket locked **6**.
3. Slacken tensioner bolt **11**. Lift tensioner against spring pressure. Tighten bolt.
4. Remove injection pump belt.
5. Slacken injection pump sprocket bolts **12**.

Installation

Injection Pump Belt

1. Ensure crankshaft web against timing pin **4**.
2. Ensure injection pump sprocket locked **6**.
3. Fit injection pump belt. Slack side of belt must be towards tensioner. Observe direction of rotation marks.
4. Slacken tensioner bolt **11**. Allow tensioner to operate. Tighten bolt to 45 Nm.
5. Tighten injection pump sprocket bolts **12**. Ensure bolts not at end of slotted holes in sprocket.
6. Fit timing belt as described previously.

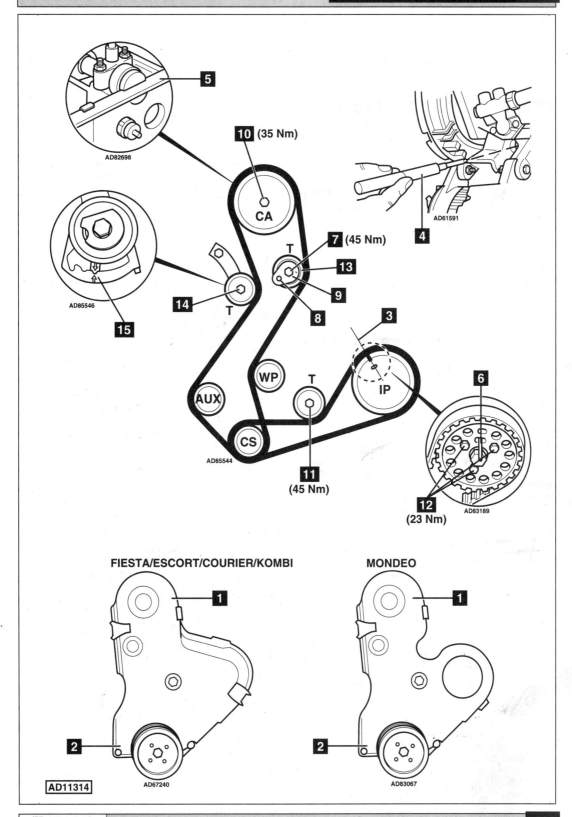

5

10 (35 Nm)

AD82698

CA

AD61591

4

7 (45 Nm)

T

13

9

14

8

T

3

AD85546

15

WP

T

IP

6

AUX

CS

AD85544

11

(45 Nm)

12

(23 Nm)

AD83189

FIESTA/ESCORT/COURIER/KOMBI

MONDEO

1

1

2

2

AD11314

AD67240

AD83067

FORD

Model:	**Fiesta 1,8D Turbo • Focus 1,8D Turbo**
Year:	**1998-03**
Engine Code:	**C9DC, F9DA/B, C9DA/B, BHDA/B**

Replacement Interval Guide

Ford recommend replacement every 100,000 miles or 10 years.

The previous use and service history of the vehicle must always be taken into account.

Refer to Timing Belt Replacement Intervals at the front of this manual.

Check For Engine Damage

CAUTION: This engine has been identified as an INTERFERENCE engine in which the possibility of valve-to-piston damage in the event of a timing belt failure is MOST LIKELY to occur. A compression check of all cylinders should be performed before removing the cylinder head(s).

Repair Times – hrs

Remove & install:

Fiesta	3,20
Focus	2,50

Special Tools

- Flywheel locking tool – Ford No.303-393 (21-168).
- Camshaft setting bar – Ford No.303-376 (21-162-B).
- Crankshaft timing pin – Ford No.303-193 (21-104).
- Camshaft sprocket puller – Ford No.303-651 (21-229).
- Camshaft sprocket holding tool – Ford No.205-072 (15-030-A).

Removal

NOTE: Vehicles built before 08/00 are fitted with manual tensioner which must be replaced with an automatic tensioner.

1. Raise and support front of vehicle.
2. Remove:
   - ❑ Intercooler cover.
   - ❑ Intercooler hoses.
   - ❑ Intercooler.
   - ❑ Coolant expansion tank. DO NOT disconnect hoses.
   - ❑ PAS pipe bracket.
   - ❑ Cylinder head cover.
   - ❑ Auxiliary drive belt cover.
   - ❑ Auxiliary drive belt.
   - ❑ Alternator drive shaft.
   - ❑ Starter motor.
3. Turn crankshaft clockwise until just before TDC on No.1 cylinder. Ensure camshaft parallel groove aligned with edge of cylinder head.
4. Remove blanking plug from cylinder block. Insert timing pin **1**. Tool No.303-193 (21-104).
5. Turn crankshaft clockwise until it stops against timing pin **1**.
6. Fit flywheel locking tool **2**.
   Tool No.303-393 (21-168).
   NOTE: Crankshaft web MUST remain locked against timing pin.
7. Fit setting bar to rear of camshaft **3**.
   Tool No.303-376 (21-162-B).
8. Support engine.

9. Remove:
   - ❑ RH engine mounting.
   - ❑ Engine mounting studs.
   - ❑ Timing belt cover **4**.
10. Slacken tensioner bolt **5**.
11. Turn tensioner pulley clockwise away from belt **6**. Use Allen key. Lightly tighten bolt **5**.
12. Hold camshaft sprocket **7**.
    Tool No.205-072 (15-030-A).
13. Slacken camshaft sprocket bolt **8**.
14. Loosen camshaft sprocket from taper **9**.
    Tool No.303-651 (21-229).
15. Ensure camshaft sprocket can turn on taper.
16. Remove timing belt.
    NOTE: DO NOT refit used belt.
    NOTE: If manual tensioner fitted it MUST be replaced with automatic tensioner.

Installation

NOTE: Refer to Ford dealer for revised tensioner.

1. Ensure crankshaft at TDC on No.1 cylinder.
2. Ensure crankshaft web against timing pin **1**.
3. Ensure flywheel locking tool located correctly **2**.
4. Ensure camshaft locked with tool **3**.
5. Ensure camshaft sprocket can turn on taper.
   NOTE: Ensure part No. of camshaft sprocket ends with AC. If not, fit new sprocket.
6. Ensure automatic tensioner arm at position **10**.
7. Fit timing belt in clockwise direction, starting at injection pump sprocket. Ensure belt is taut between sprockets on non-tensioned side.
8. Slacken tensioner bolt **5**.
9. Turn tensioner pulley anti-clockwise until pointer at position **11**. Use Allen key.
10. Tighten tensioner pulley bolt **5**.
    Tightening torque: 50 Nm.
11. Tighten camshaft sprocket bolt **8**.
    Tightening torque: 50 Nm.
12. Remove tools **1**, **2** & **3**.
13. Turn crankshaft almost 6 turns clockwise.
14. Insert timing pin in cylinder block **1**.
15. Turn crankshaft slowly clockwise to TDC on No.1 cylinder until crankshaft web against timing pin.
16. Fit flywheel locking tool **2**.
    NOTE: Crankshaft web MUST remain locked against timing pin.
17. Ensure tensioner pointer at position **11**. If not, repeat operations 1-16.
18. Ensure setting bar can be fitted to rear of camshaft **3**. If not, repeat operations 1-17.
19. Remove tools **1**, **2** & **3**.
20. Fit blanking plug to cylinder block.
    Tightening torque: 24 Nm.
21. If manual tensioner replaced with automatic tensioner, modify timing cover. File off inner corner **12**.
22. Install components in reverse order of removal.

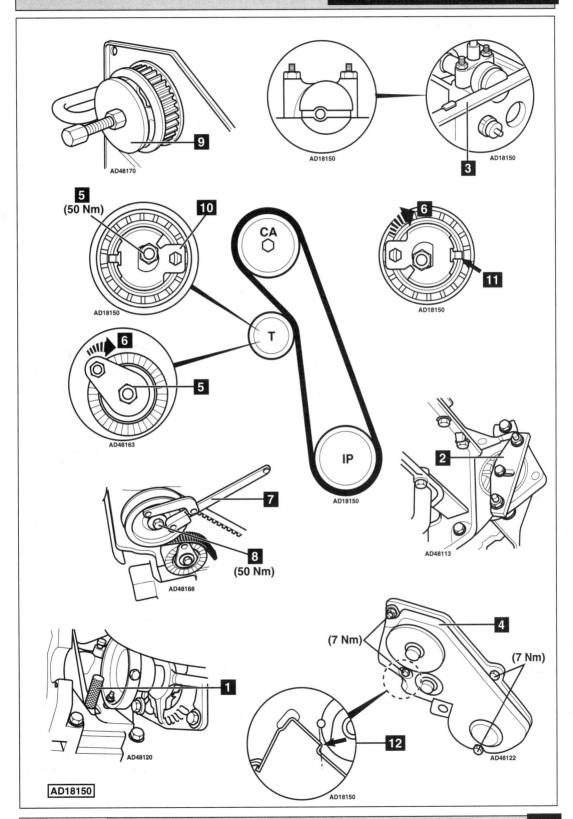

AD48170

9

AD18150

AD18150

3

5
(50 Nm)

10

AD18150

CA

6

11

AD18150

T

6

5

AD48163

IP

AD18150

7

8
(50 Nm)

AD48168

2

AD48113

4

(7 Nm)

(7 Nm)

1

AD48120

12

AD48122

AD18150

AD18150

FORD

Model:	Sierra 1,8 TD • P100 1,8 TD
Year:	1988-95
Engine Code:	RFA, RFB, RFL

Replacement Interval Guide

Ford recommend:
→08/89: Replacement every 36,000 miles or 3 years (water pump drive belt should also be replaced).
09/89→: Replacement every 40,000 miles or 5 years (water pump drive belt should also be replaced).
The previous use and service history of the vehicle must always be taken into account.
Refer to Timing Belt Replacement Intervals at the front of this manual.

Check For Engine Damage

CAUTION: This engine has been identified as an INTERFERENCE engine in which the possibility of valve-to-piston damage in the event of a timing belt failure is MOST LIKELY to occur.
A compression check of all cylinders should be performed before removing the cylinder head.

Repair Times – hrs

Check & adjust	0,90
Remove & install	1,80

Special Tools

- Crankshaft timing pin – Ford No.21-104.
- Camshaft timing pin – Ford No.23-019.

Special Precautions

- Disconnect battery earth lead.
- DO NOT turn crankshaft or camshaft when timing belt removed.
- Remove glow plugs to ease turning engine.
- Turn engine in normal direction of rotation (unless otherwise stated).
- DO NOT turn engine via camshaft or other sprockets.
- Observe all tightening torques.
- Check diesel injection pump timing after belt replacement.

Removal

Timing Belt

NOTE: Timing belt and injection pump belt should not be removed/installed until engine is cold (stopped for at least 4 hours).

1. Remove:
   - Auxiliary drive belts.
   - Water pump belt tensioner.
   - Timing belt covers **1** & **2**.
2. Turn crankshaft clockwise until slot in injection pump sprocket at 11 o'clock **3**.

3. Remove blanking plug from cylinder block. Insert timing pin in cylinder block **4**. Tool No.21-104.
4. Turn crankshaft clockwise until it stops against timing pin.
5. Insert timing pin in camshaft sprocket **5**. Tool No.23-019.
6. Insert 6 mm drill bit in injection pump sprocket **6**.
7. Slacken tensioner hexagon bolt and Torx bolt **7** & **8**.
8. Move tensioner away from belt. Tighten bolt **8**.
9. Remove timing belt.
   NOTE: DO NOT refit used belt.

Installation

Timing Belt

NOTE: Refer to Ford dealer for revised timing belt lower cover, to prevent debris entering timing belt drive area.

1. Check tensioner pulley and guide pulley/sprocket for cracks, chips or other damage. Replace if necessary.
   *NOTE: If guide sprocket **9** is plastic, it should be replaced with a steel sprocket No.7053802 and bolt No.6701537.*
2. Spin tensioner pulley and guide pulley/sprocket. Check for bearing roughness and noise. Replace if necessary.
   *NOTE: If tensioner bolt **8** has 'mushroom' head, check for evidence of removal with a chisel or self-grip wrench. If so, fit a new bolt and tensioner.*
3. Ensure timing pin located correctly in camshaft sprocket **5**. Tool No.23-019.
4. Ensure 6 mm drill bit located correctly in injection pump sprocket **6**.
5. Ensure crankshaft web against timing pin **4**. Tool No.21-104.
6. Fit timing belt. Slack side of belt must be towards tensioner. Observe direction of rotation marks on belt.
7. Slacken camshaft sprocket bolts **10**.
8. Slacken injection pump sprocket bolts **13**.
9. Slacken tensioner bolts 1/2 turn. Allow tensioner to operate.
10. Tighten camshaft sprocket bolts **10**.
11. Tighten injection pump sprocket bolts **13**.
    NOTE: Ensure bolts not at end of slotted holes in sprockets.

Autodata

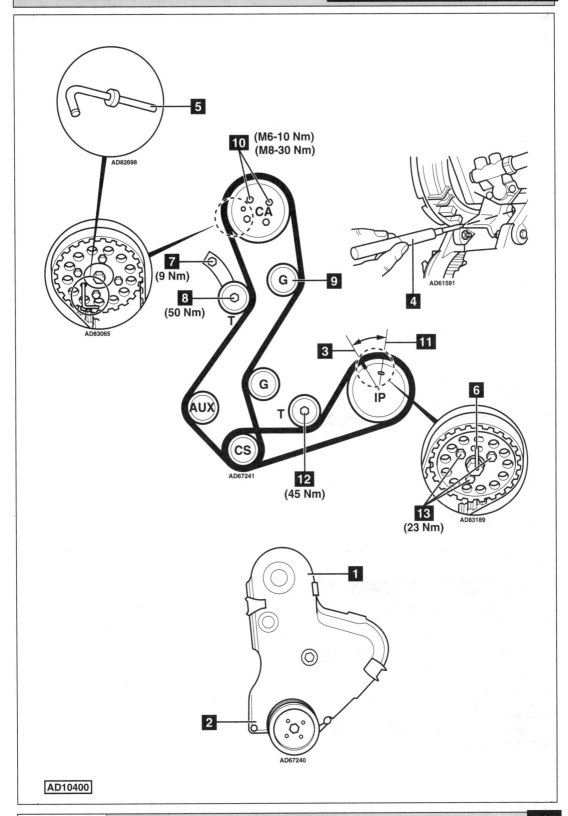

AD82698

5

10 (M6-10 Nm)
(M8-30 Nm)

CA

AD83065

7
(9 Nm)

8
(50 Nm)
T

G **9**

AD61591

4

G

AUX

T

CS

AD67241

3 **11**

IP

6

13
(23 Nm)

AD83189

12
(45 Nm)

1

2

AD67240

AD10400

12. Tighten tensioner bolts & .

13. Remove timing pins and drill bit , & . Turn crankshaft two turns clockwise. Ensure injection pump sprocket slot at 12 o'clock .

14. Turn crankshaft anti-clockwise until injection pump sprocket slot at 11 o'clock .

15. Insert timing pin in cylinder block . Tool No.21-104. Turn crankshaft clockwise until it stops against timing pin.

16. Insert timing pin in camshaft sprocket . Tool No.23-019.

17. Insert 6 mm drill bit in injection pump sprocket .

18. Slacken camshaft sprocket bolts .

19. Slacken injection pump sprocket bolts .

20. Slacken tensioner bolts & .

21. Push tensioner firmly against belt and release. Tighten tensioner bolts to specified torque & .

22. Tighten camshaft sprocket bolts to specified torque .

23. Tighten injection pump sprocket bolts to specified torque .

24. Remove timing pins and drill bit , & .

25. Install components in reverse order of removal.

Removal

Injection Pump Belt

1. Remove timing belt as described previously.

2. Ensure 6 mm drill bit located correctly in injection pump sprocket .

3. Slacken tensioner bolt . Lift tensioner against spring pressure. Tighten bolt.

4. Remove injection pump belt.

5. Slacken injection pump sprocket bolts .

Installation

Injection Pump Belt

1. Ensure crankshaft web against timing pin .

2. Ensure 6 mm drill bit located correctly in injection pump sprocket .

3. Fit injection pump belt. Slack side of belt must be towards tensioner. Observe direction of rotation marks.

4. Slacken tensioner bolt . Allow tensioner to operate. Tighten bolt to 45 Nm.

5. Tighten injection pump sprocket bolts . Ensure bolts not at end of slotted holes in sprocket.

6. Fit timing belt as described previously.

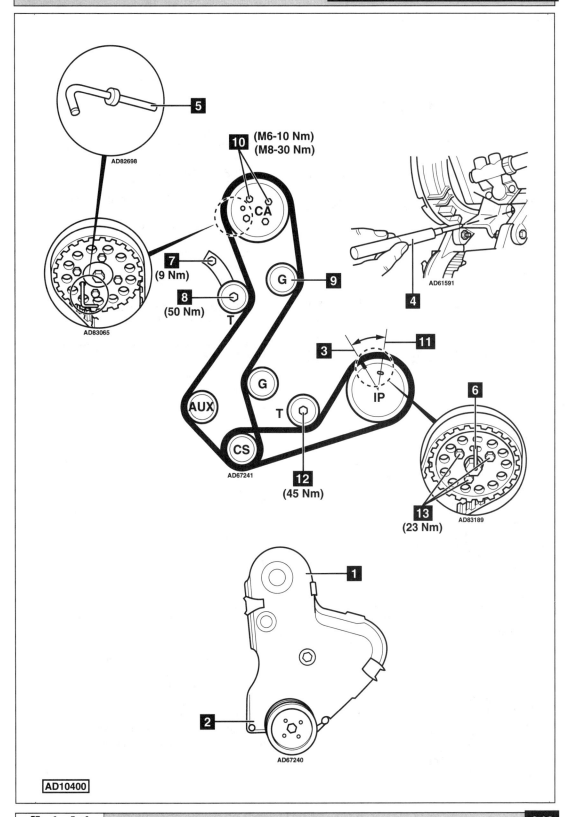

AD82698

10 (M6-10 Nm) (M8-30 Nm)

CA

5

7 (9 Nm)

8 (50 Nm) T

G

9

AD83065

AD61591

4

3 **11**

G

AUX T IP

CS **6**

AD67241

12 (45 Nm)

13 (23 Nm) AD83189

1

2

AD67240

AD10400

FORD

Model:	**Galaxy 1,9 TDI**
Year:	**1995-99**
Engine Code:	**1Z, AHU**

Replacement Interval Guide

Ford recommend replacement every
60,000 miles or 5 years, whichever occurs first.
*The previous use and service history of the vehicle
must always be taken into account.*
*Refer to Timing Belt Replacement Intervals at the front
of this manual.*

Check For Engine Damage

*CAUTION: This engine has been identified as an
INTERFERENCE engine in which the possibility of
valve-to-piston damage in the event of a timing belt
failure is MOST LIKELY to occur.*
*A compression check of all cylinders should be
performed before removing the cylinder head.*

Repair Times – hrs

Check & adjust	1,70
Remove & install	2,80

Special Tools

- Camshaft setting bar – Ford No.21-105.
- Sprocket holding tool – Ford No.15-030-A.
- Injection pump timing pin – Ford No.23-047.
- 90° angled circlip pliers.

Special Precautions

- Disconnect battery earth lead.
- DO NOT turn crankshaft or camshaft when timing belt removed.
- Remove glow plugs to ease turning engine.
- Turn engine in normal direction of rotation (unless otherwise stated).
- DO NOT turn engine via camshaft or other sprockets.
- Observe all tightening torques.
- Check diesel injection pump timing after belt replacement.

Removal

1. Raise and support front of vehicle.
2. Support engine.
3. Remove:
   - ❏ Air filter and intake hose.
   - ❏ Engine cover.
   - ❏ Breather hose.
   - ❏ Cylinder head cover.
   - ❏ RH engine mounting and bracket.
   - ❏ Engine undershield.
   - ❏ Auxiliary drive belt.
   - ❏ Timing belt upper cover **1**.
   - ❏ Water pump pulley.
   - ❏ Four crankshaft pulley bolts **11**.
   - ❏ Crankshaft pulley **2**.

4. Turn crankshaft slowly clockwise to TDC on No.1 cylinder. Ensure flywheel timing marks aligned **3**.
5. Fit setting bar No.21-105 to rear of camshaft **4**. Centralise camshaft using feeler gauges.
   NOTE: If setting bar cannot be fitted: Turn crankshaft one turn clockwise.
6. Insert timing pin in injection pump sprocket **5**. Tool No.23-047.
7. Remove timing belt lower cover **6**.
8. Slacken tensioner nut **7**. Move tensioner away from belt. Lightly tighten nut.
9. Remove:
   - ❏ Guide pulley **12**.
   - ❏ Timing belt.
   NOTE: Mark direction of rotation on belt with chalk if belt is to be reused.

Installation

1. Ensure flywheel timing marks aligned **3**.
2. Ensure setting bar fitted correctly **4**.
3. Ensure timing pin located correctly in injection pump sprocket **5**.
4. Hold camshaft sprocket. Use tool No.15-030-A. Slacken bolt **8**.
5. Tap sprocket gently to loosen it from taper. Ensure sprocket turns freely on camshaft but without free play.
6. Fit timing belt in anti-clockwise direction, starting at crankshaft sprocket. Ensure belt is taut between sprockets.
   NOTE: Observe direction of rotation marks on belt.
7. Fit guide pulley **12**. Tighten bolt to 25 Nm.
8. Remove timing pin **5**.
9. Use 90° angled circlip pliers in holes **9**: Turn tensioner back against spring force until raised mark and groove on tensioner aligned **10**.
10. Tighten tensioner nut to 20 Nm **7**.
11. Hold camshaft sprocket. Use tool No.15-030-A. Tighten bolt to 45 Nm **8**. Ensure crankshaft and camshaft do not turn.
12. Remove setting bar **4**.
13. Turn crankshaft slowly two turns clockwise. Ensure flywheel timing marks aligned **3**.
14. Ensure setting bar can be fitted **4**.
15. Ensure timing pin can be inserted in injection pump sprocket **5**.
16. Install components in reverse order of removal.
17. Tighten crankshaft pulley bolts to 25 Nm **11**.

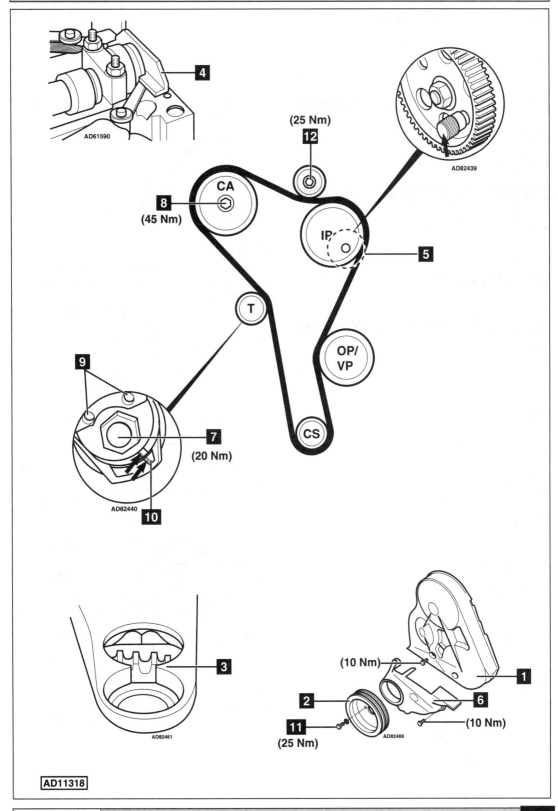

FORD

Model:	**Galaxy 1,9 TDI**
Year:	**1997-00**
Engine Code:	**AFN**

Replacement Interval Guide

Ford recommend replacement every 60,000 miles or 5 years
The previous use and service history of the vehicle must always be taken into account.
Refer to Timing Belt Replacement Intervals at the front of this manual.

Check For Engine Damage

CAUTION: This engine has been identified as an INTERFERENCE engine in which the possibility of valve-to-piston damage in the event of a timing belt failure is MOST LIKELY to occur.
A compression check of all cylinders should be performed before removing the cylinder head.

Repair Times – hrs

Check & adjust	1,70
Remove & install	2,80

Special Tools

- Injection pump sprocket locking pin – Ford No.23-047.
- Camshaft setting bar – Ford No.21-105.
- Holding tool – Ford No.15-030A.

Special Precautions

- Disconnect battery earth lead.
- DO NOT turn crankshaft or camshaft when timing belt removed.
- Remove glow plugs to ease turning engine.
- Turn engine in normal direction of rotation (unless otherwise stated).
- DO NOT turn engine via camshaft or other sprockets.
- Observe all tightening torques.
- Check diesel injection pump timing after belt replacement.

Removal

1. Remove:
   - ❏ Headlamp cover.
   - ❏ Bulkhead cover.
   - ❏ Air filter.
   - ❏ Engine cover.
2. Disconnect:
   - ❏ Mass air flow (MAF) sensor multi-plug.
   - ❏ Mass air flow (MAF) sensor intake hose.
   - ❏ Exhaust gas recirculation (EGR) solenoid hose.
   - ❏ Oil pressure switches (blue & yellow).
3. Support engine.
4. Remove engine mounting.
5. Raise and support vehicle.
6. Remove:
   - ❏ Engine undershield.
   - ❏ PAS pump hose bracket.
   - ❏ Engine torque bracket.
   - ❏ Auxiliary drive belt.
   - ❏ Timing belt upper cover **1**.
7. Turn crankshaft to TDC on No.1 cylinder **2**.
8. Fit setting bar No.21-105 to rear of camshaft **3**.
9. Centralise camshaft using feeler gauges.
10. Remove:
    - ❏ Crankshaft pulley bolts **5**.
    - ❏ Crankshaft pulley **6**.
    - ❏ Water pump pulley.
    - ❏ Timing belt lower cover **7**.
11. Lock injection pump sprocket **4**. Use tool No.23-047.
12. Remove guide pulley **9**.
13. Slacken tensioner nut **8**.
14. Turn tensioner anti-clockwise away from belt.
15. Lightly tighten nut **8**.
16. Remove timing belt.
    NOTE: Mark direction of rotation on belt.

Installation

1. Ensure flywheel timing marks aligned **2**.
2. Ensure camshaft setting bar fitted correctly **3**.
3. Ensure locking pin located correctly in injection pump sprocket **4**.
4. Hold camshaft sprocket. Use tool No.15-030A. Slacken camshaft sprocket bolt 1/2 turn **10**.
5. Loosen camshaft sprocket from taper using a drift through hole in timing belt rear cover.
6. Fit timing belt in anti-clockwise direction, starting at crankshaft sprocket. Ensure belt is taut between sprockets.
7. Fit guide pulley **9**.
8. Tighten bolt to 25 Nm **9**.
9. Remove locking pin from injection pump sprocket **4**.
10. Slacken tensioner nut **8**.
11. Turn automatic tensioner pulley clockwise until notch and raised mark on tensioner aligned **11**. Use circlip pliers.
12. Tighten tensioner nut to 20 Nm **8**.
13. Ensure flywheel timing marks aligned **2**.
14. Tighten camshaft sprocket bolt to 45 Nm **10**.
15. Remove camshaft setting bar **3**.
16. Turn crankshaft two turns clockwise until timing marks aligned **2**.
17. Ensure camshaft setting bar can be fitted **3**.
18. Ensure locking pin can be inserted in injection pump sprocket **4**.
19. Apply firm thumb pressure to belt. Tensioner marks should move out of alignment **11**.
20. Release thumb pressure from belt. Tensioner marks should realign. If not: Repeat tensioning procedure.
21. Fit timing belt lower cover **7**.
22. Fit crankshaft pulley **6**.
23. Tighten crankshaft pulley bolts to 25 Nm **5**.
24. Install components in reverse order of removal.

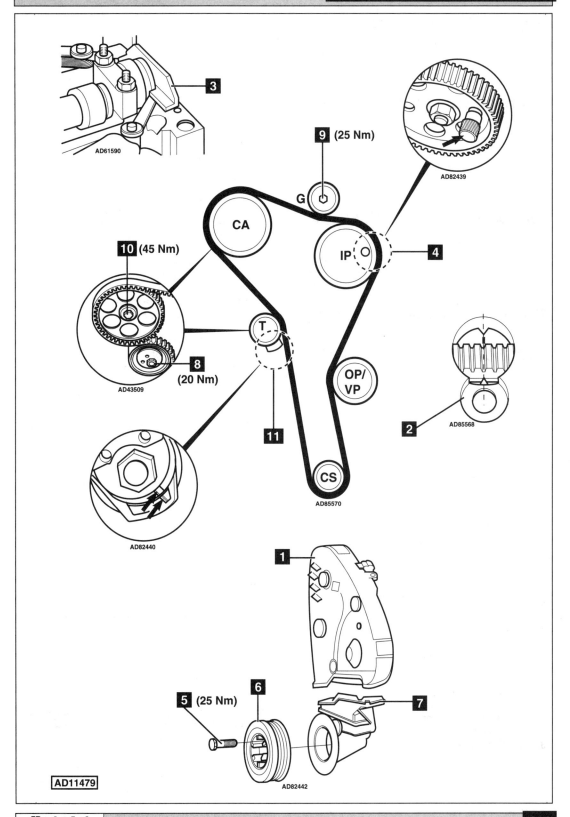

AD61590

3

9 (25 Nm)

AD82439

G

CA

10 (45 Nm)

IP

4

AD43509

8
(20 Nm)

T

11

OP/VP

2   AD85568

AD82440

CS

AD85570

1

5 (25 Nm)

6

7

AD82442

AD11479

Model:	**Galaxy 1,9 TDI**
Year:	**1999-02**
Engine Code:	**ANU, AUY**

Replacement Interval Guide

Ford recommend replacement every
40,000 miles or 5 years, whichever occurs first
(tensioner pulley must also be replaced).
*The previous use and service history of the vehicle
must always be taken into account.*
*Refer to Timing Belt Replacement Intervals at the front
of this manual.*

Check For Engine Damage

*CAUTION: This engine has been identified as an
INTERFERENCE engine in which the possibility of
valve-to-piston damage in the event of a timing belt
failure is MOST LIKELY to occur. A compression check
of all cylinders should be performed before removing
the cylinder head(s).*

Repair Times – hrs

Remove & install	2,30

Special Tools

- Crankshaft locking tool –
  Ford No.310-085 (23-059).
- Tensioner locking tool –
  Ford No.310-084 (23-058).
- Camshaft locking tool – 6 mm drill bit.
- Tensioner setting tool – 4 mm drill bit.
- Right angled circlip pliers.

Special Precautions

- Disconnect battery earth lead.
- DO NOT turn crankshaft or camshaft when timing belt removed.
- Remove glow plugs to ease turning engine.
- Turn engine in normal direction of rotation (unless otherwise stated).
- DO NOT turn engine via camshaft or other sprockets.
- Observe all tightening torques.

Removal

1. Raise and support front of vehicle.
2. Remove:
   - Engine undershield.
   - Air filter assembly.
   - Engine top cover.
   - Auxiliary drive belt.
   - Auxiliary drive belt tensioner.
   - Turbocharger air hoses.
   - Intercooler outlet hose.
3. Support engine.

4. Remove:
   - Bolts **1**.
   - Nuts **2**.
   - Bolts **3**.
5. Raise engine slightly.
6. Remove:
   - RH engine mounting **4**.
   - RH engine mounting bracket.
   - Timing belt upper cover **5**.
   - Timing belt centre cover **6**.
   - Crankshaft pulley bolts **7**.
   - Crankshaft pulley **8**.
   - Timing belt lower cover **9**.
7. Turn crankshaft clockwise to TDC on No.1 cylinder. Ensure '4Z' timing mark aligned with notch on camshaft sprocket hub **10**.
   NOTE: Notch located behind camshaft sprocket teeth.
8. Lock crankshaft **11**.
   Use tool No.310-085 (23-059).
9. Ensure timing marks aligned **12**.
10. Lock camshaft **13**. Use 6 mm drill bit.
11. Hold tensioner with right angled circlip pliers **14**.
12. Slacken tensioner nut **15**.
13. Turn tensioner pulley anti-clockwise **16**. Ensure lug and stop aligned **17** & **18**.
14. Lock automatic tensioner unit.
    Use tool No.310-084 (23-058) **19**.
15. Turn tensioner pulley clockwise **20**. Ensure lug and stop aligned **17** & **21**.
16. Remove:
    - Guide pulley **22**.
    - Automatic tensioner unit **23**.
    - Timing belt.

Installation

*NOTE: DO NOT refit used automatic tensioner
pulley **23**. Automatic tensioner pulley must always
be replaced once removed.*

1. Ensure camshaft locked with 6 mm drill bit **13**.
2. Ensure crankshaft locking tool located correctly **11**.
3. Ensure timing marks aligned **12**.
4. Slacken camshaft sprocket bolts **24**.
5. Align camshaft sprocket with bolts in centre of slotted holes. Tighten bolts finger tight **24**.
6. Fit timing belt in following order:
   - Camshaft sprocket.
   - Tensioner pulley.
   - Crankshaft sprocket.
   - Water pump sprocket.

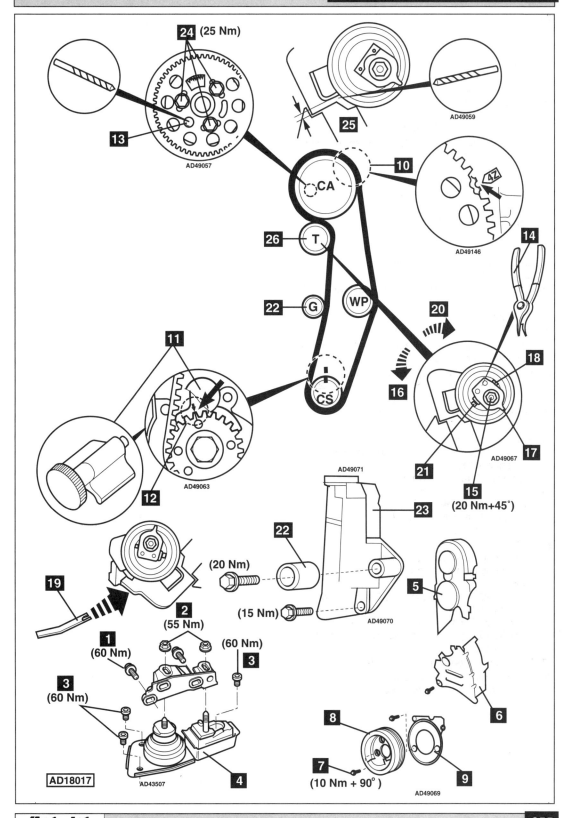

24 (25 Nm)

13

AD49057

25

AD49059

CA

10

4Z

AD49146

14

26 T

22 G

WP

20

16

18

11

CS

21

17

AD49067

12

AD49063

AD49071

15
(20 Nm+45°)

23

22

5

19

(20 Nm)

2
(55 Nm)

1
(60 Nm)

3
(60 Nm)

(15 Nm)

AD49070

3
(60 Nm)

8

6

7

9

(10 Nm + 90°)

AD49069

AD18017

AD43507

4

←

7. Install:
   ❏ Automatic tensioner unit 🔢.
   ❏ Guide pulley 🔢.
   NOTE: Ensure belt is taut between sprockets on non-tensioned side.

8. Turn tensioner pulley slowly anti-clockwise 🔢. Use right angled circlip pliers 🔢. Remove locking tool without force 🔢.

9. Turn tensioner pulley slowly clockwise 🔢. Insert 4 mm drill bit 🔢.
   NOTE: Engine must be COLD.

10. Tighten tensioner nut 🔢.
    Tightening torque: 20 Nm + 45°.

11. Tighten camshaft sprocket bolts 🔢.
    Tightening torque: 25 Nm.

12. Remove:
    ❏ 6 mm drill bit 🔢.
    ❏ Crankshaft locking tool 🔢.
    ❏ 4 mm drill bit 🔢.

13. Turn crankshaft slowly two turns clockwise to TDC on No.1 cylinder.

14. Lock crankshaft 🔢.

15. Ensure timing marks aligned 🔢.

16. Ensure 6 mm drill bit can be inserted easily into camshaft sprocket 🔢.

17. If not: Slacken camshaft sprocket bolts 🔢. Turn sprocket hub until drill bit can be inserted 🔢. Tighten bolts 🔢. Tightening torque: 25 Nm.

18. Insert 4 mm drill bit. Ensure dimension correct 🔢.

19. If not: Slacken tensioner nut 🔢. Turn tensioner pulley until dimension correct 🔢. Tighten tensioner nut 🔢.
    Tightening torque: 20 Nm + 45°.

20. Remove:
    ❏ 6 mm drill bit 🔢.
    ❏ Crankshaft locking tool 🔢.
    ❏ 4 mm drill bit 🔢.

21. Install:
    ❏ RH engine mounting bracket.
    ❏ RH engine mounting 🔢.

22. Tighten:
    ❏ Bolts 🔢. Tightening torque: 60 Nm. Lightly oil threads.
    ❏ Nuts 🔢. Tightening torque: 55 Nm.
    ❏ Bolts 🔢. Tightening torque: 60 Nm. Lightly oil threads.

23. Install components in reverse order of removal.

24. Tighten crankshaft pulley bolts 🔢.
    Tightening torque: 10 Nm + 90°.

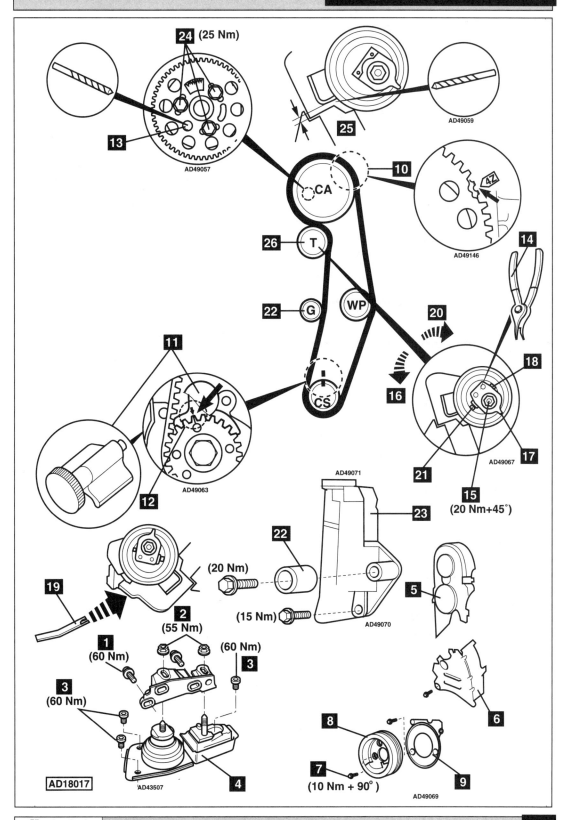

24 (25 Nm)

13

AD49057

25

AD49059

10

CA

AZ

AD49146

14

26 T

22 G

WP

20

16

18

11

12

AD49063

CS

21

15
(20 Nm+45°)

17

AD49067

AD49071

23

AD49070

5

19

2
(55 Nm)

22
(20 Nm)

1
(60 Nm)

3
(60 Nm)

3
(60 Nm)

(15 Nm)

6

AD18017

AD43507

4

8

7
(10 Nm + 90°)

9

AD49069

FORD

Model:	**Ranger 2,5D/TD**
Year:	**1999-03**
Engine Code:	**WL, WL-T**

Replacement Interval Guide

Ford recommend replacement every 60,000 miles or 5 years, whichever occurs first.
The previous use and service history of the vehicle must always be taken into account.
Refer to Timing Belt Replacement Intervals at the front of this manual.

Check For Engine Damage

CAUTION: This engine has been identified as an INTERFERENCE engine in which the possibility of valve-to-piston damage in the event of a timing belt failure is MOST LIKELY to occur. A compression check of all cylinders should be performed before removing the cylinder head(s).

Repair Times – hrs

Remove & install	1,00

Special Tools

- None required.

Special Precautions

- Disconnect battery earth lead.
- DO NOT turn crankshaft or camshaft when timing belt removed.
- Remove glow plugs to ease turning engine.
- Turn engine in normal direction of rotation (unless otherwise stated).
- DO NOT turn engine via camshaft or other sprockets.
- Observe all tightening torques.
- Check diesel injection pump timing after belt replacement.

Removal

1. Remove timing belt cover **1**.
2. Turn crankshaft clockwise to setting position.
3. Ensure timing marks aligned **2** & **3**.
4. Slacken tensioner bolt **4**.
5. Remove tensioner spring **5**.
6. Move tensioner pulley away from belt. Lightly tighten bolt **4**.
7. Remove timing belt.

Installation

1. Check free length of tensioner spring is 63 mm. Replace spring if necessary **5**.
2. Ensure timing marks aligned **2** & **3**.
3. Fit timing belt in clockwise direction, starting at injection pump sprocket. Ensure belt is taut between sprockets on non-tensioned side.
4. Slacken tensioner bolt **4**.
5. Push tensioner pulley gently against belt.
6. Fit tensioner spring **5**.
7. Lightly tighten tensioner bolt **4**.
8. Turn crankshaft slowly 2 turns clockwise.
9. Ensure timing marks aligned **2** & **3**.
10. Slacken tensioner bolt **4**.
11. Allow tensioner to operate.
12. Tighten tensioner bolt **4**.
    Tightening torque: 38-51 Nm.
13. Turn crankshaft slowly 2 turns clockwise.
14. Ensure timing marks aligned **2** & **3**.
15. Apply a load of 10 kg to belt at ▽. Belt should deflect 9-10 mm.
16. Install components in reverse order of removal.

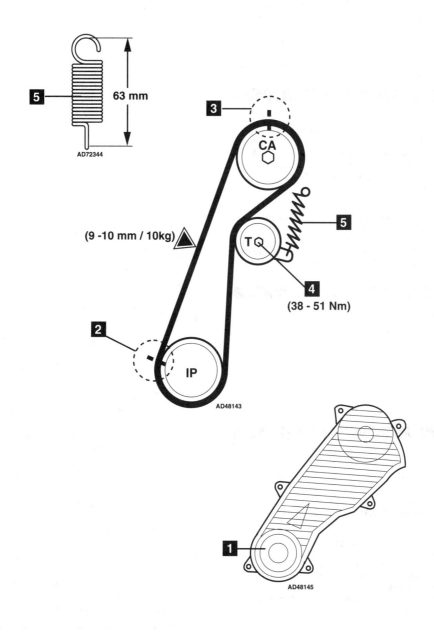

5 — 63 mm

AD72344

3

CA

(9 -10 mm / 10kg)

5

T

4

(38 - 51 Nm)

2

IP

AD48143

1

AD48145

AD11591

FORD

Model:	**Transit 2,5 DI/Turbo**
Year:	**1983-94**
Engine Code:	**EAB, 4AB, 4BC, 4CA, 4CC, 4DA, 4EA, 4FA, 4GA**

Replacement Interval Guide

Ford recommend replacement every 48,000 miles (on 6,000 mile service intervals) or every 60,000 miles or 5 years (on 10,000 mile service intervals).
The previous use and service history of the vehicle must always be taken into account.
Refer to Timing Belt Replacement Intervals at the front of this manual.

Check For Engine Damage

→1988
CAUTION: This engine has been identified as a FREEWHEELING engine in which the possibility of valve-to-piston damage in the event of a timing belt failure may be minimal or very unlikely. However, a precautionary compression check of all cylinders should be performed.

1989→
CAUTION: This engine has been identified as an INTERFERENCE engine in which the possibility of valve-to-piston damage in the event of a timing belt failure is MOST LIKELY to occur. A compression check of all cylinders should be performed before removing the cylinder head(s).

Repair Times – hrs

Check & adjust	0,40
Remove & install:	
→1986	3,50
1986-92	1,50
PAS	+0,10
EGR	+0,20
1992→	1,40
DI Turbo	1,80

Special Tools

- ■ Flywheel timing pin – Ford No.23-020.
- ■ Camshaft timing pin – Ford No.21-123.
- ■ Injection pump timing pin – Ford No.23-019.

Special Precautions

- ■ Disconnect battery earth lead.
- ■ DO NOT turn crankshaft or camshaft when timing belt removed.
- ■ Remove glow plugs to ease turning engine.
- ■ Turn engine in normal direction of rotation (unless otherwise stated).
- ■ DO NOT turn engine via camshaft or other sprockets.
- ■ Observe all tightening torques.
- ■ Check diesel injection pump timing after belt replacement.

Removal

1. Remove:
   - ❑ Engine undershield.
   - ❑ Radiator grille.
   - ❑ Radiator.
   - ❑ Auxiliary drive belt.
   - ❑ Cooling fan.
   - ❑ Viscous coupling (→1985: LH thread. 1986→: RH thread).
   - ❑ Cooling fan pulley.
   - ❑ Crankshaft pulley bolt rubber cap **1**.
   - ❑ Timing belt cover **2**.
2. Turn crankshaft until flywheel timing pin can be inserted **3**. Tool No.23-020.
3. Insert timing pin in injection pump sprocket **4**. Tool No.23-019.
4. Insert timing pin in camshaft sprocket **5**. Tool No.21-123.
5. Slacken tensioner bolts **6**. Move tensioner away from belt. Tighten clamp bolt.
6. Remove timing belt.
   NOTE: DO NOT refit used belt. Timing belt must always be renewed once it has been slackened or removed.

Installation

1. Ensure all timing pins located correctly **3**, **4** & **5**.
2. Fit timing belt in clockwise direction, starting at crankshaft sprocket.
3. Remove all timing pins **3**, **4** & **5**.
4. Slacken tensioner bolts **6**. Allow tensioner to operate. Tighten bolts.
5. Turn crankshaft two turns clockwise. Insert all timing pins **3**, **4** & **5**.
6. If timing pin cannot be inserted in injection pump sprocket: Slacken bolts **7**. Turn sprocket until timing pin can be inserted.
7. Lightly tighten bolts **7**. Remove timing pin. Fully tighten bolts.
8. Turn crankshaft two turns clockwise. Ensure all timing pins can be inserted.
9. Remove timing pins. Turn crankshaft 130° clockwise.
10. Slacken tensioner bolts **6**. Press down several times on belt at ▽.
11. Tighten tensioner bolts **6**.
12. Check and adjust injection pump timing.
13. Install components in reverse order of removal.
14. Refill cooling system.

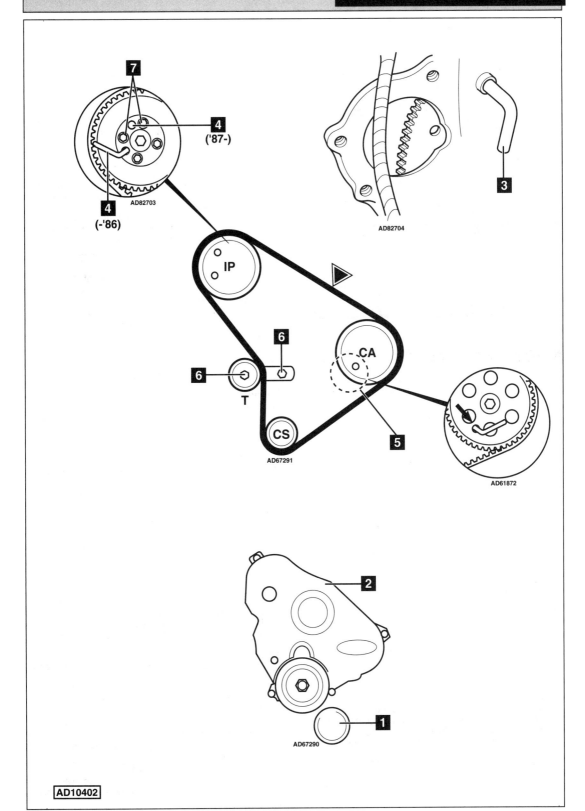

AD82703

4
('87-)

7

4
(-'86)

AD82704

3

IP

6

6

6

T

CA

CS

AD67291

5

AD61872

2

1

AD67290

AD10402

FORD

Model:	**Transit 2,5 DI/Turbo • Transit 2,5 TCI • Flareside 2,5 D**
Year:	**1994-00**
Engine Code:	**4EB, 4EC, 4ED, 4FA, 4FB, 4FC, 4FD, 4GA, 4GB, 4GC, 4GD, 4GE, 4HB, 4HC**

Replacement Interval Guide

Ford recommend:
→07/98: Replacement every 60,000 miles or 5 years, whichever occurs first.
08/98→: Replacement every 50,000 miles or 5 years, whichever occurs first.
The previous use and service history of the vehicle must always be taken into account.
Refer to Timing Belt Replacement Intervals at the front of this manual.

Check For Engine Damage

CAUTION: This engine has been identified as an INTERFERENCE engine in which the possibility of valve-to-piston damage in the event of a timing belt failure is MOST LIKELY to occur.
A compression check of all cylinders should be performed before removing the cylinder head(s).

Repair Times – hrs

Remove & install	1,80
AC	+0,20

Special Tools

- Flywheel timing pin – Ford No.23-020.
- Camshaft timing pin – Ford No.21-123.
- Injection pump timing pin – Ford No.23-019 or 23-029.

Removal

1. Drain coolant.
2. Remove:
   - ❏ Engine undershield and radiator grille.
3. Detach AC condenser from radiator (if fitted).
4. Remove:
   - ❏ Radiator.
   - ❏ Auxiliary drive belts (mark direction of rotation).
   - ❏ Cooling fan, viscous coupling and cooling fan pulley.
   - ❏ Crankshaft pulley bolt rubber cap **1**.
   - ❏ Fuel filter (turbo).
   - ❏ Crankshaft position (CKP) sensor and spacer (if fitted).
   - ❏ Timing belt covers **2**.
5. Turn crankshaft until flywheel timing pin can be inserted **3**. Tool No.23-020.
6. Insert timing pin in injection pump sprocket **4**. Tool No.23-019 or 23-029.
7. Insert timing pin in camshaft sprocket **5**. Tool No.21-123.
8. Slacken injection pump sprocket bolts **8**.
9. Slacken tensioner bolt(s) **6** & **7** or **10**.
10. Move tensioner away from belt.
11. Turbo: Tighten tensioner bolt **6**.
12. Remove timing belt. Remove tensioner pulley. Check tensioner pulley for smooth operation.
    NOTE: Non-turbo: Replace manual tensioner pulley with automatic tensioner pulley, new bolt and modified timing belt.
13. Lightly oil contact face of tensioner bracket and cylinder block.
14. Turbo: Install manual tensioner pulley. Lightly tighten bolts **6** & **7**.
15. Non-turbo: Install automatic tensioner pulley with new bolt. Lightly tighten bolt **10**.

NOTE: Turbo: DO NOT refit used belt. Timing belt must always be renewed once it has been slackened or removed.
WARNING: DO NOT fit automatic tensioner pulley or modified timing belt to turbo engines.
NOTE: Non-turbo: Original timing belt – DO NOT refit used belt. Timing belt and tensioner pulley must always be renewed once they have been slackened or removed.
NOTE: Non-turbo: Modified timing belt with automatic tensioner pulley – only replace at recommended intervals or if damaged.

Installation

1. Ensure all timing pins located correctly **3**, **4** & **5**.
2. Fit timing belt in anti-clockwise direction, starting at crankshaft sprocket.
3. Observe direction of rotation marks on belt.
4. If necessary: Turn injection pump sprocket slightly to engage timing belt teeth.

Turbo

5. Slacken tensioner bolt **6**. Allow tensioner to operate.
6. Tighten tensioner bolts **6** & **7**.

Non-turbo

7. Turn tensioner clockwise until centre punch mark on backplate aligned with lower edge of cut-out **9** & **11**. Use Allen key.
8. Tighten tensioner bolt **10**. Tightening torque: 45 Nm.

All models

9. Tighten injection pump sprocket bolts to 25 Nm **8**.
10. Remove all timing pins **3**, **4** & **5**.
11. Turn crankshaft one turn then a further 315° clockwise.

Turbo

12. Slacken tensioner bolts **6** & **7**.
13. Apply thumb pressure to belt at ▽. Allow tensioner to operate. Release thumb pressure from belt.
14. Tighten tensioner bolt **6**. Tightening torque: 24 Nm.
15. Tighten tensioner bolt **7**. Tightening torque: 58 Nm.

Non-turbo:

16. Check centre punch mark on backplate aligned with lower edge of cut-out **9** & **11**.
17. If not: Repeat tensioning procedure.
18. Tighten tensioner bolt **10**. Tightening torque: 45 Nm.

All models:

19. Turn crankshaft 45° clockwise until flywheel timing pin can be inserted **3**.
20. Insert timing pin in camshaft sprocket **5**. Tool No.21-123.
21. Insert timing pin in injection pump sprocket **4**. Tool No.23-019 or 23-029.
22. If timing pin cannot be inserted in injection pump sprocket: Slacken bolts **8**.
23. Turn sprocket until timing pin can be inserted.
24. Tighten bolts to 25 Nm **8**.
25. Remove all timing pins **3**, **4** & **5**.
26. Install components in reverse order of removal.
27. Observe direction of rotation mark on auxiliary drive belts.
28. Refill cooling system.

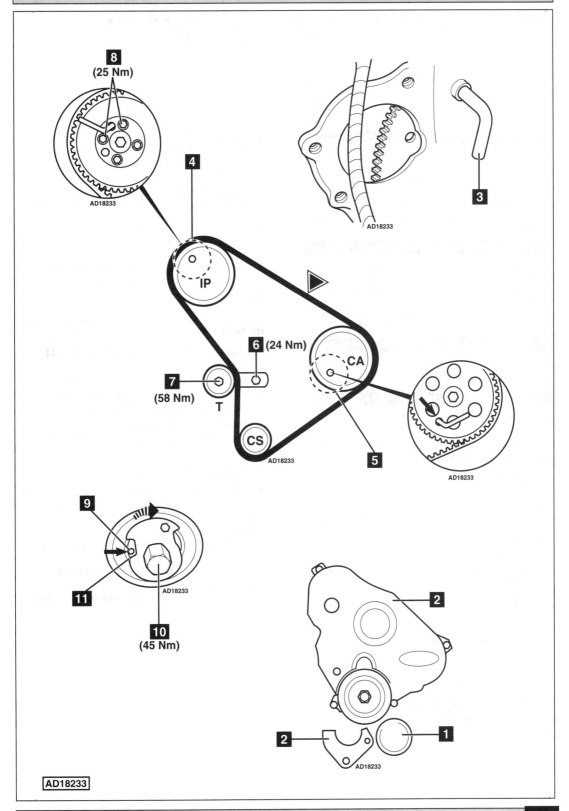

8
(25 Nm)

AD18233

4

3

AD18233

IP

6 (24 Nm)

7
(58 Nm)
T

CA

CS

AD18233

5

AD18233

9

11

10
(45 Nm)

AD18233

2

2

1

AD18233

AD18233

Model:	Civic 1,2/1,3/1,4/1,5/1,6 • Shuttle 1,4/1,5 • Civic CRX 1,5/1,6
	Concerto 1,4/1,5/1,6 SOHC • Prelude 2,0 SOHC
Year:	1988-97
Engine Code:	B20A3/4, D12/13B, D14A/1, D15B/1/2/3/4/7
	D16A6/7, D16Z1/2, D16Z6/7/9

Replacement Interval Guide

Honda recommend check & replace if necessary every 12,000 miles (→93MY) and replacement every 60,000 miles or 5 years.
The previous use and service history of the vehicle must always be taken into account.
Refer to Timing Belt Replacement Intervals at the front of this manual.

Check For Engine Damage

CAUTION: This engine has been identified as an INTERFERENCE engine in which the possibility of valve-to-piston damage in the event of a timing belt failure is MOST LIKELY to occur.
A compression check of all cylinders should be performed before removing the cylinder head.

Repair Times – hrs

| Check & adjust | 0,30 |
| Remove & install | 2,20 |

Special Tools

■ None required.

Special Precautions

■ Disconnect battery earth lead.
■ DO NOT turn crankshaft or camshaft when timing belt removed.
■ Remove spark plugs to ease turning engine.
■ Turn engine in normal direction of rotation (unless otherwise stated).
■ DO NOT turn engine via camshaft or other sprockets.
■ Observe all tightening torques.

Removal

NOTE: Normal direction of crankshaft rotation is anti-clockwise.

1. Remove:
   ❏ LH inner wing panel.
   ❏ Auxiliary drive belts.
   ❏ LH top engine mounting.
   ❏ Crankshaft pulley **1**.
   ❏ Cylinder head cover.
   ❏ Timing belt upper cover **2**.
   ❏ Timing belt lower cover **3**.
2. Temporarily fit crankshaft pulley **1**.

3. Turn crankshaft anti-clockwise to TDC on No.1 cylinder **4** or **9**. Ensure camshaft sprocket timing marks aligned with cylinder head face **5**.
   *NOTE: B20A3/4 – Align flywheel timing marks **10**. D16A/D16Z – Align camshaft sprocket timing marks **8**.*
4. Remove crankshaft pulley **1**.
5. Ensure crankshaft sprocket timing marks aligned **9**.
6. Slacken tensioner bolt **6**. Move tensioner away from belt and lightly tighten bolt.
7. Remove timing belt.

Installation

1. Ensure timing marks aligned **9** & **5** or **8**.
2. Fit timing belt.
3. Slacken and tighten tensioner bolt **6**.
4. Turn crankshaft six turns anti-clockwise. Ensure timing marks aligned **9** & **5** or **8**.
5. Slacken tensioner bolt **6**.
6. Turn crankshaft anti-clockwise for 3 teeth on camshaft sprocket.
7. Tighten tensioner bolt **6**.
   Tightening torque: 45 Nm.
8. Turn crankshaft nearly two turns anti-clockwise to TDC. Ensure timing marks aligned **9** & **5** or **8**.
9. Install components in reverse order of removal.
   NOTE: Ensure un-chamfered edge of crankshaft pulley bolt washer faces pulley.
10. Tighten crankshaft pulley bolt to 115 Nm **7**.
    Concerto: 165 Nm.
    NOTE: 1,5/1,6 1991→: Tighten bolt to 185 Nm.

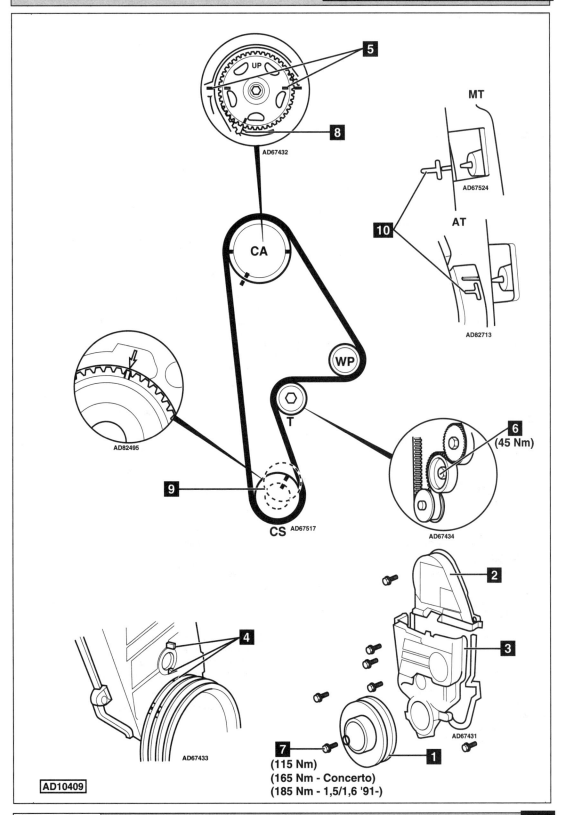

UP

5

8

AD67432

MT

AD67524

10

AT

AD82713

CA

WP

T

AD82495

6
(45 Nm)

9

CS AD67517

AD67434

2

3

4

AD67433

7
(115 Nm)
(165 Nm - Concerto)
(185 Nm - 1,5/1,6 '91-)

1

AD67431

AD10409

HONDA

Model:	Civic 1,4 • Civic 1,5 VTEC • Civic 1,6 • Civic 1,6 VTEC Civic Coupe 1,6 • CRX 1,6 • Accord 1,6 • HR-V • Logo
Year:	1995-03
Engine Code:	D13B7, D14A2, D14A3, D14A4, D14A5, D14A7, D14A8, D14Z5, D15Z6, D15Z8, D16B2, D16B6, D16W1, D16W2, D16W5, D16W7, D16Y5, D16Y7, D16Y8

Replacement Interval Guide

Honda recommend replacement as follows:
→07/98: Every 60,000 miles or 5 years, whichever occurs first.
07/98→: Every 72,000 miles or 8 years, whichever occurs first.
The previous use and service history of the vehicle must always be taken into account.
Refer to Timing Belt Replacement Intervals at the front of this manual.

Check For Engine Damage

CAUTION: This engine has been identified as an INTERFERENCE engine in which the possibility of valve-to-piston damage in the event of a timing belt failure is MOST LIKELY to occur.
A compression check of all cylinders should be performed before removing the cylinder head(s).

Repair Times – hrs

Remove & install:

Logo/Civic/CRX	2,10
Accord	2,10
HR-V	2,20
AC	+0,20

Special Tools

- None required.

Special Precautions

- Disconnect battery earth lead.
- DO NOT turn crankshaft or camshaft when timing belt removed.
- Remove spark plugs to ease turning engine.
- Turn engine in normal direction of rotation (unless otherwise stated).
- DO NOT turn engine via camshaft or other sprockets.
- Observe all tightening torques.

Removal

NOTE: Normal direction of crankshaft rotation is anti-clockwise.

1. Raise and support front of vehicle.
2. Remove:
   - LH front wheel.
   - Lower splash guard.
   - Auxiliary drive belt(s).
   - Dipstick and tube (if necessary).
3. Support engine.
4. Remove:
   - PAS pump. DO NOT disconnect hoses.
   - LH engine mounting.

5. Turn crankshaft to TDC on No.1 cylinder. Ensure timing marks aligned **1**.
6. Remove:
   - Crankshaft pulley bolt **2**.
   - Crankshaft pulley **3**.
   - Cylinder head cover.
   - Timing belt upper cover **4**.
   - Timing belt lower cover **5**.
   - Crankshaft position (CKP) sensor (if fitted).
7. Ensure camshaft sprocket timing marks aligned:
   - D13B7/D14A3/4/D15Z6/D16Y5/7/8: With arrows on cylinder head **6**.
   - Except D13B7/D14A3/4/D15Z6/D16Y5/7/8: With upper cylinder head face **7**.
8. Ensure 'UP' mark on camshaft sprocket at top.
9. Ensure crankshaft sprocket timing marks aligned:
   - D14A2/3/4/5/D15Z6/D16Y5/7/8: **8**.
   - Except D14A2/3/4/5/D15Z6/D16Y5/7/8: **9**.
10. Slacken tensioner bolt **10**. Move tensioner away from belt and lightly tighten bolt.
11. Remove timing belt.

Installation

1. Check condition of water pump.
2. Ensure crankshaft sprocket timing marks aligned **8** or **9**.
3. Ensure camshaft sprocket timing marks aligned **6** or **7**.
4. Ensure 'UP' mark on camshaft sprocket at top.
5. Fit timing belt in anti-clockwise direction, starting at crankshaft sprocket.
6. Ensure belt is taut between sprockets on non-tensioned side.
7. Slacken and tighten tensioner bolt **10**.
8. Turn crankshaft six turns anti-clockwise. Ensure crankshaft sprocket timing marks aligned **8** or **9**.
9. Ensure camshaft sprocket timing marks aligned **6** or **7**.
10. Slacken tensioner bolt **10**.
11. Turn crankshaft anti-clockwise for 3 teeth on camshaft sprocket.
12. Tighten tensioner bolt to 44 Nm **10**.
13. Turn crankshaft anti-clockwise until timing marks aligned **6** or **7** & **8** or **9**.
14. Install components in reverse order of removal.
15. Tighten crankshaft pulley bolt **2**:
    - D14A2/3/5: Tightening torque: 181-186 Nm.
    - Except D14A2/3/5:
      Tightening torque: 20 Nm + 90°.

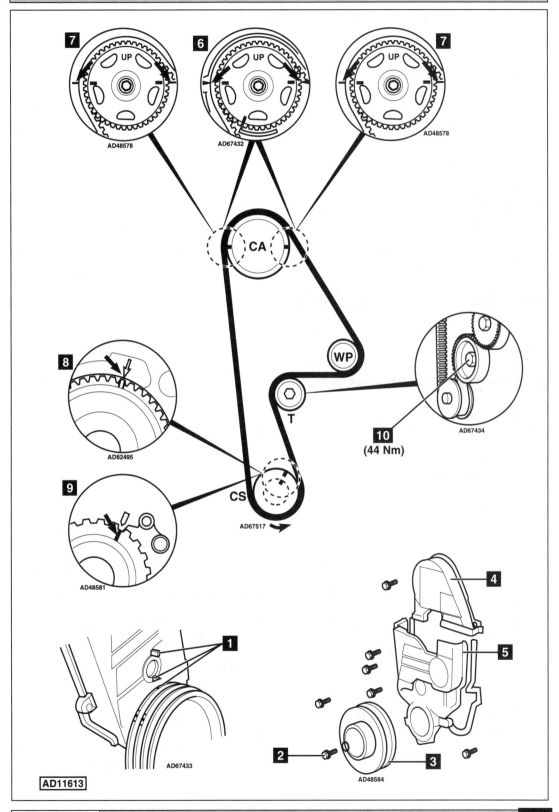

HONDA

Model:	Civic 1,4 • Civic 1,6 • Civic Coupe 1,7
Year:	2001-03
Engine Code:	D14Z5, D16W7, D17A2, D17A8, D17A9

Replacement Interval Guide

Honda recommend replacement every
72,000 miles or 8 years, whichever occurs
*The previous use and service history of the vehicle
must always be taken into account.
Refer to Timing Belt Replacement Intervals at the front
of this manual.*

Check For Engine Damage

*CAUTION: This engine has been identified as an
INTERFERENCE engine in which the possibility of
valve-to-piston damage in the event of a timing belt
failure is MOST LIKELY to occur.
A compression check of all cylinders should be
performed before removing the cylinder head(s).*

Repair Times – hrs

Remove & install:

Coupe	2,50
Civic	2,30
Hydraulic PAS	+0,20

Special Tools

- Crankshaft pulley holder (Hydraulic PAS) –
  Honda Nos.07JAB-0010200/07MAB-PY30100.
- Crankshaft pulley holder (Electronic PAS) –
  Honda Nos.07JAB-0010200/07JAB0010400.

Special Precautions

- Disconnect battery earth lead.
- DO NOT turn crankshaft or camshaft when timing belt
  removed.
- Remove spark plugs to ease turning engine.
- Turn engine in normal direction of rotation (unless
  otherwise stated).
- DO NOT turn engine via camshaft or other sprockets.
- Observe all tightening torques.

Removal

*NOTE: Normal direction of crankshaft rotation is
anti-clockwise.*

1. Raise and support front of vehicle.
2. Remove:
   ❏ LH front wheel.
   ❏ Lower splash guard.
   ❏ Auxiliary drive belt(s).
   ❏ Dipstick and tube (if necessary).
3. Support engine.
4. Remove:
   ❏ PAS pump. DO NOT disconnect hoses.
   ❏ Alternator.
   ❏ LH engine mounting & bracket.

5. Turn crankshaft to TDC on No.1 cylinder. Ensure
   timing marks aligned **1**.
   NOTE: Hydraulic PAS crankshaft pulley shown **1**.
6. Ensure 'UP' mark on camshaft sprocket at top.
7. Remove:
   ❏ Cylinder head cover.
   ❏ Timing belt upper cover **2**.
8. Ensure camshaft sprocket timing marks
   aligned **3**.
9. Remove crankshaft pulley bolt **4**:
   ❏ Hydraulic PAS –
      Tool Nos.07JAB-0010200/07MAB-PY30100.
   ❏ Electronic PAS –
      Tool Nos.07JAB-0010200/07JAB0010400.
10. Remove:
    ❏ Crankshaft pulley **5**.
    ❏ Timing belt lower cover **6**.
    ❏ Crankshaft position (CKP) sensor.
11. Ensure crankshaft sprocket timing marks
    aligned **7**.
12. Inset Allen key into tensioner and turn
    anti-clockwise to release tension on belt **8**.
13. Remove timing belt.

Installation

1. Inspect condition of water pump.
2. Ensure crankshaft sprocket timing marks
   aligned **7**.
3. Ensure camshaft sprocket timing marks
   aligned **3**.
4. Ensure 'UP' mark on camshaft sprocket at top.
5. Turn Allen key clockwise **8** until holes
   aligned **9**.
6. Insert 3mm pin **10** through holes in tensioner **9**.
7. Slacken tensioner bolt 1/2 turn **11**.
8. Fit timing belt in anti-clockwise direction, starting
   at crankshaft sprocket.
9. Ensure belt is taut between sprockets on
   non-tensioned side.
10. Turn crankshaft six turns anti-clockwise. Ensure
    crankshaft sprocket timing marks aligned **7**.
11. Ensure camshaft sprocket timing marks
    aligned **3**.
12. Tighten tensioner bolt **11**:
    Tightening torque 44 Nm.
13. Remove 3mm pin from tensioner **10**.
14. Install components in reverse order of removal.
15. Tighten crankshaft pulley bolt **4**:
    Tightening torque: 20 Nm + 90°.

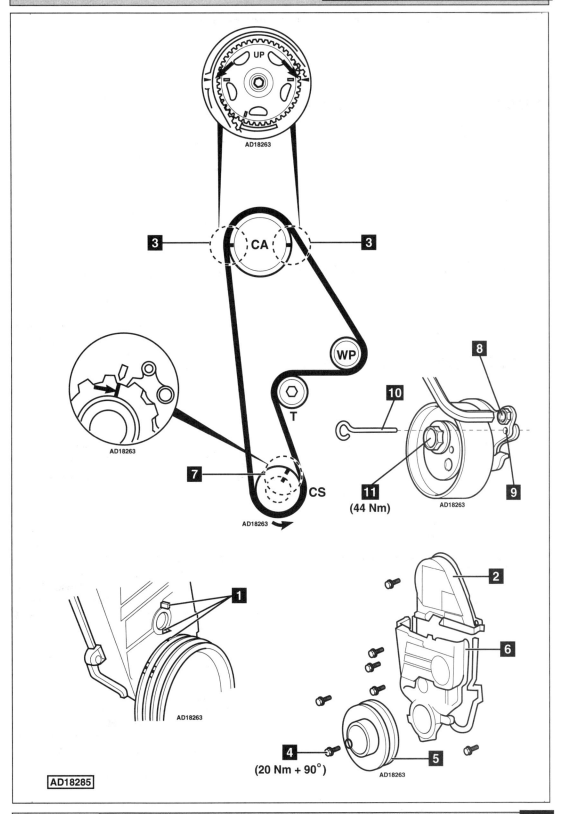

UP

AD18263

3   CA   3

WP

T

8

10

11
(44 Nm)

9

AD18263

7

CS

AD18263

AD18263

1

4
(20 Nm + 90°)

5

2

6

AD18263

AD18285

HONDA

Model:	Civic 1,5 VT/V-TEC • Civic 1,6 SOHC/V-TEC
Year:	1991-98
Engine Code:	D15Z1, D15Z3, D16Y3, D16Y2

Replacement Interval Guide

Honda recommend:
→93MY: Check & replace if necessary every 12,000 miles.
→7/98: Replacement every 60,000 miles or 5 years – replace.
7/98→: Replacement every 72,000 miles or 8 years – replace.
The previous use and service history of the vehicle must always be taken into account.
Refer to Timing Belt Replacement Intervals at the front of this manual.

Check For Engine Damage

CAUTION: This engine has been identified as an INTERFERENCE engine in which the possibility of valve-to-piston damage in the event of a timing belt failure is MOST LIKELY to occur.
A compression check of all cylinders should be performed before removing the cylinder head.

Repair Times – hrs

Remove & install:	
→1995	2,50
1995→	2,10
AC	+0,20

Special Tools

- None required.

Special Precautions

- Disconnect battery earth lead.
- DO NOT turn crankshaft or camshaft when timing belt removed.
- Remove spark plugs to ease turning engine.
- Turn engine in normal direction of rotation (unless otherwise stated).
- DO NOT turn engine via camshaft or other sprockets.
- Observe all tightening torques.

Removal

NOTE: Normal direction of crankshaft rotation is anti-clockwise.

1. Turn crankshaft to TDC on No.1 cylinder. Ensure timing marks aligned **1**.
2. Support engine.
3. Remove:
   - ❏ LH inner wing panel.
   - ❏ Auxiliary drive belt(s).
   - ❏ PAS pump. DO NOT disconnect hoses.
   - ❏ LH engine mounting.
   - ❏ Crankshaft pulley **2**.
   - ❏ Crankshaft sprocket guide washer.
   - ❏ Cylinder head cover.
   - ❏ Timing belt upper cover **3**.
   - ❏ Timing belt lower cover **4**.
4. Ensure camshaft sprocket timing marks aligned with cylinder head arrow marks (1,5 engine) **5** or pointer (1,6 engine) **6**.
5. Ensure crankshaft sprocket timing marks aligned **7**.
6. Slacken tensioner bolt **8**. Move tensioner away from belt and lightly tighten bolt.
7. Remove timing belt.

Installation

1. Ensure crankshaft sprocket timing marks aligned **7**.
2. Ensure relevant camshaft sprocket timing marks aligned **5** or **6**.
3. Fit timing belt in anti-clockwise direction, starting at crankshaft sprocket. Ensure belt is taut between sprockets on non-tensioned side.
4. Slacken and tighten tensioner bolt **8**.
5. Turn crankshaft six turns anti-clockwise. Ensure crankshaft sprocket timing marks aligned **7**.
6. Ensure relevant camshaft sprocket timing marks aligned **5** or **6**.
7. Slacken tensioner bolt **8**.
8. Turn crankshaft anti-clockwise for 3 teeth on camshaft sprocket.
9. Tighten tensioner bolt **8**.
   Tightening torque: 44 Nm.
10. Turn crankshaft anti-clockwise to TDC. Ensure timing marks aligned **7** & **5** or **6**.
11. Install components in reverse order of removal.
    NOTE: Ensure convex side of crankshaft sprocket guide washer faces belt.
12. Tighten crankshaft pulley bolt **9**.
    Tightening torque: 186 Nm.

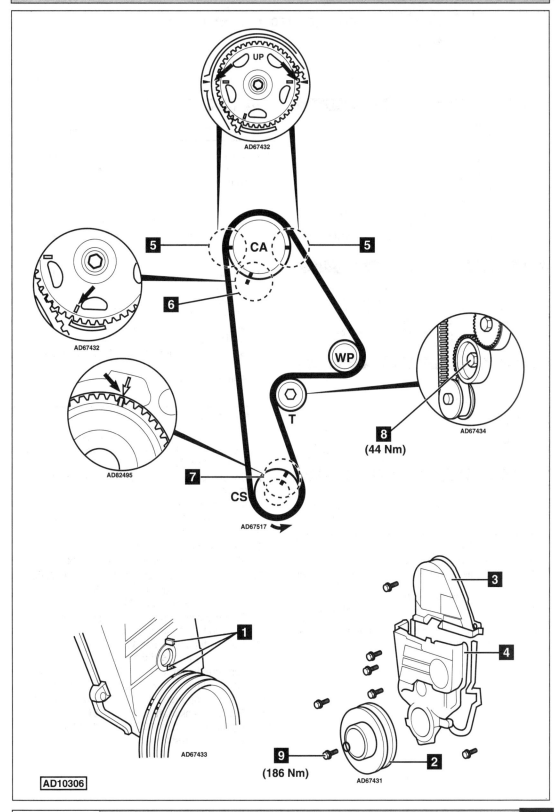

UP

AD67432

5 | CA | 5

6

AD67432

WP

T

8
(44 Nm)

AD67434

AD82495

7

CS

AD67517

3

1

4

AD67433

9
(186 Nm)

2

AD67431

AD10306

Model:	Civic 1,6 Vti • Civic 1,6 V-TEC • Integra Type R
Year:	1990-01
Engine Code:	B16A, B18C6

Replacement Interval Guide

Honda recommend:
→93MY: Check & replace if necessary every 12,000 miles.
→7/98: Replacement every 60,000 miles or 5 years – replace.
7/98→: Replacement every 72,000 miles or 8 years – replace.
The previous use and service history of the vehicle must always be taken into account.
Refer to Timing Belt Replacement Intervals at the front of this manual.

Check For Engine Damage

CAUTION: This engine has been identified as an INTERFERENCE engine in which the possibility of valve-to-piston damage in the event of a timing belt failure is MOST LIKELY to occur.
A compression check of all cylinders should be performed before removing the cylinder head.

Repair Times – hrs

Check & adjust	0,30
Remove & install:	
Civic	
→1990	2,20
1991→	2,50
Integra	2,30

Special Tools

■ Camshaft locking pins –
2 x Honda No. 07744-0010400 (5 mm pin punch).

Special Precautions

■ Disconnect battery earth lead.
■ DO NOT turn crankshaft or camshaft when timing belt removed.
■ Remove spark plugs to ease turning engine.
■ Turn engine in normal direction of rotation (unless otherwise stated).
■ DO NOT turn engine via camshaft or other sprockets.
■ Observe all tightening torques.

Removal

NOTE: Normal direction of crankshaft rotation is anti-clockwise.

1. Turn crankshaft to TDC on No.1 cylinder. Ensure timing marks aligned **1** & **2**.
2. Remove:
   ❏ LH front wheel arch liner.
   ❏ Auxiliary drive belts.
   ❏ LH engine rubber mounting.
   ❏ Spark plug lead cover and spark plug leads.
   ❏ Cylinder head cover.
   ❏ Timing belt centre cover **5**.
3. Ensure 'UP' mark on camshaft sprockets at top **3**.
4. Remove:
   ❏ Crankshaft pulley bolt **9**.
   ❏ Crankshaft pulley **6**.
   ❏ Timing belt lower cover **4**.
5. Slacken tensioner bolt **7**. Move tensioner away from belt and tighten bolt.
6. Remove timing belt.

Installation

1. Temporarily fit timing belt lower cover and crankshaft pulley **4** & **6**. Ensure timing marks aligned **1** & **2**.
2. Ensure 'UP' mark on camshaft sprockets at top **3**.
3. Move camshafts slightly if necessary and insert locking pins **8**. Tool No.07744-0010400.
4. Fit new timing belt.
5. Remove locking pins **8**.
6. Slacken and tighten tensioner bolt **7**.
7. Turn crankshaft 4 turns anti-clockwise. Ensure timing marks aligned **1**, **2** & **3**.
8. Slacken tensioner bolt **7**.
9. Turn crankshaft anti-clockwise for 3 teeth on camshaft sprocket to tension belt.
10. Tighten tensioner bolt **7**.
    Tightening torque: 55 Nm.
11. Turn crankshaft almost 2 turns anti-clockwise. Ensure timing marks aligned **1**, **2** & **3**.
12. Install components in reverse order of removal.
13. Tighten crankshaft pulley bolt **9**.
    Tightening torque: 120 Nm – 1990-91.
    180 Nm – 1992→.

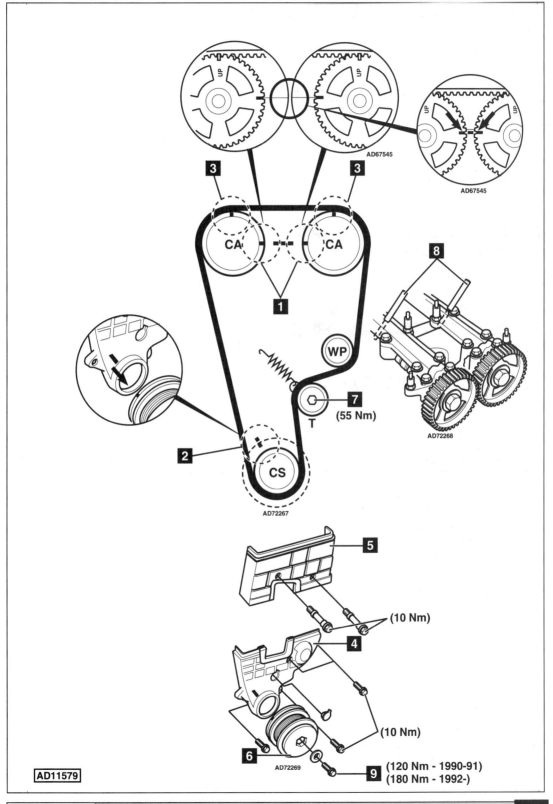

AD67545

AD67545

3

3

CA — CA

1

8

WP

7 (55 Nm)

T

AD72268

2

CS

AD72267

5

(10 Nm)

4

(10 Nm)

6

AD72269

9 (120 Nm - 1990-91)
(180 Nm - 1992-)

AD11579

HONDA

Model:	**Civic 1,6i VTEC • Civic 1,8i VTEC**
Year:	**1996-00**
Engine Code:	**B16A2, B18C4**

Replacement Interval Guide

Honda recommend replacement as follows:
→07/98: Every 60,000 miles or 5 years –
replace.
07/98→: Every 72,000 miles or 8 years –
replace.
The previous use and service history of the vehicle must always be taken into account.
Refer to Timing Belt Replacement Intervals at the front of this manual.

Check For Engine Damage

CAUTION: This engine has been identified as an INTERFERENCE engine in which the possibility of valve-to-piston damage in the event of a timing belt failure is MOST LIKELY to occur.
A compression check of all cylinders should be performed before removing the cylinder head(s).

Repair Times – hrs

Remove & install	2,30
AC	+0,20

Special Tools

■ None required.

Special Precautions

- ■ Disconnect battery earth lead.
- ■ DO NOT turn crankshaft or camshaft when timing belt removed.
- ■ Remove spark plugs to ease turning engine.
- ■ Turn engine in normal direction of rotation (unless otherwise stated).
- ■ DO NOT turn engine via camshaft or other sprockets.
- ■ Observe all tightening torques.

Removal

NOTE: Normal direction of crankshaft rotation is anti-clockwise.

1. Remove:
   ❑ LH front wheel arch liner.
   ❑ Auxiliary drive belts.
   ❑ PAS pump. DO NOT disconnect hoses.
2. Support engine.
3. Remove:
   ❑ LH engine mounting.
   ❑ PAS pump lower bracket.
   ❑ Cylinder head cover.
   ❑ Timing belt centre cover **4**.
4. Turn crankshaft to TDC on No.1 cylinder. Ensure timing marks aligned **1** & **2**.

5. Ensure 'UP' mark on camshaft sprockets at top **5**.
6. Remove:
   ❑ Crankshaft pulley **6**.
   ❑ Timing belt lower cover **7**.
7. Slacken tensioner bolt **8**. Move tensioner away from belt and tighten bolt.
8. Remove timing belt.

Installation

1. Check condition of water pump.
2. Ensure timing marks aligned **1** & **3**.
3. Ensure 'UP' mark on camshaft sprockets at top **5**.
4. Fit timing belt in anti-clockwise direction, starting at crankshaft sprocket.
5. Slacken and tighten tensioner bolt **8**.
6. Turn crankshaft six turns anti-clockwise. Ensure crankshaft sprocket timing marks aligned **3**.
7. Slacken tensioner bolt **8**.
8. Turn crankshaft anti-clockwise for 3 teeth on camshaft sprocket.
9. Tighten tensioner bolt to 54 Nm **8**.
10. Turn crankshaft nearly two turns anti-clockwise. Ensure timing marks aligned **1**, **3** & **5**.
11. Install components in reverse order of removal.
12. Tighten crankshaft pulley bolt **9**. Tightening torque: 177 Nm.

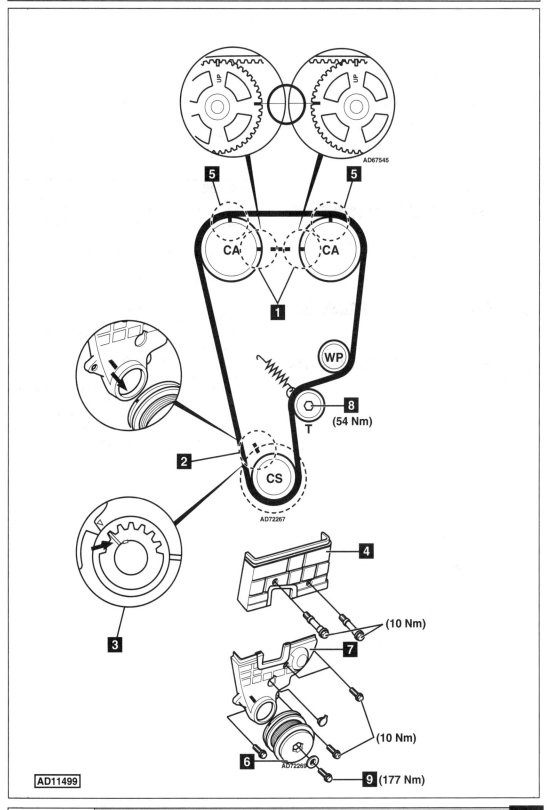

AD67545

AD72267

AD72269

(54 Nm)

(10 Nm)

(10 Nm)

(177 Nm)

Model:	Concerto 1,6 DOHC • Integra 1,6 DOHC
Year:	1988-95
Engine Code:	D16A2, D16A8, D16A9, D16Z4

Replacement Interval Guide

Honda recommend check & replace if necessary every 12,000 miles (→1993MY) and replacement every 60,000 miles or 5 years.
The previous use and service history of the vehicle must always be taken into account.
Refer to Timing Belt Replacement Intervals at the front of this manual.

Check For Engine Damage

CAUTION: This engine has been identified as an INTERFERENCE engine in which the possibility of valve-to-piston damage in the event of a timing belt failure is MOST LIKELY to occur.
A compression check of all cylinders should be performed before removing the cylinder head.

Repair Times – hrs

Check & adjust	0,30
Remove & install	2,30

Special Tools

- None required.

Special Precautions

- Disconnect battery earth lead.
- DO NOT turn crankshaft or camshaft when timing belt removed.
- Remove spark plugs to ease turning engine.
- Turn engine in normal direction of rotation (unless otherwise stated).
- DO NOT turn engine via camshaft or other sprockets.
- Observe all tightening torques.

Removal

NOTE: Normal direction of crankshaft rotation is anti-clockwise.

1. Remove:
   - Engine undershield.
   - Auxiliary drive belts.
   - Cylinder head cover.
   - Timing belt upper cover **1**.
2. Turn crankshaft anti-clockwise to TDC on No.1 cylinder. Ensure timing marks aligned **2** & **3**.
3. Remove:
   - LH top engine mounting.
   - Crankshaft pulley **6**.
   - Timing belt lower cover **4**.

4. Slacken tensioner bolt **5**. Move tensioner away from belt and lightly tighten bolt.
5. Remove:
   - Crankshaft sprocket outer guide washer.
   - Timing belt.

Installation

1. Ensure timing marks aligned **2** & **3**.
2. Fit timing belt to sprockets keeping slack on tensioner side.
3. Fit crankshaft sprocket guide washer (convex side towards belt).
4. Slacken tensioner bolt **5**.
5. Install:
   - LH engine mounting.
   - Timing belt lower cover.
6. Turn crankshaft anti-clockwise for 3 teeth on camshaft sprocket.
7. Tighten tensioner bolt **5**.
   Tightening torque: 45 Nm.
8. Turn crankshaft anti-clockwise to TDC. Ensure timing marks aligned **2** & **3**.
9. Install components in reverse order of removal.
10. Tighten crankshaft pulley bolt **7**.
    12 mm bolt: 115 Nm. 14 mm bolt: 165 Nm.

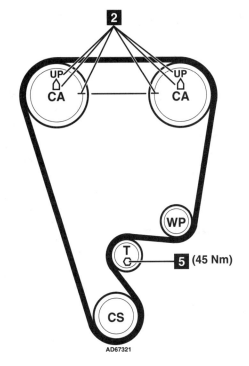

2

UP
CA

UP
CA

WP

T **5** (45 Nm)

CS

AD67321

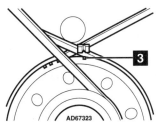

3

AD67323

1

4

6 **7** (115 Nm -12 mm)
(165 Nm -14 mm)

AD67324

AD10412

HONDA

Model:	Accord 1,8/2,0/2,2 • Prelude 2,0
Year:	1990-96
Engine Code:	F18A, F20A, F22A

Replacement Interval Guide

Honda recommend check & replace if necessary every 12,000 miles (→93MY) and replacement every 60,000 miles or 5 years.
The previous use and service history of the vehicle must always be taken into account.
Refer to Timing Belt Replacement Intervals at the front of this manual.

Check For Engine Damage

CAUTION: This engine has been identified as an INTERFERENCE engine in which the possibility of valve-to-piston damage in the event of a timing belt failure is MOST LIKELY to occur.
A compression check of all cylinders should be performed before removing the cylinder head.

Repair Times – hrs

Check & adjust	0,30
Remove & install:	
Accord	2,00
Prelude	2,30

Special Tools

- Balancer shaft locking pin – Honda No.07LAG-PT20100.

Special Precautions

- Disconnect battery earth lead.
- DO NOT turn crankshaft or camshaft when timing belt removed.
- Remove spark plugs to ease turning engine.
- Turn engine in normal direction of rotation (unless otherwise stated).
- DO NOT turn engine via camshaft or other sprockets.
- Observe all tightening torques.

Removal

NOTE: Normal direction of crankshaft rotation is anti-clockwise. Balancer shaft belt must be removed before removing timing belt.

1. Turn crankshaft to TDC on No.1 cylinder. Ensure flywheel timing marks aligned **1**.
2. Remove:
   - LH front wheel arch liner.
   - Cruise control actuator.
   - PAS drive belt.
   - PAS pump. DO NOT disconnect hoses.
   - Alternator drive belt.
   - Cylinder head cover.
   - Timing belt upper cover **3**.
   - LH engine mounting.
   - Dipstick and tube.
   - Adjusting nut **6**.
   - Crankshaft pulley **5**.
   - Timing belt lower cover **4**.
3. Ensure camshaft sprocket timing marks aligned with cylinder head face **2**.
4. Depress balancer shaft belt tensioner spring. Remove balancer shaft belt.
5. Depress timing belt tensioner spring. Remove timing belt.

Installation

1. Ensure flywheel timing marks aligned **1**.
2. Ensure 'UP' mark on camshaft sprocket at top **7**. Ensure camshaft sprocket timing marks aligned with cylinder head face **2**.
3. Fit new timing belt.
4. Align front balancer shaft timing marks **10**.
5. Remove blanking plug **8**. Insert locking pin into rear balancer shaft **9**.
   Tool No.07LAG-PT20100. Turn balancer shaft until locking pin is located in hole in shaft.
6. Fit new balancer shaft belt.
7. Install:
   - Timing belt lower cover **4**.
   - Adjusting nut **6**.
8. Remove locking pin **9**. Fit blanking plug **8**.
9. Slacken adjusting nut **6**.
10. Ensure flywheel timing marks aligned **1**. Turn crankshaft 3 teeth anti-clockwise to tension timing belt and balancer shaft belt.
11. Tighten adjusting nut to 45 Nm **6**.
12. Install components in reverse order of removal.
13. Tighten crankshaft pulley bolt **11**.
    Tightening torque: 220 Nm.

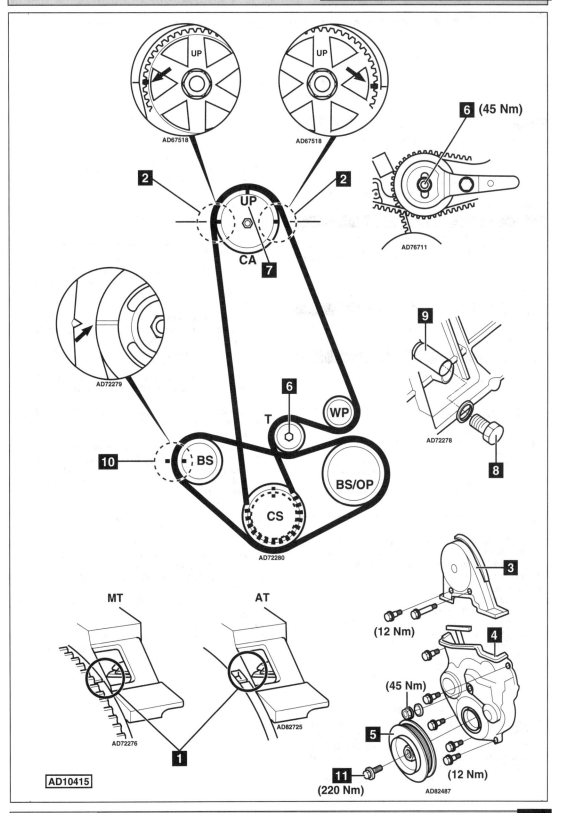

AD67518

AD67518

6 (45 Nm)

2

2

UP

CA

7

AD76711

AD72279

6

T

WP

9

AD72278

8

10

BS

BS/OP

CS

AD72280

3

(12 Nm)

4

MT

AT

(45 Nm)

5

AD72276

AD82725

1

11

(220 Nm)

(12 Nm)

AD82487

AD10415

Model:	Accord 1,8 • Accord 2,0 • Accord 2,2 VTEC • Accord Coupe 2,0/2,2 Accord Aerodeck 2,0/2,2
Year:	1993-03
Engine Code:	F18A3, F18B2, F20B3, F20B6, F20Z1, F20Z2, F20Z3, F22B5, F22B6, F22Z2

Replacement Interval Guide

Honda recommend replacement as follows:
→07/98: Every 60,000 miles or 5 years – replace.
07/98→: Every 72,000 miles or 8 years – replace.
The previous use and service history of the vehicle must always be taken into account.
Refer to Timing Belt Replacement Intervals at the front of this manual.

Check For Engine Damage

CAUTION: This engine has been identified as an INTERFERENCE engine in which the possibility of valve-to-piston damage in the event of a timing belt failure is MOST LIKELY to occur.
A compression check of all cylinders should be performed before removing the cylinder head.

Repair Times – hrs

Check & adjust	0,30
Remove & install	2,30

Special Tools

- Balancer shaft locking pin – Honda No.07LAG-PT20100.

Special Precautions

- Disconnect battery earth lead.
- DO NOT turn crankshaft or camshaft when timing belt removed.
- Remove spark plugs to ease turning engine.
- Turn engine in normal direction of rotation (unless otherwise stated).
- DO NOT turn engine via camshaft or other sprockets.
- Observe all tightening torques.

Removal

NOTE: Normal direction of crankshaft rotation is anti-clockwise. Balancer shaft belt must be removed before removing timing belt.

1. Support engine.
2. Remove:
   - ❑ Lower splash guard.
   - ❑ Auxiliary drive belts.
   - ❑ PAS pump. DO NOT disconnect hoses.
   - ❑ Cylinder head cover.
   - ❑ Timing belt upper cover **3**.
   - ❑ LH engine mounting.
   - ❑ Dipstick and tube.
   - ❑ Rubber seal for adjusting nut **6**. DO NOT slacken nut.
   - ❑ Crankshaft pulley **5**.
   - ❑ Timing belt lower cover **4**.
3. Temporarily fit crankshaft pulley. Turn crankshaft to TDC on No.1 cylinder. Ensure white flywheel timing mark and camshaft timing marks aligned **1** & **2**. Ensure 'UP' mark on camshaft sprocket at top **12**.
   *NOTE: F22Z2 engines have TDC marks on crankshaft pulley and timing belt lower cover **13**.*

4. Fit 6 mm bolt at **A** to lock tensioner arm.
5. Slacken tensioner nut **7**. Move tensioner away from belt. Lightly tighten nut.
6. Remove balancer shaft belt.
7. Slacken 6 mm bolt **A**. Slacken tensioner nut **7**. Move tensioner away from belt and lightly tighten nut.
8. Remove:
   - ❑ Crankshaft pulley **5**.
   - ❑ Timing belt.

Installation

1. Ensure timing marks aligned **1**, **2** & **13**.
2. Fit timing belt in anti-clockwise direction, starting at crankshaft sprocket. Ensure belt is taut between sprockets.
3. Ensure 6 mm bolt is slack **A**.
4. Slacken tensioner nut and tighten to 45 Nm **7**.
5. Temporarily fit crankshaft pulley. Turn crankshaft slowly six turns anti-clockwise to settle belt. Ensure timing marks aligned.
6. Turn crankshaft slowly anti-clockwise until three teeth of camshaft sprocket have passed timing marks **2**.
7. Slacken tensioner nut and tighten to 45 Nm **7**.
8. Turn crankshaft to TDC on No.1 cylinder. Ensure timing marks aligned.
9. Tighten 6 mm bolt **A** to lock tensioner arm.
10. Remove blanking plug **8**. Insert balancer shaft locking pin. Tool No.07LAG-PT20100 **9**. Turn rear balancer shaft sprocket until locking pin locates in hole in shaft.
11. Align front balancer shaft timing marks **10**.
12. Remove crankshaft pulley **5**.
13. Fit balancer shaft belt.
14. Slacken tensioner nut **7**.
15. Remove balancer shaft locking pin **9**. Fit blanking plug and tighten to 30 Nm.
16. Fit crankshaft pulley **5**. Turn crankshaft one turn anti-clockwise.
17. Tighten tensioner nut **7**. Tightening torque: 45 Nm.
18. Remove 6 mm bolt **A**.
19. Install components in reverse order of removal.
20. Tighten crankshaft pulley bolt **11**.
    Tightening torque: F18A3/F20Z1/2/3: 220 Nm.
    Except F18A3/F20Z1/2/3: 245-250 Nm.

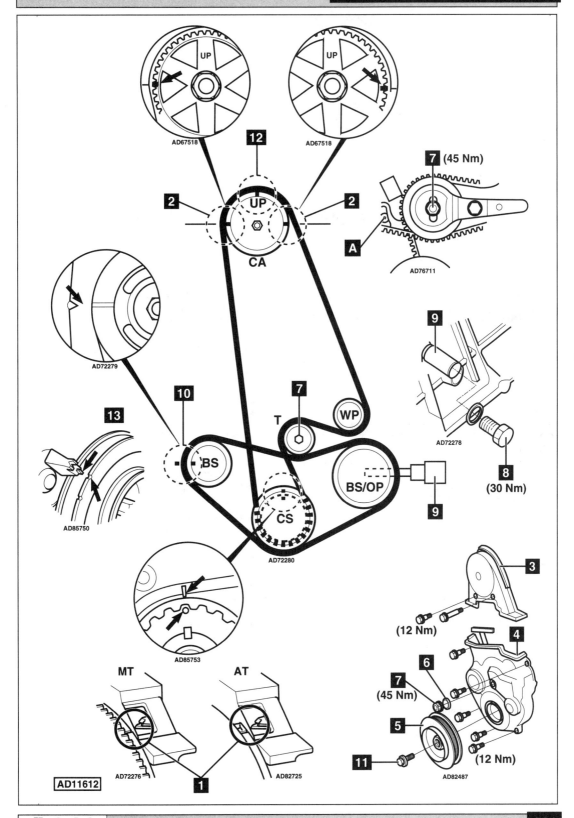

Model:	**Accord 2,0 16V • Prelude 2,0 16V**
Year:	**1987-92**
Engine Code:	**B20A5, B20A7**

Replacement Interval Guide

Honda recommend check and replace if necessary every 12,000 miles and replacement every 60,000 miles or 5 years.
The previous use and service history of the vehicle must always be taken into account.
Refer to Timing Belt Replacement Intervals at the front of this manual.

Check For Engine Damage

CAUTION: This engine has been identified as an INTERFERENCE engine in which the possibility of valve-to-piston damage in the event of a timing belt failure is MOST LIKELY to occur.
A compression check of all cylinders should be performed before removing the cylinder head.

Repair Times – hrs

Check & adjust:	
Prelude	0,40
Accord	0,30
Remove & install:	
Prelude	2,30
Accord	2,10

Special Tools

- Camshaft locking pins –
  2 x Honda No. 07744-0010400 (5 mm pin punch).

Special Precautions

- Disconnect battery earth lead.
- DO NOT turn crankshaft or camshaft when timing belt removed.
- Remove spark plugs to ease turning engine.
- Turn engine in normal direction of rotation (unless otherwise stated).
- DO NOT turn engine via camshaft or other sprockets.
- Observe all tightening torques.

Removal

NOTE: Normal direction of crankshaft rotation is anti-clockwise.

1. Remove:
   - Engine side support bracket and rubber mounting.
   - Engine undershield.
   - Auxiliary drive belts.
   - PAS drive belt adjuster.
   - PAS pump.
   - Alternator.

2. Disconnect:
   - HT cables from spark plugs.
   - Cable protector from cylinder head cover.

3. Remove:
   - Cylinder head cover.
   - Timing belt centre cover **4**.

4. Turn crankshaft to TDC on No.1 cylinder. Ensure timing marks aligned **1** & **2**.

5. Ensure 'UP' mark on camshaft sprockets at top **3**.

6. Select gear (MT) or P (AT).

7. Remove:
   - Crankshaft pulley bolt and washer **5**.
   - Crankshaft pulley **6**.
   - Timing belt lower cover **7**.

8. Push locking pins into front camshaft housings and into camshafts to hold them at TDC. Tool No.07744-0010400.

9. Slacken tensioner bolt **8**. Push tensioner against spring and tighten bolt.

10. Remove timing belt.

Installation

1. Ensure timing marks aligned **1** & **2**.

2. Ensure camshafts locked in position. Ensure 'UP' mark on camshaft sprockets at top **3**. Ensure timing marks on camshaft sprockets aligned with upper cylinder head face.

3. Fit new timing belt.

4. Slacken tensioner bolt **8**.

5. Remove locking pins.

6. Turn crankshaft 1/4 turn anti-clockwise to tension belt.

7. Tighten tensioner bolt **8**.
   Tightening torque: 43 Nm.

8. Install components in reverse order of removal.

9. Tighten crankshaft pulley bolt **5**.
   Tightening torque: 115 Nm.

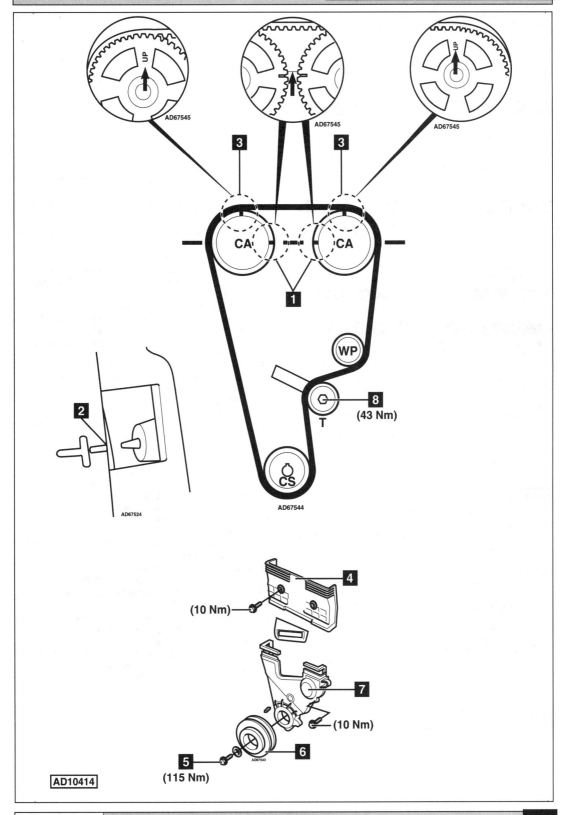

AD67545

AD67545

AD67545

3

3

CA

CA

1

WP

8
(43 Nm)

T

2

AD67524

CS

AD67544

4

(10 Nm)

7

(10 Nm)

5

6

AD67543

(115 Nm)

AD10414

HONDA

Model:	**Accord Coupe 2,0 • Prelude 2,0**
Year:	**1997-03**
Engine Code:	**F20A4, F20B5**

Check For Engine Damage

CAUTION: This engine has been identified as an INTERFERENCE engine in which the possibility of valve-to-piston damage in the event of a timing belt failure is MOST LIKELY to occur.
A compression check of all cylinders should be performed before removing the cylinder head(s).

Repair Times – hrs

Remove & install	2,30

Special Tools

- ■ Balancer shaft locking pin – Honda No.07LAG-PT20100.

Special Precautions

- ■ Disconnect battery earth lead.
- ■ DO NOT turn crankshaft or camshaft when timing belt removed.
- ■ Remove spark plugs to ease turning engine.
- ■ Turn engine in normal direction of rotation (unless otherwise stated).
- ■ DO NOT turn engine via camshaft or other sprockets.
- ■ Observe all tightening torques.

Removal

NOTE: Normal direction of crankshaft rotation is anti-clockwise.
NOTE: Balancer shaft belt must be removed before removing timing belt.

1. Support engine.
2. Remove:
   - ❑ LH splash guard.
   - ❑ Auxiliary drive belts.
   - ❑ PAS pump. DO NOT disconnect hoses.
   - ❑ Cylinder head cover.
   - ❑ Timing belt upper cover **1**.
   - ❑ LH engine mounting.
   - ❑ Dipstick and tube.
   - ❑ Rubber seal for adjusting nut **2**. DO NOT slacken nut.
3. Turn crankshaft to TDC on No.1 cylinder. Ensure timing marks aligned **5** & **6**.
4. Ensure 'UP' mark on camshaft sprocket at top **7**.
5. Remove:
   - ❑ Crankshaft pulley **4**.
   - ❑ Timing belt lower cover **3**.

6. Fit 6 mm bolt at **A** to lock tensioner arm.
7. Slacken tensioner nut **8**.
8. Move tensioner away from belt. Lightly tighten nut.
9. Remove balancer shaft belt.
10. Slacken 6 mm bolt **A**.
11. Slacken tensioner nut **8**.
12. Move tensioner away from belt. Lightly tighten nut.
13. Remove:
    - ❑ Balancer shaft belt sprocket from crankshaft **9**.
    - ❑ Timing belt.

Installation

1. Ensure timing marks aligned **6** & **10**.
2. Fit timing belt in anti-clockwise direction, starting at crankshaft sprocket. Ensure belt is taut between sprockets.
3. Ensure 6 mm bolt is slack **A**.
4. Slacken tensioner nut and tighten to 44 Nm **8**.
5. Temporarily fit crankshaft pulley **4**.
6. Turn crankshaft slowly six turns anti-clockwise to settle belt.
7. Ensure timing marks aligned **6** & **10**.
8. Turn crankshaft slowly anti-clockwise until three teeth of camshaft sprocket have passed timing marks **6**.
9. Slacken tensioner nut and tighten to 44 Nm **8**.
10. Turn crankshaft to TDC on No.1 cylinder. Ensure timing marks aligned **6** & **10**.
11. Tighten 6 mm bolt **A** to lock tensioner arm.
12. Remove blanking plug **11**.
13. Insert balancer shaft locking pin **12**. Tool No.07LAG-PT20100. Turn rear balancer shaft sprocket until locking pin locates in hole in shaft.
14. Align front balancer shaft timing marks **13**.
15. Remove crankshaft pulley **4**.
16. Install:
    - ❑ Balancer shaft belt sprocket onto crankshaft **9**.
    - ❑ Balancer shaft belt.
17. Slacken tensioner nut **8**.
18. Remove balancer shaft locking pin **12**.
19. Fit blanking plug and tighten to 29 Nm **11**.
20. Fit crankshaft pulley **4**.
21. Turn crankshaft one turn anti-clockwise.
22. Tighten tensioner nut to 44 Nm **8**.
23. Remove 6 mm bolt **A**.
24. Install components in reverse order of removal.
25. Tighten crankshaft pulley bolt **14**.
    Tightening torque: 245 Nm.

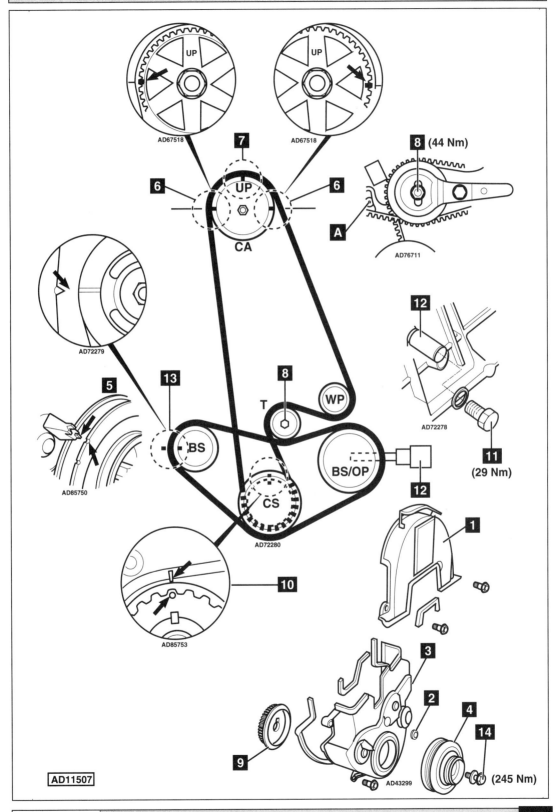

AD67518

AD67518

7

UP

8 (44 Nm)

6

6

UP

CA

A

AD76711

AD72279

12

AD72278

5

13

8

T

WP

BS

AD85750

BS/OP

12

11
(29 Nm)

CS

AD72280

1

10

AD85753

3

2

4

14

9

AD43299

(245 Nm)

AD11507

HONDA

Model:	Accord 2,3i • Prelude 2,3i
Year:	1992-96
Engine Code:	H23A2, H23A3

Replacement Interval Guide

Honda recommend replacement as follows:
Every 60,000 miles or 5 years – replace.
*The previous use and service history of the vehicle
must always be taken into account.*
*Refer to Timing Belt Replacement Intervals at the front
of this manual.*

Check For Engine Damage

*CAUTION: This engine has been identified as an
INTERFERENCE engine in which the possibility of
valve-to-piston damage in the event of a timing belt
failure is MOST LIKELY to occur.*
*A compression check of all cylinders should be
performed before removing the cylinder head.*

Repair Times – hrs

Check & adjust:

Accord	0,30
Prelude	0,40
Remove & install:	
All models	2,40

Special Tools

■ Balancer shaft locking pin –
  Honda No.07LAG-PT20100.

Special Precautions

■ Disconnect battery earth lead.
■ DO NOT turn crankshaft or camshaft when timing belt
  removed.
■ Remove spark plugs to ease turning engine.
■ Turn engine in normal direction of rotation (unless
  otherwise stated).
■ DO NOT turn engine via camshaft or other sprockets.
■ Observe all tightening torques.

Removal

*NOTE: Normal direction of crankshaft rotation is
anti-clockwise. Balancer shaft belt must be
removed before removing timing belt.*

1. Support engine.
2. Remove:
   ❑ Lower splash guard.
   ❑ Auxiliary drive belts.
   ❑ PAS pump. DO NOT disconnect hoses.
   ❑ Cylinder head cover.
   ❑ Timing belt upper cover **1**.
   ❑ LH engine mounting.
   ❑ Dipstick and tube.
   ❑ Rubber seal for adjusting nut **2**. DO NOT
     slacken nut.
   ❑ Crankshaft pulley **3**.
   ❑ Timing belt lower cover **4**.

3. Temporarily fit crankshaft pulley. Turn
   crankshaft anti-clockwise to TDC on
   No.1 cylinder. Ensure flywheel timing marks and
   timing marks on camshaft sprockets aligned **5**
   & **6**. Ensure 'UP' mark on camshaft sprockets at
   top **7**.
4. Fit 6 mm bolt at **A** to lock tensioner arm.
5. Slacken tensioner nut **8**. Move tensioner away
   from belt and lightly tighten nut.
6. Remove balancer shaft belt.
7. Slacken 6 mm bolt **A**. Slacken tensioner nut **8**.
   Move tensioner away from belt and lightly
   tighten nut.
8. Remove:
   ❑ Crankshaft pulley **3**.
   ❑ Timing belt.

Installation

1. Ensure timing marks aligned **5** & **6**.
2. Fit timing belt in clockwise direction, starting at
   crankshaft sprocket. Ensure belt is taut between
   sprockets.
3. Ensure 6 mm bolt is slack **A**.
4. Slacken tensioner nut and tighten to 45 Nm **8**.
5. Temporarily fit crankshaft pulley. Turn
   crankshaft slowly anti-clockwise until three teeth
   of camshaft sprockets have passed timing
   marks **6**.
6. Slacken tensioner nut and tighten to 45 Nm **8**.
7. Turn crankshaft anti-clockwise to TDC on
   No.1 cylinder. Ensure timing marks aligned **5**
   & **6**.
8. Tighten 6 mm bolt **A** to lock tensioner arm.
9. Remove blanking plug **9**. Insert balancer shaft
   locking pin **10**. Tool No.07LAG-PT20100. Turn
   rear balancer shaft sprocket until locking pin
   locates in hole in shaft.
10. Align front balancer shaft timing marks **11**.
11. Remove crankshaft pulley.
12. Fit balancer shaft belt.
13. Slacken tensioner nut **8**.
14. Remove balancer shaft locking pin **10**. Fit
    blanking plug and tighten to 30 Nm.
15. Fit crankshaft pulley. Turn crankshaft one turn
    anti-clockwise.
16. Tighten tensioner nut **8**.
    Tightening torque: 45 Nm.
17. Remove 6 mm bolt **A**.
18. Install components in reverse order of removal.
19. Lightly oil threads and contact face of crankshaft
    pulley bolt **12**. Tightening torque: 220 Nm.

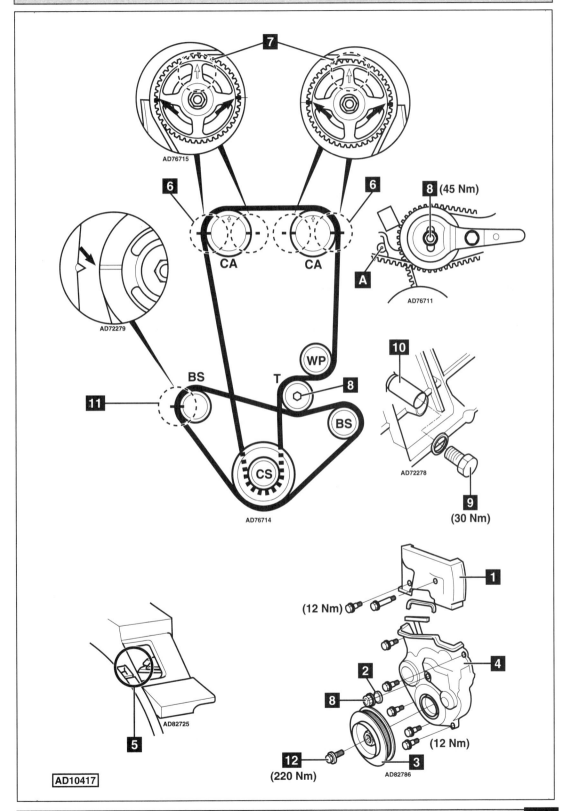

AD76715

7

6

6

8 (45 Nm)

CA

CA

A

AD76711

AD72279

WP

BS

T

8

11

BS

CS

AD76714

10

AD72278

9

(30 Nm)

1

(12 Nm)

4

2

8

(12 Nm)

5

AD82725

12

(220 Nm)

3

AD82786

AD10417

HONDA

Model:	Prelude 2,2 V-TEC • Accord Type R
Year:	1993-03
Engine Code:	H22A2, H22A5, H22A7

Replacement Interval Guide

Honda recommend replacement as follows:
→07/98: Every 60,000 miles or 5 years – replace.
07/98→: Every 72,000 miles or 8 years – replace.
The previous use and service history of the vehicle must always be taken into account.
Refer to Timing Belt Replacement Intervals at the front of this manual.

Check For Engine Damage

CAUTION: This engine has been identified as an INTERFERENCE engine in which the possibility of valve-to-piston damage in the event of a timing belt failure is MOST LIKELY to occur.
A compression check of all cylinders should be performed before removing the cylinder head.

Repair Times – hrs

Remove & install	2,40

Special Tools

- Balancer shaft locking pin – Honda No.07LAG-PT20100.
- Clamp – Honda No.14540-P13-003.

Special Precautions

- Disconnect battery earth lead.
- DO NOT turn crankshaft or camshaft when timing belt removed.
- Remove spark plugs to ease turning engine.
- Turn engine in normal direction of rotation (unless otherwise stated).
- DO NOT turn engine via camshaft or other sprockets.
- Observe all tightening torques.

Removal

NOTE: Normal direction of crankshaft rotation is anti-clockwise. Balancer shaft belt must be removed before removing timing belt.

1. Raise and support front of vehicle.
2. Support engine.
3. Remove:
   - LH front wheel.
   - LH inner wing panel.
   - Engine undershield.
   - Cruise control actuator. DO NOT disconnect control cable.
   - Auxiliary drive belts.
   - Cylinder head cover.
   - Wiring harness.
   - Timing belt upper cover **1**.
   - Dipstick and tube.
   - LH engine mounting.
4. Turn crankshaft anti-clockwise to TDC on No.1 cylinder. Ensure flywheel timing marks aligned **2**.
5. Ensure timing marks on camshaft sprockets aligned with upper cylinder head face **3**. Ensure arrows point upwards **4**.

6. Remove:
   - Crankshaft pulley bolt and washer **5**.
   - Crankshaft pulley **6**.
   - Timing belt lower cover **7**.
   - Crankshaft position (CKP) sensor (if fitted).
7. Slacken tensioner nut **8**. Move tensioner away from belts. Lightly tighten nut.
8. Remove:
   - Balancer shaft belt.
   - Timing belt.
   - Automatic tensioner unit bolts **9**.
   - Automatic tensioner unit.
9. Check tensioner body for leakage or damage.

Installation

1. Retract automatic tensioner as follows:
   - Invert automatic tensioner unit and clamp boss in vice **10**.
   - Remove blanking plug.
   - Insert screwdriver. Turn clockwise until pushrod fully retracted.
   - Fit clamp to keep pushrod retracted **11**. Tool No.14540-P13-003.
   - Fit blanking plug. Use new washer.
2. Ensure crankshaft sprocket timing marks aligned **12**.
3. Ensure timing marks on camshaft sprockets aligned with upper cylinder head face **3**. Ensure arrows point upwards **4**.
4. Fit timing belt in following order:
   - Crankshaft sprocket.
   - Exhaust camshaft sprocket.
   - Inlet camshaft sprocket.
   - Water pump sprocket.
   - Tensioner pulley.
5. Ensure belt is taut between sprockets on non-tensioned side.
6. Install automatic tensioner unit. Tighten bolts to 22 Nm **9**.
7. Remove clamp from automatic tensioner unit. Tool No.14540-P13-003.
8. Remove blanking plug **13**. Insert balancer shaft locking pin **14**. Tool No.07LAG-PT20100.
9. Turn rear balancer shaft sprocket until locking pin locates in hole in shaft.
10. Align front balancer shaft timing marks **15**.
11. Fit balancer shaft belt.
12. Slacken tensioner nut **8**. Ensure tensioner free to move.
13. Remove locking pin **14**. Turn crankshaft one turn anti-clockwise.
14. Tighten tensioner nut **8**. Tightening torque: 45 Nm.
15. Install components in reverse order of removal.
16. Lightly oil threads but NOT contact face of crankshaft pulley bolt **5**.
17. Tighten crankshaft pulley bolt **5**. Tightening torque: →1996: 220 Nm. 1997→: 245 Nm.

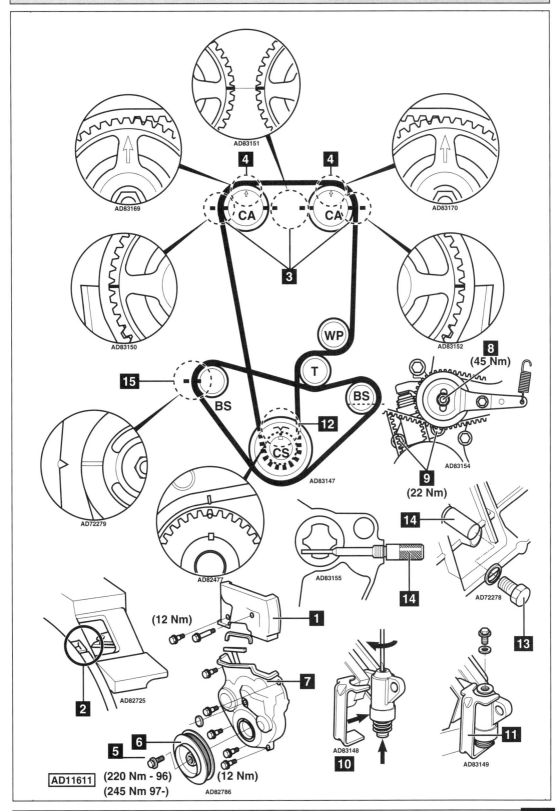

AD83151

AD83169

AD83170

4   CA   CA   **4**

3

AD83150

AD83152

WP

T

8
(45 Nm)

15 BS   BS

12

CS

AD83147

AD83154

AD72279

9
(22 Nm)

AD82477

AD83155

14

14

AD72278

13

(12 Nm) **1**

2

AD82725

7

5 **6**

10

AD83148

11

AD83149

AD11611   (220 Nm - 96)
(245 Nm 97-)

AD82786   (12 Nm)

HONDA

Model:	**Legend 3,5 V6**
Year:	**1996-02**
Engine Code:	**C35A2**

Replacement Interval Guide

Honda recommend replacement as follows:
→07/98: Every 60,000 miles or 5 years – replace.
07/98→: Every 72,000 miles or 8 years – replace.
The previous use and service history of the vehicle must always be taken into account.
Refer to Timing Belt Replacement Intervals at the front of this manual.

Check For Engine Damage

CAUTION: This engine has been identified as an INTERFERENCE engine in which the possibility of valve-to-piston damage in the event of a timing belt failure is MOST LIKELY to occur.
A compression check of all cylinders should be performed before removing the cylinder head.

Repair Times – hrs

Remove and install	2,80

Special Tools

■ Handle – Honda No.07JAB-0010200.
■ Crankshaft pulley holding tool – Honda No.07MAB-PY30100.

Removal

1. Turn crankshaft clockwise to TDC on No.1 cylinder. Ensure white timing mark on crankshaft pulley aligned **1**.
2. Remove:
   ❑ Engine top cover.
   ❑ Intake air duct.
   ❑ Auxiliary drive belts.
   ❑ Breather hoses.
   ❑ Vacuum hoses.
   ❑ Ignition amplifier bracket.
   ❑ Auxiliary drive belt tensioner.
   ❑ Dipstick and tube.
3. Hold crankshaft pulley. Use tool Nos.07JAB-0010200 and 07MAB-PY30100. Remove crankshaft pulley bolt **2**.
4. Remove:
   ❑ Crankshaft pulley **3**.
   ❑ Timing belt covers **4**, **5** & **6**.
5. Ensure timing marks aligned **7**, **8** & **9**. If not: Turn crankshaft one turn clockwise.
6. Slacken balancer shaft belt tensioner bolt 1/2 turn **10**. Move tensioner away from belt. Use screwdriver **11**. Lightly tighten bolt.
7. Remove balancer shaft belt.
8. Slacken timing belt tensioner bolt 1/2 turn **12**. Move tensioner away from belt. Use screwdriver **13**. Lightly tighten bolt.
9. Remove timing belt.
   NOTE: Mark direction of rotation on belt with chalk if belt is to be reused.

Installation

1. Remove:
   ❑ Balancer shaft belt sprocket **14**.
   ❑ Crankshaft sprocket guide washer **15**.
2. Ensure timing marks aligned **7** & **8**.
3. Fit timing belt in following order:
   ❑ Crankshaft sprocket.
   ❑ Tensioner pulley.
   ❑ Camshaft sprocket (CA1).
   ❑ Water pump pulley.
   ❑ Camshaft sprocket (CA2).
   NOTE: Ensure belt is taut between sprockets and pulleys.
   NOTE: Turn camshaft sprocket CA2 1/2 tooth width clockwise to ease installation of belt.
4. Ensure timing marks aligned **7** & **8**.
5. Slacken timing belt tensioner bolt 1/2 turn **12**. Allow tensioner to operate. Lightly tighten bolt.
6. Temporarily fit timing belt lower cover and crankshaft pulley **6** & **3**.
7. Turn crankshaft six turns clockwise to allow belt to settle. Ensure timing marks aligned **7** & **8**.
8. Turn crankshaft clockwise until camshaft sprocket CA1 has moved ten teeth. Ensure blue timing mark on crankshaft pulley aligned **1**.
9. Slacken timing belt tensioner bolt 1/2 turn **12**. Allow tensioner to operate. Tighten bolt to 42 Nm.
10. Remove:
    ❑ Crankshaft pulley **3**.
    ❑ Timing belt lower cover **6**.
11. Turn crankshaft clockwise until timing marks aligned **7** & **8**. Ensure balancer shaft timing mark aligned **9**.
12. Install:
    ❑ Crankshaft sprocket guide washer **15**.
    ❑ Balancer shaft belt sprocket **14**.
13. Fit balancer shaft belt. Ensure timing marks aligned **7** & **9**.
14. Slacken balancer shaft belt tensioner bolt 1/2 turn **10**. Allow tensioner to operate. Lightly tighten bolt.
15. Temporarily fit crankshaft pulley **3**.
16. Turn crankshaft six turns clockwise to settle belt. Ensure timing marks aligned **7**, **8** & **9**.
17. Turn crankshaft one turn clockwise. Ensure timing marks aligned **7**.
18. Slacken balancer shaft belt tensioner bolt 1/2 turn **10**. Allow tensioner to operate. Tighten bolt to 44 Nm.
19. Turn crankshaft clockwise. Ensure timing marks aligned **7**, **8** & **9**.
20. Remove crankshaft pulley **3**.
21. Install components in reverse order of removal.
22. Lightly oil threads and contact face of crankshaft pulley bolt. Hold crankshaft pulley. Use tool Nos.07JAB-0010200 & 07MAB-PY30100. Tighten crankshaft pulley bolt **2**.
    Tightening torque: 245 Nm.

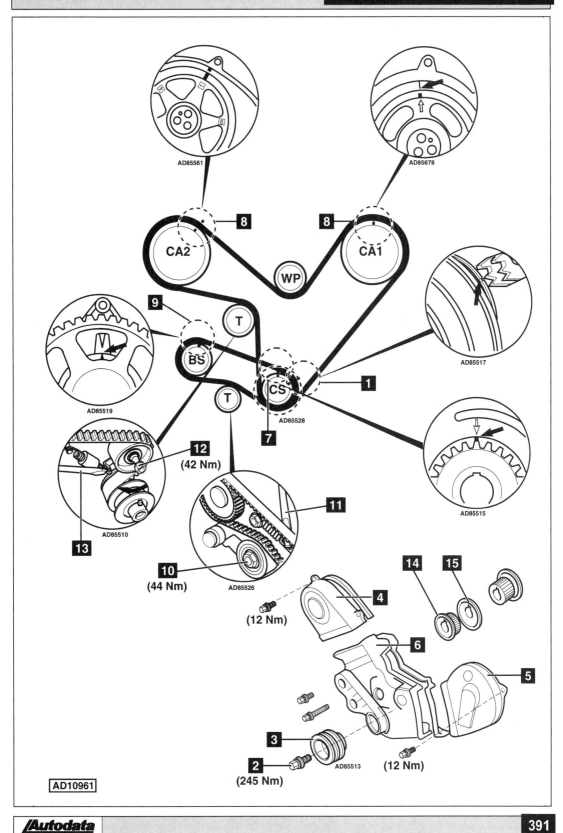

AD85561

AD85678

8

8

CA2

CA1

WP

AD85517

9

T

BS

AD85519

CS

AD85528

1

7

T

12
(42 Nm)

AD85510

13

11

10
(44 Nm)

AD85526

AD85515

14

15

4

(12 Nm)

6

5

3

2

AD85513

(12 Nm)

(245 Nm)

AD10961

Model:	**Shuttle 2,2i/2,3i**
Year:	**1995-01**
Engine Code:	**F22B8, F23A7**

Replacement Interval Guide

Honda recommend replacement as follows:
→07/98: Every 60,000 miles or 5 years, whichever occurs first.
07/98→: Every 72,000 miles or 8 years, whichever occurs first.
The previous use and service history of the vehicle must always be taken into account.
Refer to Timing Belt Replacement Intervals at the front of this manual.

Check For Engine Damage

CAUTION: This engine has been identified as an INTERFERENCE engine in which the possibility of valve-to-piston damage in the event of a timing belt failure is MOST LIKELY to occur.
A compression check of all cylinders should be performed before removing the cylinder head.

Repair Times – hrs

| Remove & install | 2,30 |

Special Tools

- Handle – Honda No.07JAB-0010200.
- Crankshaft pulley holding tool – Honda No.07MAB-PY30100.
- Balancer shaft locking pin – Honda No.07LAG-PT20100.

Special Precautions

- Disconnect battery earth lead.
- DO NOT turn crankshaft or camshaft when timing belt removed.
- Remove spark plugs to ease turning engine.
- Turn engine in normal direction of rotation (unless otherwise stated).
- DO NOT turn engine via camshaft or other sprockets.
- Observe all tightening torques.

Removal

NOTE: Normal direction of crankshaft rotation is anti-clockwise. Balancer shaft belt must be removed before removing timing belt.

1. Remove:
   - ❏ Lower splash guard.
   - ❏ Auxiliary drive belts.
   - ❏ PAS pump. DO NOT disconnect hoses.
   - ❏ Cylinder head cover.
   - ❏ Timing belt upper cover **1**.
2. Turn crankshaft to TDC on No.1 cylinder **2**. Ensure timing marks on camshaft sprocket aligned with upper cylinder head face **3**. Ensure 'UP' mark on camshaft sprocket at top **4**.
3. Hold crankshaft pulley. Use tool Nos.07JAB-0010200 and 07MAB-PY30100. Remove crankshaft pulley bolt **5**.
4. Support engine.

5. Remove:
   - ❏ Crankshaft pulley **6**.
   - ❏ LH engine mounting.
   - ❏ Dipstick and tube.
   - ❏ Rubber seal for adjusting nut **7**.
   - ❏ Timing belt lower cover **8**.
   - ❏ Crankshaft position (CKP) sensor (if fitted).
6. Ensure crankshaft sprocket timing marks aligned **9**.
7. Fit 6 mm bolt at **15**. Lightly tighten bolt.
8. Slacken tensioner nut **10**. Move tensioner away from belts. Lightly tighten nut.
9. Remove:
   - ❏ Balancer shaft belt.
   - ❏ Balancer shaft belt sprocket from crankshaft.
   - ❏ Timing belt.

Installation

1. Ensure timing marks aligned **3**, **4** & **9**.
   *NOTE: Align timing marks on camshaft sprocket with upper cylinder head face **3**.*
2. Fit timing belt in clockwise direction, starting at crankshaft sprocket. Ensure belt is taut between crankshaft sprocket and camshaft sprocket.
3. Slacken tensioner nut **10**. Allow tensioner to operate. Tighten nut.
4. Temporarily fit crankshaft pulley **6**.
5. Fit balancer shaft belt sprocket to crankshaft.
6. Turn crankshaft slowly six turns anti-clockwise to settle belt.
7. Ensure timing marks aligned **3**, **4** & **9**.
8. Slacken tensioner nut **10**.
9. Turn crankshaft slowly anti-clockwise until three teeth of camshaft sprocket have passed timing marks **3**.
10. Tighten tensioner nut **10**. Tightening torque: 44 Nm.
11. Turn crankshaft slowly anti-clockwise to TDC on No.1 cylinder. Ensure timing marks aligned **3**, **4** & **9**.
12. If not: Repeat installation and tensioning procedures.
13. Tighten 6 mm bolt **15** to lock tensioner arm.
14. Align rear balancer shaft timing marks **11**.
15. Remove blanking plug **12**. Insert balancer shaft locking pin **13**. Tool No.07LAG-PT20100.
16. Align front balancer shaft timing marks **14**.
17. Remove crankshaft pulley **6**.
18. Fit balancer shaft belt.
19. Slacken tensioner nut **10**.
20. Remove balancer shaft locking pin **13**. Fit blanking plug and tighten to 29 Nm **12**.
21. Temporarily fit crankshaft pulley **6**. Turn crankshaft one turn anti-clockwise.
22. Tighten tensioner nut **10**. Tightening torque: 44 Nm.
23. Remove 6 mm bolt **15**.
24. Remove crankshaft pulley **6**.
25. Install components in reverse order of removal.
26. Tighten crankshaft pulley bolt **5**.
    Tightening torque: 245 Nm.

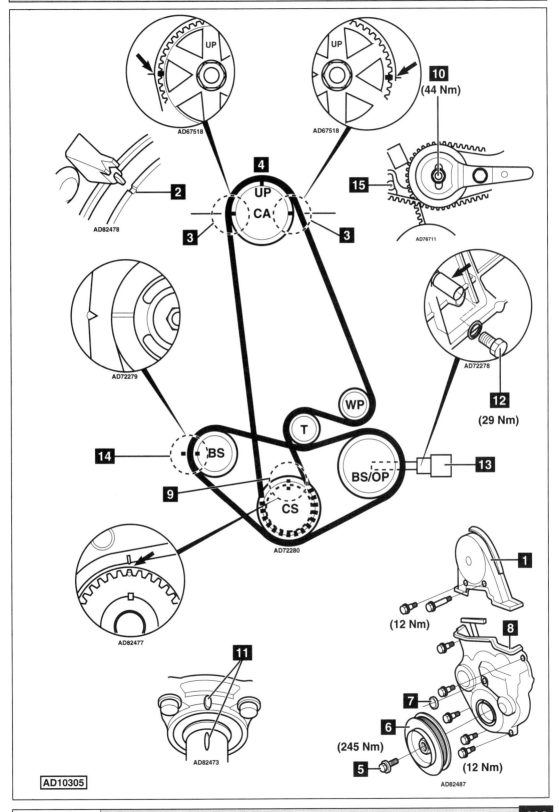

AD67518

UP

UP

AD67518

10
(44 Nm)

15

AD76711

2

AD82478

4
UP
CA

3

3

AD72279

12

(29 Nm)

AD72278

WP

T

14

BS

9

CS

AD72280

BS/OP

13

1

(12 Nm)

8

AD82477

11

7

6

(245 Nm)

5

(12 Nm)

AD82473

AD82487

AD10305

HONDA

Model:	CR-V
Year:	1997-02
Engine Code:	B20B1, B20B3

Replacement Interval Guide

Honda recommend replacement as follows:
→07/98: Every 60,000 miles or 5 years –
replace.
07/98→: Every 72,000 miles or 8 years –
replace.
*The previous use and service history of the vehicle
must always be taken into account.*
*Refer to Timing Belt Replacement Intervals at the front
of this manual.*

Check For Engine Damage

*CAUTION: This engine has been identified as an
INTERFERENCE engine in which the possibility of
valve-to-piston damage in the event of a timing belt
failure is MOST LIKELY to occur.*
*A compression check of all cylinders should be
performed before removing the cylinder head(s).*

Repair Times – hrs

Remove & install	2,30
AC	+0,20

Special Tools

■ None required.

Special Precautions

■ Disconnect battery earth lead.
■ DO NOT turn crankshaft or camshaft when timing belt
  removed.
■ Remove spark plugs to ease turning engine.
■ Turn engine in normal direction of rotation (unless
  otherwise stated).
■ DO NOT turn engine via camshaft or other sprockets.
■ Observe all tightening torques.

Removal

*NOTE: Normal direction of crankshaft rotation is
anti-clockwise.*

1. Remove:
   ❏ Undershield.
   ❏ Auxiliary drive belts.
   ❏ PAS pump. DO NOT disconnect hoses.
2. Support engine.
3. Remove:
   ❏ LH engine mounting.
   ❏ Cylinder head cover.
   ❏ Timing belt centre cover **1**.
4. Turn crankshaft to TDC on No.1 cylinder.
   Ensure timing marks aligned **2**.

5. Ensure 'UP' mark on camshaft sprockets at
   top **3**.
6. Remove:
   ❏ Crankshaft pulley bolt **9**.
   ❏ Crankshaft pulley **4**.
   ❏ Timing belt lower cover **5**.
7. Slacken tensioner bolt **6**. Move tensioner away
   from belt and tighten bolt.
8. Remove timing belt.

Installation

1. Ensure timing marks aligned **7** & **8**.
2. Ensure 'UP' mark on camshaft sprockets at
   top **3**.
3. Fit timing belt in following order:
   ❏ Crankshaft sprocket.
   ❏ Tensioner pulley.
   ❏ Water pump sprocket.
   ❏ Exhaust camshaft sprocket.
   ❏ Inlet camshaft sprocket.
4. Slacken and tighten tensioner bolt **6**.
5. Turn crankshaft six turns anti-clockwise. Ensure
   crankshaft sprocket timing marks aligned **7**.
6. Slacken tensioner bolt **6**.
7. Turn crankshaft anti-clockwise for 3 teeth on
   camshaft sprocket.
8. Tighten tensioner bolt to 54 Nm **6**.
9. Turn crankshaft two turns anti-clockwise. Ensure
   timing marks aligned **7** & **8**.
10. Install components in reverse order of removal.
11. Tighten crankshaft pulley bolt **9**.
    Tightening torque: 177 Nm.

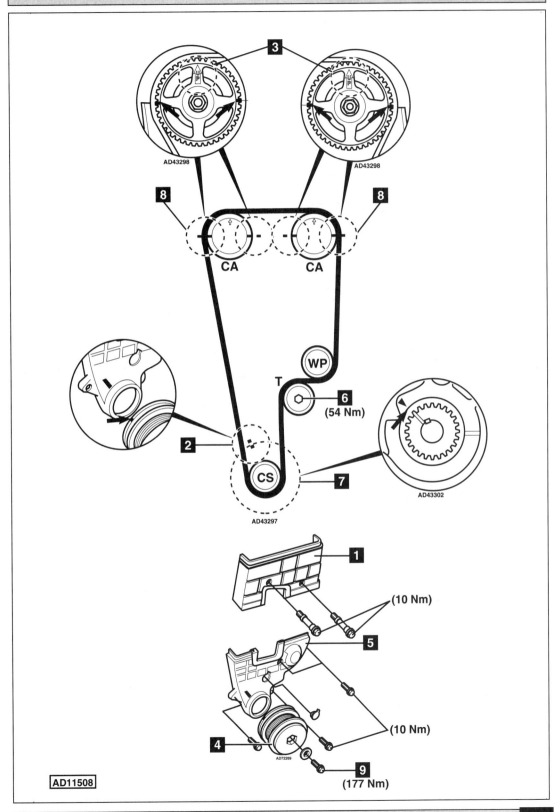

AD43298

AD43298

3

8

8

CA

CA

WP

T

6

(54 Nm)

2

AD43302

CS

7

AD43297

1

(10 Nm)

5

(10 Nm)

4

AD72269

9

(177 Nm)

AD11508

Model:	Civic 2,0 TD • Accord 2,0 TD
Year:	1996-03
Engine Code:	20T2N, 20T2R

Replacement Interval Guide

Honda recommend replacement as follows:
→07/98: Every 60,000 miles or 5 years – replace.
07/98→: Every 81,000 miles or 9 years – replace.
The previous use and service history of the vehicle must always be taken into account.
Refer to Timing Belt Replacement Intervals at the front of this manual.

Check For Engine Damage

CAUTION: This engine has been identified as an INTERFERENCE engine in which the possibility of valve-to-piston damage in the event of a timing belt failure is MOST LIKELY to occur.
A compression check of all cylinders should be performed before removing the cylinder head(s).

Repair Times – hrs

Remove & install:

Timing belt	1,60
Injection pump belt	1,30

Special Tools

- Flywheel timing pin – Honda No.18G 1523.
- Tensioning tool – Honda No.18G 1719.
- Injection pump sprocket timing pin – Honda No.18G 1717.
- Sprocket holding tool – Honda No.18G 1521.

Special Precautions

- Disconnect battery earth lead.
- DO NOT turn crankshaft or camshaft when timing belt removed.
- Remove glow plugs to ease turning engine.
- Turn engine in normal direction of rotation (unless otherwise stated).
- DO NOT turn engine via camshaft or other sprockets.
- Observe all tightening torques.
- Check diesel injection pump timing after belt replacement.

Removal

Timing Belt

1. Raise and support front of vehicle.
2. Remove:
   - ❑ RH front wheel.
   - ❑ RH engine undershield.
   - ❑ Auxiliary drive belt.
   - ❑ Timing belt upper cover **1**.
3. Turn crankshaft clockwise until camshaft timing marks aligned **2**.
4. Insert timing pin in flywheel **3**. Tool No.18G 1523.
5. Remove:
   - ❑ Crankshaft pulley bolt **4**.
   - ❑ Crankshaft pulley **5**.
   - ❑ Timing belt lower cover **6**.
   - ❑ Blanking plug **7**.

6. Slacken tensioner bolt **8**.
7. Retract automatic tensioner as follows:
   - ❑ Install special tool No.18G 1719 **9**.
   - ❑ Turn nut clockwise **22** until pushrod fully retracted.
8. Lightly tighten tensioner bolt **8**.
9. Support engine.
10. Raise engine slightly.
11. Remove:
    - ❑ Bolt **10**.
    - ❑ Nuts **11**.
    - ❑ Engine mounting support bar **12**.
    - ❑ Bolts **13**.
    - ❑ Engine mounting **14**.
    - ❑ Nuts **15**.
    - ❑ Bolts **25**.
    - ❑ Engine mounting plate **16**.
12. Remove timing belt.
    NOTE: Mark direction of rotation on belt with chalk if belt is to be reused.

Installation

Timing Belt

WARNING: If belt has been in use for more than 42,000 miles: Fit new belt.

1. Remove:
   - ❑ Special tool **9**. Tool No.18G 1719.
   - ❑ Tensioner pulley.
   - ❑ Tensioner spring and pushrod.
2. Check tensioner spring for distortion. Check free length of tensioner spring is 65 mm **24**.
3. Install:
   - ❑ Tensioner spring and pushrod.
   - ❑ Tensioner pulley.
   - ❑ Special tool **9**. Tool No.18G 1719.
4. Tighten tensioner bolt **23**. Tightening torque: 45 Nm.
5. Turn nut clockwise **22** until pushrod fully retracted.
6. Lightly tighten tensioner bolt **8**.
7. Ensure timing pin inserted in flywheel **3**.
8. Ensure timing marks aligned **2**.
   NOTE: Observe direction of rotation marks on belt.
9. Fit timing belt in anti-clockwise direction, starting at crankshaft sprocket.
10. Ensure belt is taut between sprockets on non-tensioned side.
11. Slacken tensioner bolt **8**.
12. Remove special tool **9**. Tool No.18G 1719.
13. Allow spring to push tensioner against belt.
14. Tighten tensioner bolt **8**. Tightening torque: 55 Nm.
15. Install:
    - ❑ Blanking plug **7**.
    - ❑ Engine mounting plate **16**.
    - ❑ Nuts **15**. Tightening torque: 35 Nm.
    - ❑ Bolts **25**. Tightening torque: 45 Nm.
    NOTE: If no bolts fitted, tighten all nuts to 35 Nm.

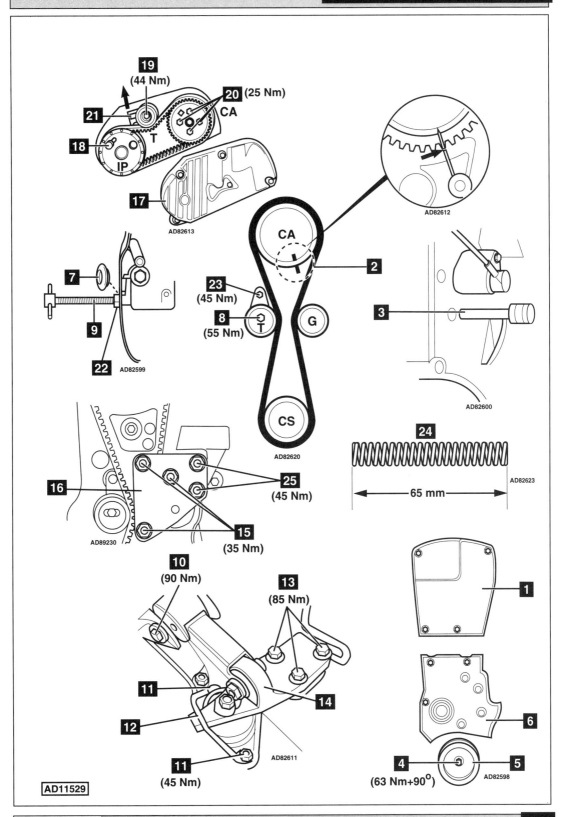

19 (44 Nm)

20 (25 Nm)

21

18

CA

T

IP

17

AD82613

AD82612

7

9

22 AD82599

23 (45 Nm)

8 (55 Nm)

CA

T

G

2

3

AD82600

CS

AD82620

24

AD82623

65 mm

16

25 (45 Nm)

15 (35 Nm)

AD89230

10 (90 Nm)

13 (85 Nm)

1

11

12

14

6

11 (45 Nm)

AD82611

4

5

(63 Nm+90°)

AD82598

AD11529

16. Install:
    - ❏ Engine mounting 14.
    - ❏ Bolts 13. Tightening torque: 85 Nm.
17. Lower engine slightly.
18. Install:
    - ❏ Engine mounting support bar 12.
    - ❏ Nuts 11. Tightening torque: 45 Nm.
    - ❏ Bolt 10. Tightening torque: 90 Nm.
    - ❏ Crankshaft pulley 5.
    - ❏ Crankshaft pulley bolt 4.
19. Temporarily tighten crankshaft pulley bolt 4.
    Tightening torque: 63 Nm.
    NOTE: DO NOT tighten further 90°.
20. Remove flywheel timing pin 3. Tool No.18G 1523.
21. Turn crankshaft two turns clockwise to setting
    position.
22. Insert timing pin in flywheel 3. Tool No.18G 1523.
23. Ensure camshaft sprocket timing marks aligned 2.
24. Slacken tensioner bolt 8.
25. Allow spring to push tensioner against belt.
26. Tighten tensioner bolt 8. Tightening torque: 55 Nm.
27. Remove:
    - ❏ Crankshaft pulley bolt 4.
    - ❏ Crankshaft pulley 5.
28. Install:
    - ❏ Timing belt lower cover 6.
    - ❏ Crankshaft pulley 5.
    - ❏ Crankshaft pulley bolt 4.
29. Tighten crankshaft pulley bolt 4.
    Tightening torque: 63 Nm + 90°.
30. Install components in reverse order of removal.

Removal

Injection Pump Belt

1. Remove:
    - ❏ Air filter housing.
    - ❏ Timing belt upper cover 1.
2. Turn crankshaft clockwise until timing marks
   aligned 2.
3. Insert timing pin in flywheel 3. Tool No.18G 1523.
4. Remove injection pump belt cover 17.
5. Insert timing pin in injection pump sprocket 18.
   Tool No.18G 1717.
6. Slacken tensioner bolt 19.
7. Hold camshaft front sprocket. Use tool
   No.18G 1521.
8. Slacken camshaft rear sprocket bolts (4 bolts) 20.
9. Move tensioner away from injection pump belt.
10. Lightly tighten tensioner bolt 19.
11. Remove injection pump belt.
    **NOTE: Mark direction of rotation on belt with chalk
    if belt is to be reused.**

Installation

Injection Pump Belt

**WARNING: If belt has been in use for more than
42,000 miles: Fit new belt.**
1. Ensure timing pin inserted in flywheel 3.
2. Ensure timing marks aligned 2.
   NOTE: Observe direction of rotation marks on belt.
3. Fit injection pump belt.
4. Slacken tensioner bolt 19.
5. Push tensioner against injection pump belt.
6. Lightly tighten tensioner bolt 19.
7. Hold camshaft front sprocket. Use tool
   No.18G 1521.
8. Tighten camshaft rear sprocket bolts (4 bolts) 20.
   Tightening torque: 25 Nm.
9. Remove:
    - ❏ Injection pump sprocket timing pin 18.
    - ❏ Flywheel timing pin 3.
10. Turn crankshaft two turns clockwise to setting
    position.
11. Ensure timing marks aligned 2.
12. Insert timing pin in flywheel 3.
13. Insert timing pin in injection pump sprocket 18.
14. Slacken tensioner bolt 19.
15. Apply clockwise torque of 6 Nm to tensioner
    pulley 21. Use dial type torque wrench.
16. Tighten tensioner bolt 19. Tightening torque: 44 Nm.
17. Remove:
    - ❏ Injection pump sprocket timing pin 18.
    - ❏ Flywheel timing pin 3.
18. Install components in reverse order of removal.

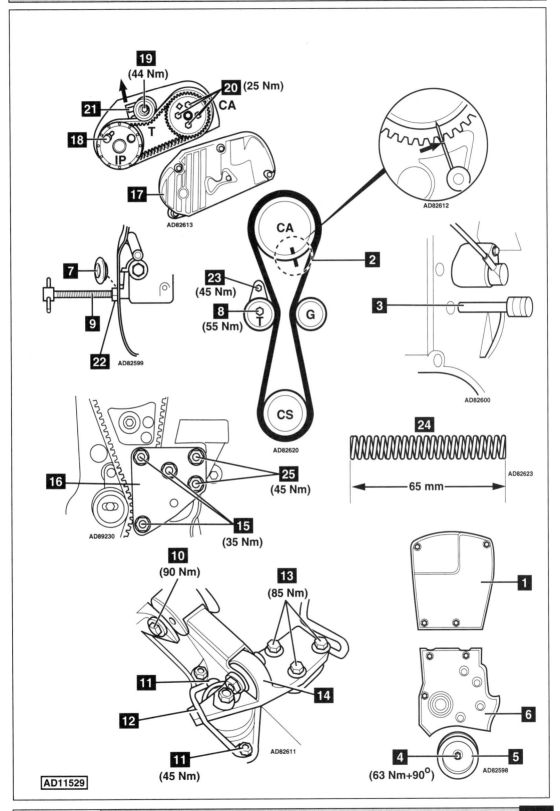

19 (44 Nm)

21

20 (25 Nm)

CA

T

18

IP

17

AD82613

AD82612

CA

2

7

23 (45 Nm)

8 (55 Nm)

T

G

3

9

22 AD82599

AD82600

CS

AD82620

24

AD82623

65 mm

16

25 (45 Nm)

15 (35 Nm)

AD89230

10 (90 Nm)

13 (85 Nm)

1

11

12

14

6

11

AD82611

4

5

AD82598

(63 Nm+90°)

(45 Nm)

AD11529

HYUNDAI

Model:	**Atoz**
Year:	**1998-03**
Engine Code:	**G4HC**

Replacement Interval Guide

Hyundai recommend replacement every
50,000 miles or 5 years.
*The previous use and service history of the vehicle
must always be taken into account.*
*Refer to Timing Belt Replacement Intervals at the front
of this manual.*

Check For Engine Damage

*CAUTION: This engine has been identified as an
INTERFERENCE engine in which the possibility of
valve-to-piston damage in the event of a timing belt
failure is MOST LIKELY to occur. A compression check
of all cylinders should be performed before removing
the cylinder head(s).*

Repair Times – hrs

Remove & install	1,40
PAS	+0,20
AC	+0,20

Special Tools

- None required.

Special Precautions

- Disconnect battery earth lead.
- DO NOT turn crankshaft or camshaft when timing belt removed.
- Remove spark plugs to ease turning engine.
- Turn engine in normal direction of rotation (unless otherwise stated).
- DO NOT turn engine via camshaft or other sprockets.
- Observe all tightening torques.

Removal

1. Remove:
   - ❏ Auxiliary drive belts.
   - ❏ Water pump pulley.
   - ❏ Dipstick and tube.
2. Raise and support front of vehicle.
3. Remove:
   - ❏ RH front wheel.
   - ❏ Crankshaft pulley bolt **1**.
   - ❏ Crankshaft pulley **2**.
   - ❏ Timing belt cover **3**.
4. Turn crankshaft clockwise to TDC on No.1 cylinder. Ensure timing marks aligned **4** & **5**.
5. Slacken tensioner bolt **6**. Push tensioner away from belt. Lightly tighten bolt.
6. Remove timing belt.

Installation

1. Ensure timing marks aligned **4** & **5**.
2. Fit timing belt in anti-clockwise direction, starting at crankshaft sprocket. Ensure belt is taut on non-tensioned side.
3. Slacken tensioner bolt **6**. Allow tensioner to operate.
4. Turn crankshaft clockwise for two camshaft sprocket teeth.
5. Tighten tensioner bolt **6**.
   Tightening torque: 22-30 Nm.
6. Turn crankshaft clockwise until timing marks aligned **4** & **5**.
7. Check belt tension as follows:
8. Apply thumb pressure to belt at ▽ and push belt away from tensioner. Check measurement between back of belt and centre of timing belt cover bolt hole is approximately 20 mm **7**.
9. Install components in reverse order of removal.
10. Tighten crankshaft pulley bolt **1**.
    Tightening torque: 140-150 Nm.

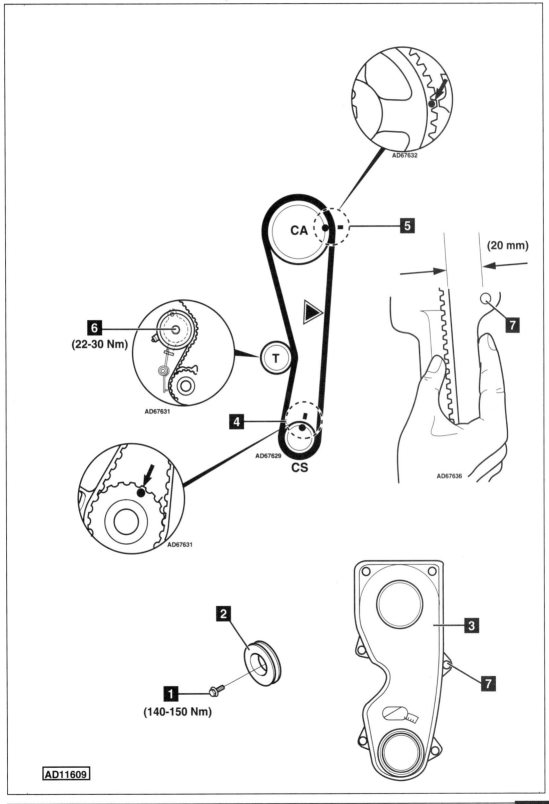

AD67632

CA **5**

(20 mm)

7

6
(22-30 Nm)

T

AD67631

4

AD67629

CS

AD67636

AD67631

2

3

7

1

(140-150 Nm)

AD11609

Model:	**Accent 1,3i/1,5i • S Coupe 1,5i (→1992) • Lantra 1,5**
Year:	**1990-02**
Engine Code:	**G4-J, G4-N, G4-K, G4EA**

Replacement Interval Guide

Hyundai recommend:

Accent:
→99MY: Replacement every 40,000 miles or 4 years.
00MY→: Replacement every 50,000 miles or 5 years.

S Coupe/Lantra:
→01/92: Replacement every 60,000 miles.
02/92→: Replacement every 40,000 miles or 4 years.
The previous use and service history of the vehicle must always be taken into account.
Refer to Timing Belt Replacement Intervals at the front of this manual.

Check For Engine Damage

CAUTION: This engine has been identified as an INTERFERENCE engine in which the possibility of valve-to-piston damage in the event of a timing belt failure is MOST LIKELY to occur.
A compression check of all cylinders should be performed before removing the cylinder head.

Repair Times – hrs

Remove & install:

Accent (→1999)	1,30
AC	+0,20
PAS	+0,20
Accent (2000→)	1,50
PAS	+0,20
AC	+0,20
S Coupe	1,60
AC	+0,20
PAS	+0,20
Lantra	1,60
AC	+0,40
PAS	+0,20

Special Tools

■ Tension gauge – Burroughs BT-33-73F.

Special Precautions

■ Disconnect battery earth lead.
■ DO NOT turn crankshaft or camshaft when timing belt removed.
■ Remove spark plugs to ease turning engine.
■ Turn engine in normal direction of rotation (unless otherwise stated).
■ DO NOT turn engine via camshaft or other sprockets.
■ Observe all tightening torques.

Removal

1. Accent: Support engine.
2. Accent: Remove RH engine mounting.
3. Remove:
   ❑ Auxiliary drive belt(s).
   ❑ Water pump pulley.
   ❑ Crankshaft pulley.
   ❑ Timing belt upper cover **1**.
   ❑ Timing belt lower cover **2**.
4. Turn crankshaft clockwise to TDC on No.1 cylinder.
5. Ensure timing marks aligned **3** & **4**.
6. Slacken tensioner bolts **5** & **6**. Move tensioner away from belt and lightly tighten bolts.
7. Remove timing belt.

Installation

NOTE: Always turn crankshaft clockwise.
1. Ensure timing marks aligned **3** & **4**.
2. Fit timing belt in anti-clockwise direction, starting at crankshaft sprocket. Ensure belt is taut between sprockets on non-tensioned side.
3. Slacken tensioner bolt **5**.
4. Slacken tensioner bolt **6**.
5. Allow tensioner to operate.
6. Tighten tensioner bolt **6**.
7. Tighten tensioner bolt **5**.
8. Turn crankshaft one turn clockwise.
9. Ensure crankshaft sprocket timing marks aligned **4**.
10. Slacken tensioner bolt **5**.
11. Slacken tensioner bolt **6**.
12. Allow tensioner to operate.
13. Tighten tensioner bolt **6**.
    Tightening torque: 20-26 Nm.
14. Tighten tensioner bolt **5**.
    Tightening torque: 20-26 Nm.
15. Apply thumb pressure to belt at ▼ (approximately 5 kg). Belt should deflect to 1/4 of tensioner bolt head width **7**.
    NOTE: Belt tension can also be set using Burroughs (BT-33-73F) tension gauge 10 at ▼. Turn crankshaft 90° anti-clockwise. Check belt tension is 9,5-16,5 kg (engine must be cold).
16. Install components in reverse order of removal.
17. Fit crankshaft pulley.
18. Tighten crankshaft pulley bolts **8**.
    ❑ G4-J: Tightening torque: 12-15 Nm.
    ❑ G4-N/G4-K: Tightening torque: 10-12 Nm.
19. G4EA: Tighten crankshaft pulley bolt **9**.
    Tightening torque: 140-150 Nm.

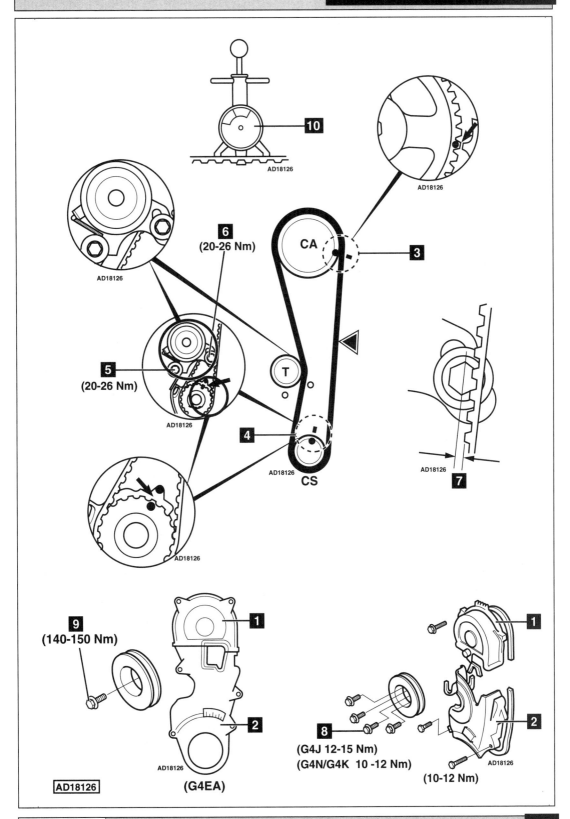

10

AD18126

AD18126

6
(20-26 Nm)

CA

3

AD18126

5
(20-26 Nm)

T

AD18126

4

AD18126

CS

7

AD18126

AD18126

9
(140-150 Nm)

1

2

AD18126

(G4EA)

1

8

(G4J 12-15 Nm)
(G4N/G4K 10 -12 Nm)

2

AD18126

(10-12 Nm)

AD18126

Model:	Accent 1,5 16V • Elantra 1,6 16V • Matrix 1,6 16V
Year:	1996-03
Engine Code:	G4K, G4GD, G4EC, G4ED

Replacement Interval Guide

→07/98: Hyundai recommend replacement every 40,000 miles or 4 years.

08/98→:

Accent/Elantra: Hyundai recommend replacement every 50,000 miles or 5 years.

Matrix: Hyundai recommend check and replacement if necessary every 30,000 miles or 3 years and replacement every 50,000 miles or 5 years.

The previous use and service history of the vehicle must always be taken into account.

Refer to Timing Belt Replacement Intervals at the front of this manual.

Check For Engine Damage

CAUTION: This engine has been identified as an INTERFERENCE engine in which the possibility of valve-to-piston damage in the event of a timing belt failure is MOST LIKELY to occur.

A compression check of all cylinders should be performed before removing the cylinder head(s).

Repair Times – hrs

Accent (→1999):	
Remove & install	1,30
AC	+0,20
PAS	+0,20
Accent (2000→):	
Remove & install	1,50
AC	+0,20
PAS	+0,20
Elantra:	
Remove & install	1,80
AC	+0,20

Special Tools

■ None required.

Special Precautions

■ Disconnect battery earth lead.
■ DO NOT turn crankshaft or camshaft when timing belt removed.
■ Remove spark plugs to ease turning engine.
■ Turn engine in normal direction of rotation (unless otherwise stated).
■ DO NOT turn engine via camshaft or other sprockets.
■ Observe all tightening torques.

Removal

1. Support engine (if necessary).
2. Remove:
   ❏ Auxiliary drive belt(s).
   ❏ Water pump pulley.
   ❏ RH engine mounting and bracket (if necessary).
   ❏ Crankshaft pulley bolt **1**.
   ❏ Crankshaft pulley **2**.
   ❏ Timing belt upper cover **3**.
   ❏ Timing belt lower cover **4**.
3. Turn crankshaft clockwise until crankshaft sprocket timing marks aligned **5**.
4. Ensure hole in camshaft sprocket aligned with timing mark on cylinder head **6**.
5. Slacken tensioner bolt **7**.
6. Slacken tensioner bolt **8**.
7. Move tensioner away from belt. Lightly tighten tensioner bolt **8**.
8. Remove timing belt.
   NOTE: Mark direction of rotation on belt with chalk if belt is to be reused.

Installation

NOTE: Observe direction of rotation marks on belt.

1. Ensure crankshaft sprocket timing marks aligned **5**.
2. Ensure hole in camshaft sprocket aligned with timing mark on cylinder head **6**.
3. Fit timing belt in following order:
   ❏ Crankshaft sprocket.
   ❏ Guide pulley.
   ❏ Camshaft sprocket.
   ❏ Tensioner pulley.
4. Ensure belt is taut between sprockets on non-tensioned side.
5. Slacken tensioner bolt **8**.
6. Allow tensioner to operate.
7. Tighten tensioner bolt **8**.
8. Tighten tensioner bolt **7**.
9. Accent/Elantra: Turn crankshaft one turn clockwise.
10. Matrix: Turn crankshaft two turns clockwise.
11. Ensure crankshaft sprocket timing marks aligned **5**.
12. Slacken tensioner bolt **7**.
13. Slacken tensioner bolt **8**.
14. Allow tensioner to operate.
15. Tighten tensioner bolt **8**.
    Tightening torque: 20-27 Nm.
16. Tighten tensioner bolt **7**.
    Tightening torque: 20-27 Nm.
17. Apply firm thumb pressure to belt at ▽ (5 kg).
18. Belt should deflect to 1/4 of tensioner bolt head width **9**.
19. Install components in reverse order of removal.
20. Tighten crankshaft pulley centre bolt **1**.
    Tightening torque: 140-150 Nm.

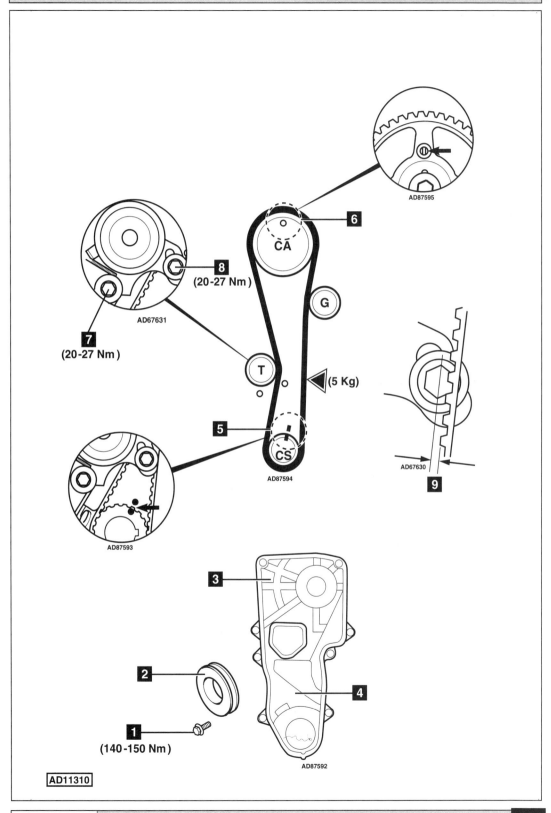

AD87595

6

CA

8
(20-27 Nm)

AD67631

G

7
(20-27 Nm)

T

(5 Kg)

5

CS

AD87594

AD67630

9

AD87593

3

2

4

AD87592

1
(140-150 Nm)

AD11310

Model:	Lantra 1,6i
Year:	1991-12/92
Engine Code:	G4-R

Replacement Interval Guide

Hyundai recommend replacement every 48,000 miles (pre 1/92) or every 40,000 miles or 4 years (01/92→).
The previous use and service history of the vehicle must always be taken into account.
Refer to Timing Belt Replacement Intervals at the front of this manual.

Check For Engine Damage

CAUTION: This engine has been identified as an INTERFERENCE engine in which the possibility of valve-to-piston damage in the event of a timing belt failure is MOST LIKELY to occur.
A compression check of all cylinders should be performed before removing the cylinder head.

Repair Times – hrs

Remove & install	2,60
AC	+0,40
PAS	+0,20

Special Tools

- Tensioner pulley tool – Hyundai No.09244-28100.
- Automatic tensioner tool – Hyundai No.09244-28000.

Special Precautions

- Disconnect battery earth lead.
- DO NOT turn crankshaft or camshaft when timing belt removed.
- Remove spark plugs to ease turning engine.
- Turn engine in normal direction of rotation (unless otherwise stated).
- DO NOT turn engine via camshaft or other sprockets.
- Observe all tightening torques.

Removal

1. Remove:
   - Auxiliary drive belts.
   - Auxiliary drive belt tensioner and bracket **1**.
   - Water pump pulley **2**.
   - Timing belt upper cover **4**.
2. Turn crankshaft clockwise until timing marks on camshaft sprockets aligned with upper cylinder head face **6**.
   NOTE: Dowel pins on camshaft sprockets should face upwards 7.
3. Remove:
   - Crankshaft pulley **3**.
   - Timing belt lower cover **5**.
4. Ensure crankshaft sprocket and oil pump sprocket timing marks aligned **8** & **9**.

5. Remove:
   - Automatic tensioner unit **10**.
   - Tensioner pulley **11**.
   - Timing belt.

Installation

NOTE: Always turn crankshaft clockwise.

1. Check tensioner body for leakage or damage **10**. Replace if necessary.
2. Check pushrod protrusion is 12 mm **12**. If not: Replace automatic tensioner unit.
3. Slowly compress pushrod into tensioner body until holes aligned. Use vice. Resistance should be felt.
4. Retain pushrod with 1,4 mm diameter pin through hole in tensioner body **13**.
5. Install automatic tensioner unit.
6. Fit tensioner pulley with spigot holes to left of bolt.
7. Ensure timing marks aligned **6**, **8** & **9**.
   NOTE: Timing marks on camshaft sprockets should align with upper cylinder head face 6. Exhaust sprocket mark is on recess, but inlet sprocket mark is on tooth.
8. Fit timing belt in anti-clockwise direction, starting at crankshaft sprocket. Ensure belt is taut between sprockets.
9. Push tensioner pulley gently against belt. Tighten tensioner pulley bolt **11**.
10. Turn crankshaft slowly 1/4 turn anti-clockwise.
11. Turn crankshaft 1/4 turn clockwise until timing marks aligned **6**, **8** & **9**.
12. Slacken tensioner pulley bolt **11**. Fit tool to tensioner pulley **16**. Tool No.09244-28100.
13. Apply clockwise torque of 2,6-2,8 Nm to tensioner pulley. Tighten bolt **11**.
14. Insert tool in hole in LH engine mounting bracket **17**. Tool No.09244-28000.
15. Screw in tool until it contacts tensioner arm.
16. Continue to screw in tool until pin can be removed from automatic tensioner unit.
17. Remove special tool **17**. Tool No.09244-28000.
18. Turn crankshaft two turns clockwise. Wait 15 minutes.
19. Check pushrod protrusion is 3,8-4,5 mm **14**.
20. If not: Repeat operations 10-17.
21. Ensure timing marks aligned **6**, **8** & **9**.
22. Install components in reverse order of removal.
23. Tighten crankshaft pulley bolts **15**.
    Tightening torque: 20-30 Nm.

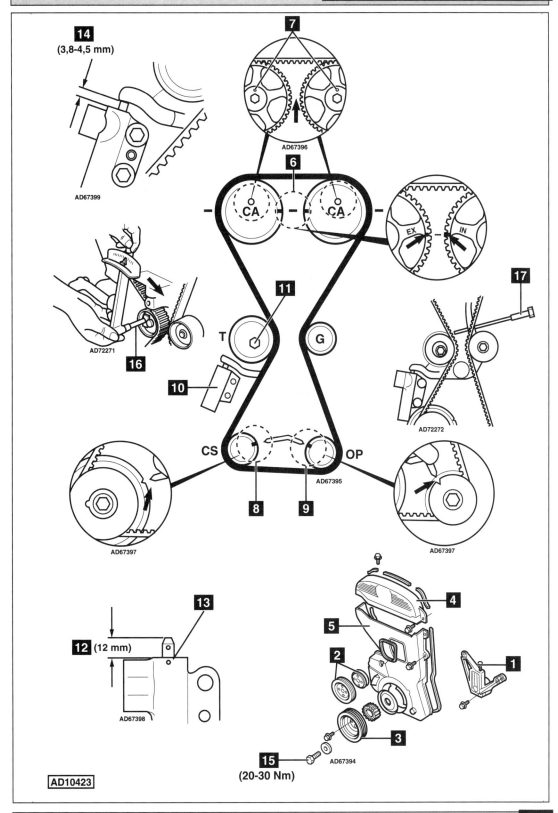

14 (3,8-4,5 mm)

AD67399

AD72271

16

10

7

AD67396

6

CA — CA

EX — IN

17

11

T   G

AD72272

8   **9**

CS   OP

AD67395

AD67397   AD67397

13

12 (12 mm)

AD67398

4

5

2

1

3

15 (20-30 Nm)   AD67394

AD10423

Model:	**Lantra 1,6/1,8 16V • Sonata 2,0 16V**
Year:	**1992-98**
Engine Code:	**G4-N, G4-P, G4-R (01/93→)**

Replacement Interval Guide

Hyundai recommend replacement every 40,000 miles or 4 years.
The previous use and service history of the vehicle must always be taken into account.
Refer to Timing Belt Replacement Intervals at the front of this manual.

Check For Engine Damage

CAUTION: This engine has been identified as an INTERFERENCE engine in which the possibility of valve-to-piston damage in the event of a timing belt failure is MOST LIKELY to occur.
A compression check of all cylinders should be performed before removing the cylinder head.

Repair Times – hrs

Lantra:	
Remove & install	2,60
AC	+0,40
PAS	+0,20
Sonata:	
Remove & install	2,80
AC	+0,40

Special Tools

- ■ Tensioner pulley tool – Hyundai No.09244-28100.
- ■ Automatic tensioner tool – Hyundai No.09244-28000.

Removal

NOTE: Record varying length of timing belt cover bolts.
1. Support engine.
2. Remove:
   - ❏ Engine mounting.
   - ❏ Auxiliary drive belts.
   - ❏ Water pump pulley **1**.
   - ❏ Crankshaft pulley bolts **21**.
   - ❏ Crankshaft pulley **2**.
   - ❏ Timing belt covers **3**.
3. Turn crankshaft to TDC on No.1 cylinder. Ensure crankshaft sprocket, camshaft sprocket, oil pump sprocket and balancer shaft sprocket timing marks aligned **4**, **5**, **6** & **7**.
   *NOTE: Timing marks on camshaft sprockets should align with upper cylinder head face **5**.*
   *NOTE: Dowel pins on camshaft sprockets should face upwards **8**.*
4. Slacken tensioner bolt **9**. Move tensioner away from belt and lightly tighten bolt.
5. Remove:
   - ❏ Automatic tensioner unit bolts **10**.
   - ❏ Automatic tensioner unit **11**.
   - ❏ Timing belt.
   - ❏ Crankshaft centre bolt **12**.
   - ❏ Crankshaft sprocket **13**.
   - ❏ Crankshaft sprocket guide washer **14**.
6. Slacken balancer shaft belt tensioner bolt **15**. Move tensioner away from belt and lightly tighten bolt.
7. Remove balancer shaft belt.

Installation

1. Ensure timing marks aligned **4**, **5**, **6** & **7**.
   *NOTE: Timing marks on camshaft sprockets should align with upper cylinder head face **5**. Exhaust sprocket mark is on recess, but inlet sprocket mark is on tooth.*
   *NOTE: To check oil pump sprocket is correctly positioned: Remove blanking plug from cylinder block **22**. Insert 8 mm diameter Phillips screwdriver in hole. Ensure screwdriver is inserted 60 mm from face of cylinder block. If screwdriver can only be inserted 20 mm: Turn oil pump sprocket 360°. Insert screwdriver again.*
2. Fit balancer shaft belt in anti-clockwise direction, starting at crankshaft sprocket.
3. Slacken balancer shaft belt tensioner bolt **15**. Turn tensioner clockwise to tension belt. Tighten bolt to 15-22 Nm.
4. Apply thumb pressure to belt at ▽. Belt should deflect 5-7 mm.
5. If not: Repeat operations 2-4.
6. Install crankshaft sprocket guide washer **14** and crankshaft sprocket **13**.
7. Tighten crankshaft centre bolt to 110-130 Nm **12**.
   NOTE: Ensure crankshaft sprocket guide washer is fitted correctly. Oil threads and face of crankshaft bolt before fitting.
8. Check tensioner body for leakage or damage **11**. Replace if necessary.
9. Check pushrod protrusion is 12 mm **16**. If not: Replace automatic tensioner unit.
10. Slowly compress pushrod into tensioner body **11** until holes aligned. Resistance should be felt.
    NOTE: Place flat washers under tensioner body to avoid damage to body end plug.
11. Retain pushrod with suitable pin through hole in tensioner body **17**.
12. Install automatic tensioner unit to cylinder block. Tighten bolts to 22-27 Nm **10**.
13. Fit timing belt in anti-clockwise direction, starting at tensioner pulley.
14. Slacken tensioner pulley bolt **9**. Turn tensioner pulley clockwise to temporarily tension belt. Tighten bolt to 43-45 Nm.
15. Turn crankshaft slowly 1/4 turn anti-clockwise.
16. Turn crankshaft clockwise until timing marks aligned **4**, **5**, **6** & **7**.
17. Slacken tensioner pulley bolt **9**. Fit tool to tensioner pulley **18**. Tool No.09244-28100.
18. Apply clockwise torque of 2,6-2,8 Nm to tensioner pulley. Tighten bolt to 43-45 Nm **9**.
19. Insert special tool in timing belt rear casing **19**. Tool No.09244-28000.
20. Screw in tool until pin can be removed from automatic tensioner unit **17**.
21. Remove special tool **19**. Tool No.09244-28000.
22. Turn crankshaft slowly two turns in direction of rotation until timing marks aligned **4**, **5**, **6** & **7**. Wait 15 minutes.
23. Check extended length of pushrod is 3,8-4,5 mm **20**.
24. Install components in reverse order of removal.
25. Tighten crankshaft pulley bolts to 20-30 Nm **21**.

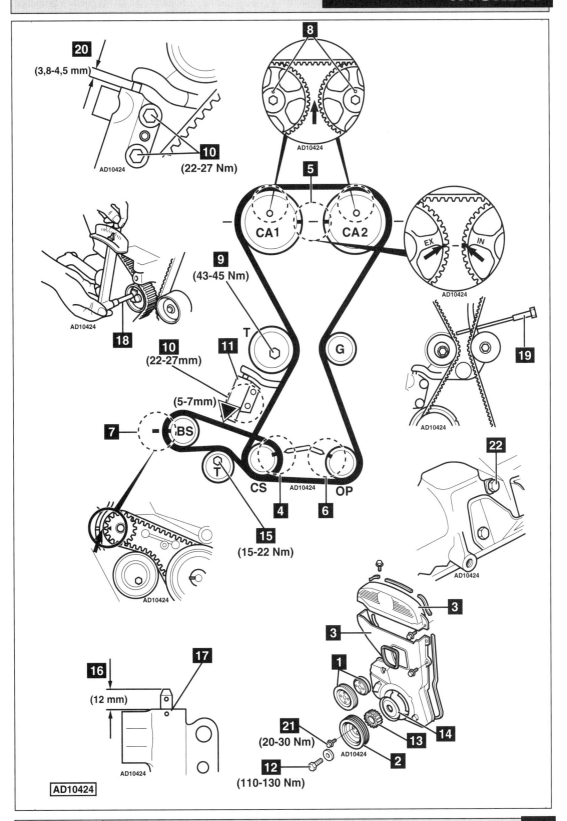

20 (3,8-4,5 mm)

10 (22-27 Nm)

8

AD10424

5

CA1 CA2

EX IN

AD10424

9 (43-45 Nm)

18 AD10424

10 (22-27mm)

11 (5-7mm)

19

AD10424

7 – BS

T G

T

CS AD10424 OP

4 **6**

22

AD10424

3

3

1

15 (15-22 Nm)

AD10424

16 (12 mm)

17

21 (20-30 Nm)

13

14

2

12 (110-130 Nm)

AD10424

AD10424

Model:	Lantra 1,6/1,8/2,0 16V • Elantra 1,8/2,0 16V • Matrix 1,8 16V
	Coupé 1,6/2,0 16V
Year:	1995-03
Engine Code:	G4GR, G4DR, G4DM, G4F, G4GB, G4GC, G4GF

Replacement Interval Guide

→07/98: Hyundai recommend replacement every 40,000 miles or 4 years.
08/98→:
Lantra/Elantra/Coupé: Hyundai recommend replacement every 50,000 miles or 5 years.
Matrix: Hyundai recommend check and replacement if necessary every 30,000 miles or 3 years and replacement every 50,000 miles or 5 years.
The previous use and service history of the vehicle must always be taken into account.
Refer to Timing Belt Replacement Intervals at the front of this manual.

Check For Engine Damage

CAUTION: This engine has been identified as an INTERFERENCE engine in which the possibility of valve-to-piston damage in the event of a timing belt failure is MOST LIKELY to occur.
A compression check of all cylinders should be performed before removing the cylinder head.

Repair Times – hrs

Remove & install:
Lantra/Coupé/Elantra	1,80
AC – Lantra/Coupé	+0,10
AC – Elantra	+0,20

Special Tools

■ None required.

Special Precautions

■ Disconnect battery earth lead.
■ DO NOT turn crankshaft or camshaft when timing belt removed.
■ Remove spark plugs to ease turning engine.
■ Turn engine in normal direction of rotation (unless otherwise stated).
■ DO NOT turn engine via camshaft or other sprockets.
■ Observe all tightening torques.

Removal

1. Support engine.
2. Remove:
   ❑ Auxiliary drive belts.
   ❑ Water pump pulley.
   ❑ RH engine mounting and bracket.
   ❑ Timing belt upper cover **1**.
   ❑ Crankshaft pulley bolt **2**.
   ❑ Crankshaft pulley **3**.
   ❑ Timing belt lower cover **4**.
   ❑ Crankshaft sprocket guide washer **5**.
3. Turn crankshaft clockwise to TDC on No.1 cylinder. Ensure small pin on crankshaft sprocket aligned with timing mark **6**.
4. Check hole in camshaft sprocket aligned with mark on camshaft bearing cap **7**. If not: Turn crankshaft one turn.
5. Slacken tensioner bolt **8**. Move tensioner away from belt and lightly tighten bolt.
6. Remove timing belt.
   NOTE: Mark direction of rotation on belt with chalk if belt is to be reused.

Installation

1. Ensure timing marks aligned **6** & **7**.
2. Fit timing belt in anti-clockwise direction, starting at crankshaft sprocket. Ensure belt is taut between sprockets.
3. Slacken tensioner bolt **8**. Push tensioner firmly against timing belt to tension belt.
4. Tighten tensioner bolt **8**.
   Tightening torque: 43-55 Nm.
5. Except Matrix: Turn crankshaft one turn clockwise.
6. Matrix: Turn crankshaft two turns clockwise.
7. Ensure timing marks aligned **6**.
8. Apply a load of 2 kg to belt at ▽. Belt should deflect 4-6 mm. If not: Adjust position of tensioner.
9. Install components in reverse order of removal.
10. Tighten crankshaft pulley bolt **2**.
    Tightening torque: 170-180 Nm.

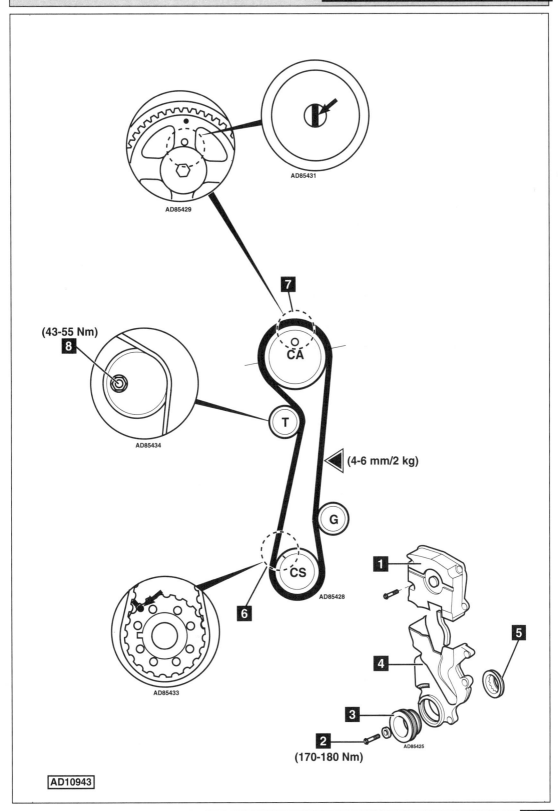

AD85431

AD85429

(43-55 Nm)

8

T

AD85434

7

CA

(4-6 mm/2 kg)

G

CS

6

AD85428

AD85433

1

4

5

3

2

(170-180 Nm)

AD85425

AD10943

HYUNDAI

Model:	Sonata 1,8/2,0/2,4
Year:	1988-96
Engine Code:	G4M, G4-P, G4-S

Replacement Interval Guide

Hyundai recommend replacement every
60,000 miles (pre 1/92) or every 40,000 miles or
4 years (01/92→).
*The previous use and service history of the vehicle
must always be taken into account.*
*Refer to Timing Belt Replacement Intervals at the
front of this manual.*

Check For Engine Damage

*CAUTION: This engine has been identified as an
INTERFERENCE engine in which the possibility of
valve-to-piston damage in the event of a timing belt
failure is MOST LIKELY to occur.*
*A compression check of all cylinders should be
performed before removing the cylinder head.*

Repair Times – hrs

Remove & install	2,00
AC	+0,40

Special Tools

■ None required.

Special Precautions

■ Disconnect battery earth lead.
■ DO NOT turn crankshaft or camshaft when timing belt removed.
■ Remove spark plugs to ease turning engine.
■ Turn engine in normal direction of rotation (unless otherwise stated).
■ DO NOT turn engine via camshaft or other sprockets.
■ Observe all tightening torques.

Removal

Timing Belt

1. Support engine.
2. Remove:
   ❑ Engine mounting bracket.
   ❑ Auxiliary drive belt(s).
   ❑ Alternator.
   ❑ Auxiliary drive belt tensioner and bracket.
   ❑ Water pump pulley **1**.
   ❑ Crankshaft pulley **2**.
   ❑ Timing belt covers **3** & **4**.
3. Turn crankshaft clockwise until camshaft sprocket, crankshaft sprocket and oil pump sprocket timing marks aligned **5**, **6** & **7**.
4. Slacken tensioner nut and bolt **8** & **9**. Move tensioner towards water pump. Lightly tighten bolt **8**.
5. Remove timing belt.

Installation

Timing Belt

1. Ensure timing marks aligned **5**, **6** & **7**.
   *NOTE: To check oil pump sprocket is correctly positioned: Remove blanking plug from cylinder block **10**. Insert 8 mm diameter Phillips screwdriver in hole. Ensure screwdriver is inserted 60 mm from face of cylinder block. If screwdriver can only be inserted 20 mm: Turn oil pump sprocket 360°. Insert screwdriver again.*
2. Fit timing belt in anti-clockwise direction, starting at crankshaft sprocket. Ensure belt is taut between sprockets on non-tensioned side.
3. Slacken tensioner bolt **8**.
4. Tighten tensioner bolt **8**.
5. Tighten tensioner nut **9**.
6. Turn crankshaft one turn clockwise. Ensure crankshaft timing marks aligned **6**.
7. Slacken and tighten tensioner nut and bolt as described previously.
8. Apply thumb pressure to belt at ▼ and push belt away from tensioner. Check measurement between back of belt and edge of timing belt rear cover is 14 mm **11**.
9. If not: Repeat tensioning procedure.
10. Install components in reverse order of removal.
11. Tighten crankshaft pulley bolts **17**.
    Tightening torque: 20-30 Nm.

Removal

Balancer Shaft Belt

1. Remove timing belt as described previously.
2. Remove:
   ❑ Crankshaft sprocket **12**.
   ❑ Crankshaft sprocket guide washer **13**.
3. Ensure balancer shaft sprocket timing marks aligned **14**.
4. Remove:
   ❑ Balancer shaft belt tensioner **15**.
   ❑ Balancer shaft belt.

Installation

Balancer Shaft Belt

1. Ensure balancer shaft sprocket timing marks aligned **14**.
2. Fit balancer shaft belt tensioner and belt.
3. Push tensioner clockwise towards belt. Tighten tensioner bolt **15**.
4. Apply thumb pressure to balancer shaft belt at ▼. Belt should deflect 5-7 mm.
5. If not: Repeat tensioning procedure.
6. Install:
   ❑ Crankshaft sprocket guide washer **13**.
   ❑ Crankshaft sprocket **12**.
7. Tighten crankshaft bolt **16**.
   Tightening torque: 108-127 Nm.

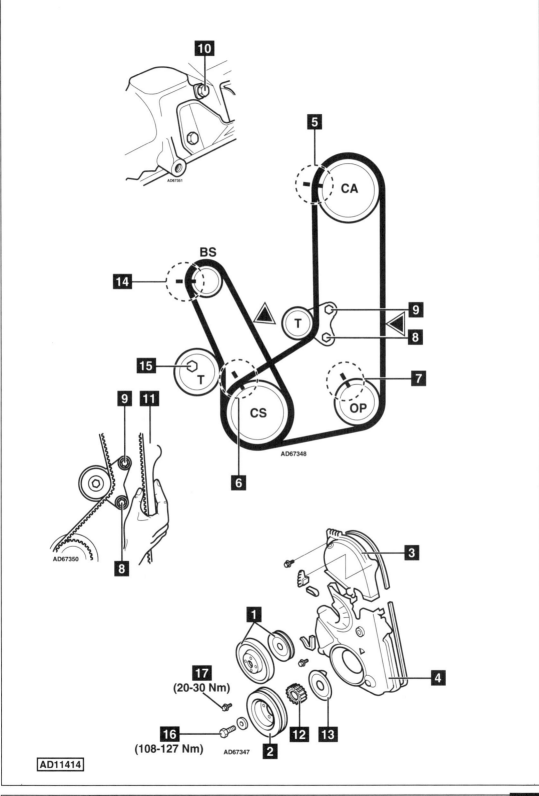

10

5

CA

BS

14

T

9

8

15 T

9 **11**

7

CS

OP

AD67348

6

AD67350

8

3

1

17
(20-30 Nm)

4

16
(108-127 Nm)

2 AD67347

12 **13**

AD67351

AD11414

HYUNDAI

Model:	**Trajet 2,0 • Santa Fe 2,4**
Year:	**2000-03**
Engine Code:	**G4JP, G4JS**

Replacement Interval Guide

Hyundai recommend replacement every 50,000 miles or 5 years.
The previous use and service history of the vehicle must always be taken into account.
Refer to Timing Belt Replacement Intervals at the front of this manual.

Check For Engine Damage

CAUTION: This engine has been identified as an INTERFERENCE engine in which the possibility of valve-to-piston damage in the event of a timing belt failure is MOST LIKELY to occur. A compression check of all cylinders should be performed before removing the cylinder head(s).

Repair Times – hrs

Remove & install	1,70

Special Tools

■ None required.

Removal

Timing Belt

1. Drain coolant.
2. Support engine.
3. Remove:
   ❏ RH engine mounting bracket.
   ❏ Auxiliary drive belt(s).
   ❏ Water pump inlet pipe.
   ❏ Water pump pulley.
   ❏ Crankshaft pulley bolts **1**.
   ❏ Crankshaft pulley **2**.
   ❏ Timing belt upper cover **3**.
   ❏ Timing belt lower cover **4**.
4. Turn crankshaft clockwise to TDC on No.1 cylinder. Ensure timing marks aligned **5**, **6**, **7** & **8**.
5. Slacken tensioner bolt **9**.
6. Remove:
   ❏ Automatic tensioner unit bolts **10**.
   ❏ Automatic tensioner unit **11**.
   NOTE: Mark direction of rotation on belt with chalk if belt is to be reused.

Installation

Timing Belt

1. Check tensioner body for leakage or damage **11**. Replace if necessary.
2. Check pushrod protrusion is 14,5 mm **12**. If not, replace automatic tensioner unit.
3. Slowly compress pushrod into tensioner body until holes aligned **13**. Use vice.
4. Retain pushrod with suitable pin through hole in tensioner body **13**.

5. Install automatic tensioner unit **11**. Tightening torque: 20-27 Nm **10**.
6. Ensure timing marks aligned **5**, **6**, **7** & **8**.
   *NOTE: To check oil pump sprocket positioned correctly: Remove blanking plug from cylinder block. Insert 8 mm diameter Phillips screwdriver in hole. Ensure screwdriver is inserted 60 mm from face of cylinder block. If screwdriver can only be inserted 20 mm: Turn oil pump sprocket 360° and reinsert screwdriver **14**.*
7. Fit timing belt in anti-clockwise direction, starting at crankshaft sprocket. Ensure belt is taut between sprockets.
8. Push tensioner pulley gently against belt. Temporarily tighten tensioner pulley bolt **9**.
9. Remove screwdriver from hole in cylinder block.
10. Ensure timing marks aligned **5**, **6**, **7** & **8**.
11. Remove pin from tensioner body to release pushrod.
12. Turn crankshaft two turns clockwise. Wait 15 minutes.
13. Check pushrod protrusion is 6-9 mm **15**.
14. Tighten tensioner pulley bolt **9**. Tightening torque: 43-55 Nm.
15. Install components in reverse order of removal.
16. Refill cooling system.

Removal

Balancer Shaft Belt

1. Remove timing belt as described previously.
2. Remove:
   ❏ Crankshaft bolt **17**.
   ❏ Crankshaft sprocket **18**.
   ❏ Crankshaft sprocket guide washer **19**.
3. Slacken balancer shaft belt tensioner bolt **16**.
4. Remove balancer shaft belt.
   NOTE: Mark direction of rotation on belt with chalk if belt is to be reused.

Installation

Balancer Shaft Belt

1. Ensure timing marks aligned **6** & **8**.
2. Fit balancer shaft belt to sprockets with no slack at ▽.
3. Turn tensioner pulley firmly clockwise against belt. Tighten bolt **16**. Tightening torque: 15-22 Nm.
4. Apply thumb pressure to belt at ▽. Belt should deflect 5-7 mm.
5. If not: Repeat tensioning procedure.
6. Install components in reverse order of removal.
7. Tighten crankshaft pulley bolt **17**. Tightening torque: 110-130 Nm.
8. Fit timing belt as described previously.

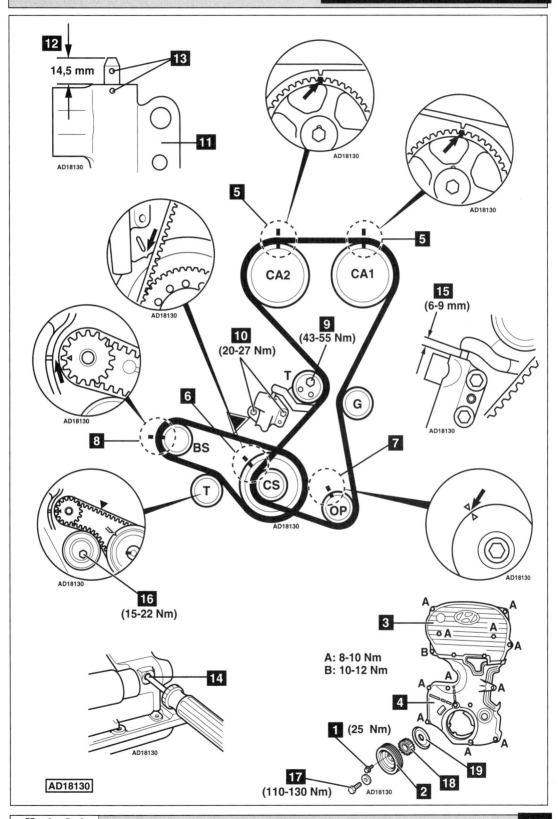

12 14,5 mm

13

11

AD18130

5

AD18130

5

AD18130

CA2 CA1

15 (6-9 mm)

AD18130

9 (43-55 Nm)

10 (20-27 Nm)

T

6

G

8

BS

7

T

CS

OP

AD18130

AD18130

16 (15-22 Nm)

AD18130

14

AD18130

A: 8-10 Nm
B: 10-12 Nm

3

4

1 (25 Nm)

17 (110-130 Nm)

AD18130

2

18

19

HYUNDAI

Model:	**H100 2,5D**
Year:	**1993-03**
Engine Code:	**D4B, D4BA**

Replacement Interval Guide

Hyundai recommend replacement every 54,000 miles or 6 years, whichever occurs first.
The previous use and service history of the vehicle must always be taken into account.
Refer to Timing Belt Replacement Intervals at the front of this manual.

Check For Engine Damage

CAUTION: This engine has been identified as an INTERFERENCE engine in which the possibility of valve-to-piston damage in the event of a timing belt failure is MOST LIKELY to occur.
A compression check of all cylinders should be performed before removing the cylinder head.

Repair Times – hrs

Remove & install	1,80

Special Tools

- Crankshaft pulley holding tool – Hyundai No.09231-43000.

Special Precautions

- Disconnect battery earth lead.
- DO NOT turn crankshaft or camshaft when timing belt removed.
- Remove glow plugs to ease turning engine.
- Turn engine in normal direction of rotation (unless otherwise stated).
- DO NOT turn engine via camshaft or other sprockets.
- Observe all tightening torques.

Removal

Timing Belt

1. Remove:
   - ❑ Viscous fan casing.
   - ❑ Viscous fan.
   - ❑ Auxiliary drive belt(s).
   - ❑ Water pump pulley.
   - ❑ Timing belt upper cover **1**.
2. Hold crankshaft pulley. Use tool No.09231-43000.
3. Remove:
   - ❑ Crankshaft pulley bolt **2**.
   - ❑ Crankshaft pulley **3**.
   - ❑ Timing belt lower cover **4**.
4. Turn crankshaft clockwise to TDC on No.1 cylinder. Ensure timing marks aligned **5**, **6**, **7** & **8**.
5. Slacken tensioner bolts **9**. Move tensioner away from belt. Lightly tighten bolts.
6. Remove timing belt.

Installation

Timing Belt

1. Ensure timing marks aligned **5**, **6**, **7** & **8**.
2. Fit timing belt. Ensure belt is taut on non-tensioned side.
3. Slacken tensioner bolts **9**. Allow tensioner to operate.
4. Turn crankshaft clockwise until camshaft sprocket timing mark **5** moves forward TWO teeth.
5. Tighten tensioner upper bolt.
   Tightening torque: 22-30 Nm.
6. Tighten tensioner lower bolt.
   Tightening torque: 22-30 Nm.
   NOTE: Observe correct tightening sequence, otherwise belt tension will not be correct.
7. Turn crankshaft anti-clockwise until timing marks aligned **5**, **6**, **7** & **8**.
8. Apply thumb pressure to timing belt at ▽ **10**. Belt should deflect 4-5 mm.
9. Install components in reverse order of removal.
10. Tighten crankshaft pulley bolt **2**.
    Tightening torque: 170-190 Nm.

Removal

Balancer Shaft Belt

1. Remove timing belt as described previously.
2. Ensure timing marks aligned **7** & **8**.
3. Slacken nut and bolt **11** & **12**.
4. Move tensioner (T2) towards water pump and tighten nut **11**.
5. Remove balancer shaft belt.

Installation

Balancer Shaft Belt

1. Ensure timing marks aligned **7** & **8**.
2. Fit balancer shaft belt to sprockets with no slack at ▽ **13**.
3. Press down several times on belt at **14**.
4. Slacken tensioner nut **11**. Allow tensioner to operate.
5. Tighten tensioner nut **11**.
   Tightening torque: 22-30 Nm.
6. Tighten tensioner bolt **12**.
   Tightening torque: 22-30 Nm.
   NOTE: Observe correct tightening sequence, otherwise belt tension will not be correct.
7. Apply thumb pressure to balancer shaft belt at ▽ **13**. Belt should deflect 4-5 mm.
8. Fit timing belt as described previously.

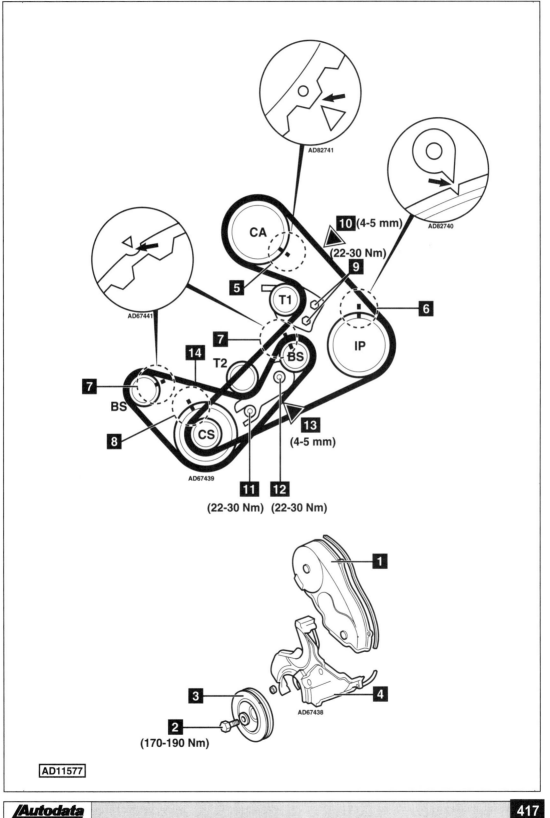

AD82741

AD82740

10 (4-5 mm)

(22-30 Nm)

CA

AD67441

5

T1

9

6

IP

7

14

T2

BS

7

BS

8

CS

13

(4-5 mm)

AD67439

11   **12**

(22-30 Nm) (22-30 Nm)

1

3

4

AD67438

2

(170-190 Nm)

AD11577

ISUZU

Model:	Trooper 3,2 V6 24V • Trooper 3,5 V6 24V
Year:	1992-03
Engine Code:	6VD1 (SOHC), 6VE1 (DOHC)

Replacement Interval Guide

→1998: Isuzu recommend replacement every 54,000 miles or 72 months.
1998→: Isuzu recommend replacement every 72,000 miles or 96 months.
The previous use and service history of the vehicle must always be taken into account.
Refer to Timing Belt Replacement Intervals at the front of this manual.

Check For Engine Damage

CAUTION: This engine has been identified as an INTERFERENCE engine in which the possibility of valve-to-piston damage in the event of a timing belt failure is MOST LIKELY to occur.
A compression check of all cylinders should be performed before removing the cylinder head.

Repair Times – hrs

Remove & install	1,60

Special Tools

■ None required.

Special Precautions

■ Disconnect battery earth lead.
■ DO NOT turn crankshaft or camshaft when timing belt removed.
■ Remove spark plugs to ease turning engine.
■ Turn engine in normal direction of rotation (unless otherwise stated).
■ DO NOT turn engine via camshaft or other sprockets.
■ Observe all tightening torques.

Removal

1. Remove:
   ❏ Air filter and hose.
   ❏ Upper radiator shroud.
   ❏ Lower radiator shroud.
   ❏ Cooling fan and viscous coupling.
   ❏ Cooling fan pulley.
   ❏ Auxiliary drive belts.
   ❏ Oil cooler hose.
   ❏ Timing belt RH upper cover **1**.
   ❏ Timing belt LH upper cover **2**.
   ❏ Crankshaft pulley **3**.
   ❏ Timing belt lower cover **4**.
2. Turn crankshaft until timing marks aligned **5** & **6**. Ensure marks on belt aligned with marks on sprockets **7** & **8**.
   NOTE: These marks DO NOT indicate TDC.
   1998→: Timing marks of camshaft sprockets in alternative location.

3. Slacken tensioner pulley bolt **9**. Turn tensioner pulley clockwise away from belt. Lightly tighten bolt.
4. Remove:
   ❏ Automatic tensioner unit bolts **10** & **11**.
   ❏ Automatic tensioner unit **12**.
   ❏ Timing belt.

Installation

1. Ensure timing marks aligned **5** & **6**.
2. Slowly compress pushrod into tensioner body **12** with a force of approximately 100 kg until holes aligned. Retain pushrod with 1,4 mm diameter pin through hole in tensioner body **13**.
3. Install automatic tensioner unit to cylinder block. Tighten bolts to 22 Nm **10** & **11**.
4. Fit timing belt in anti-clockwise direction, starting at crankshaft sprocket. Ensure belt is taut between sprockets on non-tensioned side. Ensure timing marks aligned **7** & **8**.
   *NOTE: 'ISUZU' lettering on belt should be readable from front of engine **16**.*
5. Slacken tensioner pulley bolt **9**. Turn tensioner pulley anti-clockwise to tension belt. Tighten bolt to 44 Nm.
6. Remove pin from tensioner body to release pushrod.
7. Fit crankshaft pulley **3**.
8. Turn crankshaft slowly two turns in direction of rotation until timing marks aligned **5** & **6**.
9. Check extended length of pushrod is 4,0-6,0 mm **14**.
10. If not: Repeat operations 2-8.
11. Install components in reverse order of removal.
12. Tighten crankshaft pulley bolt **15**.
    Tightening torque: 167 Nm.

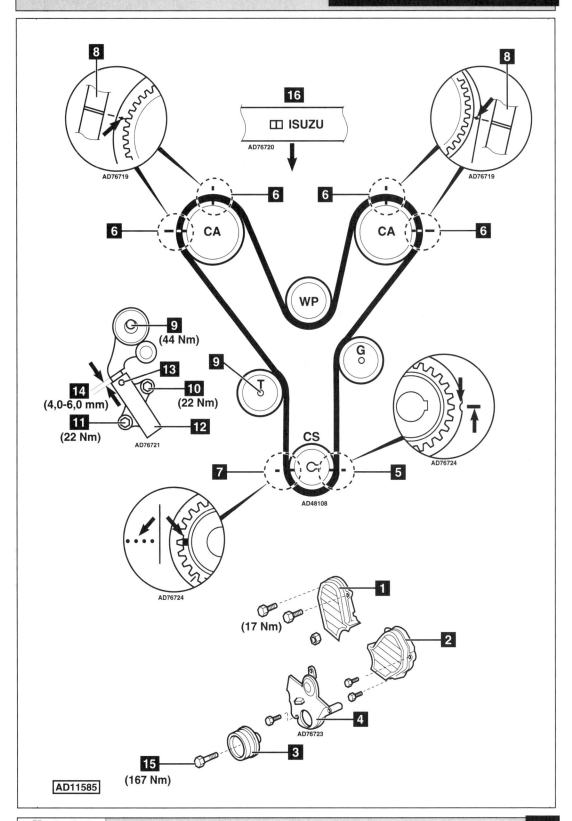

8

16

☐☐ ISUZU

AD76720

8

AD76719

AD76719

6

6

6

CA

CA

6

WP

G

9
(44 Nm)

9

T

13

14
(4,0-6,0 mm)

10
(22 Nm)

11
(22 Nm)

12

AD76721

CS

C

AD76724

7

5

AD48108

AD76724

1

(17 Nm)

2

4

AD76723

3

15
(167 Nm)

AD11585

Model:	**Trooper 2,8/3,1 TD**
Year:	**1988-98**
Engine Code:	**4JB1T, 4JG2T**

Replacement Interval Guide

Isuzu recommend replacement every
60,000 miles (pre 92) or every 54,000 miles or
72 months (1992→).
*The previous use and service history of the vehicle
must always be taken into account.*
*Refer to Timing Belt Replacement Intervals at the front
of this manual.*

Check For Engine Damage

*CAUTION: This engine has been identified as an
INTERFERENCE engine in which the possibility of
valve-to-piston damage in the event of a timing belt
failure is MOST LIKELY to occur.*
*A compression check of all cylinders should be
performed before removing the cylinder head.*

Repair Times – hrs

Remove & install	2,00

Special Tools

■ Spring balance.

Special Precautions

■ Disconnect battery earth lead.
■ DO NOT turn crankshaft or camshaft when timing belt
    removed.
■ Remove glow plugs to ease turning engine.
■ Turn engine in normal direction of rotation (unless
    otherwise stated).
■ DO NOT turn engine via camshaft or other sprockets.
■ Observe all tightening torques.
■ Check diesel injection pump timing after belt
    replacement.

Removal

1. Drain coolant.
2. Remove:
    ❏ Auxiliary drive belts.
    ❏ Cooling fan.
    ❏ Water pump pulley.
    ❏ Timing belt upper cover **1**.
    ❏ Crankshaft pulley (4 bolts) **3**.
    ❏ Timing belt lower cover **2**.
    ❏ Coolant pipe from front of engine.
    ❏ Camshaft sprocket flange **6**.
    ❏ Injection pump sprocket flange **7**.
3. Turn crankshaft clockwise to TDC on
    No.1 cylinder. Ensure timing marks aligned **10**.
4. Screw suitable locking bolts into camshaft and
    injection pump sprockets **4** & **5**.
5. Remove tensioner lever **8**.

6. Slacken tensioner pulley bolt **9**. Move tensioner
    away from belt. Lightly tighten bolt.
7. Remove timing belt.

Installation

1. Ensure locking bolts located correctly **4** & **5**.
2. Ensure timing marks aligned **10**.
3. Fit timing belt in clockwise direction, starting at
    crankshaft sprocket.
    **NOTE: 'ISUZU' lettering on belt should be
    readable from front of engine 16.**
4. Slacken tensioner pulley bolt **9**. Push tensioner
    pulley against belt.
5. Ensure belt is taut between injection pump
    sprocket and tensioner pulley as well as
    crankshaft sprocket and camshaft sprocket.
6. Position tensioner lever **8** against tensioner
    pulley housing.
7. Remove locking bolts from sprockets **4** & **5**.
8. Apply a load of 10-12 kg to tensioner lever. Use
    spring balance **11**.
    **NOTE: 4JG2T (1992→): Apply a load of 9 kg to
    tensioner lever.**
9. Tighten tensioner pulley bolt to 80 Nm.
10. Turn crankshaft slowly 45° anti-clockwise. Use
    spring balance to tension belt.
11. Repeat tensioning procedure at 45° intervals for
    one anti-clockwise turn of the crankshaft.
    **NOTE: DO NOT turn crankshaft clockwise during
    tensioning procedure.**
12. Fit tensioner lever to original position. Tighten
    nuts **12**.
13. Install:
    ❏ Injection pump sprocket flange **6**.
    ❏ Camshaft sprocket flange **7**.
    ❏ Water pump pulley.
14. Install components in reverse order of removal.
15. Refill coolant.
16. 2,8: Reset timing belt replacement warning lamp
    as follows:
    ❏ Remove speedometer and operate switch **13**.
17. 3,1: Reset timing belt replacement warning lamp
    as follows:
    ❏ Remove instrument panel.
    ❏ Remove masking tape from hole **14**.
    ❏ Remove screw from hole **15**.
    ❏ Insert screw removed from hole **15** in hole **14**
        and cover hole **15** with masking tape.
    ❏ Fit instrument panel.
    ❏ Reverse procedure if timing belt previously
        changed.

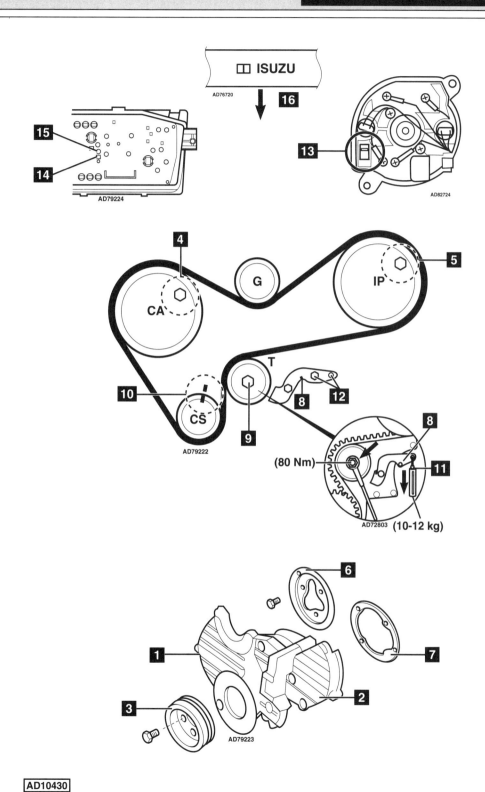

□ ISUZU

AD76720

16

15

14

AD79224

13

AD82724

4

G

5

CA

IP

T

10

CS

AD79222

9

8

12

8

(80 Nm)

11

AD72803 (10-12 kg)

6

1

7

3

2

AD79223

AD10430

Model:	**Daily 2,5 D • Turbo Daily 2,5 TD**
Year:	**1981-96**
Engine Code:	**8140.07, 8140.21, 8140.27S, 8140.47/R, 8140.61, 8140.97**

Replacement Interval Guide

Iveco recommend:
Bus and ambulance models with automatic transmission – replacement every 30,000 kilometres (18,600 miles).
Except bus and ambulance models with automatic transmission – replacement every 60,000 kilometres (37,200 miles).
The previous use and service history of the vehicle must always be taken into account.
Refer to Timing Belt Replacement Intervals at the front of this manual.

Check For Engine Damage

CAUTION: This engine has been identified as an INTERFERENCE engine in which the possibility of valve-to-piston damage in the event of a timing belt failure is MOST LIKELY to occur.
A compression check of all cylinders should be performed before removing the cylinder head.

Repair Times – hrs

Remove & install	2,00

Special Tools

■ Timing pins – Iveco No.A60617.

Special Precautions

■ Disconnect battery earth lead.
■ DO NOT turn crankshaft or camshaft when timing belt removed.
■ Remove glow plugs to ease turning engine.
■ Turn engine in normal direction of rotation (unless otherwise stated).
■ DO NOT turn engine via camshaft or other sprockets.
■ Observe all tightening torques.
■ Check diesel injection pump timing after belt replacement.

Removal

1. Remove:
   ❏ Auxiliary drive belt(s).
   ❏ Timing belt cover **1**.
2. Turn crankshaft clockwise to TDC on No.1 cylinder. Ensure camshaft timing marks aligned **2**.
3. Insert timing pins in crankshaft pulley and injection pump sprocket **3** & **4**.
   Tool No.A60617.
   *NOTE: Some engines have timing mark on crankshaft pulley **5**.*

4. Slacken tensioner nut **6**. Move tensioner pulley away from belt and lightly tighten nut.
5. Remove timing belt.

Installation

NOTE: DO NOT fit used belt.
1. Ensure timing pins located correctly **3** & **4**. Ensure timing marks aligned **2** & **5**.
2. Fit timing belt in anti-clockwise direction, starting at crankshaft sprocket. Ensure belt is taut between sprockets on non-tensioned side.
3. Slacken tensioner nut **6**. Allow tensioner to operate.
4. Remove timing pins.
5. Turn crankshaft two turns clockwise. Tighten tensioner pulley nut to 41 Nm **6**.
6. Ensure timing pins can be inserted **3** & **4**. Ensure timing marks aligned **2** & **5**.
7. Install components in reverse order of removal.

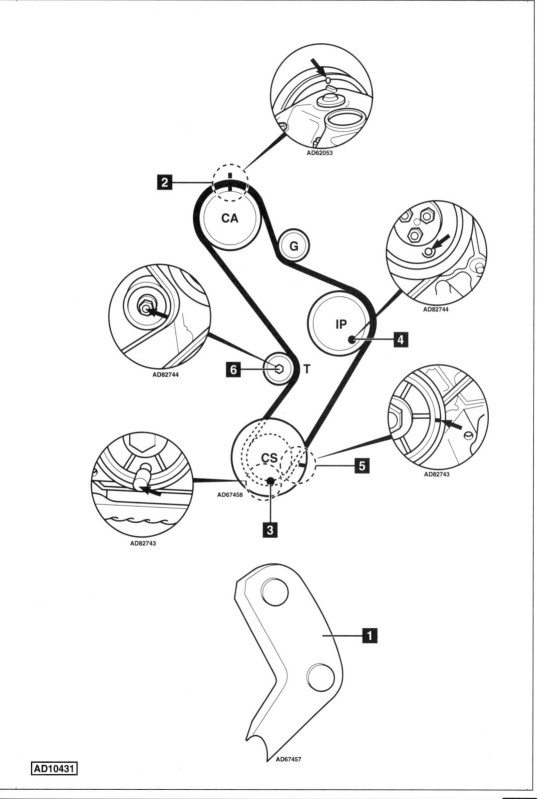

AD62053

2

CA

G

AD82744

IP

4

AD82744

6

T

AD82743

5

CS

AD67458

AD82743

3

1

AD67457

AD10431

Model:	Daily 2,5 D • Turbo Daily 2,5 TD • Turbo Daily 2,8 TD
Year:	1996-00
Engine Code:	8140.23.37, 8140.43.37, 8140.47R.2790, 8140.67F.37

Replacement Interval Guide

Iveco recommend:

→2000 MY:

Bus and ambulance models with automatic transmission – replacement every 30,000 kilometres (18,600 miles).
Except bus and ambulance models with automatic transmission – replacement every 100,000 kilometres (62,000 miles).

2000 MY→:

Replacement every 90,000 kilometres (56,000 miles).
The previous use and service history of the vehicle must always be taken into account.
Refer to Timing Belt Replacement Intervals at the front of this manual.

Check For Engine Damage

CAUTION: This engine has been identified as an INTERFERENCE engine in which the possibility of valve-to-piston damage in the event of a timing belt failure is MOST LIKELY to occur.
A compression check of all cylinders should be performed before removing the cylinder head.

Repair Times – hrs

Remove & install	2,00

Special Tools

- Flywheel timing pin – Iveco No.99360608.
- Injection pump locking pin – Iveco No.99360608.
- Clavis tension gauge.

Special Precautions

- Disconnect battery earth lead.
- DO NOT turn crankshaft or camshaft when timing belt removed.
- Remove glow plugs to ease turning engine.
- Turn engine in normal direction of rotation (unless otherwise stated).
- DO NOT turn engine via camshaft or other sprockets.
- Observe all tightening torques.
- Check diesel injection pump timing after belt replacement.

Removal

1. Remove:
   - Engine undershield.
   - Auxiliary drive belt.
2. Slacken crankshaft pulley bolt **1**.
3. Remove:
   - Water pump pulley.
   - Air intake hoses.
   - Engine top cover **2**.
   - Timing belt upper cover **3**.
   - Crankshaft pulley bolt **1**.
   - Crankshaft pulley **4**.
4. Turn crankshaft to TDC on No.1 cylinder. Insert timing pin in flywheel **5**. Tool No.99360608.
5. Ensure camshaft sprocket timing mark aligned with mark on cylinder head cover **6**.

6. Insert locking pin in injection pump sprocket **7**. Tool No.99360608.
7. Remove:
   - Bolts **8**.
   - Timing belt lower cover nut **9**.
   - Timing belt lower cover **10**.
8. Slacken tensioner nuts **11** & **12**.
9. Lever tensioner pulley away from belt. Insert spanner in tensioner plunger **13**.
10. Remove timing belt.

Installation

1. Ensure timing marks aligned **6**.
2. Ensure flywheel timing pin located correctly **5**.
3. Ensure injection pump locking pin located correctly **7**.
4. Ensure spanner fitted to tensioner plunger **13**.
5. Fit timing belt, starting at crankshaft sprocket. Ensure belt is taut on non-tensioned side.
6. Remove:
   - Spanner **13**.
   - Flywheel timing pin **5**.
   - Injection pump locking pin **7**.
7. Tension belt as follows:

→2000MY:

8. Turn crankshaft three turns clockwise. Maintain turning force on crankshaft to tension belt.
9. Ensure timing pin can be inserted in flywheel **5**.
10. Tighten tensioner nut **12**.
    Tightening torque: 37-45 Nm.
11. Tighten tensioner nut **11**.
    Tightening torque: 37-45 Nm.
12. Apply a load of 10 kg to belt at ▽. Belt should deflect 7-8 mm.

2000MY → (HSN belt):

13. Turn crankshaft two turns clockwise.
14. Ensure timing marks aligned **6**.
15. Ensure timing pin can be inserted in flywheel **5**.
16. Ensure injection pump locking pin can be inserted easily **7**.
17. Temporarily refit crankshaft pulley bolt **1**.
18. Apply clockwise torque of 28-30 Nm to crankshaft pulley bolt to tension belt.
    NOTE: Ensure camshaft sprocket timing mark 6 is still aligned.
19. Tighten tensioner nut **12**.
    Tightening torque: 37-45 Nm.
20. Tighten tensioner nut **11**.
    Tightening torque: 37-45 Nm.
21. Attach tension gauge to belt at ▽. Tension gauge should indicate 88-112 Hz.
22. Remove crankshaft pulley bolt **1**.

All models:

23. Install components in reverse order of removal.
24. Tighten crankshaft pulley bolt **1**.
    Tightening torque: 200 Nm.

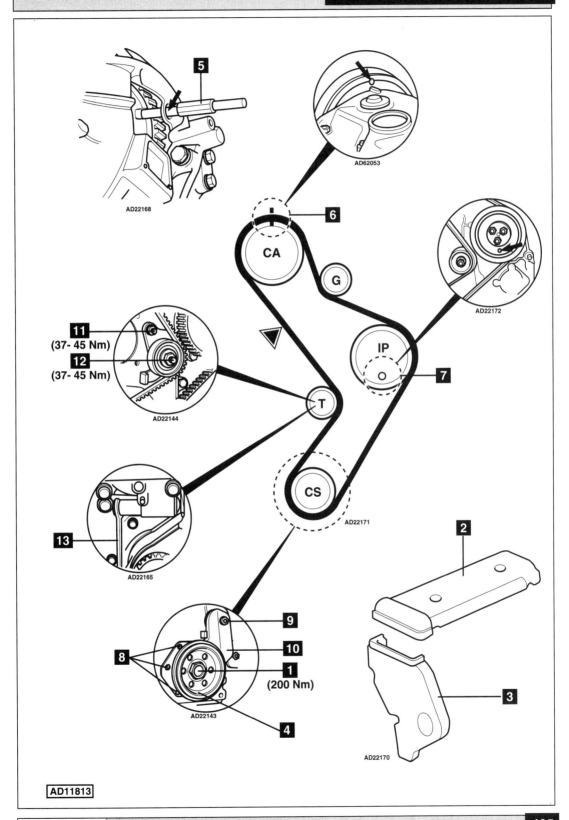

5

AD22168

AD62053

6

CA

G

AD22172

11
(37- 45 Nm)
12
(37- 45 Nm)

AD22144

IP

7

T

13

AD22165

CS

AD22171

2

3

AD22170

9

10

8

1
(200 Nm)

4

AD22143

AD11813

Model:	**Sephia 1,5/1,6 • Mentor 1,5/1,6**
Year:	**1994-99**
Engine Code:	**B5, B6**

Replacement Interval Guide

Kia recommend replacement every
48,000 miles.
*The previous use and service history of the vehicle
must always be taken into account.*
*Refer to Timing Belt Replacement Intervals at the front
of this manual.*

Check For Engine Damage

*CAUTION: This engine has been identified as an
INTERFERENCE engine in which the possibility of
valve-to-piston damage in the event of a timing belt
failure is MOST LIKELY to occur.*
*A compression check of all cylinders should be
performed before removing the cylinder head.*

Repair Times – hrs

Remove & install	1,50

Special Tools

■ None required.

Special Precautions

■ Disconnect battery earth lead.
■ DO NOT turn crankshaft or camshaft when timing belt
   removed.
■ Remove spark plugs to ease turning engine.
■ Turn engine in normal direction of rotation (unless
   otherwise stated).
■ DO NOT turn engine via camshaft or other sprockets.
■ Observe all tightening torques.

Removal

1. Remove:
   ❑ Auxiliary drive belts.
   ❑ Water pump pulley **1**.
   ❑ Crankshaft pulley bolts **2**.
   ❑ Crankshaft sprocket guide washer **3**.
   ❑ Crankshaft pulley **4**.
   ❑ Crankshaft pulley centre bolt **5**.
   ❑ Crankshaft pulley boss **6**.
   ❑ Timing belt upper cover **7**.
   ❑ Timing belt lower cover **8**.
2. Turn crankshaft clockwise to TDC on
   No.1 cylinder. Ensure timing marks aligned **9**
   & **10**.
3. Slacken tensioner bolt **11**. Disconnect tensioner
   spring **12**.
4. Remove timing belt.

Installation

1. Ensure timing marks aligned **9** & **10**.
2. Connect tensioner spring **12**. Push tensioner
   away to extend spring. Lightly tighten bolt **11**.
3. Fit timing belt in anti-clockwise direction, starting
   at crankshaft sprocket.
4. Turn crankshaft two turns clockwise.
5. Ensure timing marks aligned **9** & **10**. Allow
   spring to pull tensioner against belt.
6. Tighten tensioner bolt to 19-25 Nm **11**. Ensure
   only spring tension applied to belt.
7. Turn crankshaft two turns clockwise.
8. Ensure timing marks aligned **9** & **10**.
9. Apply a load of 10 kg to belt at ▽. Belt should
   deflect 11-13 mm.
10. If not: Repeat operations 3-9.
11. Install components in reverse order of removal.
12. Tighten crankshaft pulley centre bolt **5**.
    Tightening torque: 156-167 Nm.

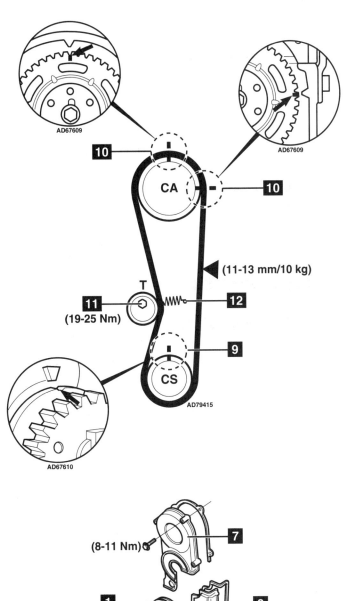

AD67609

AD67609

10

CA

10

(11-13 mm/10 kg)

T

11
(19-25 Nm)

12

9

CS

AD79415

AD67610

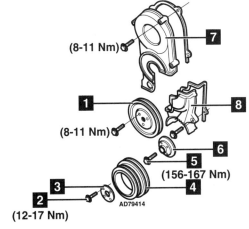

(8-11 Nm)

7

1

8

(8-11 Nm)

6

5
(156-167 Nm)

3

4

2
(12-17 Nm)

AD79414

AD10432

KIA

Model:	**Mentor 1,5 DOHC • Sephia 1,5 DOHC**
Year:	**1996-02**
Engine Code:	**B5 DOHC**

Replacement Interval Guide

Kia recommend:
→1999: Replacement every 48,000 miles.
1999→: Replacement every 54,000 miles or 6 years.
The previous use and service history of the vehicle must always be taken into account.
Refer to Timing Belt Replacement Intervals at the front of this manual.

Check For Engine Damage

CAUTION: This engine has been identified as an INTERFERENCE engine in which the possibility of valve-to-piston damage in the event of a timing belt failure is MOST LIKELY to occur.
A compression check of all cylinders should be performed before removing the cylinder head.

Repair Times – hrs

Remove & install	2,20

Special Tools

- None required.

Special Precautions

- Disconnect battery earth lead.
- DO NOT turn crankshaft or camshaft when timing belt removed.
- Remove spark plugs to ease turning engine.
- Turn engine in normal direction of rotation (unless otherwise stated).
- DO NOT turn engine via camshaft or other sprockets.
- Observe all tightening torques.

Removal

1. Raise and support front of vehicle.
2. Remove:
   - ❏ RH front wheel.
   - ❏ RH front wheel arch liner.
   - ❏ Engine undershield.
   - ❏ Auxiliary drive belts.
   - ❏ Water pump pulley.
   - ❏ Crankshaft pulley bolts **1**.
   - ❏ Crankshaft pulley **2**.
   - ❏ Crankshaft bolt **3**.
   - ❏ Crankshaft pulley boss **4**.
   - ❏ Timing belt covers **5**, **6** & **7**.
3. Turn crankshaft clockwise until timing marks aligned **8** & **9**.
4. Slacken tensioner bolt **10**.

5. Move tensioner away from belt. Lightly tighten bolt.
6. Remove timing belt.
   NOTE: Mark direction of rotation on belt with chalk if belt is to be reused.

Installation

1. Remove tensioner spring **11**.
2. Check free length of tensioner spring is 59,5 mm. Replace spring if necessary.
3. Fit tensioner spring **11**.
4. Ensure timing marks aligned **8** & **9**.
5. Fit timing belt in anti-clockwise direction, starting at crankshaft sprocket. Ensure belt is taut between sprockets.
6. Slacken tensioner bolt **10**. Allow tensioner to operate. Lightly tighten bolt.
7. Turn crankshaft two turns clockwise. Ensure timing marks aligned **8** & **9**.
8. Turn crankshaft 1 5/6 turns clockwise until crankshaft sprocket mark aligned with tension setting mark **12** (60° BTDC).
9. Slacken tensioner bolt **10**. Allow tensioner to operate. Tighten bolt to 37-52 Nm.
   *NOTE: Belt tension must only be set at tension setting mark **12**.*
10. Turn crankshaft 2 1/6 turns clockwise until timing marks aligned **8** & **9**.
11. Apply a load of 10 kg to belt at ▽. Belt should deflect 9,0-11,5 mm. If not: Repeat tensioning procedure.
12. Install components in reverse order of removal.
13. Tighten crankshaft bolt to 157-166 Nm **3**.
14. Tighten crankshaft pulley bolts to 12-17 Nm **1**.

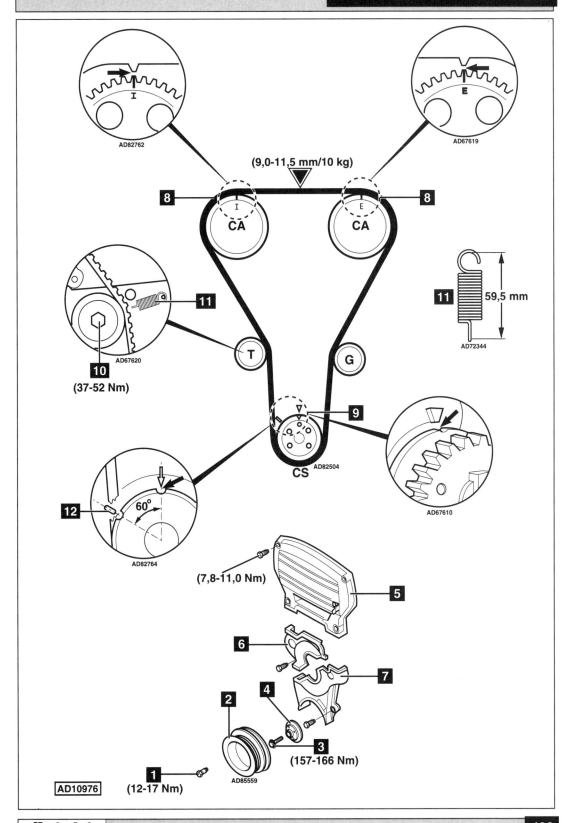

(9,0-11,5 mm/10 kg)

AD82762

AD67619

8   CA   CA   **8**

11   59,5 mm

11

AD72344

10
(37-52 Nm)

AD67620

T   G

9

CS   AD82504

AD67610

12   60°

AD82764

(7,8-11,0 Nm)

5

6

7

2

4

3
(157-166 Nm)

AD85559

1

AD10976   (12-17 Nm)

KIA

Model:	**Sportage 2,0i**
Year:	**1995-01**
Engine Code:	**FE-SOHC**

Replacement Interval Guide

Kia recommend:
→1999: Replacement every 60,000 miles.
1999→: Replacement every 54,000 miles or
6 years.
*The previous use and service history of the vehicle
must always be taken into account.*
*Refer to Timing Belt Replacement Intervals at the front
of this manual.*

Check For Engine Damage

*CAUTION: This engine has been identified as an
INTERFERENCE engine in which the possibility of
valve-to-piston damage in the event of a timing belt
failure is MOST LIKELY to occur.*
*A compression check of all cylinders should be
performed before removing the cylinder head.*

Repair Times – hrs

Remove & install	1,70

Special Tools

- None required.

Special Precautions

- Disconnect battery earth lead.
- DO NOT turn crankshaft or camshaft when timing belt
  removed.
- Remove spark plugs to ease turning engine.
- Turn engine in normal direction of rotation (unless
  otherwise stated).
- DO NOT turn engine via camshaft or other sprockets.
- Observe all tightening torques.

Removal

1. Remove:
   - ❏ Auxiliary drive belts.
   - ❏ Crankshaft pulley bolts **1**.
   - ❏ Crankshaft pulley **2**.
   - ❏ Timing belt upper cover **3**.
   - ❏ Timing belt lower cover **4**.
2. Turn crankshaft until camshaft sprocket '2'
   timing mark and crankshaft timing mark
   aligned **5** & **6**.
3. Remove crankshaft sprocket guide washer **8**.
4. Slacken tensioner bolt **7**. Move tensioner away
   from belt. Lightly tighten bolt.
5. Remove timing belt.

Installation

1. Check free length of tensioner spring **9**.
   Replace if necessary.
2. Spin tensioner pulley and guide pulley **10** & **11**.
   Check for bearing roughness and noise.
   Replace if necessary.
3. Ensure timing marks aligned **5** & **6**.
   **NOTE: If camshaft sprocket removed, install
   sprocket with dowel inserted in hole adjacent to
   '2' mark.**
4. Fit timing belt, starting at crankshaft sprocket.
   Ensure belt is taut between sprockets.
5. Slacken tensioner bolt. Turn crankshaft two
   turns clockwise. Tighten tensioner bolt to
   37-52 Nm **7**.
6. Ensure timing marks aligned **5** & **6**.
7. Apply a load of 10 kg to belt at ▽.
8. Belt should deflect 7,5-8,5 mm.
9. Fit crankshaft sprocket guide washer (convex
   side towards belt) **8**.
10. Install components in reverse order of removal.
11. Tighten crankshaft pulley bolts to 12-17 Nm **1**.

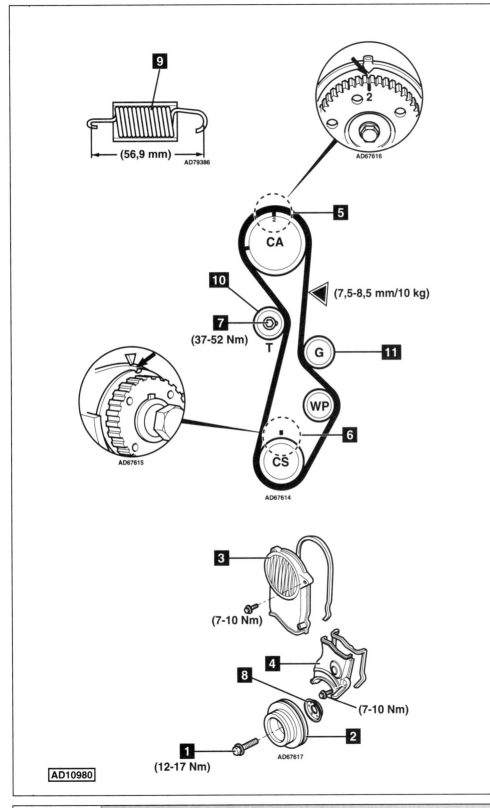

9 (56,9 mm) AD79386

AD67616

5

CA

10

7 (37-52 Nm)

T

(7,5-8,5 mm/10 kg)

G **11**

WP

6

AD67615

CS

AD67614

3 (7-10 Nm)

4 (7-10 Nm)

8

2

1 (12-17 Nm)

AD67617

AD10980

KIA

Model:	**Sportage 2,0 16V**
Year:	**1995-02**
Engine Code:	**FE DOHC**

Replacement Interval Guide

Kia recommend:
→1999: Replacement every 60,000 miles.
1999→: Replacement every 54,000 miles or 6 years.
The previous use and service history of the vehicle must always be taken into account.
Refer to Timing Belt Replacement Intervals at the front of this manual.

Check For Engine Damage

CAUTION: This engine has been identified as an INTERFERENCE engine in which the possibility of valve-to-piston damage in the event of a timing belt failure is MOST LIKELY to occur.
A compression check of all cylinders should be performed before removing the cylinder head.

Repair Times – hrs

Remove & install	2,40

Special Tools

- None required.

Special Precautions

- Disconnect battery earth lead.
- DO NOT turn crankshaft or camshaft when timing belt removed.
- Remove spark plugs to ease turning engine.
- Turn engine in normal direction of rotation (unless otherwise stated).
- DO NOT turn engine via camshaft or other sprockets.
- Observe all tightening torques.

Removal

1. Drain coolant.
2. Remove:
   - ❏ Auxiliary drive belts.
   - ❏ Cooling fan and bracket.
   - ❏ Thermostat housing **1**.
   - ❏ Crankshaft pulley bolts **2**.
   - ❏ Crankshaft pulley **3**.
   - ❏ Timing belt upper cover **4**.
   - ❏ Timing belt lower cover **5**.
3. Turn crankshaft to TDC on No.1 cylinder. Ensure timing marks aligned **6** & **7**.
4. Remove crankshaft sprocket guide washer **8**.
5. Slacken tensioner bolt **9**.
6. Move tensioner away from belt. Lightly tighten bolt **9**.
7. Remove timing belt.

Installation

1. Ensure timing marks aligned **6** & **7**.
2. Fit timing belt in anti-clockwise direction, starting at crankshaft sprocket. Ensure belt is taut between sprockets.
3. Slacken tensioner bolt **9**. Turn crankshaft two turns clockwise. Ensure timing marks aligned **6** & **7**.
4. If not: Repeat installation and tensioning procedures.
5. Turn crankshaft clockwise until 'S' mark on exhaust camshaft sprocket aligned with timing mark **10**.
6. Tighten tensioner bolt to 37-52 Nm **9**.
7. Apply a load of 10 kg to belt at ▽. Belt should deflect 7,5-8,5 mm.
8. Install components in reverse order of removal.
9. Tighten crankshaft pulley bolts to 12-17 Nm **2**.
10. Refill cooling system.

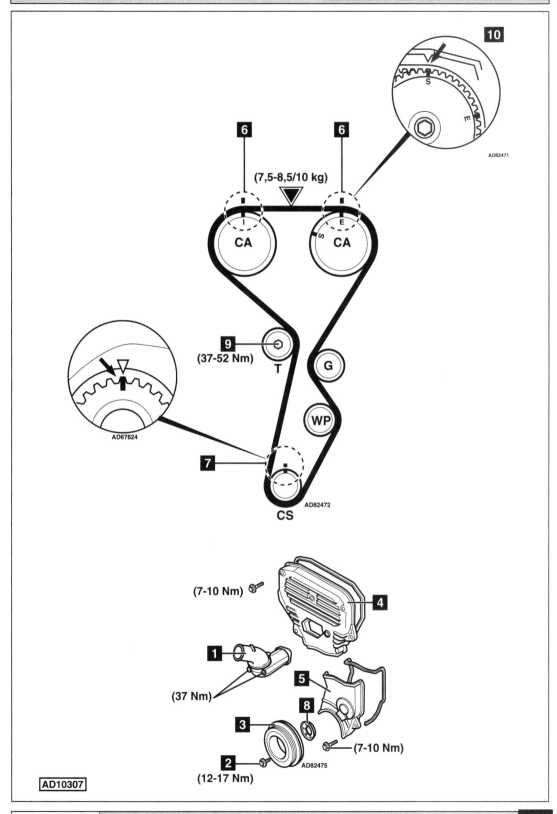

10

AD82471

6　　　　**6**

(7,5-8,5/10 kg)

CA　　CA

9 (37-52 Nm) T

G

WP

AD67624

7

CS　　AD82472

(7-10 Nm)　　**4**

1

(37 Nm)　　**5**

3　　**8**

2　　AD82475　　(7-10 Nm)

(12-17 Nm)

AD10307

Model:	**Freelander 1,8i**
Year:	**1997-00**
Engine Code:	**18F16 – manual tensioner**

Replacement Interval Guide

Land Rover recommend replacement every 72,000 miles.
The previous use and service history of the vehicle must always be taken into account.
Refer to Timing Belt Replacement Intervals at the front of this manual.

Check For Engine Damage

CAUTION: This engine has been identified as an INTERFERENCE engine in which the possibility of valve-to-piston damage in the event of a timing belt failure is MOST LIKELY to occur.
A compression check of all cylinders should be performed before removing the cylinder head(s).

Repair Times – hrs

Remove & install	2,05

Special Tools

- Camshaft sprocket locking tool – Land Rover No.LRT-12-134.
- Flywheel locking tool – Land Rover No.LRT-12-145.

Special Precautions

- Disconnect battery earth lead.
- DO NOT turn crankshaft or camshaft when timing belt removed.
- Remove spark plugs to ease turning engine.
- Turn engine in normal direction of rotation (unless otherwise stated).
- DO NOT turn engine via camshaft or other sprockets.
- Observe all tightening torques.

Removal

1. Raise and support front of vehicle.
2. Remove:
   - ❑ RH front wheel.
   - ❑ Engine undershield.
   - ❑ RH splash guard.
   - ❑ Starter motor.
   - ❑ Auxiliary drive belt.
   - ❑ Timing belt upper cover **1**.
3. Turn crankshaft clockwise until timing marks on camshaft sprockets aligned with marks on timing belt rear cover (90° BTDC) **2**.
4. Lock camshaft sprockets. Use tool No.LRT-12-134 **3**.
5. Lock flywheel. Use tool No.LRT-12-145.
6. Support engine.
7. Remove:
   - ❑ RH engine mounting.
   - ❑ Crankshaft pulley bolt **10**.
   - ❑ Crankshaft pulley **4**.
   - ❑ Timing belt lower cover and seal **5**.
   NOTE: If reusing old belt: Mark position of tensioner for refitting.
8. Slacken tensioner pulley Allen bolt 1/2 turn **6**.
9. Slacken tensioner backplate bolt 1/2 turn **7**.

10. Move tensioner away from belt. Lightly tighten tensioner backplate bolt **7**.
11. Remove timing belt.
    WARNING: If belt has been in use for more than 48,000 miles: Fit new belt.

Installation

*NOTE: If replacement timing belt is fitted: Fit new tensioner spring and pillar bolt **8**. A sleeve is not fitted to replacement spring.*

1. Ensure timing marks on camshaft sprockets aligned with marks on timing belt rear cover **2**.
2. Ensure locking tool located correctly in camshaft sprockets **3**.
3. Ensure crankshaft sprocket dots aligned with flange on oil pump **9** (90° BTDC).
4. Fit timing belt in anti-clockwise direction, starting at crankshaft sprocket. Ensure belt is taut between sprockets.
5. Install:
   - ❑ Timing belt lower cover and seal **5**.
   - ❑ Crankshaft pulley **4**.
6. Remove locking tool from camshaft sprockets **3**.

Original timing belt – proceed as follows:

7. Slacken tensioner backplate bolt **7**.
8. Turn tensioner until aligned with mark made previously.
9. Tighten tensioner backplate bolt **7**. Tightening torque: 10 Nm.
10. Tighten tensioner pulley Allen bolt to 45 Nm **6**.
11. Install components in reverse order of removal.
12. Tighten crankshaft pulley bolt to 205 Nm **10**.

Replacement timing belt – proceed as follows:

7. Slacken tensioner backplate bolt **7**.
8. Push tensioner against belt. Use finger pressure only.
9. Tighten tensioner backplate bolt **7**. Tightening torque: 10 Nm.
10. Turn crankshaft two turns clockwise.
11. Slacken tensioner backplate bolt **7**.
12. Ensure tensioner is pressing on belt.
13. Tighten tensioner pulley Allen bolt to 45 Nm **6**.
14. Tighten tensioner backplate bolt **7**. Tightening torque: 10 Nm.
15. Install components in reverse order of removal.
16. Tighten crankshaft pulley bolt to 205 Nm **10**.

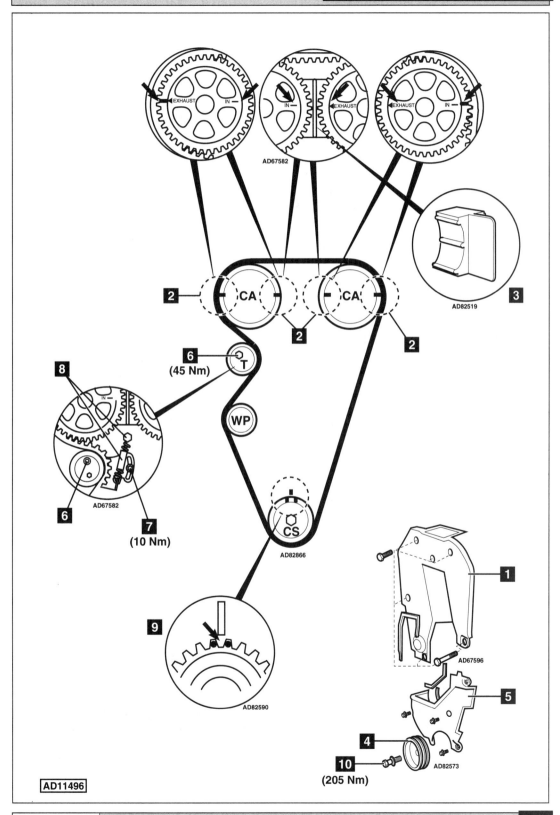

AD67582

2 | CA   CA

3 | AD82519

2

2

2

6 | T | (45 Nm)

8

6

AD67582

7 | (10 Nm)

WP

CS | AD82866

9 | AD82590

1

AD67596

5 | AD82573

4

10 | (205 Nm)

AD11496

LAND ROVER

Model:	**Freelander 1,8i**
Year:	**1997-00**
Engine Code:	**18K16 – automatic tensioner**

Replacement Interval Guide

Land Rover recommend replacement every 72,000 miles.
The previous use and service history of the vehicle must always be taken into account.
Refer to Timing Belt Replacement Intervals at the front of this manual.

Check For Engine Damage

CAUTION: This engine has been identified as an INTERFERENCE engine in which the possibility of valve-to-piston damage in the event of a timing belt failure is MOST LIKELY to occur.
A compression check of all cylinders should be performed before removing the cylinder head(s).

Repair Times – hrs

Remove & install	2,05

Special Tools

- Camshaft sprocket locking tool – Land Rover No.LRT-12-134.
- Flywheel locking tool – Land Rover No.LRT-12-145.

Special Precautions

- Disconnect battery earth lead.
- DO NOT turn crankshaft or camshaft when timing belt removed.
- Remove spark plugs to ease turning engine.
- Turn engine in normal direction of rotation (unless otherwise stated).
- DO NOT turn engine via camshaft or other sprockets.
- Observe all tightening torques.

Removal

1. Raise and support front of vehicle.
2. Remove:
   - ❏ RH front wheel.
   - ❏ Engine undershield.
   - ❏ RH splash guard.
   - ❏ Starter motor.
   - ❏ Auxiliary drive belt.
   - ❏ Timing belt upper cover **1**.
3. Turn crankshaft clockwise until timing marks on camshaft sprockets aligned with marks on timing belt rear cover (90° BTDC) **2**.
4. Lock camshaft sprockets. Use tool No.LRT-12-134 **3**.
5. Lock flywheel. Use tool No.LRT-12-145.
6. Support engine.

7. Remove:
   - ❏ RH engine mounting.
   - ❏ Crankshaft pulley bolt **13**.
   - ❏ Crankshaft pulley **4**.
   - ❏ Timing belt lower cover and seal **5**.
8. Remove tensioner bolt **6**. Fit new bolt.
9. Remove index spring and tensioner pulley **7** & **8**.
10. Remove timing belt.
    WARNING: If belt has been in use for more than 48,000 miles: Fit new belt.

Installation

1. Ensure timing marks on camshaft sprockets aligned with marks on timing belt rear cover **2**.
2. Ensure locking tool located correctly in camshaft sprockets **3**.
3. Ensure crankshaft sprocket dots aligned with flange on oil pump **9** (90° BTDC).
4. Fit tensioner pulley and index spring **8** & **7**. Use new bolt **6**.
   NOTE: Replace tensioner every 100,000 miles.
5. Ensure index spring located correctly.
6. Position tensioner lever at 9 o'clock **10**.
7. Tighten tensioner bolt until tensioner can just be moved **6**.
8. Fit timing belt in anti-clockwise direction, starting at crankshaft sprocket. Ensure belt is taut between sprockets.
9. Install:
   - ❏ Timing belt lower cover and seal **5**.
   - ❏ Crankshaft pulley **4**.
10. Remove locking tool from camshaft sprockets **3**.
11. Turn tensioner anti-clockwise until pointer **11** aligned with index spring **7**. Use Allen key (6 mm).
    NOTE: Repeat tensioning procedure if pointer passes index spring.
12. If reusing old belt: Align lower part of pointer to index spring **12**.
13. Tighten tensioner bolt to 25 Nm **6**.
14. Turn crankshaft two turns clockwise until timing marks aligned **2** & **9**.
15. Ensure pointer still aligned **11**.
16. If not: Repeat tensioning procedure.
17. Install components in reverse order of removal.
18. Tighten crankshaft pulley bolt to 205 Nm **13**.

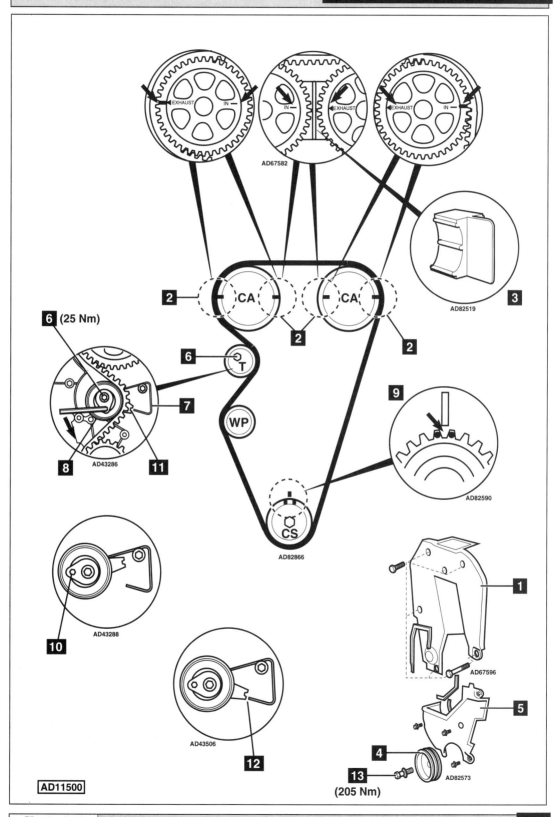

EXHAUST IN

AD67582

IN EXHAUST

EXHAUST IN

2 CA

CA

3

AD82519

2

2

2

6 (25 Nm)

6 T

9

7

8

AD43286

11

WP

AD82590

CS

AD82866

10

AD43288

1

AD67596

5

12

AD43506

4

AD82573

13

(205 Nm)

AD11500

LAND ROVER

Model:	**Freelander 2,5 V6**
Year:	**2000-03**
Engine Code:	**KV6**

Replacement Interval Guide

Land Rover recommend replacement every 72,000 miles or 6 years.
The previous use and service history of the vehicle must always be taken into account.
Refer to Timing Belt Replacement Intervals at the front of this manual.

Check For Engine Damage

CAUTION: This engine has been identified as an INTERFERENCE engine in which the possibility of valve-to-piston damage in the event of a timing belt failure is MOST LIKELY to occur. A compression check of all cylinders should be performed before removing the cylinder head(s).

Repair Times – hrs

Remove & install:

Front timing belt	3,40
LH rear timing belt	1,10
RH rear timing belt	2,00
All timing belts	5,10

Special Tools

- Crankshaft pulley holding tool – Land Rover No.LRT-12-161.
- Crankshaft pulley holding tool handle – Land Rover No.LRT-12-199.
- Locking tools for camshafts – Land Rover No.LRT-12-196.
- Alignment plate for camshaft sprockets – Land Rover No.LRT-12-175.
- Camshaft sprocket holding tool – Land Rover No.LRT-12-195.
- Camshaft alignment pins – Land Rover No.LRT-12-198.
- Camshaft aligning tool – Land Rover No.LRT-12-197.

Special Precautions

- Disconnect battery earth lead.
- DO NOT turn crankshaft or camshaft when timing belt removed.
- Remove spark plugs to ease turning engine.
- Turn engine in normal direction of rotation (unless otherwise stated).
- DO NOT turn engine via camshaft or other sprockets.
- Observe all tightening torques.

Removal

Front Timing Belt

1. Raise and support front of vehicle.
2. Remove:
   - Engine undershield.
   - RH front wheel.
   - RH splash guard.
3. Drain engine oil.
4. Support engine.

5. Remove:
   - Engine top cover.
   - RH engine mounting bracket.
   - Auxiliary drive belt.
   - Auxiliary drive belt tensioner.
   - Auxiliary drive belt guide pulley.
   - Alternator.
   - LH rear timing belt cover **1**.
6. Turn crankshaft clockwise to 'SAFE' position. Ensure timing marks aligned **2**.
7. Ensure timing marks on camshaft rear sprockets aligned **3**.
8. Remove:
   - PAS pump pulley.
   - PAS pump. DO NOT disconnect hoses.
   - Engine stabiliser bar.
   - Timing belt covers **4** & **5**.
   - Dipstick.
   - Dipstick tube.
   NOTE: Remove securing bolt. Depress dipstick tube locking collar and pull dipstick tube upwards.
9. Hold crankshaft pulley. Use tool Nos.LRT-12-161 & LRT-12-199.
10. Slacken crankshaft pulley bolt **6**.
11. Remove:
    - Crankshaft pulley bolt **6**.
    - Crankshaft pulley **7**.
    - Timing belt lower cover **8**.
    - AC compressor heat shield.
    - AC compressor. DO NOT disconnect hoses.
    - AC compressor bracket.
    - Alternator bracket.
    - Engine lifting bracket.
    - Engine front mounting plate **9**.
    - Rubber plug from around automatic tensioner unit.
12. Turn tensioner pulley away from belt. Use Allen key **10**.
13. Remove:
    - Automatic tensioner unit bolts **11**.
    - Automatic tensioner unit **12**.
14. Mark direction of rotation on belt with chalk if belt is to be reused.
15. Remove timing belt.
    NOTE: If cylinder heads are removed, water pump replaced, new camshaft sprockets fitted or if belt has been in use for more than 48,000 miles a new belt must be fitted.
16. Remove front oil seals from exhaust camshafts **13**.
17. Lock camshaft sprockets **14**. Use tool No.LRT-12-196.
18. Ensure tools located correctly on end of exhaust camshafts **15**.
19. Remove bolts from camshaft sprockets **16**. DO NOT reuse bolts.
    WARNING: Damage to camshafts may occur if locking tools are not installed when slackening or tightening camshaft sprocket bolts.
20. Remove:
    - Locking tools **14**.
    - Camshaft sprockets with hubs.

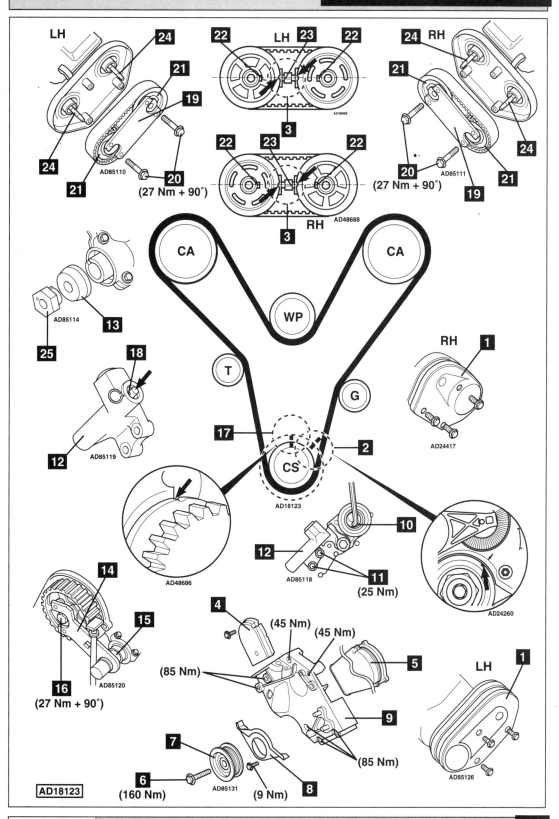

LH

24

21

19

24

21

20

(27 Nm + 90°)

AD85110

22 LH 23 22

3

22 23 22

3

RH

AD48688

AD48688

24 RH

21

20

19

24

21

(27 Nm + 90°)

AD85111

CA

CA

WP

13

25

AD85114

18

12

AD85119

T

G

17

2

CS

AD18123

RH

1

AD24417

10

12

11

AD85118

(25 Nm)

AD24260

AD48686

14

15

16

(27 Nm + 90°)

AD85120

4

(45 Nm)

(45 Nm)

5

(85 Nm)

9

1

LH

(85 Nm)

AD85126

7

6

(160 Nm)

AD85131

8

(9 Nm)

Installation

Front Timing Belt

1. Thoroughly clean camshaft sprockets and hubs.
2. Install hubs to camshaft sprockets. Then install both assemblies to camshafts.
3. Fit new bolts **16**. Tighten finger tight.
4. Ensure camshaft sprockets can turn freely but not tilt.
5. Temporarily fit timing belt to camshaft sprockets.
6. Install locking tools to camshaft sprockets **14**. Tool No.LRT-12-196.
7. Ensure tools located correctly on end of exhaust camshafts **15**.
8. Turn camshaft sprockets fully clockwise as viewed from front of engine.
9. Ensure crankshaft sprocket timing marks aligned **17**.
10. Ensure timing marks on camshaft rear sprockets aligned **3**.
11. Fit timing belt in anti-clockwise direction, starting at crankshaft sprocket. Ensure belt is taut between sprockets.
12. Use suitable wedge to hold belt in position at crankshaft sprocket.
    NOTE: Turn each camshaft sprocket anti-clockwise just enough to allow belt teeth to engage in sprocket.
13. Slowly compress pushrod into tensioner body until holes aligned **18**. Use a press. Retain pushrod with 1,5 mm diameter pin through hole in tensioner body.
14. Turn tensioner pulley against belt. Use Allen key **10**.
15. Install automatic tensioner unit **12**.
16. Coat first three threads of bolts with 'Loctite 242' or similar. Tighten bolts **11**. Tightening torque: 25 Nm.
17. Remove Allen key from tensioner pulley **10**.
18. Remove pin from tensioner body to release pushrod **18**.
19. Fit rubber plug to automatic tensioner unit.
20. Tighten camshaft sprocket bolts **16**. Tightening torque: 27 Nm + 90°.
21. Remove locking tools from camshaft sprockets **14**.
22. Remove any wedges used to hold belt in position.
23. Install new oil seals on exhaust camshafts **13**.
24. Turn crankshaft slowly two turns clockwise. Ensure timing marks aligned **3** & **17**.
25. Install components in reverse order of removal.
26. Tighten crankshaft pulley bolt **6**. Tightening torque: 160 Nm.
27. Tighten sump plug. Tightening torque: 25 Nm.
28. Refill engine oil. Capacity: 5,2 litres (includes filter).
29. Tighten wheel nuts. Tightening torque: 115 Nm.

Removal

Rear Timing Belts

NOTE: The following instructions apply to both RH and LH rear timing belts. Removal and installation should only be carried out on ONE timing belt at a time.

1. Raise and support front of vehicle.
2. Remove:
   - ❏ Engine upper cover.
   - ❏ Engine undershield.
   - ❏ RH front wheel.
   - ❏ RH splash guard.
   - ❏ Timing belt rear cover **1**.

3. Turn crankshaft clockwise to 'SAFE' position. Ensure timing marks aligned **2**.
4. Ensure timing marks on camshaft rear sprockets aligned **3**.
5. Remove and discard front oil seal from exhaust camshaft of cylinder bank being worked on (i.e. RH or LH bank) **13**.
6. Fit alignment plate to camshaft rear sprockets **19**. Tool No.LRT-12-175.
7. Remove bolts **20**.
8. Remove camshaft sprockets, timing belt and alignment plate as an assembly **21**.
9. Remove:
   - ❏ Alignment plate **19**.
   - ❏ Timing belt.
   NOTE: Mark direction of rotation on belt with chalk if belt is to be reused. If belt has been in use for more than 48,000 miles: Fit new belt.

Installation

Rear Timing Belts

1. Thoroughly clean camshaft sprockets.
2. Place sprockets on a flat surface.
3. Ensure timing marks on camshaft sprockets aligned **3**. Ensure lugs pointing towards each other **22**.
4. Fit timing belt to sprockets. Ensure timing marks aligned **3**.
5. Fit special tool between sprockets **23**. Tool No.LRT-12-195.
6. Turn centre screw to separate sprockets until alignment plate can be fitted to camshaft sprockets **19**. Tool No.LRT-12-175.
7. Remove special tool **23**. Tool No.LRT-12-195.
8. Install alignment pins to camshafts **24**. Tool No.LRT-12-198.
9. Ensure timing marks aligned **2**.
10. Install camshaft sprockets, timing belt and alignment plate as an assembly **21**.
11. Fit aligning tool to exhaust camshaft **25**. Tool No.LRT-12-197.
12. Turn exhaust camshaft and align locating slots to sprockets. Use 30 mm socket and tool No.LRT-12-197.
13. Remove alignment pins **24**. Tool No.LRT-12-198.
14. Tighten camshaft sprocket bolts **20**. Tightening torque: 27 Nm + 90°. Use new bolts.
15. Remove alignment plate **19**. Tool No.LRT-12-175.
16. Remove aligning tool **25**. Tool No.LRT-12-197.
17. Fit new oil seal **13**.
18. Turn crankshaft two turns clockwise. Ensure timing marks aligned **2** & **3**.
19. Install components in reverse order of removal.
20. Tighten wheel nuts. Tightening torque: 115 Nm.

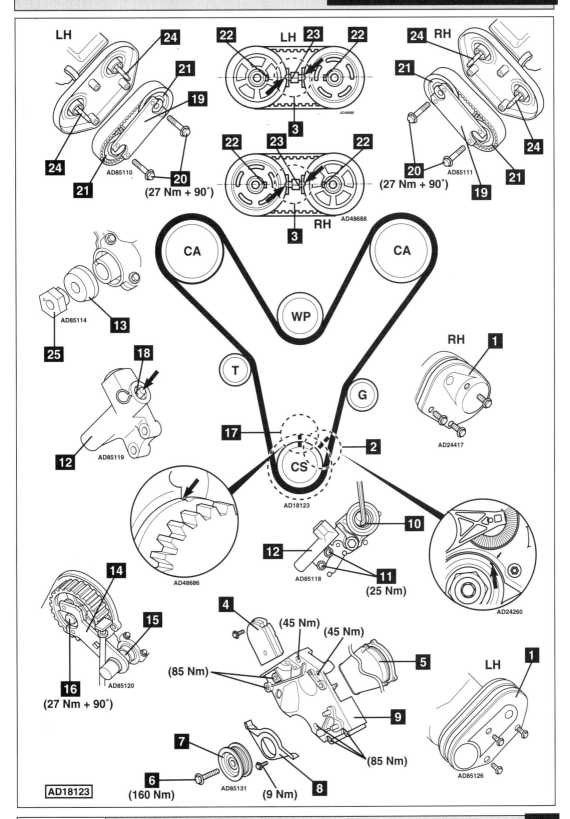

LH

24 21 19 20 (27 Nm + 90°) 21

22 LH 23 22

3

22 23 22

3 RH

24 RH 21 24 20 21 (27 Nm + 90°) 19

CA CA

WP

13 25

18 12

17 T G

2

CS AD18123

RH 1 AD24417

10 12 11 (25 Nm)

AD48686

AD24260

14 15 16 (27 Nm + 90°)

4 (45 Nm) (45 Nm) 5

(85 Nm) 9

(85 Nm)

LH 1 AD85126

7 6 (160 Nm) 8 (9 Nm)

AD18123

LAND ROVER

Model:	Freelander 2,0 TD
Year:	1997-00
Engine Code:	20T

Check For Engine Damage

CAUTION: This engine has been identified as an INTERFERENCE engine in which the possibility of valve-to-piston damage in the event of a timing belt failure is MOST LIKELY to occur.
A compression check of all cylinders should be performed before removing the cylinder head(s).

Repair Times – hrs

Remove & install:

Timing belt	4,65
Injection pump belt	1,85

Special Tools

- Flywheel timing pin – Land Rover No.LRT-12-058.
- Tensioning tool – Land Rover No.LRT-12-143.
- Sprocket holding tool – Land Rover No.LRT-12-132.
- Injection pump sprocket timing pin – Land Rover No.LRT-12-141.
- Dial type torque wrench.

Special Precautions

- Disconnect battery earth lead.
- DO NOT turn crankshaft or camshaft when timing belt removed.
- Remove glow plugs to ease turning engine.
- Turn engine in normal direction of rotation (unless otherwise stated).
- DO NOT turn engine via camshaft or other sprockets.
- Observe all tightening torques.
- Check diesel injection pump timing after belt replacement.

Removal

Timing Belt

1. Raise and support front of vehicle.
2. Remove:
   - Front drive shafts.
   - Auxiliary drive belt.
   - AC pump (if fitted). DO NOT disconnect hoses.
3. Support engine.

4. Remove:
   - Engine lower tie-bar.
   - LH engine mounting strut.
   - LH engine mounting.
5. Position engine to LH side of engine compartment.
6. Remove:
   - Timing belt upper cover **1**.
   - Camshaft sprocket damper **2**.
7. Turn crankshaft clockwise until camshaft front sprocket timing marks aligned **3**.
8. Insert timing pin in flywheel through engine backplate **4**. Tool No.LRT-12-058.
9. Remove:
   - Crankshaft pulley bolt **6**.
   - Crankshaft pulley **7**.
   - Timing belt lower cover **8**.
   - RH engine mounting plate **5**.
   - Access plug from timing belt rear cover **9**.
10. Slacken tensioner bolt **10**. Use Allen key.
11. Install special tool No.LRT-12-143 **12** until flush with end of plunger **11**.
12. Tighten nut **13** to retract plunger.
13. Lightly tighten tensioner bolt **10**.
14. Mark direction of rotation on belt with chalk if belt is to be reused.
15. Remove timing belt.
16. Remove special tool. Tool No.LRT-12-143 **12**.
    WARNING: If belt has been in use for more than 24,000 miles: Fit new belt.

Installation

Timing Belt

1. Check tensioner spring for distortion. Check free length of tensioner spring is 65 mm **14**.
2. Ensure camshaft front sprocket timing marks aligned **3**.
3. Ensure flywheel timing pin located correctly **4**.
4. Fit timing belt. Ensure belt is taut between sprockets. Observe direction of rotation marks on belt.
5. Fit engine mounting plate **5**. Tighten nuts and bolts to 30 Nm + 120°.

Original timing belt – proceed as follows:

6. Release tensioner plunger.
7. Tighten tensioner bolt **10**.
8. Temporarily fit crankshaft pulley and bolt **6** & **7**.
9. Remove flywheel timing pin **4**.

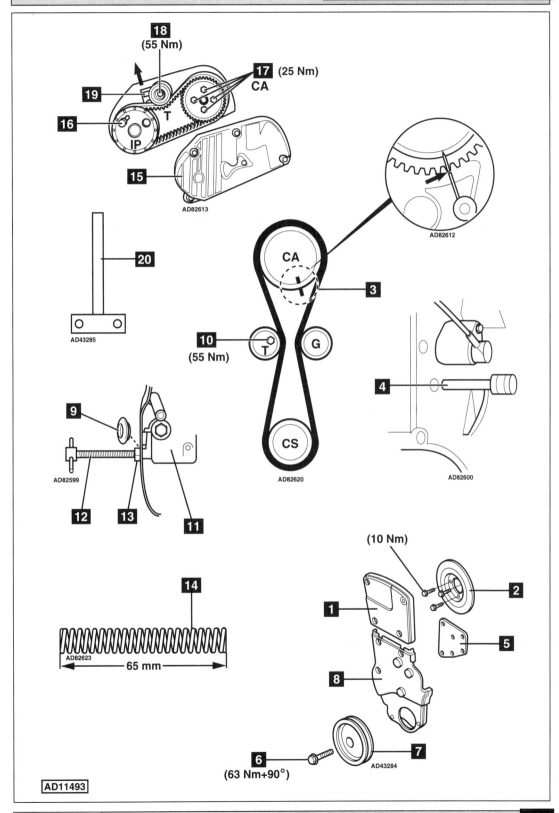

18 (55 Nm)

17 (25 Nm) CA

19

16

T

IP

15

AD82613

20

AD43285

CA

AD82612

3

10 (55 Nm) T

G

4

AD82600

9

AD82599

12 **13** **11**

CS

AD82620

14

AD82623

65 mm

(10 Nm)

1

2

5

8

6 (63 Nm+90°)

7

AD43284

AD11493

←

10. Turn crankshaft two turns clockwise until flywheel timing pin can be inserted **4**.
11. Ensure camshaft sprocket timing marks aligned **3**.
12. Remove tensioner bolt **10**. Fit new bolt. Tightening torque: 55 Nm **10**.
13. Install components in reverse order of removal.
14. Fit access plug to timing belt rear cover **9**.
15. Fit camshaft front sprocket damper **2**. Fit new bolts. Tightening torque: 10 Nm.
16. Fit new driveshaft circlips.
17. Tighten crankshaft pulley bolt **6**. Tightening torque: 63 Nm + 90°.

Replacement timing belt – proceed as follows:

6. Slacken tensioner bolt **10**.
7. Install special tool No.LRT-12-143 **12**.
8. Tighten nut **13** to fully compress plunger spring.
9. Turn centre spindle fourteen complete turns clockwise **12**.
10. Tighten tensioner bolt **10**.
11. Temporarily fit crankshaft pulley and bolt **6** & **7**.
12. Remove flywheel timing pin **4**.
13. Turn crankshaft six turns clockwise until flywheel timing pin can be inserted **4**.
14. Ensure camshaft sprocket timing marks aligned **3**.
15. Remove tensioner bolt **10**. Fit new bolt **10**. DO NOT tighten.
16. Release tensioner plunger.
17. Remove special tool **12**.
18. Tighten tensioner bolt to 55 Nm **10**.
19. Install components in reverse order of removal.
20. Fit access plug to timing belt rear cover **9**.
21. Fit camshaft front sprocket damper **2**. Fit new bolts. Tightening torque: 10 Nm.
22. Fit new driveshaft circlips.
23. Tighten crankshaft pulley bolt **6**. Tightening torque: 63 Nm + 90°.

Removal

Injection Pump Belt

1. Remove:
   ❏ Injection pump belt cover **15**.
   ❏ Timing belt upper cover **1**.
2. Turn crankshaft clockwise until camshaft front sprocket timing marks aligned **3**.
3. Insert timing pin in flywheel through engine backplate **4**. Tool No.LRT-12-058.
4. Insert timing pin into injection pump sprocket and rear cover **16**. Tool No.LRT-12-141.
5. Remove camshaft front sprocket damper **2**.

6. Hold camshaft front sprocket. Use tool No.LRT-12-132 **20**.
7. Slacken camshaft rear sprocket bolts **17**.
8. Slacken tensioner bolt **18**. Move tensioner away from belt and lightly tighten bolt.
9. Mark direction of rotation on belt with chalk if belt is to be reused.
10. Remove injection pump belt.
    WARNING: If belt has been in use for more than 24,000 miles: Fit new belt.

Installation

Injection Pump Belt

1. Turn camshaft rear sprocket fully clockwise in slotted holes.
2. Fit belt to injection pump sprocket.
3. Turn camshaft rear sprocket slowly anti-clockwise until belt teeth engage in sprocket.
4. Slacken tensioner bolt **18**.
5. Fit dial type torque wrench to square hole in tensioner plate **19**.
6. Tension belt to 6 Nm.
7. Tighten tensioner bolt **18**.
8. Hold camshaft front sprocket. Use tool No.LRT-12-132 **20**.
9. Tighten camshaft rear sprocket bolts **17**.
10. Remove:
    ❏ Flywheel timing pin **4**.
    ❏ Injection pump sprocket timing pin **16**.
11. Turn crankshaft two turns clockwise.
12. Ensure camshaft front sprocket timing marks aligned **3**.
13. Insert timing pin in flywheel **4**. Tool No.LRT-12-058.
14. Slacken camshaft rear sprocket bolts **17**.
15. Insert timing pin into injection pump sprocket and rear cover **16**. Tool No.LRT-12-141.
16. Slacken tensioner bolt **18**.
17. Fit dial type torque wrench to square hole in tensioner plate **19**.
18. Tension belt to 6 Nm.
19. Tighten tensioner bolt to 55 Nm **18**.
20. Apply anti-clockwise torque of 25 Nm to camshaft.
21. Tighten camshaft sprocket bolts to 25 Nm **17**.
22. Remove:
    ❏ Flywheel timing pin **4**.
    ❏ Injection pump sprocket timing pin **16**.
23. Install components in reverse order of removal.

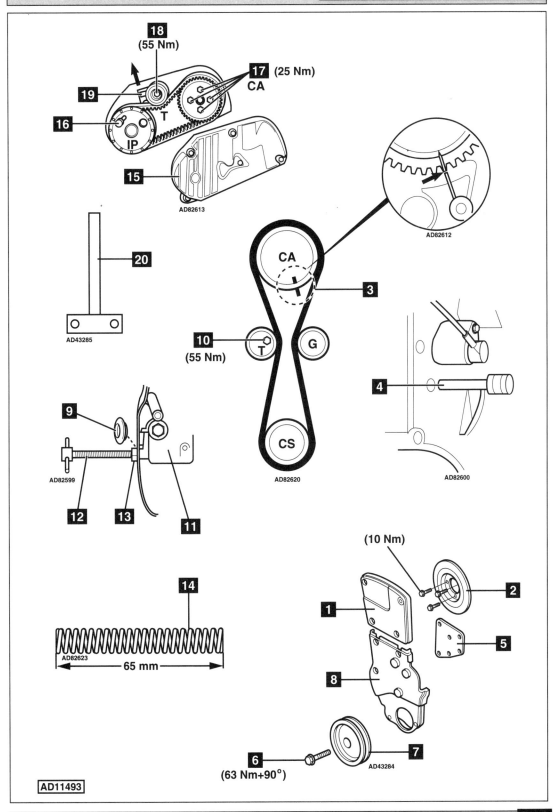

LAND ROVER

Model:	90/110/Defender 200 Tdi
Year:	1991-94
Engine Code:	200 Tdi

Replacement Interval Guide

Land Rover recommend replacement at the following intervals:
Every 60,000 miles or 5 years – normal conditions.
Every 30,000 miles or 2½ years – adverse conditions.
The previous use and service history of the vehicle must always be taken into account.
Refer to Timing Belt Replacement Intervals at the front of this manual.

Check For Engine Damage

CAUTION: This engine has been identified as an INTERFERENCE engine in which the possibility of valve-to-piston damage in the event of a timing belt failure is MOST LIKELY to occur.
A compression check of all cylinders should be performed before removing the cylinder head.

Repair Times – hrs

Check & adjust	3,80
Remove & install	4,10
AC	+0,20

Special Tools

- Crankshaft pulley/damper holding tool – Land Rover No.LST 127.
- Flywheel timing pin – Land Rover No.LST 128.
- Injection pump timing pin – Land Rover No.LST 129/2.
- Puller – Land Rover No.LST 136.
- Puller adaptor – Land Rover No.18G 1464/5.

Special Precautions

- Disconnect battery earth lead.
- DO NOT turn crankshaft or camshaft when timing belt removed.
- Remove glow plugs to ease turning engine.
- Turn engine in normal direction of rotation (unless otherwise stated).
- DO NOT turn engine via camshaft or other sprockets.
- Observe all tightening torques.
- Check diesel injection pump timing after belt replacement.

Removal

1. Drain coolant.
2. Remove:
   - ❏ Auxiliary drive belts.
   - ❏ Top hose.
   - ❏ Cooling fan and viscous coupling (LH thread).
   - ❏ Cooling fan cowling.
   - ❏ Water pump pulley.
   - ❏ Water pump.
3. Hold crankshaft pulley/damper. Use tool No.LST 127.
4. Remove:
   - ❏ Crankshaft pulley bolt **1**. Use tool Nos.LST 136 and 18G 1464/5.
   - ❏ Crankshaft damper (if fitted).
   - ❏ Crankshaft pulley **2**.
   - ❏ Timing belt cover **3**.

5. Turn crankshaft clockwise to TDC on No.1 cylinder.
6. Ensure timing marks aligned **4**.
7. Ensure crankshaft keyway aligned with 'arrow' mark **5**.
8. Lock flywheel with timing pin. Tool No.LST 128 **6**.
   NOTE: Timing pin fits into narrow slot in flywheel.
9. Insert timing pin in injection pump sprocket. Tool No.LST 129/2 **7**.
10. Slacken injection pump sprocket bolts **8**.
11. Slacken tensioner bolt **9**. Push tensioner away from belt.
12. Lightly tighten tensioner bolt **9**.
13. Remove timing belt.
    NOTE: Mark direction of rotation on belt with chalk if belt is to be reused.

Installation

1. Ensure crankshaft keyway aligned with 'arrow' mark **5**.
2. Ensure timing marks aligned **4**.
3. Ensure timing pin located correctly in injection pump sprocket **7**.
4. Ensure flywheel timing pin located correctly **6**.
5. Fit timing belt in anti-clockwise direction, starting at crankshaft sprocket.
   NOTE: Observe direction of rotation marks on belt.
6. Slacken tensioner bolt **9**.
7. Fit dial type torque wrench to square hole in tensioner plate **10**.
8. Apply clockwise torque to tensioner. New belt: 19 Nm. Used belt: 17 Nm.
   NOTE: Ensure torque wrench held in a vertical position.
9. Tighten tensioner bolt **9**. Tightening torque: 45 Nm.
10. Tighten injection pump sprocket bolts **8**. Tightening torque: 25 Nm.
11. Remove timing pins **6** & **7**.
12. Turn crankshaft two turns clockwise.
13. Ensure timing marks aligned **4**.
14. Ensure crankshaft keyway aligned with 'arrow' mark **5**.
15. Lock flywheel with timing pin **6**.
16. Insert timing pin in injection pump sprocket **7**.
17. If timing pin cannot be inserted: Slacken injection pump sprocket bolts **8**. Turn sprocket hub until timing pin can easily be inserted.
18. Tighten injection pump sprocket bolts **8**. Tightening torque: 25 Nm.
19. Repeat tensioning procedure. This is important to avoid premature timing belt failure.
20. Install components in reverse order of removal.
21. Coat threads of crankshaft pulley bolt with 'Loctite 242'.
22. Tighten crankshaft pulley bolt **1**. Tightening torque: 341 Nm.
23. Refill cooling system.

Autodata

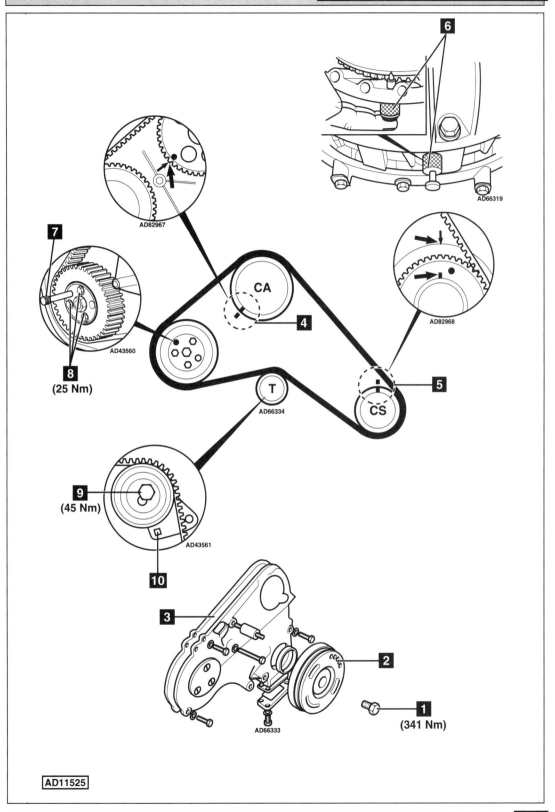

LAND ROVER

Model:	Defender 300 Tdi • Discovery 200/300 Tdi • Range Rover Tdi Range Rover Classic Tdi
Year:	1990-99
Engine Code:	200 Tdi, 300 Tdi

Replacement Interval Guide

Land Rover recommend replacement at the following intervals:

200 Tdi:
Every 60,000 miles or 5 years – normal conditions.
Every 30,000 miles or 2,5 years – adverse conditions.

300 Tdi:
Every 72,000 miles or 6 years – normal conditions.
Every 36,000 miles or 3 years – adverse conditions.
The previous use and service history of the vehicle must always be taken into account.
Refer to Timing Belt Replacement Intervals at the front of this manual.

Check For Engine Damage

CAUTION: This engine has been identified as an INTERFERENCE engine in which the possibility of valve-to-piston damage in the event of a timing belt failure is MOST LIKELY to occur.
A compression check of all cylinders should be performed before removing the cylinder head.

Repair Times – hrs

Discovery (200 Tdi):	
Check & adjust	3,80
Remove & install	4,10
AC	+0,20
All models (300 Tdi):	
Check & adjust	2,95
Remove & install	3,10

Special Tools

200 Tdi:
- Crankshaft pulley/damper holding tool – Land Rover No.LST 127.
- Flywheel timing pin – Land Rover No.LST 128.
- Injection pump timing pin – Land Rover No.LST 129.

300 Tdi:
- Crankshaft pulley/damper holding tool – Land Rover No.LRT-12-080.
- Flywheel timing pin – Land Rover No.LRT-12-044.
- Injection pump timing pin – Land Rover No.LRT-12-045.

Special Precautions

- Disconnect battery earth lead.
- DO NOT turn crankshaft or camshaft when timing belt removed.
- Remove glow plugs to ease turning engine.
- Turn engine in normal direction of rotation (unless otherwise stated).
- DO NOT turn engine via camshaft or other sprockets.
- Observe all tightening torques.
- Check diesel injection pump timing after belt replacement.

Removal

WARNING: Certain engines can suffer from premature wear of the front edge of the timing belt. Dependent on the VIN and previous repair history of the engine, the fitting of modified parts may be necessary. Refer to dealer.

1. Remove:
   - Auxiliary drive belts.
   - Top radiator hose.
   - Cooling fan and viscous coupling (LH thread).
   - Cooling fan cowling.

2. Hold crankshaft pulley/damper. Use tool No.LST 127 or LRT-12-080. Remove crankshaft pulley bolt **2**.

3. Remove:
   - Crankshaft pulley **1**.
   - Crankshaft damper (if fitted).
   - 200 Tdi: Water pump **3**.
   - Air filter/turbocharger hose.
   - Alternator.
   - PAS pump.
   - Timing belt cover **4**.

4. Turn crankshaft clockwise to TDC on No.1 cylinder. Ensure timing marks aligned **5** & **6**.

5. Lock flywheel with timing pin. MT: Remove blanking plug from lower bell housing. Insert timing pin **8**. AT: Remove two bolts and blanking plate from engine backplate. Insert timing pin in largest bolt hole. Tool No.LST 128 or LRT-12-044.

6. Insert timing pin in injection pump sprocket. Tool No.LST 129 or LRT-12-045 **7**.

7. Remove:
   - Tensioner bolt **9**.
   - Tensioner.
   - Timing belt.

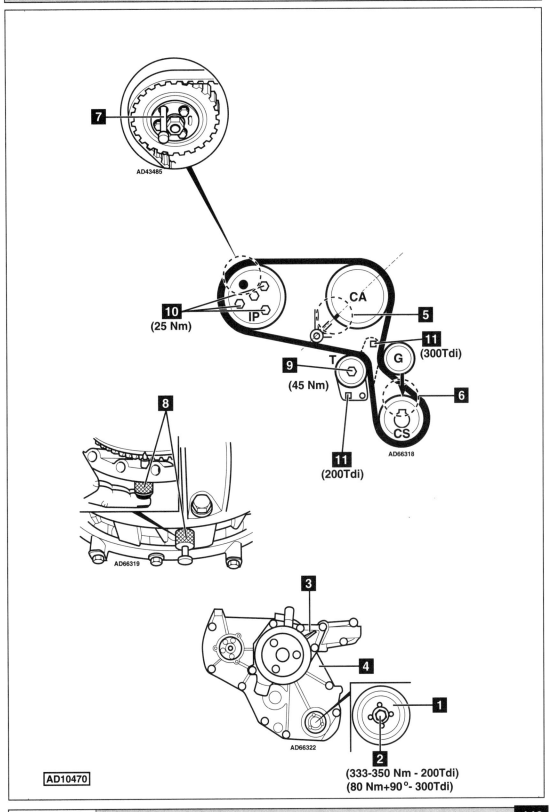

AD43485

7

10
(25 Nm)

IP

CA

5

11
(300Tdi)

9
(45 Nm)

T

G

6

CS

AD66318

11
(200Tdi)

8

AD66319

3

4

AD66322

1

2
(333-350 Nm - 200Tdi)
(80 Nm+90°- 300Tdi)

AD10470

Installation

1. Slacken injection pump sprocket bolts **⑩**.
2. Fit timing belt. Ensure belt is taut on non-tensioned side.
3. Fit tensioner ensuring it is located on dowel pin.
4. Insert torque wrench in square hole **⑪**.
   NOTE: Ensure torque wrench held in a vertical position.
5. 200 Tdi: Apply a torque of 21,7 Nm to tension belt. Used belt: 16,3 Nm.
   300 Tdi: Apply a torque of 11 Nm to tension belt.
6. Tighten tensioner bolt **⑨**.
   Tightening torque: 45 Nm.
7. Tighten injection pump sprocket bolts **⑩**.
   Tightening torque: 25 Nm.
8. Remove timing pins **⑦** & **⑧**.
9. Turn crankshaft two turns clockwise. Repeat tensioning procedure. This is important to avoid premature timing belt failure.
10. Turn crankshaft two turns clockwise. Ensure flywheel timing pin can be inserted.
11. Ensure timing marks aligned.
12. Insert timing pin in injection pump sprocket.
13. If timing pin cannot be inserted: Slacken injection pump sprocket bolts **⑩**.
14. Turn sprocket hub until timing pin can easily be inserted. Tighten injection pump sprocket bolts **⑩**. Tightening torque: 25 Nm.
15. Install components in reverse order of removal.
16. 200 Tdi: Coat threads of crankshaft pulley/damper bolt **②** and internal bore of crankshaft pulley/damper **①** with 'Loctite 242'.
17. 300 Tdi: Coat internal bore of crankshaft pulley/damper with grease **①**.
18. Tighten crankshaft pulley/damper bolt **②**.
    200 Tdi: 333-350 Nm. 300 Tdi: 80 Nm + 90°.
19. Refill cooling system.

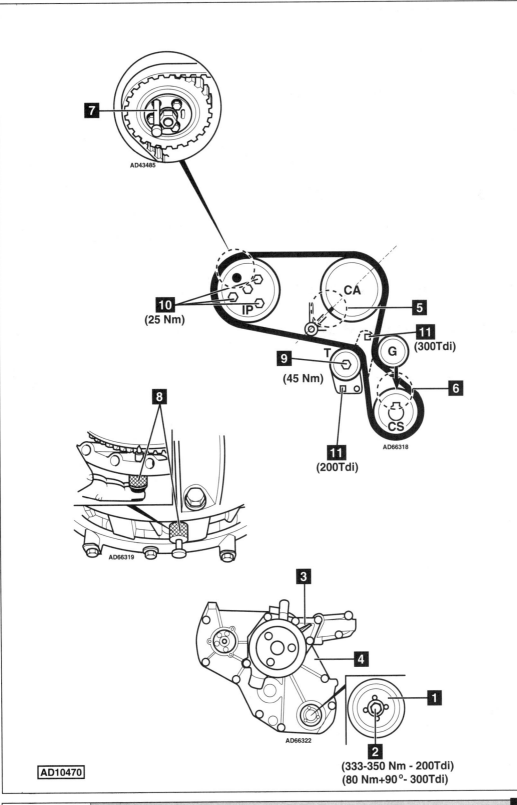

7 AD43485

10
(25 Nm)

IP

CA

5

11
(300Tdi)

9 T
(45 Nm)

G

6

11
(200Tdi)

CS

AD66318

8

AD66319

3

4

AD66322

1

2
(333-350 Nm - 200Tdi)
(80 Nm+90°- 300Tdi)

AD10470

Model:	**Cub 2,3D**
Year:	**1998-02**
Engine Code:	**LD23**

Replacement Interval Guide

LDV recommend replacement every
60,000 miles or 5 years, whichever occurs first.
*The previous use and service history of the vehicle
must always be taken into account.*
*Refer to Timing Belt Replacement Intervals at the front
of this manual.*

Check For Engine Damage

*CAUTION: This engine has been identified as an
INTERFERENCE engine in which the possibility of
valve-to-piston damage in the event of a timing belt
failure is MOST LIKELY to occur.*
*A compression check of all cylinders should be
performed before removing the cylinder head.*

Repair Times – hrs

Remove & install	1,70

Special Tools

- None required.

Special Precautions

- Disconnect battery earth lead.
- DO NOT turn crankshaft or camshaft when timing belt removed.
- Remove glow plugs to ease turning engine.
- Turn engine in normal direction of rotation (unless otherwise stated).
- DO NOT turn engine via camshaft or other sprockets.
- Observe all tightening torques.
- Check diesel injection pump timing after belt replacement.

Removal

1. Remove:
   - Engine undershield.
   - Auxiliary drive belt(s).
   - Viscous fan.
   - Water pump pulley.
   - Crankshaft pulley bolt **1**.
   - Crankshaft pulley **2**.
   - Timing belt upper cover **3**.
   - Timing belt lower cover **4**.
2. Turn crankshaft clockwise to TDC on No.1 cylinder.
   NOTE: Ensure crankshaft keyway is at 12 o'clock position.
3. Mark timing belt with chalk or paint against punch marks on sprockets **5**, **6** & **7**.
4. Slacken tensioner bolt **8**. Turn tensioner away from belt **9**. Use Allen key. Lightly tighten bolt.
5. Remove crankshaft sprocket and timing belt.

Installation

1. Ensure crankshaft at TDC on No.1 cylinder.
   NOTE: Ensure crankshaft keyway is at 12 o'clock position.
2. Install crankshaft sprocket with timing belt.
3. Fit timing belt in anti-clockwise direction, starting at crankshaft sprocket. Ensure paint marks on belt aligned with punch marks on sprockets **5**, **6** & **7**.
   NOTE: New belts are marked with white lines to ensure correct alignment with punch marks on sprockets. Ensure 'F' mark on belt faces forward.
4. Slacken tensioner bolt **8**. Allow tensioner to operate.
5. Fit crankshaft pulley bolt **1**.
6. Turn crankshaft two turns clockwise to TDC on No.1 cylinder.
   NOTE: Ensure crankshaft keyway is at 12 o'clock position.
7. Hold tensioner using Allen key **9**. Tighten tensioner bolt **8**. Tightening torque: 31-39 Nm.
   NOTE: If tensioner turns when bolt is tightened, tension may be excessive.
8. Remove crankshaft pulley bolt **1**.
9. Fit timing belt covers. Tighten bolts **10**. Tightening torque: 7-9 Nm.
10. Install components in reverse order of removal.
11. Tighten crankshaft pulley bolt **1**. Tightening torque: 137-157 Nm.

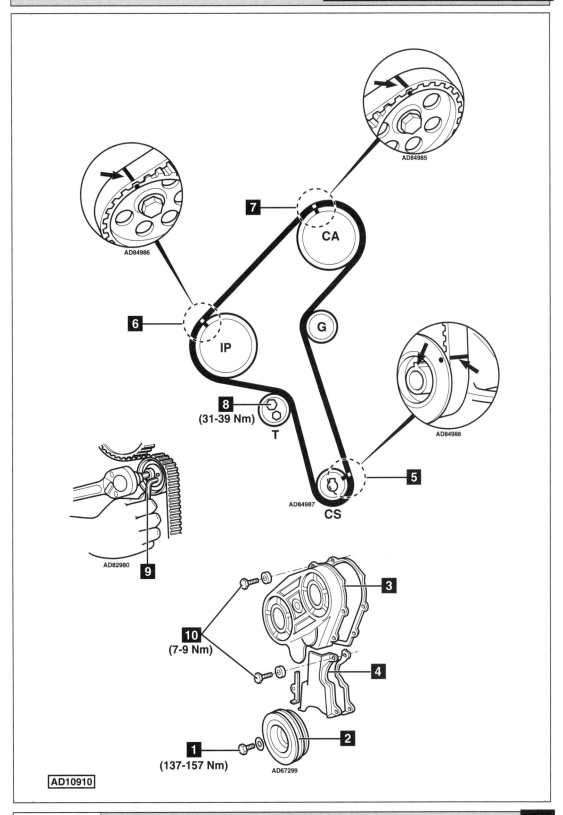

AD84985

7

CA

AD84986

6

IP

G

8
(31-39 Nm)
T

AD84988

5

AD84987

CS

AD82980

9

3

10
(7-9 Nm)

4

1
(137-157 Nm)

2

AD67299

AD10910

Model:	**Sherpa/200 Di**
Year:	**1986-94**
Engine Code:	**MDi, TN44**

Replacement Interval Guide

LDV recommend check & replace if necessary every 24,000 miles and replacement every 72,000 miles.
The previous use and service history of the vehicle must always be taken into account.
Refer to Timing Belt Replacement Intervals at the front of this manual.

Check For Engine Damage

CAUTION: This engine has been identified as an INTERFERENCE engine in which the possibility of valve-to-piston damage in the event of a timing belt failure is MOST LIKELY to occur.
A compression check of all cylinders should be performed before removing the cylinder head.

Repair Times – hrs

Check & adjust	1,65
Remove & install	1,85

Special Tools

- Crankshaft/camshaft timing pins – Leyland DAF No.18G 1523.
- Injection pump timing pins – Leyland DAF No.18G 1549.
- Tension gauge – Leyland DAF (Kent-Moore) No.KM 4088AR.

Special Precautions

- Disconnect battery earth lead.
- DO NOT turn crankshaft or camshaft when timing belt removed.
- Remove glow plugs to ease turning engine.
- Turn engine in normal direction of rotation (unless otherwise stated).
- DO NOT turn engine via camshaft or other sprockets.
- Observe all tightening torques.
- Check diesel injection pump timing after belt replacement.

Removal

1. Remove:
   - ❏ Cooling fan viscous coupling.
   - ❏ Alternator drive belt.
   - ❏ Water pump pulley.
   - ❏ Timing belt cover access panel **1**.
   - ❏ Timing belt cover **2**.
   - ❏ Cylinder head cover blanking plug.
2. Turn crankshaft until timing pin can be inserted in camshaft **3**. Tool No.18G 1523.
3. Slacken bolts **8** or centre bolt **9** of camshaft sprocket.
4. Insert timing pin in flywheel timing hole **4**. Tool No.18G 1523.
5. Ensure injection pump sprocket timing marks aligned **5**.
6. Insert timing pins in injection pump sprocket **6**. Tool No.18G 1549.
7. Remove:
   - ❏ Tensioner pulley **7**.
   - ❏ Timing belt.

Installation

*NOTE: Some injection pump sprockets have dual markings and keyways. If removed: Ensure part number (if visible) faces front. Use keyway adjacent to 'A' mark on sprocket. Align 'A' mark with pointer on engine **5**.*

1. Ensure timing pins located correctly **3**, **4** & **6**.
2. Fit timing belt.
3. Fit tensioner pulley **7**.
4. Attach tension gauge to belt at ▽. Tool No.KM 4088AR.
5. Slacken tensioner bolt **10**. Turn tensioner pulley using Allen key in tensioner pulley hole **11**.
6. Tighten tensioner bolt when tension gauge indicates 9 units (new belt) or 5 units (used belt).
7. Tighten tensioner bolt **10**.
8. Tighten bolts **8** or centre bolt **9** of camshaft sprocket.
9. Remove timing pins **3**, **4** & **6**.
10. Turn crankshaft two turns in normal direction of rotation.
11. Recheck belt tension.
12. Check injection pump timing.
13. Install components in reverse order of removal.

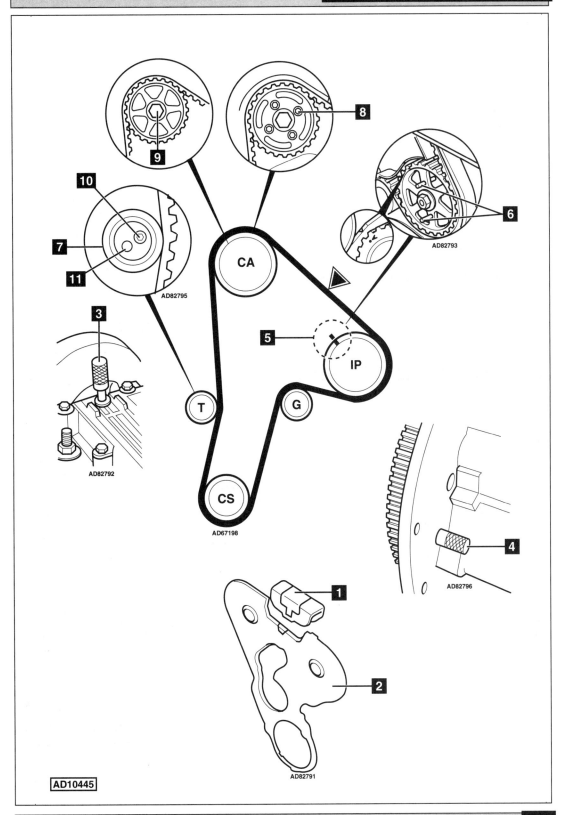

AD82793

AD82795

AD82792

AD67198

AD82796

AD82791

AD10445

LDV

Model:	**200 1,9D • Pilot 1,9D**
Year:	**1994-98**
Engine Code:	**XUD9A (10 CUA1)**

Replacement Interval Guide

LDV recommend:

→9/97:
Check & adjust every 12,000 miles.
Replacement every 48,000 miles or 4 years.

10/97→:
Check & adjust every 24,000 miles.
Replacement every 60,000 miles or 5 years.
*The previous use and service history of the vehicle
must always be taken into account.
Refer to Timing Belt Replacement Intervals at the front
of this manual.*

Check For Engine Damage

*CAUTION: This engine has been identified as an
INTERFERENCE engine in which the possibility of
valve-to-piston damage in the event of a timing belt
failure is MOST LIKELY to occur.
A compression check of all cylinders should be
performed before removing the cylinder head.*

Repair Times – hrs

Remove & install:

200	2,15
Pilot	1,90

Special Tools

- Flywheel locking tool – LDV No.LDV102.
- Flywheel timing pin – LDV No.LDV104.
- 3 x M8 bolts (for locking sprockets).

Special Precautions

- Disconnect battery earth lead.
- DO NOT turn crankshaft or camshaft when timing belt removed.
- Remove glow plugs to ease turning engine.
- Turn engine in normal direction of rotation (unless otherwise stated).
- DO NOT turn engine via camshaft or other sprockets.
- Observe all tightening torques.
- Check diesel injection pump timing after belt replacement.

Removal

1. Drain coolant.
2. Remove:
   - ❏ Intake air duct.
   - ❏ Radiator grille.
   - ❏ Bonnet shut panel.
   - ❏ Radiator (with fans).
   - ❏ Auxiliary drive belt.
3. Install flywheel locking tool **1**. Tool No.LDV102.

4. Remove:
   - ❏ Crankshaft bolt **2**.
   - ❏ Crankshaft pulleys **3**.
   - ❏ Flywheel locking tool **1**.
   - ❏ Inspection plate **4**.
   - ❏ Timing belt covers **5**, **6** & **7**.
5. Turn crankshaft clockwise to TDC on No.4 cylinder (front of engine). Insert timing pin in flywheel **8**. Tool No.LDV104.
6. Screw locking bolts into camshaft and injection pump sprockets **9** & **10**. If holes not aligned: Turn crankshaft one turn clockwise.
7. Slacken tensioner nut and bolt **11** & **12**.
8. Move tensioner away from belt. Lightly tighten bolt **12**. Lightly tighten nut **11**.
9. Remove timing belt.
   NOTE: Mark direction of rotation on belt with chalk if belt is to be reused.

Installation

1. Check tensioner spring and piston slide freely **13**.
2. Ensure timing pin and locking bolts located correctly **8**, **9** & **10**.
3. Fit timing belt in anti-clockwise direction, starting at crankshaft sprocket. Ensure belt is taut between sprockets.
   NOTE: Observe direction of rotation marks or ensure lettering on belt readable from front of engine.
4. Remove locking bolts and timing pin **8**, **9** & **10**.
5. Slacken tensioner nut and bolt two turns **11** & **12**. Allow tensioner to operate.
6. Lightly tighten bolt **12**. Lightly tighten nut **11**.
7. Fit timing belt lower cover **7**.
8. Fit crankshaft pulleys **3**. Tighten bolt finger tight **2**.
9. Turn crankshaft eight turns clockwise. Insert timing pin and locking bolts **8**, **9** & **10**.
   NOTE: If timing pin or locking bolts cannot be inserted easily: Repeat installation procedure.
10. Slacken tensioner nut and bolt two turns **11** & **12**. Allow tensioner to operate.
11. Tighten tensioner bolt and nut **12** & **11**. Tightening torque: 15 Nm.
12. Remove locking bolts and timing pin **8**, **9** & **10**.
13. Remove crankshaft bolt **2**. Degrease bolt, washer and threads in crankshaft. Coat bolt threads with 'Loctite 270' or similar thread locking compound.
14. Fit crankshaft bolt.
    Tightening torque: 40 Nm + 60° **2**.
15. Install components in reverse order of removal.
16. Refill cooling system.

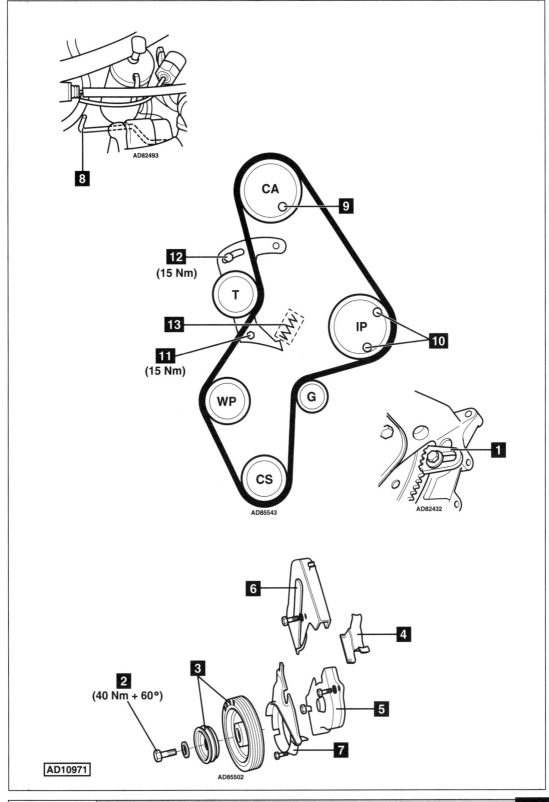

Model:	**Convoy 2,5 D/TD**
Year:	**1997-02**
Engine Code:	**4EH, 4HB**

Replacement Interval Guide

LDV recommend:

2,5 D
With manual tensioner: Replacement every 36,000 miles or 3 years.
With auto tensioner: Replacement every 60,000 miles or 5 years.

2,5 TD
Replacement every 36,000 miles or 3 years.
The previous use and service history of the vehicle must always be taken into account.
Refer to Timing Belt Replacement Intervals at the front of this manual.

Check For Engine Damage

CAUTION: This engine has been identified as an INTERFERENCE engine in which the possibility of valve-to-piston damage in the event of a timing belt failure is MOST LIKELY to occur. A compression check of all cylinders should be performed before removing the cylinder head(s).

Repair Times – hrs

Remove & install	1,05

Special Tools

- ■ Flywheel timing pin – LDV No.23-020.
- ■ Camshaft timing pin – LDV No.21-123.
- ■ Turbo: Injection pump timing pin – LDV No.23-019.
- ■ Non-turbo: Injection pump timing pin – LDV No.23-029.

Removal

1. Drain coolant.
2. Remove:
   - ❏ Bonnet lock panel.
   - ❏ Radiator.
   - ❏ Auxiliary drive belt(s).
   - ❏ Turbo: Viscous fan.
   - ❏ Water pump pulley.
   - ❏ Crankshaft pulley bolt rubber cap **1**.
   - ❏ Timing belt covers **2**.
   - ❏ Turbo: Crankshaft position (CKP) sensor and spacer.
3. Turn crankshaft clockwise to setting position (11° BTDC).
4. Insert flywheel timing pin **3**. Tool No.23-020.
5. Insert timing pin in injection pump sprocket **4**.
   - ❏ Turbo: Tool No.23-019.
   - ❏ Non-turbo: Tool No.23-029.
6. Insert timing pin in camshaft sprocket **5**. Tool No.21-123.
7. Slacken injection pump sprocket bolts **6**.
8. Slacken tensioner bolt(s) **7** & **8** or **9**. Move tensioner away from belt. Lightly tighten tensioner bolt **8** or **9**.
9. Remove timing belt.
   NOTE: Turbo: DO NOT refit used belt. Timing belt must always be renewed once it has been slackened or removed.
   WARNING: DO NOT fit automatic tensioner pulley or modified timing belt to turbo engines.
   NOTE: Non-turbo: Original timing belt – DO NOT refit used belt. Timing belt and tensioner pulley must always be renewed once they have been slackened or removed.
   NOTE: Non-turbo: Modified timing belt with automatic tensioner pulley – only replace at recommended intervals or if damaged.

Installation

1. Ensure all timing pins located correctly **3**, **4** & **5**.
2. Fit timing belt in anti-clockwise direction, starting at crankshaft sprocket.
3. Observe direction of rotation marks on belt.
4. If necessary: Turn injection pump sprocket slightly to engage timing belt teeth.
5. Ensure injection pump sprocket bolts not at end of slotted holes.

Turbo

6. Slacken tensioner bolt **8**. Allow tensioner to operate.
7. Tighten tensioner bolts **7** & **8**.

Non-turbo

8. Replace manual tensioner pulley and bolt with automatic tensioner pulley.
9. If automatic tensioner pulley already fitted: Slacken tensioner bolt **9**.
10. Turn tensioner clockwise until centre punch mark on backplate aligned with lower edge of cut-out **10** & **11**. Use Allen key.
11. Tighten tensioner bolt **9**.

All models

12. Tighten injection pump sprocket bolts **6**.
    Tightening torque: 25 Nm.
13. Remove all timing pins **3**, **4** & **5**.
14. Turn crankshaft 6 turns clockwise to setting position.
15. Insert flywheel timing pin **3**.
16. Remove flywheel timing pin **3**.
17. Turn crankshaft 130° clockwise.

Turbo

18. Slacken tensioner bolts **7** & **8**.
19. Apply thumb pressure several times to belt at ▽. Allow tensioner to operate.
20. Release thumb pressure from belt.
21. Tighten tensioner bolt **7**. Tightening torque: 58 Nm.
22. Tighten tensioner bolt **8**. Tightening torque: 24 Nm.

Non-turbo

23. Check centre punch mark on backplate aligned with lower edge of cut-out **10** & **11**. If not: Repeat tensioning procedure.
24. Tighten tensioner bolt **9**. Tightening torque: 45 Nm.

All models

25. Turn crankshaft one turn then a further 230° clockwise.
26. Insert flywheel timing pin **3**.
27. Insert timing pin in camshaft sprocket **5**.
28. Insert timing pin in injection pump sprocket **4**.
29. If timing pin cannot be inserted in injection pump sprocket: Slacken bolts **6**. Turn sprocket hub until timing pin can easily be inserted. Tighten bolts **6**.
    Tightening torque: 25 Nm.
30. Remove all timing pins **3**, **4** & **5**.
31. Install components in reverse order of removal.
32. Refill cooling system.

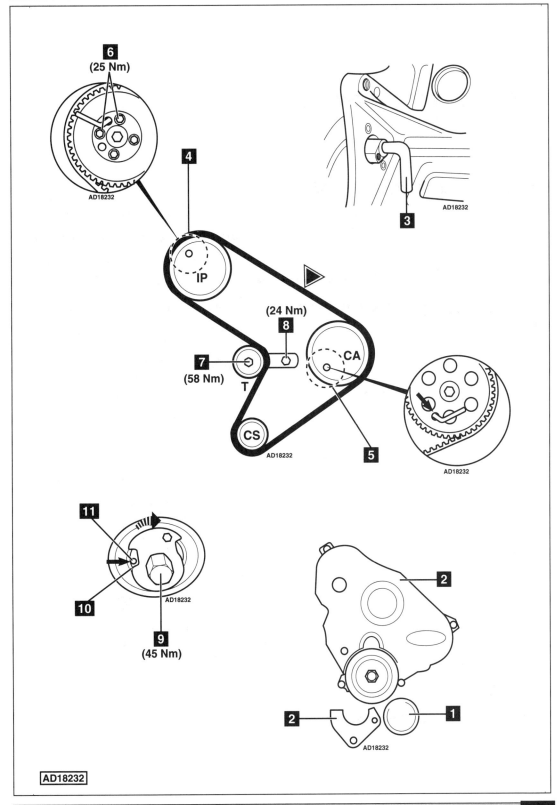

6
(25 Nm)

4

3

8
(24 Nm)

IP

CA

7
(58 Nm)
T

CS

5

11

10

9
(45 Nm)

2

1

2

AD18232

LEXUS

Model:	IS 200
Year:	1999-03
Engine Code:	1G-FE

Replacement Interval Guide

Lexus recommend replacement every 100,000 miles or 6 years.

The previous use and service history of the vehicle must always be taken into account.

Refer to Timing Belt Replacement Intervals at the front of this manual.

Check For Engine Damage

CAUTION: This engine has been identified as an INTERFERENCE engine in which the possibility of valve-to-piston damage in the event of a timing belt failure is MOST LIKELY to occur.

A compression check of all cylinders should be performed before removing the cylinder head.

Repair Times – hrs

Remove & install	2,50

Special Tools

- Crankshaft pulley holding tool – Lexus No.09213-54015.
- Crankshaft pulley puller – Lexus No.09950-50012.
- Crankshaft pulley installer – Lexus No.09316-60011.

Special Precautions

- Disconnect battery earth lead.
- DO NOT turn crankshaft or camshaft when timing belt removed.
- Remove spark plugs to ease turning engine.
- Turn engine in normal direction of rotation (unless otherwise stated).
- DO NOT turn engine via camshaft or other sprockets.
- Observe all tightening torques.

Removal

1. Raise and support front of vehicle.
2. Drain coolant.
3. Remove:
   - ❏ Air filter intake pipe.
   - ❏ Air filter assembly.
   - ❏ Engine undershield.
   - ❏ Cylinder head cover.
   - ❏ Ignition coils.
   - ❏ Auxiliary drive belts.
   - ❏ Cooling fan assembly.
   - ❏ Water pump pulley.
   - ❏ Radiator.
   - ❏ AC compressor and bracket. DO NOT disconnect pipes.
   - ❏ Timing belt upper cover **1**.
4. Turn crankshaft clockwise until timing marks aligned **2** & **3**.
   *NOTE: Timing marks **2** & **3** are at 60° BTDC.*

5. Remove:
   - ❏ Crankshaft pulley bolt **4**. Use tool No.09213-54015.
   - ❏ Crankshaft pulley **5**. Use tool No.09950-50012.
   - ❏ Timing belt lower cover **6**.
   - ❏ Crankshaft sprocket guide washer.
6. Mark crankshaft sprocket and backplate with paint **7**.
7. Insert 10 mm Allen key in tensioner pulley **8**.
8. Turn tensioner pulley slowly clockwise until pin in square hole at correct position **9**.
   NOTE: DO NOT apply a torque of more than 39 Nm to tensioner pulley.
9. Insert 5 mm Allen key in square hole of tensioner pulley **9**. Slacken tensioner pulley bolt **10**.
10. With torque maintained on Allen key **8**, remove tensioner positioning bolt **11**.
11. Turn tensioner pulley clockwise by hand **12**.
12. Remove timing belt.
    NOTE: Mark direction of rotation on belt with chalk if belt is to be reused.

Installation

1. Ensure timing marks aligned **3** & **7**.
2. Fit timing belt in following order:
   - ❏ Crankshaft sprocket.
   - ❏ Oil pump pulley.
   - ❏ Tensioner pulley.
   - ❏ Guide pulley.
   - ❏ Camshaft sprocket.
3. Ensure belt is taut between sprockets.
4. Turn tensioner pulley anti-clockwise until it touches stop **13**. Use 10 mm Allen key **8**.
5. Fit tensioner positioning bolt **11**.
   Tightening torque: 8 Nm.
6. Tighten tensioner pulley bolt **10**.
   Tightening torque: 42 Nm.
7. Remove Allen key **9**.
8. Install:
   - ❏ Crankshaft sprocket guide washer.
   - ❏ Timing belt lower cover **6**.
   - ❏ Crankshaft pulley **5**. Use tool No.09316-60011.
   - ❏ Crankshaft pulley bolt **4**.
9. Turn crankshaft two turns clockwise. Ensure timing marks aligned **2** & **3**.
10. If not: Repeat tensioning procedure.
11. Hold crankshaft pulley. Use tool No.09213-54015.
12. Tighten crankshaft pulley bolt.
    Tightening torque: 220 Nm.
13. Install components in reverse order of removal.
14. Refill cooling system.

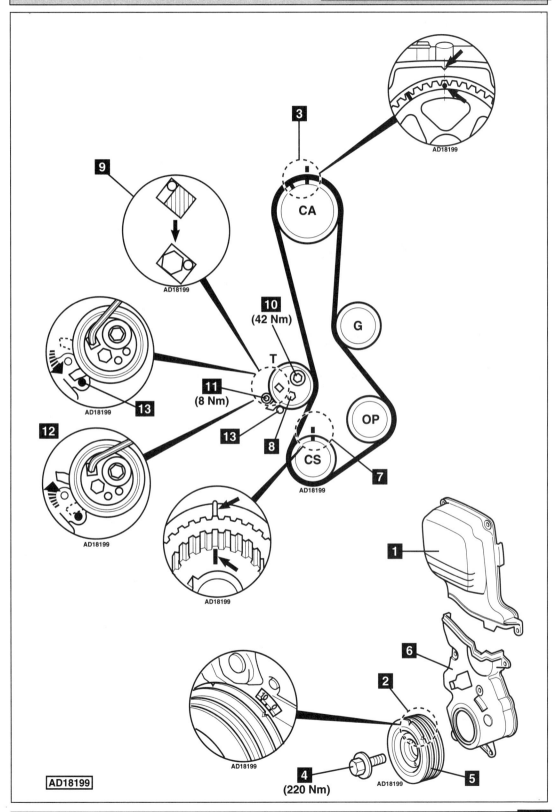

9

3

AD18199

AD18199

CA

G

10
(42 Nm)

T

OP

11
(8 Nm)

13

AD18199

8

13

12

CS

AD18199

7

AD18199

1

6

AD18199

2

4
(220 Nm)

AD18199

5

AD18199

Model:	**IS 300 • GS 300**
Year:	**1997-03**
Engine Code:	**2JZ-GE**

Replacement Interval Guide

Lexus recommend:
GS 300 →00MY: Replacement every 63,000 miles or 7 years.
GS 300 00MY→: Replacement every 100,000 miles or 10 years.
IS 300: Replacement every 100,000 miles or 10 years.
The previous use and service history of the vehicle must always be taken into account.
Refer to Timing Belt Replacement Intervals at the front of this manual.

Check For Engine Damage

CAUTION: This engine has been identified as an INTERFERENCE engine in which the possibility of valve-to-piston damage in the event of a timing belt failure is MOST LIKELY to occur.
A compression check of all cylinders should be performed before removing the cylinder head.

Repair Times – hrs

Remove & install	2,10

Special Tools

- Crankshaft pulley holding tool – Lexus No.09213-70010.
- Crankshaft pulley puller – Lexus No.09950-50012.

Special Precautions

- Disconnect battery earth lead.
- DO NOT turn crankshaft or camshaft when timing belt removed.
- Remove spark plugs to ease turning engine.
- Turn engine in normal direction of rotation (unless otherwise stated).
- DO NOT turn engine via camshaft or other sprockets.
- Observe all tightening torques.

Removal

1. Raise and support front of vehicle.
2. Remove engine undershield.
3. Drain coolant.
4. Remove:
   - Engine top cover (if necessary).
   - Air filter and air duct.
   - Radiator and cooling fan.
   - Auxiliary drive belt.
   - PAS pump and bracket. DO NOT disconnect hoses.
   - Timing belt upper cover **1**.
   - Timing belt centre cover **2**.
   - Auxiliary drive belt tensioner **3**.
5. Slacken crankshaft pulley bolt **4**. Hold crankshaft pulley. Use tool No.09213-70010.
6. Turn crankshaft clockwise until timing marks aligned **5** & **6**.
7. If timing marks not aligned: Turn crankshaft one turn clockwise.
8. Turn crankshaft 60° anti-clockwise until timing marks aligned **7** & **8**.

9. Remove:
   - Crankshaft pulley bolt **4**.
   - Crankshaft pulley **10**. Use tool No.09950-50012.
   - Timing belt lower cover **11**.
   - Crankshaft sprocket guide washer **12**.
   - Automatic tensioner unit bolts **18**.
   - Automatic tensioner unit **9**.
   - Automatic tensioner unit dust cover.
   - Timing belt.
10. Slacken tensioner pulley bolt **16**. Remove tensioner pulley **19**.
   NOTE: Mark direction of rotation on belt with chalk if belt is to be reused.

Installation

1. Check tensioner pulley for smooth operation. Replace if necessary.
2. Check tensioner body for leakage or damage **9**. Replace if necessary.
3. Check pushrod protrusion is 8,0-8,8 mm **14**. If not, replace automatic tensioner unit.
4. Slowly compress pushrod into tensioner body until holes aligned.
5. Retain pushrod with 1,5 mm Allen key through hole in tensioner body **15**.
6. Apply Loctite to tensioner pulley bolt **16**.
7. Fit tensioner pulley **13**. Tighten bolt to 35 Nm **16**.
8. Ensure timing marks aligned **8** & **17**.
9. Fit timing belt in following order:
   - Crankshaft sprocket.
   - Tensioner pulley.
10. Install:
   - Crankshaft sprocket guide washer **12**.
   - Timing belt lower cover **11**.
   - Crankshaft pulley **10**. Lightly tighten bolt **4**.
11. Ensure timing marks aligned **7** & **8**.
12. Fit timing belt to camshaft sprockets. Ensure belt is taut between sprockets.
13. Install automatic tensioner unit to cylinder block **9**. Tighten bolts to 27 Nm **18**.
14. Remove Allen key from tensioner body to release pushrod.
15. Turn crankshaft two turns clockwise. Ensure timing marks aligned **5** & **6**.
16. Install components in reverse order of removal.
17. Tighten crankshaft pulley bolt **4**.
   Tightening torque: 330 Nm.
18. Refill cooling system.

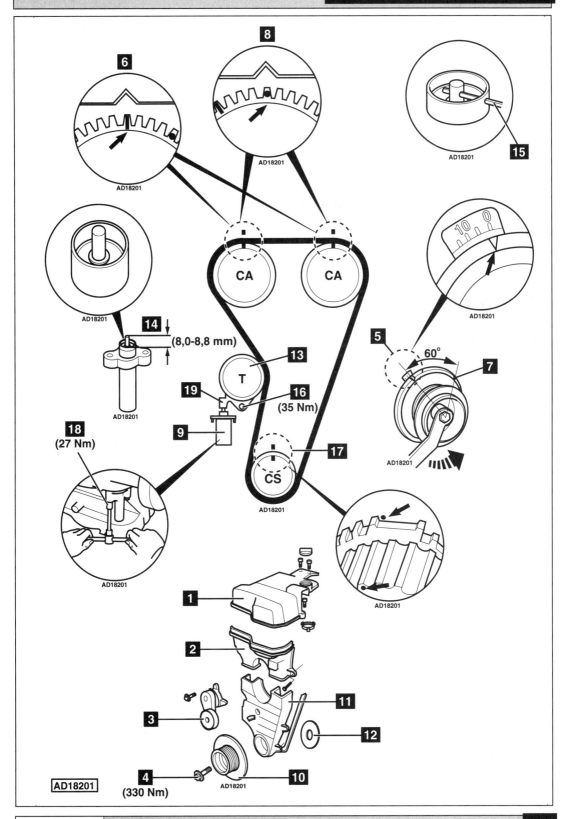

6

8

15

AD18201

AD18201

AD18201

CA

CA

AD18201

14 (8,0-8,8 mm)

AD18201

13 T

19

16 (35 Nm)

18 (27 Nm)

9

5 60° 7

AD18201

17

CS

AD18201

AD18201

1

AD18201

2

3

11

12

4 (330 Nm)

AD18201

10

AD18201

LEXUS

Model:	GS 300
Year:	1993-97
Engine Code:	2JZ-GE

Replacement Interval Guide

Lexus recommend replacement every 63,000 miles.
The previous use and service history of the vehicle must always be taken into account.
Refer to Timing Belt Replacement Intervals at the front of this manual.

Check For Engine Damage

CAUTION: This engine has been identified as an INTERFERENCE engine in which the possibility of valve-to-piston damage in the event of a timing belt failure is MOST LIKELY to occur.
A compression check of all cylinders should be performed before removing the cylinder head.

Repair Times – hrs

Remove & install	2,10

Special Tools

- Crankshaft pulley holding tool – Lexus No.09213-70010.
- Crankshaft pulley puller – Lexus No.09950-50010.

Special Precautions

- Disconnect battery earth lead.
- DO NOT turn crankshaft or camshaft when timing belt removed.
- Remove spark plugs to ease turning engine.
- Turn engine in normal direction of rotation (unless otherwise stated).
- DO NOT turn engine via camshaft or other sprockets.
- Observe all tightening torques.

Removal

1. Drain coolant.
2. Remove:
   - ❏ Air filter duct.
   - ❏ Engine undershield **1**.
   - ❏ Cooling fan assembly.
   - ❏ Water pump pulley.
3. Turn auxiliary drive belt tensioner clockwise to release tension on belt **2**. Remove auxiliary drive belt.
4. Remove:
   - ❏ Radiator.
   - ❏ Timing belt upper cover **3**.
   - ❏ Timing belt centre cover **4**.
   - ❏ Auxiliary drive belt tensioner **7**.
5. Turn crankshaft clockwise to TDC on No.1 cylinder. Ensure timing marks aligned **5** & **6**.

6. Remove:
   - ❏ Crankshaft pulley bolt **9**. Use tool No.09213-70010.
   - ❏ Crankshaft pulley **10**. Use tool No.09950-50010.
   - ❏ PAS pump front bracket.
   - ❏ Timing belt lower cover **11**.
   - ❏ Automatic tensioner unit **13**.
   - ❏ Crankshaft sprocket guide washer **12**.
   - ❏ Timing belt.
7. Slacken tensioner pulley bolt **14**. Remove tensioner pulley **8**.

Installation

1. Check tensioner pulley for smooth operation. Replace if necessary.
2. Check tensioner body for leakage or damage **13**. Replace if necessary.
3. Check pushrod protrusion is 8,0-8,8 mm **17**. If not, replace automatic tensioner unit.
4. Slowly compress pushrod into tensioner body until holes aligned.
5. Retain pushrod with 1,5 mm Allen key through hole in tensioner body **18**.
   *NOTE: Place flat washers under tensioner body to avoid damage to body end plug **19**.*
6. Apply Loctite to tensioner pulley bolt **14**.
7. Fit tensioner pulley **8**. Tighten bolt to 34 Nm **14**.
8. Ensure timing marks aligned **5** & **6**.
9. Install:
   - ❏ Timing belt.
   - ❏ Automatic tensioner unit **13**. Tighten bolts to 26 Nm.
10. Remove Allen key from tensioner body to release pushrod.
11. Turn crankshaft two turns clockwise.
12. Ensure timing marks aligned **5** & **6**.
13. Install components in reverse order of removal.
14. Tighten crankshaft pulley bolt **9**.
    Tightening torque: 324 Nm.
15. Check auxiliary drive belt tensioner. Ensure tensioner support alignment mark is in area A (used belt) or in area B (new belt). If not: Replace belt or check auxiliary drive belt tensioner.
    *NOTE: There are two types of auxiliary drive belt tensioner **15** & **16**.*
16. Refill cooling system.

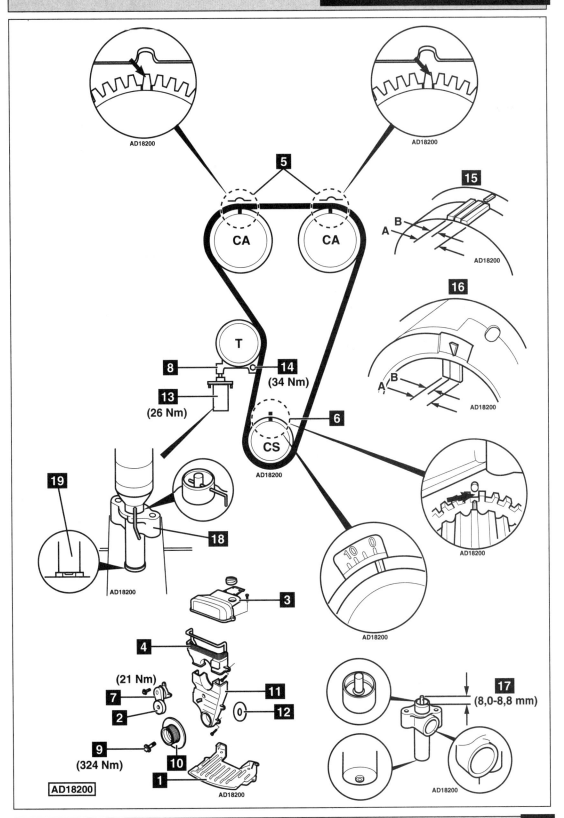

LEXUS

Model:	LS 400
Year:	1990-97
Engine Code:	1UZ-FE

Replacement Interval Guide

Lexus recommend:
→93MY: Replacement every 60,000 miles only under adverse conditions.
94MY→: Replacement every 63,000 miles.
The previous use and service history of the vehicle must always be taken into account.
Refer to Timing Belt Replacement Intervals at the front of this manual.

Check For Engine Damage

CAUTION: This engine has been identified as an INTERFERENCE engine in which the possibility of valve-to-piston damage in the event of a timing belt failure is MOST LIKELY to occur.
A compression check of all cylinders should be performed before removing the cylinder head.

Repair Times – hrs

Remove & install	2,80

Special Tools

- Crankshaft pulley puller:
  →93MY – Lexus No.09213-00050.
  94MY→ – Lexus No.09213-31021.
- Crankshaft pulley holding tool – 94MY→ – Lexus No.09213-70010.
- Crankshaft pulley installer – Lexus No.09223-46011.
- Camshaft sprocket holding tool – 94MY→ – Lexus No.09278-54012.

Removal

1. Drain coolant.
2. Remove:
   - Battery, battery tray and engine undershield.
   - Auxiliary drive belt and tensioner.
   - Cooling fan and viscous coupling.
   - Cooling fan pulley.
   - Radiator, coolant hoses, air filter and hoses.
   - AC compressor. DO NOT disconnect hoses.
   - →93MY: PAS pump. DO NOT disconnect hoses.
   - 94MY→: RH relay box cover.
   - Timing belt upper covers **1**.
   - HT leads, covers and LH ignition coil.
   - Coolant by-pass pipe.
   - Timing belt centre covers **2**.
   - Distributor caps, rotor arms and housings.
   - Alternator.
3. Turn crankshaft to TDC on No.1 cylinder. Ensure timing marks aligned **3** & **4**.
   NOTE: If belt is to be reused, check alignment marks are visible. If not: Mark timing belt with chalk against marks on sprockets.
4. Remove:
   - Automatic tensioner unit **5**.
   - Crankshaft pulley **6**. →93MY: Use tool No.09213-00050 (if necessary).
     94MY→: Use tool No.09213-70010/31021.
   - Cooling fan bracket **7**.
   - Timing belt lower cover **8**.

- Crankshaft sprocket guide washer **9**.
- Timing belt. 94MY→: Turn camshaft sprocket (CA1) slightly clockwise if necessary.

Installation

NOTE: Ensure engine is cold before installing belt. Observe alignment marks.

1. Check tensioner pulley and guide pulleys for smooth operation. Replace if necessary.
2. Check tensioner body for leakage or damage **5**. Replace if necessary.
3. Check pushrod protrusion is 10,5-11,5 mm **10**. If not, replace automatic tensioner unit.
4. Slowly compress pushrod into tensioner body until holes aligned.
5. Retain pushrod with 1,27 mm Allen key through hole in tensioner body **11**.
   *NOTE: Place flat washers under tensioner body to avoid damage to body end plug **12**.*
6. Fit dust cover **13**.
7. Ensure timing marks aligned **4** & **14**.
8. →93MY:
   - Fit timing belt in anti-clockwise direction, starting at crankshaft sprocket. Ensure belt is taut between sprockets. Ensure marks on belt aligned with marks on sprockets.
9. 94MY→:
   - Fit timing belt to crankshaft sprocket. Ensure marks on belt aligned **16**.
   - Fit timing belt around guide pulley and tensioner pulley.
10. All models – fit:
    - Crankshaft sprocket guide washer (convex side towards belt) **9** and timing belt lower cover **8**.
    - Cooling fan bracket **7** and crankshaft pulley **6**. Use tool No. 09223-46011 (if necessary).
11. 94MY→:
    - Turn camshaft sprocket (CA2) slightly clockwise until timing belt can be fitted with timing marks aligned **17**. Use tool No.09278-54012.
    - Turn camshaft sprocket (CA2) slowly anti-clockwise until timing marks aligned **4**. Ensure belt is taut between sprockets.
    - Turn camshaft sprocket (CA1) slightly clockwise until timing belt can be fitted with timing marks aligned **18**. Use tool No.09278-54012.
    - Turn camshaft sprocket (CA1) slowly anti-clockwise until timing marks aligned **4**. Ensure belt is taut between sprockets.
12. All models:
    - Install automatic tensioner unit **5**. Tighten bolts evenly. Tightening torque: →93MY: (A) 20 Nm. 94MY→: (B) 26 Nm.
    - Remove Allen key from tensioner body to release pushrod.
    - Turn crankshaft slowly two turns in direction of rotation to TDC on No.1 cylinder.
    - Ensure timing marks aligned **3** & **4**.
    - Install components in reverse order of removal.
    - Tighten crankshaft pulley bolt to 245 Nm **15**.
    - Refill cooling system.

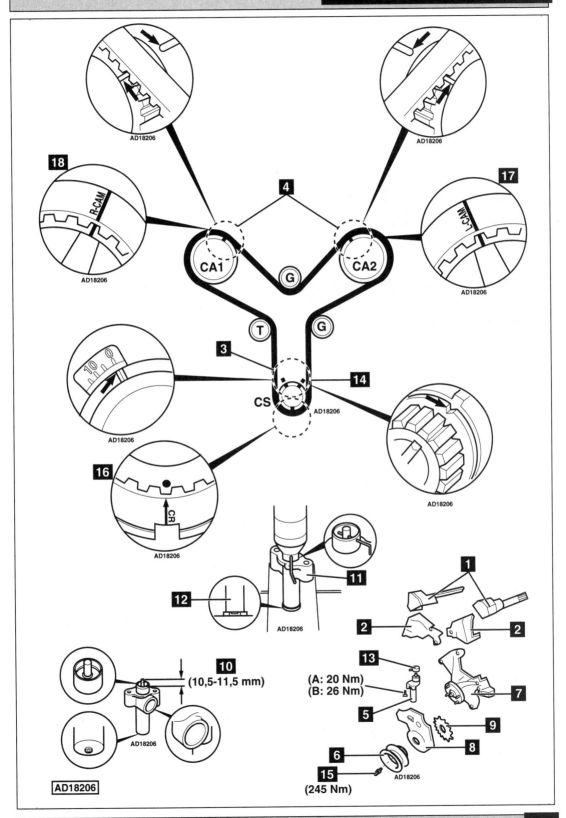

Model:	**LS 400**
Year:	**1997-00**
Engine Code:	**1UZ-FE**

Replacement Interval Guide

Lexus recommend replacement every 63,000 miles or 7 years.
The previous use and service history of the vehicle must always be taken into account.
Refer to Timing Belt Replacement Intervals at the front of this manual.

Check For Engine Damage

CAUTION: This engine has been identified as an INTERFERENCE engine in which the possibility of valve-to-piston damage in the event of a timing belt failure is MOST LIKELY to occur.
A compression check of all cylinders should be performed before removing the cylinder head.

Repair Times – hrs

Remove & install	2,80

Special Tools

- Crankshaft pulley puller – Lexus No.09950-50012.
- Crankshaft pulley holding tool – Lexus No.09213-70010.
- Crankshaft pulley installer – Lexus No.09223-46011.
- Camshaft sprocket holding tool – Lexus No.09960-10010.

Special Precautions

- Disconnect battery earth lead.
- DO NOT turn crankshaft or camshaft when timing belt removed.
- Remove spark plugs to ease turning engine.
- Turn engine in normal direction of rotation (unless otherwise stated).
- DO NOT turn engine via camshaft or other sprockets.
- Observe all tightening torques.

Removal

1. Drain coolant.
2. Remove:
   - Engine undershield.
   - Battery cover.
   - Engine covers.
   - Air filter and hoses.
   - Auxiliary drive belt.
   - Auxiliary drive belt tensioner.
   - Cooling fan and viscous coupling.
   - Cooling fan pulley.
   - Radiator and hoses.
   - Timing belt upper covers **1** & **2**.
   - Timing belt centre cover **3**.
   - AC compressor. DO NOT disconnect hoses.
   - Cooling fan bracket **4**.
3. Slacken crankshaft pulley bolt **8**. Use tool No.09213-70010.
4. Turn crankshaft clockwise to TDC on No.1 cylinder. Ensure timing marks aligned **5** & **6**.
5. Turn crankshaft clockwise a further 50° until crankshaft pulley timing mark aligned with guide pulley bolt **7**.
6. Remove:
   - Crankshaft pulley bolt **8**.

- Automatic tensioner unit bolts **9**.
- Automatic tensioner unit **10**.
- Automatic tensioner unit dust cover.
- Alternator.
- Crankshaft pulley **11**. Use tool No.09950-50012.
- Timing belt lower cover **12**.
- Crankshaft position (CKP) sensor reluctor **13**.
- Timing belt. Turn camshaft sprocket (CA1) slightly clockwise if necessary.

NOTE: If belt is to be reused, check alignment marks are visible. If not: Mark timing belt with chalk against marks on sprockets.

Installation

NOTE: Ensure engine is cold before installing belt. Observe alignment marks.

1. Check tensioner pulley and guide pulley for smooth operation. Replace if necessary.
2. Check tensioner body for leakage or damage **14**. Replace if necessary.
3. Check pushrod protrusion is 10,5-11,5 mm **15**. If not, replace automatic tensioner unit.
4. Slowly compress pushrod into tensioner body until holes aligned.
5. Retain pushrod with 1,27 mm Allen key through hole in tensioner body **16**.
   *NOTE: Place flat washers under tensioner body to avoid damage to body end plug **17**.*
6. Fit dust cover **10**.
7. Fit timing belt to crankshaft sprocket. Ensure marks on belt aligned **18**.
8. Fit timing belt around guide pulley and tensioner pulley.
9. Install:
   - Crankshaft position (CKP) sensor reluctor **13**.
   - Timing belt lower cover **12**.
   - Crankshaft pulley **11**. Use tool No.09223-46011.
   - Alternator.
10. Turn camshaft sprocket (CA2) slightly clockwise until timing belt can be fitted with timing marks aligned **19**. Use tool No.09960-10010.
11. Turn camshaft sprocket (CA2) slowly anti-clockwise until timing marks aligned **6**. Ensure belt is taut between sprockets.
12. Turn camshaft sprocket (CA1) slightly clockwise until timing belt can be fitted with timing marks aligned **20**. Use tool No.09960-10010.
13. Turn camshaft sprocket (CA1) slowly anti-clockwise until timing marks aligned **6**. Ensure belt is taut between sprockets.
14. Install automatic tensioner unit **10**. Tighten bolts evenly **9**. Tightening torque: 26 Nm.
15. Remove Allen key from tensioner body to release pushrod.
16. Turn crankshaft slowly two turns in direction of rotation to TDC on No.1 cylinder.
17. Ensure timing marks aligned **5** & **6**.
18. Install components in reverse order of removal.
19. Tighten crankshaft pulley bolt **8**. Tightening torque: 245 Nm.
20. Refill cooling system.

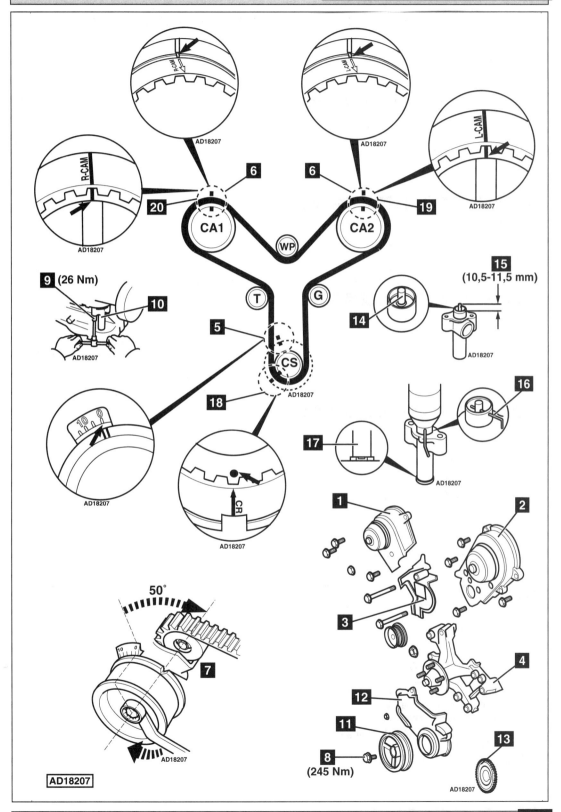

MAZDA

Model:	121 1,1/1,3 • 323 (BG) 1,3/1,6 • 323 (BF) 1,5 • 323 (BW) 1,5 Estate
Year:	1988-96
Engine Code:	B1, B3, B5, B6

Replacement Interval Guide

Mazda recommend replacement every 60,000 miles.
The previous use and service history of the vehicle must always be taken into account.
Refer to Timing Belt Replacement Intervals at the front of this manual.

Check For Engine Damage

CAUTION: This engine has been identified as an INTERFERENCE engine in which the possibility of valve-to-piston damage in the event of a timing belt failure is MOST LIKELY to occur.
A compression check of all cylinders should be performed before removing the cylinder head.

Repair Times – hrs

Remove & install:

121 (DA)	1,20
121 (DB)	1,30
323	1,50
AC	+0,10
PAS	+0,10

Special Tools

■ None required.

Special Precautions

■ Disconnect battery earth lead.
■ DO NOT turn crankshaft or camshaft when timing belt removed.
■ Remove spark plugs to ease turning engine.
■ Turn engine in normal direction of rotation (unless otherwise stated).
■ DO NOT turn engine via camshaft or other sprockets.
■ Observe all tightening torques.

Removal

1. Remove:
   ❏ RH inner wing panel.
   ❏ Alternator.
   ❏ Alternator drive belt.
   ❏ Water pump pulley **1**.
   ❏ Crankshaft pulley **2**.
   ❏ Crankshaft sprocket guide washer **3**.
   ❏ Timing belt upper cover **4**.
   ❏ Timing belt lower cover **5**.
2. Turn crankshaft clockwise until timing marks aligned **6** & **7**.
3. Slacken tensioner bolt **8**. Disconnect tensioner spring.
4. Remove timing belt.

Installation

1. Ensure timing marks aligned **6** & **7**.
2. Connect tensioner spring. Push tensioner away to extend spring. Tighten bolt.
3. Fit timing belt.
4. Turn crankshaft two turns clockwise.
5. Ensure timing marks aligned **6** & **7**.
6. Slacken tensioner bolt **8**. Allow tensioner to operate.
7. Tighten tensioner bolt **8**.
   Tightening torque: 19-25 Nm.
8. Apply a load of 10 kg to belt at ▽. Belt should deflect 12-13 mm.
9. If not: Repeat operations 4-6.
10. Turn crankshaft two turns clockwise.
11. Ensure timing marks aligned **6** & **7**.
12. Install components in reverse order of removal.
    NOTE: Install guide washer with concave side towards crankshaft pulley.
13. Tighten crankshaft pulley bolts **9**.
    Tightening torque: 12-17 Nm.

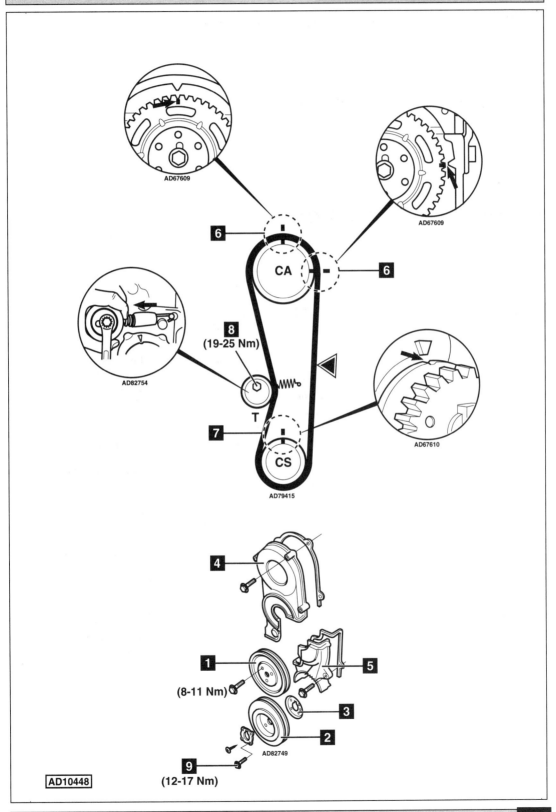

AD67609

AD67609

6

CA

6

8
(19-25 Nm)

AD82754

T

7

CS

AD79415

AD67610

4

1

(8-11 Nm)

5

3

2

AD82749

9

(12-17 Nm)

MAZDA

Model:	Demio 1,3 • 323 1,3 SOHC • MX-3 1,6 SOHC
Year:	1991-03
Engine Code:	B3, B6 SOHC

Replacement Interval Guide

Mazda recommend replacement as follows:
MX-3 − Every 60,000 miles.
Demio/323 − Every 54,000 miles.
The previous use and service history of the vehicle must always be taken into account.
Refer to Timing Belt Replacement Intervals at the front of this manual.

Check For Engine Damage

CAUTION: This engine has been identified as an INTERFERENCE engine in which the possibility of valve-to-piston damage in the event of a timing belt failure is MOST LIKELY to occur.
A compression check of all cylinders should be performed before removing the cylinder head.

Repair Times − hrs

Remove & install:
MX-3	1,50
Demio	1,30
323	1,40

Special Tools

■ Crankshaft pulley holder − Mazda No.49 D011 102.

Special Precautions

■ Disconnect battery earth lead.
■ DO NOT turn crankshaft or camshaft when timing belt removed.
■ Remove spark plugs to ease turning engine.
■ Turn engine in normal direction of rotation (unless otherwise stated).
■ DO NOT turn engine via camshaft or other sprockets.
■ Observe all tightening torques.

Removal

1. Raise and support front of vehicle.
2. Remove:
   ❏ Engine undershield.
   ❏ Auxiliary drive belt.
   ❏ Water pump pulley **1**.
3. Demio: Support engine. Remove RH engine mounting.
4. Demio: Lower engine slightly.
5. All models − remove:
   ❏ Crankshaft pulley bolts **9** and lock washer.
   ❏ Crankshaft pulley **2**.
      Use tool No.49 D011 102.
   ❏ Crankshaft position (CKP) sensor reluctor (if fitted) **12**.
   ❏ Crankshaft pulley centre bolt **10**.
   ❏ Crankshaft pulley boss **3**.
   ❏ Timing belt upper cover **4**.
   ❏ Timing belt lower cover **5**.
6. Turn crankshaft clockwise until timing marks aligned **6** & **7**.
7. Slacken tensioner bolt **8**.
8. Disconnect tensioner spring **11**.
9. Remove timing belt.

Installation

1. Check free length of tensioner spring is 64 mm **11**. Replace if necessary.
2. Ensure timing marks aligned **6** & **7**.
3. Connect tensioner spring. Push tensioner away to extend spring. Lightly tighten bolt **8**.
4. Fit timing belt.
5. Turn crankshaft two turns clockwise.
   NOTE: DO NOT allow crankshaft to turn anti-clockwise.
6. Ensure timing marks aligned **6** & **7**.
7. Slacken tensioner bolt **8**. Allow tensioner to operate. Tighten tensioner bolt **8**.
   Tightening torque: 19-25 Nm.
8. Turn crankshaft two turns clockwise.
9. Ensure timing marks aligned **6** & **7**.
10. Apply a load of 10 kg to belt at ▽. Belt should deflect 11-13 mm.
11. If not: Repeat tensioning procedure.
12. Install components in reverse order of removal.
13. Tighten crankshaft pulley centre bolt **10**.
    Tightening torque: 156-167 Nm.
14. Tighten crankshaft pulley bolts **9**.
    Tightening torque: 12-17 Nm.

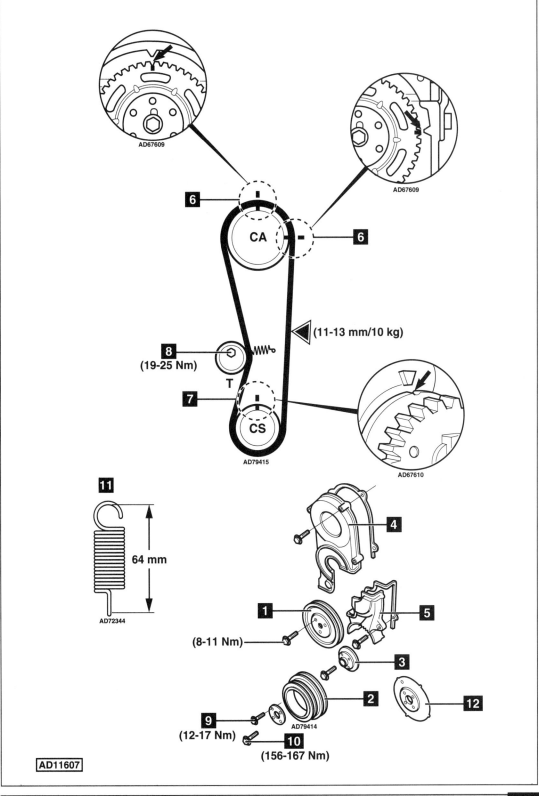

AD67609

AD67609

6

CA

6

◀ (11-13 mm/10 kg)

8
(19-25 Nm)

T

7

CS

AD79415

AD67610

11

64 mm

AD72344

4

1
(8-11 Nm)

5

3

2
AD79414

12

9
(12-17 Nm)

10
(156-167 Nm)

AD11607

MAZDA

Model:	323 (BA) 1,5i
Year:	1994-98
Engine Code:	Z5

Replacement Interval Guide

Mazda recommend replacement every
54,000 miles.
*The previous use and service history of the vehicle
must always be taken into account.
Refer to Timing Belt Replacement Intervals at the front
of this manual.*

Check For Engine Damage

*CAUTION: This engine has been identified as an
INTERFERENCE engine in which the possibility of
valve-to-piston damage in the event of a timing belt
failure is MOST LIKELY to occur.
A compression check of all cylinders should be
performed before removing the cylinder head.*

Repair Times – hrs

Remove & install	1,80

Special Tools

- None required.

Special Precautions

- Disconnect battery earth lead.
- DO NOT turn crankshaft or camshaft when timing belt removed.
- Remove spark plugs to ease turning engine.
- Turn engine in normal direction of rotation (unless otherwise stated).
- DO NOT turn engine via camshaft or other sprockets.
- Observe all tightening torques.

Removal

1. Raise and support front of vehicle.
2. Support engine.
3. Remove:
   - ❏ RH front wheel.
   - ❏ Engine undershield **1**.
   - ❏ Auxiliary drive belts.
   - ❏ Water pump pulley **2**.
   - ❏ Crankshaft pulley bolts **3**.
   - ❏ Crankshaft pulley.
   - ❏ Cylinder head cover.
   - ❏ Dipstick and tube.
   - ❏ Timing belt upper cover **4**.
   - ❏ Engine mounting **5**.
   - ❏ Timing belt centre cover **6**.
   - ❏ Timing belt lower cover **7**.
4. Remove crankshaft sprocket bolt and crankshaft pulley boss **8** & **9**.

5. Temporarily fit crankshaft sprocket bolt **8**. Turn crankshaft clockwise until camshaft sprocket 'Z' mark is at 12 o'clock position **10**.
6. Ensure timing marks aligned **11** & **12**.
7. Slacken tensioner bolt **13**. Move tensioner away from belt. Lightly tighten bolt.
8. Remove timing belt.
9. Slacken tensioner bolt **13**. Remove tensioner spring **14**. Check free length of tensioner spring is 71 mm.
10. If not: Renew tensioner spring.

Installation

1. Ensure camshaft sprocket timing marks aligned and 'Z' mark is at 12 o'clock position **10** & **12**.
   NOTE: If camshaft sprocket removed: Ensure sprocket fitted to camshaft with locating pin aligned with 'Z' mark 15.
2. Ensure crankshaft sprocket timing mark aligned **11**.
3. Fit tensioner spring **14**.
4. Push tensioner against spring tension. Lightly tighten bolt **13**.
5. Fit timing belt in anti-clockwise direction, starting at crankshaft sprocket.
6. Remove crankshaft sprocket bolt **8**.
7. Fit crankshaft pulley boss **9**. Fit crankshaft sprocket bolt **8**.
8. Turn crankshaft one turn clockwise, then a further 300° until crankshaft timing marks aligned at 60° BTDC **16**.
9. Slacken tensioner bolt **13**. Allow tensioner to operate.
10. Tighten tensioner bolt **13**.
    Tightening torque: 38-51 Nm.
11. Turn crankshaft two turns clockwise, then a further 60°. Ensure timing marks aligned **10**, **11** & **12**.
12. Apply a load of 10 kg to belt at ▽. Belt should deflect 7,0-9,0 mm.
13. If not: Repeat tensioning procedure.
14. Tighten crankshaft sprocket bolt **8**.
    Tightening torque: 157-166 Nm.
15. Install components in reverse order of removal.

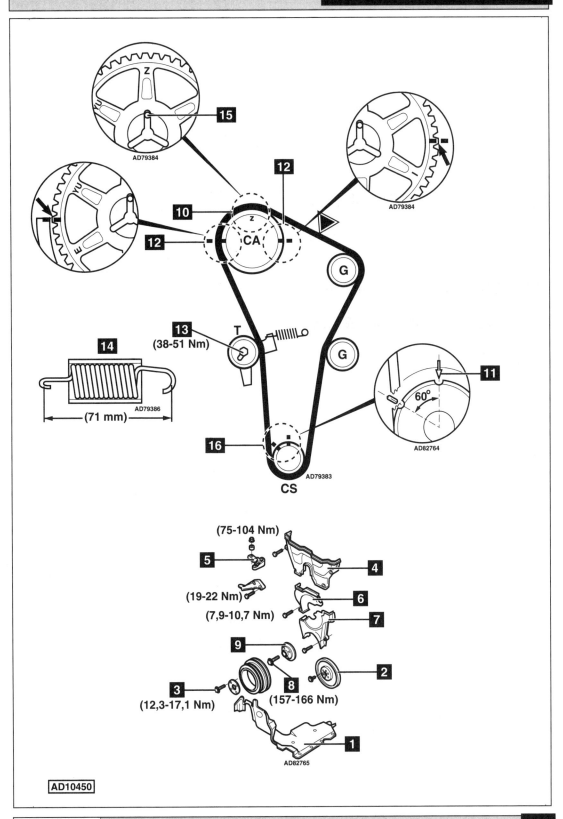

MAZDA

Model:	323 1,5 (BJ)
Year:	1998-03
Engine Code:	ZL

Replacement Interval Guide

Mazda recommend replacement every 54,000 miles.

The previous use and service history of the vehicle must always be taken into account.

Refer to Timing Belt Replacement Intervals at the front of this manual.

Check For Engine Damage

CAUTION: This engine has been identified as an INTERFERENCE engine in which the possibility of valve-to-piston damage in the event of a timing belt failure is MOST LIKELY to occur. A compression check of all cylinders should be performed before removing the cylinder head(s).

Repair Times – hrs

Remove & install	2,00

Special Tools

- Crankshaft holding tool – Mazda No.49-D011-102.

Special Precautions

- Disconnect battery earth lead.
- DO NOT turn crankshaft or camshaft when timing belt removed.
- Remove spark plugs to ease turning engine.
- Turn engine in normal direction of rotation (unless otherwise stated).
- DO NOT turn engine via camshaft or other sprockets.
- Observe all tightening torques.

Removal

1. Remove:
   - Crankshaft position (CKP) sensor.
   - Auxiliary drive belt(s).
   - Ignition coils.
   - Crankshaft pulley bolts **1**.
   - Crankshaft pulley **2**.
   - Crankshaft position sensor reluctor disc **3**.
   - Water pump pulley.
   - Cylinder head cover **4**.
2. Support engine.
3. Remove:
   - RH engine mounting.
   - Timing belt covers **5**, **6** & **7**.
4. Hold crankshaft. Use tool Nos.49-D011-102.
5. Slacken crankshaft bolt **8**.
6. Turn crankshaft clockwise until timing marks on camshaft sprockets aligned with upper cylinder head face **9**.
7. Ensure I and E marks on camshaft sprockets at 12 o'clock position **10**.
8. Ensure crankshaft pulley boss locating pin at 12 o'clock position **11**.
9. Remove:
   - Crankshaft bolt **8**.
   - Crankshaft pulley boss **12**.

10. Slacken tensioner bolt **13**.
11. Move tensioner away from belt. Lightly tighten bolt **13**.
12. Remove:
    - Timing belt.
    - Tensioner spring **14**.
    NOTE: Mark direction of rotation on belt with chalk if belt is to be reused.

Installation

1. Check free length of tensioner spring is 61,8 mm **14**. Replace if necessary.
2. Fit tensioner spring **14**.
   NOTE: Rubber sleeve on spring should point away from tensioner pulley.
3. Ensure timing marks on camshaft sprockets aligned with upper cylinder head face **9** & **10**.
4. Ensure crankshaft timing marks aligned **15**.
5. Fit timing belt in following order:
   - Crankshaft sprocket.
   - Guide pulley.
   - Exhaust camshaft sprocket.
   - Inlet camshaft sprocket.
   - Tensioner pulley.
6. Ensure belt is taut between sprockets on non-tensioned side.
   NOTE: Observe direction of rotation marks.
7. Slacken tensioner bolt **13**. Allow tensioner to operate.
8. Fit crankshaft pulley boss **12**.
9. Fit crankshaft sprocket bolt **8**.
10. Turn crankshaft slowly 1 5/6 turns clockwise until crankshaft sprocket mark aligned with tension setting mark **16**.
11. Tighten tensioner bolt **13**.
    Tightening torque: 38-51 Nm.
12. Turn crankshaft slowly 2 1/6 turns clockwise. Ensure crankshaft pulley boss locating pin at 12 o'clock position **11**.
13. Ensure timing marks on camshaft sprockets aligned **9** & **10**.
14. Apply a load of 10 kg to belt at ▽ **17**. Belt should deflect 6,0-7,5 mm. If not: Repeat tensioning procedure.
15. Install components in reverse order of removal.
16. Tighten crankshaft bolt **8**.
    Tightening torque: 157-166 Nm.
17. Tighten crankshaft bolts **1**.
    Tightening torque: 12-17 Nm.
18. Fit crankshaft position (CKP) sensor. Ensure crankshaft position (CKP) sensor air gap is 0,5-1,5 mm.

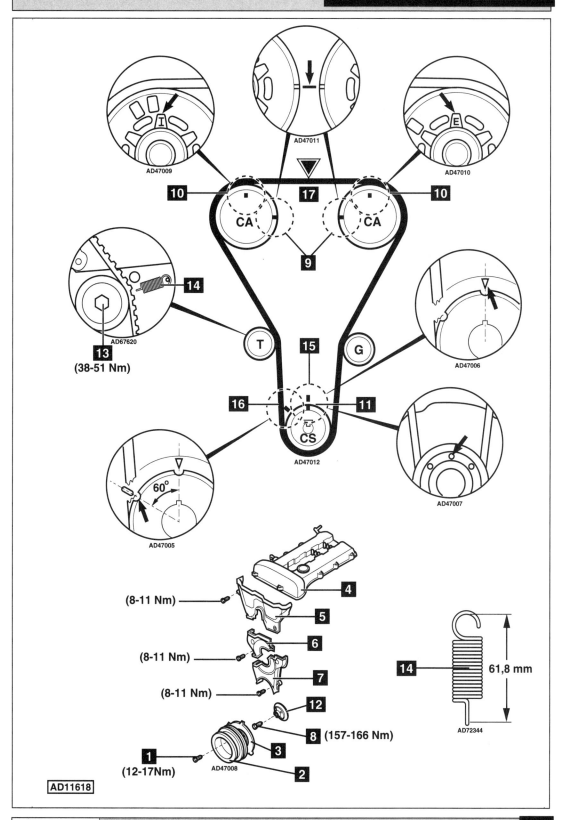

AD47011

AD47009

AD47010

10 **17** **10**

CA CA

9

14

AD67620

13
(38-51 Nm)

T **15** G

AD47006

16 **11**

CS

AD47012

AD47005

60°

AD47007

4

(8-11 Nm) **5**

6

(8-11 Nm)

7

(8-11 Nm)

12

8 (157-166 Nm)

14 61,8 mm

AD72344

1
(12-17Nm) AD47008 **3** **2**

AD11618

MAZDA

Model:	**323 (BG) 1,8 16V**
Year:	**1989-94**
Engine Code:	**BP DOHC**

Replacement Interval Guide

Mazda recommend replacement every 60,000 miles.
The previous use and service history of the vehicle must always be taken into account.
Refer to Timing Belt Replacement Intervals at the front of this manual.

Check For Engine Damage

CAUTION: This engine has been identified as an INTERFERENCE engine in which the possibility of valve-to-piston damage in the event of a timing belt failure is MOST LIKELY to occur.
A compression check of all cylinders should be performed before removing the cylinder head.

Repair Times – hrs

Remove & install	2,20

Special Tools

- None required.

Special Precautions

- Disconnect battery earth lead.
- DO NOT turn crankshaft or camshaft when timing belt removed.
- Remove spark plugs to ease turning engine.
- Turn engine in normal direction of rotation (unless otherwise stated).
- DO NOT turn engine via camshaft or other sprockets.
- Observe all tightening torques.

Removal

1. Raise and support front of vehicle.
2. Remove:
   - ❏ RH front wheel.
   - ❏ RH splash guard.
   - ❏ Engine undershield.
   - ❏ Auxiliary drive belt(s).
   - ❏ Water pump pulley.
   - ❏ Crankshaft pulley bolts **1**. Lock washer **2**.
   - ❏ Crankshaft pulley **3**.
   - ❏ Crankshaft sprocket guide washers **4**.
   - ❏ Dipstick.
   - ❏ Timing belt covers **5**, **6** & **7**.
3. Turn crankshaft clockwise to TDC on No.1 cylinder.
4. Ensure timing marks aligned **8** & **9**.
5. Slacken tensioner bolt **10**.

6. Move tensioner away from belt. Lightly tighten bolt.
7. Remove timing belt.
   NOTE: Mark direction of rotation on belt with chalk if belt is to be reused.

Installation

1. Ensure timing marks aligned **8** & **9**.
2. Fit timing belt in anti-clockwise direction, starting at crankshaft sprocket.
3. Ensure belt is taut between sprockets on non-tensioned side.
4. Slacken tensioner bolt **10**. Allow tensioner to operate. Lightly tighten bolt.
5. Turn crankshaft slowly 1 5/6 turns clockwise until crankshaft sprocket mark aligned with tension setting mark **11**.
6. Slacken tensioner bolt **10**. Allow tensioner to operate. Tighten bolt.
   Tightening torque: 38-51 Nm.
   *NOTE: Belt tension must only be set at tension setting marks **11**.*
7. Turn crankshaft slowly 2 1/6 turns clockwise. Ensure timing marks aligned **8** & **9**.
8. Apply a load of 10 kg to belt at ▽. Belt should deflect 9,0-11,5 mm. If not: Repeat tensioning procedure.
9. Install components in reverse order of removal.
10. Tighten crankshaft pulley bolts **1**.
    Tightening torque: 12-17 Nm.
    *NOTE: Ensure inner guide washer installed with concave side away from belt **4**.*

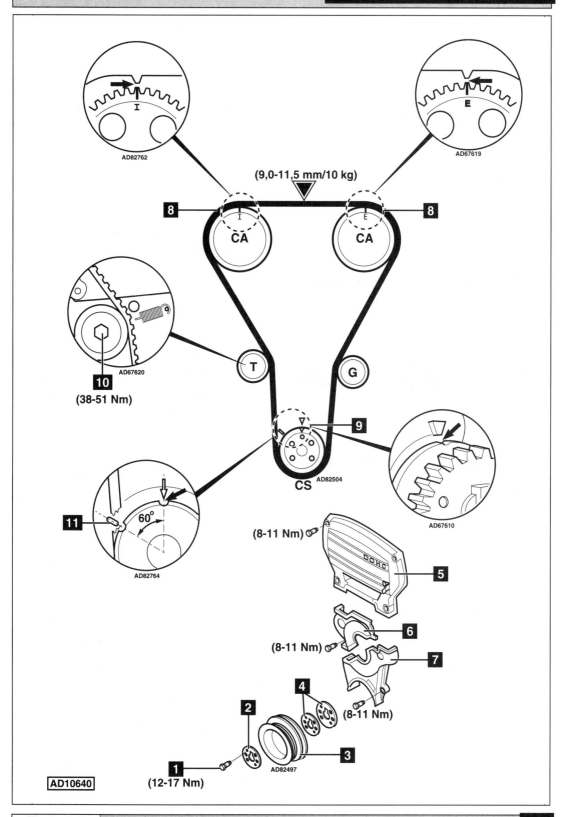

AD82762

AD67619

(9,0-11,5 mm/10 kg)

8   I   E   8

CA   CA

T   G

AD67620

10

(38-51 Nm)

9

CS   AD82504

AD67610

11   60°

AD82764

(8-11 Nm)

DOHC

5

(8-11 Nm)   **6**

7

4   (8-11 Nm)

2

1   **3**   AD82497

(12-17 Nm)

AD10640

MAZDA

Model:	**323 (BJ) 1,8 • 626 1,8/2,0 DOHC • Premacy 1,8**
Year:	**1992-03**
Engine Code:	**FP, FS**

Check For Engine Damage

CAUTION: This engine has been identified as an INTERFERENCE engine in which the possibility of valve-to-piston damage in the event of a timing belt failure is MOST LIKELY to occur.
A compression check of all cylinders should be performed before removing the cylinder head.

Repair Times – hrs

Remove & install:

323	1,80
626	1,70
Premacy	1,90

Special Tools

- Crankshaft pulley holder bolts – Mazda No.49 G011 103.
- Crankshaft pulley holder – Mazda No.49 E011 1A1.
- Crankshaft pulley holder handle – Mazda No.49 S120 710.

Special Precautions

- Disconnect battery earth lead.
- DO NOT turn crankshaft or camshaft when timing belt removed.
- Remove spark plugs to ease turning engine.
- Turn engine in normal direction of rotation (unless otherwise stated).
- DO NOT turn engine via camshaft or other sprockets.
- Observe all tightening torques.

Removal

1. Support engine.
2. Remove:
   - Engine undershield.
   - RH inner wing panel.
   - Camshaft position (CMP) sensor (if required).
   - Crankshaft position (CKP) sensor (if required).
   - RH engine mounting.
   - Auxiliary drive belts.
   - Water pump pulley.
   - PAS pump (if required). DO NOT disconnect hoses.
3. Hold crankshaft pulley. Use tool Nos. 49 G011 103/E011 1A1/S120 710.
4. Remove:
   - Crankshaft pulley bolt **1**.
   - Crankshaft pulley **2**.
   - HT leads.
   - Cylinder head cover.
   - Dipstick and tube **3**.
   - Timing belt covers **4**.
5. Temporarily fit bolt and washer to retain crankshaft sprocket **1**.
6. Turn crankshaft to TDC on No.1 cylinder. Ensure timing marks aligned **5** & **6**.
   NOTE: Timing marks on rear of camshaft sprockets align with cylinder head upper face.
7. Hold tensioner pulley **7**. Use Allen key **8**. Disconnect tensioner spring from pin **9**.
8. Turn tensioner pulley anti-clockwise to release tension on belt **7**.
9. Remove timing belt.

Installation

1. Check free length of tensioner spring is 36,6 mm **9**. Replace if necessary.
2. Ensure timing marks aligned **5** & **6**.
3. Fit timing belt in anti-clockwise direction, starting at crankshaft sprocket. Ensure belt is taut between sprockets.
4. Turn tensioner pulley clockwise **7**. Use Allen key **8**.
5. Connect tensioner spring to pin **9**. Ensure belt is tensioned.
6. Turn crankshaft slowly two turns in direction of rotation.
7. Ensure timing marks aligned **5** & **6**.
8. Remove crankshaft pulley bolt and washer **1**.
9. Install components in reverse order of removal.
10. Hold crankshaft sprocket. Use tool Nos. 49 G011 103/E011 1A1/S120 710. Tighten crankshaft pulley bolt **1**. Tightening torque: 157-166 Nm.
11. Ensure crankshaft position (CKP) sensor air gap is 0,5-1,5 mm (if removed).

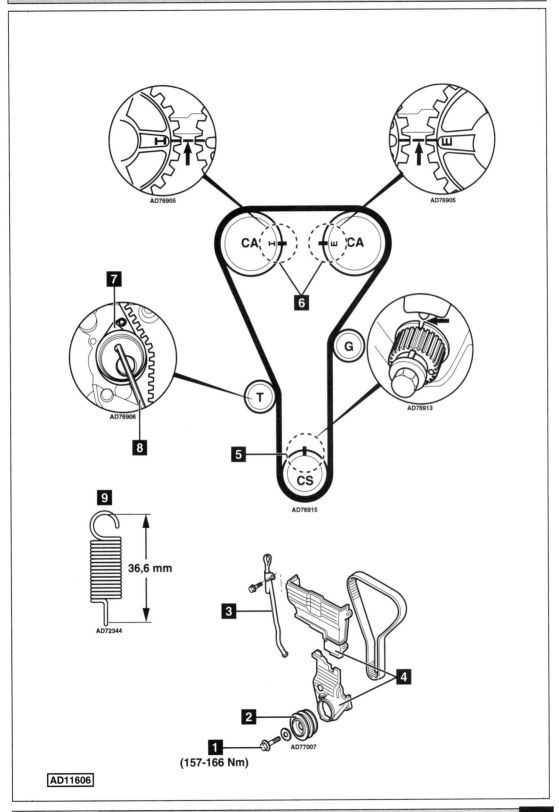

MAZDA

Model:	**323 (BA) 1,8 16V • MX-3 1,6 16V**
Year:	**1994-98**
Engine Code:	**BP DOHC, B6 DOHC**

Replacement Interval Guide

Mazda recommend:
323: Replacement every 54,000 miles.
MX-3: Replacement every 60,000 miles.
The previous use and service history of the vehicle must always be taken into account.
Refer to Timing Belt Replacement Intervals at the front of this manual.

Check For Engine Damage

CAUTION: This engine has been identified as a FREEWHEELING engine in which the possibility of valve-to-piston damage in the event of a timing belt failure may be minimal or very unlikely. However, a precautionary compression check of all cylinders should be performed.

Repair Times – hrs

Remove & install:

323	1,80
MX-3	2,20

Special Tools

- Crankshaft holding tool – Mazda No.49-D011-102.

Special Precautions

- Disconnect battery earth lead.
- DO NOT turn crankshaft or camshaft when timing belt removed.
- Remove spark plugs to ease turning engine.
- Turn engine in normal direction of rotation (unless otherwise stated).
- DO NOT turn engine via camshaft or other sprockets.
- Observe all tightening torques.

Removal

1. Raise and support front of vehicle.
2. Support engine.
3. Remove:
   - ❏ RH front wheel.
   - ❏ RH splash guard.
   - ❏ Auxiliary drive belt(s).
   - ❏ Water pump pulley.
   - ❏ Crankshaft pulley bolts **1**.
   - ❏ Crankshaft pulley **13**.
   - ❏ Dipstick tube clip.
   - ❏ RH engine mounting.
   - ❏ HT leads.
   - ❏ Cylinder head cover **2**.
   - ❏ Timing belt covers **3**, **4** & **5**.

4. Turn crankshaft clockwise to TDC on No.1 cylinder. Ensure timing marks aligned **6** & **7**.
5. Hold crankshaft sprocket. Use tool No.49-D011-102. Remove crankshaft centre bolt **8**.
6. Remove crankshaft pulley boss **9**.
7. Slacken tensioner bolt **10**. Move tensioner away from belt. Lightly tighten bolt.
8. Remove timing belt.
   NOTE: Mark direction of rotation on belt with chalk if belt is to be reused.

Installation

1. Disconnect tensioner spring **11**. Check free length of tensioner spring is 59,2 mm.
2. Fit tensioner spring.
3. Ensure timing marks aligned **6** & **7**.
4. Fit timing belt in anti-clockwise direction, starting at crankshaft sprocket. Ensure belt is taut between sprockets.
5. Slacken tensioner bolt **10**. Allow tensioner to operate. Lightly tighten bolt.
6. Fit crankshaft pulley boss and crankshaft centre bolt **8** & **9**.
7. Turn crankshaft slowly 1 5/6 turns clockwise until crankshaft sprocket mark aligned with tension setting mark **12**.
8. Slacken tensioner bolt **10**. Allow tensioner to operate. Tighten bolt.
   Tightening torque: 38-51 Nm.
   *NOTE: Belt tension must only be set at tension setting marks **12**.*
9. Turn crankshaft slowly 2 1/6 turns clockwise. Ensure timing marks aligned **6** & **7**.
10. Apply a load of 10 kg to belt at ▽. Belt should deflect 9,0-11,5 mm.
11. If not: Repeat tensioning procedure.
12. Hold crankshaft sprocket. Use tool No.49-D011-102. Tighten crankshaft centre bolt **8**. Tightening torque: 157-166 Nm.
13. Install components in reverse order of removal.
14. Tighten crankshaft pulley bolts **1**. Tightening torque: 12,5-17,0 Nm.

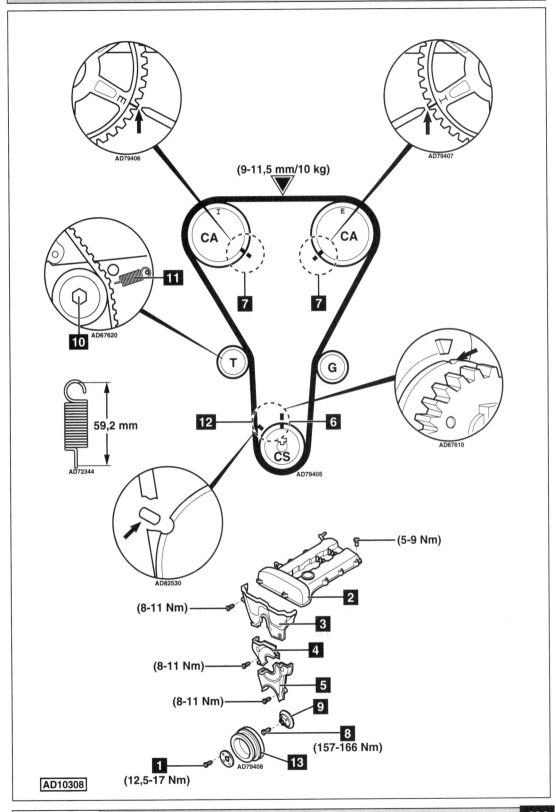

AD79406

AD79407

(9-11,5 mm/10 kg)

I — CA

E — CA

11

10 AD67620

59,2 mm

AD72344

7

7

T

G

12 — CS — **6**

AD79405

AD67610

AD82530

(5-9 Nm)

(8-11 Nm) — **2**

3

(8-11 Nm) — **4**

5

(8-11 Nm) — **9**

8
(157-166 Nm)

1 — AD79408 — **13**

(12,5-17 Nm)

AD10308

MAZDA

Model:	323 (BA) 2,0i V6
Year:	1994-98
Engine Code:	KF DOHC

Replacement Interval Guide

Mazda recommend replacement every
54,000 miles.
*The previous use and service history of the vehicle
must always be taken into account.
Refer to Timing Belt Replacement Intervals at the front
of this manual.*

Check For Engine Damage

*CAUTION: This engine has been identified as an
INTERFERENCE engine in which the possibility of
valve-to-piston damage in the event of a timing belt
failure is MOST LIKELY to occur.
A compression check of all cylinders should be
performed before removing the cylinder head.*

Repair Times – hrs

Remove & install	2,50

Special Tools

- None required.

Special Precautions

- Disconnect battery earth lead.
- DO NOT turn crankshaft or camshaft when timing belt removed.
- Remove spark plugs to ease turning engine.
- Turn engine in normal direction of rotation (unless otherwise stated).
- DO NOT turn engine via camshaft or other sprockets.
- Observe all tightening torques.

Removal

1. Raise and support front of vehicle.
2. Support engine.
3. Remove:
   - ❑ RH front wheel.
   - ❑ Engine undershield **1**.
   - ❑ Auxiliary drive belts.
   - ❑ Water pump pulley **2**.
   - ❑ Auxiliary drive belt tensioner and bracket **3**.
   - ❑ PAS pump and bracket **4**. DO NOT disconnect hoses.
   - ❑ Crankshaft pulley bolt **5**.
   - ❑ Crankshaft pulley **6**.
   - ❑ Dipstick and tube.
   - ❑ Crankshaft position (CKP) sensor.
   - ❑ Timing belt RH cover **7**.
   - ❑ Timing belt LH cover **8**.
   - ❑ Engine mounting bracket **9**.
4. Turn crankshaft to TDC on No.1 cylinder. Ensure timing marks aligned **10** & **11**.
5. Slacken automatic tensioner unit bolts **12** & **13**.

6. Remove lower bolt **13**. Allow automatic tensioner unit to swivel inwards towards belt.
7. Remove upper bolt **12** together with automatic tensioner unit **14**.
   NOTE: Hold automatic tensioner unit firmly to prevent damage to bolt threads.
8. Remove:
   - ❑ Guide pulley (G1) **15**.
   - ❑ Timing belt.

Installation

1. Ensure timing marks aligned **10** & **11**.
2. Check tensioner body for leakage or damage **14**.
3. Check pushrod protrusion is 14-16 mm. If not: Replace automatic tensioner unit.
4. Slowly compress pushrod into tensioner body **14** until holes aligned. Use suitable press. Retain pushrod with 1,6 mm diameter pin through lower hole **16**.
   NOTE: Place flat washer under tensioner body to avoid damage to body end plug. DO NOT exceed 1000 kg force.
5. Install automatic tensioner unit to cylinder block with upper bolt **12**. Lightly tighten bolt.
6. Fit timing belt in following order:
   - ❑ Crankshaft sprocket.
   - ❑ Guide pulley (G2).
   - ❑ Camshaft sprocket (CA1).
   - ❑ Tensioner pulley **17**.
   - ❑ Camshaft sprocket (CA2).
7. Fit guide pulley **15** while applying pressure on belt. Tighten bolt to 38-51 Nm.
8. Push automatic tensioner unit outwards in direction of arrow **14**. Fit lower bolt **13**. Tighten bolts to 19-25 Nm **12** & **13**.
9. Remove pin from tensioner body to release pushrod.
10. Turn crankshaft slowly two turns clockwise. Ensure timing marks aligned **10** & **11**.
11. Apply a load of 10 kg to belt at ▽. Belt should deflect 6,0-8,0 mm.
12. If not: Replace automatic tensioner unit. Repeat operations 5-11.
13. Install components in reverse order of removal.
14. Tighten crankshaft pulley bolt **5**. Tightening torque: 157-166 Nm.

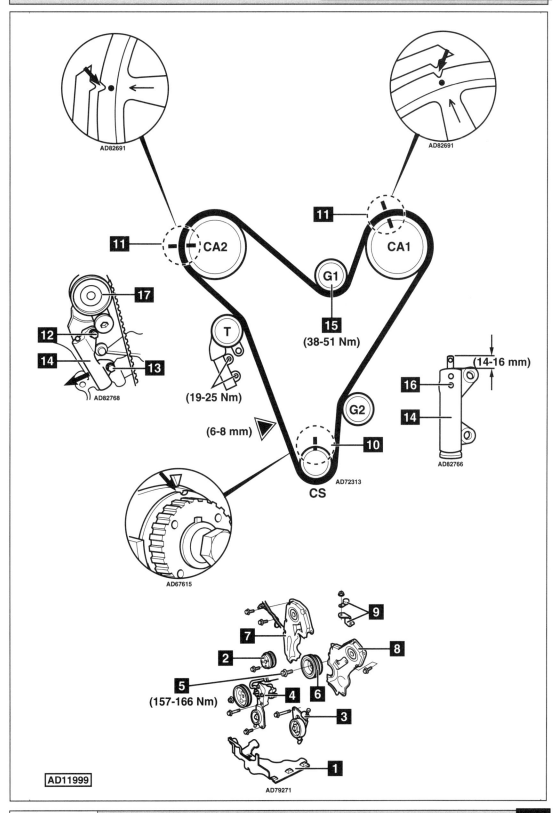

AD82691

AD82691

11 CA2

11 CA1

17

12

14

13

AD82768

T

(19-25 Nm)

G1

15

(38-51 Nm)

(14-16 mm)

16

14

AD82766

(6-8 mm)

G2

10

CS

AD72313

AD67615

9

7

2

8

5

(157-166 Nm)

4

6

3

1

AD79271

AD11999

MAZDA

Model:	626 1,6/1,8/2,0 • E1800/E2000 • B2000
Year:	1983-93
Engine Code:	F6, F8, FE – without guide pulley

Replacement Interval Guide

Mazda recommend replacement every 60,000 miles.
The previous use and service history of the vehicle must always be taken into account.
Refer to Timing Belt Replacement Intervals at the front of this manual.

Check For Engine Damage

CAUTION: This engine has been identified as an INTERFERENCE engine in which the possibility of valve-to-piston damage in the event of a timing belt failure is MOST LIKELY to occur.
A compression check of all cylinders should be performed before removing the cylinder head.

Repair Times – hrs

Remove & install:

626	1,20
AC	+0,20
E1800/E2000	1,10
B2000	1,90
AC	+0,10
PAS	+0,10

Special Tools

- None required.

Special Precautions

- Disconnect battery earth lead.
- DO NOT turn crankshaft or camshaft when timing belt removed.
- Remove spark plugs to ease turning engine.
- Turn engine in normal direction of rotation (unless otherwise stated).
- DO NOT turn engine via camshaft or other sprockets.
- Observe all tightening torques.

Removal

1. B2000 – Remove:
   - ❏ Top hose.
   - ❏ Radiator cowling.
   - ❏ Cooling fan and viscous coupling.
2. All models:
   - ❏ Raise and support front of vehicle.
3. Remove:
   - ❏ RH front wheel.
   - ❏ Inner wing panel.
   - ❏ Auxiliary drive belts.
   - ❏ Timing belt upper cover and seal **1**.

4. Turn crankshaft to TDC on No.1 cylinder. Ensure camshaft timing marks aligned **2**.
   *NOTE: Always align camshaft sprocket timing marks A, B, C or CX with mark on backplate **2** according to model: A = FE (626,E1800/2000), B = F6/F8 (626), C = F8 (E1800/2000), CX = F6 (B2000).*
5. Remove:
   - ❏ Crankshaft pulley bolts **3**.
   - ❏ Crankshaft pulley **4**.
   - ❏ Timing belt lower cover and seal **5**.
6. Ensure crankshaft timing marks aligned **6**.
7. Slacken tensioner bolt **7**. Move tensioner away from belt. Lightly tighten bolt.
8. Remove timing belt.
   NOTE: FE = blue tensioner spring. F6 and F8 = plain tensioner spring.

Installation

1. Ensure timing marks aligned **2** & **6**.
2. Clean tensioner pulley.
3. Push tensioner back and hold.
4. Fit timing belt to crankshaft sprocket and camshaft sprocket. Ensure belt is taut between sprockets.
5. Fit belt round water pump sprocket then ease it round tensioner pulley.
6. Ensure timing marks aligned **2** & **6**.
7. Slacken tensioner bolt. Turn crankshaft two turns clockwise. Tighten tensioner bolt **7**. Tightening torque: 45 Nm.
8. Ensure timing marks aligned **2** & **6**.
9. Apply a load of 10 kg to belt at ▽. Belt should deflect 12-14 mm.
10. Install components in reverse order of removal.
11. Tighten crankshaft pulley bolts **3**. Tightening torque: 15 Nm.

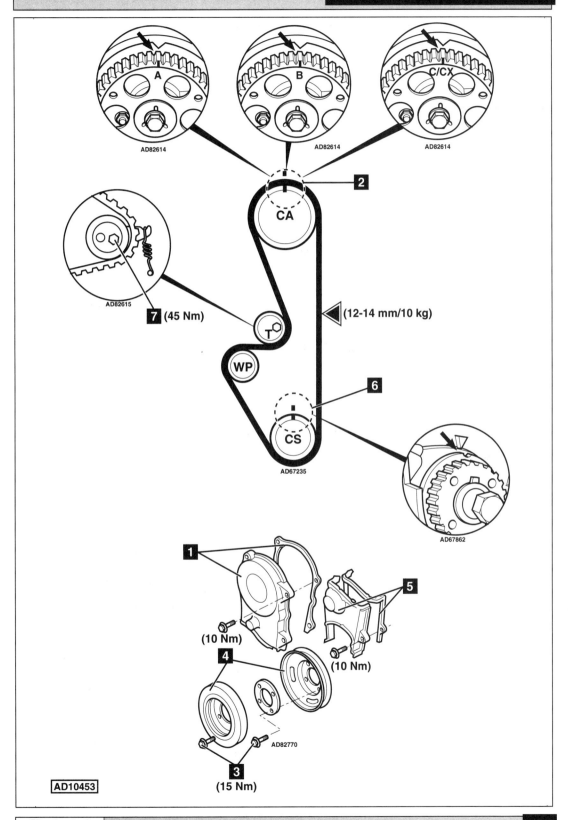

AD82614

AD82614

AD82614

2

CA

AD82615

7 (45 Nm)

T

WP

◄(12-14 mm/10 kg)

6

CS

AD67235

AD67862

1

(10 Nm)

5

(10 Nm)

4

AD82770

3

(15 Nm)

AD10453

MAZDA

Model:	**626 1,8/2,0/2,2 (→1993) • E2000 (FE EGi)**
Year:	**1987-99**
Engine Code:	**F8-1,8, FE-2,0 8V, FE-2,0 12V, F2-2,2 12V – with guide pulley**

Replacement Interval Guide

Mazda recommend:
6,000 mile service intervals: Replacement every 60,000 miles.
9,000 mile service intervals: Replacement every 54,000 miles.
The previous use and service history of the vehicle must always be taken into account.
Refer to Timing Belt Replacement Intervals at the front of this manual.

Check For Engine Damage

CAUTION: This engine has been identified as an INTERFERENCE engine in which the possibility of valve-to-piston damage in the event of a timing belt failure is MOST LIKELY to occur.
A compression check of all cylinders should be performed before removing the cylinder head.

Repair Times – hrs

Remove & install:	
626	1,20
AC	+0,20
PAS	+0,10
E2000 →10/98	1,10
E2000 11/98→	2,40

Special Tools

■ None required.

Special Precautions

■ Disconnect battery earth lead.
■ DO NOT turn crankshaft or camshaft when timing belt removed.
■ Remove spark plugs to ease turning engine.
■ Turn engine in normal direction of rotation (unless otherwise stated).
■ DO NOT turn engine via camshaft or other sprockets.
■ Observe all tightening torques.

Removal

1. Raise and support front of vehicle.
2. 626 – remove:
   ❏ RH front wheel.
   ❏ Inner wing panel.
   ❏ Auxiliary drive belts.
3. E2000 – remove:
   ❏ Front seats.
   ❏ Engine compartment cover.
   ❏ Handbrake lever assembly.
   ❏ Cooling fan and cowling.
   ❏ Auxiliary drive belts.
   ❏ Cooling fan pulley.
   ❏ Cooling fan brackets.
4. Remove crankshaft pulley bolts **1**.
5. All models – remove:
   ❏ Crankshaft pulley **2**.
   ❏ Timing belt upper cover **3**.
   ❏ Timing belt lower cover **4**.
6. Turn crankshaft to TDC on No.1 cylinder. Ensure timing marks aligned **5** & **6**.
   NOTE: F2: Align camshaft timing mark '1'.
   FE: Align camshaft timing mark '2'.
   F8: Align camshaft timing mark '3'.
7. Remove crankshaft sprocket guide washer.
8. Slacken tensioner bolt **7**. Move tensioner away from belt. Lightly tighten bolt.
9. Remove timing belt.

Installation

1. Ensure timing marks aligned **5** & **6**.
2. Clean and check tensioner pulley.
3. Fit timing belt in anti-clockwise direction, starting at crankshaft sprocket. Ensure belt is taut between sprockets.
4. Slacken tensioner bolt **7**. Turn crankshaft two turns clockwise. Tighten bolt.
   Tightening torque: 45 Nm.
5. Ensure timing marks aligned **5** & **6**.
6. Apply a load of 10 kg to belt at ▽. Belt should deflect as follows:
   ❏ F2 = 8,0-9,0 mm.
   ❏ FE = 5,5-6,5 mm.
   ❏ F8 = 4,0-5,0 mm.
7. Install components in reverse order of removal.
8. Tighten crankshaft pulley bolts **1**.
   Tightening torque: 16 Nm.

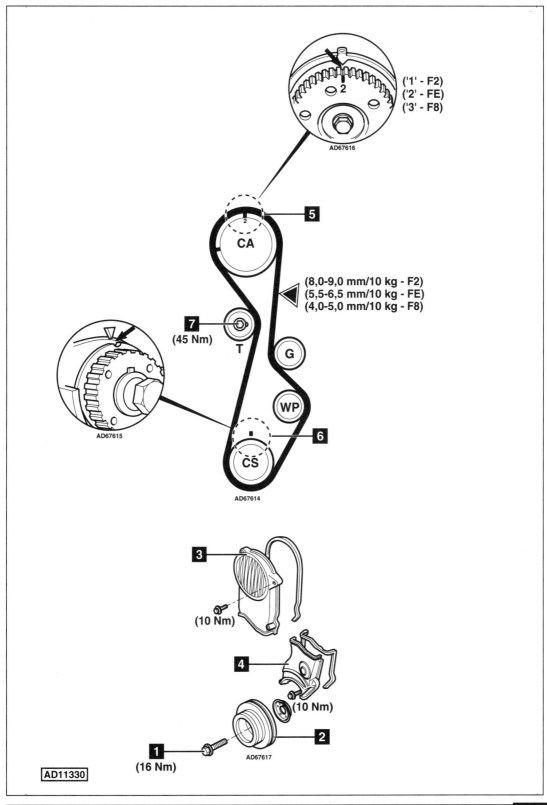

('1' - F2)
('2' - FE)
('3' - F8)

AD67616

5

CA

(8,0-9,0 mm/10 kg - F2)
(5,5-6,5 mm/10 kg - FE)
(4,0-5,0 mm/10 kg - F8)

7
(45 Nm) T

G

WP

AD67615

6

CS

AD67614

3
(10 Nm)

4
(10 Nm)

1
(16 Nm) AD67617 **2**

AD11330

Model:	**626 2,5 V6 • Xedos 6 2,0 V6 • Xedos 9 2,0/2,5 V6 • MX-6 2,5 V6**
Year:	**1992-00**
Engine Code:	**KF, KL DOHC**

Replacement Interval Guide

Mazda recommend:
6,000 mile service interval – replacement every 60,000 miles.
9,000 mile service interval – replacement every 54,000 miles.
The previous use and service history of the vehicle must always be taken into account.
Refer to Timing Belt Replacement Intervals at the front of this manual.

Check For Engine Damage

CAUTION: This engine has been identified as an INTERFERENCE engine in which the possibility of valve-to-piston damage in the event of a timing belt failure is MOST LIKELY to occur.
A compression check of all cylinders should be performed before removing the cylinder head.

Repair Times – hrs

Remove & install:
Xedos 6/626/MX-6	3,00
Xedos 9	2,50

Special Tools

■ None required.

Special Precautions

■ Disconnect battery earth lead.
■ DO NOT turn crankshaft or camshaft when timing belt removed.
■ Remove spark plugs to ease turning engine.
■ Turn engine in normal direction of rotation (unless otherwise stated).
■ DO NOT turn engine via camshaft or other sprockets.
■ Observe all tightening torques.

Removal

1. Drain coolant.
2. Support engine.
3. Remove:
   ❏ RH inner wing panel.
   ❏ Engine undershield.
   ❏ Radiator hoses.
   ❏ RH engine mounting.
   ❏ Auxiliary drive belts.
   ❏ Water pump pulley **1**.
   ❏ Auxiliary drive belt tensioner and bracket **2**.
   ❏ PAS pump and bracket **3**. DO NOT disconnect hoses.
   ❏ Crankshaft pulley **4**.
   ❏ Dipstick and tube.
   ❏ Crankshaft position (CKP) sensor bracket.

   ❏ Engine wiring harness bracket **5**.
   ❏ Timing belt RH cover **6**.
   ❏ Timing belt LH cover **7**.
4. Turn crankshaft to TDC on No.1 cylinder. Ensure timing marks aligned **8** & **9**.
5. Slacken automatic tensioner unit bolts **10** & **11**.
6. Remove bolt **10**. Allow automatic tensioner unit to swivel inwards towards belt.
7. Remove bolt **11** together with automatic tensioner unit **12**.
   NOTE: Hold automatic tensioner unit firmly to prevent damage to bolt threads.
8. Remove guide pulley (G1).
9. Remove timing belt.

Installation

1. Ensure timing marks aligned **8** & **9**.
2. Check tensioner body for leakage or damage **12**.
3. Check pushrod protrusion is 14-16 mm **13**. If not: Replace automatic tensioner unit.
4. Slowly compress pushrod **13** into tensioner body **12** until holes aligned. Use suitable press. Retain pushrod with 1,6 mm diameter pin through lower hole **14**.
   NOTE: Place flat washer under tensioner body to avoid damage to body end plug **15. DO NOT exceed 1000 kg force.**
5. Install automatic tensioner unit to cylinder block with upper bolt **11**. Lightly tighten bolt.
6. Fit timing belt in following order:
   ❏ Crankshaft sprocket.
   ❏ Guide pulley (G2).
   ❏ Camshaft sprocket (CA1).
   ❏ Tensioner pulley **16**.
   ❏ Camshaft sprocket (CA2).
7. Fit guide pulley (G1) while applying pressure on belt. Tighten bolt to 38-51 Nm.
8. Push automatic tensioner unit outwards in direction of arrow **17**. Fit lower bolt **10**. Tighten bolts to 19-25 Nm **10** & **11**.
9. Remove pin from tensioner body to release pushrod.
10. Turn crankshaft slowly two turns in direction of rotation until timing marks aligned **8** & **9**.
11. Apply a load of 10 kg to belt at ▽. Belt should deflect 6,0-8,0 mm.
12. If not: Replace automatic tensioner unit. Repeat operations 5-11.
13. Install components in reverse order of removal.
14. Tighten crankshaft pulley bolt **18**. Tightening torque: 157-166 Nm.
15. Refill cooling system.

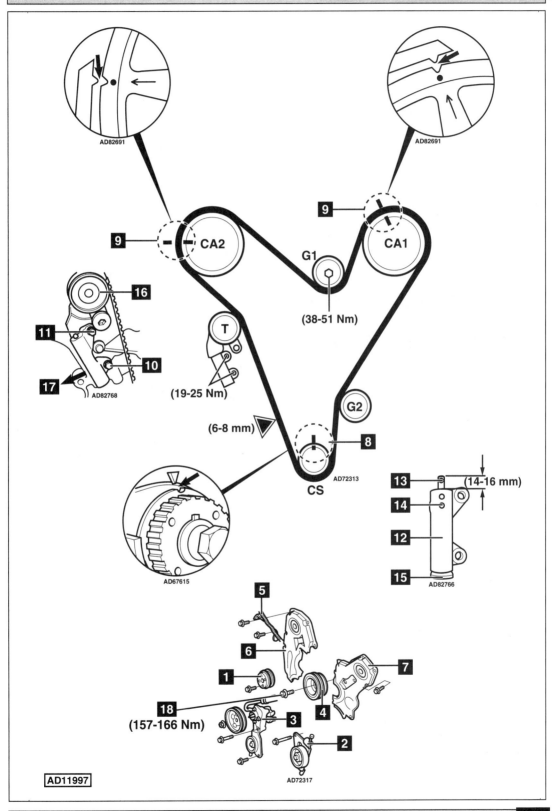

AD82691

AD82691

9 CA2

9 CA1

G1

(38-51 Nm)

16

11

10

17

AD82768

T

(19-25 Nm)

(6-8 mm)

G2

8

AD72313

CS

13 (14-16 mm)

14

12

15

AD82766

AD67615

5

6

1

7

18

(157-166 Nm)

3

4

2

AD72317

AD11997

MAZDA

Model:	**MX-3 1,8 V6**
Year:	**1991-98**
Engine Code:	**K8 DOHC**

Replacement Interval Guide

Mazda recommend replacement every 60,000 miles.
The previous use and service history of the vehicle must always be taken into account.
Refer to Timing Belt Replacement Intervals at the front of this manual.

Check For Engine Damage

CAUTION: This engine has been identified as an INTERFERENCE engine in which the possibility of valve-to-piston damage in the event of a timing belt failure is MOST LIKELY to occur.
A compression check of all cylinders should be performed before removing the cylinder head.

Repair Times – hrs

Remove & install	2,80

Special Tools

- None required.

Special Precautions

- Disconnect battery earth lead.
- DO NOT turn crankshaft or camshaft when timing belt removed.
- Remove spark plugs to ease turning engine.
- Turn engine in normal direction of rotation (unless otherwise stated).
- DO NOT turn engine via camshaft or other sprockets.
- Observe all tightening torques.

Removal

1. Remove:
   - ❑ Engine undershield.
   - ❑ Suspension strut bar.
   - ❑ Auxiliary drive belts.
   - ❑ Water pump pulley **1**.
   - ❑ Guide pulley **2**.
   - ❑ PAS pump.
   - ❑ Crankshaft pulley **3**.
   - ❑ Dipstick and tube.
   - ❑ Disconnect engine wiring harness.
   - ❑ Timing belt covers **4**.
2. Turn crankshaft clockwise until timing marks aligned **5** & **6**.
3. Remove:
   - ❑ Automatic tensioner unit **8**.
   - ❑ Tensioner pulley **7**.
   - ❑ Guide pulley (G1) **13**.
   - ❑ Timing belt.

Installation

1. Ensure timing marks aligned **5** & **6**.
2. Check pushrod protrusion is 14-16 mm **9**. If not: Replace automatic tensioner unit.
3. Slowly compress pushrod into tensioner body **15** until holes aligned. Use suitable press. Retain pushrod with 1,6 mm diameter pin **14**.
4. Install automatic tensioner unit with upper bolt loosely tightened.
5. Fit timing belt in following order:
   - ❑ Crankshaft sprocket.
   - ❑ Guide pulley (G2) **10**.
   - ❑ LH camshaft sprocket **11**.
   - ❑ Tensioner pulley.
   - ❑ RH camshaft sprocket **12**.
6. Fit guide pulley **13** while applying pressure on belt.
7. Push automatic tensioner unit outwards in direction of arrow **15**. Tighten bolts to 19-25 Nm **8**.
8. Remove pin from tensioner body to release pushrod.
9. Turn crankshaft two turns clockwise.
10. Apply a load of 10 kg to belt at ▽. Belt should deflect 6-8 mm.
11. If not: Replace automatic tensioner unit.
12. Install components in reverse order of removal.
13. Tighten crankshaft pulley bolt.
    Tightening torque: 157-167 Nm.

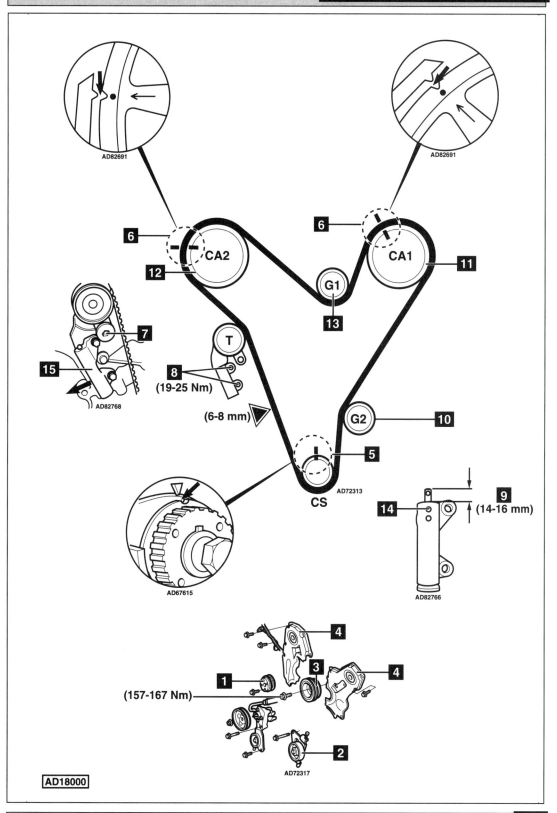

AD82691

AD82691

6 CA2

6 CA1

12

11

G1

13

7

T

15

8

(19-25 Nm)

AD82768

(6-8 mm)

G2 **10**

5

AD72313

CS

AD67615

14 **9** (14-16 mm)

AD82766

4

3 **4**

1

(157-167 Nm) **2**

AD72317

AD18000

MAZDA

Model:	MX-5 1,6/1,8
Year:	1989-03
Engine Code:	B6 DOHC, BP DOHC

Replacement Interval Guide

Mazda recommend:
6,000 mile service interval – replacement every 60,000 miles.
9,000 mile service interval – replacement every 54,000 miles.
The previous use and service history of the vehicle must always be taken into account.
Refer to Timing Belt Replacement Intervals at the front of this manual.

Check For Engine Damage

CAUTION: This engine has been identified as a FREEWHEELING engine in which the possibility of valve-to-piston damage in the event of a timing belt failure may be minimal or very unlikely. However, a precautionary compression check of all cylinders should be performed.

Repair Times – hrs

Remove & install	1,80
AC	+0,10
PAS	+0,10

Special Tools

- ■ Crankshaft pulley holder – Mazda No.49-D011-102.

Special Precautions

- ■ Disconnect battery earth lead.
- ■ DO NOT turn crankshaft or camshaft when timing belt removed.
- ■ Remove spark plugs to ease turning engine.
- ■ Turn engine in normal direction of rotation (unless otherwise stated).
- ■ DO NOT turn engine via camshaft or other sprockets.
- ■ Observe all tightening torques.

Removal

1. Drain coolant.
2. Remove:
   - ❏ Air intake pipe.
   - ❏ Camshaft position (CMP) sensor (if required).
   - ❏ Crankshaft position (CKP) sensor (if required).
   - ❏ Radiator and hoses.
   - ❏ Auxiliary drive belts.
   - ❏ Water pump pulley **1**.
   - ❏ Crankshaft pulley bolts **10**.
   - ❏ Crankshaft pulley **2**.
   - ❏ Crankshaft position (CKP) sensor reluctor (if fitted) **11**.
   - ❏ Cylinder head cover.
3. Turn crankshaft to TDC on No.1 cylinder.
4. Ensure timing marks aligned **6** & **7**.

5. Hold crankshaft sprocket. Use tool No.49-D011-102. Remove crankshaft centre bolt **12**.
6. Remove:
   - ❏ Crankshaft pulley boss (if required) **13**.
   - ❏ Timing belt covers **3**, **4** & **5**.
7. Slacken tensioner bolt **8**.
8. Move tensioner away from belt. Tighten bolt **8**.
9. Remove timing belt.
   NOTE: Mark direction of rotation on belt with chalk if belt is to be reused.

Installation

1. Check free length of tensioner spring is 59,2 mm **14**. Replace if necessary.
2. Ensure timing marks aligned **6** & **7**.
   NOTE: Align inlet camshaft 'E' mark and exhaust camshaft 'I' mark.
3. Fit timing belt in anti-clockwise direction, starting at crankshaft sprocket. Ensure belt is taut between guide pulley and camshaft sprockets.
4. Slacken tensioner bolt **8**.
5. Fit crankshaft pulley boss and crankshaft centre bolt (if removed) **12** & **13**.
   NOTE: Ensure crankshaft pulley boss dowel at 12 o'clock.
6. Turn crankshaft 1 5/6 turns clockwise. Align tension setting mark approximately 60° BTDC **9**.
   NOTE: DO NOT turn crankshaft anti-clockwise.
7. Tighten tensioner bolt **8**.
   Tightening torque: 37-52 Nm.
8. Turn crankshaft 2 1/6 turns clockwise. Ensure timing marks aligned **6** & **7**.
   NOTE: Ensure crankshaft pulley boss dowel at 12 o'clock.
9. Apply a load of 10 kg to belt at ▽. Belt should deflect 9,0-11,5 mm.
10. If not: Repeat tensioning procedure.
11. Install components in reverse order of removal.
12. Hold crankshaft sprocket. Use tool No.49-D011-102. Tighten crankshaft centre bolt **12**. Tightening torque: 157-166 Nm.
    NOTE: Fit crankshaft sprocket guide washer (convex side towards crankshaft sprocket).
13. Tighten crankshaft pulley bolts **10**.
    Tightening torque: 12-17 Nm.
14. Ensure crankshaft position (CKP) sensor air gap is 0,5-1,5 mm (if removed).
15. Refill cooling system.

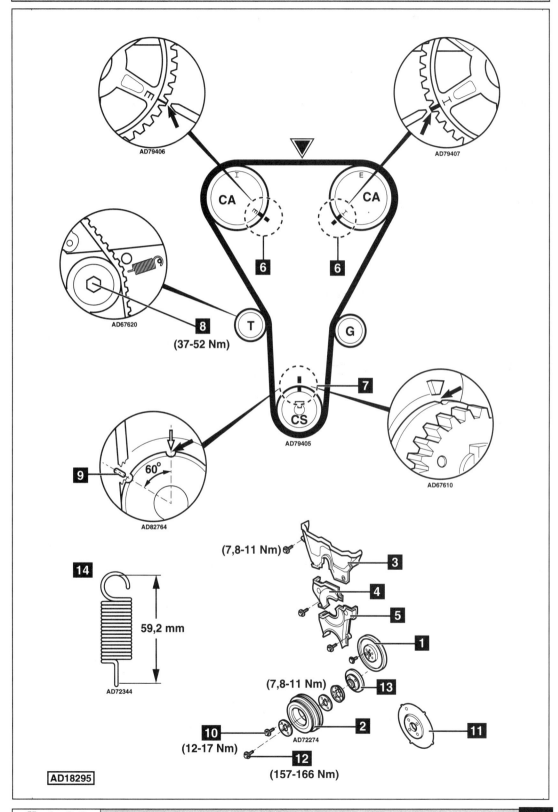

AD79406

AD79407

CA

CA

6

6

AD67620

T

G

8

(37-52 Nm)

7

AD79405

AD67610

9 60°

AD82764

(7,8-11 Nm) 3

4

5

14 1

13 (7,8-11 Nm)

59,2 mm

AD72344

11

10 2

(12-17 Nm) AD72274

12

(157-166 Nm)

AD18295

MAZDA

Model:	323 (BJ) 2,0 TD • 626 2,0 TD • Premacy 2,0 TD
Year:	1998-03
Engine Code:	RF Turbo

Replacement Interval Guide

Mazda recommend replacement every
54,000 miles.
*The previous use and service history of the vehicle
must always be taken into account.*
*Refer to Timing Belt Replacement Intervals at the front
of this manual.*

Check For Engine Damage

*CAUTION: This engine has been identified as an
INTERFERENCE engine in which the possibility of
valve-to-piston damage in the event of a timing belt
failure is MOST LIKELY to occur. A compression check
of all cylinders should be performed before removing
the cylinder head(s).*

Repair Times – hrs

Remove & install:

323/626	2,20
Premacy	2,30

Special Tools

■ 3 bolts – M8 x 1,25 mm.

Special Precautions

■ Disconnect battery earth lead.
■ DO NOT turn crankshaft or camshaft when timing belt
removed.
■ Remove glow plugs to ease turning engine.
■ Turn engine in normal direction of rotation (unless
otherwise stated).
■ DO NOT turn engine via camshaft or other sprockets.
■ Observe all tightening torques.
■ Check diesel injection pump timing after belt
replacement.

Removal

1. Remove:
   ❏ Engine top cover.
   ❏ Timing belt upper cover **1**.
   ❏ Auxiliary drive belt(s).
2. Support engine.
3. Remove:
   ❏ RH engine mounting.
4. Lower engine slightly.
5. Remove:
   ❏ Crankshaft pulley bolts **2**.
   ❏ Crankshaft pulley **3**.
   ❏ Timing belt lower cover **4**.
   ❏ Crankshaft sprocket guide washer **5**.
   ❏ Crankshaft position (CKP) sensor.

6. Turn crankshaft clockwise until timing marks
   aligned **6**, **7** & **8**.
7. Insert two M8 bolts in injection pump
   sprocket **9**.
8. Insert one M8 bolt in camshaft sprocket **10**.
9. Remove:
   ❏ Automatic tensioner unit bolts **11** & **12**.
   ❏ Automatic tensioner unit **13**.
   ❏ Timing belt.

Installation

1. Check tensioner pulley and guide pulley for
   smooth operation. Replace if necessary.
2. Check automatic tensioner unit as follows:
   ❏ Check tensioner body for leakage or
   damage.
   ❏ Check pushrod does not move when pushed
   against a firm surface.
   ❏ Check pushrod protrusion is 8 mm **14**.
3. Replace if necessary.
4. Ensure timing marks aligned **6**, **7** & **8**.
5. Ensure locking bolts located in sprockets **9**
   & **10**.
6. Fit timing belt in anti-clockwise direction, starting
   at crankshaft sprocket. Ensure belt is taut on
   non-tensioned side.
7. Slowly compress pushrod into tensioner body
   until holes aligned. Use press **15**.
8. Retain pushrod with suitable pin through hole in
   tensioner body **16**.
9. Install:
   ❏ Automatic tensioner unit **13**.
   ❏ Automatic tensioner unit bolts **11** & **12**.
10. Tighten automatic tensioner unit bolts finger
    tight **11** & **12**.
11. Remove locking bolts **9** & **10**.
12. Tighten automatic tensioner unit bolt **11**.
    Tightening torque: 19-25 Nm.
13. Tighten automatic tensioner unit bolt **12**.
    Tightening torque: 19-25 Nm.
    NOTE: Observe correct sequence.
14. Remove pin from tensioner body to release
    pushrod **16**.
15. Turn crankshaft slowly two turns clockwise.
16. Ensure timing marks aligned **6**, **7** & **8**.
17. Install components in reverse order of removal.
18. Tighten crankshaft pulley bolts **2**.
    Tightening torque: 23-32 Nm.

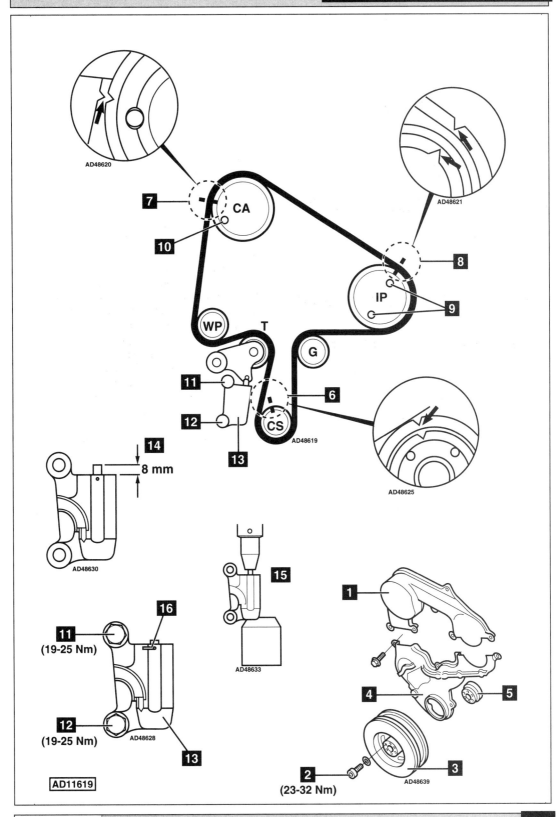

Model:	**MPV 2,5TD • B Series 2,5D/TD**
Year:	**1998-03**
Engine Code:	**WL, WL-T**

Replacement Interval Guide

Mazda recommend replacement every 60,000 miles.

The previous use and service history of the vehicle must always be taken into account.

Refer to Timing Belt Replacement Intervals at the front of this manual.

Check For Engine Damage

CAUTION: This engine has been identified as an INTERFERENCE engine in which the possibility of valve-to-piston damage in the event of a timing belt failure is MOST LIKELY to occur. A compression check of all cylinders should be performed before removing the cylinder head(s).

Repair Times – hrs

Remove & install	1,00

Special Tools

- ■ None required.

Special Precautions

- ■ Disconnect battery earth lead.
- ■ DO NOT turn crankshaft or camshaft when timing belt removed.
- ■ Remove glow plugs to ease turning engine.
- ■ Turn engine in normal direction of rotation (unless otherwise stated).
- ■ DO NOT turn engine via camshaft or other sprockets.
- ■ Observe all tightening torques.
- ■ Check diesel injection pump timing after belt replacement.

Removal

1. Remove timing belt cover **1**.
2. Turn crankshaft clockwise to setting position.
3. Ensure timing marks aligned **2** & **3**.
4. Slacken tensioner bolt **4**.
5. Remove tensioner spring **5**.
6. Move tensioner pulley away from belt. Lightly tighten bolt **4**.
7. Remove timing belt.

Installation

1. Check free length of tensioner spring is 63 mm. Replace spring if necessary **5**.
2. Ensure timing marks aligned **2** & **3**.
   NOTE: Ensure injection pump mounted securely before installing timing belt.
3. Fit timing belt in clockwise direction, starting at injection pump sprocket. Ensure belt is taut between sprockets on non-tensioned side.
4. Slacken tensioner bolt **4**.
5. Push tensioner pulley gently against belt.
6. Fit tensioner spring **5**.
7. Lightly tighten tensioner bolt **4**.
8. Turn crankshaft slowly 2 turns clockwise.
9. Ensure timing marks aligned **2** & **3**.
10. Slacken tensioner bolt **4**.
11. Allow tensioner to operate.
12. Tighten tensioner bolt **4**.
    Tightening torque: 38-51 Nm.
13. Turn crankshaft slowly 2 turns clockwise.
14. Ensure timing marks aligned **2** & **3**.
15. Apply a load of 10 kg to belt at ▽. Belt should deflect 9-10 mm.
16. Install components in reverse order of removal.

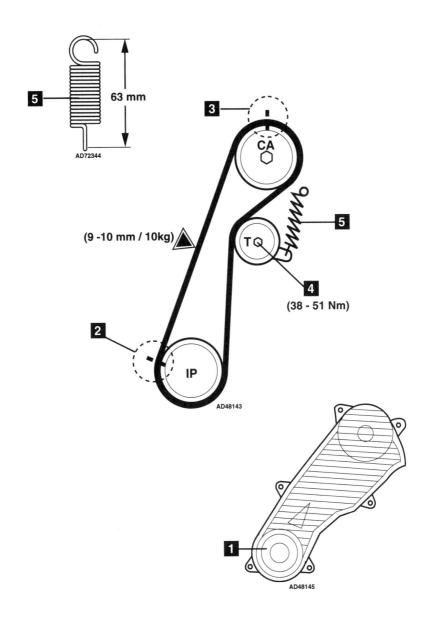

5 63 mm

AD72344

3

CA

(9 -10 mm / 10kg)

TC

5

4

(38 - 51 Nm)

2

IP

AD48143

1

AD48145

AD11591

MAZDA

Model:	**E2200D • B2200D**
Year:	**1985-03**
Engine Code:	**R2**

Replacement Interval Guide

Mazda recommend replacement every 60,000 miles.

The previous use and service history of the vehicle must always be taken into account.

Refer to Timing Belt Replacement Intervals at the front of this manual.

Check For Engine Damage

CAUTION: This engine has been identified as an INTERFERENCE engine in which the possibility of valve-to-piston damage in the event of a timing belt failure is MOST LIKELY to occur.

A compression check of all cylinders should be performed before removing the cylinder head.

Repair Times – hrs

Remove & install:

B2200	2,00
AC	+0,10
PAS	+0,10
E2200 →10/98	2,30
AC	+0,10
PAS	+0,10
E2200 11/98→	2,50
AC	+0,10

Special Tools

■ 2 bolts – M8 x 1,25.

Special Precautions

- ■ Disconnect battery earth lead.
- ■ DO NOT turn crankshaft or camshaft when timing belt removed.
- ■ Remove glow plugs to ease turning engine.
- ■ Turn engine in normal direction of rotation (unless otherwise stated).
- ■ DO NOT turn engine via camshaft or other sprockets.
- ■ Observe all tightening torques.
- ■ Check diesel injection pump timing after belt replacement.

Removal

1. Drain coolant.
2. Remove:
   - ❏ Front seats.
   - ❏ Handbrake.
   - ❏ Battery cover.
   - ❏ Radiator (if required).
   - ❏ Auxiliary drive belts.
   - ❏ Cooling fan.
   - ❏ Cooling fan pulley.
   - ❏ PAS tensioner pulley (if fitted).
   - ❏ Crankshaft pulley **1**.
   - ❏ Timing belt covers **2** & **3**.
3. Turn crankshaft to TDC on No.1 cylinder. Ensure timing marks aligned **4**, **5** & **6**.
4. 1997→: Lock injection pump sprocket with two M8 x 1,25 mm bolts **9**. Lightly tighten bolts.
   NOTE: DO NOT fully tighten injection pump sprocket locking bolts as damage will occur to pump and pulley.
5. Slacken tensioner bolt **7**. Move tensioner away from belt. Lightly tighten bolt.
6. Remove timing belt.
   NOTE: Mark direction of rotation on belt with chalk if belt to be reused.

Installation

1. 1997→: Check the free length of tensioner spring is 52,6 mm. Replace if necessary **10**.
2. Ensure timing marks aligned **4**, **5** & **6**.
3. 1997→: Ensure injection pump locked in position with bolts **9**.
4. Fit timing belt.
5. Slacken tensioner bolt **7**.
6. Turn crankshaft two turns clockwise. Ensure timing marks aligned **4**, **5** & **6**.
7. Tighten tensioner bolt **7**. Tightening torque:
   - ❏ A →1996: 31-46 Nm.
   - ❏ B 1997→: 38-51 Nm.
8. Apply a load of 10 kg to belt at ▽. Belt should deflect 10-13 mm.
9. If not: Repeat tensioning procedure.
10. Check and adjust injection pump timing.
11. Install components in reverse order of removal.
    NOTE: Ensure crankshaft pulley located correctly on dowel.
12. Tighten crankshaft pulley bolts **8**. Tightening torque: 23-32 Nm.
13. Refill cooling system.
14. Reset timing belt replacement warning lamp as follows (if fitted): Remove instrument cluster. Reverse the position of the two screws with wires attached.

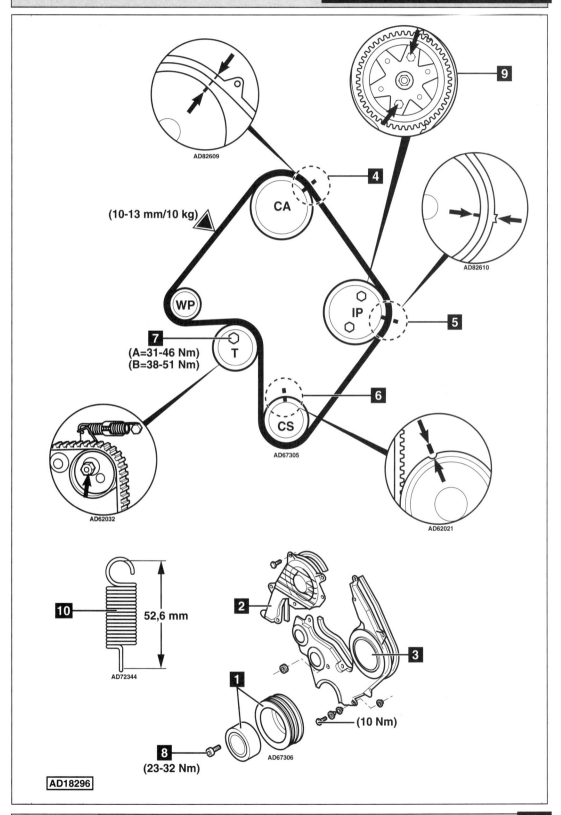

AD82609

(10-13 mm/10 kg) ⚠

CA

9

AD82610

4

WP

IP

5

7
(A=31-46 Nm)
(B=38-51 Nm)
T

6

CS

AD67305

AD62032

AD62021

10

52,6 mm

AD72344

2

3

(10 Nm)

1

8

(23-32 Nm)

AD67306

AD18296

MITSUBISHI

Model:	Colt 1200/1300/1500 • Colt/Lancer 1,3 • Lancer 1200/1300/1500
Year:	1988-96
Engine Code:	4G13, 4G15, 4G16

Replacement Interval Guide

Mitsubishi recommend replacement as follows:
To 12/90 - replacement every 60,000 miles.
1/91 on - replacement every 54,000 miles.
The previous use and service history of the vehicle must always be taken into account.
Refer to Timing Belt Replacement Intervals at the front of this manual.

Check For Engine Damage

CAUTION: This engine has been identified as an INTERFERENCE engine in which the possibility of valve-to-piston damage in the event of a timing belt failure is MOST LIKELY to occur.
A compression check of all cylinders should be performed before removing the cylinder head.

Repair Times – hrs

Remove & install	1,80
AC	+0,30
PAS	+0,20

Special Tools

■ None required.

Special Precautions

■ Disconnect battery earth lead.
■ DO NOT turn crankshaft or camshaft when timing belt removed.
■ Remove spark plugs to ease turning engine.
■ Turn engine in normal direction of rotation (unless otherwise stated).
■ DO NOT turn engine via camshaft or other sprockets.
■ Observe all tightening torques.

Removal

1. Support engine.
2. Remove:
   ❏ Engine mounting bracket.
   ❏ Auxiliary drive belt(s).
   ❏ AC drive belt tensioner pulley and bracket **1** (if fitted).
   ❏ Water pump pulley **2**.
   ❏ Crankshaft pulley bolt **12**.
   ❏ Crankshaft pulley bolts **11**.
   ❏ Crankshaft pulley **3**.
   ❏ Timing belt upper cover **4**.
   ❏ Timing belt lower cover **5**.
3. Turn crankshaft clockwise until timing marks aligned **6** & **7**.
4. Slacken tensioner bolts **8** & **9**. Release tension on belt. Lightly tighten bolt **9**.
5. Remove timing belt.

Installation

1. Ensure timing marks aligned **6** & **7**.
2. Fit timing belt in anti-clockwise direction, starting at crankshaft sprocket. Ensure belt is taut between sprockets.
3. Slacken tensioner bolt **8**.
4. Slacken tensioner bolt **9**.
5. Allow tensioner to operate.
6. Tighten tensioner bolt **9**.
7. Tighten tensioner bolt **8**.
   NOTE: If tensioner bolt 8 is tightened first, tensioner could turn causing belt to overtension.
8. Turn crankshaft two turns clockwise. Ensure timing marks aligned **6** & **7**.
9. Slacken tensioner bolt **8**.
10. Slacken tensioner bolt **9**.
11. Allow tensioner to operate.
12. Turn crankshaft clockwise for 3 teeth on camshaft sprocket.
13. Tighten tensioner bolt **9**.
    Tightening torque: 24 Nm.
14. Tighten tensioner bolt **8**.
    Tightening torque: 24 Nm.
15. Apply thumb pressure to belt at ▽. Belt should deflect to 1/4 of tensioner bolt head width **10**.
16. Install components in reverse order of removal.
17. Tighten crankshaft pulley bolts **11**.
    Tightening torque: 12-15 Nm.
18. Tighten crankshaft pulley bolt **12**.
    Tightening torque: 85 Nm.

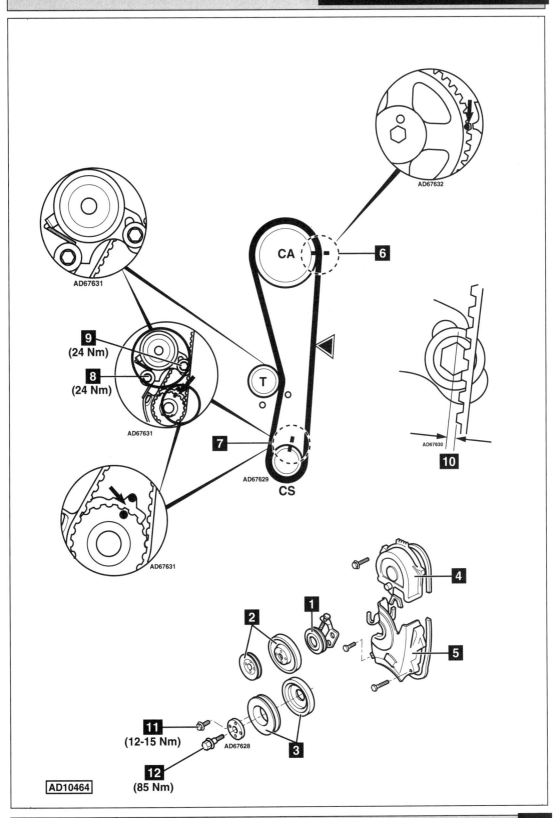

AD67632

AD67631

6

CA —

9
(24 Nm)

8
(24 Nm)

AD67631

T

7

AD67629

CS

AD67631

AD67630

10

4

2

1

5

11
(12-15 Nm)

AD67628

3

12
(85 Nm)

AD10464

Model:	Colt/Lancer 1,3 12V
Year:	1996-03
Engine Code:	4G13

Replacement Interval Guide

Mitsubishi recommend replacement every 54,000 miles.

The previous use and service history of the vehicle must always be taken into account.
Refer to Timing Belt Replacement Intervals at the front of this manual.

Check For Engine Damage

CAUTION: This engine has been identified as an INTERFERENCE engine in which the possibility of valve-to-piston damage in the event of a timing belt failure is MOST LIKELY to occur.
A compression check of all cylinders should be performed before removing the cylinder head(s).

Repair Times – hrs

Remove & install	1,10
AC	+0,20
PAS	+0,30

Special Tools

- ■ None required.

Special Precautions

- ■ Disconnect battery earth lead.
- ■ DO NOT turn crankshaft or camshaft when timing belt removed.
- ■ Remove spark plugs to ease turning engine.
- ■ Turn engine in normal direction of rotation (unless otherwise stated).
- ■ DO NOT turn engine via camshaft or other sprockets.
- ■ Observe all tightening torques.

Removal

1. Raise and support front of vehicle.
2. Remove:
   - ❏ Engine undershield.
   - ❏ Crankshaft pulley bolt **1**.
   - ❏ Crankshaft pulley **2**.
3. Remove – if fitted:
   - ❏ AC relay box. DO NOT disconnect cables.
   - ❏ AC condenser bracket bolts.
4. Support engine.
5. Remove:
   - ❏ RH engine mounting.
   - ❏ Water pump pulley.
   - ❏ PAS pump and bracket. DO NOT disconnect hoses.
   - ❏ Dipstick tube.
   - ❏ Timing belt upper cover **3**.
   - ❏ Timing belt lower cover **4**.

6. Turn crankshaft to TDC on No.1 cylinder.
7. Ensure timing marks aligned **5** & **6**.
8. If camshaft timing marks not aligned **6**: Turn crankshaft one turn clockwise.
9. Slacken tensioner bolt **7**. Move tensioner away from belt. Lightly tighten bolt.
10. Remove timing belt.
    NOTE: Mark direction of rotation on belt with chalk if belt is to be reused.

Installation

1. Ensure timing marks aligned **5** & **6**.
2. Fit timing belt in anti-clockwise direction, starting at crankshaft sprocket.
   NOTE: Ensure belt is taut between sprockets on non-tensioned side.
3. Slacken tensioner bolt 1/2 turn **7**.
4. Turn crankshaft two turns clockwise to TDC on No.1 cylinder.
5. Ensure timing marks aligned **5** & **6**.
6. Tighten tensioner bolt.
   Tightening torque: 24 Nm **7**.
7. Install:
   - ❏ Timing belt upper cover **3**.
   - ❏ Timing belt lower cover **4**.
   - ❏ Dipstick tube.
   - ❏ PAS pump and bracket.
   - ❏ Water pump pulley.
8. Install RH engine mounting.
9. Tighten engine mounting nuts **8**.
   Tightening torque: 57 Nm.
10. Tighten engine mounting bolt **9**.
    Tightening torque: 57 Nm.
11. Remove engine support.
12. Tighten engine mounting nut **10**.
    Tightening torque: 98 Nm.
13. Install components in reverse order of removal.
14. Oil crankshaft pulley bolt threads and washer face **1**.
15. Fit crankshaft pulley bolt **1**.
    Tightening torque: 125 Nm.

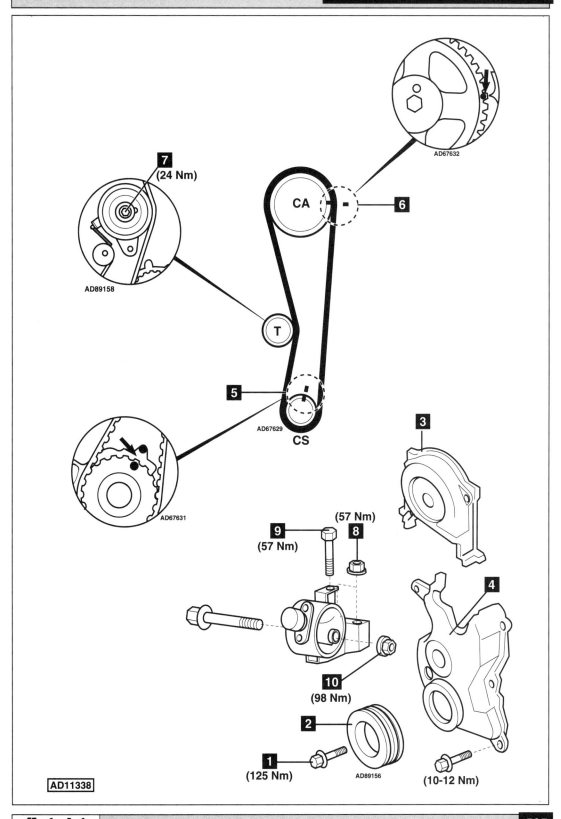

7 (24 Nm)

AD89158

AD67632

CA

6

7

T

5

AD67629

CS

AD67631

3

(57 Nm)

9 **8**

(57 Nm)

4

10

(98 Nm)

2

1

(125 Nm)

AD89156

(10-12 Nm)

AD11338

MITSUBISHI

Model:	Colt/Lancer 1,6 16V • Galant 1,8i 16V • Space Runner 1,8i 16V Space Wagon 1,8i 16V
Year:	1991-99
Engine Code:	4G92, 4G93 (SOHC)

Replacement Interval Guide

Mitsubishi recommend replacement every 54,000 miles.
The previous use and service history of the vehicle must always be taken into account.
Refer to Timing Belt Replacement Intervals at the front of this manual.

Check For Engine Damage

CAUTION: This engine has been identified as an INTERFERENCE engine in which the possibility of valve-to-piston damage in the event of a timing belt failure is MOST LIKELY to occur.
A compression check of all cylinders should be performed before removing the cylinder head.

Repair Times – hrs

Remove & install:

Space Wagon/Runner	1,80
Galant	2,00
Colt/Lancer	2,00
AC	+0,30
PAS	+0,20

Special Tools

- None required.

Special Precautions

- Disconnect battery earth lead.
- DO NOT turn crankshaft or camshaft when timing belt removed.
- Remove spark plugs to ease turning engine.
- Turn engine in normal direction of rotation (unless otherwise stated).
- DO NOT turn engine via camshaft or other sprockets.
- Observe all tightening torques.

Removal

1. Support engine.
2. Remove:
   - ❑ Engine undershield.
   - ❑ Engine mounting.
   - ❑ Dipstick and tube.
   - ❑ Auxiliary drive belt(s).
   - ❑ Crankshaft pulley bolt **8**.
   - ❑ Crankshaft pulley **1**.
   - ❑ Timing belt upper cover **2**.
   - ❑ Timing belt lower cover **3**.
   - ❑ Crankshaft sprocket guide washer **4**.
3. Turn crankshaft clockwise until timing marks aligned **5** & **6**.
4. Slacken tensioner bolt **7**. Move tensioner away from belt. Tighten bolt.
5. Remove timing belt.

Installation

1. Ensure timing marks aligned **5** & **6**.
2. Fit timing belt in anti-clockwise direction, starting at crankshaft sprocket.
3. Ensure belt is taut on non-tensioned side.
4. Slacken tensioner bolt **7**. Allow tensioner to operate.
5. Turn crankshaft two turns clockwise. Ensure timing marks aligned **5** & **6**.
   NOTE: DO NOT turn crankshaft anti-clockwise.
6. Tighten tensioner bolt to 24 Nm **7**.
7. Install components in reverse order of removal.
8. Tighten crankshaft pulley bolt to 185 Nm **8**.

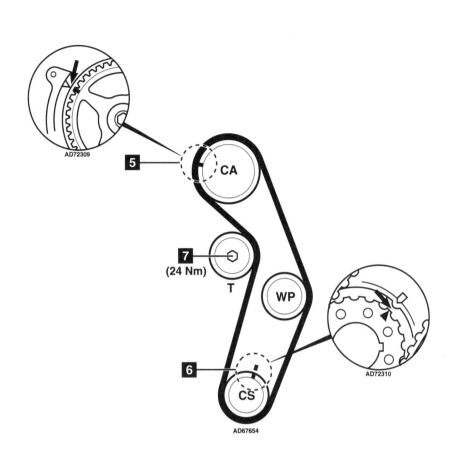

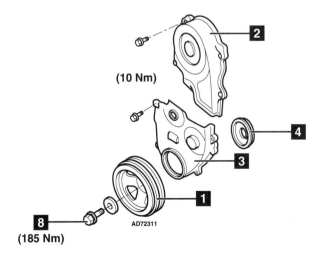

5 (CA)

AD72309

7
(24 Nm)
T

WP

6

CS

AD72310

AD67654

2

(10 Nm)

4

3

1

AD72311

8
(185 Nm)

AD10468

MITSUBISHI

Model:	**Colt/Lancer 1,6 16V • Carisma 1,6 16V • Carisma 1,8 16V**
Year:	**1996-03**
Engine Code:	**4G92, 4G93 (SOHC)**

Replacement Interval Guide

Mitsubishi recommend replacement every 54,000 miles.
The previous use and service history of the vehicle must always be taken into account.
Refer to Timing Belt Replacement Intervals at the front of this manual.

Check For Engine Damage

CAUTION: This engine has been identified as an INTERFERENCE engine in which the possibility of valve-to-piston damage in the event of a timing belt failure is MOST LIKELY to occur.
A compression check of all cylinders should be performed before removing the cylinder head.

Repair Times – hrs

Colt/Lancer:	
Remove & install	1,50
AC	+0,20
PAS	+0,30
Carisma:	
Remove & install	2,10

Special Tools

- None required.

Special Precautions

- Disconnect battery earth lead.
- DO NOT turn crankshaft or camshaft when timing belt removed.
- Remove spark plugs to ease turning engine.
- Turn engine in normal direction of rotation (unless otherwise stated).
- DO NOT turn engine via camshaft or other sprockets.
- Observe all tightening torques.

Removal

1. Raise and support front of vehicle.
2. Support engine.
3. Remove:
   - ❏ Engine undershield.
   - ❏ RH engine mounting.
   - ❏ PAS hose bracket and clip (Carisma).
   - ❏ Alternator bracket.
   - ❏ Auxiliary drive belts.
   - ❏ Engine mounting bracket **1**.
   - ❏ Crankshaft pulley bolt **2**.
   - ❏ Crankshaft pulley **3**.
   - ❏ Timing belt upper cover **4**.
   - ❏ Timing belt lower cover **5**.
   - ❏ Crankshaft sprocket guide washer **6**.

4. Turn crankshaft clockwise to TDC on No.1 cylinder. Ensure timing marks aligned **7** & **8**.
5. Slacken tensioner bolt **9**.
6. Lever tensioner bracket to right. Use screwdriver **10**.
7. Lightly tighten tensioner bolt **9**.
8. Remove timing belt.
   NOTE: Mark direction of rotation on belt with chalk if belt is to be reused.

Installation

1. Ensure timing marks aligned **7** & **8**.
2. Fit timing belt in anti-clockwise direction, starting at crankshaft sprocket. Ensure belt is taut between sprockets.
3. Slacken tensioner bolt 1/4 to 1/2 turn **9**. Allow tensioner to operate.
4. Turn crankshaft two turns clockwise. Ensure timing marks aligned **7** & **8**.
   NOTE: DO NOT allow crankshaft to turn anti-clockwise.
5. Tighten tensioner bolt to 24 Nm **9**.
6. Install components in reverse order of removal.
7. Oil crankshaft pulley bolt threads and washer face **2**.
8. Fit crankshaft pulley bolt.
   Tightening torque: 177-186 Nm **2**.

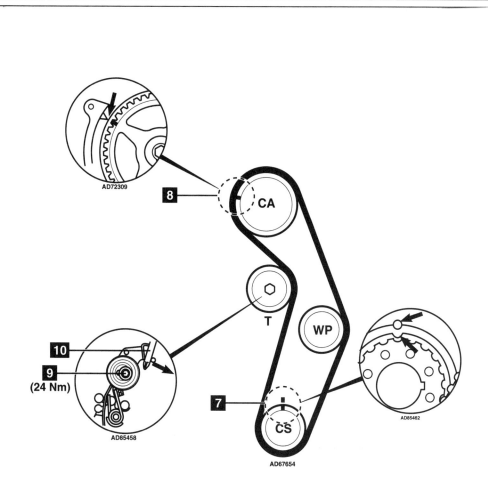

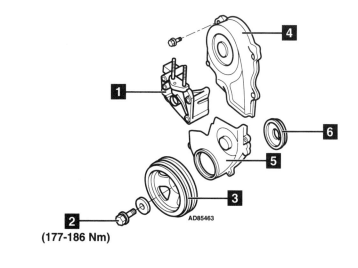

AD10948

Model:	Colt/Lancer 1,8 16V DOHC • Carisma 1,8 16V DOHC • Carisma 1,8 GDI Space Star 1,8 GDI • Shogun Pinin 1,8 GDI • Shogun Pinin 2,0 GDI
Year:	1992-03
Engine Code:	4G93

Replacement Interval Guide

Mitsubishi recommend replacement every 54,000 miles.
The previous use and service history of the vehicle must always be taken into account.
Refer to Timing Belt Replacement Intervals at the front of this manual.

Check For Engine Damage

CAUTION: This engine has been identified as an INTERFERENCE engine in which the possibility of valve-to-piston damage in the event of a timing belt failure is MOST LIKELY to occur.
A compression check of all cylinders should be performed before removing the cylinder head.

Repair Times – hrs

Colt/Lancer	1,80
AC	+0,30
PAS	+0,20
Carisma	2,30
Space Star	2,80
Shogun Pinin	1,40
AC	+0,20
Intercooler	+0,20

Special Tools

- ■ Tensioner pulley tool – Mitsubishi No.MD998767.
- ■ Crankshaft pulley holding tool – Mitsubishi No.MD990767.
- ■ Crankshaft pulley holding tool pins – Mitsubishi No.MD998719/MD998754.

Special Precautions

- ■ Disconnect battery earth lead.
- ■ DO NOT turn crankshaft or camshaft when timing belt removed.
- ■ Remove spark plugs to ease turning engine.
- ■ Turn engine in normal direction of rotation (unless otherwise stated).
- ■ DO NOT turn engine via camshaft or other sprockets.
- ■ Observe all tightening torques.

Removal

1. Raise and support front of vehicle.
2. Remove:
   - ❑ Engine undershield.
   - ❑ Auxiliary drive belts.
   - ❑ Engine upper cover (if fitted).

3. Shogun Pinin: Support engine – remove:
   - ❑ Top radiator mountings.
   - ❑ Air intake box and trunking.
   - ❑ Engine mounting.
   - ❑ Alternator bracket.
   - ❑ PAS pump bracket.
4. Carisma/Space Star: Support engine – remove:
   - ❑ RH engine mounting.
   - ❑ PAS hose bracket and clip.
   - ❑ Alternator bracket.
   - ❑ Engine mounting bracket **1**.
5. All models – remove:
   - ❑ Crankshaft pulley bolt **2**. Use tool Nos.MD990767/MD998719/MD998754.
   - ❑ Crankshaft pulley **3**.
   - ❑ Timing belt upper cover **4**.
   - ❑ Timing belt lower cover **5**.
   - ❑ Crankshaft sprocket guide washer **6**.
6. Turn crankshaft clockwise until timing marks aligned:
   - ❑ Carisma/Space Star/Shogun Pinin **7** A & **8** A.
   - ❑ Colt/Lancer **7** B & **8** A.
   - ❑ Shogun Pinin (2,0) **7** B & **8** B.
7. Slacken tensioner pulley bolt **9**. Move tensioner pulley away from belt. Lightly tighten bolt.
8. Remove timing belt.
   NOTE: Mark direction of rotation on belt with chalk if belt is to be reused.

Installation

1. Remove:
   - ❑ Automatic tensioner unit bolts **10**.
   - ❑ Automatic tensioner unit **11**.
2. Check tensioner body for leakage or damage **11**.
3. Push pushrod of automatic tensioner unit against a firm surface using a force of 10-20 kg. Pushrod should move less than 1,0 mm.
4. Slowly compress pushrod into tensioner body **11** until holes aligned. Use vice. Retain pushrod with 2 mm Allen key through hole in tensioner body **12**.
5. Install automatic tensioner unit to cylinder block. Tighten bolts to 12-15 Nm **10**.
6. Ensure timing marks aligned:
   - ❑ Carisma/Space Star/Shogun Pinin **7** A & **8** A.
   - ❑ Colt/Lancer **7** B & **8** A.
   - ❑ Shogun Pinin (2,0) **7** B & **8** B.

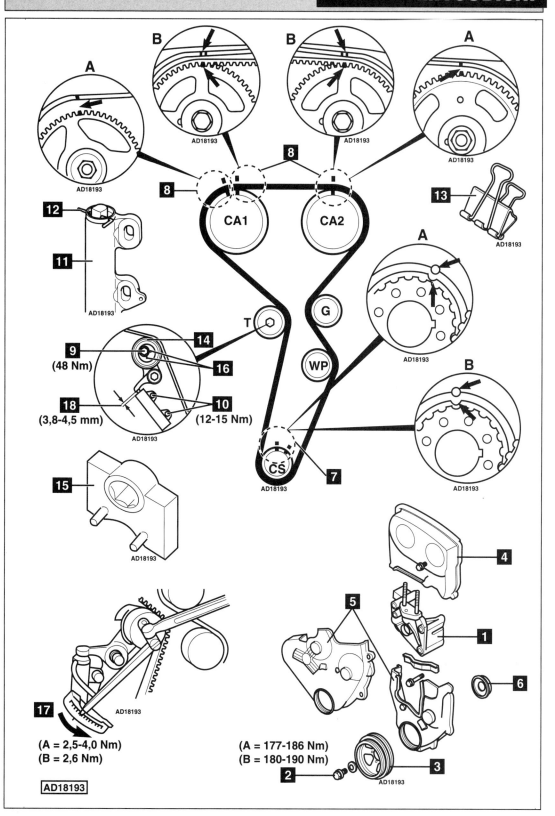

9 (48 Nm)

18 (3,8-4,5 mm)

10 (12-15 Nm)

17
(A = 2,5-4,0 Nm)
(B = 2,6 Nm)

(A = 177-186 Nm)
(B = 180-190 Nm)

AD18193

7. Slacken tensioner pulley bolt **9**.
8. Turn crankshaft sprocket 1/2 tooth width anti-clockwise.
9. Fit timing belt to exhaust camshaft sprocket (CA2). Retain in position with suitable paper clip **13**.
10. Fit timing belt to inlet camshaft sprocket (CA1). Use spanners on each sprocket bolt to align timing marks **8**. Retain in position with suitable paper clip **13**.
11. Fit timing belt in following order:
    ❏ Guide pulley.
    ❏ Water pump sprocket.
    ❏ Crankshaft sprocket.
    ❏ Tensioner pulley.
12. Adjust position of crankshaft sprocket when fitting belt to ensure timing marks aligned **7**.
13. Remove paper clips **13**.
14. Turn tensioner pulley **14** firmly anti-clockwise to temporarily tension belt. Tighten bolt to 48 Nm **9**.
15. Turn crankshaft 1/4 turn anti-clockwise.
16. Turn crankshaft clockwise until timing marks aligned:
    ❏ Carisma/Space Star/Shogun Pinin **7** A & **8** A.
    ❏ Colt/Lancer **7** B & **8** A.
    ❏ Shogun Pinin (2,0) **7** B & **8** B.
17. Slacken tensioner pulley bolt **9**. Fit tool **15** to holes in tensioner pulley **16**. Tool No.MD998767.
18. Apply anti-clockwise torque to tensioner pulley **17**:
    ❏ Carisma/Space Star/Shogun Pinin: A – 2,5-4,0 Nm.
    ❏ Colt/Lancer: B – 2,6 Nm.
19. Tighten tensioner bolt to 48 Nm **9**.
    NOTE: Ensure tensioner does not move when tightening bolt.
20. Remove Allen key from tensioner body to release pushrod.
    NOTE: Ensure Allen key can be removed easily.
21. Ensure timing marks aligned:
    ❏ Carisma/Space Star/Shogun Pinin **7** A & **8** A.
    ❏ Colt/Lancer **7** B & **8** A.
    ❏ Shogun Pinin (2,0) **7** B & **8** B.
22. Turn crankshaft slowly two turns clockwise. Ensure timing marks aligned:
    ❏ Carisma/Space Star/Shogun Pinin **7** A & **8** A.
    ❏ Colt/Lancer **7** B & **8** A.
    ❏ Shogun Pinin (2,0) **7** B & **8** B.
23. Wait 5 minutes minimum.
24. Check extended length of pushrod is 3,8-4,5 mm **18**.
25. If not: Repeat tensioning procedure.
26. Install components in reverse order of removal.

27. Carisma: Oil crankshaft pulley bolt threads and washer face.
28. Tighten crankshaft pulley bolt:
    ❏ Carisma/Space Star/Shogun Pinin **2** A. Tightening torque: 177-186 Nm.
    ❏ Colt/Lancer **2** B. Tightening torque: 180-190 Nm.

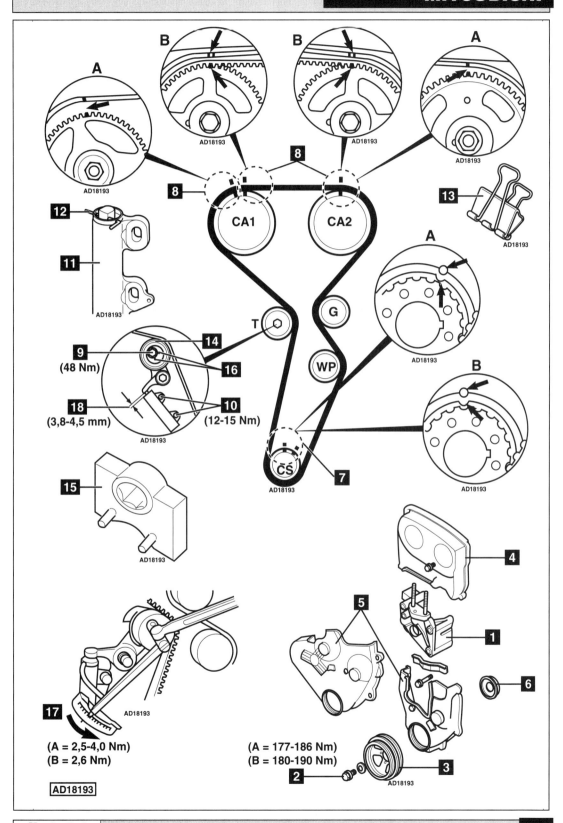

A

B

B

A

AD18193

AD18193

AD18193

AD18193

8

8

CA1

CA2

13

AD18193

12

11

A

AD18193

AD18193

G

T

14

9
(48 Nm)

16

WP

AD18193

18

10

(3,8-4,5 mm)

(12-15 Nm)

AD18193

B

15

CS

AD18193

7

AD18193

4

AD18193

5

1

17

AD18193

6

(A = 2,5-4,0 Nm)
(B = 2,6 Nm)

(A = 177-186 Nm)
(B = 180-190 Nm)

2

3

AD18193

AD18193

MITSUBISHI

Model:	**Galant 2,0 16V • Space Wagon 2,0 16V**
Year:	**1992-99**
Engine Code:	**4G63 (with balancer shaft)**

Replacement Interval Guide

Mitsubishi recommend replacement every 54,000 miles.
The previous use and service history of the vehicle must always be taken into account.
Refer to Timing Belt Replacement Intervals at the front of this manual.

Check For Engine Damage

CAUTION: This engine has been identified as an INTERFERENCE engine in which the possibility of valve-to-piston damage in the event of a timing belt failure is MOST LIKELY to occur.
A compression check of all cylinders should be performed before removing the cylinder head.

Repair Times – hrs

Remove & install:	
Galant	2,00
Space Wagon	1,80
Balancer shaft belt	+0,10
PAS	+0,20

Special Tools

■ Tensioner pulley tool – Mitsubishi No.MD998767.

Special Precautions

■ Disconnect battery earth lead.
■ DO NOT turn crankshaft or camshaft when timing belt removed.
■ Remove spark plugs to ease turning engine.
■ Turn engine in normal direction of rotation (unless otherwise stated).
■ DO NOT turn engine via camshaft or other sprockets.
■ Observe all tightening torques.

Removal

Timing Belt

1. Remove:
   ❏ Coolant expansion tank (Space Wagon).
   ❏ Engine mounting bracket (Space Wagon).
   ❏ Auxiliary drive belt(s).
   ❏ Auxiliary drive belt tensioner and bracket.
   ❏ Water pump pulley **1**.
   ❏ Crankshaft pulley **2**.
   ❏ Timing belt upper cover **3**.
   ❏ Timing belt lower cover **4**.
2. Turn crankshaft clockwise until camshaft sprocket, crankshaft sprocket and oil pump sprocket timing marks aligned **5**, **6** & **7**.

3. Remove:
   ❏ Tensioner pulley **11**.
   ❏ Timing belt.
   ❏ Automatic tensioner unit **9**.

Installation

Timing Belt

1. Ensure timing marks aligned **5**, **6** & **7**.
   *NOTE: To check oil pump sprocket is correctly positioned: Remove blanking plug from cylinder block **10**. Insert 8 mm diameter Phillips screwdriver in hole. Ensure screwdriver is inserted to a depth of 60 mm. If screwdriver can only be inserted 20 mm: Turn oil pump sprocket 360° and reinsert screwdriver.*
2. If automatic tensioner pushrod is fully extended: Slowly compress pushrod into tensioner body until holes aligned. Use vice. Retain pushrod with 1,4 mm diameter pin through hole in tensioner body **8**.
   NOTE: If end plug of tensioner body protrudes, protect with a spacer when compressing pushrod to avoid damage.
3. Install automatic tensioner unit **9**. Tighten bolts to 24 Nm.
4. Fit tensioner pulley **11**.
5. Fit timing belt in anti-clockwise direction, starting at crankshaft sprocket. Ensure belt is taut between sprockets.
6. Turn tensioner pulley anti-clockwise to pre-tension belt. Tighten tensioner pulley bolt **12**.
7. Remove screwdriver from hole in cylinder block **10**.
8. Turn crankshaft slowly 1/4 turn anti-clockwise.
9. Turn crankshaft clockwise to TDC.
10. Slacken tensioner pulley bolt **12**. Fit tool to tensioner pulley **19**. Tool No.MD998767.
11. Apply anti-clockwise torque (in direction of arrow) of 3,6 Nm to tensioner pulley. Tighten tensioner pulley bolt to 48 Nm **12**.
12. Lever tensioner arm **20** against pushrod. Remove pin from tensioner body **8**.
13. Turn crankshaft two turns clockwise. Wait 15 minutes.
14. Check pushrod protrusion is 3,8-4,5 mm **21**.
15. If not: Repeat operations 6-13.
16. Install components in reverse order of removal.
17. Tighten crankshaft pulley bolts **13**.
    Tightening torque: 20-30 Nm.

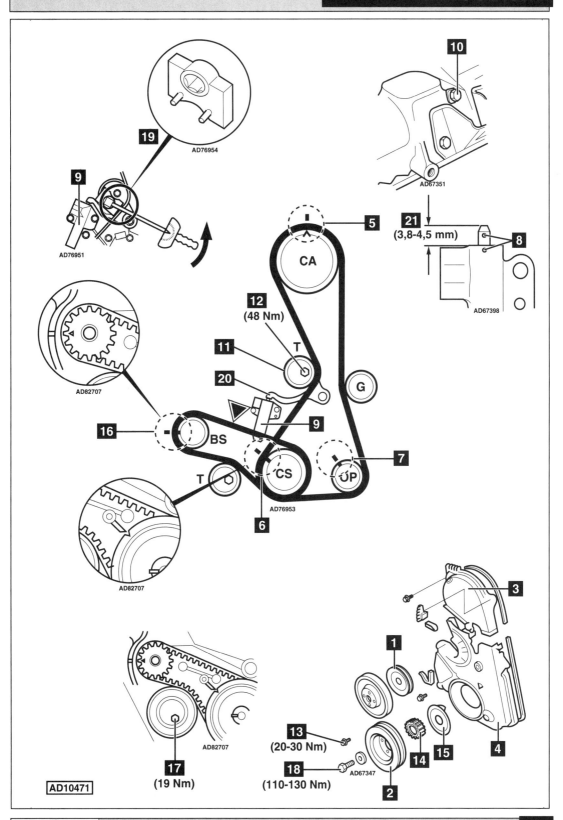

AD76954

AD76951

AD82707

AD67351

21 (3,8-4,5 mm)

AD67398

5

CA

12 (48 Nm)

11

T

20

9

G

16

BS

T

CS

UP

7

AD76953

6

AD82707

3

1

AD82707

17 (19 Nm)

13 (20-30 Nm)

18 (110-130 Nm)

AD67347

14

15

4

2

AD10471

Removal

Balancer Shaft Belt

1. Remove timing belt as described previously.
2. Remove:
   - ❏ Crankshaft sprocket bolt **18**.
   - ❏ Crankshaft sprocket **14**.
   - ❏ Crankshaft sprocket guide washer **15**.
     Ensure concave side of guide washer faces away from belt.
3. Ensure balancer shaft sprocket timing marks aligned **16**.
4. Remove:
   - ❏ Balancer shaft belt tensioner **17**.
   - ❏ Balancer shaft belt.

Installation

Balancer Shaft Belt

1. Ensure balancer shaft sprocket and crankshaft sprocket timing marks aligned **6** & **16**.
2. Install:
   - ❏ Balancer shaft belt.
   - ❏ Balancer shaft belt tensioner **17**.
3. Turn tensioner pulley firmly clockwise against belt. Tighten tensioner bolt to 19 Nm.
4. Apply thumb pressure to belt at ▽. Belt should deflect 5-7 mm.
5. If not: Repeat tensioning procedure.
6. Install:
   - ❏ Crankshaft sprocket guide washer **15**.
   - ❏ Crankshaft sprocket **14**.

   NOTE: Ensure concave side of guide washer faces away from belt.
7. Tighten crankshaft sprocket bolt **18**.
   Tightening torque: 110-130 Nm
8. Fit timing belt as described previously.

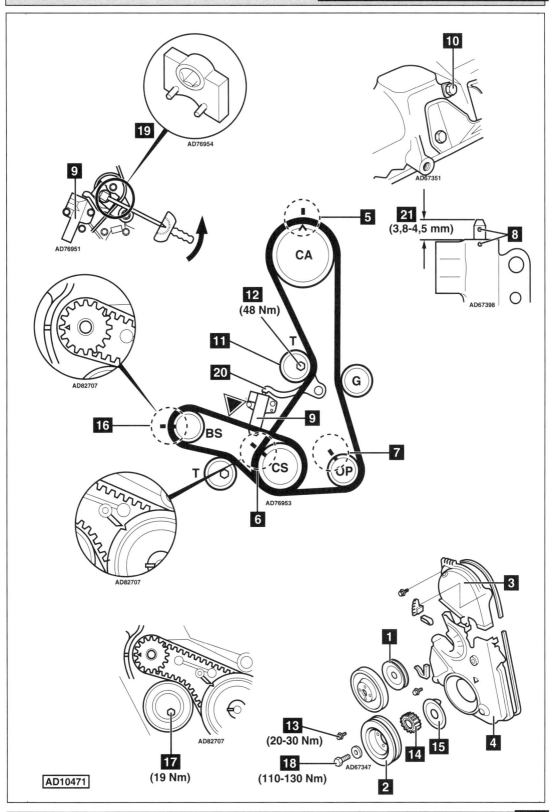

MITSUBISHI

Model:	Galant 2000/2000i/Turbo • Galant/Sapporo 2,4 • Shogun/Pajero 2,4
	Sapporo 2000/Turbo • Starion Turbo • Space Wagon 2,0 • L200/L300
Year:	1984-94
Engine Code:	4G63/T, 4G64

Replacement Interval Guide

Mitsubishi recommend replacement as follows:
To 12/90 - replacement every 60,000 miles.
1/91 on - replacement every 54,000 miles.
The previous use and service history of the vehicle must always be taken into account.
Refer to Timing Belt Replacement Intervals at the front of this manual.

Check For Engine Damage

CAUTION: This engine has been identified as an INTERFERENCE engine in which the possibility of valve-to-piston damage in the event of a timing belt failure is MOST LIKELY to occur.
A compression check of all cylinders should be performed before removing the cylinder head.

Repair Times – hrs

Remove & install:

Galant, Sapporo, Starion	1,90
Shogun/Pajero	1,60
L200/L300/Space Wagon	1,80
Balancer shaft belt	+0,10
AC	+0,30
PAS (Galant/Sapporo/Shogun/Pajero)	+0,20
PAS (Starion)	+0,30

Special Tools

■ None required.

Removal

Timing Belt

1. Remove fan and fan cowling (Shogun/Pajero).
2. Support engine.
3. Remove:
   ❏ Engine mounting bracket.
   ❏ Auxiliary drive belt(s).
   ❏ Auxiliary drive belt tensioner and bracket.
   ❏ Water pump pulley **1**.
   ❏ Crankshaft pulley **2**.
   ❏ Timing belt upper cover **3**.
   ❏ Timing belt lower cover **4**.
4. Turn crankshaft clockwise until camshaft sprocket, crankshaft sprocket and oil pump sprocket timing marks aligned **5**, **6** & **7**.
5. Slacken tensioner nut and bolt **8** & **9**. Move tensioner towards water pump. Lightly tighten bolt.
6. Remove timing belt.

Installation

Timing Belt

1. Ensure timing marks aligned **5**, **6** & **7**.
   *NOTE: To check oil pump sprocket is correctly positioned: Remove blanking plug from cylinder block **10**. Insert 8 mm diameter Phillips*

screwdriver in hole. Ensure screwdriver is inserted 60 mm from face of cylinder block. If screwdriver can only be inserted 20 mm: Turn oil pump sprocket 360° and reinsert screwdriver.

2. Fit timing belt in anti-clockwise direction, starting at crankshaft sprocket. Ensure belt is taut between sprockets on non-tensioned side.
3. Slacken tensioner bolt **9**.
4. Turn crankshaft clockwise for two camshaft sprocket teeth.
5. Pull timing belt towards tensioner **11** and release to ensure full belt contact of camshaft sprocket.
6. Tighten tensioner bolt to 49 Nm **9**.
7. Tighten tensioner nut to 49 Nm **8**.
   NOTE: Observe correct tightening sequence, otherwise belt tension will not be correct.
8. Apply thumb pressure to belt at ▽ and push belt away from tensioner. Check measurement between back of belt and edge of timing belt rear cover is 12 mm **12**.
9. If not: Repeat tensioning procedure.
10. Install components in reverse order of removal.
11. Tighten crankshaft pulley bolts **13**.
    Tightening torque: 20-30 Nm.

Removal

Balancer Shaft Belt

1. Remove timing belt as described previously.
2. Remove:
   ❏ Crankshaft sprocket bolt **18**.
   ❏ Timing belt crankshaft sprocket **14**.
   ❏ Crankshaft sprocket guide washer **15**.
3. Ensure balancer shaft sprocket timing marks aligned **16**.
4. Remove balancer shaft belt tensioner **17**.
5. Remove balancer shaft belt.

Installation

Balancer Shaft Belt

1. Ensure balancer shaft sprocket timing marks aligned **16**.
2. Fit balancer shaft belt and tensioner.
3. Push tensioner firmly clockwise towards belt. Tighten tensioner bolt to 19 Nm **17**.
4. Apply thumb pressure to balancer shaft belt at ▽. Belt should deflect 5-7 mm.
5. If not: Repeat tensioning procedure.
6. Install:
   ❏ Crankshaft sprocket guide washer **15**.
   ❏ Timing belt crankshaft sprocket **14**.
7. Tighten crankshaft sprocket bolt **18**.
   Tightening torque: 110-130 Nm.
8. Fit timing belt as described previously.

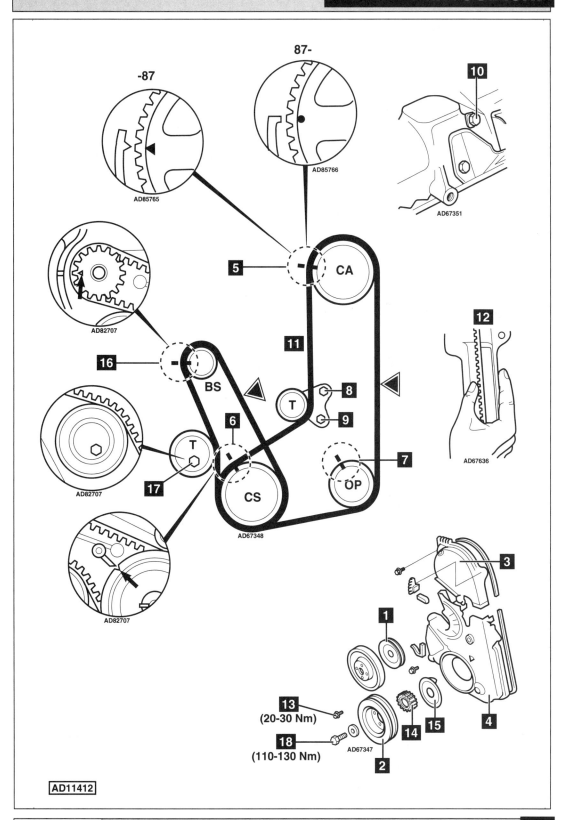

-87

AD85765

87-

AD85766

10

AD67351

5 CA

AD82707

11

16

BS

12

8

T

9

6

7

T

OP

AD67636

17

AD82707

CS

AD67348

AD82707

3

1

13
(20-30 Nm)

14 15

4

18
(110-130 Nm)

AD67347

2

AD11412

MITSUBISHI

Model:	**Galant 2,0 V6 24V**
Year:	**1993-97**
Engine Code:	**6A12**

Check For Engine Damage

CAUTION: This engine has been identified as an INTERFERENCE engine in which the possibility of valve-to-piston damage in the event of a timing belt failure is MOST LIKELY to occur.
A compression check of all cylinders should be performed before removing the cylinder head.

Repair Times – hrs

Remove & install	2,50
AC	+0,30

Special Tools

- Tensioner pulley tool – Mitsubishi No.MD998767.
- Crankshaft pulley holding tool – Mitsubishi No.MB990767.

Special Precautions

- Disconnect battery earth lead.
- DO NOT turn crankshaft or camshaft when timing belt removed.
- Remove spark plugs to ease turning engine.
- Turn engine in normal direction of rotation (unless otherwise stated).
- DO NOT turn engine via camshaft or other sprockets.
- Observe all tightening torques.

Removal

1. Support engine.
2. Remove:
   - ❏ Auxiliary drive belts.
   - ❏ Dipstick and tube.
   - ❏ Alternator.
   - ❏ Auxiliary drive belt tensioner and bracket.
   - ❏ Timing pointer.
   - ❏ AC compressor and bracket. DO NOT disconnect hoses.
   - ❏ PAS pump and bracket. DO NOT disconnect hoses.
   - ❏ Engine mounting and bracket.
   - ❏ Crankshaft pulley **1**. Use tool No.MB990767.
3. Disconnect:
   - ❏ Crankshaft position (CKP) sensor **3**.
   - ❏ Camshaft position (CMP) sensor **2**.

4. Remove timing belt covers **4**, **5** & **6**.
5. Turn crankshaft to TDC on No.1 cylinder. Ensure timing marks aligned **7** & **8**.
6. Slacken tensioner pulley bolt **9**. Turn tensioner pulley clockwise away from belt **10**. Lightly tighten bolt.
7. Remove:
   - ❏ Automatic tensioner unit bolts **11**.
   - ❏ Automatic tensioner unit **12**.
   - ❏ Timing belt.
   NOTE: Mark direction of rotation on belt with chalk if belt is to be reused.

Installation

1. Check tensioner body for leakage or damage **12**.
2. Slowly compress pushrod into tensioner body **12** until holes aligned. Retain pushrod with suitable pin through hole in tensioner body **13**.
3. Install automatic tensioner unit to cylinder block. Tighten bolts to 22 Nm **11**.
4. Ensure timing marks aligned **7** & **8**.
5. Fit timing belt in clockwise direction, starting at camshaft sprocket (CA1). Ensure belt is taut between sprockets.
   NOTE: To assist installation of belt, suitable paper clips **18 can be used to retain belt on sprockets.**
6. Slacken tensioner pulley bolt **9**. Turn tensioner pulley **10** to temporarily tension belt. Tighten bolt to 48 Nm.
7. Turn crankshaft 1/4 turn anti-clockwise.
8. Turn crankshaft clockwise until timing marks aligned **7** & **8**.
9. Slacken tensioner pulley bolt **9**. Fit tool **14** to holes in tensioner pulley **15**. Tool No.MD998767.
10. Apply anti-clockwise torque of 3 Nm to tensioner pulley. Tighten bolt to 48 Nm **9**.
11. Remove pin from tensioner body to release pushrod **13**.
12. Turn crankshaft slowly two turns in direction of rotation until timing marks aligned **7** & **8**. Wait 5 minutes minimum.
13. Check extended length of pushrod is 3,8-5,5 mm **16**.
14. If not: Repeat operations 7-13.
15. Install components in reverse order of removal.
16. Oil threads of crankshaft pulley bolt **17**. Tighten bolt to 180 Nm.

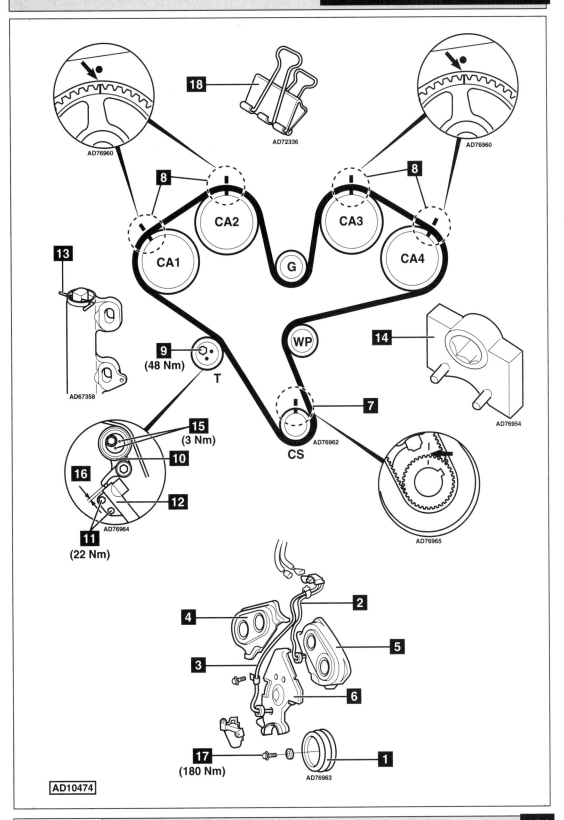

AD76960

18 AD72336

8

CA2 CA3

8

CA1 G CA4

13

AD67358

9
(48 Nm)
T

WP

14

AD76954

15
(3 Nm)

16

10

12

7

AD76962
CS

AD76965

11
(22 Nm)
AD76964

4

2

3

5

6

17
(180 Nm)

1
AD76963

AD10474

Model:	Galant 2,4 GDI • Space Wagon 2,4 GDI • Space Runner 2,4 GDI
Year:	1997-03
Engine Code:	4G64

Replacement Interval Guide

Mitsubishi recommend replacement every 54,000 miles.
The previous use and service history of the vehicle must always be taken into account.
Refer to Timing Belt Replacement Intervals at the front of this manual.

Check For Engine Damage

CAUTION: This engine has been identified as an INTERFERENCE engine in which the possibility of valve-to-piston damage in the event of a timing belt failure is MOST LIKELY to occur.
A compression check of all cylinders should be performed before removing the cylinder head.

Repair Times – hrs

Remove & install:
Galant	2,00
Except Galant	2,40

Special Tools

■ Tensioner pulley tool – Mitsubishi No.MD998767.

Special Precautions

■ Disconnect battery earth lead.
■ DO NOT turn crankshaft or camshaft when timing belt removed.
■ Remove spark plugs to ease turning engine.
■ Turn engine in normal direction of rotation (unless otherwise stated).
■ DO NOT turn engine via camshaft or other sprockets.
■ Observe all tightening torques.

Removal

Timing Belt

1. Remove:
   ❏ Engine undershield (if fitted).
   ❏ Engine top cover.
2. Support engine.
3. Remove:
   ❏ Engine mounting.
   ❏ Engine mounting bracket.
   ❏ Auxiliary drive belt(s).
   ❏ Water pump pulley.
   ❏ Crankshaft pulley bolts **1**.
   ❏ Crankshaft pulley **2**.
   ❏ Timing belt upper cover **3**.
   ❏ Timing belt lower cover **4**.
4. Turn crankshaft clockwise until timing marks aligned **5**, **6**, **7** & **8**.

5. Slacken tensioner pulley bolt **9**.
6. Remove:
   ❏ Automatic tensioner unit bolts **10**.
   ❏ Automatic tensioner unit **11**.
   ❏ Timing belt.
   NOTE: Mark direction of rotation on belt with chalk if belt is to be reused.

Installation

Timing Belt

1. Check tensioner body for leakage or damage **11**. Replace if necessary.
2. Slowly compress pushrod into tensioner body until holes aligned **12**. Use vice.
3. Retain pushrod with suitable pin through hole in tensioner body **12**.
4. Install automatic tensioner unit **11**.
5. Tighten bolts **10**. Tightening torque: 24 Nm.
6. Ensure timing marks aligned **5**, **6**, **7** & **8**.
   *NOTE: To check oil pump sprocket is correctly positioned: Remove blanking plug from cylinder block **13**. Insert 8 mm diameter Phillips screwdriver in hole. Ensure screwdriver is inserted 60 mm from face of cylinder block. If screwdriver can only be inserted 20 mm: Turn oil pump sprocket 360°. Insert screwdriver again.*
7. Ensure timing marks aligned **7**.
8. Fit timing belt in anti-clockwise direction, starting at crankshaft sprocket. Ensure belt is taut between sprockets.
9. Push tensioner pulley firmly against belt. Tighten tensioner pulley bolt **9**.
10. Remove screwdriver from hole in cylinder block.
11. Turn crankshaft 1/4 turn anti-clockwise.
12. Turn crankshaft 1/4 turn clockwise until timing marks aligned **5**, **6**, **7** & **8**.
13. Slacken tensioner pulley bolt **9**.
14. Apply anti-clockwise torque of 3,5 Nm to tensioner pulley. Use tool No.MD998767 **14**. Tighten tensioner pulley bolt **9**. Tightening torque: 48 Nm.
15. Remove pin from tensioner body to release pushrod **12**.
16. Turn crankshaft two turns clockwise. Ensure timing marks aligned **7**. Wait 15 minutes.
17. Ensure timing marks aligned **5**, **6**, **7** & **8**.
18. Check pushrod protrusion is 3,8-4,5 mm **15**. If not: Repeat tensioning procedure.

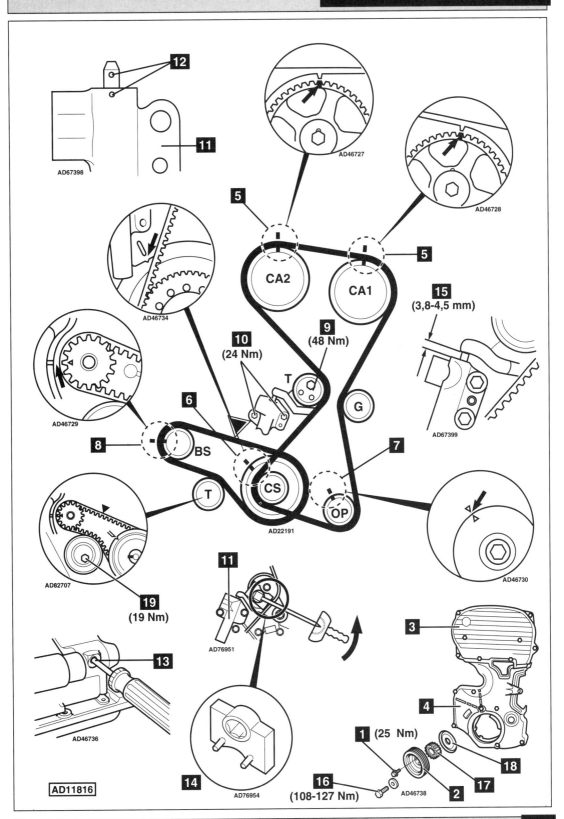

AD67398

12

11

AD46727

AD46728

5

5

AD46734

CA2

CA1

15
(3,8-4,5 mm)

AD46729

10
(24 Nm)

9
(48 Nm)

T

G

6

AD67399

8

BS

7

T

CS

OP

AD22191

AD46730

AD82707

19
(19 Nm)

11

AD76951

3

13

AD46736

4

1 (25 Nm)

16
(108-127 Nm)

14

AD76954

2

AD46738

18

17

AD11816

19. Install components in reverse order of removal.
20. Tighten crankshaft pulley bolts **1**.
    Tightening torque: 25 Nm.

Removal

Balancer Shaft Belt

1. Remove timing belt as described previously.
2. Remove:
    - ❏ Crankshaft bolt **16**.
    - ❏ Crankshaft sprocket **17**.
    - ❏ Crankshaft position (CKP) sensor reluctor **18**.
    - ❏ Crankshaft position (CKP) sensor.
3. Slacken balancer shaft belt tensioner bolt **19**.
4. Remove balancer shaft belt.
    NOTE: Mark direction of rotation on belt with chalk if belt is to be reused.

Installation

Balancer Shaft Belt

1. Ensure timing marks aligned **6** & **8**.
2. Fit balancer shaft belt.
3. Turn tensioner pulley firmly clockwise against belt. Tighten bolt **19**. Tightening torque: 19 Nm.
4. Apply thumb pressure to belt at ▽. Belt should deflect 5-7 mm.
5. If not: Repeat tensioning procedure.
6. Install components in reverse order of removal.
7. Tighten crankshaft bolt **16**.
    Tightening torque: 108-127 Nm.
8. Fit timing belt as described previously.

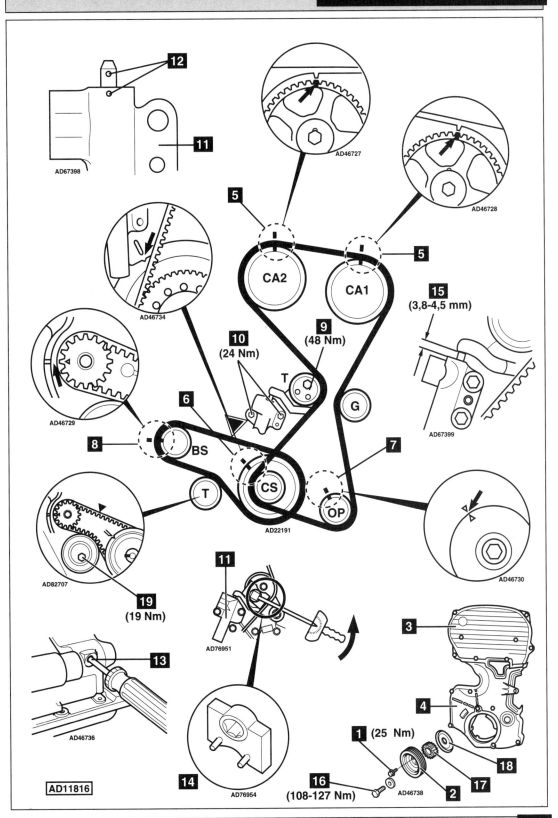

AD67398

12

11

AD46727

AD46728

5

5

AD46734

CA2

CA1

15
(3,8-4,5 mm)

10
(24 Nm)

9
(48 Nm)

AD46729

6

8

BS

T

G

AD67399

T

CS

OP

7

AD22191

AD46730

AD82707

19
(19 Nm)

11

AD76951

3

13

AD46736

AD11816

14

AD76954

1 (25 Nm)

16
(108-127 Nm)

2

AD46738

17

18

4

Model:	Galant 2,5 V6 24V DOHC • Sigma 3,0 V6 24V DOHC • 3000 GT V6 24V DOHC
Year:	1991-00
Engine Code:	6G72-DOHC, 6G73-DOHC

Replacement Interval Guide

Mitsubishi recommend replacement every 54,000 miles.
The previous use and service history of the vehicle must always be taken into account.
Refer to Timing Belt Replacement Intervals at the front of this manual.

Check For Engine Damage

CAUTION: This engine has been identified as an INTERFERENCE engine in which the possibility of valve-to-piston damage in the event of a timing belt failure is MOST LIKELY to occur.
A compression check of all cylinders should be performed before removing the cylinder head.

Repair Times – hrs

Remove & install:

Sigma	2,50
Galant	3,80
3000 GT	3,20
AC	+0,30

Special Tools

■ Tensioner pulley tool – Mitsubishi No.MD998767.

Special Precautions

■ Disconnect battery earth lead.
■ DO NOT turn crankshaft or camshaft when timing belt removed.
■ Remove spark plugs to ease turning engine.
■ Turn engine in normal direction of rotation (unless otherwise stated).
■ DO NOT turn engine via camshaft or other sprockets.
■ Observe all tightening torques.

Removal

1. Remove LH engine undershield.
2. Support engine.
3. Remove:
   ❏ Timing belt upper covers **1**.
   ❏ Auxiliary drive belts.
   ❏ Auxiliary drive belt tensioner.
   ❏ Crankshaft pulley **2**.
   ❏ Vehicles with ABS: Coolant expansion tank.
   ❏ Engine mounting bracket and guide pulley.
   ❏ Engine mounting bracket bolts in the order **10** - **7**.
   ❏ Engine mounting bracket **11**.
   ❏ 1993→: Crankshaft position (CKP) sensor.
   ❏ 1993→: Camshaft position (CMP) sensor.
   ❏ Timing belt lower cover **3**.

4. Turn crankshaft until timing marks aligned **4** & **5**.
5. Slacken tensioner pulley bolt **6**. Move tensioner away from belt. Tighten bolt.
6. Remove:
   ❏ Timing belt.
   ❏ Automatic tensioner unit.
   ❏ Tensioner pulley.

Installation

1. Check pushrod protrusion is 11,7-12,3 mm in fully extended position.
2. Slowly compress pushrod into tensioner body until holes aligned. Retain pushrod with 1,4 mm diameter pin through hole in tensioner body **16**.
3. Install automatic tensioner unit.
   NOTE: Turbo – Use Loctite or similar compound on automatic tensioner unit bolts.
4. Ensure timing marks aligned **4** & **5**.
   NOTE: If camshaft sprockets are not aligned: Turn crankshaft sprocket 3 teeth clockwise (ATDC) to avoid valves touching pistons during alignment operation. If a camshaft sprocket of one bank has been aligned and other sprocket is turned too far in either direction (approximately one turn) then valve to valve contact will occur. Always align sprockets carefully to avoid damage.
5. Fit timing belt in clockwise direction, starting at camshaft sprocket (CA1). Ensure belt is taut between sprockets.
   *NOTE: To assist installation of belt, suitable paper clips **13** can be used to retain belt on sprockets.*
6. Fit tensioner pulley with pin hole towards the top **14**.
7. Temporarily tighten tensioner pulley bolt **6**.
8. Turn crankshaft slowly 1/4 turn anti-clockwise.
9. Turn crankshaft clockwise until timing marks aligned.
10. Slacken tensioner pulley bolt **6**.
11. Fit tool to tensioner pulley **14**. Tool No.MD998767. Apply a torque of 10 Nm to tensioner pulley **15**. Tighten bolt **6**.
12. Remove pin from tensioner body.
13. Turn crankshaft two turns clockwise. Wait 5 minutes.
14. Check pushrod protrusion is 3,8-4,5 mm **12**.
15. If not: Repeat operations 8-14.
16. Install components in reverse order of removal.
17. Tighten engine mounting bracket bolts. Observe correct sequence **7** - **10**. Lubricate bolt **7** during tightening.
18. Tighten crankshaft pulley bolt to 180-190 Nm.

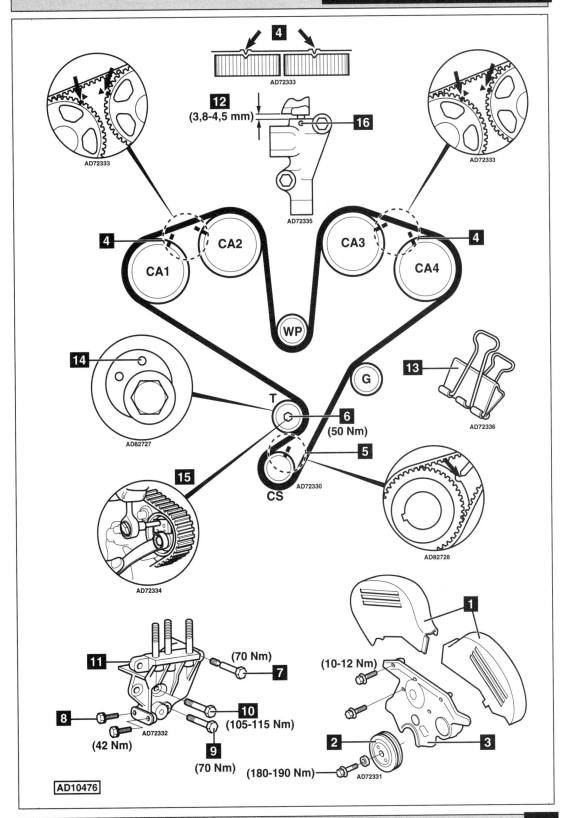

AD72333

12 (3,8-4,5 mm)

16

AD72335

4

CA2

CA1

CA3

4

CA4

WP

14

AD82727

T

G

13

AD72336

6 (50 Nm)

5

AD72330

CS

15

AD72334

AD82728

11

(70 Nm) **7**

1

(10-12 Nm)

8

10

AD72332

(105-115 Nm)

(42 Nm)

9

(70 Nm)

2

(180-190 Nm)

AD72331

3

AD10476

MITSUBISHI

Model:	Galant 2,5 V6
Year:	1997-03
Engine Code:	6A13

Replacement Interval Guide

Mitsubishi recommend replacement every 54,000 miles.
The previous use and service history of the vehicle must always be taken into account.
Refer to Timing Belt Replacement Intervals at the front of this manual.

Check For Engine Damage

CAUTION: This engine has been identified as an INTERFERENCE engine in which the possibility of valve-to-piston damage in the event of a timing belt failure is MOST LIKELY to occur.
A compression check of all cylinders should be performed before removing the cylinder head(s).

Repair Times – hrs

Remove & install	1,80

Special Tools

■ Tensioner pulley tool – Mitsubishi No.MD998767.

Special Precautions

■ Disconnect battery earth lead.
■ DO NOT turn crankshaft or camshaft when timing belt removed.
■ Remove spark plugs to ease turning engine.
■ Turn engine in normal direction of rotation (unless otherwise stated).
■ DO NOT turn engine via camshaft or other sprockets.
■ Observe all tightening torques.

Removal

1. Support engine.
2. Remove:
   ❏ Engine undershield.
   ❏ Engine cover **1**.
   ❏ Auxiliary drive belts.
   ❏ Auxiliary drive belt tensioner.
   ❏ Alternator.
   ❏ Engine mounting support bar **2**.
   ❏ Engine mounting **3**.
   ❏ Engine mounting bracket **4**.
   ❏ Timing belt LH and RH covers **5**.
3. Turn crankshaft to TDC on No.1 cylinder.
4. Ensure timing marks on camshaft sprockets aligned **6**.
5. Remove:
   ❏ Crankshaft pulley **7**.
   ❏ Timing belt lower cover **8**.
   ❏ Crankshaft sprocket outer guide washer **9**.
   ❏ Automatic tensioner unit **10**.

6. Slacken tensioner bolt **11**. Turn tensioner clockwise. Lightly tighten bolt.
7. Remove timing belt.
   NOTE: Mark direction of rotation on belt with chalk if belt is to be reused.

Installation

1. Check tensioner body for leakage or damage **12**.
2. Slowly compress pushrod into tensioner body **12** until holes aligned. Retain pushrod with suitable pin through hole in tensioner body **13**.
3. Install automatic tensioner unit to cylinder block.
4. Tighten bolts to 22 Nm **10**.
5. Ensure timing marks aligned **6** & **14**.
6. Fit timing belt in anti-clockwise direction, starting at crankshaft sprocket. Ensure belt is taut between sprockets.
7. Slacken tensioner pulley bolt **11**.
   NOTE: Ensure tensioner pulley pin holes at bottom.
8. Turn tensioner pulley anti-clockwise to temporarily tension belt. Tighten bolt.
9. Turn crankshaft 1/4 turn anti-clockwise.
10. Turn crankshaft clockwise until timing marks aligned **6** & **14**.
11. Slacken tensioner bolt **11**.
12. Fit tool **15** to holes in tensioner pulley. Tool No.MD998767.
13. Apply anti-clockwise torque of 3 Nm to tensioner pulley.
14. Tighten tensioner bolt to 48 Nm **11**.
15. Remove pin from tensioner body to release pushrod **13**.
16. Turn crankshaft slowly two turns in direction of rotation until timing marks aligned **6** & **14**.
17. Wait 5 minutes minimum.
18. Check extended length of pushrod is 3,8-4,5 mm **16**.
19. If not: Repeat operations 9-18.
20. Install components in reverse order of removal.
21. Oil crankshaft pulley bolt threads and washer face **17** & **18**.
22. Tighten crankshaft pulley bolt **17**.
    Tightening torque: 182 Nm.

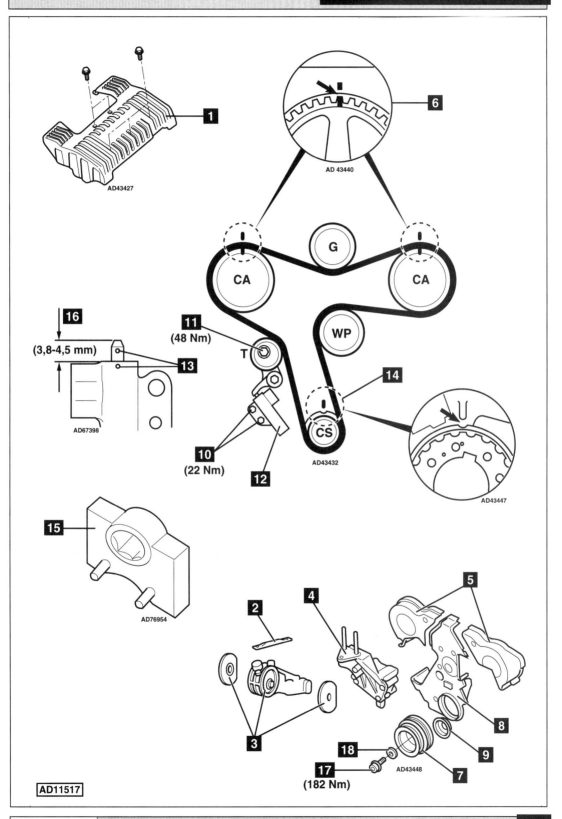

AD43427

AD 43440

6

G

CA

CA

WP

16

(3,8-4,5 mm)

11
(48 Nm)
T

13

AD67398

14

CS

AD43432

AD43447

10
(22 Nm)

12

15

AD76954

2

4

5

3

8

9

18

17
(182 Nm)

AD43448

7

AD11517

Model:	**Shogun/Pajero 3,0 12V**
Year:	**1989-94**
Engine Code:	**6G72-SOHC**

Replacement Interval Guide

Mitsubishi recommend replacement as follows:
To 12/90 - replacement every 60,000 miles.
1/91 on - replacement every 54,000 miles.
The previous use and service history of the vehicle must always be taken into account.
Refer to Timing Belt Replacement Intervals at the front of this manual.

Check For Engine Damage

CAUTION: This engine has been identified as an INTERFERENCE engine in which the possibility of valve-to-piston damage in the event of a timing belt failure is MOST LIKELY to occur.
A compression check of all cylinders should be performed before removing the cylinder head.

Repair Times – hrs

Remove & install	2,30
AC	+0,30
PAS	+0,20

Special Tools

■ None required.

Special Precautions

■ Disconnect battery earth lead.
■ DO NOT turn crankshaft or camshaft when timing belt removed.
■ Remove spark plugs to ease turning engine.
■ Turn engine in normal direction of rotation (unless otherwise stated).
■ DO NOT turn engine via camshaft or other sprockets.
■ Observe all tightening torques.

Removal

1. Drain coolant.
2. Remove:
   ❏ Top radiator hose.
   ❏ Radiator cowling.
   ❏ Cooling fan and pulley.
   ❏ Auxiliary drive belts.
   ❏ PAS pump and brackets.
   ❏ Auxiliary drive belt tensioner assembly.
   ❏ AC compressor and bracket.
   ❏ Cooling fan bracket assembly **1**.
   ❏ Timing belt covers **2** & **3**.
   ❏ Crankshaft pulley **4**.
   ❏ Timing belt cover **5**.
   ❏ Crankshaft sprocket guide washer **6**.

3. Turn crankshaft clockwise until timing marks aligned **7** & **8**.
4. Slacken tensioner bolt **9**.
5. Move tensioner away from belt. Lightly tighten bolt.
6. Remove timing belt.

Installation

1. Ensure timing marks aligned **7** & **8**.
2. Fit timing belt in anti-clockwise direction, starting at crankshaft sprocket.
3. Ensure belt is taut between sprockets on non-tensioned side.
4. Slacken tensioner bolt **9**.
5. Turn crankshaft two turns clockwise. Ensure timing marks aligned **7** & **8**.
   NOTE: DO NOT allow crankshaft to turn anti-clockwise.
6. Tighten tensioner bolt to 26 Nm **9**.
7. Install components in reverse order of removal.
8. Tighten crankshaft pulley bolt **10**.
   Tightening torque: 150-160 Nm.
9. Refill cooling system.

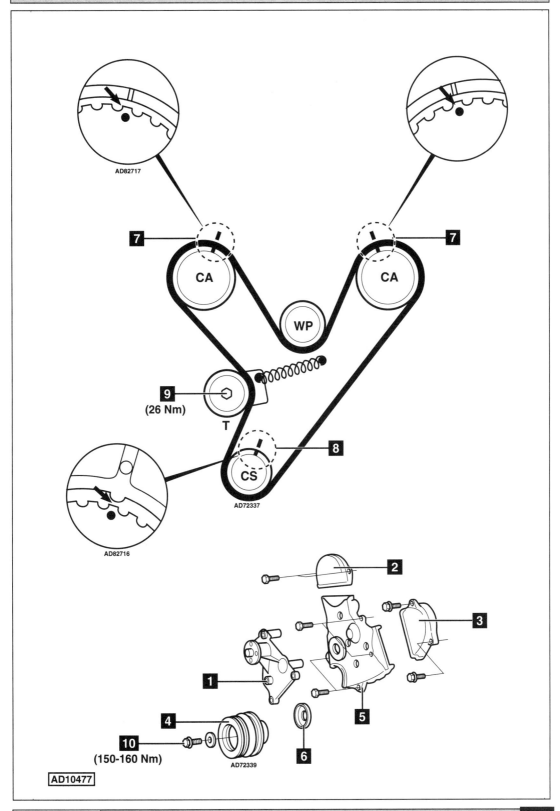

AD82717

7

CA

CA

7

WP

9
(26 Nm)

T

8

CS

AD72337

AD82716

2

3

1

5

4

10
(150-160 Nm)

6

AD72339

AD10477

MITSUBISHI

Model:	**Shogun 3,0 24V • Shogun Sport 3,0 24V**
Year:	**1994-03**
Engine Code:	**6G72-SOHC**

Replacement Interval Guide

Mitsubishi recommend replacement every 54,000 miles.
The previous use and service history of the vehicle must always be taken into account.
Refer to Timing Belt Replacement Intervals at the front of this manual.

Check For Engine Damage

CAUTION: This engine has been identified as an INTERFERENCE engine in which the possibility of valve-to-piston damage in the event of a timing belt failure is MOST LIKELY to occur. A compression check of all cylinders should be performed before removing the cylinder head(s).

Repair Times – hrs

Remove & install:

Shogun	2,30
PAS	+0,20
Shogun Sport	2,10

Special Tools

- Crankshaft pulley holding tool – Mitsubishi No.MD990767.
- Crankshaft pulley holding tool pins – Mitsubishi No.MD998715.
- Tensioner pulley tool – Mitsubishi No.MD998767.
- Crankshaft socket – Mitsubishi No.MD998716.

Special Precautions

- Disconnect battery earth lead.
- DO NOT turn crankshaft or camshaft when timing belt removed.
- Remove spark plugs to ease turning engine.
- Turn engine in normal direction of rotation (unless otherwise stated).
- DO NOT turn engine via camshaft or other sprockets.
- Observe all tightening torques.

Removal

1. Drain coolant.
2. Remove:
   - ❏ Auxiliary drive belts.
   - ❏ Auxiliary drive belt tensioner.
   - ❏ Engine undershield.
   - ❏ Top radiator hose.
   - ❏ Radiator cowling.
   - ❏ Viscous fan and pulley.
   - ❏ Viscous fan bracket.
   - ❏ Alternator.
3. Shogun Sport – Remove:
   - ❏ AC compressor and bracket.
   - ❏ PAS pump.
   - ❏ Cooling fan bracket 🔲.
4. All models – Remove:
   - ❏ Timing belt upper covers 🔲 & 🔲.
   - ❏ Crankshaft pulley bolt 🔲. Use tool Nos.MD990767/MD998715.
   - ❏ Crankshaft pulley 🔲.
   - ❏ Timing belt lower cover 🔲.
5. Turn crankshaft clockwise until timing marks aligned 🔲 & 🔲. Use tool No.MD998716.
6. Slacken tensioner pulley bolt 🔲.
7. Remove:
   - ❏ Automatic tensioner unit bolts 🔲.
   - ❏ Automatic tensioner unit 🔲.
   - ❏ Timing belt.
   NOTE: Mark direction of rotation on belt with chalk if belt is to be reused.

Installation

1. Check tensioner pulley for smooth operation.
2. Check tensioner body for leakage or damage.
3. Push pushrod of automatic tensioner unit against a firm surface using a force of 10-20 kg. Pushrod should move less than 1,0 mm.
4. If necessary: Replace automatic tensioner unit.
5. Slowly compress pushrod into tensioner body until holes aligned 🔲. Use suitable press or vice.
   NOTE: Use a suitable washer to protect tensioner body end screw when using press 🔲.
6. Retain pushrod with 1,4 mm diameter pin.
7. Install automatic tensioner unit 🔲. Tighten bolts 🔲. Tightening torque: 24 Nm.
8. Ensure timing marks aligned 🔲 & 🔲.
9. Fit timing belt in anti-clockwise direction, starting at crankshaft sprocket.
10. Ensure belt is taut between sprockets on non-tensioned side.
    NOTE: Observe direction of rotation marks on belt.
11. Push tensioner pulley by hand against belt. Lightly tighten bolt 🔲.
12. Turn crankshaft 1/4 turn anti-clockwise. Use tool No.MD998716.
13. Turn crankshaft 1/4 turn clockwise. Use tool No.MD998716.
14. Ensure timing marks aligned 🔲 & 🔲.
15. Slacken tensioner pulley bolt 🔲.
16. Apply anti-clockwise torque of 4,4 Nm to tensioner pulley 🔲. Use tool No.MD998767.
17. Hold tensioner pulley in position. Tighten tensioner pulley bolt 🔲. Tightening torque: 49 Nm.
18. Remove locking pin from automatic tensioner unit 🔲.
19. Turn crankshaft two turns clockwise. Use tool No.MD998716. Ensure timing marks aligned 🔲 & 🔲.
20. Wait 5 minutes to allow belt to settle.
21. Check pushrod protrusion is:
    - ❏ (A) Shogun: 3,8-4,5 mm 🔲.
    - ❏ (B) Shogun Sport: 3,8-5,0 mm 🔲.
22. If not: Repeat tensioning procedure.
23. Install components in reverse order of removal.
24. Tighten crankshaft pulley bolt 🔲:
    - ❏ (A) Shogun: Tightening torque: 181 Nm.
    - ❏ (B) Shogun Sport: Tightening torque: 177-186 Nm.
25. Refill coolant.

//Autodata

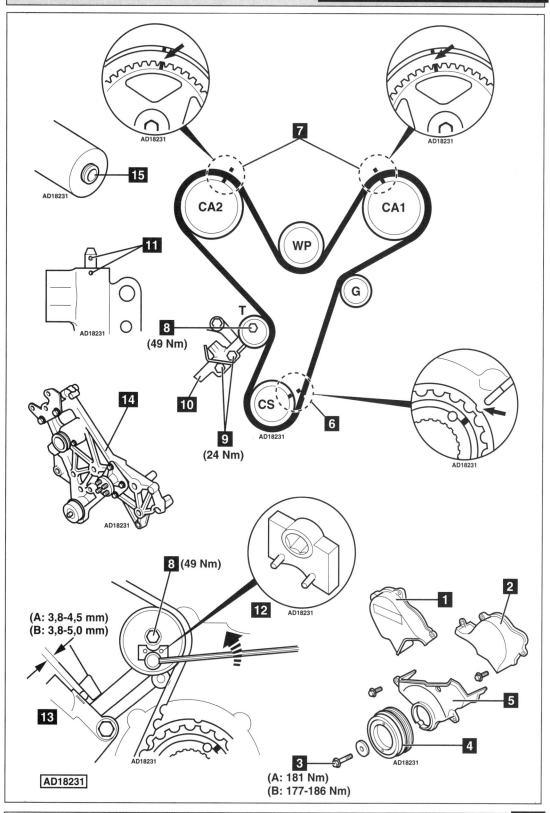

Model:	**Shogun 3,5 24V**
Year:	**1994-03**
Engine Code:	**6G74-DOHC**

Replacement Interval Guide

Mitsubishi recommend replacement every 54,000 miles.
The previous use and service history of the vehicle must always be taken into account.
Refer to Timing Belt Replacement Intervals at the front of this manual.

Check For Engine Damage

CAUTION: This engine has been identified as an INTERFERENCE engine in which the possibility of valve-to-piston damage in the event of a timing belt failure is MOST LIKELY to occur.
A compression check of all cylinders should be performed before removing the cylinder head.

Repair Times – hrs

Remove & install	3,10
AC	+0,30
PAS	+0,20

Special Tools

- Tensioner pulley tool – No.MB998767.
- Crankshaft pulley holding tool – No.MB991800.
- Crankshaft pulley holding tool pins – No.MB991802.

Removal

1. Remove:
   - ❏ Radiator & cooling fan.
   - ❏ Engine undershield.
   - ❏ Air filter & engine cover.
   - ❏ Auxiliary drive belts & tensioner.
   - ❏ RH engine lifting eye.
   - ❏ Cooling fan pulley.
   - ❏ Alternator & AC compressor.
   - ❏ PAS pump. DO NOT disconnect hoses.
   - ❏ Battery and battery tray.
   - ❏ Crankshaft position (CKP) sensor.
   - ❏ Crankshaft pulley bolt **1**. Use tool No.MB991800/MB991802.
   - ❏ Crankshaft pulley **2**.
   - ❏ Timing belt covers **3** & **4**.
2. Turn crankshaft to TDC on No.1 cylinder. Ensure crankshaft timing mark aligned **5**.
3. Ensure camshaft timing marks aligned: →2000 (A). 2000→ (B) **6**.
4. Slacken tensioner pulley bolt **7**. Move tensioner pulley away from belt. Lightly tighten bolt.
5. Remove:
   - ❏ Timing belt and tensioner unit **8**.
   - ❏ Tensioner pulley.

Installation

NOTE: If camshaft sprockets are not aligned: Turn crankshaft sprocket 3 teeth clockwise (ATDC) to avoid valves touching pistons during alignment operation. If a camshaft sprocket of one bank has been aligned and other sprocket is turned too far in either direction (approximately one turn) then valve to valve contact will occur. Always align sprockets carefully to avoid damage.

1. Check tensioner body for leakage or damage. Replace if necessary.
2. Push pushrod of automatic tensioner unit against a firm surface using a force of 10-20 kg. Pushrod should move less than 1,0 mm.
3. Slowly compress pushrod into tensioner body until holes aligned. Gently compress automatic tensioner unit **8**. Use suitable press or vice.
   NOTE: Use a suitable washer to protect tensioner body end screw when using press.
4. Retain pushrod with 1,4 mm diameter pin through hole in tensioner body **9**.
5. Install automatic tensioner unit. Tighten bolts to 24 Nm.
6. Fit tensioner pulley with pin hole towards the top **11**. Lightly tighten bolt **7**.
7. Ensure timing marks aligned: →2000 (A). 2000→ (B) **6**.
8. →2000: Align crankshaft timing mark **5**. Turn crankshaft one tooth clockwise.
9. →2000: Fit timing belt in clockwise direction, starting at camshaft sprocket (CA1). Ensure belt is taut between sprockets.
   2000→: Fit timing belt in anti-clockwise direction, starting at crankshaft sprocket. Ensure belt is taut between sprockets.
   *NOTE: To assist installation of belt, suitable paper clips **10** can be used to retain belt on sprockets.*
10. →2000: Align crankshaft timing mark **5**.
11. 2000→: Turn camshaft sprockets CA1 & CA2 anti-clockwise to tension belt.
12. Ensure timing marks aligned **5** & **6**.
13. Slacken tensioner pulley bolt **7**. Turn tensioner pulley clockwise to tension belt. Use tool No.MD998767 **12**. Lightly tighten bolt.
14. Ensure timing marks aligned **6**.
15. Remove paper clips **10**.
16. Turn crankshaft 1/4 turn anti-clockwise.
17. Turn crankshaft clockwise until timing marks aligned **5** & **6**.
18. Slacken tensioner pulley bolt **7**.
19. Fit tool to tensioner pulley. Tool No.MD998767 **12**. Apply torque to tensioner pulley:
    - ❏ →2000: 9,4 Nm. Tighten bolt to 49 Nm (A) **7**.
    - ❏ 2000→: 4,4 Nm. Tighten bolt to 42-54 Nm (B) **7**.
20. Remove pin from tensioner body. Ensure pin can be withdrawn easily.
21. Turn crankshaft two turns clockwise. Wait 5 minutes.
22. Check 1,4 mm diameter pin may be entered and withdrawn easily. Check pushrod protrusion is:
    - ❏ →2000: (A) 3,8-4,5 mm **13**.
    - ❏ 2000→: (B) 3,8-5,0 mm **13**.
23. If not: Repeat operations 13-19.
24. Install components in reverse order of removal.
25. Tighten crankshaft pulley bolt to:
    →2000: (A) 180-190 Nm.
    2000→: (B) 178-186 Nm. **1**:

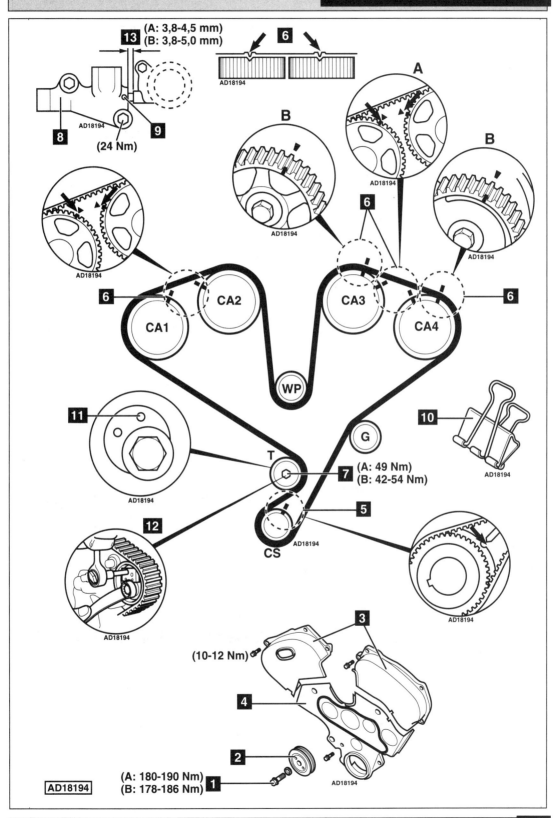

(A: 3,8-4,5 mm)
(B: 3,8-5,0 mm)

13

6

AD18194

A

AD18194

B

AD18194

B

AD18194

8

AD18194

9

(24 Nm)

6

AD18194

6

CA2

CA1

CA3

CA4

6

WP

11

AD18194

10

AD18194

G

T

7

(A: 49 Nm)
(B: 42-54 Nm)

12

AD18194

5

AD18194

CS

AD18194

3

(10-12 Nm)

4

AD18194

2

(A: 180-190 Nm)
(B: 178-186 Nm) **1**

AD18194

MITSUBISHI

Model:	Lancer 2,0 D • Galant 2,0 Turbo D • Space Wagon 2,0 Turbo D
Year:	1992-03
Engine Code:	4D68, 4D68T

Replacement Interval Guide

Mitsubishi recommend replacement every 54,000 miles.
The previous use and service history of the vehicle must always be taken into account.
Refer to Timing Belt Replacement Intervals at the front of this manual.

Check For Engine Damage

CAUTION: This engine has been identified as an INTERFERENCE engine in which the possibility of valve-to-piston damage in the event of a timing belt failure is MOST LIKELY to occur.
A compression check of all cylinders should be performed before removing the cylinder head.

Repair Times – hrs

Remove & install:
Timing belt:

Lancer	1,80
Galant →1997	2,00
Galant 1997→	2,20
Space Wagon	2,50
Balancer shaft belt:	
Lancer	2,60[1]
Galant →1997	2,20[1]
Galant 1997→	2,30[1]
Space Wagon	2,50[1]
PAS	+0,20
AC	+0,30

[1] *Includes timing belt replacement.*

Special Tools

■ None required.

Removal

Timing Belt

1. Support engine.
2. Remove:
   ❏ Front engine mounting (except Space Wagon).
   ❏ Auxiliary drive belts.
   ❏ Water pump pulley **13**.
   ❏ AC drive belt tensioner **14** (if fitted).
   ❏ Crankshaft pulley bolts **1**.
   ❏ Crankshaft pulley **2**.
   ❏ Timing belt upper cover **3**.
   ❏ Timing belt lower cover **4**.
3. Turn crankshaft clockwise to TDC on No.1 cylinder.
4. Ensure timing marks aligned **5**, **6**, **7** & **8**.
5. Slacken tensioner bolt **9**. Move tensioner away from belt and lightly tighten bolt.
6. Remove timing belt.

Installation

Timing Belt

1. Ensure timing marks aligned **5**, **6**, **7** & **8**. Ensure tensioner at bottom of slot.
   *NOTE: To check oil pump sprocket is correctly positioned: Remove blanking plug from cylinder block **10**. Insert 8 mm diameter Phillips screwdriver in hole. Ensure screwdriver is inserted to a depth of 60 mm. If screwdriver can only be inserted 20-25 mm: Turn oil pump sprocket 360° and reinsert screwdriver.*
2. Fit timing belt in clockwise direction, starting at crankshaft sprocket.
3. Remove screwdriver from oil pump shaft.
4. Turn crankshaft slowly anti-clockwise 1/2 tooth of camshaft sprocket to take up slack in belt.
5. Fit timing belt to tensioner pulley.
6. Slacken tensioner bolt **9**. Allow tensioner to operate.
7. Turn crankshaft anti-clockwise 2 1/2 teeth of camshaft sprocket.
8. Tighten tensioner bolt to 48 Nm **9**.
9. Turn crankshaft clockwise until timing marks aligned **5**, **6**, **7** & **8**.
10. Apply thumb pressure to timing belt at ▽. Belt should deflect 4-5 mm.
11. Install components in reverse order of removal.
12. Tighten crankshaft pulley bolts **1**.
    Tightening torque: 25 Nm

Removal

Balancer Shaft Belt

NOTE: It is not necessary to disturb or remove the balancer shaft belt when replacing the timing belt.
1. Remove timing belt as described previously.
2. Ensure timing marks aligned **12**.
3. Remove balancer shaft belt tensioner bolt **11** and tensioner pulley.
4. Remove balancer shaft belt.

Installation

Balancer Shaft Belt

1. Ensure timing marks aligned **12**.
2. Fit balancer shaft belt to sprockets ensuring upper side of belt kept tight.
3. Fit balancer shaft belt tensioner pulley and bolt **11**.
4. Ensure centre of tensioner pulley to left side of tensioner bolt.
5. Turn tensioner pulley clockwise to tension belt. Tighten tensioner bolt to 19 Nm **11**.
6. Apply thumb pressure to balancer shaft belt at ▽.
7. Belt should deflect 5-7 mm.
8. If not: Repeat tensioning procedure.
9. Fit timing belt as described previously.

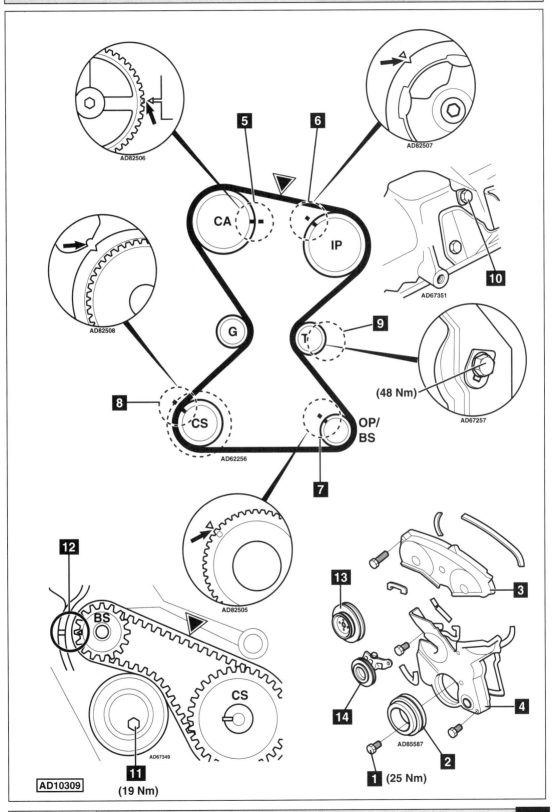

AD82506

AD82507

5

6

CA

IP

10
AD67351

9

G

T

(48 Nm)

AD67257

AD82508

8

CS

OP/
BS

AD62256

7

AD82505

12

BS

CS

13

3

AD67349

11

(19 Nm)

14

4

1 (25 Nm)

2

AD85587

AD10309

MITSUBISHI

Model:	**Carisma 1,9 TD**
Year:	**1997-02**
Engine Code:	**F8QT**

Replacement Interval Guide

Mitsubishi recommend replacement every 54,000 miles.
The previous use and service history of the vehicle must always be taken into account.
Refer to Timing Belt Replacement Intervals at the front of this manual.

Check For Engine Damage

CAUTION: This engine has been identified as an INTERFERENCE engine in which the possibility of valve-to-piston damage in the event of a timing belt failure is MOST LIKELY to occur.
A compression check of all cylinders should be performed before removing the cylinder head.

Repair Times – hrs

Remove & install	2,50
Without AC	+0,20

Special Tools

- ■ Tension gauge – Mitsubishi No.MB996032/33.

Special Precautions

- ■ Disconnect battery earth lead.
- ■ DO NOT turn crankshaft or camshaft when timing belt removed.
- ■ Remove glow plugs to ease turning engine.
- ■ Turn engine in normal direction of rotation (unless otherwise stated).
- ■ DO NOT turn engine via camshaft or other sprockets.
- ■ Observe all tightening torques.
- ■ Check diesel injection pump timing after belt replacement.

Removal

1. Raise and support front of vehicle.
2. Remove:
   - ❏ RH front wheel.
   - ❏ RH inner wing panel.
3. Support engine.
4. Remove:
   - ❏ Auxiliary drive belt.
   - ❏ RH engine mounting **1**.
   - ❏ Timing belt upper cover **2**.
5. Turn crankshaft clockwise until camshaft sprocket timing mark aligned with mark on timing belt cover **7**.
6. Insert locking pin in crankshaft **10**.
7. Rock crankshaft slightly to ensure locking pin located correctly.

8. Ensure injection pump sprocket timing marks aligned **8**.
9. Remove:
   - ❏ Timing belt centre cover **3**.
   - ❏ Crankshaft pulley bolt **6**.
   - ❏ Crankshaft pulley **5**.
   - ❏ Timing belt lower cover **4**.
10. Slacken tensioner nut to release tension on belt **9**.
11. Remove timing belt.

Installation

1. Rock crankshaft slightly to ensure locking pin located correctly **10**.
2. Observe direction of rotation marks on belt **11**.
3. Fit timing belt in following order:
   - ❏ Crankshaft sprocket.
   - ❏ Auxiliary shaft sprocket.
   - ❏ Guide pulley.
   - ❏ Injection pump sprocket.
   - ❏ Camshaft sprocket.
   - ❏ Tensioner.
4. Ensure marks on belt aligned with marks on sprockets **7**, **8** & **12**. Ensure belt is taut between sprockets.
5. Attach tension gauge to belt at ▽. Tool No.MB996032/33 **13**.
6. Use an M6 x 45 mm bolt **14** screwed into timing belt rear cover to push on tensioner pulley.
7. Screw in M6 bolt **14** until tension gauge indicates 7,5 mm.
8. Tighten tensioner nut to 50 Nm **9**.
9. Remove:
   - ❏ Crankshaft locking pin **10**.
   - ❏ Tension gauge.
10. Turn crankshaft two turns clockwise.
11. Insert locking pin in crankshaft **10**.
12. Ensure timing marks aligned **7**, **8** & **12**.
13. Recheck belt tension.
14. Remove M6 bolt and crankshaft locking pin.
15. Install components in reverse order of removal.
16. Tighten crankshaft pulley bolt to 120 Nm **6**.

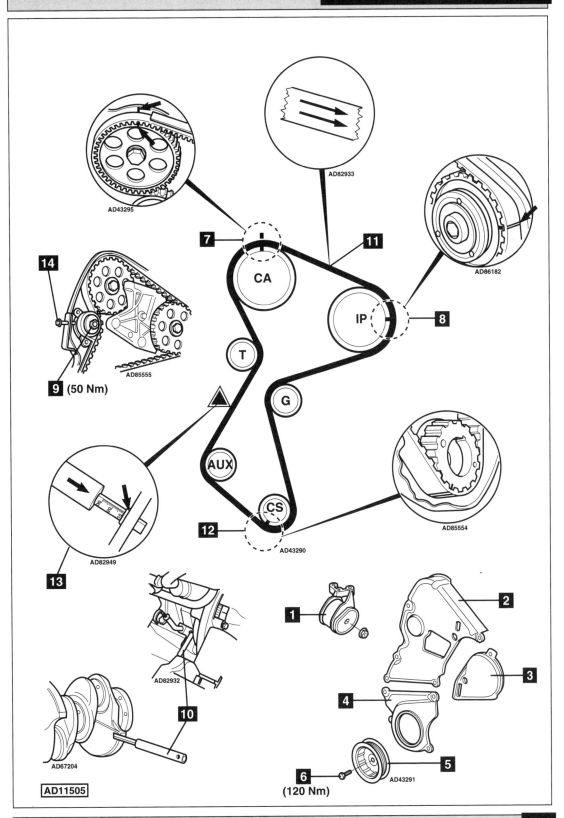

AD43295

AD82933

AD86182

7

11

8

CA

IP

14

AD85555

T

G

9 (50 Nm)

AUX

CS

AD85554

AD82949

13

12

AD43290

AD82932

10

1

2

4

3

AD67204

6

5

AD43291

(120 Nm)

AD11505

MITSUBISHI

Model:	Galant 2,3 Turbo D • Shogun 2,3/2,5 Turbo D • Shogun Sport 2,5 Turbo D Canter • L200/L300
Year:	1980-03
Engine Code:	4D55/T, 4D56/T

Replacement Interval Guide

Mitsubishi recommend replacement as follows:
→12/90 - replacement every 60,000 miles.
01/91→ - replacement every 54,000 miles.
The previous use and service history of the vehicle must always be taken into account.
Refer to Timing Belt Replacement Intervals at the front of this manual.

Check For Engine Damage

CAUTION: This engine has been identified as an INTERFERENCE engine in which the possibility of valve-to-piston damage in the event of a timing belt failure is MOST LIKELY to occur.
A compression check of all cylinders should be performed before removing the cylinder head.

Repair Times – hrs

Remove & install:
Galant:
Timing belt	2,50
Balancer shaft belt	2,70

Shogun →1991:
Timing belt	1,60
Balancer shaft belt	1,70
Intercooler	+0,20

Shogun 1991→:
Timing belt	1,90
Balancer shaft belt	2,00
Intercooler	+0,20

Shogun Sport:
Timing belt	2,00
Balancer shaft belt	2,00

L200:
Timing belt	1,60
Balancer shaft belt (→1997)	1,70
Balancer shaft belt (1997→)	1,80
Intercooler (1997→)	0,20

L300:
Timing belt	1,90
Balancer shaft belt	2,00

All models:
(except Shogun Sport) PAS	+0,20
AC	+0,30

Canter:
No information available.

Special Tools

■ L200 (97MY→), Shogun Sport: Crankshaft pulley locking tool – Mitsubishi No. MD998721.

Special Precautions

■ Disconnect battery earth lead.
■ DO NOT turn crankshaft or camshaft when timing belt removed.
■ Remove glow plugs to ease turning engine.
■ Turn engine in normal direction of rotation (unless otherwise stated).
■ DO NOT turn engine via camshaft or other sprockets.
■ Observe all tightening torques.
■ Check diesel injection pump timing after belt replacement.

Removal

Timing Belt

1. Remove:
   ❏ Oil sump protective plates (Shogun).
   ❏ Radiator.
   ❏ Auxiliary drive belt(s).
   ❏ Cooling fan.
   ❏ Timing belt upper cover **1**.
2. Turn crankshaft clockwise to TDC on No.1 cylinder. Ensure timing marks aligned **5**, **6**, **7** & **14**.
3. Remove:
   ❏ Crankshaft pulley **3** – L200 (97MY→), Shogun Sport: Use tool No.MD998721 **15**.
   ❏ Timing belt lower cover **2**.
4. Slacken tensioner bolts **8**. Move tensioner away from belt. Lightly tighten bolts.
5. Remove timing belt.
   NOTE: L200 (2000MY→): Crankshaft sprocket fitted with crankshaft position (CKP) sensor reluctor: Special care required when removing timing belt.

Installation

Timing Belt

1. Ensure timing marks aligned **5**, **6**, **7** & **14**.
2. Fit timing belt.
3. Slacken tensioner bolts **8**. Allow tensioner to operate.
4. Ensure belt is taut between sprockets on non-tensioned side.
5. Turn crankshaft clockwise TWO teeth of belt only.
6. Tighten tensioner upper bolt **8**.
   Tightening torque: 25 Nm.

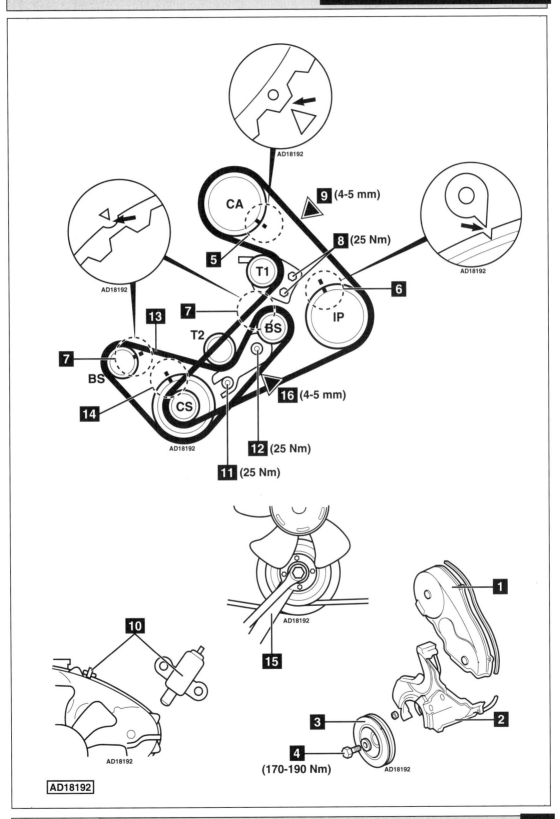

7. Tighten tensioner lower bolt **8**.
   Tightening torque: 25 Nm.
   NOTE: Observe correct tightening sequence,
   otherwise belt tension will not be correct.

8. Turn crankshaft anti-clockwise until timing marks
   aligned **5**, **6**, **7** & **14**.

9. Apply thumb pressure to timing belt at ▽ **9**. Belt
   should deflect 4-5 mm.

10. If required: Reset timing belt switch **10** by
    depressing switch until button is flush with
    switch body.

11. Check and adjust injection pump timing.

12. Install components in reverse order of removal.

13. Tighten crankshaft pulley bolt **4**.
    Tightening torque: 170-190 Nm.

Removal

Balancer Shaft Belt

1. Remove timing belt as described previously.
2. Ensure timing marks aligned **7** & **14**.
3. Slacken nut and bolt **11** & **12**.
4. Move tensioner (T2) towards water pump and
   tighten nut **11**.
5. Remove balancer shaft belt.

Installation

Balancer Shaft Belt

1. Ensure timing marks aligned.
2. Fit balancer shaft belt to sprockets with no slack
   at ▽ **16**.
3. Press down on belt at **13**. Slacken tensioner
   nut **11**.
4. Tighten tensioner nut **11**.
   Tightening torque: 25 Nm.
5. Tighten tensioner bolt **12**.
   Tightening torque: 25 Nm.
6. Apply thumb pressure to balancer shaft belt at
   ▽ **16**.
7. Belt should deflect 4-5 mm.
8. Fit timing belt as described previously.

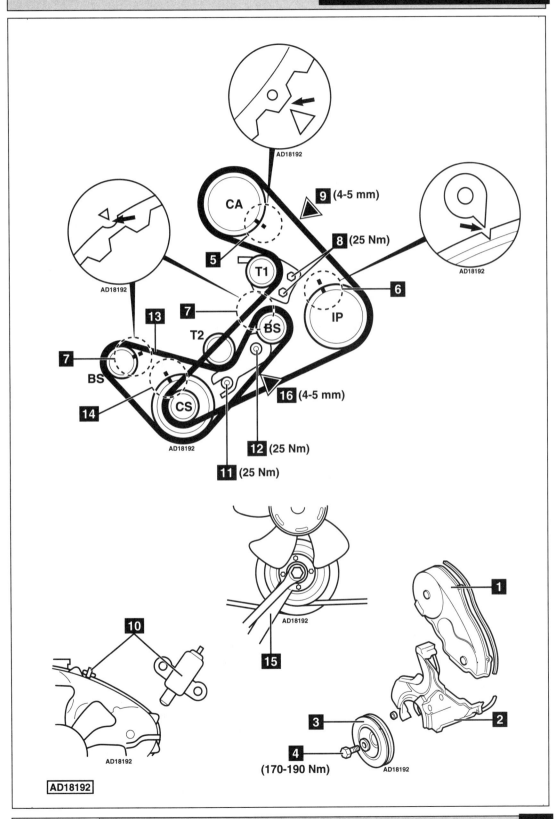

NISSAN

Model:	**Micra 1,5D**
Year:	**1998-02**
Engine Code:	**VJZ**

Replacement Interval Guide

Nissan recommend replacement as follows:
Under normal conditions − every 72,000 miles.
Under adverse conditions − every 48,000 miles.
The previous use and service history of the vehicle must always be taken into account.
Refer to Timing Belt Replacement Intervals at the front of this manual.

Check For Engine Damage

CAUTION: This engine has been identified as an INTERFERENCE engine in which the possibility of valve-to-piston damage in the event of a timing belt failure is MOST LIKELY to occur. A compression check of all cylinders should be performed before removing the cylinder head(s).

Repair Times − hrs

Remove & install	2,50

Special Tools

- Flywheel timing pin − Nissan No.KV109B0112.
- Camshaft timing pin − Nissan No.KV109B0111.
- Injection pump timing pin − Nissan No.KV109B0110.
- Tension gauge − Nissan No.KV109B0030.
- Tensioning tool − Nissan No.KV109B0080.

Special Precautions

- Disconnect battery earth lead.
- DO NOT turn crankshaft or camshaft when timing belt removed.
- Remove glow plugs to ease turning engine.
- Turn engine in normal direction of rotation (unless otherwise stated).
- DO NOT turn engine via camshaft or other sprockets.
- Observe all tightening torques.
- Check diesel injection pump timing after belt replacement.

Removal

1. Raise and support front of vehicle.
2. Remove:
   - ❏ RH front wheel.
   - ❏ Lower splash guard.
   - ❏ RH splash guard.
   - ❏ Timing belt upper cover **1**.
   - ❏ Auxiliary drive belt.
   - ❏ Crankshaft pulley bolts **2**.
   - ❏ Crankshaft pulley **3**.
3. Support engine.
4. Remove:
   - ❏ RH engine mounting and bracket.
   - ❏ Coolant expansion tank. DO NOT disconnect hoses.
   - ❏ Timing belt lower cover **4**.
5. Turn crankshaft clockwise to setting position.
6. Insert timing pin in flywheel **5**. Tool No.KV109B0112.
7. Insert timing pin in injection pump sprocket **6**. Tool No.KV109B0110.
8. Insert timing pin in camshaft sprocket **7**. Tool No.KV109B0111.
   NOTE: If locating holes for timing pins not aligned: Turn crankshaft one turn clockwise.

9. Slacken tensioner nut **8**. Move tensioner away from belt. Lightly tighten nut.
10. Remove timing belt.

Installation

NOTE: Check tensioner pulley and guide pulley for smooth operation.
1. Ensure timing pins located correctly **5**, **6** & **7**.
2. Hold sprockets. Slacken bolts **9** & **10**. Tighten bolts finger tight. Ensure sprockets can still turn.
3. Turn sprockets fully clockwise in slotted holes.
4. Fit timing belt in anti-clockwise direction, starting at crankshaft sprocket. Ensure belt is taut between sprockets.
   NOTE: Turn injection pump and camshaft sprockets slightly anti-clockwise to engage belt teeth.
5. Attach tension gauge to belt at ▽. Tool No.KV109B0030.
6. Slacken tensioner nut **8**.
7. Turn tensioner pulley anti-clockwise until tension gauge indicates the following:
   New belt: 98 SEEM units. Used belt: 75 SEEM units. Use tool No.KV109B0080 **11**.
8. Tighten tensioner nut to 20 Nm **8**.
9. Remove tension gauge. Tighten bolts of each sprocket to 25 Nm **9** & **10**.
   *NOTE: Ensure bolts of each sprocket not at end of slotted holes **9** & **10**. If bolts at end of slotted holes: Repeat installation procedure.*
10. Remove timing pins **5**, **6** & **7**.
11. Turn crankshaft two turns clockwise. Insert timing pins **5**, **6** & **7**.
12. Slacken camshaft sprocket bolts **9**.
13. Slacken injection pump sprocket bolts **10**.
14. Slacken tensioner nut **8**. Release tension on belt.
15. Attach tension gauge to belt at ▽.
16. Turn tensioner pulley anti-clockwise until tension gauge indicates the following:
    New belt: 54 SEEM units. Used belt: 44 SEEM units. Use tool No.KV109B0080 **11**.
17. Tighten tensioner nut to 20 Nm **8**.
18. Tighten bolts of each sprocket to 25 Nm **9** & **10**.
19. Remove tension gauge.
20. Remove timing pins **5**, **6** & **7**.
21. Turn crankshaft two turns clockwise. Insert flywheel timing pin **5**.
22. Ensure timing pins can be inserted easily in sprockets **6** & **7**. If not: Repeat installation procedure.
23. Attach tension gauge to belt at ▽.
24. Insert timing pins **6** & **7**.
25. Check tension gauge reading.
    New belt: 51-57 SEEM units.
    Used belt: 41-47 SEEM units. If tension not as specified: Repeat tensioning procedure.
26. Install components in reverse order of removal.
27. Tighten crankshaft pulley bolts to 15 Nm **2**.

Autodata

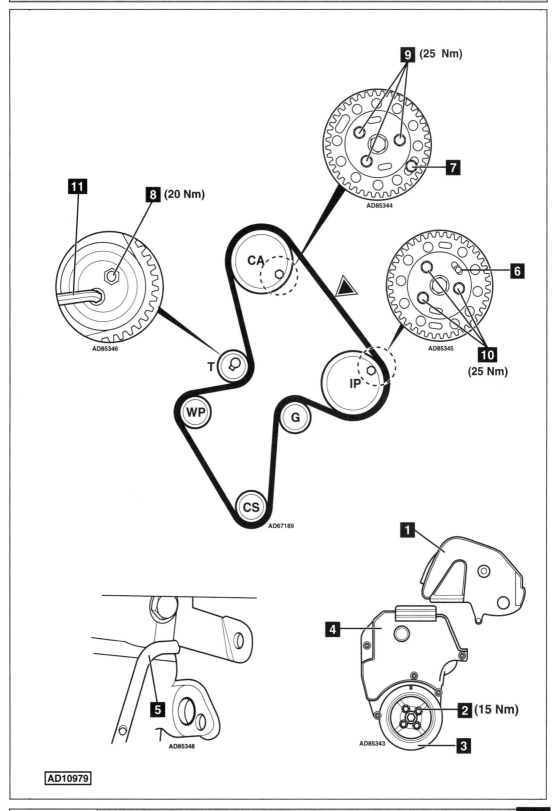

NISSAN

Model:	**Sunny 2,0D (N14)** • **Sunny Wagon 2,0D (Y10)** • **Almera 2,0D (N15)**
	Primera 2,0D (P10)
Year:	**1991-96**
Engine Code:	**CD20**

Replacement Interval Guide

Nissan recommend replacement as follows:
Pre 12/91 - replacement every 60,000 miles.
1/92 on - replacement every 54,000 miles or 60 months, whichever occurs first.
The previous use and service history of the vehicle must always be taken into account.
Refer to Timing Belt Replacement Intervals at the front of this manual.

Check For Engine Damage

CAUTION: This engine has been identified as an INTERFERENCE engine in which the possibility of valve-to-piston damage in the event of a timing belt failure is MOST LIKELY to occur.
A compression check of all cylinders should be performed before removing the cylinder head.

Repair Times – hrs

Remove & install	2,90
AC	+0,20
PAS	+0,20

Special Tools

■ None required.

Removal

Timing Belt

1. Support engine.
2. Drain coolant (Almera).
3. Remove:
   ❏ Top radiator hose (Almera).
   ❏ Auxiliary drive belts.
   ❏ Water pump pulley.
   ❏ Timing belt upper cover **1**.
   ❏ RH engine mounting.
4. Turn crankshaft to TDC on No.1 cylinder. Ensure crankshaft pulley timing marks aligned **2**.
5. Ensure lines on belt aligned with timing marks on sprockets. If not: Turn crankshaft one turn.
6. Remove:
   ❏ Crankshaft pulley.
   ❏ Timing belt lower cover **4**.
7. Slacken tensioner bolt **5**. Turn tensioner away from belt. Use Allen key. Lightly tighten bolt.
8. Remove crankshaft sprocket and timing belt.

Installation

Timing Belt

1. Temporarily fit crankshaft pulley.
2. Ensure crankshaft pulley timing marks aligned **2**.
3. Remove crankshaft pulley.
4. Install crankshaft sprocket with timing belt. Fit timing belt in anti-clockwise direction, starting at crankshaft sprocket.

5. Ensure belt is taut on non-tensioned side. Ensure lines on belt aligned with timing marks on sprockets. Arrow on belt must point away from engine.
   *NOTE: There should be 43 teeth between timing marks on sprockets **3** & **12**.*
6. Slacken tensioner bolt **5**. Allow tensioner to operate.
7. Temporarily fit crankshaft pulley.
8. Turn crankshaft slowly two turns in normal direction of rotation.
9. Except Almera: At approximately 90° BTDC insert 0,30 mm feeler gauge **6** between belt and guide pulley **7**. Continue to turn crankshaft until timing marks aligned **2** & **3**.
10. Hold tensioner using Allen key. Tighten tensioner bolt **5**. Tightening torque: 32-40 Nm.
11. Except Almera: Turn crankshaft anti-clockwise to remove feeler gauge **6**.
12. Remove crankshaft pulley.
13. Install components in reverse order of removal.
14. Tighten crankshaft pulley bolt **8**.
    Tightening torque: 142-152 Nm.

Removal

Injection Pump Belt

1. Remove air duct and resonator.
2. Turn crankshaft to TDC on No.1 cylinder. Ensure timing marks aligned **2**, **9** & **10**.
3. Slacken tensioner nut **11**. Turn tensioner clockwise to release tension on belt. Lightly tighten nut.
4. Remove injection pump belt.

Installation

Injection Pump Belt

NOTE: If injection pump sprocket removed, ensure sprocket fitted with key located in keyway marked 'A'.

1. Check tensioner pulley for smooth operation.
2. Fit injection pump belt. Align marks on belt with marks on sprockets **9** & **10**.
   *NOTE: There should be 23 teeth between timing marks on sprockets **9** & **10**.*
3. Slacken tensioner nut **11**.
4. Turn crankshaft two turns clockwise.
5. Hold tensioner with screwdriver and tighten nut **11**.
   Tightening torque: 16-21 Nm.
   NOTE: If tensioner turns when nut is tightened, tension may be excessive.

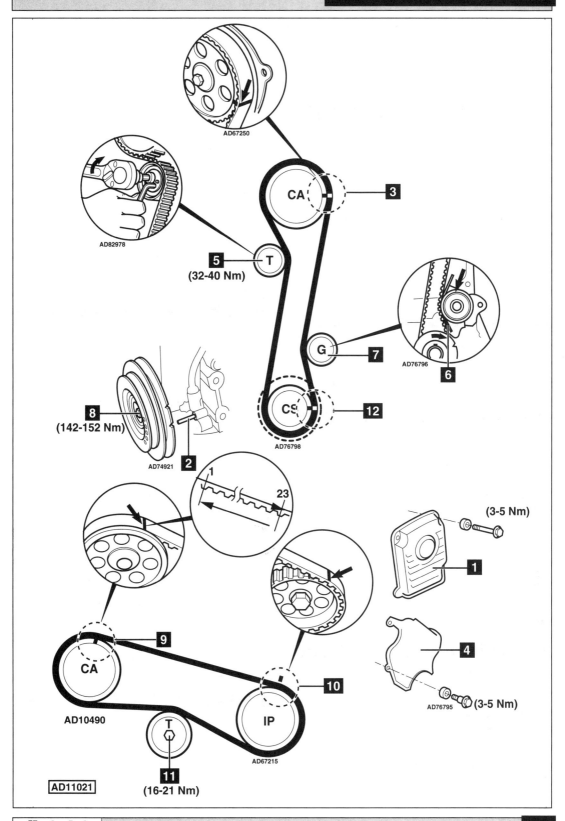

AD67250

AD82978

3

CA

5 T
(32-40 Nm)

7 G

6

AD76796

8
(142-152 Nm)

AD74921

2

CS **12**

AD76798

1

23

(3-5 Nm)

1

4

AD76795 (3-5 Nm)

9

CA

AD10490

T

11
(16-21 Nm)

IP

AD67215

10

AD11021

NISSAN

Model:	Almera 2,0D • Primera 2,0D Turbo
Year:	1996-02
Engine Code:	CD20E, CD20T

Replacement Interval Guide

Nissan recommend replacement every 54,000 miles or 60 months, whichever occurs first.
The previous use and service history of the vehicle must always be taken into account.
Refer to Timing Belt Replacement Intervals at the front of this manual.

Check For Engine Damage

CAUTION: This engine has been identified as an INTERFERENCE engine in which the possibility of valve-to-piston damage in the event of a timing belt failure is MOST LIKELY to occur.
A compression check of all cylinders should be performed before removing the cylinder head.

Repair Times – hrs

Remove & install:	
Almera	2,90
Primera	3,10
PAS	+0,20
AC	+0,20

Special Tools

- None required.

Special Precautions

- Disconnect battery earth lead.
- DO NOT turn crankshaft or camshaft when timing belt removed.
- Remove glow plugs to ease turning engine.
- Turn engine in normal direction of rotation (unless otherwise stated).
- DO NOT turn engine via camshaft or other sprockets.
- Observe all tightening torques.
- Check diesel injection pump timing after belt replacement.

Removal

Timing Belt

1. Support engine.
2. Drain coolant.
3. Remove:
   - ❑ Top radiator hose.
   - ❑ Auxiliary drive belts.
   - ❑ Water pump pulley.
   - ❑ Timing belt upper cover ■.
   - ❑ RH engine mounting.
   - ❑ Camshaft sprocket flange.

4. Turn crankshaft to TDC on No.1 cylinder. Ensure crankshaft pulley timing marks aligned ❷.
   NOTE: Crankshaft pulley TDC mark is unpainted.
   NOTE: Camshaft position can be checked using camshaft rear sprocket timing marks ❼.
5. Remove:
   - ❑ Crankshaft pulley bolt ❻.
   - ❑ Crankshaft pulley.
   - ❑ Timing belt lower cover ❹.
6. Slacken tensioner bolt ❺. Turn tensioner away from belt. Use Allen key. Lightly tighten bolt.
7. Remove crankshaft sprocket and timing belt.

Installation

Timing Belt

1. Temporarily fit crankshaft pulley.
2. Ensure crankshaft pulley timing marks aligned ❷.
3. Remove crankshaft pulley.
4. Install crankshaft sprocket with timing belt. Fit timing belt in anti-clockwise direction, starting at crankshaft sprocket.
5. Ensure belt is taut on non-tensioned side. Ensure lines on belt aligned with timing marks on sprockets ❸. Arrow on belt must point to timing belt cover.
   NOTE: There should be 43 teeth between timing marks on sprockets ❸.
6. Slacken tensioner bolt ❺. Allow tensioner to operate.
7. Turn crankshaft slowly two turns clockwise.
8. Hold tensioner using Allen key. Tighten tensioner bolt ❺. Tightening torque: 32-40 Nm.
9. Fit camshaft sprocket flange. Tighten bolts to 7 Nm.
10. Install components in reverse order of removal.
11. Tighten crankshaft pulley bolt ❻. Tightening torque: 142-152 Nm.
12. Refill cooling system.

Removal

Injection Pump Belt

1. Ensure ignition switched OFF.
2. Remove:
   - ❑ Battery.
   - ❑ Air filter, air duct and resonator.
3. Disconnect vacuum pump hoses.
4. Remove injection pump belt cover.

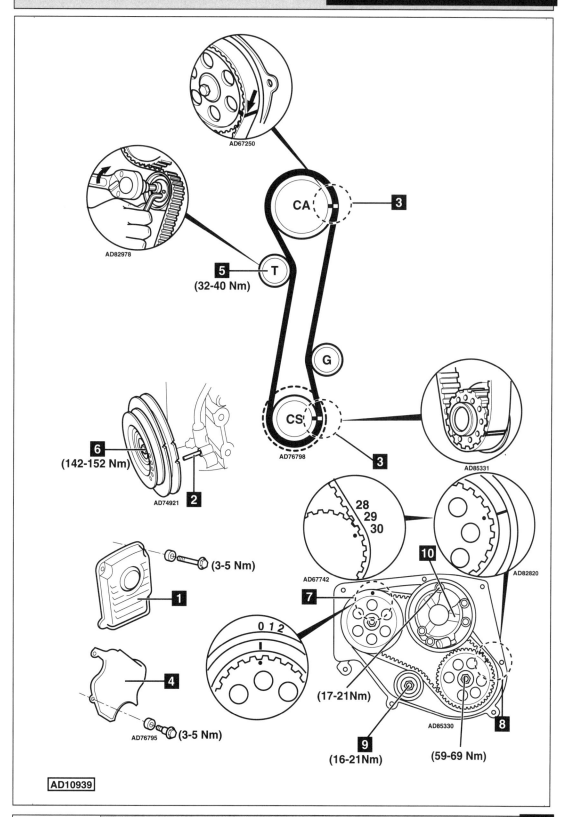

AD67250

3

CA

AD82978

5
T
(32-40 Nm)

G

6
(142-152 Nm)

AD74921

2

CS

AD76798

3

AD85331

28
29
30

AD67742

10

7

AD82820

1

(3-5 Nm)

0 1 2

4

AD76795 (3-5 Nm)

(17-21Nm)

9
(16-21Nm)

AD85330

8

(59-69 Nm)

AD10939

5. Turn crankshaft to TDC on No.1 cylinder. Ensure crankshaft pulley timing marks aligned **2**.
   NOTE: Crankshaft pulley TDC mark is unpainted.
6. Ensure camshaft rear sprocket and injection pump sprocket timing marks aligned **7** & **8**.
7. Slacken tensioner nut **9**. Turn tensioner clockwise with screwdriver. Lightly tighten nut.
8. Remove:
   ❏ Vacuum pump **10**.
   ❏ Injection pump belt.

Installation

Injection Pump Belt

NOTE: If injection pump sprocket removed, ensure sprocket fitted with key located in keyway marked 'B'.

1. Ensure timing marks aligned **2**, **7** & **8**.
2. Fit injection pump belt. Ensure belt is taut on non-tensioned side. Align marks on belt with marks on sprockets **7** & **8**. Arrow on belt must point to front of engine.
3. Install vacuum pump **10**. Tighten bolts to 17-21 Nm.
   NOTE: There should be 29 teeth between timing marks on sprockets.
4. Slacken tensioner nut **9**. Allow tensioner to operate.
5. Turn crankshaft slowly two turns clockwise.
6. Hold tensioner with screwdriver and tighten nut **9**. Tightening torque: 16-21 Nm.
7. Fit injection pump belt cover.
8. Install components in reverse order of removal.

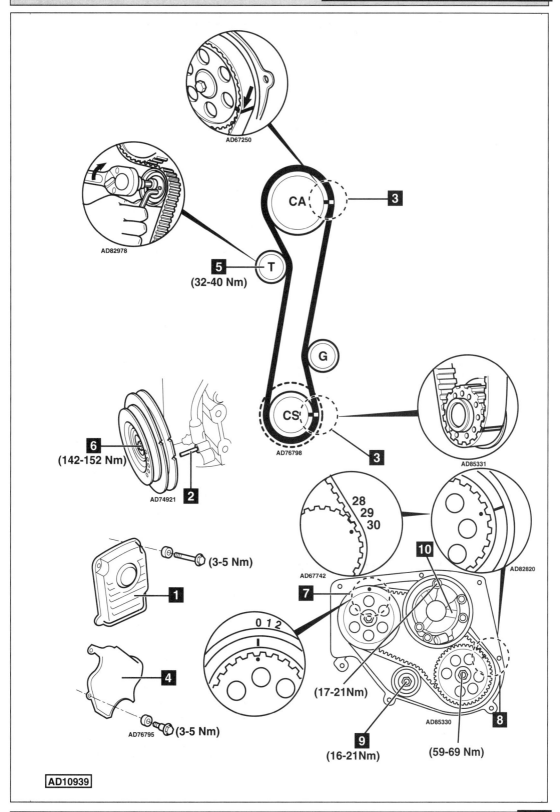

AD67250

AD82978

3

CA

5
T
(32-40 Nm)

G

CS

AD76798

3

AD85331

6
(142-152 Nm)

AD74921

2

28
29
30

AD67742

10

AD82820

(3-5 Nm)

1

7

0 1 2

4

AD76795 (3-5 Nm)

(17-21Nm)

9
(16-21Nm)

AD85330

8

(59-69 Nm)

AD10939

NISSAN

Model:	**Bluebird 2,0 D/Turbo D (U11) • Bluebird 2,0 D (T12/T72)**
	Vanette 2,0 D (C220)
Year:	**1984-96**
Engine Code:	**LD20, LD20T**

Replacement Interval Guide

Nissan recommend replacement as follows:
Check & replacement if necessary, every 36,000 miles.
Pre 12/91 - replacement every 60,000 miles.
1/92 on - replacement every 54,000 miles or 60 months, whichever occurs first.
The previous use and service history of the vehicle must always be taken into account.
Refer to Timing Belt Replacement Intervals at the front of this manual.

Check For Engine Damage

CAUTION: This engine has been identified as an INTERFERENCE engine in which the possibility of valve-to-piston damage in the event of a timing belt failure is MOST LIKELY to occur.
A compression check of all cylinders should be performed before removing the cylinder head.

Repair Times – hrs

Bluebird (U11):
Remove & install	2,20
PAS	+0,20
AC	+0,20

Bluebird (T12/T72):
Remove & install	2,10
PAS	+0,20
AC	+0,20

Vanette (C220):
Remove & install	4,80
PAS	+0,20
AC	+0,20

Special Tools

- None required.

Special Precautions

- Disconnect battery earth lead.
- DO NOT turn crankshaft or camshaft when timing belt removed.
- Remove glow plugs to ease turning engine.
- Turn engine in normal direction of rotation (unless otherwise stated).
- DO NOT turn engine via camshaft or other sprockets.
- Observe all tightening torques.
- Check diesel injection pump timing after belt replacement.

Removal

1. Raise and support front of vehicle.
2. Remove:
   - Bluebird:
     - ❏ RH front wheel.
     - ❏ Inner wing panel.
     - ❏ Upper front engine mounting.
   - Vanette:
     - ❏ Seat and floor panel.
     - ❏ Radiator and cowling.
   - All models:
     - ❏ Auxiliary drive belts.
     - ❏ Cylinder head cover.
     - ❏ Crankshaft pulley bolt **1**.
     - ❏ Crankshaft pulley **2**.
     - ❏ Water pump pulley.
     - ❏ Timing belt upper cover **3**.
     - ❏ Timing belt lower cover **4**.
3. Turn crankshaft clockwise to TDC on No.1 cylinder. Ensure timing marks aligned **5**, **6** & **7**.
4. Slacken tensioner bolt **8**. Turn tensioner clockwise with Allen key. Lightly tighten bolt.
5. Remove:
   - ❏ Crankshaft sprocket guide washer.
   - ❏ Timing belt.

Installation

1. Ensure timing marks aligned **5**, **6** & **7**.
2. Fit timing belt. Letter 'F' stamped on belt must face front of engine and NISSAN mark must be positioned between camshaft and injection pump sprockets.
3. Ensure marks on belt aligned with marks on sprockets **5**, **6** & **7**.
4. Slacken tensioner bolt **8**.
5. Turn tensioner anti-clockwise until belt is tensioned slightly.
6. Turn crankshaft two turns clockwise. Ensure timing marks aligned **5**, **6** & **7**.
7. Hold tensioner using Allen key. Tighten tensioner bolt **8**. Tightening torque: 31-39 Nm.
   NOTE: If tensioner turns when bolt is tightened, tension may be excessive.
8. Install components in reverse order of removal.
9. Tighten crankshaft pulley bolt **1**.
   Tightening torque: 137-157 Nm.

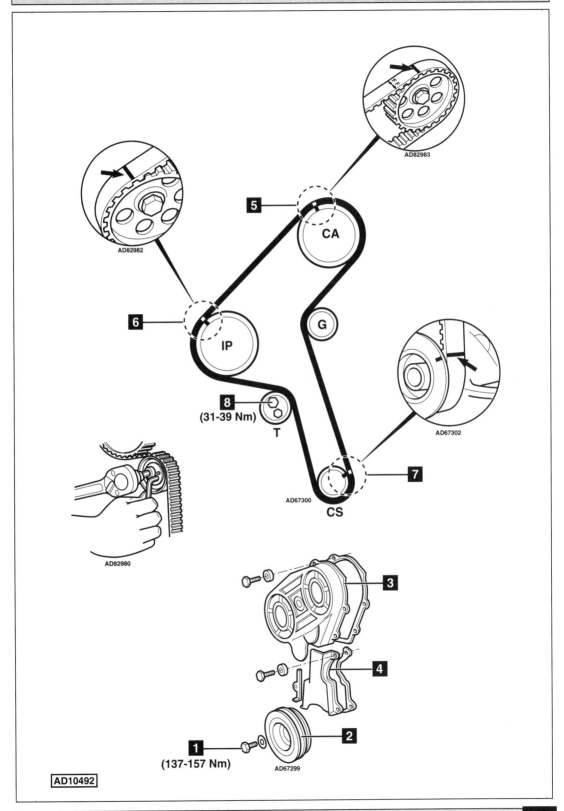

AD82983

AD82982

5

CA

6

IP

G

8
(31-39 Nm)

T

AD67302

7

AD67300

CS

AD82980

3

4

1
(137-157 Nm)

2

AD67299

AD10492

NISSAN

Model:	**Serena 2,0/2,3D • Vanette Cargo 2,3D**
Year:	**1992-02**
Engine Code:	**LD20-II, LD23**

Replacement Interval Guide

Nissan recommend replacement every 54,000 miles or 60 months, whichever occurs first.
The previous use and service history of the vehicle must always be taken into account.
Refer to Timing Belt Replacement Intervals at the front of this manual.

Check For Engine Damage

CAUTION: This engine has been identified as an INTERFERENCE engine in which the possibility of valve-to-piston damage in the event of a timing belt failure is MOST LIKELY to occur.
A compression check of all cylinders should be performed before removing the cylinder head.

Repair Times – hrs

Remove & install	1,70
PAS	+0,30
AC	+0,20

Special Tools

- None required.

Special Precautions

- Disconnect battery earth lead.
- DO NOT turn crankshaft or camshaft when timing belt removed.
- Remove glow plugs to ease turning engine.
- Turn engine in normal direction of rotation (unless otherwise stated).
- DO NOT turn engine via camshaft or other sprockets.
- Observe all tightening torques.
- Check diesel injection pump timing after belt replacement.

Removal

1. Remove:
   - ❏ Engine undershield.
   - ❏ Auxiliary drive belt(s).
   - ❏ Viscous fan.
   - ❏ Water pump pulley.
   - ❏ Crankshaft pulley bolt **1**.
   - ❏ Crankshaft pulley **2**.
   - ❏ Timing belt upper cover **3**.
   - ❏ Timing belt lower cover **4**.
2. Turn crankshaft clockwise to TDC on No.1 cylinder.
   NOTE: Check crankshaft keyway is at 12 o'clock position.

3. Mark timing belt with chalk or paint against punch marks on sprockets **5**, **6** & **7**.
4. Slacken tensioner bolt **8**. Turn tensioner away from belt **9**. Use Allen key. Lightly tighten bolt.
5. Remove crankshaft sprocket and timing belt.

Installation

1. Ensure crankshaft at TDC on No.1 cylinder.
   NOTE: Check crankshaft keyway is at 12 o'clock position.
2. Install crankshaft sprocket with timing belt.
3. Fit timing belt in anti-clockwise direction, starting at crankshaft sprocket. Ensure paint marks on belt aligned with punch marks on sprockets **5**, **6** & **7**.
   NOTE: New belts are marked with white lines to ensure correct alignment with punch marks on sprockets. Ensure 'F' mark on belt faces forward.
4. Slacken tensioner bolt **8**. Allow tensioner to operate.
5. Fit crankshaft pulley bolt **1**.
6. Turn crankshaft two turns clockwise to TDC on No.1 cylinder.
   NOTE: Ensure crankshaft keyway is at 12 o'clock position.
7. Hold tensioner using Allen key **9**. Tighten tensioner bolt **8**. Tightening torque: 31-39 Nm.
   NOTE: If tensioner turns when bolt is tightened, tension may be excessive.
8. Remove crankshaft pulley bolt **1**.
9. Fit timing belt covers. Tighten bolts **10**. Tightening torque: 7-9 Nm.
10. Install components in reverse order of removal.
11. Tighten crankshaft pulley bolt **1**. Tightening torque: 137-157 Nm.

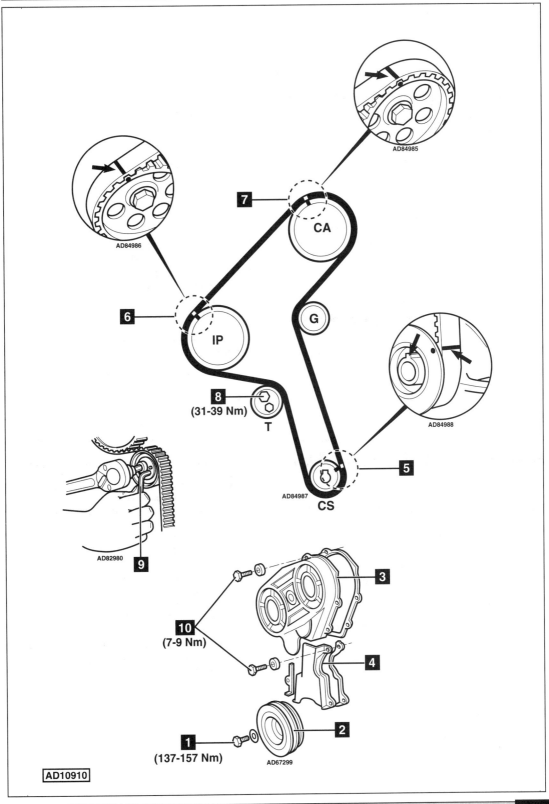

AD84985

AD84986

7

CA

6

IP

G

AD84988

8

(31-39 Nm)

T

AD84987

CS

5

AD82980

9

10

(7-9 Nm)

3

4

1

(137-157 Nm)

2

AD67299

AD10910

Model:	Patrol (260) 2,8D/TD • Patrol (Y60) 2,8TD • Patrol GR (Y61) 2,8TD
Year:	1989-00
Engine Code:	RD28, RD28T, RD28ETI

Replacement Interval Guide

Nissan recommend replacement as follows:
Pre 12/91 - replacement every 60,000 miles.
1/92 on - replacement every 54,000 miles or 60 months, whichever occurs first.
The previous use and service history of the vehicle must always be taken into account.
Refer to Timing Belt Replacement Intervals at the front of this manual.

Check For Engine Damage

CAUTION: This engine has been identified as an INTERFERENCE engine in which the possibility of valve-to-piston damage in the event of a timing belt failure is MOST LIKELY to occur.
A compression check of all cylinders should be performed before removing the cylinder head.

Repair Times – hrs

Remove & install:

Patrol (260)	4,80
Patrol (Y60)	1,30
PAS	+0,20
AC	+0,20
Patrol GR (Y61)	1,30
PAS	+0,10
AC	+0,10

Special Tools

■ None required.

Special Precautions

■ Disconnect battery earth lead.
■ DO NOT turn crankshaft or camshaft when timing belt removed.
■ Remove glow plugs to ease turning engine.
■ Turn engine in normal direction of rotation (unless otherwise stated).
■ DO NOT turn engine via camshaft or other sprockets.
■ Observe all tightening torques.
■ Check diesel injection pump timing after belt replacement.

Removal

1. Drain coolant.
2. Remove:
   ❏ Top radiator hose.
   ❏ Air hose to air filter.
   ❏ Radiator shroud.
   ❏ Cooling fan.
   ❏ Viscous coupling.
   ❏ Auxiliary drive belt(s).
   ❏ Water pump pulley.

3. Turn crankshaft clockwise to BDC on No.1 cylinder (power stroke). Ensure timing mark aligned with bolt of timing belt lower cover **1**.
4. Remove:
   ❏ Crankshaft pulley bolt **2**.
   ❏ Crankshaft pulley **3**.
   ❏ Timing belt upper cover **4**.
   ❏ Timing belt lower cover **5**.
5. Check crankshaft keyway is at 6 o'clock position **6**.
6. Mark belt with chalk or paint against timing marks on sprockets **7**, **8** & **9**.
7. Remove tensioner spring **10**.
8. Slacken tensioner nut **11**. Move tensioner away from belt and lightly tighten nut.
9. Remove timing belt.

Installation

1. Ensure crankshaft at BDC on No.1 cylinder. Check crankshaft keyway is at 6 o'clock position **6**.
2. Fit timing belt in anti-clockwise direction, starting at crankshaft sprocket. Ensure marks on belt aligned with timing marks on sprockets **7**, **8** & **9**.
   NOTE: New belts are marked with white lines to ensure correct alignment with timing marks on sprockets. Ensure belt teeth between marks on sprockets correspond with those shown in illustration. Arrow on belt must point away from engine **12**.
3. Fit tensioner spring **10**.
4. Slacken tensioner nut **11**. Allow tensioner to operate.
5. Turn crankshaft two turns clockwise to BDC on No.1 cylinder. Check crankshaft keyway is at 6 o'clock position **6**.
6. Tighten tensioner nut **11**.
   Tightening torque: 32-40 Nm.
7. Install components in reverse order of removal.
8. Tighten crankshaft pulley bolt **2**.
   Tightening torque: 142-152 Nm.

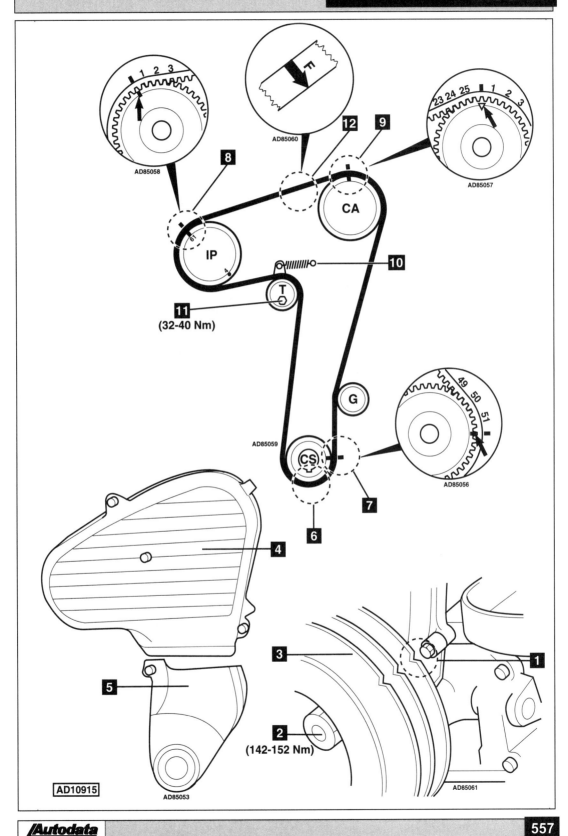

AD85058

AD85060

12 9

8

CA

AD85057

23 24 25 1 2 3

IP

10

T

11
(32-40 Nm)

G

AD85059

CS

49 50 51

AD85056

7

6

4

5

3

1

2
(142-152 Nm)

AD85053

AD85061

AD10915

PEUGEOT

Model:	106 1,0/1,1/1,4 • 205 1,0/1,1/1,4 • 205 Rallye • 309 1,1/1,4 • 405 1,4
Year:	1987-94
Engine Code:	TU1/2/3/9

Replacement Interval Guide

Peugeot recommend replacement as follows:

→1993:
Manufacturer has no recommended interval.

1993-94:
Replacement every 72,000 miles under normal conditions.
Replacement every 45,000 miles under adverse conditions.

1994→:
Replacement every 72,000 miles under normal conditions.
Replacement every 54,000 miles under adverse conditions.
The previous use and service history of the vehicle must always be taken into account.
Refer to Timing Belt Replacement Intervals at the front of this manual.

Check For Engine Damage

CAUTION: This engine has been identified as an INTERFERENCE engine in which the possibility of valve-to-piston damage in the event of a timing belt failure is MOST LIKELY to occur.
A compression check of all cylinders should be performed before removing the cylinder head.

Repair Times – hrs

Remove & install:

106/205/309	1,60
205 Rallye	1,70
405	1,20

Special Tools

- Camshaft and flywheel timing pins – Peugeot No.9767.27 (0132 R & Q).
- Tensioning tool – Peugeot No.9767.89 (0132X).
- 1,5 kg weight – for use with tensioning tool.

Special Precautions

- Disconnect battery earth lead.
- DO NOT turn crankshaft or camshaft when timing belt removed.
- Remove spark plugs to ease turning engine.
- Turn engine in normal direction of rotation (unless otherwise stated).
- DO NOT turn engine via camshaft or other sprockets.
- Observe all tightening torques.

Removal

1. Raise and support front of vehicle.
2. Remove:
   - RH front wheel.
   - Lower inner wing panel.
   - Alternator drive belt.
   - Crankshaft pulley **1**.
   - Timing belt upper cover **2**.
   - Timing belt centre cover **3**.
   - Timing belt lower cover **4**.
3. Turn crankshaft to setting position. Insert timing pin in camshaft sprocket **5**.
4. Insert timing pin in flywheel **6**.
5. Slacken tensioner nut **7**. Turn tensioner to release tension on belt.
6. Remove timing belt.

Installation

1. Ensure timing pins located correctly **5** & **6**.
2. Fit timing belt in following order:
   - Crankshaft sprocket.
   - Camshaft sprocket.
   - Water pump sprocket.
   - Tensioner pulley.
   NOTE: Ensure belt is taut between sprockets on non-tensioned side.
3. Ensure tensioner nut slackened **7**. Fit tensioning tool and weight to tensioner pulley to pre-tension belt **8**. Tool No.9767.89 (0132X).
4. Lightly tighten tensioner nut **7**.
5. Remove timing pins **5** & **6**. Turn crankshaft two turns clockwise.
6. Fit flywheel timing pin **6**.
7. Slacken tensioner nut **7**. Gently press belt at ▽.
8. Check camshaft has turned slightly by trying to insert timing pin in camshaft sprocket **5**. If timing pin cannot be inserted completely, camshaft has turned sufficiently.
9. Tighten tensioner nut **7**.
10. Remove timing pins **5** & **6**.
11. Turn crankshaft two turns clockwise. Fit flywheel timing pin **6**.
12. Gradually slacken tensioner nut **7**. Allow tensioning tool to move under influence of weight to tension belt. Tighten nut to 20 Nm.
13. Remove:
    - Flywheel timing pin **6**.
    - Tensioning tool and weight.
14. Install components in reverse order of removal.
15. Tighten crankshaft pulley bolts to 15 Nm **9**.

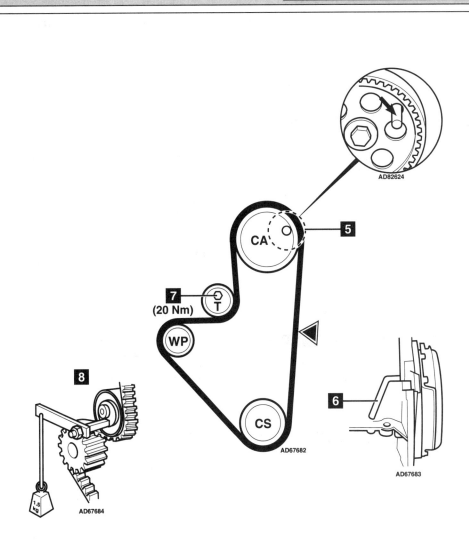

AD82624

5

CA

7
T
(20 Nm)

WP

8

AD67684

1.5 kg

6

AD67683

CS

AD67682

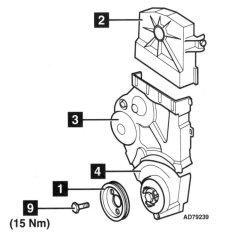

2

3

4

1

9
(15 Nm)

AD79239

AD10494

PEUGEOT

Model:	106 1,0/1,1/1,4/1,6 • 205 1,1/1,4 • 306 1,1/1,4/1,6 • 307 1,4 • 405 1,4 Partner 1,1/1,4
Year:	1993-03
Engine Code:	TU1/3/5/9

Replacement Interval Guide

Peugeot recommend:

→07/98:

Replacement every 72,000 miles under normal conditions.

Replacement every 54,000 miles under adverse conditions.

08/98→:

Replacement every 80,000 miles under normal conditions.

Replacement every 48,000 miles under adverse conditions.

The previous use and service history of the vehicle must always be taken into account.

Refer to Timing Belt Replacement Intervals at the front of this manual.

Check For Engine Damage

CAUTION: This engine has been identified as an INTERFERENCE engine in which the possibility of valve-to-piston damage in the event of a timing belt failure is MOST LIKELY to occur.

A compression check of all cylinders should be performed before removing the cylinder head.

Repair Times – hrs

Remove & install:	
106 (→07/98)	1,60
106 (08/98→)	1,50
205	1,60
306	1,20
307	1,40
Engine undershield	+0,10
AC	+0,50
PAS + AC	+1,00
405	1,20
Partner	1,40
AC	+0,30

Special Tools

- Timing pins – Peugeot No.0132Q/Z & R.
- 307: Flywheel timing pin – Peugeot No.0132Q/Y.
- Tension gauge – Peugeot No.(-).0192/ SEEM C.Tronic 105.5.
- Valve spring compressor (1997→) – Peugeot No.0132AE.

Special Precautions

- Disconnect battery earth lead.
- DO NOT turn crankshaft or camshaft when timing belt removed.
- Remove spark plugs to ease turning engine.
- Turn engine in normal direction of rotation (unless otherwise stated).
- DO NOT turn engine via camshaft or other sprockets.
- Observe all tightening torques.

Removal

1. Raise and support front of vehicle.
2. Remove:
   - ❏ RH front wheel.
   - ❏ Engine undershield (if fitted).
   - ❏ Lower inner wing panel.
   - ❏ Auxiliary drive belt(s).
   - ❏ 307: Engine tie-bars.
   - ❏ Crankshaft pulley **1**.
   - ❏ Timing belt upper cover **2**.
   - ❏ Timing belt centre cover **3**.
   - ❏ Timing belt lower cover **4**.
3. Turn crankshaft to setting position. Insert timing pins in camshaft sprocket and flywheel **5** & **6**.
4. Slacken tensioner nut to release tension on belt **7**. Lightly tighten nut.
5. Remove timing belt.

Installation

1. Ensure timing pins located correctly **5** & **6**.
2. Fit timing belt in following order:
   - ❏ Crankshaft sprocket.
   - ❏ Camshaft sprocket.
   - ❏ Water pump sprocket.
   - ❏ Tensioner pulley.

   NOTE: Ensure belt is taut between sprockets on non-tensioned side.
3. Slacken tensioner nut **7**. Turn tensioner anti-clockwise to tension belt. Temporarily tighten nut.
4. Remove timing pins **5** & **6**.
5. Attach tension gauge to belt at ▽ **8**. Tool No.(-).0192/SEEM C.Tronic 105.5.
6. Slacken tensioner nut **7**.

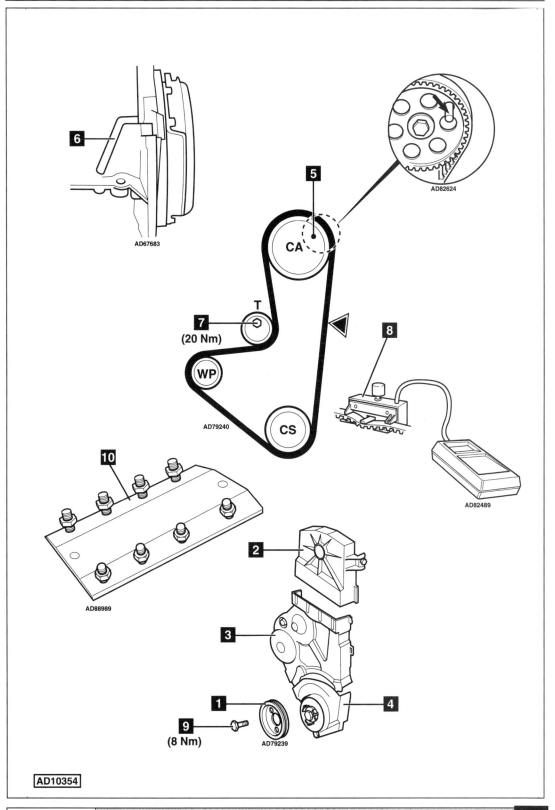

7. Tension belt as follows:

→1997:

8. Turn tensioner anti-clockwise until tension gauge indicates 45 SEEM units.
9. Tighten tensioner nut to 20 Nm **7**.

1997→:

10. Remove cylinder head cover.
11. Install valve spring compressor. Tool No.0132AE **10**.
12. Turn tensioner anti-clockwise until tension gauge indicates the following:
    ❑ Except 307: 31 SEEM units.
    ❑ 307: 44 SEEM units.
13. Tighten tensioner nut to 20 Nm **7**.
14. Remove valve spring compressor.

All models:

15. Remove tension gauge.
16. Turn crankshaft four turns clockwise to setting position.
17. Ensure timing pins can be inserted **5** & **6**.

→1997:

18. Wait 1 minute to allow belt to settle.
19. Apply firm thumb pressure to belt at ▽.
20. Attach tension gauge to belt at ▽ **8**. Tool No.(-).0192/SEEM C.Tronic 105.5.
21. Slacken tensioner nut **7**.
22. Turn tensioner anti-clockwise to tension belt. New belt: 40 SEEM units. Used belt: 36 SEEM units.
23. Tighten tensioner nut to 20 Nm **7**.

1997→:

24. Install valve spring compressor. Tool No.0132AE **10**.
25. Attach tension gauge to belt at ▽ **8**. Tool No.(-).0192/SEEM C.Tronic 105.5.
26. Tension gauge should indicate 29-33 SEEM units.
27. Remove valve spring compressor.

All models:

28. Remove tension gauge **8**.

→1997:

29. Turn crankshaft two turns clockwise.
30. Ensure flywheel timing pin can be inserted **6**.
    NOTE: DO NOT allow crankshaft to turn anti-clockwise.

31. Attach tension gauge to belt at ▽ **8**. Tool No.(-).0192/SEEM C.Tronic 105.5.
32. Check tension gauge reading. New belt: 51±3 SEEM units. Used belt: 45±3 SEEM units.

All models:

33. Install components in reverse order of removal.
34. Tighten crankshaft pulley bolts to 8 Nm **9**.

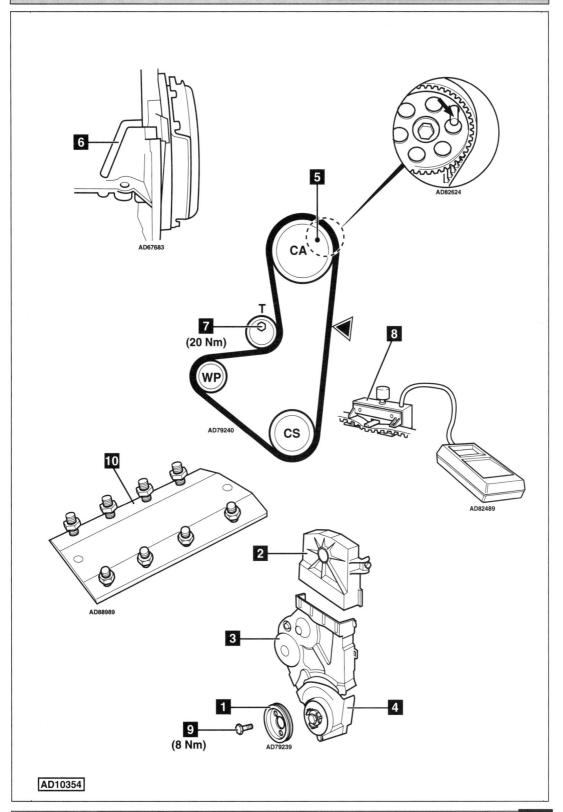

6 — AD67683

5

AD82624

CA

T

7 (20 Nm)

WP

CS

AD79240

8

AD82489

10

AD88989

2

3

1

9 (8 Nm)

AD79239

4

AD10354

PEUGEOT

Model:	**106 1,6 GTi**
Year:	**1997-03**
Engine Code:	**TU5J4/L3 (NFX)**

Replacement Interval Guide

Peugeot recommend:

→07/98:

Replacement every 72,000 miles under normal conditions.

Replacement every 54,000 miles under adverse conditions.

08/98→:

Replacement every 80,000 miles under normal conditions.

Replacement every 48,000 miles under adverse conditions.

The previous use and service history of the vehicle must always be taken into account.

Refer to Timing Belt Replacement Intervals at the front of this manual.

Check For Engine Damage

CAUTION: This engine has been identified as an INTERFERENCE engine in which the possibility of valve-to-piston damage in the event of a timing belt failure is MOST LIKELY to occur.

A compression check of all cylinders should be performed before removing the cylinder head(s).

Repair Times – hrs

Remove & install:

→07/98	2,30
08/98→	2,20

Special Tools

- Camshaft timing pin (CA1) – Peugeot No.(-).0132.AJ1.
- Camshaft timing pin (CA2) – Peugeot No.(-).0132.AJ2.
- Flywheel timing pin – Peugeot No.(-).0132.QZ.
- Tensioning tool – Peugeot No.(-).0132X1Z.
- Timing belt retaining clip – Peugeot No.(-).0132AK.
- Tension gauge – SEEM C.Tronic 105.5.

Special Precautions

- Disconnect battery earth lead.
- DO NOT turn crankshaft or camshaft when timing belt removed.
- Remove spark plugs to ease turning engine.
- Turn engine in normal direction of rotation (unless otherwise stated).
- DO NOT turn engine via camshaft or other sprockets.
- Observe all tightening torques.

Removal

1. Raise and support front of vehicle.
2. Drain coolant.
3. Remove:
   - ❏ Air filter housing.
   - ❏ Engine control module (ECM) and tray.
   - ❏ Top radiator hose.
   - ❏ Air filter intake duct.
   - ❏ Auxiliary drive belt.
   - ❏ Crankshaft pulley bolts **1**.
   - ❏ Crankshaft pulley **2**.
   - ❏ Alternator bracket.
   - ❏ Timing belt upper cover **3**.
   - ❏ Timing belt lower cover **4**.
   - ❏ Exhaust manifold heat shield.
4. Unclip AC pipe (if fitted). Push AC pipe to one side.
5. Turn crankshaft clockwise to setting position.
6. Insert timing pins in camshaft sprockets **6**. Tool Nos.(-).0132 AJ1/2.
7. Insert timing pin in flywheel **5**. Tool No.(-).0132.QZ.
8. Slacken tensioner bolt **7**. Move tensioner away from belt. Lightly tighten bolt.
9. Remove timing belt.

Installation

1. Ensure timing pins located correctly **5** & **6**.
2. Slacken bolts of each camshaft sprocket slightly **8**.
3. Turn camshaft sprockets fully clockwise in slotted holes.
   NOTE: Sprockets should turn with slight resistance.
4. Fit timing belt. Mark on belt should be aligned with mark on crankshaft **9**.
5. Secure belt to crankshaft sprocket with retaining clip. Tool No.(-).0132AK **10**.
6. Fit timing belt to guide pulley.
7. Fit timing belt to camshaft sprocket CA1 then CA2. Align marks on belt with marks on sprockets **11**.
   NOTE: Ensure belt is taut between sprockets on non-tensioned side.
8. Fit timing belt to water pump sprocket and tensioner pulley.
9. Remove retaining clip from timing belt **10**.
10. Attach tension gauge to belt at ▽. Tool No.SEEM C.Tronic 105.5 **13**.

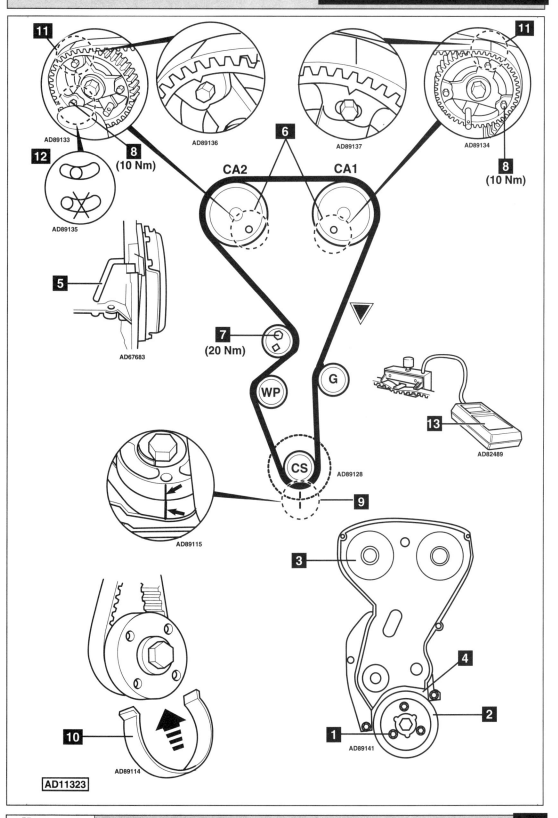

AD89133

11

12

AD89135

8
(10 Nm)

AD89136

6

CA2

CA1

AD89137

AD89134

11

8
(10 Nm)

5

AD67683

7
(20 Nm)

WP

G

CS

AD89128

13

AD82489

AD89115

9

3

4

2

10

AD89114

1

AD89141

AD11323

←

11. Insert tensioning tool in square hole of tensioner pulley. Tool No.(-).0132X1Z.

12. Turn tensioner anti-clockwise until tension gauge indicates 63 SEEM units.

13. Tighten tensioner bolt to 20 Nm .

14. Ensure bolts of each camshaft sprocket not at end of slotted holes 🖬.

15. If necessary: Repeat installation procedure.

16. Tighten bolts of each camshaft sprocket 🖪. Tightening torque: 10 Nm

17. Ensure marks on belt aligned with marks on sprockets 🖪 & 🖬.

18. Remove:
    ❏ Tension gauge 🖬.
    ❏ Timing pins 🖪 & 🖬.

19. Turn crankshaft four turns clockwise to setting position.
    NOTE: DO NOT allow crankshaft to turn anti-clockwise.

20. Insert timing pin in flywheel 🖪. Tool No.(-).0132.QZ.

21. Slacken bolts of each camshaft sprocket slightly 🖪.

22. Insert timing pins in camshaft sprockets 🖪. Tool Nos.(-).0132 AJ1/2
    NOTE: If timing pins cannot be inserted: Turn camshaft sprocket flanges (as required).

23. Attach tension gauge to belt at ▽. Tool No.SEEM C.Tronic 105.5 🖬.

24. Slacken tensioner bolt 🖪.

25. Insert tensioning tool in square hole of tensioner pulley. Tool No.(-).0132X1Z.

26. Turn tensioner anti-clockwise until tension gauge indicates 37 SEEM units.

27. Tighten tensioner bolt to 20 Nm 🖪.

28. Tighten bolts of each camshaft sprocket 🖪. Tightening torque: 10 Nm.

29. Remove tools 🖪, 🖪 & 🖬.

30. Install components in reverse order of removal.

31. Refill and bleed cooling system.

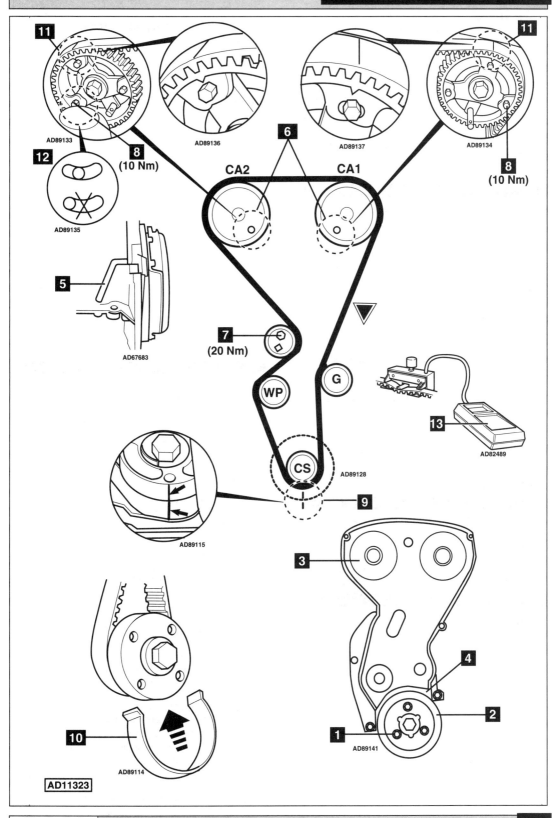

CA2　CA1

AD89133

AD89136

AD89137

AD89134

6

11

11

8
(10 Nm)

8
(10 Nm)

12

AD89135

5

AD67683

7
(20 Nm)

WP

G

CS

AD89128

AD82489

13

AD89115

9

3

4

2

1

AD89141

10

AD89114

AD11323

PEUGEOT

Model:	205/309 1,6/1,9 (1992→) • 306 1,8/2,0 • 405 1,6/1,8/1,9/2,0 (1992→) 406 1,8 • 605 2,0 • 605 2,0 Turbo • 806 1,8 • Partner 1,8 • Boxer 2,0
Year:	1989-03
Engine Code:	**Eccentric tensioner** **XU5 – B6D/E, B1A, B1E, B2A, B2B, BDY, BDZ** **XU7 – L6A, LFW, LFX, LFZ** **XU9 – D2H, D6A, D6B, D6D, DDZ, DKZ, DFZ** **XU10 – R2A, R5A, R6A, RDZ, RFW, RFX, RFZ, RGX, RGY**

Replacement Interval Guide

Peugeot recommend replacement as follows:

→1993:
Manufacturer has no recommended interval.

1993:
Replacement every 72,000 miles under normal conditions.
Replacement every 45,000 miles under adverse conditions.

1994-07/98:
Replacement every 72,000 miles under normal conditions.
Replacement every 60,000 miles under adverse conditions.

08/98→:
Replacement every 80,000 miles under normal conditions.
Replacement every 48,000 miles under adverse conditions.
The previous use and service history of the vehicle must always be taken into account.
Refer to Timing Belt Replacement Intervals at the front of this manual.

Check For Engine Damage

CAUTION: This engine has been identified as an INTERFERENCE engine in which the possibility of valve-to-piston damage in the event of a timing belt failure is MOST LIKELY to occur.
A compression check of all cylinders should be performed before removing the cylinder head.

Repair Times – hrs

Remove & install:

205/309 (carburettor)	2,60
205/309 (fuel injection)	2,90
306	2,60
Engine undershield	+0,10
AC	+0,50
405	2,60
AC	+0,10
PAS	+0,10
406	2,60
Engine undershield	+0,10
AC	+0,30

605 (carburettor)	2,40
605 (fuel injection →07/98)	2,70
605 (fuel injection 08/98→)	2,60
AC	+0,90
806 (→07/98)	2,40
806 (08/98→)	2,00
AC	+0,10
Boxer (→07/98)	2,20
Boxer (08/98→)	2,00
PAS + AC	+0,30
Partner	1,30
AC	+0,30

Special Tools

- Camshaft timing pin – Peugeot No.(-).0132.R.
- Crankshaft timing pin – Peugeot No.(-).0153.G or ZY.
- Tension gauge – SEEM No.9797.56/.70 or C.Tronic 105.5.
- Flywheel locking tool – Facom No.D.86.

Special Precautions

- Disconnect battery earth lead.
- DO NOT turn crankshaft or camshaft when timing belt removed.
- Remove spark plugs to ease turning engine.
- Turn engine in normal direction of rotation (unless otherwise stated).
- DO NOT turn engine via camshaft or other sprockets.
- Observe all tightening torques.

Removal

1. Raise and support front of vehicle.
2. Remove:
   - ❑ RH front wheel.
   - ❑ Engine undershield (if fitted).
   - ❑ RH inner wing panel.
   - ❑ Auxiliary drive belt(s).
   - ❑ Auxiliary drive belt tensioner (if fitted).
   - ❑ Timing belt upper cover **1**.
3. Turn crankshaft clockwise to setting position.
4. Insert timing pin in crankshaft pulley **2**.
   Tool No.(-).0153.G or ZY.
   NOTE: Some models have crankshaft pulley with multiple holes. Ensure correct hole used (8 mm diameter) when inserting timing pin.

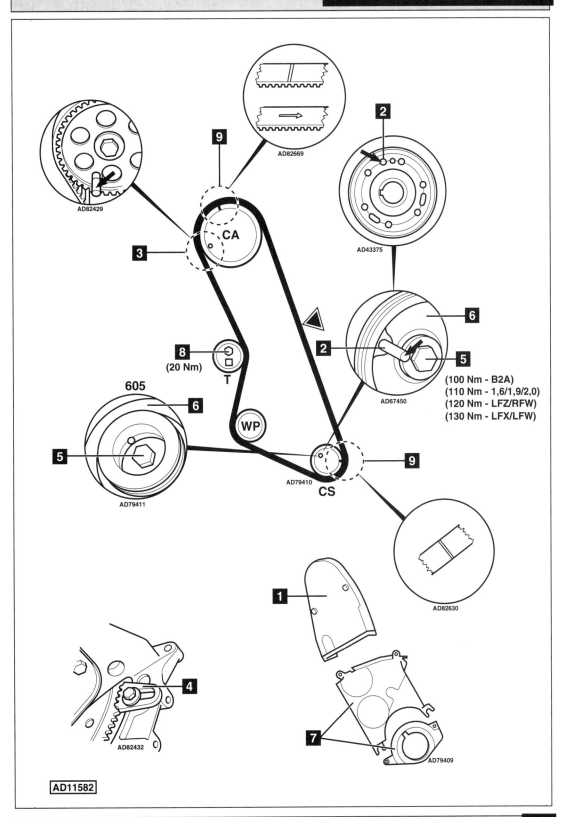

AD82669

AD82429

9

CA

3

2

AD43375

6

2

8
(20 Nm)
T

5

(100 Nm - B2A)
(110 Nm - 1,6/1,9/2,0)
(120 Nm - LFZ/RFW)
(130 Nm - LFX/LFW)

AD67450

605

6

5

AD79411

WP

AD79410

CS

9

AD82630

1

4

AD82432

7

AD79409

AD11582

5. Insert timing pin in camshaft sprocket **3**.
   Tool No. (-).0132.R.
6. Remove clutch housing lower plate. Fit flywheel locking tool **4**. Tool No.Facom D.86.
7. Remove:
   - ❑ Timing pins **2** & **3**.
   - ❑ Crankshaft pulley bolt **5**.
   - ❑ Crankshaft pulley **6**.
8. Insert timing pin in camshaft sprocket **3**.
9. Remove timing belt covers **7**.
10. Slacken tensioner bolt **8**. Move tensioner away from belt. Lightly tighten bolt.
11. Remove:
    - ❑ Timing belt.
    - ❑ Flywheel locking tool **4**.

Installation

1. Temporarily fit crankshaft pulley and timing pin **6** & **2**.
2. Ensure timing pin located correctly in camshaft sprocket **3**.
3. Remove:
   - ❑ Crankshaft pulley **6**.
   - ❑ Timing pin **2**.
4. Fit timing belt in following order:
   - ❑ Camshaft sprocket.
   - ❑ Crankshaft sprocket.
   - ❑ Tensioner pulley.
   - ❑ Water pump sprocket.
   NOTE: Ensure belt is taut between sprockets. Observe direction of rotation marks on belt. Ensure marks on belt aligned with marks on sprockets 9.
5. Slacken tensioner bolt **8**.
6. Turn tensioner anti-clockwise. Lightly tighten bolt.
7. Attach tension gauge to belt at ▽.
8. Turn tensioner anti-clockwise until tension gauge indicates 30±2 SEEM units. Use 8 mm spanner.
   NOTE: 2,0 (XU10J2 - RFW/RFX): Tension belt to 18 SEEM units.
9. Tighten tensioner bolt **8**.
   Tightening torque: 20 Nm.
10. Remove tension gauge.
11. Remove timing pin from camshaft sprocket **3**.
12. Temporarily fit crankshaft pulley.
13. Turn crankshaft two turns clockwise.
14. Ensure timing pins can be inserted easily **2** & **3**.
15. Remove timing pins. Turn crankshaft two turns clockwise.
16. Insert timing pin in camshaft sprocket **3**.
17. Attach tension gauge to belt at ▽.
18. Tension gauge should indicate 44±2 SEEM units.

19. If not: Repeat operations 5-18.
20. If tension as specified: Remove tension gauge and timing pin **3**.
21. Remove crankshaft pulley **6**.
22. Install:
    - ❑ Flywheel locking tool **4**.
    - ❑ Timing belt covers **7**.
    - ❑ Crankshaft pulley **6**.
23. Coat crankshaft pulley bolt thread with suitable thread locking compound.
24. Tighten crankshaft pulley bolt **5**.
    Tightening torque: B2A: 100 Nm, LFZ/RFW: 120 Nm, LFW/LFX: 130 Nm, other engines: 110 Nm.
25. Remove flywheel locking tool **4**.
26. Install components in reverse order of removal.

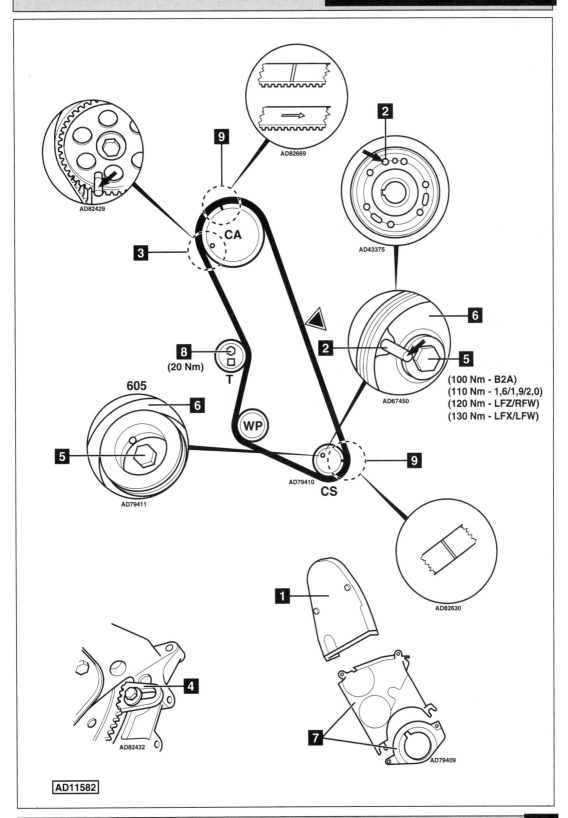

AD82669

AD82429

9

2

AD43375

CA

3

6

2

5

8
(20 Nm)
T

(100 Nm - B2A)
(110 Nm - 1,6/1,9/2,0)
(120 Nm - LFZ/RFW)
(130 Nm - LFX/LFW)

AD67450

605

6

5

AD79411

WP

AD79410

9

CS

AD82630

1

4

AD82432

7

AD79409

AD11582

Model:	**206 1,1/1,4/1,6**
Year:	**1998-03**
Engine Code:	**TU1/3/5 (HFY, HFZ, KFX, NFZ)**

Replacement Interval Guide

Peugeot recommend:
Replacement every 80,000 miles under normal conditions.
Replacement every 48,000 miles under adverse conditions.
The previous use and service history of the vehicle must always be taken into account.
Refer to Timing Belt Replacement Intervals at the front of this manual.

Check For Engine Damage

CAUTION: This engine has been identified as an INTERFERENCE engine in which the possibility of valve-to-piston damage in the event of a timing belt failure is MOST LIKELY to occur.
A compression check of all cylinders should be performed before removing the cylinder head.

Repair Times – hrs

Remove & install	1,30
AC & PAS	+0,20

Special Tools

- Flywheel timing pin – Peugeot No.(-).0132QZ.
- Camshaft timing pin – Peugeot No.(-).0132RZ.
- Valve spring compressor – Peugeot No.(-).0132AE.
- Tension gauge – SEEM C.Tronic 105.5.

Special Precautions

- Disconnect battery earth lead.
- DO NOT turn crankshaft or camshaft when timing belt removed.
- Remove spark plugs to ease turning engine.
- Turn engine in normal direction of rotation (unless otherwise stated).
- DO NOT turn engine via camshaft or other sprockets.
- Observe all tightening torques.

Removal

1. Remove:
   - ❏ Engine steady bar.
   - ❏ Auxiliary drive belt.
   - ❏ Engine control module (ECM). DO NOT disconnect cable.
   - ❏ Engine control module (ECM) bracket.
   - ❏ Crankshaft pulley bolts **1**.
   - ❏ Crankshaft pulley **2**.
   - ❏ Timing belt covers **3** & **4**.
2. Turn crankshaft clockwise to setting position.
3. Insert timing pin in camshaft sprocket **5**. Tool No.(-).0132RZ.
4. Insert flywheel timing pin **6**. Tool No.(-).0132QZ.
5. Slacken tensioner nut to release tension on belt **7**.
6. Move tensioner away from belt. Lightly tighten nut **7**.
7. Remove timing belt.

Installation

1. Ensure timing pins located correctly **5** & **6**.
2. Fit timing belt in following order:
   - ❏ Crankshaft sprocket.
   - ❏ Camshaft sprocket.
   - ❏ Water pump sprocket.
   - ❏ Tensioner pulley.
   NOTE: Ensure belt is taut between sprockets on non-tensioned side.
3. Attach tension gauge to belt at ▽. Tool No.SEEM C.Tronic 105.5 **8**.
4. Slacken tensioner nut **7**.
5. Turn tensioner anti-clockwise until tension gauge indicates 44 SEEM units.
6. Tighten tensioner nut to 22 Nm **7**.
7. Remove:
   - ❏ Timing pins **5** & **6**.
   - ❏ Tension gauge **8**.
8. Turn crankshaft four turns clockwise to setting position.
9. Insert timing pin in flywheel **6**.
   NOTE: DO NOT allow crankshaft to turn anti-clockwise.
10. Ensure timing pin can be inserted easily in camshaft sprocket **5**.
11. Remove:
    - ❏ Timing pin **5**.
    - ❏ Cylinder head cover.
12. Install valve spring compressor. Tool No.0132AE **9**.
    NOTE: Tighten bolts of tool sufficiently to free the cam followers from camshaft. Excessive tightening will cause damage to valves or pistons.
13. Attach tension gauge to belt at ▽ **8**.
14. Hold tensioner pulley in position. Slacken tensioner nut **7**.
15. Release tension on belt. Turn tensioner clockwise until tension gauge indicates 29-33 SEEM units.
16. Tighten tensioner nut to 22 Nm **7**.
17. Remove:
    - ❏ Valve spring compressor **9**.
    - ❏ Tension gauge **8**.
    - ❏ Timing pin **6**.
18. Turn crankshaft two turns clockwise.
19. Ensure timing pins can be inserted **5** & **6**.
20. Install components in reverse order of removal.

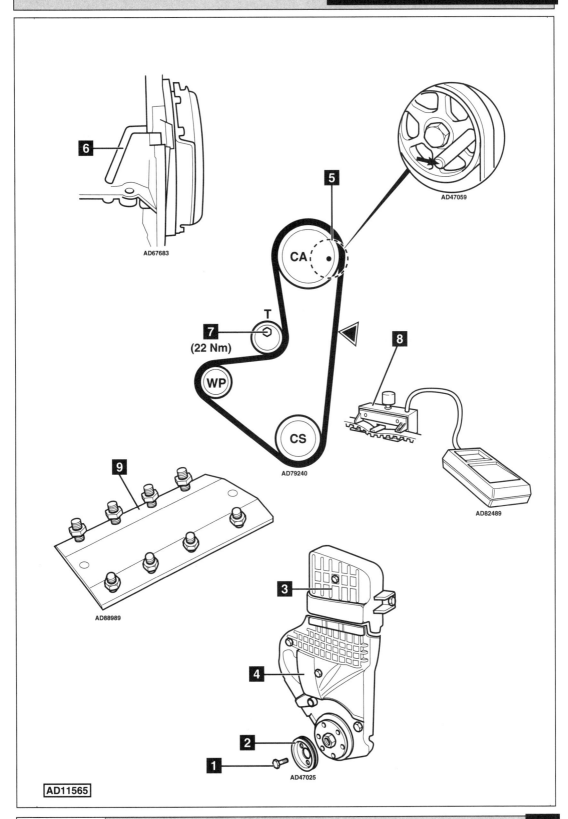

AD67683

AD47059

CA

T

(22 Nm)

WP

CS

AD79240

AD47025

AD82489

AD88989

AD11565

PEUGEOT

Model:	206 1,6 16V • 307 1,6 16V
Year:	2000-03
Engine Code:	TU5JP4 (NFU)

Replacement Interval Guide

Peugeot recommend replacement every 80,000 miles under normal conditions.
The previous use and service history of the vehicle must always be taken into account.
Refer to Timing Belt Replacement Intervals at the front of this manual.

Check For Engine Damage

CAUTION: This engine has been identified as an INTERFERENCE engine in which the possibility of valve-to-piston damage in the event of a timing belt failure is MOST LIKELY to occur.
A compression check of all cylinders should be performed before removing the cylinder head(s).

Repair Times – hrs

Remove & install	1,40

Special Tools

- Camshaft timing pin (CA1) – Peugeot No.(-).0132.AJ1.
- Camshaft timing pin (CA2) – Peugeot No.(-).0132.AJ2.
- Flywheel timing pin – Peugeot No.(-).0132.QY.
- Timing belt retaining clip – Peugeot No.(-).0132AK.

Special Precautions

- Disconnect battery earth lead.
- DO NOT turn crankshaft or camshaft when timing belt removed.
- Remove spark plugs to ease turning engine.
- Turn engine in normal direction of rotation (unless otherwise stated).
- DO NOT turn engine via camshaft or other sprockets.
- Observe all tightening torques.

Removal

1. Raise and support front of vehicle.
2. Remove:
   - ❑ RH wheel.
   - ❑ Lower inner wing panel.
   - ❑ Upper RH auxiliary drive belt cover.
   - ❑ Auxiliary drive belt.
   - ❑ Crankshaft pulley bolts **1**.
   - ❑ Crankshaft pulley **2**.
3. Support engine.
4. Unclip and reposition RH wiring harness.
5. Remove:
   - ❑ RH engine mounting.
   - ❑ Timing belt upper cover **3**.
   - ❑ Timing belt lower cover **4**.

6. Unclip AC pipe (if fitted). Push AC pipe to one side.
7. Turn crankshaft clockwise to setting position.
8. Insert timing pins in camshaft sprockets **6**. Tool Nos.(-).0132 AJ1/2.
9. Insert timing pin in flywheel **5**. Tool No.(-).0132.QY.
10. Slacken tensioner bolt **7**.
11. Turn tensioner pulley clockwise until pointer **A** at position **D**. Use Allen key **11**.
12. Remove timing belt.

Installation

1. Ensure timing pins located correctly **5** & **6**.
2. Fit timing belt. Mark on belt should be aligned with mark on crankshaft **8**.
3. Secure belt to crankshaft sprocket with retaining clip. Tool No.(-).0132AK **9**.
4. Fit timing belt to guide pulley.
5. Fit timing belt to camshaft sprocket CA1 then CA2.
   NOTE: Ensure belt is taut between sprockets on non-tensioned side.
6. Fit timing belt to water pump sprocket and tensioner pulley.
7. Remove retaining clip from timing belt **9**.
8. Turn tensioner pulley anti-clockwise **7** until pointer **A** at position **B**. Use Allen key **11**.
9. Tighten tensioner bolt to 22 Nm **7**.
10. Ensure marks on belt aligned with marks on sprockets **8** & **10**.
11. Remove timing pin **5**.
12. Turn crankshaft four turns clockwise to setting position.
    NOTE: DO NOT allow crankshaft to turn anti-clockwise.
13. Insert timing pin in flywheel **5**. Tool No.(-).0132.QY.
14. Slacken tensioner bolt **7**.
15. Align tensioner pointer **A** with position **C**.
16. Tighten tensioner bolt to 22 Nm **7**.
17. Turn crankshaft two turns clockwise.
18. Insert timing pins **5** & **6**.
19. Check tensioner position. If incorrect, repeat above operations.
20. Remove timing pins **5** & **6**.
21. Install components in reverse order of removal.
22. Tighten crankshaft pulley bolts **1**. Tightening torque: 25 Nm.

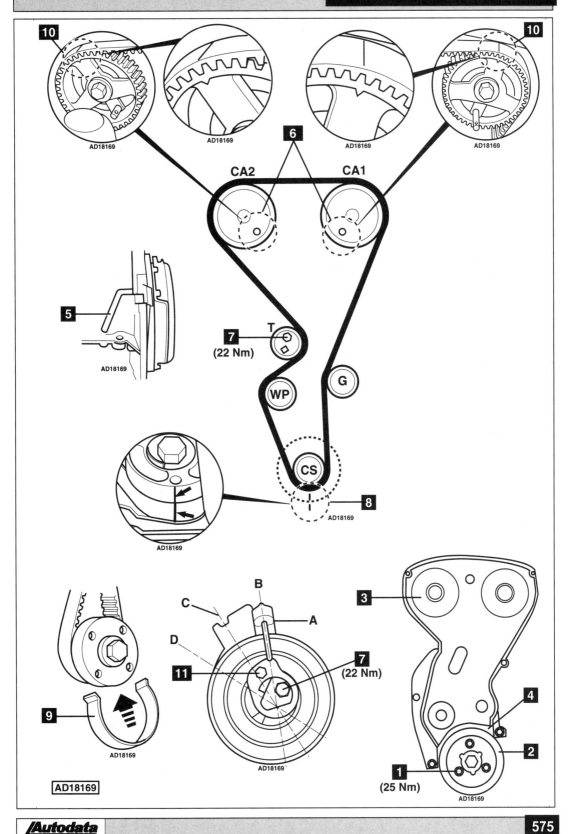

CA2 CA1

6

10

10

5

AD18169

7
T
(22 Nm)

WP

G

CS

8

AD18169

AD18169

3

B

C

A

D

11

7
(22 Nm)

9

AD18169

4

2

1
(25 Nm)

AD18169

AD18169

Model:	**206 2,0 16V • 406 2,0 16V • 406 Coupé 2,0 16V**
Year:	**1998-00**
Engine Code:	**Adjustable camshaft sprockets** **RFN (EW10J4)**

Replacement Interval Guide

Peugeot recommend:
Replacement every 80,000 miles under normal conditions.
Replacement every 48,000 miles under adverse conditions.
The previous use and service history of the vehicle must always be taken into account.
Refer to Timing Belt Replacement Intervals at the front of this manual.

Check For Engine Damage

CAUTION: This engine has been identified as an INTERFERENCE engine in which the possibility of valve-to-piston damage in the event of a timing belt failure is MOST LIKELY to occur.
A compression check of all cylinders should be performed before removing the cylinder head(s).

Repair Times – hrs

Remove & install:

206	3,90
406	2,10

Special Tools

- Camshaft timing pins – Peugeot No.(-).0189-A/Z.
- Crankshaft timing pin – Peugeot No.(-).0189-B.
- Timing belt retaining clip – Peugeot No.(-).0189-K.
- Tensioner positioning tool – Peugeot No.(-).0189-J.

Special Precautions

- Disconnect battery earth lead.
- DO NOT turn crankshaft or camshaft when timing belt removed.
- Remove spark plugs to ease turning engine.
- Turn engine in normal direction of rotation (unless otherwise stated).
- DO NOT turn engine via camshaft or other sprockets.
- Observe all tightening torques.

Removal

1. Raise and support front of vehicle.
2. Remove:
   - Engine upper cover (if fitted).
   - RH front wheel.
   - RH splash guard.
   - Auxiliary drive belt.
   - Battery and battery tray.
   - Engine lower torque reaction link.
   - 206: Engine control module (ECM) and bracket.
3. Support engine.

4. Remove:
   - RH engine mounting.
   - 206: LH gearbox mounting nut.
5. 206: Move engine/gearbox assembly towards the left.
6. Remove:
   - Timing belt upper cover **1**.
   - Crankshaft pulley bolts **2**.
   - Crankshaft pulley **3**.
   - Timing belt lower cover **4**.
7. Turn crankshaft clockwise to setting position.
8. Insert timing pin in crankshaft pulley hub **5**. Tool No.(-).0189-B.
9. Insert timing pin in hub of each camshaft sprocket **6** & **7**. Tool No.(-).0189-A/Z.
10. Slacken tensioner pulley bolt **8**.
11. Unhook tensioner pulley bracket from cylinder block rib **9**.
12. Move tensioner pulley away from belt.
13. Remove timing belt.
    NOTE: Timing belt must always be renewed once it has been removed.

Installation

NOTE: Belt adjustment must be carried out when engine is cold.

1. Ensure timing pins located correctly **5**, **6** & **7**.
2. Fit timing belt to crankshaft sprocket.
3. Secure belt to crankshaft sprocket with retaining clip. Tool No.(-).0189-K **10**.
4. Slacken bolts of each camshaft sprocket **11** & **12**.
5. Tighten bolts finger tight, then slacken 1/6 turn.
6. Turn camshaft sprockets fully clockwise in slotted holes.
   NOTE: Sprockets should turn with slight resistance.
7. Fit timing belt around guide pulley.
8. Lay belt on teeth of camshaft sprockets. Engage belt teeth by turning sprockets slightly anti-clockwise.
9. Angular movement of sprockets must not be more than one tooth space **13**.
10. Ensure belt is taut between sprockets.
11. Fit timing belt to water pump sprocket and tensioner pulley.
12. Fit tensioner pulley bracket to cylinder block rib **9**.

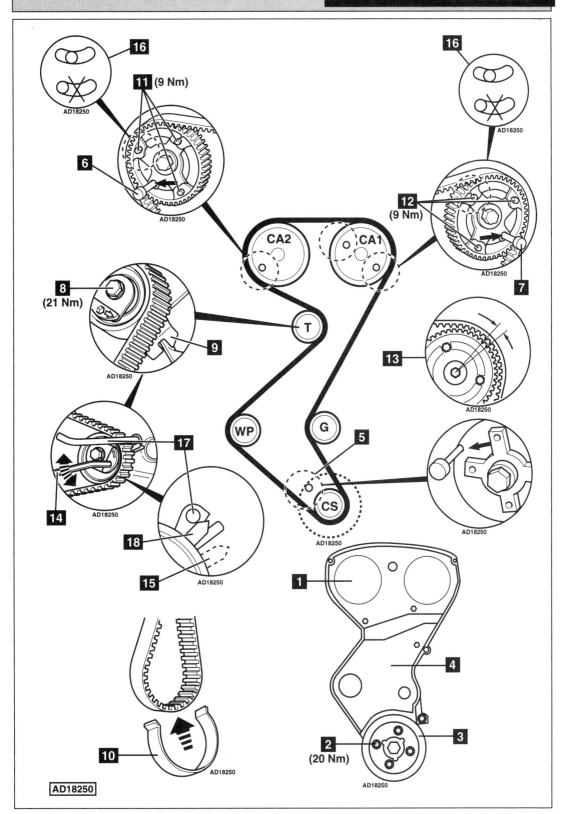

←

13. Tighten tensioner pulley bolt finger tight .
14. Turn tensioner anti-clockwise until belt tensioned to maximum. Use Allen key 🔲. Ensure pointer positioned as shown 🔲.
15. Tighten tensioner pulley bolt 🔲. Tightening torque: 21 Nm.
16. Ensure bolts of each camshaft sprocket not at end of slotted holes 🔲.
17. If necessary: Repeat installation procedure.
18. Tighten bolts of each camshaft sprocket 🔲 & 🔲. Tightening torque: 9 Nm.
19. Remove:
    ❏ Timing pins 🔲, 🔲 & 🔲.
    ❏ Timing belt retaining clip 🔲.
20. Turn crankshaft six turns clockwise to setting position.
    NOTE: DO NOT allow crankshaft to turn anti-clockwise.
21. Insert tensioner positioning tool 🔲. Tool No.(-).0189-J.
22. Hold tensioner pulley in position.
23. Slacken tensioner pulley bolt 🔲.
24. Turn tensioner clockwise until pointer and positioning tool are just touching 🔲 & 🔲. Use Allen key 🔲.
    NOTE: Pointer and notch are also aligned.
25. Tighten tensioner pulley bolt 🔲. Tightening torque: 21 Nm.
26. Remove tensioner positioning tool 🔲.
27. Insert timing pin in crankshaft pulley hub 🔲.
28. Slacken bolts of each camshaft sprocket slightly 🔲 & 🔲.
29. Insert timing pins in camshaft sprockets 🔲 & 🔲.
    NOTE: If timing pins cannot be inserted: Turn hub of of each camshaft sprocket as required.
30. Ensure bolts of each camshaft sprocket not at end of slotted holes 🔲.
31. Tighten bolts of each camshaft sprocket 🔲 & 🔲. Tightening torque: 9 Nm.
32. Remove timing pins 🔲, 🔲 & 🔲.
33. Install components in reverse order of removal.
34. Tighten crankshaft pulley bolts 🔲. Tightening torque: 20 Nm.
35. Tighten RH engine mounting:
    ❏ Three bolts securing intermediate bracket to engine bracket − 60 Nm.
    ❏ Nut securing engine mounting to intermediate bracket − 45 Nm.
    ❏ Two bolts securing strut brace to intermediate bracket − 20 Nm.
36. 206: Tighten LH gearbox mounting nut. Tightening torque: 65 Nm.
37. Tighten engine lower torque reaction link bolts. Tightening torque: 45 Nm.

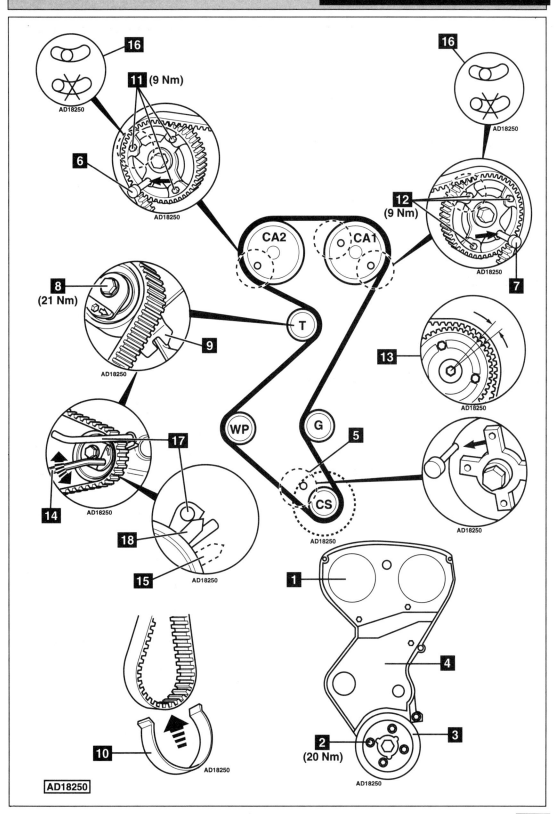

Model:	206 2,0 16V • 307 2,0 16V • 406 2,0 16V • 406 Coupé 2,0 16V • 607 2,0 16V
	806 2,0 16V
Year:	2000-03
Engine Code:	Non-adjustable camshaft sprockets
	RFR (EW10J4), RFN (EW10J4)

Replacement Interval Guide

Peugeot recommend:
Replacement every 80,000 miles under normal conditions.
Replacement every 48,000 miles under adverse conditions.
The previous use and service history of the vehicle must always be taken into account.
Refer to Timing Belt Replacement Intervals at the front of this manual.

Check For Engine Damage

CAUTION: This engine has been identified as an INTERFERENCE engine in which the possibility of valve-to-piston damage in the event of a timing belt failure is MOST LIKELY to occur.
A compression check of all cylinders should be performed before removing the cylinder head(s).

Repair Times – hrs

Remove & install:

206/307	3,90
406	2,10
607	1,80
806	No information available.

Special Tools

- Camshaft timing pins – Peugeot No.(-).0189-AZ.
- Crankshaft timing pin – Peugeot No.(-).0189-B.
- Timing belt retaining clip – Peugeot No.(-).0189-K.
- Tensioner positioning tool – Peugeot No.(-).0189-J.
- Crankshaft locking tools – Peugeot No.(-).0606-A1 & (-).0606-A2.

Special Precautions

- Disconnect battery earth lead.
- DO NOT turn crankshaft or camshaft when timing belt removed.
- Remove spark plugs to ease turning engine.
- Turn engine in normal direction of rotation (unless otherwise stated).
- DO NOT turn engine via camshaft or other sprockets.
- Observe all tightening torques.

Removal

1. Raise and support front of vehicle.
2. Remove:
   - ❑ Engine upper cover (if fitted).
   - ❑ RH front wheel.
   - ❑ RH splash guard.
   - ❑ Auxiliary drive belt.

206/406/806

3. Remove:
   - ❑ Battery and battery tray.
   - ❑ Engine lower torque reaction link.
   - ❑ 206: Engine control module (ECM) and bracket.

206/406/307

4. Support engine.
5. Remove:
   - ❑ RH engine mounting.
   - ❑ 206: LH gearbox mounting nut.
   - ❑ 307: PAS pipe bracket bolt.
6. 206: Move engine/gearbox assembly towards the left.

All models

7. Remove:
   - ❑ Timing belt upper cover **1**.
   - ❑ Crankshaft pulley bolts **2**.
   - ❑ Crankshaft pulley **3**.
   - ❑ Timing belt lower cover **4**.
8. Turn crankshaft clockwise to setting position.
9. Insert timing pin in crankshaft pulley hub **5**. Tool No.(-).0189-B.
10. Insert timing pins in camshaft sprockets **6** & **7**. Tool No.(-).0189-AZ.
11. Slacken tensioner pulley bolt **8**.
12. Unhook tensioner pulley bracket from cylinder block rib **9**.
13. Move tensioner pulley away from belt.
14. Remove timing belt.
    NOTE: Timing belt must always be renewed once it has been removed.

Installation

NOTE: Belt adjustment must be carried out when engine is cold.

1. Ensure timing pins located correctly **5**, **6** & **7**.
2. Fit timing belt to crankshaft sprocket.
   NOTE: Ensure arrow on belt faces direction of rotation.
3. Secure belt to crankshaft sprocket with retaining clip. Tool No.(-).0189-K **10**.

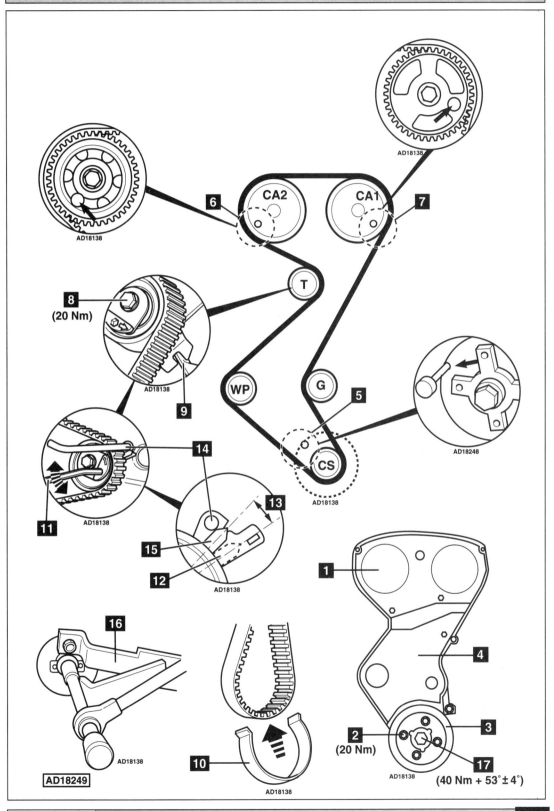

CA2 CA1 **6** **7**

8
(20 Nm)

9

14

11

15

12

13

5

WP G

CS

T

16

10

1

4

2
(20 Nm)

3

17

(40 Nm + 53° ± 4°)

AD18138
AD18138
AD18138
AD18248
AD18138
AD18138
AD18138
AD18249
AD18138
AD18138

←

4. Fit timing belt to remaining sprockets and pulleys in following order:
   - ❏ Guide pulley.
   - ❏ Inlet camshaft sprocket.
   - ❏ Exhaust camshaft sprocket.
   - ❏ Water pump sprocket.
   - ❏ Tensioner pulley.

   NOTE: Ensure belt is taut between sprockets.

5. Fit tensioner pulley bracket to cylinder block rib **9**.

6. Tighten tensioner pulley bolt finger tight **8**.

7. Remove timing belt retaining clip **10**.

8. Turn tensioner anti-clockwise until belt tensioned to maximum. Use Allen key **11**. Ensure pointer positioned as shown **12**.

9. Ensure pointer moves away from notch by at least 10° **13**. If not, replace tensioner pulley.

10. Insert tensioner positioning tool **14**.
    Tool No.(-).0189-J.

11. Turn tensioner clockwise until pointer and positioning tool are just touching **14** & **15**. Use Allen key **11**.

    NOTE: Pointer and notch are also aligned.

12. Tighten tensioner pulley bolt **8**.
    Tightening torque: 20 Nm.

13. Remove tensioner positioning tool **14**.

14. Remove timing pins **5**, **6** & **7**.

15. Turn crankshaft ten turns clockwise to setting position.

    NOTE: DO NOT allow crankshaft to turn anti-clockwise.

16. Ensure pointer and notch aligned **15**.

17. If not: Repeat tensioning procedure.

18. Insert timing pin in inlet camshaft sprocket **7**.

19. Insert timing pin in crankshaft pulley hub **5**.

20. If timing pin **5** cannot be inserted, reposition crankshaft pulley hub as follows:
    - ❏ Lock crankshaft pulley hub.
      Use tool Nos.(-).0606-A1/A2 **16**.
    - ❏ Slacken crankshaft pulley hub bolt **17**.
    - ❏ Turn crankshaft pulley hub until timing pin can be inserted **5**.
      Use tool Nos.(-).0606-A1/A2 **16**.

21. Tighten crankshaft pulley hub bolt **17**.
    Tightening torque: 40 Nm + 53±4°.

22. Remove timing pins **5** & **7**.

23. Install components in reverse order of removal.

24. Tighten crankshaft pulley bolts **2**.
    Tightening torque: 20 Nm.

206/406

25. Tighten RH engine mounting:
    - ❏ Three bolts securing intermediate bracket to engine bracket – 60 Nm.
    - ❏ Nut securing engine mounting to intermediate bracket – 45 Nm.
    - ❏ Two bolts securing strut brace to intermediate bracket – 20 Nm.

26. 206: Tighten LH gearbox mounting nut.
    Tightening torque: 65 Nm.

27. Tighten engine lower torque reaction link bolts.
    Tightening torque: 45 Nm.

307

28. Tighten RH engine mounting.
    Tightening torque: 60 Nm.

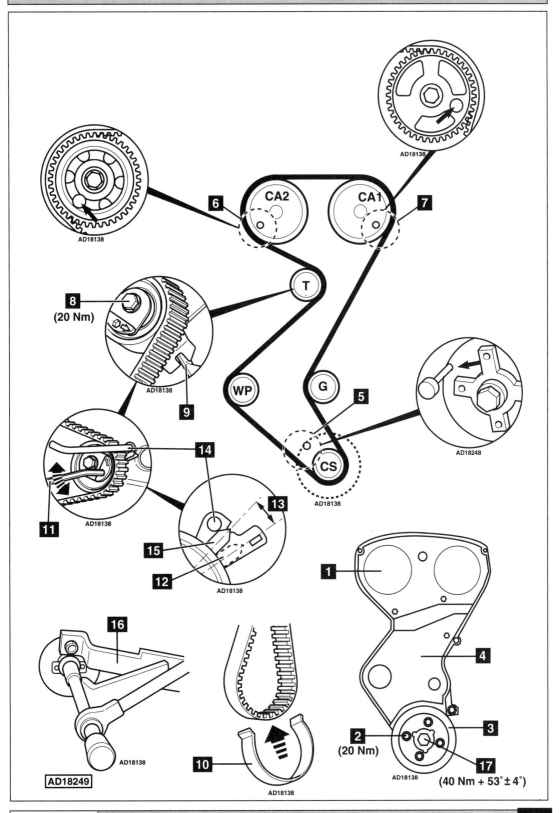

PEUGEOT

Model:	306 1,8 16V • 406 1,8 16V
Year:	1995-98
Engine Code:	XU7JP4 (LFY)

Replacement Interval Guide

Peugeot recommend replacement as follows:

→07/98:

Replacement every 72,000 miles under normal conditions.

Replacement every 54,000 miles under adverse conditions.

08/98→:

Replacement every 80,000 miles under normal conditions.

Replacement every 48,000 miles under adverse conditions.

The previous use and service history of the vehicle must always be taken into account.

Refer to Timing Belt Replacement Intervals at the front of this manual.

Check For Engine Damage

CAUTION: This engine has been identified as an INTERFERENCE engine in which the possibility of valve-to-piston damage in the event of a timing belt failure is MOST LIKELY to occur.

A compression check of all cylinders should be performed before removing the cylinder head.

Repair Times – hrs

Remove & install:	
306	2,70
Engine undershield	+0,10
AC	+0,50
406	2,90
Engine undershield	+0,10
AC	+0,30

Special Tools

- Crankshaft timing pin – Peugeot No.(-).0153G.
- Camshaft timing pins – Peugeot No.(-).0153AB.
- Tension gauge – SEEM C.Tronic 105.5.
- Flywheel locking tool – Peugeot No.0134Q.

Special Precautions

- Disconnect battery earth lead.
- DO NOT turn crankshaft or camshaft when timing belt removed.
- Remove spark plugs to ease turning engine.
- Turn engine in normal direction of rotation (unless otherwise stated).
- DO NOT turn engine via camshaft or other sprockets.
- Observe all tightening torques.

Removal

WARNING: This engine may suffer from failure of the crankshaft pulley resulting in the possible incorrect alignment of the timing pin hole. The timing belt should be removed and installed with the engine at 90° BTDC. If necessary: Fit new crankshaft pulley.

1. Raise and support front of vehicle.
2. Remove:
   - ❑ RH front wheel.
   - ❑ Engine undershield (if fitted).
   - ❑ RH wheel arch liner.
   - ❑ Auxiliary drive belt.
3. Reposition:
   - ❑ Wiring loom.
   - ❑ Evaporative emission (EVAP) canister purge valve.
4. Support engine.
5. Remove:
   - ❑ Engine mounting.
   - ❑ Timing belt upper cover **1**.
6. Turn crankshaft clockwise to setting position.
7. Insert timing pin in crankshaft pulley **2**. Tool No.(-).0153G.
8. Insert timing pins in camshaft sprockets **3**. Tool No.(-).0153AB.
9. Lock flywheel. Use tool No.0134Q.
10. Remove:
    - ❑ Crankshaft timing pin **2**.
    - ❑ Crankshaft pulley bolt **7**.
    - ❑ Crankshaft pulley **5**.

    NOTE: An extractor may be required to remove crankshaft pulley.
    - ❑ Auxiliary drive belt tensioner.
    - ❑ Timing belt lower cover **4**.
11. Slacken tensioner bolt **6**. Move tensioner away from belt. Lightly tighten bolt.
12. Remove timing belt.

Installation

1. Ensure timing pins located correctly in camshaft sprockets **3**.
2. Fit timing belt to crankshaft sprocket.
3. Install:
   - ❑ Timing belt lower cover **4**.
   - ❑ Crankshaft pulley **5**.
4. Coat crankshaft pulley bolt thread with suitable thread locking compound.
5. Tighten crankshaft pulley bolt **7**. Tightening torque: 120 Nm.

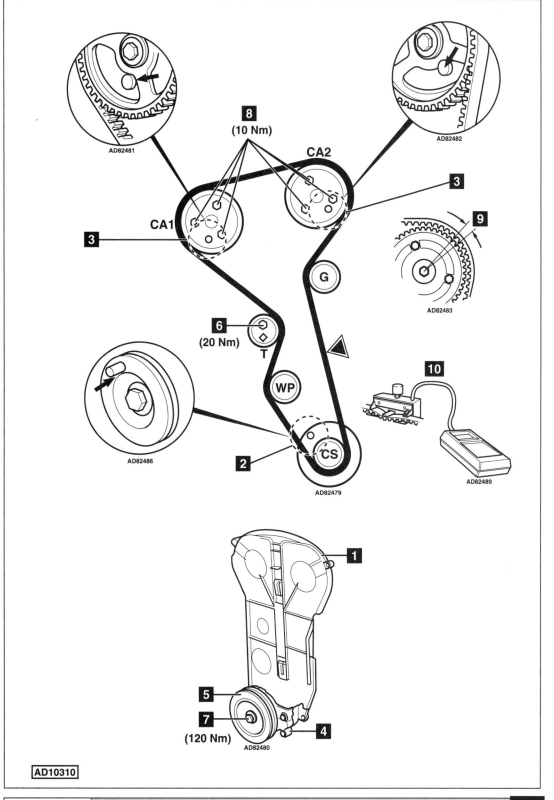

AD82481

8
(10 Nm)

CA2

CA1

3

3

9

AD82483

6
(20 Nm)
T

G

10

WP

AD82486

2

CS

AD82479

AD82489

1

5

7
(120 Nm)

4

AD82480

AD10310

←

6. Insert crankshaft timing pin **2**.
7. Ensure timing pins located correctly in camshaft sprockets **3**.
8. Slacken bolts of each camshaft sprocket **8**. Tighten bolts finger tight. Then slacken 1/6 turn.
9. Turn camshaft sprockets fully clockwise in slotted holes.
10. Fit timing belt in anti-clockwise direction. Ensure belt is taut between sprockets.
11. Lay belt on teeth of camshaft sprockets. Engage belt teeth by turning sprockets slightly anti-clockwise.
    NOTE: Angular movement of sprockets must not be more than one tooth space 9.
12. Attach tension gauge to belt at ▽. Tool No.SEEM C.Tronic 105.5 **10**.
13. Slacken tensioner bolt **6**. Turn tensioner anti-clockwise until tension gauge indicates 45 SEEM units.
14. Prevent tensioner from turning. Tighten tensioner bolt **6**. Tightening torque: 20 Nm.
15. Ensure bolts of each camshaft sprocket not at end of slotted holes **8**.
16. If bolts at end of slotted holes: Repeat installation procedure.
17. Tighten bolts of each camshaft sprocket **8**. Tightening torque: 10 Nm.
18. Remove:
    ❑ Tension gauge **10**.
    ❑ Timing pins **2** & **3**.
19. Turn crankshaft two turns clockwise. Insert crankshaft timing pin **2**.
20. Slacken bolts of each camshaft sprocket **8**. Tighten bolts finger tight. Then slacken 1/6 turn.
21. Insert timing pins in camshaft sprockets **3**.
    NOTE: If timing pins cannot be inserted: Turn camshafts slightly.
22. Attach tension gauge to belt at ▽ **10**.
23. Slacken tensioner bolt **6**. Turn tensioner until tension gauge indicates 26 SEEM units.
24. Tighten tensioner bolt **6**. Tightening torque: 20 Nm.
25. Tighten bolts of each camshaft sprocket **8**. Tightening torque: 10 Nm.
26. Remove:
    ❑ Tension gauge **10**.
    ❑ Timing pins **2** & **3**.
27. Check belt tension as follows:
28. Repeat operations 17-19.
29. Tighten bolts of each camshaft sprocket **8**. Tightening torque: 10 Nm. Remove timing pins **2** & **3**.
30. Turn crankshaft 1/4 turn clockwise. Attach tension gauge to belt at ▽.
31. Tension gauge should indicate 32-40 SEEM units.
32. Install components in reverse order of removal.

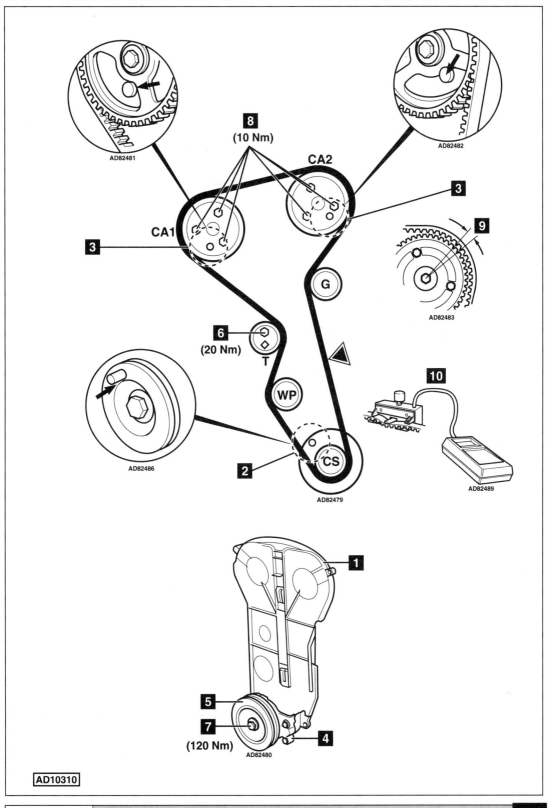

AD82481

8
(10 Nm)

CA2

CA1

3

3

AD82482

9

AD82483

6
(20 Nm)
T

G

WP

10

2

CS

AD82486

AD82479

AD82489

1

5

7
(120 Nm)

4

AD82480

AD10310

PEUGEOT

Model:	306 1,8 16V • 406 1,8 16V
Year:	1999-03
Engine Code:	LFY (XU7JP4)

Replacement Interval Guide

Peugeot recommend:
Replacement every 80,000 miles under normal conditions.
Replacement every 48,000 miles under adverse conditions.
The previous use and service history of the vehicle must always be taken into account.
Refer to Timing Belt Replacement Intervals at the front of this manual.

Check For Engine Damage

CAUTION: This engine has been identified as an INTERFERENCE engine in which the possibility of valve-to-piston damage in the event of a timing belt failure is MOST LIKELY to occur.
A compression check of all cylinders should be performed before removing the cylinder head(s).

Repair Times – hrs

306	2,70
Engine undershield	+0,10
AC	+0,50
406	2,90
Engine undershield	+0,10
AC	+0,30

Special Tools

- Flywheel locking tool – Peugeot No.(-).0134-AF.
- Crankshaft timing pin – Peugeot No.(-).0153-G.
- Camshaft timing pins – Peugeot No.(-).0153-AB.
- Tensioner retaining tool – Peugeot No.(-).0153-AL.
- Timing belt retaining clip – Peugeot No.(-).0153-AK.
- Camshaft locking tool – Peugeot No.(-).0153-AJ.

Special Precautions

- Disconnect battery earth lead.
- DO NOT turn crankshaft or camshaft when timing belt removed.
- Remove spark plugs to ease turning engine.
- Turn engine in normal direction of rotation (unless otherwise stated).
- DO NOT turn engine via camshaft or other sprockets.
- Observe all tightening torques.

Removal

WARNING: This engine may suffer from failure of the crankshaft pulley resulting in the possible incorrect alignment of the timing pin hole. The timing belt should be removed and installed with the engine at 90° BTDC. If necessary: Fit new crankshaft pulley.

1. Raise and support front of vehicle.
2. Remove:
   - ❑ RH front wheel.
   - ❑ RH splash guard.
   - ❑ Auxiliary drive belt.
3. Support engine.
4. Remove:
   - ❑ RH engine mounting.
   - ❑ Timing belt upper cover **1**.
5. Turn crankshaft clockwise to setting position. Insert timing pins in camshaft sprockets **2**. Tool No.(-).0153-AB.
6. Insert timing pin in crankshaft pulley **3**. Tool No.(-).0153-G.
7. Fit flywheel locking tool **4**. Tool No.(-).0134-AF.
8. Remove:
   - ❑ Crankshaft timing pin **3**.
   - ❑ Crankshaft pulley bolt **5**.
   - ❑ Crankshaft pulley **6**.
   - ❑ Timing belt lower covers **7**.
9. Slacken tensioner bolt **8**.
10. Turn tensioner anti-clockwise until pointer below hole **9**. Use Allen key **10**.
11. Insert tensioner retaining tool **11**. Tool No.(-).0153-AL.
12. Turn tensioner clockwise to release tension on belt.
13. Lightly tighten tensioner bolt **8**.
14. Remove timing belt.
    NOTE: Timing belt must always be renewed once it has been removed.

Installation

1. Ensure timing pins located correctly **2**.
2. Temporarily fit crankshaft pulley **6**.
3. Fit timing pin **3**.
4. Ensure flywheel locking tool located correctly **4**.
5. Remove:
   - ❑ Crankshaft timing pin **3**.
   - ❑ Crankshaft pulley **6**.
6. Fit timing belt to crankshaft sprocket.

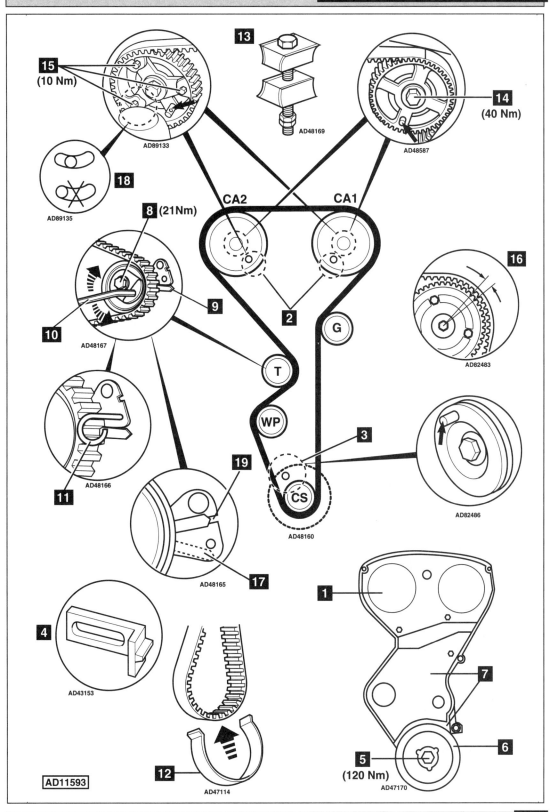

7. Secure belt to crankshaft sprocket with retaining clip **12**. Tool No.(-).0153-AK.
8. Hold camshaft sprockets.
   Use tool No.(-).0153-AJ **13**.
9. Slacken bolts of each camshaft sprocket:
   - ❏ Type 1: 1-bolt fixing **14**.
   - ❏ Type 2: 3-bolt fixing **15**.
10. Tighten bolts finger tight, then slacken 1/6 turn.
11. Remove camshaft locking tool **13**.
12. Turn camshaft sprockets fully clockwise in slotted holes.
    NOTE: Sprockets should turn with slight resistance.
13. Fit timing belt in anti-clockwise direction. Ensure belt is taut between sprockets.
14. Lay belt on teeth of camshaft sprockets. Engage belt teeth by turning sprockets slightly anti-clockwise.
    *NOTE: Angular movement of sprockets must not be more than one tooth space **16**.*
15. Fit timing belt to water pump sprocket and tensioner pulley.
16. Remove:
    - ❏ Retaining clip **12**.
17. Fit timing belt covers **7**.
18. Fit crankshaft pulley **6**.
19. Coat crankshaft pulley bolt thread with suitable thread locking compound.
20. Tighten crankshaft pulley bolt **5**.
    Tightening torque: 120 Nm.
21. Insert timing pin **3**.
22. Remove flywheel locking tool **4**.
23. Remove:
    - ❏ Tensioner retaining tool **11**.
24. Slacken tensioner bolt **8**.
25. Allow tensioner to operate.
26. Turn tensioner anti-clockwise until belt tensioned to maximum. Use Allen key **10**. Check pointer is as shown **17**.
27. Tighten tensioner bolt **8**.
    Tightening torque: 21 Nm.
28. Check camshaft sprockets. Type 2: 3-bolt fixing **15**. Ensure bolts of each camshaft sprocket not at end of slotted holes **18**.
29. If necessary: Repeat installation procedure.
30. Hold camshaft sprockets.
    Use tool No.(-).0153-AJ **13**.
31. Tighten bolts of each camshaft sprocket:
    - ❏ Type 1: 1-bolt fixing **14**.
      Tightening torque: 40 Nm.
    - ❏ Type 2: 3-bolt fixing **15**.
      Tightening torque: 10 Nm.
32. Remove:
    - ❏ Timing pins **2** & **3**.
    - ❏ Camshaft locking tool **13**.
33. Turn crankshaft four turns clockwise to setting position.
    NOTE: DO NOT allow crankshaft to turn anti-clockwise.
34. Insert timing pins **2** & **3**.
35. Hold camshaft sprockets.
    Use tool No.(-).0153-AJ **13**.
36. Slacken bolts of each camshaft sprocket:
    - ❏ Type 1: 1-bolt fixing **14**.
    - ❏ Type 2: 3-bolt fixing **15**.
37. Tighten bolts finger tight, then slacken 1/6 turn.
38. Remove camshaft locking tool **13**.
39. Hold tensioner pulley in position. Use Allen key **10**.
40. Slacken tensioner bolt **8**.
41. Turn tensioner clockwise until pointer aligned with notch **19**. Use Allen key **10**.
42. Tighten tensioner bolt **8**.
    Tightening torque: 21 Nm.
43. Hold camshaft sprockets.
    Use tool No.(-).0153-AJ **13**.
44. Tighten bolts of each camshaft sprocket:
    - ❏ Type 1: 1-bolt fixing **14**.
      Tightening torque: 40 Nm.
    - ❏ Type 2: 3-bolt fixing **15**.
      Tightening torque: 10 Nm.
45. Remove:
    - ❏ Timing pins **2** & **3**.
    - ❏ Camshaft locking tool **13**.
46. Turn crankshaft 2 turns clockwise to setting position.
47. Fit timing pin **3**.
48. Ensure timing pins can be inserted easily in sprockets **2**.
49. Ensure tensioner pointer aligned with notch **19**.
50. If not: Repeat installation and tensioning procedures.
51. Remove timing pin **3**.
52. Install components in reverse order of removal.

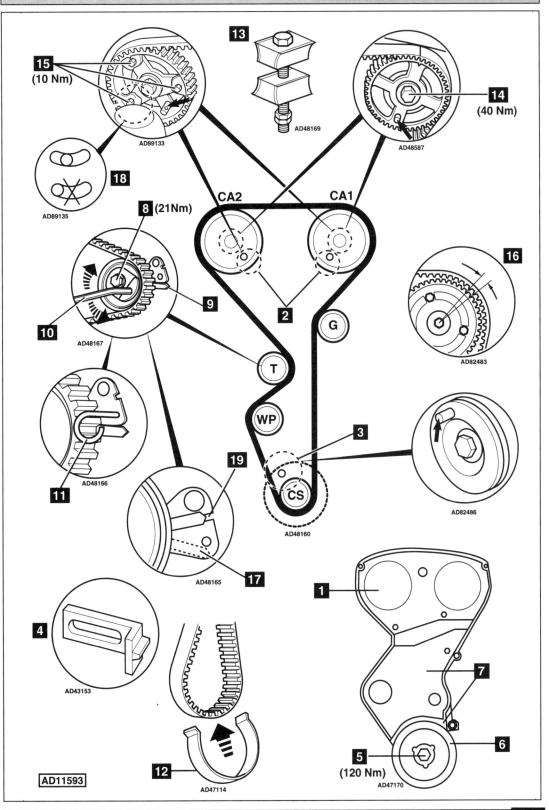

13

AD48169

15
(10 Nm)

AD89133

18

AD89135

14
(40 Nm)

AD48587

8 (21Nm)

CA2   CA1

9

10

AD48167

2

G

16

AD82483

T

11

AD48166

WP

19

3

AD48160

CS

AD82486

17

AD48165

1

4

AD43153

7

12

AD47114

5
(120 Nm)

6

AD47170

AD11593

PEUGEOT

Model:	**306 2,0 16V • 405 2,0 16V • 605 2,0 16V**
Year:	**1993-98**
Engine Code:	**XU10J4 (RFV/RFY/RFS)**

Replacement Interval Guide

Peugeot recommend replacement as folows:

→07/98:
Replacement every 72,000 miles under normal conditions.
Replacement every 54,000 miles under adverse conditions.

08/98→:
Replacement every 80,000 miles under normal conditions.
Replacement every 48,000 miles under adverse conditions.

The previous use and service history of the vehicle must always be taken into account.
Refer to Timing Belt Replacement Intervals at the front of this manual.

Check For Engine Damage

CAUTION: This engine has been identified as an INTERFERENCE engine in which the possibility of valve-to-piston damage in the event of a timing belt failure is MOST LIKELY to occur.
A compression check of all cylinders should be performed before removing the cylinder head.

Repair Times – hrs

Remove & install:

306/405	2,70
AC	+0,50
605	3,30
AC	+0,90

Special Tools

- Crankshaft timing pin – Peugeot No.9766.98.
- Camshaft timing pins – Peugeot No.9767.94.
- Tension gauge – SEEM C.Tronic 105.5.

Special Precautions

- Disconnect battery earth lead.
- DO NOT turn crankshaft or camshaft when timing belt removed.
- Remove spark plugs to ease turning engine.
- Turn engine in normal direction of rotation (unless otherwise stated).
- DO NOT turn engine via camshaft or other sprockets.
- Observe all tightening torques.

Removal

1. Support engine.
2. Remove:
   - ❏ RH wheel arch liner.
   - ❏ Engine mounting.
   - ❏ Auxiliary drive belt and tensioner.
   - ❏ Crankshaft pulley bolts **1**.
   - ❏ Crankshaft pulley **2**.
   - ❏ Timing belt upper cover **3**.
   - ❏ Timing belt lower cover **4**.
3. Turn crankshaft clockwise to setting position.
4. Insert timing pins in camshaft sprockets **5** & **6**. Tool No.9767.94.
5. Insert timing pin in crankshaft sprocket **7**. Tool No.9766.98.
6. Slacken tensioner bolts **8** & **9**.
7. Remove timing belt.

Installation

1. Ensure timing pins located correctly **5**, **6** & **7**.
2. Fit timing belt in following order:
   - ❏ Crankshaft sprocket.
   - ❏ Water pump sprocket.
   - ❏ Tensioner pulley (T1).
   - ❏ Tensioner pulley (T2).
   - ❏ Camshaft sprocket (CA2) (with no slack in belt).
   - ❏ Camshaft sprocket (CA1).
3. Apply thumb pressure to belt at each tensioner to eliminate any play in timing pins.
4. Attach tension gauge to belt at ▽ **10**.
5. Push tensioner (T1) against belt. Tension belt to 45 SEEM units.
6. Release tension on belt. Tension belt to 22±2 SEEM units.
7. Tighten tensioner bolts to 20 Nm **9**.
8. Turn tensioner (T2) anti-clockwise until tension gauge indicates 32±2 SEEM units.
9. Tighten tensioner bolt to 20 Nm **8**.
10. Remove all timing pins **5**, **6** & **7**.
11. Turn crankshaft two turns clockwise. Insert timing pin in crankshaft sprocket **7**.
12. If timing pins cannot be inserted easily in camshaft sprockets: Repeat operations 4-11.
13. Remove all timing pins **5**, **6** & **7**.
14. Attach tension gauge to belt at ▽ **10**. Tension gauge should indicate 53±2 SEEM units.
15. Install components in reverse order of removal.

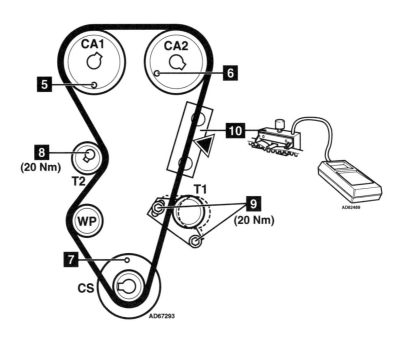

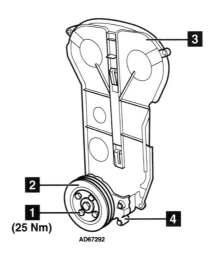

AD10499

PEUGEOT

Model:	**306 2,0 16V • 406 2,0 16V**
Year:	**1995-98**
Engine Code:	**XU10J4R/S (RFV/R6E/RFS)**

Replacement Interval Guide

Peugeot recommend replacement as follows:

→07/98:
Replacement every 72,000 miles under normal conditions.
Replacement every 54,000 miles under adverse conditions.

08/98→:
Replacement every 80,000 miles under normal conditions.
Replacement every 48,000 miles under adverse conditions.
The previous use and service history of the vehicle must always be taken into account.
Refer to Timing Belt Replacement Intervals at the front of this manual.

Check For Engine Damage

CAUTION: This engine has been identified as an INTERFERENCE engine in which the possibility of valve-to-piston damage in the event of a timing belt failure is MOST LIKELY to occur.
A compression check of all cylinders should be performed before removing the cylinder head.

Repair Times – hrs

Remove & install:

306	2,70
Engine undershield	+0,10
AC	+0,50
406	2,90
Engine undershield	+0,10
AC	+0,30

Special Tools

- Crankshaft timing pin – Peugeot No.(-).0153G.
- Camshaft timing pins – Peugeot No.(-).0153AB.
- Tension gauge – SEEM C.Tronic 105.5.
- Flywheel locking tool – Peugeot No.0134Q.

Special Precautions

- Disconnect battery earth lead.
- DO NOT turn crankshaft or camshaft when timing belt removed.
- Remove spark plugs to ease turning engine.
- Turn engine in normal direction of rotation (unless otherwise stated).
- DO NOT turn engine via camshaft or other sprockets.
- Observe all tightening torques.

Removal

WARNING: 306 GTi-16: Dependent on VIN there is a possibility of premature timing belt failure. Refer to dealer.
WARNING: This engine may suffer from failure of the crankshaft pulley resulting in the possible incorrect alignment of the timing pin hole. The timing belt should be removed and installed with the engine at 90° BTDC. If necessary: Fit new crankshaft pulley.

1. Raise and support front of vehicle.
2. Remove:
   ❑ RH front wheel.
   ❑ Engine undershield (if fitted).
   ❑ RH wheel arch liner.
   ❑ Auxiliary drive belt.
3. Reposition:
   ❑ Wiring loom.
   ❑ Evaporative emission (EVAP) canister purge valve.
4. Support engine.
5. Remove:
   ❑ Engine mounting.
   ❑ Timing belt upper cover **1**.
6. Turn crankshaft clockwise to setting position.
7. Insert timing pin in crankshaft pulley **2**.
   Tool No.(-).0153G.
8. Insert timing pins in camshaft sprockets **3**.
   Tool No.(-).0153AB.
9. Lock flywheel. Use tool No.0134Q.
10. Remove:
    ❑ Crankshaft timing pin **2**.
    ❑ Crankshaft pulley bolt **7**.
    ❑ Crankshaft pulley **5**.
    NOTE: An extractor may be required to remove crankshaft pulley.
    ❑ Auxiliary drive belt tensioner.
    ❑ Timing belt lower cover **4**.
11. Slacken tensioner bolt **6**. Move tensioner away from belt. Lightly tighten bolt.
12. Remove timing belt.

Installation

1. Ensure timing pins located correctly in camshaft sprockets **3**.
2. Fit timing belt to crankshaft sprocket.
3. Install:
   ❑ Timing belt lower cover **4**.
   ❑ Crankshaft pulley **5**.

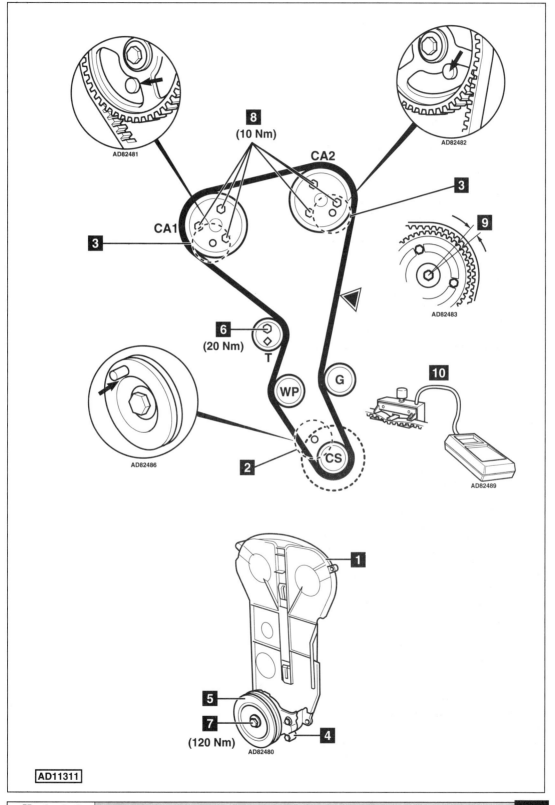

AD82481

8
(10 Nm)

CA2

AD82482

3

CA1

3

9

AD82483

6
(20 Nm)
T

G

WP

10

AD82486

CS

2

AD82489

1

5

7

(120 Nm)

4

AD82480

AD11311

←

4. Coat crankshaft pulley bolt thread with suitable thread locking compound.
5. Tighten crankshaft pulley bolt .
   Tightening torque: 120 Nm.
6. Insert crankshaft timing pin **2**.
7. Ensure timing pins located correctly in camshaft sprockets **3**.
8. Slacken bolts of each camshaft sprocket **8**.
   Tighten bolts finger tight. Then slacken 1/6 turn.
9. Turn camshaft sprockets fully clockwise in slotted holes.
10. Fit timing belt in anti-clockwise direction. Ensure belt is taut between sprockets.
11. Lay belt on teeth of camshaft sprockets.
    Engage belt teeth by turning sprockets slightly anti-clockwise.
    NOTE: Angular movement of sprockets must not be more than one tooth space **9**.
12. Attach tension gauge to belt at ▽.
    Tool No.SEEM C.Tronic 105.5 **10**.
13. Slacken tensioner bolt **6**. Turn tensioner anti-clockwise until tension gauge indicates 45 SEEM units.
14. Prevent tensioner from turning. Tighten tensioner bolt **6**. Tightening torque: 20 Nm.
15. Ensure bolts of each camshaft sprocket not at end of slotted holes **8**.
16. If bolts at end of slotted holes: Repeat installation procedure.
17. Tighten bolts of each camshaft sprocket **8**.
    Tightening torque: 10 Nm.
18. Remove:
    ❑ Tension gauge **10**.
    ❑ Timing pins **2** & **3**.
19. Turn crankshaft two turns clockwise. Insert crankshaft timing pin **2**.
20. Slacken bolts of each camshaft sprocket **8**.
    Tighten bolts finger tight. Then slacken 1/6 turn.
21. Insert timing pins in camshaft sprockets **3**.
    NOTE: If timing pins cannot be inserted: Turn camshafts slightly.
22. Attach tension gauge to belt at ▽ **10**.
23. Slacken tensioner bolt **6**. Turn tensioner until tension gauge indicates 26 SEEM units.
24. Tighten tensioner bolt **6**.
    Tightening torque: 20 Nm.
25. Tighten bolts of each camshaft sprocket **8**.
    Tightening torque: 10 Nm.
26. Remove:
    ❑ Tension gauge **10**.
    ❑ Timing pins **2** & **3**.
27. Check belt tension as follows:
28. Repeat operations 17-19.
29. Tighten bolts of each camshaft sprocket **8**.
    Tightening torque: 10 Nm. Remove timing pins **2** & **3**.

30. Turn crankshaft 1/4 turn clockwise. Attach tension gauge to belt at ▽.
31. Tension gauge should indicate 32-40 SEEM units.
32. Install components in reverse order of removal.

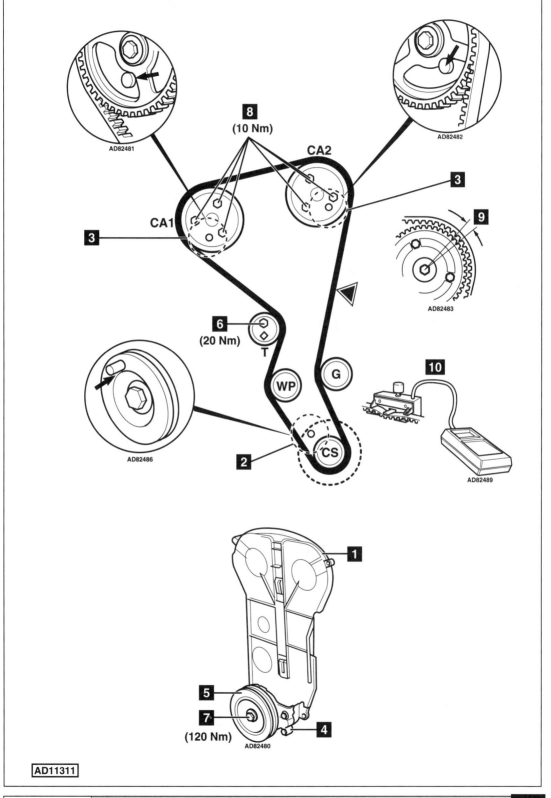

AD82481

8
(10 Nm)

CA2

3

CA1

3

AD82482

9

AD82483

6
(20 Nm)
T

AD82486

WP

G

10

AD82489

2

CS

1

5

7
(120 Nm)
AD82480

4

AD11311

PEUGEOT

Model:	406 1,6 • 406 2,0 Turbo • 806 2,0/Turbo
Year:	1995-01
Engine Code:	XU5JP (BFZ)
	XU10J2C/Z (RFU), XU10J2TE/Z (RGX), XU10J2CTE/L3 (RGX)

Replacement Interval Guide

Peugeot recommend replacement as follows:

→07/98:
Replacement every 72,000 miles under normal conditions.
Replacement every 54,000 miles under adverse conditions.

08/98→:
Replacement every 80,000 miles under normal conditions.
Replacement every 48,000 miles under adverse conditions.

The previous use and service history of the vehicle must always be taken into account.
Refer to Timing Belt Replacement Intervals at the front of this manual.

Check For Engine Damage

CAUTION: This engine has been identified as an INTERFERENCE engine in which the possibility of valve-to-piston damage in the event of a timing belt failure is MOST LIKELY to occur.
A compression check of all cylinders should be performed before removing the cylinder head.

Repair Times – hrs

Remove & install:

406	2,60
Engine undershield	+0,10
AC	+0,30
806 (→07/98)	2,10
806 (08/98→)	2,00
AC	+0,10

Special Tools

- Crankshaft timing pin – Peugeot No.(-).0153G.
- Camshaft timing pin – Peugeot No.(-).0132R.
- Tension gauge – SEEM C.Tronic 105.5.
- Flywheel locking tool – Peugeot No.9765.54 or (-).0153A.

Special Precautions

- Disconnect battery earth lead.
- DO NOT turn crankshaft or camshaft when timing belt removed.
- Remove spark plugs to ease turning engine.
- Turn engine in normal direction of rotation (unless otherwise stated).
- DO NOT turn engine via camshaft or other sprockets.
- Observe all tightening torques.

Removal

1. Raise and support front of vehicle.
2. Remove:
   - ❑ RH front wheel.
   - ❑ Engine undershield (if fitted).
   - ❑ RH wheel arch liner.
   - ❑ Auxiliary drive belt.
3. 406 – reposition:
   - ❑ Fuel pipes.
   - ❑ Evaporative emission (EVAP) canister purge valve.
   - ❑ Manifold absolute pressure (MAP) sensor.
4. Remove timing belt upper cover **1**.
5. Turn crankshaft clockwise to setting position.
6. Insert crankshaft timing pin **2**.
   Tool No.(-).0153G.
7. Insert timing pin in camshaft sprocket **3**.
   Tool No.(-)0132R.
8. Remove clutch housing plate. Lock flywheel **4**.
   Use tool No.9765.54 or (-).0153A.
9. Remove:
   - ❑ Timing pins **2** & **3**.
   - ❑ Crankshaft pulley bolt **5**.
   - ❑ Crankshaft pulley **6**.

   NOTE: An extractor may be required to remove crankshaft pulley.
   - ❑ Auxiliary drive belt tensioner.
   - ❑ Timing belt centre cover (if fitted) **7**.
   - ❑ Timing belt lower cover **8**.
   - ❑ Thrust washer **9**.
10. Insert timing pin in camshaft sprocket **3**.
11. Slacken tensioner bolt **10**. Move tensioner away from belt. Lightly tighten bolt.
12. Remove timing belt.

Installation

1. Temporarily fit crankshaft pulley and timing pin **6** & **2**.
2. Ensure timing pin located correctly **3**.
3. Remove:
   - ❑ Crankshaft pulley **6**.
   - ❑ Crankshaft timing pin **2**.
4. Fit timing belt in following order:
   - ❑ Camshaft sprocket.
   - ❑ Crankshaft sprocket.
   - ❑ Water pump sprocket.
   - ❑ Tensioner pulley.
5. Turn tensioner anti-clockwise. Lightly tension belt. Tighten tensioner bolt **10**.

➡

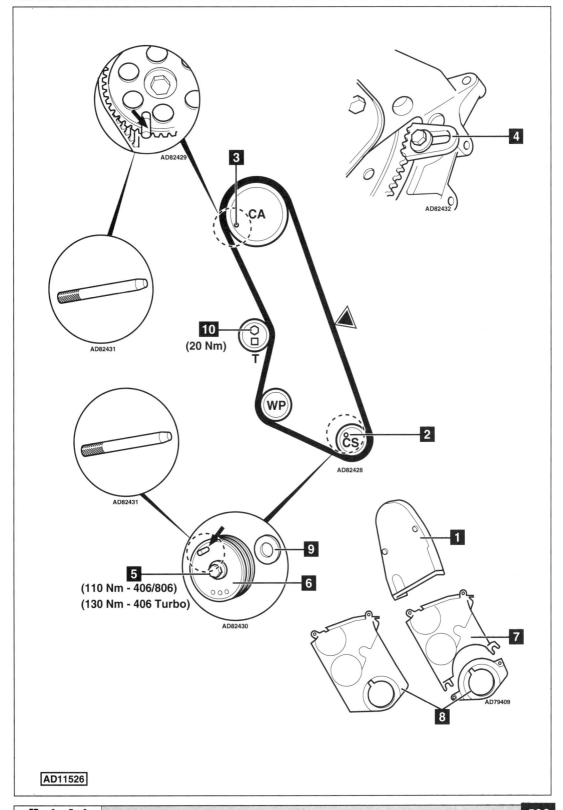

AD82429

3

CA

4

AD82432

AD82431

10
(20 Nm)
T

WP

2

CS

AD82428

AD82431

9

5
(110 Nm - 406/806)
(130 Nm - 406 Turbo)

6

AD82430

1

7

AD79409

8

AD11526

←

6. Install:
   - ❑ Thrust washer **9**.
   - ❑ Crankshaft pulley **6**.

7. Temporarily tighten crankshaft pulley bolt **5**.

8. Attach tension gauge to belt at ▽.
   Tool No.SEEM C.Tronic 105.5.

9. Slacken tensioner bolt **10**. Turn tensioner
   anti-clockwise to tension belt. 406/806 Turbo:
   16±2 SEEM units. 406 1,6: 30±2 SEEM units.

10. Tighten tensioner bolt to 20 Nm **10**.

11. Remove:
    - ❑ Tension gauge.
    - ❑ Timing pin **3**.

12. Turn crankshaft two turns clockwise.

13. Ensure timing pins can be inserted **2** & **3**.

14. Remove timing pins **2** & **3**. Turn crankshaft two
    turns clockwise.

15. Insert timing pin **3**.

16. Attach tension gauge to belt at ▽. Tension
    gauge should indicate 42-46 SEEM units.

17. If not: Repeat operations 5-9.

18. Remove:
    - ❑ Timing pin **3**.
    - ❑ Tension gauge.
    - ❑ Crankshaft pulley **6**.

19. Install components in reverse order of removal.

20. Coat crankshaft pulley bolt with suitable thread
    locking compound.

21. Tighten crankshaft pulley bolt **5**.
    Tightening torque: 406/806: 110 Nm.
    406 Turbo: 130 Nm.

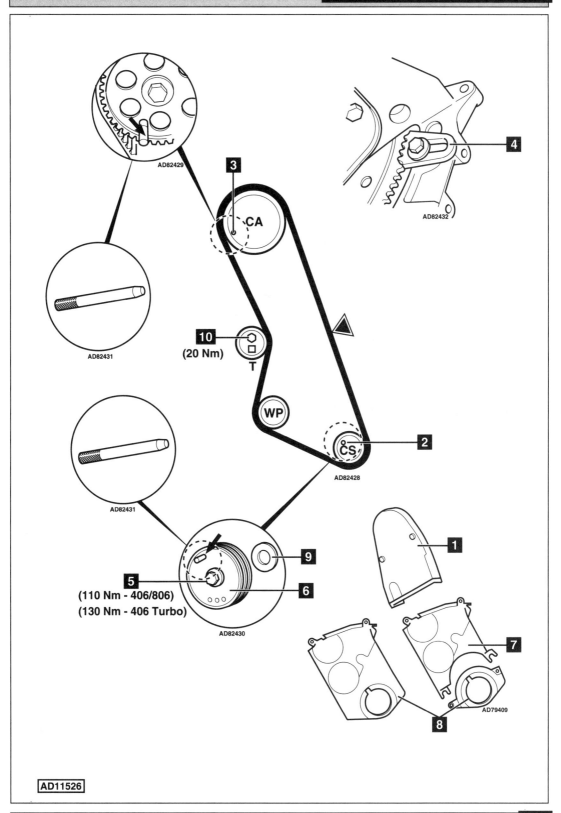

AD82429

3

CA

4

AD82432

AD82431

10
(20 Nm)
T

WP

2

CS

AD82428

AD82431

9

5
(110 Nm - 406/806)
(130 Nm - 406 Turbo)

6

AD82430

1

7

8

AD79409

AD11526

PEUGEOT

Model:	406 2,2 16V • 406 Coupé 2,2 16V • 607 2,2 16V
Year:	1999-03
Engine Code:	3FZ (EW12J4)

Replacement Interval Guide

Peugeot recommend:
Replacement every 80,000 miles under normal conditions.
Replacement every 48,000 miles under adverse conditions.
The previous use and service history of the vehicle must always be taken into account.
Refer to Timing Belt Replacement Intervals at the front of this manual.

Check For Engine Damage

CAUTION: This engine has been identified as an INTERFERENCE engine in which the possibility of valve-to-piston damage in the event of a timing belt failure is MOST LIKELY to occur.
A compression check of all cylinders should be performed before removing the cylinder head(s).

Repair Times – hrs

Remove & install:
406	2,10
607	1,80

Special Tools

- Crankshaft timing pin – Peugeot No.(-).0189-B.
- Exhaust camshaft timing pin – Peugeot No.(-).0189-AZ.
- Inlet camshaft timing pin – Peugeot No.(-).0189-L.
- Timing belt retaining clip – Peugeot No.(-).0189-K.
- Tensioner positioning tool – Peugeot No.(-).0189-J.
- Crankshaft locking tools – Peugeot No.(-).0606-A1 & (-).0606-A2.

Special Precautions

- Disconnect battery earth lead.
- DO NOT turn crankshaft or camshaft when timing belt removed.
- Remove spark plugs to ease turning engine.
- Turn engine in normal direction of rotation (unless otherwise stated).
- DO NOT turn engine via camshaft or other sprockets.
- Observe all tightening torques.

Removal

1. Raise and support front of vehicle.
2. Remove:
   - Engine upper cover (if fitted).
   - RH front wheel.
   - RH splash guard.
   - Auxiliary drive belt.
   - Engine torque reaction link.

406

3. Support engine.
   NOTE: To prevent exhaust damage, ensure engine remains in initial position.
4. Remove RH engine mounting.

All models

5. Remove:
   - Timing belt upper cover **1**.
   - Crankshaft pulley bolts **2**.
   - Crankshaft pulley **3**.
   - Timing belt lower cover **4**.
6. Turn crankshaft clockwise to setting position.
7. Insert timing pin in crankshaft pulley hub **5**. Tool No.(-).0189-B.
8. Insert timing pin in exhaust camshaft sprocket **6**. Tool No.(-).0189-AZ.
9. Insert timing pin in inlet camshaft sprocket **7**. Tool No.(-).0189-L.
10. DO NOT slacken inlet camshaft sprocket bolts **8**.
11. Slacken tensioner pulley bolt **9**.
12. Unhook tensioner pulley bracket from cylinder block rib **10**.
13. Move tensioner pulley away from belt.
14. Remove timing belt.
    NOTE: Timing belt must always be renewed once it has been removed.

Installation

NOTE: Belt adjustment must be carried out when engine is cold.
1. Ensure timing pins located correctly **5**, **6** & **7**.
2. Fit timing belt to crankshaft sprocket.
   NOTE: Ensure arrow on belt faces direction of rotation.
3. Secure belt to crankshaft sprocket with retaining clip. Tool No.(-).0189-K **11**.
4. Fit timing belt to remaining sprockets and pulleys in following order:
   - Guide pulley.
   - Inlet camshaft sprocket.
   - Exhaust camshaft sprocket.
   - Water pump sprocket.
   - Tensioner pulley.
   NOTE: Ensure belt is taut between sprockets.
5. Fit tensioner pulley bracket to cylinder block rib **10**.
6. Tighten tensioner pulley bolt finger tight **9**.
7. Remove timing belt retaining clip **11**.

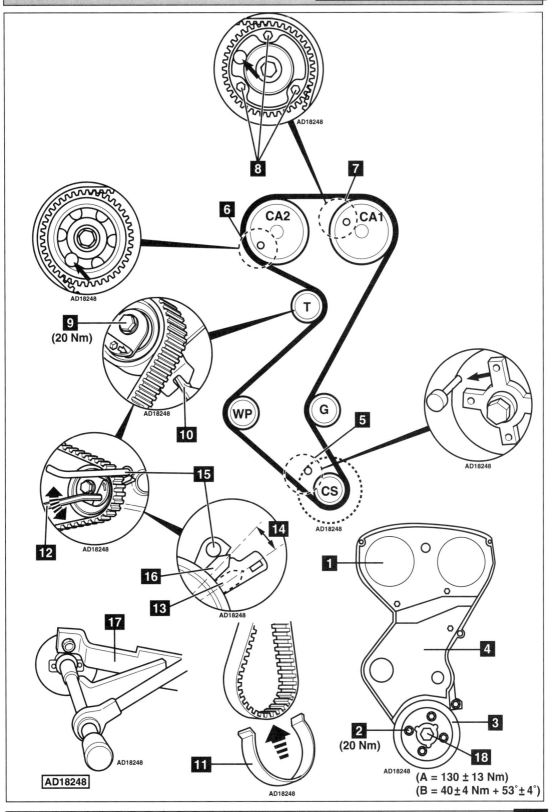

AD18248

8

7

6 CA2   CA1

9
(20 Nm)

T

10

WP   G

5

15

14

12   16

13

17

AD18248

AD18248

1

4

2
(20 Nm)   3

18

(A = 130 ± 13 Nm)
(B = 40 ± 4 Nm + 53° ± 4°)

11

8. Turn tensioner anti-clockwise until belt tensioned to maximum. Use Allen key **12**. Ensure pointer positioned as shown **13**.

9. Ensure pointer moves away from notch by at least 10° **14**. If not, replace tensioner pulley.

10. Insert tensioner positioning tool **15**.
    Tool No.(-).0189-J.

11. Turn tensioner clockwise until pointer and positioning tool are just touching **15** & **16**. Use Allen key **12**.
    NOTE: Pointer and notch are also aligned.

12. Tighten tensioner pulley bolt **9**.
    Tightening torque: 20 Nm.

13. Remove tensioner positioning tool **15**.

14. Remove timing pins **5**, **6** & **7**.

15. Turn crankshaft ten turns clockwise to setting position.
    NOTE: DO NOT allow crankshaft to turn anti-clockwise.

16. Ensure pointer and notch aligned **16**.

17. If not: Repeat tensioning procedure.

18. Insert timing pin in inlet camshaft sprocket **7**.

19. Insert timing pin in crankshaft pulley hub **5**.

20. If timing pin **5** cannot be inserted, reposition crankshaft pulley hub as follows:
    ❏ Lock crankshaft pulley hub.
      Use tool Nos.(-).0606-A1/A2 **17**.
    ❏ Slacken crankshaft pulley hub bolt **18**.
    ❏ Turn crankshaft pulley hub until timing pin can be inserted **5**.
      Use tool Nos.(-).0606-A1/A2 **17**.

21. Tighten crankshaft pulley hub bolt **18**.
    Tightening torque:
    ❏ (A) 406 – 130±13 Nm.
    ❏ (B) 607 – 40±4 Nm + 53±4°.

22. Remove timing pins **5** & **7**.

23. Install components in reverse order of removal.

24. Tighten crankshaft pulley bolts **2**.
    Tightening torque: 20 Nm.

406

25. Tighten RH engine mounting:
    ❏ Three bolts securing intermediate bracket to engine bracket – 61 Nm.
    ❏ Nut securing engine mounting to intermediate bracket – 45 Nm.

All models

26. Tighten engine torque reaction link bolts.
    Tightening torque: 45 Nm.

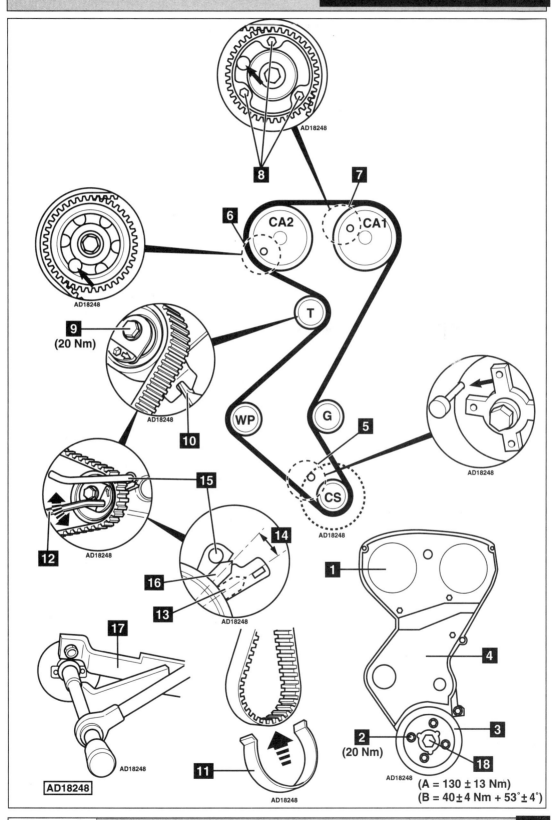

AD18248

8

7

6 CA2 CA1

9
(20 Nm)

T

10

WP **G**

5

AD18248

15

12 AD18248

16

13 AD18248

14

CS

AD18248

1

17

AD18248

AD18248

11

AD18248

4

2
(20 Nm)

3

18

AD18248

(A = 130 ± 13 Nm)
(B = 40 ± 4 Nm + 53° ± 4°)

PEUGEOT

Model:	406/Coupé 3,0 V6 • 605 3,0 V6
Year:	1997-03
Engine Code:	ES9J4 (XFZ)

Replacement Interval Guide

Peugeot recommend replacement as follows:

→07/98:
Replacement every 72,000 miles under normal conditions.
Replacement every 54,000 miles under normal conditions.

08/98→:
Replacement every 80,000 miles under normal conditions.
Replacement every 72,000 miles under normal conditions.

The previous use and service history of the vehicle must always be taken into account.
Refer to Timing Belt Replacement Intervals at the front of this manual.

Check For Engine Damage

CAUTION: This engine has been identified as an INTERFERENCE engine in which the possibility of valve-to-piston damage in the event of a timing belt failure is MOST LIKELY to occur.
A compression check of all cylinders should be performed before removing the cylinder head(s).

Repair Times – hrs

Remove & install:	
406	4,70
Engine undershield	+0,10
605	4,70

Special Tools

- Crankshaft timing pin – Peugeot No.0187A.
- Camshaft timing pins – 4 x Peugeot No.0187B.
- Timing belt retaining clip – Peugeot No.0187J.
- Camshaft holding tool – Peugeot No.0187F.
- Tensioning tool – Peugeot No.0187EZ.
- Tension gauge – SEEM C.Tronic 105.5.
- M8 x 1,25 x 75 mm bolt.
- M8 x 1,25 x 35 mm bolt.

Special Precautions

- Disconnect battery earth lead.
- DO NOT turn crankshaft or camshaft when timing belt removed.
- Remove spark plugs to ease turning engine.
- Turn engine in normal direction of rotation (unless otherwise stated).
- DO NOT turn engine via camshaft or other sprockets.
- Observe all tightening torques.

Removal

1. Remove:
   - ❑ Engine cover.
   - ❑ Engine undershield (if fitted).
   - ❑ Stabiliser bar from suspension mountings (406).
2. Disconnect engine wiring harness.
3. Move engine control module (ECM) away from engine.
4. Support engine.
5. Remove:
   - ❑ Torque reaction link.
   - ❑ RH engine mounting.
   - ❑ Engine control module (ECM) cover.
   - ❑ Engine control module (ECM) box.
   - ❑ Auxiliary drive belt.
   - ❑ PAS pump pulley.
   - ❑ Timing belt upper covers **1**.
   - ❑ Auxiliary drive belt tensioner assembly.
   - ❑ Bracket **2**.
   - ❑ Crankshaft pulley bolts **3**.
   - ❑ Crankshaft pulley **4**.
   - ❑ Timing belt lower cover **5**.
6. Turn crankshaft clockwise to setting position.
7. Insert timing pins in camshaft sprockets **7**. Tool No.0187B.
8. Slacken bolts of each camshaft sprocket **6**.
9. Insert crankshaft timing pin **8**. Tool No.0187A.
10. Insert M8 x 1,25 x 75 mm bolt into tensioner bracket **9**.
11. Tighten bolt **9** until it touches bracket **10**.
12. Slacken tensioner bolts **11**, **12** & **13**.
    NOTE: DO NOT slacken bolt 14.
13. Install special tool No.0187EZ **15**.
    NOTE: If necessary: Slacken bolt slightly 9.
14. Insert M8 x 1,25 x 35 mm bolt into tensioner bracket **16**.
15. Tighten bolt **16** until it touches bracket **17**.
16. Fully tighten bolt **16**.
17. Slacken bolt **9**.
18. Remove timing belt.
    NOTE: Mark direction of rotation on belt with chalk if belt is to be reused.

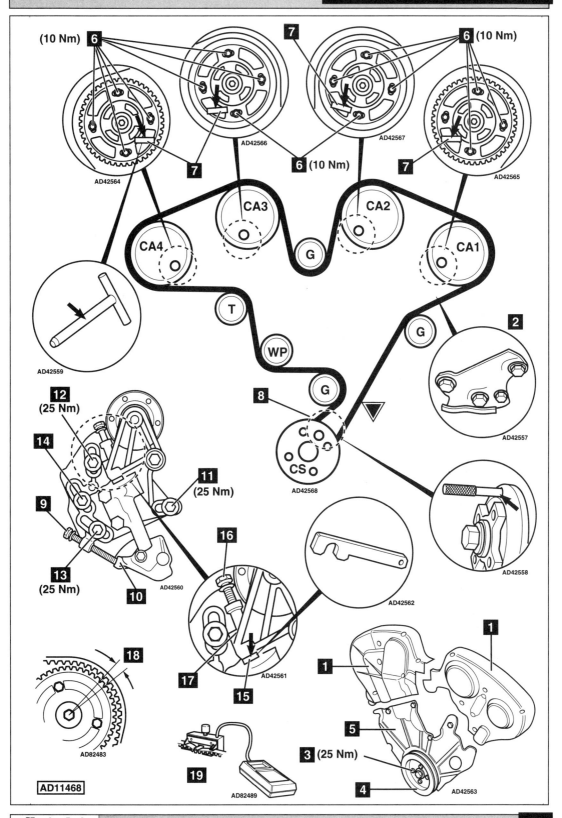

(10 Nm) **6**

7

6 (10 Nm)

AD42566

AD42567

AD42564

7

6 (10 Nm)

AD42565

7

CA3

CA2

CA4

G

CA1

AD42559

T

G

2

12
(25 Nm)

WP

AD42557

14

G

9

8

11
(25 Nm)

C

CS

AD42568

AD42558

13
(25 Nm)

16

AD42560

10

AD42562

18

AD42561

17

1

AD82483

15

1

5

3 (25 Nm)

AD11468

19

AD82489

4

AD42563

Installation

1. Ensure crankshaft timing pin located correctly **8**. Tool No.0187A.
2. Ensure timing pins located correctly in camshaft sprockets **7**. Tool No.0187B.
3. Ensure camshaft sprockets turn freely.
4. Turn camshaft sprockets fully clockwise in slotted holes.
5. Tighten bolts to 5 Nm **6**.
6. Slacken bolts 45° **6**.
7. Tighten bolts to 10 Nm **11**, **12** & **13**.
8. Slacken bolts 45° **11**, **12** & **13**.
9. Fit timing belt to crankshaft sprocket.
10. Retain in position with special tool. Tool No.0187J.
11. Fit timing belt in anti-clockwise direction.
    NOTE: If reusing old belt: Observe direction of rotation marks on belt.
12. Turn each camshaft sprocket to engage in nearest belt tooth.
    *NOTE: Ensure bolts not at end of slotted holes in sprockets **6**. Angular movement of sprockets must not be more than one tooth space **18**.*
13. Ensure belt is taut between sprockets.
14. Remove retaining clip from timing belt. Tool No.0187J.
15. Attach tension gauge to belt at ▽. Tool No.SEEM C.Tronic 105.5 **19**.
16. Screw in bolt **9** until tension gauge indicates 83±2 SEEM units.
17. Tighten bolt **11**. Tightening torque: 10 Nm.
18. Tighten bolt **12**. Tightening torque: 10 Nm.
19. Tighten bolt **13**. Tightening torque: 10 Nm.
20. Hold camshaft sprockets. Use tool No.0187F.
21. Tighten camshaft sprocket bolts (CA1) **6**. Tightening torque: 10 Nm.
22. Tighten camshaft sprocket bolts (CA2) **6**. Tightening torque: 10 Nm.
23. Tighten camshaft sprocket bolts (CA3) **6**. Tightening torque: 10 Nm.
24. Tighten camshaft sprocket bolts (CA4) **6**. Tightening torque: 10 Nm.
25. Remove tension gauge **19**.
26. Remove timing pins from camshafts **7**.
27. Remove crankshaft timing pin **8**.
28. Turn crankshaft slowly two turns clockwise.
29. Insert crankshaft timing pin **8**. Tool No.0187A.
30. Slacken bolts 45° **11**, **12** & **13**.
31. Remove bolt **16**.
32. Adjust position of bolt **9** until tool **15** slides freely without free play. Tool No.0187EZ.
33. Wait 2 minutes to allow automatic tensioner unit and belt to settle.
34. Check tool **15** slides freely without free play. Tool No.0187EZ.
35. Adjust if necessary.
36. Remove special tool **15**.
37. Tighten bolt **11**. Tightening torque: 25 Nm.
38. Tighten bolt **12**. Tightening torque: 25 Nm.
39. Tighten bolt **13**. Tightening torque: 25 Nm.
40. Remove bolt **9**.
41. Remove crankshaft timing pin **8**.
42. Turn crankshaft slowly two turns clockwise.
43. Insert crankshaft timing pin **8**. Tool No.0187A.
44. Ensure timing pins can easily be inserted in and withdrawn from camshaft sprockets **7**.
45. Adjust if necessary. Use tool No.0187F.
    *NOTE: Ensure bolts not at end of slotted holes in sprockets **6**.*
46. Remove crankshaft timing pin **8**.
47. Remove timing pins **7**.
48. Install components in reverse order of removal.
49. Tighten crankshaft pulley bolts **3**. Tightening torque: 25 Nm.

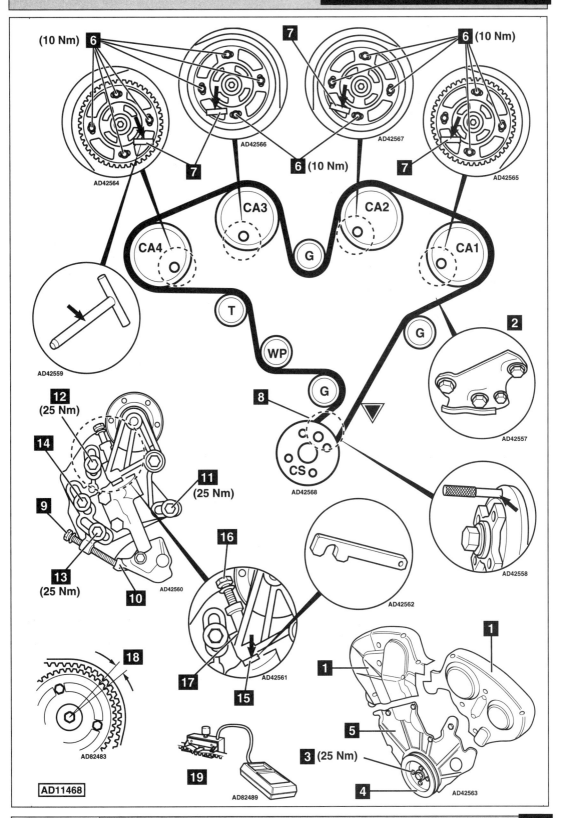

PEUGEOT

Model:	**106 1,4 D**
Year:	**1993-95**
Engine Code:	**TUD3**

Replacement Interval Guide

Peugeot recommend replacement every 72,000 miles.
The previous use and service history of the vehicle must always be taken into account.
Refer to Timing Belt Replacement Intervals at the front of this manual.

Check For Engine Damage

CAUTION: This engine has been identified as an INTERFERENCE engine in which the possibility of valve-to-piston damage in the event of a timing belt failure is MOST LIKELY to occur.
A compression check of all cylinders should be performed before removing the cylinder head.

Repair Times – hrs

Remove & install	2,00

Special Tools

- Timing pin – Peugeot No.0132Q.
- Tension gauge – Peugeot (SEEM) No.105.5.
- Tensioning tool – Peugeot No.0132X1Z/2Z.
- 1 bolt – M8 x 1,25 x 40 mm.
- 2 bolts – M8 x 1,25 x 30 mm.

Special Precautions

- Disconnect battery earth lead.
- DO NOT turn crankshaft or camshaft when timing belt removed.
- Remove glow plugs to ease turning engine.
- Turn engine in normal direction of rotation (unless otherwise stated).
- DO NOT turn engine via camshaft or other sprockets.
- Observe all tightening torques.
- Check diesel injection pump timing after belt replacement.

Removal

1. Remove:
   - ❏ Pre-heater relay. DO NOT disconnect wires.
   - ❏ Auxiliary drive belt.
   - ❏ Crankshaft pulley bolts **8**.
   - ❏ RH headlamp.
   - ❏ Timing belt covers **1** & **2**.
2. Turn crankshaft clockwise to setting position.
3. Insert timing pin in flywheel **3**. Tool No.0132Q.
4. Insert one bolt in camshaft sprocket and two bolts in injection pump sprocket **4**.
5. Slacken tensioner nut **5**. Turn tensioner clockwise to release tension on belt **6**. Use tool No.0132X1Z/2Z.
6. Lightly tighten nut.
7. Remove timing belt.

Installation

1. Ensure timing pin and locking bolts located correctly **3** & **4**.
2. Fit timing belt in following order:
   - ❏ Crankshaft sprocket.
   - ❏ Guide pulley.
   - ❏ Injection pump sprocket.
   - ❏ Camshaft sprocket.
   - ❏ Tensioner pulley.
   - ❏ Water pump sprocket.
3. Slacken tensioner nut **5**.
4. Insert tensioning tool in square hole of tensioner pulley **6**. Tool No.0132X1Z/2Z. Turn tensioner pulley anti-clockwise to pre-tension belt. Tighten tensioner nut **5**. Tightening torque: 15 Nm.
5. Remove:
   - ❏ Flywheel timing pin **3**.
   - ❏ Locking bolts **4**.
6. Turn crankshaft four turns clockwise.
7. Insert flywheel timing pin **3**.
8. Attach tension gauge to belt at ▽ **7**. Tool No.105.5. Slacken tensioner nut **5**.
9. Insert tensioning tool in square hole of tensioner pulley **6**. Tool No.0132X1Z/2Z. Tension belt to 50 SEEM units.
10. Tighten tensioner nut **5**. Tightening torque: 15 Nm.
11. Remove:
    - ❏ Flywheel timing pin **3**.
    - ❏ Tension gauge **7**.
12. Turn crankshaft one turn clockwise.
13. Insert flywheel timing pin **3**.
14. Wait 1 minute.
15. Attach tension gauge to belt at ▽. Slacken tensioner nut **5**. Release tension on belt.
16. Insert tensioning tool in square hole of tensioner pulley **6**. Tool No.0132X1Z/2Z. Tension belt to 39 SEEM units.
17. Tighten tensioner nut **5**. Tightening torque: 15 Nm.
18. Remove:
    - ❏ Flywheel timing pin **3**.
    - ❏ Tension gauge **7**.
19. Turn crankshaft two turns clockwise.
20. Insert flywheel timing pin **3**.
21. Attach tension gauge to belt at ▽. Tension gauge should indicate 48-54 SEEM units.
22. If not: Slacken tensioner nut **5**. Repeat tensioning procedure. Use tool No.0132X1Z/2Z **6**.
23. Remove flywheel timing pin **3**.
24. Install components in reverse order of removal.
25. Tighten crankshaft pulley bolts **8**. Tightening torque: 16 Nm.

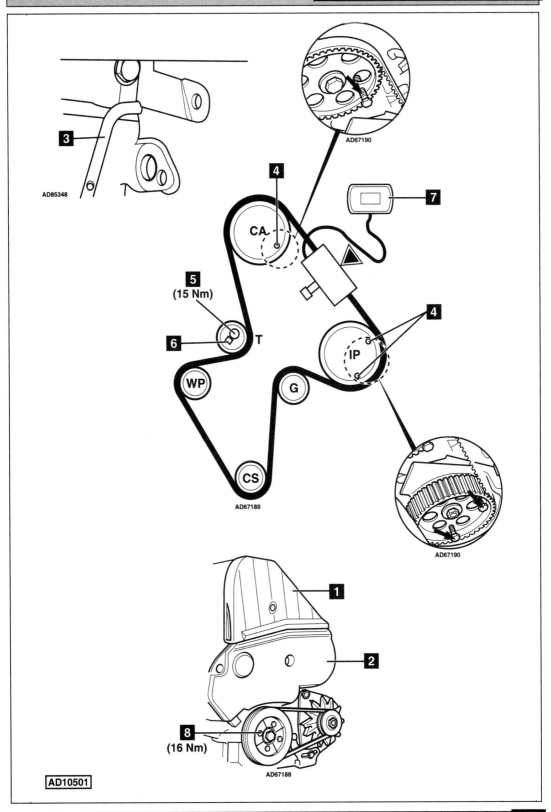

Model:	106 1,5D • 106 Van 1,5D
Year:	1994-03
Engine Code:	TUD5 (VJY/VJZ)

Replacement Interval Guide

Peugeot recommend replacement as follows:

→07/98:
Replacement every 72,000 miles under normal conditions.

08/98→:
Replacement every 80,000 miles under normal conditions.
Replacement every 48,000 miles under adverse conditions.

The previous use and service history of the vehicle must always be taken into account.
Refer to Timing Belt Replacement Intervals at the front of this manual.

Check For Engine Damage

CAUTION: This engine has been identified as an INTERFERENCE engine in which the possibility of valve-to-piston damage in the event of a timing belt failure is MOST LIKELY to occur.
A compression check of all cylinders should be performed before removing the cylinder head.

Repair Times – hrs

Remove & install:

→07/98	2,00
08/98→	1,90

Special Tools

- Flywheel timing pin – Peugeot No.0132Q.
- Injection pump timing pin – Peugeot No.0178C/D.
- Camshaft timing pin – Peugeot No.0132 AB.
- Square drive – Peugeot No.0132X.
- Tension gauge – SEEM C.Tronic 105.5.

Special Precautions

- Disconnect battery earth lead.
- DO NOT turn crankshaft or camshaft when timing belt removed.
- Remove glow plugs to ease turning engine.
- Turn engine in normal direction of rotation (unless otherwise stated).
- DO NOT turn engine via camshaft or other sprockets.
- Observe all tightening torques.
- Check diesel injection pump timing after belt replacement.

Removal

1. Raise and support front of vehicle.
2. Remove:
   - ❑ RH front wheel.
   - ❑ RH front wheel arch liner.
   - ❑ RH headlamp.
   - ❑ Glow plug relay. DO NOT disconnect wires.
   - ❑ Timing belt upper cover **1**.
   - ❑ Auxiliary drive belt.
   - ❑ Crankshaft pulley bolts **2**.
   - ❑ Crankshaft pulley **3**.
   - ❑ Timing belt lower cover **4**.
3. Turn crankshaft clockwise to setting position.
4. Insert timing pin in flywheel **5**. Tool No.0132Q.

5. Insert timing pin in injection pump sprocket **6**. Tool No.0178C/D.
6. Insert timing pin in camshaft sprocket **7**. Tool No.0132 AB.
   NOTE: If locating holes for timing pins not aligned: Turn crankshaft one turn clockwise.
7. Slacken tensioner nut **8**. Move tensioner away from belt. Lightly tighten nut.
8. Remove timing belt.

Installation

NOTE: Check tensioner pulley and guide pulley for smooth operation.

1. Ensure timing pins located correctly **5**, **6** & **7**.
2. Hold sprockets. Slacken bolts **9** & **10**. Tighten bolts finger tight. Ensure sprockets can still turn.
3. Turn sprockets fully clockwise in slotted holes.
4. Fit timing belt in anti-clockwise direction, starting at crankshaft sprocket. Ensure belt is taut between sprockets.
   NOTE: Turn injection pump and camshaft sprockets slightly anti-clockwise to engage belt teeth.
5. Attach tension gauge to belt at ▽. Tool No.SEEM C.Tronic 105.5.
6. Slacken tensioner nut **8**. Insert tool in square hole in tensioner **11**. Tool No.0132X. Tension belt. New belt: 98 SEEM units. Used belt: 75 SEEM units. Tighten tensioner nut to 20 Nm **8**.
7. Remove tension gauge. Tighten bolts of each sprocket to 25 Nm **9** & **10**.
   *NOTE: Ensure bolts of each sprocket not at end of slotted holes **9** & **10**. If bolts at end of slotted holes: Repeat installation procedure.*
8. Remove timing pins **5**, **6** & **7**.
9. Turn crankshaft ten turns clockwise. Insert timing pins **5**, **6** & **7**.
10. Slacken camshaft sprocket bolts **9**.
11. Slacken injection pump sprocket bolts **10**.
12. Attach tension gauge to belt at ▽.
13. Slacken tensioner nut **8**. Tension belt. Use tool No.0132X **11**. New belt: 54 SEEM units. Used belt: 44 SEEM units.
14. Tighten tensioner nut to 20 Nm **8**.
15. Tighten bolts of each sprocket to 25 Nm **9** & **10**. Remove tension gauge.
16. Remove timing pins **5**, **6** & **7**.
17. Turn crankshaft two turns clockwise. Insert flywheel timing pin **5**.
18. Ensure timing pins can be inserted easily in sprockets **6** & **7**. If not: Repeat installation procedure.
19. Attach tension gauge to belt at ▽.
20. Insert timing pins **6** & **7**.
21. Check tension gauge reading. New belt: 51-57 SEEM units. Used belt: 41-47 SEEM units. If tension not as specified: Repeat tensioning procedure.
22. Install components in reverse order of removal.
23. Tighten crankshaft pulley bolts to 15 Nm **2**.

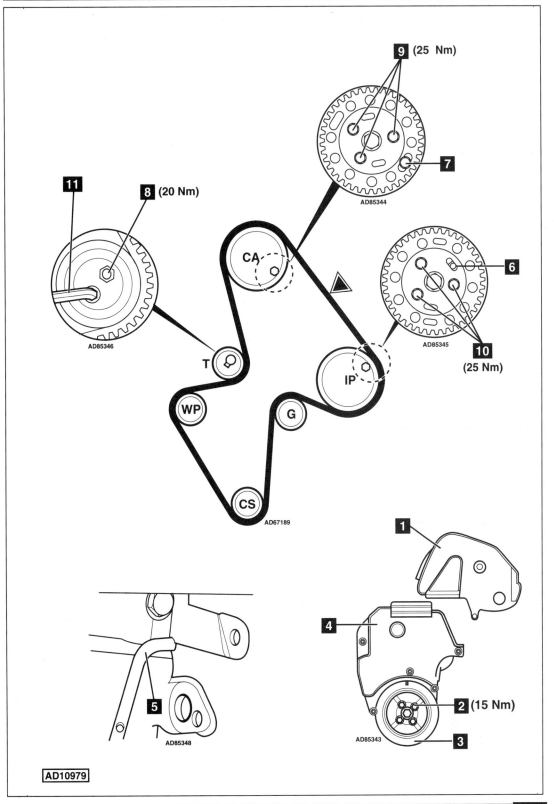

PEUGEOT

Model:	**205 1,8D/TD/1,9D • 305 1,8/1,9D • 306 1,8D/1,9D/TD • 309 1,8/1,9D/1,8TD 405 1,8TD • 405 1,9D/TD • Partner 1,8/1,9D • Talbot Horizon/Solara 1,9D J5/Talbot Express 1,9D**
Year:	**1986-01**
Engine Code:	**XUD7 (A9A), XUD9 (DHY), XUD7TE XUD9Y (DJZ), XUD9A (D9B/DJY)**

Replacement Interval Guide

Peugeot recommend replacement as follows:

→1993:

Manufacturer has no recommended interval under normal conditions.

Replacement every 45,000 miles under adverse conditions.

1993-07/98:

Replacement every 72,000 miles under normal conditions.

Replacement every 60,000 miles under adverse conditions.

08/98→:

Replacement every 80,000 miles under normal conditions.

Replacement every 48,000 miles under adverse conditions.

The previous use and service history of the vehicle must always be taken into account.
Refer to Timing Belt Replacement Intervals at the front of this manual.

Check For Engine Damage

CAUTION: This engine has been identified as an INTERFERENCE engine in which the possibility of valve-to-piston damage in the event of a timing belt failure is MOST LIKELY to occur.
A compression check of all cylinders should be performed before removing the cylinder head.

Repair Times – hrs

Remove & install:	
205/309	3,50
306	2,90
Engine undershield	+0,10
AC	+0,50
PAS + AC	+1,00
306 TD	3,40
Engine undershield	+0,10
AC	+0,50
305/405/405TD	2,90
405TD (92→)	3,40
AC	+0,10
PAS	+0,10
Partner (→07/98)	2,60
Partner (08/98→)	2,50
AC	+0,10
Horizon/Solara	3,50
J5/Express	3,20

Special Tools

- Flywheel timing pin – Peugeot No.9767.34 or 7017-TR (0.153N).
- Flywheel locking tool – Peugeot – Facom No.D86.
- Crankshaft pulley puller – Peugeot No.7015-T (0174).
- 1 bolt – M8 x 1,25 x 40 mm.
- 2 bolts – M8 x 1,25 x 35 mm.

Special Precautions

- Disconnect battery earth lead.
- DO NOT turn crankshaft or camshaft when timing belt removed.
- Remove glow plugs to ease turning engine.
- Turn engine in normal direction of rotation (unless otherwise stated).
- DO NOT turn engine via camshaft or other sprockets.
- Observe all tightening torques.
- Check diesel injection pump timing after belt replacement.

Removal

1. Raise and support front of vehicle.
2. Remove:
   - ❏ RH front wheel.
   - ❏ Engine undershield (if fitted).
   - ❏ RH wheel arch liner.
   - ❏ Auxiliary drive belts.
   - ❏ Timing belt upper covers **1** & **2**.
   - ❏ Air intake pipe (XUD7TE).
   - **NOTE: Retain rubber spacer 3.**
3. Turn crankshaft to TDC on No.1 cylinder. Insert flywheel timing pin **4**. Tool No.9767.34 or 7017-TR (0.153N).
4. Insert one bolt in camshaft sprocket and two bolts in injection pump sprocket **5** & **6**.
5. Slacken tensioner nut and bolt **7** & **8**.
6. Move tensioner away from belt. Use square drive extension and ratchet handle. Tighten bolt **7**.
7. Remove clutch housing cover plate.
8. Fit flywheel locking tool **9**. Tool No.Facom D86.
9. Remove crankshaft pulley **11**. Use puller. Tool No.7015-T (0174).
10. Remove timing belt lower cover **10**.
11. Support engine with jack or crane. Use lifting eye.

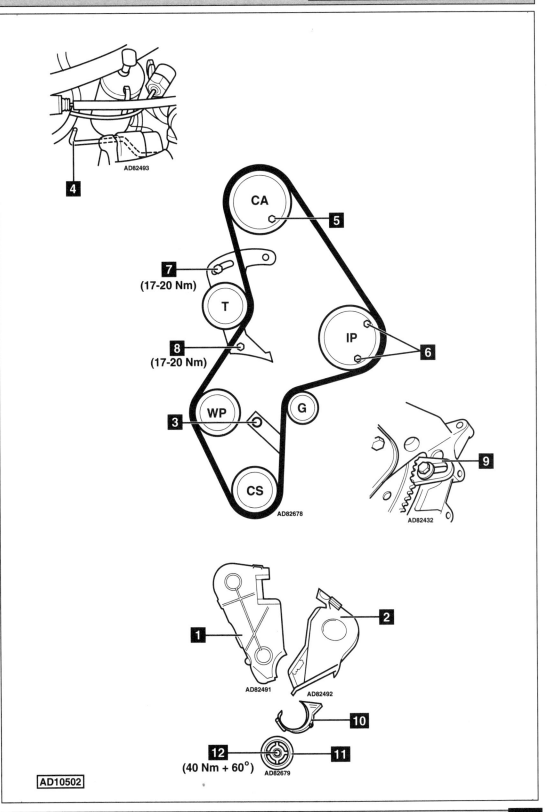

AD82493

4

CA

5

7

(17-20 Nm)

T

8

(17-20 Nm)

IP

6

WP

3

G

9

CS

AD82678

AD82432

1

AD82491

2

AD82492

10

12

11

(40 Nm + 60°)

AD82679

AD10502

12. Remove:
- ❏ Upper engine mounting.
- ❏ Timing belt.

13. Slacken tensioner bolt **7**.

14. Operate tensioner several times to ensure freedom of movement.

15. Push tensioner back and lightly tighten bolt **7**.

Installation

1. Ensure crankshaft at TDC on No.1 cylinder. Ensure timing pin and locking bolts located correctly **4**, **5** & **6**.

2. Fit timing belt in anti-clockwise direction, starting at crankshaft sprocket.

3. Slacken tensioner bolt **7**. Allow tensioner to operate. Tighten bolt.

4. Remove:
- ❏ Locking bolts **5** & **6**.
- ❏ Flywheel timing pin **4**.
- ❏ Flywheel locking tool **9**.

5. Turn crankshaft two turns clockwise.

6. Ensure timing pin and locking bolts can be inserted easily.

7. Slacken tensioner bolt **7**.

8. Tighten tensioner bolt to 17-20 Nm **7**.

9. Tighten tensioner nut to 17-20 Nm **8**.

10. Remove:
- ❏ Locking bolts **5** & **6**.
- ❏ Flywheel timing pin **4**.

11. Install components in reverse order of removal.

12. Coat crankshaft pulley bolt thread with suitable thread locking compound.

13. Fit crankshaft pulley bolt and washer **12**.
Tightening torque: 40 Nm + 60°.

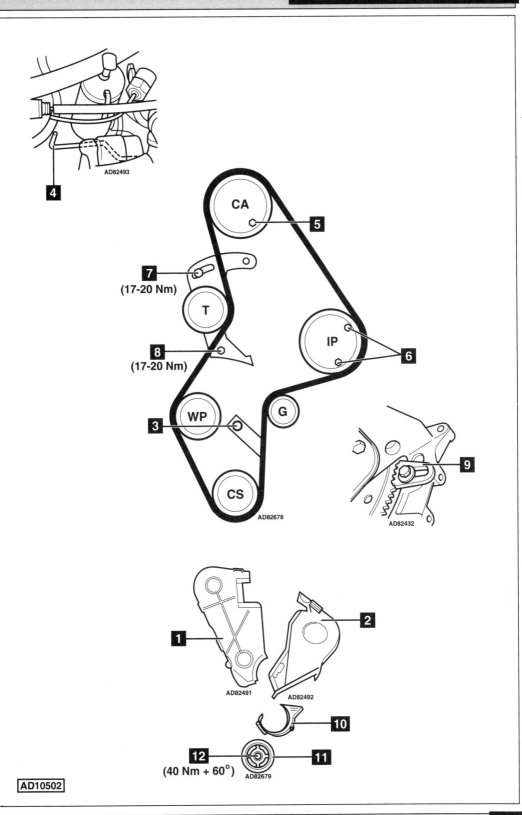

AD82493

4

CA

5

7
(17-20 Nm)

T

8
(17-20 Nm)

IP

6

WP

3

G

CS

AD82678

9

AD82432

1

AD82491

2

AD82492

10

12
(40 Nm + 60°)

11

AD82679

AD10502

PEUGEOT

Model:	**206 1,9D • 306 1,9D • Expert 1,9D • Partner 1,9D**
Year:	**1998-03**
Engine Code:	**WJY (DW8B), WJZ (DW8)**

Replacement Interval Guide

Peugeot recommend:
Replacement every 80,000 miles under normal conditions.
Replacement every 48,000 miles under adverse conditions.
The previous use and service history of the vehicle must always be taken into account.
Refer to Timing Belt Replacement Intervals at the front of this manual.

Check For Engine Damage

CAUTION: This engine has been identified as an INTERFERENCE engine in which the possibility of valve-to-piston damage in the event of a timing belt failure is MOST LIKELY to occur.
A compression check of all cylinders should be performed before removing the cylinder head(s).

Repair Times – hrs

Remove & install:

206	3,00
306	3,50
Expert	3,40
Partner	2,60

Special Tools

- Flywheel locking tool – Peugeot No.(-).0188-F.
- 206/306/Expert: Flywheel timing pin – Peugeot No.(-).0188-D.
- Partner: Flywheel timing pin – Peugeot No.(-).0188-D/0153N.
- Injection pump locking pin – Peugeot No.(-).0188-H.
- 1 bolt – M8 x 1,25 mm x 80 mm.
- Timing belt retaining clip – Peugeot No.(-).0188-K.
- Tensioning tool – Peugeot No.(-).0188-J/J1.
- Tension gauge – SEEM C.Tronic 105.5.

Special Precautions

- Disconnect battery earth lead.
- DO NOT turn crankshaft or camshaft when timing belt removed.
- Remove glow plugs to ease turning engine.
- Turn engine in normal direction of rotation (unless otherwise stated).
- DO NOT turn engine via camshaft or other sprockets.
- Observe all tightening torques.
- Check diesel injection pump timing after belt replacement.

Removal

1. Raise and support front of vehicle.
2. Disconnect and seal off fuel pipes.
3. Remove:
   - ❑ Engine upper cover.
   - ❑ RH front wheel.
   - ❑ RH splash guard.
   - ❑ Auxiliary drive belt.
   - ❑ Auxiliary drive belt tensioner pulley.
   - ❑ Flywheel housing lower cover.
4. Lock flywheel. Use tool No.(-).0188-F.
5. Remove:
   - ❑ Crankshaft pulley bolts **1**.
   - ❑ Crankshaft pulley **2**.
   - ❑ Flywheel locking tool.
6. Reposition:
   - ❑ Coolant hose from expansion tank.
   - ❑ Engine control module (ECM). DO NOT disconnect harness multi-plug.
7. Protect radiator from damage with cardboard.
8. Support engine.
9. Remove:
   - ❑ RH engine mounting and bracket.
   - ❑ Timing belt covers **3**, **4** & **5**.
10. Turn crankshaft clockwise to setting position.
11. Insert timing pin in flywheel **6**.
    Tool No.(-).0188-D/0153N.
12. Insert M8 x 1,25 mm x 80 mm bolt in camshaft sprocket **7**.
13. Insert locking pin in injection pump **8**.
    Tool No.(-).0188-H.
14. Slacken tensioner bolt **9**.
15. Turn tensioner pulley clockwise away from belt. Use tool No.(-).0188-J/J1.
16. Lightly tighten tensioner bolt **9**.
17. Remove timing belt.

Installation

1. Ensure timing pin and locking pin located correctly **6** & **8**.
2. Ensure camshaft sprocket locked with bolt **7**.
3. Slacken camshaft sprocket bolts **10**. Tighten bolts finger tight, then slacken 1/6 turn.
4. Turn camshaft sprocket fully clockwise in slotted holes.
5. Slacken injection pump sprocket bolts **11**. Tighten bolts finger tight, then slacken 1/6 turn.

 Autodata

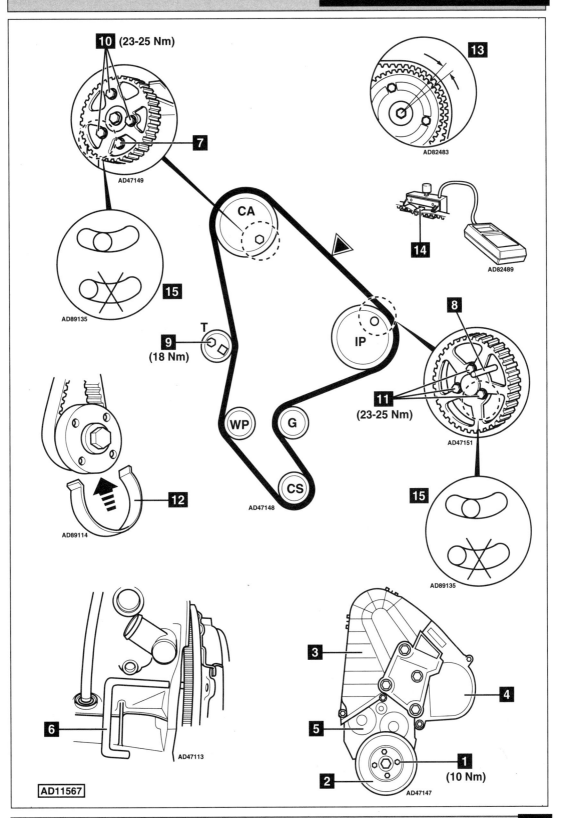

6. Turn injection pump sprocket fully clockwise in slotted holes.
   NOTE: Sprockets should turn with slight resistance.

7. Fit timing belt to crankshaft sprocket.

8. Secure belt to crankshaft sprocket with retaining clip. Tool No.(-).0188-K **12**.

9. Fit timing belt in anti-clockwise direction. Ensure belt is taut between sprockets.

10. Lay belt on injection pump sprocket teeth. Engage belt teeth by turning sprocket slightly anti-clockwise.

11. Lay belt on camshaft sprocket teeth. Engage belt teeth by turning sprocket slightly anti-clockwise.
    *NOTE: Angular movement of sprockets must not be more than one tooth space **13**.*

12. Fit timing belt to water pump sprocket and tensioner pulley.

13. Slacken tensioner bolt **9**.

14. Turn tensioner pulley anti-clockwise to temporarily tension belt.
    Use tool No.(-).0188-J/J1.

15. Lightly tighten tensioner bolt **9**.
    Tightening torque: 10 Nm.

16. Remove retaining clip **12**.

17. Attach tension gauge to belt at ▽ **14**.
    Tool No.SEEM C-Tronic 105.5

18. Turn tensioner pulley anti-clockwise until tension gauge indicates 106±2 SEEM units.

19. Tighten tensioner bolt **9**.
    Tightening torque: 18 Nm.

20. Remove tension gauge **14**.

21. Ensure bolts not at end of slotted holes in sprockets **15**.

22. If necessary: Repeat installation procedure.

23. Tighten camshaft and injection pump sprocket bolts **10** & **11**. Tightening torque: 23-25 Nm.

24. Remove timing pin and locking pin **6** & **8**.

25. Remove locking bolt from camshaft sprocket **7**.

26. Turn crankshaft eight turns clockwise to setting position.
    NOTE: DO NOT allow crankshaft to turn anti-clockwise.

27. Insert timing pin and locking pin **6** & **8**.

28. Insert locking bolt in camshaft sprocket **7**.

29. Slacken camshaft sprocket bolts **10**.

30. Slacken injection pump sprocket bolts **11**.

31. Slacken tensioner bolt to release tension on belt **9**.

32. Attach tension gauge to belt at ▽ **14**.

33. Turn tensioner pulley anti-clockwise until tension gauge indicates 42±2 SEEM units.

34. Hold tensioner pulley in position. Tighten tensioner bolt **9**. Tightening torque: 18 Nm.

35. Tighten camshaft and injection pump sprocket bolts **10** & **11**. Tightening torque: 23-25 Nm.

36. Remove tension gauge **14**.

37. Remove timing pin and locking pin **6** & **8**.

38. Remove locking bolt from camshaft sprocket **7**.

39. Turn crankshaft two turns clockwise to setting position.
    NOTE: DO NOT allow crankshaft to turn anti-clockwise.

40. Check belt tension. Attach tension gauge to belt at ▽ **14**.

41. Tension gauge should indicate the following:
    ❏ 206/306/Expert: 38-42 SEEM units.
    ❏ Partner: 38-51 SEEM units.

42. If not: Repeat tensioning procedure.

43. Remove tension gauge **14**.

44. Remove timing pin and locking pin **6** & **8**.

45. Remove locking bolt from camshaft sprocket **7**.

46. Turn crankshaft two turns clockwise to setting position.

47. Insert timing pin in flywheel **6**.

48. Ensure locking bolt can be inserted easily in camshaft sprocket **7**.

49. Ensure locking pin can be inserted easily in injection pump sprocket **8**.

50. Remove flywheel timing pin **6**.

51. Install components in reverse order of removal.

52. Tighten crankshaft pulley bolts **1**.
    Tightening torque: 10 Nm.

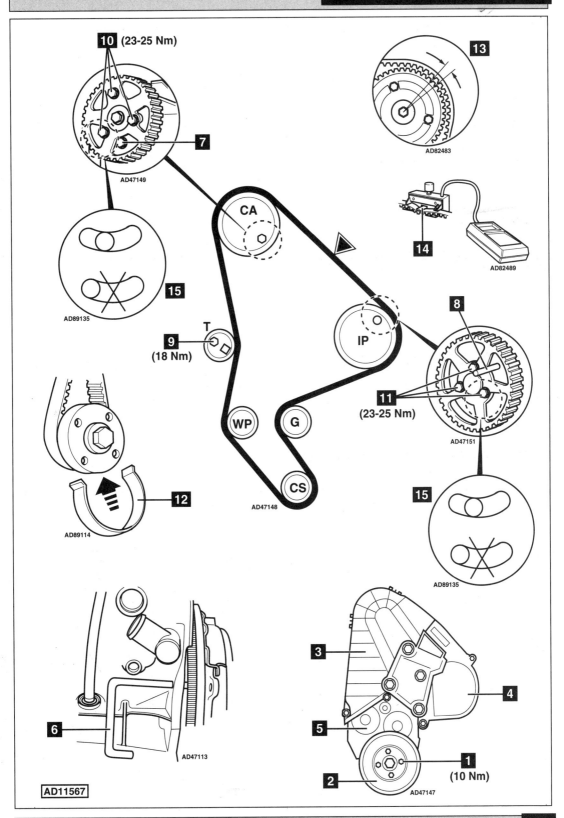

10 (23-25 Nm)

7

AD47149

15

AD89135

T

9

(18 Nm)

CA

IP

13

AD82483

14

AD82489

8

11

(23-25 Nm)

AD47151

15

AD89135

WP

G

CS

AD47148

12

AD89114

3

4

5

6

AD47113

1

(10 Nm)

2

AD47147

AD11567

PEUGEOT

Model:	206 2,0 HDi • 306 2,0 HDi • 307 2,0 HDi • 406 2,0 HDi • 607 2,0 HDi 806 2,0 HDi
Year:	1997-03
Engine Code:	RHY (DW10TD), RHZ (DW10ATED), RHW (DW10ATED4)

Replacement Interval Guide

Peugeot recommend:
Replacement every 96,000 miles under normal conditions.
Replacement every 80,000 miles under adverse conditions.
The previous use and service history of the vehicle must always be taken into account.
Refer to Timing Belt Replacement Intervals at the front of this manual.

Check For Engine Damage

CAUTION: This engine has been identified as an INTERFERENCE engine in which the possibility of valve-to-piston damage in the event of a timing belt failure is MOST LIKELY to occur.
A compression check of all cylinders should be performed before removing the cylinder head(s).

Repair Times – hrs

Remove & install:

206/306/406	3,00
307	3,30
607	3,10
806	4,50

Special Tools

- Flywheel locking tool – Peugeot No.(-).0188-F.
- Crankshaft pulley puller – Peugeot No.(-).0188-P2.
- Flywheel timing pin – Peugeot No.(-).0188-Y.
- Camshaft timing pin – Peugeot No.(-).0188-M.
- Timing belt retaining clip – Peugeot No.(-).0188-K.
- Tensioning tool – Peugeot No.(-).0188-J2.
- Tension gauge – SEEM C.Tronic 105.5.
- Set of blanking plugs – Peugeot No.(-).0188-T.

Special Precautions

- Disconnect battery earth lead.
- DO NOT turn crankshaft or camshaft when timing belt removed.
- Remove glow plugs to ease turning engine.
- Turn engine in normal direction of rotation (unless otherwise stated).
- DO NOT turn engine via camshaft or other sprockets.
- Observe all tightening torques.

Removal

NOTE: The fuel pump fitted to this engine does not require timing.

1. Raise and support front of vehicle.
2. Disconnect exhaust front pipe from manifold.
3. Remove:
   ❑ RH front wheel.
   ❑ RH splash guard.
   ❑ Engine upper cover.
   ❑ Auxiliary drive belt.
   ❑ Turbocharger air hoses.
   ❑ Flywheel housing lower cover.
4. Lock flywheel. Use tool No.(-).0188-F.
5. Remove:
   ❑ 607: Emission filter from pre-catalytic converter.
   ❑ Crankshaft pulley bolt **1**.
   ❑ Crankshaft pulley **2**.
      Use tool No.(-).0188-P2.
   ❑ Flywheel locking tool.
   ❑ Lower torque reaction link.
6. Reposition:
   ❑ Except 307/607: Coolant expansion tank. DO NOT disconnect hoses.
   ❑ Except 307/607: Engine control module (ECM) and tray (leave harness multi-plug connected).
7. Except 307: Disconnect and seal off fuel pipes. Use tool No.(-).01888-T.
8. Support engine.
9. Remove:
   ❑ RH engine mounting and bracket.
   NOTE: Protect radiator as engine may move.
   ❑ Timing belt covers **4**, **5** & **6**.
10. Turn crankshaft clockwise to setting position.
11. Insert timing pin in flywheel **3**.
    Tool No.(-).0188-Y.
12. Insert timing pin in camshaft sprocket **7**.
    Tool No.(-).0188-M.
13. Slacken tensioner bolt **8**.
14. Turn tensioner pulley clockwise away from belt. Use tool No.(-).0188-J2.
15. Lightly tighten tensioner bolt **8**.
16. Remove timing belt.

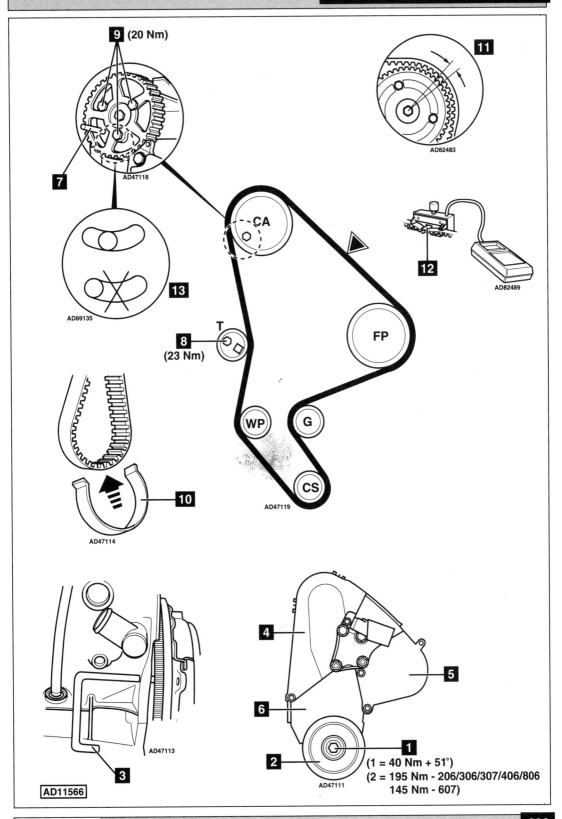

9 (20 Nm)

11

AD82483

7

AD47118

CA

12

AD82489

13

AD89135

8
(23 Nm)

T

FP

WP

G

10

AD47114

CS

AD47119

4

5

6

1

2

3

AD47113

AD11566

AD47111

(1 = 40 Nm + 51°)
(2 = 195 Nm - 206/306/307/406/806
145 Nm - 607)

Installation

1. Ensure timing pins located correctly **3** & **7**.
2. Slacken camshaft sprocket bolts **9**. Tighten bolts finger tight, then slacken 1/6 turn.
3. Turn camshaft sprocket fully clockwise in slotted holes.
   NOTE: Sprocket should turn with slight resistance.
4. Fit timing belt to crankshaft sprocket.
5. Secure belt to crankshaft sprocket with retaining clip. Tool No.(-).0188-K **10**.
6. Fit timing belt in anti-clockwise direction. Ensure belt is taut between sprockets.
7. Lay belt on camshaft sprocket teeth. Engage belt teeth by turning sprocket slightly anti-clockwise.
   *NOTE: Angular movement of sprocket must not be more than one tooth space **11**.*
8. Fit timing belt to water pump sprocket and tensioner pulley.
9. Slacken tensioner bolt **8**.
10. Turn tensioner pulley anti-clockwise to temporarily tension belt. Use tool No.(-).0188-J2.
11. Lightly tighten tensioner bolt **8**. Tightening torque: 10 Nm.
12. Remove retaining clip **10**.
13. Attach tension gauge to belt at ▽ **12**. Tool No.SEEM C-Tronic 105.5.
14. Turn tensioner pulley anti-clockwise until tension gauge indicates the following:
    ❑ Except 307/607/806: 106±2 SEEM units.
    ❑ 307/607/806: 98 SEEM units.
15. Tighten tensioner bolt **8**. Tightening torque: 23 Nm.
16. Remove tension gauge **12**.
17. Ensure sprocket bolts not at end of slotted holes **13**.
18. If necessary: Repeat installation procedure.
19. Tighten camshaft sprocket bolts **9**. Tightening torque: 20 Nm.
20. Remove timing pins **3** & **7**.
21. Turn crankshaft eight turns clockwise to setting position.
    NOTE: DO NOT allow crankshaft to turn anti-clockwise.
22. Insert timing pins **3** & **7**.
23. Slacken camshaft sprocket bolts **9**.
24. Slacken tensioner bolt to release tension on belt **8**.
25. Attach tension gauge to belt at ▽ **12**.
26. Turn tensioner pulley anti-clockwise until tension gauge indicates 54±2 SEEM units.
27. Hold tensioner pulley in position. Tighten tensioner bolt **8**. Tightening torque: 23 Nm.

28. Tighten camshaft sprocket bolts **9**. Tightening torque: 20 Nm.
29. Remove tension gauge **12**.
30. Check belt tension. Attach tension gauge to belt at ▽ **12**. Tension gauge should indicate 51-57 SEEM units.
31. If not: Repeat tensioning procedure.
32. Remove tension gauge **12**.
33. Remove timing pins **3** & **7**.
34. Turn crankshaft two turns clockwise to setting position.
35. Insert timing pin in flywheel **3**.
36. Ensure timing pin can be inserted easily **7**.
37. Remove timing pin **3**.
38. Install components in reverse order of removal.
    NOTE: Engine mounting bolts tightening torque: 60 Nm.
39. Clean crankshaft pulley bolt and crankshaft threads.
40. Coat crankshaft pulley bolt with suitable thread locking compound.
41. Tighten crankshaft pulley bolt **1**. Tightening torque: 40 Nm + 51°.
42. Check torque setting of crankshaft pulley bolt **1**.
    ❑ Except 607: 195 Nm.
    ❑ 607: 145 Nm.

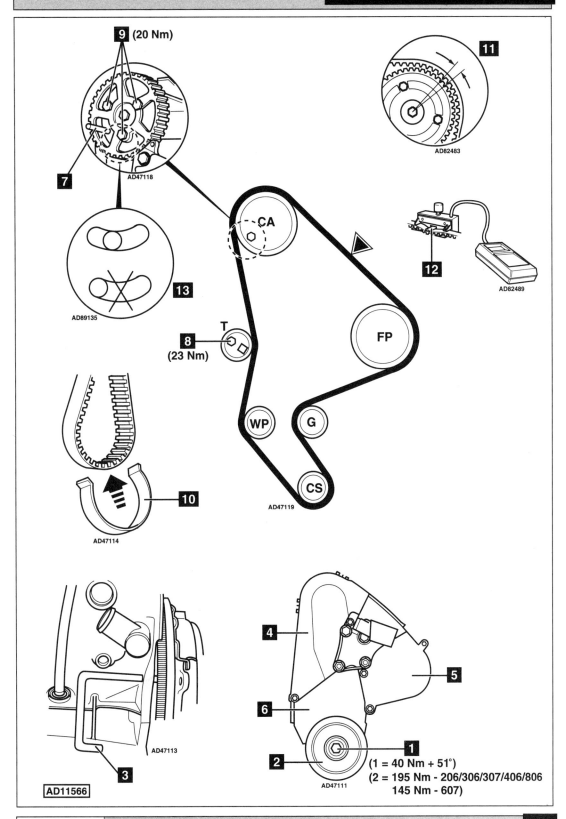

9 (20 Nm)

7

AD47118

11

AD82483

13

AD89135

CA

12

AD82489

T

8

(23 Nm)

FP

WP

G

CS

AD47119

10

AD47114

AD47113

3

AD11566

4

5

6

2

1

AD47111

(1 = 40 Nm + 51°)
(2 = 195 Nm - 206/306/307/406/806
145 Nm - 607)

PEUGEOT

Model:	**406 1,9TD • 806 1,9TD • Expert 1,9D/TD • Boxer 1,9D/TD**
Year:	**1994-01**
Engine Code:	**XUD9TF (D8B), XUD9AU (D8C), XUD9A (D9B), XUD9SD (DHW), XUD9UTF (DHX)**

Replacement Interval Guide

Peugeot recommend replacement as follows:

→07/98:
Replacement every 72,000 miles under normal conditions.
Replacement every 54,000 miles under adverse conditions.

08/98→:
Replacement every 80,000 miles under normal conditions.
Replacement every 48,000 miles under adverse conditions.
The previous use and service history of the vehicle must always be taken into account.
Refer to Timing Belt Replacement Intervals at the front of this manual.

Check For Engine Damage

CAUTION: This engine has been identified as an INTERFERENCE engine in which the possibility of valve-to-piston damage in the event of a timing belt failure is MOST LIKELY to occur.
A compression check of all cylinders should be performed before removing the cylinder head.

Repair Times – hrs

Remove & install:	
406	3,50
Engine undershield	+0,10
AC	+0,30
806 (→07/98)	3,50
806 (08/98→)	3,40
AC	+0,10
Boxer/Expert (→07/98)	3,30
Boxer/Expert (08/98→)	3,20

Special Tools

- Flywheel timing pin – Peugeot No.0.153N.
- Flywheel locking tool – Facom No.D86.
- 1 bolt – M8 x 1,25 x 40 mm.
- 2 bolts – M8 x 1,25 x 35 mm.

Removal

1. Raise and support front of vehicle.
2. Support engine.
3. Remove:
   - ❏ RH front wheel.
   - ❏ Engine undershield (if fitted).
   - ❏ RH wheel arch liner.
   - ❏ Auxiliary drive belt.
   - ❏ Upper engine mounting.
4. Reposition pipes and wiring adjacent to timing belt cover.
5. Remove timing belt covers **1** & **2**.
6. Remove retaining spacer **3** (except 406).

7. Turn crankshaft clockwise to TDC on No.1 cylinder. Insert flywheel timing pin **4**. Tool No.0.153N.
8. Insert one bolt in camshaft sprocket and two bolts in injection pump sprocket **5** & **6**.
9. Slacken tensioner nut **7**.
10. Slacken tensioner bolt **8**.
11. Turn tensioner away from belt to compress spring **9**. Use square drive in tensioner. Lightly tighten bolt **8**.
12. Remove clutch housing cover plate. Fit flywheel locking tool **10**. Tool No.Facom D86.
13. Remove:
    - ❏ Crankshaft pulley bolt and washer **11**.
    - ❏ Crankshaft pulley **12**.
    - ❏ Flywheel locking tool **10**.
    NOTE: An extractor may be required to remove crankshaft pulley.
14. 406: Remove timing belt lower cover **13**.
15. Remove timing belt.

Installation

1. Check tensioner spring and piston slide freely **9**.
2. Ensure crankshaft at TDC.
3. Ensure timing pin and locking tool located correctly **4** & **10**.
4. Ensure locking bolts located correctly **5** & **6**.
5. Fit timing belt in anti-clockwise direction, starting at crankshaft sprocket. Ensure belt is taut between sprockets.
6. Temporarily fit crankshaft pulley bolt and washer **11**.
7. Slacken tensioner bolt **8**.
8. Remove:
   - ❏ Locking bolts **5** & **6**.
   - ❏ Flywheel timing pin **4**.
   - ❏ Flywheel locking tool **10**.
9. Turn crankshaft two turns clockwise.
10. Ensure timing pin and locking bolts can be inserted easily **4**, **5** & **6**.
11. Tighten tensioner bolt to 17,5 Nm **8**.
12. Tighten tensioner nut to 17,5 Nm **7**.
13. Remove:
    - ❏ Locking bolts **5** & **6**.
    - ❏ Flywheel timing pin **4**.
    - ❏ Crankshaft pulley bolt and washer **11**.
14. Install components in reverse order of removal.
15. Coat crankshaft pulley bolt thread with suitable thread locking compound.
16. Tighten crankshaft pulley bolt **11**.
    Tightening torque: 40 Nm + 60°.

Autodata

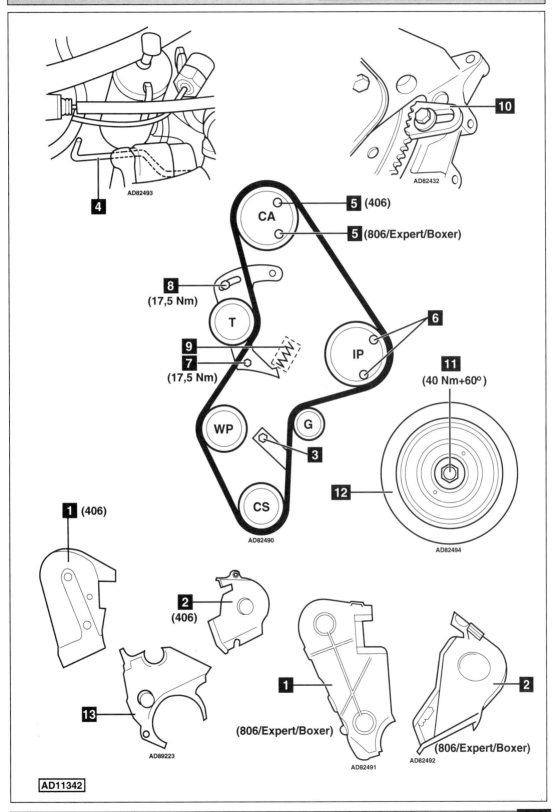

4 — AD82493

10 — AD82432

CA — 5 (406)
5 (806/Expert/Boxer)

8 — (17,5 Nm)

T

9
7 — (17,5 Nm)

IP — 6

11 (40 Nm+60°)

WP — G — 3

12

CS

AD82490

AD82494

1 (406)

2 (406)

13 — AD89223

1 (806/Expert/Boxer)
AD82491

2 (806/Expert/Boxer)
AD82492

AD11342

PEUGEOT

Model:	406 2,1D Turbo • 605 2,1D Turbo • 806 2,1D Turbo
Year:	1994-00
Engine Code:	XUD 11 BTE (P8C)

Replacement Interval Guide

Peugeot recommend replacement as follows:

→07/98:
Replacement every 72,000 miles under normal conditions.
Replacement every 60,000 miles under adverse conditions.

08/98→:
Replacement every 80,000 miles under normal conditions.
Replacement every 48,000 miles under adverse conditions.
The previous use and service history of the vehicle must always be taken into account.
Refer to Timing Belt Replacement Intervals at the front of this manual.

Check For Engine Damage

CAUTION: This engine has been identified as an INTERFERENCE engine in which the possibility of valve-to-piston damage in the event of a timing belt failure is MOST LIKELY to occur.
A compression check of all cylinders should be performed before removing the cylinder head(s).

Repair Times – hrs

Remove & install:	
406	3,50
Engine undershield	+0,10
AC	+0,30
605 (→07/98)	3,80
605 (08/98→)	3,70
AC	+0,90
806 (→07/98)	2,70
806 (08/98→)	3,40
AC	+0,10

Special Tools

- Camshaft timing pin – M8 x 1,25 x 40 mm bolt.
- Flywheel locking tool – Peugeot No.6012T.
- Flywheel timing pin – Peugeot No.7014-TJ.
- Injection pump timing pin – M8 x 1,25 x 35 mm bolt.

Special Precautions

- Disconnect battery earth lead.
- DO NOT turn crankshaft or camshaft when timing belt removed.
- Remove glow plugs to ease turning engine.
- Turn engine in normal direction of rotation (unless otherwise stated).
- DO NOT turn engine via camshaft or other sprockets.
- Observe all tightening torques.
- Check diesel injection pump timing after belt replacement.

Removal

1. Raise and support front of vehicle.
2. Remove:
   - RH front wheel.
   - Engine undershield (if fitted).
   - RH wheel arch liner.
   - Auxiliary drive belts.
3. Turn crankshaft clockwise. Insert timing pin in flywheel **1**. Tool No.7014-T.J.
4. Support engine.
5. Remove:
   - Engine torque rod.
   - RH engine mounting.
   - Timing belt upper cover **2**.
6. Insert timing pins in camshaft and injection pump sprockets **3** & **4**.
7. Remove clutch housing plate. Fit flywheel locking tool **5**. Tool No.6012T.
8. Remove:
   - Crankshaft pulley **6**.
   - Flywheel locking tool **5**.
   - Timing belt lower cover **7**.
9. Slacken tensioner nut and eccentric bolt **8** & **9**.
   NOTE: Accessible through hole in engine mounting bracket (if not removed).
10. Turn eccentric clockwise to release tensioner.
11. Tighten tensioner nut **8**.
12. Remove timing belt.

Installation

1. Ensure timing pins located correctly **1**, **3** & **4**.
2. Check tensioner pulley is held in released position. Turn eccentric anti-clockwise **9**.
3. Fit timing belt in following order:
   - Crankshaft sprocket.
   - Guide pulley.
   - Injection pump sprocket.
   - Camshaft sprocket.
   - Water pump sprocket.
   - Tensioner pulley.
   NOTE: Ensure belt is taut between sprockets. Take care not to twist belt.
4. Slacken tensioner nut **8**.
5. Remove timing pins **1**, **3** & **4**.
6. Turn crankshaft two turns clockwise.
7. Tighten tensioner nut **8**.
8. Turn crankshaft a further two turns clockwise.
9. Ensure timing pins can be inserted easily **1**, **3** & **4**.
10. Slacken tensioner nut one turn **8**.
11. Allow tensioner to operate.
12. Tighten tensioner nut to 10 Nm **8**.
13. Tighten eccentric bolt to 10 Nm **9**.
14. Install components in reverse order of removal.
15. Coat crankshaft pulley bolt thread with suitable thread locking compound.
16. Tighten crankshaft pulley bolt **10**.
    Tightening torque: 40 Nm + 60°.

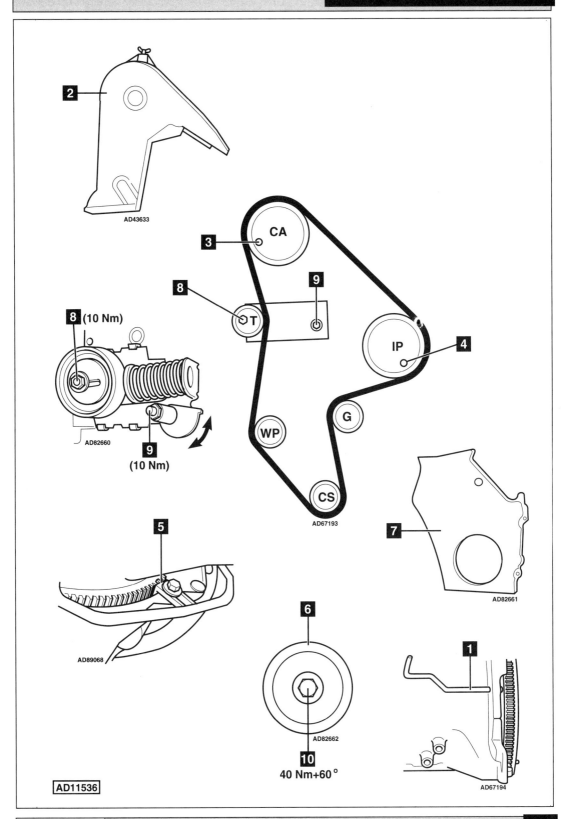

AD43633

3 — CA

8 — T

9

8 (10 Nm)

AD82660

9
(10 Nm)

IP

4

G

WP

CS

AD67193

7

AD82661

5

AD89068

6

AD82662

10
40 Nm+60°

1

AD67194

AD11536

Model:	406 2,2 HDi • 607 2,2 HDi
Year:	2000-03
Engine Code:	4HZ, 4HZ (DW12ATED)

Replacement Interval Guide

Peugeot recommend replacement every 96,000 miles or every 10 years under normal conditions or 75,000 miles or every 10 years under adverse conditions.

The previous use and service history of the vehicle must always be taken into account.

Refer to Timing Belt Replacement Intervals at the front of this manual.

Check For Engine Damage

CAUTION: This engine has been identified as an INTERFERENCE engine in which the possibility of valve-to-piston damage in the event of a timing belt failure is MOST LIKELY to occur.

A compression check of all cylinders should be performed before removing the cylinder head(s).

Repair Times – hrs

Remove & install	3,10

Special Tools

- Flywheel locking tool – Peugeot No.(-).188.F.
- Flywheel timing pin – Peugeot No.(-).0188.X.
- Camshaft timing pin – Peugeot No.(-).0188.M.
- Tensioning tool – Peugeot No.(-).0188.J2.
- Timing belt retaining clip – Peugeot No.(-).0188.K.
- Tension gauge – SEEM CTG 105.M.

Special Precautions

- Disconnect battery earth lead.
- DO NOT turn crankshaft or camshaft when timing belt removed.
- Remove glow plugs to ease turning engine.
- Turn engine in normal direction of rotation (unless otherwise stated).
- DO NOT turn engine via camshaft or other sprockets.
- Observe all tightening torques.

Removal

NOTE: The fuel pump fitted to this engine does not require timing.

1. Remove:
   - ❏ Engine upper cover.
   - ❏ Battery cover.
2. Raise and support front of vehicle.
3. Remove:
   - ❏ Engine lower cover.
   - ❏ RH front wheel.
   - ❏ RH splash guard.
   - ❏ Auxiliary drive belt.
4. Disconnect exhaust front pipe from manifold.
5. Support engine.

6. Remove:
   - ❏ Coolant expansion tank (leave hoses connected).
   - ❏ Torque reaction link.
   - ❏ RH engine mounting and bracket.
   - ❏ Flywheel housing lower cover.
7. Lock flywheel. Use tool No.(-).188.F.
8. Remove:
   - ❏ Crankshaft pulley bolt **1**.
   - ❏ Crankshaft pulley **2**.
   - ❏ Flywheel locking tool.
   - ❏ Bolt **3**.
   - ❏ Timing belt upper covers **4**.
   - ❏ Timing belt lower cover **5**.
9. Refit bolt fitted with a 17 mm thick spacer **3**. Tighten bolt to 15 Nm.
10. Turn crankshaft clockwise to setting position.
11. Insert timing pin in flywheel **6**. Tool No.(-).0188.X.
12. Insert timing pin in camshaft sprocket **7**. Tool No.(-).0188.M.
13. Slacken tensioner bolt **8**.
14. Slacken camshaft sprocket bolts **9**.
15. Turn tensioner pulley clockwise away from belt. Use tool No.(-).0188.J2.
16. Lightly tighten tensioner bolt **8**.
17. Remove timing belt.

Installation

SPECIAL NOTE: Supplementary information for engine/cylinder head overhaul.
*Align timing chain links (black) **10** & **11** with marked camshaft sprocket teeth **12** & **13**.*

1. Ensure timing pins located correctly **6** & **7**.
2. Tighten bolts finger tight **9**.
3. Turn camshaft sprocket fully clockwise in slotted holes.
4. Fit timing belt to crankshaft sprocket.
5. Secure belt to crankshaft sprocket with retaining clip. Tool No.(-).0188.K **14**.
6. Fit timing belt in anti-clockwise direction. Ensure belt is taut between sprockets.
7. Lay belt on camshaft sprocket teeth. Engage belt teeth by turning sprocket slightly anti-clockwise.

   *NOTE: Angular movement of sprocket must not be more than one tooth space **15**.*
8. Fit timing belt to water pump sprocket and tensioner pulley.
9. Slacken tensioner bolt **8**.

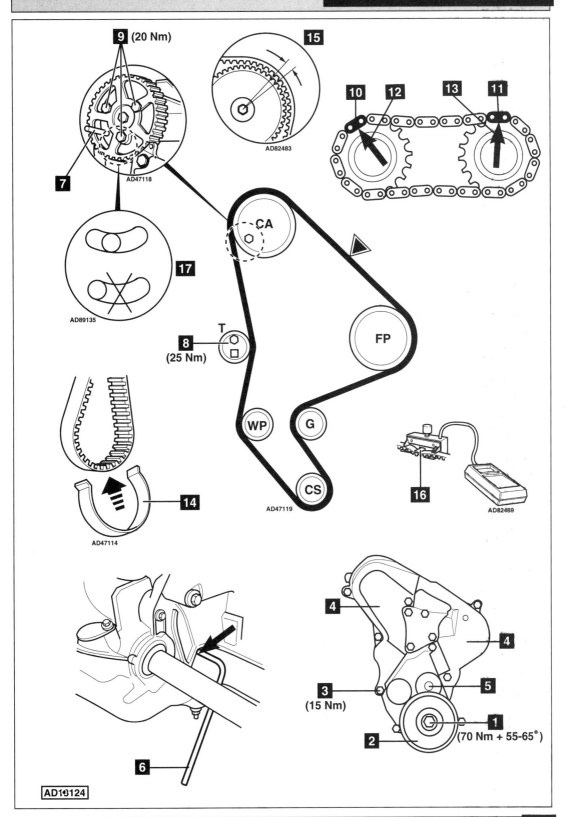

9 (20 Nm)

15

AD82483

10 **12** **13** **11**

7

AD47118

17

AD89135

CA

▶

T

8
(25 Nm)

FP

WP **G**

14

AD47114

CS

AD47119

16

AD82489

4

4

3
(15 Nm)

5

1
(70 Nm + 55-65°)

2

6

AD10124

10. Turn tensioner pulley anti-clockwise to temporarily tension belt. Use tool No.(-).0188.J2.

11. Lightly tighten tensioner bolt **8**.

12. Remove retaining clip **14**.

13. Attach tension gauge to belt at ▽ **16**. Tool No.SEEM CTG 105.M.

14. Turn tensioner pulley anti-clockwise until tension gauge indicates 106±2 SEEM units.

15. Tighten tensioner bolt **8**. Tightening torque: 25 Nm.

16. Remove tension gauge **16**.

17. Ensure sprocket bolts not at end of slotted holes **17**.

18. If necessary: Repeat installation procedure.

19. Tighten camshaft sprocket bolts **9**. Tightening torque: 20 Nm.

20. Remove timing pins **6** & **7**.

21. Turn crankshaft eight turns clockwise to setting position.
    NOTE: DO NOT allow crankshaft to turn anti-clockwise.

22. Insert timing pin **6**.

23. Slacken camshaft sprocket bolts **9**.

24. Insert timing pin **7**.

25. Slacken tensioner bolt to release tension on belt **8**.

26. Attach tension gauge to belt at ▽ **16**.

27. Turn tensioner pulley anti-clockwise until tension gauge indicates 51±3 SEEM units.

28. Hold tensioner pulley in position. Tighten tensioner bolt **8**. Tightening torque: 25 Nm.

29. Tighten camshaft sprocket bolts **9**. Tightening torque: 20 Nm.

30. Remove tension gauge **16**.

31. Check belt tension: Attach tension gauge to belt at ▽ **16**. Tension gauge should indicate 51±3 SEEM units.

32. If not: Repeat tensioning procedure.

33. Remove tension gauge **16**.

34. Remove timing pins **6** & **7**.

35. Turn crankshaft two turns clockwise to setting position.

36. Insert timing pin in flywheel **6**.

37. Ensure timing pin can be inserted easily **7**.

38. Remove timing pin **6**.

39. Install components in reverse order of removal.

40. Remove bolt and 17 mm spacer. Refit bolt **3**. Tighten bolt to 15 Nm **3**.

41. Clean crankshaft pulley bolt and crankshaft threads.

42. Coat crankshaft pulley bolt with suitable thread locking compound.

43. Tighten crankshaft pulley bolt **1**. Tightening torque: 70 Nm + 55-65°.

44. Check torque setting of crankshaft pulley bolt **1**. Tightening torque: 260 Nm.

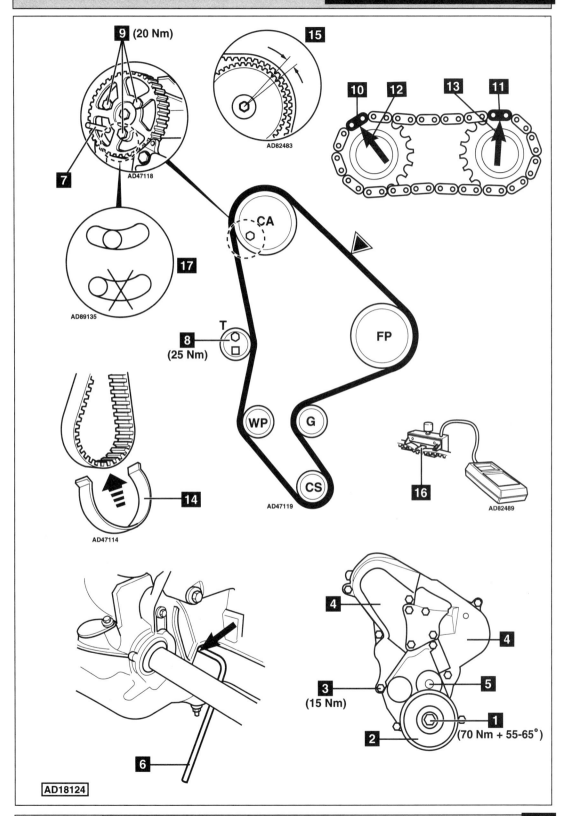

9 (20 Nm)

15

AD82483

10 **12**   **13** **11**

7

AD47118

17

AD89135

CA

FP

T
8
(25 Nm)

WP **G**

CS

AD47119

14

AD47114

16

AD82489

4

4

3
(15 Nm)

5

1
(70 Nm + 55-65°)

2

6

AD18124

PEUGEOT

Model:	**605 2,1D/TD**
Year:	**1989-98**
Engine Code:	**XUD11A, XUD11ATE**

Check For Engine Damage

CAUTION: This engine has been identified as an INTERFERENCE engine in which the possibility of valve-to-piston damage in the event of a timing belt failure is MOST LIKELY to occur.
A compression check of all cylinders should be performed before removing the cylinder head.

Repair Times – hrs

Remove & install:

→07/98	3,80
08/98→	3,70
AC	+0,90

Special Tools

- Flywheel timing pin – Peugeot No.976734 (0.153N).
- Flywheel locking tool – Peugeot-Facom No.D86.
- 1 bolt – M8 x 1,25 x 40 mm.
- 2 bolts – M8 x 1, 25 x 35 mm.

Special Precautions

- Disconnect battery earth lead.
- DO NOT turn crankshaft or camshaft when timing belt removed.
- Remove glow plugs to ease turning engine.
- Turn engine in normal direction of rotation (unless otherwise stated).
- DO NOT turn engine via camshaft or other sprockets.
- Observe all tightening torques.
- Check diesel injection pump timing after belt replacement.

Removal

1. Raise and support front of vehicle.
2. Remove:
   - RH front wheel.
   - RH wheel arch liner.
   - Auxiliary drive belts.
3. Turn crankshaft clockwise to setting position. Insert flywheel timing pin **1**. Tool No.976734 (0.153N).

4. Remove:
   - Torque reaction link.
   - Timing belt upper cover **2**.
   - Timing belt side cover **3**.
5. Insert one bolt in camshaft sprocket and two bolts in injection pump sprocket **4** & **5**. Tighten bolts finger tight.
   *NOTE: If locating holes are not aligned: Remove timing pin **1**. Turn crankshaft one turn clockwise.*
6. Slacken tensioner locknut **6**. Slacken eccentric locknut **7**.
   NOTE: Accessible through hole in engine mounting bracket.
7. Turn eccentric clockwise to release tensioner.
8. Remove clutch housing plate. Fit flywheel locking tool. Tool No.Facom D86.
9. Remove:
   - Crankshaft pulley bolt **8**.
   - Crankshaft pulley **9**.
   - Timing belt lower cover **10**.
   - Flywheel locking tool.
10. Support engine. Remove upper engine mounting.
11. Remove timing belt.

Installation

1. Ensure timing pin and locking bolts located correctly **1**, **4** & **5**.
2. Check tensioner pulley is held in released position. Lightly tighten tensioner locknut. Turn eccentric anti-clockwise **7**.
3. Fit timing belt in anti-clockwise direction, starting at crankshaft sprocket.
4. Ensure belt is taut between sprockets. Take care not to twist belt.
5. Slacken tensioner locknut **6**. Remove timing pin and locking bolts **1**, **4** & **5**.
6. Turn crankshaft two turns clockwise. Lightly tighten tensioner locknut **6**.
7. Turn crankshaft two turns clockwise. Slacken tensioner locknut **6**. Allow tensioner to operate.
8. Tighten tensioner locknut to 10 Nm **6**. Tighten eccentric locknut to 10 Nm **7**.
9. Ensure timing pin and locking bolts can be inserted easily **1**, **4** & **5**.
10. Fit timing belt covers.
11. Install:
    - Crankshaft pulley.
    - Flywheel locking tool.
12. Clean crankshaft pulley bolt. Coat threads with 'Loctite Threadlok'. Fit bolt and washer.
13. Tighten crankshaft pulley bolt **8**.
    Tightening torque: 40 Nm + 60°, then to 170 Nm.
14. Remove flywheel locking tool. Fit clutch housing plate.
15. Install components in reverse order of removal.

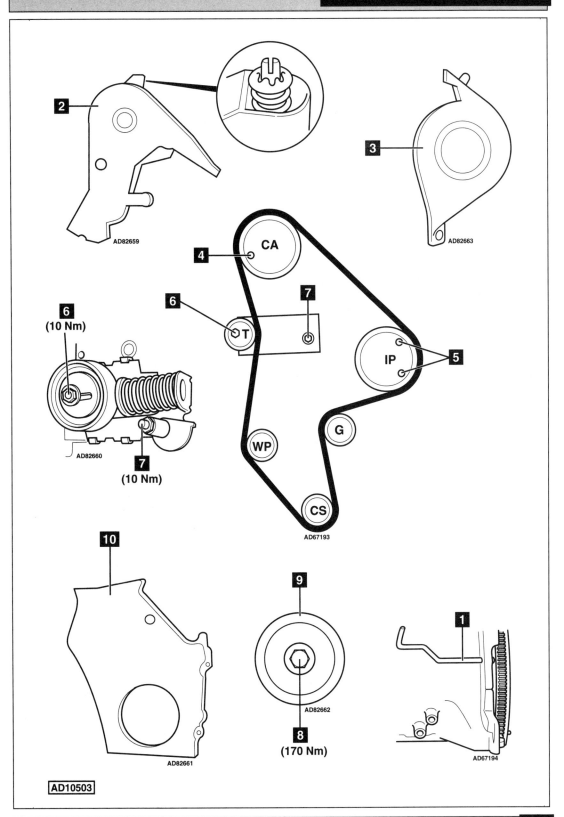

PEUGEOT

Model:	**Boxer 2,5D/TD**
Year:	**1994-02**
Engine Code:	**DJ5 (T9A), DJ5T (T8A)**

Replacement Interval Guide

Peugeot recommend:

→07/98:
Replacement every 72,000 miles under normal conditions.

08/98→:
Replacement every 80,000 miles under normal conditions.
Replacement every 48,000 miles under adverse conditions.
The previous use and service history of the vehicle must always be taken into account.
Refer to Timing Belt Replacement Intervals at the front of this manual.

Check For Engine Damage

CAUTION: This engine has been identified as an INTERFERENCE engine in which the possibility of valve-to-piston damage in the event of a timing belt failure is MOST LIKELY to occur.
A compression check of all cylinders should be performed before removing the cylinder head.

Repair Times – hrs

Remove & install:

→07/98	2,20
08/98→	2,10
AC	+0,30

Special Tools

- Flywheel timing pin – Peugeot No.(-).0153N.
- Camshaft timing pin – Peugeot No.(-).0178A.
- Injection pump timing pin – Peugeot No.(-).0178B or C.
- Tensioning lever – Peugeot No.(-).0178E.
- Tension gauge – SEEM C.Tronic 105 or 105.5.

Removal

1. Raise and support front of vehicle.
2. Disconnect battery earth lead.
3. Remove:
   - ❑ RH front wheel.
   - ❑ Engine undershield.
   - ❑ Auxiliary drive belt.
   - ❑ Timing belt upper cover **1**.
   - ❑ Crankshaft pulley **2**.
   - ❑ Timing belt lower cover **3**.
4. Turn crankshaft clockwise to setting position.
5. Insert timing pin in flywheel. Tool No.0153N.
   - ❑ Type 1 – **4**.
   - ❑ Type 2 – **12**.
6. Insert timing pin in camshaft sprocket **5**.
   Tool No.0178A.
7. Insert timing pin in injection pump sprocket **6**.
   NOTE: Bosch pump – use timing pin No.0178B (9,5 mm).
   Lucas pump – use timing pin No.0178C (6 mm).
8. Slacken camshaft sprocket bolts **7**. Tighten bolts finger tight. Then slacken 60°.

9. Slacken injection pump sprocket bolts **8**. Tighten bolts finger tight. Then slacken 60°.
10. Slacken tensioner nut **9**. Move tensioner away from belt. Lightly tighten nut.
11. Remove timing belt.
12. Turn camshaft and injection pump sprockets fully clockwise in slotted holes.

Installation

1. Ensure tensioner pulley and guide pulley in good condition.
2. Fit timing belt in anti-clockwise direction, starting at crankshaft sprocket.
   NOTE: If necessary: Turn crankshaft sprocket to engage in nearest belt tooth.
3. Attach tension gauge to belt at ▽ **10**.
   Tool No.SEEM C.Tronic 105/105.5.
4. Insert tensioning lever in square hole in tensioner. Tool No.0178E **11**.
5. Tension belt to 107 SEEM units. Tighten tensioner nut to 45 Nm **9**.
6. Remove tension gauge.
7. Tighten camshaft sprocket bolts to 10 Nm and then finally to 25 Nm **7**.
8. Tighten injection pump sprocket bolts to 10 Nm and then finally to 25 Nm **8**.
9. Remove timing pins **4** or **12**, **5** & **6**.
10. Turn crankshaft 10 turns clockwise.
11. Insert flywheel timing pin. Turn crankshaft clockwise until timing pin engages.
    NOTE: DO NOT turn crankshaft anti-clockwise.
12. Slacken camshaft sprocket bolts **7**. Tighten bolts finger tight. Then slacken 60°.
13. Slacken injection pump sprocket bolts **8**. Tighten bolts finger tight. Then slacken 60°.
14. Slacken tensioner nut to release tension on belt **9**.
15. Insert timing pin in camshaft sprocket **5**.
16. Insert timing pin in injection pump sprocket **6**.
    NOTE: Turn sprockets with a spanner if necessary.
17. Attach tension gauge to belt at ▽ **10**.
18. Insert tensioning lever in square hole in tensioner. Tool No.0178 **11**.
19. Tension belt to 58 SEEM units. Tighten tensioner nut **9**. Tightening torque: 45 Nm.
20. Remove tension gauge.
21. Tighten camshaft sprocket bolts to 10 Nm and then finally to 25 Nm **7**.
22. Tighten injection pump sprocket bolts to 10 Nm and then finally to 25 Nm **8**.
23. Remove timing pins **4** or **12**, **5** & **6**.
24. Turn crankshaft two turns clockwise.
25. Ensure timing pins can be inserted easily **4**, **5** & **6**.
26. Fit crankshaft pulley **2**. Tighten bolts to 20 Nm.
27. Check and adjust injection pump timing.
28. Install components in reverse order of removal.

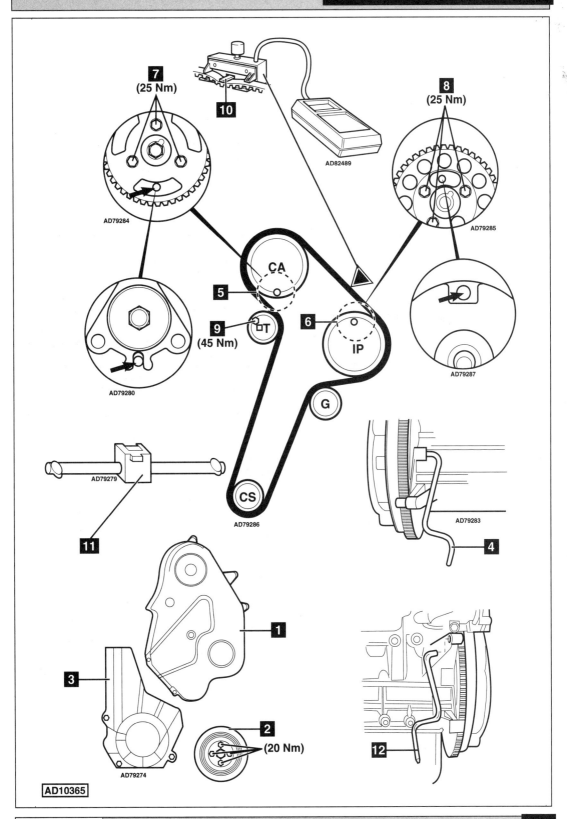

7 (25 Nm)

AD79284

10

AD82489

8 (25 Nm)

AD79285

CA

5

9 (45 Nm)

□T

6

IP

AD79287

G

AD79280

AD79279

11

CS

AD79286

AD79283

4

1

3

2 (20 Nm)

AD79274

12

AD10365

Model:	**Boxer 2,5D Turbo**
Year:	**1997-02**
Engine Code:	**THX (DJ5TED)**

Replacement Interval Guide

Peugeot recommend:
Replacement every 72,000 miles under normal conditions.
Replacement every 60,000 miles under adverse conditions.
The previous use and service history of the vehicle must always be taken into account.
Refer to Timing Belt Replacement Intervals at the front of this manual.

Check For Engine Damage

CAUTION: This engine has been identified as an INTERFERENCE engine in which the possibility of valve-to-piston damage in the event of a timing belt failure is MOST LIKELY to occur.
A compression check of all cylinders should be performed before removing the cylinder head(s).

Repair Times – hrs

Remove & install	2,20
AC	+0,30

Special Tools

- Flywheel timing pin – Peugeot No.7014-T.J.
- Camshaft timing pin – Peugeot No.5711-T.A or T.C.
- Injection pump timing pin – Peugeot No.5711-T.B.
- Tensioning lever – Peugeot No.5711-T.E.
- Tension gauge – Peugeot No.4122-T (SEEM C.Tronic 105.5).

Special Precautions

- Disconnect battery earth lead.
- DO NOT turn crankshaft or camshaft when timing belt removed.
- Remove glow plugs to ease turning engine.
- Turn engine in normal direction of rotation (unless otherwise stated).
- DO NOT turn engine via camshaft or other sprockets.
- Observe all tightening torques.
- Check diesel injection pump timing after belt replacement.

Removal

1. Raise and support front of vehicle.
2. Remove:
   - ❏ RH front wheel.
   - ❏ Engine undershield.
   - ❏ Auxiliary drive belt.
   - ❏ Timing belt upper cover **1**.
   - ❏ Crankshaft pulley bolts **2**.
   - ❏ Crankshaft pulley **3**.
   - ❏ Timing belt lower cover **4**.
3. Turn crankshaft clockwise until timing pin can be inserted in flywheel **5**. Tool No.7014-T.J.
   NOTE: Position of timing pin hole varies. Type 1: Behind starter motor. Type 2: Under turbocharger.
4. Insert timing pin in camshaft sprocket **8**. Tool No.5711-T.A/T.C.

5. Insert timing pin in injection pump sprocket **9**. Tool No.5711-T.B.
6. Turn automatic tensioner unit anti-clockwise **6**.
7. Insert suitable pin in automatic tensioner unit to retain pushrod **7**.
8. Slacken camshaft sprocket bolts **10**.
9. Slacken injection pump sprocket bolts **11**.
10. Slacken tensioner nut **12**.
11. Turn tensioner anti-clockwise. Lightly tighten nut.
12. Remove timing belt.

Installation

1. Turn camshaft and injection pump sprockets fully anti-clockwise in slotted holes.
2. Tighten bolts finger tight **10** & **11**.
3. Ensure tensioner pulleys in good condition.
4. Fit timing belt in clockwise direction, starting at crankshaft sprocket.
   NOTE: If necessary: Turn camshaft and injection pump sprockets slightly clockwise to engage in nearest belt tooth. DO NOT exceed one tooth.
5. Turn tensioner pulley **12** to ensure camshaft and injection pump sprockets turn freely.
6. Attach tension gauge to belt at ▽ **13**. Tool No.4122-T.
7. Insert tensioning lever in square hole in tensioner **14**. Tool No.5711-T.E
8. Tension belt to 41±2 SEEM units.
9. Tighten tensioner nut to 45 Nm **12**.
10. Remove tension gauge.
11. Remove locking pin from automatic tensioner unit **7**.
12. Wait 1 minute.
13. Ensure camshaft and injection pump sprocket bolts are not at end of slotted holes.
14. Tighten camshaft sprocket bolts to 5 Nm and then finally to 23 Nm **10**.
15. Tighten injection pump sprocket bolts to 5 Nm and then finally to 23 Nm **11**.
16. Remove timing pins **5**, **8** & **9**.
17. Turn crankshaft 10 turns clockwise.
18. Insert flywheel timing pin **5**.
19. Insert timing pins in camshaft and injection pump sprockets **8** & **9**.
   NOTE: DO NOT turn crankshaft anti-clockwise.
20. If timing pins cannot be inserted: Repeat installation procedure.
21. Install components in reverse order of removal.

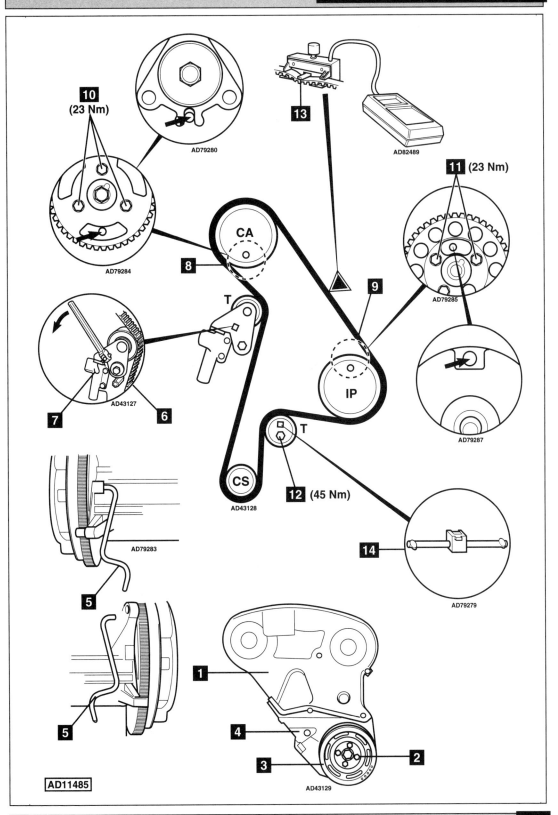

10 (23 Nm)

AD79280

AD79284

13

AD82489

11 (23 Nm)

8

CA

9

AD79285

T

7

6

AD43127

IP

AD79287

12 (45 Nm)

CS

T

AD43128

14

AD79279

AD79283

5

5

1

4

3

2

AD43129

AD11485

Model:	**Boxer 2,8 HDi**
Year:	**2000-02**
Engine Code:	**8140.43S**

Replacement Interval Guide

Peugeot recommend replacement every 75,000 miles under normal conditions or 60,000 miles under adverse conditions or every 10 years, whichever occurs first.
The previous use and service history of the vehicle must always be taken into account.
Refer to Timing Belt Replacement Intervals at the front of this manual.

Check For Engine Damage

CAUTION: This engine has been identified as an INTERFERENCE engine in which the possibility of valve-to-piston damage in the event of a timing belt failure is MOST LIKELY to occur.
A compression check of all cylinders should be performed before removing the cylinder head.

Repair Times – hrs

No information available.

Special Tools

- Tensioner locking tool – Peugeot No.1 860 638 000.

Special Precautions

- Disconnect battery earth lead.
- DO NOT turn crankshaft or camshaft when timing belt removed.
- Remove glow plugs to ease turning engine.
- Turn engine in normal direction of rotation (unless otherwise stated).
- DO NOT turn engine via camshaft or other sprockets.
- Observe all tightening torques.

Removal

NOTE: The fuel pump fitted to this engine does not require timing.

1. Raise and support front of vehicle.
2. Remove:
   - ❑ RH front wheel.
   - ❑ RH splash guard.
   - ❑ Oil filler cap **1**.
   - ❑ Engine top cover **2**.
   - ❑ Auxiliary drive belt.
   - ❑ Timing belt upper cover **3**.
   - ❑ Crankshaft position (CKP) sensor.
3. Lock flywheel. Use 8 mm diameter locking pin **4**.
4. Slacken crankshaft pulley bolt **5**.
5. Turn crankshaft to setting position. Ensure camshaft sprocket timing mark aligned with mark on cylinder head cover **6**.

6. Remove:
   - ❑ Crankshaft pulley bolt **5**.
   - ❑ Crankshaft pulley **7**.
7. Ensure crankshaft sprocket dowel pin aligned with lower cover projection **8**.
8. Remove:
   - ❑ Tensioner nut **9**.
   - ❑ Timing belt lower cover bolts **10**.
   - ❑ Timing belt lower cover **11**.
9. Lever tensioner pulley away from belt. Retain in position with tensioner locking tool **12**. Tool No.1 860 638 000.
10. Remove timing belt.

Installation

1. Temporarily fit timing belt lower cover **11**.
2. Ensure crankshaft sprocket dowel pin aligned with lower cover projection **8**.
3. Ensure camshaft sprocket timing mark aligned **6**.
4. Fit timing belt in anti-clockwise direction, starting at crankshaft sprocket. Ensure belt is taut on non-tensioned side.
5. Install:
   - ❑ Timing belt lower cover **11**.
   - ❑ Tensioner nut **9**. DO NOT tighten.
   - ❑ Crankshaft pulley **7**.
   - ❑ Crankshaft pulley bolt **5**. Hand tighten.
6. Remove tensioner locking tool **12**.
7. Turn crankshaft slowly two turns in direction of rotation.
8. Ensure crankshaft sprocket dowel pin aligned with lower cover projection **8**.
9. Ensure timing marks aligned **6**.
10. Tighten tensioner nut **9**.
    Tightening torque: 40 Nm.
11. Lock flywheel. Use 8 mm diameter locking pin **4**.
12. Tighten crankshaft pulley bolt **5**.
    Tightening torque: 200 Nm.
13. Install components in reverse order of removal.

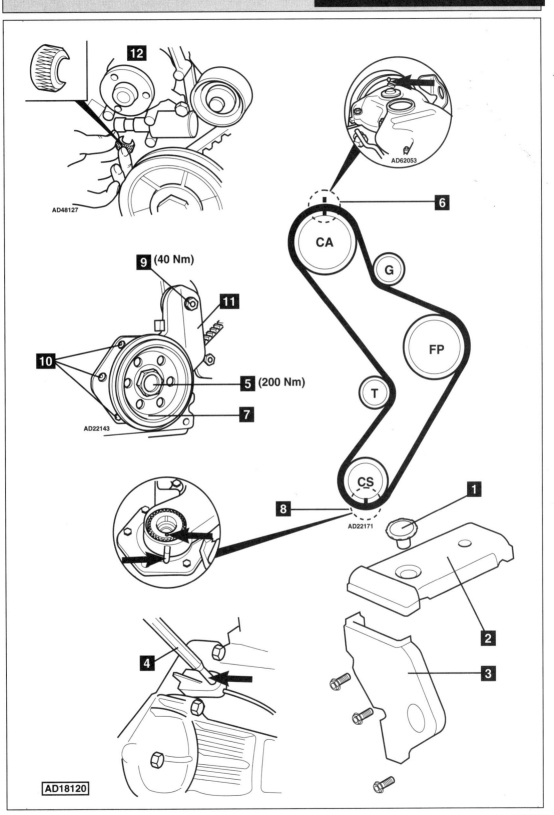

12

AD48127

AD62053

6

CA

G

9 (40 Nm)

11

10

5 (200 Nm)

7

AD22143

FP

T

CS

8

AD22171

1

2

3

4

AD18120

PROTON

Model:	1,3/1,5 • Persona/Compact 1,3/1,5 • 313/315 • 413/415
Year:	1989-02
Engine Code:	4G13, 4G15

Replacement Interval Guide

Proton recommend replacement as follows:
→1997 – Check every 24,000 miles, replace if necessary and replacement every 60,000 miles or 5 years, whichever occurs first.
1998→ – Check every 18,000 miles, replace if necessary and replacement every 54,000 miles.
The previous use and service history of the vehicle must always be taken into account.
Refer to Timing Belt Replacement Intervals at the front of this manual.

Check For Engine Damage

CAUTION: This engine has been identified as an INTERFERENCE engine in which the possibility of valve-to-piston damage in the event of a timing belt failure is MOST LIKELY to occur.
A compression check of all cylinders should be performed before removing the cylinder head.

Repair Times – hrs

Remove & install	1,80
PAS	+0,20
AC – 1,3/1,5	+0,50
AC – Persona	+0,30

Special Tools

■ None required.

Special Precautions

■ Disconnect battery earth lead.
■ DO NOT turn crankshaft or camshaft when timing belt removed.
■ Remove spark plugs to ease turning engine.
■ Turn engine in normal direction of rotation (unless otherwise stated).
■ DO NOT turn engine via camshaft or other sprockets.
■ Observe all tightening torques.

Removal

1. Remove:
   ❏ Engine undershield.
   ❏ Engine mounting bracket.
   ❏ Auxiliary drive belt(s).
   ❏ AC drive belt tensioner pulley and bracket **1**.
   ❏ Water pump pulley **2**.
   ❏ Crankshaft pulley **3**.
   ❏ Timing belt upper cover **4**.
   ❏ Timing belt lower cover **5**.
2. Turn crankshaft clockwise until timing marks aligned **6** & **7**.
   WARNING: DO NOT turn crankshaft anti-clockwise.

3. Slacken tensioner bolts **8** & **9**. Release tension on belt. Lightly tighten bolt **9**.
4. Remove timing belt.
   NOTE: Mark direction of rotation on belt with chalk if belt is to be reused.

Installation

1. Ensure timing marks aligned **6** & **7**.
2. Fit timing belt in anti-clockwise direction, starting at crankshaft sprocket. Ensure belt is taut between sprockets.
3. Slacken tensioner bolt **8**.
4. Slacken tensioner bolt **9**. Allow tensioner to operate.
5. Tighten tensioner bolt **9**.
6. Tighten tensioner bolt **8**.
7. Turn crankshaft two turns clockwise. Ensure timing marks aligned **6** & **7**.
8. Slacken tensioner bolt **8**.
9. Slacken tensioner bolt **9**. Allow tensioner to operate.
10. Tighten tensioner bolt **9**.
11. Tighten tensioner bolt **8**.
12. Apply thumb pressure to belt at ▽. Belt should deflect to 1/4 of tensioner bolt head width **11**.
13. Install components in reverse order of removal.
14. Tighten crankshaft pulley bolts **10**.
    Tightening torque: 12-15 Nm.
15. Tighten crankshaft pulley bolt **12**.
    Tightening torque: 70-100 Nm.

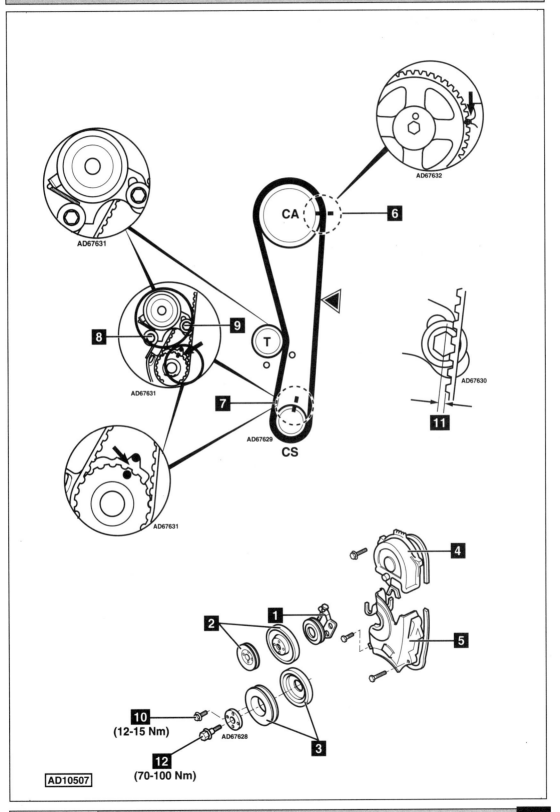

AD67632

AD67631

CA **6**

8 **9** T

AD67631

7

AD67629

CS

AD67631

AD67630

11

4

5

2 **1**

10 (12-15 Nm) AD67628

3

12 (70-100 Nm)

AD10507

PROTON

Model:	**Persona 1,6/1,8 • 416 • 418**
Year:	**1993-03**
Engine Code:	**4G92, 4G92P, 4G93**

Replacement Interval Guide

Proton recommend replacement as follows:
→1998 − Check every 24,000 miles, replace if necessary and replacement every 60,000 miles or 5 years, whichever occurs first.
1998→ − Check every 18,000 miles, replace if necessary and replacement every 54,000 miles.
The previous use and service history of the vehicle must always be taken into account.
Refer to Timing Belt Replacement Intervals at the front of this manual.

Check For Engine Damage

CAUTION: This engine has been identified as an INTERFERENCE engine in which the possibility of valve-to-piston damage in the event of a timing belt failure is MOST LIKELY to occur.
A compression check of all cylinders should be performed before removing the cylinder head.

Repair Times − hrs

Remove & install	1,80
PAS	+0,20
AC	+0,30

Special Tools

■ Crankshaft pulley holding tool − Proton No.MD998719/990767.

Special Precautions

■ Disconnect battery earth lead.
■ DO NOT turn crankshaft or camshaft when timing belt removed.
■ Remove spark plugs to ease turning engine.
■ Turn engine in normal direction of rotation (unless otherwise stated).
■ DO NOT turn engine via camshaft or other sprockets.
■ Observe all tightening torques.

Removal

1. Support engine.
2. Remove:
   ❑ Engine undershield.
   ❑ Auxiliary drive belt(s).
3. Hold crankshaft pulley.
   Use tool No.MD998719/990767.
4. Remove:
   ❑ Crankshaft pulley bolt **8**.
   ❑ Crankshaft pulley **1**.
5. Support engine. Use piece of wood between jack and sump.

6. Remove:
   ❑ Upper engine mounting.
   ❑ Timing belt upper cover **2**.
   ❑ Timing belt lower cover **3**.
   ❑ Crankshaft sprocket guide washer **4**.
7. Fit crankshaft pulley bolt. Turn crankshaft clockwise until timing marks aligned **5** & **6**.
   WARNING: DO NOT turn crankshaft anti-clockwise.
8. Slacken tensioner bolt **7**. Move tensioner away from belt and lightly tighten bolt.
9. Remove timing belt.
   NOTE: Mark direction of rotation on belt with chalk if belt is to be reused.

Installation

1. Ensure timing marks aligned **5** & **6**.
2. Fit timing belt in anti-clockwise direction, starting at crankshaft sprocket.
3. Ensure belt is taut between sprockets on non-tensioned side.
4. Slacken tensioner bolt **7**.
5. Turn crankshaft two turns clockwise. Ensure timing marks aligned **5** & **6**.
   WARNING: DO NOT turn crankshaft anti-clockwise.
6. Tighten tensioner bolt **7**.
   Tightening torque: 24 Nm.
7. Squeeze belt outwards at mid-point between camshaft sprocket and water pump sprocket.
8. Check measurement between back of belt and inner edge of timing belt rear cover is 30 mm **9**.
9. Install components in reverse order of removal.
10. Lightly oil threads of crankshaft pulley bolt.
11. Hold crankshaft pulley.
    Use tool No.MD998719/990767.
12. Tighten crankshaft pulley bolt **8**.
    Tightening torque: 185 Nm.

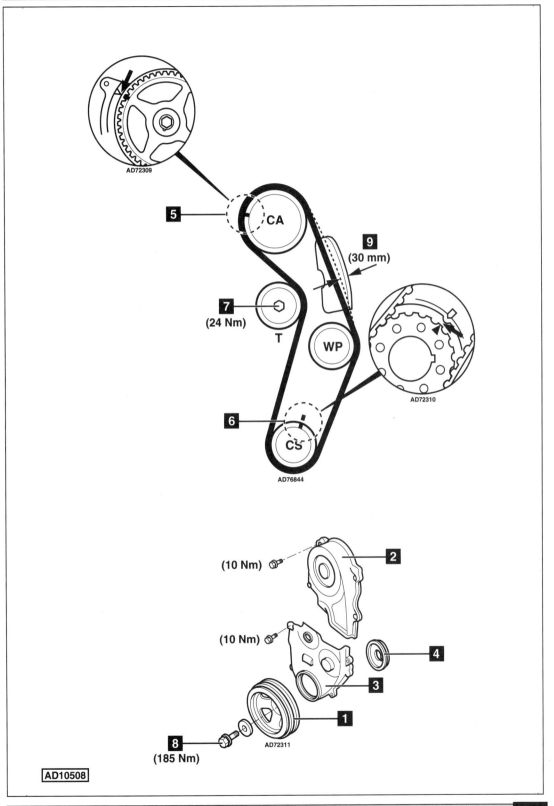

AD72309

5 CA

9 (30 mm)

7 (24 Nm) T

WP

AD72310

6 CS

AD76844

(10 Nm) **2**

(10 Nm) **4**

3

1

8 (185 Nm) AD72311

AD10508

Model:	**Persona 1,8 DOHC • 418**
Year:	**1996-03**
Engine Code:	**4G93P**

Replacement Interval Guide

Proton recommend replacement as follows:
→1998 – Check every 24,000 miles, replace if necessary and replacement every 60,000 miles or 5 years, whichever occurs first.
1998→ – Check every 18,000 miles, replace if necessary and replacement every 54,000 miles.
The previous use and service history of the vehicle must always be taken into account.
Refer to Timing Belt Replacement Intervals at the front of this manual.

Check For Engine Damage

CAUTION: This engine has been identified as an INTERFERENCE engine in which the possibility of valve-to-piston damage in the event of a timing belt failure is MOST LIKELY to occur.
A compression check of all cylinders should be performed before removing the cylinder head.

Repair Times – hrs

Remove & install	1,80
AC	+0,30
PAS	+0,20

Special Tools

- Crankshaft pulley holding tool – Proton No.MB990767.
- Holding tool adaptor pins – Proton No.MD998754.
- Tensioner pulley tool – Proton No.MD998767.

Special Precautions

- Disconnect battery earth lead.
- DO NOT turn crankshaft or camshaft when timing belt removed.
- Remove spark plugs to ease turning engine.
- Turn engine in normal direction of rotation (unless otherwise stated).
- DO NOT turn engine via camshaft or other sprockets.
- Observe all tightening torques.

Removal

1. Remove:
   - ❑ Auxiliary drive belt(s).
   - ❑ Crankshaft pulley bolt **1**.
   - ❑ Crankshaft pulley **2**.
     Use tool Nos.MB990767 & MD998754.
   - ❑ Timing belt covers **3** & **4**.
   - ❑ Crankshaft sprocket guide washer **5**.
2. Turn crankshaft clockwise until timing marks aligned **11** & **12**.
3. Slacken tensioner bolt **16**. Move tensioner away from belt and lightly tighten bolt.
4. Remove timing belt.
   NOTE: Mark direction of rotation on belt with chalk if belt is to be reused.

Installation

1. Remove automatic tensioner unit **6**.
2. Check tensioner body for leakage or damage **6**.
3. Check pushrod protrusion is 11 mm **8**. If not: Replace automatic tensioner unit.
4. Slowly compress pushrod **7** into tensioner body until holes aligned **9**. Use vice. Considerable resistance should be felt. If not: Replace automatic tensioner unit. Retain pushrod with 2 mm Allen key through hole in tensioner body.
5. Install automatic tensioner unit. Tighten bolts to 12-15 Nm **10**.
6. Ensure timing marks on camshafts aligned **11**.
7. Set crankshaft sprocket 1/2 tooth before timing mark **12**.
8. Align marks on belt with marks on sprockets.
9. Fit timing belt to inlet camshaft sprocket **13**. Retain in position with suitable paper clip **15**.
10. Fit timing belt to exhaust camshaft sprocket **14**. Retain in position with suitable paper clip **15**.
11. Fit timing belt in following order:
    - ❑ Guide pulley.
    - ❑ Water pump sprocket.
    - ❑ Crankshaft sprocket.
    - ❑ Tensioner pulley.
12. Slacken tensioner pulley bolt **16**. Turn tensioner pulley anti-clockwise to temporarily tension belt **17**.
13. Ensure timing marks aligned **11** & **12**.
14. Turn crankshaft 1/4 turn anti-clockwise.
15. Turn crankshaft clockwise until timing marks aligned **11** & **12**.
16. Slacken tensioner pulley bolt **16**. Fit tool **21** to holes in tensioner pulley **18**. Tool No.MD998767.
17. Apply anti-clockwise torque of 2,6 Nm to tensioner pulley.
18. Tighten tensioner pulley bolt **16**.
    Tightening torque: 49 Nm.
19. Turn crankshaft two turns clockwise. Ensure timing marks aligned **11** & **12**. Wait 15 minutes.
20. Remove Allen key from tensioner body to release pushrod.
    NOTE: Ensure Allen key can be removed easily. If not: Repeat tensioning procedure.
21. Check extended length of pushrod is 3,8-4,5 mm **19**.
22. If not: Repeat tensioning procedure.
23. Install components in reverse order of removal.
24. Tighten timing belt cover bolts to 10 Nm **20**.
25. Tighten crankshaft pulley bolt **1**.
    Tightening torque: 180-190 Nm.

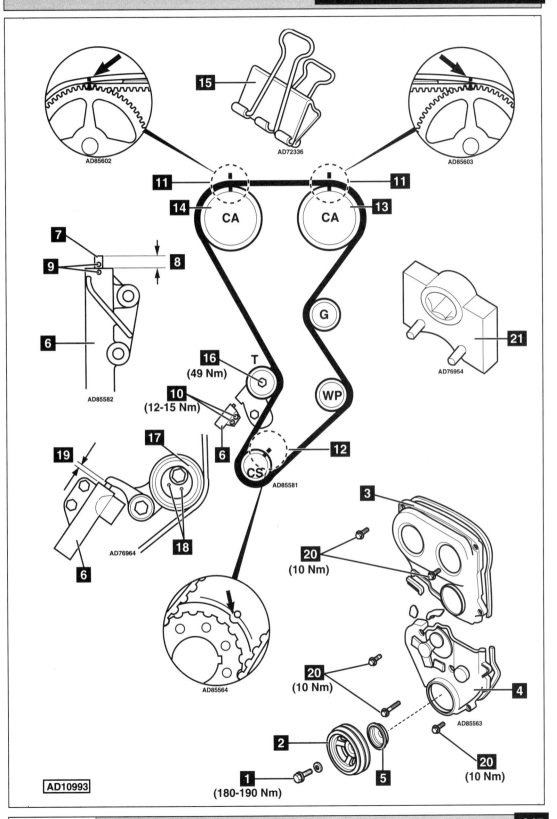

AD85602

AD72336 **15**

AD85603

11 CA **14** **13** CA **11**

7
9 **8**

6

AD85582

16
T
(49 Nm)

10
(12-15 Nm)

6

19

17

6

AD76964 **18**

G

WP

21
AD76954

12
CS
AD85581

3

20
(10 Nm)

AD85564

20
(10 Nm)

4
AD85563

20
(10 Nm)

2

5

1
(180-190 Nm)

PROTON

Model:	**Impian 1,6**
Year:	**2001-03**
Engine Code:	**4G18**

Replacement Interval Guide

Proton recommend:
Check every 18,000 or 2 years.
Replacement every 54,000 miles or 6 years.
The previous use and service history of the vehicle must always be taken into account.
Refer to Timing Belt Replacement Intervals at the front of this manual.

Check For Engine Damage

CAUTION: This engine has been identified as an INTERFERENCE engine in which the possibility of valve-to-piston damage in the event of a timing belt failure is MOST LIKELY to occur. A compression check of all cylinders should be performed before removing the cylinder head(s).

Repair Times – hrs

Remove & install	1,10
PAS	+0,30
AC	+0,30

Special Tools

- None required.

Special Precautions

- Disconnect battery earth lead.
- DO NOT turn crankshaft or camshaft when timing belt removed.
- Remove spark plugs to ease turning engine.
- Turn engine in normal direction of rotation (unless otherwise stated).
- DO NOT turn engine via camshaft or other sprockets.
- Observe all tightening torques.

Removal

1. Raise and support front of vehicle.
2. Remove:
   - ❏ Engine undershield (if fitted).
   - ❏ RH front wheel.
   - ❏ Auxiliary drive belt(s).
   - ❏ Water pump pulley.
   - ❏ Crankshaft pulley bolt **1**.
   - ❏ Crankshaft pulley **2**.
   - ❏ Crankshaft position (CKP) sensor.
   - ❏ PAS pump and bracket. DO NOT disconnect hoses.
3. Support engine.
4. Remove:
   - ❏ RH engine mounting and bracket.
   - ❏ Timing belt upper cover **3**.
   - ❏ Timing belt lower cover **4**.

5. Turn crankshaft clockwise to TDC on No.1 cylinder. Ensure timing marks aligned **5** & **6**.
6. Slacken tensioner pulley bolt **7**. Push tensioner away from belt. Lightly tighten bolt.
7. Remove timing belt.
   NOTE: Mark direction of rotation on belt with chalk if belt is to be reused.

Installation

1. Ensure timing marks aligned **5** & **6**.
2. Fit timing belt in anti-clockwise direction, starting at crankshaft sprocket.
   NOTE: Ensure belt is taut between sprockets on non-tensioned side.
3. Slacken tensioner pulley bolt 1/2 turn **7**. Allow tensioner to operate.
4. Turn crankshaft two turns clockwise to TDC on No.1 cylinder.
5. Ensure timing marks aligned **5** & **6**.
6. Tighten tensioner pulley bolt **7**.
   Tightening torque: 20-27 Nm.
7. Install components in reverse order of removal.
8. Tighten crankshaft pulley bolt **1**.
   Tightening torque: 120-135 Nm.

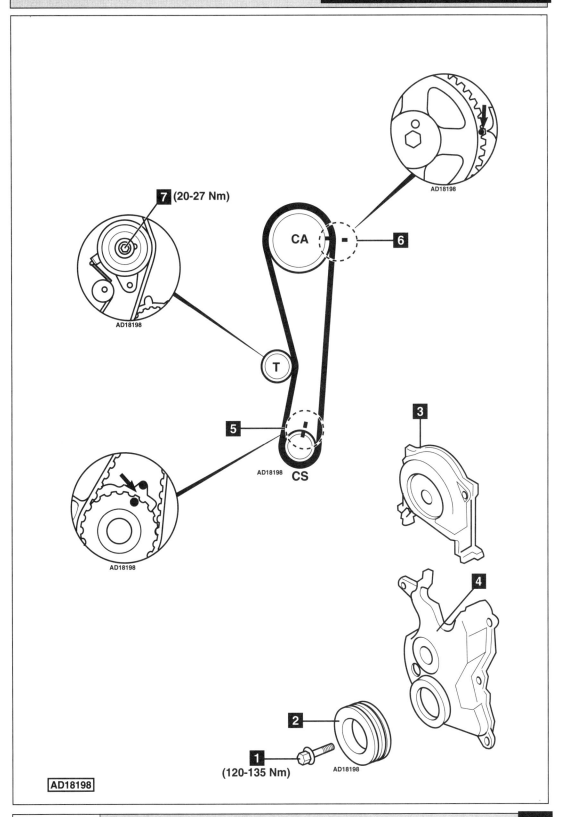

AD18198

7 (20-27 Nm)

AD18198

CA

6

T

3

5

AD18198 **CS**

AD18198

4

2

1

(120-135 Nm)

AD18198

AD18198

PROTON

Model:	**Persona 2,0 D/TD • 420 D/TD**
Year:	**1996-00**
Engine Code:	**4D68-2, 4D68TDi**

Replacement Interval Guide

Proton recommend replacement as follows:
→1998 – Check every 24,000 miles, replace if necessary and replacement every 60,000 miles or 5 years, whichever occurs first.
1998→ – Check every 18,000 miles, replace if necessary and replacement every 54,000 miles.
The previous use and service history of the vehicle must always be taken into account.
Refer to Timing Belt Replacement Intervals at the front of this manual.

Check For Engine Damage

CAUTION: This engine has been identified as an INTERFERENCE engine in which the possibility of valve-to-piston damage in the event of a timing belt failure is MOST LIKELY to occur.
A compression check of all cylinders should be performed before removing the cylinder head.

Repair Times – hrs

Remove & install	2,50
PAS	+0,20
AC	+0,30

Special Tools

■ None required.

Removal

Timing Belt

NOTE: It is not necessary to disturb or remove the balancer shaft belt when replacing the timing belt.
1. Support engine.
2. Remove:
   ❑ RH engine mounting.
   ❑ Auxiliary drive belts.
   ❑ Water pump pulley **13**.
   ❑ Crankshaft pulley bolts **1**.
   ❑ Crankshaft pulley **2**.
   ❑ Timing belt upper cover **3**.
   ❑ Timing belt lower cover **4**.
3. Turn crankshaft clockwise to TDC on No.1 cylinder.
4. Ensure timing marks aligned **5**, **6**, **7** & **8**.
5. Slacken tensioner bolt **9**. Move tensioner away from belt and lightly tighten bolt.
6. Remove timing belt.

Installation

Timing Belt

1. Ensure timing marks aligned **5**, **6**, **7** & **8**.
2. Ensure tensioner at bottom of slot **9**.
   *NOTE: To check oil pump sprocket is correctly positioned: Remove blanking plug from cylinder block **10**. Insert 8 mm diameter Phillips screwdriver in hole. Ensure screwdriver is inserted*

to a depth of 60 mm. If screwdriver can only be inserted 20-25 mm: Turn oil pump sprocket 360° and reinsert screwdriver.
3. Fit timing belt in following order:
   ❑ Crankshaft sprocket.
   ❑ Guide pulley.
   ❑ Camshaft sprocket.
   ❑ Injection pump sprocket.
   ❑ Oil pump sprocket.
4. Turn crankshaft slowly anti-clockwise 1/2 tooth of camshaft sprocket to take up slack in belt.
5. Fit timing belt to tensioner pulley.
6. Slacken tensioner bolt **9**. Allow tensioner to operate.
7. Turn crankshaft anti-clockwise for three teeth on camshaft sprocket.
8. Tighten tensioner bolt **9**. Tightening torque: 48 Nm.
9. Turn crankshaft clockwise until timing marks aligned **5**, **6**, **7** & **8**.
10. Apply thumb pressure to belt at ▽. Belt should deflect 4-5 mm.
11. If not: Repeat tensioning procedure.
12. Install components in reverse order of removal.
13. Tighten crankshaft pulley bolts **1**.
    Tightening torque: 25 Nm.

Removal

Balancer Shaft Belt

1. Remove timing belt as described previously.
2. Ensure crankshaft timing marks aligned **8**.
3. Ensure balancer shaft sprocket timing marks aligned **12**.
4. Remove:
   ❑ Balancer shaft belt tensioner bolt and tensioner pulley **11**.
   ❑ Balancer shaft belt.

Installation

Balancer Shaft Belt

1. Ensure timing marks aligned **8** & **12**.
2. Fit balancer shaft belt. Ensure belt is taut on non-tensioned side.
3. Fit balancer shaft belt tensioner pulley and bolt **11**.
4. Ensure centre of tensioner pulley to left side of tensioner bolt.
5. Turn tensioner pulley clockwise to tension belt. Tighten tensioner bolt **11**. Tightening torque: 19 Nm.
6. Apply thumb pressure to belt at ▽. Belt should deflect 5-7 mm.
7. If not: Repeat tensioning procedure.
8. Fit timing belt as described previously.

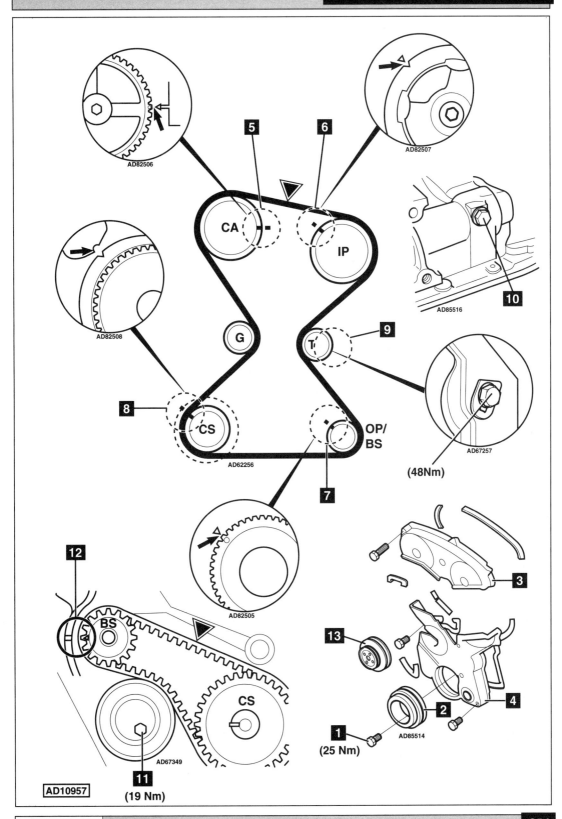

AD82506

AD82507

5

6

CA

IP

AD82508

G

T

9

10

AD85516

8

CS

AD62256

OP/
BS

AD67257

(48Nm)

7

AD82505

12

BS

CS

13

2

4

11

AD67349

1

(25 Nm)

AD85514

(19 Nm)

AD10957

RENAULT

Model:	**Twingo 1,2 • Clio 1,2 • Kangoo 1,2**
Year:	**1995-03**
Engine Code:	**D7F 700/701/710/730**

Replacement Interval Guide

Renault recommend:
6,000 mile service intervals – replacement every 72,000 miles or 5 years.
10,000 mile service intervals – replacement every 70,000 miles or 5 years.
12,000/18,000 mile service intervals – replacement every 72,000 miles or 5 years.
The previous use and service history of the vehicle must always be taken into account.
Refer to Timing Belt Replacement Intervals at the front of this manual.

Check For Engine Damage

CAUTION: This engine has been identified as an INTERFERENCE engine in which the possibility of valve-to-piston damage in the event of a timing belt failure is MOST LIKELY to occur.
A compression check of all cylinders should be performed before removing the cylinder head(s).

Repair Times – hrs

Remove & install:
Twingo	1,90
Clio (→05/98)	1,80
Clio (06/98→)	2,00
Kangoo	2,00

Special Tools

- TDC timing pin – Renault No.Mot.1054.
- Tensioner spanner – Renault No.Mot.1135-01.
- Tensioning tool – Renault No.Mot.1386.
- Tension gauge – Renault No.Mot.1273.

Special Precautions

- Disconnect battery earth lead.
- DO NOT turn crankshaft or camshaft when timing belt removed.
- Remove spark plugs to ease turning engine.
- Turn engine in normal direction of rotation (unless otherwise stated).
- DO NOT turn engine via camshaft or other sprockets.
- Observe all tightening torques.

Removal

1. Raise and support front of vehicle.
2. Support engine.
3. Remove:
   - ❏ RH engine mounting **1**.
   - ❏ RH front wheel.
   - ❏ RH splash guard.
   - ❏ Auxiliary drive belt(s).
   - ❏ Timing belt upper cover **2**.
   - ❏ Crankshaft pulley centre bolt **3**.
   - ❏ Crankshaft pulley **4**.
   - ❏ Timing belt lower cover **5**.
4. Turn crankshaft to TDC on No.1 cylinder.

5. Ensure timing marks aligned **6** & **7**.
   NOTE: Camshaft sprocket has multiple markings. Use rectangular timing mark adjacent to 'E1-E3' lettering.
6. Insert timing pin in flywheel **8**. Tool No.Mot.1054.
7. Slacken tensioner nut **9**.
8. Turn tensioner clockwise away from belt. Use tool No.Mot.1135-01 **10**.
9. Lightly tighten tensioner nut **9**.
10. Remove timing belt.

Installation

1. Ensure timing marks aligned **6** & **7**.
2. Ensure timing pin inserted in flywheel **8**.
3. Fit timing belt in anti-clockwise direction, starting at crankshaft sprocket.
4. Ensure marks on belt aligned with marks on sprockets.
5. Ensure belt is taut between sprockets.
6. Temporarily fit crankshaft pulley centre bolt **3**. Lightly tighten bolt.
7. Attach tension gauge to belt at ▽. Tool No.Mot.1273 **11**.
8. Slacken tensioner nut **9**.
9. Turn tensioner anti-clockwise until tension gauge indicates 20 units. Use tool No.Mot.1135-01 **10**.
10. Tighten tensioner nut **9**. Tightening torque: 50 Nm.
11. Remove tension gauge **11**.
12. Remove timing pin **8**.
13. Turn crankshaft two turns clockwise to TDC on No.1 cylinder.
14. Ensure timing pin can be inserted in flywheel **8**.
    NOTE: Ensure timing pin removed.
15. Ensure timing marks aligned **6** & **7**.
16. Slacken tensioner nut **9**.
17. Turn tensioner until holes are horizontal **12**. Use tool No.Mot.1135-01 **10**.
18. Tighten tensioner nut **9**. Tightening torque: 50 Nm.
19. Turn crankshaft two turns clockwise to TDC on No.1 cylinder.
20. Ensure timing pin can be inserted in flywheel **8**.
    NOTE: Ensure timing pin removed.
21. Ensure timing marks aligned **6** & **7**.
22. Fit tensioning tool **13**. Tool No.Mot.1386.
23. Apply anti-clockwise torque of 10 Nm **14**.
24. Remove tensioning tool **13**.
25. Attach tension gauge to belt at ▽ **11**.
26. Tension gauge should indicate 20±3 units.
27. If tension not as specified: Adjust tensioner manually. Use tool No.Mot.1135-01 **10**.
28. Tighten tensioner nut **9**. Tightening torque: 50 Nm.
29. Repeat operations 19-28.
30. Install components in reverse order of removal.
31. Tighten crankshaft pulley centre bolt **3**. Tightening torque: 20 Nm + 90°.

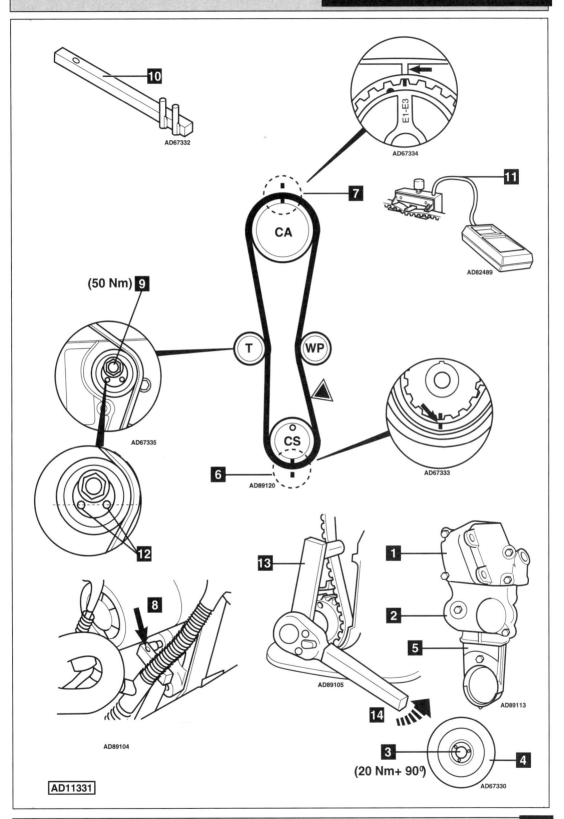

Model:	R5 1,7 • Clio 1,7 • R11 1,7 • R19 1,7 • R21 1,7 • Trafic 1,7
Year:	1986-96
Engine Code:	F1N, F2N, F3N, F2R

Replacement Interval Guide

Renault recommend replacement every 72,000 miles or 5 years.
The previous use and service history of the vehicle must always be taken into account.
Refer to Timing Belt Replacement Intervals at the front of this manual.

Check For Engine Damage

CAUTION: This engine has been identified as an INTERFERENCE engine in which the possibility of valve-to-piston damage in the event of a timing belt failure is MOST LIKELY to occur.
A compression check of all cylinders should be performed before removing the cylinder head.

Repair Times – hrs

Remove & install:

R5/Clio	3,60
R11/R19	2,10
R21	2,50
Trafic	1,80

Special Tools

- Tension gauge – Renault No.Mot.1273 (SEEM C. Tronic 105).
- Crankshaft timing pin – Renault No.Mot.861.
- M6 bolt – for adjusting tensioner.

Special Precautions

- Disconnect battery earth lead.
- DO NOT turn crankshaft or camshaft when timing belt removed.
- Remove spark plugs to ease turning engine.
- Turn engine in normal direction of rotation (unless otherwise stated).
- DO NOT turn engine via camshaft or other sprockets.
- Observe all tightening torques.

Removal

1. Remove:
   ❑ Auxiliary drive belts.
   ❑ Timing belt cover **1**.
2. Turn crankshaft clockwise until timing marks aligned **2** & **3**.
3. Insert timing pin in crankshaft **4**.
   Tool No.Mot.861.
4. Slacken tensioner nut **5**. Move tensioner away from belt. Lightly tighten nut.
5. Remove timing belt.

Installation

1. Ensure camshaft sprocket timing marks aligned **2**.
2. Ensure timing pin located correctly **4**.
3. Fit timing belt in clockwise direction, starting at crankshaft sprocket.
4. Ensure directional arrows on belt are between camshaft sprocket and auxiliary shaft sprocket.
5. Align marks on belt with marks on sprockets **2** & **6**.
6. Slacken tensioner nut **5**.
7. Screw an M6 bolt into timing belt upper rear cover until tensioner pulley contacts belt **7**.
8. Apply firm thumb pressure to belt at ▽.
9. Attach tension gauge to belt at ▽.
   Tool No.Mot.1273.
10. Screw in M6 bolt until tension gauge indicates 25 units **7**.
11. Tighten tensioner nut to 40 Nm **5**.
12. Remove:
    ❑ M6 bolt **7**.
    ❑ Tension gauge.
13. Ensure timing marks aligned.
14. Remove timing pin **4**.
15. Turn crankshaft three turns clockwise. Insert timing pin **4**.
16. Remove timing pin **4**.
17. Apply firm thumb pressure to belt at ▽.
18. Recheck belt tension. Tension gauge should indicate 22-25 units.
19. If not: Repeat tensioning procedure.
20. Install components in reverse order of removal.

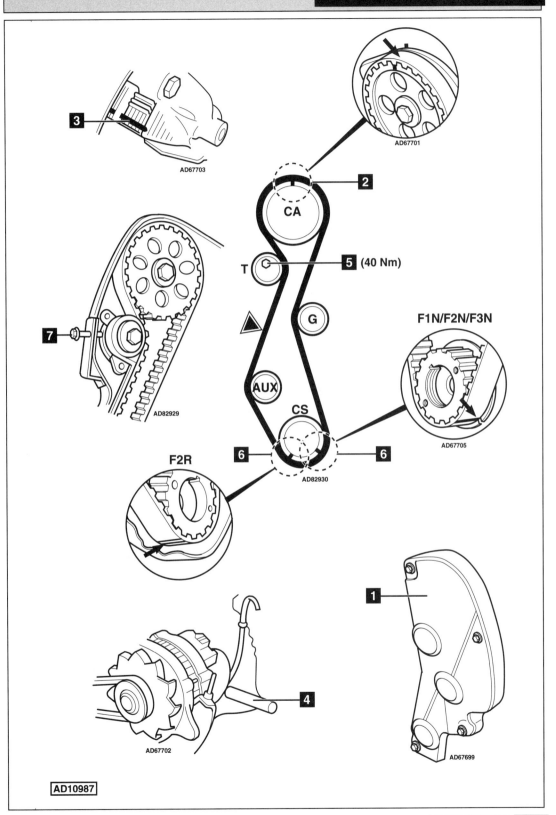

3 AD67703

AD67701

2

CA

T **5** (40 Nm)

G

△

F1N/F2N/F3N

AD67705

AUX

CS

7 AD82929

F2R

6 **6**

AD82930

1

4 AD67702

AD67699

AD10987

RENAULT

Model:	Clio 1,2/1,4 • R19 1,4 • R19 1,4 ECO (X535) • Extra/Express/Rapid 1,4 Kangoo 1,4
Year:	1988-03
Engine Code:	E5F, E6J, E7F, E7J

Replacement Interval Guide

Renault recommend replacement as follows:
6,000 mile service intervals – replacement every 72,000 miles or 5 years.
10,000 mile service intervals – replacement every 70,000 miles or 5 years.
12,000/18,000 mile service intervals – replacement every 72,000 miles or 5 years.
The previous use and service history of the vehicle must always be taken into account.
Refer to Timing Belt Replacement Intervals at the front of this manual.

Check For Engine Damage

CAUTION: This engine has been identified as an INTERFERENCE engine in which the possibility of valve-to-piston damage in the event of a timing belt failure is MOST LIKELY to occur.
A compression check of all cylinders should be performed before removing the cylinder head.

Repair Times – hrs

Remove & install:

R19	1,80
Clio (→05/98)	2,30
Clio (06/98→)	2,70
Extra/Express/Rapid	2,50
Kangoo	2,70

Special Tools

- Tension gauge – Renault No.Mot.1273 (SEEM C.Tronic 105).
- Tensioning tool – Renault No.Mot.1135.

Special Precautions

- Disconnect battery earth lead.
- DO NOT turn crankshaft or camshaft when timing belt removed.
- Remove spark plugs to ease turning engine.
- Turn engine in normal direction of rotation (unless otherwise stated).
- DO NOT turn engine via camshaft or other sprockets.
- Observe all tightening torques.

Removal

1. Raise and support front of vehicle.
2. Support engine.
3. Remove:
   - ❏ Auxiliary drive belt(s).
   - ❏ Engine mounting (Clio/Kangoo).
   - ❏ Timing belt covers **1**.
   - ❏ RH inner wing panel.
   - ❏ Crankshaft pulley **2**.
4. Turn crankshaft clockwise until camshaft sprocket timing mark **3** aligned with rib on cylinder head cover and crankshaft sprocket timing mark **4** aligned.
   NOTE: Some camshaft sprockets have multiple markings. Use rectangular timing mark adjacent to 'E1-E3' lettering.
5. Slacken tensioner pulley nut **5**. Move tensioner pulley away from belt and lightly tighten nut.
6. Remove timing belt.

Installation

NOTE: DO NOT fit used belt.
1. Ensure timing marks aligned **3** & **4**.
2. Fit timing belt. Observe direction of rotation marks on belt. Align marks on belt with marks on sprockets.
3. Attach tension gauge to belt at ▽. Tool No.Mot.1273.
4. Slacken tensioner pulley nut **5**. Turn tensioner pulley against belt until tension gauge indicates 30 units. Use tool No.Mot.1135 **6**.
5. Tighten tensioner pulley nut to 50 Nm **5**.
6. Turn crankshaft three turns clockwise. Ensure timing marks aligned **4**.
7. Recheck belt tension. Tension gauge should indicate 30 units. If not: Repeat tensioning procedure.
8. Install components in reverse order of removal.
9. Tighten crankshaft pulley bolt **7**.
   →1995: 80-90 Nm. 1996→: 20 Nm + 62-74°.

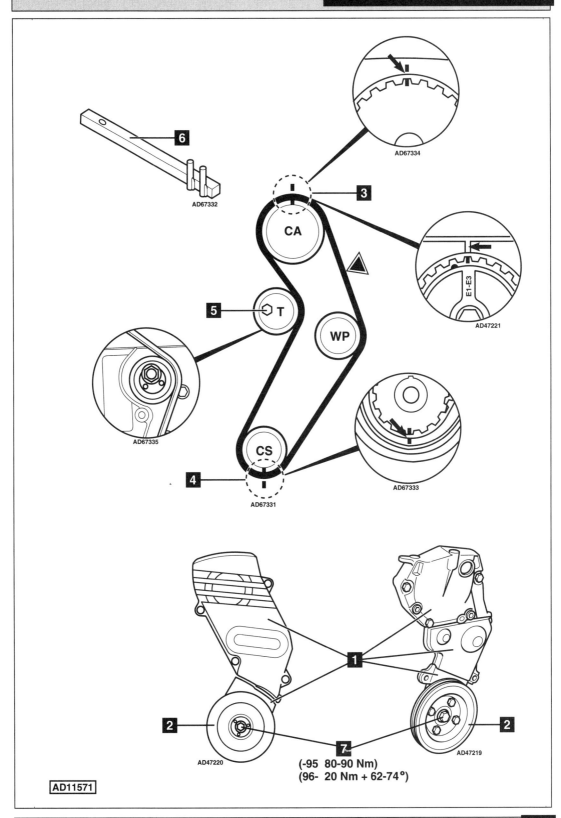

AD67334

AD67332

6

3

CA

AD47221

E1-E3

5

T

WP

AD67335

CS

AD67333

4

AD67331

1

2

2

AD47220

AD47219

7

(-95 80-90 Nm)
(96- 20 Nm + 62-74°)

AD11571

Model:	Clio 1,4/1,6 16V • Mégane/Scénic 1,4/1,6 16V • Laguna 1,6 16V
Year:	1998-03
Engine Code:	K4J 712/713/750, K4M 700/701/720/724/748

Replacement Interval Guide

Renault recommend:
→03/99: Replacement every 60,000 miles or 5 years.
04/99→: Replacement every 72,000 miles or 5 years.
The previous use and service history of the vehicle must always be taken into account.
Refer to Timing Belt Replacement Intervals at the front of this manual.

Check For Engine Damage

CAUTION: This engine has been identified as an INTERFERENCE engine in which the possibility of valve-to-piston damage in the event of a timing belt failure is MOST LIKELY to occur. A compression check of all cylinders should be performed before removing the cylinder head(s).

Repair Times – hrs

Remove & install:

Clio	2,70
Mégane	3,60
Mégane Scénic	3,90
Laguna	2,50

Special Tools

- ■ Crankshaft timing pin – Renault No.Mot.1489.
- ■ Camshaft setting bar – Renault No.Mot.1496.

Removal

1. Raise and support front of vehicle.
2. Remove:
   - ❏ Engine top cover.
   - ❏ RH splash guard.
   - ❏ Auxiliary drive belt.
3. Support engine.
4. Remove:
   - ❏ RH engine mounting.
   - ❏ Blanking plugs from rear of camshafts.
   - ❏ Blanking plug from cylinder block **1**.
5. Turn crankshaft clockwise to setting position. Ensure grooves in camshafts aligned **3**.
   NOTE: Grooves located below cylinder head upper surface.
6. Insert timing pin in cylinder block **4**. Tool No.Mot.1489.
7. Ensure crankshaft web against timing pin **5**.
8. Fit setting bar to rear of camshafts **6**. Tool No.Mot.1496.
9. Lock flywheel with large screwdriver. Slacken crankshaft pulley bolt **2**.
10. Remove:
    - ❏ Crankshaft pulley bolt **2**.
    - ❏ Crankshaft pulley **7**.
    - ❏ Timing belt covers **8** & **9**.
11. Slacken tensioner nut **10**.
12. Allow tensioner pulley to move away from belt.

13. Remove:
    - ❏ Tensioner nut **10**.
    - ❏ Tensioner pulley **11**.
    - ❏ Guide pulley **12**.
    - ❏ Timing belt.
    NOTE: DO NOT allow crankshaft sprocket to fall off crankshaft.

Installation

1. Ensure timing pin fitted **4**. Ensure crankshaft web against timing pin **5**.
2. Ensure grooves in camshafts aligned **3**.
3. Ensure setting bar fitted correctly **6**.
4. Fit new guide pulley **12**. Tighten bolt to 45 Nm.
5. Fit new tensioner pulley **11**. Temporarily tighten nut to 7 Nm **10**.
   NOTE: Ensure lug on rear of tensioner pulley located in groove in cylinder head.
6. Remove crankshaft sprocket. Degrease sprocket. Degrease end of crankshaft. Refit crankshaft sprocket.
7. Fit timing belt in anti-clockwise direction, starting at crankshaft sprocket. Ensure belt is taut on non-tensioned side.
8. Degrease crankshaft pulley **7**.
9. Fit crankshaft pulley **7**.
10. Measure length of crankshaft pulley bolt **2**. Maximum length: 49,1 mm. If longer: Use new bolt.
11. Fit crankshaft pulley bolt **2**. New bolt: DO NOT oil. Used bolt: Oil threads and contact face.
12. Temporarily tighten crankshaft pulley bolt **2**. Leave 2-3 mm clearance between bolt contact face and crankshaft pulley **7**.
13. Slacken tensioner nut **10**. Turn tensioner pulley clockwise until movable pointer **13** at right hand stop **14**. Use 6 mm Allen key **15**.
    NOTE: Movable pointer **13 should be 7-8 mm past fixed pointer **16**.**
14. Temporarily tighten tensioner nut **10**. Tightening torque: 7 Nm.
15. Lock flywheel with large screwdriver. Temporarily tighten crankshaft pulley bolt **2**. Tightening torque: 20 Nm.
16. Remove:
    - ❏ Timing pin **4**.
    - ❏ Setting bar **6**.
17. Lock flywheel with large screwdriver. Tighten crankshaft pulley bolt a further 120-150° **2**.
18. Turn crankshaft two turns clockwise to setting position.
19. Insert timing pin in cylinder block **4**.
20. Ensure crankshaft web against timing pin **5**.
21. Ensure camshaft setting bar can be fitted easily **6**.
22. Hold tensioner pulley. Use 6 mm Allen key **15**. Slacken tensioner nut **10**.
23. Turn tensioner pulley anti-clockwise until pointers aligned **13** & **16**.
24. Tighten tensioner nut **10**. Tightening torque: 27 Nm.
25. Remove timing pin **4**.
26. Refit blanking plug **1**.
27. Fit new blanking plugs to rear of camshafts.
28. Install components in reverse order of removal.

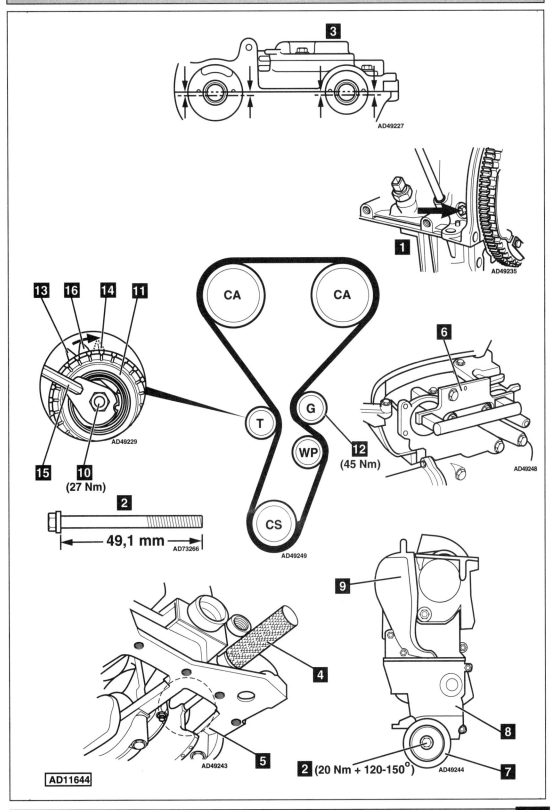

3

AD49227

1

AD49235

13 16 14 11

CA   CA

T   G

WP

12
(45 Nm)

6

AD49248

AD49229

15

10
(27 Nm)

2

◄— **49,1 mm** —►
AD73266

CS

AD49249

9

4

8

5

AD49243

2 **(20 Nm + 120-150°)**   AD49244   **7**

AD11644

RENAULT

Model:	Clio 1,8 16V • Clio Williams 2,0 16V • R19 1,8 16V
Year:	1989-97
Engine Code:	F7P 700/4, F7P 720/2, F7R 700

Replacement Interval Guide

Renault recommend replacement as follows:
6,000 mile service intervals – replacement every 72,000 miles or 5 years.
10,000 mile service intervals – replacement every 70,000 miles or 5 years.
The previous use and service history of the vehicle must always be taken into account.
Refer to Timing Belt Replacement Intervals at the front of this manual.

Check For Engine Damage

CAUTION: This engine has been identified as an INTERFERENCE engine in which the possibility of valve-to-piston damage in the event of a timing belt failure is MOST LIKELY to occur.
A compression check of all cylinders should be performed before removing the cylinder head.

Repair Times – hrs

Remove & install:

R19	3,90
Clio	4,90

Special Tools

- Crankshaft timing pin – Renault No.Mot.1054 or No.Mot.861.
- Sprocket locking tool – Renault No.Mot.1196.
- Tension gauge – Renault No.Mot.1273 (SEEM C.Tronic 105).
- M6 x 45 mm bolt – for adjusting tensioner.

Special Precautions

- Disconnect battery earth lead.
- DO NOT turn crankshaft or camshaft when timing belt removed.
- Remove spark plugs to ease turning engine.
- Turn engine in normal direction of rotation (unless otherwise stated).
- DO NOT turn engine via camshaft or other sprockets.
- Observe all tightening torques.

Removal

1. Raise and support front of vehicle.
2. Support engine (Clio).
3. Remove:
   - RH front wheel.
   - RH front wheel arch liner.
   - Engine mounting (Clio).
   - Air filter (Clio).
   - RH headlamp (Clio).
   - Fuel pipes (Clio).
   - Auxiliary drive belts.
   - Timing belt covers.
   - Blanking plug from cylinder block.
4. Use a large screwdriver to lock flywheel.
5. Remove:
   - Crankshaft pulley bolt **6**.
   - Crankshaft pulley.
6. Turn crankshaft until timing marks aligned **1**.
7. Insert timing pin in crankshaft **2**.
   Tool No.Mot.1054/No.Mot.861. Rock crankshaft slightly to ensure timing pin located correctly **2**.
8. Lock camshaft sprockets.
   Use tool No.Mot.1196 **3**.
9. Slacken tensioner nut to release tension on belt **4**.
10. Remove timing belt.

Installation

NOTE: DO NOT fit used belt.

1. Ensure timing pin located correctly **2**.
2. Ensure locking tool located correctly **3**.
3. Ensure timing marks aligned **1**.
4. Observe direction of rotation marks on belt. Align marks on belt with marks on sprockets.
5. Fit timing belt in clockwise direction, starting at crankshaft sprocket. Ensure belt is taut between sprockets.
6. Remove:
   - Timing pin **2**.
   - Locking tool **3**.
7. Screw an M6x45 mm bolt into timing belt rear cover until tensioner pulley contacts belt **5**.
8. Attach tension gauge to belt at ▽.
   Use tool No.Mot.1273.
9. Screw in M6 bolt until tension gauge indicates 32±3 units (cold) **5**.
10. Tighten tensioner nut **4**.
    Tightening torque: 48 Nm.
11. Turn crankshaft three turns clockwise. Insert timing pin **2**.
12. Remove timing pin **2**.
13. Attach tension gauge to belt at ▽. Tension gauge should indicate 32±3 units.
14. If not: Repeat tensioning procedure.
15. Install:
    - Timing belt covers.
    - Crankshaft pulley.
16. Tighten crankshaft pulley bolt to 90 Nm **6**.
17. Install components in reverse order of removal.

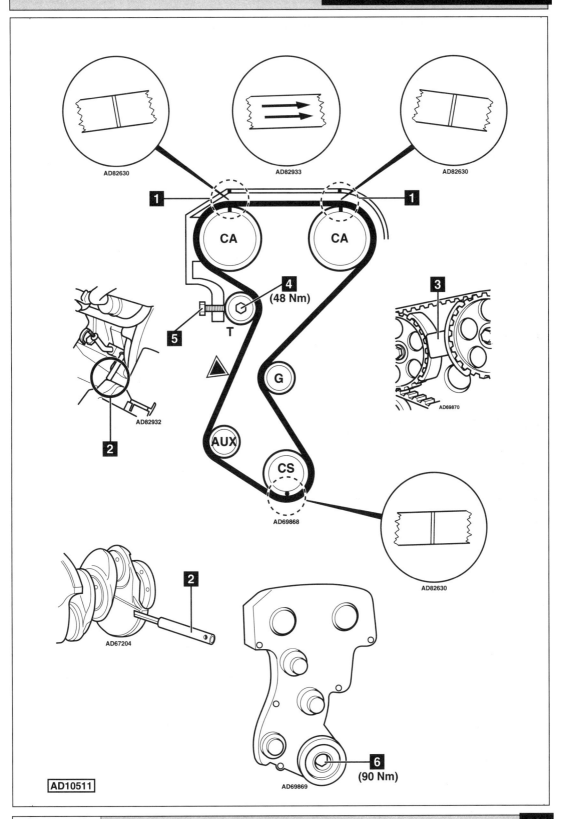

AD82630

AD82933

AD82630

1 1

CA CA

4
(48 Nm)

3

5

T

2

AD82932

G

AD69870

AUX

CS

AD69868

AD82630

2

AD67204

6
(90 Nm)

AD69869

AD10511

RENAULT

Model:	Clio 1,8 • R19 1,8 • Laguna 1,8/2,0 • Espace 2,0
Year:	1992-01
Engine Code:	F3P, F3R

Replacement Interval Guide

Renault recommend replacement as follows:
6,000 mile service interval – replacement every 72,000 miles or 5 years.
10,000 mile service interval – replacement every 70,000 miles or 5 years.
12,000/18,000 mile service intervals – replacement every 72,000 miles or 5 years.
The previous use and service history of the vehicle must always be taken into account.
Refer to Timing Belt Replacement Intervals at the front of this manual.

Check For Engine Damage

CAUTION: This engine has been identified as an INTERFERENCE engine in which the possibility of valve-to-piston damage in the event of a timing belt failure is MOST LIKELY to occur.
A compression check of all cylinders should be performed before removing the cylinder head.

Repair Times – hrs

Remove & install:

Clio	3,60
R19	1,80
Laguna	3,00
Espace	2,00

Special Tools

- Crankshaft timing pin – Renault No.Mot.1054.
- Tension gauge – Renault No.Mot.1273 (SEEM C.Tronic).
- M6 x 45 mm bolt – for adjusting tensioner.

Special Precautions

- Disconnect battery earth lead.
- DO NOT turn crankshaft or camshaft when timing belt removed.
- Remove spark plugs to ease turning engine.
- Turn engine in normal direction of rotation (unless otherwise stated).
- DO NOT turn engine via camshaft or other sprockets.
- Observe all tightening torques.

Removal

1. Raise and support front of vehicle.
2. Remove:
   - ❏ RH front wheel.
   - ❏ RH inner wing panel.
   - ❏ Engine undershield.
3. Turn crankshaft to TDC on No.1 cylinder. Ensure timing marks aligned **1**.
4. Insert timing pin in crankshaft **2**. Tool No.Mot.1054.
5. Support engine.
6. Remove:
   - ❏ RH engine mounting (2 parts).
   - ❏ Auxiliary drive belt.
   - ❏ Crankshaft pulley.
   - ❏ Timing belt cover **3**.

7. Ensure mark on belt aligned with mark on camshaft sprocket **4**.
   *NOTE: If no mark visible on backplate, make reference mark in line with camshaft sprocket mark **4**.*
8. Ensure mark on belt aligned with mark on crankshaft sprocket:
   - ❏ F3P engine: **5**.
   - ❏ F3R engine: **9**.
9. Slacken tensioner nut **6**. Move tensioner away from belt. Lightly tighten nut.
10. Remove timing belt.

Installation

1. Ensure timing pin located correctly **2**.
2. Fit timing belt in following order:
   - ❏ Crankshaft sprocket
   - ❏ Auxiliary shaft sprocket.
   - ❏ Guide pulley.
   - ❏ Camshaft sprocket.
   - ❏ Tensioner pulley.
3. Ensure marks on belt aligned with marks on sprockets. F3P: **4** & **5** or F3R: **4** & **9**.
4. Slacken tensioner nut **6**.
5. Screw an M6x45 mm bolt into timing belt upper rear cover until tensioner pulley contacts belt **7**.
6. Apply firm thumb pressure to belt at ▽.
7. Attach tension gauge to belt at ▽. Tool No.Mot.1273 **8**.
8. Screw in M6 bolt until tension gauge indicates the following **7**:
   F3P: 25 SEEM units.
   F3R: 36 SEEM units.
9. Tighten tensioner nut to 50 Nm **6**.
10. Remove:
    - ❏ Timing pin **2**.
    - ❏ Tension gauge **8**.
    - ❏ M6 bolt **7**.
11. Turn crankshaft three turns clockwise to TDC on No.1 cylinder. Insert timing pin **2**.
12. Remove timing pin.
13. Apply firm thumb pressure to belt at ▽.
14. Attach tension gauge to belt at ▽. Tool No.Mot.1273 **8**.
15. Tension gauge should indicate the following:
    F3P: 23-27 SEEM units.
    F3R: 33-39 SEEM units.
16. If not: Repeat tensioning procedure.
17. Install components in reverse order of removal.
18. Tighten crankshaft pulley bolt to 90-100 Nm.

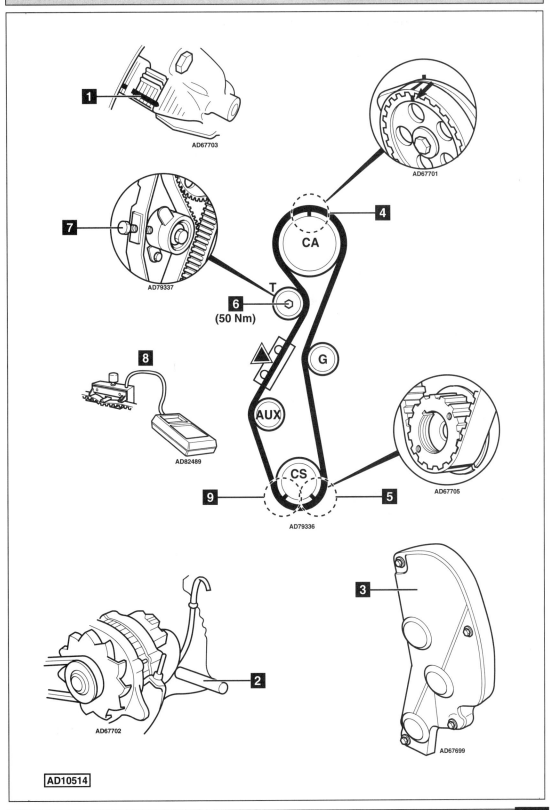

1 — AD67703

4 — AD67701

7 — AD79337

T

6 (50 Nm)

CA

G

8 — AD82489

AUX

CS

9 — AD79336

5 — AD67705

3

2 — AD67702

AD67699

AD10514

RENAULT

Model:	Clio Sport 2,0 16V • Mégane/Scénic 2,0 16V • Laguna 1,8 16V
	Laguna 2,0 16V • Espace 2,0 16V
Year:	1998-03
Engine Code:	F4P 760, F4R 700/701/730/740/741/744/780, F5R 740

Replacement Interval Guide

Renault recommend:

Except Laguna 1,8 16V
Replacement every 72,000 miles or 5 years.

Laguna 1,8 16V
→03/99: Replacement every 60,000 miles or
5 years.
04/99→: Replacement every 72,000 miles or
5 years.
*The previous use and service history of the vehicle
must always be taken into account.
Refer to Timing Belt Replacement Intervals at the front
of this manual.*

Check For Engine Damage

*CAUTION: This engine has been identified as an
INTERFERENCE engine in which the possibility of
valve-to-piston damage in the event of a timing belt
failure is MOST LIKELY to occur. A compression check
of all cylinders should be performed before removing
the cylinder head(s).*

Repair Times – hrs

Remove & install:

Clio	5,20
Mégane/Scénic	4,70
Laguna	2,50
Espace	3,50

Special Tools

- Crankshaft timing pin – Renault No.Mot.1054.
- Camshaft setting bar – Renault No.Mot.1496.

Special Precautions

- Disconnect battery earth lead.
- DO NOT turn crankshaft or camshaft when timing belt removed.
- Remove spark plugs to ease turning engine.
- Turn engine in normal direction of rotation (unless otherwise stated).
- DO NOT turn engine via camshaft or other sprockets.
- Observe all tightening torques.

Removal

*NOTE: If timing belt is excessively noisy between
idle and 2000 rpm a modified timing belt,
tensioner and guide pulleys is available, refer to
dealer.*

1. Raise and support front of vehicle.
2. Remove:
   - ❑ Engine top cover.
   - ❑ RH splash guard.
   - ❑ Auxiliary drive belt.
3. Support engine.
4. Remove:
   - ❑ RH engine mounting.
   - ❑ Blanking plugs from rear of camshafts.
   - ❑ Blanking plug from cylinder block **1**.
5. Lock flywheel with large screwdriver. Slacken crankshaft pulley bolt **2**.
6. Turn crankshaft clockwise to setting position. Ensure grooves in camshafts aligned **3**.
   NOTE: Grooves located below cylinder head upper surface.
7. Insert timing pin through hole in cylinder block and into crankshaft slot **4**. Tool No.Mot.1054.
   *NOTE: Ensure timing pin not inserted into crankshaft web balance hole **5**.*
8. Fit setting bar to rear of camshafts **6**. Tool No.Mot.1496.
9. Remove:
   - ❑ Crankshaft pulley bolt **2**.
   - ❑ Crankshaft pulley **7**.
   - ❑ Timing belt covers **8** & **9**.
10. Slacken tensioner nut **10**.
11. Allow tensioner pulley to move away from belt.
12. Remove:
    - ❑ Tensioner nut **10**.
    - ❑ Tensioner pulley **11**.
    - ❑ Guide pulley (G1) **12**.
    - ❑ Timing belt.
    NOTE: DO NOT allow crankshaft sprocket to fall off crankshaft.

Installation

1. Ensure timing pin located correctly in crankshaft **4**.
2. Ensure grooves in camshafts aligned **3**.
3. Ensure setting bar fitted correctly **6**.
4. Fit new guide pulley (G1) **12**. Tighten bolt to 45 Nm.

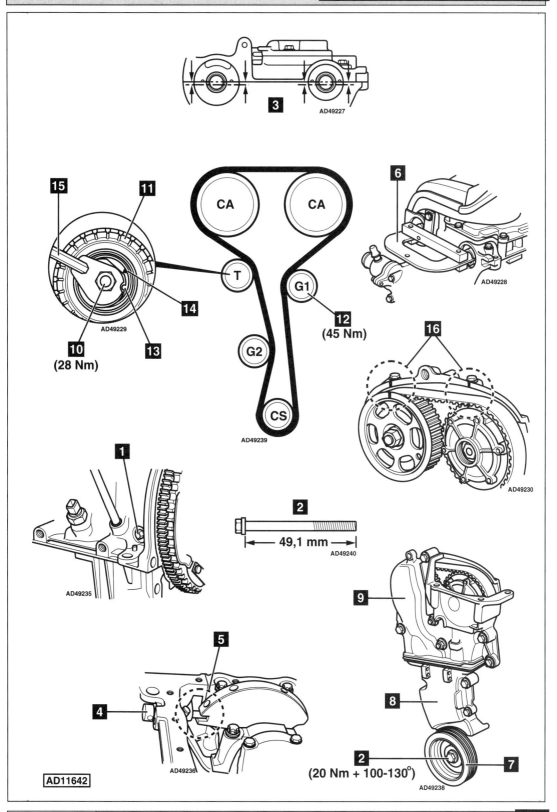

AD49227

CA   CA

T

G1

G2

CS

AD49239

12
(45 Nm)

10
(28 Nm)

AD49229

AD49228

16

AD49230

AD49235

2

← 49,1 mm →
AD49240

AD49236

AD11642

2
(20 Nm + 100-130°)

AD49238

5. Fit new tensioner pulley ⑪. Temporarily tighten nut to 7 Nm ⑩.

NOTE: Ensure lug on rear of tensioner pulley located in groove in cylinder head.

6. Remove crankshaft sprocket. Degrease sprocket. Degrease end of crankshaft. Refit crankshaft sprocket.

7. Fit timing belt in anti-clockwise direction, starting at crankshaft sprocket. Ensure belt is taut on non-tensioned side.

8. Degrease crankshaft pulley ⑦.

9. Fit crankshaft pulley ⑦.

10. Measure length of crankshaft pulley bolt ②. Maximum length: 49,1 mm. If longer: Use new bolt.

11. Fit crankshaft pulley bolt ②.
New bolt: DO NOT oil.
Used bolt: Oil threads and contact face.

12. Temporarily tighten crankshaft pulley bolt ②. Leave 2-3 mm clearance between bolt contact face and crankshaft pulley ⑦.

13. Slacken tensioner nut ⑩. Turn tensioner pulley clockwise until tensioner marks aligned ⑬ & ⑭. Use 6 mm Allen key ⑮.

14. Temporarily tighten tensioner nut ⑩.
Tightening torque: 7 Nm.

15. Lock flywheel with large screwdriver. Temporarily tighten crankshaft pulley bolt ②. Tightening torque: 20 Nm.

16. Mark camshaft sprockets and cylinder head with paint or chalk where indicated ⑯.

17. Remove:
❑ Timing pin ④.
❑ Setting bar ⑥.

18. Lock flywheel with large screwdriver. Tighten crankshaft pulley bolt a further 100-130° ②.

19. Turn crankshaft two turns clockwise to setting position. Ensure timing marks aligned ⑯.

20. Insert timing pin through hole in cylinder block and into crankshaft ④.

21. Ensure camshaft setting bar can be fitted easily ⑥.

22. Ensure tensioner pulley marks aligned ⑬ & ⑭. If not: Repeat tensioning procedure.

23. Tighten tensioner nut ⑩.
Tightening torque: 28 Nm.

24. Remove timing pin ④.

25. Refit blanking plug ①.

26. Fit new blanking plugs to rear of camshafts.

27. Install components in reverse order of removal.

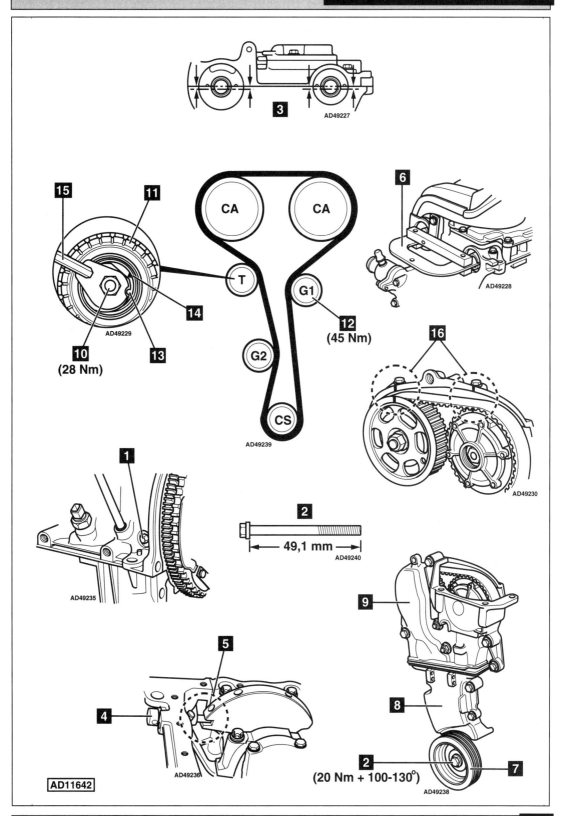

RENAULT

Model:	R18/Fuego 2,0 (8V) • R20/21 2,0 (8V/12V) • R25 2,0 (8V/12V)
	Espace 2,0/2,2 (8V) • Trafic/Master 2,0/2,2 (8V)
Year:	1980-98
Engine Code:	829, J5R, J6R
	J7R 720/721/722/723/726/740/750/751/754/755/768
	J7T, J7T 600/772/782

Replacement Interval Guide

Renault recommend replacement every
72,000 miles or 5 years.
*The previous use and service history of the vehicle
must always be taken into account.*
*Refer to Timing Belt Replacement Intervals at the front
of this manual.*

Check For Engine Damage

*CAUTION: This engine has been identified as an
INTERFERENCE engine in which the possibility of
valve-to-piston damage in the event of a timing belt
failure is MOST LIKELY to occur.*
*A compression check of all cylinders should be
performed before removing the cylinder head.*

Repair Times – hrs

Remove & install:

R18/Fuego	1,50
R20	1,50
R25 (8V)	1,50
R25 (12V)	2,80
R21	3,00
Espace (→1990)	2,80
Espace (1991→)	3,80
Trafic	1,80
Master	1,60

Special Tools

- Tension gauge – Renault No.Mot.1273 (SEEM C.Tronic 105).
- Crankshaft timing pin – Renault No.Mot.861.
- Tensioner spanner – Renault No.Mot.1384.
- Tensioner spanner – Renault No.Mot.1135.

Special Precautions

- Disconnect battery earth lead.
- DO NOT turn crankshaft or camshaft when timing belt removed.
- Remove spark plugs to ease turning engine.
- Turn engine in normal direction of rotation (unless otherwise stated).
- DO NOT turn engine via camshaft or other sprockets.
- Observe all tightening torques.

Removal

1. 1991→: Drain coolant.
2. Remove:
   - ❑ Auxiliary drive belt(s).
   - ❑ 1991→: Radiator.
3. Turn crankshaft clockwise until camshaft sprocket timing marks and flywheel TDC marks aligned **1** & **2**.
4. J7R 720: Align mark **11** with pointer in timing belt cover **1**.
   *NOTE: TDC position may also be determined using crankshaft timing pin No.Mot.861 **3**.*
   WARNING: DO NOT use tool No.Mot.861 to hold crankshaft when slackening/tightening bolts.
5. Disconnect hose from front of inlet manifold.
6. Remove timing belt cover **4**.
7. Except J7R 720: Slacken tensioner backplate nuts **5**.
8. J7R 720: Slacken tensioner nuts **12** & **13**.
9. Move tensioner away from belt. Lightly tighten nuts.
10. Remove timing belt.

Installation

*NOTE: DO NOT fit used belt. Belts for 829 engine
and for J5R and J6R engines have 116 teeth.
Sprockets may have round or square teeth.
Ensure correct replacement timing belt fitted.*

1. Ensure timing marks aligned:
   - ❑ TDC flywheel marks **2**.
   - ❑ Camshaft sprocket and centre line of rocker shaft **6**.
   - ❑ Crankshaft sprocket **7**.
   - ❑ Auxiliary shaft sprocket and web on cylinder block **8**.
2. J7R 720: Temporarily fit timing belt cover. Align mark **11** with pointer in timing belt cover **1**.
3. Fit timing belt.
   NOTE: Observe direction of rotation mark on belt.

Autodata

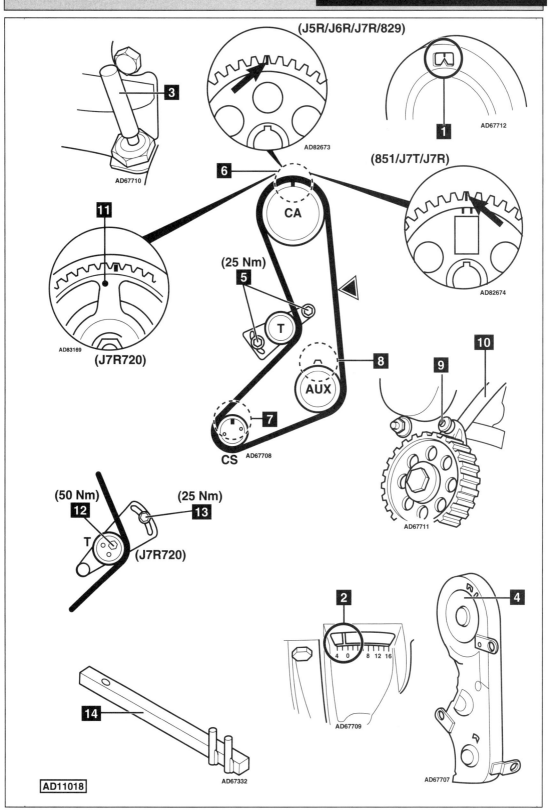

(J5R/J6R/J7R/829)

AD82673

3

AD67710

AD67712

1

(851/J7T/J7R)

6

CA

AD82674

11

AD83169

(J7R720)

(25 Nm)

5

T

8

9

10

7

CS AD67708

AUX

AD67711

(50 Nm)

12

13

(25 Nm)

T

(J7R720)

2

4

4 0   8 12 16

AD67709

14

AD67332

AD67707

AD11018

Except J7R 720: Proceed as follows:

> *NOTE: There are two types of auxiliary shaft housing. One with and one without adjusting screw* 🔟.

4. Auxiliary shaft housing without adjusting screw: Check clearance between auxiliary shaft housing and cylinder block with feeler gauge 🔟.

5. Install feeler gauge of 0,1 mm less than thickness measured in operation 4.

6. Slacken tensioner nuts 1/4 turn 🔟. Allow tensioner to operate.

7. Turn crankshaft one turn clockwise.

8. Tighten tensioner nuts. Tightening torque: 25 Nm. Remove feeler gauge.

9. Auxiliary shaft housing with adjusting screw: Adjust clearance with adjusting screw 🔟. Use a 0,1 mm feeler gauge.

10. Slacken tensioner nuts 1/4 turn. Allow tensioner to operate. Lightly tighten nuts.

11. Turn crankshaft three turns clockwise. Ensure flywheel timing marks aligned 🔟.

12. Check clearance between auxiliary shaft housing and cylinder block as described in operations 4 and 9.

13. Slacken tensioner nuts 1/4 turn, then tighten 🔟. Tightening torque: 25 Nm.

14. Attach tension gauge to belt at ▽. Tool No.Mot.1273.

15. Tension gauge should indicate 39 units. If not: Repeat tensioning procedure. Use tool No.Mot.1384 on upper nut 🔟.

J7R 720: Proceed as follows:

16. Slacken tensioner backplate nut 🔟.

17. Position tensioner backplate nut in centre of slotted hole. Tighten nut 🔟. Tightening torque: 25 Nm.

18. Attach tension gauge to belt at ▽. Tool No.Mot.1273.

19. Slacken tensioner nut 🔟.

20. Turn tensioner pulley anti-clockwise until tension gauge indicates 39±4 SEEM units. Use tool No.Mot.1135 🔟.

21. Tighten tensioner nut 🔟. Tightening torque: 50 Nm.

All models:

22. Remove tension gauge.

23. Turn crankshaft two turns clockwise.

24. Attach tension gauge to belt at ▽. Recheck belt tension.

25. Install components in reverse order of removal.

26. 1991→: Refill cooling system.

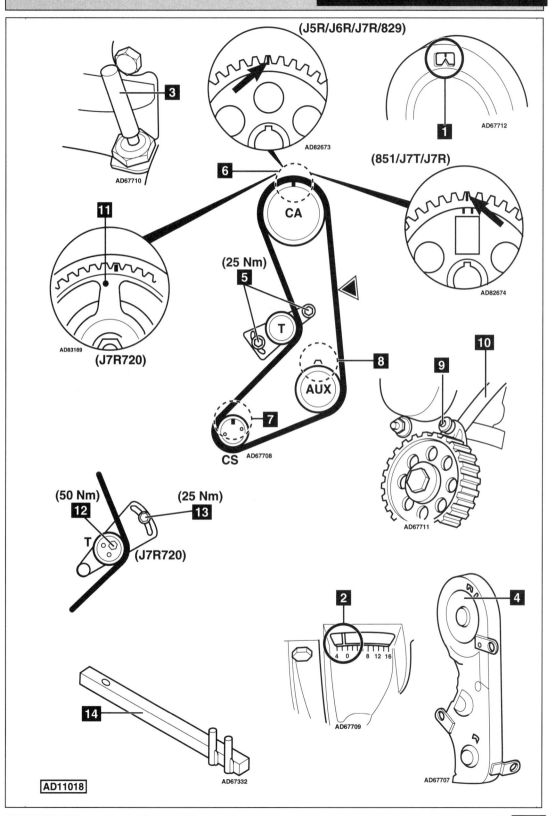

(J5R/J6R/J7R/829)

AD82673

AD67712

1

3

AD67710

6

(851/J7T/J7R)

AD82674

CA

11

(25 Nm)

5

T

AD83169

(J7R720)

8

9

10

AUX

7

CS AD67708

AD67711

(50 Nm)

12

(25 Nm)

13

T

(J7R720)

2

4

4 0   8 12 16

AD67709

14

AD67332

AD67707

AD11018

RENAULT

Model:	Mégane/Scénic 1,4/1,6
Year:	1996-02
Engine Code:	E7J 624/764, K7M 702/703/720/744/745

Replacement Interval Guide

Renault recommend replacement as follows:
10,000 mile service intervals – replacement every 70,000 miles or 5 years.
12,000 mile service intervals – replacement every 72,000 miles or 5 years.
The previous use and service history of the vehicle must always be taken into account.
Refer to Timing Belt Replacement Intervals at the front of this manual.

Check For Engine Damage

CAUTION: This engine has been identified as an INTERFERENCE engine in which the possibility of valve-to-piston damage in the event of a timing belt failure is MOST LIKELY to occur.
A compression check of all cylinders should be performed before removing the cylinder head.

Repair Times – hrs

Remove & install	1,80

Special Tools

- Tensioning tool – Renault No.Mot.1135-01.
- Tension gauge – Renault No.Mot.1273 (SEEM C.Tronic 105.6).

Special Precautions

- Disconnect battery earth lead.
- DO NOT turn crankshaft or camshaft when timing belt removed.
- Remove spark plugs to ease turning engine.
- Turn engine in normal direction of rotation (unless otherwise stated).
- DO NOT turn engine via camshaft or other sprockets.
- Observe all tightening torques.

Removal

1. Raise and support front of vehicle.
2. Remove:
   - ❏ RH wheel.
   - ❏ RH inner wing panel.
   - ❏ Auxiliary drive belts.
   - ❏ Crankshaft pulley bolt **1**.
   - ❏ Crankshaft pulley **2**.
   - ❏ Timing belt cover **3**.

3. Turn crankshaft clockwise until camshaft sprocket timing mark **4** aligned with rib on cylinder head cover and crankshaft sprocket timing mark **5** at 6 o'clock position. Ensure crankshaft sprocket keyway at 12 o'clock position.
   NOTE: Some camshaft sprockets have multiple markings. Use rectangular timing mark adjacent to 'E1-E3' lettering.
   If rib on cylinder head cover not present, mark with paint or chalk.
4. Slacken tensioner nut **6**. Turn tensioner clockwise to release tension on belt. Use tool No.Mot.1135-01 **7**. Lightly tighten nut.
5. Remove timing belt.

Installation

NOTE: DO NOT fit used belt.
1. Ensure timing marks **4** & **5** aligned at 12 o'clock and 6 o'clock positions respectively.
2. Fit timing belt in anti-clockwise direction, starting at crankshaft sprocket. Ensure belt is taut between sprockets.
   NOTE: Align marks on belt with marks on sprockets. Observe direction of rotation marks on belt.
3. Attach tension gauge to belt at ▽. Tool No.Mot.1273 **8**.
4. Slacken tensioner nut **6**. Turn tensioner anti-clockwise until tension gauge indicates 30 SEEM units. Use tool No.Mot.1135-01 **7**.
   NOTE: Engine must be COLD when installing belt.
5. Tighten tensioner nut **6**. Tightening torque: 50 Nm.
6. Turn crankshaft three turns clockwise until crankshaft timing mark **5** aligned at 6 o'clock position.
7. Recheck belt tension. Tension gauge should indicate 27-33 SEEM units. If not: Repeat tensioning procedure.
8. Install components in reverse order of removal.
9. Tighten crankshaft pulley bolt **1**. Tightening torque: 20 Nm + 62°-74°.

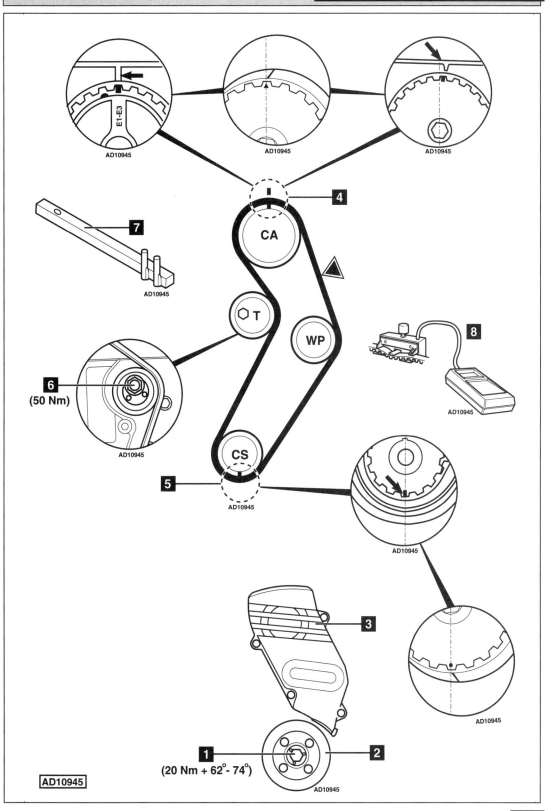

AD10945

E1-E3

CA

WP

T

CS

7

4

8

6
(50 Nm)

5

3

1
(20 Nm + 62°- 74°)

2

Model:	**Mégane/Scénic 2,0**
Year:	**1996-99**
Engine Code:	**F3R 750/751**

Replacement Interval Guide

Renault recommend replacement as follows:
10,000 mile service intervals – replacement every 70,000 miles or 5 years.
12,000 mile service intervals – replacement every 72,000 miles or 5 years.
The previous use and service history of the vehicle must always be taken into account.
Refer to Timing Belt Replacement Intervals at the front of this manual.

Check For Engine Damage

CAUTION: This engine has been identified as an INTERFERENCE engine in which the possibility of valve-to-piston damage in the event of a timing belt failure is MOST LIKELY to occur.
A compression check of all cylinders should be performed before removing the cylinder head.

Repair Times – hrs

Remove & install:	
Mégane	2,60
Mégane Scénic	3,10

Special Tools

- Crankshaft timing pin – Renault No.Mot.1054.
- Tension gauge – Renault No.Mot.1273 (SEEM C.Tronic 105.6).
- M6 x 45 mm bolt – for adjusting tensioner.

Special Precautions

- Disconnect battery earth lead.
- DO NOT turn crankshaft or camshaft when timing belt removed.
- Remove spark plugs to ease turning engine.
- Turn engine in normal direction of rotation (unless otherwise stated).
- DO NOT turn engine via camshaft or other sprockets.
- Observe all tightening torques.

Removal

1. Raise and support front of vehicle.
2. Remove engine control module (ECM). Leave multi-plug connected.
3. Hold auxiliary drive belt tensioner. Use 22 mm spanner.
4. Slacken tensioner 1/4 turn. Use 7 mm Allen key. Remove auxiliary drive belt.
   NOTE: DO NOT turn 22 mm nut as tensioner will be permanently damaged.
5. Remove:
   - RH front wheel.
   - RH front wheel arch liner.
   - Support bar from strut top mountings.
6. Support engine.

7. Remove:
   - Engine mounting.
   - Timing belt cover.
   - RH engine mounting and bracket.
   - Fuel pipe mounting plate.
   - Crankshaft pulley bolt **1**.
   - Crankshaft pulley **2**.
   - Timing belt centre cover **3**.
   - Timing belt lower cover **4**.
   - Blanking plug from cylinder block.
8. Turn crankshaft clockwise to TDC on No.1 cylinder.
9. Insert timing pin in crankshaft **5**. Tool No.Mot.1054. Rock crankshaft slightly to ensure timing pin located correctly.
10. Ensure timing marks on belt aligned with marks on sprockets **6** & **7**.
11. Slacken tensioner nut **8**. Move tensioner away from belt.
    NOTE: DO NOT slacken tensioner nut more than one turn.
12. Remove timing belt.

Installation

NOTE: DO NOT fit used belt.
1. Ensure timing pin located correctly **5**.
2. Fit timing belt in following order:
   - Crankshaft sprocket.
   - Auxiliary shaft sprocket.
   - Guide pulley.
   - Camshaft sprocket.
   - Tensioner pulley.
   NOTE: Ensure belt is taut between sprockets and pulleys.
3. Ensure marks on belt aligned with marks on sprockets **6** & **7**. Observe direction of rotation marks.
   NOTE: There should be 61 belt teeth between timing marks on crankshaft sprocket and camshaft sprocket on the guide pulley side.
4. Slacken tensioner nut **8**.
5. Screw an M6x45 mm bolt into timing belt upper rear cover until tensioner pulley contacts belt **9**.
6. Apply firm thumb pressure to belt at ▽.
7. Attach tension gauge to belt at ▽. Tool No.Mot.1273.
8. Screw in M6 bolt until tension gauge indicates 29 SEEM units **9**.
9. Tighten tensioner nut **8**. Tightening torque: 40 Nm.
10. Remove:
    - Timing pin **5**.
    - Tension gauge.
    - M6 bolt **9**.
11. Turn crankshaft three turns clockwise to TDC on No.1 cylinder. Insert timing pin **5**.
12. Remove timing pin **5**.
13. Apply firm thumb pressure to belt at ▽.
14. Recheck belt tension. Tension gauge should indicate 26-32 SEEM units.
15. If not: Repeat tensioning procedure.
16. Install components in reverse order of removal.
17. Tighten crankshaft pulley bolt to 120 Nm **1**.

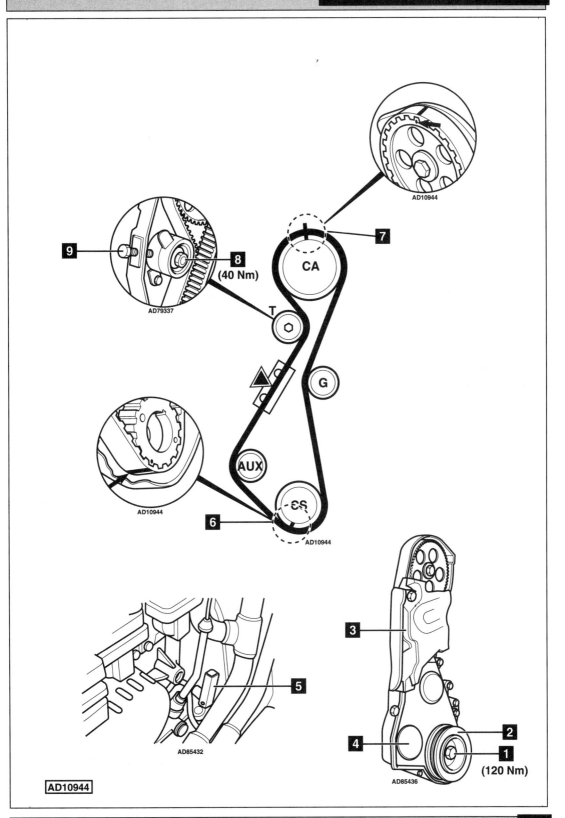

AD10944

Model:	**Mégane 2,0 16V**
Year:	**1996-99**
Engine Code:	**F7R 710**

Replacement Interval Guide

Renault recommend replacement as folows:
10,000 mile service intervals – replacement every 70,000 miles or 5 years.
12,000 mile service intervals – replacement every 72,000 miles or 5 years.
The previous use and service history of the vehicle must always be taken into account.
Refer to Timing Belt Replacement Intervals at the front of this manual.

Check For Engine Damage

CAUTION: This engine has been identified as an INTERFERENCE engine in which the possibility of valve-to-piston damage in the event of a timing belt failure is MOST LIKELY to occur.
A compression check of all cylinders should be performed before removing the cylinder head.

Repair Times – hrs

Remove & install	3,60

Special Tools

- Fuel pipe union remover – Renault No.1311-06.
- Crankshaft timing pin – Renault No.Mot.1054.
- Tension gauge – Renault No.Mot.1273 (SEEM C.Tronic 105.6).
- M6 x 45 mm bolt – for adjusting tensioner.

Special Precautions

- Disconnect battery earth lead.
- DO NOT turn crankshaft or camshaft when timing belt removed.
- Remove spark plugs to ease turning engine.
- Turn engine in normal direction of rotation (unless otherwise stated).
- DO NOT turn engine via camshaft or other sprockets.
- Observe all tightening torques.

Removal

1. Raise and support front of vehicle.
2. Remove:
   - ❑ Engine undershield.
   - ❑ RH front wheel.
   - ❑ RH front wheel arch liner.
   - ❑ Auxiliary drive belt.
   - ❑ Engine mounting cover.
   - ❑ Earth lead from engine mounting.
   - ❑ Support bar from strut top mountings.
3. Support engine.
4. Remove:
   - ❑ RH engine mounting.
   - ❑ PAS reservoir. Move to one side.
5. Disconnect fuel pipes from fuel rail.
6. Unclip fuel pipes from timing belt centre cover.
7. Remove:
   - ❑ Exhaust downpipe from manifold.
   - ❑ Crankshaft pulley bolt **1**.
   - ❑ Crankshaft pulley **2**.
   - ❑ Timing belt covers **3**, **4** & **5**.
   - ❑ Air filter.
   - ❑ Blanking plug from cylinder block.
8. Turn crankshaft clockwise until timing marks aligned **6**.
9. Insert timing pin in crankshaft **7**. Tool No.Mot.1054. Rock crankshaft slightly to ensure timing pin located correctly **8**.
10. Slacken tensioner nut **9**.
11. Remove timing belt.

Installation

NOTE: DO NOT fit used belt.

1. Ensure timing pin located correctly **7**.
2. Ensure timing marks aligned **6**.
3. Fit timing belt in anti-clockwise direction, starting at crankshaft sprocket. Ensure belt is taut between sprockets.
   NOTE: Align marks on belt with marks on sprockets. Observe direction of rotation marks on belt.
4. Screw an M6x45 mm bolt into timing belt rear cover until tensioner pulley contacts belt **10**.
5. Remove timing pin **7**.
6. Apply firm thumb pressure to belt at ▽.
7. Attach tension gauge to belt at ▽. Tool No.Mot.1273.
8. Screw in M6 bolt until tension gauge indicates 32±3 SEEM units **10**.
9. Tighten tensioner nut **9**. Tightening torque: 50 Nm.
10. Remove:
    - ❑ Tension gauge.
    - ❑ M6 bolt **10**.
11. Turn crankshaft three turns clockwise. Insert timing pin **7**.
12. Remove timing pin **7**.
13. Apply firm thumb pressure to belt at ▽. Tension gauge should indicate 32±3 SEEM units.
14. If not: Repeat tensioning procedure.
15. Install components in reverse order of removal.
16. Tighten crankshaft pulley bolt **1**. Tightening torque: 100 Nm.

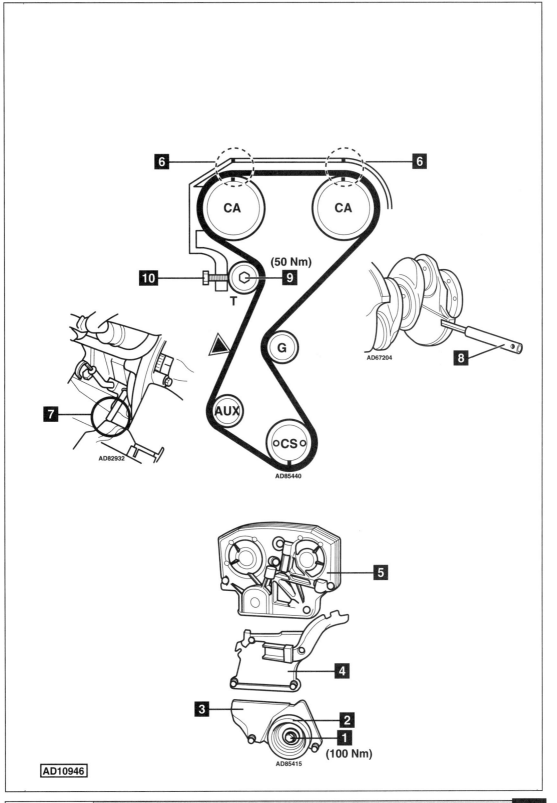

RENAULT

Model:	**Laguna 2,0 16V • Safrane 2,0 16V • Safrane 2,5 20V**
Year:	**1996-01**
Engine Code:	**N7Q 700/704/710/711, N7U 700/701**

Replacement Interval Guide

Renault recommend replacement every 70,000 miles or 5 years.
The previous use and service history of the vehicle must always be taken into account.
Refer to Timing Belt Replacement Intervals at the front of this manual.

Check For Engine Damage

CAUTION: This engine has been identified as an INTERFERENCE engine in which the possibility of valve-to-piston damage in the event of a timing belt failure is MOST LIKELY to occur.
A compression check of all cylinders should be performed before removing the cylinder head.

Repair Times – hrs

Remove & install:

Laguna	4,10
Safrane	3,60

Special Tools

- Tensioning tool (auxiliary drive belt) – Renault No.Mot.1348.
- Crankshaft TDC timing pin – Renault No.Mot.1340.
- Camshaft setting tool – Renault No.Mot.1337.
- Sprocket holding tool – Renault No.Mot.799.
- Tension gauge – Renault No.Mot.1273 (SEEM C.Tronic 105).
- Inlet camshaft plug fitting tool – Renault No.Mot.1345.

Special Precautions

- Disconnect battery earth lead.
- DO NOT turn crankshaft or camshaft when timing belt removed.
- Remove spark plugs to ease turning engine.
- Turn engine in normal direction of rotation (unless otherwise stated).
- DO NOT turn engine via camshaft or other sprockets.
- Observe all tightening torques.

Removal

1. Raise and support front of vehicle.
2. Remove:
   - ❏ RH front wheel.
   - ❏ RH front wheel arch liner.
   - ❏ Engine undershield.
   - ❏ RH engine mounting.
3. Turn auxiliary drive belt tensioner away from belt **1**. Use tool No.1348.
4. Lock auxiliary drive belt tensioner. Use 4 mm pin **2**.

5. Remove:
   - ❏ Auxiliary drive belt.
   - ❏ Auxiliary drive belt tensioner and bracket.
   - ❏ Crankshaft pulley nut and washer **4**.
   - ❏ Crankshaft pulley bolts (4 bolts) **3**.
   - ❏ Crankshaft pulley **5**.
   - ❏ Cylinder head cover and ignition coil bracket.
   - ❏ Camshaft position (CMP) sensor and rotor from rear of exhaust camshaft housing **6**.
   - ❏ Plug from rear of inlet camshaft housing **7**.
6. Turn crankshaft until grooves in camshafts are aligned **8**.
7. Fit setting tool to camshafts **9**. Tool No.Mot.1337.
8. Turn crankshaft slowly clockwise until setting tool arms meet. Fit bolt and retainer **10**.
9. Slacken bolts of each camshaft sprocket **19**.
10. Remove blanking plug from cylinder block. Ensure TDC timing pin can be inserted **11**. Tool No.Mot.1340. If necessary, turn crankshaft slightly anti-clockwise.
11. Remove:
    - ❏ Timing belt covers **12** & **13**.
    - ❏ Lower belt shield **14**.
12. Ensure crankshaft timing marks aligned **15**.
    NOTE: Engines available with either hydraulic or mechanical tensioner.

Hydraulic tensioner:

13. Remove:
    - ❏ Automatic tensioner unit bolts **16**.
    - ❏ Automatic tensioner unit.

Mechanical tensioner:

14. Slacken tensioner pulley bolt **21**.
15. Turn tensioner pulley clockwise until pointer on LH stop **22**.
16. Lightly tighten tensioner pulley bolt **21**.

All tensioners:

17. Remove timing belt.

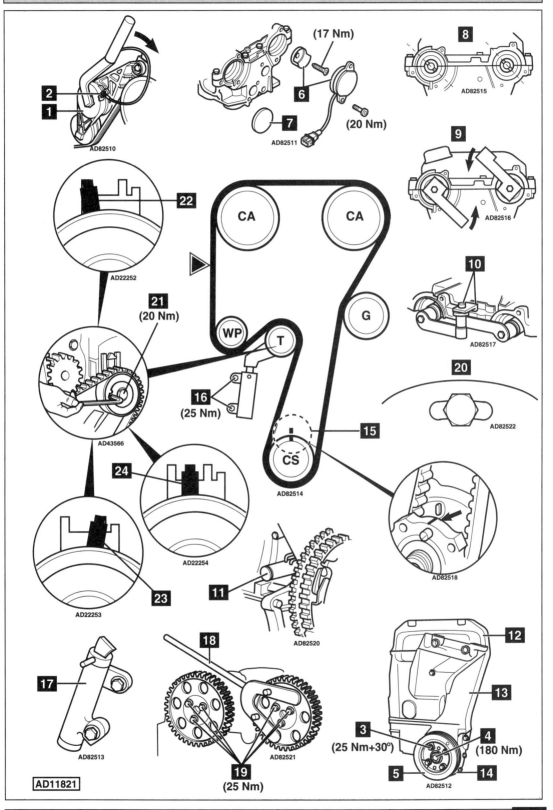

Installation

Hydraulic tensioner:

1. Slowly compress pushrod into tensioner body until holes aligned. Use press. Retain pushrod with suitable pin through hole in tensioner body **17**.

All tensioners:

2. Ensure setting tool fitted correctly **9** & **10**.
3. Ensure TDC timing pin located correctly **11**. Tool No.Mot.1340.
4. Ensure crankshaft sprocket timing marks aligned **15**.
5. Hold camshaft sprockets **18**. Use tool No.Mot.799. Slacken bolts **19**.
6. Align camshaft sprockets with bolts in centre of slotted holes **20**.
7. Fit timing belt in anti-clockwise direction, starting at crankshaft sprocket. Ensure belt is taut between sprockets.

Hydraulic tensioner:

8. Install automatic tensioner unit. Tighten bolts to 25 Nm **16**.
9. Remove pin from tensioner body to release pushrod **17**.

Mechanical tensioner:

10. Slacken tensioner pulley bolt **21**.
11. Turn tensioner pulley anti-clockwise until pointer on RH stop **23**.
12. Turn tensioner pulley clockwise until pointer in centre **24**.
13. Tighten tensioner pulley bolt **21**.
    Tightening torque: 20 Nm.

All tensioners:

14. Hold camshaft sprockets **18**.
    Use tool No.Mot.799. Tighten bolts to 25 Nm **19**.
15. Remove:
    ❏ TDC timing pin **11**.
    ❏ Setting tool **9** & **10**.
16. Turn crankshaft almost two turns clockwise.
17. Fit setting tool to camshafts **9**.
    Tool No.Mot.1337.
18. Turn crankshaft slowly clockwise until setting tool arms meet. Fit bolt and retainer **10**.
19. Ensure TDC timing pin can be inserted **11**.
    Tool No.Mot.1340.
20. Ensure crankshaft sprocket timing marks aligned **15**.

Hydraulic tensioner:

21. Attach tension gauge to belt at ▼.
    Tool No.Mot.1273. Tension gauge should indicate 36-46 SEEM units.
22. If not: Suspect faulty automatic tensioner unit.

Mechanical tensioner:

23. Ensure pointer in centre **24**. If not: Re-adjust.
24. Push on belt at ▼. Pointer should move. Release pressure from belt. Pointer should return to centre.

All tensioners:

25. Remove:
    ❏ Setting tool **9** & **10**. Tool No.Mot.1337.
    ❏ TDC timing pin **11**. Tool No.Mot.1340.
26. Fit camshaft position (CMP) sensor and rotor **6**.
27. Tighten rotor bolt to 17 Nm. Tighten sensor bolts to 20 Nm.
28. Fit new plug to rear of inlet camshaft housing **7**. Use tool No.Mot.1345.
29. Fit lower belt shield **14**.
30. Fit crankshaft pulley **5**.
31. Tighten crankshaft pulley bolts **3**.
    Tightening torque: 25 Nm + 30°.
32. Fit crankshaft pulley nut and washer **4**. Tighten nut to 180 Nm.
33. Install components in reverse order of removal.

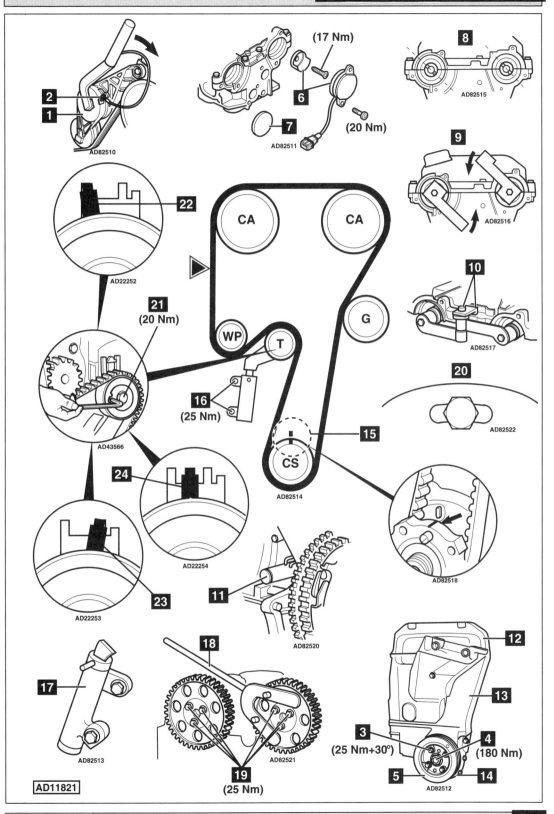

RENAULT

Model:	**Laguna 3,0 V6 • Safrane 3,0 V6 • Espace 3,0 V6**
Year:	**1997-03**
Engine Code:	**L7X 700/701/727**

Replacement Interval Guide

10,000 mile service intervals – replacement every 70,000 miles or 5 years.
12,000 mile service intervals – replacement every 72,000 miles or 5 years.
The previous use and service history of the vehicle must always be taken into account.
Refer to Timing Belt Replacement Intervals at the front of this manual.

Check For Engine Damage

CAUTION: This engine has been identified as an INTERFERENCE engine in which the possibility of valve-to-piston damage in the event of a timing belt failure is MOST LIKELY to occur.
A compression check of all cylinders should be performed before removing the cylinder head(s).

Repair Times – hrs

Remove & install:

Laguna	5,50
Safrane	4,50
Espace	10,50

Special Tools

- Crankshaft & camshaft timing pins – Renault No.Mot.1430.
- Timing belt retaining clip – Renault No.Mot.1436.
- Camshaft holding tool – Renault No.Mot.1428.
- Tensioning tool – Renault No.Mot.1429.
- Tension gauge – Renault No.Mot.1273.
- M8 x 1,25 x 75 mm bolt.
- M8 x 1,25 x 35 mm bolt.

Special Precautions

- Disconnect battery earth lead.
- DO NOT turn crankshaft or camshaft when timing belt removed.
- Remove spark plugs to ease turning engine.
- Turn engine in normal direction of rotation (unless otherwise stated).
- DO NOT turn engine via camshaft or other sprockets.
- Observe all tightening torques.

Removal

1. Espace: Remove engine and transmission assembly.
2. Laguna/Safrane: Raise and support front of vehicle.
3. Remove:
   - RH front wheel.
   - RH splash guard.
   - Auxiliary drive belt.
4. Disconnect engine wiring harness.
5. Move hoses aside.
6. Remove:
   - Engine cover.
   - Engine control module (ECM) cover.
   - Engine control module (ECM) box.
7. Move engine control module (ECM) away from engine.
8. Support engine.
9. Remove:
   - Torque reaction link.
   - RH engine mounting.
   - Auxiliary drive belt tensioner assembly.
   - PAS pump pulley.
10. All models: Remove:
    - Timing belt upper covers **1**.
    - Crankshaft pulley bolts **2**.
    - Crankshaft pulley **3**.
    - Timing belt lower cover **4**.
    - Bracket **5**.
11. Turn crankshaft clockwise to setting position.
12. Insert timing pins in camshaft sprockets **6**. Tool No.Mot.1430.
13. Slacken bolts of each camshaft sprocket **7**.
14. Insert crankshaft timing pin **8**. Tool No.Mot.1430.
15. Insert M8 x 1,25 x 75 mm bolt into tensioner bracket **9**.
16. Tighten bolt **9** until it touches bracket **10**.
17. Slacken tensioner bolts **11**, **12** & **13**.
    *NOTE: DO NOT slacken bolt **14**.*
18. Install special tool No.Mot.1429 **15**.
    *NOTE: If necessary: Slacken bolt slightly **9**.*
19. Insert M8 x 1,25 x 35 mm bolt into tensioner bracket **16**.
20. Tighten bolt **16** until it touches bracket **17**.
21. Fully tighten bolt **16**.
22. Slacken bolt **9**.
23. Remove timing belt.
    NOTE: Mark direction of rotation on belt with chalk if belt is to be reused.

Installation

1. Ensure crankshaft timing pin located correctly **8**. Tool No.Mot.1430.
2. Ensure timing pins located correctly in camshaft sprockets **6**. Tool No.Mot.1430.
3. Ensure camshaft sprockets turn freely.
4. Tighten bolts to 10 Nm **11**, **12** & **13**.
5. Slacken bolts 45° **11**, **12** & **13**.

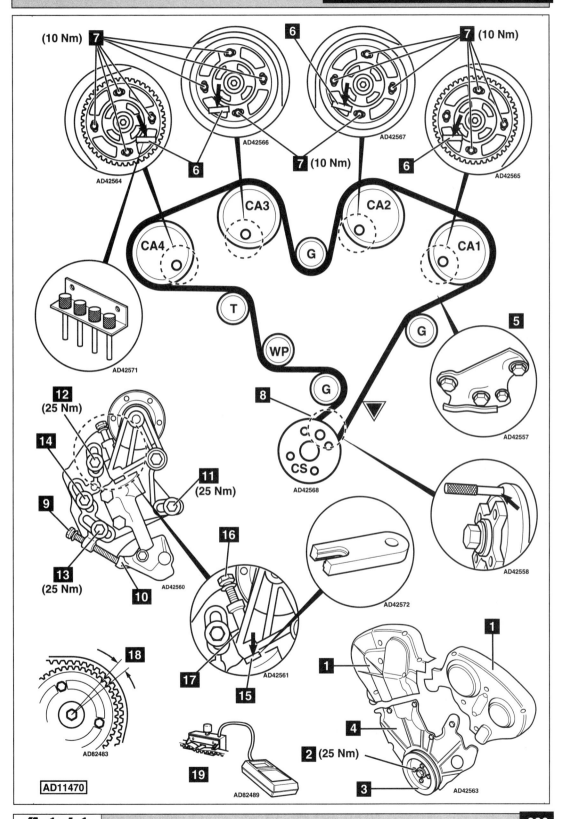

←

6. Turn camshaft sprockets fully clockwise in slotted holes.

7. Tighten bolts to 5 Nm .

8. Slacken bolts 45° .

9. Fit timing belt to crankshaft sprocket.

10. Retain in position with retaining clip.
    Tool No.Mot.1436.

11. Fit timing belt in anti-clockwise direction.
    NOTE: If reusing old belt: Observe direction of rotation marks on belt.

12. Turn each camshaft sprocket to engage in nearest belt tooth.
    NOTE: Ensure bolts not at end of slotted holes in sprockets . Angular movement of sprockets must not be more than one tooth space .

13. Ensure belt is taut between sprockets.

14. Remove retaining clip from timing belt.

15. Attach tension gauge to belt at ▽ .
    Tool No.Mot.1273.

16. Screw in bolt until tension gauge indicates 83±2 SEEM units.

17. Tighten bolt . Tightening torque: 10 Nm.

18. Tighten bolt . Tightening torque: 10 Nm.

19. Tighten bolt . Tightening torque: 10 Nm.

20. Tighten camshaft sprocket bolts (CA1) .
    Tightening torque: 10 Nm.

21. Tighten camshaft sprocket bolts (CA2) .
    Tightening torque: 10 Nm.

22. Tighten camshaft sprocket bolts (CA3) .
    Tightening torque: 10 Nm.

23. Tighten camshaft sprocket bolts (CA4) .
    Tightening torque: 10 Nm.

24. Remove tension gauge .

25. Remove timing pins .

26. Remove crankshaft timing pin .

27. Turn crankshaft slowly two turns clockwise.

28. Insert crankshaft timing pin .
    Tool No.Mot.1430.

29. Slacken bolts 45° , & .

30. Remove bolt .

31. Adjust position of bolt until tool slides freely without free play. Tool No.Mot.1429.

32. Wait 2 minutes to allow automatic tensioner unit and belt to settle.

33. Check tool slides freely without free play.
    Tool No.Mot.1429.

34. Adjust if necessary.

35. Remove special tool .

36. Tighten bolt . Tightening torque: 25 Nm.

37. Tighten bolt . Tightening torque: 25 Nm.

38. Tighten bolt . Tightening torque: 25 Nm.

39. Remove bolt .

40. Remove crankshaft timing pin .

41. Turn crankshaft slowly two turns clockwise.

42. Insert crankshaft timing pin .
    Tool No.Mot.1430.

43. Ensure timing pins can easily be inserted in and withdrawn from camshaft sprockets .

44. Adjust if necessary. Use tool No.Mot.1428.
    NOTE: Ensure bolts not at end of slotted holes in sprockets .

45. Remove crankshaft timing pin .

46. Remove timing pins .

47. Install components in reverse order of removal.

48. Tighten crankshaft pulley bolts .
    Tightening torque: 25 Nm.

49. Espace: Refit engine and transmission assembly.

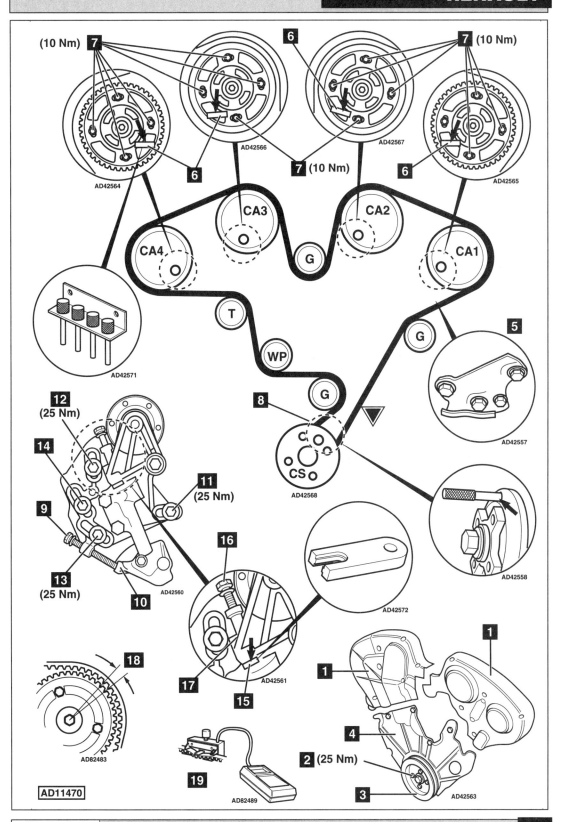

(10 Nm) **7**

6

7 (10 Nm)

AD42564

AD42566

6

7 (10 Nm)

AD42567

6

AD42565

AD42571

CA4

CA3

G

CA2

CA1

T

G

WP

5

G

AD42557

12
(25 Nm)

14

9

11
(25 Nm)

8

C

CS

AD42568

13
(25 Nm)

10

AD42560

16

AD42561

17

15

AD42572

AD42558

18

AD82483

19

AD82489

1

1

4

2 (25 Nm)

3

AD42563

AD11470

RENAULT

Model:	R5/9/11 1,6D • R5 1,9D • R5 Van 1,9D • Clio 1,9D • Clio Van 1,9D Express/Extra/Rapid 1,9D • R19/Turbo/Chamade 1,9D • R21 1,9D
Year:	1988-02
Engine Code:	F8M (4th type), F8Q – non-adjustable or three bolt adjustable injection pump sprocket

Replacement Interval Guide

Renault recommend replacement as follows:
5,000 mile service intervals – replacement every
70,000 miles or 5 years.
6,000 mile service intervals – replacement every
72,000 miles or 5 years.
*The previous use and service history of the vehicle
must always be taken into account.*
*Refer to Timing Belt Replacement Intervals at the
front of this manual.*

Check For Engine Damage

*CAUTION: This engine has been identified as an
INTERFERENCE engine in which the possibility of
valve-to-piston damage in the event of a timing belt
failure is MOST LIKELY to occur.*
*A compression check of all cylinders should be
performed before removing the cylinder head.*

Repair Times – hrs

Remove & install:

R5	3,90
R9/R11	2,10
Clio/Express/Extra/Rapid	3,60
R19/Chamade	3,50
R21	2,20

Special Tools

- Crankshaft timing pin – Renault No.Mot.1054.
- Tension gauge – Renault No.Mot.1273
  (SEEM C.Tronic 105.6).
- Injection pump sprocket locking tool –
  Renault No.Mot.1131.

Special Precautions

- Disconnect battery earth lead.
- DO NOT turn crankshaft or camshaft when timing belt
  removed.
- Remove glow plugs to ease turning engine.
- Turn engine in normal direction of rotation (unless
  otherwise stated).
- DO NOT turn engine via camshaft or other sprockets.
- Observe all tightening torques.
- Check diesel injection pump timing after belt
  replacement.

Removal

1. Remove:
   - ❑ Engine undershield.
   - ❑ RH front wheel.
   - ❑ RH front inner wing panel.
   - ❑ Diesel fuel filter assembly.
2. Support engine.
3. Remove RH engine mounting and fixing studs.
4. Turn crankshaft clockwise until camshaft sprocket
   timing mark aligned with mark on timing belt
   cover **1**.

5. Insert timing pin in crankshaft. Tool No.Mot.1054.
   Rock crankshaft slightly to ensure timing pin located
   correctly **2**.
6. Remove:
   - ❑ Auxiliary drive belt.
   - ❑ Crankshaft pulley bolt **3**.
   - ❑ Crankshaft pulley.
   - ❑ Timing belt covers.
7. Lock injection pump sprocket.
   Use tool No.Mot.1131 **4**.
8. Slacken tensioner nut to release tension on belt **5**.
9. Remove timing belt.

Installation

NOTE: DO NOT fit used belt.

1. Rock crankshaft slightly to ensure timing pin located
   correctly **2**.
2. Observe direction of rotation marks on belt.
3. Fit timing belt in anti-clockwise direction, starting at
   crankshaft sprocket. Ensure belt is taut between
   sprockets. Ensure marks on belt aligned with marks
   on sprockets **7**, **9** & **10**.
   ***NOTE: Injection pump sprocket has two timing
   marks: 'B' (Bosch), 'R' (Lucas/RotoDiesel)***
4. F8Q: Three-bolt injection pump sprocket: Align
   timing mark with belt mark **8**.
5. Remove locking tool from injection pump sprocket.
6. Attach tension gauge to belt at ▽.
   Tool No.Mot.1273.
7. Use an M6x45 mm bolt **6** screwed into timing belt
   rear cover to push on tensioner pulley.
8. Screw in M6 bolt **6** until tension gauge indicates the
   following:
   - ❑ F8M: 41 units.
   - ❑ F8Q-640/644/646/672/676/678/714/718/722/
     724/730/732/774/776/778: 28 units.
   - ❑ F8Q-610/620/706/710/740/742/744/764/766/
     768/784: 38 units.
9. Tighten tensioner nut to 50 Nm **5**.
10. Remove:
    - ❑ Crankshaft timing pin **2**.
    - ❑ Tension gauge.
11. Turn crankshaft three turns clockwise.
12. Recheck belt tension. If tension not as specified:
    Repeat tensioning procedure.
13. Turn crankshaft one turn clockwise.
14. Temporarily fit timing pin and timing belt covers to
    ensure timing marks aligned.
15. Remove M6 bolt and crankshaft timing pin.
16. Check and adjust injection pump timing.
17. Install components in reverse order of removal.
18. Tighten crankshaft pulley bolt to 95 Nm **3**.

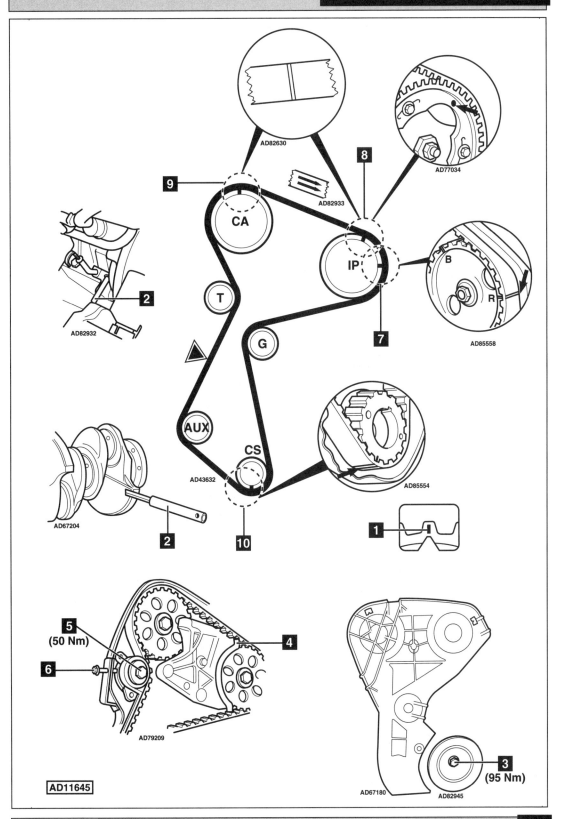

AD82630

AD77034

8

9

CA

AD82933

2

AD82932

T

IP

B

R

AD85558

7

G

AUX

CS

AD43632

AD85554

2

10

AD67204

1

5
(50 Nm)

6

4

AD79209

3
(95 Nm)

AD67180

AD82945

AD11645

RENAULT

Model:	Clio 1,9 D • Mégane/Scénic 1,9 TD • R19 1,9 D/TD
	Express/Extra/Rapid 1,9 D • Kangoo 1,9 D • Trafic 1,9 D
Year:	1991-02
Engine Code:	F8Q – Lucas pump with centrally adjustable sprocket

Replacement Interval Guide

Renault recommend replacement as follows:
5,000 mile service intervals – replacement every 70,000 miles or 5 years.
6,000 mile service intervals – replacement every 72,000 miles or 5 years.
10,000 mile service intervals – replacement every 70,000 miles or 5 years.
The previous use and service history of the vehicle must always be taken into account.
Refer to Timing Belt Replacement Intervals at the front of this manual.

Check For Engine Damage

CAUTION: This engine has been identified as an INTERFERENCE engine in which the possibility of valve-to-piston damage in the event of a timing belt failure is MOST LIKELY to occur.
A compression check of all cylinders should be performed before removing the cylinder head.

Repair Times – hrs

Remove & install:

Clio/Express/Extra/Rapid	3,60
Mégane	3,20
Mégane Scénic	3,70
R19	3,50
Kangoo	2,90
Trafic	2,00

Special Tools

- ■ Crankshaft timing pin – Renault No.Mot.1054.
- ■ Tension gauge – Renault No.Mot.1273 (SEEM C.Tronic 105.6).

Special Precautions

- ■ Disconnect battery earth lead.
- ■ DO NOT turn crankshaft or camshaft when timing belt removed.
- ■ Remove glow plugs to ease turning engine.
- ■ Turn engine in normal direction of rotation (unless otherwise stated).
- ■ DO NOT turn engine via camshaft or other sprockets.
- ■ Observe all tightening torques.
- ■ Check diesel injection pump timing after belt replacement.

Removal

1. Raise and support front of vehicle.

Except Trafic:

2. Remove:
   - ❑ Engine undershield (if fitted).
   - ❑ RH front wheel.
   - ❑ RH front inner wing panel.
   - ❑ RH engine mounting cover.
   - ❑ Diesel fuel filter assembly (if required).

Mégane:

3. Remove:
   - ❑ Strut brace.
   - ❑ Engine tie-bar.
   - ❑ Exhaust downpipe.

Except Trafic:

4. Support engine.
5. Remove RH engine mounting and fixing studs.

All models:

6. Turn crankshaft clockwise until camshaft sprocket timing mark aligned with mark on timing belt cover **1**.
7. Insert timing pin in crankshaft. Tool No.Mot.1054. Rock crankshaft slightly to ensure timing pin located correctly **2**.
8. Remove:
   - ❑ Fuel pipe mounting bolts **3**.
   - ❑ Auxiliary drive belt cover **4**.
   - ❑ Auxiliary drive belt.
   - ❑ Timing belt covers **7**.
   - ❑ Crankshaft pulley bolt **8**.
   - ❑ Crankshaft pulley **9**.
9. Slacken tensioner nut to release tension on belt **5**.
   NOTE: DO NOT slacken tensioner nut more than one turn.
10. Remove timing belt.

Installation

NOTE: DO NOT fit used belt.

1. Rock crankshaft slightly to ensure timing pin located correctly **2**.
2. Observe direction of rotation marks on belt **10**.
3. Fit timing belt. Ensure marks on belt aligned with marks on sprockets **11**, **12** & **13**. Ensure belt is taut between sprockets.
4. Attach tension gauge to belt at ▽. Tool No.Mot.1273.
5. Use an M6x45 mm bolt **6** screwed into timing belt rear cover to push on tensioner pulley.
6. Screw in M6 bolt **6** until tension gauge indicates 47 units.
7. Tighten tensioner nut to 50 Nm **5**.
8. Remove:
   - ❑ Crankshaft timing pin **2**.
   - ❑ Tension gauge.
9. Turn crankshaft three turns clockwise.
10. Recheck belt tension.
11. Tension gauge should indicate 42,3-51,7 units. If tension not as specified: Repeat tensioning procedure.
12. Turn crankshaft one turn clockwise.
13. Temporarily fit timing pin and timing belt covers to ensure timing marks aligned.
14. Remove M6 bolt and crankshaft timing pin.
15. Install components in reverse order of removal.
16. Tighten crankshaft pulley bolt **8**.
    - ❑ Except Trafic: A – 20 Nm + 120°.
    - ❑ Trafic: B: – 20 Nm + 115°.

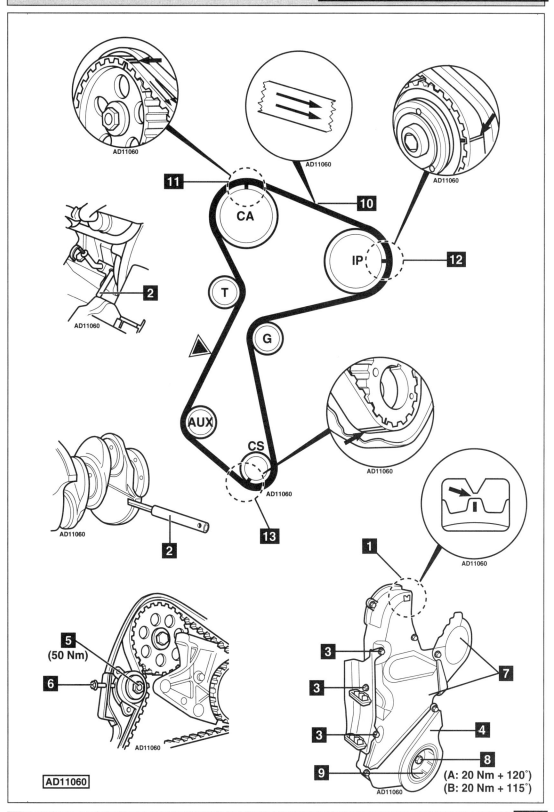

AD11060

5
(50 Nm)

6

AD11060

AD11060

1

3
3
3

7

4

8

9

(A: 20 Nm + 120°)
(B: 20 Nm + 115°)

AD11060

RENAULT

Model:	Clio 1,9 TD • Mégane 1,9 TD • Laguna 1,9 TD • Kangoo 1,9 TD
Year:	1999-03
Engine Code:	F9Q 717/731/780

Replacement Interval Guide

Renault recommend:
10,000 mile service intervals - replacement every 50,000 miles or 5 years.
18,000 mile service intervals - replacement every 45,000 miles or 5 years.
The previous use and service history of the vehicle must always be taken into account.
Refer to Timing Belt Replacement Intervals at the front of this manual.

Check For Engine Damage

CAUTION: This engine has been identified as an INTERFERENCE engine in which the possibility of valve-to-piston damage in the event of a timing belt failure is MOST LIKELY to occur. A compression check of all cylinders should be performed before removing the cylinder head(s).

Repair Times – hrs

Remove & install:

Clio	3,10
Mégane	2,60
Laguna	3,00

Special Tools

- Crankshaft timing pin – Renault No.Mot.1054.
- Injection pump sprocket locking tool – Renault No.Mot.1200-01.

Removal

1. Raise and support front of vehicle.
2. Remove:
   - ❏ RH front wheel.
   - ❏ RH splash guard.
   - ❏ Auxiliary drive belt.
   - ❏ Crankshaft bolt **1**.
   - ❏ Crankshaft pulley **2**.
   - ❏ Blanking plug from cylinder block **3**.
3. Support engine.
4. Remove:
   - ❏ RH engine mounting.
5. Turn crankshaft clockwise to setting position. Ensure timing marks aligned **4**.
6. Insert timing pin through hole in cylinder block and into crankshaft slot **5**. Tool No.Mot.1054.
   *NOTE: Ensure timing pin not inserted into crankshaft web balance hole **6**.*
7. Remove timing belt cover **7**.
   NOTE: Some engines do not have timing marks on timing belt rear cover.
8. Mark timing belt rear cover adjacent to camshaft sprocket timing mark **8**.
9. Mark timing belt rear cover adjacent to injection pump sprocket timing mark **9**.
10. Lock injection pump sprocket **10**.
    Use tool No.Mot.1200-01.
11. Slacken tensioner pulley nut **11**.
12. Undo tensioner pulley nut approximately 8 mm **11**.
13. Push tensioner pulley downwards **12**.

14. Push tensioner pulley sideways **13**.
15. Ensure bracket touches inner face of timing belt rear cover **14**.
16. Remove timing belt.

Installation

NOTE: First timing belt replacement requires new camshaft sprocket to be fitted. Second timing belt replacement requires new camshaft sprocket, tensioner pulley and guide pulley to be fitted.

1. If necessary: Fit new camshaft sprocket. Fit new tensioner pulley and guide pulley.
2. If fitting new camshaft sprocket:
   - ❏ Remove engine mounting bracket.
   - ❏ Fit new sprocket.
   - ❏ Tighten bolt **15**. Tightening torque: 60 Nm.
   - ❏ Refit engine mounting bracket.
   - ❏ Tighten bolts: Tightening torque: 35 Nm.
3. If fitting new tensioner pulley: Ensure tab located in slot **16**.
4. Ensure timing pin fitted **5**.
5. Crankshaft sprocket timing mark should be located one tooth to the left of 6 o'clock position **17**.
6. Ensure timing marks aligned **8** & **9**.
7. Ensure injection pump sprocket locking tool fitted **10**.
8. Ensure tensioner pulley pulled down and sideways **12** & **13**.
9. Ensure tensioner pulley bracket touches inner face of timing belt rear cover **14**.
10. Fit timing belt in anti-clockwise direction, starting at crankshaft sprocket. Ensure belt is taut between sprockets. Ensure marks on belt aligned with marks on sprockets **8**, **9** & **17**.
11. Fit 6 mm Allen key to tensioner bracket **18**.
12. Turn tensioner pulley anti-clockwise and lift up into normal running position **19** & **20**.
13. Ensure bracket does NOT touch inner face of timing belt rear cover **14**.
14. Remove:
    - ❏ Crankshaft timing pin **5**.
    - ❏ Injection pump sprocket locking tool **10**.
15. Turn tensioner pulley anti-clockwise until tensioner pulley marks aligned **21** & **22**. Use 6 mm Allen key **23**.
16. Temporarily tighten tensioner pulley nut **11**. Tightening torque: 10 Nm.
17. Ensure timing marks aligned **8**, **9** & **17**.
18. Turn crankshaft two turns clockwise to setting position.
19. Insert timing pin through hole in cylinder block and into crankshaft **5**.
20. Ensure timing marks aligned **8** & **9**.
21. Ensure tensioner pulley marks aligned **21** & **22**. If not: Slacken tensioner pulley nut **11**. Adjust position of tensioner pulley.
22. Tighten tensioner pulley nut **11**. Tightening torque: 20 Nm.
23. Remove timing pin **5**.
24. Refit blanking plug **3**.
25. Install components in reverse order of removal.
26. Tighten crankshaft bolt **1**.
    Tightening torque: 20 Nm + 100-130º.

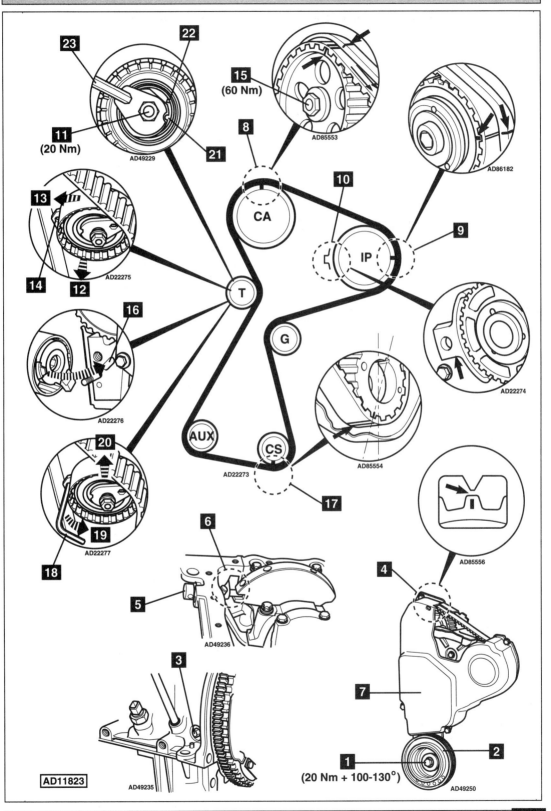

23 **22**

15
(60 Nm)

AD85553

8

AD86182

11
(20 Nm)

AD49229

21

10

9

13

CA

AD22275

14 **12**

IP

16

T

G

AD22274

AD22276

20

AUX

CS

AD85554

19

AD22273

18

AD22277

17

AD85556

6

4

5

3

7

AD49236

1
(20 Nm + 100-130°)

2

AD49250

AD11823

AD49235

RENAULT

Model:	Fuego D Turbo • R18D/Turbo • R20D/Turbo • R21D/Turbo • R25D/Turbo R30D Turbo • Espace D Turbo • Trafic/Master 2,1D
Year:	1979-00
Engine Code:	J8S

Replacement Interval Guide

5,000 mile service intervals – replacement every 70,000 miles or 5 years.
6,000 mile service intervals – replacement every 72,000 miles or 5 years.
The previous use and service history of the vehicle must always be taken into account.
Refer to Timing Belt Replacement Intervals at the front of this manual.

Check For Engine Damage

CAUTION: This engine has been identified as an INTERFERENCE engine in which the possibility of valve-to-piston damage in the event of a timing belt failure is MOST LIKELY to occur.
A compression check of all cylinders should be performed before removing the cylinder head.

Repair Times – hrs

Remove & install:

R18/Fuego	1,90
R21	3,40
R25	1,90
Espace →1991	3,00
Espace 1991→	4,00
Trafic	2,80
Master	1,80

Special Tools

- Crankshaft timing pin – Renault No.Mot.861.
- Tension gauge – Renault No.Mot.1273 (SEEM C.Tronic 105).
- Camshaft/injection pump locking plate – Renault No.Mot.854.
- Tensioner spanner – Renault No.Mot.1384.

Special Precautions

- Disconnect battery earth lead.
- DO NOT turn crankshaft or camshaft when timing belt removed.
- Remove glow plugs to ease turning engine.
- Turn engine in normal direction of rotation (unless otherwise stated).
- DO NOT turn engine via camshaft or other sprockets.
- Observe all tightening torques.
- Check diesel injection pump timing after belt replacement.

Removal

1. Remove auxiliary drive belts.
2. Turn crankshaft to TDC. Insert timing pin in crankshaft **1**. Tool No.Mot.861.
   NOTE: Timing pin hole located in cylinder block under rear of injection pump and adjacent to vacuum pump.
   Take care not to insert timing pin in crankshaft counterweight balance holes.

3. Ensure camshaft sprocket and injection pump sprocket timing marks aligned with timing belt cover marks **2** & **3**.
4. Remove timing belt cover **4**.
5. Fit locking plate **5**. Tool No.Mot.854.
6. Slacken tensioner bolts **6**. Move tensioner away from belt. Lightly tighten bolts.
   NOTE: Later engines may have two bolt tensioner mounting plate.
7. Remove timing belt.

Installation

NOTE: DO NOT fit used belt.

1. Ensure crankshaft timing pin located correctly **1**.
2. Ensure timing marks aligned **2** & **3**.
3. Check clearance between tensioner support and adjusting screw.
4. 0,1 mm feeler gauge should just be nipped **7**. Then tighten locknut.
5. Fit timing belt.
6. Remove crankshaft timing pin **1**.
7. Remove locking plate **5**.
8. Engines with spring loaded tensioner: Slacken tensioner bolts **6**. Allow tensioner to operate. Tighten bolts.
9. Engines without spring loaded tensioner: Use tool No.Mot.1384 on eccentric at top tensioner bolt to turn tensioner against belt.
10. Attach tension gauge to belt at ▽. Tool No.Mot.1273.
11. Tension gauge should indicate 45 SEEM units.
12. If tension not as specified:
    - ❏ Engines with spring loaded tensioner: Slacken and tighten tensioner bolts **6**.
    - ❏ Engines without spring loaded tensioner: Use tool No.Mot.1384 on eccentric at top tensioner bolt to adjust tensioner.
13. Tension gauge should indicate 45 SEEM units.
14. Remove tension gauge.
15. Turn crankshaft three turns clockwise. Fit crankshaft timing pin **1**.
16. Attach tension gauge to belt at ▽. Tool No.Mot.1273.
17. Tension gauge should indicate 45 SEEM units.
18. If not: Repeat tensioning procedure.
19. Install components in reverse order of removal.

Autodata

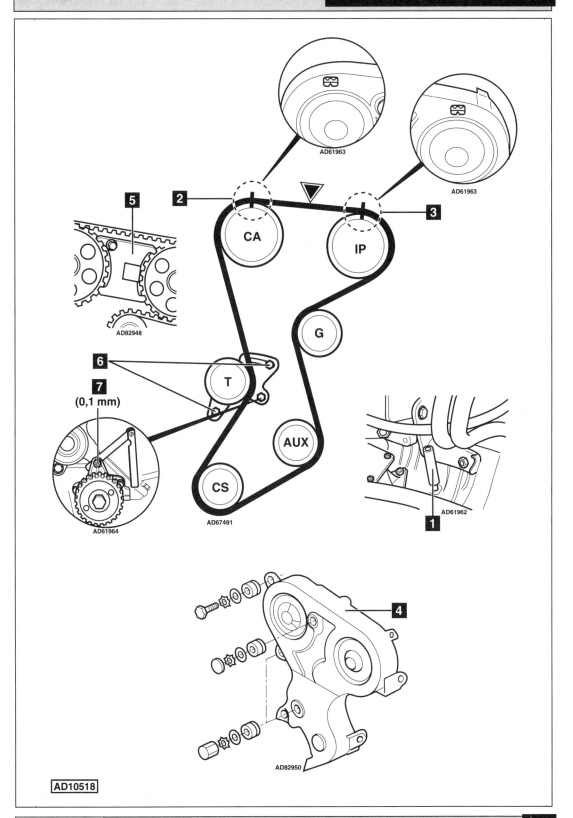

AD61963

AD61963

5

2

CA

IP

3

AD82948

6

T

7
(0,1 mm)

G

AUX

CS

AD61964

AD67491

AD61962

1

4

AD82950

AD10518

RENAULT

Model:	Mégane 1,9 D • Mégane/Scénic 1,9 TD
Year:	1995-99
Engine Code:	F8Q620, F8Q784 – non-adjustable pump sprocket

Replacement Interval Guide

Renault recommend:
6,000 mile service intervals - replacement every 72,000 miles or 5 years.
10,000 mile service intervals - replacement every 70,000 miles or 5 years.
The previous use and service history of the vehicle must always be taken into account.
Refer to Timing Belt Replacement Intervals at the front of this manual.

Check For Engine Damage

CAUTION: This engine has been identified as an INTERFERENCE engine in which the possibility of valve-to-piston damage in the event of a timing belt failure is MOST LIKELY to occur.
A compression check of all cylinders should be performed before removing the cylinder head.

Repair Times – hrs

Remove & install:

Mégane	3,20
Mégane Scénic	3,70

Special Tools

- Crankshaft timing pin – Renault No.Mot.1054.
- Tension gauge – Renault No.Mot.1273 (SEEM C.Tronic 105.6).

Special Precautions

- Disconnect battery earth lead.
- DO NOT turn crankshaft or camshaft when timing belt removed.
- Remove glow plugs to ease turning engine.
- Turn engine in normal direction of rotation (unless otherwise stated).
- DO NOT turn engine via camshaft or other sprockets.
- Observe all tightening torques.
- Check diesel injection pump timing after belt replacement.

Removal

1. Raise and support front of vehicle.
2. Remove:
   - ❏ RH front wheel.
   - ❏ RH front inner wing panel.
   - ❏ Strut brace.
   - ❏ Engine tie-bar.
   - ❏ Exhaust downpipe.
   - ❏ RH engine mounting cover.
3. Support engine.
4. Remove RH engine mounting and fixing studs.

5. Turn crankshaft clockwise until timing mark on camshaft sprocket aligned with mark on timing belt cover **1**.
6. Insert timing pin in crankshaft. Tool No.Mot.1054. Rock crankshaft slightly to ensure timing pin located correctly **2**.
7. Remove:
   - ❏ Fuel pipe mounting bolts **3**.
   - ❏ Auxiliary drive belt cover **4**.
   - ❏ Auxiliary drive belt.
   - ❏ Timing belt covers **7**.
   - ❏ Crankshaft pulley bolt **8**.
   - ❏ Crankshaft pulley **9**.
8. Slacken tensioner nut to release tension on belt **5**.
   NOTE: DO NOT slacken tensioner nut more than one turn.
9. Remove timing belt.

Installation

NOTE: DO NOT fit used belt.

1. Rock crankshaft slightly to ensure timing pin located correctly **2**.
2. Observe direction of rotation marks on belt **10**.
3. Fit timing belt. Ensure marks on belt aligned with marks on sprockets **11**, **12** & **13**. Ensure belt is taut between sprockets.
   NOTE: Injection pump sprocket has two timing marks 12. 'B' (Bosch). 'R' (Lucas).
4. Attach tension gauge to belt at ▽. Tool No.Mot.1273.
5. Use an M6x45 mm bolt **6** screwed into timing belt rear cover to push on tensioner pulley.
6. Screw in M6 bolt **6** until tension gauge indicates 38 units.
7. Tighten tensioner nut to 50 Nm **5**.
8. Remove:
   - ❏ Timing pin **2**.
   - ❏ Tension gauge.
9. Turn crankshaft three turns clockwise.
10. Recheck belt tension. Tension gauge should indicate not less than 36 units. If not: Repeat tensioning procedure.
11. Turn crankshaft one turn clockwise.
12. Temporarily fit timing pin and timing belt covers to ensure timing marks aligned.
13. Remove:
    - ❏ M6 bolt **6**.
    - ❏ Timing pin **2**.
14. Install components in reverse order of removal.
15. Tighten crankshaft pulley bolt to 120 Nm **8**.

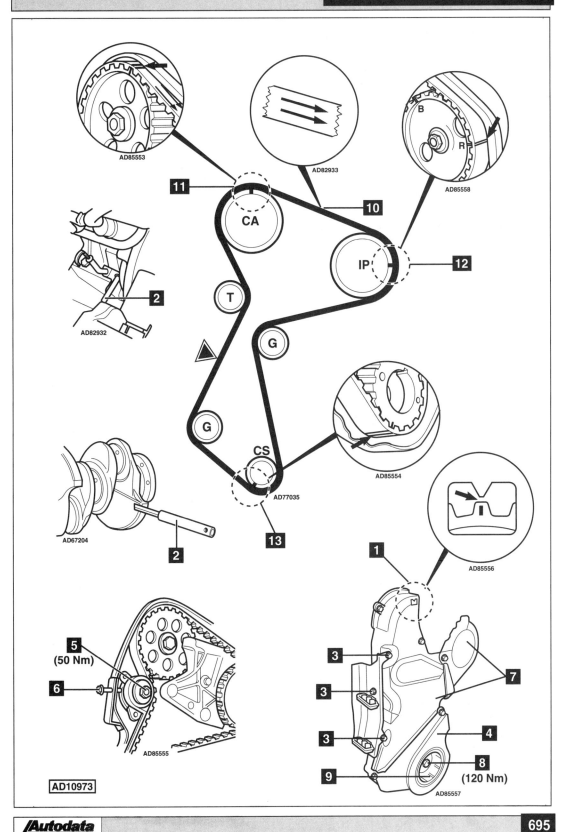

AD85553

AD82933

AD85558

11

10

12

CA

IP

2

AD82932

T

G

G

CS

AD85554

AD67204

2

AD77035

13

5
(50 Nm)

6

AD85555

AD10973

1

AD85556

3

3

3

9

7

4

8
(120 Nm)

AD85557

Model:	**Mégane/Scénic 1,9 TD • Laguna 1,9 TD/dTi • Espace 1,9 dTi**
Year:	**1997-02**
Engine Code:	**F9Q 710/716/720/722/730/734/736**

Replacement Interval Guide

Renault recommend:
10,000 mile service intervals – replacement every 50,000 miles or 5 years.
18,000 mile service intervals – replacement every 45,000 miles or 5 years.
The previous use and service history of the vehicle must always be taken into account.
Refer to Timing Belt Replacement Intervals at the front of this manual.

Check For Engine Damage

CAUTION: This engine has been identified as an INTERFERENCE engine in which the possibility of valve-to-piston damage in the event of a timing belt failure is MOST LIKELY to occur.
A compression check of all cylinders should be performed before removing the cylinder head(s).

Repair Times – hrs

Remove & install:

Mégane	2,60
Mégane Scénic	3,10
Laguna	2,60
Espace	3,80
AC	+0,40

Special Tools

- Crankshaft locking pin – Renault No.Mot.1054.
- Tension gauge – Renault No.Mot.1273 (SEEM C.Tronic 105.6).

Special Precautions

- Disconnect battery earth lead.
- DO NOT turn crankshaft or camshaft when timing belt removed.
- Remove glow plugs to ease turning engine.
- Turn engine in normal direction of rotation (unless otherwise stated).
- DO NOT turn engine via camshaft or other sprockets.
- Observe all tightening torques.
- Check diesel injection pump timing after belt replacement.

Removal

WARNING: F9Q 734 engines may require a modified tensioner pulley. Vehicles with F9Q 734 engine from No.C008270 to No.C025670 require replacement of camshaft sprocket. On certain engines the camshaft sprocket may NOT be keyed to camshaft. Special equipment and procedure required. Refer to Renault Dealer.

1. Raise and support front of vehicle.
2. Remove:
   - ❏ Engine undershield.
   - ❏ RH front wheel.
   - ❏ RH front inner wing panel.
   - ❏ Engine cover.
   - ❏ Diesel fuel filter assembly (if required).
   - ❏ Engine tie-bar.

3. Support engine.
4. Remove RH engine mounting.
5. Turn crankshaft clockwise until camshaft sprocket timing mark aligned with mark on timing belt cover **1**.
6. Remove blanking plug from cylinder block. Insert locking pin in crankshaft. Tool No.Mot.1054. Rock crankshaft slightly to ensure locking pin located correctly **2**.
7. Remove:
   - ❏ Fuel pipe mounting bolts **3**.
   - ❏ Auxiliary drive belt cover **4**.
   - ❏ Auxiliary drive belt.
   - ❏ Timing belt covers **7**.
   - ❏ Crankshaft pulley bolt **8**.
   - ❏ Crankshaft pulley **9**.
8. Slacken tensioner nut to release tension on belt **5**.
   NOTE: DO NOT slacken tensioner nut more than one turn.
9. Remove timing belt.

Installation

NOTE: DO NOT fit used belt.
1. Rock crankshaft slightly to ensure locking pin located correctly **2**.
2. Observe direction of rotation marks on belt **10**.
3. Fit timing belt. Ensure marks on belt aligned with marks on sprockets **11**, **12** & **13**. Ensure belt is taut between sprockets.
4. Attach tension gauge to belt at ▽. Tool No.Mot.1273.
5. Use an M6 x 45 mm bolt **6** screwed into timing belt rear cover to push on tensioner pulley.
6. Remove crankshaft locking pin **2**.
7. Screw in M6 bolt **6** until tension gauge indicates 42 SEEM units.
8. Tighten tensioner nut to 50 Nm **5**.
9. Remove tension gauge.
10. Turn crankshaft four turns clockwise.
11. Insert locking pin in crankshaft **2**.
12. Ensure marks on belt aligned with marks on sprockets **11**, **12** & **13**.
13. Remove crankshaft locking pin **2**.
14. Press sharply on belt at ▽.
15. Recheck belt tension.
16. Tension gauge should indicate 37 SEEM units. If tension not as specified: Repeat tensioning procedure.
17. Remove M6 bolt **6**.
18. Install components in reverse order of removal.
19. Fit new bolt. Tighten crankshaft pulley bolt **8**. Tightening torque: 20 Nm + 100-130°.

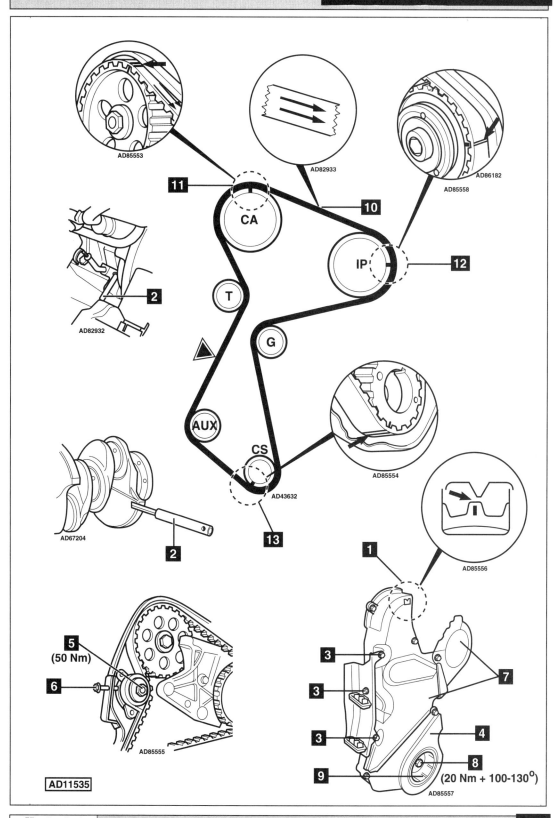

AD85553

AD82933

AD86182

AD85558

11

10

CA

2

AD82932

T

IP

12

G

AUX

CS

AD85554

AD67204

2

AD43632

13

1

AD85556

5
(50 Nm)

6

3

3

7

3

4

AD85555

9

8

(20 Nm + 100-130°)

AD85557

AD11535

RENAULT

Model:	**Mégane/Scénic 1,9 TD • Laguna 1,9 TD • Trafic 1,9 dCi**
Year:	**1999-02**
Engine Code:	**F9Q 732, F9Q 718, F9Q 760**

Replacement Interval Guide

Renault recommend replacement as follows:
Except Trafic:
10,000 mile service intervals – replacement every 50,000 miles or 5 years.
18,000 mile service intervals – replacement every 72,000 miles or 5 years.
Trafic:
Replacement every 90,000 miles or 5 years.
The previous use and service history of the vehicle must always be taken into account.
Refer to Timing Belt Replacement Intervals at the front of this manual.

Check For Engine Damage

CAUTION: This engine has been identified as an INTERFERENCE engine in which the possibility of valve-to-piston damage in the event of a timing belt failure is MOST LIKELY to occur. A compression check of all cylinders should be performed before removing the cylinder head(s).

Repair Times – hrs

Remove & install:

Laguna	2,60
AC	+0,40
Mégane	2,60
Scénic	3,10
Trafic	2,70

Special Tools

- ■ Crankshaft timing pin – Renault No.Mot.1054.
- ■ Tensioning tool – Renault No.1543.
- ■ Tension gauge – Renault No.Mot.1505.
- ■ Trafic: Engine support tool – Renault No.Mot.1367/02.
- ■ M6 bolt – for adjusting tensioner.

Removal

1. Raise and support front of vehicle.
2. Remove:
   - ❏ Engine top cover.
   - ❏ RH front wheel.
   - ❏ RH splash guard.
   - ❏ Engine undershield.
   - ❏ Auxiliary drive belt.
3. Trafic: Remove:
   - ❏ Auxiliary drive belt tensioner.
   - ❏ PAS reservoir. DO NOT disconnect pipes.
4. Mégane/Scénic/Laguna: Remove:
   - ❏ Engine control module (ECM).
   - ❏ Glow plug relay.
5. Trafic: Support engine. Use tool No.1367/02.
6. Remove:
   - ❏ RH engine mounting.
   - ❏ RH engine mounting bracket.
7. Lock flywheel with large screwdriver. Slacken crankshaft pulley bolt **1**.
8. Turn crankshaft clockwise until timing marks aligned **2**.
9. Remove blanking plug from cylinder block **3**.

10. Insert timing pin in crankshaft **4**. Tool No.Mot.1054. Rock crankshaft slightly to ensure timing pin located correctly.
11. Remove:
    - ❏ Timing belt cover **5**.
    - ❏ Crankshaft pulley bolt **1**.
    - ❏ Crankshaft pulley **6**.
12. Slacken tensioner nut **7**. Move tensioner away from belt. Lightly tighten nut.
13. Remove timing belt.

Installation

1. Ensure timing pin fitted **4**.
2. Ensure crankshaft keyway located centrally between lugs on engine front housing **8**.
3. Ensure crankshaft sprocket timing mark located one tooth to left of centre **9**.
4. Temporarily fit timing belt cover. Ensure timing marks aligned **2**.
5. Fit timing belt. Ensure marks on belt aligned with marks on sprockets **10** & **11**. Ensure belt is taut on non-tensioned side.
   NOTE: There should be 77 teeth between timing marks on crankshaft sprocket and camshaft sprocket.
6. Slacken tensioner nut **7**.
7. Screw an M6 bolt into tensioner bracket until tensioner pulley contacts belt **12**.
8. Remove timing pin **4**.
9. Temporarily fit crankshaft pulley bolt **1** with large washer.
10. Fit tensioning tool **13**. Tool No.1543.
11. Apply clockwise torque of 11 Nm **14**.
12. Remove tensioning tool.
13. Attach tension gauge to belt at ▽ **15**. Tool No.Mot.1505.
14. Screw in M6 bolt until tension gauge indicates 88-91 Hz.
15. Temporarily tighten tensioner nut **7**. Tightening torque: 10 Nm.
16. Remove tension gauge.
17. Mark timing belt rear cover with paint or chalk adjacent to timing mark **2**.
18. Turn crankshaft two turns clockwise.
19. Insert timing pin in crankshaft **4**. Rock crankshaft slightly to ensure timing pin located correctly.
20. Remove timing pin **4**.
21. Fit tensioning tool **13**.
22. Apply clockwise torque of 11 Nm **14**.
23. Remove tensioning tool.
24. Attach tension gauge to belt at ▽ **15**.
25. Tension gauge should indicate 82-88 Hz.
26. If not: Repeat tensioning procedure.
27. Remove M6 bolt **12**.
28. Remove tension gauge **15**.
29. Tighten tensioner nut **7**. Tightening torque: 50 Nm.
30. Refit blanking plug **3**.
31. Remove crankshaft pulley bolt **1**. Discard large washer. Refit bolt.
32. Tighten crankshaft pulley bolt **1**. Tightening torque: 20 Nm + 100°-130°.
33. Install components in reverse order of removal.

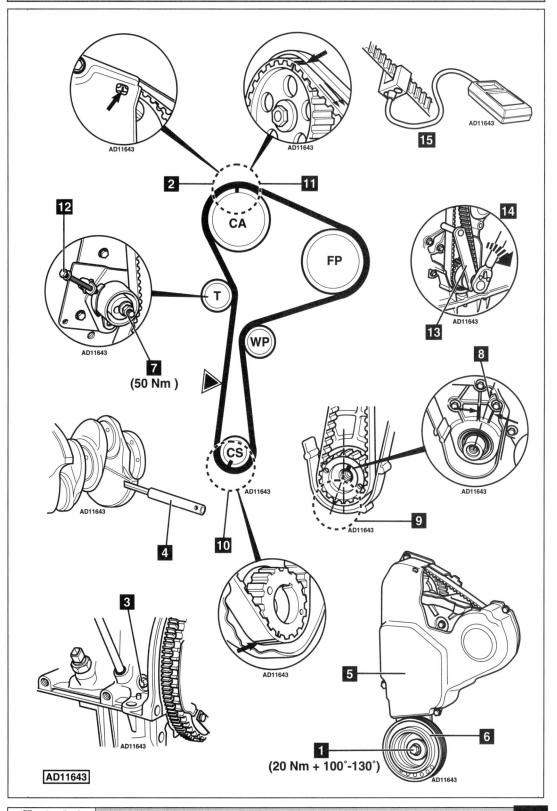

AD11643

CA

FP

T

WP

CS

7
(50 Nm)

1
(20 Nm + 100°-130°)

RENAULT

Model:	**Laguna 2,2D**
Year:	**1994-96**
Engine Code:	**G8T-706/790**

Replacement Interval Guide

Renault recommend replacement every 70,000 miles or 5 years.
NOTE: Vacuum pump drive belt MUST be renewed whenever timing belt replaced.
The previous use and service history of the vehicle must always be taken into account.
Refer to Timing Belt Replacement Intervals at the front of this manual.

Check For Engine Damage

CAUTION: This engine has been identified as an INTERFERENCE engine in which the possibility of valve-to-piston damage in the event of a timing belt failure is MOST LIKELY to occur.
A compression check of all cylinders should be performed before removing the cylinder head.

Repair Times – hrs

Remove & install	5,60

Special Tools

- Tensioning tool – No.Mot.1312.
- Crankshaft timing pin – No.Mot.1318.

Special Precautions

- Disconnect battery earth lead.
- DO NOT turn crankshaft or camshaft when timing belt removed.
- Remove glow plugs to ease turning engine.
- Turn engine in normal direction of rotation (unless otherwise stated).
- DO NOT turn engine via camshaft or other sprockets.
- Observe all tightening torques.
- Check diesel injection pump timing after belt replacement.

Removal

1. Support engine.
2. Remove:
   - ❑ Auxiliary drive belt.
   - ❑ RH engine mounting.
   - ❑ Crankshaft pulley **1**.
   - ❑ Timing belt covers.
3. Turn crankshaft to TDC on No.1 cylinder.
4. Insert timing pin in crankshaft **2**.
   Tool No.Mot.1318.
5. Rock crankshaft slightly to ensure timing pin located correctly.
6. Ensure crankshaft timing marks aligned **3**.
7. Ensure camshaft sprocket inner timing mark aligned with mark on cylinder head cover **4**.

8. Ensure camshaft sprocket outer timing mark **5** approximately 46° to left of inner timing mark **4**.
9. Ensure injection pump sprocket timing mark aligned with pump mounting boss **6**.
10. If timing marks not aligned: Remove timing pin. Turn crankshaft one turn clockwise to TDC on No.1 cylinder.
11. Slacken tensioner pulley nut to release tension on belt **7**.
12. Turn tensioner pulley clockwise. Lightly tighten nut.
13. Remove timing belt.
    NOTE: DO NOT fit used belt.

Installation

1. Ensure timing pin located correctly **2**.
2. Ensure timing marks aligned **4**, **5** & **6**.
3. Fit timing belt in anti-clockwise direction, starting at crankshaft sprocket. Ensure belt is taut between sprockets.
   *NOTE: Ensure marks on belt aligned with marks on sprockets **3**, **5** & **6**.*
4. Slacken tensioner pulley nut **7**.
5. Fit tensioning tool to cylinder block **8**.
   Tool No.Mot.1312.
6. Lever tensioner pulley anti-clockwise until it touches stop **9**. Use tool No.Mot.1312.
7. Tighten tensioner pulley nut **7**.
8. Remove timing pin **2**.
9. Turn crankshaft three turns clockwise until crankshaft timing mark aligned **3**.
10. Fit timing pin **2**.
    NOTE: DO NOT allow crankshaft to turn anti-clockwise.
11. Hold tensioning tool against tensioner pulley. Slowly slacken tensioner pulley nut **7**.
12. Allow tensioner pulley to move clockwise until minimum marks aligned **10**.
13. Tighten tensioner pulley nut **7**.
    Tightening torque: 28-34 Nm.
    NOTE: DO NOT allow crankshaft to turn anti-clockwise.
14. Remove tensioning tool.
15. Check and adjust injection pump timing.
16. Install components in reverse order of removal.
17. Tighten crankshaft pulley bolt.
    Tightening torque: 25 Nm + 58°-70°.

Autodata

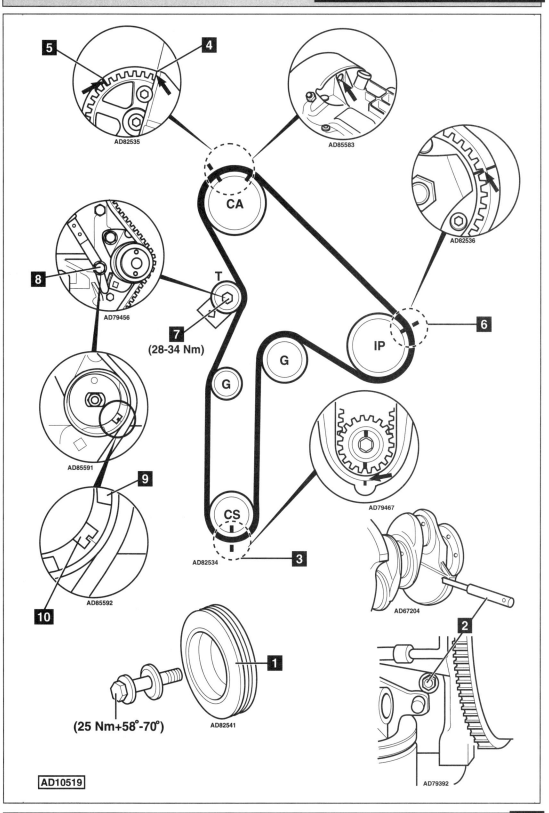

5 4

AD82535

AD85583

AD82536

CA

AD79456

8

T

7

(28-34 Nm)

IP

6

G

G

AD85591

9

CS

AD79467

AD82534

3

AD85592

10

AD67204

2

1

(25 Nm+58°-70°)

AD82541

AD79392

AD10519

Model:	Laguna 2,2 D/TD • Safrane 2,2 TD • Espace 2,2 TD
Year:	1995-00
Engine Code:	G8T 714/716/740/752/760/792/794

Replacement Interval Guide

5,000 mile service intervals – replacement every 70,000 miles or 5 years.
6,000 mile service intervals – replacement every 72,000 miles or 5 years.
10,000 mile service intervals – replacement every 70,000 miles or 5 years.
The previous use and service history of the vehicle must always be taken into account.
Refer to Timing Belt Replacement Intervals at the front of this manual.

Check For Engine Damage

CAUTION: This engine has been identified as an INTERFERENCE engine in which the possibility of valve-to-piston damage in the event of a timing belt failure is MOST LIKELY to occur.
A compression check of all cylinders should be performed before removing the cylinder head.

Repair Times – hrs

Remove & install:

Laguna	3,80
Safrane	3,10
Espace	3,80

Special Tools

■ Crankshaft timing pin – No.Mot.1318.

Special Precautions

■ Disconnect battery earth lead.
■ DO NOT turn crankshaft or camshaft when timing belt removed.
■ Remove glow plugs to ease turning engine.
■ Turn engine in normal direction of rotation (unless otherwise stated).
■ DO NOT turn engine via camshaft or other sprockets.
■ Observe all tightening torques.
■ Check diesel injection pump timing after belt replacement.

Removal

1. Raise and support front of vehicle.
2. Remove:
   ❏ RH front wheel.
   ❏ Engine undershield.
   ❏ Wheel arch liner.
3. Support engine.
4. Remove:
   ❏ Auxiliary drive belt.
   ❏ Auxiliary drive belt tensioner.
   ❏ RH engine mounting.
   ❏ Fuel filter and wiring.
   ❏ Crankshaft pulley bolt **1**.
   ❏ Crankshaft pulley **2**.
   ❏ Timing belt covers.
5. Turn crankshaft to TDC on No.1 cylinder.
6. Insert timing pin through hole in cylinder block and into crankshaft **3**. Tool No.1318.
7. Rock crankshaft slightly to ensure timing pin located correctly.
8. Ensure crankshaft timing marks aligned **4**.
9. Ensure camshaft sprocket inner timing mark aligned with mark on cylinder head cover **5**.
10. Ensure camshaft sprocket outer timing mark **6** approximately 46° to left of inner timing mark **5**.
11. Ensure injection pump sprocket timing mark aligned with pump casing mark **7**.
12. If timing marks not aligned: Remove timing pin. Turn crankshaft one turn clockwise to TDC on No.1 cylinder.
13. Slacken tensioner nut **12**.
14. Slacken tensioner bolt locknut **8**. Turn bolt **9** to release tension on belt.
15. Remove timing belt.
    NOTE: DO NOT fit used belts. If timing belt or auxiliary drive belt are removed they MUST NOT be reused.

Installation

1. Ensure timing pin located correctly **3**.
2. Ensure timing marks aligned **5**, **6** & **7**.
3. Fit timing belt in anti-clockwise direction, starting at crankshaft sprocket. Ensure belt is taut between sprockets.
   NOTE: Ensure marks on belt aligned with marks on sprockets **4, **6** & **7**.**
4. Slacken tensioner bolt locknut **8**. Turn bolt **9** until tensioner touches stop **10**.
5. Remove timing pin **3**.
6. Turn crankshaft a minimum of three turns clockwise. Ensure timing marks aligned **4**.
7. Fit timing pin **3**.
   NOTE: DO NOT allow crankshaft to turn anti-clockwise.
8. Slowly slacken tensioner bolt **9** until minimum marks **11** aligned.
9. Tighten tensioner bolt locknut **8**.
10. Tighten tensioner nut **12**.
    Tightening torque: 30 Nm.
11. Install components in reverse order of removal.
12. Tighten crankshaft pulley bolt **1**.
    Tightening torque: 25 Nm + 64°.

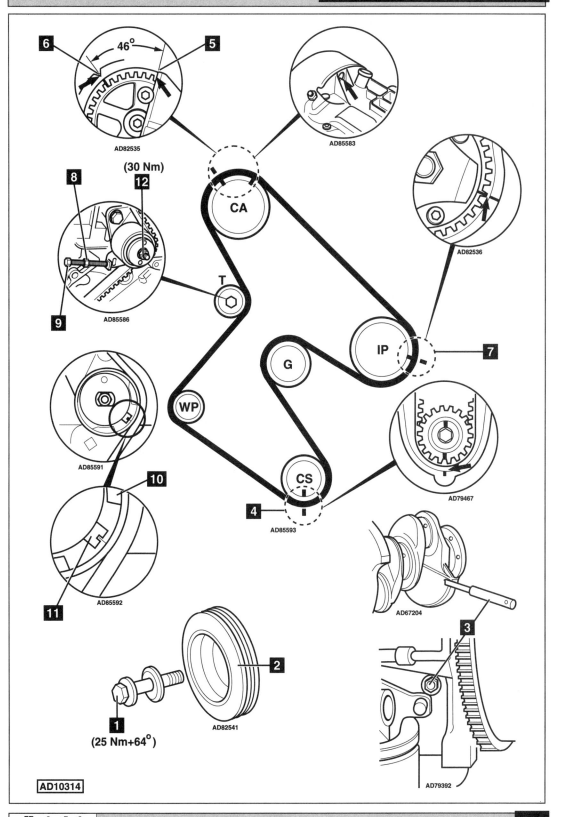

46°

6
5
AD82535

AD85583

(30 Nm)

8
12
CA

AD82536

9
AD85586

T

IP

7

G

WP

AD85591

10

CS

4
AD85593

AD79467

11
AD85592

AD67204

3

1
(25 Nm+64°)

2
AD82541

AD79392

AD10314

Model:	Trafic 2,5D • Master 2,5D/TD
Year:	1980-00
Engine Code:	S8U 720/722/730/731/742/748/750/752/758/780/782, S9U 700/702/704/714/740

Replacement Interval Guide

5,000 mile service intervals – replacement every 50,000 miles or 5 years.
6,000 mile service intervals – replacement every 72,000 miles or 5 years.
The previous use and service history of the vehicle must always be taken into account.
Refer to Timing Belt Replacement Intervals at the front of this manual.

Check For Engine Damage

CAUTION: This engine has been identified as an INTERFERENCE engine in which the possibility of valve-to-piston damage in the event of a timing belt failure is MOST LIKELY to occur.
A compression check of all cylinders should be performed before removing the cylinder head.

Repair Times – hrs

Remove & install:
Trafic	2,80
Master	1,80

Special Tools

- Tension gauge – Renault No.Mot.1273 (SEEM C.Tronic 105).
- Crankshaft timing pin – Renault No.Mot.910.
- Injection pump sprocket timing pin – Renault No.Mot.910.

Special Precautions

- Disconnect battery earth lead.
- DO NOT turn crankshaft or camshaft when timing belt removed.
- Remove glow plugs to ease turning engine.
- Turn engine in normal direction of rotation (unless otherwise stated).
- DO NOT turn engine via camshaft or other sprockets.
- Observe all tightening torques.
- Check diesel injection pump timing after belt replacement.

Removal

1. Remove:
   - ❏ Auxiliary drive belt.
   - ❏ Timing belt cover **1**.
   - ❏ Crankshaft sprocket cover (if fitted) **8**.
2. Turn crankshaft to TDC on No.1 cylinder. Ensure camshaft sprocket timing mark aligned with mark on cylinder head cover **2**.

3. Insert timing pin in crankshaft pulley **3**. Tool No.Mot.910.
   *NOTE: If no hole in crankshaft pulley: Insert timing pin in flywheel hole **5**.*
4. Insert timing pin in injection pump sprocket **4**. Tool No.Mot.910.
5. Slacken tensioner nut **6**. Move tensioner away from belt. Remove guide pulley **7**.
6. Remove timing belt.

Installation

NOTE: DO NOT fit used belt.
1. Ensure camshaft sprocket timing mark aligned **2**.
2. Ensure timing pins located correctly **3** & **4**.
3. Fit timing belt, starting at crankshaft sprocket. Ensure belt is taut on non-tensioned side.
4. Fit guide pulley **7**.
5. Slacken tensioner nut **6**.
6. Remove timing pins.
7. Turn crankshaft 1/4 turn clockwise.
8. Tighten tensioner nut **6**. Tightening torque: 45 Nm.
9. Turn crankshaft 3/4 turn clockwise.
10. Slacken and tighten tensioner nut **6**. Tightening torque: 45 Nm.
11. Attach tension gauge to belt at ▽. Tool No.Mot.1273.
12. Tension gauge should indicate 45 units. If not: Repeat tensioning procedure.
13. Turn crankshaft three turns to TDC. Ensure timing pin can be inserted **3**.
14. Attach tension gauge to belt at ▽. Tension gauge should indicate 45 units.
15. Install components in reverse order of removal.

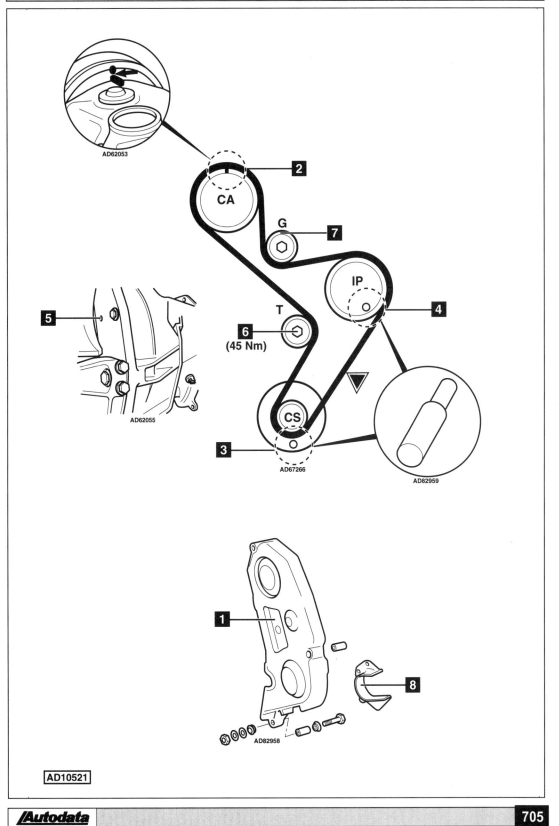

AD10521

RENAULT

Model:	Master 2,5D • Master 2,8D Turbo
Year:	1998-02
Engine Code:	S8U 770/772, S9W 700/702

Replacement Interval Guide

Renault recommend:
→10/99: Replacement every 72,000 miles or 5 years, whichever occurs first.
11/99→: Replacement every 60,000 miles or 5 years, whichever occurs first.
The previous use and service history of the vehicle must always be taken into account.
Refer to Timing Belt Replacement Intervals at the front of this manual.

Check For Engine Damage

CAUTION: This engine has been identified as an INTERFERENCE engine in which the possibility of valve-to-piston damage in the event of a timing belt failure is MOST LIKELY to occur.
A compression check of all cylinders should be performed before removing the cylinder head.

Repair Times – hrs

Remove & install	1,80

Special Tools

- Injection pump sprocket locking pin – Renault No.Mot.910.
- Flywheel timing pin – Renault No.Mot.1054.

Special Precautions

- Disconnect battery earth lead.
- DO NOT turn crankshaft or camshaft when timing belt removed.
- Remove glow plugs to ease turning engine.
- Turn engine in normal direction of rotation (unless otherwise stated).
- DO NOT turn engine via camshaft or other sprockets.
- Observe all tightening torques.
- Check diesel injection pump timing after belt replacement.

Removal

1. Raise and support front of vehicle.
2. Remove:
   - ❏ Engine undershield.
   - ❏ Auxiliary drive belt.
   - ❏ Engine top cover **1**.
   - ❏ Water pump pulley.
   - ❏ Timing belt upper cover **2**.
3. Slacken crankshaft pulley bolt **3**.
4. Turn crankshaft to TDC on No.1 cylinder. Ensure timing marks aligned **4**.
5. Remove:
   - ❏ Crankshaft pulley bolt **3**.
   - ❏ Crankshaft pulley **5**.

   - ❏ Tensioner nut and washer **6**.
   - ❏ Timing belt lower cover **7**.
6. Ensure camshaft sprocket timing mark aligned with mark on cylinder head cover **8**.
7. Insert locking pin in injection pump sprocket **9**. Tool No.Mot.910.
8. Insert timing pin in flywheel **10**. Tool No.Mot.1054.
9. Refit nut and washer to stud **6**. DO NOT tighten.
10. Lever tensioner pulley away from belt using screwdriver. Compress tensioner spring to release tension on belt. Retain in position with U-shaped tool **11**.
    NOTE: Tool can be manufactured from a 14 mm nut cut to shape.
11. Remove timing belt.

Installation

1. Ensure camshaft sprocket timing mark aligned **8**.
2. Ensure flywheel timing pin located correctly **10**.
3. Ensure locking pin located correctly in injection pump sprocket **9**.
4. Fit timing belt, starting at crankshaft sprocket. Ensure belt is taut on non-tensioned side.
5. Remove nut and washer **6**.
6. Fit timing belt lower cover **7**.
7. Refit nut and washer to stud **6**. DO NOT tighten.
8. Fit crankshaft pulley **5**.
9. Fit crankshaft pulley bolt **3**. Hand tighten.
10. Remove:
    - ❏ Tensioner tool **11**.
    - ❏ Injection pump locking pin **9**.
    - ❏ Flywheel timing pin **10**.
11. Turn crankshaft 1/4 turn clockwise.
12. Tighten tensioner nut **6**.
    Tightening torque: 41-45 Nm.
13. Turn crankshaft 3/4 turn clockwise.
14. Slacken and tighten tensioner nut **6**.
    Tightening torque: 41-45 Nm.
15. Turn crankshaft one turn clockwise to TDC. Ensure timing marks aligned **4** & **8**.
16. Ensure timing pins can be inserted **9** & **10**.
17. Install components in reverse order of removal.
18. Tighten crankshaft pulley bolt **3**.
    Tightening torque: 200 Nm.

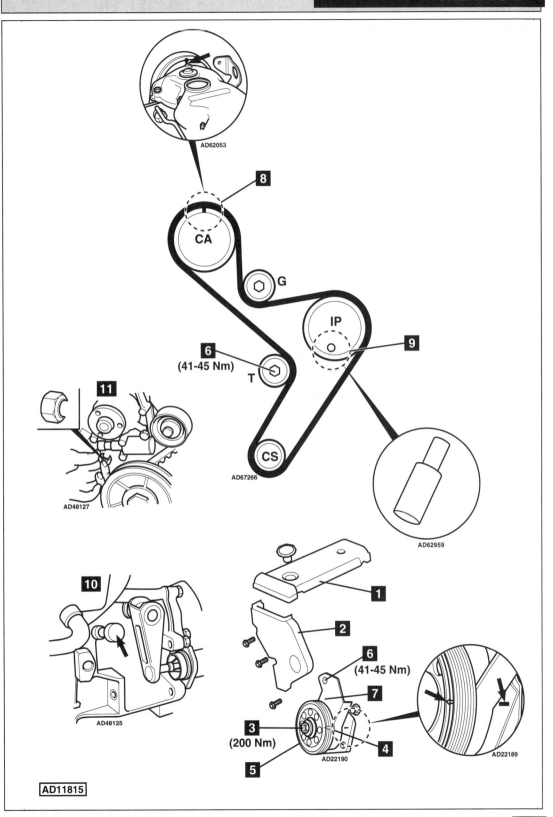

AD62053

8

CA

G

IP

9

6
(41-45 Nm)

T

11

AD48127

CS

AD67266

AD62959

10

AD48125

1

2

6
(41-45 Nm)

7

3
(200 Nm)

5

AD22190

4

AD22189

AD11815

ROVER

Model:	**Metro 1,1/1,4 • 111 • 114 • 214S**
Year:	**1990-95**
Engine Code:	**K8 (SOHC)**

Replacement Interval Guide

Rover recommend replacement every 60,000 miles or 5 years, whichever occurs first.

NOTE: Vehicles that have had the timing belt replaced at the previously recommended replacement interval of 96,000 miles, should have the belt replaced again at 120,000 miles.

The previous use and service history of the vehicle must always be taken into account.
Refer to Timing Belt Replacement Intervals at the front of this manual.

Check For Engine Damage

CAUTION: This engine has been identified as an INTERFERENCE engine in which the possibility of valve-to-piston damage in the event of a timing belt failure is MOST LIKELY to occur.
A compression check of all cylinders should be performed before removing the cylinder head.

Repair Times – hrs

Remove & install	1,90

Special Tools

- None required.

Special Precautions

- Disconnect battery earth lead.
- DO NOT turn crankshaft or camshaft when timing belt removed.
- Remove spark plugs to ease turning engine.
- Turn engine in normal direction of rotation (unless otherwise stated).
- DO NOT turn engine via camshaft or other sprockets.
- Observe all tightening torques.

Removal

1. Remove:
   - ❏ Alternator drive belt.
   - ❏ Timing belt upper cover **1**.
2. Turn crankshaft clockwise until camshaft sprocket timing mark aligned with upper cylinder head face (90° BTDC) **2**.
3. Remove:
   - ❏ Crankshaft pulley **3**.
   - ❏ Timing belt lower cover **4**.
   - ❏ Engine mounting bracket.
4. Slacken tensioner pulley bolt 1/2 turn **5**.
5. Slacken tensioner backplate bolt 1/2 turn **6**.
6. Move tensioner away from belt. Tighten tensioner backplate bolt **6**.
   Tightening torque: 10 Nm.

7. Remove timing belt.
   WARNING: If belt has been in use for more than 48,000 miles: Fit new belt.

Installation

1. Ensure camshaft sprocket timing mark aligned **2**. Ensure crankshaft sprocket dots aligned with flange on oil pump **7**.
2. Fit timing belt in anti-clockwise direction, starting at crankshaft sprocket. Ensure belt is taut between sprockets on non-tensioned side.
3. Install:
   - ❏ Engine mounting bracket.
   - ❏ Timing belt lower cover **4**.
   - ❏ Crankshaft pulley **3**.
4. Slacken tensioner backplate bolt **6**.
5. Turn crankshaft two turns clockwise. Ensure timing marks aligned.
6. Tighten tensioner pulley bolt **5**.
   Tightening torque: 45 Nm.
7. Tighten tensioner backplate bolt **6**.
   Tightening torque: 25 Nm.
8. Install components in reverse order of removal.
9. Tighten crankshaft pulley bolt **8**.
   Tightening torque: 160 Nm.

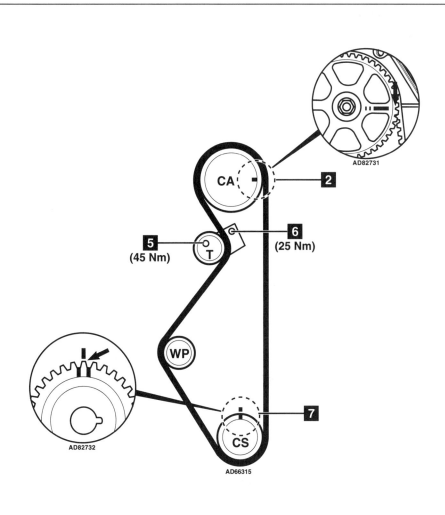

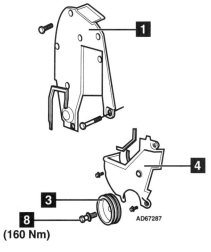

ROVER

Model:	111 • 114 • 211 • 214
Year:	1995-99
Engine Code:	K8 (SOHC)

Replacement Interval Guide

Rover recommend replacement every 60,000 miles or 5 years, whichever occurs first.

NOTE: Early vehicles that have had the timing belt replaced at the previously recommended replacement interval of 96,000 miles, should have the belt replaced again at 120,000 miles.
The previous use and service history of the vehicle must always be taken into account.
Refer to Timing Belt Replacement Intervals at the front of this manual.

Check For Engine Damage

CAUTION: This engine has been identified as an INTERFERENCE engine in which the possibility of valve-to-piston damage in the event of a timing belt failure is MOST LIKELY to occur.
A compression check of all cylinders should be performed before removing the cylinder head.

Repair Times – hrs

Remove & install:

111/114	2,00
211/214	1,70

Special Tools

- ■ Flywheel locking tool – Rover No.18G 1571.

Removal

1. Raise and support front of vehicle.
2. Remove:
   - ❏ RH front wheel.
   - ❏ Inner wing panel.
   - ❏ Auxiliary drive belt.
   - ❏ Timing belt upper cover **1**.
3. Turn crankshaft clockwise until camshaft sprocket timing mark aligned with upper cylinder head face (90° BTDC) **2**.
4. Remove starter motor. Fit locking tool in starter aperture to lock flywheel. Tool No.18G 1571.
5. Support engine.
6. Remove:
   - ❏ RH engine mounting.
   - ❏ Crankshaft pulley bolt **9**.
   - ❏ Crankshaft pulley **3**.
   - ❏ Timing belt lower cover **4**.
   - ❏ Engine mounting bracket.
7. If belt is to be reused:
   - ❏ Mark direction of rotation on belt with chalk.
   - ❏ Early models: Check scribed mark on tensioner backplate aligned with centre punch mark on cylinder head **10**. If not: Mark position of tensioner backplate to cylinder head.
   - ❏ Later models: Mark position of tensioner backplate to cylinder head.
8. Slacken tensioner pulley bolt 1/2 turn **5**.
9. Slacken tensioner backplate bolt 1/2 turn **6**.
10. Move tensioner away from belt. Lightly tighten tensioner backplate bolt **6**.
11. Remove timing belt.
    WARNING: If belt has been in use for more than 48,000 miles: Fit new belt.

Installation

1. Early models: Tensioner spring and sleeve **7** are not fitted in production on some engines. A new spring and sleeve are supplied with manufacturer's timing belt.
2. If reusing old belt: Disconnect tensioner spring (if fitted).
3. If fitting new belt:
   - ❏ Slacken tensioner backplate bolt 1/2 turn **6**.
   - ❏ Fit tensioner spring and sleeve (if missing) **7**.
   - ❏ Ensure tensioner spring and sleeve are located correctly **7**.
   - ❏ Ensure tensioner slides freely. Turn tensioner fully clockwise.
   - ❏ Lightly tighten tensioner backplate bolt **6**.
4. Ensure camshaft sprocket timing mark aligned with upper cylinder head face **2**. Ensure crankshaft sprocket dots aligned with flange on oil pump **8**.
5. Fit timing belt in anti-clockwise direction, starting at crankshaft sprocket. Ensure belt is taut between sprockets on non-tensioned side.
   NOTE: If reusing old belt: Ensure directional arrows point in direction of rotation.
6. Install:
   - ❏ Engine mounting bracket.
   - ❏ Timing belt lower cover **4**.
   - ❏ Crankshaft pulley **3**.
   - ❏ Crankshaft pulley bolt **9**.
7. Tighten crankshaft pulley bolt **9**. 111/114: 160 Nm. 211/214: 205 Nm.
8. Remove flywheel locking tool. Tool No.18G 1571.
9. Slacken tensioner backplate bolt 1/2 turn **6**.
10. If reusing old belt:
    - ❏ Align markings on tensioner backplate and cylinder head.
    - ❏ Tighten tensioner backplate bolt **6**. Tightening torque: 10 Nm.
    - ❏ Tighten tensioner pulley bolt **5**. Tightening torque: 45 Nm.
    - ❏ Reconnect tensioner spring (if fitted).
11. If new belt fitted:
    - ❏ Push tensioner against belt to pre-tension belt.
    - ❏ Tighten tensioner backplate bolt **6**. Tightening torque: 10 Nm.
    - ❏ Turn crankshaft two turns clockwise until timing marks aligned **2**.
    - ❏ Slacken tensioner backplate bolt 1/2 turn **6**.
    - ❏ Ensure only spring tension applied to belt.
    - ❏ Tighten tensioner backplate bolt **6**. Tightening torque: 10 Nm.
    - ❏ Tighten tensioner pulley bolt **5**. Tightening torque: 45 Nm.
12. Install components in reverse order of removal.

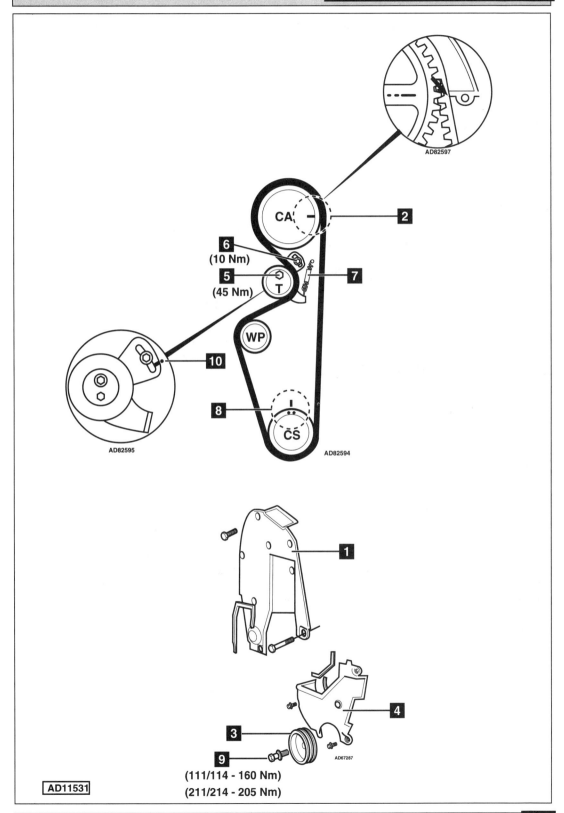

AD82597

CA

2

6
(10 Nm)

5
(45 Nm)

T

7

WP

10

8

CS

AD82595

AD82594

1

4

3

9

(111/114 - 160 Nm)

(211/214 - 205 Nm)

AD67287

AD11531

Model:	Maestro 2,0 • Montego 2,0 • MG Maestro/Montego 2,0 • LDV 200 1,7/2,0 LDV 300/400 2,0
Year:	**1984-94**
Engine Code:	**20H, 17V (C95-L), 20V (C95-L/C96-H)**

Replacement Interval Guide

Rover recommend replacement as follows:
Maestro/Montego: Replacement every
48,000 miles.
Other models:
Check every 24,000 miles - replace if necessary.
Replacement every 48,000 miles.
The previous use and service history of the vehicle must always be taken into account.
Refer to Timing Belt Replacement Intervals at the front of this manual.

Check For Engine Damage

CAUTION: This engine has been identified as an INTERFERENCE engine in which the possibility of valve-to-piston damage in the event of a timing belt failure is MOST LIKELY to occur.
A compression check of all cylinders should be performed before removing the cylinder head.

Repair Times – hrs

Check & adjust:	
Maestro/Montego	0,80
LDV	0,60
PAS	+0,40
Remove & install:	
Maestro/Montego	1,35
LDV	1,05
PAS	+0,40

Special Tools

- ■ Tensioning tool – Rover No.18G 1315.
- ■ Tension gauge – Rover No.KM 4088 AR.

Special Precautions

- ■ Disconnect battery earth lead.
- ■ DO NOT turn crankshaft or camshaft when timing belt removed.
- ■ Remove spark plugs to ease turning engine.
- ■ Turn engine in normal direction of rotation (unless otherwise stated).
- ■ DO NOT turn engine via camshaft or other sprockets.
- ■ Observe all tightening torques.

Removal

1. Remove:
   - ❑ Auxiliary drive belts.
   - ❑ LDV: Distributor cap.
   - ❑ Timing belt cover **1**.
   - ❑ Radiator bottom hose.
2. Turn crankshaft clockwise until No.1 cylinder 90° BTDC **2**. Ensure camshaft sprocket timing marks aligned **3**.
3. Slacken tensioner pulley bolt **4**.
4. Remove timing belt.

Installation

1. Ensure timing marks aligned **2** & **3**.
2. Fit timing belt.
3. Hook tensioning tool to belt at ▽.
   Tool No.18G 1315. Attach spring balance or tension gauge. Tool No. KM 4088 AR.
4. Turn tensioner pulley anti-clockwise to take up belt slack **5**. Use 8 mm Allen key.
5. Tighten tensioner pulley bolt **4**.
6. Pull spring balance until belt aligned with vertical mark on water pump inlet pipe **6**.
   NOTE: Later models: Mark one line on timing belt rear cover in line with back of belt and a second parallel line 8 mm from first line.
7. If tension required to align belt with mark is not 49 N or 5-7 units on tension gauge: Repeat tensioning procedure.
8. Turn crankshaft two turns to timing mark **2**. Ensure camshaft sprocket timing marks aligned **3**.
9. Install components in reverse order of removal.

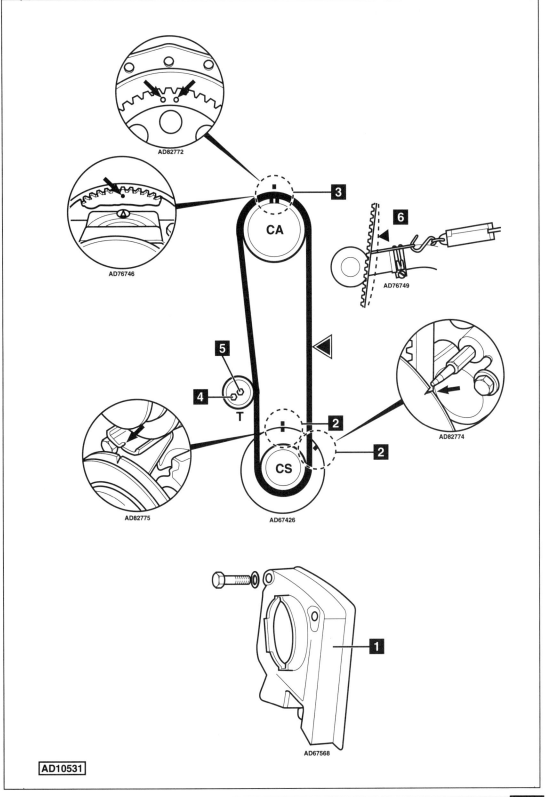

AD82772

AD76746

CA

3

6

AD76749

5

4

T

AD82774

2

2

AD82775

CS

AD67426

1

AD67568

AD10531

ROVER

Model:	Metro GTa/GTi 16V • 114 • 214 • 414
Year:	1989-95
Engine Code:	K16 (DOHC)

Replacement Interval Guide

Rover recommend replacement every 60,000 miles or 5 years, whichever occurs first.

NOTE: Vehicles that have had the timing belt replaced at the previously recommended replacement interval of 96,000 miles, should have the belt replaced again at 120,000 miles.

The previous use and service history of the vehicle must always be taken into account.

Refer to Timing Belt Replacement Intervals at the front of this manual.

Check For Engine Damage

CAUTION: This engine has been identified as an INTERFERENCE engine in which the possibility of valve-to-piston damage in the event of a timing belt failure is MOST LIKELY to occur.

A compression check of all cylinders should be performed before removing the cylinder head.

Repair Times – hrs

Remove & install:

Metro/114	1,90
214/414	2,50

Special Tools

- Camshaft sprocket locking tool – Rover No.18G 1570.
- Flywheel locking tool – Rover No.18G 1571.

Removal

214/414:

1. Raise and support front of vehicle.
2. Remove:
   - ❏ RH front wheel.
   - ❏ Front undershield.

All models:

3. Remove timing belt upper cover and seal **1**.
4. Turn crankshaft clockwise until timing marks on camshaft sprockets aligned (90° BTDC) **8**.
5. Lock camshaft sprockets **2**. Use tool No.18G 1570.
6. Fit locking tool in starter aperture to lock flywheel. Tool No.18G 1571.
7. Remove auxiliary drive belts.

214/414:

8. Remove screws retaining coolant expansion tank. Move expansion tank to one side.
9. Support engine on jack with wood block to protect sump.
10. Remove nuts and bolts retaining RH engine mounting.
11. Lower engine.
12. Remove RH engine mounting and rubber washers.
13. Raise engine to normal position.

All models:

14. Remove:
    - ❏ Crankshaft pulley bolt and washer **3**.
    - ❏ Crankshaft pulley **4**.
    - ❏ Timing belt lower cover and seal **5**.
15. Slacken tensioner pulley bolt 1/2 turn **6**.
16. Slacken tensioner backplate bolt 1/2 turn **7**.
17. Move tensioner away from belt. Lightly tighten tensioner backplate bolt **7**.
18. Remove timing belt.
    WARNING: If belt has been in use for more than 48,000 miles: Fit new belt.

Installation

1. Ensure timing marks on camshaft sprockets aligned **8**.
2. Ensure locking tool located correctly in camshaft sprockets **2**.
3. Ensure crankshaft sprocket dots aligned with flange on oil pump (90° BTDC) **9**.
   NOTE: If camshaft sprockets have been removed: Ensure dowel pin aligned with 'IN' slot of inlet camshaft sprocket and 'EX' slot of exhaust camshaft sprocket.
4. Fit timing belt in anti-clockwise direction, starting at crankshaft sprocket.
5. Install:
   - ❏ Timing belt lower cover and seal **5**.
6. Fit crankshaft pulley **4**. Ensure indent on crankshaft pulley locates over lug on crankshaft sprocket.
7. Fit crankshaft pulley bolt and washer **3**. Tightening torque: 160 Nm.
8. Remove:
   - ❏ Locking tool from camshaft sprockets **2**.
   - ❏ Flywheel locking tool.
9. Slacken tensioner backplate bolt **7**.
10. Turn crankshaft two turns clockwise. Ensure timing marks on camshaft sprockets aligned **8**.
11. Tighten tensioner backplate bolt **7**. Tightening torque: 10 Nm.
12. Tighten tensioner pulley bolt **6**. Tightening torque: 45 Nm.

214/414:

13. Lower engine slightly. Fit RH engine mounting and rubber washers.
14. Raise engine into position.
15. Fit engine mounting bolt to body. Tighten bolt finger tight.
16. Fit engine mounting nuts. Tightening torque: 100 Nm.
17. Tighten engine mounting to body bolt. Tightening torque: 80 Nm.
18. Remove jack. Refit coolant expansion tank.

All models:

19. Install components in reverse order of removal.

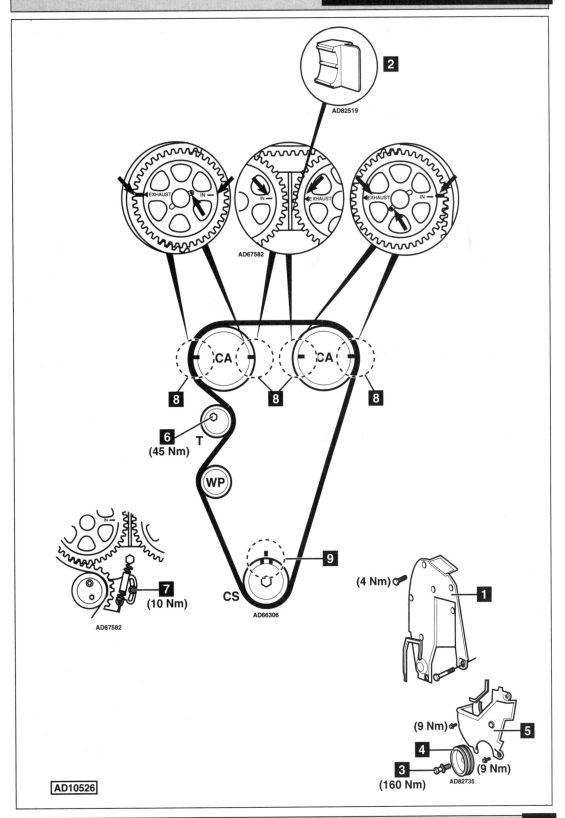

ROVER

Model:	214/216 16V • 216 Cabrio/Coupe • 218 16V • 414/416 16V • 416 Tourer
Year:	1995-99
Engine Code:	K16, 16K16, 18K16 (DOHC)

Replacement Interval Guide

Rover recommend replacement every 60,000 miles or 5 years, whichever occurs first.

NOTE: Early vehicles that have had the timing belt replaced at the previously recommended replacement interval of 96,000 miles, should have the belt replaced again at 120,000 miles.

The previous use and service history of the vehicle must always be taken into account.

Refer to Timing Belt Replacement Intervals at the front of this manual.

Check For Engine Damage

CAUTION: This engine has been identified as an INTERFERENCE engine in which the possibility of valve-to-piston damage in the event of a timing belt failure is MOST LIKELY to occur.

A compression check of all cylinders should be performed before removing the cylinder head.

Repair Times – hrs

Remove & install:

214/216/218	1,90
AC	+0,30
414/416	2,20

Special Tools

- Camshaft sprocket locking tool – Rover No.18G 1570.
- Flywheel locking tool – Rover No.18G 1571.

Special Precautions

- Disconnect battery earth lead.
- DO NOT turn crankshaft or camshaft when timing belt removed.
- Remove spark plugs to ease turning engine.
- Turn engine in normal direction of rotation (unless otherwise stated).
- DO NOT turn engine via camshaft or other sprockets.
- Observe all tightening torques.

Removal

1. Raise and support front of vehicle.
2. Remove:
   - ❑ RH front wheel.
   - ❑ Engine undershield (414/416).
   - ❑ Auxiliary drive belt(s).
   - ❑ Starter motor.
   - ❑ Timing belt upper cover **1**.
3. Turn crankshaft clockwise until timing marks on camshaft sprockets aligned with marks on backplate (90° BTDC) **2**.

4. Lock camshaft sprockets **3**.
   Use tool No.18G 1570.
5. Fit locking tool in starter aperture to lock flywheel. Tool No.18G 1571.
6. Support engine.
7. Remove:
   - ❑ RH engine mounting.
   - ❑ PAS drive belt tensioner.
   - ❑ Crankshaft pulley bolt **11**.
   - ❑ Crankshaft pulley **4**.
   - ❑ Timing belt lower cover and seal **5**.
8. If belt is to be reused:
   - ❑ Mark direction of rotation on belt with chalk.
   - ❑ Check scribed mark on tensioner backplate aligned with centre punch mark on cylinder head **10**. If tensioner backplate not scribed: Mark position of tensioner backplate to cylinder head.
9. Slacken tensioner pulley bolt 1/2 turn **6**.
10. Slacken tensioner backplate bolt 1/2 turn **7**.
11. Move tensioner away from belt. Lightly tighten tensioner backplate bolt **7**.
12. Remove timing belt.
    WARNING: If belt has been in use for more than 48,000 miles: Fit new belt.

Installation

*NOTE: If camshaft sprockets have been removed: Ensure dowel pin aligned with 'IN' slot of inlet camshaft sprocket and 'EX' slot of exhaust camshaft sprocket **12**.*

1. Early models: Tensioner spring and sleeve **9** are not fitted in production on some engines. A new spring and sleeve are supplied with manufacturer's timing belt.
2. If reusing old belt: Disconnect tensioner spring (if fitted).
3. If fitting new belt:
   - ❑ Slacken tensioner backplate bolt 1/2 turn **7**.
   - ❑ Fit tensioner spring and sleeve (if missing) **9**.
   - ❑ Ensure tensioner spring and sleeve are located correctly **9**.
   - ❑ Ensure tensioner slides freely. Turn tensioner fully clockwise.
   - ❑ Lightly tighten tensioner backplate bolt **7**.
4. Ensure timing marks on camshaft sprockets aligned with marks on backplate **2**.
5. Ensure locking tool located correctly in camshaft sprockets **3**.
6. Ensure crankshaft sprocket dots aligned with flange on oil pump **8** (90° BTDC).

➤

Autodata

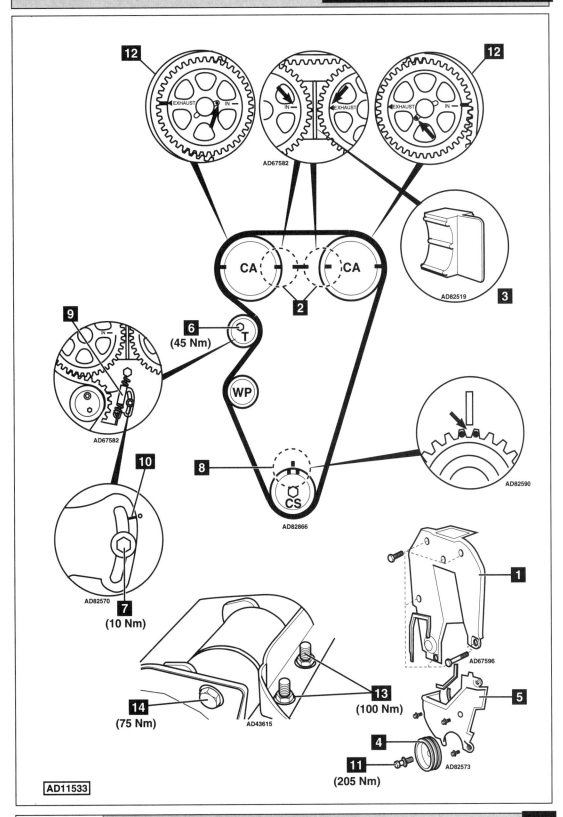

AD11533

←

7. Fit timing belt in anti-clockwise direction, starting at crankshaft sprocket. Ensure belt is taut between sprockets.

   NOTE: If reusing old belt: Ensure directional arrows point in direction of rotation.

8. Install:
   - ❏ Timing belt lower cover and seal .
   - ❏ Crankshaft pulley ◆.
   - ❏ Crankshaft pulley bolt ◆.

9. Tighten crankshaft pulley bolt ◆.
   Tightening torque: 205 Nm.

10. Remove flywheel locking tool.
    Tool No.18G 1571.

11. Fit RH engine mounting.
    - ❏ Tighten nuts ◆. Tightening torque: 100 Nm. Use new nuts.
    - ❏ Tighten bolt ◆. Tightening torque: 75 Nm.

12. Remove locking tool from camshaft sprockets ◆.

13. Slacken tensioner backplate bolt 1/2 turn ◆.

14. If reusing old belt:
    - ❏ Align markings on tensioner backplate and cylinder head.
    - ❏ Tighten tensioner backplate bolt ◆. Tightening torque: 10 Nm.
    - ❏ Tighten tensioner pulley bolt ◆. Tightening torque: 45 Nm.
    - ❏ Reconnect tensioner spring (if fitted).

15. If new belt fitted:
    - ❏ Push tensioner against belt to pre-tension belt.
    - ❏ Tighten tensioner backplate bolt ◆. Tightening torque: 10 Nm.
    - ❏ Turn crankshaft two turns clockwise until timing marks aligned ◆.
    - ❏ Slacken tensioner backplate bolt 1/2 turn ◆.
    - ❏ Ensure only spring tension applied to belt.
    - ❏ Tighten tensioner backplate bolt ◆. Tightening torque: 10 Nm.
    - ❏ Tighten tensioner pulley bolt ◆. Tightening torque: 45 Nm.

16. Install components in reverse order of removal.

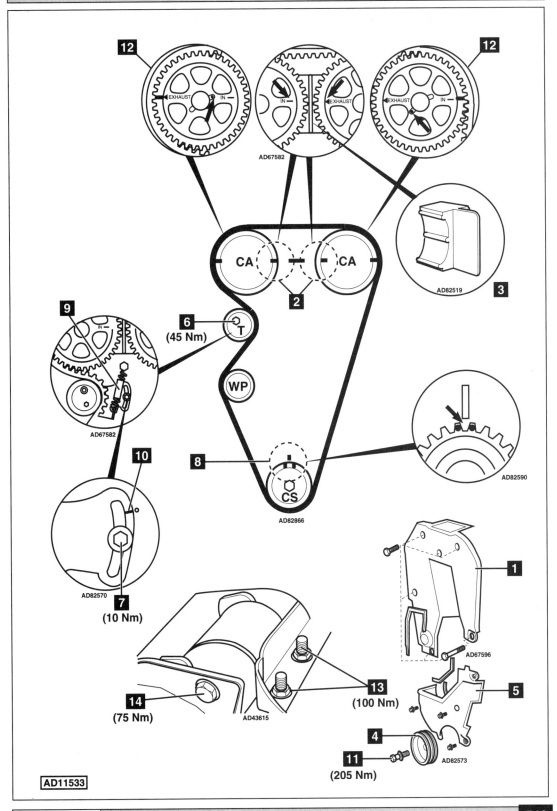

AD67582

AD82519

3

12

12

CA — CA

2

6
(45 Nm)
T

WP

9
AD67582

10

8
CS
AD82866

AD82590

7
AD82570
(10 Nm)

1
AD67596

5

14
(75 Nm)
AD43615

13
(100 Nm)

4
AD82573

11
(205 Nm)

AD11533

ROVER

Model:	25 1,1/1,4/1,6/1,8 • 45 1,4/1,6/1,8 • 75 1,8 • 214/216 16V
	216 Cabrio/Coupé • 218 16V • 414/416 16V • 416 Tourer
	MG ZR 1,4/1,8 • MG ZS 1,6/1,8
Year:	1998-03
Engine Code:	K16

Replacement Interval Guide

Rover recommend:
25/45/ZR/ZS/75 – replacement every
90,000 miles or 6 years, whichever occurs first.
200/400 – replacement every 60,000 miles or
5 years, whichever occurs first.
The previous use and service history of the vehicle
must always be taken into account.
Refer to Timing Belt Replacement Intervals at the front
of this manual.

Check For Engine Damage

CAUTION: This engine has been identified as an
INTERFERENCE engine in which the possibility of
valve-to-piston damage in the event of a timing belt
failure is MOST LIKELY to occur.
A compression check of all cylinders should be
performed before removing the cylinder head.

Repair Times – hrs

Remove & install:

214	1,90
AC	+0,20
216/218	1,90
25/ZR 1,4	1,90
AC	-0,10
25/ZR 1,8	1,90
AC (MT)	+0,10
AC (AT)	+0,20
414/416	2,30
AC	-0,20
45/ZS 1,4/1,6	2,00
45/ZS 1,8	1,90
AC	+0,10
75	2,60

Special Tools

- Camshaft sprocket locking tool –
  Rover No.18G 1570.
- Flywheel locking tool:
  214/216/400/25/45/ZR/ZS
  (except with PG1 gearbox) – Rover No.18G 1571.
  218/25/45/ZR/ZS (with PG1 gearbox) –
  Rover No.18G 1742.
  75 – Rover No.12-170.

Special Precautions

- Disconnect battery earth lead.
- DO NOT turn crankshaft or camshaft when timing belt removed.
- Remove spark plugs to ease turning engine.
- Turn engine in normal direction of rotation (unless otherwise stated).
- DO NOT turn engine via camshaft or other sprockets.
- Observe all tightening torques.

Removal

1. Raise and support front of vehicle.
2. Support engine.
3. Remove engine undershield (if fitted).
4. Unclip top coolant hose. Move to one side. DO NOT disconnect hose.
5. Remove:
   - ❏ Timing belt upper cover **1**.
   - ❏ RH engine mounting.
   - ❏ RH engine mounting bracket.
   - ❏ RH front wheel.
   - ❏ Auxiliary drive belt.
6. Turn crankshaft clockwise until timing marks on camshaft sprockets aligned **2**.
7. Ensure crankshaft timing marks aligned **3**.
8. Lock camshaft sprockets. Use tool No.18G 1570 **4**.
9. Remove starter motor.
10. Fit flywheel locking tool:
    - ❏ 214/216/400/25/45/ZR/ZS (except with PG1 gearbox) – Tool No.18G 1571.
    - ❏ 218/25/45/ZR/ZS (with PG1 gearbox) – Tool No. 18G 1742.
    - ❏ 75 – Tool No.12-170.
11. Remove:
    - ❏ Crankshaft pulley bolt **5**.
    - ❏ Crankshaft pulley **6**.
    - ❏ Auxiliary drive belt tensioner.
    - ❏ Timing belt lower cover and seal **7**.
    - ❏ Tensioner bolt **8**.
    - ❏ Tensioner spring **9**.
    - ❏ Tensioner pulley **10**.
    - ❏ Timing belt.

NOTE: Mark direction of rotation on belt with chalk if belt is to be reused.

WARNING: If belt has been in use for more than 45,000 miles (25/45/ZR/ZS/75) or 48,000 miles (200/400): Fit new belt.

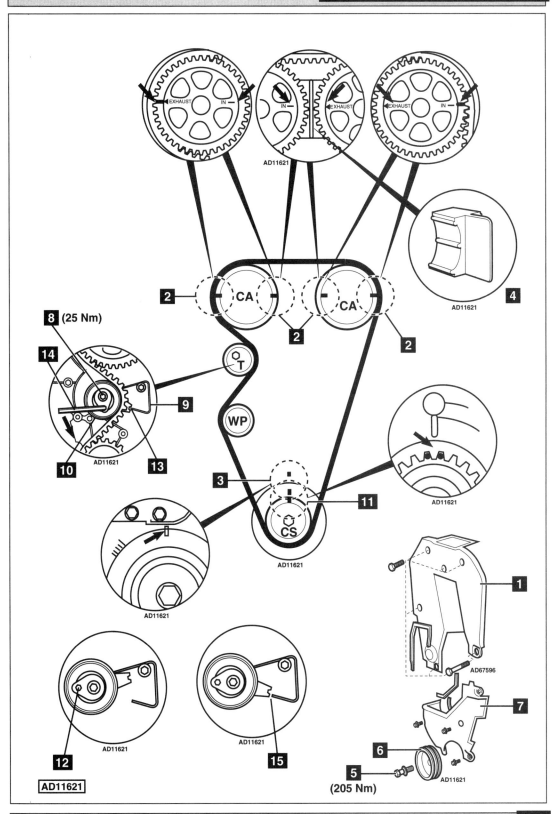

←

Installation

1. Ensure timing marks aligned **2**.
2. Ensure locking tool located correctly in camshaft sprockets **4**.
3. Ensure timing marks aligned **11**.
4. Install:
   - ❏ Tensioner pulley **10**.
   - ❏ Tensioner spring **9**.
   - ❏ Tensioner bolt **8**. Fit new bolt.
5. Position tensioner lever at 9 o'clock **12**.
6. Lightly tighten tensioner bolt **8**.
7. Fit timing belt in anti-clockwise direction, starting at crankshaft sprocket. Ensure belt is taut between sprockets.
8. Install:
   - ❏ Timing belt lower cover and seal **7**.
   - ❏ Crankshaft pulley **6**.
   - ❏ Crankshaft pulley bolt **5**.
9. Ensure flywheel locking tool located correctly:
   - ❏ 214/216/400/25/45/ZR/ZS (except with PG1 gearbox) – Tool No.18G 1571.
   - ❏ 218/25/45/ZR/ZS (with PG1 gearbox) – Tool No. 18G 1742.
   - ❏ 75 – Tool No.12-170.
10. Tighten crankshaft pulley bolt **5**. Tightening torque: 205 Nm.
11. Remove:
    - ❏ Flywheel locking tool.
    - ❏ Camshaft locking tool **4**.
12. Turn tensioner anti-clockwise until pointer aligned with index spring **13**. Use Allen key **14**.
    NOTE: Repeat tensioning procedure if pointer passes index spring.
13. If reusing old belt: Align lower part of pointer to index spring **15**.
14. Tighten tensioner bolt **8**. Tightening torque: 25 Nm.
15. Turn crankshaft two turns clockwise until timing marks aligned **2** & **3**.
16. Ensure pointer still aligned **13** or **15**.
17. If not: Repeat tensioning procedure.
18. Install components in reverse order of removal.

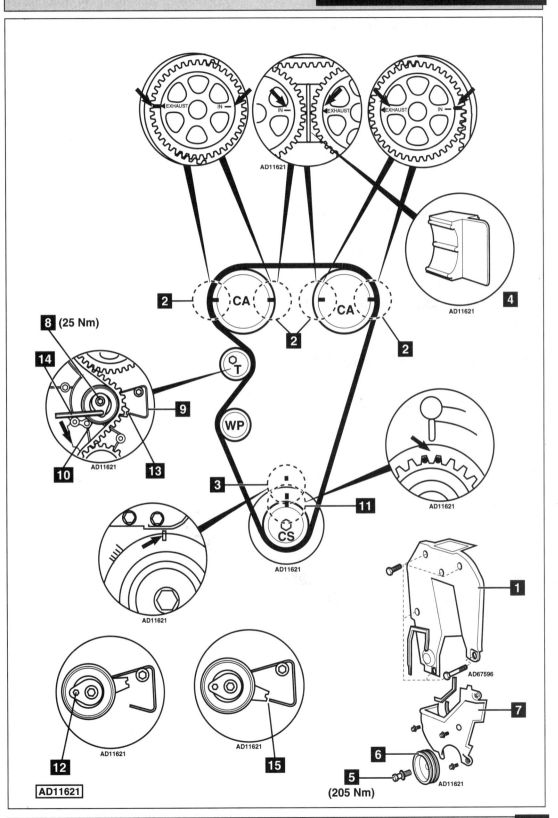

AD11621

AD11621

2

8 (25 Nm)

14

9

10 AD11621 13

3

4 AD11621

11 AD11621

AD11621

1 AD67596

7

6

5 (205 Nm) AD11621

12 AD11621

15 AD11621

AD11621

ROVER

Model:	216 16V • 416/Tourer 16V
Year:	1989-96
Engine Code:	D16A6, D16A7 (SOHC)

Replacement Interval Guide

Rover recommend replacement every
60,000 miles or 5 years, whichever occurs first.
*The previous use and service history of the vehicle
must always be taken into account.*
*Refer to Timing Belt Replacement Intervals at the front
of this manual.*

Check For Engine Damage

*CAUTION: This engine has been identified as an
INTERFERENCE engine in which the possibility of
valve-to-piston damage in the event of a timing belt
failure is MOST LIKELY to occur.*
*A compression check of all cylinders should be
performed before removing the cylinder head.*

Repair Times – hrs

Remove & install	2,10
PAS	+1,40

Special Tools

■ None required.

Special Precautions

■ Disconnect battery earth lead.
■ DO NOT turn crankshaft or camshaft when timing belt
removed.
■ Remove spark plugs to ease turning engine.
■ Turn engine in normal direction of rotation (unless
otherwise stated).
■ DO NOT turn engine via camshaft or other sprockets.
■ Observe all tightening torques.

Removal

*NOTE: Normal direction of crankshaft rotation is
anti-clockwise.*

1. Remove:
   ❏ Engine undershield.
   ❏ Auxiliary drive belts.
   ❏ Cylinder head cover.
   ❏ Timing belt upper cover **1**.
2. Turn crankshaft anti-clockwise to TDC. Ensure
   timing marks aligned **2** & **3**.
3. Remove:
   ❏ LH engine mounting.
   ❏ Crankshaft pulley **4**.
   ❏ Timing belt lower cover **5**.
4. Slacken tensioner bolt **6**. Move tensioner away
   from belt and lightly tighten bolt.
5. Remove timing belt.

Installation

1. Ensure timing marks aligned **2** & **3**.
2. Fit timing belt, starting at crankshaft sprocket.
   Ensure belt is taut on non-tensioned side.
3. Slacken tensioner bolt **6**.
4. Install:
   ❏ LH engine mounting.
   ❏ Timing belt lower cover **5**.
5. Turn crankshaft 1/4 turn anti-clockwise.
6. Tighten tensioner bolt **6**.
   Tightening torque: 45 Nm.
7. Turn crankshaft anti-clockwise to TDC. Ensure
   timing marks aligned **2** & **3**.
8. Install components in reverse order of removal.
9. Tighten crankshaft pulley bolt **7**.
   Tightening torque: 115 Nm.

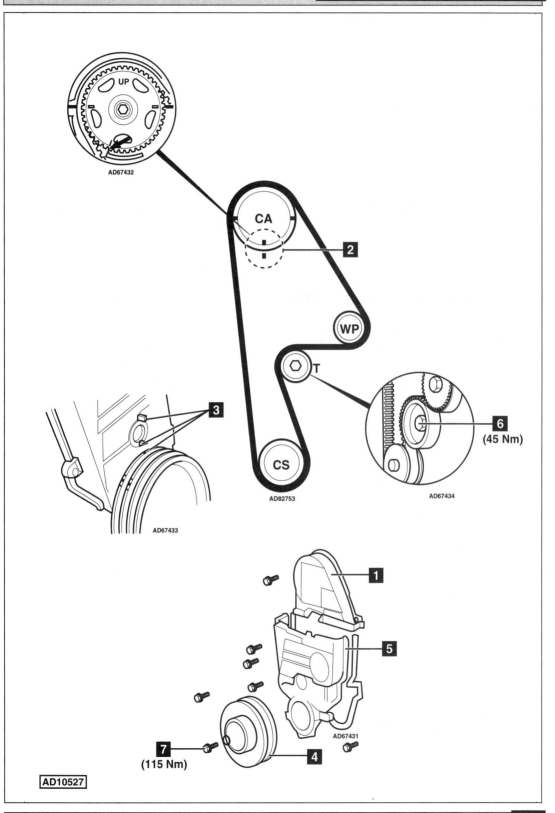

AD67432

CA

2

WP

T

6

(45 Nm)

AD67434

3

AD67433

CS

AD82753

1

5

AD67431

7

(115 Nm)

4

AD10527

ROVER

Model:	**216 GTi 16V • 416 GTi 16V**
Year:	**1990-95**
Engine Code:	**D16A8, D16A9 (DOHC)**

Replacement Interval Guide

Rover recommend replacement every
60,000 miles or 5 years, whichever occurs first.
*The previous use and service history of the vehicle
must always be taken into account.*
*Refer to Timing Belt Replacement Intervals at the front
of this manual.*

Check For Engine Damage

*CAUTION: This engine has been identified as an
INTERFERENCE engine in which the possibility of
valve-to-piston damage in the event of a timing belt
failure is MOST LIKELY to occur.*
*A compression check of all cylinders should be
performed before removing the cylinder head.*

Repair Times – hrs

Remove & install	3,20

Special Tools

■ Camshaft timing pins – Rover No.18G 1580.

Special Precautions

- ■ Disconnect battery earth lead.
- ■ DO NOT turn crankshaft or camshaft when timing belt removed.
- ■ Remove spark plugs to ease turning engine.
- ■ Turn engine in normal direction of rotation (unless otherwise stated).
- ■ DO NOT turn engine via camshaft or other sprockets.
- ■ Observe all tightening torquès.

Removal

*NOTE: Normal direction of crankshaft rotation is
anti-clockwise.*

1. Remove:
   ❏ Engine undershield.
   ❏ Auxiliary drive belts.
   ❏ Cylinder head cover.
   ❏ Timing belt upper cover **1**.
2. Turn crankshaft anti-clockwise to TDC. Ensure timing marks aligned **2** & **3**.
3. Remove:
   ❏ LH top engine mounting.
   ❏ Crankshaft pulley **8**.
   ❏ Timing belt lower cover **4**.
4. Slacken tensioner bolt **5**. Move tensioner away from belt and lightly tighten bolt.
5. Remove:
   ❏ Crankshaft sprocket outer guide washer.
   ❏ Timing belt.

Installation

1. Ensure timing marks aligned.
2. Insert timing pins in camshafts **6**.
   Tool No.18G 1580.
3. Fit timing belt, starting at crankshaft sprocket. Ensure belt is taut on non-tensioned side.
4. Remove timing pins **6**.
5. Fit crankshaft sprocket guide washer (convex side towards belt).
6. Slacken tensioner bolt **5**.
7. Install:
   ❏ LH top engine mounting.
   ❏ Timing belt lower cover.
8. Turn crankshaft 1/4 turn anti-clockwise.
9. Tighten tensioner bolt **5**.
   Tightening torque: 45 Nm.
10. Turn crankshaft anti-clockwise to TDC. Ensure timing marks aligned.
11. Install components in reverse order of removal.
12. Tighten crankshaft pulley bolt **7**.
    Tightening torque: 115 Nm.

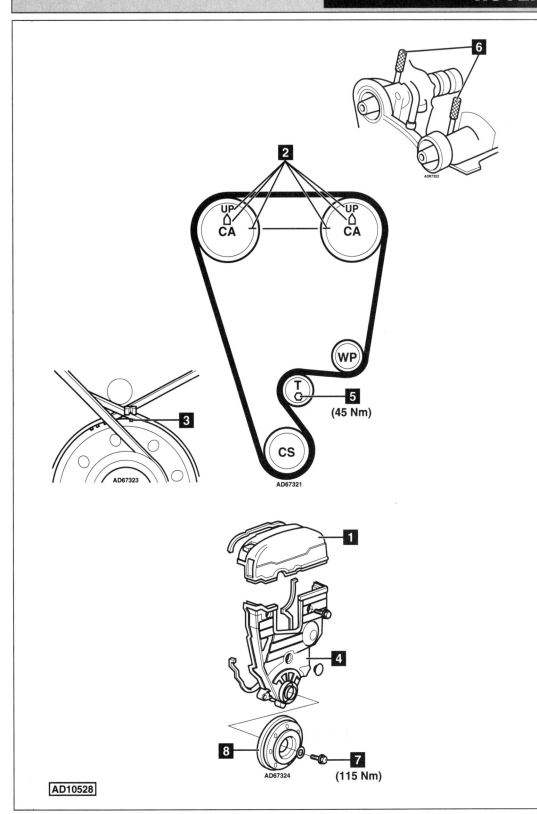

UP CA

UP CA

WP

T
5
(45 Nm)

3

AD67323

CS

AD67321

6

AD67322

1

4

8

AD67324

7
(115 Nm)

AD10528

Model:	**200 Vi • 218 Coupé • BRM 1,8 VVC • 25 1,8 VVC • ZR 1,8 VVC MGF 1,8 VVC**
Year:	**1995-03**
Engine Code:	**18K16, K16-1,8**

Replacement Interval Guide

Rover recommend:

200 Vi/218 Coupé/BRM 1,8 VVC:
Replacement every 60,000 miles or 5 years, whichever occurs first.

25/ZR:
Replacement every 60,000 miles or 4 years, whichever occurs first.

MGF:
→1999 (VIN.→RD522572): Replacement every 60,000 miles or 5 years, whichever occurs first.
1999→ (VIN.RD522573→): Replacement every 60,000 miles or 4 years, whichever occurs first.
The previous use and service history of the vehicle must always be taken into account.
Refer to Timing Belt Replacement Intervals at the front of this manual.

Check For Engine Damage

CAUTION: This engine has been identified as an INTERFERENCE engine in which the possibility of valve-to-piston damage in the event of a timing belt failure is MOST LIKELY to occur.
A compression check of all cylinders should be performed before removing the cylinder head.

Repair Times – hrs

Remove & install:

Front timing belt:	
25/ZR	1,90
AC	+0,10
200/BRM	1,90
AC	+0,30
Coupé	1,70
AC	+0,10
MGF	2,80
AC	-0,10
Rear timing belt:	
25/ZR	0,80
200/BRM	0,50
Coupé	0,60
MGF	1,10

Special Tools

- Camshaft sprocket locking tool – Rover No.18G.1570.
- Flywheel locking tool – Rover No.18G.1742.
- Holding tool – 200/BRM/Coupé/MGF: Rover No.18G.1521.
- Holding tool – 25/ZR: Rover No.12.182.

Removal

Front Timing Belt

MGF:

1. Lower windows and release hood catches. DO NOT lower hood.
2. Release and pull forward rear of hood well carpet.

3. Release 5 rear hood clips and raise rear edge of hood.
4. Remove:
   - Hood well trim.
   - Sound insulating pad.
   - Engine cover (11 bolts).
5. Raise and support rear of vehicle.
6. Remove:
   - RH rear wheel.
   - Inner wing panel.

25/ZR/200/BRM/Coupé:

1. Raise and support front of vehicle.
2. Remove:
   - RH front wheel.

All models:

1. Slacken timing belt upper cover bolt **1**.
2. Remove:
   - Timing belt upper cover bolts **2**.
   - Timing belt upper cover **3**.
3. Turn crankshaft clockwise until timing marks on camshaft front sprockets aligned with backplate **4**.
4. If fitted: Ensure crankshaft pulley timing marks aligned **5**.
   NOTE: This position is 90° BTDC.
5. Lock camshaft front sprockets. Use tool No.18G.1570 **6**.
6. Remove starter motor. Fit locking tool in starter aperture to lock flywheel. Tool No.18G.1742. Use two starter bolts.
7. Remove:
   - Auxiliary drive belt(s).
   - Auxiliary drive belt tensioner (if required).
   - Crankshaft pulley bolt **7**.
   - Crankshaft pulley **8**.
8. Support engine.

MGF:

   - Remove two engine mounting bolts **9**.
   - Unclip Hydragas pipe and move aside.
   - Remove four engine mounting bolts **10**.
   - Lower engine slightly. Remove engine mounting.
   - Raise engine to normal position.

200/BRM/Coupé:

   - Remove engine mounting nut and bolts **11**, **12** & **13**.
   - Remove engine mounting.

25/ZR:

   - Slacken engine steady bar nut **24**.
   - Slacken engine mounting bolt **25**.
   - Remove engine steady bar nut **26**.
   - Remove engine mounting nut and bolts **11**, **12** & **13**.
   - Remove engine mounting.
   - Lower engine slightly.

All models:

9. Remove timing belt lower cover **14**.
10. If belt is to be reused:
    - Mark direction of rotation on belt with chalk.
    - Mark position of tensioner backplate to cylinder head.

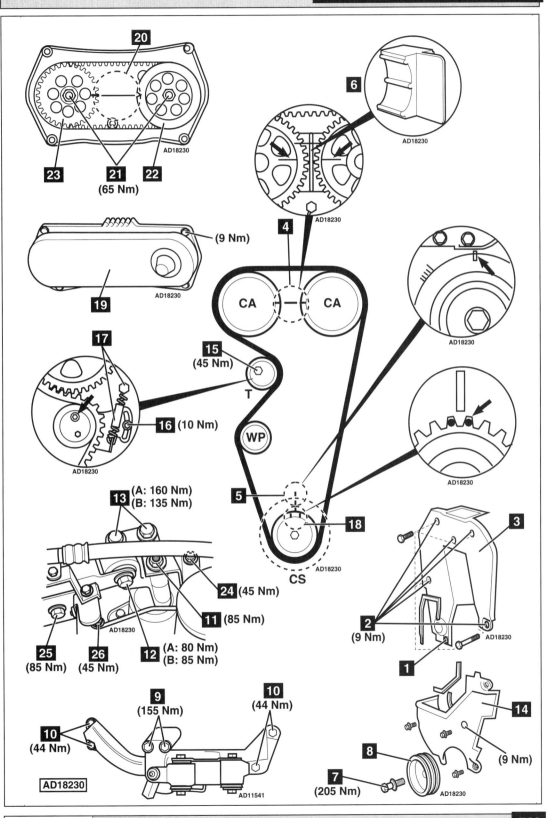

20

AD18230

23 **21** **22**
(65 Nm)

6

AD18230

AD18230

4

(9 Nm)

19

AD18230

AD18230

17

15
(45 Nm)

T

16 **(10 Nm)**

AD18230

CA — CA

WP

AD18230

5

18

CS

AD18230

13 **(A: 160 Nm)**
(B: 135 Nm)

24 **(45 Nm)**

11 **(85 Nm)**

AD18230

25 **26** **12** **(A: 80 Nm)**
(85 Nm) **(45 Nm)** **(B: 85 Nm)**

3

2
(9 Nm)

AD18230

1

14

(9 Nm)

9 **10**
(155 Nm) **(44 Nm)**

10
(44 Nm)

8

7
(205 Nm)

AD18230

AD11541

AD18230

11. Slacken tensioner bolts 1/2 turn **15** & **16**. Move tensioner away from belt. Lightly tighten bolt **16**.
12. Remove timing belt.
    WARNING: *If belt has been in use for more than 45,000 miles: Fit new belt.*

Installation

Front Timing Belt

*NOTE: Tensioner spring and bolt **17** are not fitted in production. A new spring and bolt are supplied with manufacturer's timing belt.*

1. If reusing old belt: DO NOT install tensioner spring and bolt **17**.
2. If a new belt is being fitted:
   - ❏ Slacken tensioner backplate bolt 1/2 turn **16**.
   - ❏ Install tensioner spring and bolt **17**.
   - ❏ Ensure tensioner slides freely. Turn tensioner fully clockwise.
   - ❏ Lightly tighten tensioner backplate bolt **16**.
3. Ensure crankshaft sprocket timing marks aligned with flange on oil pump **18**.
   NOTE: This position is 90° BTDC.
4. With locking tool in position **6**: Ensure timing marks on camshaft front sprockets aligned **4**.
5. Fit timing belt in anti-clockwise direction, starting at crankshaft sprocket. Ensure belt is taut between sprockets.
   NOTE: If reusing old belt: Ensure directional arrows point in direction of rotation.
6. Fit timing belt lower cover **14**.
7. Fit crankshaft pulley **8**. Ensure indent on crankshaft pulley locates over lug on crankshaft sprocket.
8. Tighten crankshaft pulley bolt **7**.
   Tightening torque: 205 Nm.
9. Remove flywheel locking tool.
10. Raise engine.

MGF:

11. Reposition mounting on engine and subframe.
12. Tighten engine mounting bolts **10**.
    Tightening torque: 44 Nm.
13. Lower engine.
14. Fit engine mounting bolts **9**.
    Tightening torque: 155 Nm.

200/BRM/Coupé:

15. Fit engine mounting.
16. Tighten bolts **13**. Tightening torque: (A) 160 Nm.
17. Tighten bolt **12**. Tightening torque: (A) 80 Nm.
18. Tighten nut **11**. Tightening torque: 85 Nm.

25/ZR:

19. Fit engine mounting.
20. Tighten bolts **13**. Tightening torque: (B) 135 Nm.
21. Tighten bolt **12**. Tightening torque: (B) 85 Nm.
22. Tighten nut **11**. Tightening torque: 85 Nm.
23. Tighten nuts **24** & **26**. Tightening torque: 45 Nm.
24. Tighten bolt **25**. Tightening torque: 85 Nm.

All models:

25. Remove locking tool from camshaft front sprockets **6**.
26. Slacken tensioner backplate bolt **16**.
27. If reusing old belt:
    - ❏ Align markings on tensioner backplate and cylinder head.
    - ❏ Tighten tensioner backplate bolt **16**.
      Tightening torque: 10 Nm.
    - ❏ Tighten tensioner pulley bolt **15**.
      Tightening torque: 45 Nm.

28. If a new belt is being fitted:
    - ❏ Push tensioner against belt to pre-tension belt.
    - ❏ Tighten tensioner backplate bolt **16**.
      Tightening torque: 10 Nm.
    - ❏ Turn crankshaft two turns clockwise until timing marks aligned **4**.
    - ❏ Slacken tensioner backplate bolt 1/2 turn **16**.
    - ❏ Ensure only spring tension applied to belt.
    - ❏ Tighten tensioner backplate bolt **16**.
      Tightening torque: 10 Nm.
    - ❏ Tighten tensioner pulley bolt **15**.
      Tightening torque: 45 Nm.
    - ❏ Remove tensioner spring and bolt **17**.
29. Install:
    - ❏ Starter motor.
    - ❏ Timing belt upper cover **3**.
    - ❏ Auxiliary drive belt tensioner.
    - ❏ Auxiliary drive belt.
30. Install components in reverse order of removal.

Removal

Rear Timing Belt

1. Disconnect HT cables and wiring harness from rear timing belt cover.
2. 25/ZR: Remove air filter.
3. All models – Remove:
   - ❏ Timing belt upper cover **3**.
   - ❏ Cover for rear timing belt **19**.
4. Turn crankshaft clockwise until timing marks on camshaft front sprockets aligned with backplate **4**.
5. Lock camshaft front sprockets. Use tool No.18G.1570 **6**.
6. Ensure timing marks on camshaft rear sprockets aligned with backplate **20**.
7. Hold camshaft rear sprockets.
   200/BRM/Coupé/MGF: Use tool No.18G.1521.
   25/ZR: Use tool No.12.182.
8. Remove:
   - ❏ Bolt of each camshaft rear sprocket **21**.
   - ❏ Camshaft rear sprockets.
   - ❏ Timing belt.

Installation

Rear Timing Belt

1. With locking tool in position **6**: Ensure timing marks on camshaft front sprockets aligned **4**.
2. Fit camshaft rear sprocket **22**.
3. Align timing marks on camshaft sprockets **20**.
4. Fit camshaft rear sprocket **23** and belt together.
   NOTE: Ensure camshaft sprocket located correctly on dowel pin.
5. Align timing marks on camshaft sprockets **20**.
6. Use straightedge between shaft bolt centres to check that timing marks on camshaft sprockets and backplate aligned.
7. Hold camshaft rear sprockets.
   200/BRM/Coupé/MGF: Use tool No.18G.1521.
   25/ZR: Use tool No.12.182.
8. Tighten bolt of each camshaft rear sprocket **21**.
   Tightening torque: 65 Nm.
9. Use straightedge between shaft bolt centres to check that timing marks on camshaft sprockets and backplate aligned.
10. Remove locking tool from camshaft front sprockets **6**.
11. Install components in reverse order of removal.

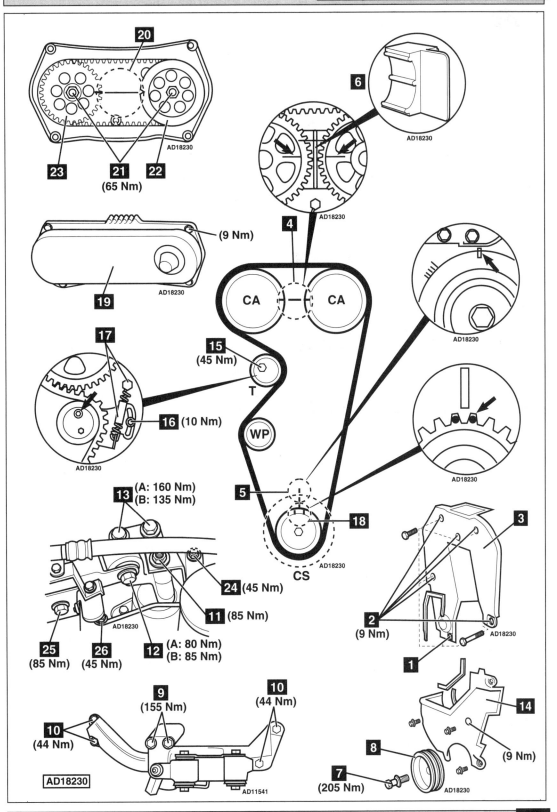

20

AD18230

23 **21** **22**
(65 Nm)

6

AD18230

4

AD18230

(9 Nm)

19

AD18230

CA — CA

AD18230

17

15
(45 Nm)

T

16 (10 Nm)

AD18230

WP

AD18230

5

18

CS

AD18230

13 **(A: 160 Nm)**
(B: 135 Nm)

3

24 (45 Nm)

11 (85 Nm)

2
(9 Nm)

1

AD18230

25
(85 Nm)

26
(45 Nm)

12 **(A: 80 Nm)**
(B: 85 Nm)

AD18230

9
(155 Nm)

10
(44 Nm)

10
(44 Nm)

14

AD18230

8

7
(205 Nm)

AD18230

AD11541

ROVER

Model:	220 2,0 16V/Turbo (10/92→) • 420i • 420 2,0 16V (10/92→) 820 2,0 16V (11/91→) • Vitesse 2,0 16V Turbo (12/91→)
Year:	1991-99
Engine Code:	20T4 (T16)

Replacement Interval Guide

Rover recommend replacement every 60,000 miles or 5 years, whichever occurs first.

NOTE: Vehicles that have had the timing belt replaced at the previously recommended replacement interval of 96,000 miles, should have the belt replaced again at 120,000 miles.

The previous use and service history of the vehicle must always be taken into account.
Refer to Timing Belt Replacement Intervals at the front of this manual.

Check For Engine Damage

CAUTION: This engine has been identified as an INTERFERENCE engine in which the possibility of valve-to-piston damage in the event of a timing belt failure is MOST LIKELY to occur.
A compression check of all cylinders should be performed before removing the cylinder head.

Repair Times – hrs

Remove & install:

220/420	1,40
820/Vitesse	1,50

Special Tools

- Camshaft sprocket locking tool – Rover No.18G 1524.
- Flywheel/drive plate timing pin – Rover No.18G 1523.
- Crankshaft pulley holding tool – Rover No.18G 1641.

Special Precautions

- Disconnect battery earth lead.
- DO NOT turn crankshaft or camshaft when timing belt removed.
- Remove spark plugs to ease turning engine.
- Turn engine in normal direction of rotation (unless otherwise stated).
- DO NOT turn engine via camshaft or other sprockets.
- Observe all tightening torques.

Removal

1. Support engine.
2. Remove:
   - ❑ Auxiliary drive belt.
   - ❑ RH engine mounting.
   - ❑ Timing belt upper cover **1**.
3. Raise engine.
4. Remove timing belt centre cover **2**.
5. Turn crankshaft until timing marks aligned **3**.
6. Insert flywheel/drive plate timing pin **5**.
   Tool No.18G 1523.
7. Lock camshaft sprockets **4**.
   Use tool No.18G 1524.
   NOTE: Crankshaft now set at 90° BTDC.
8. Slacken tensioner bolt **6**. Move tensioner away from belt and lightly tighten bolt.
9. Remove crankshaft pulley bolts **7**.
10. Hold crankshaft pulley. Use tool No.18G 1641 **8**.

11. Remove:
    - ❑ Crankshaft pulley centre bolt **9**.
    - ❑ Holding tool **8**.
    - ❑ Crankshaft pulley **10**.
    - ❑ Timing belt lower cover **11**.
    - ❑ Timing belt.
    WARNING: If belt has been in use for more than 48,000 miles: Fit new belt.

Installation

NOTE: The installation procedure will require the co-operation of an assistant.

1. Fit timing belt. Ensure belt is taut between crankshaft sprocket and exhaust camshaft sprocket.
2. Install:
   - ❑ Timing belt lower cover **11**.
   - ❑ Crankshaft pulley **10**.
3. Hold crankshaft pulley.
   Use tool No.18G 1641 **8**.
4. Tighten crankshaft pulley centre bolt **9**.
   Tightening torque: 85 Nm.
5. Slacken tensioner bolt **6**. Allow tensioner to operate.
6. Remove locking tool from camshaft sprockets **4**.

820/Vitesse:

7. Hold crankshaft pulley.
   Use tool No.18G 1641 **8**.
8. Apply anti-clockwise torque of 40 Nm to bolt of inlet camshaft sprocket to tension belt **12**.
9. Tighten tensioner bolt **6**. Tightening torque: 25 Nm.
10. Release torque loading from bolt of inlet camshaft sprocket **12**.
11. Remove holding tool **8**.
12. Fit crankshaft pulley bolts **7**.
    Tightening torque: 10 Nm.
13. Remove timing pin **5**.
14. Install components in reverse order of removal.

220/420:

7. Remove holding tool **8**.
8. Fit crankshaft pulley bolts **7**.
    Tightening torque: 10 Nm.
9. Remove timing pin **5**.
10. Turn crankshaft two turns clockwise. Ensure timing marks aligned.
11. Tighten tensioner bolt **6**. Tightening torque: 25 Nm.
12. Install components in reverse order of removal.

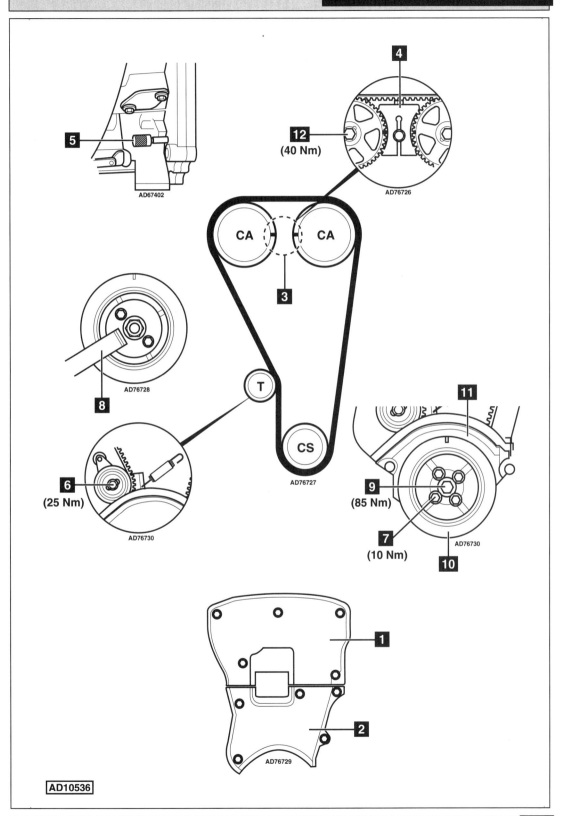

ROVER

Model:	**416i Automatic**
Year:	**1995-99**
Engine Code:	**D16Y3, D16B2**

Replacement Interval Guide

Rover recommend replacement every 60,000 miles or 5 years, whichever occurs first. *The previous use and service history of the vehicle must always be taken into account. Refer to Timing Belt Replacement Intervals at the front of this manual.*

Check For Engine Damage

*CAUTION: This engine has been identified as an INTERFERENCE engine in which the possibility of valve-to-piston damage in the event of a timing belt failure is MOST LIKELY to occur.
A compression check of all cylinders should be performed before removing the cylinder head.*

Repair Times – hrs

Remove & install:

→1997½MY	2,90
1997½MY→	3,20

Special Tools

- None required.

Special Precautions

- Disconnect battery earth lead.
- DO NOT turn crankshaft or camshaft when timing belt removed.
- Remove spark plugs to ease turning engine.
- Turn engine in normal direction of rotation (unless otherwise stated).
- DO NOT turn engine via camshaft or other sprockets.
- Observe all tightening torques.

Removal

NOTE: Normal direction of crankshaft rotation is anti-clockwise.

1. Support engine.
2. Remove:
   - ❏ Auxiliary drive belt(s).
   - ❏ PAS pump. DO NOT disconnect hoses.
   - ❏ LH engine mounting.
   - ❏ Cylinder head cover.
   - ❏ Timing belt upper cover **1**.
3. Turn crankshaft anti-clockwise to TDC on No.1 cylinder. Ensure timing marks aligned **2** & **3**.
4. Remove:
   - ❏ Crankshaft pulley **4**.
   - ❏ Crankshaft sprocket guide washer.
   - ❏ Timing belt lower cover **5**.

5. Ensure crankshaft sprocket timing marks aligned **6**.
6. Slacken tensioner bolt **7**. Move tensioner away from belt and lightly tighten bolt.
7. Remove timing belt.

Installation

1. Ensure timing marks aligned **3** & **6**.
2. Fit timing belt, starting at crankshaft sprocket. Ensure belt is taut on non-tensioned side.
3. Slacken tensioner bolt **7**.
4. Turn crankshaft one turn anti-clockwise.
5. Tighten tensioner bolt **7**.
6. Turn crankshaft five turns anti-clockwise until timing marks aligned **3** & **6**.
7. Slacken tensioner bolt **7**.
8. Turn crankshaft 30° anti-clockwise.
9. Tighten tensioner bolt **7**.
   Tightening torque: 44 Nm.
10. Turn crankshaft anti-clockwise to TDC on No.1 cylinder. Ensure timing marks aligned **3** & **6**.
11. Install components in reverse order of removal.
    NOTE: Ensure convex side of crankshaft sprocket guide washer faces belt.
12. Tighten crankshaft pulley bolt **8**.
    Tightening torque: 165 Nm.

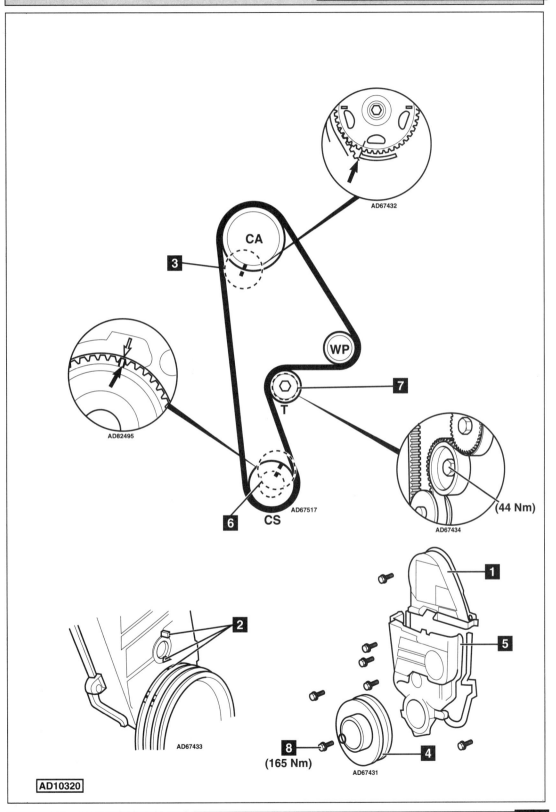

CA

3

AD67432

WP

7

T

AD82495

(44 Nm)

AD67434

6 CS

AD67517

1

2

5

AD67433

8 (165 Nm)

AD67431

4

AD10320

ROVER

Model:	45 2,0 V6 • MG ZS - 180 2,5 V6 • 75 2,0/2,5 V6
	MG ZT - 160/180/190/ZT-T - 160/180/190 2,5 V6
Year:	1999-03
Engine Code:	KV6

Replacement Interval Guide

Rover recommend replacement every 90,000 miles or 6 years, whichever occurs first.
The previous use and service history of the vehicle must always be taken into account.
Refer to Timing Belt Replacement Intervals at the front of this manual.

Check For Engine Damage

CAUTION: This engine has been identified as an INTERFERENCE engine in which the possibility of valve-to-piston damage in the event of a timing belt failure is MOST LIKELY to occur.
A compression check of all cylinders should be performed before removing the cylinder head.

Repair Times – hrs

Remove & install:

45/ZS

Front belt	5,20
LH rear belt	1,00
RH rear belt	1,70
All belts	6,90

75/ZT/ZT-T

Front belt	3,70
AC	1,30
LH rear belt	0,90
RH rear belt	1,20
All belts	4,90
AC	1,40

Special Tools

- Crankshaft pulley holding tool – Rover No.12-161.
- Crankshaft pulley holding tool handle – Rover No.18G 1672.
- Locking tools for camshafts:
  Except 2,0/2,5-190 models – Rover No.18G 1747-2.
  2,0 models – Rover No.12-187.
  2,5-190 models – Rover No.12R208.
- Alignment plate for camshaft sprockets – Rover No.12-175.
- Sprocket tool – Rover No.18G 1747-1.
- Tool for camshafts – Rover No.18G 1747-4.
- Alignment pins – Rover No.18G 1747-5.

Special Precautions

- Disconnect battery earth lead.
- DO NOT turn crankshaft or camshaft when timing belt removed.
- Remove spark plugs to ease turning engine.
- Turn engine in normal direction of rotation (unless otherwise stated).
- DO NOT turn engine via camshaft or other sprockets.
- Observe all tightening torques.

Removal

Front Timing Belt

1. Remove:
   - AC models: Oil cooler.
   - Auxiliary drive belt.
2. Raise and support front of vehicle.
3. Remove:
   - RH front wheel.
   - RH splash guard.
   - LH rear timing belt cover **1**.
4. Turn crankshaft clockwise until timing marks aligned **2**.
   NOTE: White timing mark aligned with 'SAFE'.
5. Ensure timing marks on camshaft sprockets aligned **3**.
6. Remove:
   - PAS pump pulley.
   - PAS pump. DO NOT disconnect hoses.
   - Alternator.
   - Alternator bracket.
   - Auxiliary drive belt guide pulley.
   - Timing belt upper covers **4** & **5**.
7. Hold crankshaft pulley.
   Use tool Nos.12-161/18G 1672.
8. Remove:
   - Crankshaft pulley bolt **6**.
   - Crankshaft pulley **7**.
   - Timing belt lower cover **8**.
   - Auxiliary drive belt tensioner.
9. Drain engine oil.
10. Remove:
    - Dipstick and tube.
    - AC compressor (if fitted). DO NOT disconnect hoses.
    - AC compressor bracket (if fitted).
11. Support engine.
12. Remove:
    - RH engine mounting.
    - RH engine mounting bracket **9**.
13. Turn tensioner pulley clockwise to release tension from automatic tensioner unit. Use Allen key **10**.
14. Remove:
    - Blanking cap (if fitted) **27**.
    - Automatic tensioner unit bolts **11**.
    - Automatic tensioner unit **12**.
    - Timing belt.
    NOTE: If belt has been in use for more than 45,000 miles: Fit new belt.
    - Front oil seals from exhaust camshafts **13**.
15. Except 2,0/2,5-190 models: Install locking tools to camshaft sprockets **14**. Tool No.18G 1747-2. Ensure tools located correctly on end of exhaust camshafts **15**.
16. 2,0 models: Install locking tools to camshaft sprockets **14**. Tool No.12-187. Ensure tools located correctly on end of exhaust camshafts **15**.

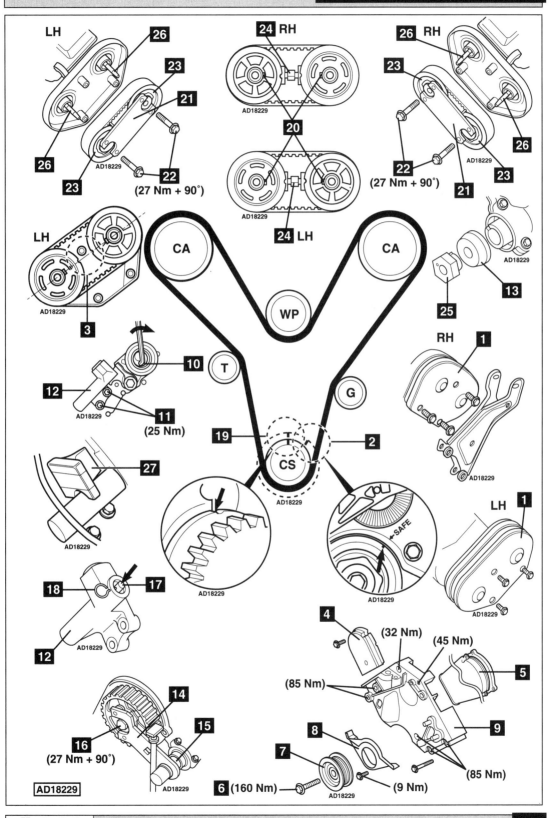

17. 2,5-190 models: Install locking tools to camshaft sprockets **14**. Tool No.12R208. Ensure tools located correctly on end of exhaust camshafts **15**.

18. Remove:
- ❏ Bolts from camshaft sprockets **16**. DO NOT reuse bolts.

WARNING: Damage to camshafts may occur if locking tools are not installed when slackening or tightening camshaft sprocket bolts.
- ❏ Locking tools **14**.
- ❏ Camshaft sprockets with hubs.

Installation

Front Timing Belt

1. Slowly compress pushrod into tensioner body until holes aligned **17**. Use a press. Retain pushrod with 1,5 mm diameter pin through hole in tensioner body **18**.

2. Thoroughly clean camshaft sprockets and hubs.

3. Install hubs to camshaft sprockets. Then install both assemblies to camshafts.

4. Fit new bolts **16**. Tighten bolts finger tight. Ensure camshaft sprockets turn freely without rocking.

5. Ensure timing marks aligned **19**.

6. Temporarily fit timing belt to camshaft sprockets.

7. Install locking tools to camshaft sprockets **14**. Tool No.18G 1747-2/12-187/12R208. Ensure tools located correctly on end of exhaust camshafts **15**.

8. Turn camshaft sprockets fully clockwise as viewed from front of engine.

9. Fit timing belt in anti-clockwise direction, starting at crankshaft sprocket. Ensure belt is taut between sprockets.

NOTE: Turn each camshaft sprocket anti-clockwise just enough to allow belt teeth to engage in sprocket.

Use suitable wedge to hold belt in position at crankshaft sprocket.

10. Turn tensioner pulley clockwise against belt **10**. Use Allen key.

11. Install automatic tensioner unit **12**. Tighten bolts to 25 Nm **11**.
Coat first three threads of bolts with 'Loctite 242' or similar.

12. Remove Allen key from tensioner pulley **10**.

13. Remove pin from tensioner body to release pushrod **18**.

14. Install blanking cap to automatic tensioner unit **27**.

15. Tighten bolts of camshaft sprockets **16**.
Tightening torque: 27 Nm + 90°.

16. Remove locking tools from camshaft sprockets **14**.

17. Remove any wedges used to hold belt in position.

18. Install new oil seals on exhaust camshafts **13**.

19. Install components in reverse order of removal.

20. Tighten crankshaft pulley bolt **6**.
Tightening torque: 160 Nm.

21. Refill engine with oil.

Removal

Rear Timing Belts

NOTE: The following instructions apply to both RH and LH rear timing belts. Removal and installation should only be carried out on ONE timing belt at a time.

1. Raise and support front of vehicle.

2. Remove:
- ❏ Covers for rear timing belts **1**.
- ❏ RH front wheel.
- ❏ RH splash guard.

3. Turn crankshaft clockwise until timing marks aligned **2**.
NOTE: White timing mark aligned with 'SAFE'.

4. Ensure timing marks on camshaft sprockets aligned **3**.

5. Ensure sprocket hubs are aligned as shown **20**.

6. Fit alignment plate to camshaft sprockets **21**. Tool No.12-175.
NOTE: Use hand force only to fit alignment plate.

7. Remove bolts **22**.

8. Remove camshaft sprockets, timing belt and alignment plate as an assembly **23**.

9. Remove alignment plate **21**.

10. Mark direction of rotation on belt with chalk if belt is to be reused.

11. Remove timing belt from sprockets.
NOTE: If belt has been in use for more than 45,000 miles: Fit new belt.

Installation

Rear Timing Belts

1. Thoroughly clean camshaft sprockets.

2. Place sprockets on a flat surface.

3. Ensure sprocket hubs are aligned as shown **20**.

4. Fit timing belt to sprockets.

5. Fit special tool between sprockets **24**. Tool No.18G 1747-1.

6. Turn centre screw to separate sprockets until alignment plate can be fitted to camshaft sprockets **21**. Tool No.12-175.
NOTE: Use hand force only to fit alignment plate.

7. Remove special tool **24**. Tool No.18G 1747-1.

8. Remove and discard front oil seal from exhaust camshaft of cylinder bank being worked on (i.e. RH or LH bank) **13**.

9. Install tool to camshaft **25**. Tool No.18G 1747-4.

10. Install alignment pins to camshafts **26**. Tool No.18G 1747-5.

11. Install camshaft sprockets, timing belt and alignment plate as an assembly **23**.

12. Use tool to align exhaust camshaft to sprocket **25**. Tool No.18G 1747-4.

13. Remove alignment pins **26**. Tool No.18G 1747-5.

14. Fit bolts to camshaft sprockets **22**.
Tightening torque: 27 Nm + 90°.

15. Remove alignment plate from sprockets **21**. Tool No.12-175.

16. Remove tool from camshaft **25**. Tool No.18G 1747-4.

17. Fit new oil seal **13**.

18. Install components in reverse order of removal.

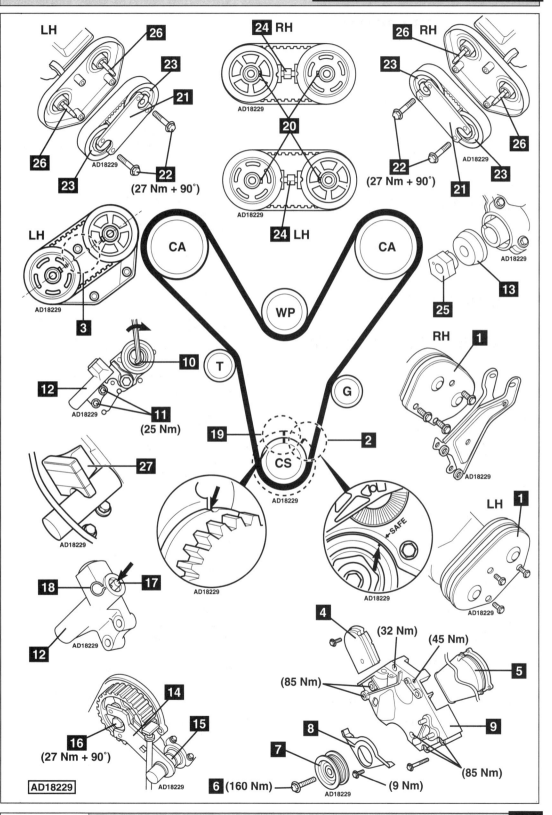

ROVER

Model:	**618 • 620 (SOHC)**
Year:	**1993-99**
Engine Code:	**F18A3, F20Z1, F20Z2**

Replacement Interval Guide

Rover recommend replacement every 60,000 miles or 5 years, whichever occurs first.
The previous use and service history of the vehicle must always be taken into account.
Refer to Timing Belt Replacement Intervals at the front of this manual.

Check For Engine Damage

CAUTION: This engine has been identified as an INTERFERENCE engine in which the possibility of valve-to-piston damage in the event of a timing belt failure is MOST LIKELY to occur.
A compression check of all cylinders should be performed before removing the cylinder head.

Repair Times – hrs

Remove & install	2,20

Special Tools

- ■ Balancer shaft locking pin – Rover No.18G 1671.

Special Precautions

- ■ Disconnect battery earth lead.
- ■ DO NOT turn crankshaft or camshaft when timing belt removed.
- ■ Remove spark plugs to ease turning engine.
- ■ Turn engine in normal direction of rotation (unless otherwise stated).
- ■ DO NOT turn engine via camshaft or other sprockets.
- ■ Observe all tightening torques.

Removal

> *NOTE: Normal direction of crankshaft rotation is anti-clockwise. Balancer shaft belt must be removed before removing timing belt.*

1. Support engine.
2. Remove:
   - ❑ Lower splash guard.
   - ❑ Auxiliary drive belt(s).
   - ❑ PAS pump. DO NOT disconnect hoses.
   - ❑ Cylinder head cover.
   - ❑ Timing belt upper cover **3**.
   - ❑ LH engine mounting.
   - ❑ Dipstick and tube.
   - ❑ Rubber seal for adjusting nut **6**. DO NOT slacken nut.
   - ❑ Crankshaft pulley **5**.
   - ❑ Timing belt lower cover **4**.

3. Temporarily fit crankshaft pulley. Turn crankshaft to TDC on No.1 cylinder. Ensure flywheel and camshaft sprocket timing marks aligned **1** & **2**. Ensure 'UP' mark on camshaft sprocket at top **12**.
4. Fit 6 mm bolt at **A** to lock tensioner arm.
5. Slacken tensioner nut **7**. Move tensioner away from belt and lightly tighten nut.
6. Remove balancer shaft belt.
7. Slacken 6 mm bolt **A**. Slacken tensioner nut **7**. Move tensioner away from belt and lightly tighten nut.
8. Remove:
   - ❑ Crankshaft pulley.
   - ❑ Timing belt.

Installation

1. Ensure flywheel and camshaft sprocket timing marks aligned **1** & **2**.
2. Fit timing belt in clockwise direction, starting at crankshaft sprocket. Ensure belt is taut between sprockets.
3. Ensure 6 mm bolt is slack **A**.
4. Slacken tensioner nut and tighten to 45 Nm **7**.
5. Temporarily fit crankshaft pulley **5**.
6. Turn crankshaft slowly anti-clockwise until three teeth of camshaft sprocket have passed timing marks **2**.
7. Slacken tensioner nut and tighten to 45 Nm **7**.
8. Turn crankshaft to TDC on No.1 cylinder. Ensure timing marks aligned **1** & **2**.
9. Tighten 6 mm bolt **A** to lock tensioner arm.
10. Remove blanking plug **8**. Insert balancer shaft locking pin **9**. Tool No.18G 1671. Turn rear balancer shaft sprocket until locking pin locates in hole in shaft.
11. Align front balancer shaft timing marks **10**.
12. Remove crankshaft pulley **5**.
13. Fit balancer shaft belt.
14. Slacken tensioner nut **7**.
15. Remove balancer shaft locking pin **9**. Fit blanking plug and tighten to 30 Nm.
16. Fit crankshaft pulley **5**. Turn crankshaft one turn anti-clockwise.
17. Tighten tensioner nut **7**.
    Tightening torque: 45 Nm.
18. Remove 6 mm bolt **A**.
19. Install components in reverse order of removal.
20. Tighten crankshaft pulley bolt **11**.
    Tightening torque: 220 Nm.

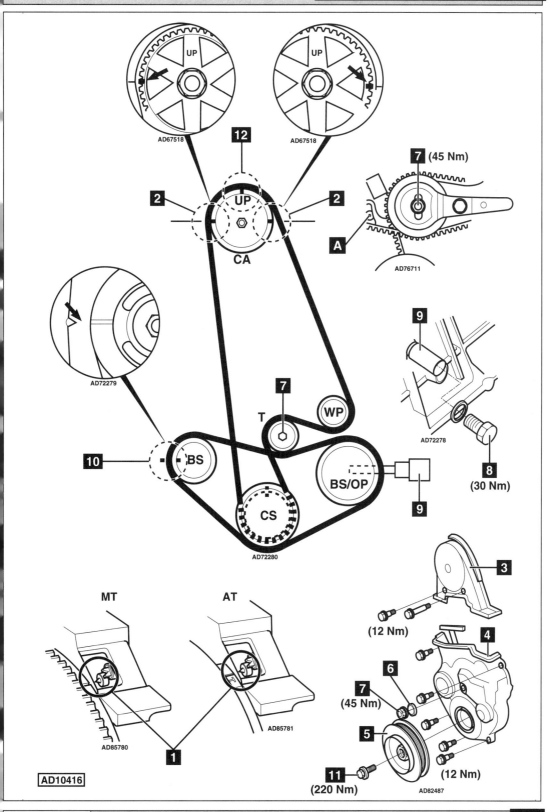

AD67518

12

AD67518

7 (45 Nm)

2

UP

2

CA

A

AD76711

9

AD72279

AD72278

8
(30 Nm)

7

T

WP

10

BS

9

CS

BS/OP

AD72280

3

(12 Nm)

4

MT

AT

6

7
(45 Nm)

5

AD85781

AD85780

1

11
(220 Nm)

(12 Nm)

AD82487

AD10416

Model:	**620 Ti**
Year:	**1994-99**
Engine Code:	**20T4**

Replacement Interval Guide

Rover recommend replacement every 60,000 miles or 5 years, whichever occurs first.

NOTE: Vehicles that have had the timing belt replaced at the previously recommended replacement interval of 96,000 miles, should have the belt replaced again at 120,000 miles.

The previous use and service history of the vehicle must always be taken into account.

Refer to Timing Belt Replacement Intervals at the front of this manual.

Check For Engine Damage

CAUTION: This engine has been identified as an INTERFERENCE engine in which the possibility of valve-to-piston damage in the event of a timing belt failure is MOST LIKELY to occur.

A compression check of all cylinders should be performed before removing the cylinder head.

Repair Times – hrs

Remove & install	1,30

Special Tools

- Camshaft sprocket locking tool – Rover No.18G 1524.
- Flywheel/drive plate timing pin – Rover No.18G 1523.
- Crankshaft pulley holding tool – Rover No.18G 1641.

Special Precautions

- Disconnect battery earth lead.
- DO NOT turn crankshaft or camshaft when timing belt removed.
- Remove spark plugs to ease turning engine.
- Turn engine in normal direction of rotation (unless otherwise stated).
- DO NOT turn engine via camshaft or other sprockets.
- Observe all tightening torques.

Removal

1. Support engine.
2. Remove:
   - ❑ Auxiliary drive belt.
   - ❑ RH engine mounting.
   - ❑ Timing belt upper cover **1**.
3. Raise engine.
4. Remove timing belt centre cover **2**.
5. Turn crankshaft until timing marks aligned **3**.
6. Lock camshaft sprockets **4**.
   Use tool No.18G 1524.

7. Insert flywheel/drive plate timing pin **5**.
   Tool No.18G 1523.
   NOTE: Crankshaft now set at 90° BTDC.
8. Slacken tensioner bolt **6**. Move tensioner away from belt and lightly tighten bolt.
9. Remove crankshaft pulley bolts **7**.
10. Hold crankshaft pulley.
    Use tool No.18G 1641 **8**.
11. Remove:
    - ❑ Crankshaft pulley centre bolt **9**.
    - ❑ Holding tool **8**.
    - ❑ Crankshaft pulley **10**.
    - ❑ Timing belt lower cover **11**.
    - ❑ Timing belt.
    WARNING: If belt has been in use for more than 48,000 miles: Fit new belt.

Installation

1. Remove tensioner spring. Check free length of tensioner spring is 57,5-58,5 mm **12**. Replace spring if necessary.
2. Fit timing belt. Ensure belt is taut between crankshaft sprocket and exhaust camshaft sprocket.
3. Install:
   - ❑ Timing belt lower cover **11**.
   - ❑ Crankshaft pulley **10**.
4. Hold crankshaft pulley.
   Use tool No.18G 1641 **8**.
5. Tighten crankshaft pulley centre bolt **9**.
   Tightening torque: 85 Nm.
6. Remove locking tool from camshaft sprockets **4**.
7. Slacken tensioner bolt **6**. Allow tensioner to operate.
8. Remove timing pin **5**.
9. Turn crankshaft two turns clockwise. Ensure timing marks aligned **3**.
10. Tighten tensioner bolt **6**.
    Tightening torque: 30 Nm.
11. Install components in reverse order of removal.
12. Ensure full weight of engine on engine mounting before final tightening of bolts.

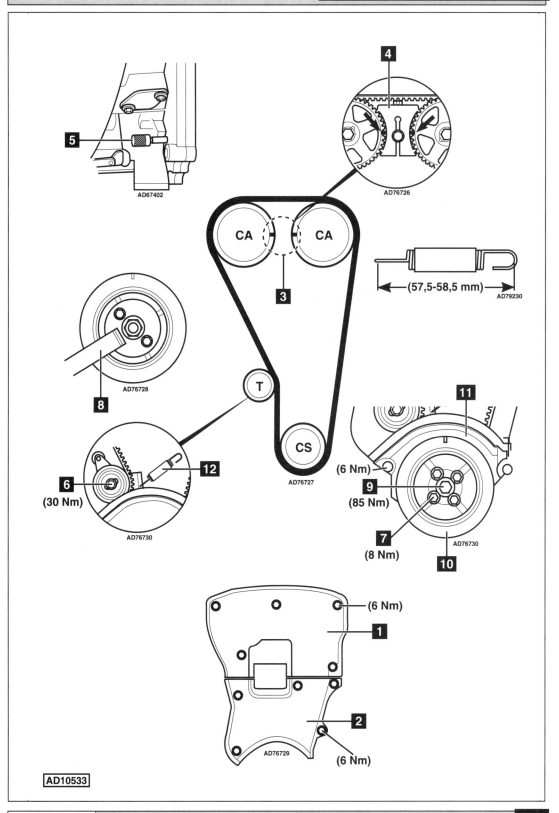

AD67402

AD76726

4

5

CA   CA

3

(57,5-58,5 mm)
AD79230

AD76728

8

T

CS
AD76727

6
(30 Nm)

12

AD76730

11

(6 Nm)

9
(85 Nm)

7
(8 Nm)

10

AD76730

(6 Nm)

1

2

AD76729

(6 Nm)

AD10533

ROVER

Model:	**623**
Year:	**1993-99**
Engine Code:	**H23A3**

Replacement Interval Guide

Rover recommend replacement every 60,000 miles or 5 years, whichever occurs first.
The previous use and service history of the vehicle must always be taken into account.
Refer to Timing Belt Replacement Intervals at the front of this manual.

Check For Engine Damage

CAUTION: This engine has been identified as an INTERFERENCE engine in which the possibility of valve-to-piston damage in the event of a timing belt failure is MOST LIKELY to occur.
A compression check of all cylinders should be performed before removing the cylinder head.

Repair Times – hrs

Remove & install   2,50

Special Tools

■ Balancer shaft locking pin – Rover No.18G 1671.

Special Precautions

■ Disconnect battery earth lead.
■ DO NOT turn crankshaft or camshaft when timing belt removed.
■ Remove spark plugs to ease turning engine.
■ Turn engine in normal direction of rotation (unless otherwise stated).
■ DO NOT turn engine via camshaft or other sprockets.
■ Observe all tightening torques.

Removal

NOTE: Normal direction of crankshaft rotation is anti-clockwise. Balancer shaft belt must be removed before removing timing belt.

1. Support engine.
2. Remove:
   ❑ Lower splash guard.
   ❑ Auxiliary drive belts.
   ❑ PAS pump. DO NOT disconnect hoses.
   ❑ Cylinder head cover.
   ❑ Timing belt upper cover **1**.
   ❑ LH engine mounting.
   ❑ Dipstick and tube.
   ❑ Rubber seal for adjusting nut **2**. DO NOT slacken nut.
   ❑ Crankshaft pulley **3**.
   ❑ Timing belt lower cover **4**.
3. Temporarily fit crankshaft pulley. Turn crankshaft anti-clockwise to TDC on No.1 cylinder.
4. Ensure flywheel timing marks and timing marks on camshaft sprockets aligned **5** & **6**. Ensure arrow mark on camshaft sprockets at top **7**.

5. Fit 6 mm bolt at **A** to lock tensioner arm. Lightly tighten bolt.
6. Slacken tensioner nut **8**. Move tensioner away from belt and lightly tighten nut.
7. Remove:
   ❑ Crankshaft pulley **3**.
   ❑ Balancer shaft belt.
8. Slacken 6 mm bolt **A**. Slacken tensioner nut **8**. Move tensioner away from belt and lightly tighten nut.
9. Remove timing belt.

Installation

1. Ensure timing marks aligned **5** & **6**.
2. Fit timing belt in clockwise direction, starting at crankshaft sprocket. Ensure belt is taut between sprockets.
3. Ensure 6 mm bolt is slack **A**.
4. Slacken tensioner nut **8**. Allow tensioner to operate.
5. Tighten tensioner nut **8**.
   Tightening torque: 45 Nm.
6. Temporarily fit crankshaft pulley.
7. Turn crankshaft slowly anti-clockwise until three teeth of camshaft sprockets have passed timing marks **6**.
8. Slacken tensioner nut and tighten to 45 Nm **8**.
9. Turn crankshaft anti-clockwise to TDC on No.1 cylinder. Ensure timing marks aligned **5** & **6**.
10. Tighten 6 mm bolt **A** to lock tensioner arm.
11. Remove blanking plug **9**. Insert balancer shaft locking pin **10**. Tool No.18G 1671. Turn rear balancer shaft sprocket until locking pin locates in hole in shaft.
12. Align front balancer shaft timing marks **11**.
13. Remove crankshaft pulley.
14. Fit balancer shaft belt.
15. Slacken tensioner nut **8**.
16. Remove balancer shaft locking pin **10**. Fit blanking plug and tighten to 30 Nm.
17. Fit crankshaft pulley. Turn crankshaft one turn anti-clockwise.
18. Tighten tensioner nut **8**.
   Tightening torque: 45 Nm.
19. Remove 6 mm bolt **A**.
20. Install components in reverse order of removal.
21. Lightly oil threads and contact face of crankshaft pulley bolt **12**. Tightening torque: 220 Nm.

Autodata

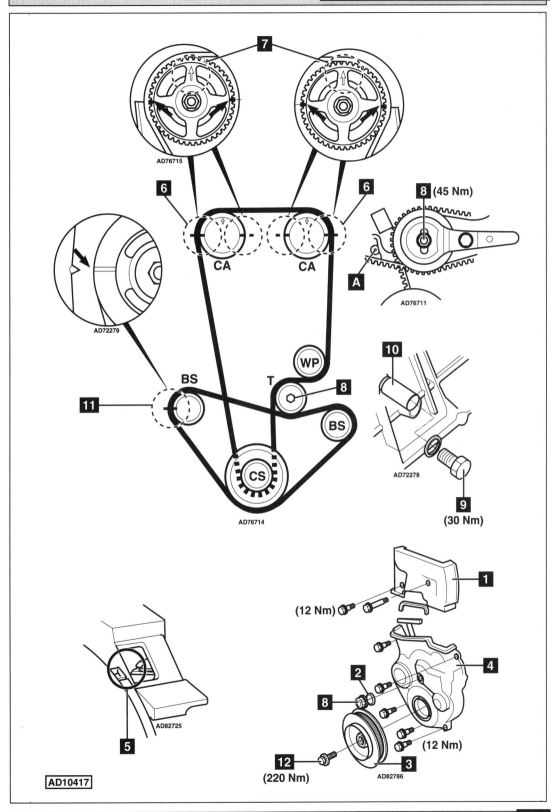

ROVER

Model:	825 • 827
Year:	1986-96
Engine Code:	V6 2,5/2,7

Replacement Interval Guide

Rover recommend replacement as follows:
→1990 - manufacturer has no recommended interval.
1990→ - replacement every 96,000 miles or 8 years.
The previous use and service history of the vehicle must always be taken into account.
Refer to Timing Belt Replacement Intervals at the front of this manual.

Check For Engine Damage

CAUTION: This engine has been identified as an INTERFERENCE engine in which the possibility of valve-to-piston damage in the event of a timing belt failure is MOST LIKELY to occur.
A compression check of all cylinders should be performed before removing the cylinder head.

Repair Times – hrs

Check & adjust	0,80
Remove & install	3,00
AC	+0,50

Special Tools

■ Flywheel locking tool – Rover No.18G 1513.

Special Precautions

■ Disconnect battery earth lead.
■ DO NOT turn crankshaft or camshaft when timing belt removed.
■ Remove spark plugs to ease turning engine.
■ Turn engine in normal direction of rotation (unless otherwise stated).
■ DO NOT turn engine via camshaft or other sprockets.
■ Observe all tightening torques.

Removal

1. Remove:
   ❏ RH inner wing panel.
   ❏ Auxiliary drive belts.
   ❏ Starter motor.
2. Lock flywheel. Use tool No.18G 1513.
3. Remove:
   ❏ Crankshaft pulley **1**.
   ❏ Timing belt covers **2** & **3**.
4. Turn crankshaft clockwise to TDC on No.1 cylinder. Ensure crankshaft sprocket timing marks aligned **4**. Ensure timing marks on camshaft sprockets aligned with marks on timing belt rear cover **5**.

5. Slacken tensioner pulley bolt **6**. Move tensioner pulley away from belt and lightly tighten bolt.
6. Remove timing belt.

Installation

1. Ensure timing marks aligned **4** & **5**.
2. Fit timing belt in following order:
   ❏ Crankshaft sprocket.
   ❏ Camshaft sprocket CA2.
   ❏ Water pump pulley.
   ❏ Tensioner pulley.
   ❏ Camshaft sprocket CA1.
   *NOTE: To ease installation: Turn camshaft rear sprocket 1/2 tooth clockwise from timing mark **5**.*
3. Slacken and tighten tensioner pulley bolt **6**.
4. Turn crankshaft 5 or 6 turns clockwise to TDC on No.1 cylinder.
5. Turn crankshaft clockwise until camshaft sprocket has moved 9 teeth.
6. Slacken tensioner pulley bolt and tighten to 43 Nm **6**.
7. Turn crankshaft clockwise to TDC on No.1 cylinder. Ensure timing marks aligned **4** & **5**.
8. Install components in reverse order of removal.
   NOTE: Ensure concave side of crankshaft pulley bolt washer faces outwards.
9. Tighten crankshaft pulley bolt **7**.
   Tightening torque: 115 Nm.

Autodata

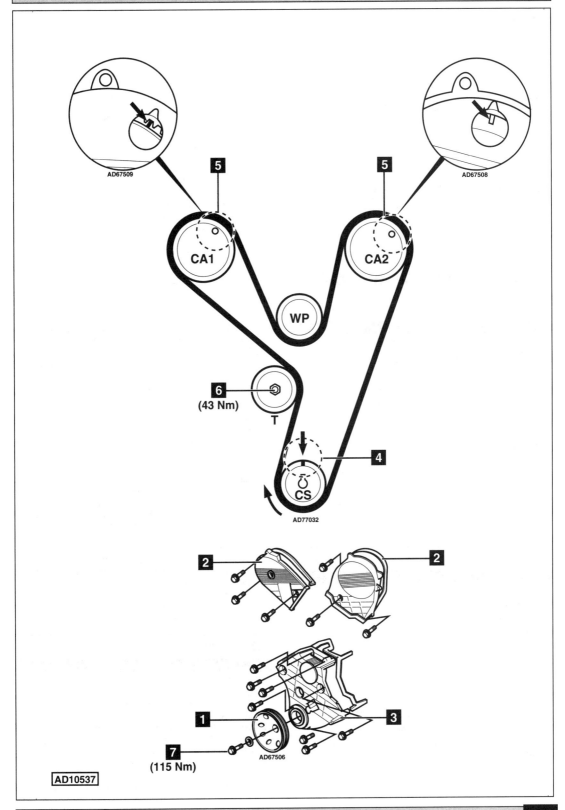

AD67509

5 **5**

AD67508

CA1 CA2

WP

6
(43 Nm)
T

4

CS

AD77032

2 **2**

1 **3**

7
(115 Nm)

AD67506

AD10537

ROVER

Model:	**825 V6**
Year:	**1996-99**
Engine Code:	**KV6**

Replacement Interval Guide

Rover recommend replacement every 96,000 miles or 8 years.

The previous use and service history of the vehicle must always be taken into account.

Refer to Timing Belt Replacement Intervals at the front of this manual.

Check For Engine Damage

CAUTION: This engine has been identified as an INTERFERENCE engine in which the possibility of valve-to-piston damage in the event of a timing belt failure is MOST LIKELY to occur.

A compression check of all cylinders should be performed before removing the cylinder head.

Repair Times – hrs

Remove & install	4,00
AC	+0,70

Special Tools

- Alignment plate for camshaft sprockets – Rover No.18G 1747.
- Sprocket tool – Rover No.18G 1747-1.
- Locking tools for camshafts – Rover No.18G 1747-2.
- Tool for camshafts – Rover No.18G 1747-4.
- Alignment pins – Rover No.18G 1747-5.
- Timing pin – Rover No.18G 1746.

Special Precautions

- Disconnect battery earth lead.
- DO NOT turn crankshaft or camshaft when timing belt removed.
- Remove spark plugs to ease turning engine.
- Turn engine in normal direction of rotation (unless otherwise stated).
- DO NOT turn engine via camshaft or other sprockets.
- Observe all tightening torques.

Removal

Front Timing Belt

1. Raise and support front of vehicle.
2. Support engine.
3. Remove:
   - RH front wheel.
   - RH splash guard.
   - Oil filter.
   - Spark plug cover.
   - Inlet manifold upper chamber.
   - Engine steady bar.
   - Timing belt rear cover of LH cylinder bank **1**.
4. Turn crankshaft clockwise to setting position (60° BTDC). Ensure hubs of LH rear camshaft sprockets aligned as shown **2**.
5. Insert timing pin into hole in lower crankcase (located under LH cylinder bank) **3**. Tool No.18G 1746.

6. Remove:
   - Timing belt LH cover **4**.
   - Auxiliary drive belt.
   - Alternator.
   - PAS pump. DO NOT disconnect hoses.
   - Engine mounting bracket spacer **5**.
   - Timing belt RH cover **6**.
   - Alternator/PAS pump mounting bracket.
   - Engine lifting bracket.
   - Crankshaft pulley bolt **7**.
   - Crankshaft pulley **8**.
   - Timing belt lower cover **9**.
   - Access caps **10**.
   - Auxiliary drive belt tensioner.
   - AC compressor (if fitted). DO NOT disconnect hoses.
   - AC compressor heat shield (if fitted).
   - AC compressor bracket (if fitted).
7. Raise engine slightly. Remove engine mounting bracket with lower bolt **11** & **12**.
   NOTE: AT – Remove transmission dipstick to prevent fouling on ABS modulator.
8. Turn tensioner pulley clockwise to release tension from automatic tensioner unit. Use Allen key **13**.
9. Remove:
   - Automatic tensioner unit bolts **15**.
   - Automatic tensioner unit **14**.
10. Mark direction of rotation on belt with chalk if belt is to be reused.
11. Remove timing belt.
    NOTE: If belt has been in use for more than 48,000 miles: Fit new belt.
12. Remove:
    - Dipstick tube securing bolt.
    - Front oil seals from exhaust camshafts **28**.
13. Install locking tools to camshaft sprockets **16**. Tool No.18G 1747-2. Ensure tools located correctly on end of exhaust camshafts **17**.
14. Remove bolts from camshaft sprockets **18**. DO NOT reuse bolts.
    WARNING: Damage to camshafts may occur if locking tools are not installed when slackening or tightening camshaft sprocket bolts.
15. Remove:
    - Locking tools **16**.
    - Camshaft sprockets with hubs.

Installation

Front Timing Belt

1. Slowly compress pushrod into tensioner body until holes aligned **19**. Use vice. Retain pushrod with 1,5 mm diameter pin through hole in tensioner body **20**.
2. Thoroughly clean camshaft sprockets and hubs.
3. Install hubs to camshaft sprockets. Then install both assemblies to camshafts.
4. Fit new bolts **18**. Tighten bolts finger tight. Ensure camshaft sprockets turn freely without rocking.
5. Temporarily fit timing belt to camshaft sprockets.

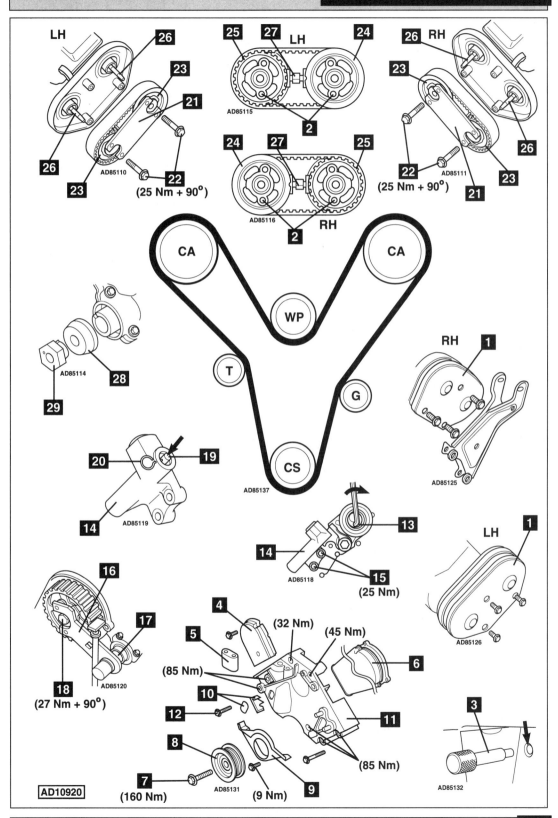

6. Install locking tools to camshaft sprockets **16**. Tool No.18G 1747-2. Ensure tools located correctly on end of exhaust camshafts **17**.
7. Turn camshaft sprockets fully clockwise as viewed from front of engine.
8. Fit timing belt in anti-clockwise direction, starting at crankshaft sprocket. Ensure belt is taut between sprockets.
   NOTE: Turn each camshaft sprocket anti-clockwise just enough to allow belt teeth to engage in sprocket.
   Use suitable wedge to hold belt in position at crankshaft sprocket.
9. Turn tensioner pulley clockwise against belt **13**. Use Allen key.
10. Install automatic tensioner unit **14**. Coat first three threads of bolts with 'Loctite 242' or similar **15**. Tighten bolts to 25 Nm.
11. Remove Allen key from tensioner pulley **13**. Remove pin from tensioner body to release pushrod **20**.
12. Tighten bolts of camshaft sprockets **18**. Tightening torque: 27 Nm + 90°.
13. Remove locking tools from camshaft sprockets **16**.
14. Remove any wedges used to hold belt in position.
15. Remove timing pin **3**. Tool No.18G 1746.
16. Install new oil seals on exhaust camshafts.
17. Install components in reverse order of removal.
    *NOTE: Ensure lower bolt **12** is located in engine mounting bracket **11** prior to installation.*
18. Tighten crankshaft pulley bolt **7**. Tightening torque: 160 Nm.

Removal

Rear Timing Belts

NOTE: The following instructions apply to both RH and LH rear timing belts. Removal and installation should only be carried out on ONE timing belt at a time.

1. Raise and support front of vehicle.
2. Remove:
   - ❏ Air filter assembly (LH belt).
   - ❏ Inlet manifold upper chamber (RH belt).
   - ❏ Rear engine steady bracket (RH belt).
   NOTE: AT models have a two part engine steady bracket.
   - ❏ Timing belt rear cover **1**.
   - ❏ RH front wheel.
   - ❏ RH splash guard.
3. Turn crankshaft clockwise until alignment plate can be fitted to camshaft sprockets **21**. Tool No.18G 1747.
4. Use hand force only to fit alignment plate.
5. Remove bolts **22**.
6. Remove camshaft sprockets, timing belt and alignment plate as an assembly **23**.
7. Remove alignment plate **21**.
8. Mark direction of rotation on belt with chalk if belt is to be reused.
9. Remove timing belt from sprockets.
   NOTE: If belt has been in use for more than 48,000 miles: Fit new belt.

Installation

Rear Timing Belts

1. Thoroughly clean camshaft sprockets and hubs.
2. Install hubs to camshaft sprockets.
3. Place sprocket and hub assemblies on a flat surface.
4. Ensure belt guide flange on inlet camshaft sprocket is facing upwards **24**.
5. Ensure belt guide flange on exhaust camshaft sprocket is facing downwards **25**.
6. Ensure sprocket hubs are aligned as shown **2**.
7. Fit timing belt to sprockets.
8. Fit special tool between sprockets **27**. Tool No.18G 1747-1. Turn centre screw to separate sprockets until alignment plate can be fitted to camshaft sprockets **21**. Tool No.18G 1747.
9. Use hand force only to fit alignment plate.
10. Remove and discard front oil seal from exhaust camshaft of cylinder bank being worked on (i.e. RH or LH bank) **28**. Install tool to camshaft **29**. Tool No.18G 1747-4.
11. Install alignment pins to camshafts **26**. Tool No.18G 1747-5.
12. Install camshaft sprockets, timing belt and alignment plate as an assembly **23**. Use tool to align exhaust camshaft to sprocket **29**. Tool No.18G 1747-4.
13. Remove alignment pins **26**. Tool No.18G 1747-5.
14. Fit bolts to camshaft sprockets **22**. Tightening torque: 25 Nm + 90°.
15. Remove alignment plate from sprockets **21**. Tool No.18G 1747.
16. Remove tool from sprockets **27**. Tool No.18G 1747-1.
17. Remove tool from camshaft **29**. Tool No.18G 1747-4. Fit new oil seal **28**.
18. Install components in reverse order of removal.

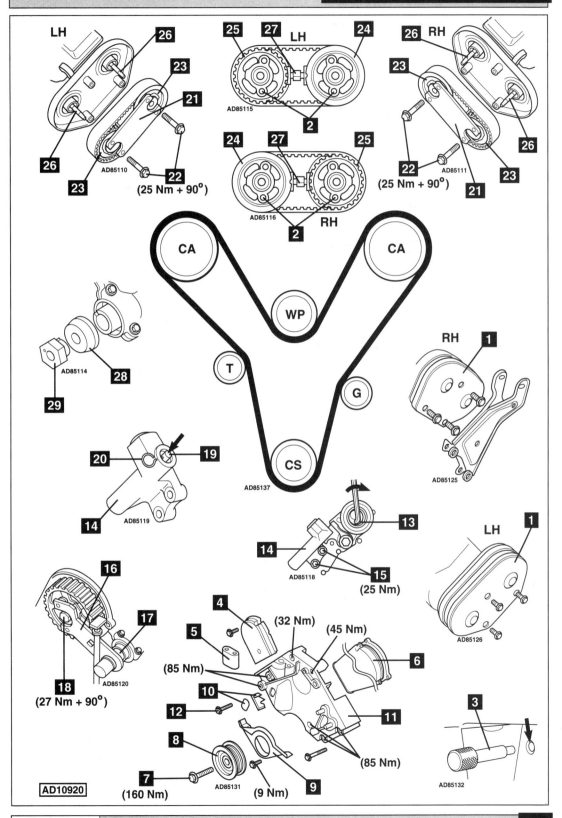

ROVER

Model:	MGF 1,6 • MGF 1,8 • MGF 1,8 Trophy
Year:	1995-02
Engine Code:	K16-1,6, K16-1,8 (G)

Replacement Interval Guide

Rover recommend:
→1999 (VIN.→RD522572): Replacement every 60,000 miles or 5 years, whichever occurs first.
1999→ (VIN.RD522573→): Replacement every 60,000 miles or 4 years, whichever occurs first.
The previous use and service history of the vehicle must always be taken into account.
Refer to Timing Belt Replacement Intervals at the front of this manual.

Check For Engine Damage

CAUTION: This engine has been identified as an INTERFERENCE engine in which the possibility of valve-to-piston damage in the event of a timing belt failure is MOST LIKELY to occur.
A compression check of all cylinders should be performed before removing the cylinder head.

Repair Times – hrs

Remove & install:

MT	3,00
AC	-0,40
AT	2,70
AC	+0,20

Special Tools

- Camshaft sprocket locking tool – Rover No.18G.1570.
- Flywheel locking tool – →2000MY: Rover No.18G.1742. 2000MY→: Rover No.18G.1571.

Special Precautions

- Disconnect battery earth lead.
- DO NOT turn crankshaft or camshaft when timing belt removed.
- Remove spark plugs to ease turning engine.
- Turn engine in normal direction of rotation (unless otherwise stated).
- DO NOT turn engine via camshaft or other sprockets.
- Observe all tightening torques.

Removal

1. Lower windows and release hood catches. DO NOT lower hood.
2. Release and pull forward rear of hood well carpet.
3. Release 5 rear hood clips and raise rear edge of hood.
4. Remove:
   - ❏ Hood well trim.
   - ❏ Sound insulating pad.
   - ❏ Engine cover (11 bolts).
   - ❏ Timing belt upper cover bolts **1**.
   - ❏ Timing belt upper cover **2**.
5. Raise and support rear of vehicle.
6. Remove:
   - ❏ RH rear wheel.
   - ❏ Inner wing panel.
7. Turn crankshaft clockwise until timing marks on camshaft sprockets aligned **3**.
8. Lock camshaft sprockets. Use tool No.18G.1570 **4**.
9. Remove:
   - ❏ Coolant hose heat shield from exhaust manifold (if fitted).
   - ❏ Flywheel cover.
10. Remove starter motor. Fit locking tool in starter aperture to lock flywheel. Tool No.18G.1742/18G.1571. Use two starter bolts.
11. Remove:
    - ❏ Auxiliary drive belt.
    - ❏ Crankshaft pulley bolt **5**.
    - ❏ Crankshaft pulley **6**.
    - ❏ AC compressor mounting bolts (if fitted). Move to one side.
12. Unclip Hydragas pipe and move aside.
13. Support engine.
14. Remove:
    - ❏ Engine mounting bolts **7** & **8**.
    - ❏ Engine mounting nut **23**.
    - ❏ Engine steady bar nut and bolt **24**.
    - ❏ Engine mounting **9**.
15. Remove timing belt lower cover **10**.

Manual tensioner

16. If belt is to be reused:
    - ❏ Mark direction of rotation on belt with chalk.
    - ❏ Mark position of tensioner backplate to cylinder head.
17. Slacken tensioner bolts 1/2 turn **11** & **12**.
18. Move tensioner away from belt. Lightly tighten bolt **12**.

Automatic tensioner

19. If belt is to be reused:
    - ❏ Mark direction of rotation on belt with chalk.
20. Remove and discard tensioner bolt **13**.
    NOTE: New bolt must be fitted.
21. Unhook index spring from tensioner **14**.
22. Remove tensioner pulley **15**.

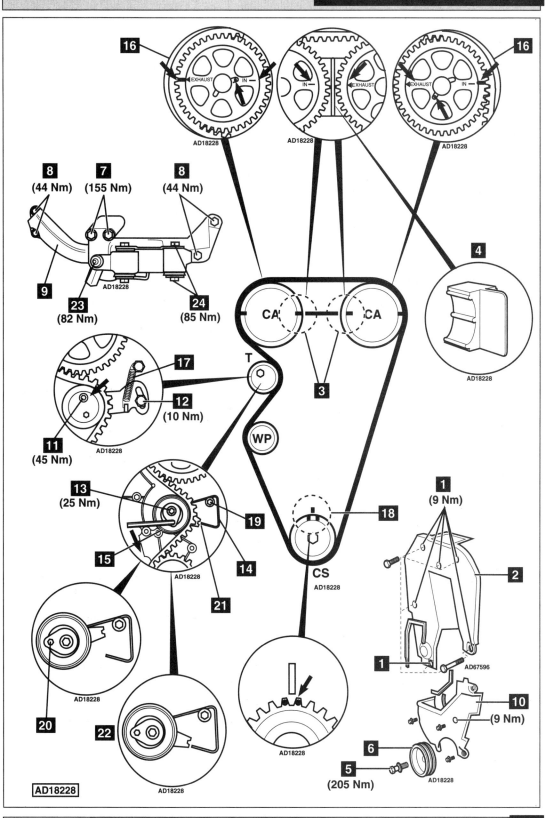

All models

23. Lower engine slightly.
24. Remove timing belt.
    WARNING: If belt has been in use for more than 48,000 miles: Fit new belt.

Installation

NOTE: If camshaft sprockets have been removed: Ensure dowel pin aligned with 'IN' slot of inlet camshaft sprocket and 'EX' slot of exhaust camshaft sprocket 16.

Manual tensioner

1. If reusing old belt: Disconnect tensioner spring 17.
2. If fitting new belt:
   - ❏ Slacken tensioner backplate bolt 1/2 turn 12.
   - ❏ Ensure tensioner spring and sleeve are located correctly 17.
   - ❏ Ensure tensioner slides freely. Turn tensioner fully clockwise.
   - ❏ Lightly tighten tensioner backplate bolt 12.

All models

3. Ensure crankshaft sprocket timing marks aligned with flange on oil pump 18.
   NOTE: This position is 90° BTDC.
4. With locking tool in position 4: Ensure timing marks on camshaft sprockets aligned 3.
5. Fit timing belt in anti-clockwise direction, starting at crankshaft sprocket. Ensure belt is taut between sprockets.
   NOTE: If reusing old belt: Ensure directional arrows point in direction of rotation.

Automatic tensioner

6. Fit tensioner pulley 15.
7. Ensure index spring hooked over retaining bolt 14 & 19.
8. Fit new tensioner bolt 13. Lightly tighten bolt 13.
9. Position tensioner at 9 o'clock position 20.

All models

10. Fit timing belt lower cover 10.
11. Fit crankshaft pulley 6. Ensure indent on crankshaft pulley locates over lug on crankshaft sprocket.
12. Tighten crankshaft pulley bolt 5.
    Tightening torque: 205 Nm.
13. Remove flywheel locking tool.
    Tool No.18G.1742/18G.1571
14. Raise engine.
15. Reposition mounting on engine and subframe.
16. Tighten engine mounting bolts 8.
    Tightening torque: 44 Nm.
17. Lower engine.

18. Fit engine mounting bolts 7.
    Tightening torque: 155 Nm.
19. Fit engine mounting nut 23.
    Tightening torque: 82 Nm.
20. Fit engine steady bar nut and bolt 24.
    Tightening torque: 85 Nm.
21. Remove locking tool from camshaft sprockets 4. Tool No.18G.1570.

Manual tensioner

22. Slacken tensioner backplate bolt 12.
23. If reusing old belt:
    - ❏ Align markings on tensioner backplate and cylinder head.
    - ❏ Tighten tensioner backplate bolt 12.
      Tightening torque: 10 Nm.
    - ❏ Tighten tensioner pulley bolt 11.
      Tightening torque: 45 Nm.
    - ❏ Reconnect tensioner spring 17.
24. If new belt fitted:
    - ❏ Push tensioner against belt to pre-tension belt.
    - ❏ Tighten tensioner backplate bolt 12.
      Tightening torque: 10 Nm.

Automatic tensioner

25. Turn tensioner pulley anti-clockwise until pointer aligned with index spring 21. Use 6 mm Allen key.
    NOTE: Repeat tensioning procedure if pointer passes index spring.
26. If reusing old belt: Align lower part of pointer to index spring 22.
27. Tighten tensioner pulley bolt 13.
    Tightening torque: 25 Nm.

All models:

28. Turn crankshaft two turns clockwise until timing marks aligned 3 & 18.

Manual tensioner

29. Slacken tensioner backplate bolt 1/2 turn 12.
30. Ensure only spring tension applied to belt.
31. Tighten tensioner backplate bolt 12.
    Tightening torque: 10 Nm.
32. Tighten tensioner pulley bolt 11.
    Tightening torque: 45 Nm.

Automatic tensioner

33. Ensure pointer still aligned 21 or 22. Adjust if necessary.

All models

34. Install components in reverse order of removal.

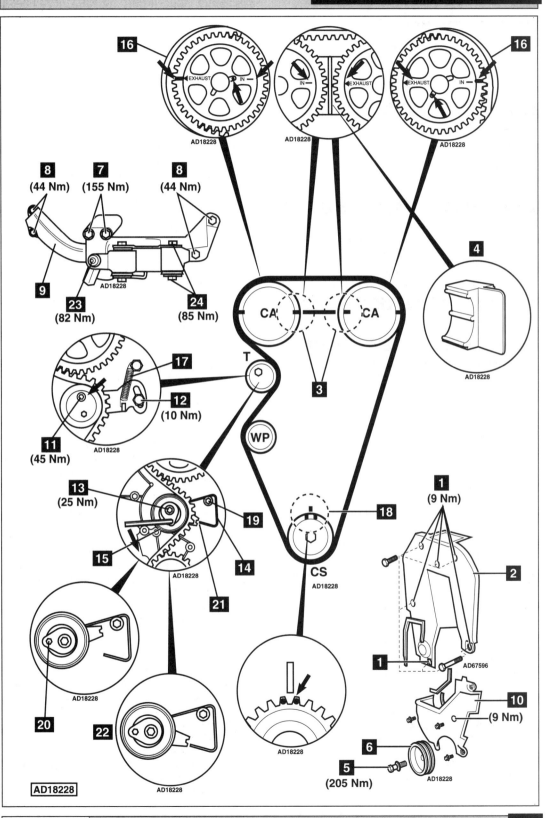

ROVER

Model:	**115 D**
Year:	**1995-98**
Engine Code:	**VJ7 (TUD5)**

Replacement Interval Guide

Rover recommend replacement every
72,000 miles or 6 years.
*The previous use and service history of the vehicle
must always be taken into account.
Refer to Timing Belt Replacement Intervals at the front
of this manual.*

Check For Engine Damage

*CAUTION: This engine has been identified as an
INTERFERENCE engine in which the possibility of
valve-to-piston damage in the event of a timing belt
failure is MOST LIKELY to occur.
A compression check of all cylinders should be
performed before removing the cylinder head.*

Repair Times – hrs

Remove & install	1,90

Special Tools

- Flywheel locking tool – Rover No.4507-T.L.
- Flywheel timing pin – Rover No.4507-T.A.
- Tension gauge – Rover No.KM 4088 AR.
- Tensioning tool – Rover No.4507-T.J.

Special Precautions

- Disconnect battery earth lead.
- DO NOT turn crankshaft or camshaft when timing belt removed.
- Remove glow plugs to ease turning engine.
- Turn engine in normal direction of rotation (unless otherwise stated).
- DO NOT turn engine via camshaft or other sprockets.
- Observe all tightening torques.
- Check diesel injection pump timing after belt replacement.

Removal

1. Raise and support front of vehicle.
2. Remove:
   - ❑ Auxiliary drive belt.
   - ❑ Flywheel blanking plug **1**.
3. Turn crankshaft clockwise to setting position.
   Insert locking tool in flywheel **2**.
   Tool No.4507-T.L.
4. Remove:
   - ❑ Crankshaft pulley bolts **3**.
   - ❑ Crankshaft pulley **4**.
   - ❑ Timing belt upper cover **5**.
   - ❑ Screen washer reservoir. DO NOT disconnect pipes.
   - ❑ Coolant expansion tank. DO NOT disconnect hoses.

5. Support engine.
6. Remove:
   - ❑ RH engine mounting bracket.
   - ❑ Timing belt lower cover **6**.
   - ❑ Flywheel locking tool **2**.
7. Turn crankshaft clockwise until camshaft sprocket timing hole aligned **7**.
8. Insert timing pin in flywheel **8**.
   Tool No.4507-T.A.
9. Insert one M8 bolt through camshaft sprocket timing hole into cylinder head **9**. Tighten bolt finger tight.
10. Insert 6 mm drill shank in injection pump sprocket and hub **10**.
11. Slacken tensioner nut **11**. Move tensioner pulley away from belt and lightly tighten nut.
12. Mark direction of rotation on belt with chalk if belt is to be reused.
13. Remove timing belt.
    WARNING: If belt has been in use for more than 36,000 miles: Fit new belt.

Installation

*NOTE: Sprockets are made from sintered steel. If
contaminated with oil, soak and wash with
suitable solvent prior to installation of timing belt.*

1. Ensure timing pin and locking bolt located correctly **8** & **9**.
2. Ensure 6 mm drill shank located correctly in injection pump sprocket and hub **10**.
3. Slacken bolts on injection pump sprocket and camshaft sprocket **12**. Tighten bolts finger tight. Ensure sprockets can still turn.
4. Turn sprockets fully clockwise in slotted holes.
5. Fit timing belt in anti-clockwise direction, starting at crankshaft sprocket. Ensure belt is taut between sprockets.
   NOTE: Turn injection pump and camshaft sprockets slightly anti-clockwise to engage belt teeth.
6. Attach tension gauge to belt at ▽.
   Tool No.KM 4088 AR.
7. Slacken tensioner nut **11**. Insert tensioning tool in square hole of tensioner pulley **13**.
   Tool No.4507-T.J. Tension belt to 15 units.
8. Tighten tensioner nut **11**.
   Tightening torque: 20 Nm.
9. Remove tension gauge.
10. Tighten bolts on injection pump sprocket and camshaft sprocket **12**.
    Tightening torque: 23 Nm.

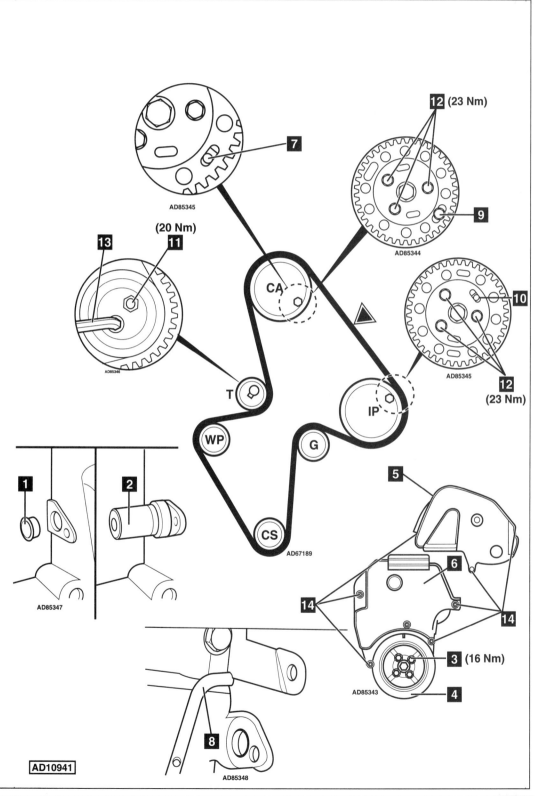

AD85345

7

12 (23 Nm)

9

AD85344

(20 Nm)

13 **11**

AD85346

CA

10

AD85345

12
(23 Nm)

T

IP

WP

G

CS

AD67189

1 **2**

AD85347

5

6

14 **14**

3 (16 Nm)

AD85343 **4**

8

AD85348

AD10941

←

11. Remove:
    - ❏ Timing pin **8**.
    - ❏ Locking bolt **9**.
    - ❏ 6 mm drill bit **10**.
12. Turn crankshaft 10 turns clockwise. Insert timing pin in flywheel **8**.
13. Insert one M8 bolt through camshaft sprocket timing hole into cylinder head **9**.
14. Insert 6 mm drill shank in injection pump sprocket and hub **10**.
15. Slacken bolts on injection pump sprocket and camshaft sprocket **12**.
16. Attach tension gauge to belt at ▽.
17. Slacken tensioner nut **11**. Tension belt.
    Use tool No.4507-T.J **13**. New belt: 10 units.
    Used belt: 8 units.
18. Tighten tensioner nut **11**.
    Tightening torque: 20 Nm.
19. Remove tension gauge.
20. Tighten bolts on injection pump sprocket and camshaft sprocket **12**.
    Tightening torque: 23 Nm.
21. Remove:
    - ❏ Timing pin **8**.
    - ❏ Locking bolt **9**.
    - ❏ 6 mm drill bit **10**.
22. Install components in reverse order of removal.
    WARNING: When installing timing belt covers, ensure metal sleeves do not fall into timing belt area **14**.
23. Insert locking tool in flywheel **2**.
    Tool No.4507-T.L.
24. Tighten crankshaft pulley bolts **3**.
    Tightening torque: 16 Nm.

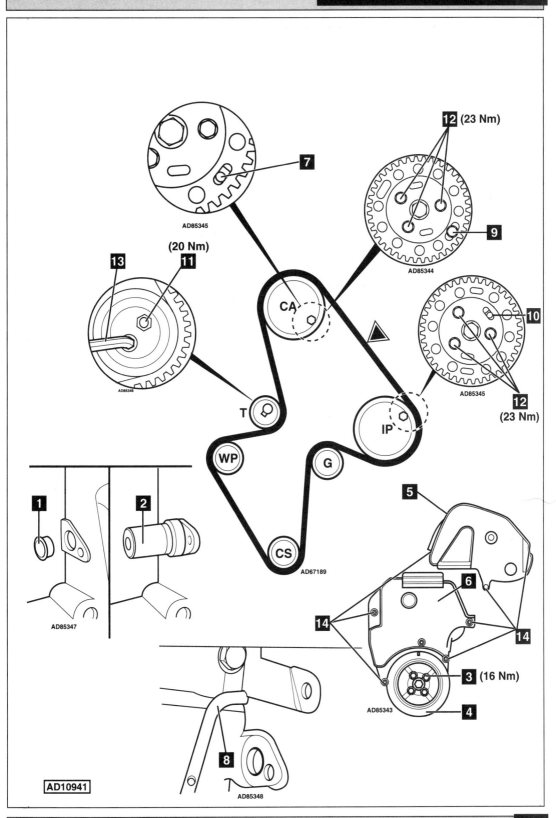

AD85345

7

12 (23 Nm)

AD85344

9

(20 Nm)

13 **11**

AD85346

CA

10

AD85345

12
(23 Nm)

T

IP

WP

G

1 **2**

AD85347

CS

AD67189

5

6

14

14

3 (16 Nm)

AD85343

4

8

AD85348

AD10941

ROVER

Model:	Maestro 2,0D/Van • Maestro 2,0D Turbo • Montego 2,0D Turbo
Year:	1986-95
Engine Code:	MDi

Replacement Interval Guide

Rover recommend replacement every 72,000 miles or 6 years.
The previous use and service history of the vehicle must always be taken into account.
Refer to Timing Belt Replacement Intervals at the front of this manual.

Check For Engine Damage

CAUTION: This engine has been identified as an INTERFERENCE engine in which the possibility of valve-to-piston damage in the event of a timing belt failure is MOST LIKELY to occur.
A compression check of all cylinders should be performed before removing the cylinder head.

Repair Times – hrs

Check & adjust:	
Maestro (→VIN 545365)	2,00
Maestro (VIN 545366→)	2,80
Maestro/Montego Turbo	2,80
Remove & install:	
Maestro (→VIN 545365)	2,45
Maestro (VIN 545366→)	4,00
Maestro/Montego Turbo	4,05

Special Tools

- Crankshaft/camshaft timing pins – Rover No.18G 1523.
- Injection pump timing pins – Rover No.18G 1549.
- Tension gauge – Rover (Kent-Moore) No.KM 4088AR.

Special Precautions

- Disconnect battery earth lead.
- DO NOT turn crankshaft or camshaft when timing belt removed.
- Remove glow plugs to ease turning engine.
- Turn engine in normal direction of rotation (unless otherwise stated).
- DO NOT turn engine via camshaft or other sprockets.
- Observe all tightening torques.
- Check diesel injection pump timing after belt replacement.

Removal

1. Raise and support front of vehicle.
2. Remove:
   - ❏ RH front wheel.
   - ❏ RH wheel arch mud shield.
   - ❏ Turbo & models 1989→: Support engine on trolley jack. Remove coolant expansion tank. Remove RH engine mounting.
   - ❏ Auxiliary drive belt.
   - ❏ Water pump pulley.
   - ❏ Timing belt cover access panel **1**.
   - ❏ Coolant hoses from water pump.
   - ❏ 1989 Maestro & Montego Turbo: Crankshaft pulley.
   - ❏ Timing belt cover **2**.
   - ❏ Cylinder head cover blanking plug.

3. Turn crankshaft until timing pin can be inserted in camshaft **3**. Tool No.18G 1523.
4. Insert timing pin in flywheel timing hole **4** or **5**. Tool No.18G 1523.
5. Insert timing pins in injection pump sprocket **6**. Tool No.18G 1549.
6. Ensure timing marks aligned **7**. Non-turbo: A. Turbo: B.
   NOTE: Early models may have incorrectly marked injection pump sprockets. Refer to Rover dealer.
7. Remove tensioner pulley **8**.
   *NOTE: 1989 Maestro & Montego Turbo: Remove guide pulley **9**.*
8. Remove timing belt.

Installation

*NOTE: Some injection pump sprockets have dual markings and keyways. If removed: Ensure part number (if visible) faces front. Non-turbo: Use keyway adjacent to 'A' mark on sprocket. Align 'A' mark with pointer on engine **7**. Turbo: Use keyway adjacent to 'B' mark on sprocket. Align 'B' mark with pointer on engine **7**.*

1. Ensure timing pins located correctly **3**, **4** or **5**, & **6**.
2. Fit timing belt.
3. Install:
   - ❏ Guide pulley **9**.
   - ❏ Tensioner pulley **8**.
4. Slacken bolts **10** or centre bolt **11** of camshaft sprocket.
5. Slacken tensioner bolt **12**. Turn tensioner pulley anti-clockwise using Allen key in tensioner pulley hole **13**.
6. Attach tension gauge to belt at ▽. Tool No.KM 4088AR.
7. Tighten tensioner bolt when tension gauge indicates 6,5-7,5 units (new belt) or 6 units (used belt).
8. Tighten bolts **10** or centre bolt **11** of camshaft sprocket.
9. Remove timing pins **3**, **4** or **5**, & **6**.
10. Turn crankshaft two turns in normal direction of rotation.
11. Recheck belt tension.
12. Check injection pump timing.
13. Install components in reverse order of removal.

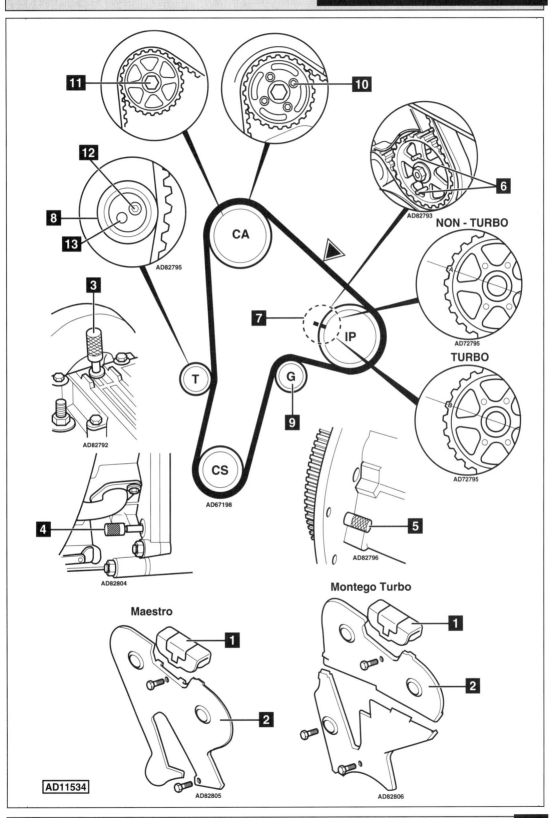

NON - TURBO

AD72795

TURBO

AD72795

Montego Turbo

Maestro

AD82805

AD82806

AD11534

ROVER

Model:	218 SD • 218 D Turbo • 418/Tourer D Turbo
Year:	1991-98
Engine Code:	A8A (XUD7TE), D9B (XUD9A)

Replacement Interval Guide

Rover recommend replacement every
72,000 miles or 6 years.
*The previous use and service history of the vehicle
must always be taken into account.
Refer to Timing Belt Replacement Intervals at the front
of this manual.*

Check For Engine Damage

*CAUTION: This engine has been identified as an
INTERFERENCE engine in which the possibility of
valve-to-piston damage in the event of a timing belt
failure is MOST LIKELY to occur.
A compression check of all cylinders should be
performed before removing the cylinder head.*

Repair Times – hrs

Remove & install   3,80

Special Tools

- Flywheel timing pin – Rover No.18G 1632.
- Flywheel locking tool – Rover No.18G 1547.
- 1 bolt – M8 x 1,25 x 40 mm.
- 2 bolts – M8 x 1,25 x 35 mm.

Special Precautions

- Disconnect battery earth lead.
- DO NOT turn crankshaft or camshaft when timing belt removed.
- Remove glow plugs to ease turning engine.
- Turn engine in normal direction of rotation (unless otherwise stated).
- DO NOT turn engine via camshaft or other sprockets.
- Observe all tightening torques.
- Check diesel injection pump timing after belt replacement.

Removal

1. Raise and support front of vehicle.
2. Support engine.
3. Remove:
   - ❏ RH front wheel.
   - ❏ Timing belt upper cover **1**.
   - ❏ RH engine mounting bracket.
   - ❏ Auxiliary drive belt.
   - ❏ Starter motor.
4. Lower engine to gain access to crankshaft pulley bolt.
5. Turn crankshaft to TDC on No.1 cylinder.
6. Insert timing pin in flywheel **2**.
   Tool No.18G 1632.
7. Lock flywheel **3**. Use tool No.18G 1547.

8. Insert one bolt in camshaft sprocket and two bolts in injection pump sprocket **4** & **5**. Tighten bolts finger tight.
9. Remove:
   - ❏ Crankshaft pulley.
   - ❏ Timing belt lower cover **6**.
   - **NOTE: Retain rubber spacer 7.**
10. Slacken tensioner nut and bolt **8** & **9**.
11. Move tensioner away from belt to release tension on belt. Tighten bolt **8**.
12. Remove timing belt.

Installation

1. Ensure crankshaft at TDC on No.1 cylinder. Ensure timing pin and locking bolts located correctly **2**, **4** & **5**.
2. Fit timing belt in anti-clockwise direction, starting at crankshaft sprocket.
3. Slacken tensioner bolt **8**. Allow tensioner to operate. Tighten tensioner nut and bolt **8** & **9**.
4. Remove:
   - ❏ Locking bolts **4** & **5**.
   - ❏ Flywheel locking tool **3**.
   - ❏ Flywheel timing pin **2**.
5. Turn crankshaft two turns clockwise.
6. Ensure timing pin and locking bolts can be inserted easily.
7. Slacken tensioner nut and bolt **8** & **9**.
8. Tighten tensioner bolt **8**.
   Tightening torque: 15 Nm.
9. Tighten tensioner nut **9**.
   Tightening torque: 15 Nm.
10. Remove:
    - ❏ Locking bolts **4** & **5**.
    - ❏ Flywheel timing pin **2**.
11. Install components in reverse order of removal.
12. Tighten crankshaft pulley bolt.
    Tightening torque: 40 Nm + 60°.
13. Check and adjust injection pump timing.

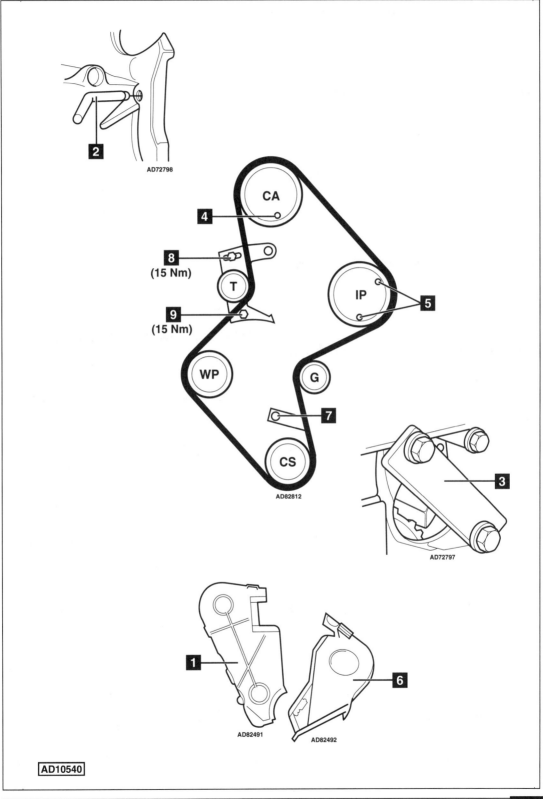

ROVER

Model:	220 D Turbo • 25 2,0 TD • 45 2,0 TD
Year:	1995-01
Engine Code:	20T, 20T2N, 20T2R

Replacement Interval Guide

Rover recommend replacement of timing belt & injection pump belt every 84,000 miles or 7 years.
The previous use and service history of the vehicle must always be taken into account.
Refer to Timing Belt Replacement Intervals at the front of this manual.

Check For Engine Damage

CAUTION: This engine has been identified as an INTERFERENCE engine in which the possibility of valve-to-piston damage in the event of a timing belt failure is MOST LIKELY to occur.
A compression check of all cylinders should be performed before removing the cylinder head.

Repair Times – hrs

Remove & install:	
Timing belt:	
220	2,10
25/45	2,20
Injection pump belt	1,20

Special Tools

- Flywheel timing pin – Rover No.18G 1523.
- Tensioning tool – Rover No.18G 1719.
- Tensioner tool – Rover No.18G 12-184.
- Sprocket holding tool – Rover No.18G 1521.
- Injection pump sprocket timing pin – Rover No.18G 1717.
- Dial type torque wrench.

Special Precautions

- Disconnect battery earth lead.
- DO NOT turn crankshaft or camshaft when timing belt removed.
- Remove glow plugs to ease turning engine.
- Turn engine in normal direction of rotation (unless otherwise stated).
- DO NOT turn engine via camshaft or other sprockets.
- Observe all tightening torques.
- Check diesel injection pump timing after belt replacement.

Removal

Timing Belt

1. Raise and support front of vehicle.
2. Remove:
   - ❏ Engine undershield.
   - ❏ RH front wheel.
   - ❏ Timing belt upper cover **1**.
   - ❏ Auxiliary drive belt.
3. Turn crankshaft clockwise until camshaft front sprocket timing marks aligned **2**.
4. Insert timing pin in flywheel through engine backplate **3**. Tool No.18G 1523.
5. Remove:
   - ❏ Crankshaft pulley bolt **4**.
   - ❏ Crankshaft pulley **5**.
   - ❏ Timing belt lower cover **6**.
   - ❏ Access plug from timing belt rear cover **7**.

6. Move PAS hoses to one side.
7. Support engine.
8. Remove:
   - ❏ Top engine tie-rod.
   - ❏ RH engine mounting.
   - ❏ RH engine mounting bracket.
   - ❏ Engine mounting plate.
9. Slacken tensioner bolt **8**. Use Allen key.
10. Retract automatic tensioner as follows:
    - ❏ Install special tool **9**. Tool No.18G 1719.
    - ❏ Turn nut clockwise **17** until pushrod fully retracted.
11. Lightly tighten tensioner bolt **8**.
12. Mark direction of rotation on belt with chalk if belt is to be reused.
13. Remove timing belt.
    WARNING: If belt has been in use for more than 42,000 miles: Fit new belt.

Installation

Timing Belt

1. Remove:
   - ❏ Special tool **9**. Tool No.18G 1719.
   - ❏ Tensioner pulley.
   - ❏ Tensioner spring and pushrod.
2. Check tensioner spring for distortion. Check free length of tensioner spring is 65 mm **10**.
3. Install:
   - ❏ Tensioner spring and pushrod.
   - ❏ Tensioner pulley.
   - ❏ Special tool **9**. Tool No.18G 1719.
4. Tighten tensioner bolt **18**. Tightening torque: 45 Nm.
5. Turn nut clockwise **17** until pushrod fully retracted.
6. Lightly tighten tensioner bolt **8**.
7. Ensure camshaft front sprocket timing marks aligned **2**.
8. Ensure flywheel timing pin located correctly **3**.
9. Fit timing belt. Ensure belt is taut between sprockets. Observe direction of rotation marks on belt.
10. Clean mating surfaces for engine mounting plate.
11. Fit engine mounting plate.
    - ❏ Tighten nuts. Tightening torque: 35 Nm.
    - ❏ Tighten bolts. Tightening torque: 45 Nm.
    NOTE: If no bolts fitted, tighten all nuts to 35 Nm.
12. Slacken tensioner bolt **8**.
13. Remove special tool **9**. Tool No.18G 1719.
14. Tighten tensioner bolt **8**. Tightening torque: 55 Nm.
15. Fit access plug to timing belt rear cover **7**.
16. Remove access plug from timing belt lower cover **19**.
17. Install:
    - ❏ Timing belt lower cover **6**. Tighten bolts to 5 Nm.
    - ❏ Crankshaft pulley **5**.
    - ❏ Crankshaft pulley bolt **4**.
18. Tighten crankshaft pulley bolt **4**.
    Tightening torque: 63 Nm + 90°.
19. Remove flywheel timing pin. Tool No.18G 1523 **3**.

➡

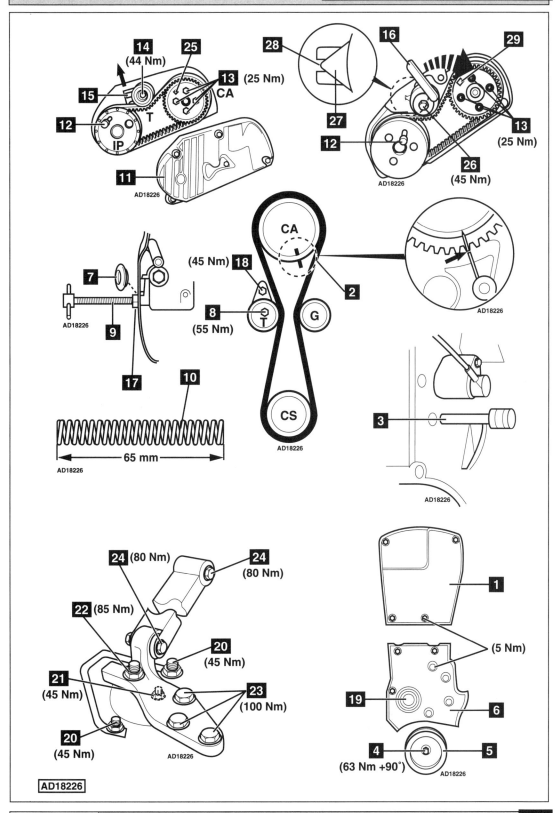

←

20. Turn crankshaft two turns clockwise. Ensure camshaft front sprocket timing marks aligned **2**.
21. Ensure flywheel timing pin can be inserted **3**. Tool No.18G 1523.
22. Slacken tensioner bolt **8**. Allow tensioner to operate.
23. Tighten tensioner bolt **8**. Tightening torque: 55 Nm.
24. Install components in reverse order of removal.
25. Tighten:
    - ❑ Nuts **20** & **21**. Tightening torque: 45 Nm.
    - ❑ Nut **22**. Tightening torque: 85 Nm.
    - ❑ Bolts **23**. Tightening torque: 100 Nm.
    - ❑ Nut and bolt **24**. Tightening torque: 80 Nm.

Removal

Injection Pump Belt

1. Remove:
   - ❑ Timing belt upper cover **1**.
   - ❑ Injection pump belt cover **11**.
   - ❑ RH wheel.
2. Turn crankshaft clockwise until camshaft front sprocket timing marks aligned **2**.
3. Insert timing pin in flywheel through engine backplate **3**. Tool No.18G 1523.
4. Slacken camshaft rear sprocket bolts **13**.
5. Insert timing pin into injection pump sprocket and mounting plate **12**. Tool No.18G 1717.
6. Ensure camshaft front sprocket timing marks aligned **2**.
7. Slacken tensioner bolt:
   - ❑ 220 – **14**.
   - ❑ 25/45 – **26**.
8. Move tensioner away from belt and lightly tighten bolt.
9. Mark direction of rotation on belt with chalk if belt is to be reused.
10. Remove injection pump belt.
    WARNING: If belt has been in use for more than 42,000 miles: Fit new belt.

Installation

Injection Pump Belt

1. Turn camshaft rear sprocket fully clockwise in slotted holes.
2. Fit belt to injection pump sprocket.
3. Turn camshaft rear sprocket slowly anti-clockwise until belt teeth engage in sprocket.
   NOTE: Fitting injection pump belt in other possible position will not allow correct belt adjustment.

220

4. Slacken tensioner bolt **14**.
5. Push tensioner against belt to take up slack.
6. Fit dial type torque wrench to square hole in tensioner plate **15**.
7. Apply clockwise torque of 6 Nm to belt.
8. Tighten tensioner bolt **14**. Tightening torque: 44 Nm.
9. Tighten camshaft rear sprocket bolts **13**.

25/45

10. Slacken tensioner bolt **26**.
11. Turn tensioner clockwise until pointer past notch **27** & **28**. Use tool No.18G12-184 **16**.

12. Turn tensioner back until pointer aligned with notch **27** & **28**.
13. Tighten tensioner bolt **26**. Tightening torque: 45 Nm.

All models

14. Remove:
    - ❑ Flywheel timing pin **3**.
    - ❑ Injection pump sprocket timing pin **12**.
15. Turn crankshaft two turns clockwise.
16. Insert timing pin in flywheel **3**. Tool No.18G 1523.
17. Ensure camshaft front sprocket timing marks aligned **2**.

220

18. Slacken camshaft rear sprocket bolts **13**.
19. Insert timing pin into injection pump sprocket and mounting plate **12**. Tool No.18G 1717.
20. Slacken tensioner bolt **14**.
21. Fit dial type torque wrench to square hole in tensioner plate **15**.
22. Apply clockwise torque of 6 Nm to belt.
23. Tighten tensioner bolt **14**. Tightening torque: 44 Nm.
24. Insert dial type torque wrench in camshaft rear sprocket **25**.
25. Apply anti-clockwise torque of 25 Nm (in normal direction of rotation).
26. Tighten camshaft rear sprocket bolts **13**. Tightening torque: 25 Nm.
27. Remove:
    - ❑ Flywheel timing pin **3**.
    - ❑ Injection pump sprocket timing pin **12**.
28. Turn crankshaft two turns clockwise.
29. Insert timing pin in flywheel **3**. Tool No.18G 1523.
30. Ensure camshaft front sprocket timing marks aligned **2**.
31. Ensure timing pin can be inserted easily in injection pump sprocket **12**. Tool No.18G 1717.
32. If not: Repeat tensioning procedure.

25/45

33. Ensure timing pin can be inserted easily in injection pump sprocket **12**. Tool No.18G 1717.
34. Ensure pointer aligned with notch **27** & **28**.
35. If not: Repeat tensioning procedure.
36. Insert dial type torque wrench in camshaft rear sprocket **29**.
37. Apply anti-clockwise torque of 25 Nm (in normal direction of rotation).
38. Tighten camshaft rear sprocket bolts **13**. Tightening torque: 25 Nm.

All models

39. Install components in reverse order of removal.

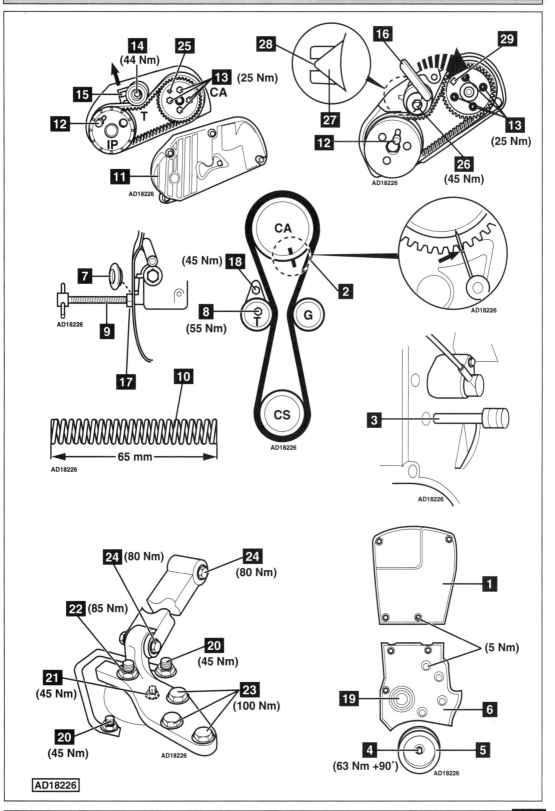

ROVER

Model:	**25 2,0 TD • 45 2,0 TD • MG ZR 2,0 TD • MG ZS 2,0 TD**
Year:	**2001-03**
Engine Code:	**20T**

Replacement Interval Guide

Rover recommend replacement of timing belt & injection pump belt every 84,000 miles or 7 years (with automatic tensioner).
Rover recommend replacement of timing belt & injection pump belt every 48,000 miles or 4 years (with manual tensioner).
The previous use and service history of the vehicle must always be taken into account.
Refer to Timing Belt Replacement Intervals at the front of this manual.

Check For Engine Damage

CAUTION: This engine has been identified as an INTERFERENCE engine in which the possibility of valve-to-piston damage in the event of a timing belt failure is MOST LIKELY to occur.
A compression check of all cylinders should be performed before removing the cylinder head.

Repair Times – hrs

Remove & install:
Timing belt and injection pump belt:

25/ZR	3,40
45/ZS	3,20
Timing belt:	
25/ZR	3,30
45/ZS	3,10
Injection pump belt:	
25/ZR	1,20
45/ZS	1,40

Special Tools

- Flywheel timing pin – Rover No.18G 1523.
- Tensioning tool – Rover No.18G 1719.
- Tensioner tool – Rover No.12-184.
- Injection pump sprocket timing pin – Rover No.12-185.
- Dial type torque wrench.

Special Precautions

- Disconnect battery earth lead.
- DO NOT turn crankshaft or camshaft when timing belt removed.
- Remove glow plugs to ease turning engine.
- Turn crankshaft in normal direction of rotation (unless otherwise stated).
- DO NOT turn engine via camshaft or other sprockets.
- Observe all tightening torques.
- Check diesel injection pump timing after belt replacement.

Removal

Timing Belt

1. Raise and support front of vehicle.
2. Remove:
   - ❏ Engine undershield.
   - ❏ RH front wheel.
   - ❏ Timing belt upper cover **1**.
   - ❏ Auxiliary drive belt.
3. Remove lower engine steady bar.
4. Turn crankshaft clockwise until camshaft front sprocket timing marks aligned **2**.
5. Insert timing pin in flywheel through engine backplate **3**. Tool No.18G 1523.
6. Remove:
   - ❏ Camshaft damper and bolts **16**.
   - ❏ Crankshaft pulley bolt **4**.
   - ❏ Crankshaft pulley **5**.
   - ❏ Timing belt lower cover **6**.
   - ❏ Access plug from timing belt rear cover **7**.
7. Support engine.
8. Remove:
   - ❏ RH engine mounting.
   - ❏ RH engine mounting bracket.
   - ❏ Engine mounting plate.
9. Slacken tensioner bolt **8**. Use Allen key.
10. Retract automatic tensioner as follows:
    - ❏ Install special tool No.18G 1719 **9**.
    - ❏ Turn nut clockwise **17** until pushrod fully retracted.
11. Lightly tighten tensioner bolt **8**.
12. Mark direction of rotation on belt with chalk if belt is to be reused.
13. Remove timing belt.
    WARNING: If belt has been in use for more than 42,000 miles: Fit new belt.

Installation

Timing Belt

1. Remove:
   - ❏ Special tool **9**. Tool No.18G 1719.
   - ❏ Tensioner pulley.
   - ❏ Tensioner spring and pushrod.
2. Check tensioner spring for distortion. Check free length of tensioner spring is 65 mm **10**.
3. Install:
   - ❏ Tensioner spring and pushrod.
   - ❏ Tensioner pulley.
   - ❏ Special tool **9**. Tool No.18G 1719.
4. Turn nut clockwise **17** until pushrod fully retracted.
5. Lightly tighten tensioner bolt **8**.
6. Ensure camshaft front sprocket timing marks aligned **2**.
7. Ensure flywheel timing pin located correctly **3**.
8. Fit timing belt. Ensure belt is taut between sprockets. Observe direction of rotation marks on belt.
9. Clean mating surfaces for engine mounting plate.
10. Fit engine mounting plate.
    - ❏ Tighten nuts. Tightening torque: 30 Nm +120°.
    - ❏ Tighten bolts. Tightening torque: 45 Nm.
11. Slacken tensioner bolt **8**.
12. Remove special tool **9**. Tool No.18G 1719.
13. Tighten tensioner bolt **8**. Tightening torque: 55 Nm.
14. Fit access plug to timing belt rear cover **7**.
15. Remove access plug from timing belt lower cover **19**.

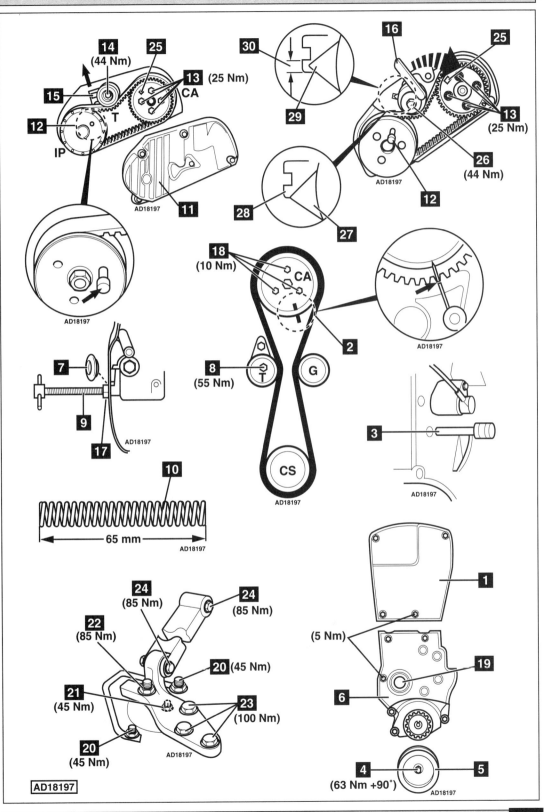

16. Install:
    - ❏ Timing belt lower cover **6**. Tighten bolts to 5 Nm.
    - ❏ Crankshaft pulley **5**.
    - ❏ Crankshaft pulley bolt **4**.
17. Tighten crankshaft pulley bolt **4**.
    Tightening torque: 63 Nm + 90°.
18. Remove flywheel timing pin. Tool No.18G 1523 **3**.
19. Turn crankshaft two turns clockwise. Ensure camshaft front sprocket timing marks aligned **2**.
20. Ensure flywheel timing pin can be inserted **3**. Tool No.18G 1523.
21. Slacken tensioner bolt **8**. Allow tensioner to operate.
22. Tighten tensioner bolt **8**. Tightening torque: 55 Nm.
23. Install camshaft damper. Tighten bolts **10**. Tightening torque: 10 Nm.
24. Install components in reverse order of removal.
25. Tighten:
    - ❏ Nuts **20** & **21**. Tightening torque: 45 Nm.
    - ❏ Nut **22**. Tightening torque: 85 Nm.
    - ❏ Bolts **23**. Tightening torque: 100 Nm.
    - ❏ Nut and bolt **24**. Tightening torque: 85 Nm.

Removal

Injection Pump Belt

1. Remove:
   - ❏ Timing belt upper cover **1**.
   - ❏ Camshaft damper and bolts **10**.
   - ❏ Injection pump belt cover **11**.
   - ❏ RH wheel.
2. Turn crankshaft clockwise until camshaft front sprocket timing marks aligned **2**.
3. Insert timing pin in flywheel through engine backplate **3**. Tool No.18G 1523.
4. Slacken camshaft rear sprocket bolts **13**.
5. Insert timing pin into injection pump sprocket and mounting plate **12**. Tool No.12-185.
6. Ensure camshaft front sprocket timing marks aligned **2**.
7. Slacken tensioner bolt:
   - ❏ Manual tensioner – **14**.
   - ❏ Automatic tensioner – **26**.
8. Move tensioner away from belt and lightly tighten bolt.
9. Mark direction of rotation on belt with chalk if belt is to be reused.
10. Remove injection pump belt.
    WARNING: If belt has been in use for more than 42,000 miles (automatic tensioner) or 24,000 miles (manual tensioner): Fit new belt.

Installation

Injection Pump Belt

1. Turn camshaft rear sprocket fully clockwise in slotted holes.
2. Fit belt to injection pump sprocket.
3. Turn camshaft rear sprocket slowly anti-clockwise until belt teeth engage in sprocket.
   NOTE: Fitting injection pump belt in other possible position will not allow correct belt adjustment.

Manual tensioner

4. Slacken tensioner bolt **14**.
5. Push tensioner against belt to take up slack.

6. Fit dial type torque wrench to square hole in tensioner plate **15**.
7. Apply clockwise torque of 6 Nm to belt.
8. Tighten tensioner bolt **14**. Tightening torque: 44 Nm.

Automatic tensioner

9. Slacken tensioner bolt **26**.
10. Turn tensioner clockwise until pointer past lower edge of baseplate **28**. Use tool No.18G12-184 **16**.
11. Turn tensioner back until pointer aligned with lower edge of baseplate **27** & **28**.
12. Tighten tensioner bolt **26**. Tightening torque: 44 Nm.

All models

13. Remove:
    - ❏ Flywheel timing pin **3**.
    - ❏ Injection pump sprocket timing pin **12**.
14. Turn crankshaft two turns clockwise.
15. Insert timing pin in flywheel **3**. Tool No.18G 1523.
16. Ensure camshaft front sprocket timing marks aligned **2**.
17. Insert timing pin into injection pump sprocket and mounting plate **12**. Tool No.12-185.

Manual tensioner

18. Insert dial type torque wrench in camshaft rear sprocket **25**.
19. Apply anti-clockwise torque of 25 Nm (in normal direction of rotation).
20. Tighten camshaft rear sprocket bolts **13**. Tightening torque: 25 Nm.
21. Remove:
    - ❏ Flywheel timing pin **3**.
    - ❏ Injection pump sprocket timing pin **12**.
22. Turn crankshaft two turns clockwise.
23. Insert timing pin in flywheel **3**. Tool No.18G 1523.
24. Ensure camshaft front sprocket timing marks aligned **2**.
25. Ensure timing pin can be inserted easily in injection pump sprocket **12**. Tool No.12-185. If not: Repeat tensioning procedure.
26. Remove:
    - ❏ Flywheel timing pin **3**.
    - ❏ Injection pump sprocket timing pin **12**.

Automatic tensioner

27. Ensure tensioner pointer aligned within lower edge of baseplate **29** & **30**. If not: Repeat tensioning procedure.
28. Insert dial type torque wrench in camshaft rear sprocket **25**.
29. Apply anti-clockwise torque of 25 Nm (in normal direction of rotation).
30. Tighten camshaft rear sprocket bolts **13**. Tightening torque: 25 Nm.
31. Remove:
    - ❏ Flywheel timing pin **3**.
    - ❏ Injection pump sprocket timing pin **12**.
32. Turn crankshaft two turns clockwise.
33. Insert timing pin in flywheel **3**. Tool No.18G 1523.
34. Ensure camshaft front sprocket timing marks aligned **2**.
35. Ensure timing pin can be inserted easily in injection pump sprocket **12**. Tool No.12-185. If not: Repeat tensioning procedure.

All models

36. Install components in reverse order of removal.

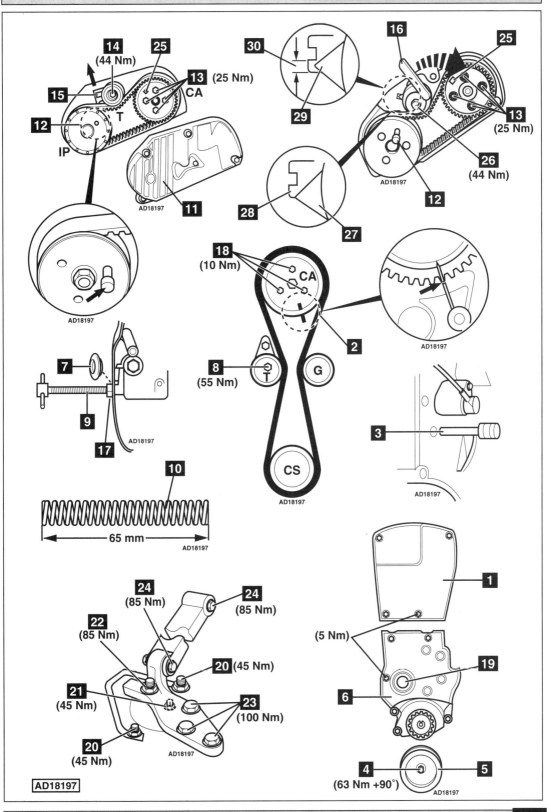

Model:	**420 D Turbo • 620 D Turbo**
Year:	**1995-99**
Engine Code:	**20T2N, 20T2R**

Replacement Interval Guide

Rover recommend replacement of timing belt & injection pump belt every 84,000 miles or 7 years.
The previous use and service history of the vehicle must always be taken into account.
Refer to Timing Belt Replacement Intervals at the front of this manual.

Check For Engine Damage

CAUTION: This engine has been identified as an INTERFERENCE engine in which the possibility of valve-to-piston damage in the event of a timing belt failure is MOST LIKELY to occur.
A compression check of all cylinders should be performed before removing the cylinder head.

Repair Times – hrs

Remove & install:

420: Timing belt	3,10
620: Timing belt	2,10
420: Injection pump belt	1,60
620: Injection pump belt	1,80

Special Tools

- Flywheel timing pin – Rover No.18G 1523.
- Tensioning tool – Rover No.18G 1719.
- Injection pump sprocket timing pin – Rover No.18G 1717.
- Dial type torque wrench.

Special Precautions

- Disconnect battery earth lead.
- DO NOT turn crankshaft or camshaft when timing belt removed.
- Remove glow plugs to ease turning engine.
- Turn engine in normal direction of rotation (unless otherwise stated).
- DO NOT turn engine via camshaft or other sprockets.
- Observe all tightening torques.
- Check diesel injection pump timing after belt replacement.

Removal

Timing Belt

1. Raise and support front of vehicle.
2. Remove:
   - ❏ Engine undershield.
   - ❏ RH front wheel.
   - ❏ Auxiliary drive belt.
   - ❏ Timing belt upper cover **1**.
3. Turn crankshaft clockwise until camshaft front sprocket timing marks aligned **2**.
4. Remove:
   - ❏ Crankshaft pulley bolt **4**.
   - ❏ Crankshaft pulley **5**.
   - ❏ Timing belt lower cover **6**.
   - ❏ Access plug from timing belt rear cover **7**.
5. Insert timing pin in flywheel through engine backplate **3**. Tool No.18G 1523.

6. Slacken tensioner bolt **8**. Use Allen key.
7. Retract automatic tensioner as follows:
   - ❏ Install special tool No.18G 1719 **9**.
   - ❏ Turn nut clockwise **24** until pushrod fully retracted.
8. Lightly tighten tensioner bolt **8**.
9. Support engine.
10. Remove top engine tie-rod **22**.
11. Slacken RH engine mounting nut **23**.
12. Remove:
    - ❏ RH engine mounting nuts **11**.
    - ❏ RH engine mounting bolts **10**.
    - ❏ RH engine mounting.
    - ❏ RH engine mounting bracket.
    - ❏ Engine mounting plate.
13. Mark direction of rotation on belt with chalk if belt is to be reused.
14. Remove timing belt.
    WARNING: If belt has been in use for more than 42,000 miles: Fit new belt.

Installation

Timing Belt

1. Remove:
   - ❏ Special tool **9**. Tool No.18G 1719.
   - ❏ Tensioner pulley.
   - ❏ Tensioner spring and pushrod.
2. Check tensioner spring for distortion. Check free length of tensioner spring is 65 mm **25**.
3. Install:
   - ❏ Tensioner spring and pushrod.
   - ❏ Tensioner pulley.
   - ❏ Special tool **9**. Tool No.18G 1719.
4. Tighten tensioner bolt **26**. Tightening torque: 45 Nm.
5. Turn nut clockwise **24** until pushrod fully retracted.
6. Lightly tighten tensioner bolt **8**.
7. Ensure timing pin inserted in flywheel **3**.
8. Ensure camshaft front sprocket timing marks aligned **2**.
9. Clean mating surfaces for engine mounting plate.
10. Fit timing belt. Ensure belt is taut between sprockets on non-tensioned side.
11. Coat engine mounting plate mating surfaces with locking compound. Use Loctite 638.
12. Fit engine mounting plate.
    - ❏ Tighten nuts. Tightening torque: 35 Nm.
    - ❏ Tighten bolts. Tightening torque: 45 Nm.
    NOTE: If no bolts fitted, tighten all nuts to 35 Nm.
13. Clean dowels and dowel holes for engine mounting.
14. Fit top engine tie-rod and RH engine mounting.
    - ❏ Tighten bolts **10**. Tightening torque: 105 Nm.
    - ❏ Tighten nuts **11**. Tightening torque: 45 Nm.
    - ❏ Tighten nut **23**. Tightening torque: 80 Nm.
    - ❏ Tighten nut **21**. Tightening torque: 80 Nm.
    - ❏ Tighten bolt **12**. Tightening torque: 80 Nm.
15. Remove engine support.

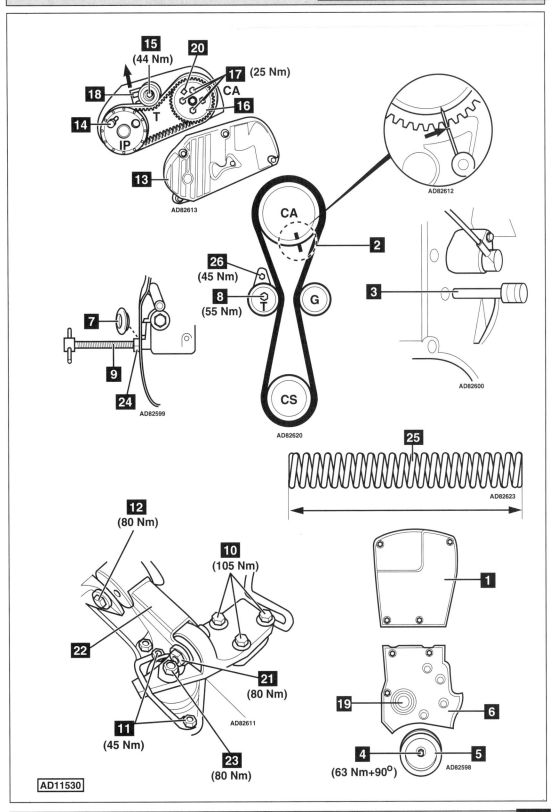

←

16. Remove access plug from timing belt lower cover 🔟.
17. Fit timing belt lower cover 🔟. Tighten bolts to 5 Nm.
18. Fit crankshaft pulley 🔟. Tighten bolt 🔟.
    Tightening torque: 63 Nm + 90°.
19. Ensure camshaft front sprocket timing marks aligned 🔟.
20. Slacken tensioner bolt 🔟.
21. Remove special tool 🔟. Tool No.18G 1719.
22. Tighten tensioner bolt 🔟. Tightening torque: 55 Nm.
23. Fit access plug to timing belt rear cover 🔟.
24. Remove flywheel timing pin 🔟.
25. Turn crankshaft two turns clockwise.
26. Ensure camshaft front sprocket timing marks aligned 🔟.
27. Insert timing pin in flywheel 🔟. Tool No.18G 1523.
28. Slacken tensioner bolt 🔟. Allow tensioner to operate.
29. Tighten tensioner bolt 🔟. Tightening torque: 55 Nm.
30. Remove flywheel timing pin 🔟.
31. Fit access plug to timing belt lower cover 🔟.
32. Install components in reverse order of removal.

Removal

Injection Pump Belt

1. Remove:
   - ❏ Engine undershield.
   - ❏ Air filter.
   - ❏ Timing belt upper cover 🔟.
   - ❏ PAS pipe bracket bolts.
   - ❏ Injection pump belt cover 🔟.
2. Turn crankshaft clockwise until camshaft front sprocket timing marks aligned 🔟.
3. Insert timing pin in flywheel through engine backplate 🔟. Tool No.18G 1523.
4. Slacken camshaft rear sprocket bolts 🔟.
5. Insert timing pin into injection pump sprocket and mounting plate 🔟. Tool No.18G 1717.
6. Ensure camshaft front sprocket timing marks aligned 🔟.
7. Slacken tensioner bolt 🔟. Move tensioner away from belt and lightly tighten bolt.
8. Mark direction of rotation on belt with chalk if belt is to be reused.
9. Remove injection pump belt.
   WARNING: If belt has been in use for more than 42,000 miles: Fit new belt.

Installation

Injection Pump Belt

1. Turn camshaft rear sprocket fully clockwise in slotted holes.
2. Fit belt to injection pump sprocket.
3. Turn camshaft rear sprocket slowly anti-clockwise until belt teeth engage in sprocket.
   NOTE: Fitting injection pump belt in other possible position will not allow correct belt adjustment.
4. Slacken tensioner bolt 🔟. Push tensioner against belt to take up slack.
5. Fit dial type torque wrench to square hole in tensioner plate 🔟.
6. Apply clockwise torque of 6 Nm to belt (against normal direction of rotation).

7. Tighten tensioner bolt 🔟. Tightening torque: 44 Nm.
8. Tighten camshaft rear sprocket bolts 🔟.
9. Remove:
   - ❏ Flywheel timing pin 🔟.
   - ❏ Injection pump sprocket timing pin 🔟.
10. Turn crankshaft two turns clockwise.
11. Insert timing pin in flywheel 🔟. Tool No.18G 1523.
12. Ensure camshaft front sprocket timing marks aligned 🔟.
13. Slacken camshaft rear sprocket bolts 🔟.
14. Insert timing pin into injection pump sprocket and mounting plate 🔟. Tool No.18G 1717.
15. Slacken tensioner bolt 🔟.
16. Fit dial type torque wrench to square hole in tensioner plate 🔟.
17. Apply clockwise torque of 6 Nm to belt.
18. Tighten tensioner bolt 🔟.
    Tightening torque: 44 Nm.
19. Insert dial type torque wrench in camshaft rear sprocket 🔟.
20. Apply anti-clockwise torque of 25 Nm (in normal direction of rotation).
21. Tighten camshaft rear sprocket bolts 🔟. Tightening torque: 25 Nm.
22. Remove:
   - ❏ Flywheel timing pin 🔟.
   - ❏ Injection pump sprocket timing pin 🔟.
23. Turn crankshaft two turns clockwise.
24. Insert timing pin in flywheel 🔟. Tool No.18G 1523.
25. Ensure camshaft front sprocket timing marks aligned 🔟.
26. Ensure timing pin can be inserted easily in injection pump sprocket 🔟. Tool No.18G 1717.
27. If not: Repeat operations 13-26.
28. Install components in reverse order of removal.

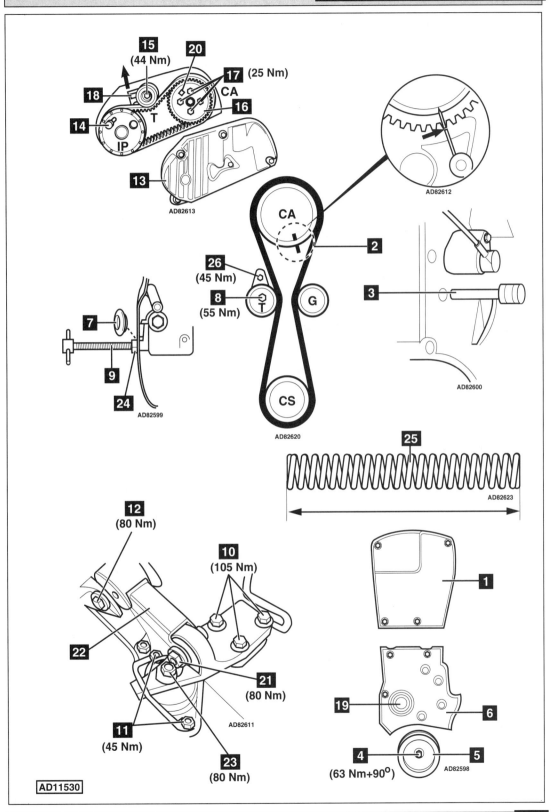

15 (44 Nm)

20

17 (25 Nm)

18

CA

14 IP

T

16

13

AD82613

AD82612

CA

2

26 (45 Nm)

8 T (55 Nm)

G

3

AD82600

7

9

24

AD82599

CS

AD82620

25

AD82623

12 (80 Nm)

10 (105 Nm)

22

21 (80 Nm)

11 (45 Nm)

AD82611

23 (80 Nm)

1

19

6

4 (63 Nm+90°)

5

AD82598

AD11530

SAAB

Model:	**900 2,5 V6 • 9000 3,0 V6**
Year:	**1993-96**
Engine Code:	**Without raised outer edge on tensioner pulley** **B258, B308**

Replacement Interval Guide

Saab recommend replacement at 30,000 miles or 3 years, then replacement every 36,000 miles or 3 years.
The previous use and service history of the vehicle must always be taken into account.
Refer to Timing Belt Replacement Intervals at the front of this manual.

Check For Engine Damage

CAUTION: This engine has been identified as an INTERFERENCE engine in which the possibility of valve-to-piston damage in the event of a timing belt failure is MOST LIKELY to occur.
A compression check of all cylinders should be performed before removing the cylinder head.

Repair Times – hrs

Remove & install	1,70

Special Tools

- Timing belt installation tool kit – SAAB No.83 95 006.
- Tension gauge – SAAB No.83 93 985.

Special Precautions

- Disconnect battery earth lead.
- DO NOT turn crankshaft or camshaft when timing belt removed.
- Remove spark plugs to ease turning engine.
- Turn engine in normal direction of rotation (unless otherwise stated).
- DO NOT turn engine via camshaft or other sprockets.
- Observe all tightening torques.

Removal

1. 900: Check campaign plate for completion of timing belt modifications **17**. Refer vehicle to main dealer if section B2 or C3 are unstamped.
   NOTE: Location of campaign plate: RH front door shut panel.
   - Section C3 stamped 7 or 8 (from chassis No.R2000001-R2047292 or No.S7000001-S7002837).
   - Section B2 stamped 7 or 8 (from chassis No.V2000001-V2007396 or No.V7000001-V7001213).
2. Raise and support front of vehicle.

3. 900 – remove:
   - RH front wheel.
   - Inner wing panel.
   - Air filter housing and hoses.
   - Auxiliary drive belt.
   - PAS pump pulley.
   - Water pump pulley.
4. 9000 – remove:
   - RH engine torque arm.
   - RH engine mounting.
   - Auxiliary drive belt.
   - Coolant hose from expansion tank.
   - Alternator air intake duct.
   - Auxiliary drive belt tensioner.
   - Water pump pulley.
   - PAS pump.
5. All models – remove:
   - Crankshaft pulley bolts **1**.
   - Crankshaft pulley **2**.
   - Timing belt cover.
6. Turn crankshaft clockwise until just before timing marks aligned **3** & **4**.
7. Fit locking tool to crankshaft **5**.
   Tool No.KM-800-10. Turn crankshaft slowly clockwise until tool arm rests against water pump flange. Secure in position.
8. Fit locking tools to camshafts **6**.
   KM-800-1: cyls. 1-3-5. KM-800-2: cyls. 2-4-6.
9. Slacken tensioner bolt **9**. Turn tensioner pulley clockwise. Lightly tighten bolt.
10. Slacken upper guide pulley bolt **7**.
11. Slacken lower guide pulley bolt **8**.
12. Remove timing belt.
    NOTE: Mark direction of rotation on belt with chalk if belt is to be reused.
13. If camshaft sprockets are to be removed: Note position of dowel pins in relation to timing marks.
    - Camshaft sprocket CA1 uses No.1 for dowel pin location and timing mark alignment **20**.
    - Camshaft sprocket CA2 uses No.2, CA3 uses No.3, CA4 uses No.4.
    - Tighten bolts **21**.
      Tightening torque: 50 Nm + 60°.

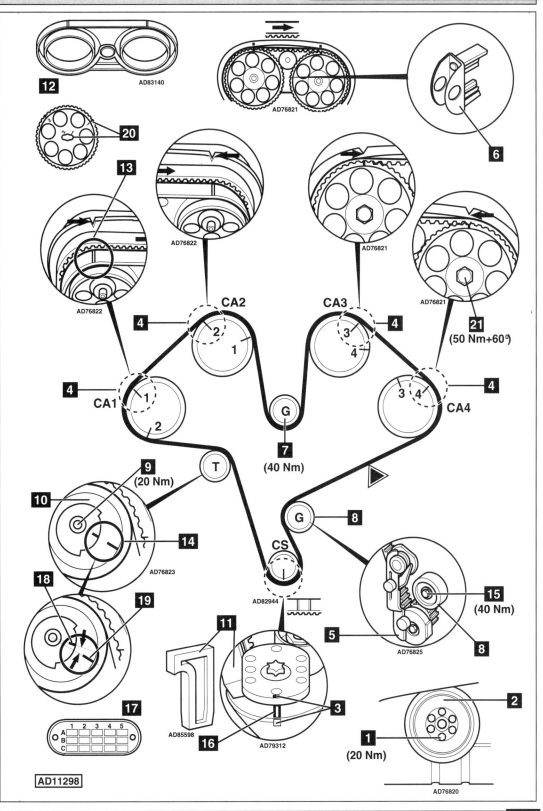

AD83140

CA2

CA3

CA1

CA4

(50 Nm+60°)

G

7

(40 Nm)

9
(20 Nm)

T

10

14

AD76823

18

19

CS

AD82944

11

G

8

15
(40 Nm)

AD76825

5

8

17

AD85598

16

AD79312

3

1
(20 Nm)

2

AD76820

AD11298

←

Installation

NOTE: Observe direction of rotation marks on belt.

1. Ensure crankshaft locking tool located correctly **5**. Tool No.KM-800-10. Ensure timing marks aligned **3**.

2. Ensure locking tools located correctly on camshafts **6**. KM-800-1: cyls. 1-3-5. KM-800-2: cyls. 2-4-6. Ensure timing marks aligned **4**.

3. Ensure double lines on belt **16** aligned with crankshaft timing marks **3**.

4. Fit timing belt in anti-clockwise direction, starting at crankshaft sprocket. Ensure belt is taut between sprockets.

5. Wedge belt into position. Use tool KM-800-30 **11**.

6. Align single marks on belt with timing marks on camshaft sprockets and timing belt rear cover **4**.

7. Turn lower guide pulley **8** anti-clockwise to remove slack from belt. Lightly tighten bolt.

8. Attach tension gauge to belt at ▽. Tool No.83 93 985.

9. Turn lower guide pulley anti-clockwise **8**. Tension belt to 275-300 N. Tighten bolt to 40 Nm **15**.

10. Remove tension gauge.

11. Turn tensioner pulley **10** anti-clockwise until marks aligned **14**. Tighten bolt to 20 Nm **9**.

12. Remove locking tool from camshaft sprockets CA1 & CA2 **6**. Tool No.KM-800-1.

13. Turn upper guide pulley **7** anti-clockwise until camshaft sprocket CA2 timing mark moves 1,0-2,0 mm clockwise **4**. Tighten bolt to 40 Nm **7**.

14. Remove locking tool from camshaft sprockets CA3 & CA4 **6**. Tool No.KM-800-2.

15. Remove:
   ❏ Crankshaft locking tool **5**.
   ❏ Belt holding tool **11**.

16. Turn crankshaft two turns clockwise until just before timing marks aligned **3**.

17. Fit locking tool to crankshaft **5**. Tool No.KM-800-10. Turn crankshaft slowly clockwise until tool arm rests against water pump flange. Secure in position.

18. Fit timing gauge to camshaft sprockets CA1 & CA2 and CA3 & CA4 in turn **12**. Tool No.KM-800-20. Ensure marks on sprockets aligned with marks on timing gauge **13**. Check edge of timing belt aligned with edge of sprockets.

19. If not: Repeat installation and tensioning procedures.

20. Check tensioner marks **14** as follows:
   NOTE: Confirm completion of timing belt modifications on 900. Refer to removal section.
   ❏ 900: Tensioner mark **18** approximately 2 mm higher than tensioner mark **19**. Tighten bolt to 20 Nm **9**.
   ❏ 9000: Tensioner marks aligned **14**. Tighten bolt to 20 Nm **9**.

21. Install components in reverse order of removal.

22. Tighten crankshaft pulley bolts to 20 Nm **1**.

23. Tighten water pump pulley bolts to 8 Nm.

24. 9000: Refill cooling system.

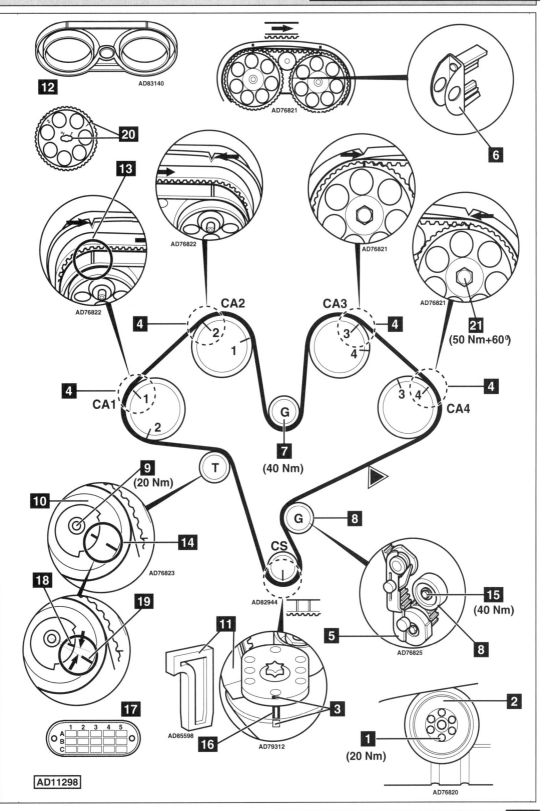

SAAB

Model:	**900 2,5 V6 • 9000 3,0 V6**
Year:	**1995-98**
Engine Code:	**With raised outer edge on tensioner pulley** **B258, B308**

Replacement Interval Guide

Saab recommend replacement as follows:
→96MY: Replacement at 30,000 miles or
3 years, then replacement every 36,000 miles or
3 years.
97MY→: Replacement at 54,000 miles or
5 years, then replacement at 114,000 miles or
5 years.
*The previous use and service history of the vehicle
must always be taken into account.*
*Refer to Timing Belt Replacement Intervals at the front
of this manual.*

Check For Engine Damage

*CAUTION: This engine has been identified as an
INTERFERENCE engine in which the possibility of
valve-to-piston damage in the event of a timing belt
failure is MOST LIKELY to occur.*
*A compression check of all cylinders should be
performed before removing the cylinder head.*

Repair Times – hrs

Remove & install   1,70

Special Tools

■ Timing belt installation tool kit –
 SAAB No.83 95 006.
■ Tension gauge – SAAB No.83 93 985.

Special Precautions

■ Disconnect battery earth lead.
■ DO NOT turn crankshaft or camshaft when timing belt
 removed.
■ Remove spark plugs to ease turning engine.
■ Turn engine in normal direction of rotation (unless
 otherwise stated).
■ DO NOT turn engine via camshaft or other sprockets.
■ Observe all tightening torques.

Removal

1. Raise and support front of vehicle.
2. 900 – remove:
   ❏ RH front wheel.
   ❏ Inner wing panel.
   ❏ Air filter housing and hoses.
   ❏ Auxiliary drive belt.
   ❏ PAS pump pulley.
   ❏ Water pump pulley.

3. 9000 – remove:
   ❏ RH engine torque arm.
   ❏ RH engine mounting.
   ❏ Auxiliary drive belt.
   ❏ Coolant hose from expansion tank.
   ❏ Alternator air intake duct.
   ❏ Auxiliary drive belt tensioner.
   ❏ Water pump pulley.
   ❏ PAS pump.
4. All models – remove:
   ❏ Crankshaft pulley bolts **1**.
   ❏ Crankshaft pulley **2**.
   ❏ Timing belt cover.
5. Turn crankshaft clockwise until just before timing
   marks aligned **4** & **5**.
6. Fit locking tool to crankshaft **3**.
   Tool No.KM-800-10. Turn crankshaft slowly
   clockwise until tool arm rests against water
   pump flange. Secure in position.
7. Fit locking tools to camshafts **6**.
   KM-800-1: cyls. 1-3-5. KM-800-2: cyls. 2-4-6.
8. Slacken tensioner bolt **7**. Turn tensioner
   clockwise. Lightly tighten bolt.
9. Slacken upper guide pulley bolt **9**.
10. Slacken lower guide pulley bolt **8**.
11. Remove timing belt.
    **NOTE: Mark direction of rotation on belt with
    chalk if belt is to be reused.**
12. If camshaft sprockets are to be removed: Note
    position of dowel pins in relation to timing marks.
    ❏ Camshaft sprocket CA1 uses No.1 for dowel
    pin location and timing mark alignment **10**.
    ❏ Camshaft sprocket CA2 uses No.2,
    CA3 uses No.3, CA4 uses No.4.
    ❏ Tighten bolts **19**.
    Tightening torque: 50 Nm + 60°.

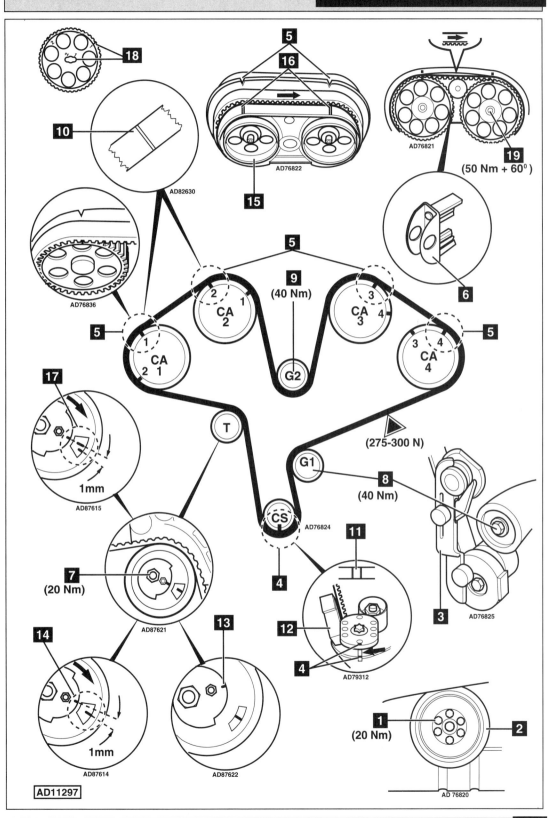

18

10

5

16

AD76822

15

AD76821

19

(50 Nm + 60°)

6

5

9

(40 Nm)

AD82630

2 1

CA 2

3

CA 3 4

5

5

AD76836

1

CA 2 1

3 4

CA 4

5

17

G2

3 4

T

(275-300 N)

1mm

AD87615

G1

8

(40 Nm)

3

AD76825

7

(20 Nm)

CS

AD76824

11

AD87621

4

12

14

13

4

AD79312

1mm

AD87614

AD87622

1

(20 Nm)

2

AD 76820

AD11297

Installation

NOTE: Observe direction of rotation marks on belt.

1. Ensure crankshaft locking tool located correctly **3**. Tool No.KM-800-10. Ensure timing marks aligned **4**.

2. Ensure locking tools located correctly on camshafts **6**. KM-800-1: cyls. 1-3-5. KM-800-2: cyls. 2-4-6. Ensure timing marks aligned **5**.

3. Fit belt to camshaft sprockets CA1 & CA2 with belt timing marks aligned **10**. Then fit around tensioner pulley **7**.

4. Ease belt around crankshaft sprocket. Ensure belt is taut between sprockets.

   *NOTE: Double lines on belt **11** will not line up with crankshaft timing marks at this time, but should be one tooth offset to the right.*

5. Wedge belt into position. Use tool KM-800-30 **12**.

6. Fit belt around lower guide pulley **8**.

7. Fit belt around upper guide pulley **9**.

8. Fit belt to camshaft sprockets CA3 & CA4. Ensure timing marks aligned **10**.

9. Feed belt slack round lower guide pulley G1 **8**. Remove belt holding tool. Tool No.KM-800-30. Pass belt slack back under crankshaft sprocket.

   *NOTE: Double lines on belt **11** should now line up with crankshaft timing marks **4**.*

10. Turn lower guide pulley **8** anti-clockwise to remove slack from belt. Lightly tighten bolt **8**.

11. Attach tension gauge to belt at ▽. Tool No.83 93 985.

12. Turn lower guide pulley anti-clockwise **8**. Tension belt to 275-300 N. Tighten bolt to 40 Nm **8**.

13. Remove tension gauge.

14. Remove locking tool from camshaft sprockets CA1 & CA2 **6**. Tool No.KM-800-1.

15. Turn upper guide pulley **9** anti-clockwise until camshaft sprocket CA2 timing mark moves 1,0-2,0 mm clockwise. Tighten bolt to 40 Nm **9**.

16. Turn tensioner pulley **7** anti-clockwise as far as it will go **13** to pre-tension belt.

17. Turn tensioner pulley clockwise until tensioner mark is 1 mm above floating mark **14**.

18. Tighten tensioner bolt to 20 Nm **7**.

19. Remove locking tool from camshaft sprockets CA3 & CA4 **6**. Tool No.KM-800-2.

20. Remove locking tool from crankshaft sprocket **3**. Tool No.KM-800-10.

21. Turn crankshaft slowly two turns clockwise until just before timing marks aligned **4** & **5**.

22. Fit locking tool to crankshaft **3**. Tool No.KM-800-10. Turn crankshaft slowly clockwise until tool arm rests against water pump flange. Secure in position.

23. Ensure timing marks aligned **4**.

24. Fit timing gauge to camshaft sprockets CA1 & CA2 and CA3 & CA4 in turn **15**. Tool No.KM-800-20.

    NOTE: Belt marks are for installation purposes only. They will no longer align once crankshaft has been turned.

25. Ensure marks on sprockets aligned with marks on timing gauge **16**. Check edge of timing belt aligned with edge of sprockets.

26. Attach tension gauge to belt at ▽. Tool No.83 93 985. Repeat tensioning procedures.

27. Turn tensioner pulley until tensioner mark is 1 mm below floating mark **17**. Tighten bolt to 20 Nm **7**.

28. Install components in reverse order of removal.

29. Tighten crankshaft pulley bolts to 20 Nm **1**.

30. Tighten water pump pulley bolts to 8 Nm.

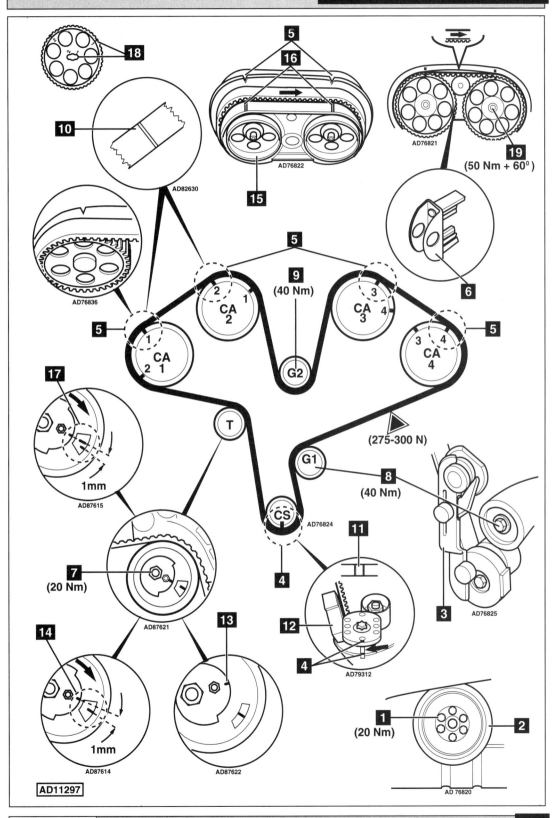

AD11297

SAAB

Model:	9-5 3,0 V6
Year:	1998-03
Engine Code:	B308E

Replacement Interval Guide

Saab recommend replacement at 54,000 miles or 5 years then every 48,000 miles or 4 years.
The previous use and service history of the vehicle must always be taken into account.
Refer to Timing Belt Replacement Intervals at the front of this manual.

Check For Engine Damage

CAUTION: This engine has been identified as an INTERFERENCE engine in which the possibility of valve-to-piston damage in the event of a timing belt failure is MOST LIKELY to occur. A compression check of all cylinders should be performed before removing the cylinder head(s).

Repair Times – hrs

Remove & install	1,80

Special Tools

- Engine support wedges – SAAB No.83 95 238.
- Crankshaft locking tool – SAAB No.KM-800-10.
- Belt holding tool (wedge) – SAAB No.KM-800-30.
- Camshaft sprocket locking tools – SAAB No.KM-800-1/2.
- Camshaft timing gauge – SAAB No.KM-800-20.
- Tensioner wrench – SAAB No.83 94 983.
- Tension gauge – SAAB No.83 93 985.

Special Precautions

- Disconnect battery earth lead.
- DO NOT turn crankshaft or camshaft when timing belt removed.
- Remove spark plugs to ease turning engine.
- Turn engine in normal direction of rotation (unless otherwise stated).
- DO NOT turn engine via camshaft or other sprockets.
- Observe all tightening torques.

Removal

1. Raise and support front of vehicle.
2. Remove lower splash guards.
3. Insert two wedges between engine sump and subframe and transmission and subframe to support engine. Tool No.83 95 238.

4. Remove:
   - ❏ Mass air flow sensor and hoses.
   - ❏ PAS hose bracket.
   - ❏ RH engine mounting and bracket.
   - ❏ Auxiliary drive belt.
   - ❏ PAS pump pulley.
   - ❏ Water pump pulley.
   - ❏ Auxiliary drive belt tensioner.
   - ❏ Crankshaft pulley bolts **1**.
   - ❏ Crankshaft pulley **2**.
   - ❏ Timing belt cover.
5. Turn crankshaft slowly clockwise until just before timing marks aligned **4** & **5**.
6. Fit locking tool to crankshaft **3**.
   Tool No.KM-800-10.
7. Turn crankshaft slowly clockwise until tool arm rests against water pump flange. Secure in position.
8. Ensure crankshaft timing marks aligned **4**.
9. Ensure timing marks on camshaft sprockets aligned **5**.
10. Fit locking tools to camshafts **6**.
    KM-800-1: cyls. 1-3-5. KM-800-2: cyls. 2-4-6.
11. Hold tensioner pulley. Use tool No.83 94 983. Slacken nut **7**.
12. Hold guide pulleys. Use tool No.83 94 983. Slacken bolts **8** & **9**.
13. Turn tensioner pulley clockwise. Use Allen key (5 mm). Lightly tighten nut **7**.
14. Remove:
    - ❏ Crankshaft locking tool **3**.
    - ❏ Guide pulley (G1) and support washer.
    - ❏ Timing belt.
    NOTE: Mark direction of rotation and position of sprockets on belt with chalk if belt is to be reused.

Installation

1. Ensure crankshaft timing marks aligned **4**.
2. Ensure locking tools located correctly on camshafts **6**. KM-800-1: cyls. 1-3-5. KM-800-2: cyls. 2-4-6.
3. Ensure timing marks on camshaft sprockets aligned **5**.
4. Fit timing belt to crankshaft sprocket. Wedge belt into position. Use tool No.KM-800-30 **10**.
5. Ensure double lines on belt **11** aligned with crankshaft timing marks.
6. Ensure single lines on belt **12** aligned with timing marks on camshaft sprockets.

➡

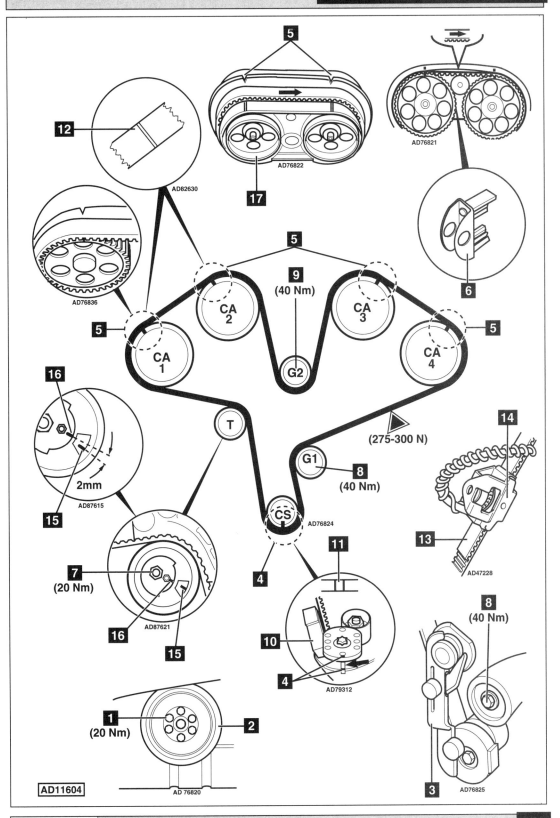

AD82630

12

AD76836

AD76822

17

AD76821

5

6

5

9
(40 Nm)

CA 2

CA 3

CA 1

G2

CA 4

5

5

16

15
2mm
AD87615

T

G1

8
(40 Nm)

▲ (275-300 N)

14

AD47228

13

7
(20 Nm)

16
AD87621

15

CS
AD76824

4

11

10

4
AD79312

8
(40 Nm)

3
AD76825

1
(20 Nm)

2

AD11604

AD 76820

←

7. Fit timing belt to remaining sprockets and pulleys. Ensure belt is taut between sprockets and pulleys.

8. Fit guide pulley (G1) and support washer.

9. Turn lower guide pulley (G1) anti-clockwise to remove slack from belt. Lightly tighten bolt **8**.

10. Remove wedge **10**.

11. Fit locking tool to crankshaft **3**.
Tool No.KM-800-10.

12. Position a spare piece of belt **13** onto timing belt. Attach tension gauge to belt at ▽ **14**.
Tool No.83 93 985.

13. Turn lower guide pulley (G1) anti-clockwise until tension gauge indicates 275-300 N. Tighten bolt to 40 Nm **8**.

14. Remove tension gauge **14**. Remove spare piece of timing belt **13**.

15. Ensure upper guide pulley (G2) bolt turns freely **9**.

16. Turn tensioner pulley until marks aligned **15** & **16**. Use Allen key (5 mm).

17. Tighten nut to 20 Nm **7**.

18. Remove locking tool from camshaft sprockets CA1 & CA2 **6**. Tool No.KM-800-1.

19. Fit timing gauge to camshaft sprockets CA1 & CA2 **17**. Tool No.KM-800-20.

20. Turn upper guide pulley (G2) anti-clockwise until marks on sprockets **5** align with marks on timing gauge **17**. Use tool No.83 94 983.

21. Tighten guide pulley (G2) bolt to 40 Nm **9**.

22. Remove:
    ❏ Crankshaft locking tool **3**.
    ❏ Camshaft timing gauge **17**.

23. Remove locking tool from camshaft sprockets CA3 & CA4 **6**.

24. Turn crankshaft slowly two turns clockwise until just before timing marks aligned **4** & **5**.

25. Fit locking tool to crankshaft **3**.
Tool No.KM-800-10.

26. Turn crankshaft slowly clockwise until tool arm rests against water pump flange.

27. Ensure crankshaft timing marks aligned **4**.

28. Fit timing gauge to camshaft sprockets CA1 & CA2 **17**. Tool No.KM-800-20.

29. Ensure marks on sprockets aligned with marks on timing gauge **5**. Check edge of timing belt aligned with edge of sprockets.
    NOTE: Belt marks are for installation purposes only. They will no longer align once crankshaft has been turned.

30. Check tensioner mark **16** is now approximately 2 mm above mark **15**.

31. If mark **16** is less than 2 mm above mark **15**: Repeat installation and tensioning procedures.

32. Install components in reverse order of removal.

33. Tighten crankshaft pulley bolts to 20 Nm **1**.

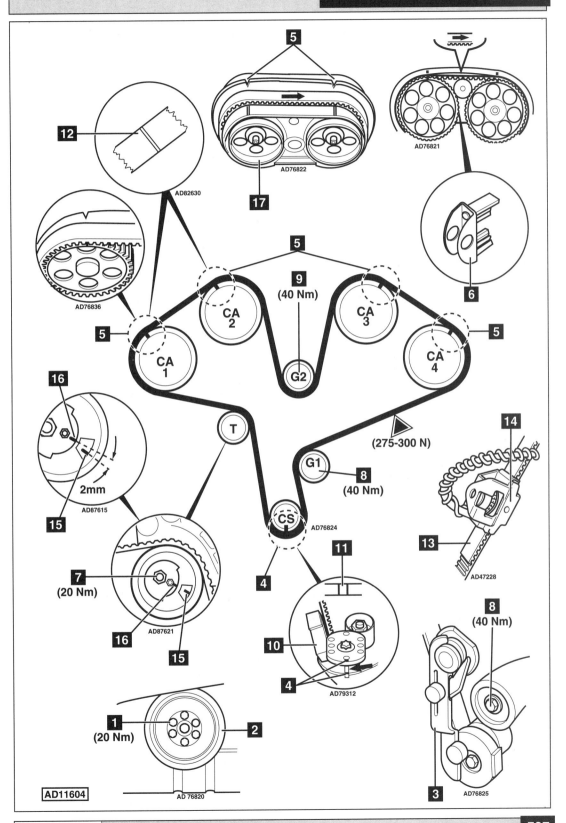

SEAT

Model:	Arosa 1,0/1,4 • Ibiza 1,0/1,4/1,6 • Cordoba 1,0/1,4/1,6 • Inca 1,4/1,6
Year:	1996-03
Engine Code:	AEE, AER, AEX, AKK, AKV, ALD, ALL, ALM, ANV, ANW, ANX, APQ, AUC, AUD

Replacement Interval Guide

Seat recommend check at first 60,000 miles and then check every 20,000 miles, replace if necessary.

No manufacturer's recommended replacement interval.

The previous use and service history of the vehicle must always be taken into account.

Refer to Timing Belt Replacement Intervals at the front of this manual.

Check For Engine Damage

CAUTION: This engine has been identified as an INTERFERENCE engine in which the possibility of valve-to-piston damage in the event of a timing belt failure is MOST LIKELY to occur.

A compression check of all cylinders should be performed before removing the cylinder head(s).

Repair Times – hrs

Remove & install	
Arosa	1,60
Ibiza/Cordoba/Inca:	
→1998	0,80
1999→	1,60
AC	+0,20

Special Tools

- ■ None required.

Special Precautions

- ■ Disconnect battery earth lead.
- ■ DO NOT turn crankshaft or camshaft when timing belt removed.
- ■ Remove spark plugs to ease turning engine.
- ■ Turn engine in normal direction of rotation (unless otherwise stated).
- ■ DO NOT turn engine via camshaft or other sprockets.
- ■ Observe all tightening torques.

Removal

1. Raise and support front of vehicle.
2. Remove:
   - ❑ Air filter intake duct.
   - ❑ RH wheel arch mud shield (if fitted).
   - ❑ Auxiliary drive belt.
3. Arosa:
   - ❑ Support engine.
   - ❑ Remove RH engine mounting.
   - ❑ Lower engine slightly.

4. Remove:
   - ❑ Crankshaft pulley bolts **1**.
   - ❑ Crankshaft pulley **2**.
   - ❑ Timing belt upper cover **3**.
   - ❑ Timing belt lower cover **4**.
5. Turn crankshaft to TDC on No.1 cylinder. Ensure timing marks aligned **5** & **6**.
6. If timing mark on camshaft not aligned **6**: Turn crankshaft one turn clockwise.
7. Slacken tensioner nut **7**. Turn tensioner anti-clockwise away from belt. Lightly tighten nut **7**.
8. Remove timing belt.
   NOTE: Mark direction of rotation on belt with chalk if belt is to be reused.

Installation

1. Ensure timing marks aligned **5** & **6**.
2. Fit timing belt.
3. Ensure tensioner baseplate is supported by bolt **8**.
4. Slacken tensioner nut **7**.
5. Turn tensioner clockwise until pointer aligned with notch **9**.
6. Tighten tensioner nut to 20 Nm **7**.
7. Turn crankshaft two turns clockwise to TDC on No.1 cylinder.
8. Ensure timing marks aligned **5** & **6**.
9. Ensure tensioner pointer aligned with notch **9**.
10. If not: Repeat tensioning procedure.
11. Apply firm thumb pressure to belt at ▽.
12. Turn crankshaft two turns clockwise to TDC on No.1 cylinder.
13. Ensure tensioner pointer aligned with notch **9**.
14. Install:
    - ❑ Timing belt lower cover **4**.
    - ❑ Timing belt upper cover **3**.
    - ❑ Crankshaft pulley **2**.
    - ❑ Crankshaft pulley bolts **1**.
15. Tighten crankshaft pulley bolts to 20 Nm **1**.
16. Arosa:
    - ❑ Raise engine slightly.
    - ❑ Renew RH engine mounting bolts **10**, **11** & **12**.
    - ❑ Fit RH engine mounting.
17. Arosa – tighten:
    - ❑ Engine mounting bolts **10**.
      Tightening torque: 25 Nm + 45°.
    - ❑ Engine mounting bolts **11**.
      Tightening torque: 50 Nm.
    - ❑ Engine mounting bolts **12**.
      Tightening torque: 25 Nm.
18. Install components in reverse order of removal.

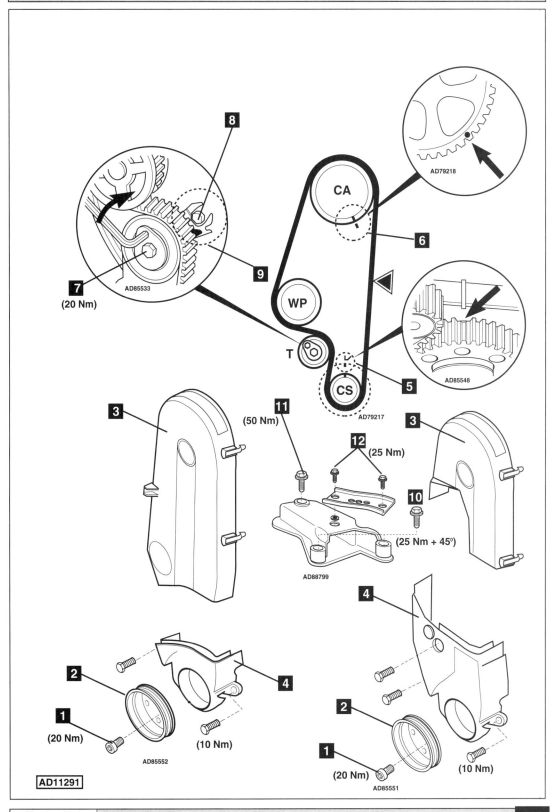

CA

WP

T

CS

8

7
(20 Nm)
AD85533

9

6

AD79218

5
AD85548

AD79217

3

11
(50 Nm)

12 (25 Nm)

3

10

(25 Nm + 45°)

AD88799

4

2

4

1
(20 Nm)
AD85552

(10 Nm)

2

1
(20 Nm)
AD85551

(10 Nm)

AD11291

SEAT

Model:	**Arosa 1,4 16V • Ibiza 1,4 16V • Cordoba 1,4 16V • Toledo 1,4 16V Leon 1,4 16V**
Year:	**1999-03**
Engine Code:	**AFK, AHW, APE, AQQ, AUA, AUB, AXP**

Replacement Interval Guide

Seat recommend check at first 60,000 miles and then check every 20,000 miles, replace if necessary. No manufacturer's recommended replacement interval.
The previous use and service history of the vehicle must always be taken into account.
Refer to Timing Belt Replacement Intervals at the front of this manual.

Check For Engine Damage

CAUTION: This engine has been identified as an INTERFERENCE engine in which the possibility of valve-to-piston damage in the event of a timing belt failure is MOST LIKELY to occur.
A compression check of all cylinders should be performed before removing the cylinder head(s).

Repair Times – hrs

Remove & install:

Arosa	1,90
Ibiza/Cordoba	1,60
Toledo/Leon	2,50

Special Tools

- ■ Camshaft locking tool – No.T10016.
- ■ Crankshaft pulley holding tool – No.T20018/A.

Special Precautions

- ■ Disconnect battery earth lead.
- ■ DO NOT turn crankshaft or camshaft when timing belt removed.
- ■ Remove spark plugs to ease turning engine.
- ■ Turn engine in normal direction of rotation (unless otherwise stated).
- ■ DO NOT turn engine via camshaft or other sprockets.
- ■ Observe all tightening torques.

Removal

Timing Belt

1. Remove:
   - ❑ Upper engine cover.
   - ❑ Air filter assembly.
   - ❑ PAS reservoir. DO NOT disconnect hose(s) (Leon/Toledo).
   - ❑ Timing belt upper cover **1**.
2. Turn crankshaft clockwise to TDC on No.1 cylinder. Ensure crankshaft pulley timing marks aligned **2**.
3. Ensure camshaft sprocket locating holes aligned **3**.
4. If locating holes are not aligned: Turn crankshaft one turn clockwise.
5. Fit locking tool to camshaft sprockets **4**. Tool No.T10016 **5**.
   NOTE: Ensure locking tool located correctly in cylinder head.

6. Arosa/Toledo/Leon:
   - ❑ Support engine.
   - ❑ Remove RH engine mounting and bracket.
7. Remove:
   - ❑ RH engine undershield.
   - ❑ Auxiliary drive belt.
8. Raise and support front of vehicle.
9. Lower engine until crankshaft pulley bolt accessible.
10. Fit crankshaft pulley holding tool. Tool No.T20018/A.
11. Slacken crankshaft pulley bolt **6**.
12. Remove:
    - ❑ Holding tool. Tool No.T20018/A.
    - ❑ Crankshaft pulley bolt **6**.
    - ❑ Crankshaft pulley **7**.
13. Fit two washers to crankshaft pulley bolt **6**.
14. Fit crankshaft pulley bolt **6**. Lightly tighten bolt.
15. Remove:
    - ❑ Auxiliary drive belt tensioner.
    - ❑ Auxiliary drive belt guide pulley (models with AC).
    - ❑ Timing belt lower cover **8**.
16. Slacken tensioner pulley bolt **9**.
17. Turn tensioner pulley anti-clockwise to release tension on belt.
18. Remove timing belt.
    NOTE: Mark direction of rotation on belt with chalk if belt is to be reused.

Installation

Timing Belt

1. Ensure locking tool fitted to camshaft sprockets **4**. Tool No.T10016 **5**.
2. Ensure crankshaft sprocket timing marks aligned **10**.
   NOTE: Ground tooth on crankshaft sprocket aligned with timing mark labelled '4V' on oil pump housing.
3. Fit timing belt in anti-clockwise direction, starting at water pump sprocket.
4. Tighten tensioner pulley bolt finger tight **9**. Ensure baseplate is supported by bolt **11**.
5. Turn tensioner pulley clockwise **12** until pointer **13** aligned with notch in baseplate **14**.
6. Tighten tensioner pulley bolt to 20 Nm **9**.
7. Remove locking tool from camshaft sprockets **4**. Tool No.T10016 **5**.
8. Turn crankshaft two turns clockwise to TDC on No.1 cylinder. Ensure crankshaft sprocket timing marks aligned **10**.
9. Ensure locking tool can be inserted into camshaft sprockets **4**. Tool No.T10016 **5**.
10. Ensure pointer **13** aligned with notch in baseplate **14**.
11. If not: Repeat tensioning procedure.
12. Apply firm thumb pressure to belt at ▽. Pointer **13** and notch in baseplate **14** must move apart.

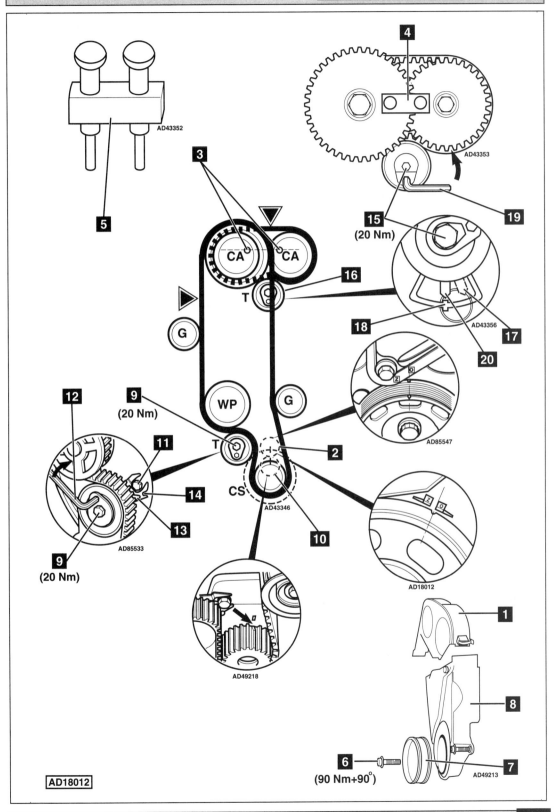

13. Release thumb pressure from belt at ▽.
14. Turn crankshaft two turns clockwise to TDC on No.1 cylinder.
15. Ensure pointer **13** aligned with notch in baseplate **14**.
16. Remove crankshaft pulley bolt **6**.
17. Install:
    - ❏ Timing belt lower cover **8**.
    - ❏ Crankshaft pulley **7**.
    - ❏ New oiled crankshaft pulley bolt **6**.
18. Tighten crankshaft pulley bolt **6**.
    Tightening torque: 90 Nm + 90°.
19. Install components in reverse order of removal.
20. Check engine control module (ECM) fault memory.
21. Arosa:
    - ❏ Tighten bolts securing engine mounting bracket to engine. Use new bolts.
      Tightening torque: 50 Nm.
    - ❏ Tighten long engine mounting bolts. Use new bolts. Tightening torque: 40 Nm + 90°.
    - ❏ Tighten short engine mounting bolt.
      Tightening torque: 50 Nm.
22. Toledo/Leon:
    - ❏ Tighten bolts securing engine mounting bracket to engine. Tightening torque: 50 Nm.
    - ❏ Tighten long bolts securing engine mounting to body. Use new bolts.
      Tightening torque: 40 Nm + 90°.
    - ❏ Tighten short bolts securing engine mounting to body. Use new bolts. Tightening torque: 25 Nm.
    - ❏ Tighten bolts securing engine mounting to bracket. Use new bolts.
      Tightening torque: 60 Nm + 90°.

Removal

Exhaust Camshaft Drive Belt

1. Remove timing belt as described previously.
2. Slacken tensioner pulley bolt **15**.
3. Turn tensioner pulley clockwise to release tension on belt.
4. Remove:
    - ❏ Tensioner pulley bolt **15**.
    - ❏ Tensioner pulley **16**.
    - ❏ Drive belt.
    NOTE: Mark direction of rotation on belt with chalk if belt is to be reused.

Installation

Exhaust Camshaft Drive Belt

1. Ensure crankshaft sprocket secured to crankshaft.
2. Ensure crankshaft sprocket timing marks aligned **10**.
   NOTE: Ground tooth on crankshaft sprocket aligned with timing mark labelled '4V' on oil pump housing.
3. Ensure locking tool fitted to camshaft sprockets **4**. Tool No.T10016 **5**.
4. Fit drive belt in clockwise direction, starting at top of inlet camshaft sprocket.
5. Ensure belt is taut between sprockets on non-tensioned side.
6. Turn tensioner pulley clockwise until pointer in position as shown **17**.

7. Install:
    - ❏ Tensioner pulley **16**.
    - ❏ Tensioner pulley bolt **15**.
8. Tighten tensioner pulley bolt finger tight **15**.
   *NOTE: Ensure lug in baseplate **18** is located in cylinder head hole.*
9. Turn tensioner anti-clockwise **19** until pointer **20** aligned with lug in baseplate **18**.
10. Tighten tensioner pulley bolt to 20 Nm **15**.
11. Fit timing belt as described previously.
12. Remove locking tool from camshaft sprockets **4**. Tool No.T10016 **5**.
13. Turn crankshaft two turns clockwise to TDC on No.1 cylinder. Ensure crankshaft sprocket timing marks aligned **10**.
14. Ensure locking tool can be inserted into camshaft sprockets **4**. Tool No.T10016 **5**.
15. Ensure pointer **20** aligned with lug in baseplate **18**.
16. If not: Repeat tensioning procedure.
17. Apply firm thumb pressure to belt at ▽. Pointer **20** and lug in baseplate **18** must move apart.
18. Release thumb pressure from belt at ▽.
19. Turn crankshaft two turns clockwise to TDC on No.1 cylinder.
20. Ensure pointer **20** aligned with lug in baseplate **18**.
21. Install components in reverse order of removal.

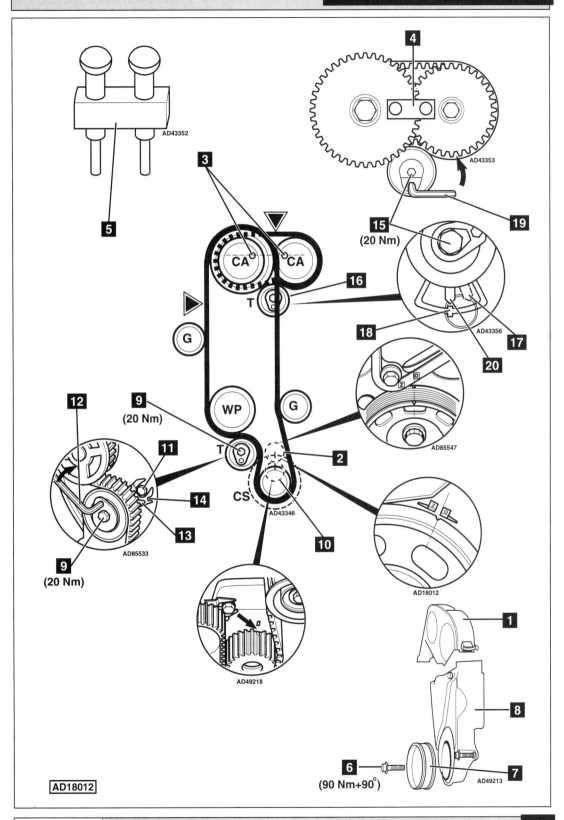

AD43352

AD43353

5

4

3

CA CA

T

G

16

15 (20 Nm)

18

19

AD43356

17

20

12

9 (20 Nm)

11

T

WP

G

2

AD85547

14

CS

13

AD85533

9 (20 Nm)

10

AD43346

AD18012

AD49218

1

8

6 (90 Nm+90°)

7

AD49213

SEAT

Model:	Ibiza/Cordoba 1,05/1,3/1,4/1,6
Year:	1993-99
Engine Code:	Water pump tensioner AAV, ABU, AAU, ABD, 2G

Replacement Interval Guide

Seat recommend check every 20,000 miles, and replacement if necessary.
No manufacturer's recommended replacement interval.
The previous use and service history of the vehicle must always be taken into account.
Refer to Timing Belt Replacement Intervals at the front of this manual.

Check For Engine Damage

CAUTION: This engine has been identified as an INTERFERENCE engine in which the possibility of valve-to-piston damage in the event of a timing belt failure is MOST LIKELY to occur.
A compression check of all cylinders should be performed before removing the cylinder head.

Repair Times – hrs

Check & adjust	0,20
Remove & install	0,80

Special Tools

■ None required.

Special Precautions

■ Disconnect battery earth lead.
■ DO NOT turn crankshaft or camshaft when timing belt removed.
■ Remove spark plugs to ease turning engine.
■ Turn engine in normal direction of rotation (unless otherwise stated).
■ DO NOT turn engine via camshaft or other sprockets.
■ Observe all tightening torques.

Removal

1. Disconnect air intake pipe.
2. Remove:
   ❑ Timing belt upper cover **1**.
   ❑ Auxiliary drive belt.
   NOTE: Models with Poly-V belt: Fit ring spanner to tensioner nut. Turn clockwise to release tension on belt.
3. Turn crankshaft to TDC on No.1 cylinder. Ensure timing marks aligned **2** & **3**.
4. Remove:
   ❑ Inner wing panel.
   ❑ Crankshaft damper bolts (4 bolts) **5**.
   ❑ Crankshaft damper.
   ❑ Timing belt lower cover.

5. Slacken water pump bolts **4**.
6. Turn water pump anti-clockwise to release tension on belt.
7. Remove timing belt.

Installation

1. Ensure timing marks aligned **2** & **3**.
2. Fit timing belt in anti-clockwise direction. Ensure belt is taut between sprockets on non-tensioned side.
3. Turn water pump clockwise to tension belt. Use screwdriver.
   *NOTE: Belt is correctly tensioned when it can just be twisted with finger and thumb through 90° at ▽ **6**.*
4. Tighten water pump bolts to 10 Nm.
5. Install components in reverse order of removal.

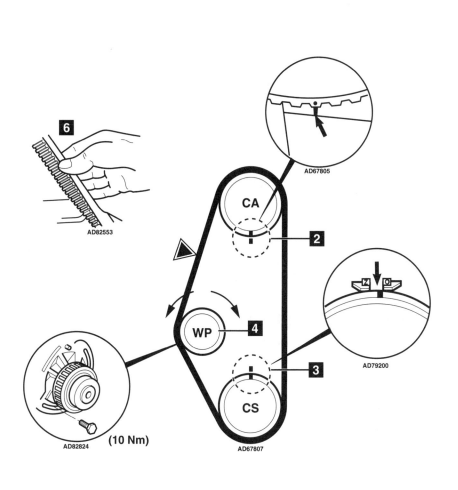

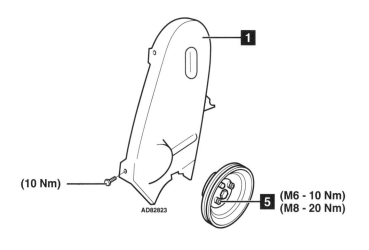

AD10543

Model:	Ibiza 1,4 16V • Cordoba 1,4 16V
Year:	1995-99
Engine Code:	AFH

Replacement Interval Guide

The vehicle manufacturer has not recommended a timing belt replacement interval for this engine. *The previous use and service history of the vehicle must always be taken into account.*
Refer to Timing Belt Replacement Intervals at the front of this manual.

Check For Engine Damage

CAUTION: This engine has been identified as an INTERFERENCE engine in which the possibility of valve-to-piston damage in the event of a timing belt failure is MOST LIKELY to occur.
A compression check of all cylinders should be performed before removing the cylinder head.

Repair Times – hrs

Information not available.

Special Tools

■ Crankshaft pulley holding tool – No.T20018.

Special Precautions

■ Disconnect battery earth lead.
■ DO NOT turn crankshaft or camshaft when timing belt removed.
■ Remove spark plugs to ease turning engine.
■ Turn engine in normal direction of rotation (unless otherwise stated).
■ DO NOT turn engine via camshaft or other sprockets.
■ Observe all tightening torques.

Removal

1. Remove:
   ❏ Auxiliary drive belt cover.
   ❏ Auxiliary drive belt.
2. Hold crankshaft pulley **1**. Use tool No.T20018.
3. Remove:
   ❏ Crankshaft pulley bolt **2**.
   ❏ Crankshaft pulley **1**.
4. Fit crankshaft pulley bolt **2** with two large washers and tighten.
5. Remove:
   ❏ Timing belt upper cover **3**.
   ❏ Timing belt lower cover **4**.
6. Turn crankshaft clockwise until timing marks aligned **5** & **6**.
   NOTE: Crankshaft sprocket has one tooth ground at an angle.

7. Slacken tensioner bolt **7**.
8. Remove timing belt.
   NOTE: Mark direction of rotation on belt with chalk if belt is to be reused.

Installation

1. Ensure timing marks aligned **5** & **6**.
2. Fit timing belt in anti-clockwise direction, starting at crankshaft sprocket. Ensure belt is taut between sprockets.
3. Tighten tensioner bolt finger tight **7**. Ensure baseplate is supported by bolt **8**.
4. Turn tensioner clockwise **9** until pointer **10** aligned with notch in baseplate **11**.
5. Tighten tensioner bolt to 20 Nm **7**.
6. Turn crankshaft two turns clockwise to TDC on No.1 cylinder. Ensure timing marks aligned **5** & **6**.
7. Ensure pointer **10** aligned with notch in baseplate **11**.
8. Remove crankshaft pulley bolt **2**. Discard two large washers.
9. Install components in reverse order of removal.
10. Hold crankshaft pulley **1**. Use tool No.T20018.
11. Fit new oiled crankshaft pulley bolt **2**.
    Tightening torque: 90 Nm + 90°.

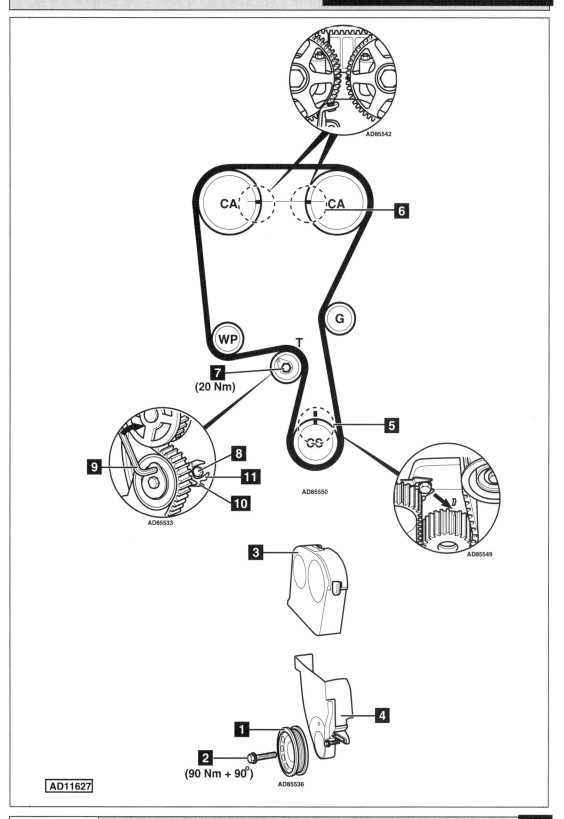

AD85542

CA   CA   6

G

WP   T

7
(20 Nm)

5

9

8
11
10

AD85533

AD85550

AD85549

3

4

1

2
(90 Nm + 90°)   AD85536

SEAT

Model:	**Ibiza/Cordoba 1,6/1,8/2,0 • Toledo 1,8/2,0 • Alhambra 2,0 • Inca 1,6**
Year:	**1993-02**
Engine Code:	**Timing mark on front of camshaft sprocket** **ABS, ADY, ADZ, AGG, 1F, 2E**

Replacement Interval Guide

Seat recommend:

Ibiza/Cordoba/Toledo:
Check every 20,000 miles (replace if necessary).

Alhambra:
Check at the first 60,000 miles and then every 20,000 miles (replace if necessary).

Inca:
→1999: Check every 20,000 miles (replace if necessary).
2000→: Check at the first 60,000 miles and then every 20,000 miles (replace if necessary).
The previous use and service history of the vehicle must always be taken into account.
Refer to Timing Belt Replacement Intervals at the front of this manual.

Check For Engine Damage

CAUTION: This engine has been identified as an INTERFERENCE engine in which the possibility of valve-to-piston damage in the event of a timing belt failure is MOST LIKELY to occur.
A compression check of all cylinders should be performed before removing the cylinder head.

Repair Times – hrs

Except Alhambra:	
Check & adjust	0,70
Remove & install	1,20
AC	+0,20
Alhambra:	
Check & adjust	0,90
Remove & install	1,90

Special Tools

■ Two-pin wrench – SEAT No.U-30009/Matra V.159.

Special Precautions

■ Disconnect battery earth lead.
■ DO NOT turn crankshaft or camshaft when timing belt removed.
■ Remove spark plugs to ease turning engine.
■ Turn engine in normal direction of rotation (unless otherwise stated).
■ DO NOT turn engine via camshaft or other sprockets.
■ Observe all tightening torques.

Removal

1. Raise and support front of vehicle.
2. Disconnect air intake trunking for access.
3. Remove:
   ❑ Timing belt upper cover **1**.
   ❑ Auxiliary drive belts.
   ❑ Auxiliary drive belt tensioner.
   ❑ Crankshaft pulley bolts **2**.
   ❑ Crankshaft pulley **3**.
   ❑ Water pump pulley (if required).
   ❑ Timing belt lower cover **4**.
4. Temporarily fit crankshaft pulley. Retain with two bolts lightly tightened.
5. Turn crankshaft clockwise until timing marks aligned **5** & **6**.
   NOTE: Alhambra: Use flywheel timing marks 8.
6. Slacken tensioner nut **7**.
7. Turn tensioner anti-clockwise away from belt. Use tool No.U-30009/Matra V.159.
8. Lightly tighten tensioner nut.
9. Remove:
   ❑ Crankshaft pulley **3**.
   ❑ Timing belt.
   NOTE: Mark direction of rotation on belt with chalk if belt is to be reused.

Installation

1. Ensure timing marks aligned. All models: **6**. Alhambra: **8**.
2. Fit timing belt to crankshaft sprocket and auxiliary shaft sprocket.
3. Temporarily fit crankshaft pulley. Retain with two bolts lightly tightened.
4. Ensure timing mark on crankshaft pulley aligned with mark on auxiliary shaft sprocket **5**.
5. Fit timing belt to camshaft sprocket and tensioner pulley.
6. Ensure timing marks aligned **6** & **5** or **8**.
7. Turn tensioner clockwise until belt can just be twisted with finger and thumb through 90° at ▽. Use tool No.U-30009/Matra V.159.
8. Tighten tensioner nut to 45 Nm **7**.
9. Turn crankshaft two turns clockwise. Ensure timing marks aligned **6** & **5** or **8**.
10. Recheck belt tension.
11. Install components in reverse order of removal.
12. Tighten crankshaft pulley bolts to 20 Nm **2**.

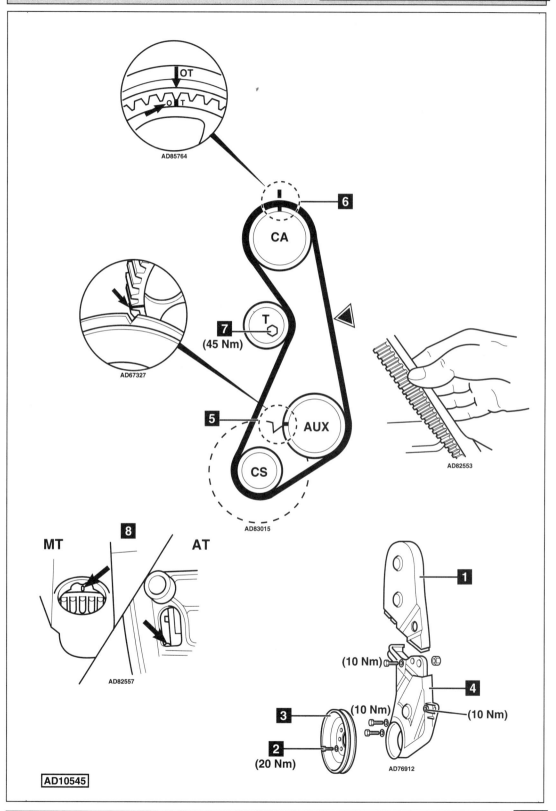

AD85764

6

CA

AD67327

7 T
(45 Nm)

5 AUX

CS

AD82553

AD83015

MT 8 AT

AD82557

1

(10 Nm)

4

(10 Nm)

3 (10 Nm)

2
(20 Nm)

AD76912

AD10545

SEAT

Model:	Ibiza/Cordoba/Toledo 1,6
Year:	1996-00
Engine Code:	AFT, AKS

Replacement Interval Guide

Seat recommend check every 20,000 miles, and replacement if necessary.
The previous use and service history of the vehicle must always be taken into account.
Refer to Timing Belt Replacement Intervals at the front of this manual.

Check For Engine Damage

CAUTION: This engine has been identified as an INTERFERENCE engine in which the possibility of valve-to-piston damage in the event of a timing belt failure is MOST LIKELY to occur.
A compression check of all cylinders should be performed before removing the cylinder head(s).

Repair Times – hrs

Check & adjust	0,70
Remove & install	1,20

Special Tools

- Two-pin wrench – SEAT No.U-30009.

Special Precautions

- Disconnect battery earth lead.
- DO NOT turn crankshaft or camshaft when timing belt removed.
- Remove spark plugs to ease turning engine.
- Turn engine in normal direction of rotation (unless otherwise stated).
- DO NOT turn engine via camshaft or other sprockets.
- Observe all tightening torques.

Removal

1. Raise and support front of vehicle.
2. Remove:
   - ❏ Air filter intake duct.
   - ❏ Timing belt upper cover **1**.
3. Turn crankshaft to TDC on No.1 cylinder.
4. Ensure timing marks aligned **2** or **3** & **4**.
5. If timing mark on camshaft sprocket not aligned **4**: Turn crankshaft one turn clockwise.
6. Remove:
   - ❏ Auxiliary drive belts.
   - ❏ Crankshaft pulley bolts **5**.
   - ❏ Crankshaft pulleys **6**.
   - ❏ Water pump pulley **7**.
   - ❏ Timing belt lower cover **8**.
   - ❏ Distributor cap.

7. Ensure timing marks aligned **3** & **4**.
8. Ensure distributor rotor arm aligned with mark for cylinder No.1 on distributor body **9**.
9. Slacken tensioner nut **10**. Turn tensioner anti-clockwise away from belt. Lightly tighten nut **10**.
10. Remove timing belt.
    NOTE: Mark direction of rotation on belt with chalk if belt is to be reused.

Installation

1. Ensure timing marks aligned **3** & **4**.
2. Ensure distributor rotor arm aligned with mark for cylinder No.1 on distributor body **9**.
3. Fit timing belt in anti-clockwise direction, starting at crankshaft sprocket.
4. Slacken tensioner nut **10**.
5. Turn tensioner clockwise until belt can just be twisted with finger and thumb through 90° at ▽. Use tool No.U-30009.
6. Tighten tensioner nut to 45 Nm **10**.
7. Turn crankshaft two turns clockwise to TDC on No.1 cylinder. Ensure timing marks aligned **3** & **4**.
8. Ensure distributor rotor arm aligned with mark for cylinder No.1 on distributor body **9**.
9. Recheck belt tension.
10. Install components in reverse order of removal.
11. Tighten crankshaft pulley bolts to 25 Nm **5**.

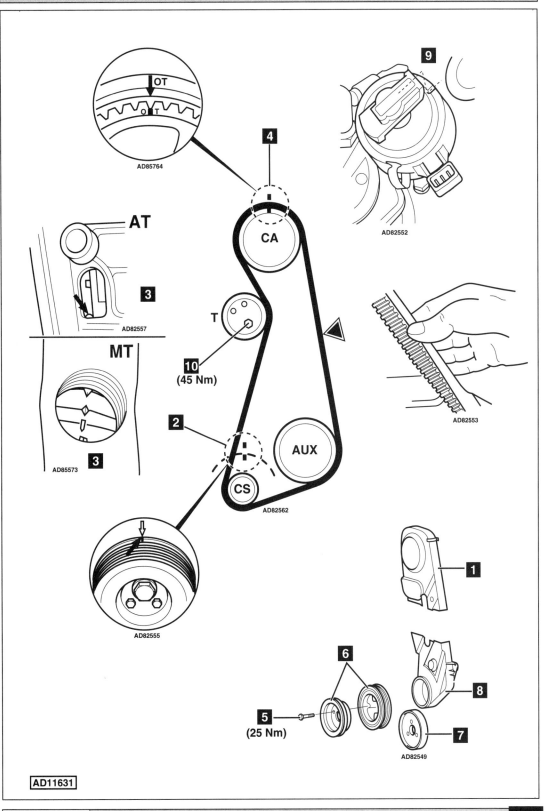

AD85764

AT

3

AD82557

MT

3

AD85573

4

CA

T

10
(45 Nm)

2

AUX

CS

AD82562

AD82555

9

AD82552

AD82553

1

6

5
(25 Nm)

8

7

AD82549

AD11631

SEAT

Model:	Ibiza 1,6 • Cordoba 1,6 • Leon 1,6 • Toledo 1,6
Year:	1999-02
Engine Code:	AEH, AKL, APF, AUR

Replacement Interval Guide

Seat recommend check at first 60,000 miles and then check every 20,000 miles, replace if necessary.
No manufacturer's recommended replacement interval.
The previous use and service history of the vehicle must always be taken into account.
Refer to Timing Belt Replacement Intervals at the front of this manual.

Check For Engine Damage

CAUTION: This engine has been identified as an INTERFERENCE engine in which the possibility of valve-to-piston damage in the event of a timing belt failure is MOST LIKELY to occur.
A compression check of all cylinders should be performed before removing the cylinder head(s).

Repair Times – hrs

Remove & install	2,50

Special Tools

■ Two-pin wrench – No.U-30009/A or T10020.

Removal

1. Raise and support front of vehicle.
2. Remove:
   ❏ RH engine undershield.
   ❏ Auxiliary drive belt.
   ❏ Auxiliary drive belt tensioner **1**.
   ❏ Timing belt upper cover **2**.
3. Turn crankshaft to TDC on No.1 cylinder.
4. Ensure timing marks aligned **3** or **4**.
5. Ensure camshaft sprocket timing mark aligned **5**.
6. Support engine.
7. Remove:
   ❏ Coolant expansion tank. DO NOT disconnect hoses.
   ❏ PAS reservoir. DO NOT disconnect hoses.
   ❏ RH engine mounting bolts **6**, **7** & **8**.
   ❏ RH engine mounting.
   ❏ Crankshaft pulley bolts **9**.
   ❏ Crankshaft pulley **10**.
   ❏ Timing belt centre cover **11**.
   ❏ Timing belt lower cover **12**.
   ❏ RH engine mounting bracket bolts **13**.
   ❏ RH engine mounting bracket.
8. Slacken tensioner nut **14**. Turn tensioner clockwise away from belt. Lightly tighten nut.
9. Remove timing belt.
   NOTE: Mark direction of rotation on belt with chalk if belt is to be reused.

Installation

1. Ensure timing marks aligned **4** & **5**.
2. Fit timing belt to crankshaft sprocket and water pump sprocket.
3. Install:
   ❏ RH engine mounting bracket.
   ❏ RH engine mounting bracket bolts **13**.
4. Tighten RH engine mounting bracket bolts to 45 Nm **13**.
5. Install:
   ❏ Timing belt lower cover **12**.
   ❏ Timing belt centre cover **11**.
   ❏ Crankshaft pulley **10**.
   ❏ Crankshaft pulley bolts **9**.
6. Tighten crankshaft pulley bolts **9**. Tightening torque: 25 Nm.
7. Fit engine mounting.
8. Tighten:
   ❏ Engine mounting bolts **8**. Tightening torque: 40 Nm + 90°. Use new bolts.
   ❏ Engine mounting bolts **7**. Tightening torque: 25 Nm. Use new bolts.
   ❏ Engine mounting bolts **6**. Tightening torque: 60 Nm + 90°. Use new bolts.
9. Check engine mounting alignment:
   ❏ Engine mounting clearance: 14 mm **15**.
   ❏ Engine mounting clearance: 10 mm minimum **16**.
   ❏ Ensure engine mounting bolts **6** aligned with edge of mounting **17**.
10. Install:
    ❏ Coolant expansion tank.
    ❏ PAS reservoir.
11. Fit timing belt to tensioner pulley and camshaft sprocket.
12. Ensure timing marks aligned **3** or **4** & **5**.
    NOTE: Ensure belt is taut between sprockets on non-tensioned side.
13. Slacken tensioner nut **14**.
14. Ensure tensioner retaining lug is properly engaged **18**.
15. Turn tensioner 5 times fully anti-clockwise and clockwise from stop to stop. Use tool No.U-30009/A or T10020.
16. Turn tensioner fully anti-clockwise then slowly clockwise until pointer **19** aligned with notch **20** in baseplate. Use tool No.U-30009/A or T10020.
17. Tighten tensioner nut **14**. Tightening torque: 20 Nm.
18. Turn crankshaft two turns clockwise to TDC on No.1 cylinder.
    NOTE: Turn crankshaft last 45° smoothly without stopping.
19. Ensure timing marks aligned **3** or **4** & **5**.
20. Ensure pointer **19** aligned with notch **20** in baseplate.
21. If not: Repeat tensioning procedure.
22. Apply firm thumb pressure to belt at ▽. Pointer **19** and notch **20** must move apart.
23. Release thumb pressure from belt at ▽.
24. Turn crankshaft two turns clockwise to TDC on No.1 cylinder.
    NOTE: Turn crankshaft last 45° smoothly without stopping.
25. Ensure pointer **19** aligned with notch **20** in baseplate.
26. Install components in reverse order of removal.

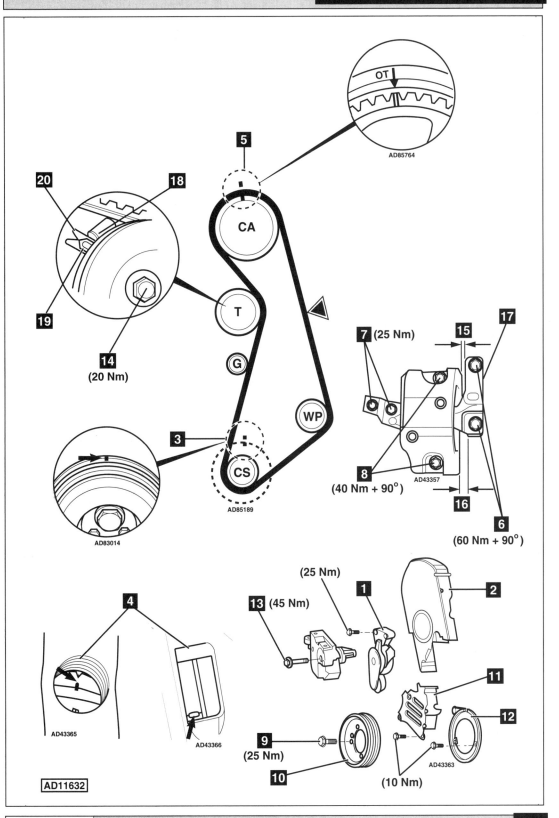

AD85764

5

20 **18**

CA

T

G

WP

14
(20 Nm)

19

3

CS

AD85189

AD83014

7 **(25 Nm)**

15 **17**

8

AD43357

(40 Nm + 90°)

16

6

(60 Nm + 90°)

4

AD43365

AD43366

(25 Nm)

1

2

13 **(45 Nm)**

11

12

AD43363

9
(25 Nm)

10

(10 Nm)

AD11632

Model:	Ibiza 1,8 16V • Cordoba 1,8 16V • Toledo 1,8/2,0 16V
Year:	1991-99
Engine Code:	KR, PL, ABF, ADL

Replacement Interval Guide

Seat recommend check every 20,000 miles, and replacement if necessary.
No manufacturer's recommended replacement interval.
The previous use and service history of the vehicle must always be taken into account.
Refer to Timing Belt Replacement Intervals at the front of this manual.

Check For Engine Damage

CAUTION: This engine has been identified as an INTERFERENCE engine in which the possibility of valve-to-piston damage in the event of a timing belt failure is MOST LIKELY to occur.
A compression check of all cylinders should be performed before removing the cylinder head.

Repair Times – hrs

Check & adjust	0,70
Remove & install	1,20
AC	+0,20

Special Tools

- Tension gauge – SEAT No.U-10028.
- Two-pin wrench – SEAT No.U-30009.

Special Precautions

- Disconnect battery earth lead.
- DO NOT turn crankshaft or camshaft when timing belt removed.
- Remove spark plugs to ease turning engine.
- Turn engine in normal direction of rotation (unless otherwise stated).
- DO NOT turn engine via camshaft or other sprockets.
- Observe all tightening torques.

Removal

1. Remove:
   - Air filter and intake pipe.
   - Auxiliary drive belt(s). Mark direction of rotation.
   - Auxiliary drive belt tensioner (if fitted) **1**.
   - Water pump pulley.
   - Timing belt upper cover **3**.
2. Turn crankshaft to TDC on No.1 cylinder. Ensure timing marks aligned **5**, **7** & **8**.
3. If cylinder head cover removed: Use camshaft sprocket rear timing mark **6**.
   NOTE: Align notch with upper cylinder head face.

4. Remove:
   - Crankshaft pulley bolts (4 bolts).
   - Crankshaft pulley **2**.
   - Crankshaft damper **4**.
   - Timing belt lower cover **9**.
5. Slacken tensioner nut **10**. Turn tensioner anti-clockwise to release tension on belt. Use tool No.U-30009 **12**. Lightly tighten nut.
6. Remove timing belt.

Installation

1. Ensure timing marks aligned **5** & **7**.
2. If cylinder head cover removed: Ensure camshaft sprocket rear timing mark aligned **6**.
   NOTE: Notch aligned with upper cylinder head face.
3. Fit timing belt in anti-clockwise direction, starting at crankshaft sprocket.
4. Attach tension gauge to belt at ▽. Tool No.U-10028 **11**.
5. Slacken tensioner nut **10**.
6. Turn tensioner clockwise until tension gauge indicates 13-14 units. Use tool No.U-30009 **12**.
   NOTE: 2,0 16V: Turn tensioner until belt can just be twisted with finger and thumb through 90° at ▽ 13.
7. Tighten tensioner nut to 45 Nm **10**.
8. Remove tension gauge.
9. Turn crankshaft two turns clockwise to TDC on No.1 cylinder. Ensure timing marks aligned **5**, **6** & **7**.
10. Recheck belt tension.
11. Install components in reverse order of removal.
12. Tighten crankshaft pulley bolts to 20 Nm.

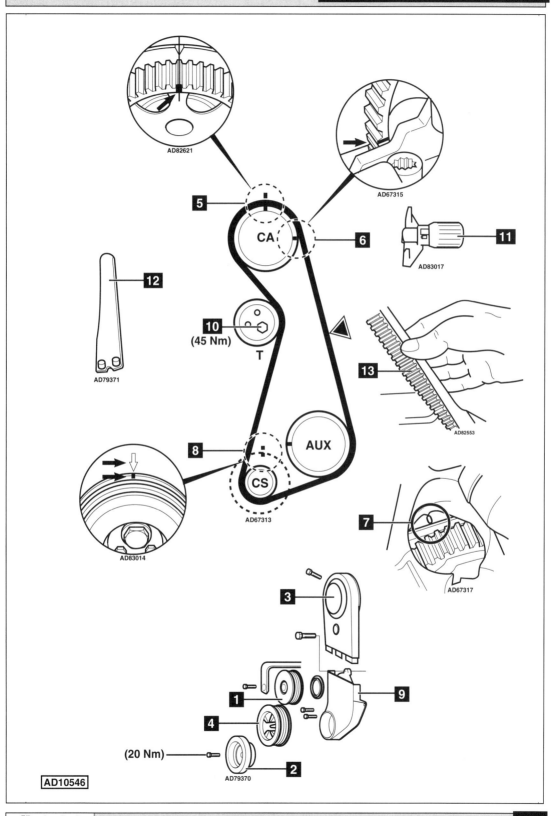

AD82621

AD67315

5

CA

6

11

AD83017

12

AD79371

10
(45 Nm)

T

13

AD82553

8

AUX

CS

AD67313

AD83014

7

AD67317

3

9

1

4

2

AD79370

(20 Nm)

AD10546

SEAT

Model:	**Ibiza/Cordoba 1,8 Turbo • Toledo 1,8 • Leon 1,8/Turbo**
Year:	**1999-03**
Engine Code:	**AGN, APG, APP, APT, AQX, AYP**

Replacement Interval Guide

Seat recommend:
Check at the first 60,000 miles and then every 20,000 miles (replace if necessary).
Replacement every 120,000 miles (tensioner pulley must also be replaced).
The previous use and service history of the vehicle must always be taken into account.
Refer to Timing Belt Replacement Intervals at the front of this manual.

Check For Engine Damage

CAUTION: This engine has been identified as an INTERFERENCE engine in which the possibility of valve-to-piston damage in the event of a timing belt failure is MOST LIKELY to occur.
A compression check of all cylinders should be performed before removing the cylinder head.

Repair Times – hrs

Remove & install:	
Ibiza/Cordoba	1,60
Toledo/Leon	2,20

Special Tools

■ Tensioner bolt/locking pin – SEAT No.T20046.

Removal

1. Raise and support front of vehicle.
2. Remove:
   ❏ Engine top cover (if fitted).
   ❏ Intercooler hoses (Leon).

Toledo/Leon

3. Remove:
   ❏ RH headlamp (Toledo).
   ❏ Coolant expansion tank. DO NOT disconnect hoses.
   ❏ PAS reservoir. DO NOT disconnect hoses.

All models

4. Remove:
   ❏ RH engine undershield.
   ❏ Auxiliary drive belt.
   ❏ Auxiliary drive belt tensioner **1**.
   ❏ Timing belt upper cover **2**.

Toledo/Leon

5. Disconnect vacuum hoses from:
   ❏ Evaporative emission (EVAP) canister.
   ❏ Throttle body.
6. Support engine.
7. Remove:
   ❏ RH engine mounting bolts **3**, **4** & **5**.
   ❏ RH engine mounting.
   ❏ Engine mounting bracket bolts **6**.
   ❏ Engine mounting bracket.

All models

8. Turn crankshaft to TDC on No.1 cylinder.
9. Ensure timing marks aligned **7** or **18**.
10. Ensure camshaft sprocket timing marks aligned **8**.
11. If not: Turn crankshaft one turn clockwise.
12. Remove:
    ❏ Crankshaft pulley bolts **9**.
    ❏ Crankshaft pulley **10**.
    ❏ Timing belt centre cover **11**.
    ❏ Timing belt lower cover **12**.
13. Separate tensioner bolt and locking pin **13** & **14**. Tool No.T20046.
14. Fit tensioner bolt to tensioner **13**.
15. Tighten bolt sufficiently to allow locking pin to be inserted **14**.
16. Remove timing belt.
    NOTE: Mark direction of rotation on belt with chalk if belt is to be reused.

Installation

1. Ensure timing marks aligned **8**.
2. Ensure tensioner bolt and locking pin fitted **13** & **14**. Tool No.T20046.
3. Fit timing belt to crankshaft sprocket.
4. Install:
   ❏ Timing belt lower cover **12**.
   ❏ Crankshaft pulley **10**.
   ❏ Crankshaft pulley bolts **9**.
5. Lightly tighten crankshaft pulley bolts **9**.
6. Ensure timing marks aligned **7**.
7. Fit timing belt in following order:
   ❏ Water pump sprocket.
   ❏ Tensioner pulley.
   ❏ Camshaft sprocket.
   NOTE: Ensure belt is taut between sprockets on non-tensioned side.
8. Remove locking pin **14**.
9. Remove tensioner bolt **13**.
10. Turn crankshaft two turns clockwise. Ensure timing marks aligned **7** & **8**.
11. Tighten crankshaft pulley bolts **9**.
    Tightening torque: 25 Nm.
12. Install components in reverse order of removal.

Toledo/Leon

13. Tighten engine mounting bracket bolts to 45 Nm **6**.
14. Tighten:
    ❏ Engine mounting bolts **5**.
      Tightening torque: 40 Nm + 90°. Use new bolts.
    ❏ Engine mounting bolts **4**.
      Tightening torque: 25 Nm.
    ❏ Engine mounting bolts **3**.
      Tightening torque: 60 Nm + 90°. Use new bolts.
15. Check engine mounting alignment:
    ❏ Engine mounting clearance: 14 mm **15**.
    ❏ Engine mounting clearance: 10 mm minimum **16**.
    ❏ Ensure engine mounting bolts **3** aligned with edge of mounting **17**.

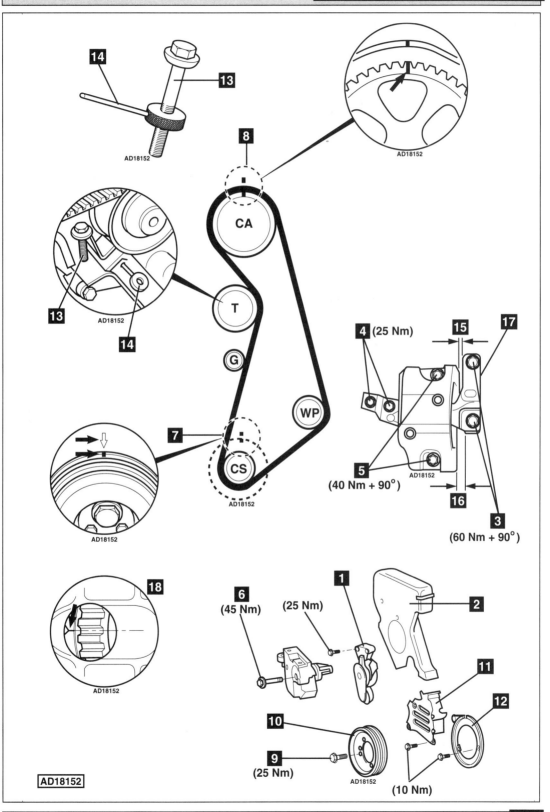

AD18152

14

13

AD18152

8

CA

13

T

14

AD18152

G

4 (25 Nm)

15

17

WP

5

AD18152

7

CS

AD18152

AD18152

(40 Nm + 90°)

16

3

(60 Nm + 90°)

18

AD18152

6
(45 Nm)

1
(25 Nm)

2

11

12

10

9
(25 Nm)

AD18152

(10 Nm)

AD18152

Model:	**Ibiza 2,0 16V • Cordoba 2,0 16V**
Year:	**1996-99**
Engine Code:	**ABF**

Replacement Interval Guide

The vehicle manufacturer has not recommended
a timing belt replacement interval for this engine.
*The previous use and service history of the vehicle
must always be taken into account.*
*Refer to Timing Belt Replacement Intervals at the front
of this manual.*

Check For Engine Damage

*CAUTION: This engine has been identified as an
INTERFERENCE engine in which the possibility of
valve-to-piston damage in the event of a timing belt
failure is MOST LIKELY to occur.*
*A compression check of all cylinders should be
performed before removing the cylinder head(s).*

Repair Times – hrs

Remove & install	1,20

Special Tools

■ Two-pin wrench – SEAT No.U-30009.

Special Precautions

■ Disconnect battery earth lead.
■ DO NOT turn crankshaft or camshaft when timing belt
  removed.
■ Remove spark plugs to ease turning engine.
■ Turn engine in normal direction of rotation (unless
  otherwise stated).
■ DO NOT turn engine via camshaft or other sprockets.
■ Observe all tightening torques.

Removal

1. Raise and support front of vehicle.
2. Remove:
   ❏ Engine undershield.
   ❏ Air intake hose.
   ❏ Air filter housing.
   ❏ Timing belt upper cover **1**.
   ❏ Auxiliary drive belts.
   ❏ Auxiliary drive belt tensioner pulley **2**.
3. Turn crankshaft to TDC on No.1 cylinder.
4. Ensure timing marks aligned **3**, **4** & **5**.
5. If cylinder head cover removed: Use camshaft
   sprocket rear timing mark **6**.
   NOTE: Align notch with upper cylinder head face.
6. Remove:
   ❏ Crankshaft pulley bolts **7**.
   ❏ Crankshaft pulleys **8**.
   ❏ Water pump pulley (if required).
   ❏ Timing belt lower cover **9**.

7. Ensure timing marks aligned **4** & **5** or **6**.
8. Slacken tensioner nut **10**.
9. Turn tensioner anti-clockwise away from belt.
   Lightly tighten nut.
10. Remove timing belt.
    *NOTE: Mark direction of rotation on belt with
    chalk if belt is to be reused.*

Installation

1. Ensure timing marks aligned **4** & **5**.
2. If cylinder head cover removed: Ensure
   camshaft sprocket rear timing mark aligned **6**.
   *NOTE: Notch aligned with upper cylinder head
   face.*
3. Fit timing belt in anti-clockwise direction, starting
   at crankshaft sprocket.
4. Slacken tensioner nut **10**.
5. Turn tensioner clockwise until marks aligned **11**.
   Use tool No.U-30009.
   NOTE: Engine must be cold.
6. Tighten tensioner nut to 25 Nm **10**.
7. Turn crankshaft two turns clockwise to TDC on
   No.1 cylinder. Ensure timing marks aligned
   4 & **5** or **6**.
8. Check tensioner marks aligned **11**.
9. If not: Repeat tensioning procedure.
10. Apply firm thumb pressure to belt at ▽.
    Tensioner marks must move apart.
11. Release thumb pressure from belt at ▽.
12. Tensioner marks should realign **11**.
13. Install components in reverse order of removal.
14. Tighten crankshaft pulley bolts to 20 Nm **7**.

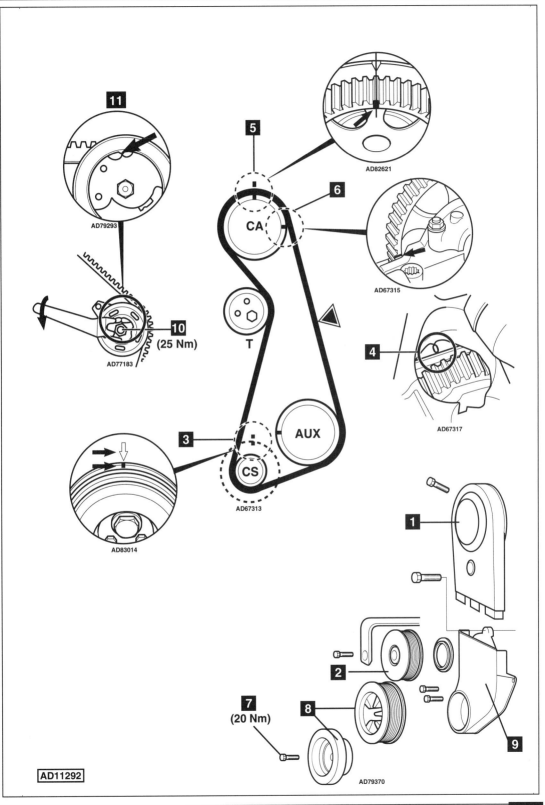

11

AD79293

5

AD82621

6

AD67315

CA

10
(25 Nm)

AD77183

T

4

AD67317

AUX

3

CS

AD67313

AD83014

1

2

7
(20 Nm)

8

9

AD79370

AD11292

SEAT

Model:	Toledo 1,6/1,8
Year:	1991-97
Engine Code:	Timing mark on rear of camshaft sprocket EZ, 1F, RP

Replacement Interval Guide

Seat recommend check every 20,000 miles, and replacement if necessary.
No manufacturer's recommended replacement interval.
The previous use and service history of the vehicle must always be taken into account.
Refer to Timing Belt Replacement Intervals at the front of this manual.

Check For Engine Damage

CAUTION: This engine has been identified as an INTERFERENCE engine in which the possibility of valve-to-piston damage in the event of a timing belt failure is MOST LIKELY to occur.
A compression check of all cylinders should be performed before removing the cylinder head.

Repair Times – hrs

Check & adjust	0,70
Remove & install	1,20
AC	+0,20

Special Tools

- Two-pin wrench – SEAT No.U-30009.

Special Precautions

- Disconnect battery earth lead.
- DO NOT turn crankshaft or camshaft when timing belt removed.
- Remove spark plugs to ease turning engine.
- Turn engine in normal direction of rotation (unless otherwise stated).
- DO NOT turn engine via camshaft or other sprockets.
- Observe all tightening torques.

Removal

1. Remove:
   - ❏ Auxiliary drive belts.
   - ❏ Crankshaft pulley bolts **4**.
   - ❏ Crankshaft pulley **5**.
   - ❏ Timing belt upper cover **6**.
   - ❏ Timing belt lower cover **7**.
2. Temporarily fit crankshaft pulley.
3. Turn crankshaft to TDC on No.1 cylinder. Ensure timing marks on crankshaft pulley and camshaft sprocket aligned **1** & **3**.
4. Ensure distributor rotor arm aligned with mark on distributor body **2**.
5. Slacken tensioner nut **8**. Turn tensioner anti-clockwise.
6. Remove crankshaft pulley.
7. Remove timing belt.

Installation

1. Ensure timing marks aligned **1** & **2**.
2. Fit timing belt to crankshaft sprocket and auxiliary shaft sprocket.
3. Temporarily fit crankshaft pulley **5**.
4. Ensure timing mark on crankshaft pulley aligned with mark on auxiliary shaft sprocket **3**.
5. Fit timing belt to camshaft sprocket and tensioner pulley.
6. Ensure timing marks aligned **1**, **2** & **3**.
7. Turn tensioner clockwise until belt can just be twisted with finger and thumb through 90° at ▽. Use tool No.U-30009.
8. Tighten tensioner nut to 45 Nm **8**.
9. Turn crankshaft two turns clockwise.
10. Ensure timing marks aligned **1**, **2** & **3**.
11. Recheck belt tension.
12. Install components in reverse order of removal.
13. Tighten crankshaft pulley bolts to 20 Nm **4**.

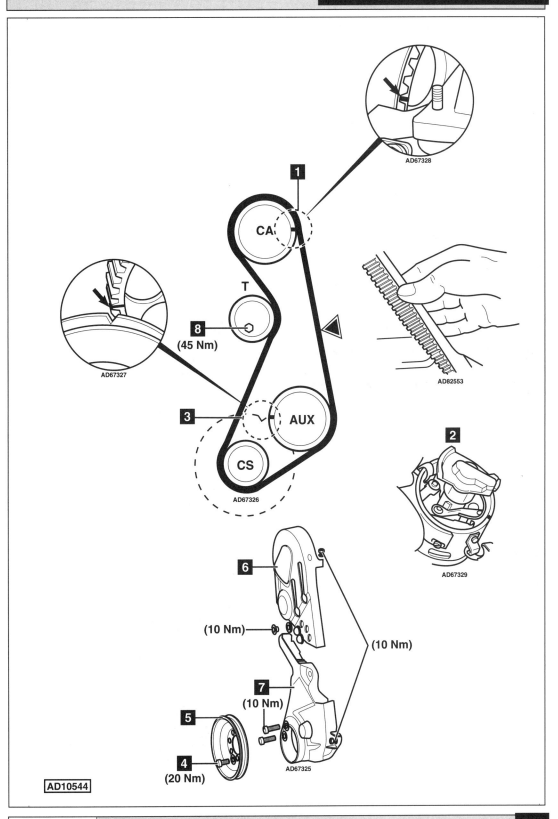

AD67328

1

CA

T

8
(45 Nm)

AD67327

AD82553

3

AUX

CS

AD67326

2

AD67329

6

(10 Nm)

(10 Nm)

7
(10 Nm)

5

4
(20 Nm)

AD67325

AD10544

Model:	**Alhambra 1,8 Turbo**
Year:	**1997-03**
Engine Code:	**AJH, AWC**

Replacement Interval Guide

Seat recommend:
Check at 60,000 miles and then every 20,000 miles
(replace if necessary).
AJH →1999: Replacement every 80,000 miles
(tensioner pulley must also be replaced).
AJH 2000→ and AWC: Replacement every
120,000 miles (tensioner pulley must also be
replaced).
*The previous use and service history of the vehicle
must always be taken into account.*
*Refer to Timing Belt Replacement Intervals at the
front of this manual.*

Check For Engine Damage

*CAUTION: This engine has been identified as an
INTERFERENCE engine in which the possibility of
valve-to-piston damage in the event of a timing belt
failure is MOST LIKELY to occur.*
*A compression check of all cylinders should be
performed before removing the cylinder head.*

Repair Times – hrs

Remove & install	2,20

Special Tools

- Tensioner tool – SEAT No.T10092.
- Pin – SEAT No.T40011.

Special Precautions

- Disconnect battery earth lead.
- DO NOT turn crankshaft or camshaft when timing belt removed.
- Remove spark plugs to ease turning engine.
- Turn engine in normal direction of rotation (unless otherwise stated).
- DO NOT turn engine via camshaft or other sprockets.
- Observe all tightening torques.

Removal

1. Raise and support front of vehicle.
2. Remove:
   - ❏ Air filter assembly.
   - ❏ Engine undershield.
   - ❏ Intercooler to turbocharger hose.
   - ❏ Intercooler to turbocharger hose bracket.
   - ❏ Lower engine torque rod.
   - ❏ Auxiliary drive belt.
   - ❏ Auxiliary drive belt tensioner **1**.
   - ❏ Timing belt upper cover **2**.
3. Turn crankshaft to TDC on No.1 cylinder.
4. Ensure timing marks aligned **3** & **4**.
5. Support engine.

6. Remove:
   - ❏ RH engine mounting bolts **5** & **6**.
   - ❏ RH engine mounting nuts **7**.
   - ❏ RH engine mounting.
   - ❏ Engine mounting bracket bolts **8**.
   - ❏ Engine mounting bracket.
   - ❏ Crankshaft pulley bolts **9**.
   - ❏ Crankshaft pulley **10**.
   - ❏ Timing belt centre cover **11**.
   - ❏ Timing belt lower cover **12**.
7. Screw tensioner tool into tensioner **13** & **14**.
   Tool No.T10092.
8. Tighten nut gradually **14**.
   NOTE: DO NOT overtighten.
9. Fit locking pin **15**. Tool No.T40011.
10. Remove timing belt.
    **NOTE: Mark direction of rotation on belt with chalk
    if belt is to be reused.**

Installation

1. Ensure timing marks aligned **4**.
2. Ensure tensioner tool fitted **13** & **14**.
3. Ensure locking pin fitted **15**.
4. Fit timing belt to crankshaft sprocket.
5. Install:
   - ❏ Timing belt lower cover **12**.
   - ❏ Crankshaft pulley **10**.
   - ❏ Crankshaft pulley bolts **9**.
6. Tighten crankshaft pulley bolts **9**.
   Tightening torque: 25 Nm.
7. Ensure timing marks aligned **3**.
8. Fit timing belt to remaining sprockets and pulleys in following order:
   - ❏ Water pump sprocket.
   - ❏ Tensioner pulley.
   - ❏ Camshaft sprocket.
   **NOTE: Ensure belt is taut between sprockets on
   non-tensioned side.**
9. Remove locking pin **15**.
10. Remove tensioner tool **13** & **14**.
11. Turn crankshaft two turns clockwise.
12. Ensure timing marks aligned **3** & **4**.
13. Fit engine mounting bracket.
14. Tighten engine mounting bracket bolts **8**.
    Tightening torque: 45 Nm.
15. Install:
    - ❏ Timing belt centre cover **11**.
    - ❏ Engine mounting.
16. Tighten:
    - ❏ Engine mounting bolts **6**.
      Tightening torque: 60 Nm. Lightly oil threads.
    - ❏ Engine mounting nuts **7**.
      Tightening torque: 55 Nm.
    - ❏ Engine mounting bolts **5**.
      Tightening torque: 60 Nm. Lightly oil threads.
17. Install components in reverse order of removal.

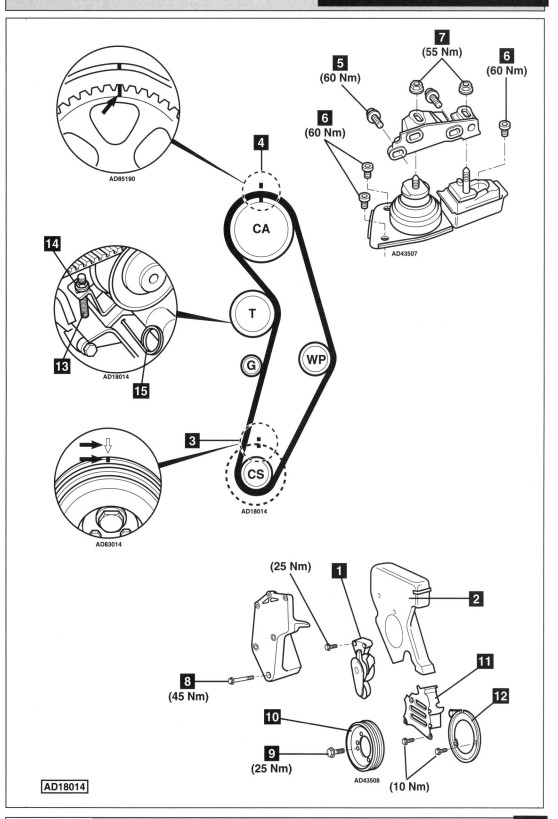

AD85190

7
(55 Nm)

5
(60 Nm)

6
(60 Nm)

6
(60 Nm)

4

CA

AD43507

14

T

WP

13

G

AD18014

15

3

CS

AD18014

AD83014

(25 Nm)

1

2

8
(45 Nm)

11

12

10

9
(25 Nm)

AD43508

(10 Nm)

AD18014

SEAT

Model:	**Alhambra 2,0**
Year:	**2000-03**
Engine Code:	**ATM**

Replacement Interval Guide

Seat recommend check at the first 60,000 miles and then every 20,000 miles (replace if necessary).
The previous use and service history of the vehicle must always be taken into account.
Refer to Timing Belt Replacement Intervals at the front of this manual.

Check For Engine Damage

CAUTION: This engine has been identified as an INTERFERENCE engine in which the possibility of valve-to-piston damage in the event of a timing belt failure is MOST LIKELY to occur. A compression check of all cylinders should be performed before removing the cylinder head(s).

Repair Times – hrs

Remove & install	1,90

Special Tools

- ■ Tensioner wrench – SEAT No.T10020.

Removal

1. Raise and support front of vehicle.
2. Remove:
   - ❏ RH splash guard.
   - ❏ Engine cover.
   - ❏ Air filter assembly.
   - ❏ Auxiliary drive belt.
   - ❏ Auxiliary drive belt tensioner **1**.
   - ❏ Timing belt upper cover **2**.
3. Turn crankshaft clockwise to TDC on No.1 cylinder.
4. Ensure timing marks aligned **3** & **4**.
5. Support engine.
6. Remove:
   - ❏ RH engine mounting bolts **5** & **6**.
   - ❏ RH engine mounting nuts **7**.
   - ❏ RH engine mounting.
   - ❏ Engine mounting bracket bolts **8**.
   - ❏ Engine mounting bracket **9**.
   - ❏ Crankshaft pulley bolts **10**.
   - ❏ Crankshaft pulley **11**.
   - ❏ Timing belt centre cover **12**.
   - ❏ Timing belt lower cover **13**.
7. Slacken tensioner nut **14**.
8. Turn tensioner clockwise away from belt. Use tool No.T10020 **15**.
9. Lightly tighten nut **14**.
10. Remove timing belt.
    NOTE: Mark direction of rotation on belt with chalk if belt is to be reused.

Installation

1. Ensure timing marks aligned **4**.
2. Fit timing belt loosely to crankshaft sprocket and water pump sprocket.
   NOTE: Engine must be COLD when installing belt. If belt is to be reused: Observe direction of rotation marks on belt.
3. Install:
   - ❏ Timing belt lower cover **13**.
   - ❏ Timing belt centre cover **12**.
   - ❏ Crankshaft pulley **11**.
   - ❏ Crankshaft pulley bolts **10**.
4. Tighten crankshaft pulley bolts **10**.
   Tightening torque: 25 Nm.
5. Ensure timing marks aligned **3**.
6. Fit timing belt to tensioner pulley, then camshaft sprocket.
7. Ensure belt is taut between sprockets on non-tensioned side.
8. Ensure tensioner retaining lug **16** is properly engaged.
9. Slacken tensioner nut **14**.
10. Turn tensioner 5 times fully anti-clockwise and clockwise from stop to stop. Use tool No.T10020 **15**.
11. Turn tensioner fully anti-clockwise then slowly clockwise until pointer **17** aligned with notch **18** in baseplate. Use tool No.T10020.
12. Tighten tensioner nut **14**. Tightening torque: 20 Nm.
13. Turn crankshaft slowly two turns clockwise until timing marks aligned **3** & **4**.
    NOTE: Turn crankshaft last 45° smoothly without stopping.
14. Ensure pointer **17** aligned with notch **18** in baseplate.
15. If not: Repeat tensioning procedure.
16. Apply firm thumb pressure to belt at ▽. Pointer **17** and notch **18** must move apart.
17. Release thumb pressure from belt at ▽.
18. Turn crankshaft slowly two turns clockwise until timing marks aligned **3** & **4**.
    NOTE: Turn crankshaft last 45° smoothly without stopping.
19. Ensure pointer **17** aligned with notch **18** in baseplate.
20. Install:
    - ❏ RH engine mounting bracket **9**.
21. Tighten engine mounting bracket bolts **8**.
    Tightening torque: 45 Nm.
22. Install:
    - ❏ Engine mounting.
23. Tighten:
    - ❏ Engine mounting bolts **6**.
      Tightening torque: 60 Nm. Lightly oil threads.
    - ❏ Engine mounting nuts **7**.
      Tightening torque: 55 Nm.
    - ❏ Engine mounting bolts **5**.
      Tightening torque: 60 Nm. Lightly oil threads.
24. Install components in reverse order of removal.

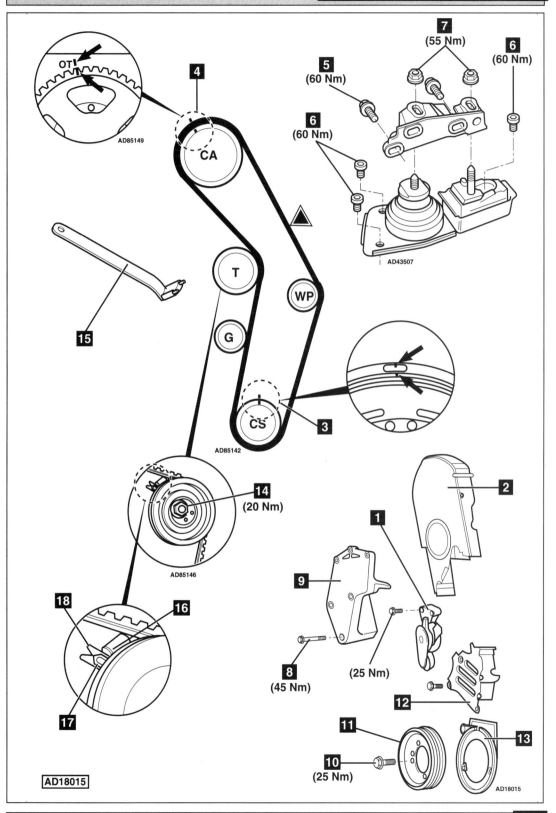

OT

AD85149

4

CA

5
(60 Nm)

7
(55 Nm)

6
(60 Nm)

6
(60 Nm)

AD43507

T

WP

G

15

CS

3

AD85142

14
(20 Nm)

AD85146

18

16

17

2

1

9

8
(45 Nm)

(25 Nm)

12

11

13

10
(25 Nm)

AD18015

AD18015

SEAT

Model:	Arosa 1,4 TDI PD
Year:	1999-03
Engine Code:	AMF

Check For Engine Damage

CAUTION: This engine has been identified as an INTERFERENCE engine in which the possibility of valve-to-piston damage in the event of a timing belt failure is MOST LIKELY to occur. A compression check of all cylinders should be performed before removing the cylinder head(s).

Repair Times – hrs

Remove & install: 2,90

Special Tools

- Crankshaft sprocket locking tool – SEAT No.T10050.
- Camshaft locking tool – SEAT No.T20102.
- Two-pin wrench – SEAT No.U-30009A.
- Tensioner locking tool – SEAT No.T10008.
- 4 mm drill bit.

Special Precautions

- Disconnect battery earth lead.
- DO NOT turn crankshaft or camshaft when timing belt removed.
- Remove glow plugs to ease turning engine.
- Turn engine in normal direction of rotation (unless otherwise stated).
- DO NOT turn engine via camshaft or other sprockets.
- Observe all tightening torques.

Removal

1. Raise and support front of vehicle.
2. Remove:
   - ❏ Engine top cover.
   - ❏ Turbocharger air hoses.
   - ❏ Intercooler outlet hose.
   - ❏ Mass air flow (MAF) sensor.
   - ❏ Engine undershield.
   - ❏ Lower engine steady bar (located on gearbox).
   - ❏ Auxiliary drive belt.
   - ❏ Auxiliary drive belt tensioner.
   - ❏ Timing belt upper cover **1**.
3. Support engine.
4. Remove:
   - ❏ RH engine mounting.
   - ❏ RH engine mounting bracket **2**.
   - ❏ Timing belt centre cover **3**.
5. Lower engine slightly.
6. Remove:
   - ❏ Crankshaft pulley bolts **4**.
   - ❏ Crankshaft pulley **5**.
   - ❏ Timing belt lower cover **6**.
7. Turn crankshaft clockwise to TDC on No.1 cylinder.
8. Ensure timing mark aligned with notch on camshaft sprocket hub **7**.
   NOTE: Notch located behind camshaft sprocket teeth.
9. Lock crankshaft sprocket **8**.
   Use tool No.T10050.
10. Ensure timing marks aligned **9**.
11. Lock camshaft **10**. Use tool No.T20102.
12. Fully insert Allen key into tensioner pulley **11**.
13. Turn tensioner pulley slowly anti-clockwise until locking tool can be inserted **12**. Tool No.T10008.
14. Slacken tensioner nut **13**.
15. Remove:
    - ❏ Automatic tensioner unit **14**.
    - ❏ Timing belt.

Installation

1. Ensure camshaft locked with tool **10**.
2. Ensure crankshaft sprocket locking tool located correctly **8**.
3. Ensure timing marks aligned **9**.
4. Ensure automatic tensioner unit locked with tool **12**. Tool No.T10008.
5. Slacken camshaft sprocket bolts **15**.
6. Turn camshaft sprocket fully clockwise in slotted holes. Tighten bolts finger tight **15**.
7. Turn tensioner pulley slowly clockwise **16** until lug **17** just reaches stop **18**.
   Use tool No.U-30009A **19**.
8. Fit timing belt in following order:
   - ❏ Camshaft sprocket.
   - ❏ Tensioner pulley.
   - ❏ Crankshaft sprocket.
   - ❏ Water pump sprocket.
9. Install:
   - ❏ Automatic tensioner unit **14**.
   NOTE: Ensure belt is taut between sprockets on non-tensioned side.

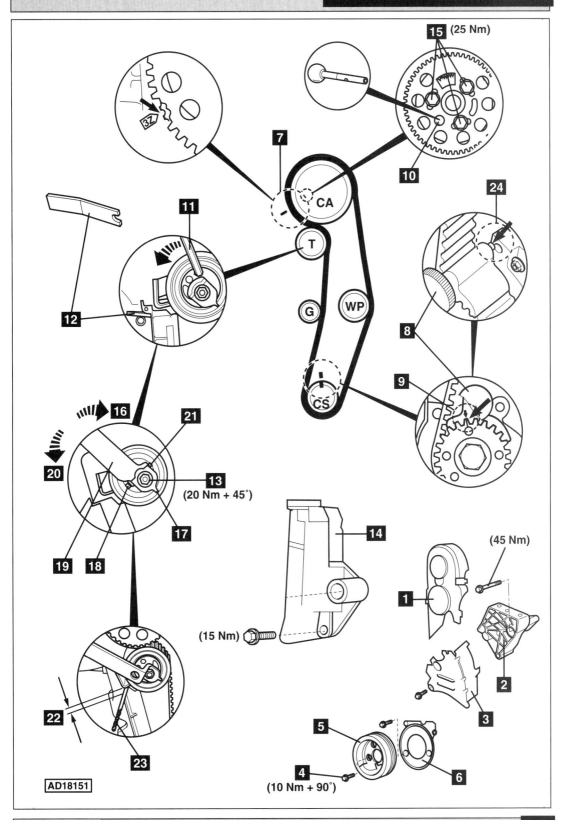

15 (25 Nm)

7

10

24

11

12

T

CA

G

WP

8

9

CS

16

21

20

13
(20 Nm + 45°)

17

19 18

14

22

23

(15 Nm)

1

(45 Nm)

2

3

5

6

4
(10 Nm + 90°)

AD18151

10. Turn tensioner pulley slowly anti-clockwise **20** (lug **17** moves towards stop **21**). Use tool No.U-30009A **19**.

11. Remove locking tool without force **12**.

12. Turn tensioner pulley slowly clockwise **16** (lug **17** moves towards stop **18**) until dimension **22** is 4±1 mm. Use drill bit **23**.
    NOTE: Engine must be COLD.

13. Tighten tensioner nut **13**.
    Tightening torque: 20 Nm + 45°.

14. Tighten camshaft sprocket bolts **15**.
    Tightening torque: 25 Nm.

15. Remove:
    - ❏ Camshaft locking tool **10**.
    - ❏ Crankshaft sprocket locking tool **8**.
    - ❏ Drill bit **23**.

16. Turn crankshaft slowly two turns clockwise to TDC on No.1 cylinder.

17. Ensure dimension **22** is 4±1 mm. Use drill bit **23**.

18. If not: Slacken tensioner nut **13**. Turn tensioner pulley until dimension correct **22**. Tighten tensioner nut **13**.
    Tightening torque: 20 Nm + 45°.

19. Lock crankshaft sprocket **8**.
    Use tool No.T10050.

20. Ensure timing marks aligned **9**.

21. Ensure camshaft locking tool can be inserted easily **10**. Tool No.T20102.

22. If not:
    - ❏ Remove lug of crankshaft sprocket locking tool from hole in oil seal housing.
    - ❏ Turn crankshaft until camshaft locking tool can be inserted **10**. Tool No.T20102.
    - ❏ Slacken camshaft sprocket bolts **15**.
    - ❏ Turn crankshaft anti-clockwise until lug of locking tool just passes hole in oil seal housing **24**.
    - ❏ Turn crankshaft clockwise until lug and hole aligned.
    - ❏ Lock crankshaft sprocket **8**.
      Use tool No.T10050.
    - ❏ Tighten camshaft sprocket bolts **15**.
      Tightening torque: 25 Nm.
    - ❏ Remove locking tools **8** & **10**.
    - ❏ Turn crankshaft slowly two turns clockwise to TDC on No.1 cylinder.
    - ❏ Ensure locking tools can be fitted correctly **8** & **10**.

23. Remove:
    - ❏ Camshaft locking tool **10**.
    - ❏ Crankshaft sprocket locking tool **8**.
    - ❏ Drill bit **23**.

24. Install components in reverse order of removal.

25. Tighten crankshaft pulley bolts **4**.
    Tightening torque: 10 Nm + 90°.

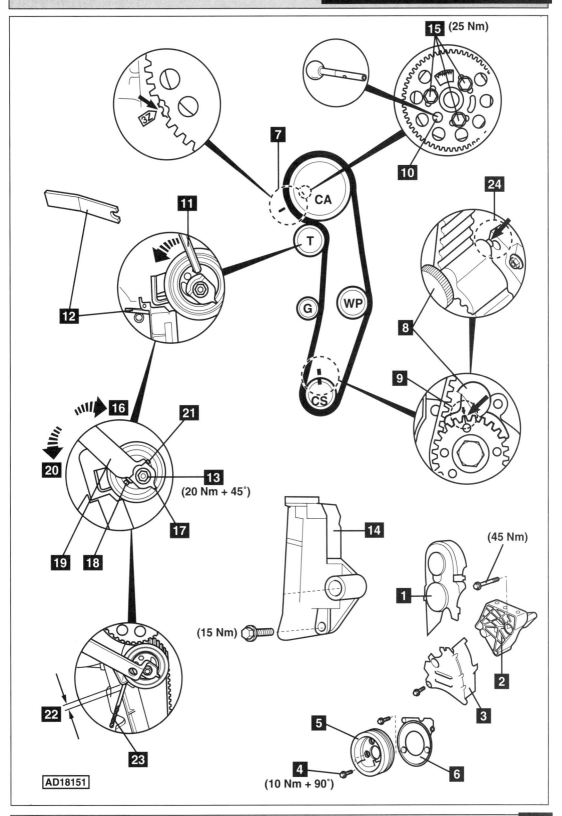

15 (25 Nm)

7

CA

10

24

11

T

8

12

G   WP

9

CS

16

20

21

13
(20 Nm + 45°)

17

19 **18**

14

(45 Nm)

1

2

22

3

(15 Nm)

23

5

6

4
(10 Nm + 90°)

AD18151

SEAT

Model:	**Arosa 1,7 SDI**
Year:	**1998-03**
Engine Code:	**AKU**

Replacement Interval Guide

Seat recommend:
Check condition and width every 10,000 miles or 12 months, whichever occurs first (replacement width 21 mm).
Replacement every 60,000 miles.
The previous use and service history of the vehicle must always be taken into account.
Refer to Timing Belt Replacement Intervals at the front of this manual.

Check For Engine Damage

CAUTION: This engine has been identified as an INTERFERENCE engine in which the possibility of valve-to-piston damage in the event of a timing belt failure is MOST LIKELY to occur.
A compression check of all cylinders should be performed before removing the cylinder head.

Repair Times – hrs

Remove & install	1,90

Special Tools

- Camshaft setting bar – SEAT No.U-40021.
- Injection pump sprocket locking pin – SEAT No.U-20003.
- Two-pin wrench – SEAT No.30009/A.
- Holding tool – SEAT No.T20018.

Removal

1. Raise and support front of vehicle.
2. Remove:
   - ❑ Engine undershield.
   - ❑ Auxiliary drive belt.
   - ❑ Air intake pipes.
   - ❑ Engine top cover.
   - ❑ Timing belt upper cover **1**.
   - ❑ Cylinder head cover.
3. Turn crankshaft to TDC on No.1 cylinder. Ensure flywheel timing marks aligned **2**.
4. Fit setting bar No.U-40021 to rear of camshaft **3**. Centralise camshaft using feeler gauges.
5. Support engine.
6. Remove:
   - ❑ Lower engine steady bar.
   - ❑ RH engine mounting.
   - ❑ RH engine mounting bracket **4**.
7. Lock injection pump sprocket **5**. Use tool No.U-20003.
8. Slacken tensioner nut **6**. Turn tensioner anti-clockwise away from belt. Lightly tighten nut.
9. Remove timing belt from camshaft and injection pump sprockets.

10. Lower engine slightly.
11. Remove:
    - ❑ Water pump pulley.
12. Hold crankshaft pulley. Use tool No.T20018.
13. Remove:
    - ❑ Crankshaft pulley bolts **7**.
    - ❑ Crankshaft pulley **8**.
    - ❑ Auxiliary drive belt tensioner (LH thread).
    - ❑ Timing belt lower cover **9**.
    - ❑ Timing belt.
    NOTE: Mark direction of rotation on belt with chalk if belt is to be reused.

Installation

1. Ensure flywheel timing marks aligned **2**.
2. Ensure camshaft setting bar fitted correctly **3**.
3. Ensure locking pin located correctly in injection pump sprocket **5**.
4. Hold camshaft sprocket. Use tool No.T20018.
5. Slacken camshaft sprocket bolt 1/2 turn **10**.
6. Loosen sprocket from taper using a drift through hole in timing belt rear cover **11**. Ensure sprocket can turn on taper.
7. Fit timing belt in anti-clockwise direction, starting at crankshaft sprocket. Ensure belt is taut between sprockets.
8. Slacken tensioner nut **6**.
9. Turn automatic tensioner pulley clockwise until notch and raised mark on tensioner aligned **12**. Use tool No.30009/A **13**.
10. If tensioner pulley turned too far: Turn fully anti-clockwise and repeat tensioning procedure.
11. Tighten tensioner nut **6**. Tightening torque: 20 Nm.
12. Ensure flywheel timing marks aligned **2**.
13. Hold camshaft sprocket. Use tool No.T20018.
14. Tighten camshaft sprocket bolt **10**. Tightening torque: 45 Nm.
15. Remove camshaft setting bar **3**.
16. Remove locking pin from injection pump sprocket **5**.
17. Turn crankshaft two turns clockwise until timing marks aligned **2**.
18. Ensure camshaft setting bar can be fitted **3**.
19. Ensure locking pin can be inserted in injection pump sprocket **5**.
20. Ensure notch and raised mark on tensioner aligned **12**. If not: Repeat tensioning procedure.
21. Install components in reverse order of removal.
22. Hold crankshaft pulley. Use tool No.T20018.
23. Tighten crankshaft pulley bolts **7**. Tightening torque: 25 Nm.

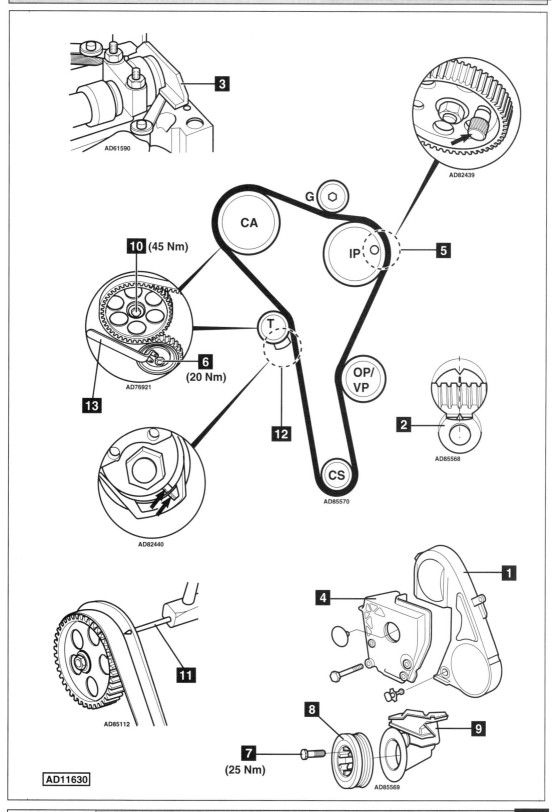

AD61590

3

AD82439

G

CA

10 (45 Nm)

IP

5

6

(20 Nm)

AD76921

13

T

12

OP/
VP

2

AD85568

CS

AD85570

AD82440

11

AD85112

1

4

9

8

7

(25 Nm)

AD85569

AD11630

SEAT

Model:	Ibiza/Cordoba 1,7 SDI • Ibiza/Cordoba 1,9 D/TD • Ibiza/Cordoba 1,9 SDI/TDI Toledo 1,9 D/TD • Toledo 1,9 TDI • Alhambra 1,9 TDI • Inca 1,9 SDI
Year:	1994-00
Engine Code:	Automatic tensioner AAZ, AEY, AFN, AHU, AKW, AVG, 1Y, 1Z

Replacement Interval Guide

Seat recommend:

Except AAZ/1Y: Check condition and width every 10,000 miles or 12 months, whichever occurs first (replacement width 22 mm).

Alhambra:

Replacement every 60,000 miles.

Ibiza/Cordoba/Toledo/Inca:

AAZ/1Y: Check and replacement if necessary every 60,000 miles.

Except AAZ/1Y: Replacement every 60,000 miles (tensioner pulley must also be replaced).

The previous use and service history of the vehicle must always be taken into account.

Refer to Timing Belt Replacement Intervals at the front of this manual.

Check For Engine Damage

CAUTION: This engine has been identified as an INTERFERENCE engine in which the possibility of valve-to-piston damage in the event of a timing belt failure is MOST LIKELY to occur.

A compression check of all cylinders should be performed before removing the cylinder head.

Repair Times – hrs

Remove & install:

Except Alhambra	1,80
Alhambra	2,90

Special Tools

■ Injection pump sprocket locking pin – SEAT No.U-20003 or No.2064.

■ Camshaft setting bar – SEAT No.U-40021 or No.2065A.

■ Two-pin wrench – SEAT No.U-30009 or Matra V.159.

Removal

1. Raise and support front of vehicle.
2. Remove:
   - ❏ Engine cover (if fitted).
   - ❏ Air filter assembly.
   - ❏ Engine undershield (if fitted).
   - ❏ Auxiliary drive belt(s).
   - ❏ Auxiliary drive belt tensioner (if fitted).
3. Alhambra: Support engine. Remove RH engine mounting.
4. Remove:
   - ❏ Timing belt upper cover **1**.
   - ❏ Breather hose/pipe.
   - ❏ Cylinder head cover.
5. Turn crankshaft to TDC on No.1 cylinder. Ensure flywheel timing marks aligned **2**.
6. Fit setting bar to rear of camshaft **3**. Tool No.U-40021 or No.2065A.
7. Centralise camshaft using feeler gauges.

8. Lock injection pump sprocket. Use tool No.U-20003 or No.2064 **4**.
9. Remove:
   - ❏ Crankshaft pulley bolts **5**.
   - ❏ Crankshaft pulley **6**.
   - ❏ Water pump pulley (if required).
   - ❏ Timing belt lower cover **7**.
10. Slacken tensioner nut **8**. Turn tensioner anti-clockwise away from belt. Lightly tighten nut.
11. Remove:
    - ❏ Guide pulley bolt **9**.
    - ❏ Guide pulley.
    - ❏ Timing belt.
    - **NOTE: Mark direction of rotation on belt with chalk if belt is to be reused.**

Installation

1. Ensure flywheel timing marks aligned **2**.
2. Ensure camshaft setting bar fitted correctly **3**.
3. Ensure locking pin located correctly in injection pump sprocket **4**.
4. Slacken camshaft sprocket bolt 1/2 turn **10**.
   NOTE: DO NOT use setting bar to hold camshaft when slackening sprocket bolt.
5. Loosen sprocket from taper using a drift through hole in timing belt rear cover.
6. Fit timing belt in anti-clockwise direction, starting at crankshaft sprocket. Ensure belt is taut between sprockets.
7. Remove locking pin from injection pump sprocket **4**.
8. Fit guide pulley. Tighten bolt to 25 Nm **9**.
9. Slacken tensioner nut **8**.
10. Turn automatic tensioner pulley clockwise until notch and raised mark on tensioner aligned **11**. Use tool No.U-30009 or Matra V.159 **12**.
11. Tighten tensioner nut to 20 Nm **8**.
12. Ensure flywheel timing marks aligned **2**. Tighten camshaft sprocket bolt to 45 Nm **10**.
13. Remove camshaft setting bar **3**.
14. Turn crankshaft two turns clockwise until timing marks aligned **2**.
15. Ensure camshaft setting bar can be fitted **3**.
16. Ensure locking pin can be inserted in injection pump sprocket **4**.
17. Apply firm thumb pressure to belt. Tensioner marks should move out of alignment **11**.
18. Release thumb pressure from belt. Tensioner marks should realign.
19. Install:
    - ❏ Timing belt lower cover **7**.
    - ❏ Crankshaft pulley **6**.
20. Tighten crankshaft pulley bolts to 25 Nm **5**.
21. Install components in reverse order of removal.

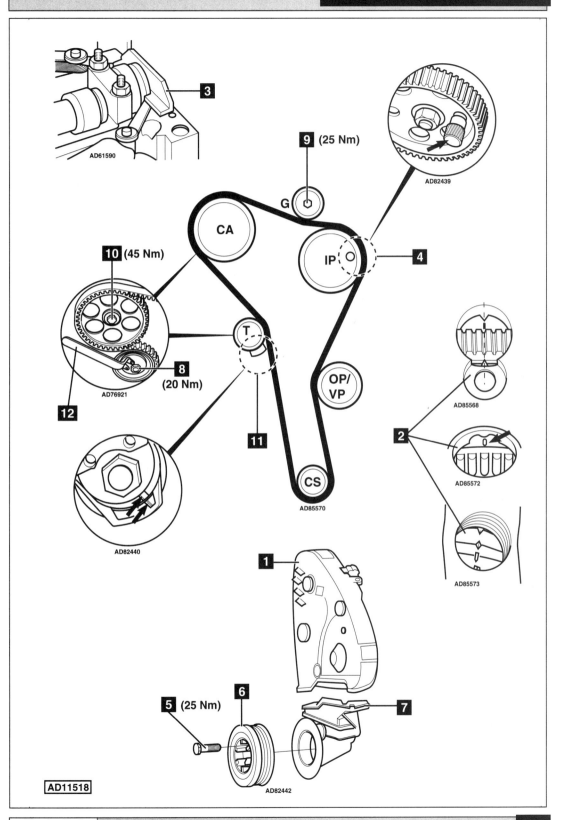

Model:	Ibiza/Cordoba 1,9 D/TD • Toledo 1,9 D/TD
Year:	1991-95
Engine Code:	Manual tensioner
	1Y, AAZ

Replacement Interval Guide

Seat recommend check every 20,000 miles, and replacement if necessary.
The previous use and service history of the vehicle must always be taken into account.
Refer to Timing Belt Replacement Intervals at the front of this manual.

Check For Engine Damage

CAUTION: This engine has been identified as an INTERFERENCE engine in which the possibility of valve-to-piston damage in the event of a timing belt failure is MOST LIKELY to occur.
A compression check of all cylinders should be performed before removing the cylinder head.

Repair Times – hrs

Check & adjust	0,70
Remove & install	1,80

Special Tools

- Tension gauge – SEAT No.U-10028.
- Injection pump sprocket locking pin – SEAT No.U-20003.
- Camshaft setting bar – SEAT No.U-40021 or No.U-20006.
- Two-pin wrench – SEAT No.U-30009.
- Sprocket holding tool – SEAT No.U-20002.

Special Precautions

- Disconnect battery earth lead.
- DO NOT turn crankshaft or camshaft when timing belt removed.
- Remove glow plugs to ease turning engine.
- Turn engine in normal direction of rotation (unless otherwise stated).
- DO NOT turn engine via camshaft or other sprockets.
- Observe all tightening torques.
- Check diesel injection pump timing after belt replacement.

Removal

1. Remove:
   - ❑ Air filter.
   - ❑ Auxiliary drive belts.
   - ❑ Water pump pulley.
   - ❑ PAS pump pulley (if required).
   - ❑ Cylinder head cover.
   - ❑ Timing belt upper cover **1**.
   - ❑ Crankshaft pulley bolts (4 bolts) **2**.
   - ❑ Crankshaft pulley **3**.
   - ❑ Timing belt lower cover **4**.

2. Turn crankshaft to TDC on No.1 cylinder. Ensure flywheel timing marks aligned **5**.
3. Fit setting bar to rear of camshaft **6**. Tool No.U-40021 or No.U-20006.
4. Centralise camshaft using feeler gauges.
5. Lock injection pump sprocket. Use tool No.U-20003 **7**.
6. Slacken tensioner nut **8**. Turn tensioner anti-clockwise away from belt. Use tool No.U-30009. Lightly tighten nut.
7. Remove timing belt.
   NOTE: Mark direction of rotation on belt with chalk if belt is to be reused.

Installation

1. Ensure flywheel timing marks aligned **5**.
2. Ensure camshaft setting bar fitted correctly **6**.
3. Ensure locking pin located correctly in injection pump sprocket **7**.
4. Hold camshaft sprocket. Use tool No.U-20002. Slacken bolt **9**.
5. Loosen sprocket from taper using a drift through hole in timing belt rear cover.
6. Fit timing belt in anti-clockwise direction, starting at crankshaft sprocket.
7. Attach tension gauge to belt at ▽ **10**. Tool No.U-10028.
8. Turn tensioner clockwise until tension gauge indicates 12-13 units. Use tool No.U-30009. Tighten nut to 45 Nm **8**.
9. Tighten camshaft sprocket bolt to 45 Nm **9**.
10. Remove locking pin from injection pump sprocket **7**.
11. Remove camshaft setting bar **6**.
12. Turn crankshaft two turns clockwise.
13. Recheck belt tension.
14. Install:
    - ❑ Timing belt lower cover **4**.
    - ❑ Crankshaft pulley **3**.
15. Tighten crankshaft pulley bolts to 20 Nm **2**.
16. Install components in reverse order of removal.

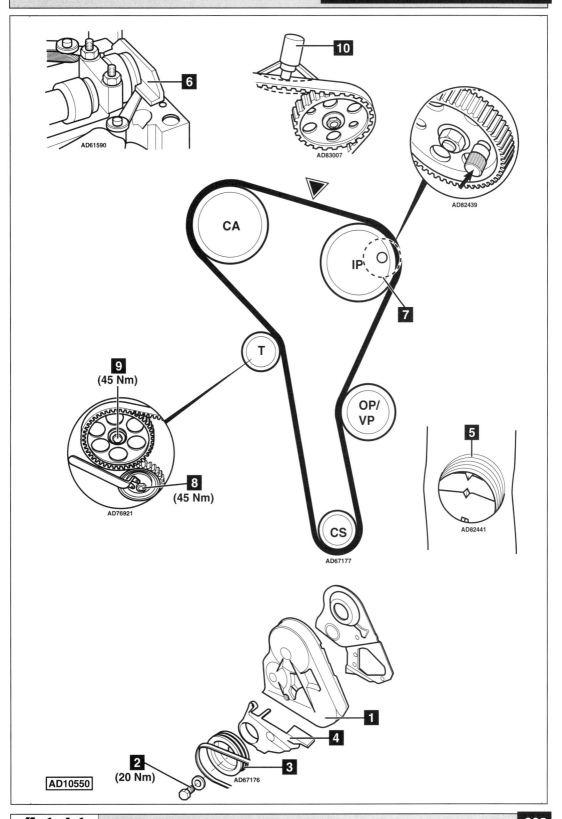

AD61590

10

AD83007

AD82439

CA

IP

7

T

9
(45 Nm)

8
(45 Nm)

AD76921

OP/
VP

5

AD82441

CS

AD67177

1

4

3

2
(20 Nm)

AD67176

AD10550

SEAT

Model:	**Ibiza/Cordoba 1,9 D/TD • Toledo 1,9 D/TD • Inca 1,9 D**
Year:	**1994-00**
Engine Code:	**Two-piece injection pump sprocket** **1Y, AAZ**

Replacement Interval Guide

Seat recommend check and replacement if necessary every 60,000 miles.
The previous use and service history of the vehicle must always be taken into account.
Refer to Timing Belt Replacement Intervals at the front of this manual.

Check For Engine Damage

CAUTION: This engine has been identified as an INTERFERENCE engine in which the possibility of valve-to-piston damage in the event of a timing belt failure is MOST LIKELY to occur.
A compression check of all cylinders should be performed before removing the cylinder head.

Repair Times – hrs

Remove & install	1,80

Special Tools

- Injection pump locking pin – SEAT No.U-40074.
- Camshaft setting bar – SEAT No.U-40021.
- Two-pin wrench – SEAT No.U-30009.

Removal

1. Raise and support front of vehicle.
2. Remove:
   - ❏ Air filter assembly.
   - ❏ Auxiliary drive belt(s).
   - ❏ Auxiliary drive belt tensioner (if fitted).
   - ❏ Timing belt upper cover **1**.
   - ❏ Breather hose/pipe.
   - ❏ Cylinder head cover.
   - ❏ Crankshaft pulley bolts **2**.
   - ❏ Crankshaft pulley **3**.
   - ❏ Water pump pulley.
   - ❏ Timing belt lower cover **4**.
3. Turn crankshaft to TDC on No.1 cylinder. Ensure flywheel timing marks aligned **5**.
4. Fit setting bar No.U-40021 to rear of camshaft **6**.
5. Centralise camshaft using feeler gauges.
6. Insert locking pin in injection pump **7**. Tool No.U-40074.
7. Slacken injection pump sprocket bolts **8**.
8. Slacken tensioner nut **9**. Move tensioner away from belt. Lightly tighten nut.
9. Remove timing belt.
   NOTE: Mark direction of rotation on belt with chalk if belt is to be reused.

Installation

1. Ensure flywheel timing marks aligned **5**.
2. Ensure camshaft setting bar fitted correctly **6**.
3. Ensure locking pin located correctly in injection pump **7**.
4. Slacken camshaft sprocket bolt 1/2 turn **10**.
   NOTE: DO NOT use setting bar to hold camshaft when slackening sprocket bolt.
5. Loosen sprocket from taper using a drift through hole in timing belt rear cover.
6. Fit timing belt in anti-clockwise direction, starting at crankshaft sprocket. Ensure belt is taut between sprockets.
7. Turn automatic tensioner pulley clockwise until notch and raised mark on tensioner aligned **11**. Use tool No.U-30009.
8. Tighten tensioner nut to 20 Nm **9**.
9. Ensure flywheel timing marks aligned **5**.
10. Tighten camshaft sprocket bolt to 45 Nm **10**.
11. Tighten injection pump sprocket bolts to 25 Nm **8**.
12. Remove:
    - ❏ Injection pump locking pin **7**.
    - ❏ Camshaft setting bar **6**.
13. Turn crankshaft two turns clockwise to TDC on No.1 cylinder.
14. Ensure flywheel timing marks aligned **5**.
15. Ensure camshaft setting bar can be fitted **6**. Tool No.U-40021.
16. Ensure locking pin can be inserted in injection pump **7**. Tool No.U-40074.
17. If not: Slacken injection pump sprocket bolts **8**. Turn sprocket hub until locking pin can be inserted **7**. Tighten bolts to 25 Nm **8**.
18. Apply firm thumb pressure to belt. Tensioner marks should move out of alignment **11**.
19. Release thumb pressure from belt. Tensioner marks should realign.
20. Install components in reverse order of removal.
21. Tighten crankshaft pulley bolts to 25 Nm **2**.

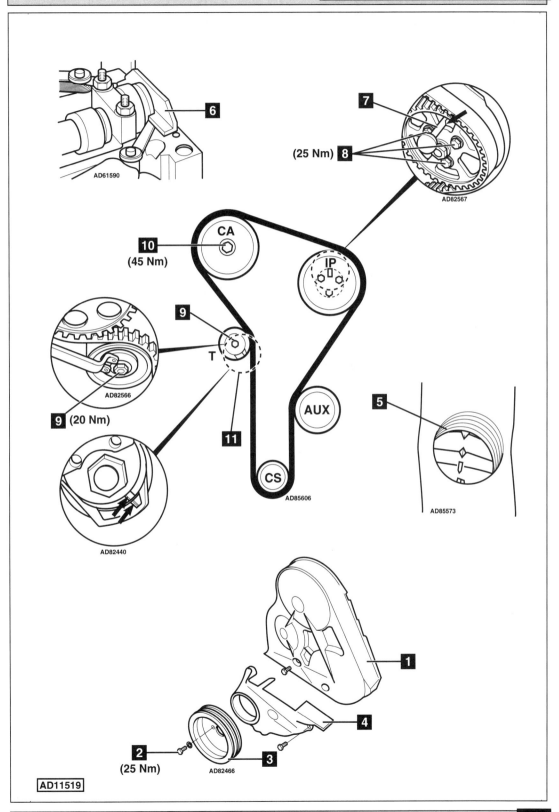

SEAT

Model:	Ibiza 1,9 SDI/TDI • Cordoba 1,9 SDI/TDI • Leon 1,9 SDI/TDI Toledo 1,9 TDI
Year:	1999-03
Engine Code:	AGP, AGR, AHF, ALH, AQM, ASV

Replacement Interval Guide

Seat recommend:
Check condition and width every 10,000 miles or 12 months, whichever occurs first (replacement width less than 22 mm).
Replacement every 60,000 miles (tensioner pulley must also be replaced).
The previous use and service history of the vehicle must always be taken into account.
Refer to Timing Belt Replacement Intervals at the front of this manual.

Check For Engine Damage

CAUTION: This engine has been identified as an INTERFERENCE engine in which the possibility of valve-to-piston damage in the event of a timing belt failure is MOST LIKELY to occur.
A compression check of all cylinders should be performed before removing the cylinder head(s).

Repair Times – hrs

Remove & install:

Ibiza/Cordoba (SDI)	1,90
Ibiza/Cordoba (TDI)	2,20
Leon/Toledo	2,90

Special Tools

- Camshaft setting bar – SEAT No.T20038.
- Injection pump locking pin – SEAT No.U-40074.
- Two-pin wrench – SEAT No.U-30009.

Special Precautions

- Disconnect battery earth lead.
- DO NOT turn crankshaft or camshaft when timing belt removed.
- Remove glow plugs to ease turning engine.
- Turn engine in normal direction of rotation (unless otherwise stated).
- DO NOT turn engine via camshaft or other sprockets.
- Observe all tightening torques.
- Check diesel injection pump timing after belt replacement.

Removal

1. Remove:
   - ❏ Engine top cover.
   - ❏ TDI: Intercooler to intake manifold hose.
   - ❏ SDI: Air hose to air filter.
   - ❏ SDI: Inlet manifold upper chamber.
   - ❏ Leon/Toledo: Coolant expansion tank. DO NOT disconnect hoses.
   - ❏ Leon/Toledo: PAS reservoir. DO NOT disconnect hoses.
2. Raise and support front of vehicle.
3. Remove:
   - ❏ Leon/Toledo: Front bumper.
   - ❏ Leon/Toledo: RH headlamp.
   - ❏ Timing belt upper cover **1**.
   - ❏ Vacuum pump.
   - ❏ Engine undershield (if fitted).
   - ❏ Auxiliary drive belt.
4. Turn crankshaft clockwise to TDC on No.1 cylinder.
5. Ensure timing marks aligned **2**.
6. Fit setting bar to rear of camshaft **3**. Tool No.T20038.
7. Centralise camshaft using feeler gauges **4**.
8. Lock injection pump. Use tool No.U-40074 **5**.
9. Remove bolts one at a time from injection pump sprocket **6**. Fit new bolts. Tighten bolts finger tight.
   NOTE: DO NOT slacken injection pump hub nut.

Leon/Toledo:

10. Support engine.
11. Remove:
    - ❏ RH engine mounting bolts **7**, **8** & **9**.
    - ❏ RH engine mounting.
12. Raise engine slightly.
13. Remove:
    - ❏ Engine mounting bracket bolts **10**.
    - ❏ Engine mounting bracket.

All models:

   - ❏ Crankshaft pulley bolts **11**.
   - ❏ Crankshaft pulley **12**.
   - ❏ Timing belt centre cover **13**.
   - ❏ Timing belt lower cover **14**.
14. Slacken tensioner nut **15**.
15. Turn tensioner anti-clockwise away from belt:
    - ❏ SDI: Use Allen key.
    - ❏ TDI: Use tool No.U-30009.

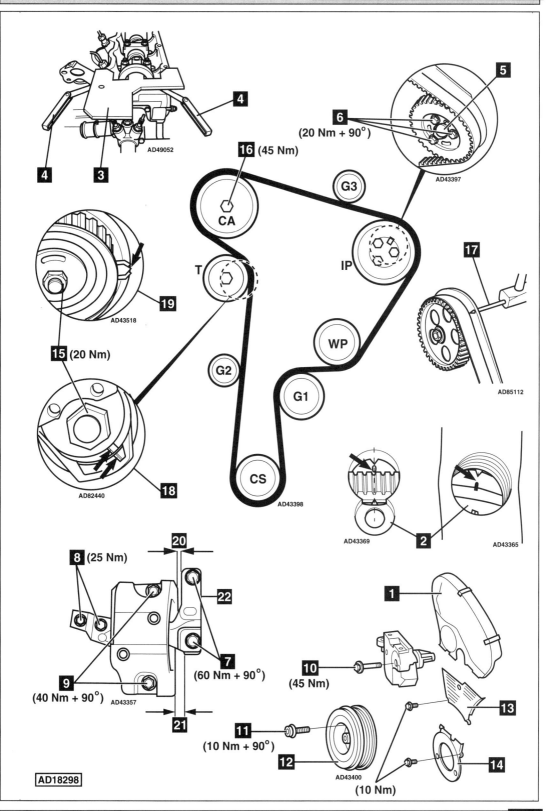

4

4

3

AD49052

5

6 (20 Nm + 90°)

AD43397

16 (45 Nm)

G3

CA

IP

17

AD85112

19

AD43518

15 (20 Nm)

T

G2

G1

WP

18

AD82440

CS

AD43398

AD43369

2

AD43365

8 (25 Nm)

20

22

1

9

(40 Nm + 90°) AD43357

7

(60 Nm + 90°)

10

(45 Nm)

13

21

11

(10 Nm + 90°)

12

AD43400

14

(10 Nm)

AD18298

16. Lightly tighten tensioner nut **15**.
17. Remove timing belt.

    NOTE: Mark direction of rotation on belt with chalk if belt is to be reused.

Installation

1. Ensure timing marks aligned **2**.
2. Ensure camshaft setting bar fitted correctly **3**.
3. Ensure locking pin located correctly in injection pump **5**.
4. Slacken camshaft sprocket bolt 1/2 turn **16**.

    NOTE: DO NOT use setting bar to hold camshaft when slackening sprocket bolt.

5. Loosen sprocket from taper using a drift through hole in timing belt rear cover **17**.
6. Remove:
    - ❏ Camshaft sprocket bolt **16**.
    - ❏ Camshaft sprocket.
7. Align injection pump sprocket with bolts in centre of slotted holes **6**.
8. Fit timing belt in following order:
    - ❏ Crankshaft sprocket.
    - ❏ Guide pulley G1.
    - ❏ Water pump sprocket.
    - ❏ Injection pump sprocket.
    - ❏ Guide pulley G2.
    - ❏ Tensioner pulley.
    - ❏ Guide pulley G3.

    NOTE: Turn injection pump sprocket slightly to engage timing belt teeth.

9. Ensure belt is taut between sprockets.
10. Fit camshaft sprocket to belt, then install camshaft sprocket with belt onto end of camshaft.
11. Fit camshaft sprocket bolt **16**.
12. Lightly tighten camshaft sprocket bolt **16**. Sprocket should turn freely on taper but not tilt.
13. Slacken tensioner nut **15**.
14. Turn tensioner pulley clockwise as follows:
    - ❏ SDI: Until pointer and notch in backplate aligned **19**. Use Allen key.
    - ❏ TDI: Until notch and raised mark on tensioner aligned **18**. Use tool No.U-30009.
15. If tensioner pulley turned too far: Turn fully anti-clockwise and repeat tensioning procedure.
16. Tighten tensioner nut to 20 Nm **15**.
17. Ensure timing marks aligned **2**.
18. Tighten camshaft sprocket bolt **16**. Tightening torque: 45 Nm.

    NOTE: DO NOT use setting bar to hold camshaft when tightening sprocket bolt.

19. Temporarily tighten injection pump sprocket bolts **6**. Tightening torque: 20 Nm.

    NOTE: DO NOT tighten further 90°.

20. Remove:
    - ❏ Camshaft setting bar **3**.
    - ❏ Injection pump locking pin **5**.
21. Turn crankshaft two turns clockwise until timing marks aligned **2**.
22. Ensure camshaft setting bar can be fitted **3**.
23. Ensure locking pin can be inserted in injection pump **5**. Tighten bolts further 90° **6**.
24. If not: Slacken injection pump sprocket bolts **6**. Turn sprocket hub until locking pin can be inserted **5**. Tighten bolts to 20 Nm + 90° **6**.
25. Ensure notch and raised mark on tensioner aligned **18**.
26. If not: Repeat tensioning procedure.
27. Apply firm thumb pressure to belt. Tensioner marks should move out of alignment **18** or **19**.
28. Release thumb pressure from belt:
    - ❏ SDI: Tensioner marks should move back following movement of belt.
    - ❏ TDI: Tensioner marks should realign.

    NOTE: SDI: Tensioner marks will not realign.

Leon/Toledo:

29. Install:
    - ❏ RH engine mounting bracket.
    - ❏ RH engine mounting bracket bolts **10**.
30. Tighten RH engine mounting bracket bolts **10**. Tightening torque: 45 Nm.
31. Fit and align RH engine mounting:
    - ❏ Engine mounting clearance: 14 mm **20**.
    - ❏ Engine mounting clearance: 10 mm minimum **21**.
    - ❏ Ensure engine mounting bolts **7** aligned with edge of mounting **22**.
32. Tighten:
    - ❏ Engine mounting bolts **9**. Tightening torque: 40 Nm + 90°. Use new bolts.
    - ❏ Engine mounting bolts **7**. Tightening torque: 60 Nm + 90°. Use new bolts.
    - ❏ Engine mounting bolts **8**. Tightening torque: 25 Nm.

All models:

33. Tighten crankshaft pulley bolts **11**. Tightening torque: 10 Nm + 90°.
34. Install components in reverse order of removal.

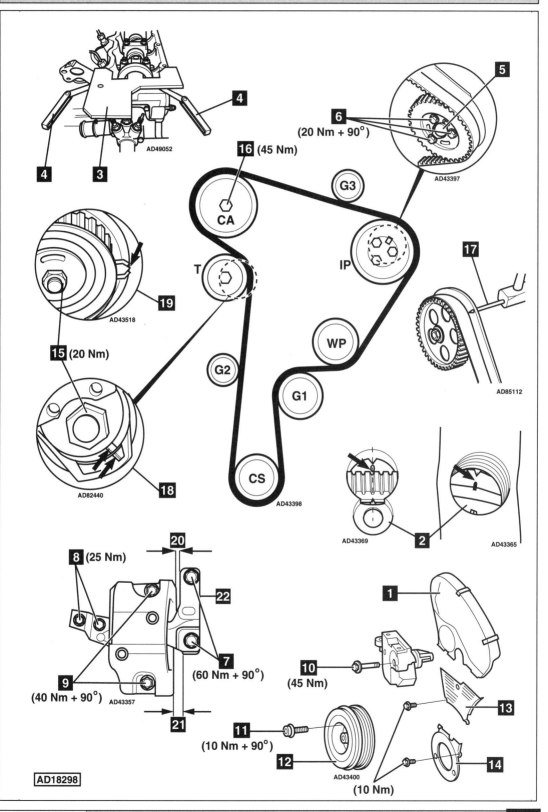

SEAT

Model:	**Alhambra 1,9 TDI PD**
Year:	**1999-03**
Engine Code:	**ANU, AUY**

Replacement Interval Guide

Seat recommend:
Check condition and width every 10,000 miles or 12 months, whichever occurs first (replacement width 27 mm).
Replacement every 40,000 miles (tensioner pulley must also be replaced).
The previous use and service history of the vehicle must always be taken into account.
Refer to Timing Belt Replacement Intervals at the front of this manual.

Check For Engine Damage

CAUTION: This engine has been identified as an INTERFERENCE engine in which the possibility of valve-to-piston damage in the event of a timing belt failure is MOST LIKELY to occur. A compression check of all cylinders should be performed before removing the cylinder head(s).

Repair Times – hrs

Remove & install	2,50

Special Tools

- Crankshaft locking tool – SEAT No.T10050.
- Camshaft locking tool – SEAT No.3359.
- Two-pin wrench – SEAT No.3387.
- Tensioner locking tool – SEAT No.T10008.
- 4 mm drill bit.

Special Precautions

- Disconnect battery earth lead.
- DO NOT turn crankshaft or camshaft when timing belt removed.
- Remove glow plugs to ease turning engine.
- Turn engine in normal direction of rotation (unless otherwise stated).
- DO NOT turn engine via camshaft or other sprockets.
- Observe all tightening torques.

Removal

1. Raise and support front of vehicle.
2. Remove:
   - ❏ Engine undershield.
   - ❏ Engine top cover.
   - ❏ Air filter assembly.
   - ❏ Turbocharger air hoses.
   - ❏ Intercooler outlet hose.
   - ❏ Auxiliary drive belt.
   - ❏ Auxiliary drive belt tensioner.
3. Support engine.
4. Raise engine slightly.

5. Remove:
   - ❏ Bolts **1**.
   - ❏ Nuts **2**.
   - ❏ Bolts **3**.
   - ❏ RH engine mounting **4**.
   - ❏ RH engine mounting bracket.
   - ❏ Timing belt upper cover **5**.
   - ❏ Timing belt centre cover **6**.
   - ❏ Crankshaft pulley bolts **7**.
   - ❏ Crankshaft pulley **8**.
   - ❏ Timing belt lower cover **9**.
6. Turn crankshaft clockwise to TDC on No.1 cylinder. Ensure '4Z' timing mark aligned with notch on camshaft sprocket hub **10**.
   NOTE: Notch located behind camshaft sprocket teeth.
7. Lock crankshaft **11**. Use tool No.T10050.
8. Ensure timing marks aligned **12**.
9. Lock camshaft **13**. Use tool No.3359.
10. Hold tensioner with two-pin wrench **14**. Tool No.3387.
11. Slacken tensioner nut **15**.
12. Turn tensioner pulley anti-clockwise **16**. Ensure lug and stop aligned **17** & **18**.
13. Lock automatic tensioner unit. Use tool No.T10008 **19**.
14. Turn tensioner pulley clockwise **20**. Ensure lug and stop aligned **17** & **21**.
15. Remove:
    - ❏ Guide pulley **22**.
    - ❏ Automatic tensioner unit **23**.
    - ❏ Timing belt.

Installation

1. Ensure camshaft locked with tool **13**.
2. Ensure crankshaft locking tool located correctly **11**.
3. Ensure timing marks aligned **12**.
4. Slacken camshaft sprocket bolts **24**.
5. Align camshaft sprocket with bolts in centre of slotted holes. Tighten bolts finger tight **24**.
6. Fit timing belt in following order:
   - ❏ Camshaft sprocket.
   - ❏ Tensioner pulley.
   - ❏ Crankshaft sprocket.
   - ❏ Water pump sprocket.
7. Install:
   - ❏ Automatic tensioner unit **23**.
   - ❏ Guide pulley **22**.
   NOTE: Ensure belt is taut between sprockets on non-tensioned side.

➡

Autodata

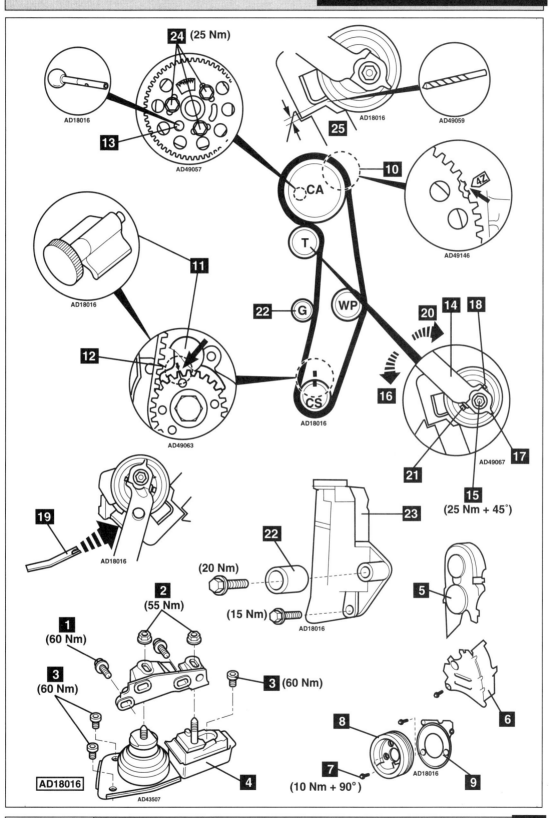

24 (25 Nm)

AD18016

13

AD49057

25

AD18016

AD49059

10

4Z

AD49146

11

AD18016

12

AD49063

22 G

T

WP

CS

AD18016

20

14 **18**

16

21

AD49067 **17**

15
(25 Nm + 45°)

19

AD18016

23

22

(20 Nm)

5

2
(55 Nm)

1
(60 Nm)

3
(60 Nm)

3 (60 Nm)

(15 Nm)

AD18016

6

8

7

(10 Nm + 90°)

AD18016

9

AD18016

4

AD43507

8. Turn tensioner pulley slowly anti-clockwise 🔟.
   Use tool No.3387 🔢. Remove locking tool
   without force 🔢.

9. Turn tensioner pulley slowly clockwise 🔢. Insert
   4 mm drill bit 🔢.
   NOTE: Engine must be COLD.

10. Tighten tensioner nut 🔢.
    Tightening torque: 20 Nm + 45°.

11. Tighten camshaft sprocket bolts 🔢.
    Tightening torque: 25 Nm.

12. Remove:
    ❏ Camshaft locking tool 🔢.
    ❏ Crankshaft locking tool 🔢.
    ❏ Drill bit.

13. Turn crankshaft slowly two turns clockwise to
    TDC on No.1 cylinder.

14. Lock crankshaft 🔢.

15. Ensure timing marks aligned 🔢.

16. Ensure locking tool can be inserted easily into
    camshaft sprocket 🔢.

17. If not: Slacken camshaft sprocket bolts 🔢. Turn
    sprocket hub until locking tool can be
    inserted 🔢. Tighten bolts 🔢.
    Tightening torque: 25 Nm.

18. Insert 4 mm drill bit. Ensure dimension
    correct 🔢.

19. If not: Slacken tensioner nut 🔢. Turn tensioner
    pulley until dimension correct 🔢. Tighten
    tensioner nut 🔢.
    Tightening torque: 20 Nm + 45°.

20. Remove:
    ❏ Camshaft locking tool 🔢.
    ❏ Crankshaft locking tool 🔢.
    ❏ Drill bit.

21. Install:
    ❏ RH engine mounting bracket.
    ❏ RH engine mounting 🔢.

22. Tighten:
    ❏ Bolts 🔢. Tightening torque: 60 Nm. Lightly
      oil threads.
    ❏ Nuts 🔢. Tightening torque: 55 Nm.
    ❏ Bolts 🔢. Tightening torque: 60 Nm. Lightly
      oil threads.

23. Install components in reverse order of removal.

24. Tighten crankshaft pulley bolts 🔢.
    Tightening torque: 10 Nm + 90°.

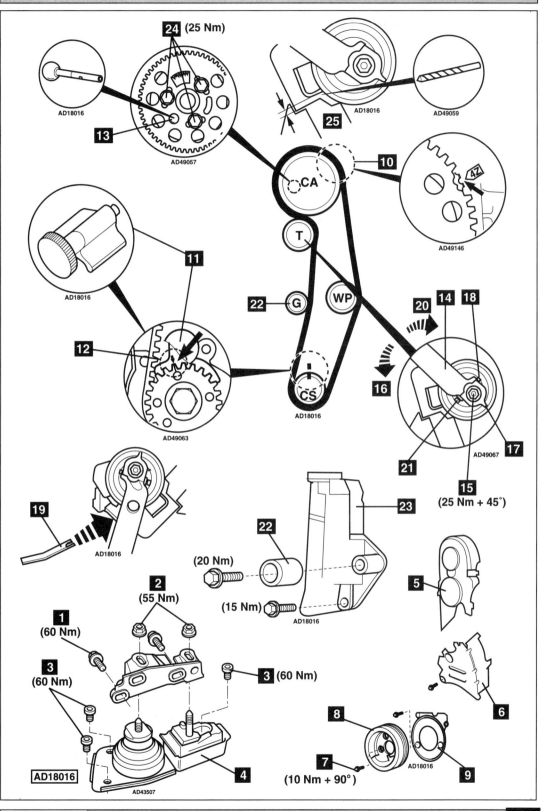

SKODA

Model:	**Fabia 1,4**
Year:	**1999-03**
Engine Code:	**AUA, AUB**

Replacement Interval Guide

Skoda recommend check at the first 60,000 miles and then every 20,000 miles (replace if necessary).

The previous use and service history of the vehicle must always be taken into account.
Refer to Timing Belt Replacement Intervals at the front of this manual.

Check For Engine Damage

CAUTION: This engine has been identified as an INTERFERENCE engine in which the possibility of valve-to-piston damage in the event of a timing belt failure is MOST LIKELY to occur.
A compression check of all cylinders should be performed before removing the cylinder head(s).

Repair Times – hrs

Remove & install	1,30

Special Tools

■ Camshaft locking tool – No.T10016.

Special Precautions

■ Disconnect battery earth lead.
■ DO NOT turn crankshaft or camshaft when timing belt removed.
■ Remove spark plugs to ease turning engine.
■ Turn engine in normal direction of rotation (unless otherwise stated).
■ DO NOT turn engine via camshaft or other sprockets.
■ Observe all tightening torques.

Removal

Timing Belt

1. Remove:
   ❏ Upper engine cover.
   ❏ Air filter assembly.
   ❏ Timing belt upper cover **1**.
2. Turn crankshaft clockwise to TDC on No.1 cylinder. Ensure timing marks on crankshaft pulley aligned **2**.
3. Ensure locating holes aligned **3**.
4. If locating holes are not aligned: Turn crankshaft one turn clockwise.
5. Fit locking tool to camshaft sprockets **4**. Tool No.T10016 **5**.
   NOTE: Ensure locking tool located correctly in cylinder head.
6. Support engine.

7. Remove:
   ❏ RH engine mounting.
   ❏ RH engine mounting bracket.
   ❏ Auxiliary drive belt.
8. Raise and support front of vehicle.
9. Lower engine until crankshaft pulley bolt accessible.
10. Remove:
    ❏ Crankshaft pulley bolt **6**.
    ❏ Crankshaft pulley **7**.
11. Fit two washers to crankshaft pulley bolt **6**.
12. Fit crankshaft pulley bolt **6**. Lightly tighten bolt.
13. Remove:
    ❏ Auxiliary drive belt tensioner.
    ❏ Timing belt lower cover **8**.
14. Slacken tensioner pulley bolt **9**.
15. Turn tensioner pulley anti-clockwise to release tension on belt.
16. Remove timing belt.
    NOTE: Mark direction of rotation on belt with chalk if belt is to be reused.

Installation

Timing Belt

1. Ensure locking tool fitted to camshaft sprockets **4**. Tool No.T10016 **5**.
2. Ensure timing mark on crankshaft sprocket aligned **10**.
   NOTE: Align ground tooth on crankshaft sprocket.
3. Fit timing belt in anti-clockwise direction, starting at water pump sprocket.
4. Tighten tensioner pulley bolt finger tight **9**. Ensure baseplate is supported by bolt **11**.
5. Turn tensioner pulley clockwise **12** until pointer **13** aligned with notch in baseplate **14**.
6. Remove locking tool from camshaft sprockets **4**. Tool No.T10016 **5**.
7. Tighten tensioner pulley bolt to 20 Nm **9**.
8. Turn crankshaft two turns clockwise to TDC on No.1 cylinder. Ensure timing mark on crankshaft sprocket aligned **10**.
9. Ensure locking tool can be inserted into camshaft sprockets **4**. Tool No.T10016 **5**.
10. Ensure pointer **13** aligned with notch in baseplate **14**.
11. If not: Repeat tensioning procedure.
12. Apply firm thumb pressure to belt at ▽. Pointer **13** and notch in baseplate **14** must move apart.

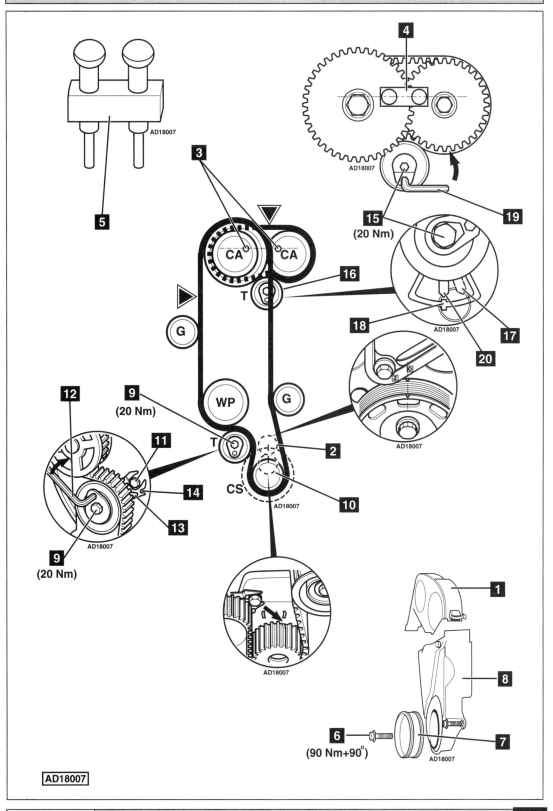

13. Release thumb pressure from belt at ▽.
14. Turn crankshaft two turns clockwise to TDC on No.1 cylinder.
15. Ensure pointer 13 aligned with notch in baseplate 14.
16. Remove crankshaft pulley bolt 6.
17. Install:
    ❏ Timing belt lower cover 8.
    ❏ Crankshaft pulley 7.
    ❏ New oiled crankshaft pulley bolt 6.
18. Tighten crankshaft pulley bolt 6. Tightening torque: 90 Nm + 90°.
19. Install components in reverse order of removal.

Removal

Exhaust Camshaft Drive Belt

1. Remove timing belt as described previously.
2. Slacken tensioner pulley bolt 15.
3. Turn tensioner pulley clockwise to release tension on belt.
4. Remove:
    ❏ Tensioner pulley bolt 15.
    ❏ Tensioner pulley 16.
    ❏ Drive belt.
    NOTE: Mark direction of rotation on belt with chalk if belt is to be reused.

Installation

Exhaust Camshaft Drive Belt

1. Ensure locking tool fitted to camshaft sprockets 4. Tool No.T10016 5.
2. Fit drive belt in clockwise direction, starting at top of inlet camshaft sprocket.
3. Ensure belt is taut between sprockets on non-tensioned side.
4. Turn tensioner pulley clockwise until pointer in position as shown 17.
5. Install:
    ❏ Tensioner pulley 16.
    ❏ Tensioner pulley bolt 15.
6. Tighten tensioner pulley bolt finger tight 15.
    NOTE: Ensure lug in baseplate 18 is located in cylinder head hole.
7. Turn tensioner anti-clockwise 19 until pointer 20 aligned with lug in baseplate 18.
8. Tighten tensioner pulley bolt to 20 Nm 15.
9. Fit timing belt as described previously.
10. Remove locking tool from camshaft sprockets 4. Tool No.T10016 5.
11. Turn crankshaft two turns clockwise to TDC on No.1 cylinder. Ensure timing marks on crankshaft sprocket aligned 10.

12. Ensure locking tool can be inserted into camshaft sprockets 4. Tool No.T10016 5.
13. Ensure pointer 20 aligned with lug in baseplate 18.
14. If not: Repeat tensioning procedure.
15. Apply firm thumb pressure to belt at ▽. Pointer 20 and lug in baseplate 18 must move apart.
16. Release thumb pressure from belt at ▽.
17. Turn crankshaft two turns clockwise to TDC on No.1 cylinder.
18. Ensure pointer 20 aligned with lug in baseplate 18.
19. Install components in reverse order of removal.

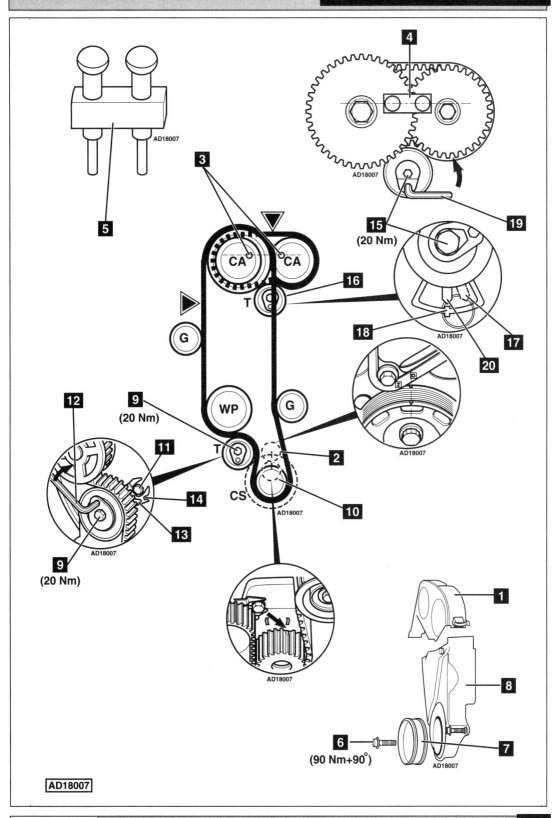

AD18007

SKODA

Model:	Felicia 1,6 • Octavia 1,6
Year:	1996-03
Engine Code:	AEE

Replacement Interval Guide

Skoda recommend:

Felicia:
Check at 60,000 miles (replace if necessary).
No manufacturer's recommended replacement interval.

Octavia:
Check at the first 60,000 miles and then every 20,000 miles (replace if necessary).
No manufacturer's recommended replacement interval.
The previous use and service history of the vehicle must always be taken into account.
Refer to Timing Belt Replacement Intervals at the front of this manual.

Check For Engine Damage

CAUTION: This engine has been identified as an INTERFERENCE engine in which the possibility of valve-to-piston damage in the event of a timing belt failure is MOST LIKELY to occur.
A compression check of all cylinders should be performed before removing the cylinder head.

Repair Times – hrs

Remove & install	1,30

Special Tools

- None required.

Special Precautions

- Disconnect battery earth lead.
- DO NOT turn crankshaft or camshaft when timing belt removed.
- Remove spark plugs to ease turning engine.
- Turn engine in normal direction of rotation (unless otherwise stated).
- DO NOT turn engine via camshaft or other sprockets.
- Observe all tightening torques.

Removal

1. Remove:
   - ❏ Auxiliary drive belt.
   - ❏ Crankshaft pulley bolts **1**.
   - ❏ Crankshaft pulley **2**.
   - ❏ Timing belt upper cover **3**.
   - ❏ Timing belt lower cover **4**.
2. Turn crankshaft clockwise to TDC on No.1 cylinder.
3. Ensure crankshaft sprocket timing mark aligned **5**.
   NOTE: Crankshaft sprocket has one tooth ground at an angle.

4. Ensure camshaft sprocket timing mark aligned **6**.
5. Slacken tensioner nut **7**.
6. Remove timing belt.
   NOTE: Mark direction of rotation on belt with chalk if belt is to be reused.

Installation

NOTE: If camshaft turned with engine at TDC, valves will interfere with pistons.
1. Ensure timing marks aligned **5** & **6**.
2. Fit timing belt in anti-clockwise direction, starting at crankshaft sprocket. Ensure belt is taut between sprockets on non-tensioned side.
3. Tighten tensioner nut finger tight **7**. Ensure baseplate is supported by bolt **8** & **9**.
4. Turn tensioner clockwise **10** until pointer **11** aligned with notch **12** in baseplate. Use Allen key.
5. Tighten tensioner nut to 20 Nm **7**.
6. Turn crankshaft two turns clockwise to TDC on No.1 cylinder. Ensure timing marks aligned **5** & **6**.
7. Apply firm thumb pressure to belt at ▽. Pointer **11** and notch **12** must move apart.
8. Release thumb pressure from belt at ▽. Ensure pointer **11** aligned with notch **12** in baseplate.
9. Install components in reverse order of removal.
10. Tighten crankshaft pulley bolts to 20 Nm **1**.

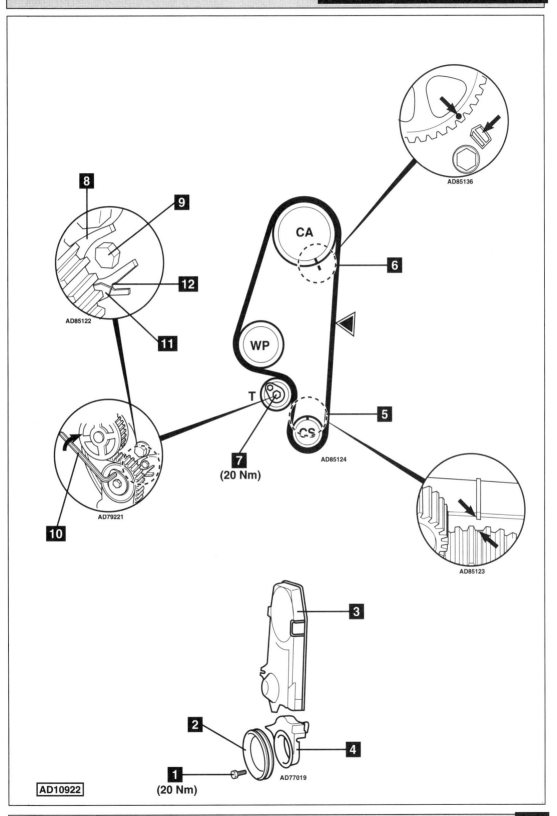

AD85136

8

9

CA

6

12

AD85122

WP

11

5

T

CS

7
(20 Nm)

AD85124

10

AD79221

AD85123

3

2

4

1
(20 Nm)

AD77019

AD10922

SKODA

Model:	**Octavia 1,6**
Year:	**1997-03**
Engine Code:	**AEH, AKL**

Replacement Interval Guide

Skoda recommend check at the first 60,000 miles and then every 20,000 miles (replace if necessary). The vehicle manufacturer has not recommended a timing belt replacement interval for this engine.
The previous use and service history of the vehicle must always be taken into account.
Refer to Timing Belt Replacement Intervals at the front of this manual.

Check For Engine Damage

CAUTION: This engine has been identified as an INTERFERENCE engine in which the possibility of valve-to-piston damage in the event of a timing belt failure is MOST LIKELY to occur.
A compression check of all cylinders should be performed before removing the cylinder head(s).

Repair Times – hrs

Remove & install	2,50

Special Tools

■ Two-pin wrench – No.MP 1-302/T10020.

Special Precautions

■ Disconnect battery earth lead.
■ DO NOT turn crankshaft or camshaft when timing belt removed.
■ Remove spark plugs to ease turning engine.
■ Turn engine in normal direction of rotation (unless otherwise stated).
■ DO NOT turn engine via camshaft or other sprockets.
■ Observe all tightening torques.

Removal

1. Raise and support front of vehicle.
2. Remove:
   ❏ Engine cover.
   ❏ Engine undershield.
   ❏ RH lower inner wing panel.
   ❏ Auxiliary drive belt.
   ❏ Auxiliary drive belt tensioner **1**.
3. Turn crankshaft to TDC on No.1 cylinder.
4. Ensure timing marks aligned **2** or **3**.
5. Ensure camshaft sprocket timing marks aligned **4**.
6. Remove:
   ❏ Crankshaft pulley bolts **5**.
   ❏ Crankshaft pulley **6**.
   ❏ Timing belt upper cover **7**.
   ❏ Timing belt centre cover **8**.
   ❏ Timing belt lower cover **9**.
   ❏ PAS reservoir. DO NOT disconnect hoses.
7. Support engine. Remove:
   ❏ RH engine mounting bolts **10**, **11** & **12**.
   ❏ RH engine mounting.
   ❏ RH engine mounting bracket bolts **13**.
   ❏ RH engine mounting bracket.
8. Slacken tensioner nut **14**. Turn tensioner clockwise away from belt. Lightly tighten nut.

9. Remove timing belt.
   NOTE: Mark direction of rotation on belt with chalk if belt is to be reused.

Installation

1. Ensure timing marks aligned **3** & **4**.
2. Fit timing belt in following order:
   ❏ Crankshaft sprocket.
   ❏ Water pump sprocket.
   ❏ Tensioner pulley.
   ❏ Camshaft sprocket.
   NOTE: Ensure belt is taut between sprockets on non-tensioned side.
3. Ensure tensioner retaining lug is properly engaged **15**.
4. Turn tensioner fully anti-clockwise.
   Use tool No.MP 1-302/T10020.
5. Turn tensioner clockwise until pointer **16** aligned with notch in baseplate **17**.
   Use tool No.MP 1-302/T10020.
6. Tighten tensioner nut **14**. Tightening torque: 25 Nm.
7. Turn crankshaft two turns clockwise to TDC on No.1 cylinder.
8. Ensure timing marks aligned **3** & **4**.
9. Ensure pointer **16** aligned with notch **17** in baseplate.
10. If not: Repeat tensioning procedure.
11. Apply firm thumb pressure to belt at ▽. Pointer **16** and notch **17** must move apart.
12. Release thumb pressure from belt at ▽. Pointer must realign with notch.
13. Install:
    ❏ Timing belt lower cover **9**.
    ❏ Crankshaft pulley **6**.
    ❏ Crankshaft pulley bolts **5**.
14. Tighten crankshaft pulley bolts **5**.
    Tightening torque: 25 Nm.
15. Install:
    ❏ RH engine mounting bracket.
    ❏ RH engine mounting bracket bolts **13**.
16. Tighten RH engine mounting bracket bolts to 45 Nm **13**.
17. Fit and align RH engine mounting.
    ❏ Engine mounting clearance: 14 mm **18**.
    ❏ Engine mounting clearance: 10 mm minimum **19**.
    ❏ Ensure engine mounting bolts **10** aligned with edge of mounting **20**.
18. Tighten:
    ❏ Engine mounting bolts **12**.
       Tightening torque: 40 Nm + 90°. Use new bolts.
    ❏ Engine mounting bolts **11**.
       Tightening torque: 25 Nm.
    ❏ Engine mounting bolts **10**.
       Tightening torque: 100 Nm. Use new bolts.
19. Install remainder of components in reverse order of removal.

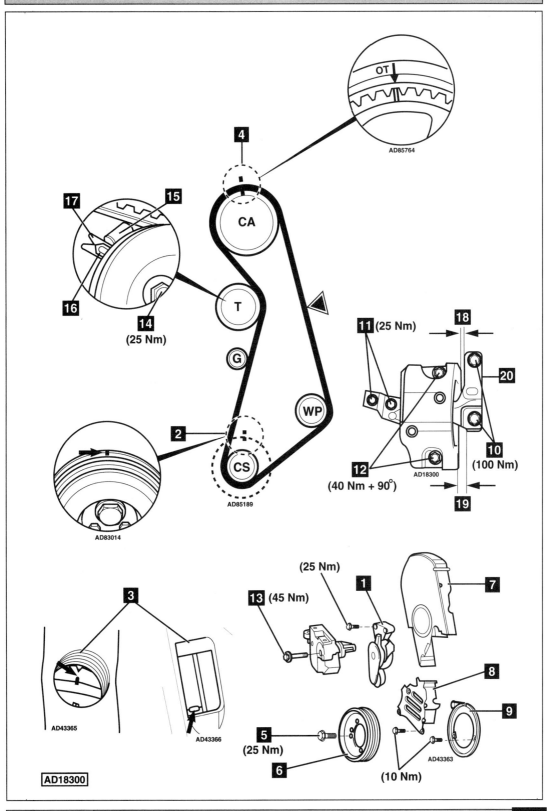

OT

AD85764

4

CA

17 15

T

16

14
(25 Nm)

G

11 (25 Nm) 18

20

2

CS

WP

12

10
(100 Nm)

(40 Nm + 90°)

AD18300

19

AD85189

AD83014

(25 Nm)

3

13 (45 Nm)

1

7

8

9

5
(25 Nm)

6

AD43365

AD43366

AD43363

(10 Nm)

AD18300

SKODA

Model:	**Octavia 1,8/Turbo**
Year:	**1996-03**
Engine Code:	**AGN, AGU**

Check For Engine Damage

CAUTION: This engine has been identified as an INTERFERENCE engine in which the possibility of valve-to-piston damage in the event of a timing belt failure is MOST LIKELY to occur.
A compression check of all cylinders should be performed before removing the cylinder head(s).

Repair Times – hrs

Remove & install	2,20

Special Tools

- Tensioner locking pin – No.T40011.
- 1 x M5 x 55 mm stud, nut and washer.

Special Precautions

- Disconnect battery earth lead.
- DO NOT turn crankshaft or camshaft when timing belt removed.
- Remove spark plugs to ease turning engine.
- Turn engine in normal direction of rotation (unless otherwise stated).
- DO NOT turn engine via camshaft or other sprockets.
- Observe all tightening torques.

Removal

1. Remove:
   - ❏ Auxiliary drive belt.
   - ❏ Auxiliary drive belt tensioner.
   - ❏ Coolant expansion tank. DO NOT disconnect hoses.
   - ❏ PAS reservoir. DO NOT disconnect hoses.
   - ❏ Timing belt upper cover **1**.
2. Raise and support front of vehicle.
3. Remove:
   - ❏ Engine undershield.
   - ❏ Turbo: Intercooler hoses.
4. Support engine.
5. Remove:
   - ❏ RH engine mounting **2**.
   - ❏ Engine mounting bracket **3**.
   - ❏ Timing belt centre cover **4**.
6. Turn crankshaft clockwise to TDC on No.1 cylinder.
7. Ensure timing marks aligned **5** or **6**.
8. Ensure camshaft sprocket timing marks aligned **7**.
9. Remove:
   - ❏ Crankshaft pulley bolts (4 bolts) **8**.
   - ❏ Crankshaft pulley **9**.
   - ❏ Timing belt lower cover **10**.

10. Insert M5 stud into tensioner **11**.
11. Fit nut and washer **12** to stud. Tighten nut sufficiently to allow locking pin to be inserted **13**.
    Tool No.T40011.
    NOTE: DO NOT overtighten nut.
12. Remove timing belt.
    NOTE: Mark direction of rotation on belt with chalk if belt is to be reused.

Installation

1. Ensure timing marks aligned **7**.
2. Fit timing belt to crankshaft sprocket.
3. Fit timing belt lower cover **10**.
4. Install:
   - ❏ Crankshaft pulley **9**.
   - ❏ Crankshaft pulley bolts **8**.
5. Tighten crankshaft pulley bolts **8**.
   Tightening torque: 25 Nm.
6. Ensure timing marks aligned **5** & **6**.
7. Fit timing belt in following order:
   - ❏ Water pump sprocket.
   - ❏ Tensioner pulley.
   - ❏ Camshaft sprocket.
   NOTE: Ensure belt is taut between sprockets on non-tensioned side.
8. Remove locking pin **13**.
9. Remove nut, washer and stud **11** & **12**.
10. Turn crankshaft two turns clockwise to TDC on No.1 cylinder.
11. Ensure timing marks aligned **5**, **6** & **7**.
12. Install:
    - ❏ Timing belt centre cover **4**.
    - ❏ Engine mounting bracket **3**.
13. Tighten engine mounting bracket bolts **14**.
    Tightening torque: 45 Nm.
14. Fit and align RH engine mounting **2**:
    - ❏ Engine mounting clearance: 14 mm **15**.
    - ❏ Engine mounting clearance: 10 mm minimum **16**.
    - ❏ Ensure engine mounting bolts **17** aligned with edge of mounting **18**.
15. Tighten:
    - ❏ Engine mounting bolts **19**.
      Tightening torque: 40 Nm + 90°. Use new bolts.
    - ❏ Engine mounting bolts **20**.
      Tightening torque: 25 Nm.
    - ❏ Engine mounting bolts **17**.
      Tightening torque: 100 Nm. Use new bolts.
16. Install remainder of components in reverse order of removal.

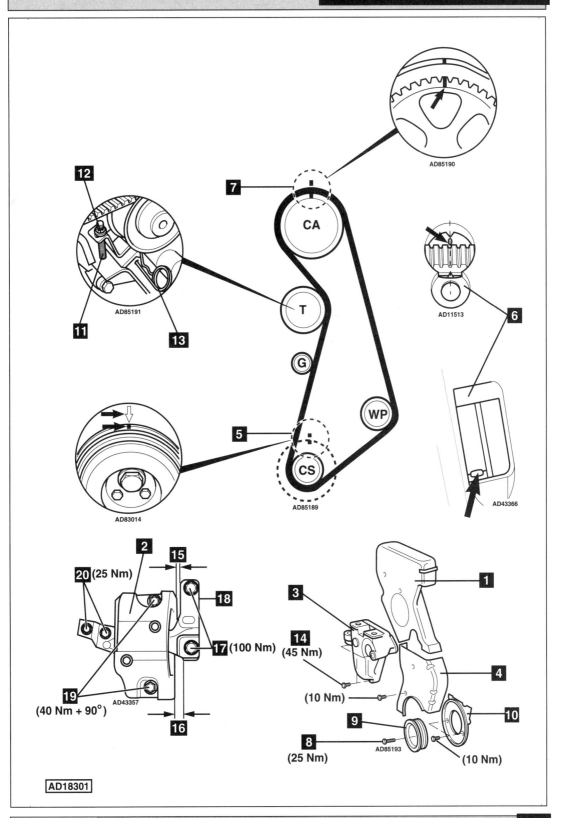

AD85190

12

7

CA

AD11513

6

T

11

13

AD85191

G

WP

5

CS

AD85189

AD83014

AD43366

2

15

20 (25 Nm)

18

17 (100 Nm)

19

(40 Nm + 90°)

AD43357

16

3

14
(45 Nm)

(10 Nm)

9

8
(25 Nm)

AD85193

(10 Nm)

1

4

10

AD18301

SKODA

Model:	**Octavia 2,0**
Year:	**1999-03**
Engine Code:	**AQY**

Replacement Interval Guide

Skoda recommend check at the first 60,000 miles and then every 20,000 miles (replace if necessary). The vehicle manufacturer has not recommended a timing belt replacement interval for this engine.
The previous use and service history of the vehicle must always be taken into account.
Refer to Timing Belt Replacement Intervals at the front of this manual.

Check For Engine Damage

CAUTION: This engine has been identified as an INTERFERENCE engine in which the possibility of valve-to-piston damage in the event of a timing belt failure is MOST LIKELY to occur. A compression check of all cylinders should be performed before removing the cylinder head(s).

Repair Times – hrs

Remove & install	1,30

Special Tools

■ Tensioner wrench – Skoda No.T10020.

Special Precautions

■ Disconnect battery earth lead.
■ DO NOT turn crankshaft or camshaft when timing belt removed.
■ Remove spark plugs to ease turning engine.
■ Turn engine in normal direction of rotation (unless otherwise stated).
■ DO NOT turn engine via camshaft or other sprockets.
■ Observe all tightening torques.

Removal

1. Remove:
   ❑ Dipstick.
   ❑ Engine top cover.
   ❑ Auxiliary drive belt.
   ❑ Auxiliary drive belt tensioner.
   ❑ Coolant expansion tank. DO NOT disconnect hoses.
   ❑ PAS reservoir. DO NOT disconnect hoses.
   ❑ Timing belt upper cover **1**.
2. Raise and support front of vehicle.
3. Remove:
   ❑ Engine lower cover.
   ❑ RH splash guard.
4. Turn crankshaft clockwise to TDC on No.1 cylinder.
5. Ensure timing marks aligned **2** or **3**.
6. Ensure timing marks aligned **4**.
7. Support engine.
8. Remove:
   ❑ RH engine mounting **5**.
   ❑ Crankshaft pulley bolts **6**.
   ❑ Crankshaft pulley **7**.
   ❑ Timing belt centre cover **8**.
   ❑ Timing belt lower cover **9**.
9. Raise engine slightly.
10. Remove RH engine mounting bracket.

11. Slacken tensioner nut **10**. Turn tensioner clockwise away from belt. Use tool No.T10020 **11**.
12. Lightly tighten nut **10**.
13. Remove timing belt.
    NOTE: Mark direction of rotation on belt with chalk if belt is to be reused.

Installation

1. Ensure timing marks aligned **4**.
2. Fit timing belt loosely to crankshaft sprocket and water pump sprocket.
   NOTE: Engine must be COLD when installing belt. If belt is to be reused: Observe direction of rotation marks on belt.
3. Install:
   ❑ RH engine mounting bracket.
   ❑ Timing belt lower cover **9**.
   ❑ Timing belt centre cover **8**.
   ❑ Crankshaft pulley **7**.
   ❑ Crankshaft pulley bolts **6**.
4. Tighten crankshaft pulley bolts **6**.
   Tightening torque: 40 Nm.
5. Ensure timing marks aligned **2** or **3**.
6. Install:
   ❑ RH engine mounting **5**.
   ❑ Coolant expansion tank.
   ❑ PAS reservoir.
7. Fit timing belt to tensioner pulley, then camshaft sprocket.
8. Ensure belt is taut between sprockets on non-tensioned side.
9. Ensure tensioner retaining lug is properly engaged **12**.
10. Slacken tensioner nut **10**.
11. Turn tensioner 5 times fully anti-clockwise and clockwise from stop to stop. Use tool No.T10020 **11**.
12. Turn tensioner fully anti-clockwise then slowly clockwise until pointer **13** aligned with notch **14** in baseplate. Use tool No.T10020.
13. Tighten tensioner nut **10**. Tightening torque: 25 Nm.
14. Turn crankshaft slowly two turns clockwise until timing marks aligned **2** or **3**.
    NOTE: Turn crankshaft last 45° smoothly without stopping.
15. Ensure timing marks aligned **4**.
16. Ensure pointer **13** aligned with notch **14** in baseplate.
17. If not: Repeat tensioning procedure.
18. Apply firm thumb pressure to belt at ▽. Pointer **13** and notch **14** must move apart.
19. Release thumb pressure from belt at ▽.
20. Turn crankshaft slowly two turns clockwise until timing marks aligned **2** or **3**.
    NOTE: Turn crankshaft last 45° smoothly without stopping.
21. Ensure timing marks aligned **4**.
22. Ensure pointer **13** aligned with notch **14** in baseplate.
23. Install components in reverse order of removal.

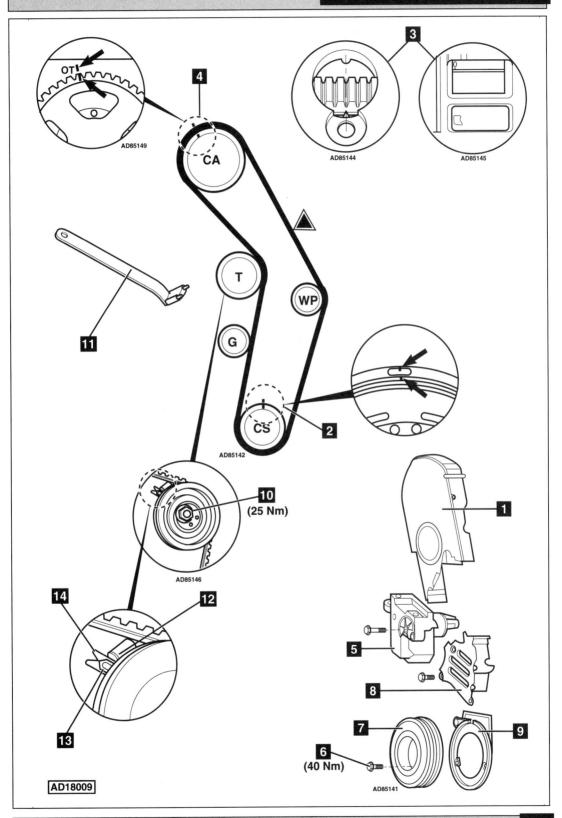

OTI · AD85149

3

AD85144

AD85145

4

CA

⚠

T

WP

11

G

2

CS

AD85142

10
(25 Nm)

AD85146

14

12

13

1

5

8

7

9

6
(40 Nm)

AD85141

AD18009

SKODA

Model:	Felicia 1,9D • Felicia Pick-up 1,9D • Felicia Cube Van 1,9D
Year:	1996-01
Engine Code:	AEF

Replacement Interval Guide

Skoda recommend check every 60,000 miles, and replacement if necessary.
The previous use and service history of the vehicle must always be taken into account.
Refer to Timing Belt Replacement Intervals at the front of this manual.

Check For Engine Damage

CAUTION: This engine has been identified as an INTERFERENCE engine in which the possibility of valve-to-piston damage in the event of a timing belt failure is MOST LIKELY to occur.
A compression check of all cylinders should be performed before removing the cylinder head.

Repair Times – hrs

Remove & install	1,90

Special Tools

- Sprocket holding tool – Skoda No.MP 1-216.
- Camshaft setting bar – Skoda No.MP 1-300.
- Injection pump locking pin – Skoda No.MP 1-301.
- Two-pin wrench – Skoda No.MP 1-302.

Special Precautions

- Disconnect battery earth lead.
- DO NOT turn crankshaft or camshaft when timing belt removed.
- Remove glow plugs to ease turning engine.
- Turn engine in normal direction of rotation (unless otherwise stated).
- DO NOT turn engine via camshaft or other sprockets.
- Observe all tightening torques.
- Check diesel injection pump timing after belt replacement.

Removal

1. Raise and support front of vehicle.
2. Remove:
   - ❏ Air filter cover and air intake hose.
   - ❏ Timing belt upper cover **1**.
   - ❏ Cylinder head cover.
   - ❏ Engine undershield.
   - ❏ Auxiliary drive belt.
   - ❏ Water pump pulley.
   - ❏ Crankshaft pulley bolts **3**.
   - ❏ Crankshaft pulley **4**.
   - ❏ Timing belt lower cover **5**.
3. Slacken engine mounting bracket bolts **17**.
4. Support engine.

5. Remove:
   - ❏ RH engine mounting bracket bolts **2**.
   - ❏ RH engine mounting bracket spacer **6**.
6. Turn crankshaft clockwise to TDC on No.1 cylinder. Ensure flywheel timing mark aligned in centre of hole **7**.
7. Fit setting bar to rear of camshaft **8**. Tool No.MP 1-300. Centralise camshaft using feeler gauges **9**.
   NOTE: If setting bar cannot be fitted: Turn crankshaft one turn clockwise.
8. Lock injection pump with locking pin in top hole **10**. Tool No.MP 1-301. Ensure groove on locking pin is 1,0 mm away from sprocket hub.
9. Slacken tensioner nut **11**. Move tensioner away from belt. Lightly tighten nut.
10. Hold camshaft sprocket. Use tool No.MP 1-216. Remove bolt **12**.
11. Loosen sprocket from taper using a drift through hole in timing belt rear cover **13**.
12. Remove:
    - ❏ Camshaft sprocket.
    - ❏ Timing belt through gap for engine mounting bracket spacer **6**.
    NOTE: Mark direction of rotation on belt with chalk if belt is to be reused.

Installation

1. Ensure crankshaft at TDC on No.1 cylinder. Ensure flywheel timing mark aligned in centre of hole **7**.
2. Ensure camshaft setting bar fitted correctly **8**.
3. Ensure locking pin located correctly in injection pump **10**.
4. Hold injection pump sprocket. Use tool No.MP 1-216. Slacken bolts 180° **14**.
5. Ensure injection pump sprocket is located with bolts in centre of holes **14**.
6. Fit timing belt in following order:
   - ❏ Crankshaft sprocket.
   - ❏ Oil/vacuum pump pulley.
   - ❏ Injection pump sprocket.
   - ❏ Tensioner pulley.
   NOTE: Ensure belt is taut between sprockets. Observe direction of rotation marks on belt.
7. Fit camshaft sprocket to belt, then install camshaft sprocket with belt onto end of camshaft.
8. Fit camshaft sprocket bolt **12**. Tighten bolt finger tight.

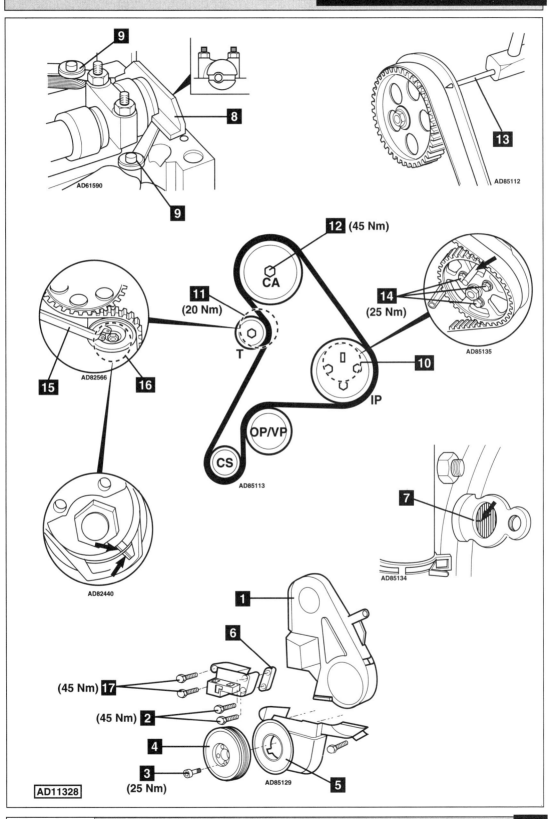

9. Turn automatic tensioner pulley clockwise until notch and raised mark on tensioner aligned 🔟. Use tool No.MP 1-302 🔟.

10. Tighten tensioner nut to 20 Nm 🔟.

11. Ensure flywheel timing mark aligned in centre of hole 🔟.

12. Hold camshaft sprocket. Use tool No.MP 1-216. Tighten bolt to 45 Nm 🔟.

13. Hold injection pump sprocket. Use tool No.MP 1-216. Tighten bolts to 25 Nm 🔟.

14. Remove:
    ❏ Setting bar 🔟.
    ❏ Feeler gauges 🔟.
    ❏ Injection pump locking pin 🔟.

15. Turn crankshaft two turns clockwise to TDC on No.1 cylinder. Ensure flywheel timing mark aligned in centre of hole 🔟.

16. Ensure camshaft setting bar can be fitted 🔟.

17. Ensure locking pin can be inserted in injection pump 🔟. Ensure groove on locking pin is 1,0 mm away from sprocket hub.

18. If not: Repeat installation and tensioning procedures.

19. Install components in reverse order of removal.

20. Tighten RH engine mounting bracket bolts to 45 Nm 🔟 & 🔟.

21. Tighten crankshaft pulley bolts to 25 Nm 🔟.

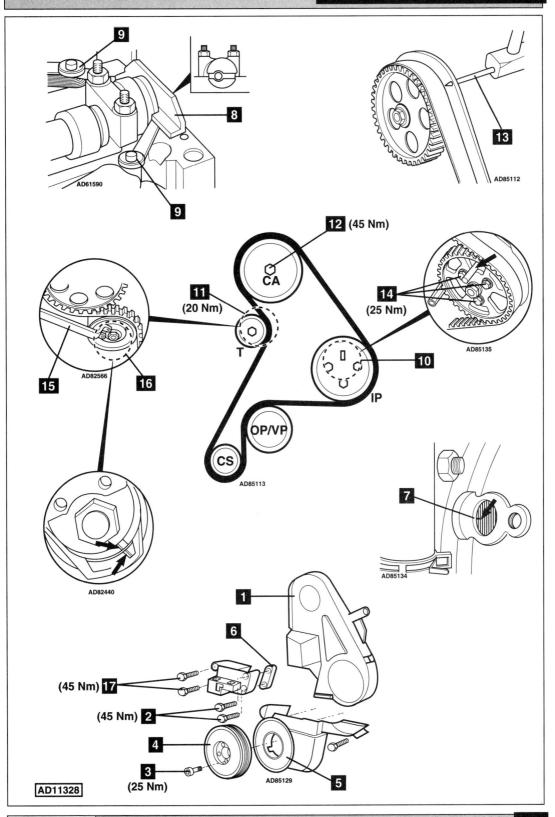

AD61590

AD85112

13

9

8

9

12 (45 Nm)

CA

11 (20 Nm)

T

14 (25 Nm)

AD85135

10

IP

OP/VP

CS

AD85113

15

AD82566

16

AD82440

7

AD85134

1

6

(45 Nm) **17**

(45 Nm) **2**

4

3 (25 Nm)

AD85129 **5**

AD11328

SKODA

Model:	**Octavia 1,9 SDI**
Year:	**1997-02**
Engine Code:	**AGP, AQM**

Replacement Interval Guide

Skoda recommend:
Check condition and width every 10,000 miles or 12 months, whichever occurs first (replacement width 22 mm).
Check timing belt part No. If fitted with:
Part No. 038 109 119 – replacement every 40,000 miles.
Part No. 038 109 119D – replacement every 60,000 miles.
The previous use and service history of the vehicle must always be taken into account.
Refer to Timing Belt Replacement Intervals at the front of this manual.

Check For Engine Damage

CAUTION: This engine has been identified as an INTERFERENCE engine in which the possibility of valve-to-piston damage in the event of a timing belt failure is MOST LIKELY to occur.
A compression check of all cylinders should be performed before removing the cylinder head(s).

Repair Times – hrs

Remove & install	3,30

Special Tools

- Sprocket holding tool – No.MP 1-216.
- Camshaft setting bar – No.MP 1-312.
- Injection pump locking pin – No.MP 1-301/3359.
- Puller – No.T40001.

Special Precautions

- Disconnect battery earth lead.
- DO NOT turn crankshaft or camshaft when timing belt removed.
- Remove glow plugs to ease turning engine.
- Turn engine in normal direction of rotation (unless otherwise stated).
- DO NOT turn engine via camshaft or other sprockets.
- Observe all tightening torques.
- Check diesel injection pump timing after belt replacement.

Removal

1. Raise and support front of vehicle.
2. Remove:
   - ❏ Engine cover.
   - ❏ Windscreen washer reservoir. DO NOT disconnect hoses.
   - ❏ Coolant expansion tank. DO NOT disconnect hoses.
   - ❏ PAS reservoir. DO NOT disconnect hoses.
   - ❏ Fuel filter. DO NOT disconnect hoses.
   - ❏ Engine undershield.
   - ❏ RH lower inner wing panel.
   - ❏ Auxiliary drive belt.
   - ❏ Auxiliary drive belt tensioner.
   - ❏ Air intake pipe.
   - ❏ Timing belt upper cover **1**.
   - ❏ Cylinder head cover.
   - ❏ Vacuum pump.
3. Turn crankshaft to TDC on No.1 cylinder.
4. Ensure timing marks aligned **2**.
5. Fit setting bar No.MP 1-312 to rear of camshaft **4**.
6. Centralise camshaft using feeler gauges.
7. Lock injection pump.
   Use tool No.MP 1-301/3359 **5**.
8. Slacken injection pump sprocket bolts **6**.
   NOTE: DO NOT slacken injection pump centre hub nut.
9. Support engine.
10. Remove:
    - ❏ RH engine mounting bolts **7**, **8** & **16**.
    - ❏ RH engine mounting.
    - ❏ RH engine mounting bracket bolts **9**.
    - ❏ RH engine mounting bracket.
    - ❏ Crankshaft pulley bolts **10**.
    - ❏ Crankshaft pulley **11**.
    - ❏ Timing belt centre cover **12**.
    - ❏ Timing belt lower cover **13**.
11. Slacken tensioner nut **14**.
12. Turn tensioner anti-clockwise away from belt. Use Allen key.
13. Lightly tighten tensioner nut **14**.
14. Remove timing belt.
    NOTE: Mark direction of rotation on belt with chalk if belt is to be reused.

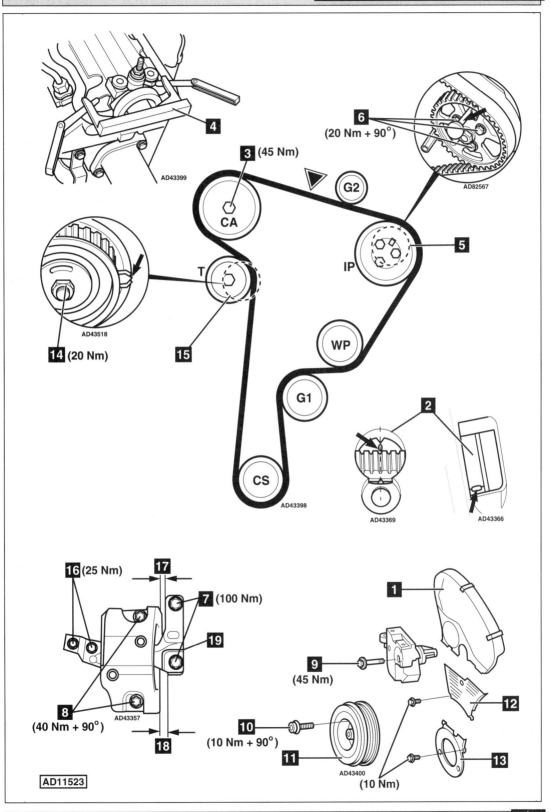

4 AD43399

3 (45 Nm)

6 (20 Nm + 90°) AD82567

G2

CA

IP **5**

T

AD43518

14 (20 Nm)

15

WP

G1

CS AD43398

2 AD43369 AD43366

16 (25 Nm) **17**

7 (100 Nm)

19

8 AD43357

(40 Nm + 90°)

18

1

9 (45 Nm)

12

10 (10 Nm + 90°)

11 AD43400

13

(10 Nm)

AD11523

←

Installation

1. Ensure timing marks aligned **2**.
2. Ensure camshaft setting bar fitted correctly. Tool No.MP 1-312 **4**.
3. Ensure locking pin located correctly in injection pump **5**. Tool No. MP 1-301/3359.
4. Hold camshaft sprocket. Use tool No.MP 1-216.
5. Slacken camshaft sprocket bolt 1/2 turn **3**.
   NOTE: DO NOT use setting bar to hold camshaft when slackening sprocket bolt.
6. Loosen camshaft sprocket from taper:
   ❑ Method 1: Use drift through hole in timing belt rear cover.
   ❑ Method 2: Use puller. Tool No.T40001.
7. Remove:
   ❑ Camshaft sprocket bolt **3**.
   ❑ Camshaft sprocket.
8. Remove bolts from injection pump sprocket **6**. Fit new bolts.
9. Align injection pump sprocket with bolts in centre of slotted holes.
10. Fit timing belt in following order:
    ❑ Crankshaft sprocket.
    ❑ Guide pulley G1.
    ❑ Water pump sprocket.
    ❑ Injection pump sprocket.
    ❑ Tensioner pulley.
    ❑ Guide pulley G2.
    NOTE: Turn injection pump sprocket slightly to engage timing belt teeth.
11. Ensure belt is taut between sprockets.
12. Fit camshaft sprocket to belt, then install camshaft sprocket with belt onto end of camshaft.
13. Fit camshaft sprocket bolt **3**.
14. Lightly tighten camshaft sprocket bolt **3**. Sprocket should turn freely on taper but not tilt.
15. Slacken tensioner nut **14**.
16. Turn tensioner pulley clockwise until pointer and notch in backplate aligned **15**. Use Allen key.
17. If tensioner pulley turned too far: Turn fully anti-clockwise and repeat tensioning procedure.
18. Tighten tensioner nut to 20 Nm **14**.
    NOTE: Ensure tensioner retaining lug is properly engaged.
19. Ensure timing marks aligned **2**.
20. Hold camshaft sprocket. Use tool No.MP 1-216.
21. Tighten camshaft sprocket bolt to 45 Nm **3**.
22. Temporarily tighten injection pump sprocket bolts to 20 Nm **6**.
    NOTE: DO NOT tighten further 90°.
23. Remove:
    ❑ Camshaft setting bar **4**. Tool No.MP 1-312.
    ❑ Injection pump locking pin **5**. Tool No. MP 1-301/3359.

24. Turn crankshaft two turns clockwise until timing marks aligned **2**.
25. Ensure camshaft setting bar can be fitted **4**.
26. Ensure locking pin can be inserted in injection pump **5**. Tighten bolts further 90° **6**.
27. If not:
    ❑ Turn crankshaft until locking pin can be inserted **5**. Tool No.MP 1-301/3359.
    ❑ Slacken injection pump sprocket bolts **6**.
    ❑ Turn crankshaft to TDC on No.1 cylinder.
    ❑ Ensure timing marks aligned **2**.
    ❑ Tighten bolts to 20 Nm + 90° **6**.
28. Ensure tensioner pointer aligned with notch in backplate **15**.
29. If not: Repeat tensioning procedure.
30. Apply firm thumb pressure to belt at ▽. Tensioner marks should move out of alignment **15**.
31. Release thumb pressure from belt. Pointer must realign with notch.
32. Install:
    ❑ Timing belt lower cover **13**.
    ❑ Timing belt centre cover **12**.
    ❑ Crankshaft pulley **11**.
    ❑ Crankshaft pulley bolts **10**.
33. Tighten crankshaft pulley bolts **10**. Tightening torque: 10 Nm + 90°.
34. Install:
    ❑ RH engine mounting bracket.
    ❑ RH engine mounting bracket bolts **9**.
35. Tighten RH engine mounting bracket bolts to 45 Nm **9**.
36. Fit and align RH engine mounting:
    ❑ Engine mounting clearance: 14 mm **17**.
    ❑ Engine mounting clearance: 10 mm minimum **18**.
    ❑ Ensure engine mounting bolts **7** aligned with edge of mounting **19**.
37. Tighten:
    ❑ Engine mounting bolts **8**. Tightening torque: 40 Nm + 90°. Use new bolts.
    ❑ Engine mounting bolts **16**. Tightening torque: 25 Nm.
    ❑ Engine mounting bolts **7**. Tightening torque: 100 Nm. Use new bolts.
38. Install remainder of components in reverse order of removal.

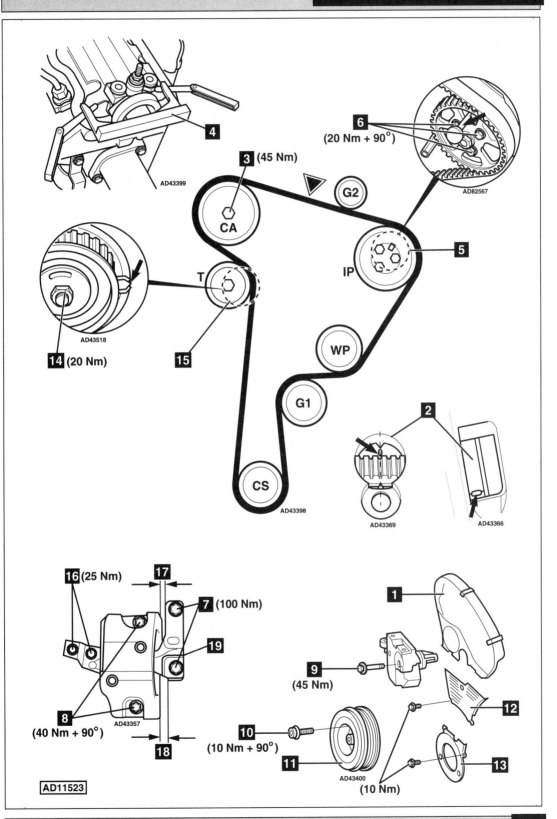

4

AD43399

3 (45 Nm)

G2

CA

6
(20 Nm + 90°)

AD82567

5

IP

AD43518

14 (20 Nm)

T

15

WP

G1

CS

AD43398

2

AD43369

AD43366

16 (25 Nm)

17

7 (100 Nm)

19

8
(40 Nm + 90°)

AD43357

18

10
(10 Nm + 90°)

11

AD43400

9
(45 Nm)

1

12

13

(10 Nm)

AD11523

SKODA

Model:	**Octavia 1,9 TDI**
Year:	**1996-03**
Engine Code:	**AGR, AHF, ALH**

Replacement Interval Guide

Skoda recommend:

→06/99:

Check condition and width every 10,000 miles or 12 months, whichever occurs first (replacement width = less than 22 mm).
Replacement at the first 40,000 miles using:
Part No.038 109 119 – replacement then every 40,000 miles.
Part No.038 109 119D – replacement then every 60,000 miles.

07/99-04/01:

→2000MY: Check condition and width every 10,000 miles or 12 months, whichever occurs first (replacement width = less than 22 mm).
Part No.038 109 119D – replacement every 60,000 miles.

05/01→:

Replacement every 80,000 miles (guide pulley must also be replaced – part No.038 109 244H).
The previous use and service history of the vehicle must always be taken into account.
Refer to Timing Belt Replacement Intervals at the front of this manual.

Check For Engine Damage

CAUTION: This engine has been identified as an INTERFERENCE engine in which the possibility of valve-to-piston damage in the event of a timing belt failure is MOST LIKELY to occur.
A compression check of all cylinders should be performed before removing the cylinder head(s).

Repair Times – hrs

Remove & install   3,30

Special Tools

- Camshaft setting bar – No.MP 1-312.
- Injection pump locking pin – No.3359.
- Sprocket holding tool – No.MP 1-216.
- Two-pin wrench – Matra V.159.
- Puller – No.T40001.

Special Precautions

- Disconnect battery earth lead.
- DO NOT turn crankshaft or camshaft when timing belt removed.
- Remove glow plugs to ease turning engine.
- Turn engine in normal direction of rotation (unless otherwise stated).
- DO NOT turn engine via camshaft or other sprockets.
- Observe all tightening torques.
- Check diesel injection pump timing after belt replacement.

Removal

1. Raise and support front of vehicle.
2. Disconnect wiring (as required).
3. Remove:
   - ❏ RH headlamp.
   - ❏ Engine cover.
   - ❏ Intercooler to intake manifold hose.
   - ❏ Coolant expansion tank. DO NOT disconnect hoses.
   - ❏ PAS reservoir. DO NOT disconnect hoses.
   - ❏ Windscreen washer reservoir. DO NOT disconnect hoses.
   - ❏ Fuel filter. DO NOT disconnect hoses.
   - ❏ Turbocharger to intercooler pipe.
   - ❏ Engine undershield.
   - ❏ RH lower inner wing panel.
   - ❏ Auxiliary drive belt.
   - ❏ Timing belt upper cover **1**.
   - ❏ Cylinder head cover.
   - ❏ Vacuum pump.
4. Turn crankshaft to TDC on No.1 cylinder.
5. Ensure timing marks aligned **2**.
6. Fit setting bar No.MP 1-312 to rear of camshaft **3**.
7. Centralise camshaft using feeler gauges.
8. Lock injection pump. Use tool No.3359 **4**.
9. Slacken injection pump sprocket bolts **5**.
   NOTE: DO NOT slacken injection pump centre hub nut.
10. Support engine.
11. Remove:
    - ❏ RH engine mounting bolts **6**, **7** & **20**.
    - ❏ RH engine mounting.
    - ❏ RH engine mounting bracket bolts **8**.
    - ❏ RH engine mounting bracket.
    - ❏ Crankshaft pulley bolts **9**.
    - ❏ Crankshaft pulley **10**.
    - ❏ Timing belt centre cover **11**.
    - ❏ Timing belt lower cover **12**.

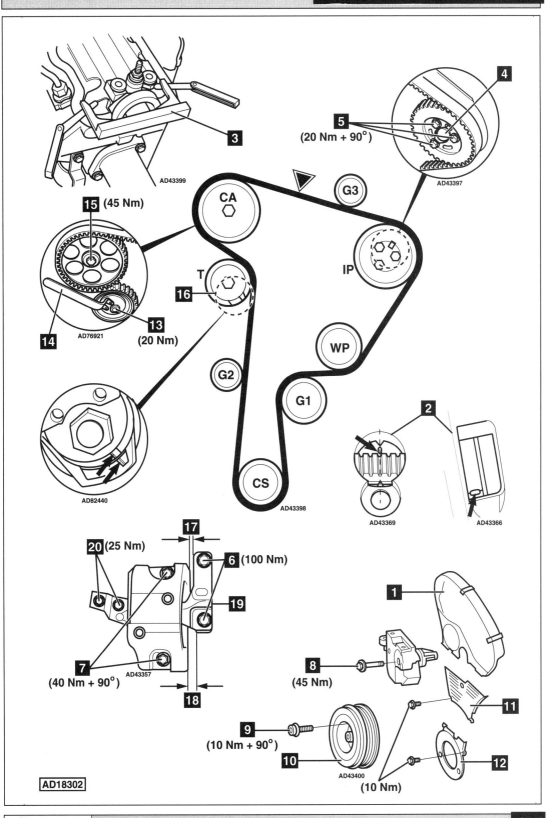

AD43399

3

4

5 (20 Nm + 90°)

AD43397

15 (45 Nm)

CA

G3

IP

T

16

13 (20 Nm)

14

AD76921

AD82440

G2

G1

WP

CS

AD43398

2

AD43369

AD43366

17

20 (25 Nm)

6 (100 Nm)

19

7 (40 Nm + 90°)

AD43357

18

1

8 (45 Nm)

11

9 (10 Nm + 90°)

10

AD43400

12

(10 Nm)

AD18302

←

12. Slacken tensioner nut **13**.
13. Turn tensioner anti-clockwise away from belt. Use wrench Matra V.159 **14**.
14. Lightly tighten tensioner nut **13**.
15. Remove timing belt.

NOTE: Mark direction of rotation on belt with chalk if belt is to be reused.

Installation

1. Ensure timing marks aligned **2**.
2. Ensure camshaft setting bar fitted correctly. Tool No.MP 1-312 **3**.
3. Ensure locking pin located correctly in injection pump **4**. Tool No.3359.
4. Hold camshaft sprocket. Use tool No.MP 1-216.
5. Slacken camshaft sprocket bolt 1/2 turn **15**.

NOTE: DO NOT use setting bar to hold camshaft when slackening sprocket bolt.

6. Loosen camshaft sprocket from taper:
   ❏ Method 1: Use drift through hole in timing belt rear cover.
   ❏ Method 2: Use puller. Tool No.T40001.
7. Remove:
   ❏ Camshaft sprocket bolt **15**.
   ❏ Camshaft sprocket.
8. Remove bolts from injection pump sprocket **5**. Fit new bolts.
9. Align injection pump sprocket with bolts in centre of slotted holes.
10. Fit timing belt in following order:
    ❏ Crankshaft sprocket.
    ❏ Guide pulley G1.
    ❏ Water pump sprocket.
    ❏ Injection pump sprocket.
    ❏ Guide pulley G2.
    ❏ Tensioner pulley.
    ❏ Guide pulley G3.

NOTE: Turn injection pump sprocket slightly to engage timing belt teeth.

11. Ensure belt is taut between sprockets.
12. Fit camshaft sprocket to belt, then install camshaft sprocket with belt onto end of camshaft.
13. Fit camshaft sprocket bolt **15**.
14. Lightly tighten camshaft sprocket bolt **15**. Sprocket should turn freely on taper but not tilt.
15. Slacken tensioner nut **13**.
16. Turn tensioner pulley clockwise until notch and raised mark on tensioner aligned **16**. Use wrench Matra V.159 **14**.
17. If tensioner pulley turned too far: Turn fully anti-clockwise and repeat tensioning procedure.
18. Tighten tensioner nut to 20 Nm **13**.

NOTE: Ensure tensioner retaining lug is properly engaged.

19. Ensure timing marks aligned **2**.
20. Hold camshaft sprocket. Use tool No.MP 1-216.
21. Tighten camshaft sprocket bolt to 45 Nm **15**.
22. Temporarily tighten injection pump sprocket bolts to 20 Nm **5**.

NOTE: DO NOT tighten further 90°.

23. Remove:
    ❏ Camshaft setting bar **3**. Tool No.MP 1-312.
    ❏ Injection pump locking pin **4**. Tool No.3359.
24. Turn crankshaft two turns clockwise until timing marks aligned **2**.
25. Ensure camshaft setting bar can be fitted **3**.
26. Ensure locking pin can be inserted in injection pump **4**. Tighten bolts further 90° **5**.
27. If not:
    ❏ Turn crankshaft until locking pin can be inserted **4**. Tool No.3359.
    ❏ Slacken injection pump sprocket bolts **5**.
    ❏ Turn crankshaft to TDC on No.1 cylinder.
    ❏ Ensure timing marks aligned **2**.
    ❏ Tighten bolts to 20 Nm + 90° **5**.
28. Ensure notch and raised mark on tensioner aligned **16**.
29. If not: Repeat tensioning procedure.
30. Apply firm thumb pressure to belt at ▽. Tensioner marks should move out of alignment **16**.
31. Release thumb pressure from belt. Tensioner marks should realign.
32. Install:
    ❏ Timing belt lower cover **12**.
    ❏ Timing belt centre cover **11**.
    ❏ Crankshaft pulley **10**.
    ❏ Crankshaft pulley bolts **9**.
33. Tighten crankshaft pulley bolts **9**. Tightening torque: 10 Nm + 90°.
34. Install:
    ❏ RH engine mounting bracket.
    ❏ RH engine mounting bracket bolts **8**.
35. Tighten RH engine mounting bracket bolts to 45 Nm **8**.
36. Fit and align RH engine mounting:
    ❏ Engine mounting clearance: 14 mm **17**.
    ❏ Engine mounting clearance: 10 mm minimum **18**.
    ❏ Ensure engine mounting bolts **6** aligned with edge of mounting **19**.
37. Tighten:
    ❏ Engine mounting bolts **7**. Tightening torque: 40 Nm + 90°. Use new bolts.
    ❏ Engine mounting bolts **20**. Tightening torque: 25 Nm.
    ❏ Engine mounting bolts **6**. Tightening torque: 100 Nm. Use new bolts.
38. Install remainder of components in reverse order of removal.

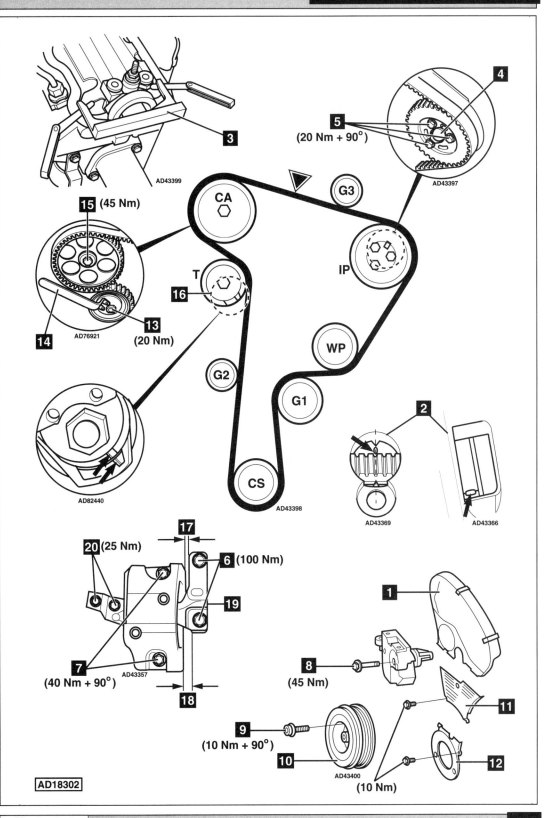

3

AD43399

5 (20 Nm + 90°)

4

AD43397

CA

G3

15 (45 Nm)

T

IP

16

13 (20 Nm)

14

AD76921

WP

G2

G1

AD82440

2

CS

AD43398

AD43369

AD43366

17

20 (25 Nm)

6 (100 Nm)

19

7

AD43357

18

(40 Nm + 90°)

1

8 (45 Nm)

11

9 (10 Nm + 90°)

10

AD43400

12

(10 Nm)

AD18302

Model:	Justy • Sumo (E10/E12)
Year:	1987-97
Engine Code:	EF10, EF12

Replacement Interval Guide

Subaru recommend:
Justy: Replacement every 45,000 miles or 36 months, whichever occurs first.
Sumo: Replacement every 24,000 miles or 24 months, whichever occurs first.
The previous use and service history of the vehicle must always be taken into account.
Refer to Timing Belt Replacement Intervals at the front of this manual.

Check For Engine Damage

CAUTION: This engine has been identified as a FREEWHEELING engine in which the possibility of valve-to-piston damage in the event of a timing belt failure may be minimal or very unlikely. However, a precautionary compression check of all cylinders should be performed.

Repair Times – hrs

Remove & install:

Justy	1,40
Sumo (E10/E12)	0,80[1]
Sumo (E12)	5,40[2]

[1] Early models (engine installed).
[2] Later models (includes removal and installation of engine).

Special Tools

- ■ Crankshaft pulley holding tool (EF10) – Subaru No.499205500.
- ■ Crankshaft pulley holding tool (EF12) – Subaru No.498715600.

Special Precautions

- ■ Disconnect battery earth lead.
- ■ DO NOT turn crankshaft or camshaft when timing belt removed.
- ■ Remove spark plugs to ease turning engine.
- ■ Turn engine in normal direction of rotation (unless otherwise stated).
- ■ DO NOT turn engine via camshaft or other sprockets.
- ■ Observe all tightening torques.

Removal

1. Remove auxiliary drive belt.
2. Hold crankshaft pulley **2**. Use appropriate tool. Slacken crankshaft pulley bolt **1**.
3. Turn crankshaft to TDC on No.3 cylinder exhaust stroke.
4. Remove:
   - ❑ Crankshaft pulley bolt **1**.
   - ❑ Crankshaft pulley **2**.
   - ❑ Timing belt cover and seal **3** & **4**.
   - ❑ Crankshaft sprocket guide washer **7**.
5. Ensure timing marks aligned **5** & **6**.
6. Slacken tensioner bolts **8**. Move tensioner away from belt and lightly tighten bolts.
7. Remove timing belt.

Installation

1. Ensure timing marks aligned **5** & **6**.
2. Fit timing belt in following order:
   - ❑ Crankshaft sprocket
   - ❑ Camshaft sprocket.
   - ❑ Tensioner pulley.
4. Slacken tensioner bolts **8**. Allow tensioner to operate and tighten bolts.
5. Turn crankshaft two turns clockwise. Ensure timing marks aligned **5** & **6**.
6. Fit crankshaft sprocket guide washer **7**.
7. Install components in reverse order of removal.
8. Hold crankshaft pulley **2**. Use appropriate tool. Tighten crankshaft pulley bolt **1**. EF10: 70 Nm. EF12: 88 Nm.

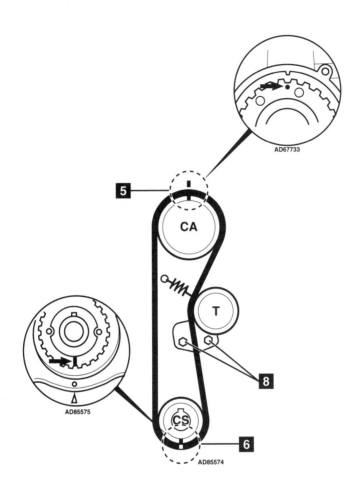

AD67733

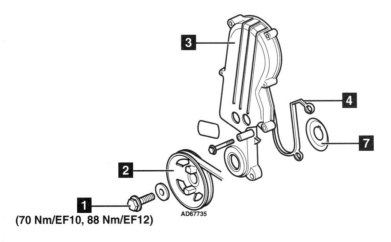

(70 Nm/EF10, 88 Nm/EF12)

AD10988

SUBARU

Model:	Impreza 1,6/2,0 • Legacy 2,0/2,2 • Forester 2,0/Turbo • Legacy 2,5 Legacy Outback 2,5
Year:	1997-03
Engine Code:	EJ16/EJ20/EJ20J2/EJ205/EJ22/EJ25

Replacement Interval Guide

Subaru recommend:

→1999MY:
Replacement every 45,000 miles or 3 years, whichever occurs first.

2000-01MY:
Replacement every 60,000 miles or 4 years, whichever occurs first.

2002MY→:
Except Forester 2,0 Turbo: Replacement every 60,000 miles or 5 years, whichever occurs first.
Forester 2,0 Turbo: Replacement every 50,000 miles or 5 years, whichever occurs first.
The previous use and service history of the vehicle must always be taken into account.
Refer to Timing Belt Replacement Intervals at the front of this manual.

Check For Engine Damage

CAUTION: This engine has been identified as a FREEWHEELING engine in which the possibility of valve-to-piston damage in the event of a timing belt failure may be minimal or very unlikely. However, a precautionary compression check of all cylinders should be performed.

Repair Times – hrs

Remove & install:

Impreza/Forester	1,20
Forester Turbo	2,10
Legacy 2,0/2,5	1,20
Legacy 2,2	1,30

Special Tools

- Crankshaft pulley holding tool – Subaru No.499977 000/300.
- Crankshaft sprocket tool – Subaru No.499987500.

Removal

1. Remove:
   - ❑ Coolant expansion tank.
   - ❑ Cooling fan.
   - ❑ AC condenser blower.
   - ❑ Auxiliary drive belt cover.
   - ❑ Auxiliary drive belts.
   - ❑ AC drive belt tensioner.
2. Hold crankshaft pulley **1**. Use tool No.499977 000/300.
3. Remove:
   - ❑ Crankshaft pulley bolt **2**.
   - ❑ Crankshaft pulley **1**.
   - ❑ Timing belt covers **3**, **4** & **5**.
   - ❑ MT: Timing belt guide bolts **6**.
   - ❑ MT: Timing belt guide **7**.
4. Turn crankshaft clockwise until timing marks aligned **8**, **9** & **10**. Use tool No.499987500.

5. Remove:
   - ❑ Guide pulley (G1 or G2) (to allow belt to slacken).
   - ❑ Timing belt.
   - ❑ Automatic tensioner unit bolt **11**.
   - ❑ Automatic tensioner unit **12**.
   NOTE: Mark direction of rotation on belt with chalk if belt is to be reused.

Installation

1. Check tensioner pulley and guide pulleys for smooth operation. Replace if necessary.
2. Check tensioner body for leakage or damage. Replace if necessary.
3. Check pushrod protrusion is 5,5-6,5 mm **13**. If not: Replace automatic tensioner unit.
   NOTE: Ensure automatic tensioner unit remains in the normal upright position during the following reset procedure.
4. Hold tensioner body. Check pushrod does not move when pushed against a firm surface with a force of less than 30 kg.
5. If pushrod moves easily, proceed as follows:
   - ❑ Slowly press pushrod until flush with upper surface of tensioner body, then release. Repeat 3 times.
   - ❑ Press pushrod with a force of 30 kg. If pushrod moves easily: Replace automatic tensioner unit.
6. Slowly compress pushrod into tensioner body until holes aligned **14**. Use a press. Retain pushrod with 2 mm diameter pin through hole in tensioner body **15**.
7. Install automatic tensioner unit **12**. Tighten bolt to 35-43 Nm **11**.
8. Ensure timing marks aligned **8**, **9** & **10**.
9. Fit timing belt in anti-clockwise direction, starting at crankshaft sprocket. Ensure belt is taut on non-tensioned side.
10. Fit guide pulley (G1 or G2). Tighten bolt to 35-43 Nm.
11. Remove pin from tensioner body to release pushrod **15**.
12. Turn crankshaft two turns clockwise. Use tool No.499987500. Ensure timing marks aligned **8**, **9** & **10**.
13. MT: Fit timing belt guide **7**. Tighten bolts finger tight **6**.
14. MT: Adjust clearance between back of timing belt and belt guide to 0,5-1,5 mm **16**. Tighten bolts to 10 Nm **6**.
15. Install components in reverse order of removal.
16. Lightly oil threads and contact face of crankshaft pulley bolt **2**.
17. Hold crankshaft pulley **1**. Use tool No.499977 000/300.
18. Tighten crankshaft pulley bolt **2**. Stage 1: 44 Nm. Stage 2: 122-137 Nm.
19. Check bolt has turned 45°. If not: Fit new bolt. Repeat tightening procedure.

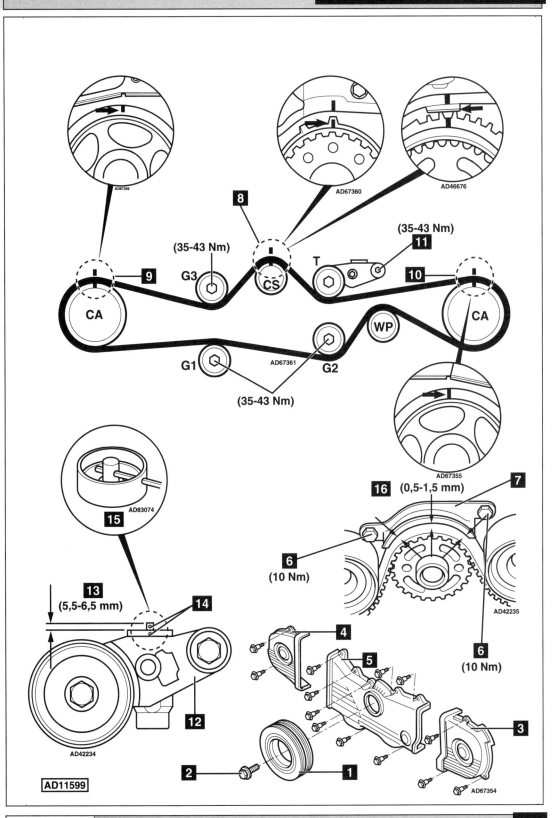

SUBARU

Model:	**Impreza 1,6/1,8 16V**
Year:	**1993-00**
Engine Code:	**1600/1800 OHC**

Replacement Interval Guide

Subaru recommend replacement every 45,000 miles or 36 months, whichever occurs first.
The previous use and service history of the vehicle must always be taken into account.
Refer to Timing Belt Replacement Intervals at the front of this manual.

Check For Engine Damage

CAUTION: This engine has been identified as a FREEWHEELING engine in which the possibility of valve-to-piston damage in the event of a timing belt failure may be minimal or very unlikely. However, a precautionary compression check of all cylinders should be performed.

Repair Times – hrs

Remove & install	1,20
PAS	+0,10
AC	+0,10

Special Tools

- None required.

Special Precautions

- Disconnect battery earth lead.
- DO NOT turn crankshaft or camshaft when timing belt removed.
- Remove spark plugs to ease turning engine.
- Turn engine in normal direction of rotation (unless otherwise stated).
- DO NOT turn engine via camshaft or other sprockets.
- Observe all tightening torques.

Removal

1. Remove:
   - ❑ Coolant expansion tank.
   - ❑ Cooling fan assembly.
   - ❑ Auxiliary drive belt cover.
   - ❑ Auxiliary drive belts.
   - ❑ Crankshaft pulley **1**.
   - ❑ Timing belt covers **2**, **3** & **4**.
2. Turn crankshaft until timing marks on belt aligned with timing marks on sprockets **5** & **6**.
3. Slacken automatic tensioner unit bolts **7**. Move automatic tensioner unit away from tensioner pulley.

4. Remove:
   - ❑ Guide pulleys (G2 & G3).
   - ❑ Tensioner pulley and spacer.
   - ❑ Automatic tensioner unit **8**.
   NOTE: Mark direction of rotation and position of sprockets on belt with chalk if belt is to be reused.
5. Remove timing belt.

Installation

1. Check tensioner body for leakage or damage **8**.
2. Check extended length of pushrod is 15,4-16,4 mm **10**. If not: Replace automatic tensioner unit **8**.
3. Slowly compress pushrod into tensioner body until holes aligned. Retain pushrod with 1,50 mm diameter pin through hole in tensioner body **11**.
4. Install:
   - ❑ Automatic tensioner unit **8**. Lightly tighten bolts **7**.
   - ❑ Guide pulleys (G2 & G3). Tighten bolts to 40 Nm.
   - ❑ Tensioner pulley and spacer. Tighten bolt to 40 Nm **9**.
5. Ensure timing marks aligned **5** & **6**.
6. Fit timing belt in anti-clockwise direction, starting at crankshaft sprocket. Ensure belt is taut between sprockets.
7. Ensure timing marks on belt aligned with timing marks on sprockets **5** & **6**. Observe direction of rotation mark on belt.
8. Slacken automatic tensioner unit bolts **7**. Move automatic tensioner unit fully towards tensioner pulley. Tighten bolts to 25 Nm **7**.
9. Ensure timing marks aligned **5** & **6**.
10. Remove pin from tensioner body to release pushrod **11**.
11. Install components in reverse order of removal.
12. Tighten crankshaft pulley bolt **12**.
    Tightening torque: 95 Nm.

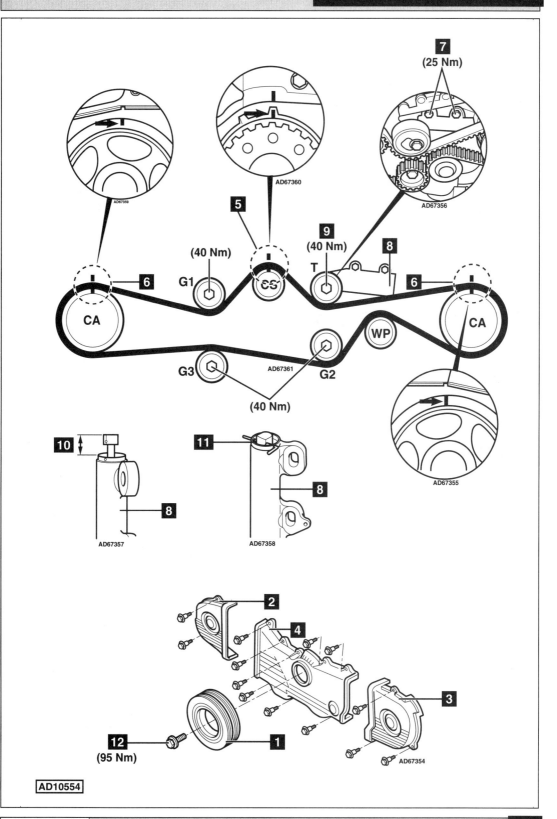

SUBARU

Model:	**Impreza 2,0 • Legacy 1,8/2,0/2,2**
Year:	**1989-96**
Engine Code:	**AY/EJ 18, AY/EJ 20-EN, 2200 SOHC**

Replacement Interval Guide

Subaru recommend replacement every 45,000 miles or 36 months, whichever occurs first.
The previous use and service history of the vehicle must always be taken into account.
Refer to Timing Belt Replacement Intervals at the front of this manual.

Check For Engine Damage

CAUTION: This engine has been identified as a FREEWHEELING engine in which the possibility of valve-to-piston damage in the event of a timing belt failure may be minimal or very unlikely. However, a precautionary compression check of all cylinders should be performed.

Repair Times – hrs

Remove & install	1,20
PAS	+0,10
AC	+0,10

Special Tools

- Crankshaft pulley holding tool – Subaru No.499977000.

Special Precautions

- Disconnect battery earth lead.
- DO NOT turn crankshaft or camshaft when timing belt removed.
- Remove spark plugs to ease turning engine.
- Turn engine in normal direction of rotation (unless otherwise stated).
- DO NOT turn engine via camshaft or other sprockets.
- Observe all tightening torques.

Removal

1. Remove:
   - ❑ Coolant expansion tank.
   - ❑ Cooling fan cowling upper bolts. Slacken lower bolts.
   - ❑ Cooling fan and cowling.
   - ❑ AC fan (if fitted).
   - ❑ Auxiliary drive belt cover.
   - ❑ Auxiliary drive belts.
2. Hold crankshaft pulley **1**.
   Use tool No.499977000.
3. Remove:
   - ❑ Crankshaft pulley bolt **2**.
   - ❑ Crankshaft pulley **1**.
   - ❑ Timing belt covers and seals **3**, **4** & **5**.

4. Turn crankshaft until timing marks aligned **6**, **7** & **8**.
5. Slacken automatic tensioner unit bolts **9**.
6. Remove:
   - ❑ Guide pulleys (G2 & G3).
   - ❑ Timing belt.
   - ❑ Guide pulley (G1).
   - ❑ Tensioner pulley and spacer **10**.
   - ❑ Automatic tensioner unit **11**.

Installation

1. Check tensioner body for leakage or damage.
2. Check pushrod protrusion is 15,4-16,4 mm **12**. If not: Replace automatic tensioner unit.
3. Hold tensioner body. Check pushrod does not move when pushed against a firm surface with a force of 15-50 kgf. If pushrod moves: Replace automatic tensioner unit.
4. Slowly compress pushrod into tensioner body until holes aligned. Use a press. Pressure must not exceed 1,000 kgf. Retain pushrod with 1,5 mm diameter pin through hole in tensioner body **13**.
5. Install automatic tensioner unit, push to right (looking at front of engine) and lightly tighten bolts.
6. Install:
   - ❑ Guide pulley (G1). Tighten bolt to 40 Nm.
   - ❑ Tensioner pulley and spacer **10**. Tighten bolt to 40 Nm.
7. Ensure timing marks aligned **6**, **7** & **8**.
8. Fit timing belt, starting at crankshaft sprocket.
9. Fit guide pulleys (G2 & G3). Tighten bolts to 40 Nm.
10. Slacken automatic tensioner unit bolts **9**. Move tensioner unit fully to left and tighten bolts to 25 Nm.
11. Ensure timing marks aligned **6**, **7** & **8**.
12. Remove pin from tensioner body to release pushrod.
13. Install:
    - ❑ Timing belt covers and seals **3**, **4** & **5**.
    - ❑ Crankshaft pulley **1**.
    - ❑ Crankshaft pulley bolt **2**.
14. Hold crankshaft pulley. Use tool No.499977000. Tighten crankshaft pulley bolt **2**.
    Tightening torque: 95 Nm.
15. Install components in reverse order of removal.

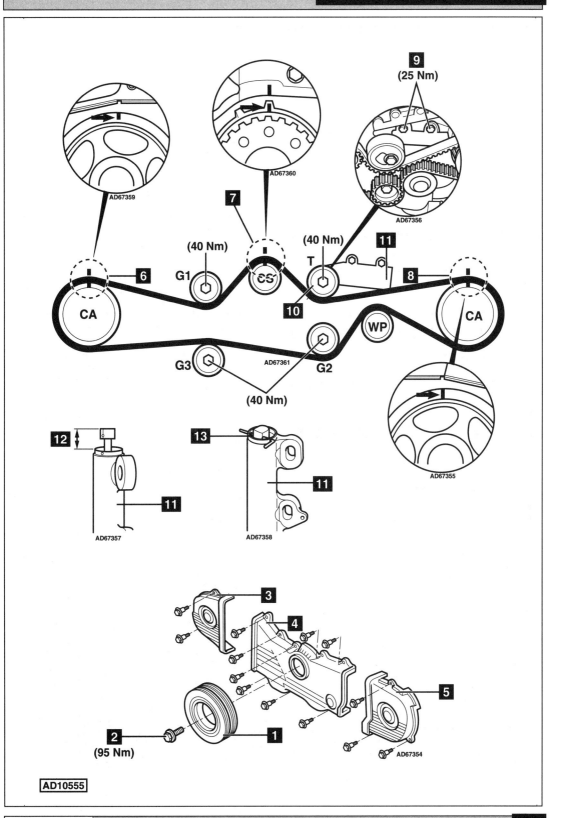

SUBARU

Model:	Impreza 2,0 Turbo • Legacy 2,0 Turbo • Legacy 2,5 Legacy Outback 2,5
Year:	1991-99
Engine Code:	EJ20/EJ20-G/EJ25

Replacement Interval Guide

Subaru recommend replacement every 45,000 miles or 36 months, whichever occurs first.
The previous use and service history of the vehicle must always be taken into account.
Refer to Timing Belt Replacement Intervals at the front of this manual.

Check For Engine Damage

CAUTION: This engine has been identified as an INTERFERENCE engine in which the possibility of valve-to-piston damage in the event of a timing belt failure is MOST LIKELY to occur.
A compression check of all cylinders should be performed before removing the cylinder head.

Repair Times – hrs

Remove & install:

Impreza	2,10
Legacy	1,50
PAS	+0,10
AC	+0,10

Special Tools

- None required.

Special Precautions

- Disconnect battery earth lead.
- DO NOT turn crankshaft or camshaft when timing belt removed.
- Remove spark plugs to ease turning engine.
- Turn engine in normal direction of rotation (unless otherwise stated).
- DO NOT turn engine via camshaft or other sprockets.
- Observe all tightening torques.

Removal

1. Remove:
   - ❏ Auxiliary drive belt.
   - ❏ Crankshaft pulley **1**.
   - ❏ Timing belt covers **2**, **3** & **4**.
   - ❏ Temporarily fit crankshaft pulley.
2. Turn crankshaft until timing marks aligned **5**, **6**, **7**, **8**, **9**, **10** & **11**.
   *NOTE: Ensure double lines on inlet camshaft sprockets and exhaust camshaft sprockets aligned **7** & **10**. Ensure crankshaft key at 6 o'clock position.*
3. Slacken automatic tensioner unit bolts **12**.

4. Remove:
   - ❏ Guide pulley (G1).
   - ❏ Timing belt.
   - ❏ Guide pulleys (G2 & G3).
   - ❏ Automatic tensioner unit.
   - ❏ Tensioner pulley and spacer.

Installation

1. Check tensioner body for leakage or damage.
2. Check pushrod protrusion is 15,4-16,4 mm **13**. If not: Replace automatic tensioner unit.
3. Hold tensioner body. Check pushrod does not move when pushed against a firm surface with a force of 15-50 kg. If pushrod moves: Replace automatic tensioner unit.
4. Slowly compress pushrod into tensioner body until holes aligned **14**. Use a press. Force must not exceed 1,000 kg. Retain pushrod with 1,5 mm diameter pin through hole in tensioner body **15**.
5. Install automatic tensioner unit, push to right (looking at front of engine) and lightly tighten bolts.
6. Ensure timing marks aligned **5**, **6**, **7**, **8**, **9**, **10** & **11**.
   NOTE: If it is necessary to turn camshafts, turn them separately by small amounts to prevent contact and possible damage to inlet and exhaust valves. Turn LH camshafts in direction shown by arrows.
7. Fit timing belt, starting at crankshaft sprocket.
8. Install:
   - ❏ Tensioner pulley and spacer. Tighten bolt to 40 Nm.
   - ❏ Guide pulleys (G1, G2 & G3). Tighten bolts to 40 Nm.
9. Slacken automatic tensioner unit bolts **12**. Move tensioner unit fully to left and tighten bolts to 25 Nm.
10. Ensure timing marks aligned **5**, **6**, **7**, **8**, **9**, **10** & **11**.
11. Remove pin from tensioner body to release pushrod.
12. Turn crankshaft two turns. Ensure timing marks aligned.
13. Install components in reverse order of removal.
14. Tighten crankshaft pulley bolt **1**.
    Tightening torque: 103-118 Nm.

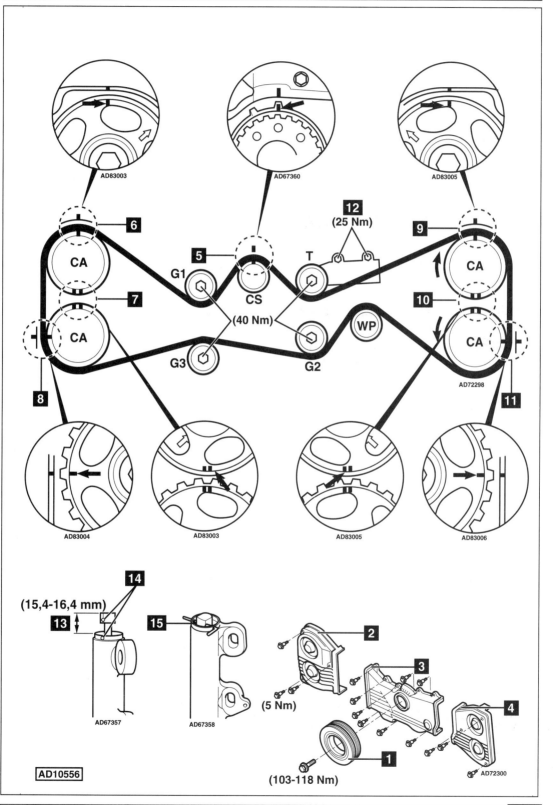

SUBARU

Model:	**Impreza 2,0 Turbo**
Year:	**1997-00**
Engine Code:	**EJ20**

Replacement Interval Guide

Subaru recommend:
→1999MY: Replacement every 45,000 miles or 3 years, whichever occurs first.
2000MY: Replacement every 60,000 miles or 4 years, whichever occurs first.
The previous use and service history of the vehicle must always be taken into account.
Refer to Timing Belt Replacement Intervals at the front of this manual.

Check For Engine Damage

CAUTION: This engine has been identified as an INTERFERENCE engine in which the possibility of valve-to-piston damage in the event of a timing belt failure is MOST LIKELY to occur. A compression check of all cylinders should be performed before removing the cylinder head(s).

Repair Times – hrs

Remove & install	2,10

Special Tools

■ Crankshaft pulley holding tool – Subaru No.499977100.

Special Precautions

■ Disconnect battery earth lead.
■ DO NOT turn crankshaft or camshaft when timing belt removed.
■ Remove spark plugs to ease turning engine.
■ Turn engine in normal direction of rotation (unless otherwise stated).
■ DO NOT turn engine via camshaft or other sprockets.
■ Observe all tightening torques.

Removal

1. Raise and support front of vehicle.
2. Remove:
   ❏ Lower splash guard.
   ❏ Coolant expansion tank. Move to one side.
   ❏ Cooling fan assembly.
   ❏ Auxiliary drive belt cover.
   ❏ Auxiliary drive belt(s).
   ❏ AC drive belt tensioner.
3. Hold crankshaft pulley **2**. Use tool No.499977100.
4. Remove:
   ❏ Crankshaft pulley bolt **1**.
   ❏ Crankshaft pulley **2**.
   ❏ Timing belt covers **3**, **4** & **5**.
   ❏ MT: Timing belt guide bolts **6**.
   ❏ MT: Timing belt guide **7**.
5. Temporarily fit crankshaft pulley **2**.
6. Turn crankshaft clockwise until timing marks aligned **8**, **9**, **10**, **11**, **12**, **13** & **14**.
   NOTE: Ensure crankshaft sprocket keyway at 6 o'clock position.
7. Remove:
   ❏ Crankshaft pulley **2**.
   ❏ Guide pulley (G3).
   ❏ Automatic tensioner unit bolt **15**.
   ❏ Automatic tensioner unit **16**.
   ❏ Timing belt.
   NOTE: Mark direction of rotation on belt with chalk if belt is to be reused.

Installation

1. Check tensioner pulley and guide pulleys for smooth operation. Replace if necessary.
2. Check tensioner body for leakage or damage **16**. Replace if necessary.
3. Check pushrod protrusion is 5,5-6,5 mm **17**. If not: Replace automatic tensioner unit.
   NOTE: Ensure automatic tensioner unit remains in the normal upright position during the following reset procedure.
4. Hold tensioner body. Check pushrod does not move when pushed against a firm surface with a force of less than 30 kg.
5. If pushrod moves easily, proceed as follows:
   ❏ Slowly press pushrod until flush with upper surface of tensioner body, then release. Repeat 3 times.
   ❏ Press pushrod with a force of 30 kg. If pushrod moves easily: Replace automatic tensioner unit.
6. Slowly compress pushrod into tensioner body until holes aligned **18**. Use a press. Retain pushrod with 2 mm diameter pin through hole in tensioner body **19**.
   NOTE: Ensure automatic tensioner unit remains upright.
7. Install automatic tensioner unit **16**. Tighten bolt to 35-43 Nm **15**.
8. Ensure timing marks aligned **8**, **9**, **10**, **11**, **12**, **13** & **14**.
   NOTE: If it is necessary to turn camshafts, turn them separately by small amounts to prevent contact and possible damage to inlet and exhaust valves. Turn LH camshafts in direction shown by arrows.
9. Fit timing belt in anti-clockwise direction, starting at crankshaft sprocket. Ensure belt is taut on non-tensioned side.
10. Align marks on belt with marks on sprockets **8**, **9**, **11**, **12** & **14**.
11. Fit guide pulley (G3). Tighten bolt to 35-43 Nm.
12. Ensure timing marks aligned **8**, **9**, **10**, **11**, **12**, **13** & **14**.
13. Remove pin from tensioner body to release pushrod **19**.
14. Turn crankshaft two turns clockwise. Ensure all timing marks aligned **8**, **9**, **10**, **11**, **12**, **13** & **14**.
15. MT: Fit timing belt guide **7**. Tighten bolts finger tight **6**.
16. MT: Adjust clearance between back of timing belt and belt guide to 0,5-1,5 mm **20**. Tighten bolts to 10 Nm **6**.
17. Install components in reverse order of removal.
18. Hold crankshaft pulley **2**. Use tool No.499977100.
19. Tighten crankshaft pulley bolt **1**.
    Tightening torque: 167-177 Nm.

Autodata

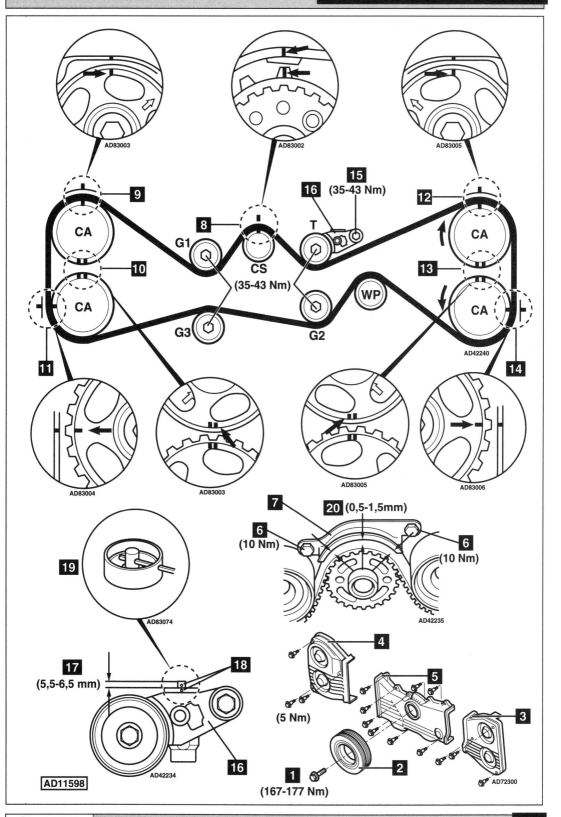

SUZUKI

Model:	Swift 1,0/1,3 • SJ 413/Samurai • Alto 1,0
Year:	1985-03
Engine Code:	G10, G10B, G13

Replacement Interval Guide

Suzuki recommend:
→08/99: Replacement every 60,000 miles or 5 years.
09/99→: Replacement every 54,000 miles or 6 years.
The previous use and service history of the vehicle must always be taken into account.
Refer to Timing Belt Replacement Intervals at the front of this manual.

Check For Engine Damage

CAUTION: This engine has been identified as an INTERFERENCE engine in which the possibility of valve-to-piston damage in the event of a timing belt failure is MOST LIKELY to occur.
A compression check of all cylinders should be performed before removing the cylinder head.

Repair Times – hrs

Remove & install:

Swift 1,0	1,80
Alto/Swift 1,3	1,90
SJ 413	2,20

Special Tools

■ None required.

Special Precautions

■ Disconnect battery earth lead.
■ DO NOT turn crankshaft or camshaft when timing belt removed.
■ Remove spark plugs to ease turning engine.
■ Turn engine in normal direction of rotation (unless otherwise stated).
■ DO NOT turn engine via camshaft or other sprockets.
■ Observe all tightening torques.

Removal

1. Raise and support front of vehicle.
2. Remove:
   ❏ RH front wheel.
   ❏ Air filter.
   ❏ Engines with adjustable tappets:
      Cylinder head cover.
   ❏ Auxiliary drive belt.
   ❏ Water pump pulley.
   ❏ Crankshaft pulley **1**.
   ❏ Timing belt cover and seal **2** & **3**.
3. Slacken tensioner locknut **5**. Slacken tensioner bolt **4**.

4. Move tensioner lever upwards. Lightly tighten locknut.
5. Remove timing belt.

Installation

1. Engines with adjustable tappets: Slacken all tappet adjusting screws to allow camshaft to rotate freely.
2. Ensure timing marks aligned **6** & **7**.
3. Fit timing belt, starting at crankshaft sprocket. Ensure belt is taut between sprockets on non-tensioned side.
4. Slacken tensioner locknut **5**. Allow tensioner to operate.
5. Turn crankshaft two turns clockwise.
6. Ensure timing marks aligned **6** & **7**.
7. Tighten tensioner locknut to 11 Nm **5**.
8. Tighten tensioner bolt to 26 Nm **4**.
9. Fit timing belt cover and seal **2** & **3**.
10. Fit crankshaft pulley **1**. Tighten bolts to 12 Nm **8**.
11. Engines with adjustable tappets: Adjust tappet clearances (cold). Inlet: 0,13-0,17 mm. Exhaust: 0,16-0,20 mm.
12. Install components in reverse order of removal.

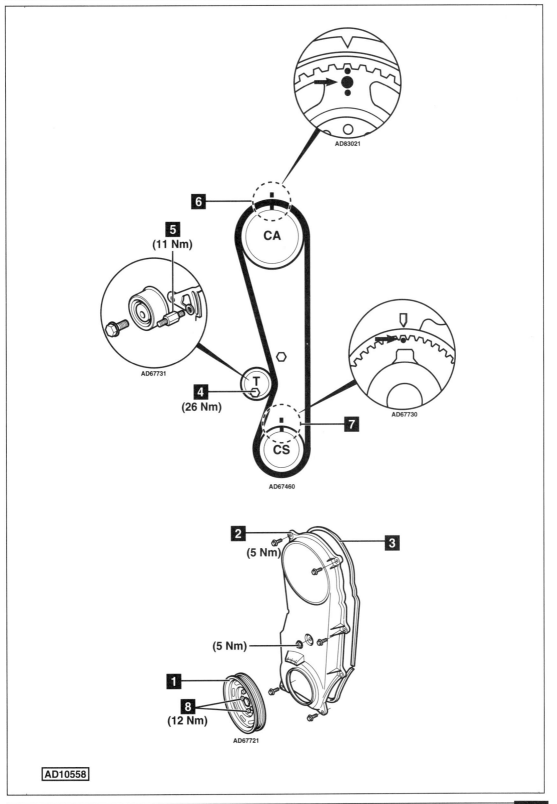

AD83021

6

5
(11 Nm)

CA

AD67731

4
(26 Nm)

T

AD67730

7

CS

AD67460

2
(5 Nm)

3

(5 Nm)

1

8
(12 Nm)

AD67721

AD10558

SUZUKI

Model:	**Swift GTi**
Year:	**1986-97**
Engine Code:	**G13B DOHC**

Replacement Interval Guide

Suzuki recommend replacement every 60,000 miles or 5 years.
The previous use and service history of the vehicle must always be taken into account.
Refer to Timing Belt Replacement Intervals at the front of this manual.

Check For Engine Damage

CAUTION: This engine has been identified as an INTERFERENCE engine in which the possibility of valve-to-piston damage in the event of a timing belt failure is MOST LIKELY to occur.
A compression check of all cylinders should be performed before removing the cylinder head.

Repair Times – hrs

Remove & install	1,90

Special Tools

■ None required.

Special Precautions

■ Disconnect battery earth lead.
■ DO NOT turn crankshaft or camshaft when timing belt removed.
■ Remove spark plugs to ease turning engine.
■ Turn engine in normal direction of rotation (unless otherwise stated).
■ DO NOT turn engine via camshaft or other sprockets.
■ Observe all tightening torques.

Removal

1. Remove:
   ❏ Air filter.
   ❏ Auxiliary drive belt.
   ❏ Water pump pulley.
   ❏ RH engine mounting bracket.
2. Raise and support front of vehicle.
3. Remove:
   ❏ Later models: Lower inner wing liner.
   ❏ Early models: Access hole cover.
   ❏ Crankshaft pulley **1**.
   ❏ Timing belt covers **2** & **3**.
4. Turn crankshaft to TDC on No.1 cylinder. Ensure timing marks aligned **4**, **5** & **6**.
5. Slacken tensioner bolt **7**. Slacken tensioner locknut **8**.
6. Move tensioner lever upwards. Lightly tighten locknut.
7. Remove timing belt.

Installation

1. Ensure timing marks aligned **4**, **5** & **6**.
   NOTE: If necessary use small wooden wedges to hold camshaft sprockets on timing marks.
2. Fit timing belt, starting at crankshaft sprocket. Ensure belt is taut between sprockets on non-tensioned side.
3. Slacken tensioner locknut **8**. Allow tensioner to operate.
4. Turn crankshaft two turns clockwise.
5. Ensure timing marks aligned **4**, **5** & **6**.
6. Tighten tensioner locknut to 11 Nm **8**.
7. Tighten tensioner bolt to 26 Nm **7**.
8. Fit timing belt covers and seals **2** & **3**.
9. Fit crankshaft pulley **1**. Tighten bolts to 11 Nm **9**.
10. Fit water pump pulley. Tighten bolts to 10 Nm.
11. Install components in reverse order of removal.

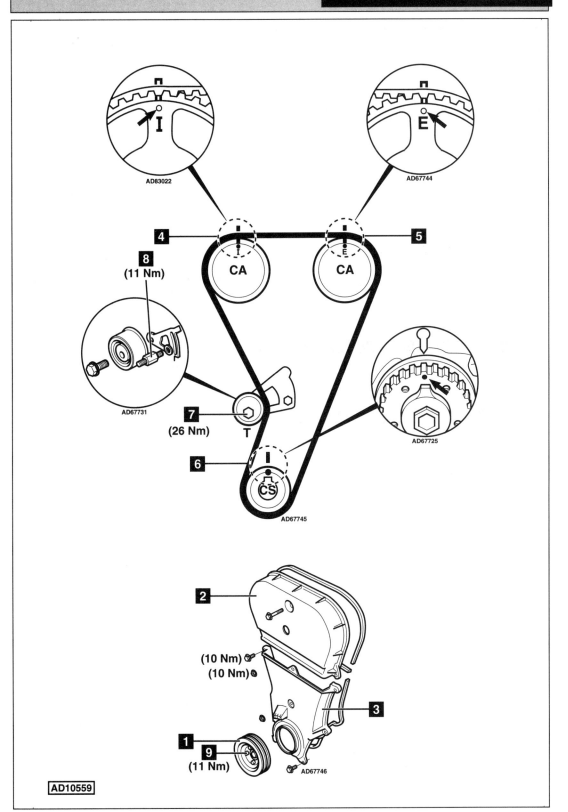

AD83022

AD67744

4

5

8
(11 Nm)

CA

CA

AD67731

7
(26 Nm) T

AD67725

6 I

CS

AD67745

2

(10 Nm)
(10 Nm)

3

1
9
(11 Nm)

AD67746

AD10559

SUZUKI

Model:	**Baleno 1,3/1,6 16V • Swift 1,6 16V**
Year:	**1990-02**
Engine Code:	**G13B, G16B**

Replacement Interval Guide

Suzuki recommend:
→08/99: Replacement every 60,000 miles or
5 years.
09/99→: Replacement every 54,000 miles or
6 years.
*The previous use and service history of the vehicle
must always be taken into account.*
*Refer to Timing Belt Replacement Intervals at the front
of this manual.*

Check For Engine Damage

*CAUTION: This engine has been identified as an
INTERFERENCE engine in which the possibility of
valve-to-piston damage in the event of a timing belt
failure is MOST LIKELY to occur.*
*A compression check of all cylinders should be
performed before removing the cylinder head.*

Repair Times – hrs

Remove & install:
Baleno	1,70
Swift	2,30

Special Tools

- ■ None required.

Special Precautions

- ■ Disconnect battery earth lead.
- ■ DO NOT turn crankshaft or camshaft when timing belt removed.
- ■ Remove spark plugs to ease turning engine.
- ■ Turn engine in normal direction of rotation (unless otherwise stated).
- ■ DO NOT turn engine via camshaft or other sprockets.
- ■ Observe all tightening torques.

Removal

1. Raise and support front of vehicle.
2. Remove:
   - ❏ RH front wheel.
   - ❏ RH engine undershield.
   - ❏ Air filter.
   - ❏ Auxiliary drive belt(s).
   - ❏ PAS pump. DO NOT disconnect hoses.
   - ❏ Water pump pulley.
3. Support engine.
4. Remove:
   - ❏ RH engine mounting.
   - ❏ Crankshaft pulley bolts **1**.
   - ❏ Crankshaft pulley **2**.
   - ❏ Timing belt cover and seal **3**.

5. Turn crankshaft clockwise until timing marks aligned **4** & **5**.
6. Slacken tensioner nut and bolt **6** & **7**. Move tensioner away from belt. Lightly tighten nut **6**.
7. Remove timing belt.

Installation

1. Check tensioner pulley and tensioner plate for smooth operation.
2. Ensure timing marks aligned **4** & **5**.
3. Fit timing belt in anti-clockwise direction, starting at crankshaft sprocket. Ensure belt is taut between sprockets.
4. Slacken tensioner nut **6**. Allow tensioner to operate.
5. Turn crankshaft two turns clockwise.
6. Ensure timing marks aligned **4** & **5**.
7. Tighten tensioner nut to 11 Nm **6**.
8. Tighten tensioner bolt to 25 Nm **7**.
9. Install components in reverse order of removal.
10. Tighten crankshaft pulley bolts to 16 Nm **1**.

Autodata

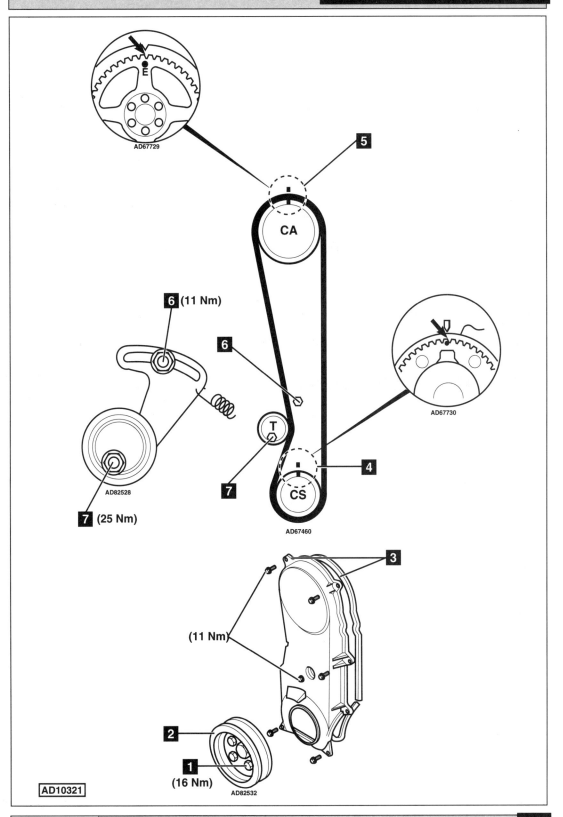

AD67729

5

CA

6 (11 Nm)

6

AD67730

T

7

4

CS

AD67460

AD82528

7 (25 Nm)

3

(11 Nm)

2

1

(16 Nm)

AD82532

AD10321

Model:	**Vitara 1,6 8V**
Year:	**1989-03**
Engine Code:	**G16A**

Replacement Interval Guide

Suzuki recommend:
→08/99: Replacement every 60,000 miles or 5 years.
09/99→: Replacement every 54,000 miles or 6 years.
The previous use and service history of the vehicle must always be taken into account.
Refer to Timing Belt Replacement Intervals at the front of this manual.

Check For Engine Damage

CAUTION: This engine has been identified as an INTERFERENCE engine in which the possibility of valve-to-piston damage in the event of a timing belt failure is MOST LIKELY to occur.
A compression check of all cylinders should be performed before removing the cylinder head.

Repair Times – hrs

Remove & install	2,40

Special Tools

■ None required.

Special Precautions

■ Disconnect battery earth lead.
■ DO NOT turn crankshaft or camshaft when timing belt removed.
■ Remove spark plugs to ease turning engine.
■ Turn engine in normal direction of rotation (unless otherwise stated).
■ DO NOT turn engine via camshaft or other sprockets.
■ Observe all tightening torques.

Removal

1. Remove:
   ❏ Cooling fan and cowling.
   ❏ Auxiliary drive belt.
   ❏ Water pump pulley.
   ❏ Crankshaft pulley **1**.
   ❏ Timing belt cover **2**.
2. Turn crankshaft clockwise until timing marks aligned **5** & **6**.
3. Slacken tensioner bolt **7**.
4. Remove tensioner lever nut **3**.
5. Remove tensioner spring **4**. Move tensioner lever upwards. Lightly tighten bolt.
6. Remove timing belt.

Installation

1. Remove:
   ❏ Air filter.
   ❏ Cylinder head cover.
2. Slacken all tappet adjusting screws to allow camshaft to rotate freely.
3. Ensure crankshaft timing marks aligned **5**.
4. Ensure camshaft timing marks aligned **6**.
5. Fit timing belt. Ensure belt is taut between sprockets on non-tensioned side.
6. Slacken tensioner bolt **7**.
7. Fit tensioner spring and tensioner lever nut **4** & **3**.
8. Turn crankshaft two turns clockwise. Ensure timing marks aligned **5** & **6**.
9. Tighten tensioner lever nut to 10 Nm **3**.
10. Tighten tensioner bolt to 26 Nm **7**.
11. Fit timing belt cover.
12. Fit crankshaft pulley **1**. Tighten bolts to 11 Nm **8**.
13. Adjust tappet clearances to specification – see Autodata Technical Data Manual.
14. Install components in reverse order of removal.

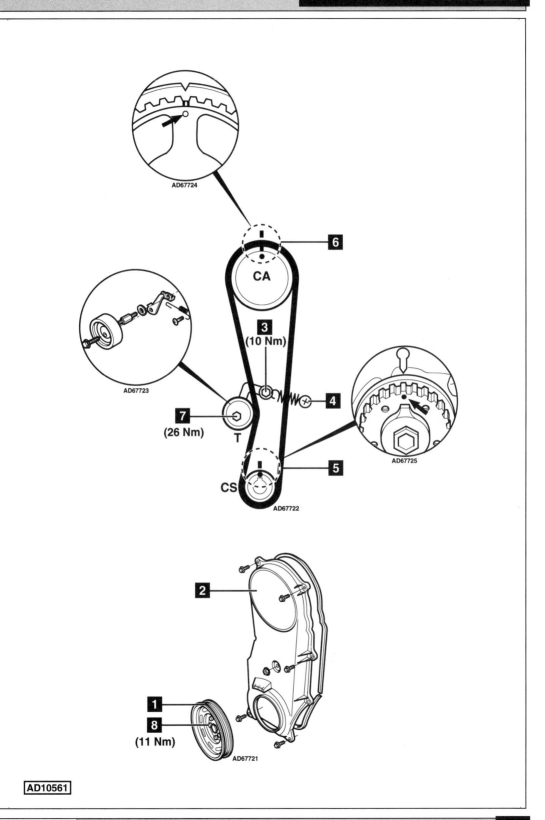

AD67724

6

CA

3
(10 Nm)

AD67723

7
(26 Nm)
T

4

AD67725

5

CS

AD67722

2

1

8
(11 Nm)

AD67721

AD10561

Model:	**Jimny 1,3 16V • Vitara 1,6 16V • X-90**
Year:	**1991-03**
Engine Code:	**G13B, G16B**

Replacement Interval Guide

Suzuki recommend:
→08/99: Replacement every 60,000 miles or 5 years.
09/99→: Replacement every 54,000 miles or 6 years.
The previous use and service history of the vehicle must always be taken into account.
Refer to Timing Belt Replacement Intervals at the front of this manual.

Check For Engine Damage

CAUTION: This engine has been identified as an INTERFERENCE engine in which the possibility of valve-to-piston damage in the event of a timing belt failure is MOST LIKELY to occur.
A compression check of all cylinders should be performed before removing the cylinder head.

Repair Times – hrs

Remove & install:

Jimny	1,90
X-90/Vitara	2,40

Special Tools

■ None required.

Special Precautions

■ Disconnect battery earth lead.
■ DO NOT turn crankshaft or camshaft when timing belt removed.
■ Remove spark plugs to ease turning engine.
■ Turn engine in normal direction of rotation (unless otherwise stated).
■ DO NOT turn engine via camshaft or other sprockets.
■ Observe all tightening torques.

Removal

1. Jimny: Drain coolant.
2. Remove:
   ❏ Jimny: Top hose.
   ❏ Auxiliary drive belts.
   ❏ Cooling fan and cowling.
   ❏ Water pump pulley.
   ❏ Crankshaft pulley bolts **1**.
   ❏ Crankshaft pulley **2**.
   ❏ Timing belt cover and seal **3**.
3. Turn crankshaft clockwise until timing marks aligned **4** & **5**.
   NOTE: Use 'E' mark on camshaft sprocket.
4. Remove tensioner spring.

5. Slacken tensioner nut and bolt **6** & **7**. Move tensioner away from belt. Lightly tighten nut **6**.
6. Remove timing belt.

Installation

1. Check tensioner pulley for smooth operation.
2. Ensure timing marks aligned **4** & **5**.
3. Fit timing belt in anti-clockwise direction, starting at crankshaft sprocket. Ensure belt is taut between sprockets.
4. Fit tensioner spring.
5. Slacken tensioner nut **6**. Allow tensioner to operate.
6. Turn crankshaft two turns clockwise.
7. Ensure timing marks aligned **4** & **5**.
8. Tighten tensioner nut to 11 Nm **6**.
9. Tighten tensioner bolt to 25 Nm **7**.
10. Install components in reverse order of removal.
11. Tighten crankshaft pulley bolts to 16 Nm **1**.

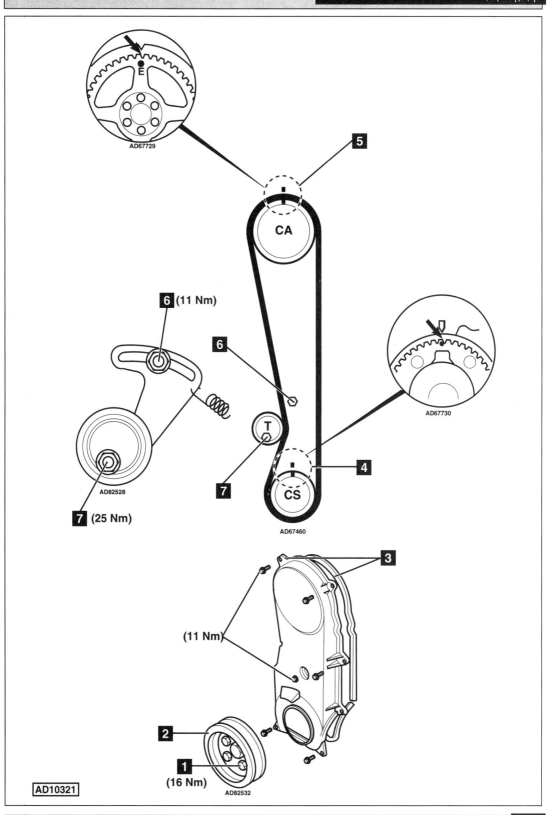

AD67729

5

CA

6 (11 Nm)

6

AD67730

T

4

7

CS

AD67460

AD82528

7 (25 Nm)

3

(11 Nm)

2

1
(16 Nm)

AD82532

AD10321

Model:	**Vitara 2,0TD • Grand Vitara 2,0TD**
Year:	**1995-03**
Engine Code:	**RF-Turbo**

Replacement Interval Guide

Suzuki recommend replacement every 60,000 miles or 5 years, whichever occurs first.
The previous use and service history of the vehicle must always be taken into account.
Refer to Timing Belt Replacement Intervals at the front of this manual.

Check For Engine Damage

CAUTION: This engine has been identified as an INTERFERENCE engine in which the possibility of valve-to-piston damage in the event of a timing belt failure is MOST LIKELY to occur.
A compression check of all cylinders should be performed before removing the cylinder head.

Repair Times – hrs

Remove & install	2,70

Special Tools

■ Two M8 x 1,25 x 30 mm bolts (for locking sprocket).

Special Precautions

■ Disconnect battery earth lead.
■ DO NOT turn crankshaft or camshaft when timing belt removed.
■ Remove glow plugs to ease turning engine.
■ Turn engine in normal direction of rotation (unless otherwise stated).
■ DO NOT turn engine via camshaft or other sprockets.
■ Observe all tightening torques.
■ Check diesel injection pump timing after belt replacement.

Removal

1. Drain coolant.
2. Remove:
   ❏ Top radiator hose.
   ❏ Cooling fan cowling.
   ❏ Cooling fan and coupling.
   ❏ Auxiliary drive belts.
   ❏ Auxiliary drive belt tensioner bracket.
   ❏ Crankshaft pulley bolts **10**.
   ❏ Crankshaft pulley **1**.
   ❏ Timing belt covers **2** & **3**.
3. Turn crankshaft to TDC on No.1 cylinder. Ensure timing marks aligned **4**, **5** & **6**.
4. Lock injection pump sprocket. Use two M8 bolts **7**.

5. Slacken tensioner bolt **8**. Move tensioner away from belt. Lightly tighten bolt.
6. Remove timing belt.
7. Slacken tensioner bolt **8**. Remove tensioner spring **9**.

Installation

1. Check free length of tensioner spring is 52,6 mm **9**. Replace if necessary.
2. Check tensioner pulley for smooth operation. Replace if necessary.
3. Fit tensioner spring. Push tensioner against spring tension. Lightly tighten bolt **8**.
4. Ensure timing marks aligned **4**, **5** & **6**.
5. Fit timing belt in anti-clockwise direction, starting at crankshaft sprocket. Ensure belt is taut on non-tensioned side.
6. Remove locking bolts **7**.
7. Slacken tensioner bolt **8**. Allow tensioner to operate.
8. Turn crankshaft two turns clockwise. Ensure timing marks aligned **4**, **5** & **6**.
9. Tighten tensioner bolt to 40 Nm **8**.
10. Apply a load of 10 kg to belt at ▽. Belt should deflect 9-11,5 mm.
11. If not: Repeat operations 6-10.
12. Install components in reverse order of removal.
    NOTE: Ensure crankshaft pulley located correctly on dowel.
13. Tighten crankshaft pulley bolts to 28 Nm **10**.
14. Refill cooling system.

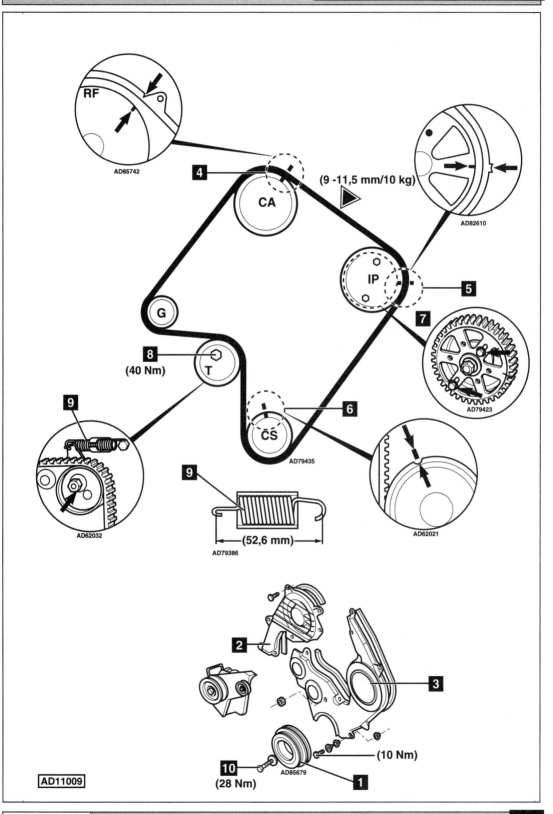

RF

AD85742

4

CA

(9 -11,5 mm/10 kg) ▷

AD82610

IP

5

7

AD79423

G

8
(40 Nm)

T

9

6

CS

AD79435

AD62032

9

(52,6 mm)

AD79386

AD62021

2

3

10
(28 Nm)

AD85679

1

(10 Nm)

AD11009

Model:	**Starlet 1,0/1,3 • Corolla 1,3**
Year:	**1985-99**
Engine Code:	**1E, 2E, 2E-E, 2E-C**

Replacement Interval Guide

Toyota recommend replacement as follows:
→1993 - replacement every 48,000 miles or
4 years, whichever occurs first.
1994→ - replacement every 63,000 miles or
5 years, whichever occurs first.
The previous use and service history of the vehicle must always be taken into account.
Refer to Timing Belt Replacement Intervals at the front of this manual.

Check For Engine Damage

CAUTION: This engine has been identified as an INTERFERENCE engine in which the possibility of valve-to-piston damage in the event of a timing belt failure is MOST LIKELY to occur.
A compression check of all cylinders should be performed before removing the cylinder head.

Repair Times – hrs

Remove & install:

Starlet (EP70)	1,20
Starlet (EP80/81/90)	1,70
PAS	+0,30
AC	+0,10
Corolla	1,30
PAS	+0,20
AC	+0,10

Special Tools

- Puller – Toyota No.09213-31021.

Special Precautions

- Disconnect battery earth lead.
- DO NOT turn crankshaft or camshaft when timing belt removed.
- Remove spark plugs to ease turning engine.
- Turn engine in normal direction of rotation (unless otherwise stated).
- DO NOT turn engine via camshaft or other sprockets.
- Observe all tightening torques.

Removal

1. Raise and support front of vehicle.
2. Support engine.
3. Remove:
   - ❏ RH front wheel.
   - ❏ Inner wing panel.
   - ❏ Engine mounting and bracket.
   - ❏ Auxiliary drive belts.
   - ❏ Timing belt upper cover **1**.

4. Turn crankshaft clockwise to TDC on No.1 cylinder. Ensure timing marks aligned **2**.
5. Remove:
   - ❏ Crankshaft pulley bolt **3**.
   - ❏ Crankshaft pulley **4**. Use tool No.09213-31021.
   - ❏ Timing belt lower cover **5**.
6. Fit crankshaft pulley bolt **3**.
7. Ensure timing marks aligned **2** & **6**.
8. Slacken tensioner bolt **7**. Move tensioner away from belt and lightly tighten bolt.
9. Remove:
   - ❏ Timing belt.
   - ❏ Tensioner pulley.

Installation

1. Check tensioner pulley for smooth operation.
2. Check free length of tensioner spring is 38,4 mm. Replace spring if necessary.
3. Fit tensioner and spring. Ensure spring is connected correctly.
4. Push tensioner to left and lightly tighten bolt **7**.
5. Ensure timing marks aligned **2** & **6**.
6. Fit timing belt.
7. Slacken tensioner bolt **7**.
8. Turn crankshaft two turns clockwise to TDC on No.1 cylinder.
9. Ensure timing marks aligned **2** & **6**.
10. Tighten tensioner bolt **7**.
    Tightening torque: 18 Nm.
11. Check belt is under tension at ▽.
12. Remove crankshaft pulley bolt **3**.
13. Install:
    - ❏ Timing belt lower cover.
    - ❏ Crankshaft pulley **4**.
    - ❏ Crankshaft pulley bolt **3**.
      Tightening torque: 147-152 Nm.
14. Install components in reverse order of removal.

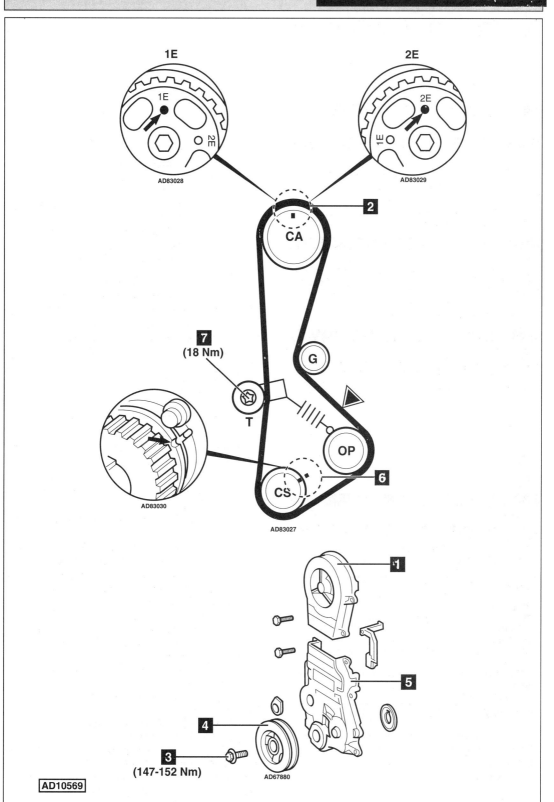

1E

AD83028

2E

AD83029

2

CA

7
(18 Nm)

G

T

AD83030

OP

CS

6

AD83027

1

5

4

3
(147-152 Nm)

AD67880

AD10569

TOYOTA

Model:	**Corolla 1,3i 16V • Starlet 1,3i 16V**
Year:	**1992-00**
Engine Code:	**4E-FE**

Replacement Interval Guide

Toyota recommend replacement as follows:
→1993 - replacement every 60,000 miles or
5 years, whichever occurs first.
1994→ - replacement every 63,000 miles or
5 years, whichever occurs first.
*The previous use and service history of the vehicle
must always be taken into account.*
*Refer to Timing Belt Replacement Intervals at the front
of this manual.*

Check For Engine Damage

*CAUTION: This engine has been identified as an
INTERFERENCE engine in which the possibility of
valve-to-piston damage in the event of a timing belt
failure is MOST LIKELY to occur.*
*A compression check of all cylinders should be
performed before removing the cylinder head.*

Repair Times – hrs

Remove & install:

Corolla →1997	1,40
AC	+0,20
PAS	+0,20
Corolla 1997→	1,80
AC	+0,10
PAS	+0,30
Starlet	1,50
AC	+0,10
PAS	+0,20

Special Tools

- ■ Puller – Toyota No.09213-31021.
- ■ Crankshaft pulley holding tool –
  Toyota No.09213-14010.
- ■ Handle – Toyota No.09330-00021.

Special Precautions

- ■ Disconnect battery earth lead.
- ■ DO NOT turn crankshaft or camshaft when timing belt removed.
- ■ Remove spark plugs to ease turning engine.
- ■ Turn engine in normal direction of rotation (unless otherwise stated).
- ■ DO NOT turn engine via camshaft or other sprockets.
- ■ Observe all tightening torques.

Removal

1. Raise and support front of vehicle.
2. Support engine.
3. Remove:
   - ❏ RH front wheel.
   - ❏ Inner wing panel.
   - ❏ Engine mounting and bracket.
   - ❏ Auxiliary drive belt(s).
   - ❏ Cylinder head cover.
   - ❏ Timing belt upper cover **1**.
4. Turn crankshaft clockwise to TDC on No.1 cylinder. Ensure crankshaft pulley timing mark aligned **2**.
5. Check hole in camshaft sprocket aligned with mark on camshaft bearing cap **3**.
6. If not: Turn crankshaft one turn clockwise.
7. Remove:
   - ❏ Crankshaft pulley bolt **4**. Use tool Nos.09213-14010 & 09330-00021.
   - ❏ Crankshaft pulley **5**. Use tool No.09213-31021.
   - ❏ Timing belt lower cover **6**.
8. Temporarily fit crankshaft pulley bolt **4**.
9. Ensure timing marks aligned **3** & **7**.
10. Slacken tensioner bolt **8**. Move tensioner away from belt and lightly tighten bolt.
11. Remove:
    - ❏ Timing belt.
    - ❏ Tensioner pulley and spring **9** & **10**.

Installation

1. Check tensioner pulley for smooth operation **9**.
2. Check free length of tensioner spring is 38,4 mm **10**.
3. Fit tensioner and spring. Ensure spring is connected correctly.
4. Push tensioner to left and lightly tighten bolt **8**.
5. Ensure timing marks aligned **3** & **7**.
6. Fit timing belt in anti-clockwise direction, starting at crankshaft sprocket. Ensure belt is taut between sprockets.
7. Slacken tensioner bolt **8**.
8. Turn crankshaft two turns clockwise to TDC on No.1 cylinder. Ensure timing marks aligned **3** & **7**.
9. If not: Repeat installation and tensioning procedures.
10. Tighten tensioner bolt **8**.
    Tightening torque: 19 Nm.
11. Remove crankshaft pulley bolt **4**.
12. Install:
    - ❏ Timing belt lower cover **6**.
    - ❏ Crankshaft pulley **5**.
13. Install components in reverse order of removal.
14. Tighten crankshaft pulley bolt **4**.
    Tightening torque: 152 Nm. Use tool Nos.09213-14010 & 09330-00021.

Autodata

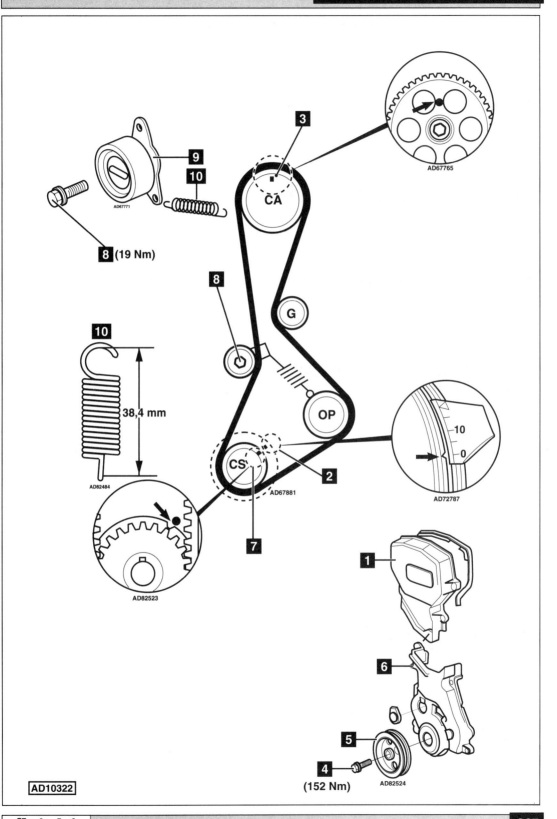

9

AD67771

10

8 (19 Nm)

3

CA

AD67765

8

G

10

38,4 mm

AD82484

OP

10
0

AD72787

CS

AD67881

2

7

AD82523

1

6

5

AD82524

4

(152 Nm)

AD10322

Model:	Corolla 1,3/1,6 • Tercel 1,3/1,5 • Carina II 1,6
Year:	1982-92
Engine Code:	2A, 3A, 4A, 4A-LC, 4A-F, 4A-FE

Replacement Interval Guide

Toyota recommend replacement every 60,000 miles or 5 years, whichever occurs first.
The previous use and service history of the vehicle must always be taken into account.
Refer to Timing Belt Replacement Intervals at the front of this manual.

Check For Engine Damage - except 4A engines

CAUTION: This engine has been identified as an INTERFERENCE engine in which the possibility of valve-to-piston damage in the event of a timing belt failure is MOST LIKELY to occur.
A compression check of all cylinders should be performed before removing the cylinder head.

Check For Engine Damage - 4A engine

CAUTION: This engine has been identified as a FREEWHEELING engine in which the possibility of valve-to-piston damage in the event of a timing belt failure may be minimal or very unlikely. However, a precautionary compression check of all cylinders should be performed.

Repair Times – hrs

Corolla (AE82) – 4A/4A-LC:	
Remove & install	2,00
PAS	+0,10
AC	+0,10
Corolla (AE92/95) – 4A-F:	
Remove & install	2,30
PAS	+0,20
AC	+0,10
Corolla/Tercel – 2A/3A:	
Remove & install	1,20
AC	+0,10
Carina (AT151) – 4A:	
Remove & install	1,90
AC	+0,40
Carina (AT171) – 4AF:	
Remove & install	2,00
PAS	+0,20
AC	+0,20

Special Tools

- Puller – Toyota No.09213-31021.

Special Precautions

- Disconnect battery earth lead.
- DO NOT turn crankshaft or camshaft when timing belt removed.
- Remove spark plugs to ease turning engine.
- Turn engine in normal direction of rotation (unless otherwise stated).
- DO NOT turn engine via camshaft or other sprockets.
- Observe all tightening torques.

Removal

1. Tercel: Remove cooling fan and coupling.
2. Carina/Corolla: Support engine. Remove engine mounting and bracket.
3. All models – remove:
   - ❑ Auxiliary drive belts.
   - ❑ Air filter.
   - ❑ Timing belt upper cover **1**.
4. Turn crankshaft clockwise to TDC on No.1 cylinder. Ensure timing marks aligned **2**.
5. Remove:
   - ❑ Crankshaft pulley bolt **3**.
   - ❑ Crankshaft pulley **4**. Use tool No.09213-31021.
   - ❑ Timing belt lower covers **5**.
   - ❑ Crankshaft sprocket guide washer **9**.
6. Ensure timing marks aligned **2** & **6**.
7. Slacken tensioner bolt **7**. Move tensioner away from belt and lightly tighten bolt.
8. Remove timing belt.

Installation

1. Remove tensioner and spring **8** & **10**. Check tensioner pulley for smooth operation.
2. Check free length of tensioner spring is 38,4 mm **8**. 4A-F: 43,3 mm. Replace spring if necessary.
3. Fit tensioner and spring. Ensure spring is connected correctly.
4. Push tensioner to left and lightly tighten bolt **7**.
5. Ensure timing marks aligned **2** & **6**.
6. Install:
   - ❑ Timing belt.
   - ❑ Crankshaft sprocket guide washer **9**.
7. Temporarily fit crankshaft pulley **4**.
8. Slacken tensioner bolt **7**. Turn crankshaft two turns clockwise to TDC on No.1 cylinder.
9. Ensure timing marks aligned **2** & **6**.
10. Tighten tensioner bolt **7**. Tightening torque: 37 Nm.
11. Apply a load of 2 kg to belt at ▽. Belt should deflect 6-7 mm. 4A-F: 5-6 mm.
12. If not: Repeat tensioning procedure.
13. Remove crankshaft pulley **4**.
14. Install:
    - ❑ Timing belt lower covers **5**.
    - ❑ Crankshaft pulley **4**.
15. Tighten crankshaft pulley bolt **3**. Tightening torque: 118 Nm.
16. Install components in reverse order of removal.

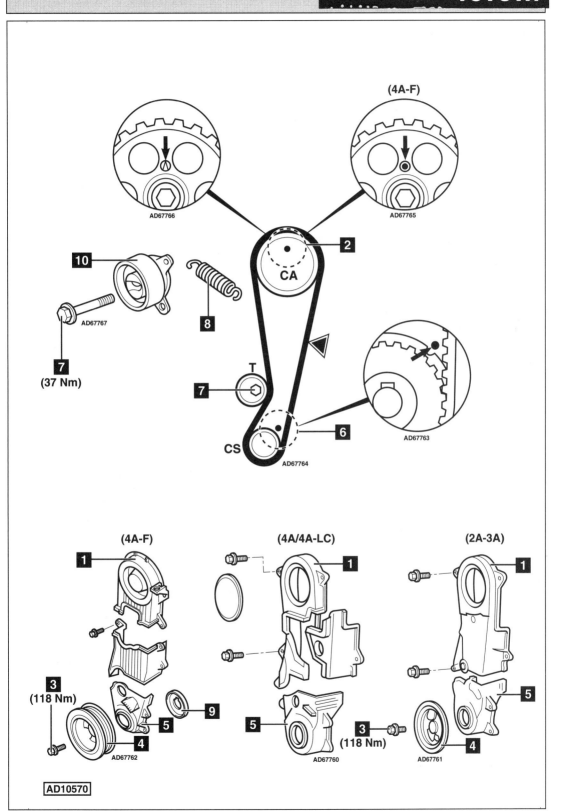

(4A-F)

AD67766

AD67765

2

CA

10

AD67767

8

7

(37 Nm)

T

7

AD67763

CS

6

AD67764

(4A-F)

1

3
(118 Nm)

4

5

9

AD67762

(4A/4A-LC)

1

5

3
(118 Nm)

AD67760

(2A-3A)

1

5

4

AD67761

AD10570

TOYOTA

Model:	Corolla 1,6/1,8 16V • Carina E 1,6/1,8 • Avensis 1,6/1,8 • Celica GT 1,8 16V
Year:	1992-00
Engine Code:	4A-FE, 7A-FE

Replacement Interval Guide

Toyota recommend replacement as follows:
→1993 - replacement every 60,000 miles or
5 years, whichever occurs first.
1994→ - replacement every 63,000 miles or
5 years, whichever occurs first.
*The previous use and service history of the vehicle
must always be taken into account.*
*Refer to Timing Belt Replacement Intervals at the
front of this manual.*

Check For Engine Damage

*CAUTION: This engine has been identified as an
INTERFERENCE engine in which the possibility of
valve-to-piston damage in the event of a timing belt
failure is MOST LIKELY to occur.*
*A compression check of all cylinders should be
performed before removing the cylinder head.*

Repair Times – hrs

Corolla →1997/Carina E:
Remove & install	1,60
PAS	+0,10
AC	+0,20

Corolla 1997→:
Remove & install	1,70
PAS	+0,10
AC	+0,20

Avensis:
Remove & install	2,00
AC	+0,20

Celica:
Remove & install	2,00
AC	+0,30

Special Tools

■ Puller – Toyota No.09213-31021.

Special Precautions

■ Disconnect battery earth lead.
■ DO NOT turn crankshaft or camshaft when timing belt
removed.
■ Remove spark plugs to ease turning engine.
■ Turn engine in normal direction of rotation (unless
otherwise stated).
■ DO NOT turn engine via camshaft or other sprockets.
■ Observe all tightening torques.

Removal

1. Support engine.
2. Remove:
   ❏ Engine mounting.
   ❏ Auxiliary drive belts.
   ❏ Water pump pulley (if necessary).
   ❏ Cylinder head cover.
   ❏ Timing belt upper cover **1**.

3. Turn crankshaft clockwise to TDC on
   No.1 cylinder **10**.
4. Check timing mark on camshaft sprocket aligned
   with mark on camshaft bearing cap **2**.
5. Remove:
   ❏ Crankshaft pulley bolt **3**.
   ❏ Crankshaft pulley **4**. Use tool No.09213-31021.
   ❏ Timing belt lower covers **5**.
   ❏ Crankshaft sprocket guide washer **9**.
6. Ensure timing marks aligned **6** & **2**.
7. Slacken tensioner bolt **7**. Move tensioner away from
   belt and lightly tighten bolt.
8. Remove:
   ❏ Timing belt.
   ❏ Tensioner pulley.

Installation

1. Check tensioner pulley for smooth operation.
2. Check free length of tensioner spring **8**.
   ❏ 4A-FE: Except Corolla 1997→ – 35,3 mm.
   ❏ 4A-FE: Corolla 1997→ – 36,9 mm.
   ❏ 7A-FE: 31,8 mm.
3. Replace spring if necessary.
4. Fit tensioner and spring. Ensure spring is connected
   correctly.
5. Push tensioner to left and lightly tighten bolt **7**.
6. Ensure timing marks aligned **6** & **2**.
7. Install:
   ❏ Timing belt.
   ❏ Crankshaft sprocket guide washer **9**.
8. Temporarily fit crankshaft pulley.
9. Slacken tensioner bolt **7**.
10. Turn crankshaft two turns clockwise to TDC on
    No.1 cylinder.
11. Ensure timing marks aligned **6** & **2**.
12. Tighten tensioner bolt to 37 Nm **7**.
13. Apply a load of 2 kg to belt at ▽. Belt should deflect
    5-6 mm.
14. If not: Repeat tensioning procedure.
15. Remove crankshaft pulley **4**.
16. Install:
    ❏ Timing belt lower covers **5**.
    ❏ Crankshaft pulley **4**.
    ❏ Crankshaft pulley bolt **3**.
       Tightening torque: 118 Nm.
17. Install components in reverse order of removal.

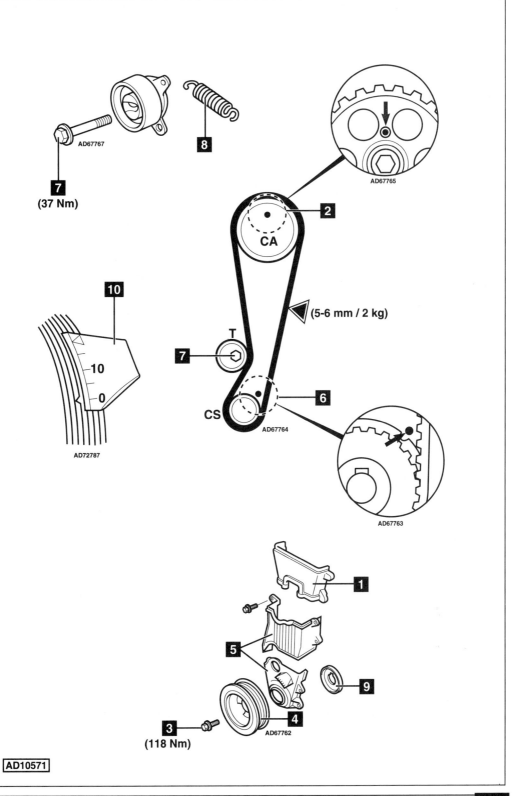

AD67767

7
(37 Nm)

8

AD67765

2

CA

(5-6 mm / 2 kg)

T

7

6

CS

AD67764

AD67763

10

10

0

AD72787

1

5

9

3
(118 Nm)

4

AD67762

AD10571

TOYOTA

Model:	Corolla GT Coupe 1,6 • Corolla GTi 1,6 • Celica (AT160) 1,6 • MR2 1,6
Year:	1984-92
Engine Code:	4A-GE, 4A-GEL

Replacement Interval Guide

Toyota recommend replacement every 60,000 miles or 5 years, whichever occurs first.
The previous use and service history of the vehicle must always be taken into account.
Refer to Timing Belt Replacement Intervals at the front of this manual.

Check For Engine Damage

CAUTION: This engine has been identified as an INTERFERENCE engine in which the possibility of valve-to-piston damage in the event of a timing belt failure is MOST LIKELY to occur.
A compression check of all cylinders should be performed before removing the cylinder head.

Repair Times – hrs

Remove & install:

Corolla (AE86)	0,90
AC	+0,10
PAS	+0,10
Corolla (AE92)	2,20
AC	+0,10
PAS	+0,20
Celica (AT160)	1,40
AC	+0,40
PAS	+0,10
MR2	1,60
AC	+0,40

Special Tools

■ Puller – Toyota No.09213-31021.

Special Precautions

■ Disconnect battery earth lead.
■ DO NOT turn crankshaft or camshaft when timing belt removed.
■ Remove spark plugs to ease turning engine.
■ Turn engine in normal direction of rotation (unless otherwise stated).
■ DO NOT turn engine via camshaft or other sprockets.
■ Observe all tightening torques.

Removal

1. Except AE86: Support front of engine.
2. Remove:
   ❏ Except AE86: Engine mounting.
   ❏ AE86: Air intake pipe.
   ❏ Auxiliary drive belts.
   ❏ AE86: Cooling fan and viscous coupling.
   ❏ Water pump pulley **1**.
   ❏ Timing belt upper cover **2**.

3. Turn crankshaft clockwise to TDC on No.1 cylinder. Ensure timing marks aligned **3** & **4**.
4. Check camshaft notch is visible through cylinder head cover oil filler hole **5**.
5. Remove:
   ❏ Crankshaft pulley bolt **6**.
   ❏ Crankshaft pulley **7**.
      Use tool No.09213-31021.
   ❏ Timing belt lower covers **8** & **9**.
   ❏ Crankshaft sprocket guide washer **10**.
6. Slacken tensioner pulley bolt **11**. Push tensioner pulley to left and lightly tighten bolt.
7. Remove:
   ❏ Timing belt.
   ❏ Tensioner pulley.

Installation

NOTE: Ensure engine is cold before installing belt.

1. Check tensioner pulley for smooth operation.
2. Check free length of tensioner spring is 43,5 mm. Replace spring if necessary.
3. Fit tensioner and spring. Ensure spring is connected correctly.
4. Push tensioner to left and lightly tighten bolt **11**.
5. Ensure timing marks aligned **3**, **4** & **12**.
6. Fit timing belt. Ensure belt is taut between sprockets.
7. Slacken tensioner bolt **11**.
8. Temporarily fit crankshaft pulley **7**.
9. Turn crankshaft two turns clockwise to TDC on No.1 cylinder.
10. Ensure timing marks aligned.
11. Tighten tensioner bolt **11**.
    Tightening torque: 37 Nm.
12. Apply a load of 2 kg to belt at ▽. Belt should deflect 4 mm.
13. If not: Repeat tensioning procedure.
14. Remove crankshaft pulley **7**.
15. Fit crankshaft sprocket guide washer **10**.
16. Fit timing belt lower cover **9**. Note length of bolts **13**.
17. Fit crankshaft pulley **7**. Except AE86: Tighten bolt to 137 Nm **6**. AE86: Tighten bolt to 158 Nm **6**.
18. Install components in reverse order of removal.

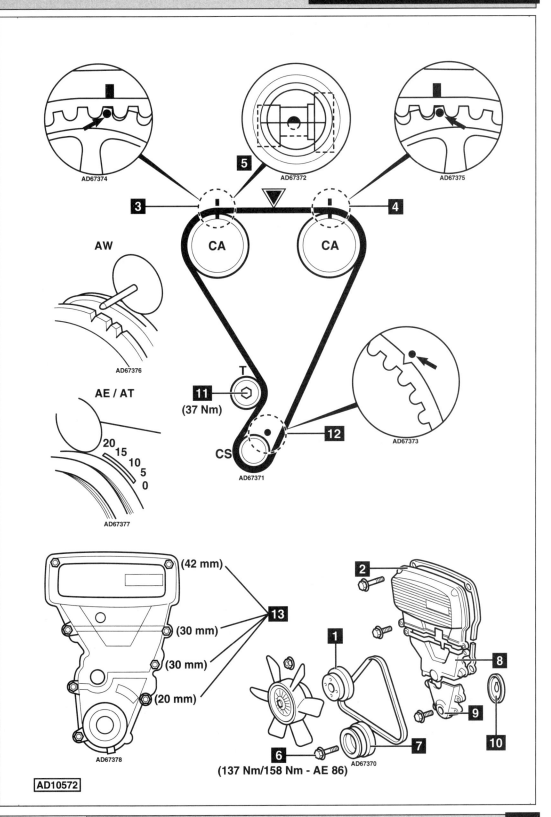

AD67374

5

AD67372

AD67375

3 | **4**

CA | CA

AW

AD67376

AE / AT

20
15
10
5
0

AD67377

T

11

(37 Nm)

CS

AD67371

12

AD67373

(42 mm)

13

(30 mm)

(30 mm)

(20 mm)

AD67378

2

1

8

9

7

10

6

(137 Nm/158 Nm - AE 86)   AD67370

AD10572

Model:	Carina II 1,8/2,0i • Camry 1,8/2,0i • MR2 2,0
Year:	1983-94
Engine Code:	1S, 1S-E, 2S-E, 3S-FE

Replacement Interval Guide

Toyota recommend replacement as follows:
→1993 - replacement every 60,000 miles or
5 years, whichever occurs first.
1994→ - replacement every 63,000 miles or
5 years, whichever occurs first.
*The previous use and service history of the vehicle
must always be taken into account.*
*Refer to Timing Belt Replacement Intervals at the front
of this manual.*

Check For Engine Damage - 3S-FE engine

*CAUTION: This engine has been identified as an
INTERFERENCE engine in which the possibility of
valve-to-piston damage in the event of a timing belt
failure is MOST LIKELY to occur.*
*A compression check of all cylinders should be
performed before removing the cylinder head.*

Check For Engine Damage - 1S, 1SE & 2S-E engines

*CAUTION: This engine has been identified as a
FREEWHEELING engine in which the possibility of
valve-to-piston damage in the event of a timing belt
failure may be minimal or very unlikely. However, a
precautionary compression check of all cylinders
should be performed.*

Repair Times – hrs

Remove & install:

Carina	2,00
AC	+0,10
Camry (1SE/2SE)	1,90
PAS	+0,50
Camry (3S-FE)	2,10
PAS	+0,20
MR2	2,50

Special Tools

- ■ Puller – Toyota No.09213-31021 (1S/2S).
- ■ Puller – Toyota No.09213-60017 (3S-FE).

Special Precautions

- ■ Disconnect battery earth lead.
- ■ DO NOT turn crankshaft or camshaft when timing belt removed.
- ■ Remove spark plugs to ease turning engine.
- ■ Turn engine in normal direction of rotation (unless otherwise stated).
- ■ DO NOT turn engine via camshaft or other sprockets.
- ■ Observe all tightening torques.

Removal

1. Remove auxiliary drive belts.
2. Support engine.
3. Remove:
   - ❏ Engine mounting and bracket.
   - ❏ Timing belt upper cover **1**.
4. Turn crankshaft clockwise to TDC on No.1 cylinder. Ensure timing marks aligned **2** & **3**.
5. Remove:
   - ❏ Crankshaft pulley bolt **4**.
   - ❏ Crankshaft pulley **5**. Use tool Nos.09213-31021 (1S/2S) or 09213-60017 (3S-FE).
6. Ensure timing marks aligned **2**.
7. Slacken tensioner pulley bolt **6**. Move tensioner pulley away from belt and lightly tighten bolt.
8. Remove:
   - ❏ Timing belt lower cover **7**.
   - ❏ Crankshaft sprocket guide washer **9**.
   - ❏ Timing belt.

Installation

1. Remove tensioner pulley and spring. Check tensioner pulley for smooth operation.
2. Check free length of tensioner spring **8**. 1S/2S: 51 mm. 3S-FE: 46 mm. Replace spring if necessary.
3. Fit tensioner pulley and spring. Ensure spring is connected correctly.
4. Push tensioner pulley to left and lightly tighten bolt **6**.
5. Ensure timing marks aligned **2**.
6. Fit timing belt. Ensure belt is taut between sprockets.
7. Install:
   - ❏ Crankshaft sprocket guide washer **9**.
   - ❏ Crankshaft pulley **5**.
8. Slacken tensioner pulley bolt 1/2 turn **6**.
9. Turn crankshaft two turns clockwise to TDC on No.1 cylinder.
10. Ensure timing marks aligned **2** & **3**.
11. Tighten tensioner pulley bolt **6**. Tightening torque: 42 Nm.
12. Check belt deflects slightly at ▽.
13. Remove crankshaft pulley. Fit timing belt lower cover.
14. Fit crankshaft pulley **5**. Ensure crankshaft pulley mark aligned with TDC mark on timing belt cover **10**.
15. Tighten crankshaft pulley bolt **4**. Tightening torque: 108 Nm.
16. Install components in reverse order of removal.

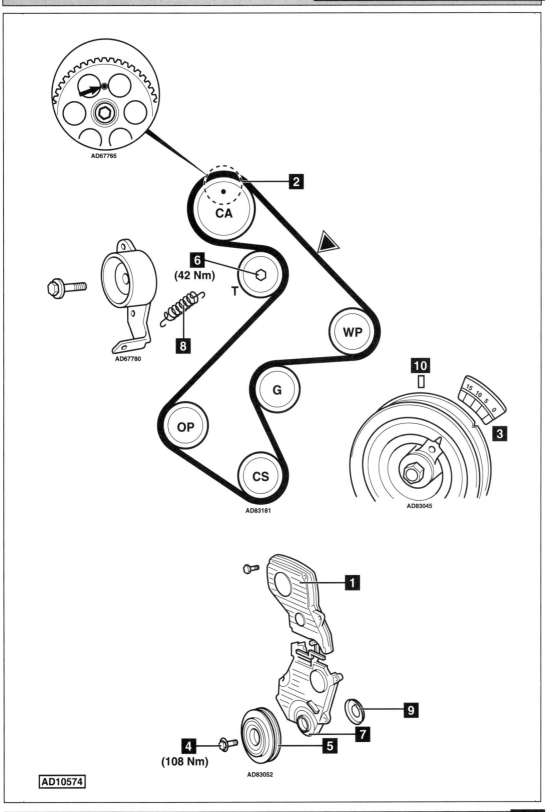

AD67765

2

CA

6
(42 Nm)

T

8

AD67780

WP

10

3

15 10 5 0

G

OP

CS

AD83181

AD83045

1

9

7

4
(108 Nm)

5

AD83052

AD10574

TOYOTA

Model:	Carina E 2,0i • Camry 2,2 16V
Year:	**1992-01**
Engine Code:	**3S-FE, 5S-FE**

Replacement Interval Guide

Toyota recommend replacement as follows:
→1993 - replacement every 60,000 miles or 5 years, whichever occurs first.
1994→ - replacement every 63,000 miles or 5 years, whichever occurs first.
The previous use and service history of the vehicle must always be taken into account.
Refer to Timing Belt Replacement Intervals at the front of this manual.

Check For Engine Damage

CAUTION: This engine has been identified as an INTERFERENCE engine in which the possibility of valve-to-piston damage in the event of a timing belt failure is MOST LIKELY to occur.
A compression check of all cylinders should be performed before removing the cylinder head.

Repair Times – hrs

Remove & install	2,10

Special Tools

- Puller – Toyota No. 09213-60017.

Special Precautions

- Disconnect battery earth lead.
- DO NOT turn crankshaft or camshaft when timing belt removed.
- Remove spark plugs to ease turning engine.
- Turn engine in normal direction of rotation (unless otherwise stated).
- DO NOT turn engine via camshaft or other sprockets.
- Observe all tightening torques.

Removal

1. Remove auxiliary drive belts.
2. Support engine.
3. Remove:
   - ❏ Engine mounting and bracket.
   - ❏ Timing belt upper cover **1**.
4. Turn crankshaft clockwise to TDC on No.1 cylinder. Ensure timing marks aligned **2** & **3**.
5. Remove:
   - ❏ Crankshaft pulley bolt **4**.
   - ❏ Crankshaft pulley **5**.
     Use tool No.09213-60017.
6. Ensure timing marks aligned **2** & **3**.
7. Slacken tensioner bolt **6**. Move tensioner away from belt and lightly tighten bolt.

8. Remove:
   - ❏ Timing belt lower cover **7**.
   - ❏ Crankshaft sprocket guide washer **9**.
   - ❏ Timing belt.
   - ❏ Tensioner pulley.

Installation

1. Check tensioner pulley for smooth operation.
2. Check free length of tensioner spring is 46,0 mm **8**. Replace spring if necessary.
3. Fit tensioner and spring. Ensure spring is connected correctly.
4. Push tensioner to left and lightly tighten bolt **6**.
5. Ensure timing marks aligned **2** & **3**.
6. Fit timing belt. Ensure belt is taut between sprockets.
7. Install:
   - ❏ Crankshaft sprocket guide washer **9**.
   - ❏ Timing belt lower cover **7**.
   - ❏ Crankshaft pulley **5**.
8. Slacken tensioner bolt 1/2 turn **6**.
9. Turn crankshaft two turns clockwise to TDC on No.1 cylinder. Ensure timing marks aligned **2** & **3**.
10. Turn crankshaft slowly 1 7/8 turns clockwise until crankshaft pulley timing mark **3** aligned with 45° BTDC mark **10** on timing belt lower cover **7**.
11. Tighten tensioner bolt **6**.
    Tightening torque: 42 Nm.
12. Install components in reverse order of removal.
13. Tighten crankshaft pulley bolt **4**.
    Tightening torque: 108 Nm.

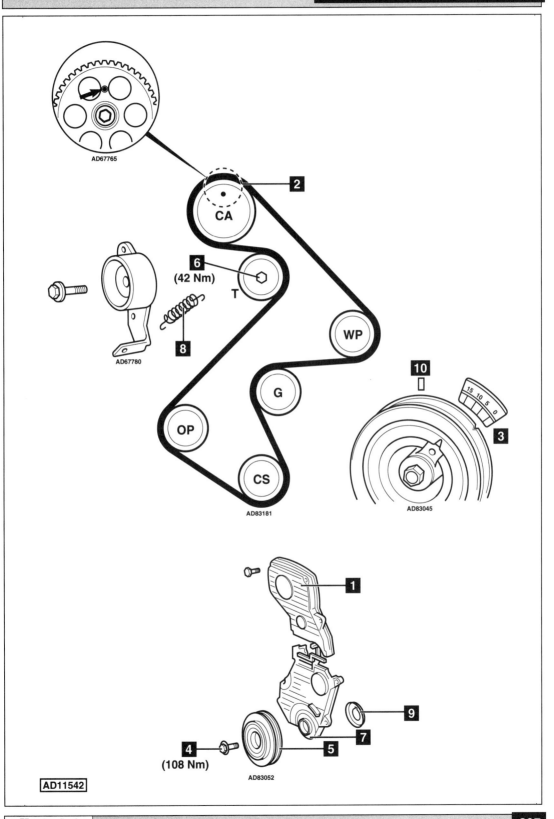

AD67765

2

CA

6
(42 Nm)

T

8

AD67780

WP

10

G

15 10 5 0

3

OP

CS

AD83181

AD83045

1

9

7

4
(108 Nm)

5

AD83052

AD11542

TOYOTA

Model:	Carina E 2,0 GTi • Celica GT 2,0 • Celica GT-4 2,0 • MR2 2,0 GT
Year:	1990-00
Engine Code:	3S-GE, 3S-GTE

Replacement Interval Guide

Toyota recommend replacement as follows:
→1993 - replacement every 60,000 miles or 5 years, whichever occurs first.
1994→ - replacement every 63,000 miles or 5 years, whichever occurs first.
The previous use and service history of the vehicle must always be taken into account.
Refer to Timing Belt Replacement Intervals at the front of this manual.

Check For Engine Damage

CAUTION: This engine has been identified as an INTERFERENCE engine in which the possibility of valve-to-piston damage in the event of a timing belt failure is MOST LIKELY to occur.
A compression check of all cylinders should be performed before removing the cylinder head.

Repair Times – hrs

Remove & install:

Carina E (ST191)	2,30
Celica GT (ST182)	2,50
Celica GT (ST202)	2,00
Celica GT-4 (ST185)	3,20
Celica GT-4 (ST205)	3,10
AC	+0,10
MR2 (SW20)	3,20
AC	+0,10

Special Tools

■ Puller – Toyota No.09213-31021.

Special Precautions

■ Disconnect battery earth lead.
■ DO NOT turn crankshaft or camshaft when timing belt removed.
■ Remove spark plugs to ease turning engine.
■ Turn engine in normal direction of rotation (unless otherwise stated).
■ DO NOT turn engine via camshaft or other sprockets.
■ Observe all tightening torques.

Removal

1. Support engine.
2. Remove:
   ❏ Auxiliary drive belts.
   ❏ Engine mounting and bracket.
   ❏ Alternator.
   ❏ Timing belt upper cover **1**.

3. Turn crankshaft clockwise to TDC on No.1 cylinder. Ensure timing marks aligned **2** & **3**.
4. Remove automatic tensioner unit **5**.
5. Remove timing belt from camshaft sprockets.
6. Prevent crankshaft from turning. Remove crankshaft pulley bolt **6**.
7. Remove:
   ❏ Crankshaft pulley **7**.
      Use tool No.09213-31021.
   ❏ Timing belt lower cover **8**.
   ❏ Crankshaft sprocket guide washer **9**.
   ❏ Timing belt.
   ❏ Tensioner pulley bolt **10**.
   ❏ Tensioner pulley **11**.

Installation

NOTE: Ensure engine is cold before installing belt.

1. Check tensioner pulley for smooth operation **11**.
2. Fit tensioner pulley. Apply locking compound to bolt threads. Tighten bolt to 43 Nm **10**.
3. Fit timing belt in anti-clockwise direction. Ensure belt is taut between sprockets.
4. Install:
   ❏ Timing belt lower cover **8**.
   ❏ Crankshaft pulley **7**.
5. Ensure timing marks aligned **2** & **3**.
6. Check tensioner body for leakage or damage **5**. Replace if necessary.
7. Check pushrod does not move when pushed against a firm surface. If pushrod moves: Replace automatic tensioner unit.
8. Check pushrod protrusion is: →1994 – 8,5-9,5 mm. 1994→ – 10-11 mm. **12**. If not: Replace automatic tensioner unit.
9. Slowly compress pushrod into tensioner body until holes aligned. Use press.
10. Retain pushrod with 1,27 mm Allen key through hole in tensioner body **13**.
11. Install automatic tensioner unit **5**.

→1994:

12. Lightly tighten bolts **4**.
13. Turn tensioner pulley bolt anti-clockwise with a torque of 18 Nm **14**.
14. Turn crankshaft 300° clockwise to align crankshaft mark with 60° BTDC mark on timing belt cover **15**.

Autodata

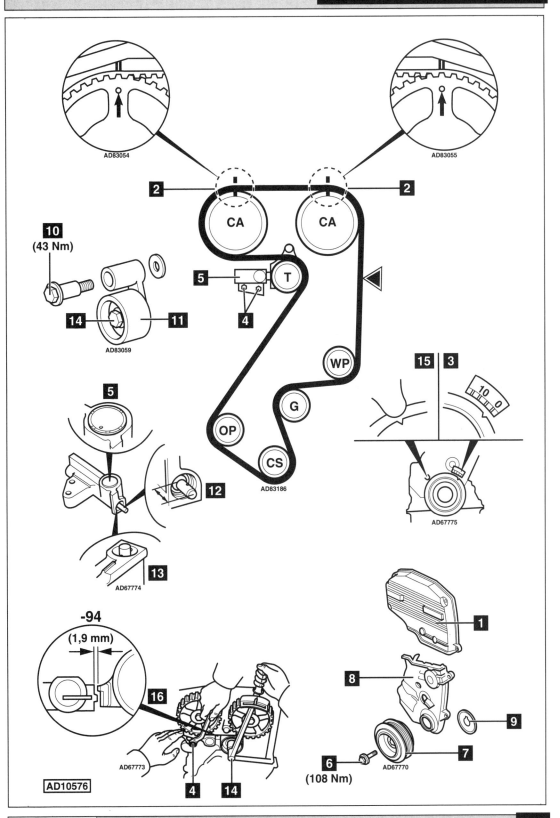

AD83054

AD83055

2

2

CA

CA

10
(43 Nm)

5

T

14 **11**

4

AD83059

WP

15 **3**

10 0

G

5

OP

CS

AD83186

12

13

AD67774

AD67775

1

8

-94
(1,9 mm)

16

9

7

AD67773

6
(108 Nm)

AD67770

4 **14**

AD10576

←

15. Slacken bolts . Insert 1,9 mm feeler gauge between tensioner body and tensioner pulley bracket ⑯.

16. Turn tensioner pulley bolt anti-clockwise with a torque of 18 Nm ⑭.

17. Hold tensioner pulley in position. Tighten automatic tensioner unit bolts to 21 Nm ④. Remove feeler gauge.

18. Remove Allen key from tensioner body to release pushrod ⑬.

19. Turn crankshaft one turn clockwise to 60° BTDC mark ⑮.

20. Insert feeler gauges between tensioner body and tensioner pulley bracket. Check gap is 1,8-2,2 mm ⑯.

21. If not: Remove automatic tensioner unit. Repeat operations 9-20.

22. Turn crankshaft 2 1/6 turns clockwise to TDC on No.1 cylinder. Ensure timing marks aligned ② & ③.

1994→:

23. Tighten automatic tensioner unit bolts ④. Tightening torque: 21 Nm.

24. Turn tensioner pulley bolt anti-clockwise with a torque of 69 Nm ⑭.

25. Remove Allen key from tensioner body to release pushrod ⑬.

26. Remove torque wrench ⑭.

27. Turn crankshaft 2 turns clockwise to TDC on No.1 cylinder. Ensure timing marks aligned ② & ③.

28. If not: Repeat installation and tensioning procedures.

All models:

29. Install components in reverse order of removal.

30. Tighten crankshaft pulley bolt ⑥. Tightening torque: 108 Nm.

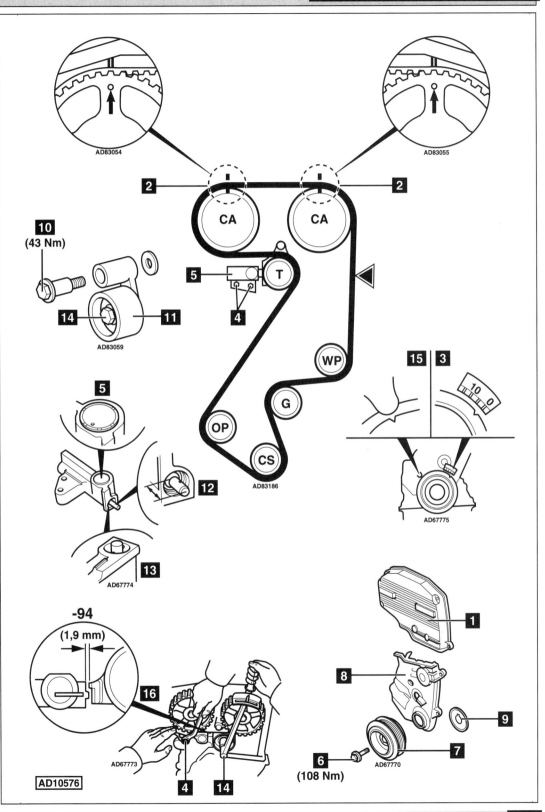

AD83054

AD83055

2 | 2

CA | CA

10
(43 Nm)

5 | T

14 | 11

4

AD83059

WP

5

15 | 3

10 0

G

OP

CS

12

AD83186

13

AD67774

AD67775

-94
(1,9 mm)

16

1

8

AD67773

9

4 | 14

6
(108 Nm)

7

AD67770

AD10576

Model:	Avensis 2,0 • Picnic • RAV-4
Year:	1994-01
Engine Code:	3S-FE

Replacement Interval Guide

Toyota recommend replacement every
63,000 miles or 5 years, whichever occurs first.
*The previous use and service history of the vehicle
must always be taken into account.*
*Refer to Timing Belt Replacement Intervals at the front
of this manual.*

Check For Engine Damage

*CAUTION: This engine has been identified as an
INTERFERENCE engine in which the possibility of
valve-to-piston damage in the event of a timing belt
failure is MOST LIKELY to occur.*
*A compression check of all cylinders should be
performed before removing the cylinder head.*

Repair Times – hrs

Remove & install:

RAV-4	2,40
ABS	+0,90
Picnic	1,50
Avensis	1,40
AC	+0,10

Special Tools

- Puller – Toyota No.09950-50010.

Special Precautions

- Disconnect battery earth lead.
- DO NOT turn crankshaft or camshaft when timing belt removed.
- Remove spark plugs to ease turning engine.
- Turn engine in normal direction of rotation (unless otherwise stated).
- DO NOT turn engine via camshaft or other sprockets.
- Observe all tightening torques.

Removal

1. Support engine.
2. Remove:
   - ❏ RH engine mounting and bracket **1**.
   - ❏ Auxiliary drive belt(s).
   - ❏ Timing belt upper cover **2**.
3. Turn crankshaft clockwise to TDC on No.1 cylinder. Ensure timing marks aligned **3**.
4. Check hole in camshaft sprocket aligned with mark on camshaft bearing cap **4**. If not: Turn crankshaft one turn.

5. Remove:
   - ❏ Crankshaft pulley bolt **5**.
   - ❏ Crankshaft pulley **6**. Use tool No.09950-50010.
   - ❏ Timing belt lower cover **7**.
6. Ensure timing marks aligned **4** & **8**.
7. Slacken tensioner bolt **9**. Move tensioner away from belt and lightly tighten bolt.
8. Remove:
   - ❏ Crankshaft sprocket guide washer **10**.
   - ❏ Timing belt.
   - ❏ Tensioner **11**.

Installation

1. Check tensioner pulley for smooth operation.
2. Check free length of tensioner spring is 46 mm **12**. Replace spring if necessary.
3. Fit tensioner and spring. Ensure spring is connected correctly.
4. Push tensioner to left and lightly tighten bolt **9**.
5. Ensure timing marks aligned **4** & **8**.
6. Fit timing belt. Ensure belt is taut between sprockets.
7. Fit crankshaft sprocket guide washer **10**.
8. Temporarily fit crankshaft pulley **6**.
9. Slacken tensioner bolt 1/2 turn **9**. Allow tensioner to operate.
10. Turn crankshaft two turns clockwise to TDC on No.1 cylinder.
11. Ensure timing marks aligned **4** & **8**.
12. If not: Repeat installation procedure.
13. Remove crankshaft pulley **6**.
14. Install:
    - ❏ Timing belt lower cover **7**.
    - ❏ Crankshaft pulley **6**.
    - ❏ Crankshaft pulley bolt **5**.
15. Ensure crankshaft pulley and timing belt cover TDC marks aligned **3**.
16. Turn crankshaft slowly 1 7/8 turns clockwise until crankshaft pulley timing mark **3** aligned with 45° BTDC mark **13** on timing belt lower cover **7**.
17. Tighten tensioner bolt **9**. Tightening torque: 42 Nm.
18. Tighten crankshaft pulley bolt **5**. Tightening torque: 108 Nm.
19. Install components in reverse order of removal.

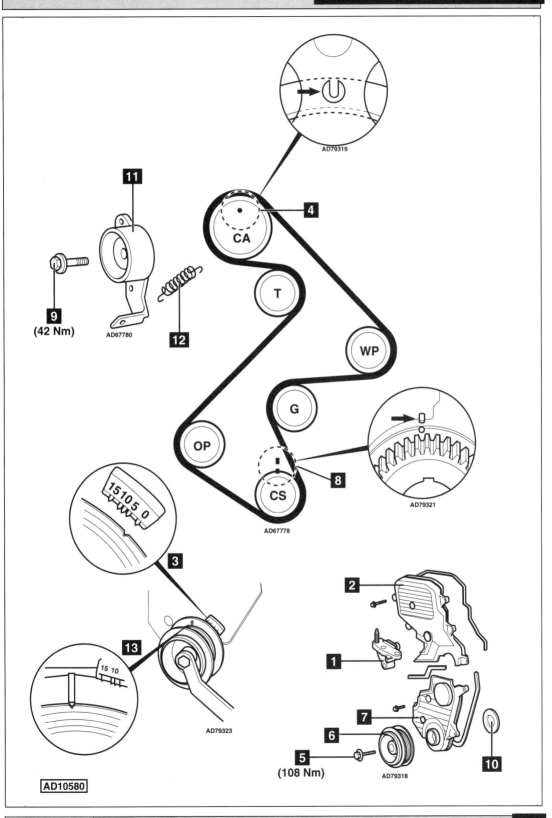

AD79319

11

4

CA

T

WP

G

OP

CS

8

AD79321

9
(42 Nm)

AD67780

12

3

15 10 5 0

AD67778

13

15 10

AD79323

2

1

7

6

5
(108 Nm)

AD79318

10

AD10580

TOYOTA

Model:	Camry 3,0 V6
Year:	1991-96
Engine Code:	3VZ-FE

Replacement Interval Guide

Toyota recommend replacement as follows:
→1993 - replacement every 60,000 miles or
5 years, whichever occurs first.
1994→ - replacement every 63,000 miles or
5 years, whichever occurs first.
The previous use and service history of the vehicle must always be taken into account.
Refer to Timing Belt Replacement Intervals at the front of this manual.

Check For Engine Damage

CAUTION: This engine has been identified as an INTERFERENCE engine in which the possibility of valve-to-piston damage in the event of a timing belt failure is MOST LIKELY to occur.
A compression check of all cylinders should be performed before removing the cylinder head.

Repair Times – hrs

Remove & install	2,20

Special Tools

■ Puller – Toyota No.09213-60017.

Special Precautions

■ Disconnect battery earth lead.
■ DO NOT turn crankshaft or camshaft when timing belt removed.
■ Remove spark plugs to ease turning engine.
■ Turn engine in normal direction of rotation (unless otherwise stated).
■ DO NOT turn engine via camshaft or other sprockets.
■ Observe all tightening torques.

Removal

1. Raise and support front of vehicle.
2. Support front of engine.
3. Remove:
   ❏ Auxiliary drive belts.
   ❏ Engine mounting.
   ❏ Timing belt upper cover **1**.
   ❏ RH engine mounting bracket **2**.
4. Turn crankshaft clockwise to TDC on No.1 cylinder. Ensure timing marks aligned **3**, **4** & **5**.
5. From under front of engine remove automatic tensioner unit **6**.
6. Remove:
   ❏ Crankshaft pulley bolt **7**.
   ❏ Crankshaft pulley **8**.
      Use tool No.09213-60017.

❏ Timing belt lower cover **9**.
❏ Crankshaft sprocket guide washer **10**.
NOTE: If belt is to be reused, check alignment marks are visible. If not: Mark timing belt with chalk against marks on sprockets.

7. Turn rear camshaft sprocket slightly clockwise to release tension on belt.
8. Remove timing belt.

Installation

NOTE: Ensure engine is cold before installing belt. Observe alignment marks.

1. Check tensioner pulley and guide pulley for smooth operation.
2. Check tensioner body for leakage or damage. Replace if necessary.
3. Check pushrod does not move when pushed against a firm surface. Replace if necessary.
4. Check pushrod protrusion is 10,0-10,5 mm **11**. Replace if necessary.
5. Ensure timing marks aligned **3**, **4** & **5**.
6. Install:
   ❏ Timing belt.
   ❏ Crankshaft sprocket guide washer **10**.
7. Temporarily fit timing belt lower cover and crankshaft pulley.
8. Slowly compress pushrod into tensioner body until holes aligned. Use press.
   *NOTE: Use a suitable washer to protect tensioner body end screw when using press **12**.*
9. Retain pushrod with 1,5 mm Allen key through hole in tensioner body.
10. Fit dust cover.
11. Install automatic tensioner unit. Tighten bolts to 26 Nm.
12. Remove Allen key from tensioner body to release pushrod.
13. Turn crankshaft two turns clockwise to TDC on No.1 cylinder.
14. Ensure timing marks aligned **3**, **4** & **5**.
15. Remove crankshaft pulley and timing belt lower cover.
16. Install:
   ❏ Timing belt lower cover with new seal.
   ❏ Crankshaft pulley **8**.
   ❏ Crankshaft pulley bolt **7**.
      Tightening torque: 245 Nm.
17. Install components in reverse order of removal.
   *NOTE: Timing belt upper cover bolts must be fitted in correct locations **13**.*

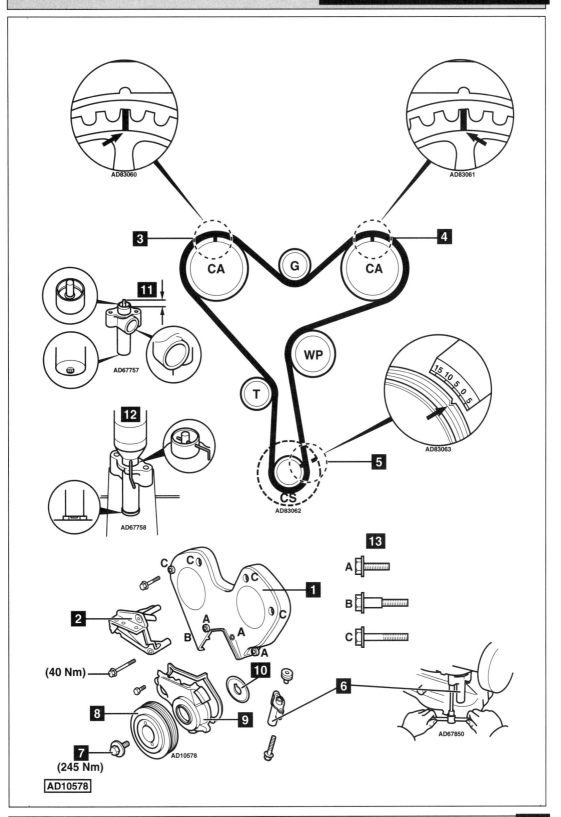

AD83060

AD83061

3

4

CA G CA

11

AD67757

WP

T

AD83063

12

5

AD67758

CS

AD83062

13

A

1

2

B

(40 Nm)

10

6

8

9

AD10578

7

(245 Nm)

AD10578

AD67850

TOYOTA

Model:	**Camry 3,0 V6 24V**
Year:	**1996-01**
Engine Code:	**1MZ-FE**

Replacement Interval Guide

Toyota recommend replacement every 63,000 miles or 5 years, whichever occurs first.
The previous use and service history of the vehicle must always be taken into account.
Refer to Timing Belt Replacement Intervals at the front of this manual.

Check For Engine Damage

CAUTION: This engine has been identified as an INTERFERENCE engine in which the possibility of valve-to-piston damage in the event of a timing belt failure is MOST LIKELY to occur.
A compression check of all cylinders should be performed before removing the cylinder head.

Repair Times – hrs

Remove & install	2,10

Special Tools

- Puller – Toyota No.09950-50010.
- Crankshaft pulley holding tool – Toyota No.09213-54015.

Special Precautions

- Disconnect battery earth lead.
- DO NOT turn crankshaft or camshaft when timing belt removed.
- Remove spark plugs to ease turning engine.
- Turn engine in normal direction of rotation (unless otherwise stated).
- DO NOT turn engine via camshaft or other sprockets.
- Observe all tightening torques.

Removal

1. Raise and support front of vehicle.
2. Support engine.
3. Remove:
   - ❏ RH front wheel.
   - ❏ RH inner wing panel.
   - ❏ Auxiliary drive belts.
   - ❏ RH engine mounting.
   - ❏ Alternator bracket **1**.
   - ❏ Crankshaft bolt **2**.
     Use tool No.09213-54015.
   - ❏ Crankshaft pulley **3**.
     Use tool No.09950-50010.
   - ❏ Timing belt covers **4** & **5**.
   - ❏ Engine mounting bracket **6**.
   - ❏ Crankshaft sprocket guide washer **7**.
4. Temporarily fit crankshaft bolt **2**.

5. Turn crankshaft clockwise to TDC on No.1 cylinder. Ensure timing marks aligned **8**, **9** & **10**.
6. From under front of engine remove automatic tensioner unit and dust cover.
7. Remove timing belt.
   NOTE: If belt is to be reused, check alignment marks are visible. If not: Mark timing belt with chalk against marks on sprockets.

Installation

NOTE: Ensure engine is cold before installing belt. Observe alignment marks.

1. Check tensioner pulley and guide pulley for smooth operation. Replace if necessary.
2. Check tensioner body for leakage or damage. Replace if necessary.
3. Check pushrod does not move when pushed against a firm surface **12**. Replace if necessary.
4. Check pushrod protrusion is 10,0-10,8 mm **13**. Replace if necessary.
5. Ensure timing marks aligned **8**, **9** & **10**.
6. Fit timing belt in anti-clockwise direction, starting at crankshaft sprocket. Ensure belt is taut between sprockets.
7. Slowly compress pushrod into tensioner body until holes aligned **14**. Use press.
8. Retain pushrod with 1,27 mm Allen key through hole in tensioner body **15**.
9. Install automatic tensioner unit **11**. Fit dust cover. Tighten bolts evenly to 27 Nm **16**.
10. Remove Allen key from tensioner body to release pushrod **15**.
11. Turn crankshaft slowly two turns clockwise to TDC on No.1 cylinder. Ensure timing marks aligned **8**, **9** & **10**.
12. Remove crankshaft bolt **2**.
13. Install components in reverse order of removal.
14. Tighten crankshaft bolt **2**.
    Tightening torque: 216 Nm.

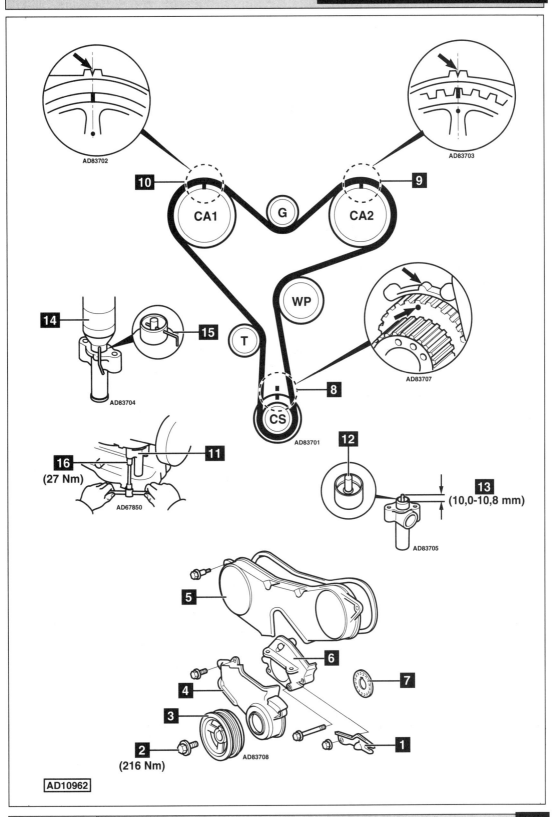

AD83702

AD83703

10

9

CA1

G

CA2

14

15

T

WP

AD83707

AD83704

8

CS

AD83701

16
(27 Nm)

11

12

13
(10,0-10,8 mm)

AD67850

AD83705

5

6

4

7

3

1

2
(216 Nm)

AD83708

AD10962

TOYOTA

Model:	**Supra 3,0 Turbo**
Year:	**1993-97**
Engine Code:	**2JZ-GTE**

Replacement Interval Guide

Toyota recommend replacement as follows:
→1993 - replacement every 60,000 miles or
5 years, whichever occurs first.
1994→ - replacement every 63,000 miles or
5 years, whichever occurs first.
*The previous use and service history of the vehicle
must always be taken into account.*
*Refer to Timing Belt Replacement Intervals at the
front of this manual.*

Check For Engine Damage

*CAUTION: This engine has been identified as an
INTERFERENCE engine in which the possibility of
valve-to-piston damage in the event of a timing belt
failure is MOST LIKELY to occur.*
*A compression check of all cylinders should be
performed before removing the cylinder head.*

Repair Times – hrs

Remove & install	2,40

Special Tools

■ Puller – Toyota No.09213-31021.

Special Precautions

■ Disconnect battery earth lead.
■ DO NOT turn crankshaft or camshaft when timing belt
removed.
■ Remove spark plugs to ease turning engine.
■ Turn engine in normal direction of rotation (unless
otherwise stated).
■ DO NOT turn engine via camshaft or other sprockets.
■ Observe all tightening torques.

Removal

1. Remove:
   ❑ Battery and battery tray.
   ❑ Engine undershield.
   ❑ Air filter duct and air hoses.
   ❑ LH headlight beam angle gauge.
   ❑ Upper fan cowling.
2. Drain coolant. Disconnect coolant hoses. Remove
   radiator.
3. Remove auxiliary drive belt tensioner damper.
4. Slacken viscous fan nuts.
5. Remove:
   ❑ Auxiliary drive belt.
   ❑ Viscous fan nuts and fan.
   ❑ Timing belt upper cover **1**.
   ❑ Timing belt centre cover **2**.
   ❑ Auxiliary drive belt tensioner **3**.
6. Turn crankshaft to TDC on No.1 cylinder **4**.
7. Ensure timing marks on camshaft sprockets
   aligned **5**.
8. Disconnect oil cooler pipe clamps and reposition
   pipes.
9. Remove:
   ❑ Crankshaft pulley bolt **6**.
   ❑ Crankshaft pulley **7**. Use tool No.09213-31021.

10. Ensure timing mark on crankshaft sprocket
    aligned **11**.
11. Remove:
    ❑ Automatic tensioner unit **8**.
    ❑ Timing belt lower cover **9**.
    ❑ Crankshaft sprocket guide washer **10**.
    ❑ Timing belt.
    ❑ Tensioner pulley **12**.

Installation

1. Check tensioner pulley for smooth operation.
   Replace if necessary.
2. Fit tensioner pulley **12**. Tighten bolt to 34 Nm.
3. Remove dust cover. Check tensioner body for
   leakage or damage.
4. Check pushrod protrusion is 8,0-8,8 mm **13**.
5. Check pushrod does not move when pushed against
   a firm surface. If pushrod moves: Replace automatic
   tensioner unit.
6. Slowly compress pushrod into tensioner body until
   holes aligned. Use press.
7. Retain pushrod with 1,5 mm Allen key through hole
   in tensioner body **14**. Fit dust cover.
8. Ensure timing marks aligned **5** & **11**.
9. Fit timing belt in anti-clockwise direction, starting at
   crankshaft sprocket. Ensure belt is taut between
   sprockets.
10. Fit crankshaft sprocket guide washer **10**.
11. Install automatic tensioner unit **8**. Tighten bolts
    evenly to 26 Nm.
12. Remove Allen key from tensioner body to release
    pushrod. Use pliers.
13. Temporarily fit crankshaft pulley.
14. Turn crankshaft slowly two turns clockwise to TDC
    on No.1 cylinder.
15. Ensure timing marks aligned **5** & **11**.
16. Remove crankshaft pulley **7**.
17. Install:
    ❑ Timing belt lower cover **9**.
    ❑ Crankshaft pulley **7**.
    ❑ Crankshaft pulley bolt **6**.
18. Ensure timing marks on crankshaft pulley and timing
    belt lower cover aligned **4**.
19. Fit auxiliary drive belt tensioner **3**. Ensure bolts do
    not drop into timing belt lower cover.
20. Tighten tensioner bolts evenly to 21 Nm.
21. Install:
    ❑ Timing belt centre cover **2**.
    ❑ Timing belt upper cover **1**.
    ❑ Auxiliary drive belt
    ❑ Viscous fan. Tighten nuts to 16 Nm.
22. Check auxiliary drive belt damper unit for leakage.
    Tighten nuts to 20 Nm.
23. Tighten crankshaft pulley bolt **6**.
    Tightening torque: 324 Nm.
24. Fit radiator and reconnect coolant hoses. Refill
    cooling system.
25. Install components in reverse order of removal.

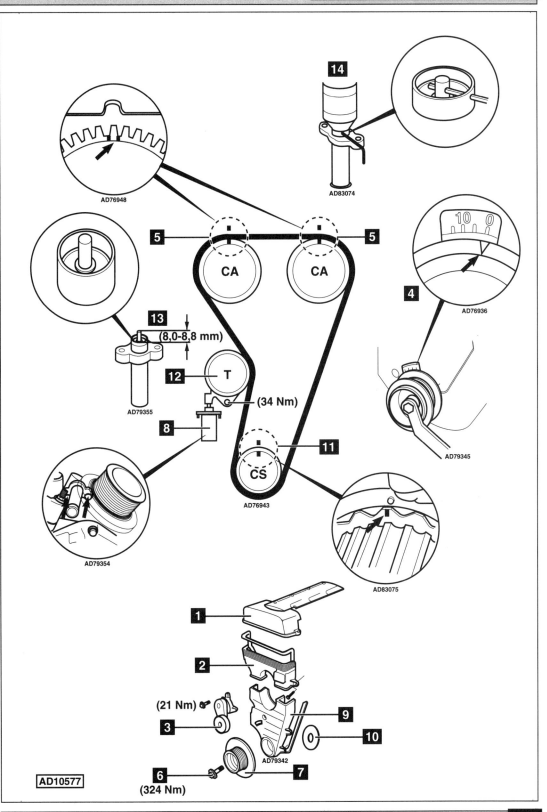

AD76948

14
AD83074

5   5

CA   CA

10 0

4
AD76936

13
(8,0-8,8 mm)

12   T

AD79355

8

(34 Nm)

AD79345

11

CS
AD76943

AD79354

AD83075

1

2

(21 Nm)

3   9

AD79342

6   10

(324 Nm)   7

AD10577

TOYOTA

Model:	4-Runner 3,0 V6
Year:	1989-96
Engine Code:	3VZ-E

Replacement Interval Guide

Toyota recommend replacement as follows:
→1993 - replacement every 60,000 miles or
5 years, whichever occurs first.
1994→ - replacement every 63,000 miles or
5 years, whichever occurs first.
*The previous use and service history of the vehicle
must always be taken into account.*
*Refer to Timing Belt Replacement Intervals at the
front of this manual.*

Check For Engine Damage

*CAUTION: This engine has been identified as a
FREEWHEELING engine in which the possibility of
valve-to-piston damage in the event of a timing belt
failure may be minimal or very unlikely. However, a
precautionary compression check of all cylinders
should be performed.*

Repair Times – hrs

Remove & install	2,90
AC	+0,10

Special Tools

■ Puller – Toyota No.09213-31021.

Special Precautions

■ Disconnect battery earth lead.
■ DO NOT turn crankshaft or camshaft when timing belt removed.
■ Remove spark plugs to ease turning engine.
■ Turn engine in normal direction of rotation (unless otherwise stated).
■ DO NOT turn engine via camshaft or other sprockets.
■ Observe all tightening torques.

Removal

1. Drain coolant.
2. Remove top hose and outlet pipe.
3. Remove:
   ❑ Air intake hoses.
   ❑ Auxiliary drive belts.
   ❑ Cooling fan.
   ❑ Timing belt upper cover **1**.
   *NOTE: Mark position of cooling fan to viscous
   coupling before removal.*
4. Turn crankshaft clockwise to TDC on No.1 cylinder.
   Ensure timing marks aligned **2**, **3** & **4**.
5. Remove:
   ❑ Crankshaft pulley bolt **5**.
   ❑ Crankshaft pulley **6**. Use tool No.09213-31021.
   ❑ Viscous coupling and bracket **7**.
   ❑ Timing belt lower cover **8**.
6. Slacken tensioner bolt **9**. Move tensioner away from belt **10**. Lightly tighten bolt.

7. Remove:
   ❑ Crankshaft sprocket guide washer **11**.
   ❑ Timing belt.
   *NOTE: Mark direction of rotation on belt with chalk
   if belt is to be reused.*

Installation

1. Check tensioner pulley and guide pulley for smooth operation.
2. Slacken tensioner bolt **9**. Remove tensioner spring **12**.
3. Check free length of tensioner spring is 56,27 mm. Replace spring if necessary.
4. Fit tensioner spring. Push tensioner against spring tension. Lightly tighten bolt **9**.
5. Remove bolts from camshaft sprockets **13**.
6. Remove dowel pins from camshaft sprockets **14**.
   *NOTE: DO NOT turn camshafts. Dowel pin holes
   should remain at 12 o'clock position **14**.*
7. Fit bolts to camshaft sprockets **13**. Ensure bolt heads do NOT touch sprockets and sprockets can turn freely.
8. Ensure timing marks aligned **3**, **4** & **15**.
9. Fit timing belt, starting at camshaft sprockets. Ensure belt is taut between sprockets.
10. Temporarily fit crankshaft pulley bolt **5**.
11. Slacken tensioner bolt **9**. Allow tensioner to operate.
12. Turn crankshaft slowly two turns clockwise to TDC on No.1 cylinder.
13. Ensure timing marks aligned **3**, **4** & **15**.
14. Tighten tensioner bolt **9**. Tightening torque: 37 Nm.
15. Remove bolts from camshaft sprockets **13**. Check dowel pin holes are aligned and insert dowel pins **14**.
16. Fit bolts to camshaft sprockets **13**.
    Tightening torque: 108 Nm.
17. Remove crankshaft pulley bolt **5**.
18. Fit crankshaft sprocket guide washer **11**.
19. Install:
    ❑ Timing belt lower cover **8**.
    ❑ Crankshaft pulley **6**.
    ❑ Crankshaft pulley bolt **5**.
       Tightening torque: 245 Nm.
20. Install viscous coupling and bracket. Tighten bolts to 30 Nm.
    *NOTE: Ensure viscous coupling and cooling fan
    locating marks aligned.*
21. Install components in reverse order of removal.
22. Refill cooling system.

Autodata

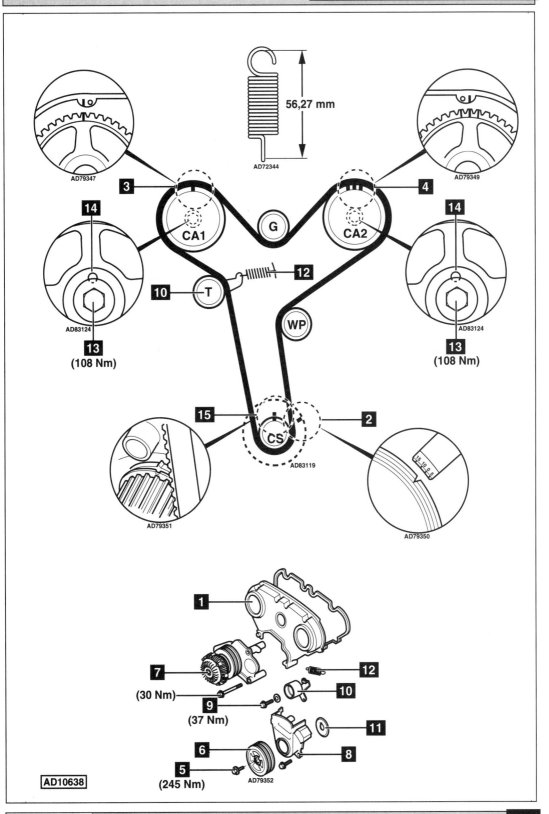

56,27 mm

AD72344

AD79347

AD79349

3

4

14

14

G

CA1

CA2

10 T

12

WP

13
(108 Nm)

13
(108 Nm)

15

2

CS

AD83119

AD79351

AD79350

1

7

12

(30 Nm)

9
(37 Nm)

10

11

6

8

5
(245 Nm)

AD79352

AD10638

TOYOTA

Model:	**Landcruiser Colorado 3,4 V6**
Year:	**1996-02**
Engine Code:	**5VZ-FE**

Replacement Interval Guide

Toyota recommend replacement every
63,000 miles or 5 years, whichever occurs first.
*The previous use and service history of the vehicle
must always be taken into account.*
*Refer to Timing Belt Replacement Intervals at the front
of this manual.*

Check For Engine Damage

*CAUTION: This engine has been identified as an
INTERFERENCE engine in which the possibility of
valve-to-piston damage in the event of a timing belt
failure is MOST LIKELY to occur.
A compression check of all cylinders should be
performed before removing the cylinder head.*

Repair Times – hrs

Remove & install	2,70
AC	+0,30

Special Tools

- Puller – Toyota No.09950-50010.
- Crankshaft pulley holding tool –
  Toyota No.09213-54015.

Special Precautions

- Disconnect battery earth lead.
- DO NOT turn crankshaft or camshaft when timing belt removed.
- Remove spark plugs to ease turning engine.
- Turn engine in normal direction of rotation (unless otherwise stated).
- DO NOT turn engine via camshaft or other sprockets.
- Observe all tightening torques.

Removal

1. Drain coolant.
2. Remove:
   - ❑ Top radiator hose.
   - ❑ Auxiliary drive belt.
   - ❑ Cooling fan and viscous coupling.
   - ❑ Dipstick and tube.
   - ❑ Timing belt upper cover **1**.
   - ❑ Cooling fan bracket **2**.
   - ❑ Crankshaft bolt **3**.
     Use tool No.09213-54015.
   - ❑ Crankshaft pulley **4**.
     Use tool No.09950-50010.
   - ❑ Timing belt lower cover **5**.
   - ❑ Crankshaft sprocket guide washer **6**.
3. Temporarily fit crankshaft bolt **3**.

4. Turn crankshaft clockwise to TDC on No.1 cylinder. Ensure timing marks aligned **7**, **8** & **9**.
5. From under front of engine remove automatic tensioner unit and dust cover.
6. Remove timing belt.
   NOTE: If belt is to be reused, check alignment marks are visible. If not: Mark timing belt with chalk against marks on sprockets.

Installation

NOTE: Ensure engine is cold before installing belt. Observe alignment marks.

1. Check tensioner pulley and guide pulley for smooth operation. Replace if necessary.
2. Check tensioner body for leakage or damage **10**. Replace if necessary.
3. Check pushrod does not move when pushed against a firm surface **11**. Replace if necessary.
4. Check pushrod protrusion is 10,0-10,8 mm **12**. Replace if necessary.
5. Ensure timing marks aligned **7**, **8** & **9**.
6. Fit timing belt in following order:
   - ❑ Camshaft sprocket (CA2).
   - ❑ Guide pulley.
   - ❑ Camshaft sprocket (CA1).
   - ❑ Water pump pulley.
   - ❑ Crankshaft sprocket.
   - ❑ Tensioner pulley.
   NOTE: Ensure belt is taut between sprockets.
7. Slowly compress pushrod into tensioner body until holes aligned. Use press.
8. Retain pushrod with 1,27 mm Allen key **14** through hole in tensioner body **13**.
9. Install automatic tensioner unit **10**. Fit dust cover. Tighten bolts evenly to 27 Nm **15**.
10. Remove Allen key from tensioner body to release pushrod.
11. Turn crankshaft slowly two turns clockwise to TDC on No.1 cylinder. Ensure timing marks aligned **7**, **8** & **9**.
12. Remove crankshaft bolt **3**.
13. Install components in reverse order of removal.
14. Tighten crankshaft bolt **3**.
    Tightening torque: 250 Nm.
15. Refill cooling system.

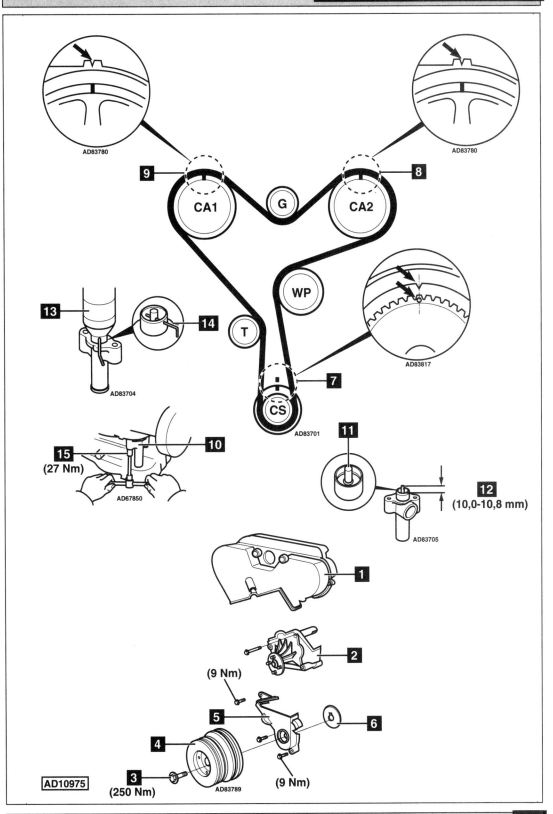

AD83780

AD83780

9

8

CA1

G

CA2

13

14

WP

AD83704

T

AD83817

7

CS

AD83701

15
(27 Nm)

10

AD67850

11

12
(10,0-10,8 mm)

AD83705

1

2
(9 Nm)

5

6

4

3
(250 Nm)

AD83789

(9 Nm)

AD10975

TOYOTA

Model:	Corolla D • Carina II D • Carina E D/TD • Camry Turbo D Avensis 2,0 TD • Lite-Ace D
Year:	1984-99
Engine Code:	1C, 1C-T, 2C, 2C-T, 2C-E, 2C-TE

Replacement Interval Guide

Toyota recommend replacement as follows:
→1993: Replacement every 60,000 miles or
5 years, whichever occurs first.
1994→: Replacement every 63,000 miles or
5 years, whichever occurs first.
The previous use and service history of the vehicle must always be taken into account.
Refer to Timing Belt Replacement Intervals at the front of this manual.

Check For Engine Damage

CAUTION: This engine has been identified as an INTERFERENCE engine in which the possibility of valve-to-piston damage in the event of a timing belt failure is MOST LIKELY to occur.
A compression check of all cylinders should be performed before removing the cylinder head.

Repair Times – hrs

Remove & install:

Corolla (CE80)	1,50
Corolla (CE90)	2,20
AC	+0,10
PAS	+0,40
Corolla (CE100)	2,10
AC	+0,10
PAS	+0,40
Corolla (CE110/2C-E)	2,10
AC	+0,10
PAS	+0,40
Camry II (CV10)	1,70
Camry II (CV20)	1,90
Carina (CT150) →1988	1,80
Carina (CT170) 1989→	2,00
Carina E (CT190) →1995	2,30
Carina E (CT190) 1996→	2,40
AC	+0,10
Avensis	2,40
AC	+0,10
Lite-Ace D	1,80
AC	+0,10

Special Tools

■ None required.

Special Precautions

- ■ Disconnect battery earth lead.
- ■ DO NOT turn crankshaft or camshaft when timing belt removed.
- ■ Remove glow plugs to ease turning engine.
- ■ Turn engine in normal direction of rotation (unless otherwise stated).
- ■ DO NOT turn engine via camshaft or other sprockets.
- ■ Observe all tightening torques.
- ■ Check diesel injection pump timing after belt replacement.

Removal

1. Remove:
   - ❏ Auxiliary drive belt(s).
   - ❏ PAS pump.
   - ❏ Cooling fan (RWD).
   - ❏ Timing belt upper cover **1**.
2. Turn crankshaft clockwise until timing marks aligned **2** & **3**.
3. Remove:
   - ❏ Crankshaft pulley bolt **9**.
   - ❏ Crankshaft pulley **5**.
   - ❏ Timing belt lower cover **6**.
4. Ensure timing marks aligned **4**.
5. Support engine (FWD).
6. Remove:
   - ❏ RH engine mounting (FWD).
   - ❏ Cooling fan bracket (RWD).
   - ❏ Engine mounting bracket (FWD).
   - ❏ Tensioner spring **7**.
7. Slacken tensioner bolt **8**.
8. Remove timing belt.

Installation

1. Check free length of tensioner spring is 51,93 mm. Replace spring if necessary.
2. Ensure timing marks aligned **2** & **3**.
3. Fit timing belt.
   NOTE: Ensure numbers/letters on belt can be read from rear of engine.
4. Fit tensioner spring **7**.
5. Turn crankshaft two turns clockwise until timing marks aligned **2** & **3**.
6. Tighten tensioner bolt **8**. Tightening torque: 37 Nm.
7. Install components in reverse order of removal.
8. Tighten crankshaft pulley bolt **9**.
   - ❏ Except Carina E/Corolla (2C-E)/Avensis: Tightening torque: 98 Nm.
   - ❏ Carina E/Corolla (2C-E)/Avensis: Tightening torque: 196 Nm.

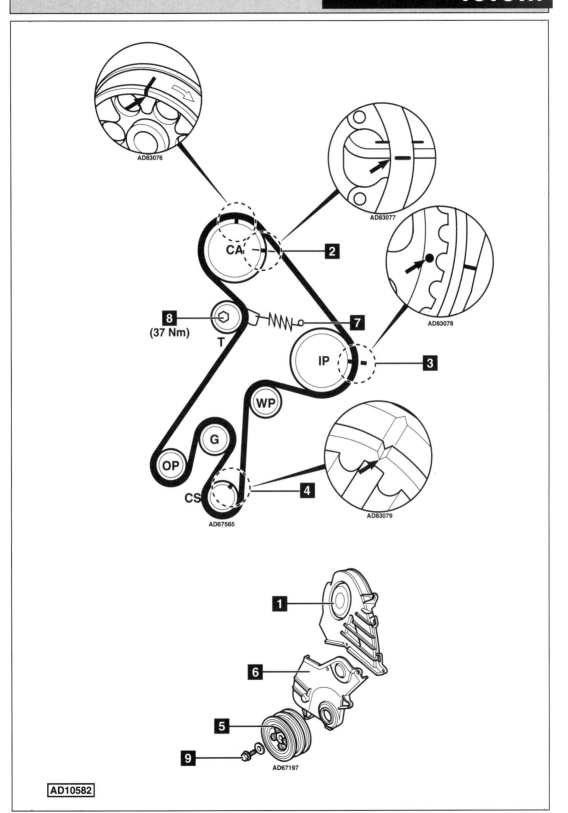

AD83076

AD83077

AD83078

CA

2

8
(37 Nm)
T

7

3

IP

WP

G

OP

CS

AD67565

AD83079

4

1

6

5

9

AD67197

AD10582

Model:	Avensis 2,0 TD • Avensis Verso 2,0 TD • Corolla 2,0 TD Corolla Verso 2,0 TD
Year:	1992-03
Engine Code:	1CD-FTV

Replacement Interval Guide

Toyota recommend replacement every 60,000 miles or 5 years.
The previous use and service history of the vehicle must always be taken into account.
Refer to Timing Belt Replacement Intervals at the front of this manual.

Check For Engine Damage

CAUTION: This engine has been identified as an INTERFERENCE engine in which the possibility of valve-to-piston damage in the event of a timing belt failure is MOST LIKELY to occur. A compression check of all cylinders should be performed before removing the cylinder head(s).

Repair Times – hrs

Remove & install:

Corolla	2,00
With PAS	+0,30
Avensis	2,40

Special Tools

- Crankshaft pulley holding tool – Toyota No.09213-54015.
- Puller – Toyota No.09950-50012.

Special Precautions

- Disconnect battery earth lead.
- DO NOT turn crankshaft or camshaft when timing belt removed.
- Remove glow plugs to ease turning engine.
- Turn engine in normal direction of rotation (unless otherwise stated).
- DO NOT turn engine via camshaft or other sprockets.
- Observe all tightening torques.
- Check diesel injection pump timing after belt replacement.

Removal

1. Remove auxiliary drive belt.
2. Hold crankshaft pulley.
   Use tool No.09213-54015.
3. Remove crankshaft pulley bolt **1**.
4. Remove crankshaft pulley **2**.
   Use tool No.09950-50012.
5. Remove:
   - ❏ Timing belt covers **3** & **4**.
   - ❏ Crankshaft sprocket guide washer **5**.
6. Support engine.

7. Remove:
   - ❏ RH engine mounting.
   - ❏ RH engine mounting bracket **6**.
8. Temporarily fit crankshaft pulley bolt **1**.
9. Turn crankshaft clockwise to TDC on No.1 cylinder. Ensure timing marks aligned **7**, **8** & **9**.
10. Remove:
    - ❏ Automatic tensioner unit bolts **10** & **11**.
    - ❏ Automatic tensioner unit **12**.
    - ❏ Timing belt.
    NOTE: Mark direction of rotation on belt with chalk if belt is to be reused.

Installation

1. Check and reset automatic tensioner unit as follows:
   - ❏ Check tensioner body for leakage or damage. Replace if necessary.
   - ❏ Keep automatic tensioner unit upright. Push pushrod against a firm surface **13**. If pushrod moves: Replace automatic tensioner unit.
   - ❏ Check pushrod protrusion is 9-10,6 mm **14**. If not: Replace automatic tensioner unit.
   - ❏ Slowly compress pushrod into tensioner body until holes aligned **15**. Use a press **16**.
   - ❏ Retain pushrod with suitable pin through hole in tensioner body **17**.
   NOTE: DO NOT exceed 1000 kg force.
2. Ensure timing marks aligned **7**, **8** & **9**.
3. Fit timing belt in clockwise direction, starting at camshaft sprocket. Ensure belt is taut on non-tensioned side.
4. Install automatic tensioner unit to cylinder block **12**.
5. Fit automatic tensioner unit bolt **11**. Tighten bolt finger tight.
6. Turn automatic tensioner unit clockwise. Fit automatic tensioner unit bolt **10**. Tighten bolt finger tight.
7. Tighten bolts evenly to 21 Nm **10** & **11**.
8. Remove pin from tensioner body **17**.
9. Turn crankshaft slowly two turns clockwise to TDC on No.1 cylinder. Ensure timing marks aligned **7**, **8** & **9**.
10. If not: Repeat installation and tensioning procedures.
11. Remove crankshaft pulley bolt **1**.
12. Install components in reverse order of removal.
13. Tighten crankshaft pulley bolt **1**.
    Tightening torque: 180 Nm.

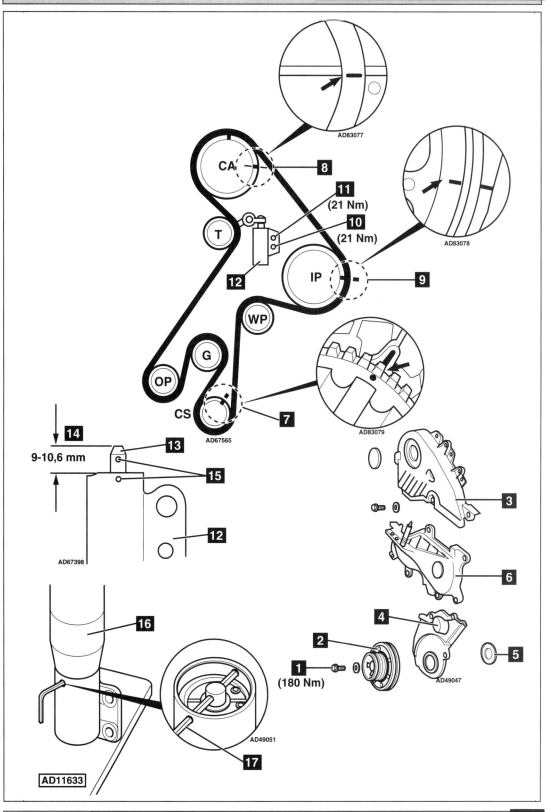

AD83077

CA

8

11
(21 Nm)

10
(21 Nm)

AD83078

T

12

IP

9

WP

G

OP

CS

7

AD67565

AD83079

14

9-10,6 mm

13

15

12

AD67398

16

4

5

2

1
(180 Nm)

AD49047

AD49051

17

AD11633

TOYOTA

Model:	Picnic 2,2 TD
Year:	1998-01
Engine Code:	3C-TE

Replacement Interval Guide

Toyota recommend replacement every 63,000 miles or 5 years.
The previous use and service history of the vehicle must always be taken into account.
Refer to Timing Belt Replacement Intervals at the front of this manual.

Check For Engine Damage

CAUTION: This engine has been identified as an INTERFERENCE engine in which the possibility of valve-to-piston damage in the event of a timing belt failure is MOST LIKELY to occur. A compression check of all cylinders should be performed before removing the cylinder head(s).

Repair Times – hrs

Remove & install	1,70
AC	+0,10

Special Tools

- Crankshaft pulley holding tool – Toyota No.09213-54015.
- Handle – Toyota No.09330-00021.
- Puller – Toyota No.09950-50010.

Special Precautions

- Disconnect battery earth lead.
- DO NOT turn crankshaft or camshaft when timing belt removed.
- Remove glow plugs to ease turning engine.
- Turn engine in normal direction of rotation (unless otherwise stated).
- DO NOT turn engine via camshaft or other sprockets.
- Observe all tightening torques.
- Check diesel injection pump timing after belt replacement.

Removal

1. Remove:
   - ❏ PAS pump hoses (collect fluid).
   - ❏ Auxiliary drive belt(s).
2. Slacken crankshaft pulley bolt **1**. Use tool Nos.09213-54015 and 09330-00021.
3. Remove:
   - ❏ Crankshaft pulley bolt **1**.
   - ❏ Crankshaft pulley **2**.
     Use tool No.09950-50010.
   - ❏ Timing belt covers **3** & **4**.
   - ❏ Crankshaft sprocket guide washer **5**.

4. Support engine.
5. Remove:
   - ❏ RH engine mounting.
   - ❏ RH engine mounting bracket **6**.
6. Turn crankshaft clockwise until timing marks aligned **7**.
7. Ensure camshaft sprocket timing marks aligned with cylinder head face **8**.
8. Ensure injection pump timing marks aligned **9**.
9. Slacken tensioner bolt **10**.
10. Remove:
    - ❏ Tensioner spring **11**.
    - ❏ Timing belt.

Installation

1. Check free length of tensioner spring is 52 mm **11**. Replace spring if necessary.
2. Ensure timing marks aligned **7**, **8** & **9**.
3. Fit timing belt in anti-clockwise direction, starting at crankshaft sprocket. Ensure belt is taut on non-tensioned side.
4. Fit tensioner spring **11**. Ensure tensioner free to move.
5. Turn crankshaft two turns clockwise until timing marks aligned **7**, **8** & **9**.
6. Tighten tensioner bolt **10**.
   Tightening torque: 37 Nm.
7. Install components in reverse order of removal.
8. Tighten crankshaft pulley bolt **1**.
   Tightening torque: 196 Nm.

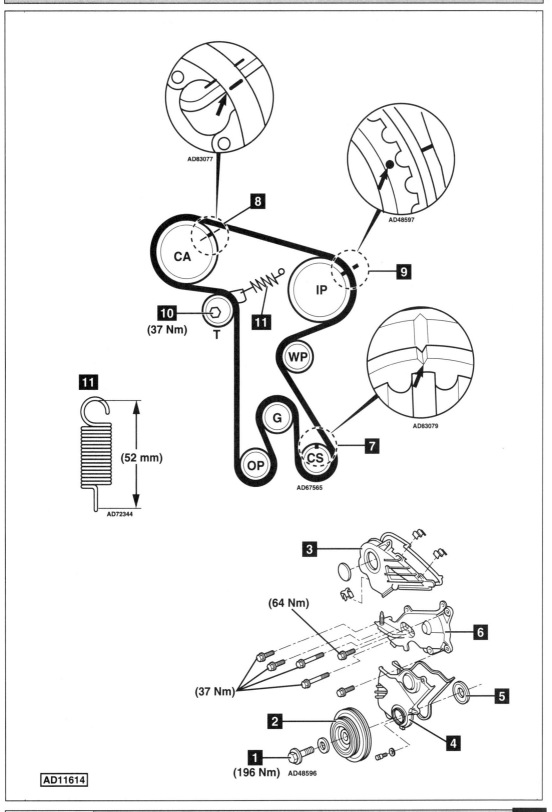

AD83077

AD48597

8

CA

9

AD48597

IP

10
(37 Nm)

11

T

WP

AD83079

11

(52 mm)

G

AD72344

7

OP

CS

AD67565

3

(64 Nm)

6

(37 Nm)

5

2

4

1

(196 Nm)   AD48596

AD11614

Model:	**4-Runner 3,0 TD • Landcruiser 3,0 TD • Landcruiser Colorado 3,0 TD** **Landcruiser 4,2 TD • Landcruiser Amazon 4,2 TD**
Year:	**1993-03**
Engine Code:	**1KZ-T, 1KZ-TE, 1HD-FT, 1HD-FTE**

Replacement Interval Guide

Toyota recommend:
→1993: Replacement every 60,000 miles or 5 years, whichever occurs first.
1994→: Replacement every 63,000 miles or 5 years, whichever occurs first.
1996→: Replacement every 60,000 miles or 5 years, whichever occurs first (Colorado only).
The previous use and service history of the vehicle must always be taken into account.
Refer to Timing Belt Replacement Intervals at the front of this manual.

Check For Engine Damage

CAUTION: This engine has been identified as an INTERFERENCE engine in which the possibility of valve-to-piston damage in the event of a timing belt failure is MOST LIKELY to occur.
A compression check of all cylinders should be performed before removing the cylinder head.

Repair Times – hrs

Remove & install:	
4-Runner	1,00
Landcruiser (1KZ-T/TE)	1,20
AC	0,20
Landcruiser (1HD-FT/1HD-FTE)	1,10

Special Tools

■ None required.

Special Precautions

■ Disconnect battery earth lead.
■ DO NOT turn crankshaft or camshaft when timing belt removed.
■ Remove glow plugs to ease turning engine.
■ Turn engine in normal direction of rotation (unless otherwise stated).
■ DO NOT turn engine via camshaft or other sprockets.
■ Observe all tightening torques.
■ Check diesel injection pump timing after belt replacement.

Removal

1. Remove timing belt cover **1**.
2. Turn crankshaft clockwise until timing marks aligned **2**.
3. Ensure timing marks aligned:
   ❏ 1KZ-T/TE: **3**.
   ❏ 1HD-FT/1HD-FTE: **4**.
4. Evenly slacken and remove automatic tensioner unit bolts **5**.

5. Remove:
   ❏ Automatic tensioner unit **6**.
   ❏ Tensioner pulley bolt **7**.
   ❏ Tensioner pulley **8**.
   ❏ Timing belt.

Installation

1. Check tensioner pulley for smooth operation **8**. Replace if necessary.
2. Remove dust cover. Check tensioner body for leakage or damage **6**. Check pushrod protrusion is 9,0-9,8 mm **9**.
3. Check pushrod does not move when pushed against a firm surface. If pushrod moves: Replace automatic tensioner unit.
4. Slowly compress pushrod into tensioner body until holes aligned. Use press.
5. Retain pushrod with 1,5 mm Allen key through hole in tensioner body **10**. Fit dust cover.
6. Ensure timing marks aligned **2**.
7. Ensure timing marks aligned:
   ❏ 1KZ-T/TE: **3**.
   ❏ 1HD-FT/1HD-FTE: **4**.
8. Fit timing belt in anti-clockwise direction, starting at lower sprocket. Ensure belt is taut between sprockets.
9. Fit tensioner pulley **8**. Tighten bolt to 34 Nm **7**.
10. Push tensioner pulley against belt. Install automatic tensioner unit **6**. Tighten bolts evenly to 13 Nm **5**.
11. Remove Allen key from tensioner body to release pushrod.
12. Turn crankshaft two turns clockwise. Ensure timing marks aligned **2** & **3** or **4**.
13. If not: Repeat installation and tensioning procedures.
14. Install components in reverse order of removal.
15. Reset timing belt replacement warning lamp as follows:

4-Runner/Landcruiser:
   ❏ Remove grommet in speedometer housing. Depress switch using a thin screwdriver **11**.
Colorado:
   ❏ Remove speedometer housing. Remove screw indicated and insert in alternative location **12**.
Amazon:
   ❏ Lamp illuminates at 90,000 miles. Return vehicle to dealer.

Autodata

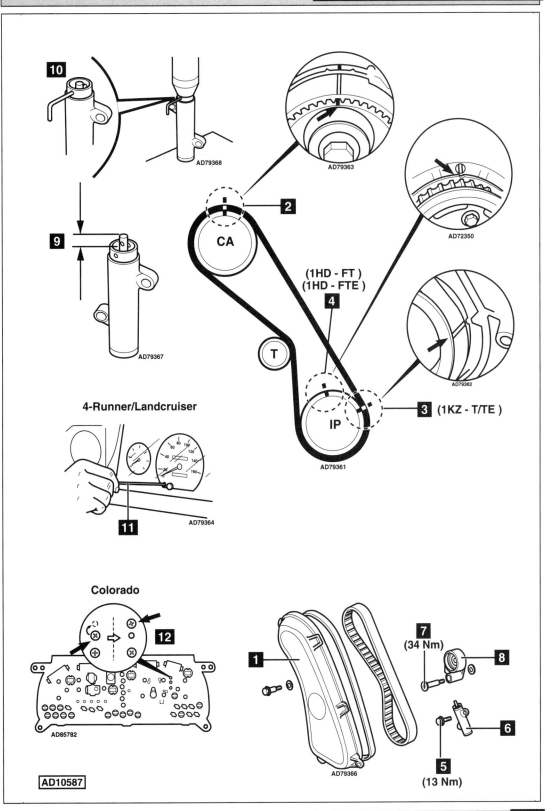

10

AD79368

9

AD79367

2

CA

AD79363

AD72350

(1HD - FT)
(1HD - FTE)

4

AD79362

3 (1KZ - T/TE)

T

IP

AD79361

4-Runner/Landcruiser

11

AD79364

Colorado

12

AD85782

AD10587

1

7
(34 Nm)

8

6

5
(13 Nm)

AD79366

TOYOTA

Model:	Hi-Lux 2,4D/TD • Hi-Ace 2,4D (LH61) • Hi-Ace PowerVan 2,4D/TD (LXH12) Dyna 100/150 • Landcruiser 2,4 TD
Year:	1985-02
Engine Code:	2L, 2L-T, 3L

Replacement Interval Guide

Toyota recommend:
6,000 mile service intervals – replacement every 60,000 miles or 5 years, whichever occurs first.
9,000 mile service intervals – replacement every 63,000 miles or 5 years, whichever occurs first.
The previous use and service history of the vehicle must always be taken into account.
Refer to Timing Belt Replacement Intervals at the front of this manual.

Check For Engine Damage

CAUTION: This engine has been identified as an INTERFERENCE engine in which the possibility of valve-to-piston damage in the event of a timing belt failure is MOST LIKELY to occur.
A compression check of all cylinders should be performed before removing the cylinder head.

Repair Times – hrs

Remove & install:

Dyna →1995	2,00
Dyna 100 1996→	2,60
Dyna 150 1996→	2,70
PAS	+0,20
AC	+0,30
Hi-Lux D	1,80
Hi-Lux TD	2,00
PAS	+0,20
AC	+0,50
Hi-Ace →1995	2,40
PAS	+0,20
Hi-Ace D 1995→	2,70
Hi-Ace TD 1995→	2,90
Landcruiser	2,00
PAS	+0,10
AC	+0,40

Special Tools

- Crankshaft pulley drift – Toyota No.09223-63010.
- Crankshaft pulley holding tool – Toyota No.09213-54015 & 09330-00021.
- Crankshaft pulley puller – Toyota No.09213-60017.

Special Precautions

- Disconnect battery earth lead.
- DO NOT turn crankshaft or camshaft when timing belt removed.
- Remove glow plugs to ease turning engine.
- Turn engine in normal direction of rotation (unless otherwise stated).
- DO NOT turn engine via camshaft or other sprockets.
- Observe all tightening torques.
- Check diesel injection pump timing after belt replacement.

Removal

1. Drain coolant.
2. Remove:
   - ❑ Radiator.
   - ❑ Cooling fan and pulley.
   - ❑ Auxiliary drive belts.
   - ❑ Crankshaft pulley bolt and washer **1**.
     Use tool Nos.09213-54015 & 09330-00021.
   - ❑ Crankshaft pulley **2**.
     Use tool No.09213-60017 or suitable puller.
   - ❑ Timing belt cover bolts **3**.
   - ❑ Timing belt cover and seals **4** & **5**.
3. Fit crankshaft pulley bolt.
4. Turn crankshaft to TDC on No.1 cylinder. Ensure timing marks aligned **6**, **7** & **8**.
5. Slacken tensioner bolt **9**.
6. Lever tensioner away from belt using screwdriver and lightly tighten bolt.
7. Remove:
   - ❑ Crankshaft sprocket guide washer.
   - ❑ Timing belt.

Installation

1. Ensure timing marks aligned **6**, **7** & **8**.
2. Fit timing belt in following order:
   - ❑ Crankshaft sprocket.
   - ❑ Tensioner pulley.
   - ❑ Injection pump sprocket.
   - ❑ Camshaft sprocket.
   - ❑ Guide sprocket.
   NOTE: Ensure belt is taut between sprockets.
3. Slacken tensioner bolt **9**.
4. Turn crankshaft four turns clockwise. Ensure timing marks aligned **6**, **7** & **8**.
5. Tighten tensioner bolt **9**.
   Tightening torque: 44 Nm.

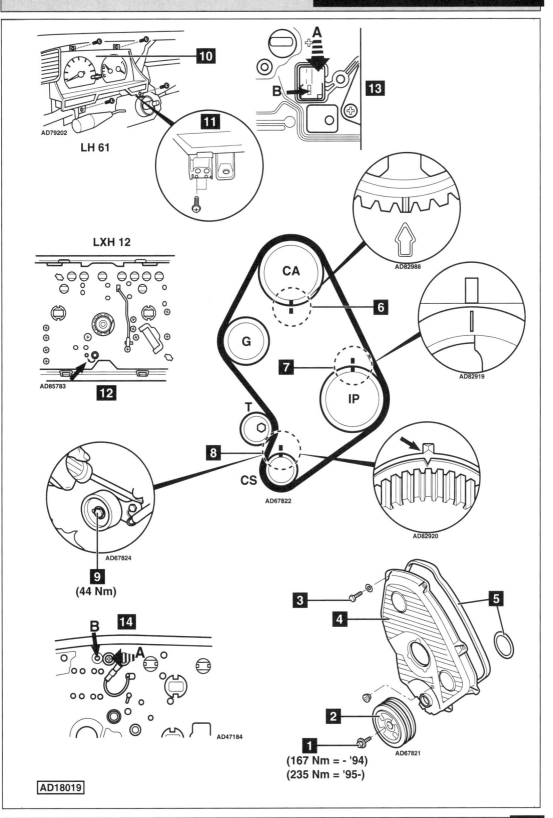

LH 61

AD79202

10

11

LXH 12

AD85783

12

9
(44 Nm)

AD67824

B

A

14

AD47184

A

B

13

CA

AD82988

6

7

G

IP

AD82919

T

8

CS

AD67822

AD82920

3

4

5

2

AD67821

1
(167 Nm = - '94)
(235 Nm = '95-)

AD18019

←

6. Install:
   - ❑ Crankshaft sprocket guide washer.
   - ❑ Timing belt cover and seal **4** & **5**.

7. Remove crankshaft pulley bolt **1**.

8. Fit crankshaft pulley **2**.
   Use tool No.09223-63010. Tap crankshaft pulley into place.

9. Fit crankshaft pulley bolt and washer **1**.

10. Hold crankshaft pulley **2**.
    Use tool Nos.09213-54015 & 09330-00021.
    Tighten crankshaft pulley bolt **1**.
    →1994: 167 Nm. 1995→: 235 Nm.

11. Install components in reverse order of removal.

12. Refill cooling system.

13. Timing belt replacement warning lamp will illuminate at approximately 60,000 miles. After illumination. Reset as follows:
    - ❑ Landcruiser/Dyna: Remove grommet from front of speedometer. Press reset switch.
    - ❑ Hi-Ace (LH61): Remove instrument panel **10**. Remove warning lamp switch screw **11** and transfer to next hole. Check warning lamp goes out. Fit instrument panel.
    - ❑ Hi-Ace (LXH12): Remove speedometer housing. Remove screw indicated and insert in alternative location **12**. Fit speedometer housing.
    - ❑ Hi-Lux:
    - ❑ Remove instrument panel.
    - ❑ Without tachometer: Move connector from position 'A' to 'B' or 'B' to 'A' **13**.
    - ❑ With tachometer: Remove screw indicated. Move screw from position 'A' to 'B' or 'B' to 'A' **14**.
    - ❑ Fit instrument panel.

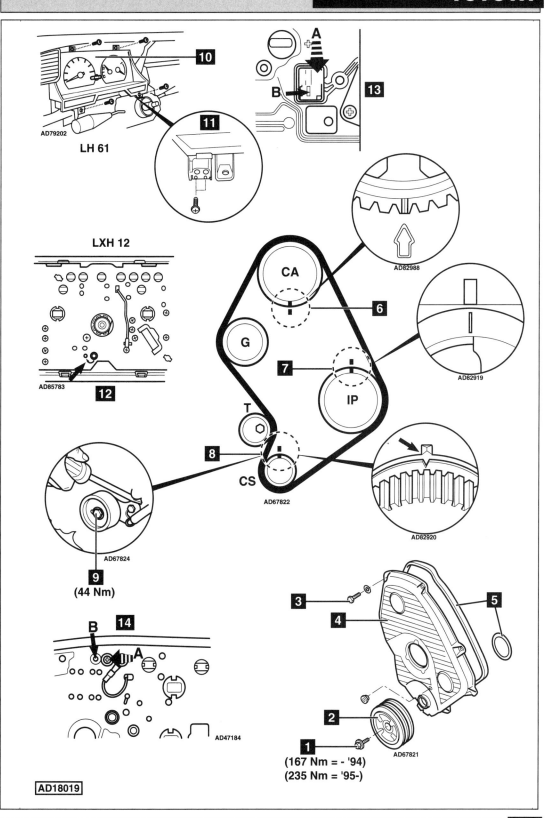

AD79202

LH 61

10

11

13

AD82988

LXH 12

AD85783

12

CA

6

G

7

AD82919

IP

T

8

CS

AD67822

AD82920

AD67824

9
(44 Nm)

B

14

A

AD47184

3

4

5

2

1

AD67821

(167 Nm = - '94)
(235 Nm = '95-)

AD18019

Model:	Nova 1,2/1,4/1,6 • Nova GTE • Corsa-A 1,2/1,4/1,6 • Corsa-A GSi Corsa-B/Tigra 1,4/1,6 • Astra/Belmont 1,4/1,6 • Kadett-E 1,4/1,6 Astra Van 1,4/1,6 • Cavalier 1,4/1,6 • Vectra-A 1,4/1,6
Year:	1991-00
Engine Code:	Additional automatic tensioner C12NZ, 14NV, C14NZ, C14SE/L, 16SV, E16NZ, E16SE, C16NZ, C16SE/L

Replacement Interval Guide

Vauxhall recommend replacement as follows:

Nova/Corsa-A:
Check and replacement if necessary every 18,000 miles.

All models:
→94 MY - replacement every 36,000 miles or 4 years, whichever occurs first.
95-96 MY - replacement every 40,000 miles or 4 years, whichever occurs first.
97 MY→ - replacement every 80,000 miles or 8 years, whichever occurs first.
The previous use and service history of the vehicle must always be taken into account.
Refer to Timing Belt Replacement Intervals at the front of this manual.

Check For Engine Damage – 1,2

CAUTION: This engine has been identified as an INTERFERENCE engine in which the possibility of valve-to-piston damage in the event of a timing belt failure is MOST LIKELY to occur.
A compression check of all cylinders should be performed before removing the cylinder head.

Check For Engine Damage – 1,4/1,6

CAUTION: This engine has been identified as a FREEWHEELING engine in which the possibility of valve-to-piston damage in the event of a timing belt failure may be minimal or very unlikely. However, a precautionary compression check of all cylinders should be performed.

Repair Times – hrs

Nova/Corsa-A	
Remove & install	0,90
1,6i	+0,30
Corsa-B/Tigra	
Remove & install	0,90
Astra/Astra Van/Belmont/Kadett	
Remove & install	0,90
PAS	+0,20
Cavalier/Vectra-A	
Remove & install	1,00
Fuel injection models	+0,20
PAS	+0,20

Special Tools

■ Water pump wrench – Kent Moore No.KM-421-A.

Removal

1. Remove:
   ❏ Air filter and air intake hose (if fitted).
   ❏ Auxiliary drive belt(s).
   ❏ Timing belt upper cover **1**.
2. Turn crankshaft clockwise until timing marks aligned **3** & **4**.
   *NOTE: C14SE/L, X14SZ & C16SE/L: Timing mark is on toothed reluctor disc **5**. 14NV: Align crankshaft pulley with first mark at 10° BTDC **6**.*
3. Remove:
   ❏ Crankshaft pulley **13**.
   ❏ Timing belt lower cover **2**.
4. Slacken water pump bolts.
5. Turn movable part of tensioner until holes aligned **7**. Insert suitable pin to hold tensioner.
6. Remove timing belt.

Installation

NOTE: Ensure engine is cold before adjusting timing belt.

1. Temporarily fit timing belt lower cover and crankshaft pulley **2** & **13**.
   NOTE: C14SE/L, X14SZ & C16SE/L: Temporarily fit toothed reluctor disc.
2. Ensure timing marks aligned **3** & **4**, **5** or **6**.
3. Install:
   ❏ Timing belt.
   ❏ Crankshaft pulley.
   ❏ Crankshaft pulley bolt **8**.
4. Tighten crankshaft pulley bolt **8**.
   Tightening torque: M10 x 23 mm: 55 Nm.
   M10 x 30 mm: 55 Nm + 45-60°.
   M12: 95 Nm + 30° + 15°.
5. Remove locking pin from tensioner.
6. Turn water pump clockwise to tension belt **9**.
   Use tool No.KM-421-A.
7. Movable part of tensioner must be against stop **10**.
8. Turn crankshaft two turns clockwise until timing marks aligned **3** & **4**, **5** or **6**.
   NOTE: Position of water pump must not alter when turning crankshaft.
9. Turn water pump anti-clockwise **11** until tensioner pointer aligned with notch in support plate **12**.
   Use tool No.KM-421-A.
10. Tighten water pump bolts. Tightening torque: 8 Nm.
11. If tensioner pointer not aligned: Repeat tensioning procedure.
12. Install components in reverse order of removal.

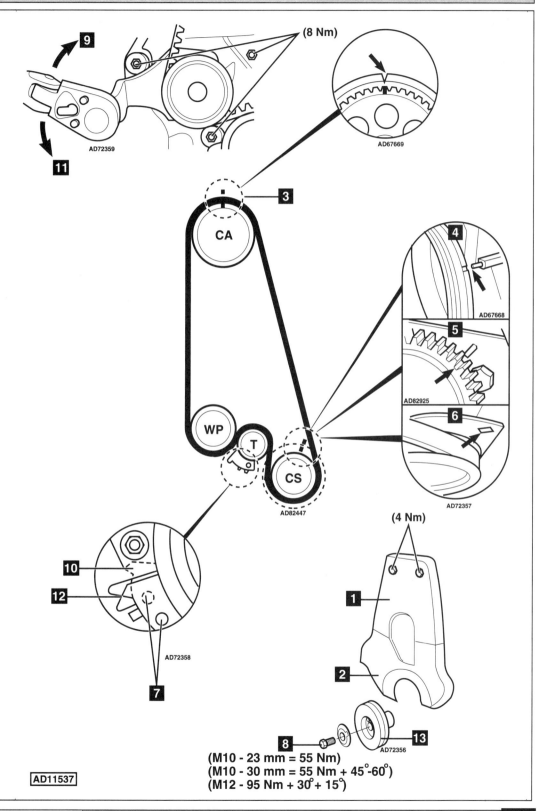

(8 Nm)

AD72359

AD67669

3

CA

4

AD67668

5

AD82925

6

AD72357

WP

T

CS

AD82447

AD72358

10

12

7

(4 Nm)

1

2

8

13

AD72356

(M10 - 23 mm = 55 Nm)
(M10 - 30 mm = 55 Nm + 45°-60°)
(M12 - 95 Nm + 30° + 15°)

AD11537

Model:	Corsa-B 1,2/1,4 • Corsa Van/Combo 1,4 • Astra-F 1,4/1,6 • Astra Van 1,6
Year:	1993-00
Engine Code:	12NZ, C12NZ, C14NZ, C14SE, X12SZ, X14NZ, X14SZ, X16SZ, X16SZR

Replacement Interval Guide

Vauxhall recommend replacement as follows:
→94MY - replacement every 36,000 miles or 4 years, whichever occurs first.
95-96MY - replacement every 40,000 miles or 4 years, whichever occurs first.
97-98MY - replacement every 80,000 miles or 8 years, whichever occurs first.
99MY→ - replacement every 80,000 miles or 8 years, whichever occurs first (X16SZR - tensioner pulley must also be replaced).
The previous use and service history of the vehicle must always be taken into account.
Refer to Timing Belt Replacement Intervals at the front of this manual.

Check For Engine Damage - 1,2/1,6 engines

CAUTION: This engine has been identified as an INTERFERENCE engine in which the possibility of valve-to-piston damage in the event of a timing belt failure is MOST LIKELY to occur.
A compression check of all cylinders should be performed before removing the cylinder head.

Check For Engine Damage - 1,4 engines

CAUTION: This engine has been identified as a FREEWHEELING engine in which the possibility of valve-to-piston damage in the event of a timing belt failure may be minimal or very unlikely. However, a precautionary compression check of all cylinders should be performed.

Repair Times – hrs

Remove & install:
Corsa-B	0,90
Astra-F/Van	1,10
Corsa Van/Combo	0,90

Special Tools

■ Water pump wrench – Kent Moore No.KM-421-A.

Special Precautions

■ Disconnect battery earth lead.
■ DO NOT turn crankshaft or camshaft when timing belt removed.
■ Remove spark plugs to ease turning engine.
■ Turn engine in normal direction of rotation (unless otherwise stated).
■ DO NOT turn engine via camshaft or other sprockets.
■ Observe all tightening torques.

Removal

1. Remove:
   ❏ Air filter, air duct and air intake hose.
   ❏ C14SE: Disconnect intake air temperature (IAT) sensor multi-plug.
   ❏ Auxiliary drive belt.
   ❏ Timing belt upper cover **1**.
   ❏ Crankshaft pulley bolt **2**.
   ❏ Crankshaft pulley **3**.
   ❏ Timing belt lower cover **4**.
2. Fit crankshaft pulley bolt. Turn crankshaft to TDC on No.1 cylinder.
3. Ensure timing marks aligned **5** & **6**.
4. Turn movable part of tensioner until holes aligned **7**. Insert suitable pin to hold tensioner.
5. Remove timing belt.

Installation

1. Ensure timing marks aligned **5** & **6**.
2. Fit timing belt in anti-clockwise direction, starting at crankshaft sprocket. Ensure belt is taut between sprockets.
3. Remove locking pin from tensioner **7**.
4. Slacken water pump bolts **8**.
5. Turn water pump clockwise to tension belt **9**. Use tool No.KM-421-A.
6. Movable part of tensioner must be against stop **10**.
7. Turn crankshaft two turns clockwise to TDC on No.1 cylinder. Ensure timing marks aligned **5** & **6**.
   NOTE: Position of water pump must not alter when turning crankshaft.
8. Turn water pump anti-clockwise **11** until tensioner pointer aligned with notch in support plate **12**. Use tool No.KM-421-A.
9. Tighten water pump bolts to 8 Nm **8**.
10. If tensioner pointer not aligned: Repeat tensioning procedure.
11. Install components in reverse order of removal.
12. Fit new crankshaft pulley bolt **2**.
    Tightening torque:
    ❏ Corsa-B/Combo: 95 Nm +30° +15°.
    ❏ Astra-F/Van: 95 Nm +45° +15°.

Autodata

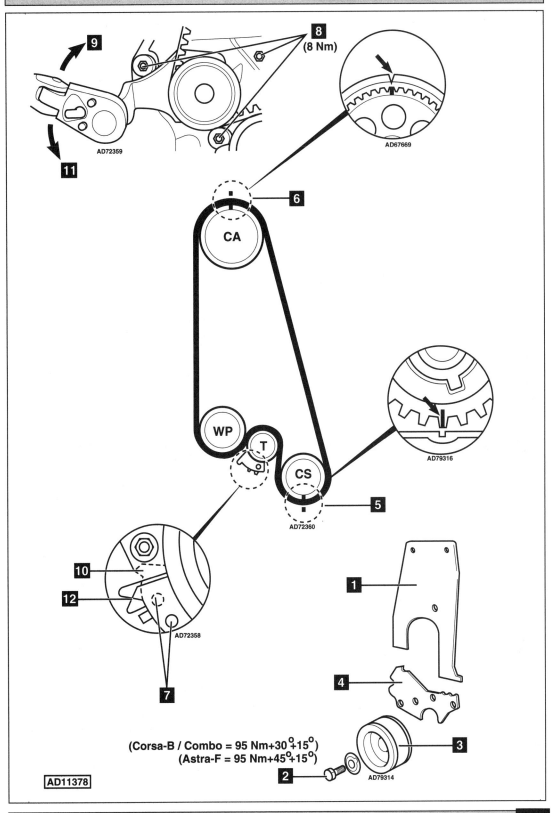

8 (8 Nm)

AD72359

AD67669

9

11

6

CA

AD79316

WP

T

CS

5

AD72360

10

12

AD72358

7

1

4

(Corsa-B / Combo = 95 Nm+30°+15°)
(Astra-F = 95 Nm+45°+15°)

2

AD79314

3

AD11378

VAUXHALL-OPEL

Model:	Corsa-B 1,4/1,6 16V • Astra-F 1,4/1,6 16V • Tigra 1,4/1,6 16V
Year:	1993-00
Engine Code:	X14XE, X16XE, X16XEL, C16XE

Replacement Interval Guide

Vauxhall recommend replacement as follows:
→94MY - replacement every 36,000 miles or
4 years, whichever occurs first.
95-96MY - replacement every 40,000 miles or
4 years, whichever occurs first.
97MY→ - replacement every 40,000 miles or
4 years, whichever occurs first (tensioner pulley
must also be replaced).
*The previous use and service history of the vehicle
must always be taken into account.*
*Refer to Timing Belt Replacement Intervals at the front
of this manual.*

Check For Engine Damage

*CAUTION: This engine has been identified as an
INTERFERENCE engine in which the possibility of
valve-to-piston damage in the event of a timing belt
failure is MOST LIKELY to occur.*
*A compression check of all cylinders should be
performed before removing the cylinder head.*

Repair Times – hrs

Remove & install:

Corsa-B/Tigra	1,30
Astra-F	1,10

Special Tools

- None required.

Special Precautions

- Disconnect battery earth lead.
- DO NOT turn crankshaft or camshaft when timing belt removed.
- Remove spark plugs to ease turning engine.
- Turn engine in normal direction of rotation (unless otherwise stated).
- DO NOT turn engine via camshaft or other sprockets.
- Observe all tightening torques.

Removal

*WARNING: Certain engines require modification
to tensioner pulley/guide pulleys due to possible
failure. Refer to dealer.*

1. Mark direction of rotation on auxiliary drive belt with chalk.
2. Turn auxiliary drive belt tensioner away from belt to release tension on belt. Use ring spanner.
3. Insert 4 mm pin through hole in tensioner and into mounting bracket **1**.
4. Remove auxiliary drive belt.
5. C16XE: Disconnect mass air flow (MAF) sensor multi-plug.
6. Remove:
   - ❏ Air intake trunking.
   - ❏ Air filter and housing.
   - ❏ C16XE: Mass air flow (MAF) sensor.
   - ❏ Timing belt upper cover **2**.
   - ❏ Crankshaft pulley bolt **3**.
   - ❏ Crankshaft pulley.
   - ❏ Timing belt lower cover **4**.
7. Temporarily fit crankshaft pulley bolt **3**.
8. Turn crankshaft clockwise until timing marks aligned **5** & **6**.
9. Ensure timing marks on camshaft sprockets aligned with upper edge of cylinder head.
10. Slacken tensioner bolt **7**. Turn tensioner clockwise until pointer on LH stop **10**.
11. Tighten tensioner bolt **7**.
12. Remove timing belt.
    NOTE: Tensioner should be carefully examined for wear or damage. Replace if necessary.

Installation

1. Ensure lug on water pump aligned with corresponding lug on cylinder block **8**.
2. Ensure timing marks aligned **5** & **6**.
3. Fit timing belt in anti-clockwise direction, starting at crankshaft sprocket. Ensure belt is taut between sprockets.
4. Slacken tensioner bolt **7**.
5. Turn tensioner anti-clockwise until pointer on RH stop **11**. Use Allen key **9**. Lightly tighten bolt **7**.
6. Turn crankshaft two turns clockwise. Ensure timing marks aligned **5** & **6**.
7. Slacken tensioner bolt **7**.
8. Turn tensioner clockwise until pointer aligned with:
   'V' notch in bracket (new belt) **12**.
   LH edge of 'V' notch (used belt) **13**.
   NOTE: Drive side of timing belt must remain taut during adjustment procedure.
9. Tighten tensioner bolt to 20 Nm **7**.
10. Remove crankshaft pulley bolt **3**.
11. Install components in reverse order of removal.
    NOTE: Observe direction of rotation marks on auxiliary drive belt.
12. Fit new crankshaft pulley bolt **3**.
    Tightening torque: 95 Nm + 30° + 15°.

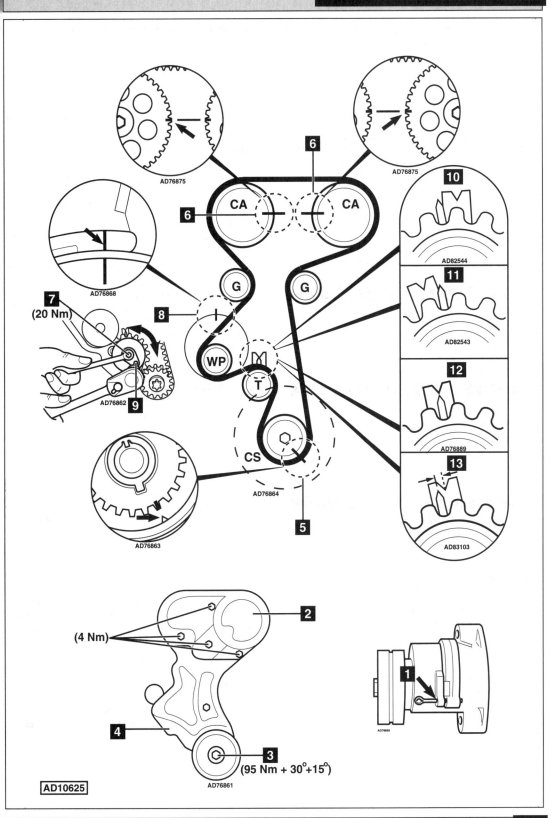

AD76875

AD76875

6

6

CA

CA

10

AD82544

11

AD82543

12

AD76889

13

AD83103

AD76868

G

G

7
(20 Nm)

8

WP

T

AD76862

9

CS

AD76864

5

AD76863

2

(4 Nm)

4

3
(95 Nm + 30° + 15°)

AD76861

1

VAUXHALL-OPEL

Model:	Corsa-C 1,4/1,8 16V • Astra-G 1,4/1,6/1,8 16V • Vectra 1,6/1,8 16V Zafira 1,6/1,8 16V
Year:	1995-03
Engine Code:	X14XE, X16XEL, X16SEJ, X18E1, X18XE1, Y16XE, Z14XE, Z16XE, Z18XE, Z18XEL

Replacement Interval Guide

Vauxhall recommend replacement as follows:
→96MY - replacement every 40,000 miles or 4 years, whichever occurs first.
97MY→ - replacement every 40,000 miles or 4 years, whichever occurs first (tensioner pulley must also be replaced).
The previous use and service history of the vehicle must always be taken into account.
Refer to Timing Belt Replacement Intervals at the front of this manual.

Check For Engine Damage

CAUTION: This engine has been identified as an INTERFERENCE engine in which the possibility of valve-to-piston damage in the event of a timing belt failure is MOST LIKELY to occur.
A compression check of all cylinders should be performed before removing the cylinder head.

Repair Times – hrs

Remove & install:

Corsa-C 1,4	1,30
Corsa-C 1,8	1,40
Astra/Zafira	1,10
Vectra 1,6	0,90
Vectra 1,8	1,00

Special Tools

- Flywheel locking tool – Kent Moore No.KM-911.
- Camshaft sprocket locking tool – Kent Moore No.KM-852.
- Engine support tool (→2000 Astra/Zafira) – Kent Moore No.KM-6001.
- Engine alignment tool (→2000 Astra/Zafira) – Kent Moore No.KM-909-B.
- Engine support tool (2000→ Astra/Zafira) – Kent Moore No.KM-6001-A.
- Engine alignment tool (2000→ Astra/Zafira) – Kent Moore No.KM-6173.
- Engine support tool (Corsa) – Kent Moore No.6169-3.
- Engine alignment tool (Corsa) – Kent Moore No.6169.
- Auxiliary drive belt tensioner locking tool – Kent Moore No.KM-6130.

Special Precautions

- Disconnect battery earth lead.
- DO NOT turn crankshaft or camshaft when timing belt removed.
- Remove spark plugs to ease turning engine.
- Turn engine in normal direction of rotation (unless otherwise stated).
- DO NOT turn engine via camshaft or other sprockets.
- Observe all tightening torques.

Removal

WARNING: Certain engines require modification to tensioner pulley/guide pulleys due to possible failure. Refer to dealer.

1. Remove air filter housing.
2. Raise and support front of vehicle.
3. Remove:
   - ❑ RH front wheel.
   - ❑ RH inner wing panel.
   - ❑ Engine top cover (if fitted).
4. Mark direction of rotation on auxiliary drive belt with chalk.
5. Except Corsa/Astra: Turn auxiliary drive belt tensioner clockwise to release tension on belt. Use ring spanner.
6. Corsa/Astra: Turn auxiliary drive belt tensioner anti-clockwise to release tension on belt. Use ring spanner.
7. Insert 4 mm pin or locking tool No.KM-6130 through hole in tensioner and into mounting bracket **1**.
8. Remove:
   - ❑ Auxiliary drive belt.
   - ❑ Timing belt upper cover **2**.
9. →2000 Astra/Zafira: Fit engine support and alignment tools. Tool Nos.KM-6001/KM-909-B.
10. 2000→ Astra/Zafira: Fit engine support and alignment tools. Tool Nos.KM-6001-A/6173.
11. Corsa: Fit engine support and alignment tools. Tool Nos.KM-6169-3/6169.
12. Corsa/Astra/Zafira – remove:
    - ❑ Auxiliary drive belt tensioner.
    - ❑ RH engine mounting and bracket.
13. Insert tool in bell housing to lock flywheel **3**. Tool No.KM-911.

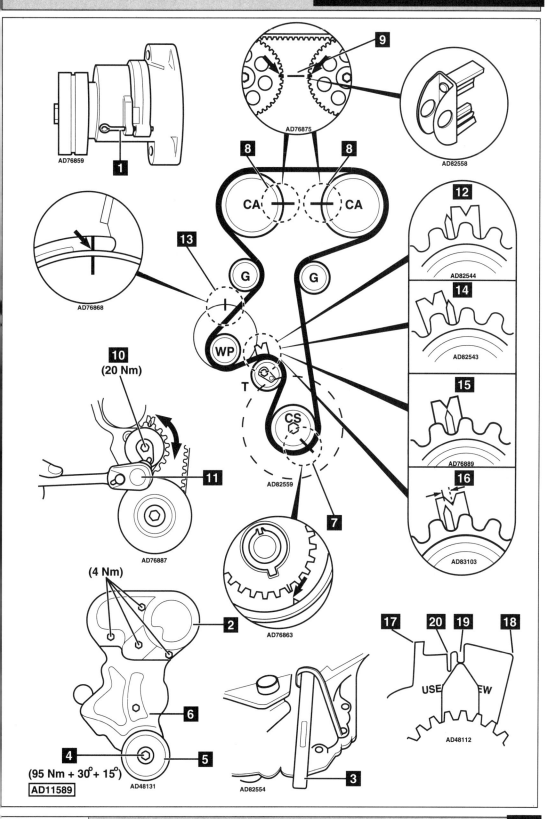

←

14. Remove:
    ❑ Crankshaft pulley bolt **4**.
    ❑ Crankshaft pulley **5**.
    ❑ Timing belt lower cover **6**.
    ❑ Flywheel locking tool **3**.
    ❑ Camshaft position (CMP) sensor (if fitted).

15. Temporarily fit crankshaft pulley bolt **4**.

16. Turn crankshaft clockwise to TDC on No.1 cylinder. Ensure crankshaft timing marks aligned **7**.

17. Timing marks on camshaft sprockets must be aligned with upper edge of cylinder head **8**.

18. Lock camshaft sprockets. Use tool No.KM-852 **9**.

19. Slacken tensioner bolt **10**.

20. Turn tensioner clockwise until pointer on LH stop: Use Allen key **11**.
    ❑ Type 1 – **12**.
    ❑ Type 2 – **17**.

21. Lightly tighten tensioner bolt **10**.

22. Remove timing belt.

Installation

1. Ensure mark on water pump aligned with mark on cylinder block **13**.

2. Ensure timing marks aligned **7** & **8**.

3. Fit timing belt in anti-clockwise direction, starting at crankshaft sprocket. Ensure belt is taut between sprockets.

4. Slacken tensioner bolt **10**.

5. Turn tensioner anti-clockwise until pointer on RH stop: Use Allen key **11**.
    ❑ Type 1 – **14**.
    ❑ Type 2 – **18**.

6. Lightly tighten tensioner bolt **10**.

7. Remove locking tool from sprockets **9**.

8. Turn crankshaft two turns clockwise. Ensure timing marks aligned **7** & **8**.

9. Lock camshaft sprockets. Use tool No.KM-852 **9**.

10. Slacken tensioner bolt **10**.

11. Turn tensioner clockwise until pointer aligned as follows:
    ❑ Type 1:
    ❑ New belt – 'V' notch in bracket **15**.
    ❑ Used belt – LH edge of 'V' notch **16**.
    ❑ Type 2:
    ❑ New belt – 'NEW' notch in bracket **19**.
    ❑ Used belt – 'USED' notch in bracket **20**.

12. Lightly tighten tensioner bolt **10**.

13. Remove locking tool from sprockets **9**.

14. Turn crankshaft slowly two turns clockwise. Ensure timing marks aligned **7** & **8**.

15. Check pointer aligned as follows:
    ❑ Type 1:
    ❑ New belt – 'V' notch in bracket **15**.
    ❑ Used belt – LH edge of 'V' notch **16**.
    ❑ Type 2:
    ❑ New belt – 'NEW' notch in bracket **19**.
    ❑ Used belt – 'USED' notch in bracket **20**.

16. Tighten tensioner bolt to 20 Nm **10**.

17. Insert tool in bell housing to lock flywheel **3**. Tool No.KM-911.

18. Remove crankshaft pulley bolt **4**.

19. Install components in reverse order of removal.
    NOTE: Observe direction of rotation marks on auxiliary drive belt.

20. Fit new crankshaft pulley bolt **4**. Tightening torque: 95 Nm + 30° + 15°.

21. Remove flywheel locking tool **3**.

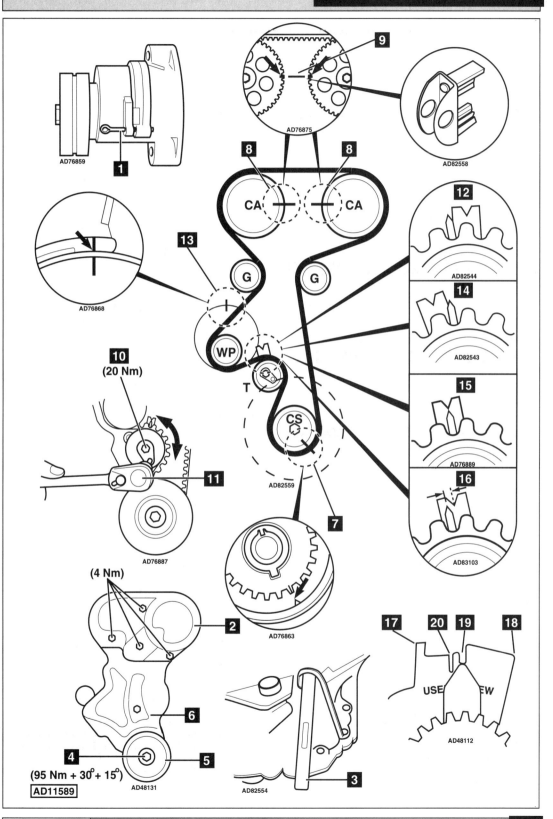

AD76859

9

AD76875

AD82558

8　8

CA — CA

13

G　G

AD76868

WP

T

CS

AD82559

10
(20 Nm)

11

AD76887

7

AD76863

12

AD82544

14

AD82543

15

AD76889

16

AD83103

(4 Nm)

2

6

17　20　19　18

USE ... EW

AD48112

4

5

(95 Nm + 30° + 15°)

AD48131

AD11589

3

AD82554

VAUXHALL-OPEL

Model:	Corsa-C 1,6 • Astra-G 1,6 • Vectra 1,6
Year:	1995-03
Engine Code:	X16SZR, Z16SE

Replacement Interval Guide

Vauxhall recommend replacement as follows:
→96MY - replacement every 40,000 miles or
4 years, whichever occurs first.
97-98MY - replacement every 80,000 miles or
8 years, whichever occurs first.
99MY→ - replacement every 80,000 miles or
8 years, whichever occurs first (tensioner pulley
must also be replaced).
*The previous use and service history of the vehicle
must always be taken into account.*
*Refer to Timing Belt Replacement Intervals at the
front of this manual.*

Check For Engine Damage

*CAUTION: This engine has been identified as an
INTERFERENCE engine in which the possibility of
valve-to-piston damage in the event of a timing belt
failure is MOST LIKELY to occur.
A compression check of all cylinders should be
performed before removing the cylinder head.*

Repair Times – hrs

Remove & install:

Corsa-C	No information available.
Astra-G	0,80
Vectra	0,90

Special Tools

- Flywheel locking tool – Kent Moore No.KM-911.
- Water pump wrench – Kent Moore No.KM-421-A.
- Water pump wrench (Astra 2000→) –
  Kent Moore No.KM-421-B.
- Engine support tool (Astra →2000) –
  Kent Moore No.KM-6001.
- Engine support tool (Astra 2000→) –
  Kent Moore No.KM-6001-A.
- Engine alignment tool (Astra →2000) –
  Kent Moore No.KM-909-B.
- Engine alignment tool (Astra 2000→) –
  Kent Moore No.KM-6173.
- Auxiliary drive belt tensioner locking tool –
  Kent Moore No.KM-6130.

Removal

1. Remove air filter housing.
2. Raise and support front of vehicle.
3. Remove:
   - ❑ RH front wheel.
   - ❑ RH inner wing panel.
4. Mark direction of rotation on auxiliary drive belt with chalk.
5. Except Corsa/Astra 2000→: Turn auxiliary drive belt tensioner clockwise to release tension on belt. Use ring spanner.
6. Corsa/Astra 2000→: Turn auxiliary drive belt tensioner anti-clockwise to release tension on belt. Use ring spanner.
7. Lock tensioner pulley. Use tool No.KM-6130.

8. Remove:
   - ❑ Auxiliary drive belt.
   - ❑ Timing belt upper cover **1**.
9. Astra →2000: Fit engine support and alignment tools. Tool Nos.KM-6001/KM-909-B.
10. Astra 2000→: Fit engine support and alignment tools. Tool Nos.KM-6001-A/KM-6173.
11. Corsa/Astra: Remove:
    - ❑ Auxiliary drive belt tensioner.
12. Astra: Remove:
    - ❑ RH engine mounting and bracket.
13. Insert tool in bell housing to lock flywheel **2**. Tool No.KM-911.
14. Remove:
    - ❑ Crankshaft pulley bolt **3**.
    - ❑ Crankshaft pulley **4**.
    - ❑ Timing belt lower cover **5**.
    - ❑ Flywheel locking tool **2**.
15. Temporarily fit crankshaft pulley bolt.
16. Turn crankshaft to TDC on No.1 cylinder. Ensure timing marks aligned **6**.
17. Ensure camshaft timing mark aligned with notch in timing belt rear cover **7**.
18. If not: Turn crankshaft one turn.
19. Turn movable part of tensioner until holes aligned **8**. Insert suitable pin to hold tensioner.
20. Remove timing belt.

Installation

1. Ensure timing marks aligned **6** & **7**.
2. Fit timing belt in anti-clockwise direction, starting at crankshaft sprocket. Ensure belt is taut between sprockets.
3. Remove locking pin from tensioner **8**.
4. Slacken water pump bolts **9**.
5. Turn water pump clockwise to tension belt **10**. Use tool No.KM-421-A/KM-421-B.
6. Movable part of tensioner must be against stop **11**.
7. Lightly tighten water pump bolts **9**.
8. Turn crankshaft two turns clockwise until timing marks aligned **6** & **7**.
9. Slacken water pump bolts **9**.
10. Turn water pump anti-clockwise **12** until tensioner pointer aligned with 'V' notch in support plate **13**. Use tool No.KM-421-A/KM-421-B.
11. Tighten water pump bolts to 8 Nm **9**.
12. Turn crankshaft two turns clockwise to TDC on No.1 cylinder. Ensure timing marks aligned **6** & **7**.
13. If not: Repeat tensioning procedure.
14. Insert tool in bell housing to lock flywheel **2**. Tool No.KM-911.
15. Remove crankshaft pulley bolt **3**.
16. Install components in reverse order of removal.
17. Fit new crankshaft pulley bolt **3**. Tightening torque:
    - ❑ Corsa/Astra – 95 Nm + 30° + 15°.
    - ❑ Vectra – 55 Nm + 45° + 15°.
18. Remove flywheel locking tool **2**.

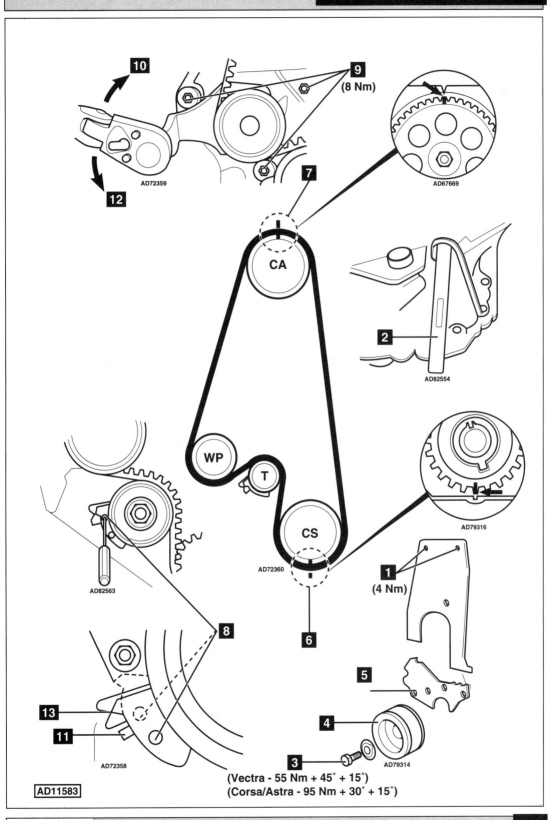

10

9
(8 Nm)

AD72359

12

AD67669

7

CA

2

AD82554

WP

T

AD79316

AD82563

CS

AD72360

1
(4 Nm)

8

13

11

AD72358

6

5

4

AD79314

3

(Vectra - 55 Nm + 45˚ + 15˚)
(Corsa/Astra - 95 Nm + 30˚ + 15˚)

AD11583

Model:	Astra/Belmont/Kadett-E 1,3/1,4/1,6/1,8/2,0 • Astra-F 1,8/2,0
	Astravan/Astramax 1,3/1,4/1,6
	Cavalier/Ascona/Vectra-A 1,3/1,4/1,6/1,8/2,0 • Manta-B 1,8 • Calibra 2,0
	Carlton/Rekord/Omega-A 1,8/2,0 • Frontera 2,0
Year:	1982-94
Engine Code:	Water pump tensioner
	13, 14, 16, 18, 20

Replacement Interval Guide

Vauxhall recommend check every 36,000 miles or 4 years (replace if necessary). The vehicle manufacturer has not recommended a timing belt replacement interval for this engine.
The previous use and service history of the vehicle must always be taken into account.
Refer to Timing Belt Replacement Intervals at the front of this manual.

Check For Engine Damage

CAUTION: This engines has been identified as a FREEWHEELING engine in which the possibility of valve-to-piston damage in the event of a timing belt failure may be minimal or very unlikely. However, a precautionary compression check of all cylinders should be performed.

Repair Times – hrs

Astra/Belmont/Astravan/Astramax/Kadett:	
Check & adjust	0,70
Remove & install:	
1,3/1,4/1,6	0,90
1,8/2,0	1,10
Astra-F:	
Check & adjust	0,80
Remove & install	1,40
Cavalier/Ascona (→1987):	
Check & adjust:	
1,3/1,6	0,70
1,8/2,0	0,80
Remove & install:	
1,3/1,6	0,90
1,8/2,0	1,10
Cavalier/Vectra-A (1988→):	
Check & adjust:	
1,4/1,6	0,60
1,6i	0,70
1,8/2,0	0,80
Remove & install:	
1,4/1,6	1,00
1,6i	1,20
1,8/2,0	1,30
Calibra:	
Check & adjust	0,80
Remove & install	1,40
Carlton/Rekord/Omega-A:	
Check & adjust	0,80
Remove & install	1,30
PAS	+0,20
Frontera:	
Check & adjust	0,80
Remove & install	1,40

Special Tools

- Tension gauge – Kent Moore No.KM-510-A.
- 1,3/1,4/1,6: Water pump wrench – Kent Moore No.KM-421-A.
- 1,8/2,0: Water pump wrench – Kent Moore No.KM-637-A.

Removal

1. Remove:
   - ❑ Air intake hose (if fitted).
   - ❑ Viscous fan.
   - ❑ Auxiliary drive belt.
   - ❑ Viscous fan pulley **1**.
   - ❑ Timing belt front cover **2**.
   NOTE: Viscous fan coupling has LH thread.
2. Turn crankshaft until timing marks aligned **3** & **4**.
   NOTE: 14NV: Align crankshaft pulley with first mark at 10° BTDC.
3. Remove:
   - ❑ 1,8/2,0: Crankshaft pulley bolts **6**.
   - ❑ Crankshaft pulley bolt **5** (if required).
   - ❑ Crankshaft pulley **7**.
4. Slacken water pump bolts.
5. Turn water pump anti-clockwise to release tension on belt. Use water pump wrench.
   Tool No.KM-421-A (1,3/1,4/1,6) or KM-637-A (1,8/2,0).
6. Remove timing belt.

Installation

1. Fit timing belt to crankshaft sprocket.
2. Fit crankshaft pulley.
3. Fit and tighten crankshaft pulley bolt **5**. 23 mm bolt: 55 Nm. 30 mm bolt: 55 Nm + 45° + 15°. **6**: 20 Nm.
4. Ensure timing marks aligned **3** & **4**.
5. Fit belt to remaining sprockets.
6. Turn water pump clockwise to take up belt slack. Use water pump wrench. Tool No.KM-421-A (1,3/1,4/1,6) or KM-637-A (1,8/2,0).
7. Turn crankshaft two turns clockwise.
8. Attach tension gauge to belt at ▽. Tool KM-510-A.
9. Lightly tap tension gauge with fingers. Check tension gauge reading.
10. Adjust water pump position to set tension. Use water pump wrench. Tool No.KM-421-A (1,3/1,4/1,6) or KM-637-A (1,8/2,0).
    NOTE: For a cold new belt tension gauge should indicate the following: 1,3/1,4: 6,0 units. 1,6: 5,5 units. 1,8/2,0: 4,5 units.
11. Tighten water pump bolts:
    - ❑ 1,3/1,4/1,6: M6 bolts – 8 Nm.
    - ❑ 1,8/2,0: M8 bolts – 25 Nm.
12. Remove tension gauge. Turn crankshaft one turn clockwise. Recheck belt tension.
13. Install components in reverse order of removal.

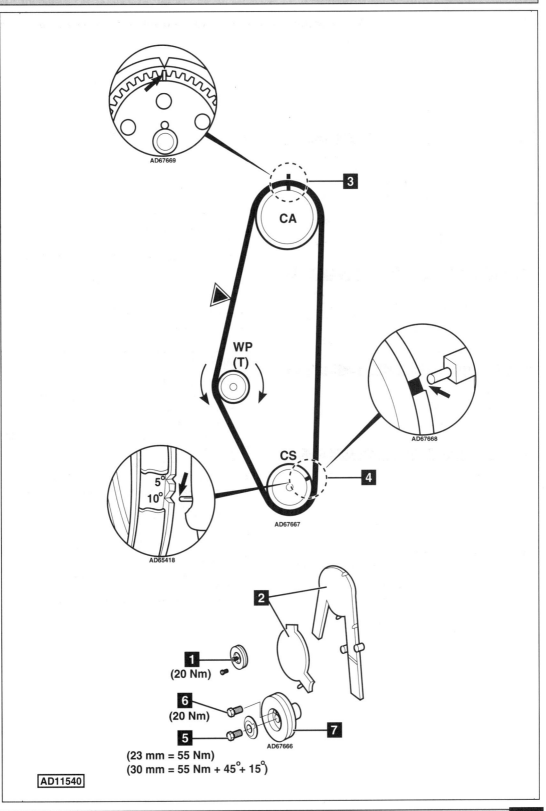

AD67669

3

CA

WP
(T)

AD67668

CS

5°
10°

AD65418

AD67667

4

2

1
(20 Nm)

6
(20 Nm)

5
(23 mm = 55 Nm)
(30 mm = 55 Nm + 45°+ 15°)

AD67666

7

AD11540

Model:	Astra 2,0 16V • Kadett-E 2,0 16V • Astra-F 2,0 16V • Cavalier 2,0 16V
	Vectra-A 2,0 16V • Calibra 2,0 16V
Year:	1988-92
Engine Code:	Manual tensioner
	20XE, C20XE

Replacement Interval Guide

Vauxhall recommend replacement as follows:
To 90MY - replacement every 63,000 miles or 7 years, whichever occurs first.
91MY on - replacement every 36,000 miles or 4 years, whichever occurs first.
The previous use and service history of the vehicle must always be taken into account.
Refer to Timing Belt Replacement Intervals at the front of this manual.

Check For Engine Damage

CAUTION: This engine has been identified as an INTERFERENCE engine in which the possibility of valve-to-piston damage in the event of a timing belt failure is MOST LIKELY to occur.
A compression check of all cylinders should be performed before removing the cylinder head.

Repair Times – hrs

Remove & install:

Astra/Kadett-E/Astra-F	1,00
Cavalier/Vectra-A	1,20
Calibra	1,50

Special Tools

- Tensioning tool – Kent Moore No.KM-666.
- Crankshaft pulley bolt socket – Kent Moore No.KM-321-A.

Special Precautions

- Disconnect battery earth lead.
- DO NOT turn crankshaft or camshaft when timing belt removed.
- Remove spark plugs to ease turning engine.
- Turn engine in normal direction of rotation (unless otherwise stated).
- DO NOT turn engine via camshaft or other sprockets.
- Observe all tightening torques.

Removal

1. Remove:
   - ❑ Air filter.
   - ❑ Air intake hose.
   - ❑ Auxiliary drive belt(s).
   - ❑ Timing belt cover **1**.
2. Turn crankshaft until timing marks aligned **2** & **3**. Use tool KM-321-A.
3. Slacken oil cooler hose clip. Move hoses aside.
4. Remove:
   - ❑ Crankshaft pulley bolts **5**.
   - ❑ Crankshaft pulley **6**.
5. Slacken tensioner bolt **4**.
6. Remove timing belt.

Installation

NOTE: DO NOT refit used belt.

1. Fit timing belt, starting at crankshaft sprocket. Ensure belt is taut between sprockets on non-tensioned side.
2. Fit crankshaft pulley. Tighten bolts to 20 Nm **5**.
3. Mark 8th tooth anti-clockwise from TDC mark on camshaft sprocket **8**.
4. Tension timing belt. Use tool KM-666 **7**.
5. Turn crankshaft two turns clockwise until mark on 8th tooth aligns with timing mark on timing belt rear cover **3**.
6. Tighten tensioner bolt **4**.
   Tightening torque: 25 Nm + 45-60°.
7. Remove tensioning tool.
8. Turn crankshaft until timing marks aligned **2** & **3**.
9. Install components in reverse order of removal.

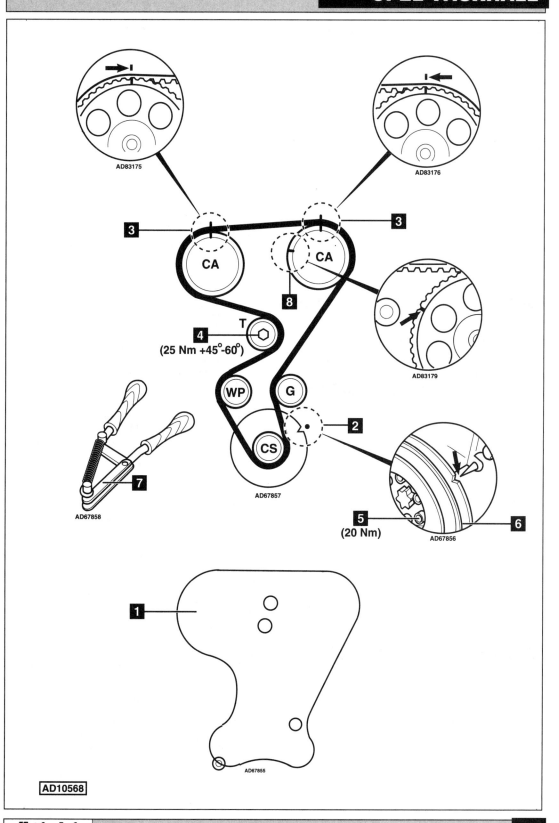

AD83175

AD83176

3

CA

3

CA

8

AD83179

T
4
(25 Nm +45°-60°)

WP

G

CS

AD67857

2

7

AD67858

5
(20 Nm)

6

AD67856

1

AD67855

AD10568

VAUXHALL-OPEL

Model:	Astra-F 1,6/1,8/2,0 • Cavalier 1,6/1,8/2,0 • Vectra-A 1,6/1,8/2,0 Vectra-B 1,6/2,0 • Calibra 2,0 • Omega-B 2,0 • Frontera 2,0
Year:	1990-00
Engine Code:	Automatic tensioner 16LZ2, C16NZ2, C18NZ, C18SEL, 20NE, 20NEJ, C20NE, X20SE

Replacement Interval Guide

Vauxhall recommend replacement as follows:
→94MY - replacement every 36,000 miles or 4 years, whichever occurs first.
95-96MY - replacement every 40,000 miles or 4 years, whichever occurs first.
97MY→ - replacement every 80,000 miles or 8 years, whichever occurs first (X20SE - tensioner pulley must also be replaced).
The previous use and service history of the vehicle must always be taken into account.
Refer to Timing Belt Replacement Intervals at the front of this manual.

Check For Engine Damage

CAUTION: This engine has been identified as an INTERFERENCE engine in which the possibility of valve-to-piston damage in the event of a timing belt failure is MOST LIKELY to occur.
A compression check of all cylinders should be performed before removing the cylinder head.

Repair Times – hrs

Astra-F
Remove & install:

1,6	1,20
1,8/2,0	1,40

Cavalier/Vectra

Remove & install	1,30
AC (→1993 MY)	+0,20
PAS (→1993 MY)	+0,30

Omega-B

Remove & install	0,90

Calibra

Remove & install	1,40
AC	+0,20

Frontera
Remove & install:

C20NE	1,40
X20SE	0,60
AC (X20SE)	+0,80

Special Tools

■ None required.

Removal

1. Raise and support front of vehicle.
2. Turn auxiliary drive belt tensioner clockwise to release tension on belt. Use ring spanner.

3. 2,0 – except Frontera:
   ❏ Disconnect intake air temperature (IAT) sensor multi-plug.
   ❏ Disconnect mass air flow (MAF) sensor multi-plug.
4. Remove:
   ❏ RH front wheel.
   ❏ RH splash guard.
   ❏ Air intake trunking.
   ❏ Air filter housing.
   ❏ RH engine mounting torque reaction link.
5. All models – remove:
   ❏ Auxiliary drive belt.
   ❏ Frontera: Cooling fan viscous coupling bolt (LH thread).
   ❏ Frontera: Cooling fan, viscous coupling and pulley.
   ❏ Crankshaft pulley bolts (4 bolts) **7**.
   ❏ Crankshaft pulley.
   ❏ Timing belt upper cover.
   ❏ Timing belt lower cover.
6. Turn crankshaft clockwise until TDC mark on crankshaft pulley aligned with mark on oil pump housing **1**.
7. Mark on camshaft sprocket must be aligned with mark on timing belt cover backplate **2**.
8. Slacken tensioner bolt **3**.
9. Turn tensioner clockwise until pointer on LH stop **5**. Use Allen key **4**.
10. Lightly tighten tensioner bolt **3**.
11. Remove timing belt.

Installation

1. Ensure timing marks aligned **1** & **2**.
2. Fit timing belt. Ensure belt is taut between crankshaft sprocket and camshaft sprocket.
3. Ensure lug on water pump aligned with corresponding lug on cylinder block **6**.
4. Slacken tensioner bolt **3**.
5. Turn tensioner anti-clockwise until pointer on RH stop **8**.
6. Tighten tensioner bolt **3**.
7. Turn crankshaft two turns clockwise. Ensure timing marks aligned **1** & **2**.
8. Slacken tensioner bolt **3**.
9. Turn tensioner clockwise until pointer aligned with:
   'V' notch in bracket (new belt) **9**.
   LH edge of 'V' notch (used belt) **10**.
10. Turn crankshaft two turns clockwise. Ensure timing marks aligned **1** & **2**.
11. Check pointer aligned with 'V' notch in bracket **9**.
12. If not: Repeat tensioning procedure.
13. Tighten tensioner bolt to 20 Nm **3**.
14. Install components in reverse order of removal.
15. Tighten crankshaft pulley bolts to 20 Nm **7**.
16. Frontera: Tighten cooling fan viscous coupling bolt (LH thread). Tightening torque: 26 Nm.

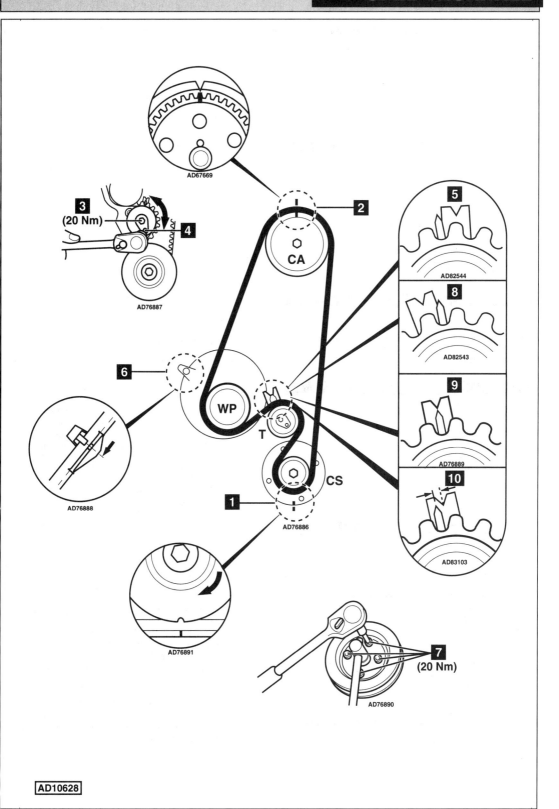

AD67669

3
(20 Nm)

4

AD76887

2

CA

5

AD82544

8

AD82543

9

AD76889

10

AD83103

6

AD76888

WP

T

CS

1

AD76886

AD76891

7
(20 Nm)

AD76890

AD10628

Model:	Astra-F 1,8/2,0 16V • Cavalier 2,0 16V/Turbo • Vectra-A 2,0 16V/Turbo
	Calibra 2,0 16V/Turbo • Omega-B 2,0 16V
Year:	1992-00
Engine Code:	Automatic tensioner
	C18XE, C18XEL, X18XE, C20XE, C20LET, X20XEV, 20XEJ

Replacement Interval Guide

Vauxhall recommend replacement as follows:
→94MY - replacement every 36,000 miles or 4 years, whichever occurs first.
95-96MY - replacement every 40,000 miles or 4 years, whichever occurs first.
97MY→ - replacement every 40,000 miles or 4 years, whichever occurs first (tensioner pulley must also be replaced).
The previous use and service history of the vehicle must always be taken into account.
Refer to Timing Belt Replacement Intervals at the front of this manual.

Check For Engine Damage

CAUTION: This engine has been identified as an INTERFERENCE engine in which the possibility of valve-to-piston damage in the event of a timing belt failure is MOST LIKELY to occur.
A compression check of all cylinders should be performed before removing the cylinder head.

Repair Times – hrs

Remove & install:

Astra-F – 1,8	0,90
Astra-F – 2,0 (C20XE)	1,50
Astra-F – 2,0 (X20XEV)	0,90
Vectra/Cavalier	1,20
Calibra (C20XE)	1,50
Calibra (X20XEV)	1,10
Calibra Turbo	1,40
Omega-B	1,00

Special Tools

■ None required.

Removal

WARNING: Certain engines require modification to tensioner pulley/guide pulleys due to possible failure. Refer to dealer.

1. Remove:
   ❑ Air filter and housing.
   ❑ Air intake hose.
   ❑ RH engine mounting (PAS and AC) **1**.
   ❑ C18XE/XEL: Mass air flow (MAF) sensor.
2. Mark direction of rotation on auxiliary drive belt with chalk.
3. Turn auxiliary drive belt tensioner clockwise to release tension on belt **2**. Use suitable spanner.
4. Remove:
   ❑ Auxiliary drive belt.
   ❑ Timing belt cover.

5. Turn crankshaft clockwise until crankshaft pulley timing mark aligned **3**.
6. Camshaft sprockets with single timing marks:
   ❑ Ensure camshaft sprocket timing marks aligned **4**.
7. Camshaft sprockets with dual timing marks:
   ❑ Ensure 'INTAKE' timing marks aligned **14**.
   ❑ Ensure 'EXHAUST' timing marks aligned **15**.
8. If timing marks on camshafts not aligned: Turn crankshaft one turn clockwise.
9. Remove:
   ❑ Crankshaft pulley bolts **5**.
   ❑ Crankshaft pulley **6**.
10. Slacken tensioner bolt **7**.
11. Turn tensioner clockwise until pointer on LH stop **10**. Use Allen key **9**.
12. Lightly tighten tensioner bolt **7**.
13. Remove timing belt.

Installation

1. Ensure lug on water pump aligned with corresponding lug on cylinder block **8**.
2. Camshaft sprockets with single timing marks:
   ❑ Ensure timing marks aligned **4** & **16**.
3. Camshaft sprockets with dual timing marks:
   ❑ Ensure timing marks aligned **14**, **15** & **16**.
   NOTE: If camshaft sprockets removed: Ensure correct dowel location of sprockets.
4. Fit timing belt.
   NOTE: Belt adjustment must be carried out when engine is cold.
5. Slacken tensioner bolt **7**.
6. Turn tensioner anti-clockwise until pointer on RH stop **11**. Use Allen key.
7. Lightly tighten tensioner bolt **7**.
8. Fit crankshaft pulley. Tighten crankshaft pulley bolts to 20 Nm **5**.
9. Turn crankshaft two turns in direction of rotation until timing marks aligned **3**.
10. Camshaft sprockets with single timing marks:
    ❑ Ensure timing marks aligned **4**.
11. Camshaft sprockets with dual timing marks:
    ❑ Ensure timing marks aligned **14** & **15**.
12. Slacken tensioner bolt **7**.
13. Turn tensioner clockwise until pointer aligned with:
    'V' notch in bracket (new belt) **12**.
    LH edge of 'V' notch (used belt) **13**.
14. Tighten tensioner bolt to 20 Nm **7**.
15. Install components in reverse order of removal.
    NOTE: Observe direction of rotation marks on auxiliary drive belt.

Autodata

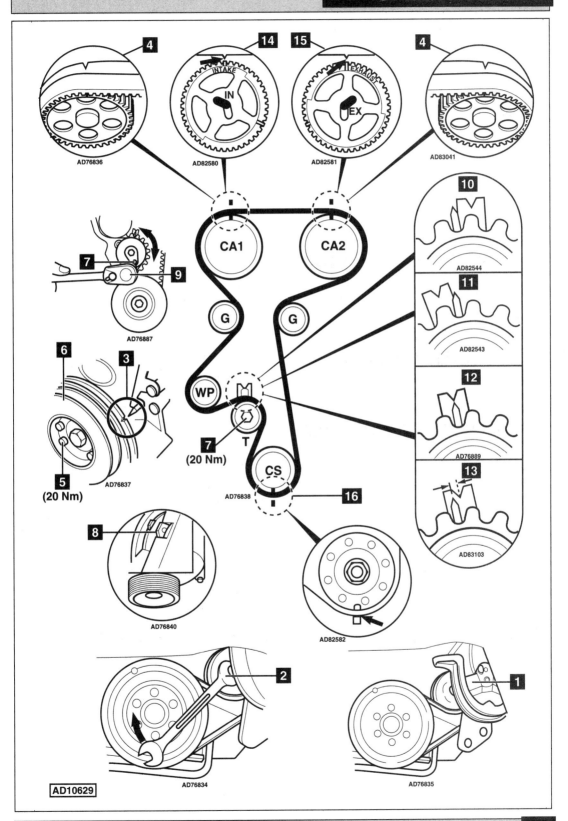

AD76836

4

AD82580

14

AD82581

15 EX

AD83041

4

10 AD82544

11 AD82543

12 AD76889

13 AD83103

CA1

CA2

7 **9**

AD76887

G

G

6

3

WP

M

7 T
(20 Nm)

AD76837

5
(20 Nm)

CS

AD76838

16

8

AD76840

AD82582

2

AD76834

1

AD76835

AD10629

Model:	**Astra-G 2,0 16V • Vectra 1,8/2,0 16V**
Year:	**1995-00**
Engine Code:	**X18XE, C20SEL, X20XEV**

Replacement Interval Guide

Vauxhall recommend replacement as follows:
→96MY - replacement every 40,000 miles or 4 years, whichever occurs first.
97MY→ - replacement every 40,000 miles or 4 years, whichever occurs first (tensioner pulley must also be replaced).
The previous use and service history of the vehicle must always be taken into account.
Refer to Timing Belt Replacement Intervals at the front of this manual.

Check For Engine Damage

CAUTION: This engine has been identified as an INTERFERENCE engine in which the possibility of valve-to-piston damage in the event of a timing belt failure is MOST LIKELY to occur.
A compression check of all cylinders should be performed before removing the cylinder head.

Repair Times – hrs

Remove & install:

Vectra	1,20
Astra-G	1,70

Special Tools

- Camshaft sprocket locking tool – Kent Moore No.KM-853.
- Engine support tool (Astra) – Kent Moore No.KM-6001.
- Engine alignment tool (Astra) – Kent Moore No.KM-909-B.

Special Precautions

- Disconnect battery earth lead.
- DO NOT turn crankshaft or camshaft when timing belt removed.
- Remove spark plugs to ease turning engine.
- Turn engine in normal direction of rotation (unless otherwise stated).
- DO NOT turn engine via camshaft or other sprockets.
- Observe all tightening torques.

Removal

WARNING: Certain engines require modification to tensioner pulley/guide pulleys due to possible failure. Refer to dealer.

1. Remove:
   ❏ Air filter housing.
2. Raise and support front of vehicle.
3. Remove RH front wheel.
4. Mark direction of rotation on auxiliary drive belt with chalk.
5. Turn auxiliary drive belt tensioner clockwise to release tension on belt. Use ring spanner.
6. Insert 4 mm pin through hole in tensioner and into mounting bracket **1**.
7. Remove:
   ❏ Auxiliary drive belt.
   ❏ Vectra: Engine torque bracket **2**.
   ❏ Timing belt cover(s) **3**.
8. Astra: Fit engine support and alignment tools. Tool Nos.KM-6001/KM-909-B.
9. Astra – remove:
   ❏ Auxiliary drive belt tensioner.
   ❏ RH engine mounting and bracket.
10. Turn crankshaft clockwise until timing marks aligned **4**.
11. Remove:
    ❏ Crankshaft pulley bolts **5**.
    ❏ Crankshaft pulley **6**.
12. Ensure crankshaft timing marks aligned **7**.
13. Timing marks on camshaft sprockets must be aligned with notches in timing belt rear cover **8**.
14. Lock camshaft sprockets. Use tool No.KM-853 **9**.
15. Slacken tensioner bolt **10**.
16. Turn tensioner clockwise until pointer on LH stop: Use Allen key **11**.
    ❏ Type 1 – **12**.
    ❏ Type 2 – **16**.
17. Lightly tighten tensioner bolt **10**.
18. Remove timing belt.

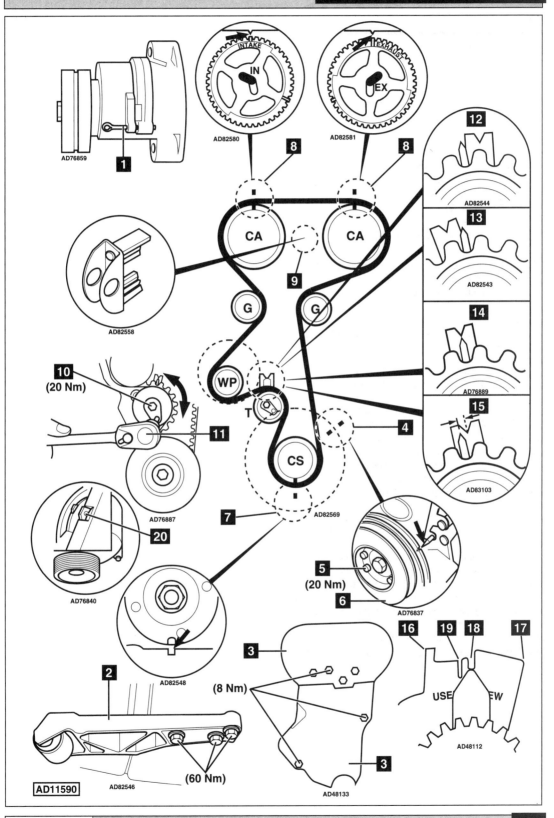

AD76859 **1**

AD82580 INTAKE IN **8**

AD82581 EXHAUST EX **8**

AD82544 **12**

AD82543 **13**

AD76889 **14**

AD83103 **15**

9

CA CA

AD82558

G G

WP M T **4**

10 (20 Nm)

11

AD76887

AD76840 **20**

CS

7 AD82569

AD82548 **2**

5 (20 Nm) **6**

AD76837

3

(8 Nm)

16 **19** **18** **17**

USE NEW

AD48112

AD11590 AD82546 (60 Nm)

3 AD48133

Installation

NOTE: Ensure lug on water pump aligned with corresponding lug on cylinder block ⅜.

1. Ensure timing marks aligned ⅞ & ⅜.

   NOTE: If camshaft sprockets removed: Ensure correct dowel location of sprockets.

2. Fit timing belt in anti-clockwise direction, starting at crankshaft sprocket. Ensure belt is taut between sprockets.

3. Slacken tensioner bolt ⑩.

4. Turn tensioner anti-clockwise until pointer on RH stop: Use Allen key ⑪.
   - ❑ Type 1 – ⑬.
   - ❑ Type 2 – ⑰.

5. Lightly tighten tensioner bolt ⑩.

6. Remove locking tool ⑨.

7. Turn crankshaft two turns clockwise. Ensure timing marks aligned ⑦ & ⑧.

8. Lock camshaft sprockets.
   Use tool No.KM-853 ⑨.

9. Slacken tensioner bolt ⑩.

10. Turn tensioner clockwise until pointer aligned as follows:
    - ❑ Type 1:
    - ❑ New belt – 'V' notch in bracket ⑭.
    - ❑ Used belt – LH edge of 'V' notch ⑮.
    - ❑ Type 2:
    - ❑ New belt – 'NEW' notch in bracket ⑱.
    - ❑ Used belt – 'USED' notch in bracket ⑲.

11. Tighten tensioner bolt to 20 Nm ⑩.

12. Remove locking tool ⑨.

13. Turn crankshaft two turns clockwise. Ensure timing marks aligned ⑦ & ⑧.

14. Check pointer aligned as follows:
    - ❑ Type 1:
    - ❑ New belt – 'V' notch in bracket ⑭.
    - ❑ Used belt – LH edge of 'V' notch ⑮.
    - ❑ Type 2:
    - ❑ New belt – 'NEW' notch in bracket ⑱.
    - ❑ Used belt – 'USED' notch in bracket ⑲.

15. If not: Repeat tensioning procedure.

16. Install components in reverse order of removal.

    NOTE: Observe direction of rotation marks on auxiliary drive belt.

17. Tighten crankshaft pulley bolts to 20 Nm ⑤.

18. Tighten engine torque bracket bolts to 60 Nm.

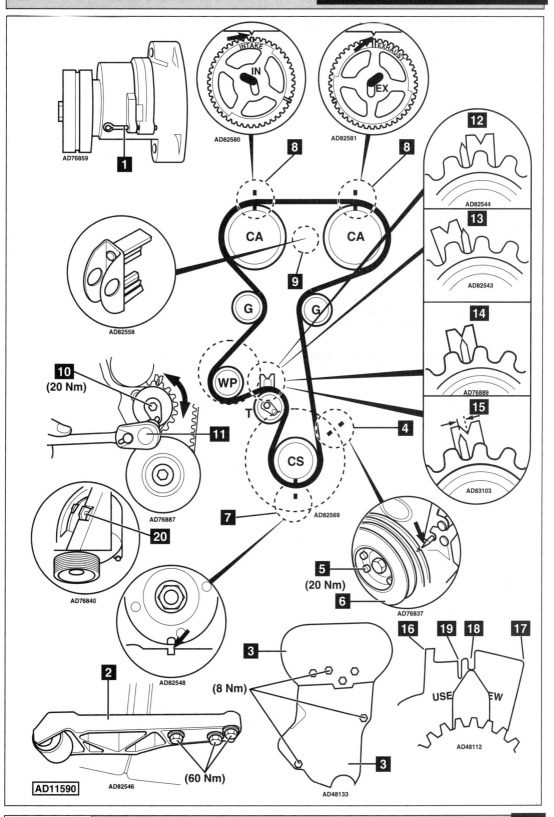

Model:	**Astra-G 2,0 Turbo • Zafira 2,0 Turbo**
Year:	**2000-03**
Engine Code:	**Z20LET**

Replacement Interval Guide

Vauxhall recommend replacement every 40,000 miles or 4 years (tensioner pulley must also be replaced).

The previous use and service history of the vehicle must always be taken into account.

Refer to Timing Belt Replacement Intervals at the front of this manual.

Check For Engine Damage

CAUTION: This engine has been identified as an INTERFERENCE engine in which the possibility of valve-to-piston damage in the event of a timing belt failure is MOST LIKELY to occur.

A compression check of all cylinders should be performed before removing the cylinder head.

Repair Times – hrs

Remove & install	1,30

Special Tools

- Camshaft sprocket locking tool – Kent Moore No.KM-853.
- Engine support tool – Kent Moore No.KM-6001-A.
- Engine alignment tool – Kent Moore No.KM-6173.

Special Precautions

- Disconnect battery earth lead.
- DO NOT turn crankshaft or camshaft when timing belt removed.
- Remove spark plugs to ease turning engine.
- Turn engine in normal direction of rotation (unless otherwise stated).
- DO NOT turn engine via camshaft or other sprockets.
- Observe all tightening torques.

Removal

1. Remove:
   ❑ Air filter housing.
2. Raise and support front of vehicle.
3. Remove:
   ❑ RH front wheel.
   ❑ RH splash guard.
4. Mark direction of rotation on auxiliary drive belt with chalk.
5. Turn auxiliary drive belt tensioner anti-clockwise to release tension on belt.
6. Remove:
   ❑ Auxiliary drive belt.
   ❑ Timing belt upper cover **1**.

7. Fit engine support and alignment tools. Tool Nos.KM-6001-A/KM-6173.
8. Remove:
   ❑ Auxiliary drive belt tensioner.
   ❑ RH engine mounting and bracket.
9. Turn crankshaft clockwise until timing marks aligned **2**.
10. Remove:
    ❑ Crankshaft pulley bolts **3**.
    ❑ Crankshaft pulley **4**.
    ❑ Timing belt lower cover **5**.
11. Ensure crankshaft timing marks aligned **6**.
12. Timing marks on camshaft sprockets must be aligned with notches in timing belt rear cover **7**.
13. Lock camshaft sprockets. Use tool No.KM-853 **8**.
14. Slacken tensioner bolt **11**.
15. Turn tensioner clockwise until pointer on LH stop. Use Allen key **10**.
    ❑ Type 1: **9**.
    ❑ Type 2: **17**.
16. Lightly tighten tensioner bolt **11**.
17. Remove timing belt.

Installation

*NOTE: Ensure lug on water pump aligned with corresponding lug on cylinder block **19**.*

1. Ensure timing marks aligned **6** & **7**.
   NOTE: If camshaft sprockets removed: Ensure correct dowel location of sprockets.
2. Fit timing belt in anti-clockwise direction, starting at crankshaft sprocket. Ensure belt is taut between sprockets.
3. Slacken tensioner bolt **11**.
4. Turn tensioner anti-clockwise until pointer on RH stop. Use Allen key **10**.
   ❑ Type 1: **12**.
   ❑ Type 2: **18**.
5. Lightly tighten tensioner bolt **11**.
6. Remove locking tool **8**.
7. Turn crankshaft two turns clockwise. Ensure timing marks aligned **6** & **7**.
8. Lock camshaft sprockets. Use tool No.KM-853 **8**.
9. Slacken tensioner bolt **11**.

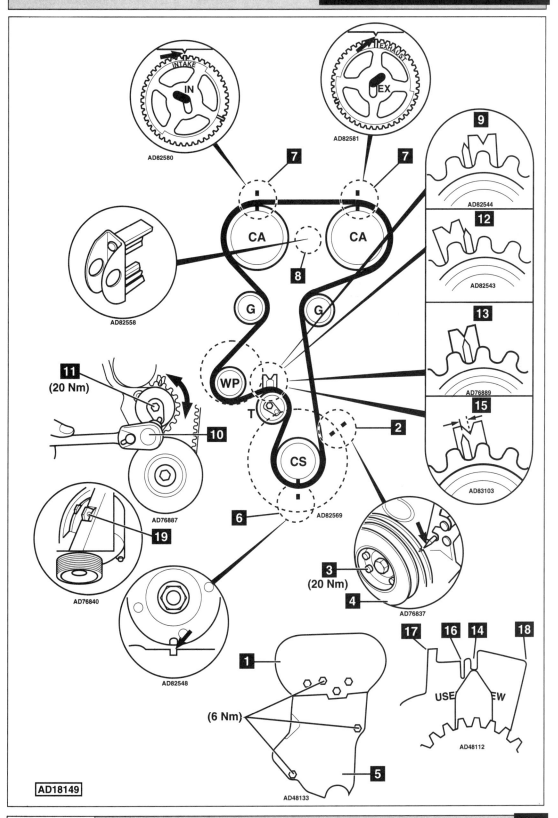

AD82580

AD82581

7

7

IN

EX

CA

CA

9

AD82544

12

AD82543

13

AD76889

15

AD83103

AD82558

8

G

G

11
(20 Nm)

WP

T

10

AD76887

19

AD76840

2

CS

AD82569

6

3
(20 Nm)

4

AD76837

AD82548

1

17 **16** **14** **18**

USE NEW

AD48112

(6 Nm)

5

AD48133

AD18149

←

10. Turn tensioner clockwise until pointer aligned as follows:
    - ❏ Type 1:
    - ❏ New belt – 'V' notch in bracket 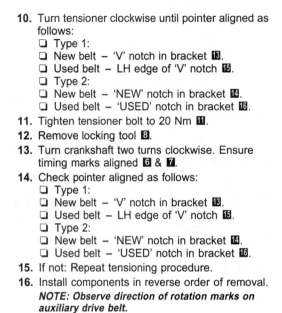.
    - ❏ Used belt – LH edge of 'V' notch 🔢.
    - ❏ Type 2:
    - ❏ New belt – 'NEW' notch in bracket 🔢.
    - ❏ Used belt – 'USED' notch in bracket 🔢.

11. Tighten tensioner bolt to 20 Nm 🔢.

12. Remove locking tool 🔢.

13. Turn crankshaft two turns clockwise. Ensure timing marks aligned 🔢 & 🔢.

14. Check pointer aligned as follows:
    - ❏ Type 1:
    - ❏ New belt – 'V' notch in bracket 🔢.
    - ❏ Used belt – LH edge of 'V' notch 🔢.
    - ❏ Type 2:
    - ❏ New belt – 'NEW' notch in bracket 🔢.
    - ❏ Used belt – 'USED' notch in bracket 🔢.

15. If not: Repeat tensioning procedure.

16. Install components in reverse order of removal.

    NOTE: Observe direction of rotation marks on auxiliary drive belt.

17. Tighten crankshaft pulley bolts to 20 Nm 🔢.

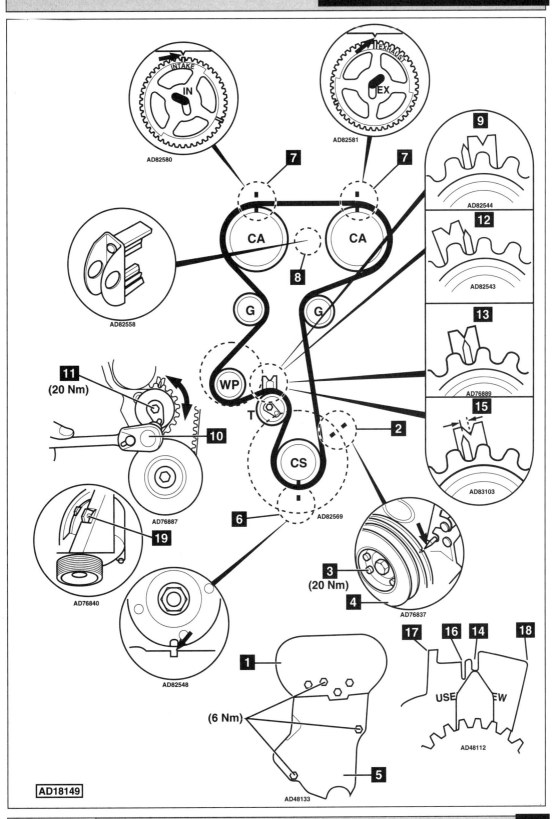

AD82580

AD82581

7

7

9

AD82544

12

AD82543

13

AD76889

15

AD83103

CA

CA

8

G

G

AD82558

WP

2

11
(20 Nm)

T

10

CS

AD76887

19

6

AD82569

AD76840

3
(20 Nm)

4

AD76837

1

17 16 14 18

USE   NEW

AD82548

(6 Nm)

AD48112

5

AD48133

VAUXHALL-OPEL

Model:	Cavalier 2,5i V6 • Vectra-A 2,5i V6 • Calibra 2,5i V6
Year:	1993-96
Engine Code:	C25XE

Replacement Interval Guide

Vauxhall recommend replacement as follows:
→94 MY - replacement every 36,000 miles or 4 years, whichever occurs first.
95 MY→ - replacement every 40,000 miles or 4 years, whichever occurs first.
The previous use and service history of the vehicle must always be taken into account.
Refer to Timing Belt Replacement Intervals at the front of this manual.

Check For Engine Damage

CAUTION: This engine has been identified as an INTERFERENCE engine in which the possibility of valve-to-piston damage in the event of a timing belt failure is MOST LIKELY to occur.
A compression check of all cylinders should be performed before removing the cylinder head.

Repair Times – hrs

Remove & install:

Cavalier/Vectra	1,40
Calibra	1,80
MT	+0,20

Special Tools

- ■ RH camshaft locking tool (red) – Kent Moore No.KM-800-1.
- ■ LH camshaft locking tool (green) – Kent Moore No.KM-800-2.
- ■ Crankshaft locking tool – Kent Moore No.KM-800-10.
- ■ Camshaft timing gauge – Kent Moore No.KM-800-20.
- ■ Belt holding tool (wedge) – Kent Moore No.KM-800-30.

Special Precautions

- ■ Disconnect battery earth lead.
- ■ DO NOT turn crankshaft or camshaft when timing belt removed.
- ■ Remove spark plugs to ease turning engine.
- ■ Turn engine in normal direction of rotation (unless otherwise stated).
- ■ DO NOT turn engine via camshaft or other sprockets.
- ■ Observe all tightening torques.

Removal

1. Remove:
   - ❑ Air filter housing and hoses.
   - ❑ Engine stabiliser (if fitted).
   - ❑ Auxiliary drive belt.
   - ❑ PAS pump pulley.
   - ❑ Water pump pulley.
   - ❑ Crankshaft pulley bolts **1**.
   - ❑ Crankshaft pulley **2**.
   - ❑ Timing belt cover.

2. Turn crankshaft until just before timing marks aligned **3** & **4**.
   *NOTE: Camshaft sprockets fitted to early models may have single timing marks 1, 2, 3 and 4. Ensure sprockets are fitted to their respective camshafts CA1, CA2, CA3 & CA4. On later models (illustrated), camshaft sprockets have dual timing marks. Ensure CA1 sprocket timing mark 1 aligned with timing mark on exhaust camshaft CA1 **4** etc.*

3. Fit locking tool to crankshaft **5**.
   Tool No. KM-800-10. Turn crankshaft slowly clockwise until tool arm rests against water pump flange. Secure in position.

4. Fit locking tool between camshaft sprockets CA1 & CA2 **6**. Tool No. KM-800-1 (red).
   *NOTE: If locking tool will not fit between camshaft sprockets CA1 & CA2: Release belt tension slightly by turning upper guide pulley **7**.*

5. Fit locking tool between camshaft sprockets CA3 & CA4 **6**. Tool No. KM-800-2 (green).
   *NOTE: If locking tool will not fit between camshaft sprockets CA3 & CA4: Release belt tension slightly by turning lower guide pulley **8**.*

6. Slacken tensioner bolt **9**. Turn tensioner clockwise to release tension on belt **10**.

7. Slacken upper guide pulley bolt **7**.

8. Slacken lower guide pulley bolt **8**.

9. Remove timing belt.

10. If camshaft sprockets are to be removed: Note position of dowel pins in relation to timing marks.
    - ❑ Camshaft sprocket CA1 uses No.1 for dowel pin location and timing mark alignment **15**.
    - ❑ Camshaft sprocket CA2 uses No.2, CA3 uses No.3, CA4 uses No.4.
    - ❑ Fit new bolts **17**.
      Tightening torque: 50 Nm + 60° + 15°.

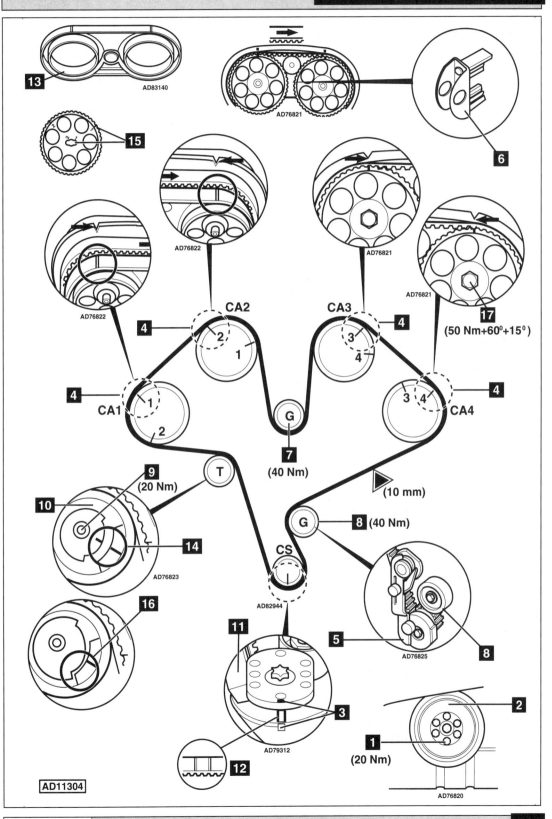

AD83140

CA2

CA3

4

4

4

4

G

7
(40 Nm)

CA1

CA4

9
(20 Nm)

T

(10 mm)

10

G

8 (40 Nm)

14

AD76823

CS

AD82944

16

11

5

8

AD76825

3

12

AD79312

1
(20 Nm)

2

AD76820

AD11304

Installation

NOTE: Renew timing belt if it has been disturbed or removed from engine.

1. Ensure all locking tools located correctly **5** & **6**.
2. Fit timing belt in anti-clockwise direction, starting at crankshaft sprocket. Ensure double lines on belt **12** aligned with crankshaft timing marks **3**. Observe direction of rotation marks on belt.
3. Wedge belt into position.
   Use tool No. KM-800-30 **11**. Ensure belt is taut between sprockets.
4. Fit belt to camshaft sprockets CA3 & CA4. Ensure timing marks aligned **4**. Ensure slack at ▽ is less than 10 mm.
5. Fit belt around upper guide pulley **7** and camshaft sprockets CA1 & CA2. Ensure timing marks aligned **4**.

   *NOTE: Align single marks on belt with timing marks on camshaft sprockets and timing belt rear cover **4**.*

6. Fit belt around tensioner pulley **10**.
7. Turn lower guide pulley **8** anti-clockwise until belt taut between lower guide pulley **8** and camshaft sprocket CA4. Tighten bolt to 40 Nm **8**.
8. Turn upper guide pulley **7** anti-clockwise until belt taut between camshaft sprockets CA2 & CA3. Tighten bolt to 40 Nm **7**.
9. Turn tensioner pulley **10** anti-clockwise as far as it will go **16** to pre-tension belt. Tighten bolt to 20 Nm **9**.
10. Remove:
    ❏ Locking tools **5** & **6**.
    ❏ Belt holding tool **11**.
11. Turn crankshaft slowly two turns clockwise until just before timing marks aligned **3** & **4**.
12. Slacken tensioner bolt slightly **9**. Turn tensioner pulley clockwise until tensioner marks aligned **14**. Tighten bolt to 20 Nm **9**.
13. Fit locking tool to crankshaft **5**.
    Tool No. KM-800-10. Turn crankshaft slowly clockwise until tool arm rests against water pump flange. Secure in position.
14. Fit timing gauge to camshaft sprockets CA1 & CA2 and CA3 & CA4 in turn **13**.
    Tool No. KM-800-20.
15. Ensure timing marks aligned **4**.

    NOTE: Belt marks are for installation purposes only. They will no longer align once crankshaft has been turned.

16. If necessary: Repeat tensioning and adjustment procedures. Start at lower guide pulley **8**.
17. Remove crankshaft locking tool **5**.

18. Install components in reverse order of removal.

    NOTE: During removal of timing belt covers the sealing tape may become damaged or detached, resulting in severe engine damage if it comes into contact with timing belt. Ensure sealing tape is secured to timing belt covers with a suitable adhesive and renewed if necessary.

19. Tighten crankshaft pulley bolts to 20 Nm **1**.

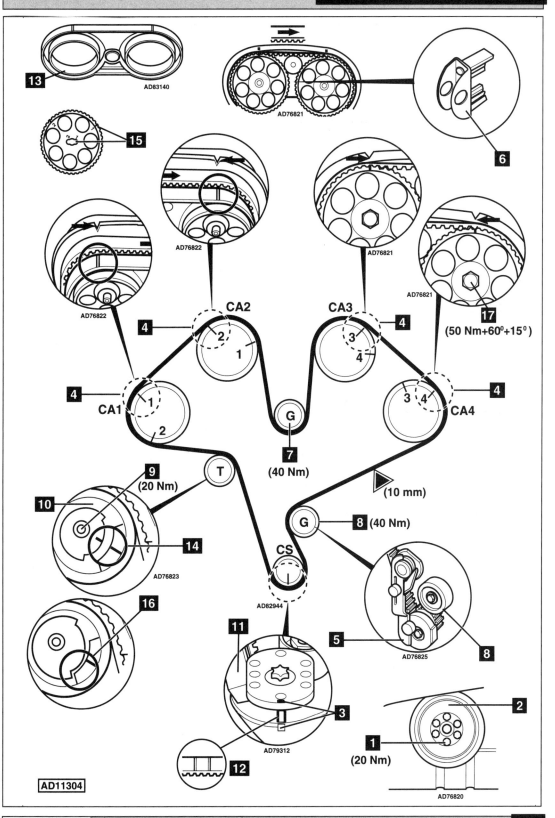

13 AD83140

AD76821

6

15

AD76822

AD76822

AD76821

AD76821

CA2

CA3

4

4

17

(50 Nm+60°+15°)

4

CA1

4

4

CA4

9
(20 Nm)

T

10

14

AD76823

G

7
(40 Nm)

(10 mm)

G

8 (40 Nm)

16

CS

AD82944

11

5

8

AD76825

3

AD79312

12

2

1
(20 Nm)

AD76820

AD11304

Model:	Vectra 2,5 V6 • Omega-B 2,5/3,0 V6 • Calibra 2,5 V6 • Sintra 3,0 V6
Year:	1994-1996
Engine Code:	Without raised outer edge on tensioner pulley X25XE, X30XE

Replacement Interval Guide

Vauxhall recommend replacement as follows:
Vectra/Sintra/Calibra:
Every 40,000 miles or 4 years - replace.

Omega-B:
94MY:
Every 36,000 miles or 4 years - replace.
95MY→:
Every 40,000 miles or 4 years - replace.
The previous use and service history of the vehicle must always be taken into account.
Refer to Timing Belt Replacement Intervals at the front of this manual.

Check For Engine Damage

CAUTION: This engine has been identified as an INTERFERENCE engine in which the possibility of valve-to-piston damage in the event of a timing belt failure is MOST LIKELY to occur.
A compression check of all cylinders should be performed before removing the cylinder head.

Repair Times – hrs

Remove & install:

Vectra-B	2,50
Omega-B	2,40
Sintra	3,10
Calibra (AT)	1,80
Calibra (MT)	2,00

Special Tools

- RH camshaft locking tool (red) – Kent Moore No.KM-800-1.
- LH camshaft locking tool (green) – Kent Moore No.KM-800-2.
- Crankshaft locking tool – Kent Moore No.KM-800-10.
- Camshaft timing gauge – Kent Moore No.KM-800-20.
- Belt holding tool (wedge) – Kent Moore No.KM-800-30.

Special Precautions

- Disconnect battery earth lead.
- DO NOT turn crankshaft or camshaft when timing belt removed.
- Remove spark plugs to ease turning engine.
- Turn engine in normal direction of rotation (unless otherwise stated).
- DO NOT turn engine via camshaft or other sprockets.
- Observe all tightening torques.

Removal

1. Remove:
   - ❑ Air filter housing and hoses.
   - ❑ Engine undershield (if fitted).
   - ❑ Engine stabiliser (if fitted).
   - ❑ Auxiliary drive belt.
   NOTE: Mark direction of rotation on belt.
   - ❑ Water pump pulley.
2. Slacken retaining clips and remove cable duct.
3. Remove:
   - ❑ PAS pump pulley.
   - ❑ Auxiliary drive belt tensioner.
   - ❑ Crankshaft pulley bolts **1**.
   - ❑ Crankshaft pulley **2**.
   - ❑ Timing belt cover.
4. Turn crankshaft clockwise until just before timing marks aligned **3** & **4**.
5. Fit locking tool to crankshaft **5**.
   Tool No. KM-800-10. Turn crankshaft slowly clockwise until tool arm rests against water pump flange. Secure in position.
6. Ensure timing marks on camshaft sprockets aligned **4**.
   *NOTE: Camshaft sprockets have dual timing marks. Camshaft sprocket CA1 timing mark 1 must align with timing mark **4**, camshaft sprocket CA2 timing mark 2 must align with timing mark **4**, etc.*
7. Fit locking tool between camshaft sprockets CA1 & CA2 **6**. Tool No. KM-800-1 (red).
   *NOTE: If locking tool will not fit between camshaft sprockets CA1 & CA2: Release belt tension slightly by turning upper guide pulley G2 **7**.*
8. Fit locking tool between camshaft sprockets CA3 & CA4 **6**. Tool No. KM-800-2 (green).
   *NOTE: If locking tool will not fit between camshaft sprockets CA3 & CA4: Release belt tension slightly by turning lower guide pulley clockwise **8**.*
9. Slacken tensioner bolt **9**. Turn tensioner clockwise to release tension on belt **10**. Lightly tighten bolt.
10. Slacken upper guide pulley bolt **7**.
11. Slacken lower guide pulley bolt **8**.
12. Remove timing belt.

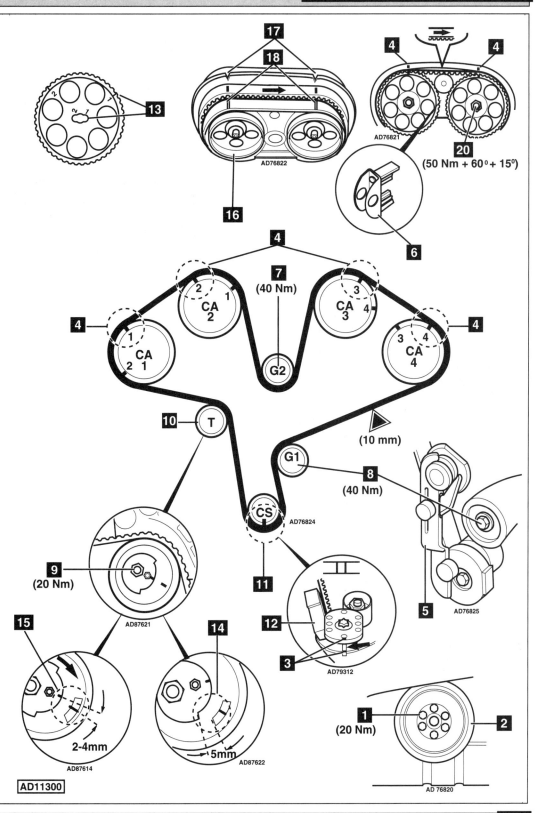

13

17 **18**

AD76822

16

4 **4**

AD76821

20
(50 Nm + 60° + 15°)

6

4

7
(40 Nm)

CA
2
CA
3
CA
4

4

CA
2 1

4

4

G2

10 T

G1

(10 mm)

8
(40 Nm)

5

AD76825

CS

AD76824

11

12

3

AD79312

9
(20 Nm)

AD87621

15

2-4mm

AD87614

14

5mm AD87622

1
(20 Nm)

2

AD 76820

AD11300

←

13. If camshaft sprockets are to be removed: Note position of dowel pins in relation to timing marks.
- ❏ Camshaft sprocket CA1 uses No.1 for dowel pin location and timing mark alignment **⓳**.
- ❏ Camshaft sprocket CA2 uses No.2, CA3 uses No.3, CA4 uses No.4.
- ❏ Fit new bolts **⓴**. Tightening torque: 50 Nm + 60° + 15°.

Installation

NOTE: Renew timing belt if it has been disturbed or removed from engine.

1. Ensure all locking tools located correctly **❺** & **❻**.

2. Fit timing belt in anti-clockwise direction, starting at crankshaft sprocket. Ensure double lines on belt **⓫** aligned with crankshaft timing marks **❸**. Observe direction of rotation marks on belt.

3. Wedge belt into position. Use tool No. KM-800-30 **⓬**. Ensure belt is taut between sprockets.

4. Fit belt to camshaft sprockets CA3 & CA4. Ensure timing marks aligned **❹** & **⓱**. Ensure slack at ▽ is less than 10 mm.

5. Fit belt around upper guide pulley **❼** and camshaft sprockets CA1 & CA2. Ensure timing marks aligned **❹** & **⓱**.

6. Fit belt around tensioner pulley **❿**.

7. Turn lower guide pulley **❽** anti-clockwise until belt taut between lower guide pulley **❽** and camshaft sprocket CA4. Tighten bolt to 40 Nm **❽**.

8. Turn upper guide pulley **❼** anti-clockwise until belt taut between camshaft sprockets CA2 & CA3. Tighten bolt to 40 Nm **❼**.

9. Turn tensioner pulley **❿** anti-clockwise until marks are 5 mm apart **⓮** to tension belt. Tighten bolt to 20 Nm **❾**.

10. Remove:
- ❏ Locking tools **❺** & **❻**.
- ❏ Belt holding tool **⓬**.

11. Turn crankshaft slowly two turns clockwise until just before timing marks aligned **❸** & **❹**.

12. Slacken tensioner bolt slightly **❾**. Turn tensioner pulley clockwise until marks are 2-4 mm apart **⓯**. Tighten bolt to 20 Nm **❾**.

13. Fit locking tool to crankshaft **❺**. Tool No. KM-800-10. Turn crankshaft slowly clockwise until tool arm rests against water pump flange. Secure in position.

14. Ensure timing marks aligned **❸**.

15. Fit timing gauge to camshaft sprockets CA1 & CA2 and CA3 & CA4 in turn **⓰**. Tool No. KM-800-20. Ensure timing marks aligned **❹** & **⓲**.
NOTE: Belt marks are for installation purposes only. They will no longer align once crankshaft has been turned.

16. If necessary: Repeat tensioning and adjustment procedures. Start at lower guide pulley **❽**.

17. Remove crankshaft locking tool **❺**.

18. Install components in reverse order of removal.
NOTE: During removal of timing belt covers the sealing tape may become damaged or detached, resulting in severe engine damage if it comes into contact with timing belt. Ensure sealing tape is secured to timing belt covers with a suitable adhesive and renewed if necessary.

19. Tighten crankshaft pulley bolts to 20 Nm **❶**.

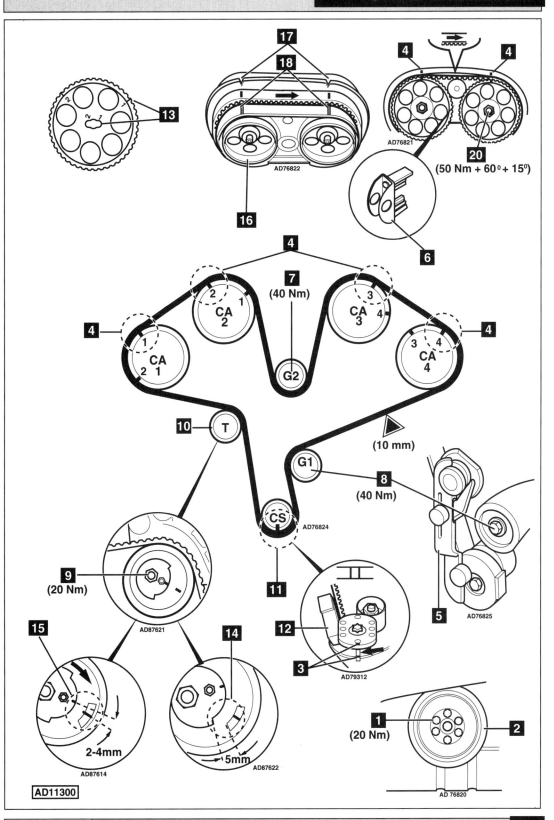

VAUXHALL-OPEL

Model:	**Vectra-B 2,5/2,6 V6 • Omega-B 2,5/2,6/3,0/3,2 V6 • Sintra 3,0 V6** **Calibra 2,5 V6**
Year:	**1996-03**
Engine Code:	**X25XE, X30XE, Y26SE, Y32SE**

Replacement Interval Guide

Vauxhall recommend replacement as follows:
96MY - replacement every 80,000 miles or
8 years, whichever occurs first (tensioner pulley
must also be replaced).
97MY→ - replacement every 40,000 miles or
4 years, whichever occurs first (tensioner pulley
must also be replaced).
*The previous use and service history of the vehicle
must always be taken into account.
Refer to Timing Belt Replacement Intervals at the front
of this manual.*

Check For Engine Damage

*CAUTION: This engine has been identified as an
INTERFERENCE engine in which the possibility of
valve-to-piston damage in the event of a timing belt
failure is MOST LIKELY to occur.
A compression check of all cylinders should be
performed before removing the cylinder head.*

Repair Times – hrs

Remove & install:

Vectra-B (→2001 MY)	2,50
Vectra-B (2001 MY→)	2,30
Omega-B 2,5	2,20
Omega-B 3,0	2,90
Omega-B 2,6/3,2	2,70
Sintra	3,10
Calibra (AT)	1,80
Calibra (MT)	2,00

Special Tools

- RH camshaft locking tool (red) –
  Kent Moore No.KM-800-1.
- LH camshaft locking tool (green) –
  Kent Moore No.KM-800-2.
- Crankshaft locking tool –
  Kent Moore No.KM-800-10.
- Camshaft timing gauge –
  Kent Moore No.KM-800-20.
- Belt holding tool (wedge) –
  Kent Moore No.KM-800-30.
- Engine support – Kent Moore No.KM-883-1.

Special Precautions

- Disconnect battery earth lead.
- DO NOT turn crankshaft or camshaft when timing belt removed.
- Remove spark plugs to ease turning engine.
- Turn engine in normal direction of rotation (unless otherwise stated).
- DO NOT turn engine via camshaft or other sprockets.
- Observe all tightening torques.

Removal

1. Remove:
   - ❏ Air filter housing and hoses.
   - ❏ Engine undershield.
   - ❏ RH inner wing panel.
   - ❏ Engine stabiliser (if fitted).
2. Vectra-B 2001 MY→ with AC: Support engine. Use tool No.883-1.
3. Vectra-B 2001 MY→ with AC: Loosen RH engine mounting and move aside.
4. Remove:
   - ❏ Auxiliary drive belt. Mark direction of rotation.
   - ❏ Water pump pulley.
   - ❏ Slacken retaining clips and remove cable duct.
   - ❏ PAS pump pulley.
   - ❏ Auxiliary drive belt tensioner.
   - ❏ Crankshaft pulley bolts **1**.
   - ❏ Crankshaft pulley **2**.
   - ❏ Timing belt cover.
5. Turn crankshaft clockwise until just before timing marks aligned **4** & **5**.
6. Fit locking tool to crankshaft **3**. Tool No.KM-800-10. Turn crankshaft slowly clockwise until tool arm rests against water pump flange. Secure in position.
7. Ensure timing marks on camshaft sprockets aligned **5**.
   *NOTE: Camshaft sprockets have dual timing marks. Camshaft sprocket CA1 timing mark 1 must align with timing mark **5**, camshaft sprocket CA2 timing mark 2 must align with timing mark **5** etc.*
8. Fit locking tool between camshaft sprockets CA1 & CA2 **6**. Tool No.KM-800-1 (red).
   *NOTE: If locking tool will not fit between camshaft sprockets CA1 & CA2: Release belt tension slightly by turning upper guide pulley **9**.*

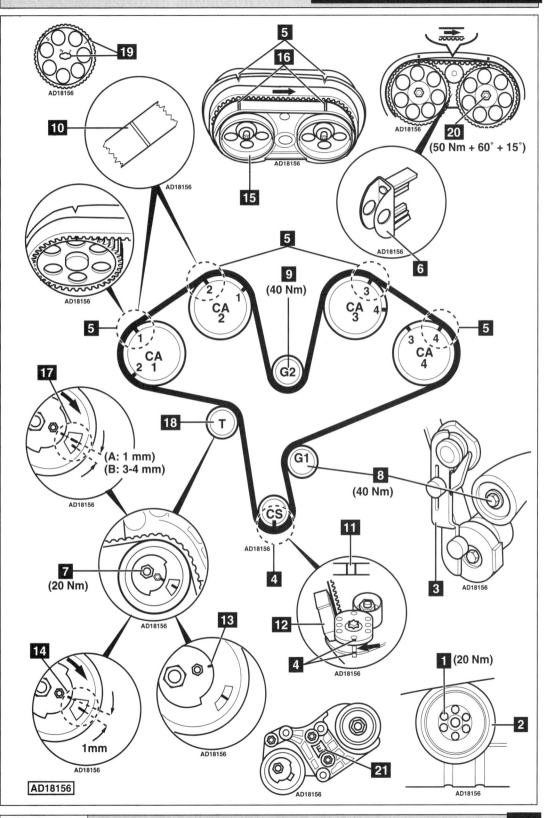

19 AD18156

10

5 **16**

15 AD18156

AD18156

AD18156

20 (50 Nm + 60° + 15°)

6 AD18156

5 **9** (40 Nm)

2 1 CA 2

3 CA 3 4

1 CA 2 1

5 3 4 CA 4 **5**

G2

17 (A: 1 mm) (B: 3-4 mm) AD18156

18 T

G1 **8** (40 Nm)

CS AD18156

3 AD18156

7 (20 Nm) AD18156

11

4

12 **4** AD18156

14 1mm AD18156

13 AD18156

1 (20 Nm)

2 AD18156

21 AD18156

AD18156

9. Fit locking tool between camshaft sprockets CA3 & CA4 **6**. Tool No.KM-800-2 (green).

   *NOTE: If locking tool will not fit between camshaft sprockets CA3 & CA4: Release belt tension slightly by turning lower guide pulley clockwise **8**.*

10. Slacken tensioner bolt **7**. Turn tensioner clockwise to release tension on belt **18**. Lightly tighten bolt.

11. Slacken upper guide pulley bolt **9**.

12. Slacken lower guide pulley bolt **8**.

13. Remove timing belt.

14. If camshaft sprockets are to be removed: Note position of dowel pins in relation to timing marks.
    - ❑ Camshaft sprocket CA1 uses No.1 for dowel pin location and timing mark alignment **19**.
    - ❑ Camshaft sprocket CA2 uses No.2, CA3 uses No.3, CA4 uses No.4.
    - ❑ Fit new bolts **20**.
      Tightening torque: 50 Nm + 60° + 15°.

Installation

NOTE: Renew timing belt if it has been disturbed or removed from engine.

1. Ensure all locking tools located correctly **3** & **6**.

2. Fit timing belt in clockwise direction, starting at crankshaft sprocket. Ensure double lines on belt **11** aligned with crankshaft timing marks **4**. Observe direction of rotation marks on belt.

3. Wedge belt into position.
   Use tool No.KM-800-30 **12**. Ensure belt is taut between sprockets.

   *NOTE: Base plate must also be replaced with tensioner: Base plate with code 'D' replace every 40,000 miles or 4 years: Base plate with code 'E'/'EA' or '01' replace every 80,000 miles or 8 years **21**.*

4. Fit belt around tensioner pulley **18**.

5. Fit belt to camshaft sprockets CA1 & CA2. Ensure belt timing marks aligned **10**. Fit belt around upper guide pulley **9**.

6. Fit belt around lower guide pulley **8**. Then fit around camshaft sprockets CA3 & CA4. Ensure belt timing marks aligned **10**.

7. Turn lower guide pulley **8** anti-clockwise until belt taut between lower guide pulley and camshaft sprocket CA4. Tighten bolt to 40 Nm **8**.

8. Turn upper guide pulley **9** anti-clockwise until belt taut between camshaft sprockets CA2 & CA3. Tighten bolt to 40 Nm **9**.

9. Turn tensioner pulley **18** anti-clockwise as far as it will go **13** to pre-tension belt.

10. Turn tensioner pulley clockwise until tensioner mark is 1 mm above floating mark **14**.

11. Tighten tensioner bolt to 20 Nm **7**.

12. Remove:
    - ❑ Locking tools **3** & **6**.
    - ❑ Belt holding tool **12**.

13. Turn crankshaft slowly two turns clockwise until just before timing marks aligned **4** & **5**.

14. Fit locking tool to crankshaft **3**.
    Tool No.KM-800-10. Turn crankshaft slowly clockwise until tool arm rests against water pump flange. Secure in position.

15. Ensure timing marks aligned **4**.

16. Fit timing gauge to camshaft sprockets CA1 & CA2 and CA3 & CA4 in turn **15**.
    Tool No.KM-800-20.

    NOTE: Belt marks are for installation purposes only. They will no longer align once crankshaft has been turned.

17. Ensure marks on sprockets aligned with marks on timing gauge **16**. Check edge of timing belt aligned with edge of sprockets.

18. If not: Repeat tensioning and adjustment procedures. Start at lower guide pulley **8**.

19. Turn tensioner pulley until tensioner mark is as follows:
    - ❑ (A) →2001 MY = 1 mm below floating mark.
    - ❑ (B) 2001 MY→ = 3-4 mm below floating mark **17**.

20. Tighten bolt to 20 Nm **7**.

21. Install components in reverse order of removal.

    NOTE: During removal of timing belt covers the sealing tape may become damaged or detached, resulting in severe engine damage if it comes into contact with timing belt. Ensure sealing tape is secured to timing belt covers with a suitable adhesive and renewed if necessary.

22. Tighten crankshaft pulley bolts **1**.
    Tightening torque: 20 Nm.

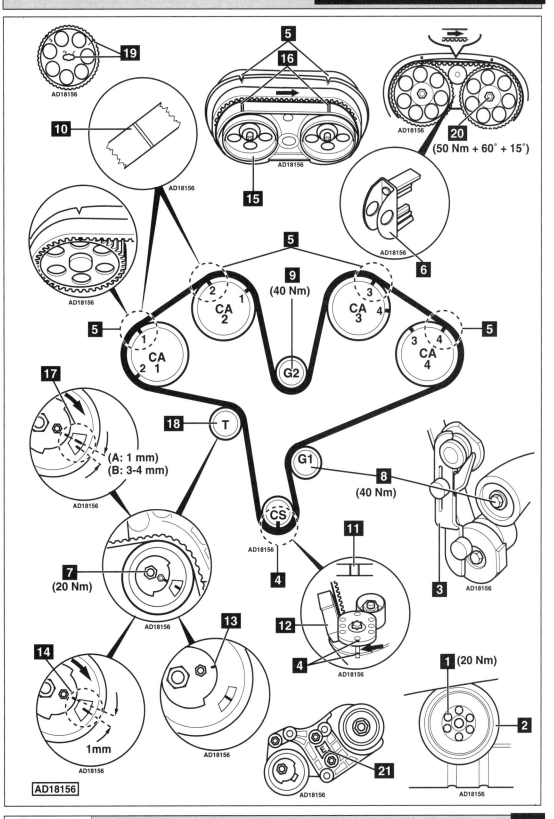

Model:	**Omega-B 2,2 16V • Frontera-B 2,2 16V • Sintra-A 2,2 16V**
Year:	**1995-03**
Engine Code:	**X22XE, Y22XE, Z22XE**

Replacement Interval Guide

Vauxhall recommend replacement as follows:
95MY - replacement every 36,000 miles or 4 years, whichever occurs first.
96MY - replacement every 40,000 miles or 4 years, whichever occurs first.
97MY→ - replacement every 40,000 miles or 4 years, whichever occurs first (tensioner pulley must also be replaced).
The previous use and service history of the vehicle must always be taken into account.
Refer to Timing Belt Replacement Intervals at the front of this manual.

Check For Engine Damage

CAUTION: This engine has been identified as an INTERFERENCE engine in which the possibility of valve-to-piston damage in the event of a timing belt failure is MOST LIKELY to occur.
A compression check of all cylinders should be performed before removing the cylinder head.

Repair Times – hrs

Remove & install:

Omega-B	0,80
Sintra	1,70
Frontera	0,80
AC	+0,20
Frontera-B	0,70

Special Tools

- Frontera/Sintra: Camshaft sprocket locking tool – Kent Moore No.KM-832.
- Omega-B: Camshaft sprocket locking tool – Kent Moore No.KM-853.

Special Precautions

- Disconnect battery earth lead.
- DO NOT turn crankshaft or camshaft when timing belt removed.
- Remove spark plugs to ease turning engine.
- Turn engine in normal direction of rotation (unless otherwise stated).
- DO NOT turn engine via camshaft or other sprockets.
- Observe all tightening torques.

Removal

WARNING: Certain engines require modification to tensioner pulley/guide pulleys due to possible failure. Refer to dealer.

1. Mark direction of rotation on auxiliary drive belt with chalk.
2. Turn auxiliary drive belt tensioner clockwise to release tension on belt **1**.
3. Remove auxiliary drive belt.
4. Turn crankshaft clockwise until timing marks aligned **2**.
5. Omega: Reposition pipes and wiring adjacent to timing belt cover.

6. Remove:
   - Engine undershield.
   - Crankshaft pulley bolts **8**.
   - Crankshaft pulley **3**.
   - Timing belt cover.
7. Turn crankshaft clockwise until timing marks aligned **4**, **5** & **6**.
8. Lock camshaft sprockets. Use tool No.KM-832/853.
9. Slacken tensioner bolt **7**.
10. Turn tensioner clockwise until pointer on LH stop. Use Allen key.
    - Type 1 – **10**.
    - Type 2 – **14**.
11. Lightly tighten tensioner bolt.
12. Remove timing belt.

Installation

1. Ensure lug on water pump aligned with corresponding lug on cylinder block **9**.
2. Ensure timing marks aligned **4**, **5** & **6**.
3. Ensure locking tool located correctly.
   *NOTE: Sprockets fitted to inlet camshaft (CA1) and exhaust camshaft (CA2) have similar timing marks. If removed, ensure correct dowel location of sprockets on camshafts and timing marks aligned with 'INTAKE' **5** and 'EXHAUST' **6** marks respectively.*
4. Fit timing belt in anti-clockwise direction, starting at crankshaft sprocket. Ensure belt is taut between crankshaft sprocket and exhaust camshaft sprocket (CA2).
   NOTE: Belt adjustment must be carried out when engine is cold.
5. Slacken tensioner bolt **7**.
6. Turn tensioner anti-clockwise until pointer on RH stop.
   - Type 1 – **11**.
   - Type 2 – **15**.
7. Lightly tighten tensioner bolt **7**.
8. Remove locking tool.
9. Turn crankshaft two turns clockwise. Ensure timing marks aligned **4**, **5** & **6**.
10. Slacken tensioner bolt **7**.
11. Turn tensioner clockwise until pointer aligned as follows:
    - Type 1:
    - New belt – 'V' notch in bracket **12**.
    - Used belt – LH edge of 'V' notch **13**.
    - Type 2:
    - New belt – 'NEW' notch in bracket **16**.
    - Used belt – 'USED' notch in bracket **17**.
12. Tighten tensioner bolt.
    - (A) →2000: 25 Nm **7**.
    - (B) 2000→: 20 Nm **7**.
13. Install components in reverse order of removal.
14. Tighten crankshaft pulley bolts to 20 Nm **8**.
    NOTE: Observe direction of rotation marks on auxiliary drive belt.

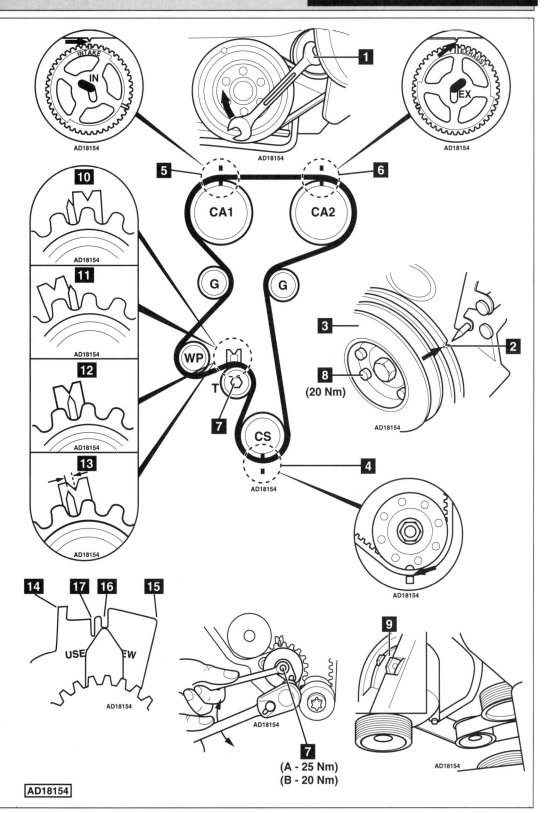

AD18154

Model:	Monterey 3,2 V6 24V • Frontera 3,2 V6
Year:	1994-03
Engine Code:	6VD1, 6VD1-W

Replacement Interval Guide

Vauxhall recommend replacement as follows:
→95MY - replacement every 54,000 miles or
6 years, whichever occurs first.
96-98MY - replacement every 60,000 miles or
6 years, whichever occurs first.
98½MY→ - replacement every 80,000 miles.
*The previous use and service history of the vehicle
must always be taken into account.*
*Refer to Timing Belt Replacement Intervals at the front
of this manual.*

Check For Engine Damage

*CAUTION: This engine has been identified as an
INTERFERENCE engine in which the possibility of
valve-to-piston damage in the event of a timing belt
failure is MOST LIKELY to occur.*
*A compression check of all cylinders should be
performed before removing the cylinder head.*

Repair Times – hrs

Remove & install:

Monterey	2,00
AC	+0,20
Frontera	1,90

Special Tools

■ None required.

Special Precautions

■ Disconnect battery earth lead.
■ DO NOT turn crankshaft or camshaft when timing belt removed.
■ Remove spark plugs to ease turning engine.
■ Turn engine in normal direction of rotation (unless otherwise stated).
■ DO NOT turn engine via camshaft or other sprockets.
■ Observe all tightening torques.

Removal

1. Remove:
   ❏ Air filter and hose.
   ❏ Upper radiator shroud.
   ❏ Lower radiator shroud.
   ❏ Cooling fan and viscous coupling.
   ❏ Auxiliary drive belt(s).
   ❏ Cooling fan pulley.
   ❏ PAS pump pulley.
   ❏ Auxiliary drive belt tensioner.
   ❏ Oil cooler hose.
   ❏ Crankshaft pulley bolt **1**.
   ❏ Crankshaft pulley **2**.
   ❏ Timing belt RH upper cover **3**.
   ❏ Timing belt LH upper cover **4**.
   ❏ Timing belt lower cover **5**.

2. Temporarily fit crankshaft pulley bolt **1**. Turn crankshaft clockwise until timing marks aligned **6** & **7**.
   *NOTE: Some engines have alternative location for timing marks of camshaft sprockets **7**.*

3. Early engines with eccentric tensioner pulley:
   ❏ Slacken tensioner pulley bolt **10**.
   ❏ Turn tensioner clockwise away from belt.
   ❏ Lightly tighten bolt.

4. Remove:
   ❏ Automatic tensioner unit bolts **11**.
   ❏ Automatic tensioner unit **12**.
   ❏ Timing belt.

Installation

1. Ensure timing marks aligned **6** & **7**.

2. Fit timing belt in anti-clockwise direction, starting at crankshaft sprocket. Ensure belt is taut between sprockets on non-tensioned side.

3. Ensure marks on belt **8** & **9** aligned with marks on sprockets **6** & **7**.
   *NOTE: Letters on belt should be readable from front of engine **13**.*

4. Slowly compress pushrod into tensioner body **12** with a force of approximately 100 kg until holes aligned.

5. Retain pushrod with 1,4 mm diameter pin through hole in tensioner body **14**.

6. Install automatic tensioner unit to cylinder block. Tighten bolts **11**. Monterey: 19 Nm. Frontera: 25 Nm.

7. Early engines with eccentric tensioner pulley:
   ❏ Slacken tensioner pulley bolt **10**.
   ❏ Turn tensioner anti-clockwise to tension belt.
   ❏ Tighten bolt to 42 Nm.

8. Remove pin from tensioner body to release pushrod.

9. Temporarily fit crankshaft pulley **2**.

10. Turn crankshaft slowly two turns clockwise. Ensure timing marks aligned **6** & **7**.

11. Check extended length of pushrod is 4,0-6,0 mm **15**.

12. If not: Repeat operations 4-11.

13. Install components in reverse order of removal.

14. Tighten crankshaft pulley bolt **1**. Tightening torque: 167 Nm.

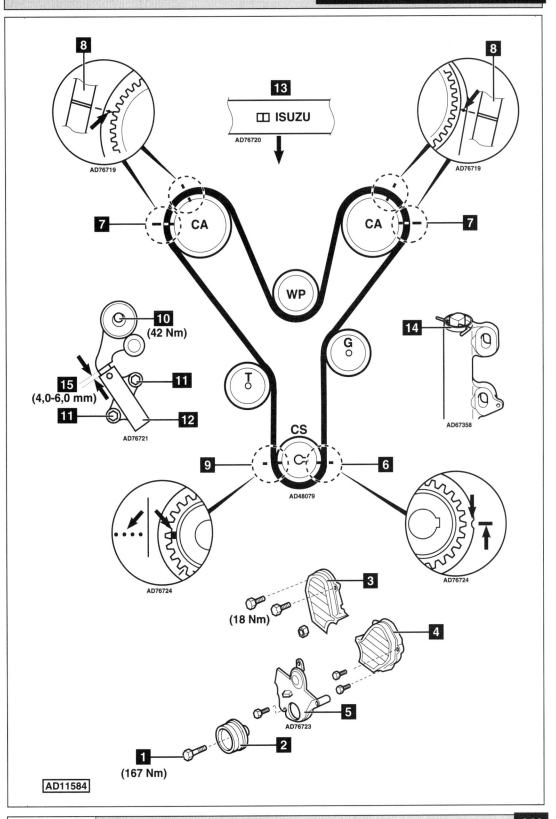

8

13

□ **ISUZU**

AD76720

AD76719

AD76719

7 CA CA **7**

WP

10
(42 Nm)

G

15
(4,0-6,0 mm)

11

11

T

14

12

AD76721

AD67358

CS

9

6

AD48079

AD76724

AD76724

3

(18 Nm)

4

5

AD76723

1

2

(167 Nm)

AD11584

VAUXHALL-OPEL

Model:	Nova 1,5 TD • Corsa-A 1,5 TD • Corsa-B 1,5D/TD
Year:	1988-00
Engine Code:	15D (4EC1), 15DT (T-4EC1)

Replacement Interval Guide

Vauxhall recommend:

→ 94MY:

Check every 36,000 miles or 4 years, whichever occurs first.

→91MY – replacement every 63,000 miles or 7 years, whichever occurs first.

92-94MY – replacement every 72,000 miles or 8 years, whichever occurs first.

95MY→:

Replacement every 80,000 miles or 8 years, whichever occurs first.

The previous use and service history of the vehicle must always be taken into account.

Refer to Timing Belt Replacement Intervals at the front of this manual.

Check For Engine Damage

CAUTION: This engine has been identified as an INTERFERENCE engine in which the possibility of valve-to-piston damage in the event of a timing belt failure is MOST LIKELY to occur.

A compression check of all cylinders should be performed before removing the cylinder head.

Repair Times – hrs

Remove & install:	
Nova/Corsa-A	1,60
Corsa-B	1,30
PAS	+0,20

Special Tools

- 1 bolt – M6 x 30 x 1,00 – to lock camshaft sprocket.
- 1 bolt – M8 x 40 x 1,25 – to lock injection pump sprocket.

Special Precautions

- Disconnect battery earth lead.
- DO NOT turn crankshaft or camshaft when timing belt removed.
- Remove glow plugs to ease turning engine.
- Turn engine in normal direction of rotation (unless otherwise stated).
- DO NOT turn engine via camshaft or other sprockets.
- Observe all tightening torques.
- Check diesel injection pump timing after belt replacement.

Removal

1. Remove auxiliary drive belt.
2. Turn crankshaft clockwise to TDC on No.1 cylinder. Ensure timing marks aligned **1**.
3. Release vacuum pipe, alternator cable and coolant hoses from timing belt upper cover **2** and remove cover.
4. Screw locking bolts into camshaft and injection pump sprockets **3** & **4**. Tighten bolts finger tight.
5. Slacken tensioner nut and bolt **5** & **6**.
6. Remove tensioner spring to release tension on belt.
7. Remove:
   - ❑ Crankshaft pulley bolts **7**.
   - ❑ Crankshaft pulley.
   - ❑ Timing belt lower cover **8**.
   - ❑ Timing belt.

   NOTE: Mark direction of rotation on belt with chalk if belt is to be reused.

Installation

1. Ensure timing marks aligned **1**.
2. Ensure locking bolts located in sprockets **3** & **4**.
3. Fit timing belt, starting at crankshaft sprocket. Ensure belt is taut between sprockets.
4. Fit tensioner spring to tension belt. Tighten nut to 25 Nm **5**. Tighten bolt to 25 Nm **6**.
5. Install:
   - ❑ Timing belt lower cover.
   - ❑ Crankshaft pulley.
6. Ensure timing marks aligned **1**.
7. Tighten crankshaft pulley bolts to 25 Nm **7**.
8. Remove locking bolts from sprockets **3** & **4**.
9. Turn crankshaft two turns clockwise. Ensure timing marks aligned **1**.
10. Install components in reverse order of removal.

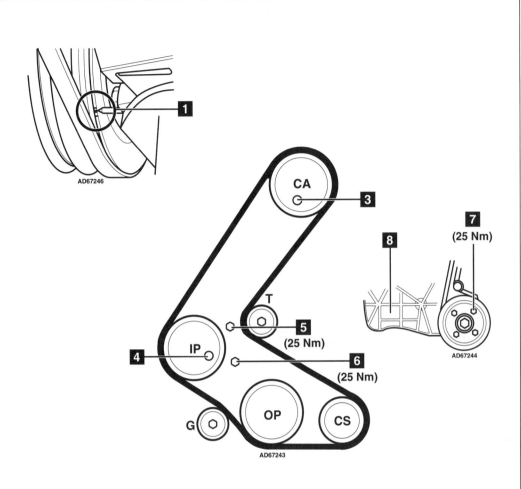

AD67246

CA

3

7
(25 Nm)

8

T

5
(25 Nm)

IP

4

6
(25 Nm)

G

OP

CS

AD67243

AD67244

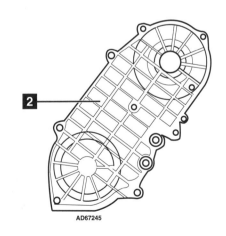

2

AD67245

AD10563

VAUXHALL-OPEL

Model:	Corsa-B 1,7D • Corsavan/Combo 1,7D • Astra-F 1,7 TD Cavalier 1,7 TD • Vectra-A 1,7 TD
Year:	1992-00
Engine Code:	17D/X17D (4EE1), 17DT (TC4EE1), X17DT (TC4EE1)

Replacement Interval Guide

Vauxhall recommend replacement as follows:
→94MY - replacement every 72,000 miles or 8 years, whichever occurs first.
95MY→ - replacement every 80,000 miles or 8 years, whichever occurs first.
The previous use and service history of the vehicle must always be taken into account.
Refer to Timing Belt Replacement Intervals at the front of this manual.

Check For Engine Damage

CAUTION: This engine has been identified as an INTERFERENCE engine in which the possibility of valve-to-piston damage in the event of a timing belt failure is MOST LIKELY to occur.
A compression check of all cylinders should be performed before removing the cylinder head.

Repair Times – hrs

Remove & install:

Corsa-B/Corsavan/Combo	1,70
PAS	+0,20
Astra-F	1,50
Cavalier/Vectra	3,00

Special Tools

- None required.

Special Precautions

- Disconnect battery earth lead.
- DO NOT turn crankshaft or camshaft when timing belt removed.
- Remove glow plugs to ease turning engine.
- Turn engine in normal direction of rotation (unless otherwise stated).
- DO NOT turn engine via camshaft or other sprockets.
- Observe all tightening torques.
- Check diesel injection pump timing after belt replacement.

Removal

1. Support engine.
2. Remove:
   - ❏ Air filter and air intake hoses.
   - ❏ Auxiliary drive belts.
   - ❏ RH engine mounting.
   - ❏ Vacuum pipe bracket.
   - ❏ Timing belt upper cover **1**.

3. Turn crankshaft clockwise to TDC on No.1 cylinder. Ensure crankshaft sprocket timing marks aligned **2**.
4. Insert M6 x 1,00 bolt in camshaft sprocket and M8 x 1,25 bolt in injection pump sprocket to lock sprockets in position **3**.
   NOTE: If locating holes not aligned: Turn crankshaft one turn.
5. Remove:
   - ❏ Crankshaft pulley (4 bolts).
   - ❏ Timing belt lower cover **4**.
6. Slacken tensioner bolts **6**. Remove tensioner spring **5**.
7. Move tensioner pulley away from belt. Lightly tighten bolts.
8. Remove timing belt.

Installation

1. Use crankshaft pulley to check TDC marks aligned **2**. Ensure locking bolts located correctly **3**.
2. Fit timing belt in clockwise direction, starting at crankshaft sprocket. Ensure belt is taut between sprockets.
3. Temporarily fit crankshaft pulley. Ensure timing marks aligned **2**.
4. Fit tensioner spring **5**.
5. Corsa-B/Corsavan/Combo:
   - ❏ Slacken tensioner bolts **6**. Allow tensioner to operate.
   - ❏ Remove locking bolts **3**.
   - ❏ Tighten tensioner bolts to 19 Nm **6**.
   - ❏ Turn crankshaft slowly two turns clockwise. Ensure timing marks aligned **2**.
6. Astra-F/Vectra/Cavalier:
   - ❏ Slacken tensioner bolts **6**.
   - ❏ Temporarily fit crankshaft pulley.
   - ❏ Remove locking bolts **3**.
   - ❏ Turn crankshaft 60° anti-clockwise (against normal direction of rotation) **7**.
   - ❏ Tighten tensioner bolts to 19 Nm **6**.
   - ❏ Turn crankshaft slowly two turns + 60° clockwise until timing marks aligned **2**.
7. Ensure timing marks aligned **2**. Ensure locking bolts can be inserted in sprockets **3**.
8. If not: Repeat tensioning procedure.
9. Remove crankshaft pulley. Fit timing belt lower cover **4**.
10. Fit crankshaft pulley. Tighten bolts to 20 Nm.
11. Install components in reverse order of removal.

Autodata

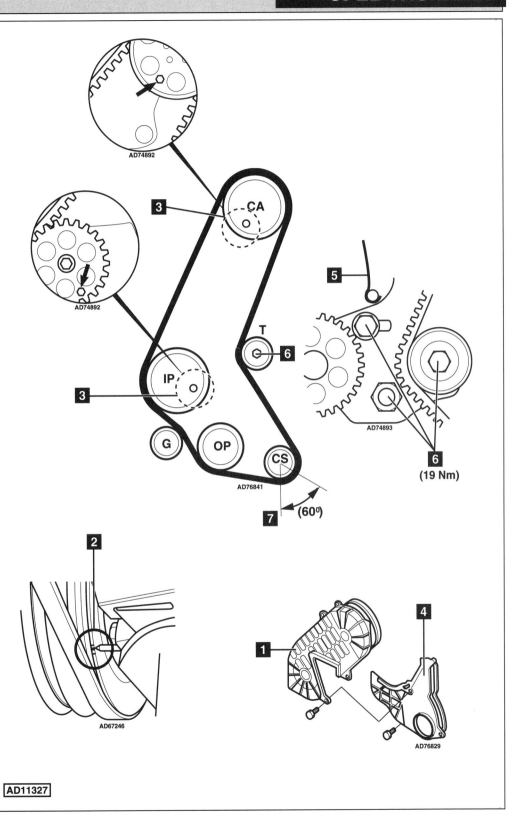

AD74892

3

CA

AD74892

5

T

6

IP

3

G

OP

CS

AD74893

6

(19 Nm)

AD76841

7 (60°)

2

AD67246

1

4

AD76829

AD11327

Model:	Corsa-C 1,7 TD • Astra-G 1,7 TD • Vectra-B 1,7 TD • Zafira 1,7 TD
Year:	1995-02
Engine Code:	X17DT, Y17DT, Y17DTI, Y17DTL

Replacement Interval Guide

Vauxhall recommend replacement as follows:

Corsa/Astra/Zafira:
Replacement every 100,000 miles or 10 years, whichever occurs first.

Vectra:
Replacement every 80,000 miles or 8 years, whichever occurs first.
The previous use and service history of the vehicle must always be taken into account.
Refer to Timing Belt Replacement Intervals at the front of this manual.

Check For Engine Damage

CAUTION: This engine has been identified as an INTERFERENCE engine in which the possibility of valve-to-piston damage in the event of a timing belt failure is MOST LIKELY to occur.
A compression check of all cylinders should be performed before removing the cylinder head.

Repair Times – hrs

Remove & install:

Corsa/Astra/Vectra	1,70
Zafira	No information available.

Special Tools

■ Engine support and alignment tool –
Kent Moore No.KM-6169.

Special Precautions

■ Disconnect battery earth lead.
■ DO NOT turn crankshaft or camshaft when timing belt removed.
■ Remove glow plugs to ease turning engine.
■ Turn engine in normal direction of rotation (unless otherwise stated).
■ DO NOT turn engine via camshaft or other sprockets.
■ Observe all tightening torques.
■ Check diesel injection pump timing after belt replacement.

Removal

1. Raise and support vehicle.
2. Remove:
   ❑ RH wheel.
   ❑ RH inner wing panel.
   ❑ Air filter housing.
3. Corsa: Fit engine support and alignment tool. Tool No.KM-6169.
4. Support engine.
5. Remove RH engine mounting and bracket.

6. Corsa/Astra/Zafira: Refit engine mounting bolt and tighten finger tight **11**.
7. Remove:
   ❑ Auxiliary drive belt(s).
   ❑ AC drive belt tensioner (if fitted).
8. Corsa: Move wiring harness to one side.
9. Remove timing belt upper cover **1**.
10. Turn crankshaft to TDC on No.1 cylinder. Ensure timing marks aligned **2**.
11. Insert M6 x 1,00 bolt in camshaft sprocket and M8 x 1,25 bolt in injection pump sprocket to lock sprockets in position **3**.
12. If locating holes not aligned: Turn crankshaft one turn clockwise.
13. Remove:
    ❑ Crankshaft pulley bolts **4**.
    ❑ Crankshaft pulley **5**.
    ❑ Timing belt lower cover **6**.
14. Slacken tensioner bolt(s) **7** or **8**. Remove tensioner spring **9**.
15. Move tensioner away from belt. Lightly tighten bolts.
16. Remove timing belt.

Installation

1. Ensure timing marks aligned **2**. Ensure locking bolts located correctly **3**.
2. Check tensioner for smooth operation.
3. Fit timing belt in anti-clockwise direction, starting at crankshaft sprocket. Ensure belt is taut between sprockets.
4. Fit tensioner spring **9**. Slacken tensioner bolts **7** or **8**.
5. Temporarily fit crankshaft pulley **5**.
6. Remove locking bolts **3**.
7. Turn crankshaft 60° anti-clockwise (against normal direction of rotation) **10**.
8. Tighten tensioner bolt(s). Vectra: 19 Nm **7**. Corsa/Astra/Zafira: 38 Nm **8**.
9. Remove crankshaft pulley. Fit timing belt lower cover **6**.
10. Fit crankshaft pulley. Tighten bolts to 20 Nm **4**.
11. Turn crankshaft two turns +60° clockwise until timing marks aligned **2**. Ensure locking bolts can be installed **3**.
12. Install components in reverse order of removal.
13. Tighten engine mounting bolt to 40 Nm **11**.

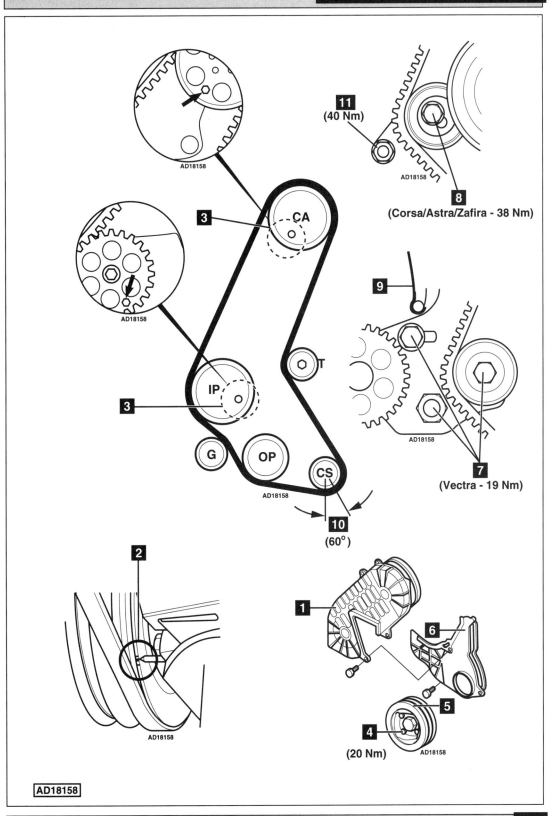

AD18158

11 (40 Nm)

8 (Corsa/Astra/Zafira - 38 Nm)

3 CA

3 IP

T

G   OP   CS

9

7 (Vectra - 19 Nm)

10 (60°)

2

1

6

4 (20 Nm)   **5**

AD18158

VAUXHALL-OPEL

Model:	**Astra/Belmont 1,6/1,7 D • Kadett-E 1,6/1,7 D • Cavalier 1,6/1,7 D** **Ascona-C 1,6 D • Vectra-A 1,7 D • Astramax 1,6/1,7 D • Astravan 1,6/1,7 D**
Year:	**1986-96**
Engine Code:	**16DA, 17D**

Replacement Interval Guide

Vauxhall recommend:

→90MY:
Check every 36,000 miles or 4 years, whichever occurs first.
Replacement every 54,000 miles or 8 years, whichever occurs first.

91MY→:
Replacement every 36,000 miles or 4 years, whichever occurs first.
The previous use and service history of the vehicle must always be taken into account.
Refer to Timing Belt Replacement Intervals at the front of this manual.

Check For Engine Damage

CAUTION: This engine has been identified as an INTERFERENCE engine in which the possibility of valve-to-piston damage in the event of a timing belt failure is MOST LIKELY to occur.
A compression check of all cylinders should be performed before removing the cylinder head.

Repair Times – hrs

Remove & install:

Except Ascona/Vectra/Cavalier	2,70
PAS	+0,20
Ascona/Vectra/Cavalier	2,30

Special Tools

- ■ Flywheel locking tool – Kent Moore No.KM-517-B.
- ■ Tension gauge – 16DA: Kent Moore No.KM-510-2.
- ■ Tension gauge – 17D: Kent Moore No.KM-510-A.
- ■ Measuring bar – Kent Moore No.KM-661-1.
- ■ Holding tool – Kent Moore No.KM-661-2.
- ■ Water pump spanner – Kent Moore No.KM-509-A.
- ■ Dial gauge – Kent Moore No.MKM-571-A.

Removal

1. Remove:
   - ❏ Bell housing lower inspection plate.
   - ❏ Auxiliary drive belt.
   - ❏ Crankshaft pulley bolts **1**.
   - ❏ Crankshaft pulley **2**. Use tool KM-517-B.
   - ❏ Timing belt covers **3** & **4**.
   - ❏ Cylinder head cover.
2. Turn crankshaft clockwise to TDC on No.1 cylinder. Ensure timing marks aligned **5** & **6**.
3. Hold camshaft. Use 22 mm spanner on flats of camshaft at No.4 cylinder. Slacken bolt **7**.
4. Slacken water pump bolts. Turn water pump to release tension on belt **8**. Use tool KM-509-A.
5. Undo RH front engine mounting from sidemember. Move engine away from sidemember.
6. Remove timing belt from sprockets and slip belt between sidemember and engine mounting.

Installation

1. Ensure timing marks aligned **5** & **6**.
2. Slide timing belt between sidemember and engine mounting and fit to sprockets.
3. Fit engine mounting to sidemember.
4. Attach tension gauge to belt at ▽.
   Tool KM-510-2 (16DA)/KM-510-A (17D).
5. Turn water pump clockwise to tension belt.
   Use tool KM-509-A.
   Engine COLD: 9,5 units (new belt), 9,0 units (used belt).
   Engine HOT: 7,5 units (new belt), 6,0 units (used belt).
6. Tighten water pump bolts to 25 Nm.
7. Ensure timing marks aligned **5** & **6**.
8. Temporarily tighten camshaft sprocket bolt **7**.
9. Check valve timing as follows:
   - ❏ Turn crankshaft to 90° BTDC.
   - ❏ Install measuring bar KM-661-1. Ensure left-hand locating pins are positioned in bores of camshaft housing **9**.
   - ❏ Install dial gauge MKM-571-A with foot of dial gauge on base circle of No.1 cylinder inlet valve cam lobe (2nd lobe from front) **10**.
   - ❏ Set dial gauge to zero with measuring foot on cam lobe base circle.
   - ❏ Move measuring bar KM-661-1 to left, so that right-hand locating pins are positioned in bores of camshaft housing and dial gauge foot is over cam lobe **11**.
   - ❏ Turn crankshaft clockwise until TDC marks aligned **5** & **6**.
   - ❏ Dial gauge should indicate 0,55±0,03 mm.
10. If not, adjust as follows:
    - ❏ Slacken camshaft sprocket bolt **7**. Ensure position of crankshaft and camshaft remain unaltered.
    - ❏ Install holding tool KM-661-2 over No.4 cylinder **12**.
    - ❏ Adjust position of camshaft until dial gauge indicates approximately 0,80 mm **14**. Use spanner **13**.
    - ❏ Carry out fine adjustment of camshaft using screw **15** until dial gauge indicates 0,60-0,64 mm.
    - ❏ Fit new camshaft sprocket bolt **7**.
      Tightening torque: 75 Nm + 60-65°.
11. Remove holding tool **12**.
12. Carefully remove measuring bar KM-661-1 without altering dial gauge position.
13. Turn crankshaft two turns clockwise. Ensure TDC marks aligned **5** & **6**.
14. Install measuring bar KM-661-1. Ensure right-hand locating pins are positioned in bores of camshaft housing **11**. Dial gauge should indicate 0,55±0,03 mm.
15. Install components in reverse order of removal.
16. Tighten crankshaft pulley bolts to 20 Nm **1**.

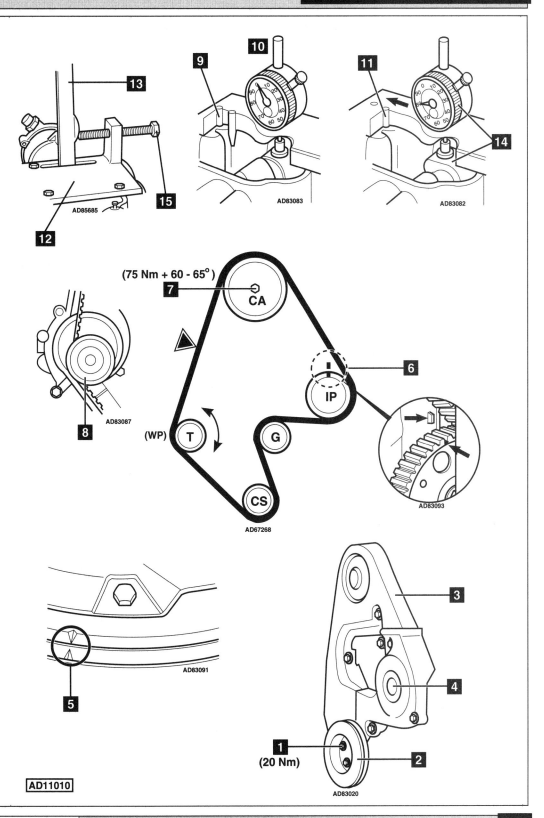

(75 Nm + 60 - 65°)

AD83083

AD83082

AD85685

AD83087

(WP)

AD67268

AD83093

AD83091

AD83020

(20 Nm)

AD11010

Model:	**Astra-F 1,7 D/TD • Astravan 1,7 D/TD • Cavalier 1,7 D • Vectra-A 1,7 D**
Year:	**1992-98**
Engine Code:	**17DR, X17DTL**

Replacement Interval Guide

Vauxhall recommend replacement as follows:

Astra-F/Astravan:

→94MY - replacement every 36,000 miles or 4 years, whichever occurs first.

95-96MY - replacement every 40,000 miles or 4 years, whichever occurs first.

97MY→ - replacement every 80,000 miles or 8 years, whichever occurs first.

Cavalier/Vectra:

→94MY - replacement every 36,000 miles or 4 years, whichever occurs first.

95MY - replacement every 40,000 miles or 4 years, whichever occurs first.

The previous use and service history of the vehicle must always be taken into account.

Refer to Timing Belt Replacement Intervals at the front of this manual.

Check For Engine Damage

CAUTION: This engine has been identified as an INTERFERENCE engine in which the possibility of valve-to-piston damage in the event of a timing belt failure is MOST LIKELY to occur.

A compression check of all cylinders should be performed before removing the cylinder head.

Repair Times – hrs

Remove & install:

Astra-F	2,20
Vectra/Cavalier	2,40

Special Tools

- Flywheel locking tool – Kent Moore No.KM-517-B.
- TDC position tool – Kent Moore No.KM-851.
- Measuring bar – Kent Moore No.KM-661-1.
- Holding tool – Kent Moore No.KM-661-2.
- Dial gauge – Kent Moore No.KM-571-B.
- Flywheel timing pin (X17DTL 97MY→) – Kent Moore No.KM-951.

Special Precautions

- Disconnect battery earth lead.
- DO NOT turn crankshaft or camshaft when timing belt removed.
- Remove glow plugs to ease turning engine.
- Turn engine in normal direction of rotation (unless otherwise stated).
- DO NOT turn engine via camshaft or other sprockets.
- Observe all tightening torques.
- Check diesel injection pump timing after belt replacement.

Removal

1. Remove:
   - Auxiliary drive belts.
   - Cylinder head cover.
   - 17DR: Clutch cover plate.
   - (X17DTL →96MY: Flywheel cover.
   - Crankshaft pulley **18**. Use tool No.KM-517-B.
   - Timing belt upper cover **2**.
   - Timing belt lower cover **3**.
2. 17DR: Turn crankshaft clockwise until timing marks aligned **1** & **6**.
3. X17DTL →96MY: Fit tool No.KM-851 to flywheel housing **10**. Turn crankshaft clockwise until timing marks aligned **6** & **11**.
4. X17DTL 97MY→: Turn crankshaft clockwise until timing marks aligned **6**. Insert timing pin in flywheel housing **20**. Tool No.KM-951.
   *NOTE: Hold in position by attaching timing pin spring to bolt **21**.*
5. Slacken tensioner bolt **5**.
6. Turn tensioner clockwise until pointer on LH stop **9**. Use Allen key **8**.
7. Lightly tighten tensioner bolt **5**.
8. Remove timing belt.
   NOTE: Mark direction of rotation on belt with chalk if belt is to be reused.

Installation

1. 17DR: Ensure timing marks aligned **1** & **6**.
2. X17DTL →96MY: Ensure timing marks aligned **6** & **11**.
3. X17DTL 97MY→: Ensure flywheel timing pin located correctly **20**. Tool No.KM-951.
4. Fit timing belt. Ensure belt is taut between sprockets.
5. Slacken tensioner bolt **5**.
6. Turn tensioner fully anti-clockwise until pointer on RH stop **12**. Use Allen key.
7. Tighten tensioner bolt to 25 Nm **5**.
8. X17DTL 97MY→: Remove flywheel timing pin **20**. Tool No.KM-951.
9. Turn crankshaft two turns clockwise.
10. 17DR: Ensure timing marks aligned **1** & **6**.
11. X17DTL →96MY: Ensure timing marks aligned **6** & **11**.
12. X17DTL 97MY→: Ensure flywheel timing pin can be inserted **20**. Tool No.KM-951.

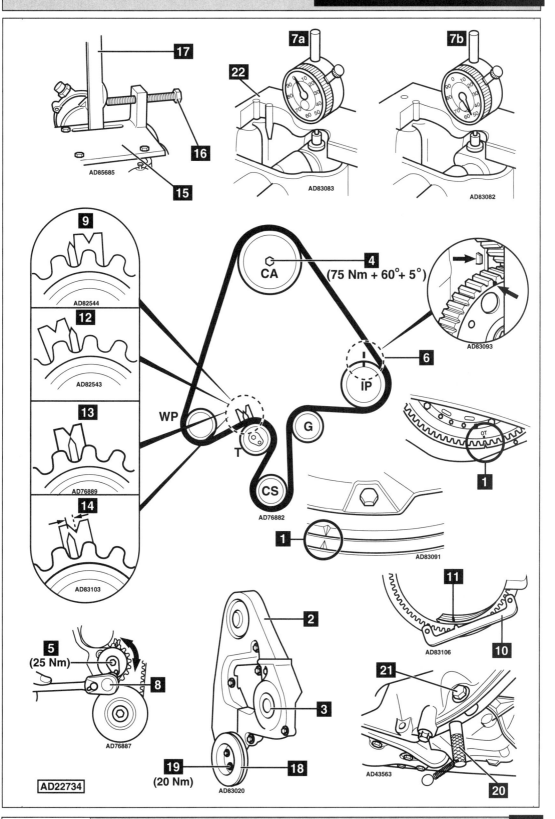

17

22

7a

7b

AD85685

16

15

AD83083

AD83082

9

AD82544

12

AD82543

13

AD76889

14

AD83103

4
(75 Nm + 60° + 5°)

CA

AD83093

WP

6

IP

T

G

1

CS

AD76882

1

AD83091

11

10

AD83106

5
(25 Nm)

8

2

21

AD76887

3

AD22734

19
(20 Nm)

18

AD83020

AD43563

20

←

13. Slacken tensioner bolt **5**. Turn tensioner clockwise until pointer aligned with: 'V' notch in bracket (new belt) **13**. LH edge of 'V' notch (used belt) **14**.

14. Tighten tensioner bolt to 25 Nm **5**.

15. Turn crankshaft to 90° BTDC.

16. Install measuring bar KM-661-1 **22**. Ensure left-hand locating pins are positioned in bores of camshaft housing **7a**.

17. Install dial gauge KM-571-B with foot of dial gauge on base circle of No.1 cylinder inlet valve cam lobe (2nd lobe from front) **7a**.
    NOTE: Dial gauge must be fitted with 10 mm diameter flat measuring foot. Pre-tension must not exceed 0,50 mm.

18. Set dial gauge to zero with measuring foot on cam lobe base circle.

19. Move measuring bar KM-661-1 **22** to left, so that right-hand locating pins are positioned in bores of camshaft housing and dial gauge foot is over cam lobe **7b**.

20. 17DR: Turn crankshaft clockwise until timing marks aligned **1** & **6**.

21. X17DTL →96MY: Turn crankshaft clockwise until timing marks aligned **6** & **11**.

22. X17DTL 97MY→: Turn crankshaft clockwise until timing marks aligned **6**. Insert timing pin in flywheel housing **20**. Tool No.KM-951.
    *NOTE: Hold in position by attaching timing pin spring to bolt **21**.*

23. Dial gauge should indicate 0,55±0,03 mm.

24. If not: Slacken camshaft sprocket bolt **4**. Ensure position of crankshaft and camshaft remain unaltered.

25. Install holding tool KM-661-2 over No.4 cylinder **15**.

26. Adjust position of camshaft until dial gauge indicates approximately 0,80 mm. Use spanner **17**.

27. Carry out fine adjustment of camshaft using screw **16** until dial gauge indicates 0,60-0,64 mm.

28. Fit new camshaft sprocket bolt **4**. Tightening torque: 75 Nm + 60° + 5°.

29. Remove holding tool **15**.

30. Carefully remove measuring bar KM-661-1 **22** without altering dial gauge position.

31. X17DTL 97MY→: Remove flywheel timing pin **20**.

32. Turn crankshaft two turns clockwise to TDC.

33. 17DR: Ensure timing marks aligned **1** & **6**.

34. X17DTL →96MY: Ensure timing marks aligned **6** & **11**.

35. X17DTL 97MY→: Ensure timing marks aligned **6**. Insert timing pin in flywheel housing **20**. Tool No.KM-951.
    *NOTE: Hold in position by attaching timing pin spring to bolt **21**.*

36. Install measuring bar KM-661-1 **22**. Ensure right-hand locating pins are positioned in bores of camshaft housing **7b**. Dial gauge should indicate 0,55±0,03 mm.

37. Install components in reverse order of removal.

38. Tighten crankshaft pulley bolts to 20 Nm **19**.

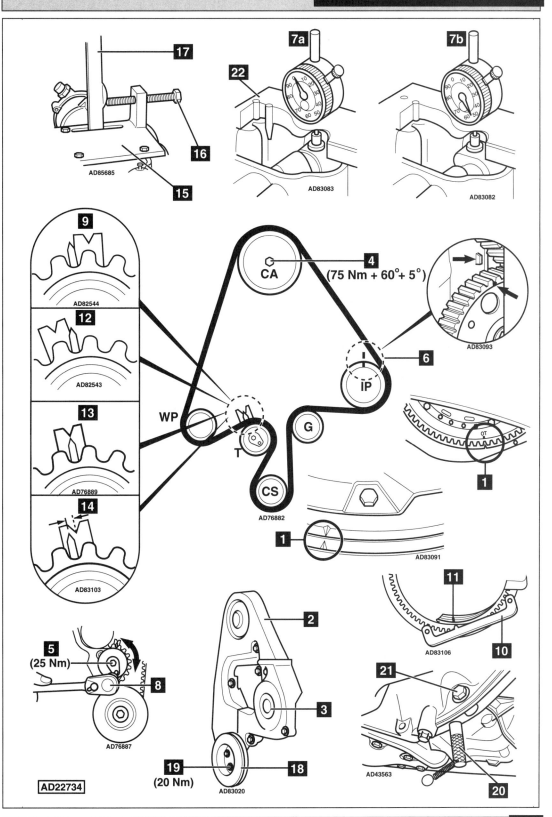

17

16

15

AD85685

22

7a

7b

AD83083

AD83082

9

AD82544

12

AD82543

13

AD76889

14

AD83103

CA

4

(75 Nm + 60° + 5°)

AD83093

6

IP

I

WP

G

T

CS

AD76882

OT

1

1

AD83091

11

10

AD83106

5

(25 Nm)

8

AD76887

2

3

19

(20 Nm)

18

AD83020

21

20

AD43563

AD22734

VAUXHALL-OPEL

Model:	**Astra-G 1,7 TD**
Year:	**1998-00**
Engine Code:	**X17DTL**

Replacement Interval Guide

Vauxhall recommend replacement every
80,000 miles or 8 years, whichever occurs first.
*The previous use and service history of the vehicle
must always be taken into account.*
*Refer to Timing Belt Replacement Intervals at the front
of this manual.*

Check For Engine Damage

*CAUTION: This engine has been identified as an
INTERFERENCE engine in which the possibility of
valve-to-piston damage in the event of a timing belt
failure is MOST LIKELY to occur.*
*A compression check of all cylinders should be
performed before removing the cylinder head.*

Repair Times – hrs

Remove & install	1,60

Special Tools

- Engine support tool – Kent Moore No.KM-6001.
- Engine alignment tool – Kent Moore No.KM-909-B.
- Dial gauge – Kent Moore No.MKM-571-B.
- Measuring bar – Kent Moore No.KM-661-1.
- Holding tool – Kent Moore No.KM-661-2.
- Injection pump locking pin –
  Kent Moore No.KM-6011.
- Flywheel timing pin – Kent Moore No.KM-951.

Special Precautions

- Disconnect battery earth lead.
- DO NOT turn crankshaft or camshaft when timing belt
  removed.
- Remove glow plugs to ease turning engine.
- Turn engine in normal direction of rotation (unless
  otherwise stated).
- DO NOT turn engine via camshaft or other sprockets.
- Observe all tightening torques.

Removal

1. Remove:
   - ❑ Air filter and intake pipe.
   - ❑ Mass air flow (MAF) sensor.
   - ❑ Timing belt upper cover **1**.
   - ❑ Auxiliary drive belt(s).
2. Raise and support front of vehicle.
3. Remove:
   - ❑ RH front wheel.
   - ❑ Engine undershield.
   - ❑ Crankshaft pulley bolts **2**.
   - ❑ Crankshaft pulley **3**.
   - ❑ Timing belt lower cover **4**.

4. Support engine. Use tool Nos.KM-6001/KM-909-B.
5. Remove:
   - ❑ RH engine mounting and bracket.
   - ❑ Auxiliary drive belt tensioner.
   - ❑ Cylinder head cover.
6. Turn crankshaft to 90° BTDC on
   No.1 cylinder **5**.
7. Ensure No.1 cylinder exhaust valve cam lobe
   points vertically upwards **6**.
8. Slacken injection pump sprocket bolts **7**.
   Tighten bolts finger tight.
9. Slacken tensioner bolt **8**.
10. Turn tensioner clockwise until pointer to left of
    LH stop **9**. Use Allen key **10**.
11. Lightly tighten tensioner bolt **8**.
12. Remove timing belt.
13. Remove camshaft sprocket bolt **11**. DO NOT
    reuse bolt.
14. Fit new camshaft sprocket bolt **11**. Tighten bolt
    finger tight. Ensure sprocket can still turn.
    **NOTE: Mark direction of rotation on belt with
    chalk if belt is to be reused.**

Installation

1. Ensure No.1 cylinder exhaust valve cam lobe
   points vertically upwards **6**.
2. Ensure crankshaft at 90° BTDC on
   No.1 cylinder **5**.
3. Install measuring bar KM-661-1 **12**. Ensure
   left-hand locating pins are positioned in bores of
   camshaft housing.
4. Install dial gauge MKM-571-B with foot of dial
   gauge on base circle of No.1 cylinder inlet valve
   cam lobe (2nd lobe from front) **13a**.
   **NOTE: Dial gauge must be fitted with 10 mm
   diameter flat measuring foot. Pre-load must not
   exceed 0,50 mm.**
5. Set dial gauge to zero with measuring foot on
   cam lobe base circle.
6. Move measuring bar KM-661-1 **12** to left, so that
   right-hand locating pins are positioned in bores
   of camshaft housing and dial gauge foot is over
   cam lobe **13b**.
7. Install holding tool KM-661-2 over
   No.4 cylinder **14**.
8. Adjust position of camshaft until dial gauge
   indicates approximately 0,80 mm. Use
   spanner **15**.

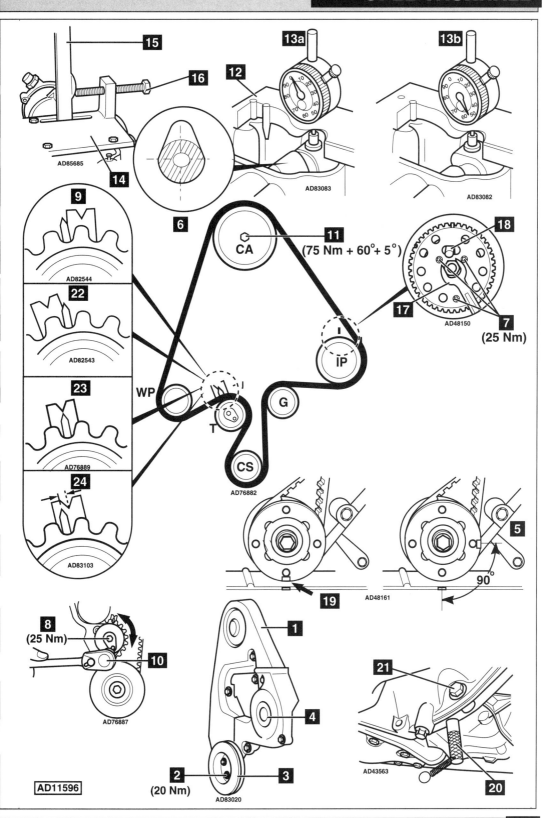

15

16

AD85685

14

12

13a

AD83083

13b

AD83082

9

AD82544

22

AD82543

23

AD76889

24

AD83103

6

11
(75 Nm + 60° + 5°)

CA

WP

T

G

CS

IP

18

17

7
(25 Nm)

AD48150

AD76882

19

5

90°

AD48161

8
(25 Nm)

10

AD76887

1

4

2
(20 Nm)

3

AD83020

21

20

AD43563

AD11596

←

9. Carry out fine adjustment of camshaft using screw until dial gauge indicates 0,60-0,64 mm.

10. Turn injection pump sprocket slowly clockwise. Use spanner ⟨17⟩. Insert locking pin in injection pump ⟨18⟩. Tool No.KM-6011.

11. Turn crankshaft clockwise to TDC on No.1 cylinder. Ensure timing marks aligned ⟨19⟩.

12. Insert timing pin in flywheel housing ⟨20⟩. Tool No.KM-951.
    NOTE: Hold in position by attaching timing pin spring to bolt ⟨21⟩.

13. Fit timing belt in clockwise direction, starting at crankshaft sprocket. Ensure belt is taut between sprockets on non-tensioned side.

14. Slacken tensioner bolt ⟨8⟩.

15. Turn tensioner anti-clockwise until pointer near to RH stop ⟨22⟩. Use Allen key ⟨10⟩.

16. Tighten tensioner bolt to 25 Nm ⟨8⟩.

17. Hold camshaft sprocket. Tighten bolt ⟨11⟩. Tightening torque: 75 Nm + 60° + 5°.

18. Hold injection pump sprocket. Tighten bolts ⟨7⟩. Tightening torque: 25 Nm.

19. Remove holding tool ⟨14⟩.

20. Carefully remove measuring bar KM-661-1 ⟨12⟩ without altering dial gauge position.

21. Remove:
    ❑ Flywheel timing pin ⟨20⟩.
    ❑ Locking pin ⟨18⟩.

22. Turn crankshaft slowly two turns clockwise to TDC on No.1 cylinder. Ensure timing marks aligned ⟨19⟩.

23. Slacken tensioner bolt ⟨8⟩. Turn tensioner clockwise until pointer aligned as follows:
    ❑ New belt – 'V' notch in bracket ⟨23⟩.
    ❑ Used belt – 4 mm left of 'V' notch ⟨24⟩.

24. Tighten tensioner bolt to 25 Nm ⟨8⟩.

25. Turn crankshaft slowly two turns clockwise to TDC on No.1 cylinder. Ensure timing marks aligned ⟨19⟩.

26. Check tensioner marks as follows:
    ❑ New belt – 'V' notch in bracket ⟨23⟩.
    ❑ Used belt – 4 mm left of 'V' notch ⟨24⟩.

27. If not: Repeat tensioning procedure.

28. Turn crankshaft two turns clockwise. Ensure timing marks aligned ⟨19⟩. Insert timing pin in flywheel housing ⟨20⟩. Tool No.KM-951.

29. Insert locking pin in injection pump ⟨18⟩. Tool No.KM-6011.

30. Install measuring bar KM-661-1 ⟨12⟩. Ensure right-hand locating pins are positioned in bores of camshaft housing [13b].

31. Dial gauge should indicate 0,55±0,03 mm.

32. If not: Repeat installation procedure.

33. Install components in reverse order of removal.

34. Tighten crankshaft pulley bolts to 20 Nm ⟨2⟩.

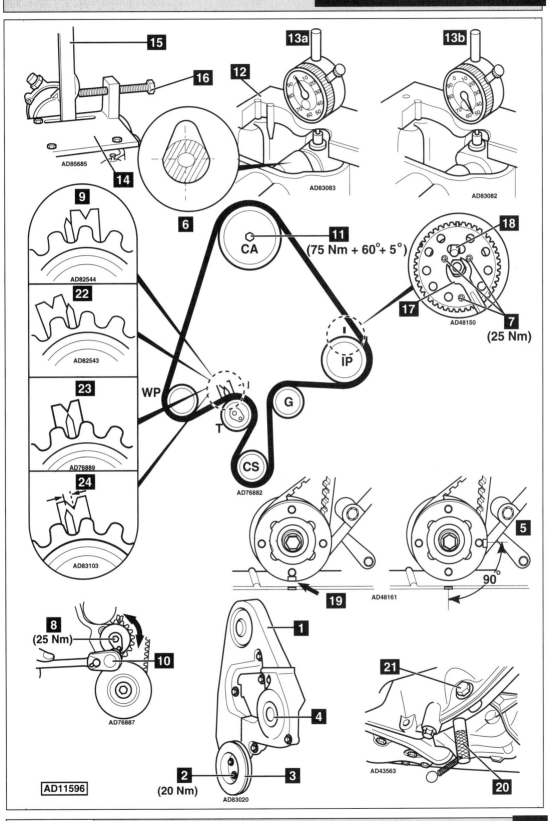

VAUXHALL-OPEL

Model:	**Frontera 2,8 TD**
Year:	**1995-96**
Engine Code:	**4JB1-TC**

Replacement Interval Guide

Vauxhall recommend replacement every 60,000 miles or 6 years, whichever occurs first.
The previous use and service history of the vehicle must always be taken into account.
Refer to Timing Belt Replacement Intervals at the front of this manual.

Check For Engine Damage

CAUTION: This engine has been identified as an INTERFERENCE engine in which the possibility of valve-to-piston damage in the event of a timing belt failure is MOST LIKELY to occur.
A compression check of all cylinders should be performed before removing the cylinder head.

Repair Times – hrs

Remove & install	1,90

Special Tools

■ Spring balance.

Special Precautions

■ Disconnect battery earth lead.
■ DO NOT turn crankshaft or camshaft when timing belt removed.
■ Remove glow plugs to ease turning engine.
■ Turn engine in normal direction of rotation (unless otherwise stated).
■ DO NOT turn engine via camshaft or other sprockets.
■ Observe all tightening torques.
■ Check diesel injection pump timing after belt replacement.

Removal

1. Remove:
   ❏ Cooling fan.
   ❏ Auxiliary drive belt(s).
   ❏ Crankshaft pulley (4 bolts) **1**.
   ❏ Timing belt upper cover **2**.
   ❏ Timing belt lower cover **3**.
   ❏ Camshaft sprocket flange **4**.
   ❏ Injection pump sprocket flange **5**.
2. Turn crankshaft clockwise to TDC on No.1 cylinder. Ensure timing marks aligned **6**.
3. Screw suitable locking bolts into camshaft and injection pump sprockets **7** & **8**.
4. Remove tensioner lever **9**.
5. Slacken tensioner bolt **10**. Move tensioner pulley away from belt. Lightly tighten bolt.
6. Remove timing belt.

Installation

1. Ensure locking bolts located in sprockets **7** & **8**.
2. Ensure timing marks aligned **6**.
3. Fit timing belt in clockwise direction, starting at crankshaft sprocket.
4. Slacken tensioner pulley bolt **10**. Push tensioner pulley against belt. Tighten bolt finger tight.
5. Ensure belt is taut between sprockets on non-tensioned side.
6. Position tensioner lever **9** against tensioner pulley housing.
7. Remove locking bolts from sprockets **7** & **8**.
8. Apply a load of 9 kg to tensioner lever **11**. Use spring balance.
9. Temporarily tighten tensioner pulley bolt **10**.
10. Turn crankshaft 45° anti-clockwise. Slacken tensioner pulley bolt and tighten to 76 Nm **10**.
    NOTE: DO NOT turn crankshaft clockwise during tensioning procedure.
11. Fit tensioner lever to original position **9**. Tighten nuts **12**.
12. Install:
    ❏ Camshaft sprocket flange **4**.
    ❏ Injection pump sprocket flange **5**.
13. Install components in reverse order of removal.

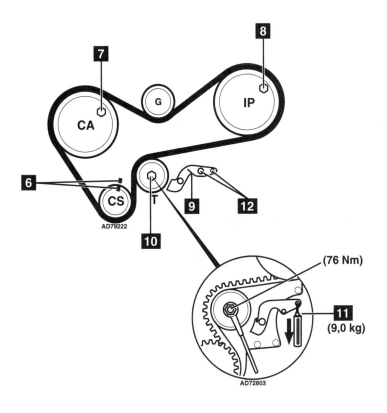

AD79222

(76 Nm)

(9,0 kg)

AD72803

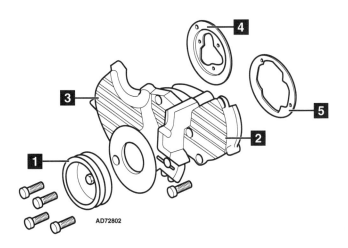

AD72802

AD10330

VAUXHALL-OPEL

Model:	**Monterey 3,0 TD**
Year:	**1998-02**
Engine Code:	**4JX1**

Replacement Interval Guide

Vauxhall recommend replacement every 120,000 miles.
The previous use and service history of the vehicle must always be taken into account.
Refer to Timing Belt Replacement Intervals at the front of this manual.

Check For Engine Damage

CAUTION: This engine has been identified as an INTERFERENCE engine in which the possibility of valve-to-piston damage in the event of a timing belt failure is MOST LIKELY to occur. A compression check of all cylinders should be performed before removing the cylinder head(s).

Repair Times – hrs

Remove & install	1,30

Special Tools

- 1 x M8 bolt.

Special Precautions

- Disconnect battery earth lead.
- DO NOT turn crankshaft or camshaft when timing belt removed.
- Remove glow plugs to ease turning engine.
- Turn engine in normal direction of rotation (unless otherwise stated).
- DO NOT turn engine via camshaft or other sprockets.
- Observe all tightening torques.

Removal

1. Remove:
   - ❏ Air filter and air intake hoses.
   - ❏ Viscous fan.
   - ❏ Auxiliary drive belts.
   - ❏ Water pump pulley.
2. Disconnect:
   - ❏ Camshaft position (CMP) sensor.
3. Turn crankshaft to TDC on No.1 cylinder. Ensure timing marks aligned **1**.
4. Remove:
   - ❏ Timing belt cover **2**.
   - ❏ Camshaft position (CMP) sensor bracket **3**.
5. Insert one M8 bolt through camshaft sprocket timing hole into cylinder head **4**.
6. Slacken tensioner bolts **5** & **6**.
7. Push tensioner pulley against belt.
8. Remove:
   - ❏ Tensioner spring **7**.
   - ❏ Timing belt.

Installation

1. Ensure timing marks aligned **1**.
2. Ensure locking bolt located in camshaft sprocket **4**.
3. Fit timing belt in clockwise direction, starting at camshaft sprocket. Ensure belt is taut on non-tensioned side.
4. Push tensioner pulley against belt.
5. Fit tensioner spring **7**.
6. Remove locking bolt from camshaft sprocket **4**.
7. Turn crankshaft slowly two turns clockwise until timing marks aligned **1**.
8. Ensure locking bolt can be inserted in camshaft sprocket **4**.
9. Tighten tensioner bolt **5**.
   Tightening torque: 50 Nm.
10. Tighten tensioner bolt **6**.
    Tightening torque: 20 Nm.
    NOTE: Observe correct sequence.
11. Remove locking bolt from camshaft sprocket **4**.
12. Install components in reverse order of removal.

Autodata

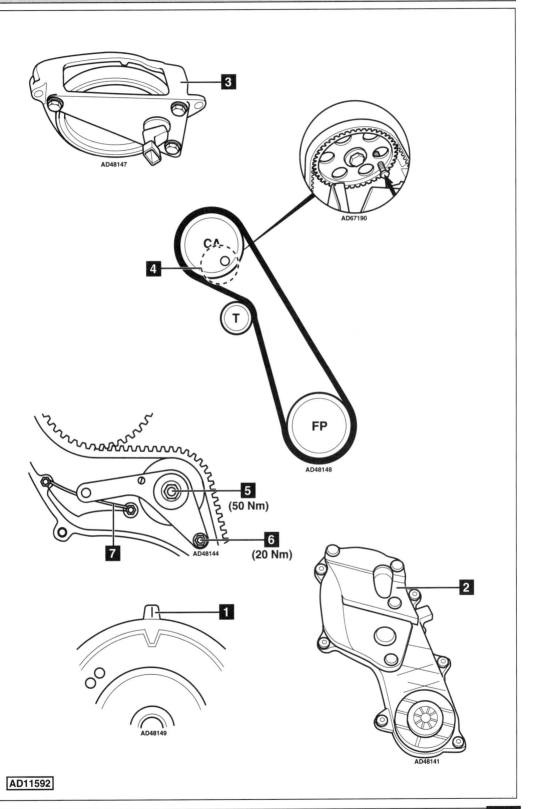

AD48147

AD67190

3

CA

4

T

FP

AD48148

5
(50 Nm)

6
(20 Nm)

7

AD48144

2

1

AD48149

AD48141

AD11592

VAUXHALL-OPEL

Model:	**Monterey 3,1 Turbo D**
Year:	**1992-98**
Engine Code:	**4JG2T/C**

Replacement Interval Guide

Vauxhall recommend replacement as follows:
→95MY - replacement every 54,000 miles or
6 years, whichever occurs first.
96MY→ - replacement every 60,000 miles or
6 years, whichever occurs first.
*The previous use and service history of the vehicle
must always be taken into account.*
*Refer to Timing Belt Replacement Intervals at the front
of this manual.*

Check For Engine Damage

*CAUTION: This engine has been identified as an
INTERFERENCE engine in which the possibility of
valve-to-piston damage in the event of a timing belt
failure is MOST LIKELY to occur.*
*A compression check of all cylinders should be
performed before removing the cylinder head.*

Repair Times – hrs

Remove & install	2,00
AC	+0,10

Special Tools

- Spring balance.
- 2 bolts – M8 x 1,25 x 40 mm.

Special Precautions

- Disconnect battery earth lead.
- DO NOT turn crankshaft or camshaft when timing belt removed.
- Remove glow plugs to ease turning engine.
- Turn engine in normal direction of rotation (unless otherwise stated).
- DO NOT turn engine via camshaft or other sprockets.
- Observe all tightening torques.
- Check diesel injection pump timing after belt replacement.

Removal

1. Remove:
   - ❏ Auxiliary drive belts.
   - ❏ Cooling fan and pulley.
   - ❏ Crankshaft pulley (4 bolts) **1**.
   - ❏ Timing belt upper cover **2**.
   - ❏ Timing belt lower cover **3**.
   - ❏ Camshaft sprocket flange **4**.
   - ❏ Injection pump sprocket flange **5**.
2. Turn crankshaft clockwise to TDC on No.1 cylinder. Ensure timing marks aligned **6**.
3. Lock camshaft sprocket **7**. Use M8 x 40 bolt.
4. Lock injection pump sprocket **8**. Use M8 x 40 bolt.
5. Slacken tensioner pulley bolt **9**. Move tensioner pulley away from belt. Lightly tighten bolt.
6. Remove timing belt.

Installation

1. Ensure locking bolts located correctly **7** & **8**.
2. Ensure timing marks aligned **6**.
3. Fit timing belt in clockwise direction, starting at crankshaft sprocket. Ensure belt is taut between sprockets.
   *NOTE: 'ISUZU' lettering on belt should be readable from front of engine **17**.*
4. Slacken tensioner pulley bolt **9**. Push tensioner pulley against belt.
5. Ensure belt is taut between injection pump sprocket and tensioner pulley as well as crankshaft sprocket and camshaft sprocket.
6. Remove tensioner lever bolt **10**. Slacken tensioner lever nut **11**.
7. Position tensioner lever **12** against tensioner pulley housing.
8. Remove locking bolts from sprockets **7** & **8**.
9. Apply a load of 9 kg to tensioner lever **13**. Use spring balance.
10. Tighten tensioner bolt to 76 Nm **9**.
11. Turn crankshaft 45° anti-clockwise. Repeat tensioning procedure.
    NOTE: DO NOT turn crankshaft clockwise during tensioning procedure.
12. Position tensioner lever. Fit bolt **10**. Tighten bolt and nut **10** & **11**.
13. Install:
    - ❏ Injection pump sprocket flange **5**. Tighten bolts to 19 Nm.
    - ❏ Camshaft sprocket flange **4**. Tighten bolts to 19 Nm.
14. Install components in reverse order of removal.
15. Tighten crankshaft pulley bolts to 19 Nm **16**.
16. Check and adjust injection pump timing.
17. Reset timing belt replacement warning lamp as follows:
    - ❏ Remove instrument panel.
    - ❏ Remove masking tape from hole **15**.
    - ❏ Remove screw from hole **14**.
    - ❏ Insert screw removed from hole **14** in hole **15** and cover hole **14** with masking tape.
    - ❏ Fit instrument panel.
    - ❏ Reverse procedure if timing belt previously changed.

Autodata

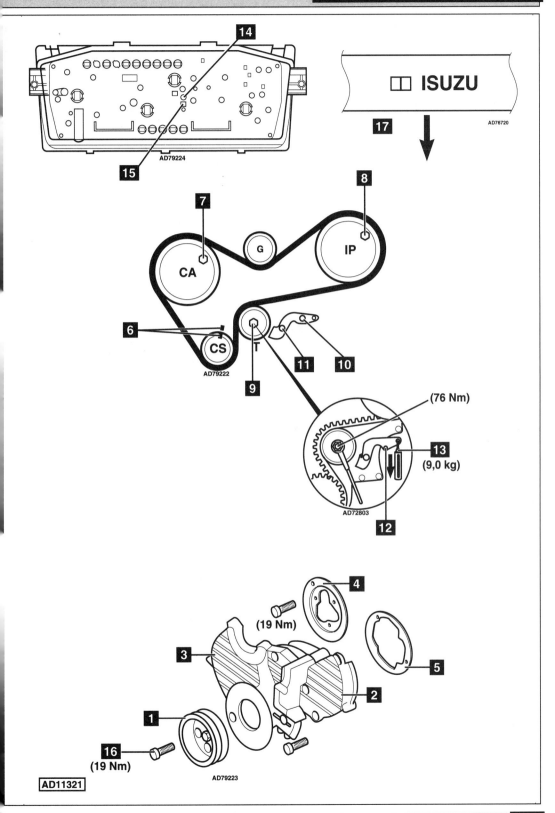

AD79224

ISUZU

AD76720

(76 Nm)

(9,0 kg)

AD72803

(19 Nm)

(19 Nm)

AD79222

AD79223

AD11321

VAUXHALL-OPEL

Model:	**Midi 2,0/2,4 TD • Midi 2,2 D**
Year:	**1988-97**
Engine Code:	**4FC1-T, 4FD1, 4FG1-T**

Replacement Interval Guide

Vauxhall recommend:
2,0 TD/2,2 D - replacement every 48,000 miles.
NOTE: Timing belt warning lamp will illuminate at 50,000 miles.
2,4 TD - replacement every 60,000 miles or 6 years, whichever occurs first.
The previous use and service history of the vehicle must always be taken into account.
Refer to Timing Belt Replacement Intervals at the front of this manual.

Check For Engine Damage

CAUTION: This engine has been identified as an INTERFERENCE engine in which the possibility of valve-to-piston damage in the event of a timing belt failure is MOST LIKELY to occur.
A compression check of all cylinders should be performed before removing the cylinder head.

Repair Times – hrs

Remove & install:	
2,0 TD	2,10
2,2 D	1,90
PAS	+0,10
2,4 TD	1,70
PAS	+0,30

Special Tools

- Camshaft setting bar – Kent Moore No.KM-8070.
- Tension gauge – Kent Moore No.KM-4088.
- M8 bolt – to lock injection pump sprocket.
- Puller – for camshaft sprocket.

Special Precautions

- Disconnect battery earth lead.
- DO NOT turn crankshaft or camshaft when timing belt removed.
- Remove glow plugs to ease turning engine.
- Turn engine in normal direction of rotation (unless otherwise stated).
- DO NOT turn engine via camshaft or other sprockets.
- Observe all tightening torques.
- Check diesel injection pump timing after belt replacement.

Removal

1. Drain coolant.
2. Remove:
   - ❏ Auxiliary drive belt.
   - ❏ Cooling fan and pulley.
   - ❏ Timing belt upper cover (10 bolts) **1**.
   - ❏ Cooling fan cowling.
   - ❏ Front undershield.
   - ❏ Coolant by-pass pipe **2**.
3. Turn crankshaft clockwise to TDC on No.1 cylinder. Ensure timing marks aligned **4** & **5**.

4. Screw M8 bolt into injection pump sprocket **11**.
5. Remove cylinder head cover.
6. Slacken all the valve adjusting screws.
7. Fit setting bar KM-8070 to rear of camshaft **9**.
8. Remove:
   - ❏ Crankshaft pulley.
   - ❏ Timing belt lower cover **3**.
   - ❏ Timing belt holder **10**.
   - ❏ Three tensioner bolts **6**, **7** & **8**.
   - ❏ Tensioner spring.
   - ❏ Timing belt.

Installation

1. Remove camshaft sprocket bolt.
2. Withdraw camshaft sprocket approximately 3 mm. Use suitable puller.
3. Fit timing belt in anti-clockwise direction, starting at crankshaft sprocket.
4. Depress tensioner by hand. Fit tensioner spring.
5. Fit tensioner bolts. Lightly tighten bolts in following sequence: **6**, **7** & **8**.
6. Tighten camshaft sprocket bolt to 60-70 Nm.
7. Remove:
   - ❏ Locking bolt **11**.
   - ❏ Camshaft setting bar **9**.
8. Temporarily fit crankshaft pulley. Ensure crankshaft at TDC **5**.
9. Turn crankshaft two turns clockwise. Ensure timing marks aligned **4** & **5**.
10. Ensure camshaft setting bar can be fitted **9**.
11. Slacken tensioner bolts **6**, **7** & **8**.
12. Push tensioner against belt. Tighten bolts in following sequence: **6**, **7** & **8**.
13. Attach tension gauge to belt at ▽. Tool No. KM-4088.
14. Tension gauge should indicate 21-29 kg.
15. Adjust valve clearances. Inlet: 0,25 mm. Exhaust: 0,35 mm.
16. Remove crankshaft pulley.
17. Fit timing belt holder **10**. Ensure timing belt holder does not contact belt.
18. Install:
   - ❏ Timing belt lower cover.
   - ❏ Crankshaft pulley.
19. Check and adjust injection pump timing.
20. Install components in reverse order of removal.
21. Refill cooling system.
22. Reset timing belt replacement warning lamp as follows: Slide switch under fascia to opposite position **12**.

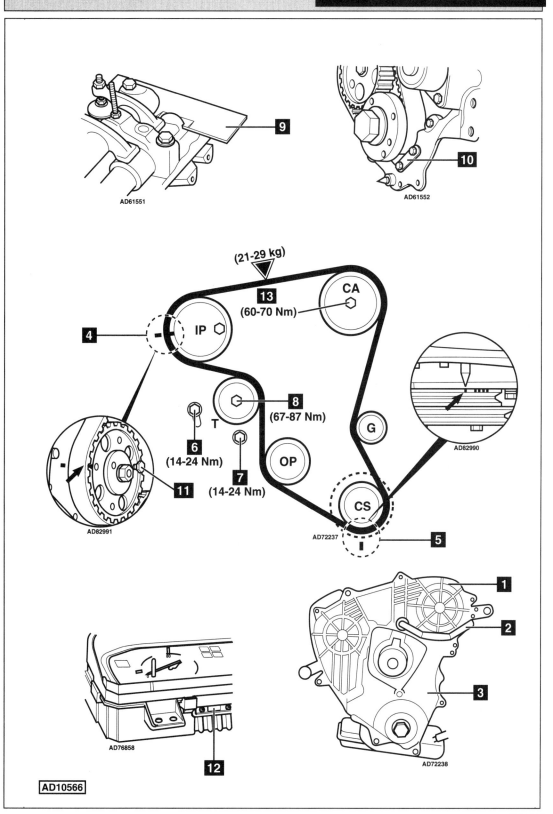

AD61551

9

AD61552

10

(21-29 kg)

13

(60-70 Nm)

CA

IP

4

8

(67-87 Nm)

T

6

(14-24 Nm)

11

7

(14-24 Nm)

OP

G

AD82990

AD82991

CS

AD72237

5

1

2

3

AD72238

AD76858

12

AD10566

Model:	**Brava 2,5D/TD**
Year:	**1990-01**
Engine Code:	**4JA1, 4JA1T, 4JG2T**

Replacement Interval Guide

Vauxhall recommend replacement as follows:
4500 mile service intervals – replacement every 54,000 miles or 6 years, whichever occurs first.
5000 mile service intervals – replacement every 60,000 miles or 6 years, whichever occurs first.
The previous use and service history of the vehicle must always be taken into account.
Refer to Timing Belt Replacement Intervals at the front of this manual.

Check For Engine Damage

CAUTION: This engine has been identified as an INTERFERENCE engine in which the possibility of valve-to-piston damage in the event of a timing belt failure is MOST LIKELY to occur.
A compression check of all cylinders should be performed before removing the cylinder head.

Repair Times – hrs

Remove & install:	
2,5D	2,20
2,5TD	2,40
PAS	+0,20
AC	+0,20

Special Tools

- Crankshaft pulley puller – No.5-8840-0086-0.
- Spring balance (0-20 kg).

Special Precautions

- Disconnect battery earth lead.
- DO NOT turn crankshaft or camshaft when timing belt removed.
- Remove glow plugs to ease turning engine.
- Turn engine in normal direction of rotation (unless otherwise stated).
- DO NOT turn engine via camshaft or other sprockets.
- Observe all tightening torques.
- Check diesel injection pump timing after belt replacement.

Removal

1. Drain coolant.
2. Remove:
   - ❏ Radiator.
   - ❏ Auxiliary drive belt.
   - ❏ Cooling fan.
   - ❏ Water pump pulley.
   - ❏ Crankshaft pulley bolt **13**.
   - ❏ Crankshaft pulley **3**.
   - ❏ Timing belt upper cover **1**.
   - ❏ Timing belt lower cover **2**.
   - ❏ Camshaft sprocket flange **6**.
   - ❏ Injection pump sprocket flange **7**.
3. Turn crankshaft to TDC on No.1 cylinder. Ensure timing marks aligned **10**.
4. Screw suitable locking bolts into camshaft and injection pump sprockets **4** & **5**.

5. Remove tensioner lever **8**.
6. Slacken tensioner pulley bolt **9**. Move tensioner pulley away from belt. Lightly tighten bolt.
7. Remove timing belt.

Installation

1. Ensure locking bolts located correctly **4** & **5**.
2. Ensure timing marks aligned **10**.
3. Fit timing belt in clockwise direction, starting at crankshaft sprocket. Ensure belt is taut between sprockets.
4. Slacken tensioner pulley bolt **9**. Push tensioner pulley against belt.
5. Fit and position tensioner lever **8** against tensioner bracket.
6. Remove locking bolts from sprockets **4** & **5**.
7. Apply a load of 10-12 kg to tensioner lever. Use spring balance **11**.
8. Tighten tensioner pulley bolt to 80 Nm **9**.
9. Turn crankshaft slowly 45° anti-clockwise. Repeat tensioning procedure.
10. Repeat tensioning procedure at 45° intervals for one anti-clockwise turn of the crankshaft.
    NOTE: DO NOT turn crankshaft clockwise during tensioning procedure.
11. Fit tensioner lever to original position. Tighten nut and bolt **12**.
12. Install:
    - ❏ Injection pump sprocket flange **6**.
    - ❏ Camshaft sprocket flange **7**.
13. Install components in reverse order of removal.
14. Tighten crankshaft pulley bolt **13**.
    Up to engine No.868393: 200 Nm.
    From engine No.868394: 186-226 Nm.
15. Check injection pump timing.
16. Refill cooling system.
17. Reset timing belt replacement warning lamp as follows:
    - ❏ Remove instrument panel.
    - ❏ Remove masking tape from hole **15**.
    - ❏ Remove screw from hole **14**.
    - ❏ Insert screw removed from hole **14** in hole **15** and cover hole **14** with masking tape.
    - ❏ Fit instrument panel.
    - ❏ Reverse procedure if timing belt previously changed.
    NOTE: The timing belt warning lamp will illuminate at 62,000 miles. Check that the manufacturer's recommended replacement interval has not been amended.

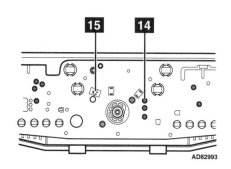

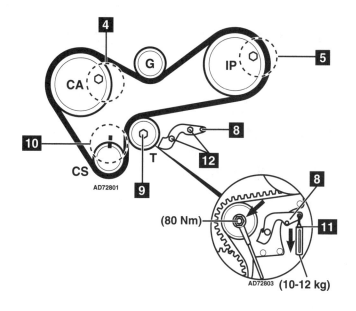

15 **14**

AD82993

4

G

CA

IP

5

10

CS

T

8

12

AD72801

9

(80 Nm)

8

11

AD72803 **(10-12 kg)**

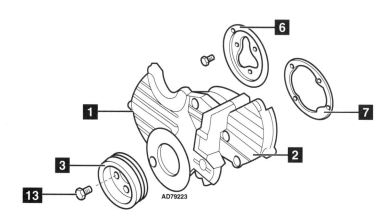

6

1

7

3

2

13

AD79223

AD10597

VAUXHALL-OPEL

Model:	**Arena 1,9 D**
Year:	**1997-01**
Engine Code:	**F8Q 606**

Replacement Interval Guide

Vauxhall recommend:
98MY: Replacement every 70,000 miles or
5 years.
99MY→: Replacement every 72,000 miles or
5 years.
The previous use and service history of the vehicle must always be taken into account.
Refer to Timing Belt Replacement Intervals at the front of this manual.

Check For Engine Damage

CAUTION: This engine has been identified as an INTERFERENCE engine in which the possibility of valve-to-piston damage in the event of a timing belt failure is MOST LIKELY to occur.
A compression check of all cylinders should be performed before removing the cylinder head.

Repair Times – hrs

Remove & install   0,90

Special Tools

- Crankshaft locking pin – No.Mot.861.
- Injection pump sprocket locking tool – No.Mot.1131.

Special Precautions

- Disconnect battery earth lead.
- DO NOT turn crankshaft or camshaft when timing belt removed.
- Remove glow plugs to ease turning engine.
- Turn engine in normal direction of rotation (unless otherwise stated).
- DO NOT turn engine via camshaft or other sprockets.
- Observe all tightening torques.
- Check diesel injection pump timing after belt replacement.

Removal

NOTE: No.1 cylinder is at flywheel end.

1. Remove:
   - ❏ Heater fan intake ducting.
   - ❏ Air filter intake hose.
   - ❏ Fuel return pipe clip.
   - ❏ Timing belt covers **1**.
2. Turn crankshaft clockwise until camshaft sprocket timing mark aligned with mark on timing belt cover **2**.
3. Ensure crankshaft at TDC on No.1 cylinder. Insert crankshaft locking pin **3**. Tool No.Mot.861.

4. Rock crankshaft slightly to ensure locking pin located correctly.
5. Ensure marks on belt aligned with marks on sprockets **6** & **7**.
6. Lock injection pump sprocket **8**. Use tool No.Mot.1131.
7. Remove:
   - ❏ Auxiliary drive belt.
   - ❏ Crankshaft pulley bolt **4**.
   - ❏ Crankshaft pulley **5**.
8. Slacken tensioner nut to release tension on belt **9**.
9. Remove timing belt.

Installation

1. Rock crankshaft slightly to ensure locking pin located correctly **3**.
2. Observe direction of rotation marks on belt **10**.
3. Fit timing belt. Ensure marks on belt aligned with marks on sprockets **6** & **7**. Ensure belt is taut between sprockets.
4. Remove:
   - ❏ Crankshaft locking pin **3**.
   - ❏ Injection pump sprocket locking tool **8**.
5. Use an M6 bolt screwed into timing belt rear cover to push on tensioner pulley **11**. Adjust position of tensioner pulley.
6. Apply a load of 30 N to belt at ▽. Belt should deflect 7,8 mm.
7. Screw in M6 bolt **11** until belt tension is correct.
8. Tighten tensioner nut to 50 Nm **9**.
9. Remove M6 bolt **11**.
10. Turn crankshaft two turns clockwise to TDC on No.1 cylinder.
11. Insert crankshaft locking pin **3**. Tool No.Mot.861.
12. Ensure marks on belt aligned with marks on sprockets **6** & **7**.
13. Recheck belt tension.
14. Install components in reverse order of removal.
15. Tighten crankshaft pulley bolt **4**. Tightening torque: 20 Nm + 115°.

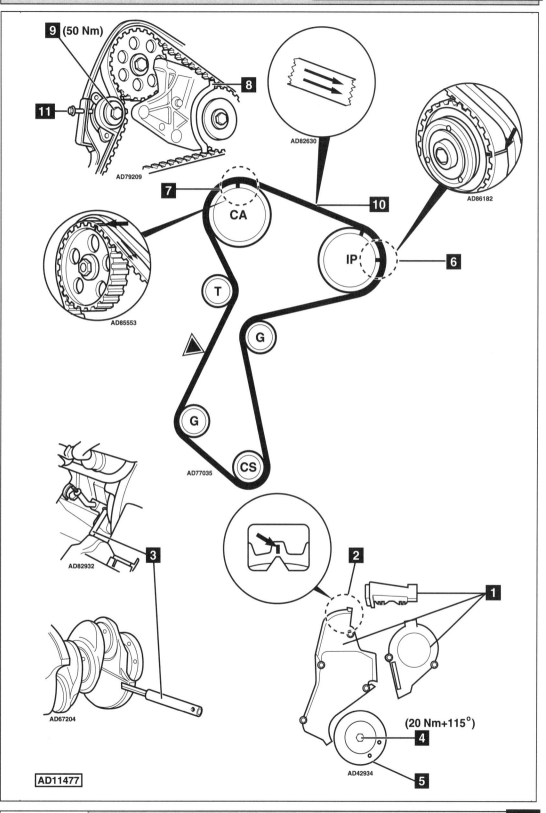

9 (50 Nm)

11

AD79209

8

AD82630

10

AD86182

7

CA

IP

6

AD85553

T

G

G

AD77035

CS

AD82932

3

AD67204

2

1

4 (20 Nm+115°)

AD42934

5

AD11477

VAUXHALL-OPEL

Model:	**Arena 2,5 D • Movano 2,5 D • Movano 2,8 D/TD**
Year:	**1997-01**
Engine Code:	**S8U-758, S8U-770, S9W-700/702**

Replacement Interval Guide

Vauxhall recommend:

Arena/Movano 2,5 D:
→98MY: Replacement every 50,000 miles or 5 years.
99MY→: Replacement every 72,000 miles or 5 years.

Movano 2,8 D/TD:
Replacement every 60,000 miles or 5 years.
The previous use and service history of the vehicle must always be taken into account.
Refer to Timing Belt Replacement Intervals at the front of this manual.

Check For Engine Damage

CAUTION: This engine has been identified as an INTERFERENCE engine in which the possibility of valve-to-piston damage in the event of a timing belt failure is MOST LIKELY to occur.
A compression check of all cylinders should be performed before removing the cylinder head.

Repair Times – hrs

Remove & install:

Arena	1,90
Movano	1,00

Special Tools

- Crankshaft pulley timing pin (Arena) – Kent Moore No.KM-966.
- Flywheel timing pin (Movano) – Kent Moore No.KM-966.
- Injection pump sprocket locking pin – Kent Moore No.KM-966.

Special Precautions

- Disconnect battery earth lead.
- DO NOT turn crankshaft or camshaft when timing belt removed.
- Remove glow plugs to ease turning engine.
- Turn engine in normal direction of rotation (unless otherwise stated).
- DO NOT turn engine via camshaft or other sprockets.
- Observe all tightening torques.
- Check diesel injection pump timing after belt replacement.

Removal

1. Remove:
   - ❑ Engine undershield.
   - ❑ Air filter assembly.
   - ❑ Heater fan intake ducting.
   - ❑ Radiator fan(s).
   - ❑ Arena: Radiator.
2. Disconnect brake servo hose and throttle cable.
3. Slacken crankshaft pulley bolt **1**.
4. Turn crankshaft to TDC on No.1 cylinder. Ensure timing marks aligned **2**.
5. Remove:
   - ❑ Auxiliary drive belt.
   - ❑ Movano: Water pump pulley.
   - ❑ Oil filler cap.
   - ❑ Engine top cover **3**.
   - ❑ Timing belt upper cover **4**.
6. Ensure camshaft sprocket timing mark aligned with mark on cylinder head cover **5**.
7. Insert locking pin in injection pump sprocket **6**. Tool No.KM-966.
8. Movano: Insert timing pin in flywheel **7**. Tool No.KM-966.
9. Remove:
   - ❑ Crankshaft pulley bolt **1**.
   - ❑ Crankshaft pulley **8**.
   - ❑ Tensioner nut and washer **9**.
   - ❑ Timing belt lower cover **10**.
10. Refit nut and washer to stud **9**. DO NOT tighten.
11. Compress tensioner spring to release tension on belt. Retain in position with U-shaped tool **11**.
    NOTE: Tool can be manufactured from a 14 mm nut cut to shape.
12. Move tensioner pulley away from belt.
13. Remove timing belt.

Installation

1. Ensure camshaft sprocket timing mark aligned **5**.
2. Movano: Ensure flywheel timing pin located correctly **7**.
3. Arena: Temporarily fit crankshaft pulley **8**. Insert crankshaft pulley timing pin **12**. Tool No.KM-966.
4. Ensure injection pump sprocket locking pin located correctly **6**.
5. Arena: Remove crankshaft pulley **8**.
6. Fit timing belt, starting at crankshaft sprocket. Ensure belt is taut on non-tensioned side.
7. Remove nut and washer **9**.
8. Fit timing belt lower cover **10**.
9. Refit nut and washer to stud **9**. DO NOT tighten.
10. Fit crankshaft pulley **8**.
11. Fit crankshaft pulley bolt **1**. Hand tighten.
12. Remove:
    - ❑ Tensioner tool **11**.
    - ❑ Injection pump sprocket locking pin **6**.
    - ❑ Movano: Flywheel timing pin **7**.
13. Tighten crankshaft pulley bolt **1**.
    Tightening torque: 200 Nm.
14. Turn crankshaft 1/4 turn clockwise.
15. Tighten tensioner nut **9**. Tightening torque: 45 Nm.
16. Turn crankshaft 3/4 turn clockwise.
17. Slacken and tighten tensioner nut **9**.
    Tightening torque: 45 Nm.
18. Turn crankshaft one turn clockwise to TDC. Ensure timing marks aligned **2**.
19. Ensure camshaft sprocket timing mark aligned **5**.
20. Ensure timing pin and locking pin can be inserted:
    - ❑ Arena: **6** & **12**.
    - ❑ Movano: **6** & **7**.
21. Install components in reverse order of removal.

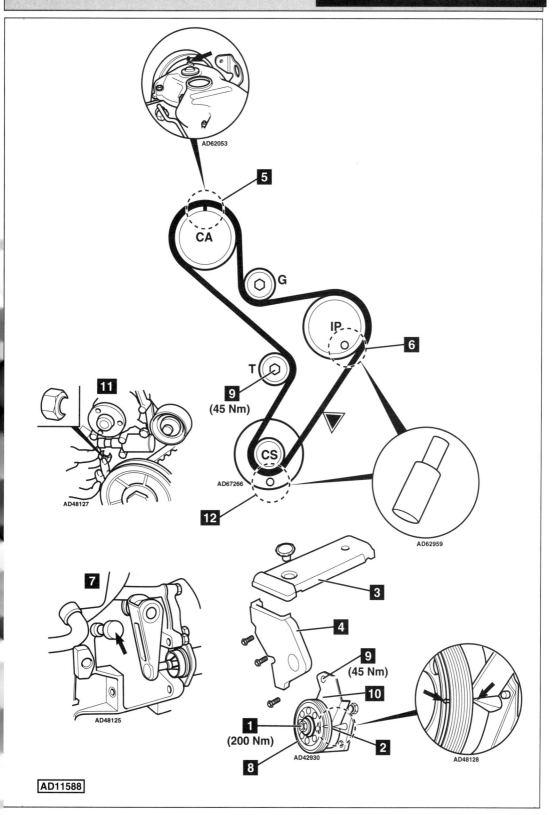

AD62053

5

CA

G

IP

6

T

9
(45 Nm)

11

AD48127

CS

AD67266

12

AD62959

7

AD48125

3

4

9
(45 Nm)

10

1
(200 Nm)

8

AD42930

2

AD48128

AD11588

VAUXHALL-OPEL

Model:	**Movano 2,2DTi**
Year:	**2000-03**
Engine Code:	**G9T 720**

Replacement Interval Guide

Vauxhall recommend replacement every 72,000 miles or 5 years (plastic pulleys must also be replaced).
The previous use and service history of the vehicle must always be taken into account.
Refer to Timing Belt Replacement Intervals at the front of this manual.

Check For Engine Damage

CAUTION: This engine has been identified as an INTERFERENCE engine in which the possibility of valve-to-piston damage in the event of a timing belt failure is MOST LIKELY to occur.
A compression check of all cylinders should be performed before removing the cylinder head.

Repair Times – hrs

Remove & install   2,20

Special Tools

- Engine support – Kent Moore No.6210.
- Crankshaft locking pin – Kent Moore No.6203.
- Camshaft locking tools – Kent Moore No.6204/6205.

Special Precautions

- Disconnect battery earth lead.
- DO NOT turn crankshaft or camshaft when timing belt removed.
- Remove glow plugs to ease turning engine.
- Turn engine in normal direction of rotation (unless otherwise stated).
- DO NOT turn engine via camshaft or other sprockets.
- Observe all tightening torques.

Removal

1. Raise and support vehicle.
2. Remove:
   - ❑ Engine undershield.
   - ❑ RH wheel.
   - ❑ RH inner wing panel.
3. Support engine. Use tool No.6210.
4. Unclip fuel feed pipe from RH engine mounting. Move to one side.
5. Remove:
   - ❑ RH engine mounting.
   - ❑ RH lower support bracket.
   - ❑ Auxiliary drive belt.
   - ❑ Timing belt upper and lower covers **1** & **2**.
   - ❑ Oil filter housing bolt **3**.
6. Turn crankshaft pulley clockwise until mark at lower position **4**. Ensure grooves in camshafts vertically aligned **5**.
7. Fit crankshaft locking pin **6**. Tool No.6203.
8. Rock crankshaft slightly to ensure locking pin located correctly.
9. Install camshaft locking tools **7** & **8**. Tool Nos.6204/6205.

10. If tools cannot be fitted correctly: Turn crankshaft one turn clockwise.
11. Slacken bolts of camshaft sprockets **9** & **15**.
12. Slacken tensioner bolt **10**.
13. Remove:
    - ❑ Exhaust camshaft sprocket bolts **9**.
    - ❑ Exhaust camshaft sprocket **16**.
    - ❑ Timing belt.

Installation

1. Rock crankshaft slightly to ensure locking pin located correctly **6**.
2. Ensure camshaft locking tools correctly located **7** & **8**.
   *NOTE: Ensure inlet camshaft sprocket bolts **15** not at end of slotted holes **14**.*
3. Fit timing belt in following order:
   - ❑ Intermediate shaft sprocket.
   - ❑ Guide pulley.
   - ❑ Inlet camshaft sprocket.
   - ❑ Tensioner pulley.
4. Fit exhaust camshaft sprocket **16** to belt, then install camshaft sprocket with belt onto end of camshaft. Ensure sprocket bolts not at end of slotted holes **14**.
5. Turn tensioner anti-clockwise until top of lever arm **11** aligns with upper edge of tool **7**. Use 6 mm Allen key.
   *NOTE: Ensure exhaust camshaft sprocket bolts not at end of slotted holes **14**.*
6. Tighten tensioner nut **10**. Tightening torque: 25 Nm.
7. Tighten bolts of camshaft sprockets **9** & **15**. Tightening torque: 10 Nm.
8. Remove:
   - ❑ Locking tools **7** & **8**.
   - ❑ Locking pin **6**.
9. Turn crankshaft two turns clockwise.
10. Fit crankshaft locking pin **6**.
11. Rock crankshaft slightly to ensure locking pin located correctly.
12. Install camshaft locking tools **7** & **8**.
13. Slacken bolts of camshaft sprockets **9** & **15**.
14. Slacken tensioner bolt **10**.
15. Turn tensioner until raised part of lever arm **12** aligns with upper edge of tool **7**. Use 6 mm Allen key.
    *NOTE: Tensioner pointer should be aligned with groove **13**. If not: Repeat tensioning procedures.*
16. Tighten tensioner nut **10**. Tightening torque: 25 Nm.
17. Tighten bolts of camshaft sprockets **9** & **15**. Tightening torque: 10 Nm.
18. Remove:
    - ❑ Locking tools **7** & **8**.
    - ❑ Locking pin **6**.
19. Install components in reverse order of removal.
NOTE: Auxiliary drive belt and plastic pulleys must also be replaced.

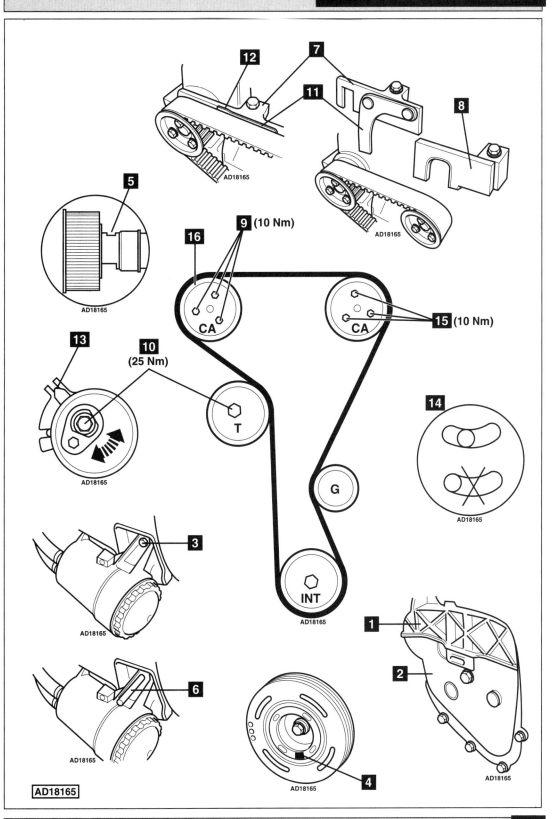

AD18165

VOLKSWAGEN

Model:	Lupo 1,0/1,4 • Polo 1,0/1,1/1,3/1,4/1,6 • Polo Classic 1,4/1,6 Caddy 1,4/1,6 • Golf/Vento 1,4/1,6
Year:	1994-03
Engine Code:	Automatic tensioner ADX, AEA, AEE, AER, AEV, AEX, AKK, AKP, AKV, ALD, ALL, ALM, ANV, ANX, APQ, AUC, AUD

Replacement Interval Guide

Volkswagen recommend check at the first 60,000 miles and then every 20,000 miles (replace if necessary).
The vehicle manufacturer has not recommended a timing belt replacement interval for this engine.
The previous use and service history of the vehicle must always be taken into account.
Refer to Timing Belt Replacement Intervals at the front of this manual.

Check For Engine Damage

CAUTION: This engine has been identified as an INTERFERENCE engine in which the possibility of valve-to-piston damage in the event of a timing belt failure is MOST LIKELY to occur.
A compression check of all cylinders should be performed before removing the cylinder head.

Repair Times – hrs

Remove & install:

Lupo/Polo	1,60
Polo Classic	1,30
Golf/Vento	1,10
Caddy 1,4	1,10
Caddy 1,6	1,30

Special Tools

- None required.

Special Precautions

- Disconnect battery earth lead.
- DO NOT turn crankshaft or camshaft when timing belt removed.
- Remove spark plugs to ease turning engine.
- Turn engine in normal direction of rotation (unless otherwise stated).
- DO NOT turn engine via camshaft or other sprockets.
- Observe all tightening torques.

Removal

1. Lupo/Polo:
   - Support engine.
   - Remove RH engine mounting bolts **14** & **15**.
   - Remove RH engine mounting.
   - Lower engine slightly.
2. All models – remove:
   - Auxiliary drive belt cover (if fitted).
   - Auxiliary drive belt.
   - Timing belt upper cover **1**.
3. Turn crankshaft to TDC on No.1 cylinder.
4. Ensure '0' timing mark aligned **2**.
5. Ensure camshaft sprocket timing marks aligned **3**.

6. Remove:
   - Crankshaft pulley bolts (4 bolts) **4**.
   - Crankshaft pulley **5**.
   - Timing belt lower cover **6**.
7. Slacken tensioner nut **7**.
8. Remove timing belt.
   NOTE: Mark direction of rotation on belt with chalk if belt is to be reused.

Installation

1. Ensure timing marks aligned **3** & **13**.
   NOTE: Align ground tooth on crankshaft sprocket with mark on sealing flange.
2. Fit timing belt in anti-clockwise direction. Ensure belt is taut between sprockets on non-tensioned side.
3. Tighten tensioner nut finger tight **7**. Ensure baseplate is supported by bolt **8**.
4. Turn tensioner clockwise **9** until pointer **11** aligned with notch in baseplate **10**.
5. Tighten tensioner nut to 20 Nm **7**.
6. Turn crankshaft two turns clockwise to TDC on No.1 cylinder. Ensure timing marks aligned **3** & **13**.
7. Ensure pointer **11** aligned with notch in baseplate **10**.
8. Apply firm thumb pressure to belt at ▽. Pointer **11** and notch **10** must move apart.
9. Release thumb pressure from belt at ▽.
10. Turn crankshaft two turns clockwise to TDC on No.1 cylinder.
11. Ensure pointer **11** aligned with notch in baseplate **10**.
12. Fit timing belt lower cover **6**. Tighten bolts to 10 Nm **12**.
13. Install:
    - Crankshaft pulley **5**.
    - Crankshaft pulley bolts **4**.
14. Tighten crankshaft pulley bolts to 20 Nm **4**.
15. Install components in reverse order of removal.
16. Fit engine mounting. Use new bolts.
17. Tighten engine mounting:
    - Bolts **14**. Tightening torque: 25 Nm + 45°.
    - Bolt **15**. Tightening torque: 50 Nm.
    - Bolts **16**. Tightening torque: 25 Nm.

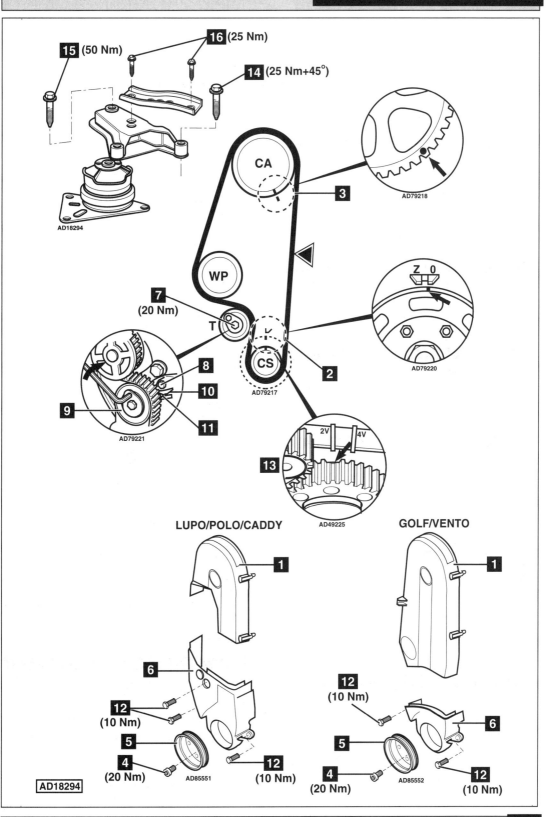

15 (50 Nm)

16 (25 Nm)

14 (25 Nm+45°)

3

AD79218

CA

WP

7
(20 Nm)

T

Z 0

AD79220

8

10

9

11

AD79221

2

CS

AD79217

2V 4V

13

AD49225

LUPO/POLO/CADDY

1

6

12
(10 Nm)

5

4
(20 Nm)

AD85551

12
(10 Nm)

GOLF/VENTO

1

12
(10 Nm)

6

5

4
(20 Nm)

AD85552

12
(10 Nm)

AD18294

Model:	Lupo 1,4 16V • Polo 1,4/1,6 16V • Polo Classic 1,4 16V Golf/Bora 1,4/1,6 16V • Caddy 1,4 16V
Year:	1997-03
Engine Code:	AFK, AHW, AJV, AKQ, APE, AQQ, ARC, AUA, AUB, ATN, AUS, AVY, AXP, AZD

Replacement Interval Guide

Volkswagen recommend check at the first 60,000 miles and then every 20,000 miles (replace if necessary).
The vehicle manufacturer has not recommended a timing belt replacement interval for this engine.
The previous use and service history of the vehicle must always be taken into account.
Refer to Timing Belt Replacement Intervals at the front of this manual.

Check For Engine Damage

CAUTION: This engine has been identified as an INTERFERENCE engine in which the possibility of valve-to-piston damage in the event of a timing belt failure is MOST LIKELY to occur.
A compression check of all cylinders should be performed before removing the cylinder head(s).

Repair Times – hrs

Remove & install:

Golf/Bora	2,50
Lupo/Polo	1,60
Polo Classic	1,30
Caddy	1,10

Special Tools

■ Camshaft locking tool (except ARC/AVY) – No.T10016.
■ Camshaft locking tools (ARC/AVY) – No.T10074.
■ Crankshaft pulley holding tool (except ARC/ATN/AUS/AVY/AZD) – No.3415.
■ Crankshaft pulley holding tool (ARC/ATN/AUS/AVY/AZD) – No.T10028.

Special Precautions

■ Disconnect battery earth lead.
■ DO NOT turn crankshaft or camshaft when timing belt removed.
■ Remove spark plugs to ease turning engine.
■ Turn engine in normal direction of rotation (unless otherwise stated).
■ DO NOT turn engine via camshaft or other sprockets.
■ Observe all tightening torques.

Removal

Timing Belt

1. Raise and support front of vehicle.
2. Remove:
   ❏ Upper engine cover.
   ❏ Air filter assembly.
   ❏ Timing belt upper cover **1**.
3. Turn crankshaft clockwise to TDC on No.1 cylinder. Ensure timing marks on crankshaft pulley aligned **2**.
4. Ensure camshaft sprocket locating holes aligned:
   ❏ Except ARC/AVY: **3**
   ❏ ARC/AVY: **4**

5. If locating holes are not aligned: Turn crankshaft one turn clockwise.
6. Fit locking tool(s) to camshaft sprockets:
   ❏ Except ARC/AVY: **5**. Tool No.T10016.
   ❏ ARC/AVY: **21**. Tool No.T10074.
   NOTE: Ensure locking tool(s) located correctly in cylinder head.
7. Remove:
   ❏ PAS reservoir. DO NOT disconnect hoses.
   ❏ RH engine undershield.

Except Polo Classic:

8. Support engine.
9. Remove:
   ❏ RH engine mounting.
   ❏ RH engine mounting bracket.
10. Lower engine until crankshaft pulley bolt accessible.

All models:

11. Remove auxiliary drive belt.
12. Fit crankshaft pulley holding tool:
    ❏ Except ARC/ATN/AUS/AVY/AZD – tool No.3415.
    ❏ ARC/ATN/AUS/AVY/AZD – tool No.T10028.
13. Slacken crankshaft pulley bolt **6**.
14. Remove:
    ❏ Holding tool. Tool No.3415 or T10028.
    ❏ Crankshaft pulley bolt **6**.
    ❏ Crankshaft pulley **7**.
15. Fit two washers to crankshaft pulley bolt **6**.
16. Fit crankshaft pulley bolt **6**. Lightly tighten bolt.
17. Remove:
    ❏ Auxiliary drive belt guide pulley (models with AC).
    ❏ Auxiliary drive belt tensioner.
    ❏ Timing belt lower cover **8**.
18. Slacken tensioner pulley bolt **9**.
19. Turn tensioner pulley anti-clockwise to release tension on belt.
20. Remove timing belt.
    NOTE: Mark direction of rotation on belt with chalk if belt is to be reused.

Installation

Timing Belt

1. Ensure locking tool(s) fitted to camshaft sprockets:
   ❏ Except ARC/AVY: **5**. Tool No.T10016.
   ❏ ARC/AVY: **21**. Tool No.T10074.
2. Ensure timing mark on crankshaft sprocket aligned **10**.
   NOTE: Align ground tooth on crankshaft sprocket.
3. Remove – ARC/AVY:
   ❏ Guide pulley bolt **22**.
   ❏ Guide pulley **23**.
4. Tighten tensioner pulley bolt finger tight **9**. Ensure baseplate is supported by bolt **11**.

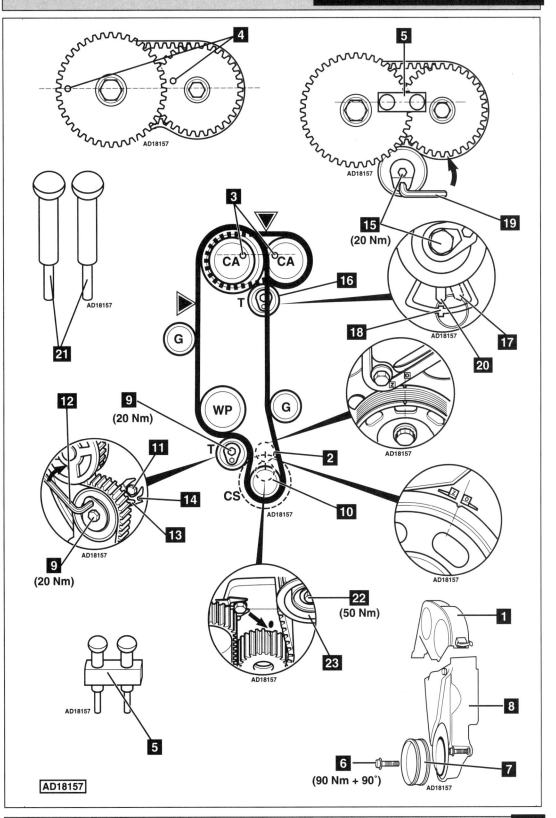

4

5

AD18157

AD18157

15
(20 Nm)

19

3

CA   CA

16

T

18

9
(20 Nm)

WP

G

G

17

20

AD18157

21

12

11

T

2

AD18157

9
(20 Nm)

14

CS

AD18157

10

13

Z   O

AD18157

AD18157

22
(50 Nm)

23

AD18157

1

5

8

6
(90 Nm + 90°)

7

AD18157

AD18157

←

5. Fit timing belt in anti-clockwise direction, starting at water pump sprocket.
6. Fit – ARC/AVY:
   ❑ Guide pulley 🔢.
   ❑ Guide pulley bolt 🔢.
7. Tighten guide pulley bolt 🔢.
   Tightening torque: 50 Nm.
8. Turn tensioner pulley clockwise 🔢 until pointer 🔢 aligned with notch in baseplate 🔢.
9. Tighten tensioner pulley bolt to 20 Nm 🔢.
10. Remove locking tool(s) from camshaft sprockets:
    ❑ Except ARC/AVY: 🔢. Tool No.T10016.
    ❑ ARC/AVY: 🔢. Tool No.T10074.
11. Turn crankshaft two turns clockwise to TDC on No.1 cylinder. Ensure timing marks on crankshaft sprocket aligned 🔢.
12. Ensure locking tool(s) can be inserted into camshaft sprockets:
    ❑ Except ARC/AVY: 🔢. Tool No.T10016.
    ❑ ARC/AVY: 🔢. Tool No.T10074.
13. Ensure pointer 🔢 aligned with notch in baseplate 🔢.
14. If not: Repeat tensioning procedure.
15. Apply firm thumb pressure to belt at ▽. Pointer 🔢 and notch in baseplate 🔢 must move apart.
16. Release thumb pressure from belt at ▽.
17. Turn crankshaft two turns clockwise to TDC on No.1 cylinder.
18. Ensure pointer 🔢 aligned with notch in baseplate 🔢.
19. Remove crankshaft pulley bolt 🔢.
20. Install:
    ❑ Timing belt lower cover 🔢.
    ❑ Crankshaft pulley 🔢.
    ❑ New oiled crankshaft pulley bolt 🔢.
21. Fit crankshaft pulley holding tool:
    ❑ Except ARC/ATN/AUS/AVY/AZD – tool No.3415.
    ❑ ARC/ATN/AUS/AVY/AZD – tool No.T10028.
22. Tighten crankshaft pulley bolt 🔢.
    Tightening torque: 90 Nm + 90°.
23. Remove holding tool. Tool No.3415 or T10028.
24. Install components in reverse order of removal.
25. Except Polo Classic: Tighten bolts securing engine mounting bracket to engine.
    Tightening torque: 50 Nm.
26. Lupo/Polo: Tighten engine mounting:
    ❑ Bolts securing engine mounting to body – 20 Nm + 45°. Use new bolts.
    ❑ Bolts securing intermediate bracket to engine mounting bracket – 40 Nm + 90°. Use new bolts.
    ❑ Bolt securing intermediate bracket to engine mounting – 50 Nm.
27. Golf/Bora: Tighten engine mounting:
    ❑ Long bolts securing engine mounting to body – 40 Nm + 90°. Use new bolts.
    ❑ Short bolts securing engine mounting to body – 25 Nm.
    ❑ Bolts securing engine mounting to engine mounting bracket – 60 Nm + 90°. Use new bolts.

Removal

Exhaust Camshaft Drive Belt

1. Remove timing belt as described previously.
2. Slacken tensioner pulley bolt 🔢.
3. Turn tensioner pulley clockwise to release tension on belt.
4. Remove:
   ❑ Tensioner pulley bolt 🔢.
   ❑ Tensioner pulley 🔢.
   ❑ Drive belt.
   NOTE: Mark direction of rotation on belt with chalk if belt is to be reused.

Installation

Exhaust Camshaft Drive Belt

1. Ensure locking tool(s) fitted to camshaft sprockets:
   ❑ Except ARC/AVY: 🔢. Tool No.T10016.
   ❑ ARC/AVY: 🔢. Tool No.T10074.
2. Fit drive belt in clockwise direction, starting at top of inlet camshaft sprocket.
3. Ensure belt is taut between sprockets on non-tensioned side.
4. Turn tensioner pulley clockwise until pointer in position as shown 🔢.
5. Install:
   ❑ Tensioner pulley 🔢.
   ❑ Tensioner pulley bolt 🔢.
6. Tighten tensioner pulley bolt finger tight 🔢.
   NOTE: Ensure lug in baseplate 🔢 is located in cylinder head hole.
7. Turn tensioner anti-clockwise 🔢 until pointer 🔢 aligned with lug in baseplate 🔢.
8. Tighten tensioner pulley bolt to 20 Nm 🔢.
9. Fit timing belt as described previously.
10. Remove locking tool(s) from camshaft sprockets:
    ❑ Except ARC/AVY: 🔢. Tool No.T10016.
    ❑ ARC/AVY: 🔢. Tool No.T10074.
11. Turn crankshaft two turns clockwise to TDC on No.1 cylinder. Ensure timing mark on crankshaft sprocket aligned 🔢.
12. Ensure locking tool(s) can be inserted into camshaft sprockets:
    ❑ Except ARC/AVY: 🔢. Tool No.T10016.
    ❑ ARC/AVY: 🔢. Tool No.T10074.
13. Ensure pointer 🔢 aligned with lug in baseplate 🔢.
14. If not: Repeat tensioning procedure.
15. Apply firm thumb pressure to belt at ▽. Pointer 🔢 and lug in baseplate 🔢 must move apart.
16. Release thumb pressure from belt at ▽.
17. Turn crankshaft two turns clockwise to TDC on No.1 cylinder.
18. Ensure pointer 🔢 aligned with lug in baseplate 🔢.
19. Install components in reverse order of removal.

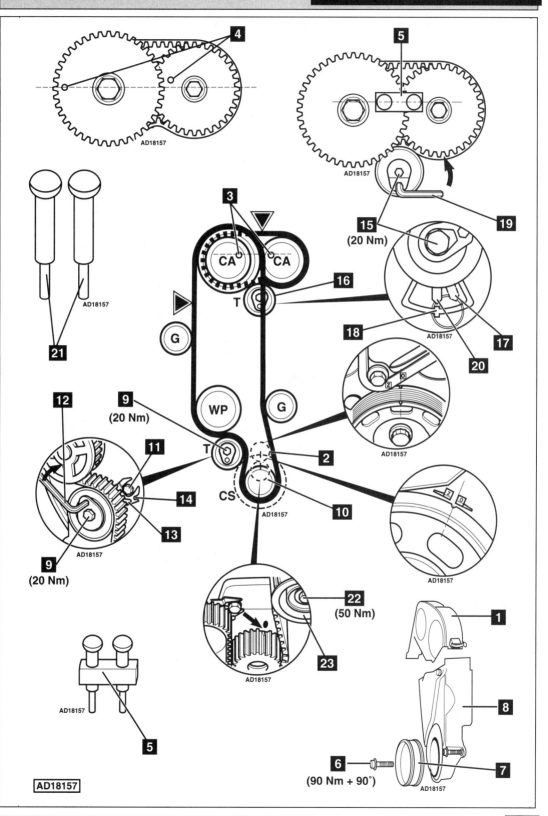

Model:	**Polo 1,1/1,3 • Polo G40 • Golf/Jetta 1,1/1,3/1,4/1,6 • Golf/Vento 1,4/1,6 Scirocco 1,3 • Passat/Santana 1,3**
Year:	**1980-95**
Engine Code:	**Water pump tensioner FA, FJ, FY, FZ, EP, EU, GF, GG, GK, GN, GL, GS, GT, HB, HH, HJ, HK, HW, HZ, MH, NU, NZ, PY, SC, AAK, AAU, AAV, ABD, ABU, ACM, 2C, 2G, 3F**

Replacement Interval Guide

Volkswagen recommend check & adjust every 20,000 miles (replace if necessary).
No manufacturer's recommended replacement interval.
The previous use and service history of the vehicle must always be taken into account.
Refer to Timing Belt Replacement Intervals at the front of this manual.

Check For Engine Damage

CAUTION: This engine has been identified as an INTERFERENCE engine in which the possibility of valve-to-piston damage in the event of a timing belt failure is MOST LIKELY to occur.
A compression check of all cylinders should be performed before removing the cylinder head.

Repair Times – hrs

Polo/Golf/Jetta/Scirocco:
Bucket tappets:

Check & adjust	0,50
Remove & install	1,10

Hydraulic tappets:

Check & adjust	0,30
Remove & install	0,50

Passat/Santana:
Bucket tappets:

Check & adjust	0,50
Remove & install	1,10

Hydraulic tappets:

Check & adjust	0,30
Remove & install	0,70

Special Tools

■ None required.

Special Precautions

■ Disconnect battery earth lead.
■ DO NOT turn crankshaft or camshaft when timing belt removed.
■ Remove spark plugs to ease turning engine.
■ Turn engine in normal direction of rotation (unless otherwise stated).
■ DO NOT turn engine via camshaft or other sprockets.
■ Observe all tightening torques.

Removal

1. Remove:
   ❏ Alternator drive belt.
   ❏ Timing belt front cover **1**.
2. Turn crankshaft to TDC on No.1 cylinder. Ensure timing marks aligned **2** & **3**.
   NOTE: Engine code PY – align crankshaft pulley 'O' mark with edge of bracket.
3. Later models – remove:
   ❏ Four crankshaft damper bolts **5**.
   ❏ Crankshaft damper **6**.
   ❏ Timing belt lower cover **7**.
4. All models: Slacken water pump bolts **4**.
5. Turn water pump anti-clockwise to release tension on belt.
6. Remove timing belt.

Installation

1. Ensure timing marks aligned **2** & **3**.
   NOTE: Engine code PY – align crankshaft pulley 'O' mark with edge of bracket.
2. Fit timing belt.
   NOTE: Ensure belt is taut between sprockets on non-tensioned side.
3. Turn water pump clockwise to tension belt. Use screwdriver.
   NOTE: Belt is correctly tensioned when it can just be twisted with finger and thumb through 90° at ▽.
4. Tighten water pump bolts to 10 Nm.
5. Turn crankshaft two turns clockwise.
6. Ensure timing marks aligned **2** & **3**.
7. Check belt tension.
8. Install components in reverse order of removal.
9. Later models: Tighten crankshaft damper bolts to 20 Nm **5**.

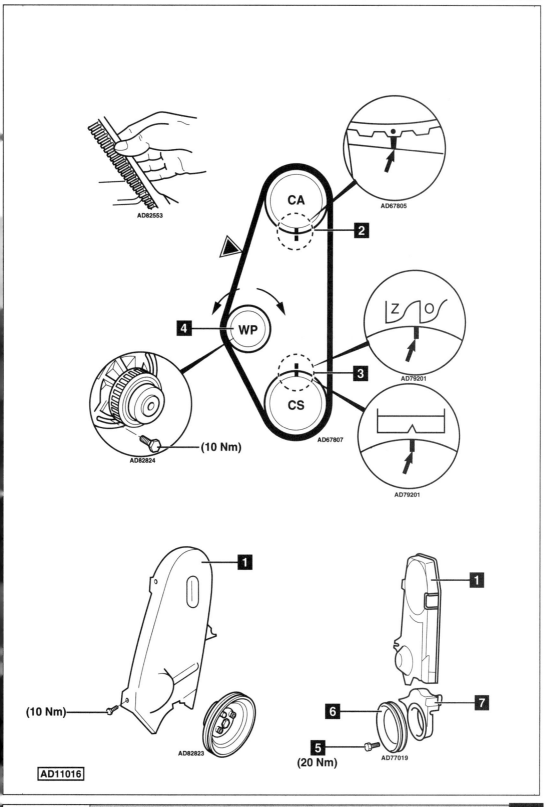

AD82553

AD67805

CA

2

4 WP

AD79201

3

CS

AD67807

(10 Nm)
AD82824

AD79201

1

1

(10 Nm)

AD82823

6 **7**

5
(20 Nm) AD77019

AD11016

VOLKSWAGEN

Model:	Polo 1,4 16V
Year:	1995-00
Engine Code:	AFH

Replacement Interval Guide

The vehicle manufacturer has not recommended a timing belt replacement interval for this engine. *The previous use and service history of the vehicle must always be taken into account. Refer to Timing Belt Replacement Intervals at the front of this manual.*

Check For Engine Damage

CAUTION: This engine has been identified as an INTERFERENCE engine in which the possibility of valve-to-piston damage in the event of a timing belt failure is MOST LIKELY to occur. A compression check of all cylinders should be performed before removing the cylinder head.

Repair Times – hrs

Remove & install	1,60

Special Tools

- ■ Crankshaft pulley holding tool – No.3415.

Special Precautions

- ■ Disconnect battery earth lead.
- ■ DO NOT turn crankshaft or camshaft when timing belt removed.
- ■ Remove spark plugs to ease turning engine.
- ■ Turn engine in normal direction of rotation (unless otherwise stated).
- ■ DO NOT turn engine via camshaft or other sprockets.
- ■ Observe all tightening torques.

Removal

1. Support engine.
2. Remove:
   - ❏ RH engine mounting.
   - ❏ Auxiliary drive belt cover.
3. Lower engine until crankshaft pulley bolt accessible.
4. Remove auxiliary drive belt.
5. Hold crankshaft pulley **2**. Use tool No.3415. Remove crankshaft pulley bolt **1**.
6. Remove crankshaft pulley **2**.
7. Fit crankshaft pulley bolt **1** with two large washers and tighten.
8. Remove:
   - ❏ Timing belt front cover **3**.
   - ❏ Timing belt upper cover **4**.
   - ❏ Timing belt lower cover **5**.
9. Turn crankshaft clockwise until timing marks aligned **6** & **7**.
10. Remove engine mounting bracket **16**.
11. Slacken tensioner nut **8**.
12. Remove timing belt.
    NOTE: Mark direction of rotation on belt with chalk if belt is to be reused.

Installation

1. Ensure timing marks aligned **6** & **7**.
2. Fit timing belt in anti-clockwise direction, starting at crankshaft sprocket. Ensure belt is taut between sprockets.
3. Tighten tensioner nut finger tight **8**. Ensure baseplate is supported by bolt **9**.
4. Turn tensioner clockwise **10** until pointer **12** aligned with notch in baseplate **11**.
5. Tighten tensioner nut to 20 Nm **8**.
6. Turn crankshaft two turns clockwise to TDC on No.1 cylinder. Ensure timing marks aligned **6** & **7**.
7. Ensure pointer **12** aligned with notch in baseplate **11**.
8. Fit timing belt covers **3**, **4** & **5**. Tighten bolts to 10 Nm **13**.
9. Fit engine mounting bracket **16**. Tighten bolt to 40 Nm + 90° **17**.
10. Remove crankshaft pulley bolt **1**.
11. Fit crankshaft pulley **2**.
12. Fit new oiled crankshaft pulley bolt **1**. Tightening torque: 90 Nm + 90°.
13. Install components in reverse order of removal.
14. Fit engine mounting. Use new bolts.
15. Tighten engine mounting bolts **14** & **15** (refer to illustration).

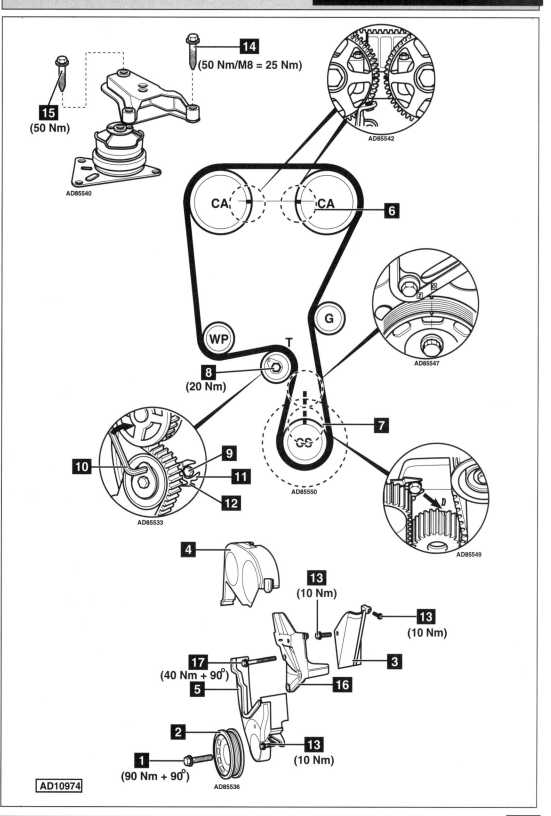

14 (50 Nm/M8 = 25 Nm)

15 (50 Nm)

AD85540

AD85542

CA **CA** **6**

AD85547

WP T **G**

8 (20 Nm)

10 **9** **11** **12**

AD85533

GS **7**

AD85550

AD85549

4

13 (10 Nm)

13 (10 Nm)

3

17 (40 Nm + 90°)

5 **16**

2

1 (90 Nm + 90°)

13 (10 Nm)

AD85536

AD10974

Model:	Polo Classic 1,6/1,8 (1995→) • Golf/Jetta/Vento/Caddy 1,6/1,8/2,0 Scirocco 1,6/1,8 • Passat 1,8/2,0 • Corrado 2,0 • Caddy 1,6/1,8 (1996→)
Year:	1986-02
Engine Code:	Timing mark on front of camshaft sprocket DX, EG, EM, EV, EW, EX, EZ, FB, FD, FH, FK, FN, FP, FR, FT, FV, GH, GU, GX, GZ, HM, HN, HT, HV, JB, JH, JJ, KT, PB, PF, PN, RD, RE, RF, RG, RH, RP, RV, AAM, ABN, ABS, ADY, ADZ, AGG, AKR, ANN, ANP, EZA, 1F, 1P, 2E, 2H

Replacement Interval Guide

Volkswagen recommend:
→1995: Check & adjust every 20,000 miles (replace if necessary).
1996→: Check at first 60,000 miles and then every 20,000 miles (replace if necessary).
No manufacturer's recommended replacement interval.
The previous use and service history of the vehicle must always be taken into account.
Refer to Timing Belt Replacement Intervals at the front of this manual.

Check For Engine Damage

CAUTION: This engine has been identified as an INTERFERENCE engine in which the possibility of valve-to-piston damage in the event of a timing belt failure is MOST LIKELY to occur.
A compression check of all cylinders should be performed before removing the cylinder head.

Repair Times – hrs

Polo Classic:	
Check & adjust	0,90
Remove & install	1,30
Golf/Jetta/Scirocco/Caddy (→1991):	
Check & adjust	0,50
Remove & install	1,10
Golf/Vento (1992→):	
Check & adjust – 1,6	0,90
Check & adjust – 1,8/2,0	0,70
Remove & install – 1,6	1,10
Remove & install – 1,8/2,0	1,30
Passat:	
Check & adjust	0,50
Remove & install	1,30
Caddy (1996→):	
Check & adjust	0,70
Remove & install	1,30

Special Tools

■ None required.

Special Precautions

■ Disconnect battery earth lead.
■ DO NOT turn crankshaft or camshaft when timing belt removed.
■ Remove spark plugs to ease turning engine.
■ Turn engine in normal direction of rotation (unless otherwise stated).
■ DO NOT turn engine via camshaft or other sprockets.
■ Observe all tightening torques.

Removal

1. Raise and support front of vehicle.
2. Remove:
   ❑ Auxiliary drive belt(s).
   ❑ Auxiliary drive belt tensioner (if fitted).
   ❑ Crankshaft pulley bolts **3**.
   ❑ Crankshaft pulley **4**.
   ❑ Water pump pulley (if required).
   ❑ Timing belt upper cover **6**.
   ❑ Timing belt lower cover **5**.
3. Temporarily fit crankshaft pulley. Lightly tighten one bolt.
4. Turn crankshaft until timing marks aligned **1** & **2**.
5. Ensure distributor rotor arm aligned with mark on distributor body **8**.
6. Remove:
   ❑ Crankshaft pulley bolt **3**.
   ❑ Crankshaft pulley **4**.
7. Slacken tensioner bolt **7**. Turn tensioner anti-clockwise away from belt. Lightly tighten bolt.
8. Remove timing belt.

Installation

1. Ensure timing mark on camshaft sprocket aligned **1**.
2. Fit timing belt to crankshaft sprocket and auxiliary shaft sprocket.
3. Temporarily fit crankshaft pulley. Lightly tighten one bolt.
4. Ensure timing mark on crankshaft pulley aligned with mark on auxiliary shaft sprocket **2**.
5. Ensure distributor rotor arm aligned with mark on distributor body **8**.
6. Fit timing belt to camshaft sprocket and tensioner pulley.
7. Ensure timing marks aligned **1** & **2**.
8. Turn tensioner clockwise until belt can just be twisted with finger and thumb through 90° at ▽.
9. Tighten tensioner bolt to 45 Nm **7**.
10. Turn crankshaft two turns clockwise.
11. Ensure timing marks aligned **1** & **2**.
12. Ensure distributor rotor arm aligned with mark on distributor body **8**.
13. Check belt tension.
14. Install components in reverse order of removal.
15. Tighten crankshaft pulley bolts **3**. Tightening torque: 20-25 Nm.
16. Tighten auxiliary drive belt tensioner bolt to 20-25 Nm (if fitted).
17. Tighten water pump pulley bolts to 20 Nm (if removed).

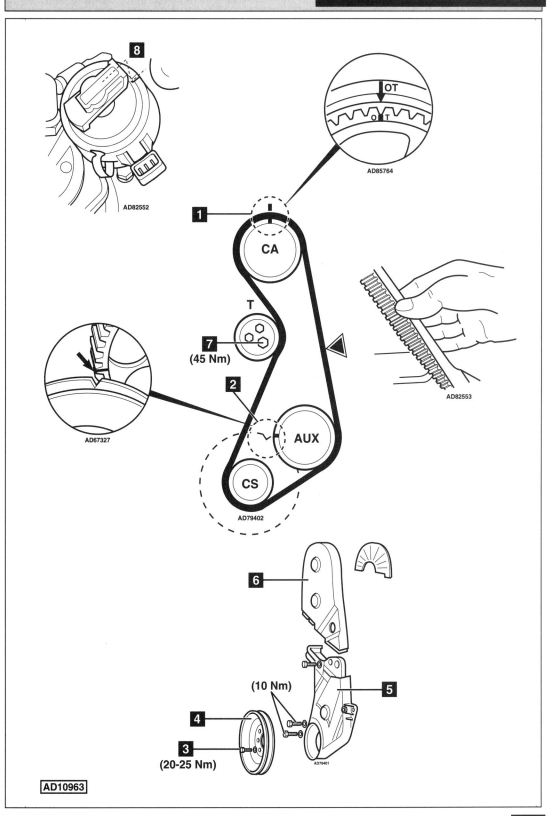

AD82552

AD85764

1

CA

8

T

7
(45 Nm)

2

AUX

AD67327

CS

AD79402

AD82553

6

(10 Nm)

5

4

3
(20-25 Nm)

AD79401

AD10963

VOLKSWAGEN

Model:	Polo Classic 1,6 • Golf/Vento 1,6 • Passat 1,6 • Sharan 2,0
Year:	1994-00
Engine Code:	ADY, AEK, AFT, AKS

Replacement Interval Guide

Volkswagen recommend:
→1995: Check & adjust every 20,000 miles (replace if necessary).
1996→: Check at first 60,000 miles and then every 20,000 miles (replace if necessary).
No manufacturer's recommended replacement interval.
The previous use and service history of the vehicle must always be taken into account.
Refer to Timing Belt Replacement Intervals at the front of this manual.

Check For Engine Damage

CAUTION: This engine has been identified as an INTERFERENCE engine in which the possibility of valve-to-piston damage in the event of a timing belt failure is MOST LIKELY to occur.
A compression check of all cylinders should be performed before removing the cylinder head.

Repair Times – hrs

Check & adjust:	
Except Passat	0,90
Passat	0,50
Remove & install:	
Polo Classic	1,30
Golf/Vento/Passat	1,10
Sharan	1,90

Special Tools

- Two-pin wrench – Matra V.159.

Special Precautions

- Disconnect battery earth lead.
- DO NOT turn crankshaft or camshaft when timing belt removed.
- Remove spark plugs to ease turning engine.
- Turn engine in normal direction of rotation (unless otherwise stated).
- DO NOT turn engine via camshaft or other sprockets.
- Observe all tightening torques.

Removal

Sharan:

1. Support front of engine.
2. Remove engine mounting.

All models:

3. Turn crankshaft clockwise to TDC on No.1 cylinder.

4. Ensure crankshaft timing marks aligned **1** or **10**. Ensure distributor rotor arm aligned with mark on distributor body **2**.
5. Remove:
   - ❑ Auxiliary drive belts.
   - ❑ Auxiliary drive belt tensioner (if fitted).
   - ❑ Crankshaft pulley bolts **3**.
   - ❑ Crankshaft pulleys **4**.
   - ❑ Water pump pulley **5**.
   - ❑ Timing belt upper cover **6**.
   - ❑ Timing belt lower cover **7**.
6. Ensure camshaft sprocket timing mark aligned **8**.
7. Slacken tensioner bolt **9**.
8. Turn tensioner anti-clockwise to release tension on belt. Lightly tighten bolt.
9. Remove timing belt.
   NOTE: Mark direction of rotation on belt with chalk if belt is to be reused.

Installation

1. Ensure camshaft sprocket timing mark aligned **8**.
2. Fit timing belt to crankshaft sprocket and auxiliary shaft sprocket.
   NOTE: Disregard mark on auxiliary shaft sprocket.
3. Install:
   - ❑ Timing belt lower cover **7**.
   - ❑ Crankshaft pulleys **4**.
   - ❑ Crankshaft pulley bolts **3**.
   NOTE: Crankshaft pulley bolt holes are offset.
4. Tighten crankshaft pulley bolts to 25 Nm **3**.
5. Ensure crankshaft timing marks aligned **1** or **10**.
6. Ensure distributor rotor arm aligned with mark on distributor body **2**.
7. Fit timing belt to tensioner pulley and camshaft sprocket.
8. Turn tensioner clockwise until belt can just be twisted with finger and thumb through 90° at ▽. Use wrench Matra V.159.
9. Tighten tensioner bolt to 45 Nm **9**.
10. Turn crankshaft two turns clockwise.
11. Ensure timing marks aligned **8** & **1** or **10**.
12. Recheck belt tension.
    NOTE: Belt is correctly tensioned when it can just be twisted with finger and thumb through 90° at ▽.
13. Install components in reverse order of removal.
14. Tighten auxiliary drive belt tensioner bolt to 20 Nm (if fitted).

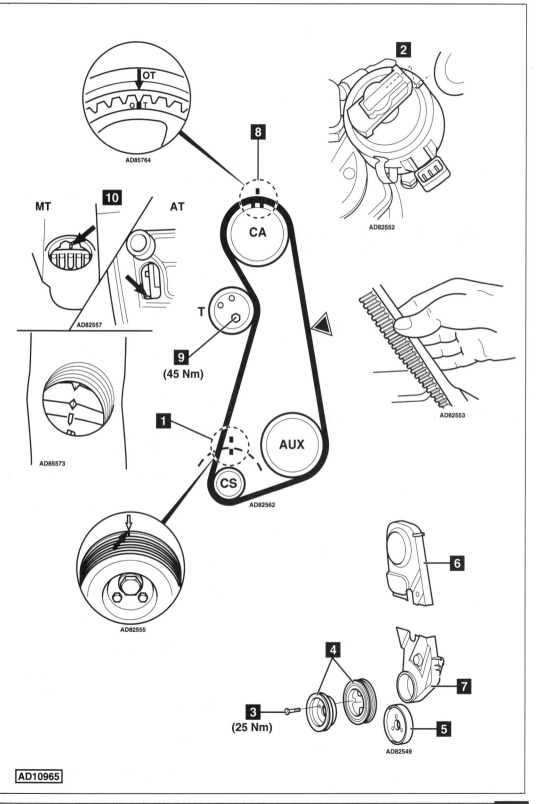

OT

AD85764

8

2

AD82552

MT 10 AT

AD82557

CA

T

9

(45 Nm)

AD82553

AD85573

1

AUX

CS

AD82562

AD82555

6

4

3

(25 Nm)

7

5

AD82549

AD10965

VOLKSWAGEN

Model:	Polo Classic 1,6 • Golf/Bora 1,6 • Beetle 1,6
Year:	1997-03
Engine Code:	AEH, AKL, APF, AUR, AVU, AYD

Replacement Interval Guide

Volkswagen recommend check at the first 60,000 miles and then every 20,000 miles (replace if necessary).
No manufacturer's recommended replacement interval.
The previous use and service history of the vehicle must always be taken into account.
Refer to Timing Belt Replacement Intervals at the front of this manual.

Check For Engine Damage

CAUTION: This engine has been identified as an INTERFERENCE engine in which the possibility of valve-to-piston damage in the event of a timing belt failure is MOST LIKELY to occur.
A compression check of all cylinders should be performed before removing the cylinder head(s).

Repair Times – hrs

Checking	0,20
Remove & install:	
Polo Classic	1,30
Golf/Bora/Beetle	2,50

Special Tools

- Auxiliary drive belt tensioner locking pin – No.T10060.
- Two-pin wrench – No.T10020.

Special Precautions

- Disconnect battery earth lead.
- DO NOT turn crankshaft or camshaft when timing belt removed.
- Remove spark plugs to ease turning engine.
- Turn engine in normal direction of rotation (unless otherwise stated).
- DO NOT turn engine via camshaft or other sprockets.
- Observe all tightening torques.

Removal

1. Raise and support front of vehicle.
2. Remove:
   - ❑ Engine cover.
   - ❑ RH engine undershield.
   - ❑ Auxiliary drive belt. Use tool No.T10060.
   - ❑ Auxiliary drive belt tensioner **1**.
   - ❑ Golf/Bora/Beetle: Coolant expansion tank. DO NOT disconnect hoses.
   - ❑ Golf/Bora: PAS reservoir. DO NOT disconnect hoses.
   - ❑ Timing belt upper cover **2**.
3. Turn crankshaft to TDC on No.1 cylinder.

4. Ensure timing marks aligned **3** or **4**.
5. Ensure camshaft sprocket timing mark aligned **5**.

Golf/Bora/Beetle:

6. Support engine.
7. Remove:
   - ❑ RH engine mounting bolts **6**, **7** & **8**.
   - ❑ RH engine mounting.

All models:

8. Remove:
   - ❑ Crankshaft pulley bolts **9**.
   - ❑ Crankshaft pulley **10**.
   - ❑ Timing belt centre cover **11**.
   - ❑ Timing belt lower cover **12**.

Golf/Bora/Beetle:

9. Remove:
   - ❑ RH engine mounting bracket bolts **13**.
   - ❑ RH engine mounting bracket.

All models:

10. Slacken tensioner nut **14**. Turn tensioner clockwise away from belt. Lightly tighten nut.
11. Remove timing belt.
    NOTE: Mark direction of rotation on belt with chalk if belt is to be reused.

Installation

1. Ensure camshaft sprocket timing marks aligned **5**.
2. Except APF/Polo Classic: Ensure timing marks aligned **4**.
3. Fit timing belt to crankshaft sprocket and water pump sprocket.

Polo Classic:

4. Install:
   - ❑ Timing belt lower cover **12**.
   - ❑ Timing belt centre cover **11**.
   - ❑ Crankshaft pulley **10**.
   - ❑ Crankshaft pulley bolts **9**.
5. Lightly tighten crankshaft pulley bolts **9**.
6. Ensure crankshaft pulley timing marks aligned **3**.

All models:

7. Fit timing belt to tensioner pulley and camshaft sprocket.
   NOTE: Ensure belt is taut between sprockets on non-tensioned side.

➡

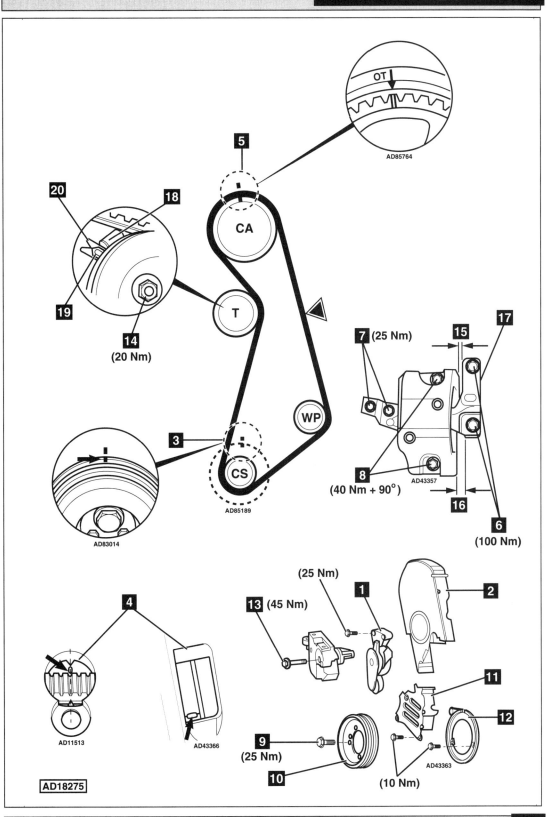

AD85764

5

20 **18**

19

14
(20 Nm)

CA

T

7 (25 Nm)

15 **17**

8

AD43357

(40 Nm + 90°)

16

6
(100 Nm)

WP

3

CS

AD85189

AD83014

4

AD11513

AD43366

13 (45 Nm)

(25 Nm)

1 **2**

11

12

AD43363

9
(25 Nm)

10

(10 Nm)

AD18275

8. Slacken tensioner nut **14**.
9. Ensure tensioner retaining lug is properly engaged **18**.
10. Turn tensioner 5 times fully anti-clockwise and clockwise from stop to stop.
Use tool No.T10020.
11. Turn tensioner fully anti-clockwise then slowly clockwise until pointer **19** aligned with notch **20** in baseplate. Use tool No.T10020.
NOTE: *Engine must be COLD.*
12. Tighten tensioner nut **14**.
Tightening torque: 20 Nm.
13. Turn crankshaft two turns clockwise to TDC on No.1 cylinder.
NOTE: *Turn crankshaft last 45° smoothly without stopping.*
14. Ensure timing marks aligned **3** or **4**.
15. Ensure camshaft sprocket timing marks aligned **5**.
16. Ensure pointer **19** aligned with notch **20** in baseplate.
17. If not: Repeat tensioning procedure.

Golf/Bora/Beetle:

18. Install:
    ❏ Timing belt lower cover **12**.
    ❏ Timing belt centre cover **11**.
    ❏ Crankshaft pulley **10**.
    ❏ Crankshaft pulley bolts **9**.
19. Install:
    ❏ RH engine mounting bracket.
    ❏ RH engine mounting bracket bolts **13**.
20. Tighten RH engine mounting bracket bolts to 45 Nm **13**.
21. Fit engine mounting.
22. Tighten:
    ❏ Engine mounting bolts **8**.
    Tightening torque: 40 Nm + 90°. Use new bolts.
    ❏ Engine mounting bolts **7**.
    Tightening torque: 25 Nm.
    ❏ Engine mounting bolts **6**.
    Tightening torque: 100 Nm.
23. Check engine mounting alignment:
    ❏ Engine mounting clearance: 14 mm **15**.
    ❏ Engine mounting clearance: 10 mm minimum **16**.
    ❏ Ensure engine mounting bolts **6** aligned with edge of mounting **17**.

All models:

24. Tighten crankshaft pulley bolts **9**.
Tightening torque: 25 Nm.
25. Apply firm thumb pressure to belt at ▽.
Pointer **19** and notch **20** must move apart.
26. Release thumb pressure from belt at ▽.
27. Turn crankshaft two turns clockwise to TDC on No.1 cylinder.
NOTE: *Turn crankshaft last 45° smoothly without stopping.*
28. Ensure pointer **19** aligned with notch **20** in baseplate.
29. Install remaining components in reverse order of removal.

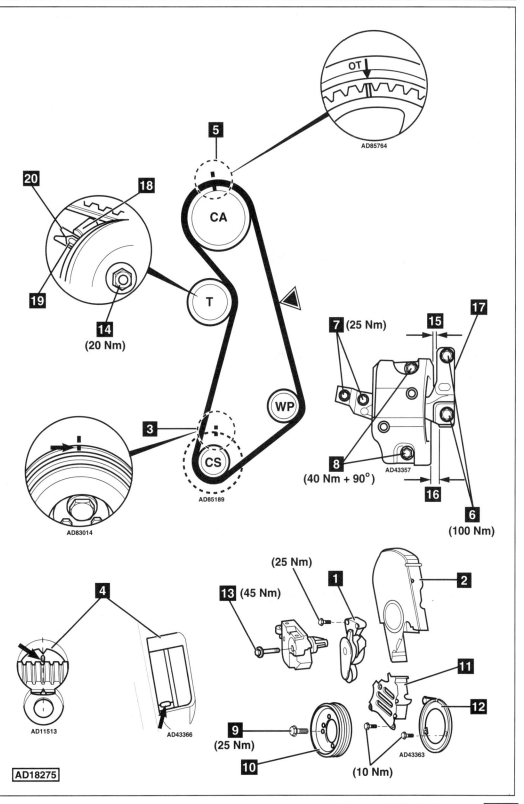

OT

AD85764

5

20 **18**

CA

19

14
(20 Nm)

T

7 (25 Nm) **15** **17**

WP

8
(40 Nm + 90°) AD43357

16

6
(100 Nm)

3

CS

AD85189

AD83014

4

AD11513

AD43366

(25 Nm)

13 (45 Nm) **1** **2**

11

12

9
(25 Nm)

10

AD43363

(10 Nm)

AD18275

VOLKSWAGEN

Model:	**Golf/Jetta/Vento 1,8/2,0 16V • Scirocco 1,8 16V • Passat 1,8/2,0 16V**
	Corrado 1,8/2,0 16V
Year:	**1985-94**
Engine Code:	**Manual tensioner**
	KR, PL, ABF, 9A

Replacement Interval Guide

Volkswagen recommend check & adjust every 20,000 miles (replace if necessary).
No manufacturer's recommended replacement interval.
The previous use and service history of the vehicle must always be taken into account.
Refer to Timing Belt Replacement Intervals at the front of this manual.

Check For Engine Damage

CAUTION: This engine has been identified as an INTERFERENCE engine in which the possibility of valve-to-piston damage in the event of a timing belt failure is MOST LIKELY to occur.
A compression check of all cylinders should be performed before removing the cylinder head.

Repair Times – hrs

Golf/Jetta →1992:
Check & adjust	0,50
Remove & install	1,10

Golf/Vento 1992→:
Check & adjust	0,90
Remove & install	1,60

Passat:
Check & adjust	0,70
Remove & install	1,30

Scirocco:
Check & adjust	0,50
Remove & install	1,10

Corrado:
Check & adjust	0,40
Remove & install	1,60

Special Tools

- Tension gauge – VAG No.210.
- Two-pin wrench – Matra V.159.

Special Precautions

- Disconnect battery earth lead.
- DO NOT turn crankshaft or camshaft when timing belt removed.
- Remove spark plugs to ease turning engine.
- Turn engine in normal direction of rotation (unless otherwise stated).
- DO NOT turn engine via camshaft or other sprockets.
- Observe all tightening torques.

Removal

1. Remove:
   - ❏ Auxiliary drive belt(s).
   - ❏ Auxiliary drive belt tensioner pulley (if fitted).
   - ❏ Timing belt upper cover **1**.
2. Turn crankshaft to TDC on No.1 cylinder. Ensure timing marks aligned **2**, **3** & **4**.
3. If cylinder head cover removed: Use camshaft sprocket rear timing mark **9**.
   NOTE: Align notch with upper cylinder head face.
4. Remove:
   - ❏ Crankshaft pulley bolts **5**.
   - ❏ Crankshaft pulley **6**.
   - ❏ Water pump pulley (if required).
   - ❏ Timing belt lower cover **7**.
5. Slacken tensioner bolt **8**. Turn tensioner anti-clockwise to release tension on belt. Lightly tighten bolt.
6. Remove timing belt.

Installation

1. Ensure timing marks aligned **2** & **4**.
2. If cylinder head cover removed: Ensure camshaft sprocket rear timing mark aligned **9**.
   NOTE: Notch aligned with upper cylinder head face.
3. Fit timing belt in anti-clockwise direction, starting at crankshaft sprocket.
4. Attach tension gauge to belt at ▽. Tool No.210.
5. Turn tensioner clockwise. Use wrench Matra V.159. Tension gauge should indicate 13-14 units.
6. Tighten tensioner bolt **8**. M8: 25 Nm. M10: 45 Nm.
7. Remove tension gauge.
8. Turn crankshaft two turns clockwise.
9. Ensure timing marks aligned **2** & **4**.
10. Recheck belt tension.
11. Install components in reverse order of removal.
12. Tighten crankshaft pulley bolts **5**. Tightening torque: 20 Nm.
    NOTE: Crankshaft pulley bolt holes are offset.

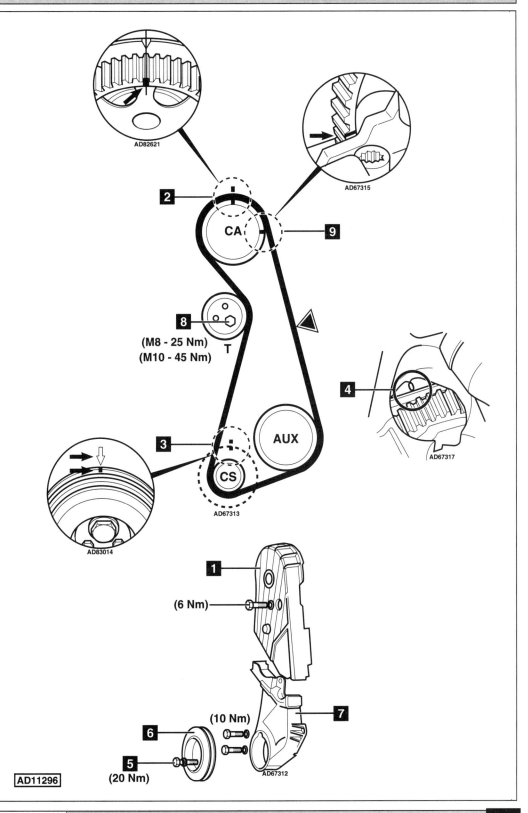

AD82621

AD67315

2

CA

9

8

(M8 - 25 Nm)
(M10 - 45 Nm)

T

4

AD67317

3

AUX

CS

AD67313

AD83014

1

(6 Nm)

(10 Nm)

7

6

5

(20 Nm)

AD67312

AD11296

VOLKSWAGEN

Model:	Golf/Bora 1,8/Turbo • Beetle 1,8 Turbo
Year:	1997-03
Engine Code:	AGN, AGU, AQA, AVC, ARZ

Check For Engine Damage

CAUTION: This engine has been identified as an INTERFERENCE engine in which the possibility of valve-to-piston damage in the event of a timing belt failure is MOST LIKELY to occur.
A compression check of all cylinders should be performed before removing the cylinder head.

Repair Times – hrs

Remove & install	2,20

Special Tools

- 1 x M5 x 55 mm stud and nut.

Special Precautions

- Disconnect battery earth lead.
- DO NOT turn crankshaft or camshaft when timing belt removed.
- Remove spark plugs to ease turning engine.
- Turn engine in normal direction of rotation (unless otherwise stated).
- DO NOT turn engine via camshaft or other sprockets.
- Observe all tightening torques.

Removal

1. Raise and support front of vehicle.
2. Remove:
   - Upper engine cover.
   - RH engine undershield.
   - Golf/Bora: RH headlamp.
   - Intercooler to turbocharger hose.
   - Auxiliary drive belt.
   - Auxiliary drive belt tensioner **1**.
3. Turn crankshaft to TDC on No.1 cylinder.
4. Ensure timing marks aligned **2** or **3**.
5. Remove:
   - Coolant expansion tank. DO NOT disconnect hoses.
   - PAS reservoir. DO NOT disconnect hoses.
   - Fuel vapour hose from charcoal canister to throttle body.
   - Timing belt upper cover **5**.
6. Ensure camshaft sprocket timing marks aligned **4**.
7. Support engine.

8. Remove:
   - RH engine mounting bolts **6**, **7** & **8**.
   - RH engine mounting.
   - Engine mounting bracket bolts **9**.
   - Engine mounting bracket.
   - Crankshaft pulley bolts **10**.
   - Crankshaft pulley **11**.
   - Timing belt centre cover **12**.
   - Timing belt lower cover **13**.
9. Insert M5 stud into tensioner **14**.
10. Fit nut and washer to stud **15**. Tighten nut sufficiently to allow a suitable locking pin to be inserted **16**.
    NOTE: DO NOT overtighten nut.
11. Remove timing belt.
    NOTE: Mark direction of rotation on belt with chalk if belt is to be reused.

Installation

1. Ensure timing marks aligned **3** & **4**.
2. Fit timing belt in following order:
   - Crankshaft sprocket.
   - Water pump sprocket.
   - Tensioner pulley.
   - Camshaft sprocket.
   NOTE: Ensure belt is taut between sprockets on non-tensioned side.
3. Remove locking pin **16**.
4. Remove nut and stud **14** & **15**.
5. Turn crankshaft two turns clockwise. Ensure timing marks aligned **3** & **4**.
6. Fit engine mounting bracket.
7. Tighten engine mounting bracket bolts to 45 Nm **9**.
8. Install:
   - Timing belt lower cover **13**.
   - Crankshaft pulley **11**.
   - Crankshaft pulley bolts **10**.
9. Tighten crankshaft pulley bolts **10**.
   Tightening torque: 25 Nm.
10. Fit engine mounting.
11. Tighten:
    - Engine mounting bolts **8**.
      Tightening torque: 40 Nm + 90°. Use new bolts.
    - Engine mounting bolts **7**.
      Tightening torque: 25 Nm.
    - Engine mounting bolts **6**.
      Tightening torque: 60 Nm + 90°. Use new bolts.
12. Check engine mounting alignment:
    - Engine mounting clearance: 14 mm **17**.
    - Engine mounting clearance: 10 mm minimum **18**.
    - Ensure engine mounting bolts **6** aligned with edge of mounting **19**.
13. Install components in reverse order of removal.

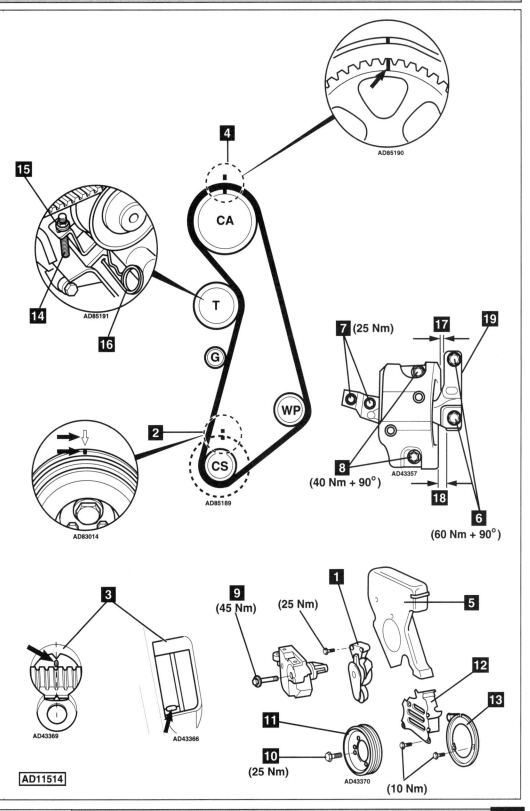

AD85190

AD85191

AD83014

AD85189

4

15

14

16

CA

T

G

WP

2

CS

7 (25 Nm)

17

19

8

AD43357

(40 Nm + 90°)

18

6

(60 Nm + 90°)

3

AD43369

AD43366

9
(45 Nm)

(25 Nm)

1

5

12

11

13

10
(25 Nm)

AD43370

(10 Nm)

AD11514

Model:	**Golf/Bora 2,0 • Beetle 2,0**
Year:	**1998-03**
Engine Code:	**AQY, APK**

Replacement Interval Guide

Volkswagen recommend check at the first 60,000 miles and then every 20,000 miles (replace if necessary).

No manufacturer's recommended replacement interval.

The previous use and service history of the vehicle must always be taken into account.
Refer to Timing Belt Replacement Intervals at the front of this manual.

Check For Engine Damage

CAUTION: This engine has been identified as an INTERFERENCE engine in which the possibility of valve-to-piston damage in the event of a timing belt failure is MOST LIKELY to occur. A compression check of all cylinders should be performed before removing the cylinder head(s).

Repair Times – hrs

Remove & install	1,90

Special Tools

- Engine support – No.10-222A.
- Engine support adaptor (Beetle) – No.10-222A/8.
- Engine support adaptor (Golf/Bora) – No.10-222A/1.
- Tensioner wrench – No.T10020.

Special Precautions

- Disconnect battery earth lead.
- DO NOT turn crankshaft or camshaft when timing belt removed.
- Remove spark plugs to ease turning engine.
- Turn engine in normal direction of rotation (unless otherwise stated).
- DO NOT turn engine via camshaft or other sprockets.
- Observe all tightening torques.

Removal

1. Raise and support front of vehicle.
2. Remove:
   - ❑ RH splash guard.
   - ❑ Engine cover.
   - ❑ Auxiliary drive belt.
   - ❑ Auxiliary drive belt tensioner.
   - ❑ Coolant expansion tank. DO NOT disconnect hoses.
   - ❑ Golf/Bora: PAS reservoir. DO NOT disconnect hoses.
   - ❑ Timing belt upper cover **1**.
3. Turn crankshaft clockwise to TDC on No.1 cylinder.
4. Ensure timing marks aligned **2** or **3**.

5. Ensure camshaft sprocket timing marks aligned **4**.
6. Install engine support and adaptor. Tool Nos.10-222A/A1/A8.
7. Remove:
   - ❑ RH engine mounting **5**.
   - ❑ Crankshaft pulley bolts **6**.
   - ❑ Crankshaft pulley **7**.
   - ❑ Timing belt centre cover **8**.
   - ❑ Timing belt lower cover **9**.
8. Raise engine slightly.
9. Remove:
   - ❑ RH engine mounting bracket bolts **10**.
   - ❑ RH engine mounting bracket **11**.
10. Slacken tensioner nut **12**. Turn tensioner clockwise away from belt. Use tool No.T10020 **13**. Lightly tighten nut **12**.
11. Remove timing belt.
    NOTE: Mark direction of rotation on belt with chalk if belt is to be reused.

Installation

1. Ensure timing marks aligned **3** & **4**.
2. Fit timing belt in following order:
   - ❑ Crankshaft sprocket.
   - ❑ Water pump sprocket.
   - ❑ Guide pulley.
   - ❑ Tensioner pulley.
   - ❑ Camshaft sprocket.
   NOTE: If belt is to be reused: Observe direction of rotation marks on belt. Ensure belt is taut between sprockets on non-tensioned side.
3. Slacken tensioner nut **12**.
4. Ensure tensioner retaining lug **14** is properly engaged.
5. Turn tensioner 5 times fully anti-clockwise and clockwise from stop to stop. Use tool No.T10020 **13**.
6. Turn tensioner fully anti-clockwise then slowly clockwise until pointer **15** aligned with notch **16** in baseplate. Use tool No.T10020 **13**.
   NOTE: Engine must be COLD.
7. Tighten tensioner nut **12**. Tightening torque: 20 Nm.
8. Turn crankshaft slowly two turns clockwise until timing marks aligned **3** & **4**.
   NOTE: Turn crankshaft last 45° smoothly without stopping.
9. Ensure pointer **15** aligned with notch **16** in baseplate.
10. If not: Repeat tensioning procedure.

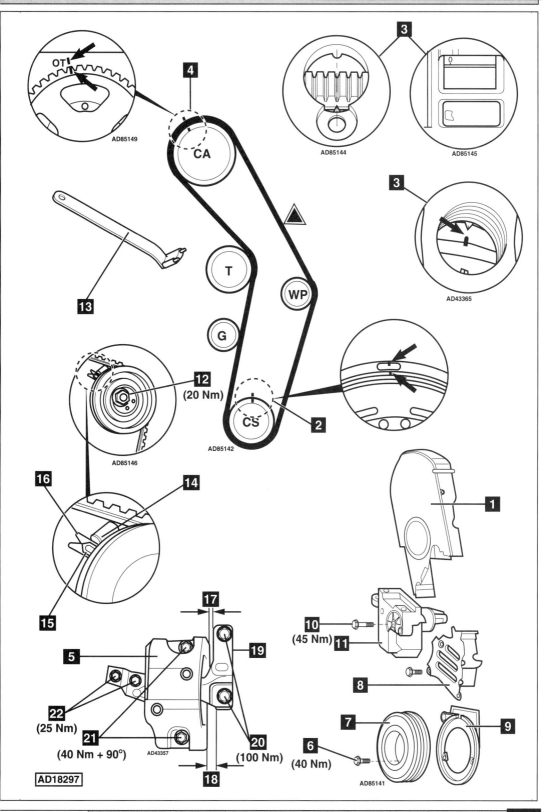

OTI AD85149

4 CA

AD85144

AD85145

3

AD43365

3

13

T

WP

G

12 (20 Nm)

CS

2

AD85146

AD85142

16

14

15

1

17

10 (45 Nm) **11**

5

19

8

22 (25 Nm)

21

20 (100 Nm)

7

9

(40 Nm + 90°) AD43357

18

6 (40 Nm)

AD85141

AD18297

11. Apply firm thumb pressure to belt at ▽.
    Pointer ⓐ and notch ⓑ must move apart.
12. Release thumb pressure from belt at ▽.
13. Turn crankshaft slowly two turns clockwise until
    timing marks aligned ❸ & ❹.
    *NOTE: Turn crankshaft last 45° smoothly without
    stopping.*
14. Ensure pointer ⓐ aligned with notch ⓑ in
    baseplate.
15. Install:
    ❑ Timing belt lower cover ❾.
    ❑ Timing belt centre cover ❽.
    ❑ Crankshaft pulley ❼.
    ❑ Crankshaft pulley bolts ❻.
16. Tighten crankshaft pulley bolts ❻.
    Tightening torque: 40 Nm.
17. Install:
    ❑ RH engine mounting bracket ⓫.
    ❑ RH engine mounting bracket bolts ⓾.
18. Tighten RH engine mounting bolts ⓾.
    Tightening torque: 45 Nm.
19. Fit and align RH engine mounting ❺.
    ❑ Engine mounting clearance: 14 mm ⓱.
    ❑ Engine mounting clearance: 10 mm ⓲.
    ❑ Ensure engine mounting bolts ⓴ aligned with
      edge of mounting ⓳.
20. Tighten:
    ❑ Engine mounting bolts ㉑.
      Tightening torque: 40 Nm + 90°. Use new
      bolts.
    ❑ Engine mounting bolts ㉒.
      Tightening torque: 25 Nm.
    ❑ Engine mounting bolts ⓴.
      Tightening torque: 100 Nm.
21. Remove engine support and adaptor.
22. Install components in reverse order of removal.

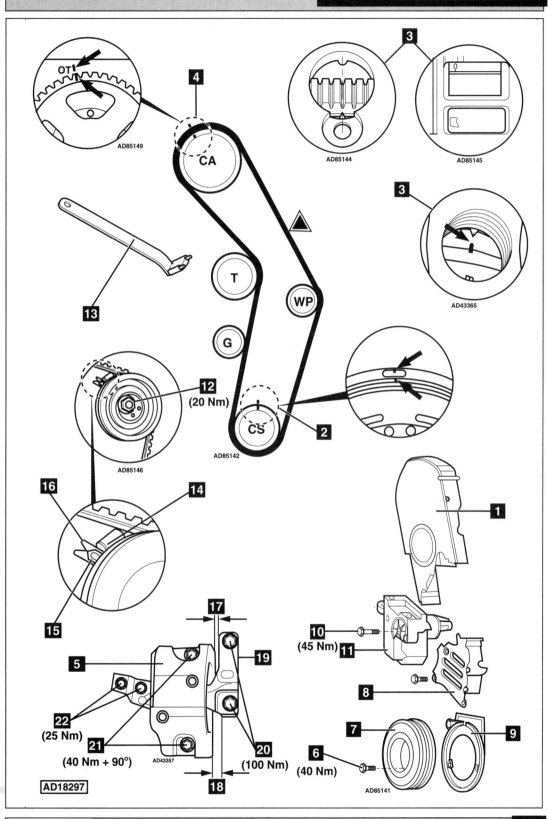

AD85149

OT

4

CA

3

AD85144

AD85145

3

AD43365

T

WP

13

G

12
(20 Nm)

AD85146

16

14

15

CS

2

AD85142

1

17

10
(45 Nm)

11

5

19

8

22
(25 Nm)

21

(40 Nm + 90°)

AD43357

20
(100 Nm)

18

7

9

6
(40 Nm)

AD85141

AD18297

VOLKSWAGEN

Model:	Golf/Vento 2,0 16V • Passat 2,0 16V • Corrado 2,0 16V
Year:	1994-98
Engine Code:	Automatic tensioner 9A, ABF

Replacement Interval Guide

Volkswagen recommend check & adjust every 20,000 miles (replace if necessary).
No manufacturer's recommended replacement interval.
The previous use and service history of the vehicle must always be taken into account.
Refer to Timing Belt Replacement Intervals at the front of this manual.

Check For Engine Damage

CAUTION: This engine has been identified as an INTERFERENCE engine in which the possibility of valve-to-piston damage in the event of a timing belt failure is MOST LIKELY to occur.
A compression check of all cylinders should be performed before removing the cylinder head.

Repair Times – hrs

Golf/Vento:	
Check & adjust	0,90
Remove & install	1,60
Passat:	
Check & adjust	0,70
Remove & install	1,30
Corrado:	
Check & adjust	0,40
Remove & install	1,60

Special Tools

■ Two-pin wrench – Matra V.159.

Special Precautions

■ Disconnect battery earth lead.
■ DO NOT turn crankshaft or camshaft when timing belt removed.
■ Remove spark plugs to ease turning engine.
■ Turn engine in normal direction of rotation (unless otherwise stated).
■ DO NOT turn engine via camshaft or other sprockets.
■ Observe all tightening torques.

Removal

1. Remove:
   ❑ Auxiliary drive belts.
   ❑ Auxiliary drive belt tensioner (if fitted).
   ❑ Timing belt upper cover **1**.
2. Turn crankshaft to TDC on No.1 cylinder. Ensure timing marks aligned **2**, **3** & **4**.
3. If cylinder head cover removed: Use camshaft sprocket rear timing mark **11**.
   NOTE: Align notch with upper cylinder head face.
4. Remove:
   ❑ Crankshaft pulley bolts **5**.
   ❑ Crankshaft pulley **6**.
   ❑ Water pump pulley (if required).
   ❑ Timing belt lower cover **7**.
5. Slacken tensioner nut **8**. Turn tensioner clockwise to release tension on belt. Lightly tighten nut.
6. Remove timing belt.

Installation

1. Ensure timing marks aligned **2** & **4**.
2. If cylinder head cover removed: Ensure camshaft sprocket rear timing mark aligned **11**.
   NOTE: Notch aligned with upper cylinder head face.
3. Fit timing belt in anti-clockwise direction, starting at crankshaft sprocket.
4. Turn tensioner anti-clockwise until it touches stop. Use wrench Matra V.159 **9**.
5. Release tension until marks align exactly **10**.
   NOTE: Engine must be cold.
6. Tighten tensioner nut **8**. Tightening torque: 25 Nm.
7. Turn crankshaft two turns clockwise. Ensure timing marks aligned **2** & **4**.
8. Check tensioner marks aligned **10**.
9. If not: Repeat tensioning procedure.
10. Apply firm thumb pressure to belt at ▽. Tensioner marks must move apart **10**.
11. Release thumb pressure from belt at ▽.
12. Turn crankshaft two turns clockwise. Ensure timing marks aligned **2** & **4**.
13. Tensioner marks should realign **10**.
14. Install components in reverse order of removal.
15. Tighten crankshaft pulley bolts **5**. Tightening torque: 20 Nm.
    NOTE: Crankshaft pulley bolt holes are offset.

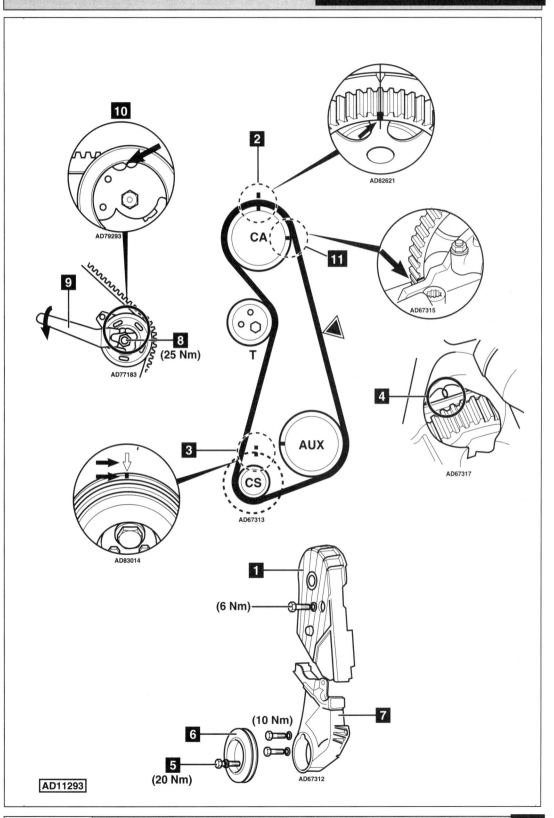

10 AD79293

2 AD82621

CA

11 AD67315

9

8 (25 Nm) AD77183

T

4 AD67317

3 AD67313

AUX

CS

AD83014

1

(6 Nm)

7

(10 Nm)

6

5 (20 Nm) AD67312

AD11293

VOLKSWAGEN

Model:	**Golf Cabriolet 2,0 • Passat 1,6**
Year:	**1996-03**
Engine Code:	**ADP, ATU, AWF, AWG**

Replacement Interval Guide

Volkswagen recommend check at the first 60,000 miles and then every 20,000 miles (replace if necessary).
No manufacturer's recommended replacement interval.
The previous use and service history of the vehicle must always be taken into account.
Refer to Timing Belt Replacement Intervals at the front of this manual.

Check For Engine Damage

CAUTION: This engine has been identified as an INTERFERENCE engine in which the possibility of valve-to-piston damage in the event of a timing belt failure is MOST LIKELY to occur.
A compression check of all cylinders should be performed before removing the cylinder head(s).

Repair Times – hrs

Remove & install:

Passat	2,50
Golf	1,30

Special Tools

■ Support guides – No.3369.
■ Two-pin wrench – Matra V.159.

Special Precautions

■ Disconnect battery earth lead.
■ DO NOT turn crankshaft or camshaft when timing belt removed.
■ Remove spark plugs to ease turning engine.
■ Turn engine in normal direction of rotation (unless otherwise stated).
■ DO NOT turn engine via camshaft or other sprockets.
■ Observe all tightening torques.

Removal

1. Raise and support front of vehicle.
2. Remove:
   ❑ Engine undershield.
   ❑ Passat: Front bumper.
   ❑ Air filter intake duct.
   ❑ Passat: Front panel bolt **1**.
3. Passat: Install support guides No.3369 in front panel **2**.
4. Passat: Remove front panel bolts **3** & **4**. Slide front panel forward.
5. Remove:
   ❑ Auxiliary drive belt.
   ❑ Auxiliary drive belt tensioner **5**.
   ❑ Golf: Water pump pulley.
   ❑ Timing belt upper cover **6**.
6. Turn crankshaft to TDC on No.1 cylinder. Ensure timing marks aligned:
   ❑ Passat: **7** & **8**.
   ❑ Golf: **8** & **16**.

7. Remove:
   ❑ Crankshaft pulley bolts **9**.
   ❑ Crankshaft pulley **10**.
   ❑ Timing belt lower cover **11**.
8. Slacken tensioner bolt **12**. Move tensioner away from belt. Lightly tighten bolt.
9. Remove timing belt.
   NOTE: Mark direction of rotation on belt with chalk if belt is to be reused.

Installation

1. Ensure timing marks aligned **8**.
2. Fit timing belt to crankshaft sprocket and auxiliary shaft sprocket.
3. Temporarily fit crankshaft pulley (with one bolt).
4. Passat: Ensure timing mark on crankshaft pulley aligned with mark on auxiliary shaft sprocket **13**.
   NOTE: DO NOT use auxiliary shaft sprocket timing mark labelled 'OT'.
5. Golf: Ensure timing marks aligned **16**.
6. Fit timing belt to tensioner pulley and camshaft sprocket.
   NOTE: Ensure belt is taut between sprockets on non-tensioned side.
7. Slacken tensioner bolt **12**.
8. Fit two-pin wrench. Use wrench Matra V.159.
   *NOTE: Two-pin wrench locates in hole **17** and against lug **18**. When turning two-pin wrench clockwise, the tensioner pulley will turn anti-clockwise and apply tension to belt.*
9. Turn two-pin wrench clockwise **19** until piston A is fully extended and piston B has risen approximately 1 mm **15**. Use wrench Matra V.159.
10. Tighten tensioner nut **12**. Tightening torque: 25 Nm.
11. Remove:
    ❑ Crankshaft pulley bolt **9**.
    ❑ Crankshaft pulley **10**.
12. Install:
    ❑ Timing belt lower cover **11**.
    ❑ Crankshaft pulley **10**.
    ❑ Crankshaft pulley bolts **9**.
13. Lightly tighten crankshaft pulley bolts **9**.
14. Turn crankshaft two turns clockwise to TDC on No.1 cylinder.
15. Ensure timing marks aligned:
    ❑ Passat: **7** & **8**.
    ❑ Golf: **8** & **16**.
16. Passat: Remove distributor cap. Ensure distributor rotor arm aligned with mark for cylinder No.1 on distributor body **14**.
17. Ensure tensioner piston B upper edge aligned with section C or dimension 'X' is 25-29 mm **15**.
18. Tighten crankshaft pulley bolts **9**.
    Tightening torque: 25 Nm.
19. Install components in reverse order of removal.

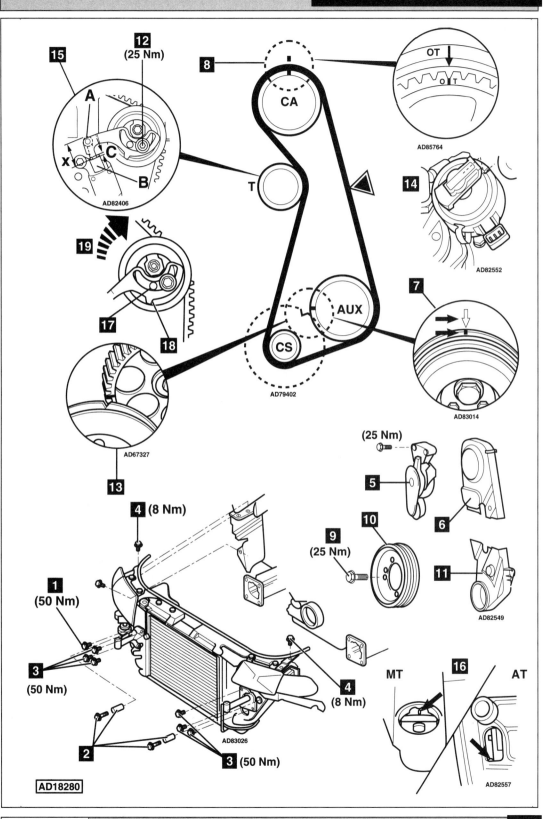

AD85764

AD82552

AD82406

AD79402

AD83014

AD67327

AD83026

AD82549

AD82557

AD18280

15 **12** (25 Nm)

8

OT

14

T **CA**

19

17 **18**

13

7 AUX

CS

X **A** **C** **B**

(25 Nm)

5 **6**

9 (25 Nm) **10**

11

4 (8 Nm)

1 (50 Nm)

3 (50 Nm)

2

3 (50 Nm)

4 (8 Nm)

16 MT AT

VOLKSWAGEN

Model:	**Passat 1,6**
Year:	**1996-02**
Engine Code:	**AHL, ALZ, ANA, ARM**

Replacement Interval Guide

Volkswagen recommend check at the first 60,000 miles and then every 20,000 miles (replace if necessary).
No manufacturer's recommended replacement interval.
The previous use and service history of the vehicle must always be taken into account.
Refer to Timing Belt Replacement Intervals at the front of this manual.

Check For Engine Damage

CAUTION: This engine has been identified as an INTERFERENCE engine in which the possibility of valve-to-piston damage in the event of a timing belt failure is MOST LIKELY to occur.
A compression check of all cylinders should be performed before removing the cylinder head(s).

Repair Times – hrs

Check	0,20
Remove & install	2,50

Special Tools

- Alternator drive belt tensioner locking pin – Volkswagen No.3204.
- Support guides (→09/00) – Volkswagen No.3369.
- Support guides (10/00→) – Volkswagen No.3411.
- Two-pin wrench – Matra V.159.

Special Precautions

- Disconnect battery earth lead.
- DO NOT turn crankshaft or camshaft when timing belt removed.
- Remove spark plugs to ease turning engine.
- Turn engine in normal direction of rotation (unless otherwise stated).
- DO NOT turn engine via camshaft or other sprockets.
- Observe all tightening torques.

Removal

1. Raise and support front of vehicle.
2. Remove:
   - ❏ Engine undershield.
   - ❏ Front bumper.
   - ❏ Air intake pipe between front panel and air filter.
   - ❏ Front panel bolt **1**.
   - ❏ Front panel bolts **2** (10/00→).
   - ❏ Front bumper carrier – not shown (10/00→).
3. Install support guides in front panel:
   - ❏ →09/00 **3**. Tool No.3369.
   - ❏ 10/00→ **4**. Tool No.3411.
4. Remove:
   - ❏ Front panel bolts **2** & **5** (→09/00)
   - ❏ Front panel bolts **5** (10/00→)
5. Slide front panel forward into service position.
6. Remove:
   - ❏ AC drive belt (if fitted).
   - ❏ Alternator drive belt. Use tool No.3204.
   - ❏ Alternator drive belt tensioner **6**.
   - ❏ Timing belt upper cover **7**.

7. Turn crankshaft to TDC on No.1 cylinder.
8. Ensure timing marks aligned **8** & **9**.
9. Remove:
   - ❏ Crankshaft pulley bolts **10**.
   - ❏ Crankshaft pulley **11**.
   - ❏ Timing belt centre cover **12**.
   - ❏ Timing belt lower cover **13**.
10. Slacken tensioner nut **14**. Turn tensioner clockwise away from belt. Lightly tighten nut.
11. Remove timing belt.
    NOTE: Mark direction of rotation on belt with chalk if belt is to be reused.

Installation

1. Ensure camshaft sprocket timing marks aligned **9**.
2. Fit timing belt to crankshaft sprocket and water pump sprocket.
3. Install:
   - ❏ Timing belt centre cover **12**.
   - ❏ Timing belt lower cover **13**.
   - ❏ Crankshaft pulley **11**.
   - ❏ Crankshaft pulley bolts **10**.
4. Tighten crankshaft pulley bolts **10**.
   Tightening torque: 25 Nm.
5. Ensure crankshaft pulley timing marks aligned **8**.
6. Fit timing belt to tensioner pulley and camshaft sprocket.
   NOTE: Ensure belt is taut between sprockets on non-tensioned side.
7. Slacken tensioner nut **14**.
8. Ensure tensioner retaining lug is properly engaged **15**.
9. Turn tensioner 5 times fully anti-clockwise and clockwise from stop to stop. Use wrench Matra V.159.
10. Turn tensioner fully anti-clockwise then slowly clockwise until pointer **16** aligned with notch **17** in baseplate. Use wrench Matra V.159.
    NOTE: Engine must be COLD.
11. Tighten tensioner nut **14**. Tightening torque: 20 Nm.
12. Turn crankshaft two turns clockwise to TDC on No.1 cylinder.
    NOTE: Turn crankshaft last 45° smoothly without stopping.
13. Ensure timing marks aligned **8** & **9**.
14. Ensure pointer **16** aligned with notch **17** in baseplate.
15. Apply firm thumb pressure to belt at ▽. Pointer **16** and notch **17** must move apart.
16. Release thumb pressure from belt at ▽.
17. Turn crankshaft two turns clockwise to TDC on No.1 cylinder.
    NOTE: Turn crankshaft last 45° smoothly without stopping.
18. Ensure pointer **16** aligned with notch **17** in baseplate.
19. Install components in reverse order of removal.

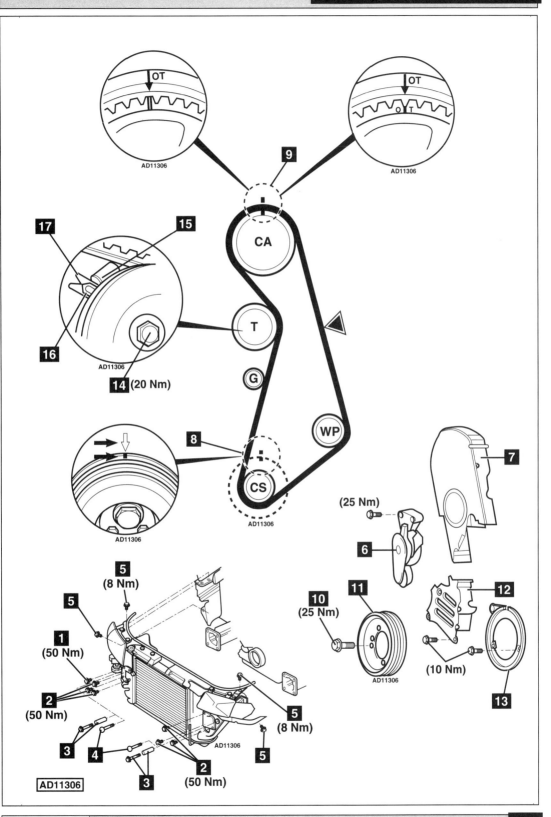

VOLKSWAGEN

Model:	**Passat 1,8**
Year:	**1996-02**
Engine Code:	**ADR, APT, ARG**

Special Precautions

- Disconnect battery earth lead.
- DO NOT turn crankshaft or camshaft when timing belt removed.
- Remove spark plugs to ease turning engine.
- Turn engine in normal direction of rotation (unless otherwise stated).
- DO NOT turn engine via camshaft or other sprockets.
- Observe all tightening torques.

Removal

1. Ensure coolant temperature above 30°C.
   NOTE: To allow tensioner pushrod to be compressed into tensioner body.
2. Raise and support front of vehicle.
3. Remove:
   - ❑ Engine undershield.
   - ❑ Front bumper.
   - ❑ Air filter intake duct.
   - ❑ Front panel bolt **1**.
4. Install support guides No.3369 in front panel **2**.
5. Remove front panel bolts **3** & **4**.
6. Slide front panel forward.

7. Remove:
   - ❑ Auxiliary drive belt.
   - ❑ Auxiliary drive belt tensioner **5**.
   - ❑ Timing belt upper cover **6**.
8. Turn crankshaft clockwise to TDC on No.1 cylinder.
9. Ensure timing marks aligned **7** & **8**.
10. Remove:
    - ❑ Crankshaft pulley bolts **9**.
    - ❑ Crankshaft pulley **10**.
    - ❑ Timing belt lower cover **11**.
11. Turn tensioner pulley slowly anti-clockwise until holes in pushrod and tensioner body aligned **13**. Use 8 mm Allen key **12**.
12. Secure in position. Use 1,5 mm pin.
    NOTE: DO NOT slacken tensioner bolts 14.
13. Remove timing belt.
    NOTE: Mark direction of rotation on belt with chalk if belt is to be reused.

Installation

1. Ensure timing marks aligned **8**.
2. Fit timing belt to crankshaft sprocket and auxiliary shaft sprocket.
3. Install:
   - ❑ Timing belt lower cover **11**.
   - ❑ Crankshaft pulley **10**.
   - ❑ Crankshaft pulley bolts **9**.
4. Lightly tighten crankshaft pulley bolts **9**.
5. Ensure crankshaft pulley timing marks aligned **7**.
6. Fit timing belt to tensioner pulley and camshaft sprocket.
   NOTE: Ensure belt is taut between sprockets on non-tensioned side.
7. Turn tensioner pulley anti-clockwise until locking pin can be removed **13**. Use 8 mm Allen key **12**.
8. Turn tensioner pulley clockwise until pushrod fully extended **15**.
9. Turn crankshaft two turns clockwise to TDC on No.1 cylinder.
10. Ensure timing marks aligned **7** & **8**.
11. Tighten crankshaft pulley bolts **9**.
    Tightening torque: 25 Nm.
12. Install components in reverse order of removal.

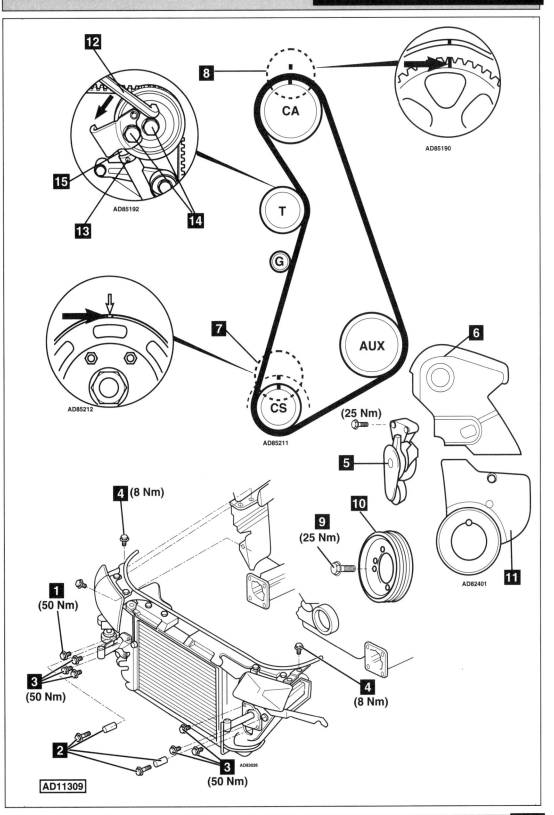

12

8

CA

AD85190

15

13

AD85192

14

T

G

AD85212

7

AUX

6

CS

AD85211

(25 Nm)

5

4 (8 Nm)

9
(25 Nm)

10

AD82401

11

1
(50 Nm)

3
(50 Nm)

2

3 AD83026
(50 Nm)

4
(8 Nm)

AD11309

VOLKSWAGEN

Model:	**Passat 1,8 Turbo**
Year:	**1996-02**
Engine Code:	**AEB, ANB, APU**

Replacement Interval Guide

Volkswagen recommend:
Check at the first 60,000 miles and then every 20,000 miles (replace if necessary) and replacement as follows:
→1999MY: Replacement every 80,000 miles.
2000MY→: Replacement every 120,000 miles.
The previous use and service history of the vehicle must always be taken into account.
Refer to Timing Belt Replacement Intervals at the front of this manual.

Check For Engine Damage

CAUTION: This engine has been identified as an INTERFERENCE engine in which the possibility of valve-to-piston damage in the event of a timing belt failure is MOST LIKELY to occur.
A compression check of all cylinders should be performed before removing the cylinder head(s).

Repair Times – hrs

Remove & install	2,50

Special Tools

- Support guides – No.3369.
- Two-pin wrench – Matra V.159.

Special Precautions

- Disconnect battery earth lead.
- DO NOT turn crankshaft or camshaft when timing belt removed.
- Remove spark plugs to ease turning engine.
- Turn engine in normal direction of rotation (unless otherwise stated).
- DO NOT turn engine via camshaft or other sprockets.
- Observe all tightening torques.

Removal

1. Raise and support front of vehicle.
2. Remove:
   - ❑ Engine undershield.
   - ❑ Front bumper.
   - ❑ Air filter intake duct.
   - ❑ Front panel bolt **1**.
3. Install support guides No.3369 in front panel **2**.
4. Remove front panel bolts **3** & **4**.
5. Slide front panel forward.
6. Remove:
   - ❑ Auxiliary drive belt.
   - ❑ Auxiliary drive belt tensioner **5**.
   - ❑ Timing belt upper cover **6**.

7. Turn crankshaft clockwise to TDC on No.1 cylinder. Ensure timing marks aligned **7** & **8**.
8. Remove:
   - ❑ Crankshaft pulley bolts **9**.
   - ❑ Crankshaft pulley **10**.
   - ❑ Timing belt lower cover **11**.
9. Slacken tensioner bolt **12**. Move tensioner away from belt. Lightly tighten bolt.
10. Remove timing belt.
    NOTE: Mark direction of rotation on belt with chalk if belt is to be reused.

Installation

1. Ensure timing marks aligned **8**.
2. Fit timing belt to crankshaft sprocket and auxiliary shaft sprocket.
3. Install:
   - ❑ Timing belt lower cover **11**.
   - ❑ Crankshaft pulley **10**.
   - ❑ Crankshaft pulley bolts **9**.
4. Lightly tighten crankshaft pulley bolts **9**.
5. Ensure crankshaft pulley timing marks aligned **7**.
6. Fit timing belt to tensioner pulley and camshaft sprocket.
   NOTE: Ensure belt is taut between sprockets on non-tensioned side.
7. Slacken tensioner bolt **12**.
8. Fit two-pin wrench. Use wrench Matra V.159.
   *NOTE: Two-pin wrench locates in hole **14** and against lug **15**. When turning two-pin wrench clockwise, the tensioner pulley will turn anti-clockwise and apply tension to belt.*
9. Turn two-pin wrench clockwise **16** until piston A is fully extended and piston B has risen approximately 1 mm **13**. Use wrench Matra V.159.
10. Tighten tensioner bolt **12**.
    Tightening torque: 25 Nm.
11. Turn crankshaft two turns clockwise to TDC on No.1 cylinder.
12. Ensure timing marks aligned **7** & **8**.
13. Ensure tensioner piston B upper edge aligned with section C or dimension 'X' is 25-29 mm **13**.
14. Tighten crankshaft pulley bolts **9**.
    Tightening torque: 25 Nm.
15. Install components in reverse order of removal.

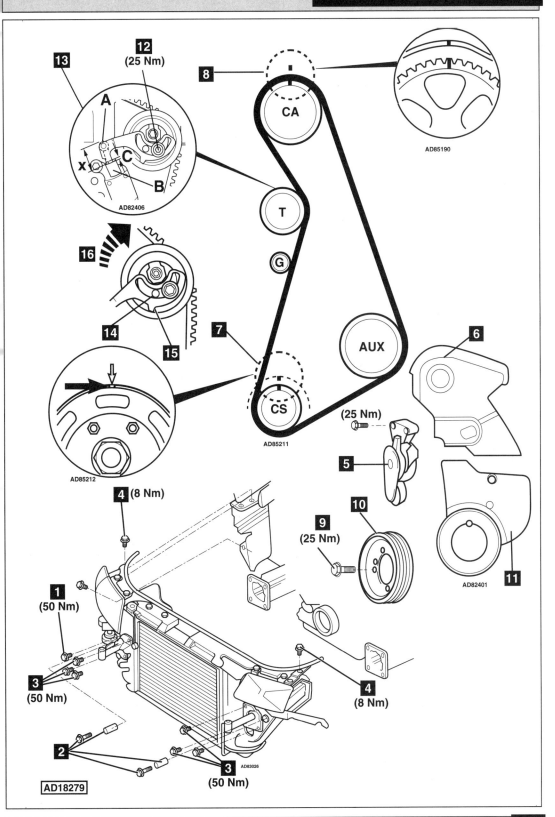

12 (25 Nm)

13

A

C

X

B

AD82406

8

CA

AD85190

T

G

16

14

15

7

AUX

CS

AD85211

(25 Nm)

5

10

9 (25 Nm)

6

11

AD82401

4 (8 Nm)

AD85212

1 (50 Nm)

3 (50 Nm)

4 (8 Nm)

2

3 (50 Nm)

AD83026

AD18279

VOLKSWAGEN

Model:	**Passat 2,8 30V**
Year:	**1996-02**
Engine Code:	**ACK, ALG, APR, AQD**

Replacement Interval Guide

Volkswagen recommend replacement every 80,000 miles (tensioner pulley must also be replaced).
The previous use and service history of the vehicle must always be taken into account.
Refer to Timing Belt Replacement Intervals at the front of this manual.

Check For Engine Damage

CAUTION: This engine has been identified as an INTERFERENCE engine in which the possibility of valve-to-piston damage in the event of a timing belt failure is MOST LIKELY to occur.
A compression check of all cylinders should be performed before removing the cylinder head(s).

Repair Times – hrs

Remove & install	3,30

Special Tools

- Crankshaft locking tool – No.3242.
- Camshaft locking tool – No.3391.
- Camshaft sprocket puller – No.3032.

Special Precautions

- Disconnect battery earth lead.
- DO NOT turn crankshaft or camshaft when timing belt removed.
- Remove spark plugs to ease turning engine.
- Turn engine in normal direction of rotation (unless otherwise stated).
- DO NOT turn engine via camshaft or other sprockets.
- Observe all tightening torques.

Removal

1. Remove:
   - ❑ Viscous fan (LH thread).
   - ❑ Auxiliary drive belt.
   - ❑ Auxiliary drive belt tensioner.
   - ❑ Timing belt covers **1**, **2** & **3**.
2. Turn crankshaft to TDC on No.3 cylinder.
3. Ensure crankshaft pulley timing marks aligned **4**.
4. Ensure large holes in locking plates of camshaft sprockets face in towards each other **5**.
5. If not: Turn crankshaft one turn clockwise.
6. Remove blanking plug from crankcase. Screw in crankshaft locking tool. Tool No.3242 **6**.
   NOTE: TDC hole in crankshaft web must be aligned with blanking plug hole.

7. Turn tensioner pulley clockwise until holes in pushrod and tensioner body aligned. Use 8 mm Allen key **7**. Retain pushrod with 2 mm diameter pin through hole in tensioner body **8**.
8. Remove:
   - ❑ Crankshaft pulley bolts **9**.
   - ❑ Crankshaft pulley **10**.
   - ❑ Viscous fan mounting bracket.
   - ❑ Timing belt lower cover **11**.
   - ❑ Timing belt.
   NOTE: Mark direction of rotation on belt with chalk if belt is to be reused.

Installation

1. Remove:
   - ❑ Bolt of each camshaft sprocket **12**.
   - ❑ Locking plate of each camshaft sprocket **5**.
2. Screw M10 bolt into camshaft to act as support for puller.
3. Loosen camshaft sprockets from taper. Use puller No.3032 **13**.
4. Install:
   - ❑ Locking plates **5**.
   - ❑ Bolts **12**.
5. Lightly tighten bolt of each camshaft sprocket **12**.
6. Ensure camshaft sprockets can turn freely but not tilt.
7. Fit timing belt.
8. Fit locking tool to camshafts. Tool No.3391.
9. Turn tensioner pulley slightly clockwise. Use 8 mm Allen key **7**. Remove pin from tensioner body to release pushrod **8**.
10. Install torque wrench to hexagon of tensioner **14**.
11. Tension timing belt in anti-clockwise direction to 15 Nm **14**.
12. Remove torque wrench.
13. Tighten bolt of each camshaft sprocket to 55 Nm **12**.
14. Fit crankshaft pulley **10**.
15. Remove:
    - ❑ Camshaft locking tool.
    - ❑ Crankshaft locking tool **6**.
16. Fit blanking plug.
17. Tighten crankshaft pulley bolts **9**.
    Tightening torque: 25 Nm.
18. Install components in reverse order of removal.

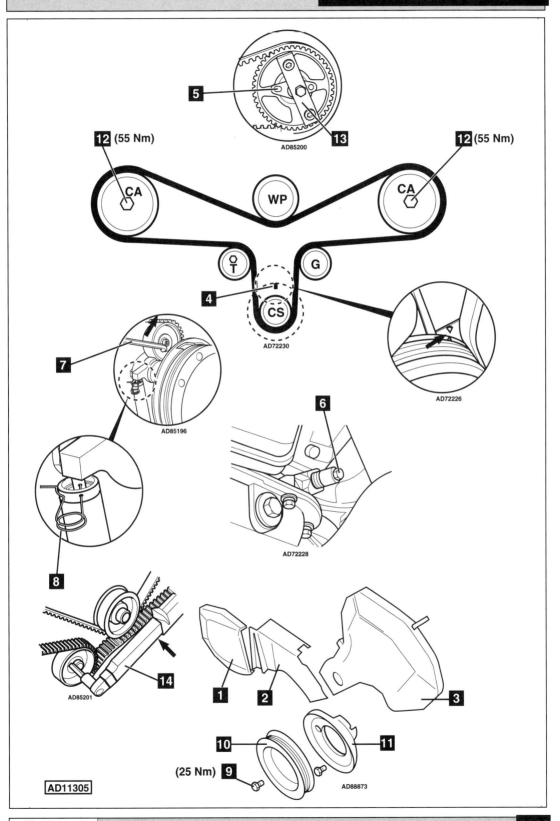

5

12 (55 Nm)   **12** (55 Nm)

13   AD85200

CA   WP   CA

T   G

4

CS

AD72230

7   AD85196

AD72226

6

AD72228

8

14   AD85201

1   **2**   **3**

10   **11**

(25 Nm) **9**   AD88873

AD11305

Model:	**Sharan 1,8 Turbo**
Year:	**1997-00**
Engine Code:	**AJH**

Replacement Interval Guide

Volkswagen recommend:
Check at the first 60,000 miles and then every 20,000 miles (replace if necessary) and replacement every 120,000 miles (tensioner pulley must also be replaced).
The previous use and service history of the vehicle must always be taken into account.
Refer to Timing Belt Replacement Intervals at the front of this manual.

Check For Engine Damage

CAUTION: This engine has been identified as an INTERFERENCE engine in which the possibility of valve-to-piston damage in the event of a timing belt failure is MOST LIKELY to occur.
A compression check of all cylinders should be performed before removing the cylinder head.

Repair Times – hrs

Remove & install	2,20

Special Tools

■ 1 x M5 x 55 mm stud and nut.

Special Precautions

■ Disconnect battery earth lead.
■ DO NOT turn crankshaft or camshaft when timing belt removed.
■ Remove spark plugs to ease turning engine.
■ Turn engine in normal direction of rotation (unless otherwise stated).
■ DO NOT turn engine via camshaft or other sprockets.
■ Observe all tightening torques.

Removal

1. Raise and support front of vehicle.
2. Remove:
   ❏ Air filter assembly.
   ❏ Engine undershield.
   ❏ Intercooler to turbocharger hose.
   ❏ Intercooler to turbocharger hose bracket.
   ❏ Auxiliary drive belt.
   ❏ Auxiliary drive belt tensioner **1**.
   ❏ Timing belt upper cover **2**.
3. Turn crankshaft to TDC on No.1 cylinder.
4. Ensure crankshaft pulley timing marks aligned **3**.
5. Ensure camshaft sprocket timing marks aligned **4**.
6. Support engine.
7. Remove:
   ❏ RH engine mounting bolts **5** & **6**.
   ❏ RH engine mounting nuts **7**.
   ❏ RH engine mounting.
   ❏ Engine mounting bracket bolts **8**.
   ❏ Engine mounting bracket.
   ❏ Crankshaft pulley bolts **9**.
   ❏ Crankshaft pulley **10**.
   ❏ Timing belt centre cover **11**.
   ❏ Timing belt lower cover **12**.

8. Insert M5 stud into tensioner **13**.
9. Fit nut and washer to stud **14**. Tighten nut sufficiently to allow a suitable locking pin to be inserted **15**.
   NOTE: DO NOT overtighten nut.
10. Remove timing belt.
    NOTE: Mark direction of rotation on belt with chalk if belt is to be reused.

Installation

1. Ensure camshaft sprocket timing marks aligned **4**.
2. Fit timing belt to crankshaft sprocket.
3. Install:
   ❏ Timing belt lower cover **12**.
   ❏ Crankshaft pulley **10**.
   ❏ Crankshaft pulley bolts **9**.
4. Lightly tighten crankshaft pulley bolts **9**.
5. Ensure crankshaft pulley timing marks aligned **3**.
6. Fit timing belt to remaining sprockets in following order:
   ❏ Water pump sprocket.
   ❏ Tensioner pulley.
   ❏ Camshaft sprocket.
   NOTE: Ensure belt is taut between sprockets on non-tensioned side.
7. Remove locking pin **15**.
8. Remove nut and stud **13** & **14**.
9. Turn crankshaft two turns clockwise.
10. Ensure timing marks aligned **3** & **4**.
11. Fit engine mounting bracket.
12. Tighten engine mounting bracket bolts to 45 Nm **8**.
13. Tighten crankshaft pulley bolts **9**.
    Tightening torque: 25 Nm.
14. Install:
    ❏ Timing belt centre cover **11**.
    ❏ Engine mounting.
15. Tighten:
    ❏ Engine mounting bolts **6**.
       Tightening torque: 60 Nm. Lightly oil threads.
    ❏ Engine mounting nuts **7**.
       Tightening torque: 55 Nm.
    ❏ Engine mounting bolts **5**.
       Tightening torque: 60 Nm. Lightly oil threads.
16. Install components in reverse order of removal.

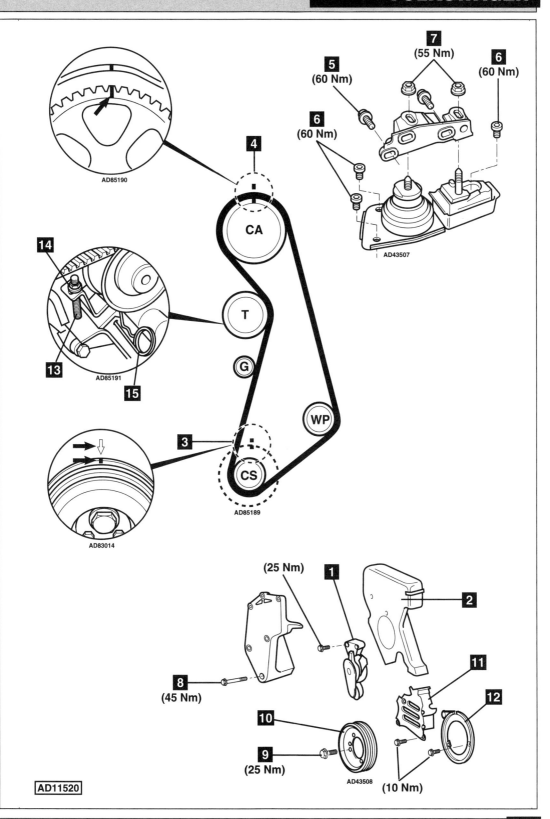

AD85190

4

CA

5 (60 Nm)

7 (55 Nm)

6 (60 Nm)

6 (60 Nm)

6 (60 Nm)

AD43507

14

13

AD85191

15

T

G

WP

3

CS

AD85189

AD83014

(25 Nm)

1

2

8 (45 Nm)

10

9 (25 Nm)

AD43508

11

12

(10 Nm)

AD11520

VOLKSWAGEN

Model:	**Transporter 2,0**
Year:	**1990-03**
Engine Code:	**AAC**

Replacement Interval Guide

Volkswagen recommend check & adjust every 20,000 miles (replace if necessary).
No manufacturer's recommended replacement interval.

The previous use and service history of the vehicle must always be taken into account.
Refer to Timing Belt Replacement Intervals at the front of this manual.

Check For Engine Damage

CAUTION: This engine has been identified as an INTERFERENCE engine in which the possibility of valve-to-piston damage in the event of a timing belt failure is MOST LIKELY to occur.
A compression check of all cylinders should be performed before removing the cylinder head.

Repair Times – hrs

Check & adjust	0,90
Remove & install	1,90

Special Tools

■ Two-pin wrench – Matra V.159.

Special Precautions

■ Disconnect battery earth lead.
■ DO NOT turn crankshaft or camshaft when timing belt removed.
■ Remove spark plugs to ease turning engine.
■ Turn engine in normal direction of rotation (unless otherwise stated).
■ DO NOT turn engine via camshaft or other sprockets.
■ Observe all tightening torques.

Removal

1. Remove timing belt upper cover **1**.
2. Turn crankshaft to TDC on No.1 cylinder. Ensure flywheel timing marks aligned **4**. Ensure camshaft timing marks aligned **5** or **6**.
3. Remove:
   ❑ Auxiliary drive belts.
   ❑ Crankshaft damper bolts **7**.
   ❑ Crankshaft damper **3**.
   ❑ Timing belt lower cover **2**.
4. Slacken tensioner bolt **8**. Turn tensioner anti-clockwise to release tension on belt. Lightly tighten bolt.
5. Remove timing belt.

Installation

1. Ensure timing marks aligned **4** & **5** or **6**.
2. Ensure distributor rotor arm aligned with mark on distributor body **9**.
3. Fit timing belt in following order:
   ❑ Crankshaft sprocket.
   ❑ Auxiliary shaft sprocket.
   ❑ Camshaft sprocket.
4. Ensure timing marks aligned **4** & **5** or **6**. Slacken tensioner bolt **8**.
5. Turn tensioner clockwise until belt can just be twisted with finger and thumb through 90° at ▽. Use wrench Matra V.159.
6. Tighten tensioner bolt to 45 Nm **8**.
7. Turn crankshaft two turns clockwise.
8. Ensure timing marks aligned **4** & **5** or **6**.
9. Check belt tension.
10. Install components in reverse order of removal.
11. Tighten crankshaft damper bolts to 25 Nm **7**.

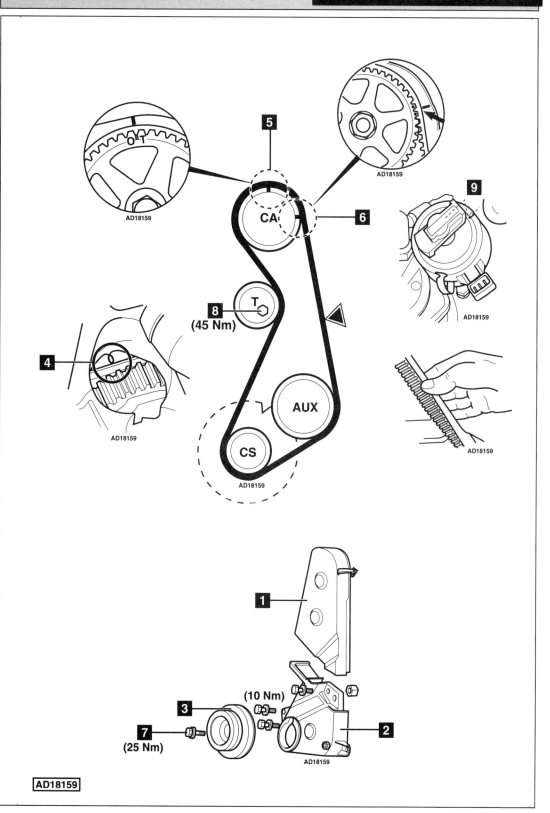

5

CA

6

9

8
(45 Nm)

T

4

AUX

CS

AD18159

1

(10 Nm)

3

7
(25 Nm)

2

AD18159

VOLKSWAGEN

Model:	Transporter 2,5
Year:	10/91-7/95
Engine Code:	AAF, ACU

Replacement Interval Guide

Volkswagen recommend check & adjust or replacement if necessary every 20,000 miles. No manufacturer's recommended replacement interval.
The previous use and service history of the vehicle must always be taken into account.
Refer to Timing Belt Replacement Intervals at the front of this manual.

Check For Engine Damage

CAUTION: This engine has been identified as an INTERFERENCE engine in which the possibility of valve-to-piston damage in the event of a timing belt failure is MOST LIKELY to occur.
A compression check of all cylinders should be performed before removing the cylinder head.

Repair Times – hrs

Check & adjust	1,30
Remove & install	2,50

Special Tools

- ■ Crankshaft damper locking tool – VAG No.3419.

Special Precautions

- ■ Disconnect battery earth lead.
- ■ DO NOT turn crankshaft or camshaft when timing belt removed.
- ■ Remove spark plugs to ease turning engine.
- ■ Turn engine in normal direction of rotation (unless otherwise stated).
- ■ DO NOT turn engine via camshaft or other sprockets.
- ■ Observe all tightening torques.

Removal

1. Remove engine undershield.
2. Mark direction of rotation on auxiliary drive belt with chalk. Remove auxiliary drive belt.
3. Remove timing belt upper cover **1**.
4. Turn crankshaft clockwise to TDC on No.1 cylinder.
5. Ensure TDC marks aligned **4** or **5**.
6. Ensure camshaft sprocket timing mark aligned with mark on timing belt rear cover **6**.
7. Lock crankshaft damper. Use tool No.3419 **9**. Remove centre bolt **10**.
8. Remove:
   - ❑ Four crankshaft damper bolts.
   - ❑ Crankshaft damper **3**.
   - ❑ Timing belt lower cover **2**.

9. Slacken tensioner nut **8**. Move tensioner away from belt.
10. Remove timing belt.

Installation

1. Ensure crankshaft at TDC on No.1 cylinder.
2. Ensure timing marks aligned **5** & **6**.
3. Fit timing belt, starting at crankshaft sprocket.
4. Install:
   - ❑ Timing belt lower cover **2**.
   - ❑ Crankshaft damper **3**.
   - ❑ Four crankshaft damper bolts **7**.
5. Tighten crankshaft damper bolts **7**. Tightening torque: 20 Nm + 90°.
6. Fit crankshaft damper centre bolt **10**. Lock crankshaft damper. Use tool No.3419 **9**.
7. Tighten centre bolt to specified torque:
   - ❑ Bolt length 'X' 65 mm = 460 Nm **10**.
   - ❑ Bolt length 'X' 110 mm = 160 Nm + 180° **10**.
8. Remove locking tool **9**.
9. Turn crankshaft two turns clockwise until timing mark just before TDC **4**.
10. Turn crankshaft slowly to TDC **4**. Tighten tensioner nut to 15 Nm **8** (while crankshaft is turning).
11. Turn crankshaft two turns clockwise. Ensure timing marks aligned **5** & **6**.
12. Install components in reverse order of removal.
    NOTE: Observe direction of rotation marks on auxiliary drive belt.

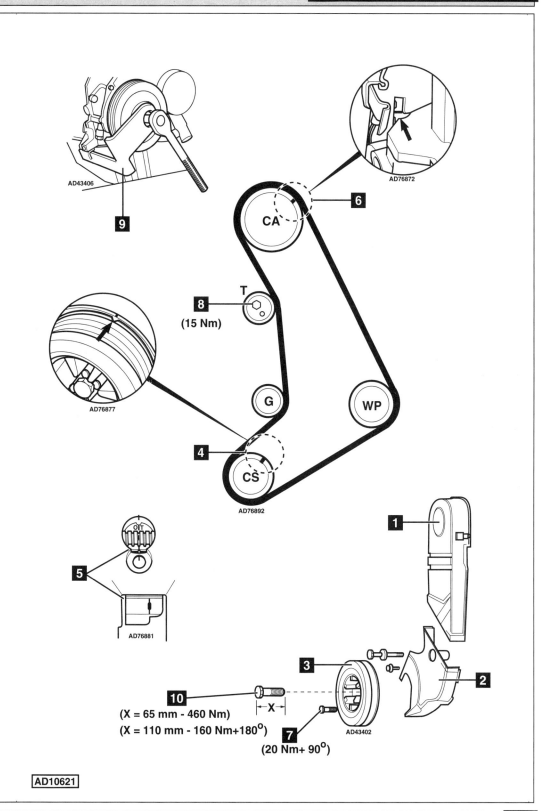

AD43406

9

AD76872

6

CA

T

8 (15 Nm)

AD76877

G

WP

4

CS

AD76892

5

AD76881

1

3

2

10

(X = 65 mm - 460 Nm)
(X = 110 mm - 160 Nm+180°)

|←—X—→|

7

AD43402

(20 Nm+ 90°)

AD10621

VOLKSWAGEN

Model:	**Transporter 2,5**
Year:	**08/95-03**
Engine Code:	**ACU, AET, APL, AVT**

Replacement Interval Guide

The vehicle manufacturer has not recommended a timing belt replacement interval for this engine.
The previous use and service history of the vehicle must always be taken into account.
Refer to Timing Belt Replacement Intervals at the front of this manual.

Check For Engine Damage

CAUTION: This engine has been identified as an INTERFERENCE engine in which the possibility of valve-to-piston damage in the event of a timing belt failure is MOST LIKELY to occur.
A compression check of all cylinders should be performed before removing the cylinder head.

Repair Times – hrs

Check & adjust	1,30
Remove & install	2,50

Special Tools

- Crankshaft damper locking tool – No.3419.
- Tensioner pulley spanner – No.3355.

Special Precautions

- Disconnect battery earth lead.
- DO NOT turn crankshaft or camshaft when timing belt removed.
- Remove spark plugs to ease turning engine.
- Turn engine in normal direction of rotation (unless otherwise stated).
- DO NOT turn engine via camshaft or other sprockets.
- Observe all tightening torques.

Removal

1. Remove engine undershield.
2. Remove:
   - ❏ Auxiliary drive belt.
   - ❏ Timing belt upper cover **1**.
3. Turn crankshaft clockwise to TDC on No.1 cylinder.
4. Ensure TDC marks aligned **4** or **5**.
5. Ensure camshaft sprocket timing mark aligned with mark on timing belt rear cover **6**.
6. Lock crankshaft damper. Use tool No.3419 **9**. Remove centre bolt **10**.
7. Remove:
   - ❏ Four crankshaft damper bolts **7**.
   - ❏ Crankshaft damper **3**.
   - ❏ Timing belt lower cover **2**.

8. Slacken tensioner bolt **8**. Move tensioner away from belt.
9. Remove timing belt.

Installation

1. Ensure crankshaft at TDC on No.1 cylinder.
2. Ensure timing marks aligned **5** & **6**.
3. Fit timing belt, starting at crankshaft sprocket.
4. Install:
   - ❏ Timing belt lower cover **2**.
   - ❏ Crankshaft damper **3**.
   - ❏ Four crankshaft damper bolts **7**.
5. Tighten crankshaft damper bolts **7**. Tightening torque: 20 Nm + 90°.
6. Fit crankshaft damper centre bolt **10**. Lock crankshaft damper. Use tool No.3419 **9**.
7. Tighten centre bolt to specified torque:
   - ❏ Bolt length 'X' 65 mm = 460 Nm **10**.
   - ❏ Bolt length 'X' 110 mm = 160 Nm + 180° **10**.
8. Remove locking tool **9**.
9. Turn tensioner pulley slowly clockwise until right edge of pointer **11** passes right edge of pointer **12**. Use tool No.3355 **13**.
10. Turn tensioner pulley slowly anti-clockwise until right edge of pointers aligned **11** & **12**. Use tool No.3355 **13**.
11. Tighten tensioner bolt to 20 Nm **8**.
12. Turn crankshaft two turns clockwise to TDC on No.1 cylinder.
13. Ensure timing marks aligned **5** & **6**.
14. Ensure right edge of tensioner pointers aligned **11** & **12**. If not: Repeat tensioning procedures.
15. Install components in reverse order of removal.

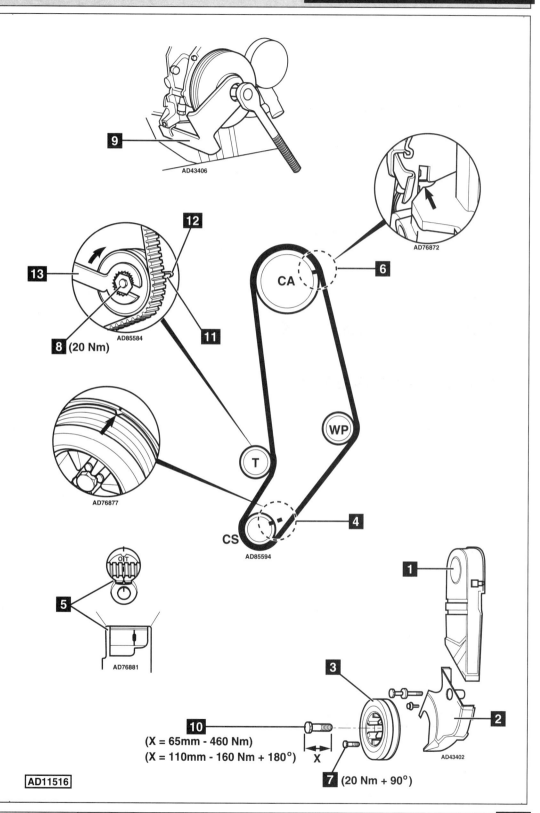

AD43406

CA

AD76872

6

13

12

11

8 (20 Nm)

AD85584

9

AD76877

T

WP

5

AD76881

CS

AD85594

4

1

3

2

AD43402

10

(X = 65mm - 460 Nm)
(X = 110mm - 160 Nm + 180°)

X

7 (20 Nm + 90°)

AD11516

VOLKSWAGEN

Model:	**LT 2,4**
Year:	**1983-96**
Engine Code:	**DL, HS, 1E**

Replacement Interval Guide

Volkswagen recommend check & adjust every 20,000 miles (replace if necessary).
No manufacturer's recommended replacement interval.
The previous use and service history of the vehicle must always be taken into account.
Refer to Timing Belt Replacement Intervals at the front of this manual.

Check For Engine Damage

CAUTION: This engine has been identified as an INTERFERENCE engine in which the possibility of valve-to-piston damage in the event of a timing belt failure is MOST LIKELY to occur.
A compression check of all cylinders should be performed before removing the cylinder head.

Repair Times – hrs

Check & adjust	0,50
Remove & install	1,30

Special Tools

- Crankshaft pulley locking tool (rigid fan) – VAG No.3037.
- Crankshaft pulley locking tool (viscous fan) – VAG No.3181.

Special Precautions

- Disconnect battery earth lead.
- DO NOT turn crankshaft or camshaft when timing belt removed.
- Remove spark plugs to ease turning engine.
- Turn engine in normal direction of rotation (unless otherwise stated).
- DO NOT turn engine via camshaft or other sprockets.
- Observe all tightening torques.

Removal

1. Remove:
   - ❏ Timing belt upper cover **1**.
   - ❏ Timing belt lower cover **2**.
   - ❏ Lower air duct.
   - ❏ Cooling fan cowling.
   - ❏ Cooling fan **3**.
   - ❏ Auxiliary drive belt.
   *NOTE: Vehicles with viscous coupling – remove fan and coupling **4** as an assembly from intermediate flange **5**.*
2. Fit appropriate crankshaft pulley locking tool. Undo bolt **6**.

3. Turn crankshaft to TDC on No.1 cylinder. Ensure camshaft and flywheel timing marks aligned **7** & **8**.
   NOTE: Timing mark on camshaft sprocket aligned with upper edge of cylinder head cover gasket.
4. Slacken water pump bolts to release tension on belt **11**.
5. Remove crankshaft pulley **10** and crankshaft sprocket with belt attached.

Installation

1. Ensure timing marks aligned **7** & **8**.
   NOTE: Timing mark on camshaft sprocket aligned with upper edge of cylinder head cover gasket.
2. Fit crankshaft pulley and crankshaft sprocket with belt attached.
   NOTE: Ensure belt is not trapped between sprocket and cylinder block.
3. Clean crankshaft pulley bolt **6**. Coat threads with corrosion inhibitor (AMV 188 000 02 or similar).
4. Fit crankshaft pulley bolt **6**. Lightly tighten bolt.
5. Fit timing belt to other sprockets, then ease it round guide pulley. Ensure belt is taut between sprockets.
6. Turn water pump anti-clockwise to tension belt **9**.
   NOTE: Belt is correctly tensioned when it can just be twisted with finger and thumb through 90° at ▼.
7. Tighten water pump bolts to 25 Nm **11**.
8. Fit appropriate crankshaft pulley locking tool. Tighten bolt to 460 Nm **6**.
9. Turn crankshaft two turns clockwise.
10. Ensure timing marks aligned **7** & **8**.
11. Check belt tension.
12. Install components in reverse order of removal.

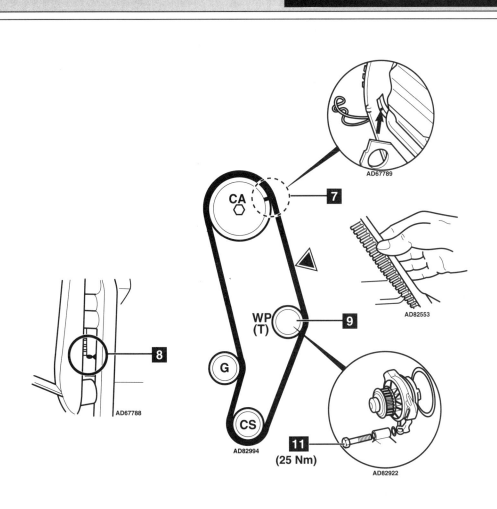

AD67789

CA

7

AD82553

WP
(T)

9

8

AD67788

G

CS

AD82994

11
(25 Nm)

AD82922

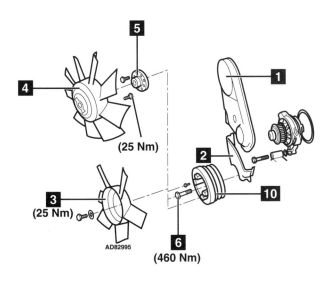

5

4

(25 Nm)

1

2

3
(25 Nm)

10

6
(460 Nm)

AD82995

AD10607

VOLKSWAGEN

Model:	**Lupo 3L 1,2 TDI PD • Lupo 1,4 TDI PD • Polo 1,4 TDI PD**
	Golf/Bora 1,9 TDI PD • Beetle 1,9 TDI PD • Sharan 1,9 TDI PD
Year:	**1998-03**
Engine Code:	**AJM, AMF, ANU, ANY, ASZ, ATD, AUY, AXR, AYZ**

Replacement Interval Guide

Volkswagen recommend:

All models:
→2000MY: Check condition and width every 10,000 miles or 12 months, whichever occurs first (replacement width 27 mm).

Lupo/Polo/Beetle:
→2000MY: Replacement every 40,000 miles.
2001MY→: Replacement every 60,000 miles.

Golf/Bora:
AJM engine →2000MY: Replacement every 40,000 miles.
All other engines →2000MY: Replacement every 60,000 miles.
2001MY→: Replacement every 60,000 miles.

Sharan:
Replacement every 40,000 miles.

All models:
Tensioner pulley must also be replaced.
The previous use and service history of the vehicle must always be taken into account.
Refer to Timing Belt Replacement Intervals at the front of this manual.

Check For Engine Damage

CAUTION: This engine has been identified as an INTERFERENCE engine in which the possibility of valve-to-piston damage in the event of a timing belt failure is MOST LIKELY to occur. A compression check of all cylinders should be performed before removing the cylinder head(s).

Repair Times – hrs

Checking	0,20
Remove & install:	
Lupo	2,20
Polo	2,50
Golf/Bora/Sharan	2,90
Beetle	3,30

Special Tools

- Auxiliary drive belt tensioner locking pin – Volkswagen No.T10060.
- Camshaft locking tool – Volkswagen No.3359.
- Crankshaft sprocket locking tool – Volkswagen No.T10050.
- Tensioner locking tool – Volkswagen No.T10008.
- Two-pin wrench – Volkswagen No.3387.
- ANY/AYZ engines – 7 mm drill bit.
- All other engines – 4 mm drill bit.

Special Precautions

- Disconnect battery earth lead.
- DO NOT turn crankshaft or camshaft when timing belt removed.
- Remove glow plugs to ease turning engine.
- Turn engine in normal direction of rotation (unless otherwise stated).
- DO NOT turn engine via camshaft or other sprockets.
- Observe all tightening torques.

Removal

1. Raise and support front of vehicle.
2. Remove engine undershield.
3. Remove:
   - ❑ Engine top cover.
   - ❑ Golf/Bora: RH headlamp.
   - ❑ Turbocharger air hoses.
   - ❑ Intercooler outlet hose.
   - ❑ Mass air flow (MAF) sensor (if necessary).
   - ❑ Lupo/Polo: Lower engine steady bar.
   - ❑ Auxiliary drive belt. Use tool No.T10060.
   - ❑ Auxiliary drive belt tensioner.
   - ❑ Golf/Bora/Beetle: Coolant expansion tank. DO NOT disconnect hoses.
   - ❑ Golf/Bora: PAS reservoir. DO NOT disconnect hoses.
   - ❑ Beetle: Fuel filter bracket.
   - ❑ Timing belt upper cover **1**.
4. Lupo/Polo/Golf/Bora: Detach fuel pipes from cylinder head cover.
   WARNING: Fuel may be under high pressure and very hot.
5. Support engine.
6. Remove:
   - ❑ RH engine mounting bolts **2**, **3** & **4**.
   - ❑ RH engine mounting.
   - ❑ RH engine mounting bracket **5**.
   NOTE: Golf/Bora/Beetle shown.
7. Lower engine slightly.
8. Remove:
   - ❑ Timing belt centre cover **6**.
   - ❑ Crankshaft pulley bolts **7**.
   - ❑ Crankshaft pulley **8**.
   - ❑ Timing belt lower cover **9**.
9. Turn crankshaft clockwise to TDC on No.1 cylinder.
10. Ensure timing mark aligned with notch on camshaft sprocket hub:
    - ❑ ANY/AMF engines: 3Z **10**.
    - ❑ All other engines: 4Z **11**.
    NOTE: Notch located behind camshaft sprocket teeth.
11. Lock crankshaft sprocket **12**. Use tool No.T10050.
12. Ensure timing marks aligned **13**.
13. Lock camshaft **14**. Use tool No.3359.
14. Fully insert Allen key into tensioner pulley **15**.
15. Turn tensioner pulley slowly anti-clockwise until locking tool can be inserted **16**. Tool No.T10008.

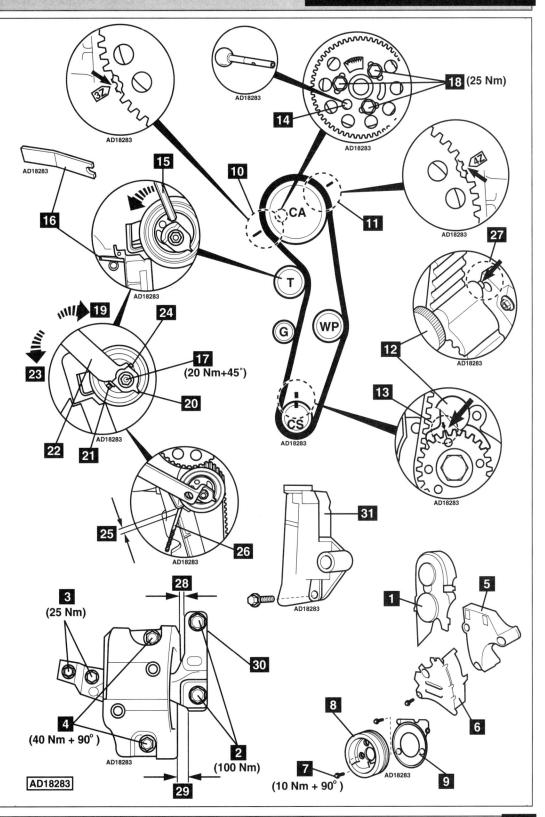

AD18283

3 (25 Nm)

18 (25 Nm)

17 (20 Nm+45°)

4 (40 Nm + 90°)

2 (100 Nm)

7 (10 Nm + 90°)

16. Slacken tensioner nut **17**.

17. Remove:
- ❏ Automatic tensioner unit **31**.
- ❏ Timing belt.

Installation

1. Ensure camshaft locked with tool **14**.

2. Ensure crankshaft sprocket locking tool located correctly **12**.

3. Ensure timing marks aligned **13**.

4. Ensure automatic tensioner unit locked with tool **16**. Tool No.T10008.

5. Slacken camshaft sprocket bolts **18**.

6. Turn camshaft sprocket fully clockwise in slotted holes. Tighten bolts finger tight **18**.

7. Turn tensioner pulley slowly clockwise **19** until lug **20** just reaches stop **21**. Use tool No.3387 **22**.

8. Fit timing belt in following order:
- ❏ Camshaft sprocket.
- ❏ Tensioner pulley.
- ❏ Crankshaft sprocket.
- ❏ Water pump sprocket.

9. Install:
- ❏ Automatic tensioner unit **31**.

NOTE: Ensure belt is taut between sprockets on non-tensioned side.

10. Turn tensioner pulley slowly anti-clockwise **23** (lug **20** moves towards stop **24**). Use tool No.3387 **22**.

11. Remove locking tool without force **16**.

12. Turn tensioner pulley slowly clockwise **19** (lug **20** moves towards stop **21**) until dimension **25** as specified:
- ❏ ANY/AYZ engines: 7±1 mm. Use drill bit **26**.
- ❏ All other engines: 4±1 mm. Use drill bit **26**.

NOTE: Engine must be COLD.

13. Tighten tensioner nut **17**.
Tightening torque: 20 Nm + 45°.

14. Tighten camshaft sprocket bolts **18**.
Tightening torque: 25 Nm.

15. Remove:
- ❏ Camshaft locking tool **14**.
- ❏ Crankshaft sprocket locking tool **12**.
- ❏ Drill bit **26**.

16. Turn crankshaft slowly two turns clockwise to TDC on No.1 cylinder.

17. Ensure dimension **25** as specified:
- ❏ ANY/AYZ engines: 7±1 mm. Use drill bit **26**.
- ❏ All other engines: 4±1 mm. Use drill bit **26**.

18. If not: Slacken tensioner nut **17**. Turn tensioner pulley until dimension as specified **25**. Tighten tensioner nut **17**. Tightening torque: 20 Nm + 45°.

19. Lock crankshaft sprocket **12**. Use tool No.T10050.

20. Ensure timing marks aligned **13**.

21. Ensure camshaft locking tool can be inserted easily **14**. Tool No.3359.

22. If not:
- ❏ Remove lug of crankshaft sprocket locking tool from hole in oil seal housing.
- ❏ Turn crankshaft until camshaft locking tool can be inserted **14**. Tool No.3359.
- ❏ Slacken camshaft sprocket bolts **18**.
- ❏ Turn crankshaft anti-clockwise until lug of locking tool just passes hole in oil seal housing **27**.

- ❏ Turn crankshaft clockwise until lug and hole aligned.
- ❏ Lock crankshaft sprocket **12**. Use tool No.T10050.
- ❏ Tighten camshaft sprocket bolts **18**. Tightening torque: 25 Nm.
- ❏ Remove locking tools **12** & **14**.
- ❏ Turn crankshaft slowly two turns clockwise to TDC on No.1 cylinder.
- ❏ Ensure locking tools can be fitted correctly **12** & **14**.

23. Remove:
- ❏ Camshaft locking tool **14**.
- ❏ Crankshaft sprocket locking tool **12**.
- ❏ Drill bit **26**.

24. Install components in reverse order of removal.

25. Tighten crankshaft pulley bolts **7**.
Tightening torque: 10 Nm + 90°.

26. Tighten bolts securing engine mounting bracket to engine **5**:
- ❏ Lupo/Polo/Golf/Bora/Beetle: 45 Nm.
- ❏ Sharan: M8 bolts – 30 Nm.
M10 bolts – 45 Nm.

27. Lupo/Polo – AMF: Tighten RH engine mounting:
- ❏ Bolts securing engine mounting to body – 20 Nm + 90°. Use new bolts.
- ❏ Bolts securing engine mounting to engine bracket – 40 Nm + 90°. Use new bolts.

28. Lupo – ANY/AYZ: Tighten RH engine mounting:
- ❏ Bolts securing engine mounting to body – 20 Nm + 45°. Use new bolts.
- ❏ Bolts securing intermediate bracket to engine bracket – 40 Nm + 90°. Use new bolts.
- ❏ Bolt securing intermediate bracket to engine mounting – 40 Nm + 90°. Use new bolt.

29. Golf/Bora/Beetle: Fit and align RH engine mounting:
- ❏ Engine mounting clearance: 14 mm. **28**.
- ❏ Engine mounting clearance: 10 mm minimum **29**.

30. Golf/Bora/Beetle: Tighten RH engine mounting:
- ❏ Long bolts securing engine mounting to body **4**. Tightening torque: 40 Nm + 90°. Use new bolts.
- ❏ Short bolts securing engine mounting to body **3**. Tightening torque: 25 Nm.
- ❏ Bolts securing engine mounting to engine bracket **2**. Tightening torque: 100 Nm.

*NOTE: Ensure bolts **2** aligned with edge of mounting **30**.*

31. Sharan: Tighten RH engine mounting:
- ❏ Bolts – 60 Nm. Lubricate bolts.
- ❏ Nuts – 55 Nm. Lubricate nuts.

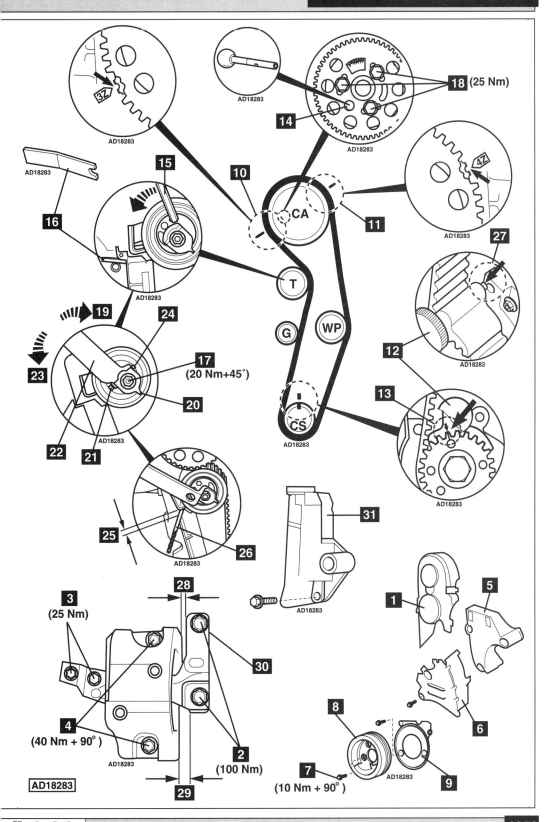

Model:	Lupo 1,7 SDI • Polo 1,7 SDI • Polo 1,9 SDI
Year:	1995-03
Engine Code:	AGD, AHG, AKU, ASX

Replacement Interval Guide

Volkswagen recommend:

All models
→2000MY: Check condition and width every 10,000 miles (replacement width – 21 mm), replace if necessary.

Lupo
→2000MY: Replacement every 40,000 miles.
2001MY→: Replacement every 60,000 miles.

Polo – AHG/AKU engines
→2000MY: Replacement every 40,000 miles.
2001MY→: Replacement every 60,000 miles.

Polo – AGD/ASX engines
Replacement every 60,000 miles.
The previous use and service history of the vehicle must always be taken into account.
Refer to Timing Belt Replacement Intervals at the front of this manual.

Check For Engine Damage

CAUTION: This engine has been identified as an INTERFERENCE engine in which the possibility of valve-to-piston damage in the event of a timing belt failure is MOST LIKELY to occur.
A compression check of all cylinders should be performed before removing the cylinder head.

Repair Times – hrs

Remove & install:
Lupo	2,50
Polo	2,20

Special Tools

- Injection pump sprocket locking pin – No.2064.
- Camshaft setting bar – No.2065A.
- Two-pin wrench – No.3387.

Special Precautions

- Disconnect battery earth lead.
- DO NOT turn crankshaft or camshaft when timing belt removed.
- Remove glow plugs to ease turning engine.
- Turn engine in normal direction of rotation (unless otherwise stated).
- DO NOT turn engine via camshaft or other sprockets.
- Observe all tightening torques.
- Check diesel injection pump timing after belt replacement.

Removal

1. Support engine.
2. Remove:
   - ❑ Insulation shield.
   - ❑ Air filter.
   - ❑ Auxiliary drive belt.
   - ❑ Lower engine steady bar.
   - ❑ Timing belt upper cover **1**.
   - ❑ Cylinder head cover.
   - ❑ Engine mounting.
   - ❑ Bracket **2**.

3. Turn crankshaft to TDC on No.1 cylinder. Ensure flywheel timing marks aligned **3**.
4. Fit setting bar No.2065A to rear of camshaft **4**. Centralise camshaft using feeler gauges.
5. Lock injection pump sprocket **5**. Use tool No.2064.
6. Slacken tensioner nut **6**. Turn tensioner anti-clockwise away from belt. Lightly tighten nut.
7. Lower engine slightly.
8. Remove:
   - ❑ Water pump pulley.
   - ❑ Crankshaft pulley bolts **7**.
   - ❑ Crankshaft pulley **8**.
   - ❑ Auxiliary drive belt tensioner.
   - ❑ Timing belt lower cover **9**.
   - ❑ Guide pulley **10**.
   - ❑ Timing belt.
   NOTE: Mark direction of rotation on belt with chalk if belt is to be reused.

Installation

1. Ensure flywheel timing marks aligned **3**.
2. Ensure camshaft setting bar fitted correctly **4**.
3. Ensure locking pin located correctly in injection pump sprocket **5**.
4. Slacken camshaft sprocket bolt 1/2 turn **11**.
5. Loosen sprocket from taper using a drift through hole in timing belt rear cover. Ensure sprocket can turn on taper.
6. Fit timing belt in anti-clockwise direction, starting at crankshaft sprocket. Ensure belt is taut between sprockets.
7. Remove locking pin from injection pump sprocket **5**.
8. Fit guide pulley **10**. Tighten bolt to 25 Nm.
9. Slacken tensioner nut **6**.
10. Turn automatic tensioner pulley clockwise until notch and raised mark on tensioner aligned **13**. Use tool No.3387 **12**.
11. Tighten tensioner nut to 20 Nm **6**.
12. Ensure flywheel timing marks aligned **3**.
13. Tighten camshaft sprocket bolt to 45 Nm **11**.
14. Remove camshaft setting bar **4**.
15. Turn crankshaft two turns clockwise until timing marks aligned **3**.
16. Ensure camshaft setting bar can be fitted **4**.
17. Ensure locking pin can be inserted in injection pump sprocket **5**.
18. Apply firm thumb pressure to belt. Tensioner marks should move out of alignment **13**.
19. Release thumb pressure from belt. Tensioner marks should realign **13**. If not: Repeat tensioning procedure.
20. Install:
    - ❑ Timing belt lower cover **9**.
    - ❑ Crankshaft pulley **8**.
21. Tighten crankshaft pulley bolts **7**:
    - ❑ (A) Except Lupo – 25 Nm.
    - ❑ (B) Lupo – 10 Nm + 90°.
22. Install components in reverse order of removal.

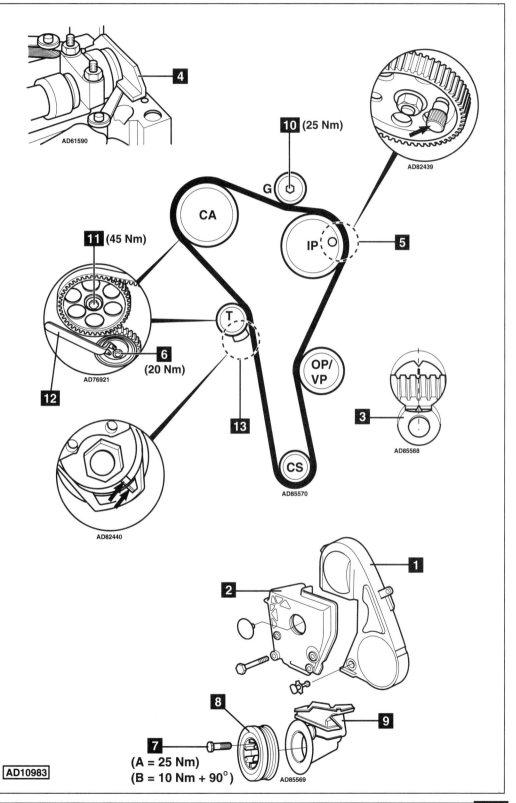

4 AD61590

10 (25 Nm)

AD82439

G

CA

11 (45 Nm)

IP **5**

AD76921

6 (20 Nm)

12

T

13

OP/ VP

3

AD85568

AD82440

CS

AD85570

1

2

8

9

7 (A = 25 Nm) (B = 10 Nm + 90°)

AD85569

AD10983

VOLKSWAGEN

Model:	**Polo 1,3/1,4 D**
Year:	**1986-94**
Engine Code:	**MN, 1W**

Replacement Interval Guide

The vehicle manufacturer has not recommended a timing belt replacement interval for this engine. *The previous use and service history of the vehicle must always be taken into account.*
Refer to Timing Belt Replacement Intervals at the front of this manual.

Check For Engine Damage

CAUTION: This engine has been identified as an INTERFERENCE engine in which the possibility of valve-to-piston damage in the event of a timing belt failure is MOST LIKELY to occur.
A compression check of all cylinders should be performed before removing the cylinder head.

Repair Times – hrs

Check & adjust	0,70
Remove & install	1,90

Special Tools

- Injection pump sprocket locking pin – VAG No.2064.
- Tension gauge – VAG No.210.
- Camshaft setting bar – VAG No.2065A.
- Sprocket holding tool – VAG No.3036.

Special Precautions

- Disconnect battery earth lead.
- DO NOT turn crankshaft or camshaft when timing belt removed.
- Remove glow plugs to ease turning engine.
- Turn engine in normal direction of rotation (unless otherwise stated).
- DO NOT turn engine via camshaft or other sprockets.
- Observe all tightening torques.
- Check diesel injection pump timing after belt replacement.

Removal

1. Turn crankshaft to TDC on No.1 cylinder. Ensure flywheel timing marks aligned **1**.
2. Remove:
   - ❏ Air filter.
   - ❏ Auxiliary drive belt.
   - ❏ Crankshaft damper bolts (4 bolts) **2**.
   - ❏ Crankshaft damper **3**.
   - ❏ Timing belt covers **4**.
   - ❏ Cylinder head cover.
   - ❏ Injection pump belt cover **5**.
3. Ensure injection pump sprocket timing mark aligned **6**.
4. Slacken injection pump belt by releasing thermostat housing/vacuum pump bolts **7**.

5. Lock injection pump sprocket **8**. Use tool No.2064.
6. Hold camshaft rear sprocket. Use tool No.3036. Remove bolt **12**.
7. Remove camshaft rear sprocket and injection pump belt.
8. Fit setting bar No.2065A to rear of camshaft **9**. Centralise camshaft using feeler gauges.
9. Slacken water pump bolts. Turn water pump to release tension on belt **10**.
10. Remove timing belt.
   NOTE: Mark direction of rotation on belt with chalk if belt is to be reused.

Installation

1. Hold camshaft sprocket. Use tool No.3036. Slacken camshaft sprocket bolt one turn **11**.
2. Loosen sprocket from taper using a drift through hole in timing belt rear cover.
   NOTE: If there is no hole in timing belt rear cover, drill 6 mm hole through cover.
3. Fit timing belt in anti-clockwise direction, starting at crankshaft sprocket.
4. Attach tension gauge to belt at ▽. Tool No.210.
5. Turn water pump until tension gauge indicates 14-15 units **10**.
6. Ensure flywheel timing marks aligned **1**.
7. Tighten water pump bolts to 20 Nm.
8. Hold camshaft sprocket. Use tool No.3036. Tighten bolt to 45 Nm **11**.
9. Remove camshaft setting bar **9**.
10. Fit camshaft rear sprocket and injection pump belt.
11. Tighten bolt **12** until sprocket can just be turned by hand.
12. Attach tension gauge to belt at ▽. Tool No.210.
13. Turn thermostat housing/vacuum pump until tension gauge indicates 17-18 units. Tighten bolts.
14. Ensure flywheel timing marks aligned **1**.
15. Hold camshaft rear sprocket. Use tool No.3036. Tighten bolt to 100 Nm **12**.
16. Remove locking pin from injection pump sprocket **8**.
17. Install components in reverse order of removal.
18. Tighten crankshaft damper bolts to 20 Nm **2**.
19. Check injection pump timing.

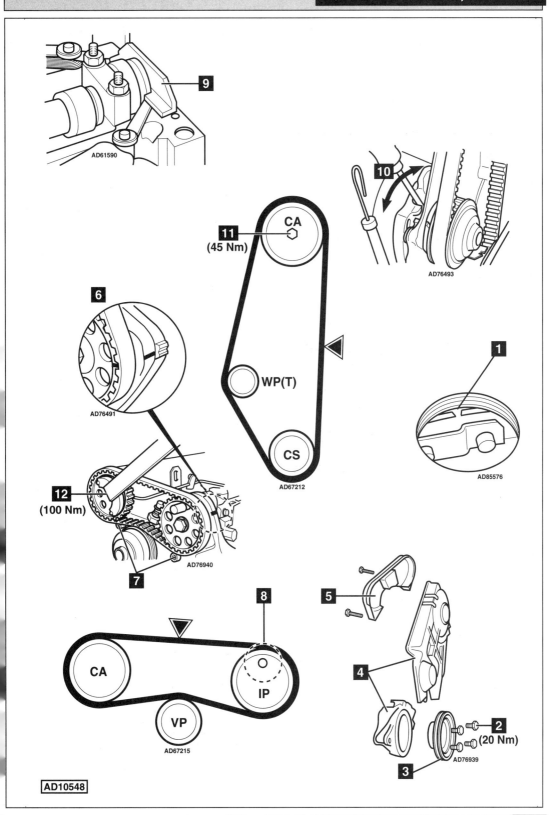

AD61590

9

10

AD76493

11
(45 Nm)

CA

WP(T)

CS

AD67212

6

AD76491

1

AD85576

12
(100 Nm)

AD76940

7

8

5

4

CA

IP

VP

AD67215

2
(20 Nm)

3

AD76939

AD10548

VOLKSWAGEN

Model:	Polo Classic 1,7/1,9 SDI • Polo Classic 1,9 TDI • Golf/Vento 1,9 SDI Caddy 1,7/1,9 SDI
Year:	1995-03
Engine Code:	AEY, AFN, AHB, AHU, AKW, ALE

Replacement Interval Guide

Volkswagen recommend:

All models:
→2000MY: Check condition and width every 10,000 miles or 12 months, whichever occurs first (replacement width 22 mm).

Polo Classic/Caddy:
Engine code AKW: Replacement every 40,000 miles.
Except engine code AKW: Replacement every 55,000 miles.

Golf/Vento:
Replacement every 60,000 miles (tensioner pulley must also be replaced).
The previous use and service history of the vehicle must always be taken into account.
Refer to Timing Belt Replacement Intervals at the front of this manual.

Check For Engine Damage

CAUTION: This engine has been identified as an INTERFERENCE engine in which the possibility of valve-to-piston damage in the event of a timing belt failure is MOST LIKELY to occur.
A compression check of all cylinders should be performed before removing the cylinder head.

Repair Times – hrs

Remove & install	1,90

Special Tools

- Injection pump sprocket locking pin – No.2064.
- Camshaft setting bar – No.2065A.
- Two-pin wrench – Matra V.159.

Special Precautions

- Disconnect battery earth lead.
- DO NOT turn crankshaft or camshaft when timing belt removed.
- Remove glow plugs to ease turning engine.
- Turn engine in normal direction of rotation (unless otherwise stated).
- DO NOT turn engine via camshaft or other sprockets.
- Observe all tightening torques.
- Check diesel injection pump timing after belt replacement.

Removal

1. Remove:
   - ❏ Air filter.
   - ❏ Auxiliary drive belt(s).
   - ❏ Auxiliary drive belt tensioner.
   - ❏ Timing belt upper cover **1**.
   - ❏ Cylinder head cover.
2. Turn crankshaft to TDC on No.1 cylinder. Ensure flywheel timing marks aligned **2**.
3. Fit setting bar No.2065A to rear of camshaft **3**. Centralise camshaft using feeler gauges.

4. Lock injection pump sprocket. Use tool No.2064 **4**.
5. Remove:
   - ❏ Water pump pulley.
   - ❏ Crankshaft pulley bolts **5**.
   - ❏ Crankshaft pulley **6**.
   - ❏ Timing belt lower cover **7**.
6. Slacken tensioner nut **8**. Turn tensioner anti-clockwise away from belt. Lightly tighten nut.
7. Remove:
   - ❏ Guide pulley **9**.
   - ❏ Timing belt.
   NOTE: Mark direction of rotation on belt with chalk if belt is to be reused.

Installation

1. Ensure flywheel timing marks aligned **2**.
2. Ensure camshaft setting bar fitted correctly **3**.
3. Ensure locking pin located correctly in injection pump sprocket **4**.
4. Slacken camshaft sprocket bolt 1/2 turn **10**.
5. Loosen sprocket from taper using a drift through hole in timing belt rear cover. Ensure sprocket can turn on taper.
6. Fit timing belt in anti-clockwise direction, starting at crankshaft sprocket. Ensure belt is taut between sprockets.
7. Fit guide pulley **9**. Tighten bolt to 25 Nm.
8. Remove locking pin from injection pump sprocket.
9. Slacken tensioner nut **8**.
10. Turn automatic tensioner pulley clockwise until notch and raised mark on tensioner aligned **11**. Use wrench Matra V.159 **12**.
11. Tighten tensioner nut to 20 Nm **8**.
12. Ensure flywheel timing marks aligned **2**. Tighten camshaft sprocket bolt to 45 Nm **10**.
13. Remove camshaft setting bar **3**.
14. Turn crankshaft two turns clockwise until timing marks aligned **2**.
15. Ensure camshaft setting bar can be fitted **3**.
16. Ensure locking pin can be inserted in injection pump sprocket **4**.
17. Ensure notch and raised mark on tensioner aligned **11**.
18. If not: Repeat tensioning procedure.
19. Apply firm thumb pressure to belt. Tensioner marks should move out of alignment **11**.
20. Release thumb pressure from belt. Tensioner marks should realign **11**.
21. Install:
    - ❏ Timing belt lower cover **7**.
    - ❏ Crankshaft pulley **6**.
    - ❏ Crankshaft pulley bolts **5**.
22. Tighten crankshaft pulley bolts **5**. Tightening torque: 25 Nm.
23. Install components in reverse order of removal.

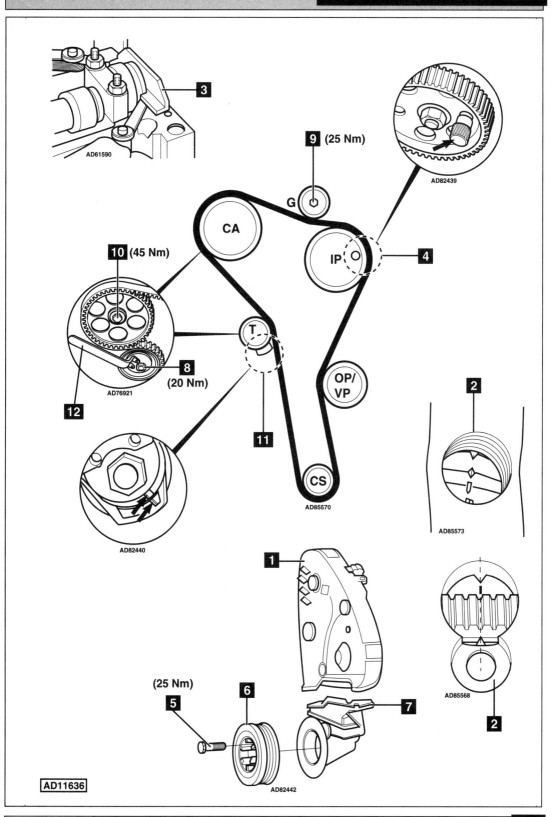

VOLKSWAGEN

Model:	Polo 1,9D • Caddy Pick-up 1,9D
Year:	1995-03
Engine Code:	AEF

Replacement Interval Guide

Volkswagen recommend:

Polo
→1996MY: Check and adjust every 40,000 miles.
1997MY→: Check and adjust every 60,000 miles.
The vehicle manufacturer has not recommended a timing belt replacement interval for this engine.

Caddy Pick-up
Check and adjust every 60,000 miles.
The vehicle manufacturer has not recommended a timing belt replacement interval for this engine.
The previous use and service history of the vehicle must always be taken into account.
Refer to Timing Belt Replacement Intervals at the front of this manual.

Check For Engine Damage

CAUTION: This engine has been identified as an INTERFERENCE engine in which the possibility of valve-to-piston damage in the event of a timing belt failure is MOST LIKELY to occur.
A compression check of all cylinders should be performed before removing the cylinder head.

Repair Times – hrs

Check & adjust	0,90
Remove & install:	
Polo	2,20
Caddy Pick-up	1,90

Special Tools

- Injection pump locking pin – No.3359.
- Camshaft setting bar – No.2065A.
- Two-pin wrench – No.3387.

Special Precautions

- Disconnect battery earth lead.
- DO NOT turn crankshaft or camshaft when timing belt removed.
- Remove glow plugs to ease turning engine.
- Turn engine in normal direction of rotation (unless otherwise stated).
- DO NOT turn engine via camshaft or other sprockets.
- Observe all tightening torques.
- Check diesel injection pump timing after belt replacement.

Removal

1. Support engine.
2. Remove:
   ❑ Insulation shield.
   ❑ Air filter.
   ❑ Auxiliary drive belt.
   ❑ Lower engine steady bar.
   ❑ Timing belt upper cover **1**.
   ❑ Cylinder head cover.
   ❑ Engine mounting.
   ❑ Bracket **2**.

3. Turn crankshaft to TDC on No.1 cylinder. Ensure flywheel timing marks aligned **3**.
4. Fit setting bar No.2065A to rear of camshaft **4**. Centralise camshaft using feeler gauges.
5. Insert locking pin in injection pump **5**. Tool No.3359.
6. Slacken injection pump sprocket bolts **6**.
   NOTE: DO NOT slacken injection pump hub nut.
7. Slacken tensioner nut **7**. Turn tensioner anti-clockwise away from belt. Lightly tighten nut.
8. Lower engine slightly.
9. Remove:
   ❑ Water pump pulley.
   ❑ Crankshaft pulley bolts **8**.
   ❑ Crankshaft pulley **9**.
   ❑ Auxiliary drive belt tensioner.
   ❑ Timing belt lower cover **10**.
   ❑ Timing belt.
   NOTE: Mark direction of rotation on belt with chalk if belt is to be reused.

Installation

1. Ensure flywheel timing marks aligned **3**.
2. Ensure camshaft setting bar fitted correctly **4**.
3. Ensure locking pin located correctly in injection pump **5**.
4. Slacken camshaft sprocket bolt 1/2 turn **11**.
5. Loosen sprocket from taper using a drift through hole in timing belt rear cover. Ensure sprocket can turn on taper.
6. Fit timing belt in anti-clockwise direction, starting at crankshaft sprocket. Ensure belt is taut between sprockets.
7. Turn automatic tensioner pulley clockwise until notch and raised mark on tensioner aligned **12**. Use tool No.3387.
8. Tighten tensioner nut to 20 Nm **7**.
9. Tighten camshaft sprocket bolt to 45 Nm **11**.
10. Tighten injection pump sprocket bolts to 25 Nm **6**.
11. Remove:
    ❑ Injection pump locking pin **5**.
    ❑ Camshaft setting bar **4**.
12. Turn crankshaft two turns clockwise to TDC on No.1 cylinder.
13. Ensure flywheel timing marks aligned **3**. Ensure camshaft setting bar can be fitted **4**.
14. Ensure locking pin can be inserted in injection pump **5**.
15. If not: Slacken injection pump sprocket bolts **6**. Turn sprocket hub until locking pin can be inserted. Tighten bolts to 25 Nm.
16. Install components in reverse order of removal.
17. Tighten crankshaft pulley bolts to 25 Nm **8**.

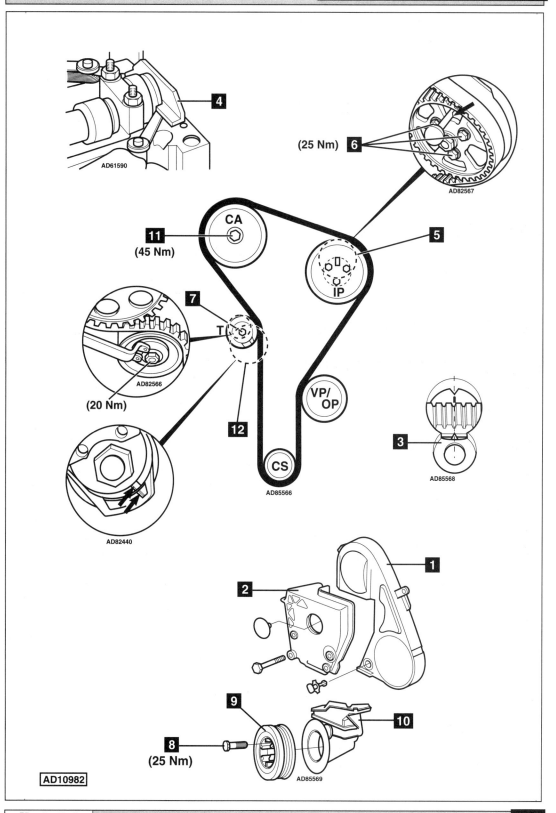

Model:	**Polo Classic 1,9 SD • Caddy 1,9 SD**
Year:	**1996-03**
Engine Code:	**1Y**

Replacement Interval Guide

Volkswagen recommend check every
55,000 miles (replace if necessary).
The vehicle manufacturer has not recommended
a timing belt replacement interval for this engine.
*The previous use and service history of the vehicle
must always be taken into account.
Refer to Timing Belt Replacement Intervals at the front
of this manual.*

Check For Engine Damage

*CAUTION: This engine has been identified as an
INTERFERENCE engine in which the possibility of
valve-to-piston damage in the event of a timing belt
failure is MOST LIKELY to occur.
A compression check of all cylinders should be
performed before removing the cylinder head.*

Repair Times – hrs

Remove & install	1,90

Special Tools

- Injection pump locking pin – No.3359.
- Camshaft setting bar – No.2065A.
- Two-pin wrench – No.3387.

Special Precautions

- Disconnect battery earth lead.
- DO NOT turn crankshaft or camshaft when timing belt removed.
- Remove glow plugs to ease turning engine.
- Turn engine in normal direction of rotation (unless otherwise stated).
- DO NOT turn engine via camshaft or other sprockets.
- Observe all tightening torques.
- Check diesel injection pump timing after belt replacement.

Removal

1. Remove:
   - ❏ Auxiliary drive belt.
   - ❏ Timing belt upper cover **1**.
   - ❏ Cylinder head cover.
2. Turn crankshaft to TDC on No.1 cylinder. Ensure flywheel timing marks aligned **2**.
3. Fit setting bar No.2065A to rear of camshaft **3**. Centralise camshaft using feeler gauges.
4. Insert locking pin in injection pump **4**. Tool No.3359.
5. Slacken injection pump sprocket bolts **5**.
   NOTE: DO NOT slacken injection pump hub nut.

6. Slacken tensioner nut **6**. Turn tensioner anti-clockwise away from belt. Lightly tighten nut.
7. Remove:
   - ❏ Crankshaft pulley bolts **7**.
   - ❏ Crankshaft pulley **8**.
   - ❏ Timing belt lower cover **9**.
   - ❏ Timing belt.
   NOTE: Mark direction of rotation on belt with chalk if belt is to be reused.

Installation

1. Ensure flywheel timing marks aligned **2**.
2. Ensure camshaft setting bar fitted correctly **3**.
3. Ensure locking pin located correctly in injection pump **4**.
4. Slacken camshaft sprocket bolt 1/2 turn **10**.
5. Loosen sprocket from taper using a drift through hole in timing belt rear cover. Ensure sprocket can turn on taper.
6. Fit timing belt in anti-clockwise direction, starting at crankshaft sprocket. Ensure belt is taut between sprockets.
7. Turn automatic tensioner pulley clockwise until notch and raised mark on tensioner aligned **11**. Use tool No.3387.
8. Tighten tensioner nut to 20 Nm **6**.
9. Tighten camshaft sprocket bolt to 45 Nm **10**.
10. Tighten injection pump sprocket bolts to 25 Nm **5**.
11. Remove locking pin from injection pump **4**.
12. Remove camshaft setting bar **3**.
13. Turn crankshaft two turns clockwise to TDC on No.1 cylinder.
14. Ensure flywheel timing marks aligned **2**.
15. Ensure camshaft setting bar can be fitted **3**.
16. Ensure locking pin can be inserted in injection pump **4**.
17. If not: Slacken injection pump sprocket bolts **5**. Turn sprocket hub until locking pin can be inserted. Tighten bolts to 25 Nm.
18. Install components in reverse order of removal.
19. Tighten crankshaft pulley bolts to 25 Nm **7**.

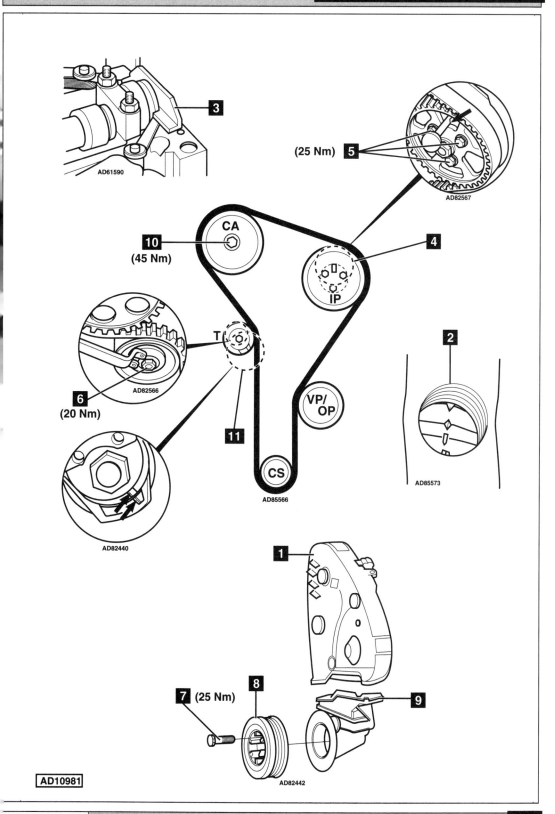

AD61590

3

(25 Nm) **5**

AD82567

10 (45 Nm)

CA

4

IP

AD82566

6 (20 Nm)

T

2

VP/OP

11

CS

AD85566

AD85573

AD82440

1

7 (25 Nm)

8

9

AD82442

AD10981

VOLKSWAGEN

Model:	Polo 1,9 TDI • Polo Classic 1,9 SDI/TDI • Golf/Bora 1,9 SDI/TDI Beetle 1,9 TDI • Caddy 1,9 TDI
Year:	1997-03
Engine Code:	AGP, AGR, AHF, ALH, AQM, ASV

Replacement Interval Guide

Volkswagen recommend:

All models
→2000MY: Check condition and width every 10,000 miles or 12 months, whichever occurs first (replacement width 22 mm).

Polo/Classic/Caddy
ALH, AT: Check timing belt and tensioner pulley every 20,000 miles.
Replacement every 60,000 miles.

Golf/Bora
ALH →1999MY: Check every 20,000 miles.
→2001MY: Replacement every 60,000 miles.
2002MY→: Replacement every 80,000 miles (guide pulley must also be replaced –
part No.038109244H).

Beetle →1999MY
Replacement at the first 40,000 miles using:
Part No.038 109 119 – replacement then every 40,000 miles.
Part No.038 109 119D – replacement then every 60,000 miles.
ALH, AT: Check every 20,000 miles and replacement every 40,000 miles.

Beetle 2000MY→
Part No.038 109 119D – replacement every 60,000 miles.
ALH, AT: Replacement every 40,000 miles.
The previous use and service history of the vehicle must always be taken into account.
Refer to Timing Belt Replacement Intervals at the front of this manual.

Check For Engine Damage

CAUTION: This engine has been identified as an INTERFERENCE engine in which the possibility of valve-to-piston damage in the event of a timing belt failure is MOST LIKELY to occur.
A compression check of all cylinders should be performed before removing the cylinder head(s).

Repair Times – hrs

Remove & install:
Golf/Bora/Beetle	3,30
Polo/Classic/Caddy	2,50

Special Tools

- Camshaft setting bar – No.3418/T10098.
- Injection pump locking pin – No.3359.
- Sprocket holding tool – No.3036.
- Two-pin wrench – Matra V.159 – TDI.
- Puller – No.T40001.

Special Precautions

- Disconnect battery earth lead.
- DO NOT turn crankshaft or camshaft when timing belt removed.
- Remove glow plugs to ease turning engine.
- Turn engine in normal direction of rotation (unless otherwise stated).
- DO NOT turn engine via camshaft or other sprockets.
- Observe all tightening torques.
- Check diesel injection pump timing after belt replacement.

Removal

1. Raise and support front of vehicle.

Golf/Bora:
2. Remove:
   - RH headlamp.
   - Coolant expansion tank. DO NOT disconnect hoses.
   - PAS reservoir. DO NOT disconnect hoses.
   - RH engine undershield.

Beetle:
3. Remove:
   - Fuel filter and bracket.
   - Engine undershield.

All models:
4. Remove:
   - Intercooler to intake manifold hose.
   - Timing belt upper cover **1**.
   - Vacuum pump.
   - Polo/Classic: Engine undershield.
   - Auxiliary drive belt.
5. Remove – when using camshaft setting bar No.3418:
   - Polo Classic AGP/AQM: Inlet manifold upper chamber.
   - Cylinder head cover.
6. Turn crankshaft to TDC on No.1 cylinder.
7. Ensure timing marks aligned. MT: **2**. AT: **3**.
8. Fit camshaft setting bar to rear of camshaft:
   - Tool No.3418 **4**. Centralise camshaft using feeler gauges.
   - Or tool No.T10098 **20**.
9. Lock injection pump. Use tool No.3359 **5**.
10. Slacken injection pump sprocket bolts **6**.
    NOTE: DO NOT slacken injection pump hub nut.

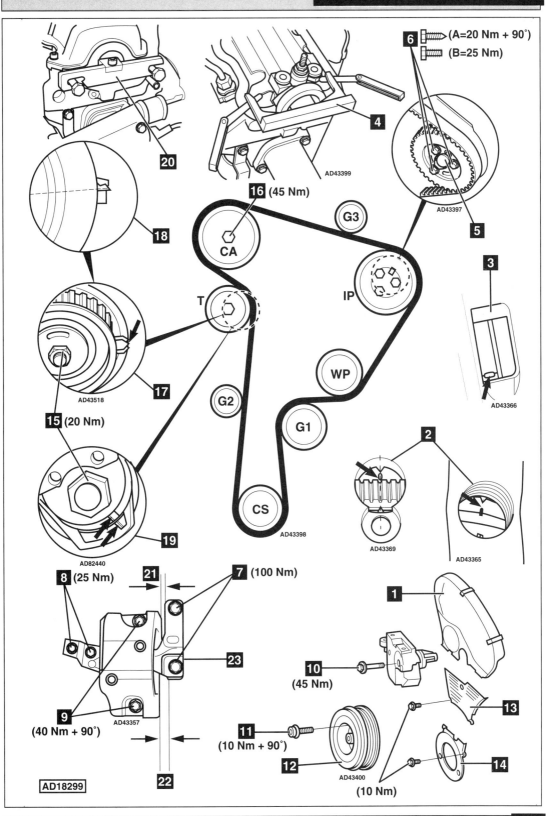

6 ⟩ (A=20 Nm + 90°)
⟩ (B=25 Nm)

4

AD43399

20

16 (45 Nm)

G3

AD43397

5

18

CA

IP

3

T

AD43518

17

WP

AD43366

15 (20 Nm)

G2

2

G1

AD43369

19

AD82440

CS

AD43398

AD43365

8 (25 Nm)

21

7 (100 Nm)

1

23

10
(45 Nm)

9

AD43357

13

(40 Nm + 90°)

11
(10 Nm + 90°)

12

AD43400

14

22

(10 Nm)

AD18299

Golf/Bora/Beetle:

11. Support engine.
12. Remove:
    - ❏ RH engine mounting bolts **7**, **8** & **9**.
    - ❏ RH engine mounting.
    - ❏ Engine mounting bracket bolts **10**.
    - ❏ Engine mounting bracket.

All models:

13. Remove:
    - ❏ Crankshaft pulley bolts **11**.
    - ❏ Crankshaft pulley **12**.
    - ❏ Timing belt centre cover **13**.
    - ❏ Timing belt lower cover **14**.
14. Slacken tensioner nut **15**.
15. SDI: Turn tensioner anti-clockwise away from belt. Use Allen key.
16. TDI: Turn tensioner anti-clockwise away from belt. Use wrench Matra V.159.
17. Lightly tighten tensioner nut **15**.
18. Remove timing belt.
    NOTE: Mark direction of rotation on belt with chalk if belt is to be reused.

Installation

1. Ensure timing marks aligned. MT: **2**. AT: **3**.
2. Ensure camshaft setting bar fitted correctly **4** or **20**.
3. Ensure locking pin located correctly in injection pump **5**.
4. Hold camshaft sprocket. Use tool No.3036.
5. Slacken camshaft sprocket bolt one turn **16**.
   NOTE: DO NOT use setting bar to hold camshaft when slackening sprocket bolt.
6. Loosen sprocket from taper.
   - ❏ Method 1: Use drift through hole in timing belt rear cover.
   - ❏ Method 2: Use puller. Tool No.T40001.
7. Remove:
   - ❏ Camshaft sprocket bolt **16**.
   - ❏ Camshaft sprocket.
8. Remove bolts from injection pump sprocket **6**. Proceed as follows:
   - ❏ Injection pump sprocket 038 130 111A: Bolts with pointed ends – type A. Fit new bolts.
   - ❏ Injection pump sprocket 038 130 111B: Bolts with flat ends – type B. Bolts can be re-used.
9. Align injection pump sprocket with bolts in centre of slotted holes.
10. Fit timing belt in following order:
    - ❏ Crankshaft sprocket.
    - ❏ Guide pulley G1.
    - ❏ Water pump sprocket.
    - ❏ Injection pump sprocket.
    - ❏ Guide pulley G2 – TDI.
    - ❏ Tensioner pulley.
    - ❏ Guide pulley G3.
    NOTE: Turn injection pump sprocket slightly to engage timing belt teeth.
11. Ensure belt is taut between sprockets.
12. Fit camshaft sprocket to belt, then install camshaft sprocket with belt onto end of camshaft.
13. Fit camshaft sprocket bolt **16**.
14. Lightly tighten camshaft sprocket bolt **16**. Sprocket should turn freely on taper but not tilt.
15. Slacken tensioner nut **15**.

16. SDI: Turn tensioner pulley clockwise until pointer and notch in backplate aligned **17**. Use Allen key.
17. TDI: Turn tensioner pulley clockwise as follows:
    - ❏ ALH, AT: Pointer and raised section of tensioner backplate aligned **18**. Use Allen key.
    - ❏ Except ALH, AT: Notch and raised mark on tensioner aligned **19**. Use wrench Matra V.159.
18. If tensioner pulley turned too far: Turn fully anti-clockwise and repeat tensioning procedure.
19. Tighten tensioner nut to 20 Nm **15**.
    NOTE: Ensure tensioner retaining lug is properly engaged.
20. Ensure timing marks aligned. MT: **2**. AT: **3**.
21. Hold camshaft sprocket. Use tool No.3036.
22. Tighten camshaft sprocket bolt to 45 Nm **16**.
23. Tighten injection pump sprocket bolts **6**:
    - ❏ Type A: Temporarily tighten to 20 Nm.
    NOTE: Tighten further 90° after checking pump timing using diagnostic equipment.
    - ❏ Type B: Tighten to 25 Nm.
24. Remove:
    - ❏ Camshaft setting bar **4** or **20**. Tool No.3418/T10098.
    - ❏ Injection pump locking pin **5**. Tool No.3359.
25. Turn crankshaft two turns clockwise until timing marks aligned. MT: **2**. AT: **3**.
26. Ensure camshaft setting bar can be fitted **4** or **20**.
27. Ensure locking pin can be inserted in injection pump **5**.
28. SDI: Ensure tensioner pointer aligned with notch in backplate **17**.
29. TDI: ALH, AT: Pointer and raised section of tensioner backplate aligned **18**.
    Except ALH, AT: Ensure notch and raised mark on tensioner aligned **19**.
30. If not: Repeat tensioning procedure.
31. Apply firm thumb pressure to belt. Tensioner marks should move out of alignment **17**, **18** or **19**.
32. Release thumb pressure from belt. Tensioner marks should follow movement of belt.
33. Install components in reverse order of removal.
34. Tighten crankshaft pulley bolts **11**.
    Tightening torque: 10 Nm + 90°.

Golf/Bora/Beetle:

35. Tighten RH engine mounting bracket bolts to 45 Nm **10**.
36. Fit and align RH engine mounting:
    - ❏ Engine mounting clearance: 14 mm **21**.
    - ❏ Engine mounting clearance: 10 mm **22**.
    - ❏ Ensure engine mounting bolts **7** aligned with edge of mounting **23**.
37. Tighten:
    - ❏ Engine mounting bolts **9**.
      Tightening torque: 40 Nm + 90°. Use new bolts.
    - ❏ Engine mounting bolts **8**.
      Tightening torque: 25 Nm.
    - ❏ Engine mounting bolts **7**.
      Tightening torque: 100 Nm.

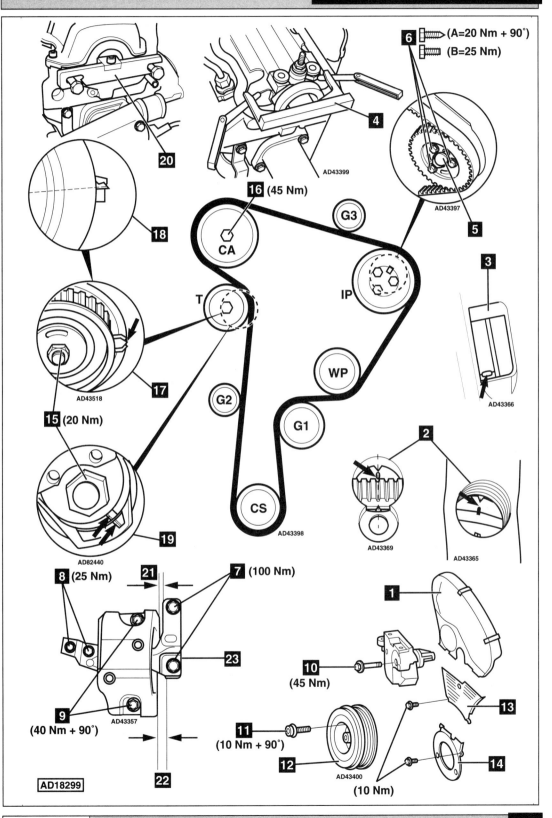

6 ▸ (A=20 Nm + 90°)
▸ (B=25 Nm)

4

AD43399

20

AD43397

16 (45 Nm)

G3

CA

5

3

AD43366

18

T

IP

AD43518

17

WP

15 (20 Nm)

G2

G1

2

CS

AD43398

AD43369

AD82440

19

AD43365

8 (25 Nm)

21

7 (100 Nm)

1

23

10
(45 Nm)

9

AD43357

13

(40 Nm + 90°)

11
(10 Nm + 90°)

22

12

AD43400

14

(10 Nm)

AD18299

Model:	Golf/Jetta/Caddy 1,5/1,6 D/Turbo • Golf/Vento 1,9 D/TD/TDI
	Passat/Santana D/Turbo • Transporter 1,6 D/1,7 D/1,9 D/Turbo
Year:	1976-95
Engine Code:	Manual tensioner
	CK, CR, CS, CY, JK, JP, JX, JR, KY
	RA, ME, SB, AAZ, ABL, 1V, 1X, 1Y, 1Z

Replacement Interval Guide

All except AAZ/ABL/1X/1Y/1Z engine:
The vehicle manufacturer has not recommended a timing belt replacement interval for these engines.
Volkswagen recommend:
AAZ/1Y engine:
Check and adjust or replacement if necessary every 20,000 miles.
ABL/1X engine:
Check and adjust or replacement if necessary every 20,000 miles.
Replacement every 80,000 miles.
1Z engine:
Check condition and width every 10,000 miles or 12 months, whichever occurs first (replacement width – 22 mm).
Check adjustment every 20,000 miles.
Replacement every 60,000 miles (tensioner pulley must also be replaced).
The previous use and service history of the vehicle must always be taken into account.
Refer to Timing Belt Replacement Intervals at the front of this manual.

Check For Engine Damage

CAUTION: This engine has been identified as an INTERFERENCE engine in which the possibility of valve-to-piston damage in the event of a timing belt failure is MOST LIKELY to occur.
A compression check of all cylinders should be performed before removing the cylinder head.

Repair Times – hrs

Check & adjust:	
Except Transporter	0,40
Transporter (→1989)	0,50
Transporter (1990→)	1,60
Remove & install:	
Golf/Jetta/Caddy	1,60
Golf/Vento (1992→)	1,90
Passat/Santana (→1987)	1,90
Passat (1988→)	1,60
Transporter (→1989)	1,90
Transporter (1990→)	2,50

Special Tools

- Tension gauge – VAG No.210.
- Injection pump sprocket locking pin – VAG No.2064.
- Camshaft setting bar – VAG No.2065A.
- Two-pin wrench – Matra V.159.

Removal

1. Hinge radiator grille forward (Transporter 1991→).
2. Remove:
   - ❏ Engine undershield (if fitted).
   - ❏ Auxiliary drive belts.
   - ❏ Water pump pulley.
   - ❏ Cylinder head cover.
   - ❏ →01/84: Timing belt cover **1**.
3. Remove (02/84→):
   - ❏ Timing belt upper cover **2**.
   - ❏ Crankshaft pulley bolts **3**.
   - ❏ Crankshaft pulley **4**.
   - ❏ Timing belt lower cover **5**.
4. Turn crankshaft to TDC on No.1 cylinder. Ensure flywheel timing marks aligned **6**.
5. Fit setting bar No.2065A to rear of camshaft **10**. Centralise camshaft using feeler gauges.
6. Lock injection pump sprocket **8**. Use tool No.2064.
7. Slacken tensioner pulley nut **9**. Turn tensioner pulley anti-clockwise away from belt. Lightly tighten nut.
8. Remove timing belt.

Installation

1. Ensure flywheel timing marks aligned **6**.
2. Ensure camshaft setting bar fitted correctly **10**.
3. Ensure locking pin located correctly in injection pump sprocket **8**.
4. Slacken camshaft sprocket bolt **7**.
5. Loosen sprocket from taper using a drift through hole in timing belt rear cover. Ensure sprocket can turn on taper.
6. Fit timing belt, starting at crankshaft sprocket. Ensure belt is taut between sprockets.
7. Attach tension gauge to belt at ▽. Tool No.210.
8. Slacken tensioner pulley nut **9**.
9. Turn tensioner pulley clockwise until tension gauge indicates 12-13 units. Use wrench Matra V.159.
10. Tighten tensioner pulley nut to 45 Nm **9**.
11. Tighten camshaft sprocket bolt to 45 Nm **7**.
12. Remove:
    - ❏ Injection pump sprocket locking pin **8**.
    - ❏ Camshaft setting bar **10**.
13. Turn crankshaft two turns clockwise. Strike belt once with a rubber faced mallet at ▽. Recheck belt tension.
14. 02/84→: Fit timing belt lower cover **5**. Fit crankshaft pulley **4**. Tighten crankshaft pulley bolts to 20-25 Nm **3**.
15. Install components in reverse order of removal.

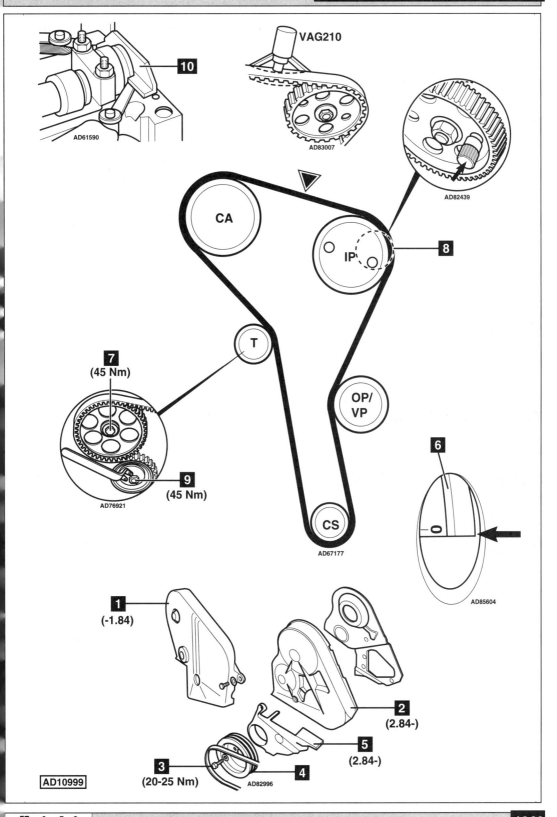

VAG210

AD61590

10

AD83007

AD82439

8

CA

IP

7
(45 Nm)

T

OP/
VP

6

9
(45 Nm)

AD76921

CS

AD67177

AD85604

1
(-1.84)

2
(2.84-)

5
(2.84-)

AD10999

3
(20-25 Nm)

4

AD82996

VOLKSWAGEN

Model:	Golf/Vento 1,9 TD/TDI • Golf Cabrio 1,9 TDI • Passat 1,9 TD/TDI (→1996)
Year:	1991-01
Engine Code:	Automatic tensioner AAZ, AFN, AHU, ALE, 1Z, AVG

Replacement Interval Guide

Volkswagen recommend:

Engine code AAZ:
Check & adjust or replacement if necessary every 20,000 miles.
No manufacturer's recommended replacement interval.

Engine code 1Z/AFN/AHU/ALE/AVG:
Check condition and width every 10,000 miles or 12 months, whichever occurs first (replacement width 22 mm).
Replacement every 60,000 miles (tensioner pulley must also be replaced).
The previous use and service history of the vehicle must always be taken into account.
Refer to Timing Belt Replacement Intervals at the front of this manual.

Check For Engine Damage

CAUTION: This engine has been identified as an INTERFERENCE engine in which the possibility of valve-to-piston damage in the event of a timing belt failure is MOST LIKELY to occur.
A compression check of all cylinders should be performed before removing the cylinder head.

Repair Times – hrs

Remove & install:
Golf/Vento	1,90
Passat	1,60

Special Tools

- Injection pump sprocket locking pin – VAG No.2064.
- Camshaft setting bar – VAG No.2065A.
- Two-pin wrench – Matra V.159.

Special Precautions

- Disconnect battery earth lead.
- DO NOT turn crankshaft or camshaft when timing belt removed.
- Remove glow plugs to ease turning engine.
- Turn engine in normal direction of rotation (unless otherwise stated).
- DO NOT turn engine via camshaft or other sprockets.
- Observe all tightening torques.
- Check diesel injection pump timing after belt replacement.

Removal

1. Remove:
   - Auxiliary drive belts.
   - Auxiliary drive belt tensioner (if fitted).
   - Timing belt upper cover **1**.
   - Cylinder head cover.
2. Turn crankshaft to TDC on No.1 cylinder. Ensure flywheel timing marks aligned **2**.
3. Fit setting bar No.2065A to rear of camshaft **3**. Centralise camshaft using feeler gauges.

4. Lock injection pump sprocket **4**. Use tool No.2064.
5. Remove:
   - Crankshaft pulley bolts **5**.
   - Crankshaft pulley **6**.
   - Timing belt lower cover **7**.
6. Slacken tensioner nut **8**. Turn tensioner anti-clockwise away from belt. Lightly tighten nut.
7. Except AAZ: Remove guide pulley **9**.
8. Remove timing belt.
   NOTE: Mark direction of rotation on belt with chalk if belt is to be reused.

Installation

1. Ensure flywheel timing marks aligned **2**.
2. Ensure camshaft setting bar fitted correctly **3**.
3. Ensure locking pin located correctly in injection pump sprocket **4**.
4. Slacken camshaft sprocket bolt 1/2 turn **10**.
5. Loosen sprocket from taper using a drift through hole in timing belt rear cover. Ensure sprocket can turn on taper.
6. Fit timing belt in anti-clockwise direction, starting at crankshaft sprocket. Ensure belt is taut between sprockets.
7. Except AAZ: Fit guide pulley. Tighten bolt to 25 Nm **9**.
8. Remove locking pin from injection pump sprocket **4**.
9. Slacken tensioner nut **8**.
10. Turn automatic tensioner pulley clockwise until notch and raised mark on tensioner aligned **11**. Use wrench Matra V.159 **12**.
11. Tighten tensioner nut to 20 Nm **8**.
12. Ensure flywheel timing marks aligned **2**. Tighten camshaft sprocket bolt to 45 Nm **10**.
13. Remove camshaft setting bar **3**.
14. Turn crankshaft two turns clockwise until timing marks aligned **2**.
15. Ensure camshaft setting bar can be fitted **3**.
16. Ensure locking pin can be inserted in injection pump sprocket **4**.
17. Apply firm thumb pressure to belt. Tensioner marks should move out of alignment **11**.
18. Release thumb pressure from belt. Tensioner marks should realign. If not: Repeat tensioning procedure.
19. Install:
   - Timing belt lower cover **7**.
   - Crankshaft pulley **6**.
20. Tighten crankshaft pulley bolts to 25 Nm **5**.
21. Install components in reverse order of removal.

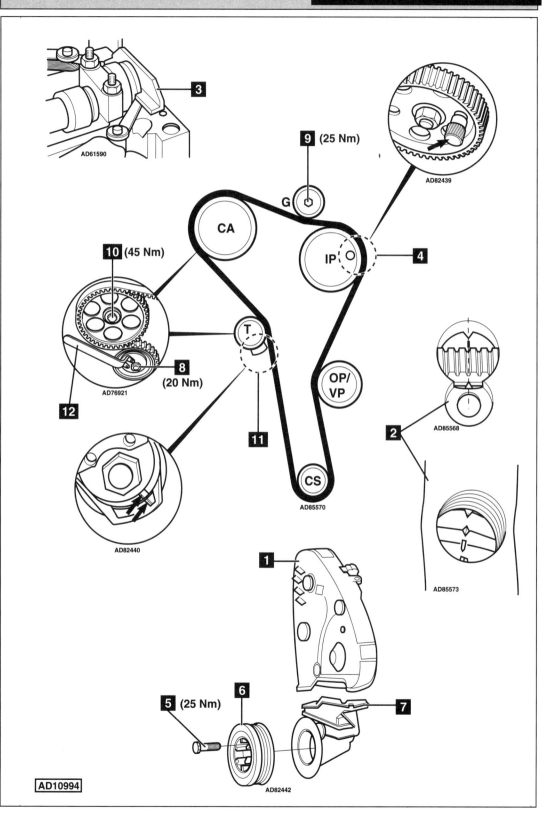

3 AD61590

9 (25 Nm)

AD82439

G

CA

10 (45 Nm)

IP

4

T

8 (20 Nm)

AD76921

12

OP/VP

AD85568

2

11

AD82440

CS

AD85570

AD85573

1

5 (25 Nm)

6

7

AD82442

AD10994

VOLKSWAGEN

Model:	Golf/Vento 1,9 D/TD • Passat 1,9 TD (-96) • Transporter 1,9TD
Year:	1994-03
Engine Code:	Two-piece injection pump sprocket 1Y, 1X, AAZ, ABL

Replacement Interval Guide

Volkswagen recommend replacement as follows:

1Y/AAZ:
Check & adjust every 20,000 miles (or replace if necessary).
No manufacturer's recommended replacement interval.

1X/ABL:
Check and adjust every 40,000 miles.
Replacement every 80,000 miles.
The previous use and service history of the vehicle must always be taken into account.
Refer to Timing Belt Replacement Intervals at the front of this manual.

Check For Engine Damage

CAUTION: This engine has been identified as an INTERFERENCE engine in which the possibility of valve-to-piston damage in the event of a timing belt failure is MOST LIKELY to occur.
A compression check of all cylinders should be performed before removing the cylinder head.

Repair Times – hrs

Check and adjust:	
Transporter	1,60
Remove & install:	
Golf/Vento	1,90
Passat	1,60
Transporter	2,50

Special Tools

- Injection pump locking pin – No.3359.
- Camshaft setting bar – No.2065A.
- Two-pin wrench – Matra V.159.

Special Precautions

- Disconnect battery earth lead.
- DO NOT turn crankshaft or camshaft when timing belt removed.
- Remove glow plugs to ease turning engine.
- Turn engine in normal direction of rotation (unless otherwise stated).
- DO NOT turn engine via camshaft or other sprockets.
- Observe all tightening torques.
- Check diesel injection pump timing after belt replacement.

Removal

1. Remove:
   - Engine undershield.
   - Radiator grille (Transporter).
   - Auxiliary drive belt.
   - Water pump pulley (if required).
   - Cylinder head cover.
   - Timing belt upper cover **1**.
   - Crankshaft pulley bolts **2**.
   - Crankshaft pulley **3**.
   - Timing belt lower cover **4**.
2. Turn crankshaft to TDC on No.1 cylinder. Ensure flywheel timing marks aligned **5**.
3. Fit setting bar No.2065A to rear of camshaft **6**. Centralise camshaft using feeler gauges.
4. Insert locking pin in injection pump **7**. Tool No.3359.
5. Slacken injection pump sprocket bolts **8**.
6. Slacken tensioner nut **9**. Move tensioner away from belt. Lightly tighten nut.
7. Remove timing belt.

Installation

1. Ensure flywheel timing marks aligned **5**.
2. Ensure camshaft setting bar fitted correctly **6**.
3. Ensure locking pin located correctly in injection pump **7**.
4. Slacken camshaft sprocket bolt 1/2 turn **10**. Loosen sprocket from taper using a drift through hole in timing belt rear cover.
5. Fit timing belt in anti-clockwise direction, starting at crankshaft sprocket. Ensure belt is taut between sprockets.
6. Turn automatic tensioner pulley clockwise until notch and raised mark on tensioner aligned **11**. Use wrench Matra V.159.
7. Tighten tensioner nut to 20 Nm **9**.
8. Ensure flywheel timing marks aligned **5**.
9. Tighten camshaft sprocket bolt to 45 Nm **10**.
10. Tighten injection pump sprocket bolts to 25 Nm **8**.
11. Remove:
    - Injection pump locking pin **7**.
    - Camshaft setting bar **6**.
12. Turn crankshaft two turns clockwise to TDC on No.1 cylinder.
13. Ensure flywheel timing marks aligned **5**.
14. Ensure camshaft setting bar can be fitted **6**.
15. Ensure locking pin can be inserted in injection pump **7**.
16. If not: Slacken injection pump sprocket bolts **8**. Turn sprocket hub until locking pin can be inserted **7**. Tighten bolts to 25 Nm **8**.
17. Install components in reverse order of removal.
18. Tighten crankshaft pulley bolts to 25 Nm **2**.

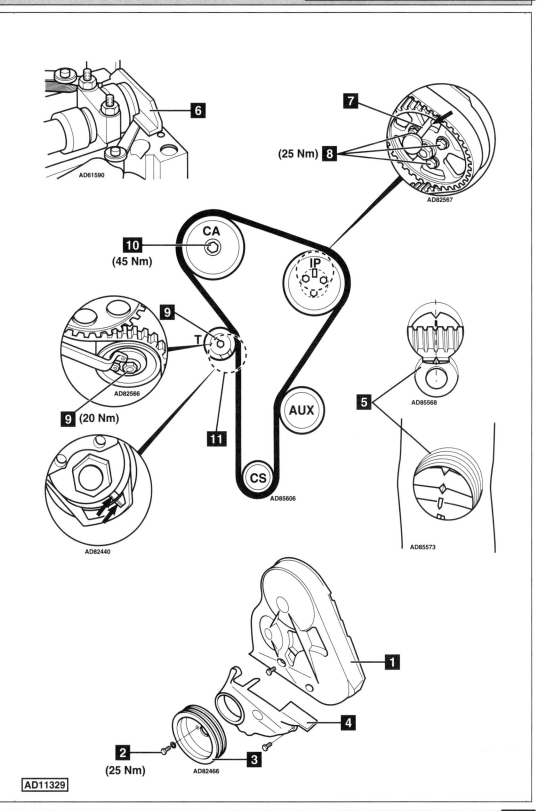

6

AD61590

7

(25 Nm) **8**

AD82567

10
(45 Nm)

CA

IP

9

T

AD82566

9 (20 Nm)

AD82440

11

AUX

CS

AD85606

5

AD85568

AD85573

1

4

2

3

(25 Nm)

AD82466

AD11329

VOLKSWAGEN

Model:	**Passat 1,9 TDI • Sharan 1,9 TDI**
Year:	**1995-00**
Engine Code:	**Single-piece injection pump sprocket** **1Z, AFN, AHH, AHU, AVG**

Replacement Interval Guide

Volkswagen recommend:

Except AHH engine
→2000MY: Check condition and width every 10,000 miles or 12 months, whichever occurs first (replacement width – 22 mm).
Replacement every 60,000 miles.

AHH engine
→2000MY: Check condition and width every 10,000 miles or 12 months, whichever occurs first (replacement width – 22 mm).
2000MY→: Check condition and tension every 20,000 miles.
Replacement every 60,000 miles.
The previous use and service history of the vehicle must always be taken into account.
Refer to Timing Belt Replacement Intervals at the front of this manual.

Check For Engine Damage

CAUTION: This engine has been identified as an INTERFERENCE engine in which the possibility of valve-to-piston damage in the event of a timing belt failure is MOST LIKELY to occur.
A compression check of all cylinders should be performed before removing the cylinder head.

Repair Times – hrs

Remove & install	3,30

Special Tools

- ■ Support guides – No.3369.
- ■ Injection pump sprocket locking pin – No.2064.
- ■ Camshaft setting bar – No.2065A.
- ■ Two-pin wrench – Matra V.159.

Removal

1. Passat – remove:
   - ❏ Front bumper.
   - ❏ Air intake pipe between front panel and air filter.
   - ❏ Front panel bolts 🔞 & 🔞.
2. Install support guides No.3369 in front panel 🔞.
3. Slide front panel forward.
4. Sharan:
   - ❏ Support engine.
   - ❏ Remove engine mounting.
5. All models – remove:
   - ❏ Auxiliary drive belt.
   - ❏ Auxiliary drive belt tensioner.
   - ❏ Timing belt upper cover 🔟.
   - ❏ Cylinder head cover.
6. Turn crankshaft to TDC on No.1 cylinder. Ensure flywheel timing marks aligned 🟦 – Passat or 🔟 – Sharan.
7. Fit setting bar No.2065A to rear of camshaft 🟥. Centralise camshaft using feeler gauges.
8. Lock injection pump sprocket 🟦. Use tool No.2064.

9. Remove:
   - ❏ Crankshaft pulley bolts 🟦.
   - ❏ Crankshaft pulley 🔟.
   - ❏ Timing belt lower cover 🟥.
10. Slacken tensioner nut 🟦. Turn tensioner anti-clockwise away from belt. Lightly tighten nut.
11. Remove:
    - ❏ Guide pulley 🟥.
    - ❏ Timing belt.
    NOTE: Mark direction of rotation on belt with chalk if belt is to be reused.

Installation

1. Ensure flywheel timing marks aligned 🟦 – Passat or 🔟 – Sharan.
2. Ensure camshaft setting bar fitted correctly 🟥.
3. Ensure locking pin located correctly in injection pump sprocket 🟦.
4. Slacken camshaft sprocket bolt 1/2 turn 🔟.
5. Loosen sprocket from taper using a drift through hole in timing belt rear cover. Ensure sprocket can turn on taper.
6. Fit timing belt in anti-clockwise direction, starting at crankshaft sprocket. Ensure belt is taut between sprockets.
7. Fit guide pulley 🟥. Tighten bolt to 25 Nm.
8. Remove locking pin from injection pump sprocket 🟦.
9. Slacken tensioner nut 🟦.
10. Turn automatic tensioner pulley clockwise until notch and raised mark on tensioner aligned 🔟. Use wrench Matra V.159 🔟.
11. Tighten tensioner nut to 20 Nm 🟦.
12. Ensure flywheel timing marks aligned 🟦 – Passat or 🔟 – Sharan.
13. Tighten camshaft sprocket bolt to 45 Nm 🔟.
14. Remove camshaft setting bar 🟥.
15. Turn crankshaft two turns clockwise until timing marks aligned 🟦 – Passat or 🔟 – Sharan.
16. Ensure camshaft setting bar can be fitted 🟥.
17. Ensure locking pin can be inserted in injection pump sprocket 🟦.
18. Apply firm thumb pressure to belt. Tensioner marks should move out of alignment 🔟.
19. Release thumb pressure from belt. Tensioner marks should realign. If not: Repeat tensioning procedure.
20. Install:
    - ❏ Timing belt lower cover 🟥.
    - ❏ Crankshaft pulley 🔟.
21. Tighten crankshaft pulley bolts to 25 Nm 🟦.
22. Install components in reverse order of removal.

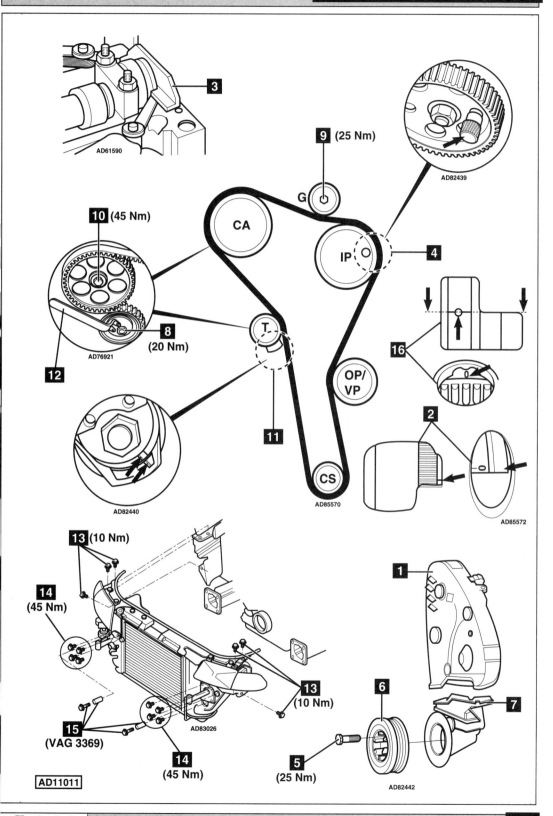

3 AD61590

9 (25 Nm)

AD82439

10 (45 Nm)

G

CA

IP **4**

16

8 (20 Nm)

AD76921

12

T

11

OP/ VP

2

AD82440

CS

AD85570

AD85572

13 (10 Nm)

14 (45 Nm)

13 (10 Nm)

1

15 (VAG 3369)

14 (45 Nm)

AD83026

6

7

5 (25 Nm)

AD82442

AD11011

VOLKSWAGEN

Model:	Passat 1,9 TDI
Year:	1997-00
Engine Code:	Two-piece injection pump sprocket AFN, AHH, AHU

Replacement Interval Guide

Volkswagen recommend:

Engine code AFN/AHU:
→2000MY: Check condition and width every 10,000 miles or 12 months, whichever occurs first (replacement width – 22 mm).
Replacement every 60,000 miles.

Engine code AHH:
→2000MY: Check width every 10,000 miles or 12 months, whichever occurs first (replacement width – 22 mm).
Check condition and tension every 20,000 miles.
Replacement every 60,000 miles.
The previous use and service history of the vehicle must always be taken into account.
Refer to Timing Belt Replacement Intervals at the front of this manual.

Check For Engine Damage

CAUTION: This engine has been identified as an INTERFERENCE engine in which the possibility of valve-to-piston damage in the event of a timing belt failure is MOST LIKELY to occur.
A compression check of all cylinders should be performed before removing the cylinder head.

Repair Times – hrs

Remove & install	3,30

Special Tools

- Support guides – No.3369.
- Camshaft setting bar – No.2065A.
- Injection pump locking pin – No.3359.
- Sprocket holding tool – No.3036.
- Two-pin wrench – Matra V.159.

Special Precautions

- Disconnect battery earth lead.
- DO NOT turn crankshaft or camshaft when timing belt removed.
- Remove glow plugs to ease turning engine.
- Turn engine in normal direction of rotation (unless otherwise stated).
- DO NOT turn engine via camshaft or other sprockets.
- Observe all tightening torques.
- Check diesel injection pump timing after belt replacement.

Removal

1. Remove:
   - ❑ Front bumper.
   - ❑ Air intake pipe between front panel and air filter.
   - ❑ Front panel bolts **1** & **2**.
2. Install support guides No.3369 in front panel **3**.
3. Slide front panel forward.
4. Remove:
   - ❑ Auxiliary drive belt.
   - ❑ Auxiliary drive belt tensioner.
   - ❑ Timing belt upper cover **4**.
   - ❑ Cylinder head cover.
5. Turn crankshaft to TDC on No.1 cylinder. Ensure flywheel timing marks aligned **5**.
6. Fit setting bar No.2065A to rear of camshaft **6**. Centralise camshaft using feeler gauges.
7. Insert locking pin in injection pump **7**. Tool No.3359.
8. Slacken injection pump sprocket locking bolts **8**.
9. Remove:
   - ❑ Crankshaft pulley bolts **9**.
   - ❑ Crankshaft pulley **10**.
   - ❑ Timing belt lower cover **11**.
10. Slacken tensioner nut **12**. Turn tensioner anti-clockwise away from belt. Lightly tighten nut.
11. Remove:
    - ❑ Guide pulley bolt **13**.
    - ❑ Guide pulley **14**.
    - ❑ Timing belt.
    NOTE: Mark direction of rotation on belt with chalk if belt is to be reused.

Installation

1. Ensure flywheel timing marks aligned **5**.
2. Ensure camshaft setting bar fitted correctly **6**.
3. Ensure locking pin located correctly in injection pump **7**.
4. Hold camshaft sprocket. Use tool No.3036.
5. Slacken camshaft sprocket bolt 1/2 turn **15**.
6. Loosen sprocket from taper using a drift through hole in timing belt rear cover.
7. Remove:
   - ❑ Camshaft sprocket bolt **15**.
   - ❑ Camshaft sprocket.
8. Remove locking bolts from injection pump sprocket **8**. Fit new bolts.

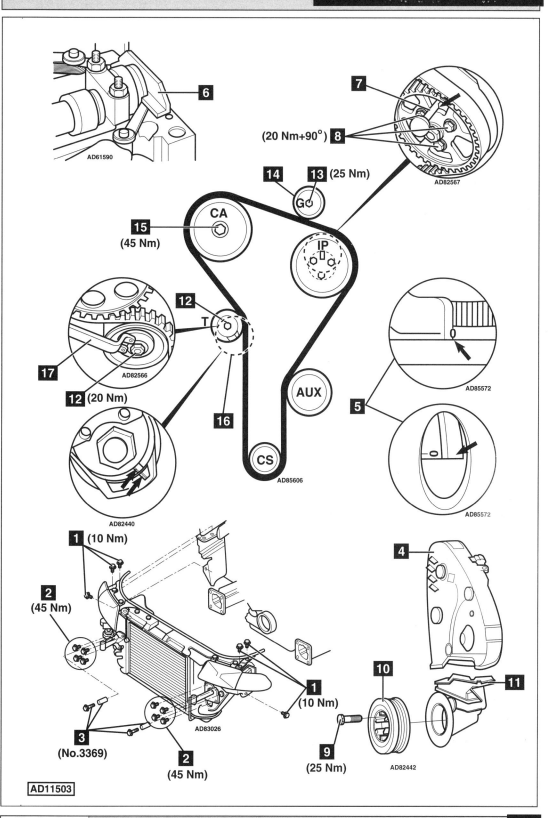

6 AD61590

7

(20 Nm+90°) **8**

AD82567

14 **13** (25 Nm)

15 CA (45 Nm)

GO

IP

17 **12** T

AD82566

12 (20 Nm)

16

AD82440

AUX

CS

AD85606

5

AD85572

AD85572

1 (10 Nm)

2 (45 Nm)

3 (No.3369)

2 (45 Nm)

AD83026

1 (10 Nm)

4

10

9 (25 Nm)

AD82442

11

AD11503

←

9. Align injection pump sprocket with bolts in centre of slotted holes.

10. Fit timing belt in following order:
    ❏ Crankshaft sprocket.
    ❏ Auxiliary shaft sprocket.
    ❏ Injection pump sprocket.
    ❏ Tensioner pulley.
    NOTE: Turn injection pump sprocket slightly to engage timing belt teeth.

11. Ensure belt is taut between sprockets.

12. Fit camshaft sprocket to belt, then install camshaft sprocket with belt onto end of camshaft.

13. Fit camshaft sprocket bolt **15**.

14. Lightly tighten camshaft sprocket bolt **15**. Sprocket should turn freely on taper but not tilt.

15. Fit guide pulley **14**. Tighten bolt to 25 Nm **13**.

16. Slacken tensioner nut **12**.

17. Turn automatic tensioner pulley clockwise until notch and raised mark on tensioner aligned **16**. Use wrench Matra V.159 **17**.

18. If tensioner pulley turned too far: Turn fully anti-clockwise and repeat tensioning procedure.

19. Tighten tensioner nut to 20 Nm **12**.

20. Ensure flywheel timing marks aligned **5**.

21. Hold camshaft sprocket. Use tool No.3036.

22. Tighten camshaft sprocket bolt to 45 Nm **15**.

23. Temporarily tighten injection pump sprocket locking bolts to 20 Nm **8**.
    NOTE: DO NOT tighten further 90°.

24. Remove:
    ❏ Camshaft setting bar **6**.
    ❏ Injection pump locking pin **7**.

25. Turn crankshaft two turns clockwise until timing marks aligned **5**.

26. Ensure camshaft setting bar can be fitted **6**.

27. Ensure locking pin can be inserted in injection pump **7**.

28. If not: Slacken injection pump sprocket locking bolts **8**. Turn sprocket hub until locking pin can be inserted **7**.

29. Tighten bolts to 20 Nm + 90° **8**.

30. Ensure notch and raised mark on tensioner aligned **16**.

31. If not: Slacken tensioner nut **12**. Turn automatic tensioner pulley clockwise until notch and raised mark on tensioner aligned **16**. Tighten tensioner nut to 20 Nm **12**.

32. Install:
    ❏ Timing belt lower cover **11**.
    ❏ Crankshaft pulley **10**.
    ❏ Crankshaft pulley bolts **9**.

33. Tighten crankshaft pulley bolts to 25 Nm **9**.

34. Install components in reverse order of removal.

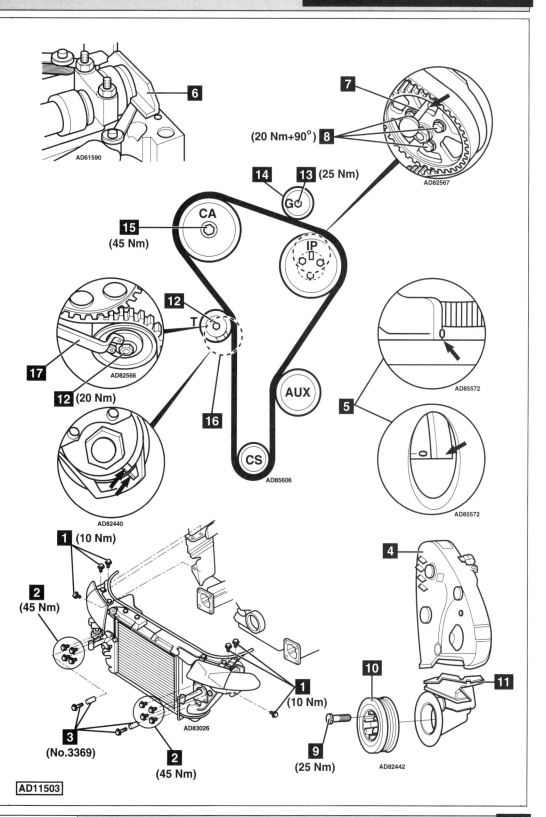

6

AD61590

7

(20 Nm+90°) 8

AD82567

14 13 (25 Nm)

GC

CA

15
(45 Nm)

IP

12

T

17

AD82566

12 (20 Nm)

AD82440

AUX

16

CS

AD85606

AD85572

5

AD85572

4

1 (10 Nm)

2
(45 Nm)

1
(10 Nm)

3
(No.3369)

2
(45 Nm)

AD83026

10

9
(25 Nm)

AD82442

11

AD11503

Model:	**Passat 1,9 TDI PD**
Year:	**1998-03**
Engine Code:	**AJM, ATJ, AVB, AVF, AWX**

Replacement Interval Guide

Volkswagen recommend:

All models:

→2000MY: Check condition and width every 10,000 miles or 12 months, whichever occurs first (replacement width 27 mm).

AJM →2000MY: Replacement every 40,000 miles.

All other engines →2000MY: Replacement every 60,000 miles.

2001MY→: Replacement every 60,000 miles. Tensioner pulley must also be replaced.

The previous use and service history of the vehicle must always be taken into account.

Refer to Timing Belt Replacement Intervals at the front of this manual.

Check For Engine Damage

CAUTION: This engine has been identified as an INTERFERENCE engine in which the possibility of valve-to-piston damage in the event of a timing belt failure is MOST LIKELY to occur. A compression check of all cylinders should be performed before removing the cylinder head(s).

Repair Times – hrs

Checking	0,20
Remove & install	2,90

Special Tools

- Camshaft locking tool – Volkswagen No.3359.
- Crankshaft sprocket locking tool – Volkswagen No.T10050.
- Support guides (→09/00) – Volkswagen No.3369.
- Support guides (10/00→) – Volkswagen No.3411.
- Tensioner locking tool – Volkswagen No.T10008.
- Two-pin wrench – Volkswagen No.3387.
- 4 mm drill bit.

Special Precautions

- Disconnect battery earth lead.
- DO NOT turn crankshaft or camshaft when timing belt removed.
- Remove glow plugs to ease turning engine.
- Turn engine in normal direction of rotation (unless otherwise stated).
- DO NOT turn engine via camshaft or other sprockets.
- Observe all tightening torques.

Removal

1. Raise and support front of vehicle.
2. Remove engine undershield.
3. Remove:
   - ❑ Front bumper.
   - ❑ Air intake pipe between front panel and air filter.
   - ❑ Front panel bolt **1**.
   - ❑ Front panel bolts **2** (10/00→).
   - ❑ Front bumper carrier – not shown (10/00→).
4. Install support guides in front panel:
   - ❑ →09/00 **3**. Tool No.3369.
   - ❑ 10/00→ **4** & **5**. Tool No.3411.
5. Remove:
   - ❑ Front panel bolts **2** & **6** (→09/00).
   - ❑ Front panel bolts **6** (10/00→).
6. Slide front panel forward into service position.
7. Remove:
   - ❑ Engine top cover.
   - ❑ Turbocharger air hoses.
   - ❑ Intercooler outlet hose.
   - ❑ Mass air flow (MAF) sensor.
   - ❑ Auxiliary drive belt.
   - ❑ Auxiliary drive belt tensioner.
   - ❑ Timing belt upper cover **7**.
   - ❑ Cooling fan assembly.
   - ❑ Timing belt centre cover **8**.
   - ❑ Crankshaft pulley bolts **9**.
   - ❑ Crankshaft pulley **10**.
   - ❑ Timing belt lower cover **11**.
8. Release coolant pipes for fuel cooler from engine
9. Turn crankshaft clockwise to TDC on No.1 cylinder.
10. Ensure timing mark aligned with notch on camshaft sprocket hub (4Z) **12**.

    NOTE: Notch located behind camshaft sprocket teeth.
11. Lock crankshaft sprocket **13**. Use tool No.T10050
12. Ensure timing marks aligned **14**.
13. Lock camshaft **15**. Use tool No.3359.
14. Fully insert Allen key **16** into tensioner pulley **17**
15. Turn tensioner pulley slowly anti-clockwise until locking tool can be inserted **18**. Tool No.T10008
16. Slacken tensioner nut **19**.
17. Remove:
    - ❑ Automatic tensioner unit **20**.
    - ❑ Timing belt.

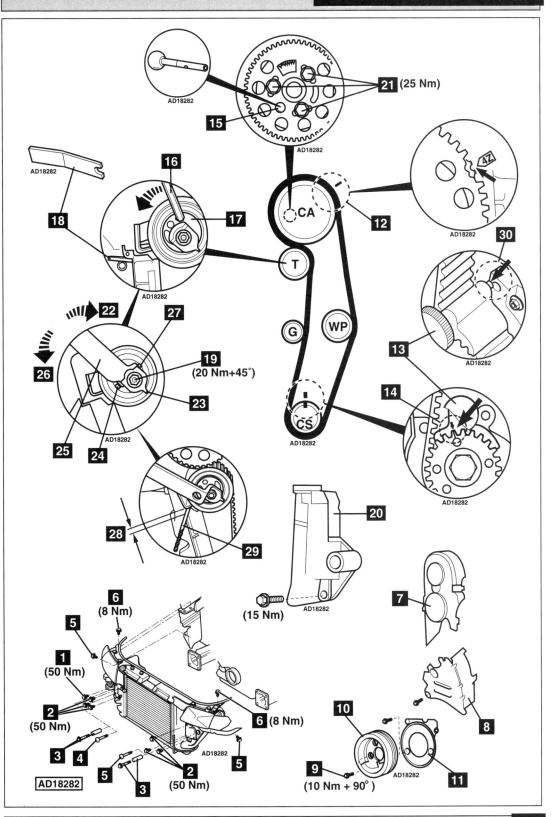

21 (25 Nm)

15

AD18282

AD18282

16

AD18282

18

17

12

CA

4Z

AD18282

30

13

AD18282

T

22

27

26

19
(20 Nm+45°)

G

WP

23

25 **24**

AD18282

14

CS

AD18282

AD18282

28

29

AD18282

20

(15 Nm)

AD18282

7

6
(8 Nm)

5

1
(50 Nm)

2
(50 Nm)

3 **4**

5

5

AD18282

2
(50 Nm)

6 (8 Nm)

10

9
(10 Nm + 90°)

11

8

AD18282

Installation

1. Ensure camshaft locked with tool .
2. Ensure crankshaft sprocket locking tool located correctly █.
3. Ensure timing marks aligned █.
4. Ensure automatic tensioner unit locked with tool █. Tool No.T10008.
5. Slacken camshaft sprocket bolts █.
6. Turn camshaft sprocket fully clockwise in slotted holes. Tighten bolts finger tight █.
7. Turn tensioner pulley slowly clockwise █ until lug █ just reaches stop █. Use tool No.3387 █.
8. Fit timing belt in following order:
   ❏ Camshaft sprocket.
   ❏ Tensioner pulley.
   ❏ Crankshaft sprocket.
   ❏ Water pump sprocket.
9. Install:
   ❏ Automatic tensioner unit █.
   NOTE: Ensure belt is taut between sprockets on non-tensioned side.
10. Turn tensioner pulley slowly anti-clockwise █ (lug █ moves towards stop █). Use tool No.3387 █.
11. Remove locking tool without force █.
12. Turn tensioner pulley slowly clockwise █ (lug █ moves towards stop █) until dimension █ is 4±1 mm. Use drill bit █.
    NOTE: Engine must be COLD.
13. Tighten tensioner nut █.
    Tightening torque: 20 Nm + 45°.
14. Tighten camshaft sprocket bolts █.
    Tightening torque: 25 Nm.
15. Remove:
    ❏ Camshaft locking tool █.
    ❏ Crankshaft sprocket locking tool █.
    ❏ Drill bit █.
16. Turn crankshaft slowly two turns clockwise to TDC on No.1 cylinder.
17. Ensure dimension █ is 4±1 mm. Use drill bit █.
18. If not: Slacken tensioner nut █. Turn tensioner pulley until dimension as specified █. Tighten tensioner nut █.
    Tightening torque: 20 Nm + 45°.
19. Lock crankshaft sprocket █. Use tool No.T10050.
20. Ensure timing marks aligned █.
21. Ensure camshaft locking tool can be inserted easily █. Tool No.3359.

22. If not:
    ❏ Remove lug of crankshaft sprocket locking tool from hole in oil seal housing.
    ❏ Turn crankshaft until camshaft locking tool can be inserted █. Tool No.3359.
    ❏ Slacken camshaft sprocket bolts █.
    ❏ Turn crankshaft anti-clockwise until lug of locking tool just passes hole in oil seal housing █.
    ❏ Turn crankshaft clockwise until lug and hole aligned.
    ❏ Lock crankshaft sprocket █. Use tool No.T10050.
    ❏ Tighten camshaft sprocket bolts █.
      Tightening torque: 25 Nm.
    ❏ Remove locking tools █ & █.
    ❏ Turn crankshaft slowly two turns clockwise to TDC on No.1 cylinder.
    ❏ Ensure locking tools can be fitted correctly █ & █.
23. Remove:
    ❏ Camshaft locking tool █.
    ❏ Crankshaft sprocket locking tool █.
    ❏ Drill bit █.
24. Install components in reverse order of removal.
25. Tighten crankshaft pulley bolts █.
    Tightening torque: 10 Nm + 90°.

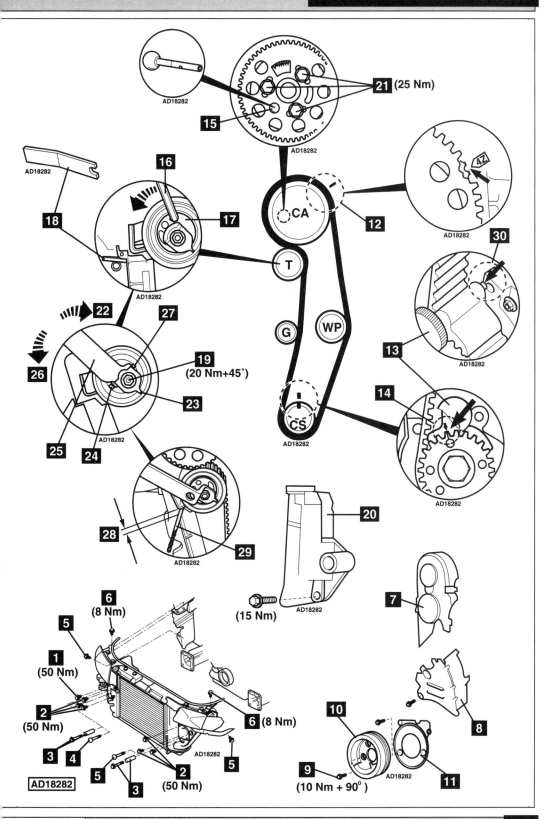

21 (25 Nm)

AD18282

15

AD18282

16

AD18282

18

AD18282

17

12

AD18282

30

13

AD18282

14

AD18282

22

27

19 (20 Nm+45°)

23

25

24 AD18282

26

28

29 AD18282

20

(15 Nm) AD18282

7

8

6 (8 Nm)

5

1 (50 Nm)

2 (50 Nm)

3 **4**

6 (8 Nm)

5

2 (50 Nm)

AD18282

5

3

AD18282

10

9 (10 Nm + 90°)

AD18282

11

CA

T

G

WP

CS

AD18282

Model:	**Passat 2,5 V6 TDI**
Year:	**1998-03**
Engine Code:	**AFB, AKN**

Replacement Interval Guide

Volkswagen recommend replacement every 80,000 miles.
The previous use and service history of the vehicle must always be taken into account.
Refer to Timing Belt Replacement Intervals at the front of this manual.

Check For Engine Damage

CAUTION: This engine has been identified as an INTERFERENCE engine in which the possibility of valve-to-piston damage in the event of a timing belt failure is MOST LIKELY to occur.
A compression check of all cylinders should be performed before removing the cylinder head(s).

Repair Times – hrs

Remove & install	3,70

Special Tools

- Support guides – No.3369.
- Crankshaft locking tool – No.3242.
- Camshaft aligning tool – 2 x No.3458.
- Injection pump locking pin – No.3359.
- Tensioner socket – No.3078.
- Holding tool – No.3036.
- Puller – No.T40001.

Special Precautions

- Disconnect battery earth lead.
- DO NOT turn crankshaft or camshaft when timing belt removed.
- Remove glow plugs to ease turning engine.
- Turn engine in normal direction of rotation (unless otherwise stated).
- DO NOT turn engine via camshaft or other sprockets.
- Observe all tightening torques.
- Check diesel injection pump timing after belt replacement.

Removal

Injection Pump Belt

1. Remove:
   - ❏ Front bumper.
   - ❏ Engine undershield.
   - ❏ Intercooler hoses.
   - ❏ Air filter intake duct.
   - ❏ Front panel bolts **1** & **2**.
2. Install support guides No.3369 in front panel **3**.
3. Move radiator support panel into service position.
4. Remove:
   - ❏ Viscous fan.
   - ❏ Viscous fan air ducting.
   - ❏ Upper engine cover.
   - ❏ Timing belt upper covers.
   - ❏ Auxiliary drive belt cover.
   - ❏ Auxiliary drive belt(s).
   NOTE: Mark direction of rotation on belt(s).
   - ❏ Oil filler cap.
5. Turn crankshaft clockwise to TDC.

6. Ensure camshaft timing mark aligned with centre of oil filler cap hole **4**.
7. Remove blanking plug from cylinder block.
8. Screw in crankshaft locking tool. Tool No.3242 **5**.
   NOTE: TDC hole in crankshaft web must be aligned with blanking plug hole.
9. Remove:
   - ❏ Coolant expansion tank. DO NOT disconnect hoses.
   - ❏ Vacuum pump. DO NOT disconnect hoses.
   - ❏ Air filter housing.
   - ❏ Turbocharger air hoses.
   - ❏ Cover plate from rear of RH camshaft **6**.
10. Install camshaft aligning tools to rear of camshafts. Tool No.3458 **7**.
    NOTE: Retain in position with chains provided, to prevent tools falling out.
11. Remove:
    - ❏ Injection pump vibration damper bolts **8**.
    - ❏ Injection pump vibration damper **9**.
    *NOTE: DO NOT slacken injection pump centre hub nut **10**.*
12. Insert locking pin in injection pump sprocket. Tool No.3359 **11**.
13. Slacken injection pump belt tensioner nut **12**. Tool No.3078.
14. Turn tensioner pulley clockwise away from belt **13**. Use Allen key.
15. Lightly tighten injection pump belt tensioner nut **12**.
16. Remove injection pump belt.
17. Remove:
    - ❏ Camshaft sprocket bolts **16**.
    - ❏ Camshaft outer sprocket (for injection pump belt) **17**.
    NOTE: Mark direction of rotation on belt with chalk if belt is to be reused.

Installation

Injection Pump Belt

1. Ensure camshaft timing mark aligned with centre of oil filler cap hole **4**.
2. Ensure crankshaft locking tool located correctly. Tool No.3242 **5**.
3. Ensure camshaft aligning tools fitted to rear of camshafts. Tool No.3458 **7**.
4. Ensure locking pin located correctly in injection pump. Tool No.3359 **11**.
5. Fit camshaft outer sprocket **17**.
6. Ensure camshaft sprocket bolts not at end of slotted holes **16**.
7. Fit camshaft sprocket bolts **16**. Lightly tighten bolts Ensure sprocket can turn.
8. Fit injection pump belt.
   NOTE: Observe direction of rotation marks.
9. Slacken injection pump belt tensioner nut **12**. Use tool No.3078.

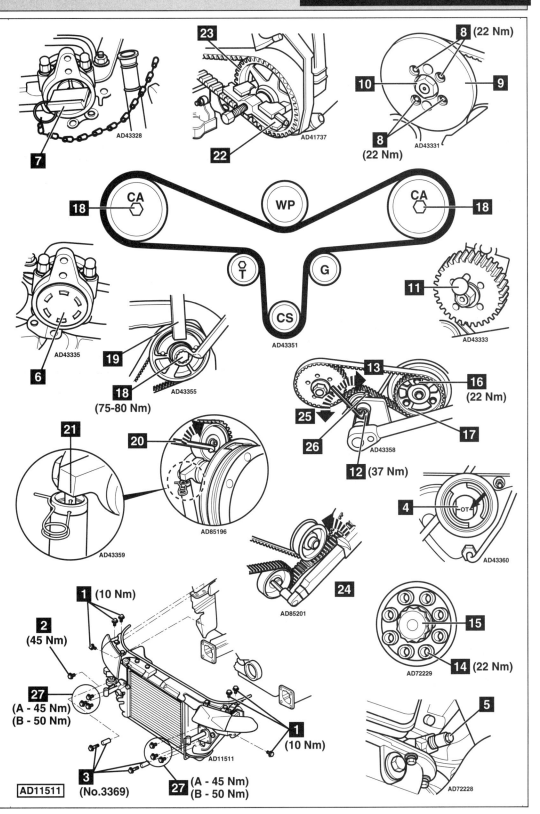

AD43328

AD41737

8 (22 Nm)

10

9

8 (22 Nm)

AD43331

7

23

22

18 CA

WP

CA 18

AD43351

6

AD43335

19

18 (75-80 Nm)

AD43355

OT

G

CS

11

AD43333

13

16 (22 Nm)

25

26

17

AD43358

12 (37 Nm)

21

20

AD85196

AD43359

4

OT

AD43360

24

AD85201

1 (10 Nm)

2 (45 Nm)

27 (A - 45 Nm) (B - 50 Nm)

15

14 (22 Nm)

AD72229

5

1 (10 Nm)

AD11511

3 (No.3369)

27 (A - 45 Nm) (B - 50 Nm)

AD72228

←

10. Turn tensioner pulley anti-clockwise **25**. Use Allen key. Check tensioner pointers align **26**.
11. Tighten tensioner nut to 37 Nm **12**.
12. Tighten sprocket bolts to 22 Nm **16**.
13. Remove injection pump locking pin **11**.
14. Remove camshaft aligning tools **7**.
15. Remove crankshaft locking tool **5**. Fit blanking plug.
16. Turn crankshaft slowly two turns clockwise.
17. Screw in crankshaft locking tool. Tool No.3242 **5**.
18. Check camshaft aligning tools can be fitted correctly **7**.
19. Check locking pin can be fitted correctly in injection pump sprocket **11**.
20. Check tensioner pointers align **26**. Adjust if necessary.
21. Remove tools **5**, **7** & **11**.
22. Install injection pump vibration damper **9**.
23. Tighten injection pump vibration damper bolts **8**. Tightening torque: 22 Nm.
24. Install components in reverse order of removal.

Removal

Timing Belt

1. Remove injection pump belt as described previously.
2. Remove:
   - ❑ Crankshaft pulley bolts **14**.
   - ❑ Crankshaft pulley.
   - *NOTE: DO NOT remove crankshaft centre bolt **15**.*
   - ❑ Timing belt lower cover.
   - ❑ Viscous fan pulley.
   - ❑ Viscous fan bracket.
3. Remove:
   - ❑ Camshaft sprocket bolts **16**.
   - ❑ Camshaft outer sprocket (for injection pump belt) **17**.
4. Hold sprockets. Use tool No.3036 **19**. Slacken bolt of each camshaft sprocket **18**.
5. Turn tensioner pulley clockwise until holes in pushrod and tensioner body aligned. Use 8 mm Allen key **20**. Retain pushrod with 2 mm diameter pin through hole in tensioner body **21**.
6. Loosen camshaft sprockets from camshafts. Use tool No.T40001 **22**.
7. Remove:
   - ❑ LH camshaft sprocket bolt **18**.
   - ❑ LH camshaft sprocket **23**.
   - ❑ Timing belt.
   - *NOTE: Mark direction of rotation on belt with chalk if belt is to be reused.*

Installation

Timing Belt

1. Ensure camshaft timing mark aligned with centre of oil filler cap hole **4**.
2. Ensure crankshaft locking tool located correctly. Tool No.3242 **5**.
3. Ensure camshaft aligning tools fitted to rear of camshafts. Tool No.3458 **7**.
4. Ensure RH camshaft sprocket can turn on taper but not tilt.

5. Fit timing belt in following order:
   - *NOTE: Observe direction of rotation marks.*
   - ❑ Crankshaft sprocket.
   - ❑ RH camshaft sprocket.
   - ❑ Tensioner pulley.
   - ❑ Guide pulley.
   - ❑ Water pump pulley.
6. Fit LH camshaft sprocket to belt, then install camshaft sprocket with belt onto end of camshaft.
7. Fit camshaft sprocket bolt **18**.
8. Ensure LH camshaft sprocket can turn on taper but not tilt.
9. Turn tensioner pulley slightly clockwise. Use 8 mm Allen key **20**. Remove pin from tensioner body to release pushrod **21**.
10. Remove Allen key from tensioner pulley **20**.
11. Install torque wrench to hexagon of tensioner.
12. Apply anti-clockwise torque (in direction of arrow) of 15 Nm to tensioner pulley **24**.
13. Remove torque wrench.
14. Hold sprockets. Use tool No.3036 **19**. Tighten bolt of each camshaft sprocket to 75-80 Nm **18**.
15. Install:
   - ❑ Timing belt lower cover.
   - ❑ Viscous fan bracket.
   - ❑ Viscous fan pulley.
   - ❑ Crankshaft pulley.
   - ❑ Crankshaft pulley bolts **14**.
   - *NOTE: Ensure notches aligned with tabs on crankshaft sprocket.*
16. Tighten crankshaft pulley bolts to 22 Nm **14**.
17. Fit injection pump belt as described previously.
18. Install components in reverse order of removal.

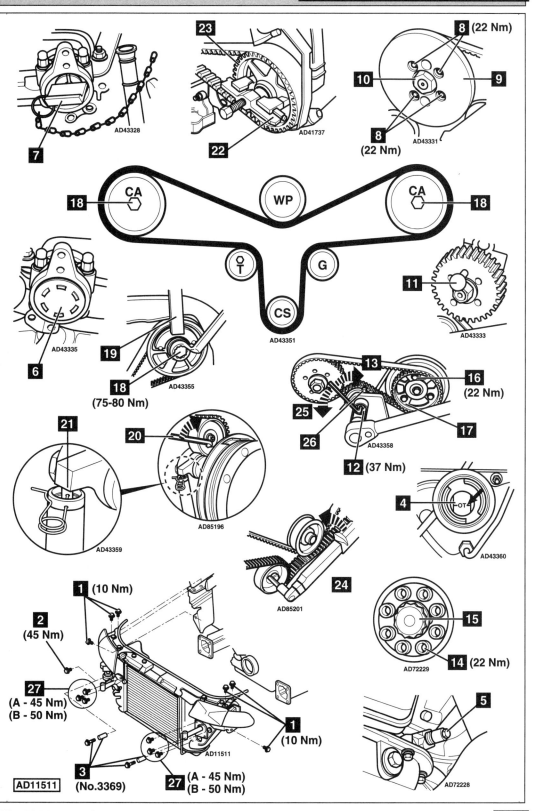

VOLKSWAGEN

Model:	**Transporter 2,4D**
Year:	**1990-01/95**
Engine Code:	**AAB**

Special Precautions

■ Disconnect battery earth lead.
■ DO NOT turn crankshaft or camshaft when timing belt removed.
■ Remove glow plugs to ease turning engine.
■ Turn engine in normal direction of rotation (unless otherwise stated).
■ DO NOT turn engine via camshaft or other sprockets.
■ Observe all tightening torques.
■ Check diesel injection pump timing after belt replacement.

Removal

1. Hinge radiator grille forward.
2. Remove:
   ❏ Auxiliary drive belt(s).
   ❏ Injection pump belt cover.
   ❏ Cylinder head cover.
   ❏ Timing belt upper cover **1**.
3. Turn crankshaft clockwise to TDC on No.1 cylinder.
4. Ensure flywheel timing marks aligned **4**. Ensure injection pump sprocket timing mark aligned **3**.
5. Lock injection pump sprocket. Use tool No.2064 **5**.
6. Hold camshaft rear sprocket. Use tool No.3036. Remove bolt **6**.

7. Remove camshaft rear sprocket and injection pump belt.
8. Lock crankshaft damper. Use tool No.3248 **11**. Remove crankshaft damper centre bolt **7**.
9. Ensure crankshaft at TDC on No.1 cylinder **4**.
10. Remove:
    ❏ Crankshaft damper bolts (4 bolts).
    ❏ Crankshaft damper **9**.
11. Fit setting bar No.2065A to rear of camshaft **8**. Centralise camshaft using feeler gauges.
12. Slacken water pump bolts **12**. Turn water pump clockwise to release tension on belt.
13. Remove timing belt lower cover **2**. Remove timing belt.

Installation

1. Ensure crankshaft at TDC on No.1 cylinder **4**.
2. Ensure camshaft setting bar fitted correctly **8**.
3. Slacken camshaft sprocket bolt **10**. Tap sprocket from behind to loosen it from taper.
4. Fit timing belt in clockwise direction, starting at crankshaft sprocket.
5. Fit timing belt lower cover **2**.
6. Attach tension gauge to belt at ▽. Tool No.210.
7. Turn water pump anti-clockwise until tension gauge indicates 12-13 units. Tighten water pump bolts to 20 Nm **12**.
8. Ensure crankshaft at TDC on No.1 cylinder.
9. Hold camshaft sprocket. Use tool No.3036. Tighten bolt to 85 Nm **10**.
10. Remove camshaft setting bar **8**.
11. Fit crankshaft damper **9**. Fit four bolts. Tighten bolts to 20 Nm.
12. Fit crankshaft damper centre bolt **7**. Lock crankshaft damper. Use tool No.3248 **11**. Tighten bolt to 460 Nm.
13. Remove locking tool.
14. Fit camshaft rear sprocket and injection pump belt. Ensure timing marks aligned **3**.
15. Tighten bolt **6** until sprocket can just be turned by hand.
16. Attach tension gauge to belt at ▽. Tool No.210.
17. Move injection pump on mountings until tension gauge indicates 12-13 units.
18. Hold camshaft rear sprocket. Use tool No.3036. Tighten bolt to 100 Nm **6**.
19. Remove locking pin from injection pump sprocket **5**.
20. Check injection pump timing.
21. Install components in reverse order of removal.

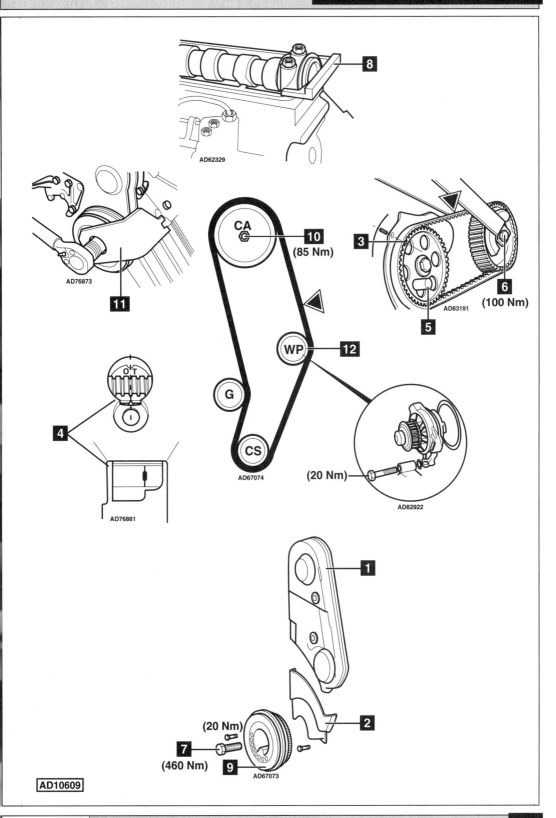

AD62329

AD76873

11

CA

10

(85 Nm)

3

6

(100 Nm)

5

AD83191

WP **12**

G

CS

AD67074

(20 Nm)

AD82922

4

AD76881

1

2

(20 Nm)

7

(460 Nm)

9

AD67073

AD10609

VOLKSWAGEN

Model:	**Transporter 2,4D**
Year:	**2/95-03**
Engine Code:	**AAB, AJA**

Replacement Interval Guide

Volkswagen recommend check & adjust every 40,000 miles (replace if necessary) and replacement every 80,000 miles.
The previous use and service history of the vehicle must always be taken into account.
Refer to Timing Belt Replacement Intervals at the front of this manual.

Check For Engine Damage

CAUTION: This engine has been identified as an INTERFERENCE engine in which the possibility of valve-to-piston damage in the event of a timing belt failure is MOST LIKELY to occur.
A compression check of all cylinders should be performed before removing the cylinder head.

Repair Times – hrs

Remove & install	3,70

Special Tools

- Injection pump sprocket locking pin – No.2064.
- Camshaft setting bar – No.2065A.
- Crankshaft damper holding tool – No.T10025.
- Sprocket holding tool – No.3036
- Tension gauge – No.210.
- Tensioner pulley spanner – No.3355.

Special Precautions

- Disconnect battery earth lead.
- DO NOT turn crankshaft or camshaft when timing belt removed.
- Remove glow plugs to ease turning engine.
- Turn engine in normal direction of rotation (unless otherwise stated).
- DO NOT turn engine via camshaft or other sprockets.
- Observe all tightening torques.
- Check diesel injection pump timing after belt replacement.

Removal

1. Remove:
   - Engine undershield.
   - Auxiliary drive belt.
   - Radiator grille.
2. Remove bonnet lock platform bolts and hinge platform and radiator forward.
3. Remove:
   - Auxiliary drive belt.
   - Injection pump belt cover.
   - Cylinder head cover.
   - Timing belt upper cover **1**.
4. Turn crankshaft clockwise to TDC on No.1 cylinder.
5. Ensure timing marks aligned **4**. Ensure injection pump sprocket timing mark aligned **3**.
6. Lock injection pump sprocket. Use tool No.2064 **5**.
7. Hold camshaft rear sprocket. Use tool No.3036. Remove bolt **6**.
8. Remove camshaft rear sprocket and injection pump belt.
9. Lock crankshaft damper. Use tool No.T10025 **7**.
10. Remove crankshaft damper centre bolt **8**.
11. Ensure crankshaft at TDC on No.1 cylinder **4**.
12. Remove:
    - Crankshaft damper bolts (4 bolts) **9**.
    - Crankshaft damper **10**.
    - Timing belt lower cover **2**.
13. Fit setting bar No.2065A to rear of camshaft **11**. Centralise camshaft using feeler gauges.
14. Slacken tensioner bolt **12**. Turn tensioner away from belt. Use tool No.3355 **13**. Lightly tighten bolt.
15. Remove timing belt.
    NOTE: Mark direction of rotation on belt with chalk if belt is to be reused.

Installation

1. Ensure crankshaft at TDC on No.1 cylinder **4**.
2. Ensure camshaft setting bar fitted correctly **11**.
3. Slacken camshaft sprocket bolt 1/2 turn **14**. Tap sprocket from behind to loosen it from taper.
   NOTE: Sprocket should turn freely on taper but not tilt.
4. Fit timing belt in anti-clockwise direction, starting at crankshaft sprocket. Ensure belt is taut between sprockets.
5. Slacken tensioner bolt **12**.

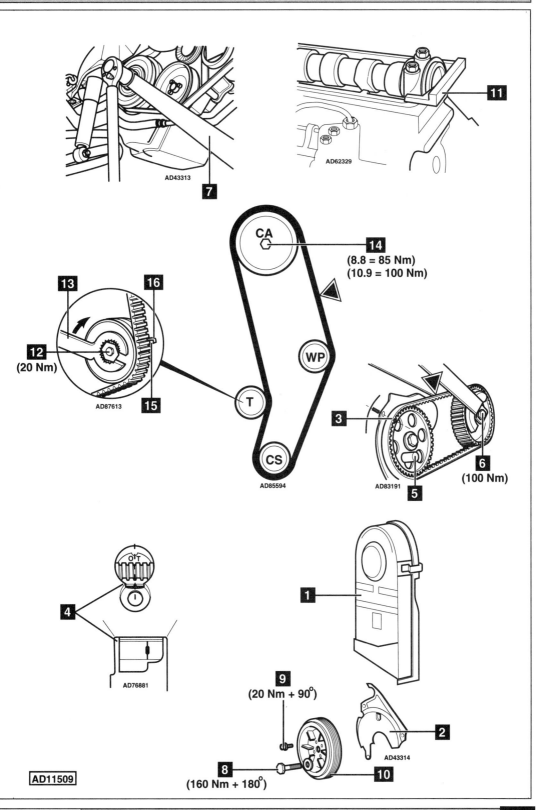

AD43313

7

AD62329

11

CA

14
(8.8 = 85 Nm)
(10.9 = 100 Nm)

13   **16**

12
(20 Nm)

AD87613   **15**

WP

T

CS

AD85594

3

AD83191

5

6
(100 Nm)

4

AD76881

1

9
(20 Nm + 90°)

2

AD43314

8
(160 Nm + 180°)

10

AD11509

←

6. Turn tensioner pulley slowly clockwise until right edge of pointers aligned **15** & **16**. Use tool No.3355 **13**.

7. If tensioner pulley turned too far: Turn fully anti-clockwise and repeat tensioning procedure.

   *NOTE: To prevent tensioner damage, right edge of pointer **15** should not go past right edge of pointer **16** during adjustment.*

8. Tighten tensioner bolt to 20 Nm **12**.

9. Ensure timing marks aligned **4**.

10. Hold camshaft front sprocket. Use tool No.3036. Tighten bolt to specified torque **14**.

    NOTE: Check marking on camshaft sprocket bolt head for correct torque setting.
    8.8 - 85 Nm.
    10.9 - 100 Nm.

11. Remove camshaft setting bar **11**.

12. Install:
    - ❑ Timing belt lower cover **2**.
    - ❑ Crankshaft damper **10**.
    - ❑ Crankshaft damper bolts (4 bolts) **9**.

13. Tighten crankshaft damper bolts **9**. Tightening torque: 20 Nm + 90°.

14. Fit crankshaft damper centre bolt **8**. Lock crankshaft damper. Use tool No.T10025 **7**. Tighten bolt to 160 Nm + 180°.

15. Fit camshaft rear sprocket and injection pump belt. Ensure timing marks aligned **3**.

16. Tighten bolt **6** until sprocket can just be turned by hand.

17. Attach tension gauge to belt at ▽. Tool No.210.

18. Move injection pump on mountings until tension gauge indicates 12-13 units.

19. Hold camshaft rear sprocket. Use tool No.3036. Tighten bolt to 100 Nm **6**.

20. Remove locking pin from injection pump sprocket **5**.

21. Install components in reverse order of removal.

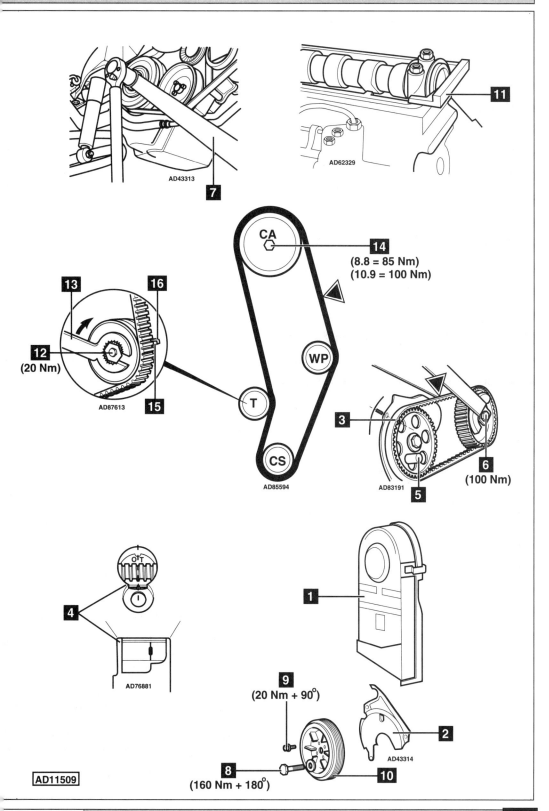

AD43313

7

AD62329

11

CA

14

(8.8 = 85 Nm)
(10.9 = 100 Nm)

13

16

12

(20 Nm)

AD87613

15

T

WP

3

6

(100 Nm)

5

AD83191

AD85594

CS

4

AD76881

1

9

(20 Nm + 90°)

2

AD43314

8

(160 Nm + 180°)

10

VOLKSWAGEN

Model:	**Transporter 2,5 TDI**
Year:	**1995-03**
Engine Code:	**ACV, AHY, AJT**

Replacement Interval Guide

Volkswagen recommend check and adjust if necessary every 40,000 miles (timing belt). Replacement every 80,000 miles – timing belt tensioner pulley must also be replaced (timing belt and injection pump belt).

The previous use and service history of the vehicle must always be taken into account.

Refer to Timing Belt Replacement Intervals at the front of this manual.

Check For Engine Damage

CAUTION: This engine has been identified as an INTERFERENCE engine in which the possibility of valve-to-piston damage in the event of a timing belt failure is MOST LIKELY to occur.

A compression check of all cylinders should be performed before removing the cylinder head.

Repair Times – hrs

Remove & install	4,20

Special Tools

- Crankshaft pulley holding tool – No.3248/A.
- Tensioner pulley spanner – No.3355.
- Sprocket holding tool – No.3036.
- Camshaft setting bar – No.2065A.
- Dial gauge adaptor – No.3313.

Special Precautions

- Disconnect battery earth lead.
- DO NOT turn crankshaft or camshaft when timing belt removed.
- Remove glow plugs to ease turning engine.
- Turn engine in normal direction of rotation (unless otherwise stated).
- DO NOT turn engine via camshaft or other sprockets.
- Observe all tightening torques.
- Check diesel injection pump timing after belt replacement.

Removal

1. Remove:
   - ❏ Auxiliary drive belt.
   - ❏ Radiator grille.
   - ❏ Intercooler bracket.
2. Remove bonnet lock platform bolts and hinge platform and radiator forward.
3. Remove:
   - ❏ Timing belt upper cover **1**.
   - ❏ Cylinder head cover.
   - ❏ Injection pump belt cover.

4. Hold crankshaft pulley. Use tool No.3248/A. Slacken centre bolt **3**.
5. Turn crankshaft to TDC on No.1 cylinder. Ensure timing marks aligned **4**. Ensure injection pump timing marks aligned **5**.
6. If not: Turn crankshaft one turn clockwise.
7. Hold camshaft rear sprocket. Use tool No.3036. Remove bolt **6**.
8. Remove:
   - ❏ Injection pump belt tensioner **7**.
   - ❏ Camshaft rear sprocket and injection pump belt.
9. Fit setting bar No.2065A to rear of camshaft **8**. Centralise camshaft using feeler gauges.
10. Remove:
    - ❏ Crankshaft pulley centre bolt **3**.
    - ❏ Crankshaft pulley bolts **9**.
    - ❏ Crankshaft pulley **10**.
    - ❏ Timing belt lower cover **2**.
11. Slacken tensioner bolt **11**. Turn tensioner away from belt. Use tool No.3355 **12**. Lightly tighten bolt.
12. Remove timing belt.

Installation

1. Ensure crankshaft at TDC on No.1 cylinder **4**.
2. Slacken camshaft sprocket bolt 1/2 turn **13**. Tap sprocket gently to loosen it from taper.
   NOTE: Sprocket should turn freely on taper but not tilt.
3. Fit timing belt in anti-clockwise direction, starting at crankshaft sprocket. Ensure belt is taut between sprockets.
4. Slacken tensioner bolt **11**.
5. Turn tensioner pulley slowly clockwise until right edge of pointers aligned **14** & **23**. Use tool No.3355 **12**.
6. If tensioner pulley turned too far: Turn fully anti-clockwise and repeat tensioning procedure.
   NOTE: To prevent tensioner damage, right edge of pointer **23** should not go past right edge of pointer **14** during adjustment.
7. Tighten tensioner bolt to 20 Nm **11**.
8. Ensure timing marks aligned **4**.
9. Hold camshaft front sprocket. Use tool No.3036. Tighten bolt to specified torque **13**.
   NOTE: Check marking on camshaft sprocket bolt head for correct torque setting. 8.8: 85 Nm. 10.9: 100 Nm.
10. Remove camshaft setting bar **8**.

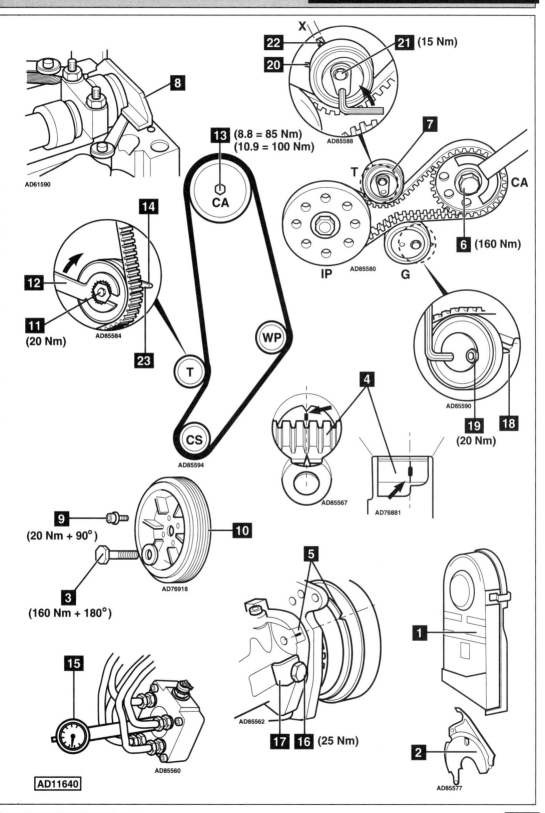

←

11. Install:
    - ❏ Timing belt lower cover **2**.
    - ❏ Crankshaft pulley **10**.

    NOTE: Fit new crankshaft pulley centre bolt 3.

12. Lightly oil new centre bolt threads and contact surfaces.

13. Hold crankshaft pulley. Use tool No.3248/A. Tighten bolt to 160 Nm + 180° **3**.

14. Tighten crankshaft pulley bolts to 20 Nm + 90° (4 bolts) **9**.

15. Install dial gauge with adaptor No.3313 in injection pump **15**.

16. Ensure injection pump timing marks aligned **5**.

17. Slacken injection pump locking bolt **16**. Remove keeper plate **17**.

18. Set dial gauge to zero.

19. Turn injection pump sprocket slowly clockwise (against normal direction of rotation). Use tool No.3036.

20. If dial gauge reading decreases: Reset to zero once dial gauge movement stops.

    NOTE: If dial gauge reading increases: Turn injection pump sprocket anti-clockwise until dial gauge movement stops and timing marks are approximately aligned 5.

21. Turn injection pump sprocket anti-clockwise (in normal direction of rotation) until dial gauge indicates 0,55 mm.

22. Tighten injection pump locking bolt to 25 Nm **16**.

23. Ensure TDC marks aligned **4**.

24. Fit injection pump belt with camshaft rear sprocket.

25. Lightly tighten sprocket bolt **6**. Ensure sprocket can just be turned by hand.

26. Ensure guide pulley pointer aligned with cylinder head flange contour **18**.

27. If not: Slacken guide pulley nut **19**. Turn guide pulley until pointer aligned. Use Allen key. Tighten nut to 20 Nm **19**.

28. Fit tensioner **7**. Ensure tensioner tab engaged in cut-out **20**. Tighten bolt finger tight **21**.

29. Turn tensioner pulley anti-clockwise until pointers aligned **22**. Tighten bolt to 15 Nm **21**.

    NOTE: To prevent tensioner damage, front pointer should not go past rear pointer during adjustment.

30. Hold camshaft rear sprocket. Use tool No.3036. Tighten bolt to 160 Nm **6**.

31. Slacken injection pump locking bolt **16**. Insert keeper plate **17**. Tighten locking bolt to 25 Nm **16**.

32. Turn crankshaft two turns clockwise to TDC on No.1 cylinder.

33. Ensure right edge of timing belt tensioner pointers aligned **14** & **23**. If not: Repeat tensioning procedures.

34. Check injection pump belt tensioner pointers are either aligned **22**, or front pointer is within dimension 'X'. If not: Repeat tensioning procedures.

35. Install components in reverse order of removal.

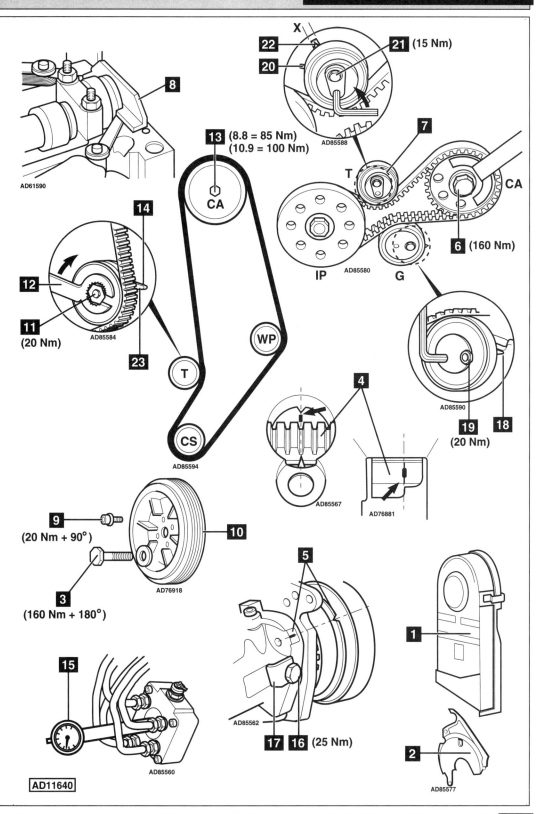

8

AD61590

22 X

20

21 (15 Nm)

AD85588

13 (8.8 = 85 Nm)
(10.9 = 100 Nm)

CA

7

T

CA

6 (160 Nm)

IP AD85580 **G**

14

12

11

(20 Nm)

AD85584

23

T

WP

CS

AD85594

4

AD85567

AD76881

AD85590

19 **18**

(20 Nm)

9

(20 Nm + 90°)

3

(160 Nm + 180°)

10

AD76918

5

1

15

17 **16** (25 Nm)

AD85562

AD85560

2

AD85577

AD11640

VOLKSWAGEN

Model:	**LT 2,4D/Turbo D**
Year:	**1992-96**
Engine Code:	**Manual tensioner** **ACL, ACT**

Replacement Interval Guide

Volkswagen recommend check & adjust every 20,000 miles (replace if necessary).
No manufacturer's recommended replacement interval.
The previous use and service history of the vehicle must always be taken into account.
Refer to Timing Belt Replacement Intervals at the front of this manual.

Check For Engine Damage

CAUTION: This engine has been identified as an INTERFERENCE engine in which the possibility of valve-to-piston damage in the event of a timing belt failure is MOST LIKELY to occur.
A compression check of all cylinders should be performed before removing the cylinder head.

Repair Times – hrs

Remove & install	5,20

Special Tools

- Injection pump sprocket locking pin – VAG No.2064.
- Camshaft setting bar – VAG No.2065A.
- Sprocket holding tool – VAG No.3036.
- Tension gauge – VAG No.210.

Special Precautions

- Disconnect battery earth lead.
- DO NOT turn crankshaft or camshaft when timing belt removed.
- Remove glow plugs to ease turning engine.
- Turn engine in normal direction of rotation (unless otherwise stated).
- DO NOT turn engine via camshaft or other sprockets.
- Observe all tightening torques.
- Check diesel injection pump timing after belt replacement.

Removal

1. Remove:
   - ❏ Coolant expansion tank bolts. Move expansion tank to one side.
   - ❏ Upper and lower fan cowling.
   - ❏ Auxiliary drive belt.
   - ❏ Viscous fan coupling (if fitted).
   - ❏ Timing belt upper cover **9**.
   - ❏ Timing belt lower cover **10**.
   - ❏ Injection pump belt cover.
   - ❏ Cylinder head cover.
2. Turn crankshaft to TDC on No.1 cylinder **1**. Ensure injection pump sprocket mark aligned **2**.
3. Lock injection pump sprocket. Use tool No.2064 **3**.
4. Hold camshaft rear sprocket. Use tool No.3036 **4**. Remove sprocket bolt **13**.
5. Remove camshaft rear sprocket and injection pump belt.
6. Remove crankshaft damper bolt **11**.

7. Turn crankshaft to TDC on No.1 cylinder.
8. Fit setting bar No.2065A to rear of camshaft **5**. Centralise camshaft using feeler gauges.
9. Slacken tensioner nut **6**.
10. Remove crankshaft damper and timing belt together. **NOTE: Mark direction of rotation on belt with chalk if belt is to be reused.**

Installation

1. Check timing belt for damage or oil contamination. Replace if necessary.
2. Ensure crankshaft at TDC on No.1 cylinder **1**.
3. Ensure camshaft setting bar fitted correctly **5**.
4. Ensure locking pin located correctly in injection pump sprocket **3**.
5. Slacken camshaft sprocket bolt approximately one turn **12**.
6. Tap sprocket from behind to loosen it from taper.
7. Ensure crankshaft at TDC on No.1 cylinder **1**.
8. Fit crankshaft damper and timing belt together.
9. Fit timing belt in anti-clockwise direction. Ensure belt is taut between sprockets.
10. Tighten crankshaft damper bolt to 460 Nm **11**.
11. Install torque wrench to hexagon of tensioner **7**. Apply a torque of 5 Nm to tensioner pulley to pre-tension belt. Remove torque wrench.
12. Tighten tensioner nut to 20 Nm **6**.
13. Ensure timing marks aligned **1**.
14. Tighten camshaft sprocket bolt to 85 Nm **12**.
15. Remove camshaft setting bar **5**.
16. Turn crankshaft 90° clockwise. DO NOT allow crankshaft to spring back.
17. Slacken tensioner nut **6**. Allow tensioner to operate. Tighten nut to 20 Nm.
18. Turn crankshaft 1³/₄ turns clockwise to TDC on No.1 cylinder **1**.
19. Fit camshaft rear sprocket and injection pump belt.
20. Tighten bolt **13** until sprocket can just be turned by hand.
21. Slacken injection pump mounting plate bolts.
22. Attach tension gauge to belt at ▽ **8**. Tool No.210.
23. Move injection pump on mounting plate until tension gauge indicates 12-13 units.
24. Tighten injection pump mounting plate bolts.
25. Hold camshaft rear sprocket. Tighten bolt to 100 Nm **13**.
26. Remove locking pin from injection pump sprocket **3**.
27. Check injection pump timing.
28. Install components in reverse order of removal.

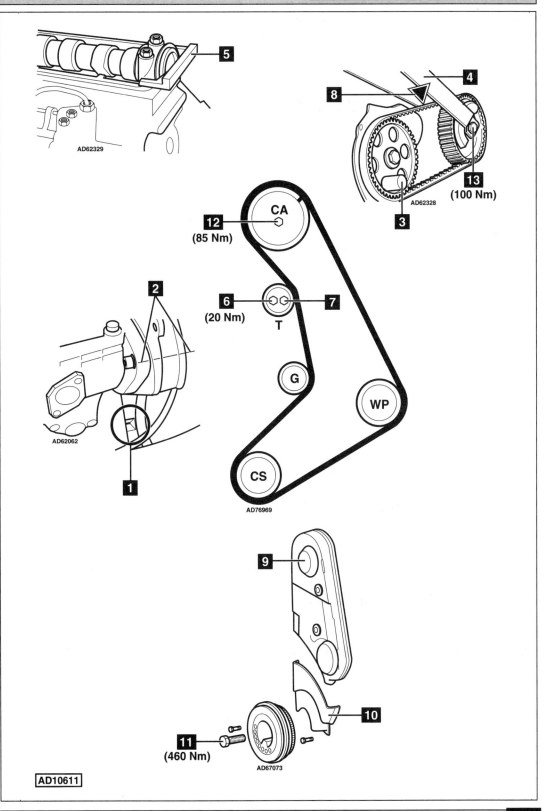

AD62329

5

4

8

13
(100 Nm)

3

AD62328

CA

12
(85 Nm)

2

6
(20 Nm)

T

7

G

WP

1

AD62062

CS

AD76969

9

10

11
(460 Nm)

AD67073

AD10611

VOLKSWAGEN

Model:	LT 2,5 SDI • LT 2,5 TDI
Year:	1996-03
Engine Code:	AGX, AHD, ANJ, APA

Replacement Interval Guide

Volkswagen recommend:

Engine code AGX →1999MY:
Check condition every 45,000 kilometres (27,961 miles) (timing belt).
Replacement every 135,000 kilometres (83,885 miles) – timing belt tensioner pulley must also be replaced (timing belt and injection pump belt).

Engine code AGX 2000MY→:
Check condition every 45,000 kilometres (27,961 miles) (timing belt).
Replacement every 90,000 kilometres (55,923 miles) – timing belt tensioner pulley must also be replaced (timing belt and injection pump belt).

Engine code AHD/ANJ/APA:
Check condition every 45,000 kilometres (27,961 miles) (timing belt).
Replacement every 135,000 kilometres (83,885 miles) – timing belt tensioner pulley must also be replaced (timing belt and injection pump belt).

NOTE: The vehicle manufacturer publishes this information only in kilometres. The conversion to miles is included for reference purposes only.
The previous use and service history of the vehicle must always be taken into account.
Refer to Timing Belt Replacement Intervals at the front of this manual.

Check For Engine Damage

CAUTION: This engine has been identified as an INTERFERENCE engine in which the possibility of valve-to-piston damage in the event of a timing belt failure is MOST LIKELY to occur.
A compression check of all cylinders should be performed before removing the cylinder head.

Repair Times – hrs

Check & adjust	2,20
Remove & install	4,20

Special Tools

- Crankshaft pulley holding tool – No.3419.
- Tensioner pulley spanner – No.3355.
- Sprocket holding tool – No.3036.
- Camshaft setting bar – No.2065A.
- Dial gauge adaptor – No.3313.

Special Precautions

- Disconnect battery earth lead.
- DO NOT turn crankshaft or camshaft when timing belt removed.
- Remove glow plugs to ease turning engine.
- Turn engine in normal direction of rotation (unless otherwise stated).
- DO NOT turn engine via camshaft or other sprockets.
- Observe all tightening torques.
- Check diesel injection pump timing after belt replacement.

Removal

1. Remove:
   - Auxiliary drive belt.
   - Viscous fan.
   - Viscous fan pulley.
   - Guide pulley from auxiliary drive belt tensioner bracket.
   - Viscous fan bracket.
   - Radiator cowling.
   - Centre console and inspection cover.
   - Injection pump belt cover.
   - Timing belt upper cover **1**.
   - Cylinder head cover.
2. Hold crankshaft pulley. Use tool No.3419. Slacken centre bolt **3**.
3. Turn crankshaft to TDC on No.1 cylinder. Ensure flywheel timing marks aligned **4**. Ensure injection pump and crankshaft pulley timing marks aligned **5** & **6**.
4. If not: Turn crankshaft one turn clockwise.
5. Hold camshaft rear sprocket. Use tool No.3036. Remove bolt **7**.
6. Remove injection pump belt tensioner **8**.
7. Remove camshaft rear sprocket and injection pump belt.
8. Fit setting bar No.2065A to rear of camshaft **9**. Centralise camshaft using feeler gauges.
9. Remove:
   - Crankshaft pulley bolts **10**.
   - Crankshaft pulley centre bolt **3**.
   - Crankshaft pulley **11**.
   - Timing belt lower cover **2**.
10. Slacken tensioner bolt **12**. Turn tensioner pulley away from belt. Use tool No.3355 **13**. Lightly tighten bolt.
11. Remove timing belt.

➡

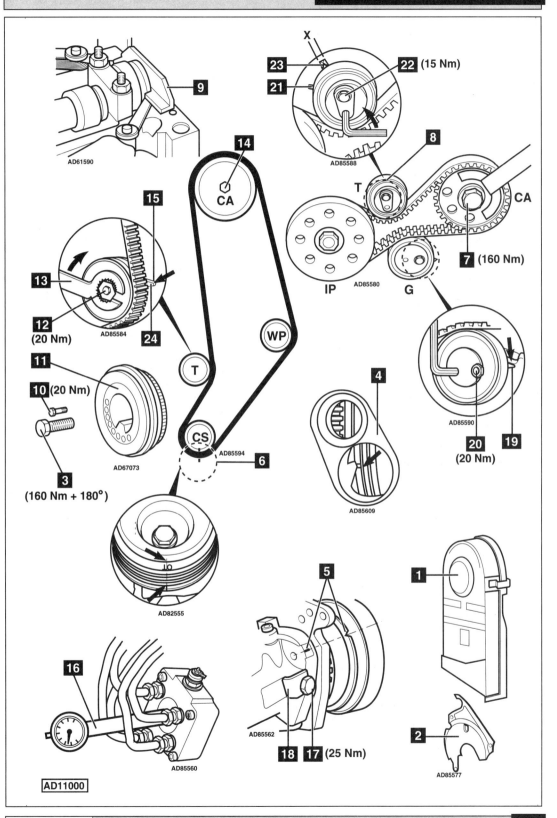

AD61590

9

X

23
21
22 (15 Nm)

AD85588

14
CA

8

T

CA

7 (160 Nm)

15

13
12 (20 Nm)
AD85584
24

IP
AD85580
G

AD85590

11
10 (20 Nm)

T

WP

4

20 (20 Nm)
19

3
(160 Nm + 180°)

AD67073

CS
AD85594
6

AD85609

AD82555

5

1

16

AD85562

AD85560

18 **17** (25 Nm)

2

AD85577

AD11000

Installation

1. Ensure crankshaft at TDC on No.1 cylinder ▪.
2. Slacken camshaft sprocket bolt 1/2 turn ▪. Tap sprocket gently to loosen it from taper.
   NOTE: Sprocket should turn freely on taper but not tilt.
3. Fit timing belt in anti-clockwise direction, starting at crankshaft sprocket. Ensure belt is taut between sprockets.
4. Slacken tensioner bolt ▪.
5. Turn tensioner pulley slowly clockwise until right edge of pointers aligned ▪ & ▪. Use tool No.3355 ▪.
6. If tensioner pulley turned too far: Turn fully anti-clockwise and repeat tensioning procedure.
   NOTE: To prevent tensioner damage, right edge of pointer ▪ should not go past right edge of pointer ▪ during adjustment.
7. Tighten tensioner bolt to 20 Nm ▪.
8. Ensure flywheel timing marks aligned ▪.
9. Hold camshaft sprocket. Use tool No.3036. Tighten bolt to specified torque ▪.
   NOTE: Check marking on camshaft sprocket bolt head for correct torque setting.
   8.8: 85 Nm.
   10.9: 100 Nm.
10. Remove camshaft setting bar ▪.
11. Install:
    ❑ Timing belt lower cover ▪.
    ❑ Crankshaft pulley ▪.
    NOTE: Fit new crankshaft pulley centre bolt ▪.
12. Lightly oil new centre bolt threads and contact surfaces.
13. Hold crankshaft pulley. Use tool No.3419. Tighten bolt to 160 Nm + 180° ▪.
14. Tighten crankshaft pulley bolts to 20 Nm (4 bolts) ▪.
15. Install dial gauge with adaptor No.3313 in injection pump ▪.
16. Ensure injection pump timing marks aligned ▪.
17. Slacken injection pump locking bolt ▪. Remove keeper plate ▪.
18. Set dial gauge to zero.
19. Turn injection pump sprocket slowly clockwise (against normal direction of rotation). Use tool No.3036.
20. If dial gauge reading decreases: Reset to zero once dial gauge movement stops.
    NOTE: If dial gauge reading increases: Turn injection pump sprocket anti-clockwise until dial gauge movement stops and timing marks are approximately aligned ▪.

21. Turn injection pump sprocket anti-clockwise (in normal direction of rotation) until dial gauge indicates the following:
    ❑ AHD: 0,55 mm.
    ❑ AGX →01/99: 0,35 mm.
    ❑ AGX 02/99→ & ANJ/APA: 0,55 mm.
22. Tighten injection pump locking bolt to 25 Nm ▪.
23. Ensure TDC marks aligned ▪.
24. Fit injection pump belt with camshaft rear sprocket.
25. Lightly tighten sprocket bolt ▪. Ensure sprocket can just be turned by hand.
26. Ensure guide pulley pointer aligned with cylinder head flange contour ▪.
27. If not: Slacken guide pulley nut ▪. Turn guide pulley until pointer aligned. Use Allen key. Tighten nut to 20 Nm ▪.
28. Fit tensioner ▪. Ensure tensioner tab engaged in cut-out ▪. Tighten bolt finger tight ▪.
29. Turn tensioner pulley anti-clockwise until pointers aligned ▪. Tighten bolt to 15 Nm ▪.
    NOTE: To prevent tensioner damage, front pointer should not go past rear pointer during adjustment.
30. Hold camshaft rear sprocket. Use tool No.3036. Tighten bolt to 160 Nm ▪.
31. Slacken injection pump locking bolt ▪. Insert keeper plate ▪. Tighten locking bolt to 25 Nm ▪.
32. Turn crankshaft two turns clockwise to TDC on No.1 cylinder.
33. Ensure right edge of timing belt tensioner pointers aligned ▪ & ▪. If not: Repeat tensioning procedures.
34. Check injection pump belt tensioner pointers are either aligned ▪, or front pointer is within dimension 'X'. If not: Repeat tensioning procedures.
35. Install components in reverse order of removal.

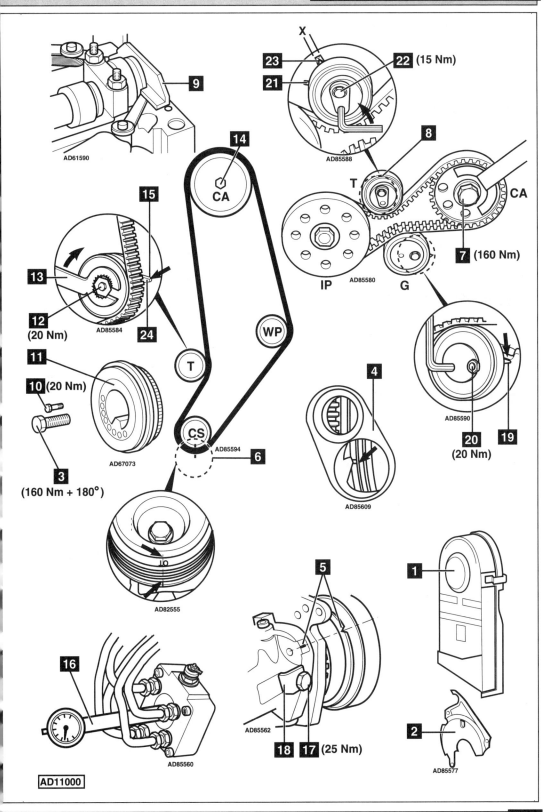

9

AD61590

X

23

22 (15 Nm)

21

AD85588

8

T

CA

7 (160 Nm)

IP   AD85580   G

14
CA

15

13

12
(20 Nm)   AD85584   **24**

11

10 (20 Nm)

T

WP

3
(160 Nm + 180°)

AD67073

CS   AD85594   **6**

AD82555

4

AD85609

AD85590

20
(20 Nm)   **19**

16

5

1

AD85562

AD85560

18   **17** (25 Nm)

2

AD85577

AD11000

Model:	340 • 440 • 460 • 480
Year:	1986-97
Engine Code:	B16, B18, B20, B172

Replacement Interval Guide

Volvo recommend:

340:
Replacement every 48,000 miles or 48 months, whichever occurs first.

440/460/480:
→1990: Replacement every 50,000 miles or 60 months, whichever occurs first.
1991→: Replacement every 60,000 miles or 72 months, whichever occurs first.
The previous use and service history of the vehicle must always be taken into account.
Refer to Timing Belt Replacement Intervals at the front of this manual.

Check For Engine Damage

CAUTION: This engine has been identified as an INTERFERENCE engine in which the possibility of valve-to-piston damage in the event of a timing belt failure is MOST LIKELY to occur.
A compression check of all cylinders should be performed before removing the cylinder head.

Repair Times – hrs

Remove & install:
340	1,00
440/460/480	1,40

Special Tools

■ Timing pin (8 mm diameter) – for setting crankshaft at TDC.
■ Tension gauge – Volvo No.5197.
■ M6 bolt – for adjusting tensioner.

Special Precautions

■ Disconnect battery earth lead.
■ DO NOT turn crankshaft or camshaft when timing belt removed.
■ Remove spark plugs to ease turning engine.
■ Turn engine in normal direction of rotation (unless otherwise stated).
■ DO NOT turn engine via camshaft or other sprockets.
■ Observe all tightening torques.

Removal

1. Remove:
   ❑ Auxiliary drive belt(s).
   ❑ Timing belt front cover **1**.
2. Turn crankshaft to TDC on No.1 cylinder. Ensure timing marks aligned **2** & **3**.
3. Remove blanking plug (located next to dipstick tube). Insert 8 mm timing pin **4**.
4. Rock crankshaft slightly to ensure timing pin located correctly.
5. Slacken tensioner nut to release tension on belt **5**.
6. Move tensioner away from belt. Lightly tighten nut.
7. Remove timing belt. Start at auxiliary shaft sprocket.

Installation

NOTE: From 1988 the positions of guide pulley and tensioner pulley were interchanged.

1. Ensure timing marks aligned **2** & **3**.
2. Rock crankshaft slightly to ensure timing pin located correctly.
3. Fit timing belt, starting at crankshaft sprocket. Align marks on belt with marks on sprockets.
4. Set tension gauge to 13 units.
5. Attach tension gauge to belt at ▽ **6**.
6. Slacken tensioner nut **5**.
7. →1988: Turn tensioner pulley until tension is correct, with mark on plunger flush with bottom of handle.
8. Tighten tensioner nut to 40 Nm **5**.
9. 1988→: Screw an M6 bolt into timing belt rear cover **7**.
10. Screw in M6 bolt until belt tension is correct.
11. Tighten tensioner nut to 40 Nm **5**.
12. Remove:
    ❑ M6 bolt.
    ❑ Tension gauge.
    All models:
13. Ensure timing marks aligned **2** & **3**.
14. Remove timing pin **4**.
15. Turn crankshaft two turns clockwise. Ensure timing marks aligned **2** & **3**.
16. Insert timing pin. Recheck belt tension.
17. Remove timing pin. Fit blanking plug.
18. Install components in reverse order of removal.

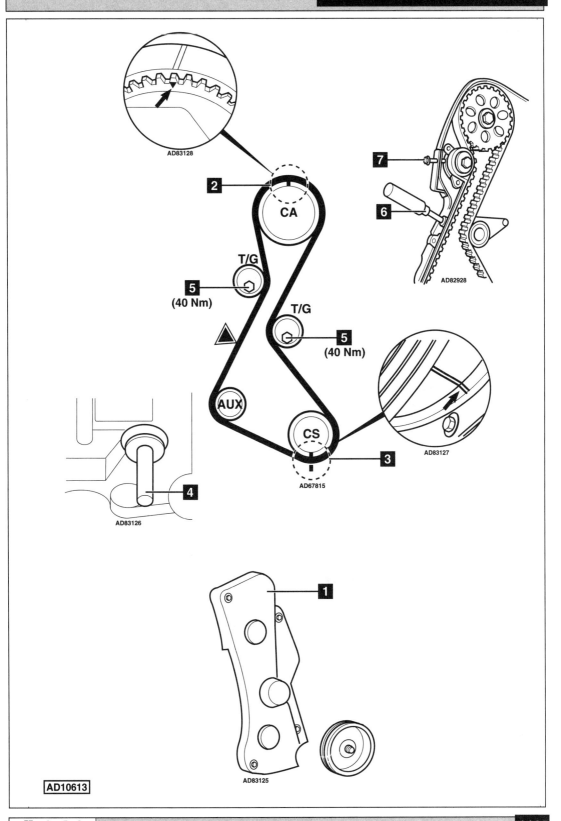

AD83128

7

6

AD82928

2

CA

T/G

5

(40 Nm)

T/G

5

(40 Nm)

AUX

AD83127

CS

3

AD67815

4

AD83126

1

AD83125

VOLVO

Model:	360 • 240 2,0/2,3 • 740/760/Turbo 2,0/2,3 • 940/960/Turbo 2,0/2,3
Year:	1983-97
Engine Code:	B200, B230

Replacement Interval Guide

Volvo recommend:

360:
Replacement every 48,000 miles or 48 months, whichever occurs first.
Check and adjust 6,000 miles after replacing belt.

240/740/760:
→1990: Replacement every 48,000 miles or 48 months, whichever occurs first.
→1990: Check and adjust 6,000 miles after replacing belt.
1991→: Replacement every 50,000 miles or 60 months, whichever occurs first.
1991→: Check and adjust 10,000 miles after replacing belt.

940/960:
Replacement every 50,000 miles or 60 months, whichever occurs first.
Check and adjust 10,000 miles after replacing belt.
The previous use and service history of the vehicle must always be taken into account.
Refer to Timing Belt Replacement Intervals at the front of this manual.

Check For Engine Damage

CAUTION: This engine has been identified as an INTERFERENCE engine in which the possibility of valve-to-piston damage in the event of a timing belt failure is MOST LIKELY to occur.
A compression check of all cylinders should be performed before removing the cylinder head.

Repair Times – hrs

Check & adjust	0,10
Remove & install:	
360	2,00
240/740/760 (except B200F/B230F)	1,60
240/740/760 (B200F/B230F)	1,40
940/960	1,60

Special Tools

■ Crankshaft pulley locking tool – Volvo No.5284.

Special Precautions

■ Disconnect battery earth lead.
■ DO NOT turn crankshaft or camshaft when timing belt removed.
■ Remove spark plugs to ease turning engine.
■ Turn engine in normal direction of rotation (unless otherwise stated).
■ DO NOT turn engine via camshaft or other sprockets.
■ Observe all tightening torques.

Removal

1. Remove:
   ❑ Auxiliary drive belts.
   ❑ Cooling fan cowling.
   ❑ Timing belt upper cover **1**.

2. Turn crankshaft to TDC on No.1 cylinder. Ensure timing marks aligned **2**, **3** & **5**.
3. Remove tensioner pulley nut and washer **6**.
4. Fit locking tool to crankshaft pulley. Tool No.5284. Hold tool in place with tensioner pulley nut.
5. Remove:
   ❑ Crankshaft pulley bolt **7**.
   ❑ Tensioner pulley nut.
   ❑ Locking tool.
6. Ensure timing marks aligned **2**, **3** & **5**.
7. Fit tensioner pulley nut.
8. Remove:
   ❑ Crankshaft pulley **8**.
   ❑ Timing belt lower cover **9**.
9. Stretch belt on non-tensioned side by pulling gently.
10. Lock tensioner pulley in depressed position with 3 mm drill bit.
11. Remove timing belt.

Installation

1. Ensure timing marks aligned **2**, **3** & **5**.
2. Fit timing belt to crankshaft sprocket.
   *NOTE: Ensure two lines on belt aligned with crankshaft timing marks by wrapping belt round crankshaft sprocket **4**.*
3. Stretch belt and fit to auxiliary shaft sprocket, camshaft sprocket and tensioner pulley.
4. Ensure timing marks aligned **2**, **3** & **5**.
5. Pull on belt to depress tensioner pulley. Remove 3 mm drill bit.
6. Install:
   ❑ Timing belt lower cover **9**.
   ❑ Crankshaft pulley **8**.
   ❑ Crankshaft pulley bolt **7**.
   ❑ Locking tool. Tool No.5284. Hold tool in place with tensioner pulley nut.
7. Tighten crankshaft pulley bolt **7**.
   240/740/760/940/960: 60 Nm + 60°.
   360: 250 Nm.
8. Remove:
   ❑ Tensioner pulley nut.
   ❑ Locking tool.
9. Fit tensioner pulley nut and washer **6**.
10. Turn crankshaft two turns clockwise.
11. Ensure timing marks aligned **2**, **3** & **5**.
12. Slacken tensioner pulley nut **6**. Allow tensioner to operate. Fully tighten nut.
13. Install components in reverse order of removal.
14. Run engine until it reaches normal operating temperature.
15. Remove rubber plug from timing belt cover **10**.
16. Turn crankshaft to TDC on No.1 cylinder.
17. Slacken tensioner pulley nut **6**. Allow tensioner to operate. Tighten nut.
18. Fit rubber plug **10**.

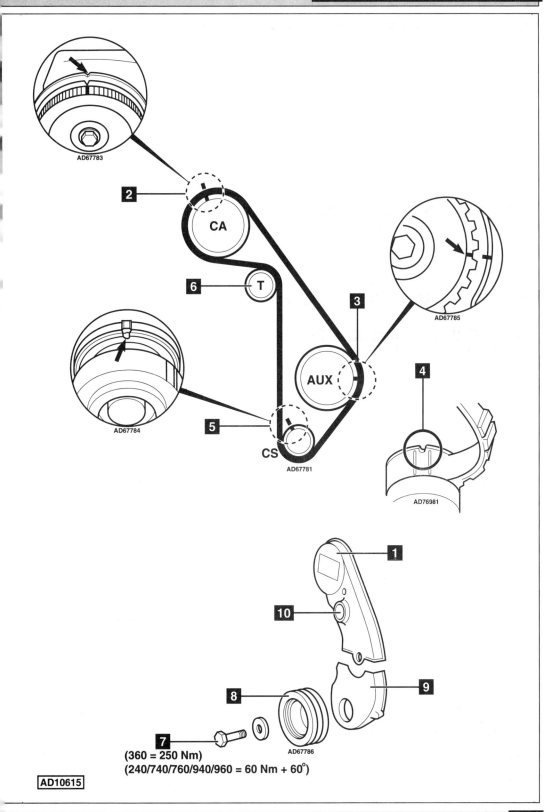

AD67783

CA

2

6 T

3

AD67785

AD67784

AUX

5

CS

AD67781

4

AD76981

1

10

8

9

7

AD67786

(360 = 250 Nm)
(240/740/760/940/960 = 60 Nm + 60°)

AD10615

VOLVO

Model:	S40/V40 1,6/1,8/2,0 • S40/V40 1,9/2,0 Turbo
Year:	1996-99
Engine Code:	Hydraulic tensioner B4164S, B4184S, B4194T, B4204S, B4204T

Replacement Interval Guide

Volvo recommend replacement every 80,000 miles or 96 months.
The previous use and service history of the vehicle must always be taken into account.
Refer to Timing Belt Replacement Intervals at the front of this manual.

Check For Engine Damage

CAUTION: This engine has been identified as an INTERFERENCE engine in which the possibility of valve-to-piston damage in the event of a timing belt failure is MOST LIKELY to occur.
A compression check of all cylinders should be performed before removing the cylinder head(s).

Repair Times – hrs

Remove & install	3,00

Special Tools

- ■ Tension gauge – Volvo No.998 8500.
- ■ Crankshaft pulley locking tool – Volvo No.999 5433.
- ■ Press for automatic tensioner unit – Volvo No.999 5456.

Special Precautions

- ■ Disconnect battery earth lead.
- ■ DO NOT turn crankshaft or camshaft when timing belt removed.
- ■ Remove spark plugs to ease turning engine.
- ■ Turn engine in normal direction of rotation (unless otherwise stated).
- ■ DO NOT turn engine via camshaft or other sprockets.
- ■ Observe all tightening torques.

Removal

1. Remove:
   - ❏ Engine top cover.
   - ❏ RH engine compartment cover (Turbo).
   - ❏ Auxiliary drive belt.
   - ❏ Auxiliary drive belt guide pulley.
   - ❏ Timing belt upper cover **1**.
2. Raise engine slightly.
3. Support engine.
4. Remove:
   - ❏ RH engine mounting.
5. Unclip AC pipe.
6. Raise and support front of vehicle.
7. Remove:
   - ❏ RH front wheel.
   - ❏ Timing belt lower cover **2**.
   - ❏ Crankshaft pulley bolts (4 bolts) **3**.
8. Lock crankshaft pulley. Use tool No.999 5433.
9. Remove crankshaft pulley centre nut **4**. If necessary: Lower engine slightly.
10. Remove:
    - ❏ Locking tool.
    - ❏ Crankshaft pulley **5**.

11. Turn crankshaft clockwise until timing marks aligned **6** & **7**.
12. Wait 5 minutes.
13. Attach tension gauge to belt at ▽ **8**. Tool No.998 8500 **9**.
14. Tension gauge should indicate 2,5-4,0 units.
15. If not: Replace automatic tensioner unit.
    NOTE: Tension value only applies to used belt.
16. Remove tension gauge **9**.
17. Remove upper bolt of automatic tensioner unit **10**.
18. Slacken lower bolt of automatic tensioner unit **11**.
19. Turn automatic tensioner unit slightly.
20. Remove:
    - ❏ Lower bolt of automatic tensioner unit **11**.
    - ❏ Automatic tensioner unit **12**.
    - ❏ Timing belt.

Installation

1. Check tensioner body for leakage or damage **12**. Replace if necessary.
2. Lubricate tensioner pulley lever pivot **13**.
3. Remove plastic washer from automatic tensioner unit **14**.
4. Slowly compress pushrod into tensioner body until holes aligned **15**. Use tool No.999 5456.
5. Retain pushrod with 2 mm diameter pin through hole in tensioner body **16**.
6. Install:
   - ❏ Automatic tensioner unit **12**.
   - ❏ Automatic tensioner unit bolts **10** & **11**. Tightening torque: 25 Nm.
7. Ensure timing marks aligned **6** & **7**.
8. Fit timing belt in anti-clockwise direction, starting at crankshaft sprocket. Ensure belt is taut between sprockets.
9. Remove pin from tensioner body to release pushrod **16**.
10. Apply firm thumb pressure to belt at ▽ **17**.
11. Apply firm thumb pressure to belt at ▽ **8**.
12. Fit new plastic washer to automatic tensioner unit **14**.
    NOTE: Countersunk side of washer should face upward.
13. Turn crankshaft two turns clockwise to setting position.
14. Ensure timing marks aligned **6** & **7**.
15. Install:
    - ❏ Crankshaft pulley **5**.
    - ❏ Crankshaft pulley bolts **3**. Tightening torque: 25 Nm + 30°.
    - ❏ Crankshaft pulley centre nut **4**. Tightening torque: 180 Nm.
16. Install components in reverse order of removal.

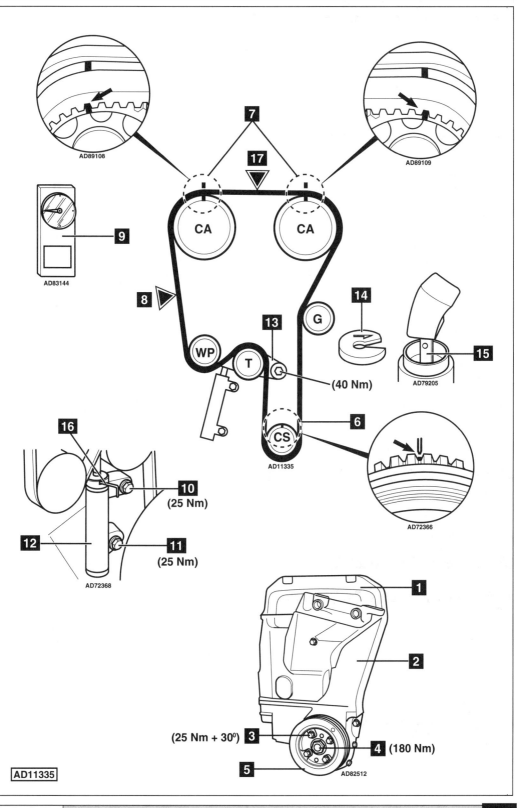

VOLVO

Model:	S40/V40 1,6/1,8/2,0 • S40/V40 1,9/2,0 Turbo
Year:	1999-03
Engine Code:	Mechanical tensioner B4164S/2, B4184S/2/3, B4184S9/10, B4194T/2, B4204S/2, B4204T/2/3, B4204T5

Replacement Interval Guide

Volvo recommend:

All models – 1999MY:
Replacement of timing belt and tensioner every 80,000 miles or 8 years.

1,6/1,8/2,0 – 2000-01MY:
Replacement of timing belt and tensioner at the following intervals:
90,000 miles/8 years.
186,000 miles/16 years.

1,9T/2,0T – 2000MY:
Replacement of timing belt and tensioner at the following intervals:
66,000 miles/6 years.
138,000 miles/12 years.
210,000 miles/18 years.

2,0T – 2001MY:
Replacement of timing belt and tensioner at the following intervals:
90,000 miles/8 years.
186,000 miles/16 years.

All models – 2002MY→:
Replacement of timing belt and tensioner every 96,000 miles or 8 years.
The previous use and service history of the vehicle must always be taken into account.
Refer to Timing Belt Replacement Intervals at the front of this manual.

Check For Engine Damage

CAUTION: This engine has been identified as an INTERFERENCE engine in which the possibility of valve-to-piston damage in the event of a timing belt failure is MOST LIKELY to occur.
A compression check of all cylinders should be performed before removing the cylinder head(s).

Repair Times – hrs

Remove & install	2,20

Special Tools

- ■ None required.

Removal

1. Remove:
   - ❏ Engine top cover.
   - ❏ RH engine compartment cover (Turbo).
   - ❏ RH headlamp cover (Turbo).
   - ❏ Auxiliary drive belt.
   - ❏ Timing belt upper cover **1**.
   - ❏ PAS hose bracket (2001→).
2. Raise engine slightly.
3. Support engine.
4. Remove RH engine mounting.
5. Raise and support front of vehicle.

6. Remove:
   - ❏ RH front wheel.
   - ❏ RH lower splash guard.
   - ❏ Engine undershield.
   - ❏ Auxiliary drive belt tensioner **2** (if required).
   - ❏ Timing belt front lower cover **3**.
   - ❏ Engine mounting bracket **4**.
7. Turn crankshaft clockwise until timing marks aligned **5** & **6**.
8. Turn crankshaft a further ¼ turn clockwise, then anti-clockwise until timing marks aligned **5** & **6**.
9. Remove:
   - ❏ Timing belt rear upper cover.
   - ❏ Tensioner bolt **7**.
   - ❏ Tensioner pulley.
   - ❏ Timing belt rear lower cover **8**.
   - ❏ Timing belt.

Installation

1. Check guide pulley. Replace if necessary.
2. Fit new tensioner pulley with Allen key hole at 10 o'clock position.
3. Tighten tensioner bolt to 3 Nm **7**.
4. Ensure timing marks aligned **5** & **6**.
5. Fit timing belt in anti-clockwise direction, starting at crankshaft sprocket. Ensure belt is taut between sprockets.
6. Fit timing belt rear upper cover.
7. Turn tensioner pulley anti-clockwise until pointer **9** passes alignment marks **10**.
   Use Allen key (6 mm) **11**.
8. Turn tensioner pulley slowly clockwise until pointer is aligned according to engine temperature:
   - ❏ -20°C – left of alignment marks **12**.
   - ❏ 20°C – centre of alignment marks **13**.
   - ❏ 50°C – right of alignment marks **14**.
9. Tighten tensioner bolt to 20 Nm **7**.
10. Press down on belt and ensure tensioner pointer moves freely **9**.
11. Turn crankshaft two turns clockwise.
12. Ensure timing marks aligned **5** & **6**.
13. Ensure tensioner pointer correctly aligned:
    - ❏ -20°C – left of alignment marks **12**.
    - ❏ 20°C – centre of alignment marks **13**.
    - ❏ 50°C – right of alignment marks **14**.
14. If not: Repeat tensioning procedure.
15. Install components in reverse order of removal.

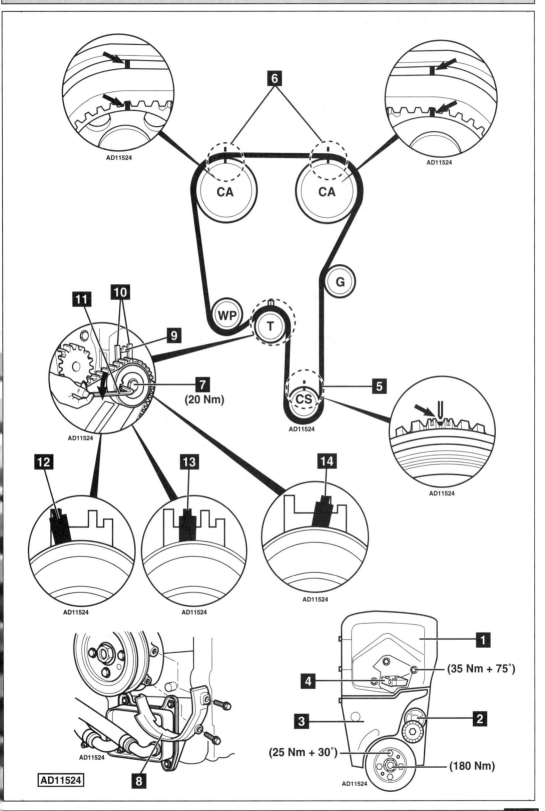

AD11524

6

AD11524

AD11524

CA

CA

G

11

10

9

WP

T

7
(20 Nm)

AD11524

5

CS

AD11524

AD11524

12

13

14

AD11524

AD11524

AD11524

1

4 — (35 Nm + 75°)

3

2

(25 Nm + 30°)

(180 Nm)

AD11524

AD11524

8

VOLVO

Model:	S40/V40 1,8 GDI
Year:	1998-02
Engine Code:	B4184SJ, B4184SM

Replacement Interval Guide

Volvo recommend replacement every
60,000 miles or 6 years, whichever occurs first.
*The previous use and service history of the vehicle
must always be taken into account.
Refer to Timing Belt Replacement Intervals at the front
of this manual.*

Check For Engine Damage

*CAUTION: This engine has been identified as an
INTERFERENCE engine in which the possibility of
valve-to-piston damage in the event of a timing belt
failure is MOST LIKELY to occur.
A compression check of all cylinders should be
performed before removing the cylinder head.*

Repair Times – hrs

Remove & install	3,10

Special Tools

- Camshaft sprocket locking tool –
  Volvo No.999 5714.
- Crankshaft pulley holding tool – Volvo No.999 5705.
- Tensioner pulley tool – Volvo No.999 5709.

Special Precautions

- Disconnect battery earth lead.
- DO NOT turn crankshaft or camshaft when timing belt removed.
- Remove spark plugs to ease turning engine.
- Turn engine in normal direction of rotation (unless otherwise stated).
- DO NOT turn engine via camshaft or other sprockets.
- Observe all tightening torques.

Removal

1. Raise and support front of vehicle.
2. Remove:
   ❑ Engine upper cover.
3. Disconnect crankshaft position (CKP) sensor multi-plug.
4. Remove:
   ❑ Engine undershields.
   ❑ Auxiliary drive belts.
   ❑ Timing belt upper cover **1**.
   ❑ Alternator bracket.
   ❑ PAS hose bracket and clip.
   ❑ Crankshaft pulley bolt **2**.
     Use tool No.999 5705.
   ❑ Crankshaft pulley **3**.
   ❑ Timing belt lower cover **4**.
   ❑ Coolant pipe support bolt (B4184SM).
   ❑ Crankshaft sprocket guide washer **5**.
   ❑ PAS pump support bracket.
5. Support engine – remove:
   ❑ RH engine mounting.
   ❑ Engine mounting bracket **6**.
6. Turn crankshaft clockwise until timing marks aligned **7** & **8**.
7. Lock camshaft sprockets.
   Use tool No.999 5714 **9**.
8. Slacken tensioner pulley bolt **10**.
9. Remove:
   ❑ Automatic tensioner unit bolts **11**.
   ❑ Automatic tensioner unit **12**.
   ❑ Timing belt.
   **NOTE: Mark direction of rotation on belt with
   chalk if belt is to be reused.**

Installation

1. Check tensioner body for leakage or damage **12**.
2. Ensure extended length of pushrod is 10,5-11,5 mm **13**.
3. Push pushrod of automatic tensioner unit against a firm surface using a force of 100-200 N. Pushrod should move less than 1,0 mm.
4. Slowly compress pushrod into tensioner body **12** until holes aligned. Use vice. Retain pushrod with 2 mm Allen key through hole in tensioner body **14**.
5. Install automatic tensioner unit to cylinder block. Tighten bolts to 13 Nm **11**.
6. Ensure timing marks aligned **7** & **8**.

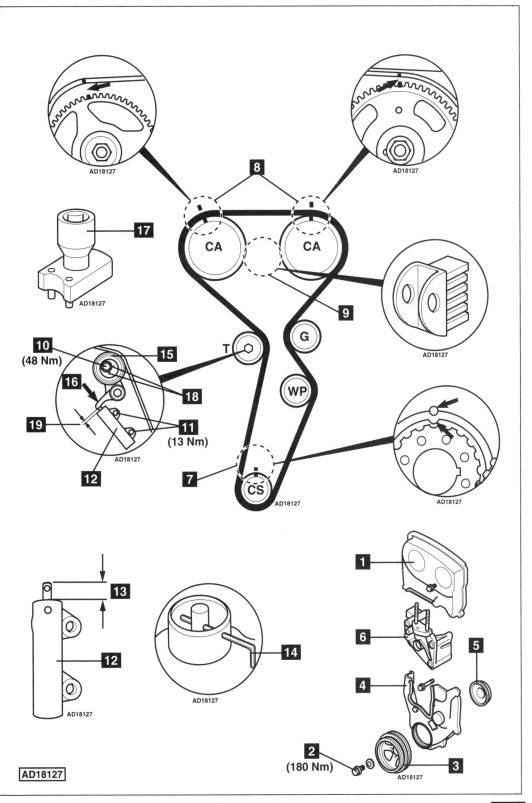

7. Ensure locking tool located correctly in camshaft sprockets **9**.

8. Turn crankshaft sprocket ½ tooth width anti-clockwise.

9. Fit timing belt in anti-clockwise direction, starting at crankshaft sprocket.

10. Remove locking tool from camshaft sprockets **9**.

11. Turn tensioner pulley **15** firmly anti-clockwise to temporarily tension belt. Tighten bolt to 48 Nm **10**.

12. Turn crankshaft ¼ turn anti-clockwise.

13. Press tensioner pulley arm towards automatic tensioner unit **16**.

14. Turn tensioner pulley anti-clockwise until timing belt correctly located on all sprockets.

15. Slacken tensioner pulley bolt **10**.

16. Fit tool **17** to holes in tensioner pulley **18**. Tool No.999 5709.

17. Apply anti-clockwise torque of 2,6 Nm to tensioner pulley **15**.

18. Tighten tensioner pulley bolt to 48 Nm **10**.
   NOTE: Ensure tensioner does not move when tightening bolt.

19. Turn crankshaft slowly 2¼ turns clockwise.

20. Ensure timing marks aligned **7** & **8**.

21. Remove Allen key from tensioner body to release pushrod.
   NOTE: Ensure Allen key can be removed easily.

22. Turn crankshaft slowly two turns clockwise.

23. Wait 2 minutes minimum.

24. Ensure Allen key can be inserted and removed easily **14**.
   *NOTE: If Allen key cannot be removed easily, ensure extended length of pushrod is 3,8-4,5 mm **19**.*

25. If not: Repeat tensioning procedure.

26. Install components in reverse order of removal.

27. Oil crankshaft pulley bolt threads and washer face.

28. Tighten crankshaft pulley bolt to 180 Nm **2**.

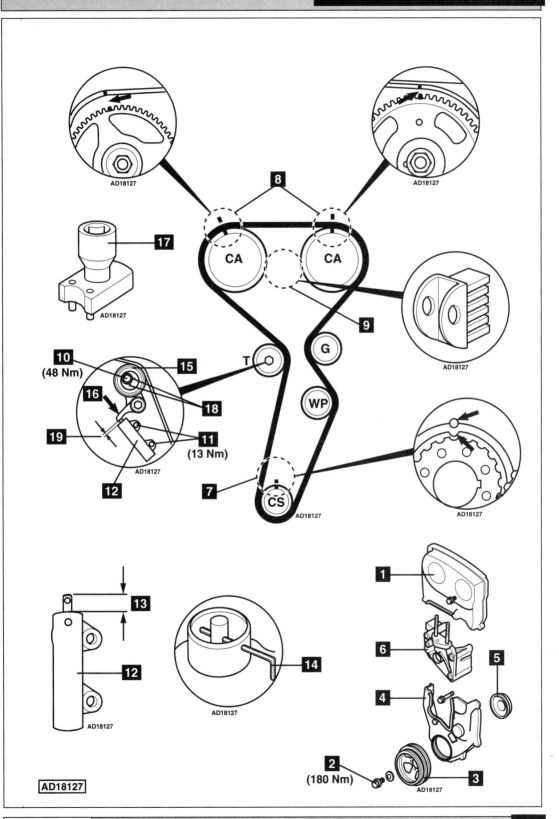

AD18127

VOLVO

Model:	**S60 2,0/2,3 Turbo • S60 2,4/Turbo • S70/V70/C70 2,0 Turbo** **S70/V70 2,3 20V • S70/V70/C70 2,3 Turbo • S70/V70/C70 2,4/Turbo** **S70/V70/C70 2,5 20V • S70/V70/C70 2,5 Turbo • S80 2,0/2,3/2,4 Turbo** **S80 2,4/2,9 • S80 2,8/2,9 Turbo • V70 XC (Cross Country) 2,0/2,3/2,4 Turbo**
Year:	**1997-03**
Engine Code:	**B5204T3/4/5, B5234FS, B5234T3, B5234T7/8, B5244S/2, B5244SG/2,** **B5244T, B5244T2/3, B5254FS, B5254T, B6284T, B6294S/2, B6294T**

Replacement Interval Guide

Volvo recommend:
→98MY: Replacement of timing belt and tensioner every 80,000 miles or 8 years.
99MY: Replacement of timing belt and tensioner every 100,000 miles or 10 years.
00-01MY: Replacement of timing belt and tensioner at 90,000 miles or 8 years and then at 186,000 miles or 8 years.
02MY→: Replacement of timing belt and tensioner every 96,000 miles or 8 years.
The previous use and service history of the vehicle must always be taken into account.
Refer to Timing Belt Replacement Intervals at the front of this manual.

Check For Engine Damage

CAUTION: This engine has been identified as an INTERFERENCE engine in which the possibility of valve-to-piston damage in the event of a timing belt failure is MOST LIKELY to occur.
A compression check of all cylinders should be performed before removing the cylinder head(s).

Repair Times – hrs

Remove & install:

S70/V70/XC →2000 MY	1,30
C70	1,30
V70/XC 2000 MY→	1,50
S60	1,50
S80 – 2,0/2,3/2,4	1,60
S80 – 2,8	2,40
S80 – 2,9	1,90

Special Tools

- Crankshaft pulley holding tool – Volvo No.999 5433.

Special Precautions

- Disconnect battery earth lead.
- DO NOT turn crankshaft or camshaft when timing belt removed.
- Remove spark plugs to ease turning engine.
- Turn engine in normal direction of rotation (unless otherwise stated).
- DO NOT turn engine via camshaft or other sprockets.
- Observe all tightening torques.

Removal

1. Raise and support front of vehicle.
2. Remove:
   - Strut brace (2000 MY→).
   - Engine steady bar (S80 2,8/2,9).
   - Air intake hoses (S80 2,8/2,9 Turbo).
   - PAS reservoir. DO NOT disconnect hose(s).
   - Coolant expansion tank. DO NOT disconnect hose(s) (except S80 2,8/2,9).
   - Coolant expansion tank. Disconnect hose(s) (S80 2,8/2,9).
   - Auxiliary drive belt.
   - Timing belt upper cover **1**.
   - Timing belt front cover **2**.
   - RH front wheel.
   - RH lower splash guard.
3. Fit timing belt upper cover **1**.
4. Turn crankshaft clockwise until timing marks aligned **3** & **4**.
5. Turn crankshaft a further ¼ turn clockwise, then anti-clockwise until timing marks aligned **3** & **4**.
6. Remove:
   - Timing belt upper cover **1**.
   - Crankshaft pulley bolts **5**.

Except S80 2,8/2,9:

7. Fit crankshaft pulley holding tool **6**. Tool No.999 5433.
8. Remove crankshaft pulley centre bolt **7**.

All models:

9. Remove:
   - Crankshaft pulley **8**.
   - Auxiliary drive belt cover (S80 2,8/2,9).
   - Tensioner bolt **9**.
   - Tensioner pulley.
   - Timing belt.

Autodata

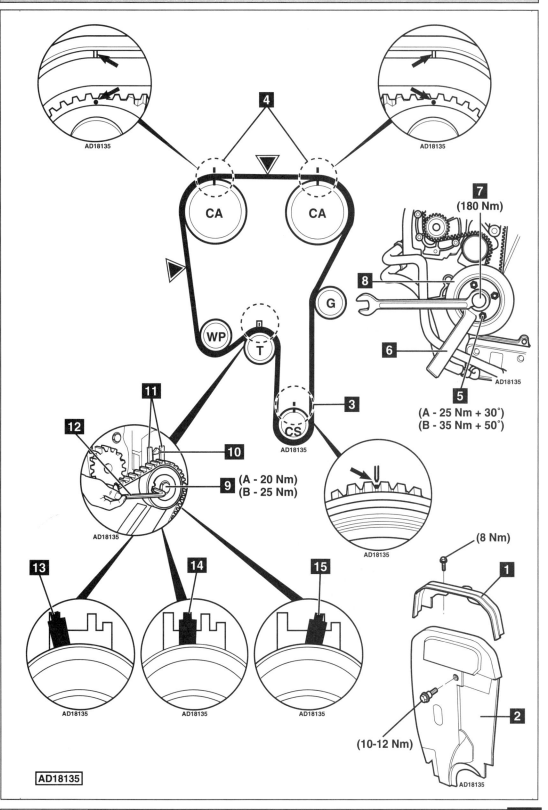

AD18135

Installation

1. Check guide pulley. Replace if necessary.
2. Fit new tensioner pulley with Allen key hole at 10 o'clock position.
3. Tighten tensioner bolt to 5 Nm .
4. Fit timing belt upper cover .
5. Ensure timing marks aligned ▇ & ▇.
6. Remove timing belt upper cover .
7. Fit timing belt in anti-clockwise direction, starting at crankshaft sprocket. Ensure belt is taut between sprockets.

Except S80 2,8/2,9:

8. Fit crankshaft pulley and centre bolt ▇ & ▇.
9. Tighten crankshaft pulley centre bolt to 180 Nm. Use tool No.999 5433 ▇.

All models:

10. Turn crankshaft sprocket slightly in clockwise direction to tension timing belt between inlet camshaft sprocket and crankshaft sprocket.
11. Turn tensioner pulley anti-clockwise until pointer ▇ passes alignment marks ▇. Use Allen key (6 mm) ▇.
12. Turn tensioner pulley slowly clockwise until pointer is aligned according to engine temperature:
    ❏ -20°C – left of alignment marks ▇.
    ❏ 20°C – centre of alignment marks ▇.
    ❏ 50°C – right of alignment marks ▇.
13. Tighten tensioner bolt:
    ❏ (A) Except S80 2,8/2,9 – 20 Nm ▇.
    ❏ (B) S80 2,8/2,9 – 25 Nm ▇.
14. Press down on belt at two positions ▽. Ensure tensioner pointer moves freely ▇.
15. Fit timing belt upper cover ▇.
16. Turn crankshaft two turns clockwise.
17. Ensure timing marks aligned ▇ & ▇.
18. Ensure tensioner pointer correctly aligned:
    ❏ -20°C – left of alignment marks ▇.
    ❏ 20°C – centre of alignment marks ▇.
    ❏ 50°C – right of alignment marks ▇.
19. If not: Repeat tensioning procedure.
20. S80 2,8/2,9: Fit crankshaft pulley ▇.
21. Fit crankshaft pulley bolts ▇. Use new bolts.
22. Tighten crankshaft pulley bolts ▇:
    ❏ (A) Except S80 2,8/2,9 – 25 Nm + 30°.
    ❏ (B) S80 2,8/2,9 – 35 Nm + 50°.
23. Install components in reverse order of removal.

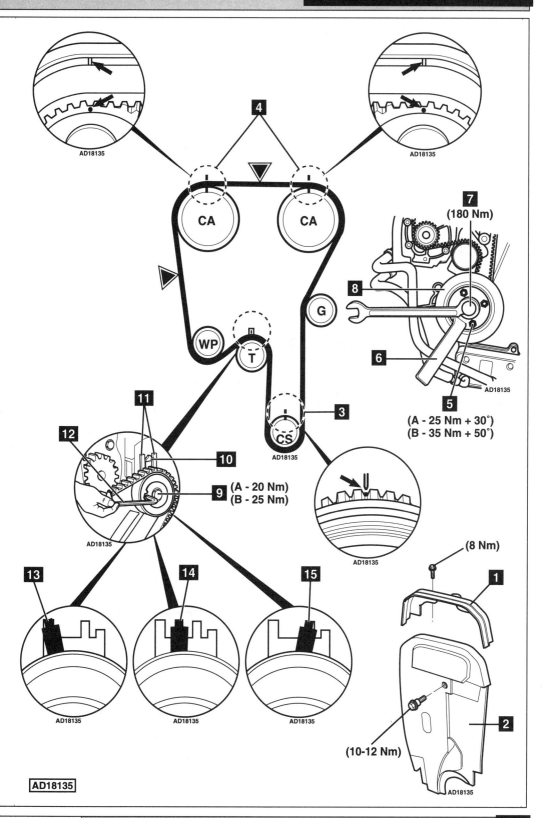

VOLVO

Model:	740 16V • 940 16V
Year:	1990-95
Engine Code:	B204FT, B234F

Replacement Interval Guide

Volvo recommend:
→1990: Replacement every 48,000 miles or 48 months, whichever occurs first.
1991→: Replacement every 50,000 miles or 60 months, whichever occurs first.
The previous use and service history of the vehicle must always be taken into account.
Refer to Timing Belt Replacement Intervals at the front of this manual.

Check For Engine Damage

CAUTION: This engine has been identified as an INTERFERENCE engine in which the possibility of valve-to-piston damage in the event of a timing belt failure is MOST LIKELY to occur.
A compression check of all cylinders should be performed before removing the cylinder head.

Repair Times – hrs

Remove & install	1,90
Balancer shaft belt	+0,50

Special Tools

- Sprocket locking plate – Volvo No.5416.
- Tension gauge (balancer shaft belt) – Volvo No.998 8500.
- 2 mm diameter pin – for locking automatic tensioner unit.

Special Precautions

- Disconnect battery earth lead.
- DO NOT turn crankshaft or camshaft when timing belt removed.
- Remove spark plugs to ease turning engine.
- Turn engine in normal direction of rotation (unless otherwise stated).
- DO NOT turn engine via camshaft or other sprockets.
- Observe all tightening torques.

Removal

Timing Belt

1. Remove:
   - Auxiliary drive belts.
   - Cooling fan cowling.
   - Timing belt upper cover **1**.
   - Timing belt centre cover **2**.
   - Timing belt lower cover **3**.
   - Engine undershield.
2. Turn crankshaft to TDC on No.1 cylinder. Ensure timing marks aligned **4**, **5** & **6**.
3. Lock camshaft sprockets. Use tool No.5416.
4. Remove bolts from automatic tensioner unit **11** & **12**.
5. Remove:
   - Automatic tensioner unit.
   - Timing belt.

Installation

Timing Belt

1. Check automatic tensioner unit as follows:
   - Check tensioner body for leakage or damage.
   - Put automatic tensioner unit between vice jaws.
   - Turn vice handle 20° every 5 secs. Resistance should be felt.
   - Insert 2 mm diameter pin to lock pushrod fully depressed.
   NOTE: If any signs of leakage, lack of resistance or if pushrod cannot be depressed, automatic tensioner unit must be replaced.
2. Ensure timing marks aligned **4**, **5** & **6**.
3. Fit timing belt to crankshaft sprocket **6**.
   NOTE: Ensure two lines on belt align with crankshaft timing marks.
4. Fit belt around oil pump sprocket and upper guide pulley (G2).
5. Fit timing belt to camshaft sprockets. Ensure marks on belt aligned with marks on sprockets.
6. Fit timing belt around lower guide pulley (G1), then ease around tensioner pulley.
7. Ensure timing marks aligned **4**, **5** & **6**.
8. Install automatic tensioner unit. Tighten upper bolt to 20 Nm **12**. Tighten lower bolt to 50 Nm **11**.
9. Remove 2 mm diameter pin from tensioner body.

Autodata

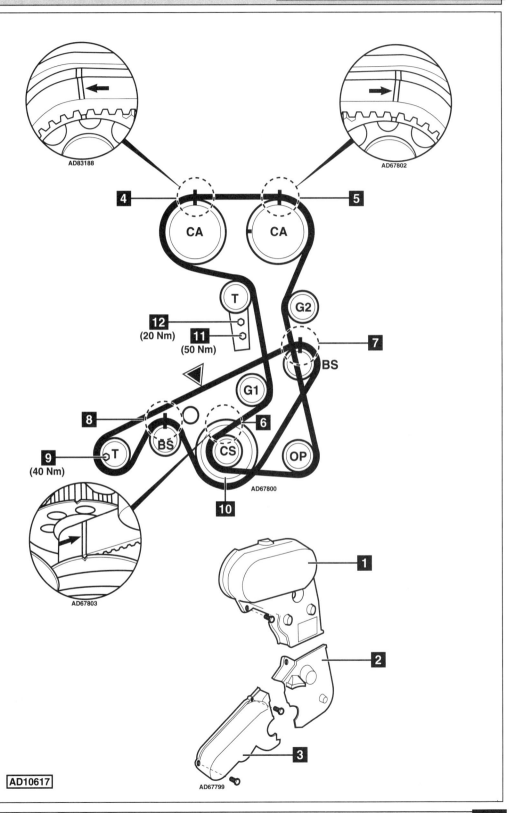

AD83188

AD67802

4

5

CA

CA

T

G2

12
(20 Nm)

11
(50 Nm)

7

BS

G1

8

6

9
(40 Nm)

T

BS

CS

OP

AD67800

10

AD67803

1

2

3

AD67799

AD10617

10. Remove locking tool.
11. Turn crankshaft two turns clockwise.
12. Ensure timing marks aligned **4**, **5** & **6**.
13. Install components in reverse order of removal.

Removal

Balancer Shaft Belt

1. Remove timing belt as described previously.
2. Ensure all timing marks aligned **4**, **5**, **6**, **7** & **8**.
3. Slacken balancer shaft belt tensioner sprocket locknut to release tension on belt **9**.
4. Remove balancer shaft belt.

Installation

Balancer Shaft Belt

1. Ensure all timing marks aligned **4**, **5**, **6**, **7** & **8**.
2. Carefully ease belt under crankshaft sprocket. Ensure blue spot marking on belt is directly opposite crankshaft TDC mark **10**.
3. Fit belt to upper balancer shaft sprocket **7** with yellow or white dot aligning with mark on sprocket.
4. Fit belt to lower balancer shaft sprocket **8** aligning white or yellow mark on belt with mark on sprocket.
5. Fit belt round tensioner sprocket **9**.
6. Ensure all timing marks aligned **4**, **5**, **6**, **7** & **8**.
7. Turn tensioner sprocket clockwise to tension belt. Use Allen key. Lightly tighten locknut **9**.
8. Turn crankshaft a small amount in each direction to ensure belt properly engaged in sprockets.
9. Ensure all timing marks aligned **4**, **5**, **6**, **7** & **8**.
10. Attach tension gauge to belt at ▽. Tool No.998 8500. Tension gauge should indicate 3,6-4,0 units (cold).
11. If not: Repeat operations 6-10.
12. Tighten tensioner sprocket locknut to 40 Nm **9**.
13. Fit timing belt as described previously.

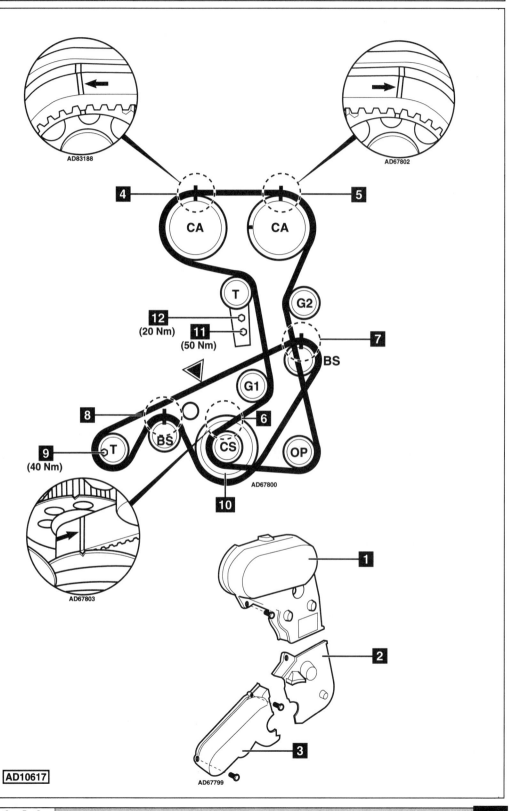

AD83188

AD67802

4

5

CA

CA

T

G2

12
(20 Nm)

11
(50 Nm)

7

BS

G1

8

BS

6

9
(40 Nm)

T

CS

OP

AD67800

AD67803

10

1

2

3

AD67799

AD10617

VOLVO

Model:	850 2,0/2,5 10V/20V • 850 2,0 Turbo • 850 2,3/2,5 Turbo • 850 R 960 2,5/3,0 24V • S70/V70 2,0 10V • S70/V70/C70 2,0 Turbo S70/V70 2,3 20V • S70/V70/C70 2,3 Turbo • S70/V70/C70 2,5 10V/20V S70/V70/C70 2,5 Turbo • S90/V90 3,0 24V • V70 XC (Cross Country) 2,5 Turbo
Year:	1990-98
Engine Code:	**Hydraulic tensioner** B5202S/FS, B5204S/FS, B5204T/FT, B5204T2/3, B5234FS, B5234T/FT, B5234T2/3, B5234T4/5, B5234T6/7, B5252S/FS, B5254S/FS, B5254T, B6254/F, B6304F/FS, B6304FS2, B6304G/GS

Replacement Interval Guide

Volvo recommend replacement as follows:
850/S70/V70/C70:
→93MY (21 mm belt) - every 50,000 miles or 60 months, whichever occurs first.
94MY→ (23 mm belt) - every 80,000 miles or 96 months, whichever occurs first.

960:
→93MY (21 mm belt) - every 30,000 miles or 36 months, whichever occurs first.
94MY (23 mm belt) - every 50,000 miles or 60 months, whichever occurs first.
95MY→ (28 mm belt) - every 80,000 miles or 96 months, whichever occurs first.

S90/V90:
Every 80,000 miles or 96 months, whichever occurs first.
The previous use and service history of the vehicle must always be taken into account.
Refer to Timing Belt Replacement Intervals at the front of this manual.

Check For Engine Damage

CAUTION: This engine has been identified as an INTERFERENCE engine in which the possibility of valve-to-piston damage in the event of a timing belt failure is MOST LIKELY to occur.
A compression check of all cylinders should be performed before removing the cylinder head.

Repair Times – hrs

Remove & install:	
850/S70/V70/C70	1,60
AC	+0,10
960/S90/V90	1,40

Special Tools

■ Tension gauge – Volvo No.998 8500.

Removal

1. Raise and support front of vehicle.
2. 850/S90/V90 – remove:
   ❏ Ignition coil cover.
   ❏ Fuel line clips.
   ❏ Coolant expansion tank. DO NOT disconnect hoses.
3. 960/S90/V90 – remove:
   ❏ Front engine splashguard.

4. All models – remove:
   ❏ Auxiliary drive belt.
   ❏ Timing belt cover **1**.
5. Turn crankshaft clockwise until timing marks aligned **2** & **3**.
6. Remove upper bolt of automatic tensioner unit **4**.
7. Remove lower bolt of automatic tensioner unit **5**. Twist tensioner body to free pushrod.
8. Remove:
   ❏ Automatic tensioner unit.
   ❏ Crankshaft pulley cover.
   ❏ Timing belt.

Installation

*NOTE: Whenever timing belt is replaced, tensioner pulley lever pivot **8** must be lubricated. If tensioner body has been leaking or pushrod has no resistance or is seized, renew automatic tensioner unit.*
*NOTE: A restricting washer (Part No.1271851-6 for engine codes B5204S, B5254S & B5234 and Part No.1275047-7 for engine codes B5202S & B5252S) should be fitted (when engine is cold) to automatic tensioner pushrod **9**, countersunk side upwards and centred on automatic tensioner body. This washer should be replaced whenever a new belt is fitted.*

1. Check timing belt guide pulley. Replace if necessary.
2. Slowly compress pushrod into tensioner body until holes aligned. Retain pushrod with 2 mm diameter pin through hole in tensioner body **6**.
3. Ensure timing marks aligned **2** & **3**.
4. Fit timing belt in anti-clockwise direction, starting at crankshaft sprocket.
5. Install automatic tensioner unit. Tighten bolts to 25 Nm **4** & **5**.
6. Remove 2 mm diameter pin from tensioner body to release pushrod **6**.
7. Turn crankshaft two turns clockwise until timing marks aligned **2** & **3**. Wait 5 minutes.
8. Attach tension gauge to belt at ▽. Tool No.998 8500 **7**.
9. Check tension gauge reading. 21 mm belts: 3,5-4,6 units. 23 mm belts: 2,5-4,0 units.
   NOTE: B5252: 2,7-4,2 units.
   B6304GS: 2,5-3,5 units.
10. If tension not as specified: Replace automatic tensioner unit.
11. Install components in reverse order of removal.

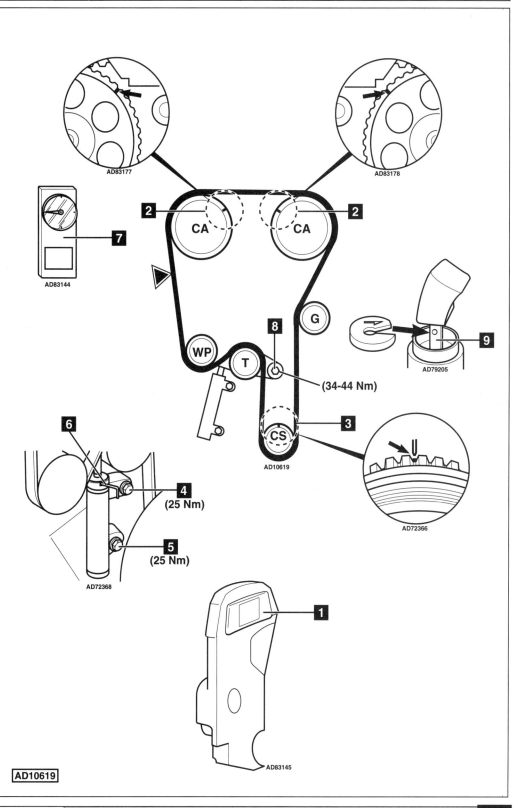

AD83177

AD83178

2

CA

CA

2

7

AD83144

G

8

9

AD79205

WP

T

(34-44 Nm)

6

CS

3

AD10619

4
(25 Nm)

AD72366

5
(25 Nm)

AD72368

1

AD83145

VOLVO

Model:	S70/V70 2,0 10V • S70/V70/C70 2,0 Turbo • S70/V70 2,3 20V
	S70/V70/C70 2,3 Turbo • S70/V70/C70 2,5 10V/20V
	S70/V70/C70 2,5 Turbo • V70 XC (Cross Country) 2,5 Turbo
Year:	1998-99
Engine Code:	**Mechanical tensioner**
	B5202FS, B5204T2/3, B5234FS, B5234T2/3, B5234T4, B5234T6/7, B5252FS, B5254FS, B5254T

Replacement Interval Guide

Volvo recommend:
98MY: Replacement of timing belt and tensioner every 80,000 miles or 8 years.
99MY: Replacement of timing belt and tensioner every 100,000 miles or 10 years.
The previous use and service history of the vehicle must always be taken into account.
Refer to Timing Belt Replacement Intervals at the front of this manual.

Check For Engine Damage

CAUTION: This engine has been identified as an INTERFERENCE engine in which the possibility of valve-to-piston damage in the event of a timing belt failure is MOST LIKELY to occur.
A compression check of all cylinders should be performed before removing the cylinder head(s).

Repair Times – hrs

Remove & install:

→98MY	1,60
AC	+0,10
99MY→	1,30

Special Tools

■ None required.

Special Precautions

■ Disconnect battery earth lead.
■ DO NOT turn crankshaft or camshaft when timing belt removed.
■ Remove spark plugs to ease turning engine.
■ Turn engine in normal direction of rotation (unless otherwise stated).
■ DO NOT turn engine via camshaft or other sprockets.
■ Observe all tightening torques.

Removal

1. Raise and support front of vehicle.
2. Remove:
   ❑ Ignition coil cover.
   ❑ Fuel line clips.
   ❑ Coolant expansion tank. DO NOT disconnect hoses.
   ❑ Auxiliary drive belt.
   ❑ Timing belt cover **1**.

3. Turn crankshaft clockwise until timing marks aligned **2** & **3**.
4. Remove:
   ❑ Tensioner bolt **4**.
   ❑ Tensioner.
   ❑ Crankshaft pulley cover.
   ❑ Timing belt.

Installation

1. Check timing belt guide pulley. Replace if necessary.
2. Fit new tensioner pulley with Allen key hole at 10 o'clock position.
3. Tighten tensioner bolt to 5 Nm **4**.
4. Ensure timing marks aligned **2** & **3**.
5. Fit timing belt in anti-clockwise direction, starting at crankshaft sprocket. Ensure belt is taut between sprockets.
6. Turn tensioner pulley anti-clockwise until pointer **6** reaches stop to right of alignment mark. Use Allen key (6 mm) **5**.
7. Turn tensioner pulley slowly clockwise until pointer is aligned according to engine temperature:
   ❑ -20°C – left of alignment marks **7**.
   ❑ 20°C – centre of alignment marks **8**.
   ❑ 50°C – right of alignment marks **9**.
8. Tighten tensioner bolt to 20 Nm **4**.
9. Press down on belt and ensure tensioner pointer moves freely **6**.
10. Turn crankshaft two turns clockwise.
11. Ensure timing marks aligned **2** & **3**.
12. Ensure tensioner pointer correctly aligned:
   ❑ -20°C – left of alignment marks **7**.
   ❑ 20°C – centre of alignment marks **8**.
   ❑ 50°C – right of alignment marks **9**.
13. If not: Repeat tensioning procedure.
14. Install components in reverse order of removal.

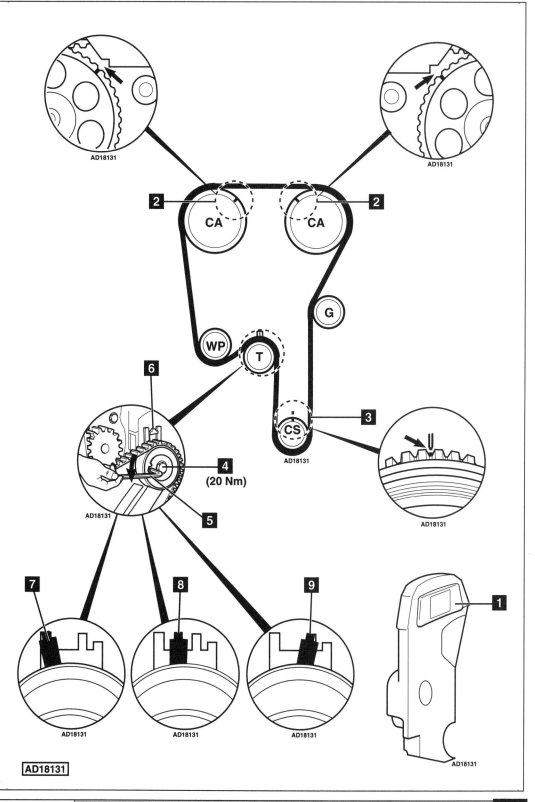

VOLVO

Model:	S80 2,8 Turbo • S80 2,9
Year:	1998-99
Engine Code:	B6284T, B6304S3

Replacement Interval Guide

Volvo recommend replacement every
100,000 miles or 10 years, whichever occurs first.
*The previous use and service history of the vehicle
must always be taken into account.
Refer to Timing Belt Replacement Intervals at the front
of this manual.*

Check For Engine Damage

*CAUTION: This engine has been identified as an
INTERFERENCE engine in which the possibility of
valve-to-piston damage in the event of a timing belt
failure is MOST LIKELY to occur.
A compression check of all cylinders should be
performed before removing the cylinder head.*

Repair Times – hrs

Remove & install:

2,8	3,10
2,9	2,60

Special Tools

■ Crankshaft pulley holding tool – Volvo No.999 5433.

Special Precautions

■ Disconnect battery earth lead.
■ DO NOT turn crankshaft or camshaft when timing belt removed.
■ Remove spark plugs to ease turning engine.
■ Turn engine in normal direction of rotation (unless otherwise stated).
■ DO NOT turn engine via camshaft or other sprockets.
■ Observe all tightening torques.

Removal

1. Raise and support front of vehicle.
2. Remove:
   ❏ Strut brace.
   ❏ Bolt securing steady bar to engine bracket.
   ❏ Air intake hoses (Turbo).
   ❏ PAS reservoir. DO NOT disconnect hoses.
   ❏ Coolant expansion tank. Disconnect hose(s).
   ❏ RH front wheel.
   ❏ RH lower splash guard.
   ❏ Auxiliary drive belt.
   ❏ Timing belt upper cover **1**.
   ❏ Timing belt front cover **2**.
3. Fit timing belt upper cover **1**.
4. Turn crankshaft clockwise until timing marks aligned **3** & **4**.
5. Turn crankshaft a further ¼ turn clockwise, then anti-clockwise until timing marks aligned **3** & **4**.

6. Remove:
   ❏ Timing belt upper cover **1**.
   ❏ Crankshaft pulley bolts **5**.
7. Fit crankshaft pulley holding tool **6**. Tool No.999 5433.
8. Remove:
   ❏ Crankshaft pulley centre nut **7**.
   ❏ Crankshaft pulley **8**.
   ❏ Upper bolt of automatic tensioner unit **9**.
   ❏ Lower bolt of automatic tensioner unit **10**.
9. Twist tensioner body to free pushrod.
10. Remove:
    ❏ Automatic tensioner unit.
    ❏ Timing belt.

Installation

*NOTE: Whenever timing belt is replaced, tensioner pulley lever pivot **11** must be lubricated. If tensioner body has been leaking or pushrod has no resistance or is seized, renew automatic tensioner unit.*

1. Check timing belt guide pulley. Replace if necessary.
2. Slowly compress pushrod into tensioner body until holes aligned. Retain pushrod with 2 mm diameter pin through hole in tensioner body **12**.
3. Install automatic tensioner unit. Tighten bolts to 25 Nm **9** & **10**.
4. Ensure timing marks aligned **3** & **4**.
5. Fit timing belt in anti-clockwise direction, starting at crankshaft sprocket. Ensure belt is taut between sprockets on non-tensioned side.
   NOTE: Ensure camshaft position actuator does not move during timing belt installation – B6304S3.
6. Remove 2 mm diameter pin from tensioner body to release pushrod **12**.
7. Fit crankshaft pulley and centre nut **7** & **8**.
8. Turn crankshaft two turns clockwise.
9. Ensure timing marks aligned **3** & **4**.
10. Tighten crankshaft pulley centre nut to 300 Nm **7**. Use tool No.999 5433 **6**.
11. Fit crankshaft pulley bolts **5**. Use new bolts.
12. Tighten crankshaft pulley bolts to 35 Nm + 50° **5**.
13. Install components in reverse order of removal.
14. Fit strut brace. Use new bolts.
    Tightening torque: 50 Nm.
15. Fit engine steady bar bolt. Use new bolt.
    Tightening torque: 80 Nm.

Autodata

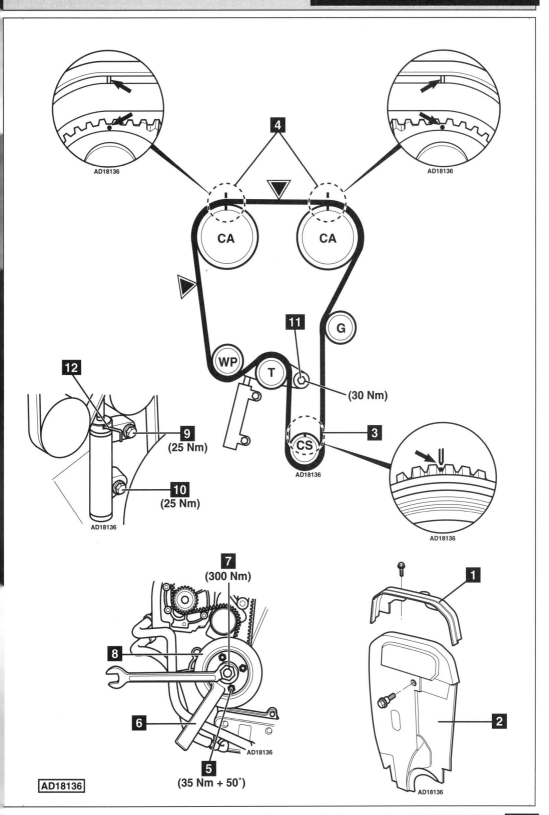

VOLVO

Model:	440 1,9TD • 460 1,9TD
Year:	1994-97
Engine Code:	D19T

Replacement Interval Guide

Volvo recommend replacement every 60,000 miles or 72 months, whichever occurs first.
The previous use and service history of the vehicle must always be taken into account.
Refer to Timing Belt Replacement Intervals at the front of this manual.

Check For Engine Damage

CAUTION: This engine has been identified as an INTERFERENCE engine in which the possibility of valve-to-piston damage in the event of a timing belt failure is MOST LIKELY to occur.
A compression check of all cylinders should be performed before removing the cylinder head.

Repair Times – hrs

Remove & install	1,90

Special Tools

- Timing pin (8 mm diameter) – for setting crankshaft at TDC.
- Tension gauge – Volvo No.999-5506/5434.
- M6 bolt – for adjusting tensioner pulley.

Special Precautions

- Disconnect battery earth lead.
- DO NOT turn crankshaft or camshaft when timing belt removed.
- Remove glow plugs to ease turning engine.
- Turn engine in normal direction of rotation (unless otherwise stated).
- DO NOT turn engine via camshaft or other sprockets.
- Observe all tightening torques.
- Check diesel injection pump timing after belt replacement.

Removal

1. Support engine.
2. Remove:
   - ❏ Engine mounting **1**.
   - ❏ Lower engine side cover.
   - ❏ Timing belt upper cover **2**.
   - ❏ Diesel fuel filter assembly.
   - ❏ Timing belt covers **3** & **4**.
   - ❏ Auxiliary drive belt.
3. Turn crankshaft clockwise until flywheel No.1 cylinder TDC mark aligned **5**.
4. Remove blanking plug (located next to dipstick tube). Insert 8 mm timing pin **6**.
5. Rock crankshaft until timing pin locates in crankshaft web **6**.

6. Ensure timing mark on camshaft sprocket aligned with mark on timing belt rear cover **7**.
7. Scribe mark on timing belt rear cover opposite mark on injection pump sprocket **8**.
8. Remove:
   - ❏ Crankshaft pulley bolt **9**.
   - ❏ Crankshaft pulley **10**.
9. Slacken tensioner pulley nut to release tension on belt **11**.
10. Remove timing belt.

Installation

1. Rock crankshaft slightly to ensure timing pin located correctly **6**.
2. Ensure timing marks aligned **7** & **8**.
3. Observe direction of rotation marks on belt. Align marks on belt with marks on sprockets **12** & **13**.
4. Ensure tensioner pulley nut fully slackened **11**.
5. Fit timing belt in following order:
   - ❏ Crankshaft sprocket.
   - ❏ Auxiliary shaft sprocket.
   - ❏ Guide pulley.
   - ❏ Injection pump sprocket.
   - ❏ Camshaft sprocket.

   NOTE: Ensure belt is taut between sprockets.
6. Fit tool No.999-5506 to belt **14**.
7. Screw M6 bolt into timing belt rear cover to push on tool at tensioner pulley **15**.
8. Screw in M6 bolt and tension belt to 7,5 mm (cold). Use tension gauge **16**.
9. Tighten tensioner pulley nut to 50 Nm **11**.
10. Remove:
    - ❏ Tensioning tool **14**.
    - ❏ Timing pin **6**.
11. Turn crankshaft two turns clockwise. Insert timing pin **6**.
12. Ensure timing marks aligned **7** & **8**.
13. Recheck belt tension.
14. Remove:
    - ❏ M6 bolt **15**.
    - ❏ Timing pin **6**.
15. Fit blanking plug.
16. Check and adjust injection pump timing.
17. Install components in reverse order of removal.
18. Bleed fuel system.
19. Tighten crankshaft pulley bolt to 95 Nm **9**.

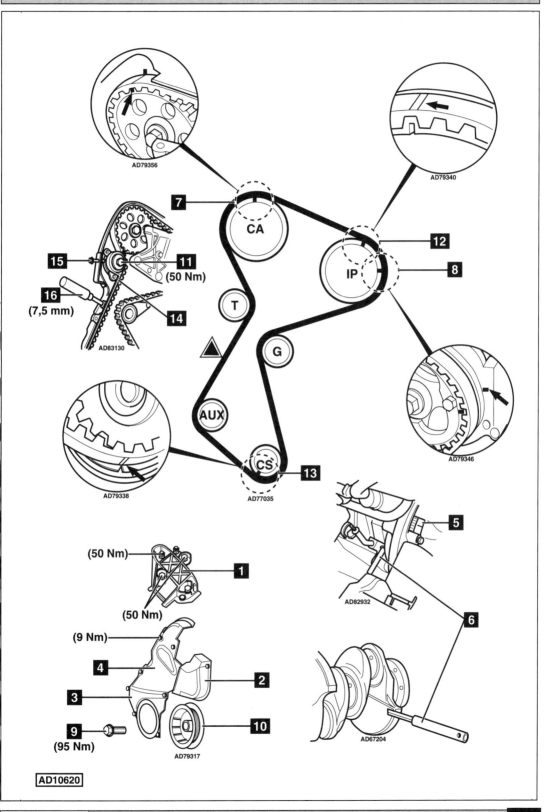

AD79356

AD79340

7

CA

12

8

IP

15

11
(50 Nm)

16
(7,5 mm)

14

AD83130

T

G

AUX

AD79346

AD79338

CS

13

AD77035

5

AD82932

(50 Nm)

1

(50 Nm)

6

(9 Nm)

4

2

3

10

9
(95 Nm)

AD79317

AD67204

AD10620

VOLVO

Model:	**S40/V40 1,9 TD**
Year:	**1996-99**
Engine Code:	**D4192T**

Replacement Interval Guide

Volvo recommend replacement every 80,000 miles or 8 years.
The previous use and service history of the vehicle must always be taken into account.
Refer to Timing Belt Replacement Intervals at the front of this manual.

Check For Engine Damage

CAUTION: This engine has been identified as an INTERFERENCE engine in which the possibility of valve-to-piston damage in the event of a timing belt failure is MOST LIKELY to occur.
A compression check of all cylinders should be performed before removing the cylinder head(s).

Repair Times – hrs

Remove & install	2,40

Special Tools

- Tension gauge – Volvo No.999-5434/5506.
- Timing pin – 8 mm diameter.
- Locking fluid – Volvo No.1161053-2.

Removal

1. Remove:
   - Engine top cover.
   - Auxiliary drive belt cover **1**.
   - Auxiliary drive belt.
2. Raise engine slightly.
3. Support engine.
4. Move PAS hose to one side.
5. Remove:
   - RH engine mounting.
   - Injection pump sprocket cover **2**.
   - Timing belt upper cover **3**.
6. Turn crankshaft clockwise until timing marks aligned **4** & **5**.
7. Remove blanking plug **6**.
8. Insert 8 mm timing pin **7**.
9. Mark injection pump bracket adjacent to sprocket timing mark **8**.
10. Remove:
    - Crankshaft pulley centre bolt **9**.
    - Crankshaft pulley **10**.
11. Slacken tensioner nut **11**. Move tensioner away from belt. Lightly tighten nut.
12. Remove timing belt.

Installation

NOTE: If injection pump sprocket marked 'HTD2', a timing belt with corresponding lettering must be fitted.

1. Ensure timing pin located correctly **7**.
2. Ensure timing marks aligned **4**, **5** & **8**.
3. Fit timing belt in following order:
   - Crankshaft sprocket.
   - Auxiliary shaft sprocket.
   - Guide pulley.
   - Injection pump sprocket.
   - Camshaft sprocket.
   NOTE: Observe direction of rotation marks on belt.
   - Tensioner pulley.
4. Ensure belt is taut between sprockets.
5. Ensure marks on belt aligned with marks on sprockets **12** & **13**.
6. Slacken tensioner nut **11**.
7. Insert M6 bolt **14**.
8. Attach tension gauge to belt at ▽ **15**. Tool No.999-5434/5506.
9. Screw in M6 bolt and tension belt to 7,5 mm **14**.
10. Tighten tensioner nut to 50 Nm **11**.
11. Ensure timing marks aligned **5**, **8** & **12**.
12. Remove:
    - Tension gauge **15**.
    - M6 bolt **14**.
13. Install:
    - Crankshaft pulley **10**.
    - Crankshaft pulley centre bolt **9**. Use locking fluid (1161053-2).
14. Tighten crankshaft pulley centre bolt **9**.
    Tightening torque: 95 Nm (to engine No.003534).
    Tightening torque: 20 Nm + 115°
    (from engine No.003535).
15. Remove timing pin **7**.
16. Turn crankshaft two turns clockwise to setting position.
17. Insert timing pin **7**.
18. Ensure timing marks aligned **4**, **5** & **8**.
19. Recheck belt tension. If necessary: Repeat tensioning procedure.
20. Remove timing pin **7**.
21. Fit blanking plug and tighten to 20 Nm **6**.
22. Install components in reverse order of removal.

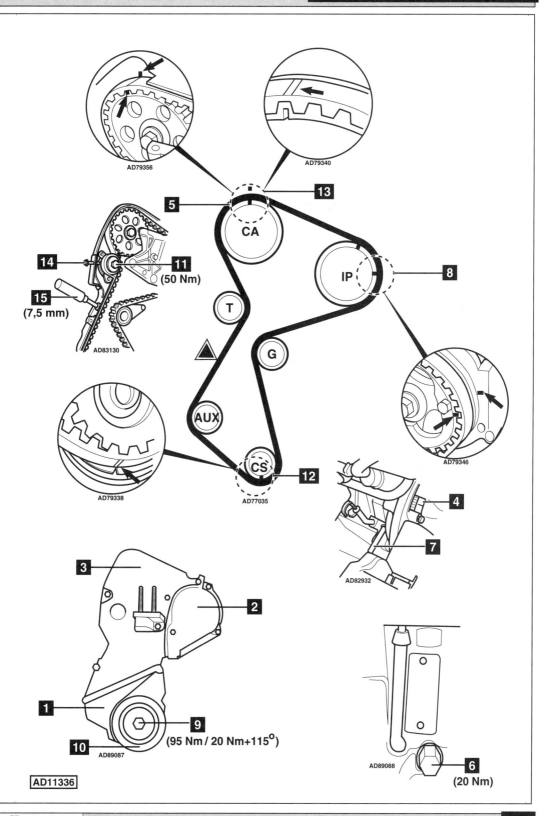

AD79356

AD79340

13

5

CA

14

11
(50 Nm)

15
(7,5 mm)

AD83130

IP

8

T

G

AUX

AD79346

CS

AD79338

12

AD77035

4

7

AD82932

3

2

1

9
(95 Nm / 20 Nm+115°)

10   AD89087

AD89088

6
(20 Nm)

AD11336

VOLVO

Model:	**S40/V40 1,9 TD**
Year:	**1999-00**
Engine Code:	**D4192T2**

Replacement Interval Guide

Volvo recommend replacement every 50,000 miles or 5 years.
The previous use and service history of the vehicle must always be taken into account.
Refer to Timing Belt Replacement Intervals at the front of this manual.

Check For Engine Damage

CAUTION: This engine has been identified as an INTERFERENCE engine in which the possibility of valve-to-piston damage in the event of a timing belt failure is MOST LIKELY to occur.
A compression check of all cylinders should be performed before removing the cylinder head(s).

Repair Times – hrs

Remove & install	2,40
AC	+0,10

Special Tools

- M6 bolt – for adjusting tensioner.
- Tension gauge – Volvo No.951 2797.
- Timing pin – 8 mm diameter.

Special Precautions

- Disconnect battery earth lead.
- DO NOT turn crankshaft or camshaft when timing belt removed.
- Remove glow plugs to ease turning engine.
- Turn engine in normal direction of rotation (unless otherwise stated).
- DO NOT turn engine via camshaft or other sprockets.
- Observe all tightening torques.
- Check diesel injection pump timing after belt replacement.

Removal

1. Ensure engine is cold before replacing belt.
2. Raise and support front of vehicle.
3. Remove:
   - ❏ Engine upper cover and bracket.
   - ❏ Auxiliary drive belt.
   - ❏ PAS pipe bracket.
   - ❏ Injection pump sprocket cover **1**.
4. Raise engine slightly.
5. Support engine.
6. Remove:
   - ❏ AC pipe bracket.
   - ❏ RH engine mounting.
7. Turn crankshaft clockwise to TDC on No.1 cylinder until timing marks aligned **2** & **3**.
8. Remove blanking plug **4**.
9. Insert 8 mm timing pin **5**.
10. Rock crankshaft slightly to ensure timing pin located correctly.
11. Remove:
    - ❏ Engine mounting stud **6**.
    - ❏ Timing belt cover **7**.
12. Mark timing belt rear cover adjacent to camshaft sprocket timing mark **8**.
13. Mark injection pump bracket adjacent to sprocket timing mark **9**.
14. Remove:
    - ❏ Crankshaft pulley centre bolt **10**.
    - ❏ Crankshaft pulley **11**.
15. Slacken tensioner nut **12**. Move tensioner away from belt. Lightly tighten nut.
16. Remove timing belt.

Installation

1. Ensure timing pin located correctly **5**.
2. Ensure timing marks aligned **2**, **8** & **9**.
3. Fit timing belt in following order:
   - ❏ Crankshaft sprocket.
   - ❏ Auxiliary shaft sprocket.
   - ❏ Guide pulley.
   - ❏ Injection pump sprocket.
   - ❏ Camshaft sprocket.
   - ❏ Tensioner pulley.

   NOTE: Observe direction of rotation marks on belt.
4. Ensure belt is taut between sprockets.
5. Ensure marks on belt aligned with marks on sprockets **13** & **14**.

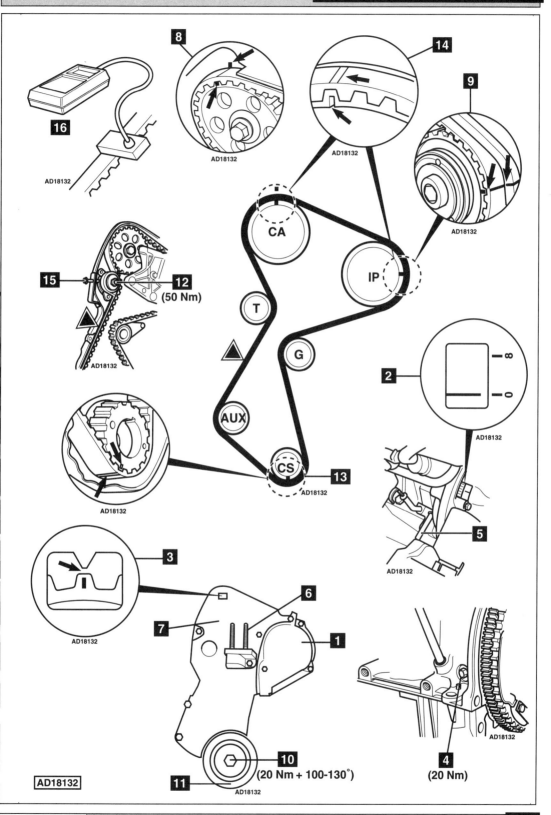

←

6. Install:
   - ❏ Crankshaft pulley **11**.
   - ❏ Crankshaft pulley centre bolt **10**.
     Use new bolt.
7. Ensure timing pin located correctly **5**.
8. Tighten crankshaft pulley centre bolt **10**.
   Tightening torque: 20 Nm + 100-130°.
9. Remove timing pin **5**.
10. Slacken tensioner nut **12**.
11. Insert M6 bolt **15**.
12. Attach tension gauge to belt at ▽.
    Tool No.951 2797 **16**.
13. Screw in M6 bolt until tension gauge indicates
    68±5 Hz.
14. Remove tension gauge **16**.
15. Temporarily tighten tensioner pulley bolt **12**.
16. Turn crankshaft 4 turns clockwise to TDC on
    No.1 cylinder.
17. Insert 8 mm timing pin **5**.
18. Rock crankshaft slightly to ensure timing pin
    located correctly.
19. Ensure timing marks aligned **2**, **8** & **9**.
20. Remove timing pin **5**.
21. Attach tension gauge to belt at ▽.
    Tool No.951 2797 **16**.
22. Tension gauge should indicate 61±5 Hz.
23. If not: Screw in M6 bolt until tension gauge
    indicates 61±5 Hz.
24. Remove tension gauge **16**.
25. Tighten tensioner nut to 50 Nm **12**.
    **NOTE: DO NOT turn crankshaft again to check
    timing belt tension.**
26. Remove M6 bolt **15**.
27. Fit blanking plug and tighten to 20 Nm **4**.
28. Install components in reverse order of removal.

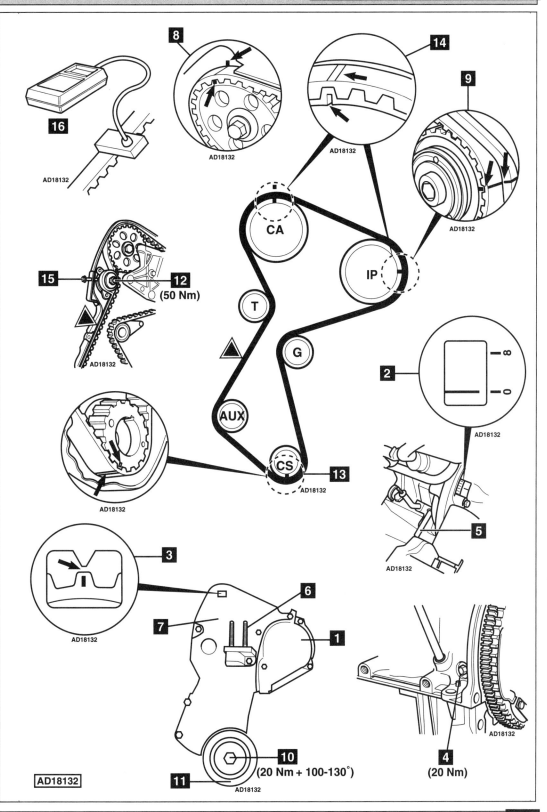

8

AD18132

14

AD18132

9

AD18132

16

AD18132

CA

15 — **12** (50 Nm)

AD18132

T

G

2

AD18132

IP

AUX

CS — **13**

AD18132

AD18132

5

AD18132

3

AD18132

7

6

1

4 (20 Nm)

AD18132

10 (20 Nm + 100-130°)

11

AD18132

AD18132

VOLVO

Model:	**S40/V40 1,9 TD**
Year:	**2001-03**
Engine Code:	**D4192T3/4**

Replacement Interval Guide

Volvo recommend:
01MY/10,000 mile service intervals –
replacement every 80,000 miles or 8 years.
02MY→/12,000 mile service intervals –
replacement every 72,000 miles or 6 years.
The previous use and service history of the vehicle must always be taken into account.
Refer to Timing Belt Replacement Intervals at the front of this manual.

Check For Engine Damage

CAUTION: This engine has been identified as an INTERFERENCE engine in which the possibility of valve-to-piston damage in the event of a timing belt failure is MOST LIKELY to occur. A compression check of all cylinders should be performed before removing the cylinder head(s).

Repair Times – hrs

Remove & install	2,00

Special Tools

- M6 bolt – for adjusting tensioner.
- Tension gauge – Volvo No.951 2797.
- Timing pin – 8 mm diameter.

Special Precautions

- Disconnect battery earth lead.
- DO NOT turn crankshaft or camshaft when timing belt removed.
- Remove glow plugs to ease turning engine.
- Turn engine in normal direction of rotation (unless otherwise stated).
- DO NOT turn engine via camshaft or other sprockets.
- Observe all tightening torques.

Removal

NOTE: The fuel pump fitted to this engine does not require timing.

1. Ensure engine is cold before replacing belt.
2. Raise and support front of vehicle.
3. Remove:
   - ❏ Engine top cover.
   - ❏ Engine control module (ECM) and bracket. Leave multi-plug(s) connected.
   - ❏ RH splash guard.
   - ❏ Engine undershield.
   - ❏ Auxiliary drive belt.
   - ❏ Holder for PAS pipes.
4. Raise engine slightly.
5. Support engine.
6. Remove:
   - ❏ RH engine mounting.
   - ❏ RH engine mounting bracket studs.
7. Turn crankshaft clockwise to TDC on No.1 cylinder (flywheel end).
8. Check camshaft alignment:

Type 1
   - ❏ Ensure camshaft sprocket timing marks aligned **1**.

Type 2
   - ❏ Remove cylinder head cover.
   - ❏ Ensure camshaft lobes for No.1 cylinder angled upwards.
9. Remove timing belt cover **2**.
10. Mark timing belt rear cover adjacent to camshaft sprocket timing mark **3**.
11. Remove blanking plug from cylinder block **4**.
12. Insert 8 mm timing pin **5**.
13. Rock crankshaft slightly to ensure timing pin located correctly.
14. Remove:
    - ❏ Crankshaft pulley bolt **6**.
    - ❏ Crankshaft pulley **7**.
15. Slacken tensioner nut **8**. Move tensioner away from belt. Lightly tighten nut.
16. Remove timing belt.
    NOTE: DO NOT refit used belt.

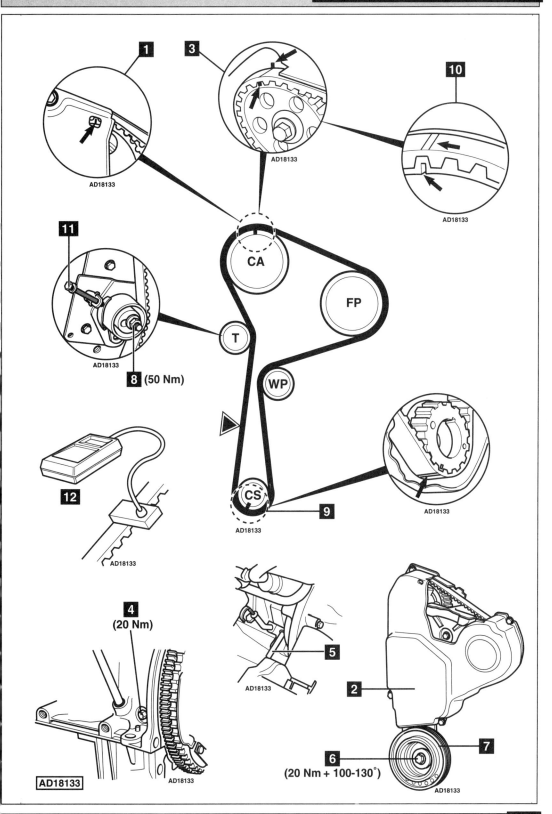

1

AD18133

3

AD18133

10

AD18133

11

AD18133

8 (50 Nm)

CA

FP

T

WP

12

AD18133

AD18133

CS

AD18133

9

AD18133

4
(20 Nm)

5

AD18133

2

7

6
(20 Nm + 100-130°)

AD18133

AD18133

←

Installation

1. Ensure timing pin fitted .
2. Ensure camshaft sprocket timing marks aligned 🔳.
3. Fit timing belt in following order:
   - ❏ Crankshaft sprocket.
   - ❏ Tensioner pulley.
   - ❏ Camshaft sprocket.
   - ❏ Water pump pulley.

   NOTE: Observe direction of rotation marks on belt.
4. Ensure belt is taut between sprockets.
5. Ensure marks on belt aligned with marks on sprockets 🔳 & 🔟.
6. Turn camshaft sprocket slightly in clockwise direction to tension timing belt against tensioner pulley.
7. Fit timing belt to fuel pump sprocket.
8. Turn camshaft sprocket slightly in anti-clockwise direction to tension timing belt against water pump pulley.
9. Install:
   - ❏ Crankshaft pulley 🔳.
   - ❏ Crankshaft pulley centre bolt 🔳. Use new bolt.
10. Ensure timing pin located correctly 🔳.
11. Tighten crankshaft pulley centre bolt 🔳. Tightening torque: 20 Nm + 100-130°.
12. Remove timing pin 🔳.
13. Slacken tensioner nut 🔳.
14. Insert M6 bolt 🔳.
15. Attach tension gauge to belt at ▽. Tool No.951 2797 🔳.
16. Screw in M6 bolt until tension gauge indicates 88±3 Hz.
17. Remove tension gauge 🔳.
18. Temporarily tighten tensioner pulley bolt 🔳.
19. Turn crankshaft 4 turns clockwise to TDC on No.1 cylinder.
20. Insert 8 mm timing pin 🔳.
21. Rock crankshaft slightly to ensure timing pin located correctly.
22. Ensure camshaft sprocket timing marks aligned 🔳.
23. Remove timing pin 🔳.
24. Attach tension gauge to belt at ▽. Tool No.951 2797 🔳.
25. Tension gauge should indicate 85±3 Hz.
26. If not: Screw in M6 bolt until tension gauge indicates 85±3 Hz.
27. Remove tension gauge 🔳.
28. Tighten tensioner nut to 50 Nm 🔳.

    NOTE: DO NOT turn crankshaft again to check timing belt tension.
29. Remove M6 bolt 🔳.
30. Fit blanking plug and tighten to 20 Nm 🔳. Use locking fluid.
31. Install components in reverse order of removal.
32. Check engine control module (ECM) fault memory.

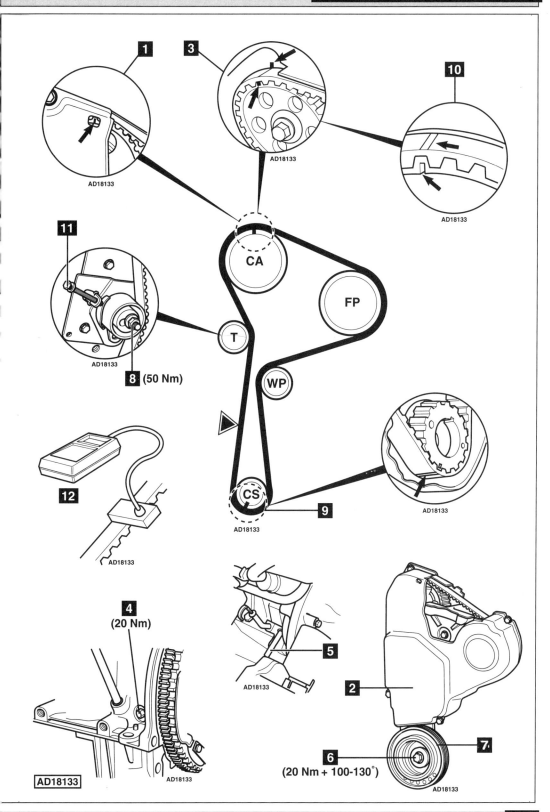

1

3

10

AD18133

AD18133

AD18133

11

CA

FP

T

WP

AD18133

8 (50 Nm)

12

CS

9

AD18133

AD18133

AD18133

4
(20 Nm)

5

AD18133

2

AD18133

6
(20 Nm + 100-130°)

7

AD18133

Model:	240D • 740D/Turbo • 760D Turbo • 940D/Turbo • 960D Turbo
Year:	1979-94
Engine Code:	D24, D24T, D24TIC (without EGR)

Replacement Interval Guide

Volvo recommend replacement of timing belt, injection pump belt and guide pulley as follows:
→12/90 - every 72,000 miles.
01/91→ - every 80,000 miles or 96 months.
The previous use and service history of the vehicle must always be taken into account.
Refer to Timing Belt Replacement Intervals at the front of this manual.

Check For Engine Damage

CAUTION: This engine has been identified as an INTERFERENCE engine in which the possibility of valve-to-piston damage in the event of a timing belt failure is MOST LIKELY to occur.
A compression check of all cylinders should be performed before removing the cylinder head.

Repair Times – hrs

Timing belt:	
Check & adjust (D24/T)	0,30
Check & adjust (D24TIC)	0,50
Remove & install (D24/T)	2,70
Remove & install (D24TIC)	2,90
Injection pump belt:	
Check & adjust	0,40
Remove & install (D24/T)	1,40
Remove & install (D24TIC)	1,50

Special Tools

■ Tension gauge – Volvo No.5197.
■ Injection pump sprocket locking pin – Volvo No.5193.
■ Camshaft setting bar – Volvo No.5190.
■ Crankshaft pulley holding tool – Volvo No.5187.
■ Camshaft sprocket holding tool – Volvo No.5199.

Special Precautions

■ Disconnect battery earth lead.
■ DO NOT turn crankshaft or camshaft when timing belt removed.
■ Remove glow plugs to ease turning engine.
■ Turn engine in normal direction of rotation (unless otherwise stated).
■ DO NOT turn engine via camshaft or other sprockets.
■ Observe all tightening torques.
■ Check diesel injection pump timing after belt replacement.

Removal

1. Remove:
   ❑ Radiator and hoses.
   ❑ Splashguard.
   ❑ Cooling fan with spacer and pulley.
   ❑ Auxiliary drive belts.
2. Turn crankshaft to TDC on No.1 cylinder. Ensure flywheel and injection pump sprocket timing marks aligned **1** & **2**.
3. Hold crankshaft pulley. Use tool No.5187. Use 22 mm socket to remove bolt. Ensure timing marks aligned **1**.

4. Remove:
   ❑ Crankshaft pulley/damper (four 6 mm Allen screws).
   ❑ Timing belt upper cover **3**.
   ❑ Cylinder head cover.
   ❑ Injection pump belt cover **4**.
5. Slacken injection pump mounting bolts.
6. Hold camshaft rear sprocket. Use tool No.5199 **5**. Remove bolt and sprocket **9** & **10**.
7. Remove injection pump belt.
8. Fit setting bar to rear of camshaft **6**. Tool No. 5190.
9. Lock injection pump sprocket. Use tool No.5193 **7**.
10. Remove timing belt lower cover **8**.
11. Slacken water pump bolts. Turn water pump to release tension on belt.
12. Remove timing belt.

Installation

NOTE: When fitting a new timing belt the guide pulley must also be renewed.

1. Slacken camshaft sprocket bolt one turn **12**. Tap sprocket from behind to loosen it from taper.
2. Install:
   ❑ Timing belt.
   ❑ Timing belt lower cover **8**.
   ❑ Crankshaft pulley/damper. Tighten Allen screws to 20 Nm.
3. Clean crankshaft pulley bolt. Coat threads with locking compound.
4. Hold crankshaft pulley. Use tool No.5187. Tighten bolt to 350 Nm.
5. Insert 0,2 mm feeler gauge under left-hand side of camshaft setting bar **11**.
6. Set tension gauge to 12,5 units. Attach tension gauge to belt at ▽. Turn water pump to tension belt. Tighten water pump bolts.
7. Ensure crankshaft at TDC **1**.
8. Tighten camshaft sprocket bolt to 45 Nm **12**.
9. Remove setting bar and feeler gauge from camshaft.
10. Ensure injection pump timing marks aligned **2**. Ensure locking pin located correctly in injection pump sprocket **7**.
11. Fit camshaft rear sprocket and injection pump belt.
12. Fit sprocket bolt and tighten until sprocket can just be turned by hand **10**.
13. Set tension gauge to 12,5 units. Attach tension gauge to belt at ▽.
14. Move injection pump mounting to tension belt. Tighten bolts.
15. Apply firm thumb pressure to belt. Recheck belt tension.
16. Hold camshaft rear sprocket. Use tool No.5199 **5**. Tighten bolt to 100 Nm **9**.
    NOTE: Ensure camshaft does not move.
17. Install components in reverse order of removal.

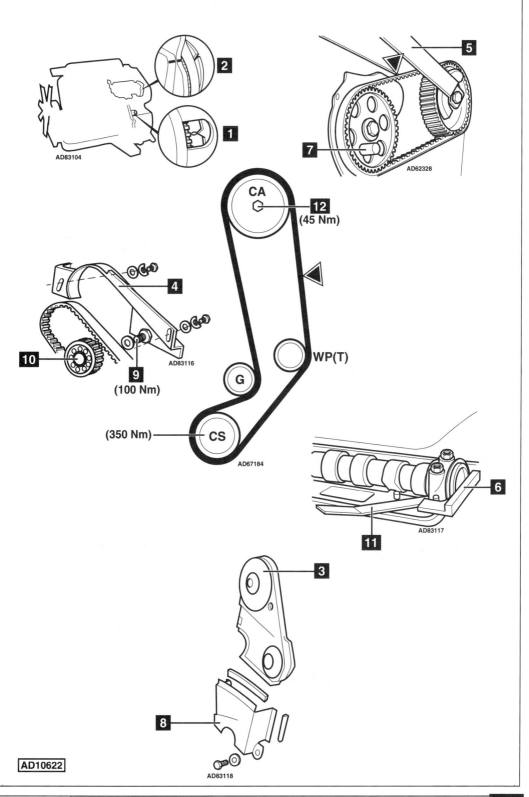

VOLVO

Model:	850 2,5 TDI • S70/V70 2,5 TDI • S80 2,5 TDI
Year:	1996-01
Engine Code:	D5252T

Replacement Interval Guide

Volvo recommend:

Timing belt:
All years: Check and adjust every 20,000 miles or
2 years.
96/97MY additions: Check at 10,000 miles or 1 year,
50,000 miles or 5 years and 90,000 miles or
9 years.
98MY additions: Check at 10,000 miles or 1 year
and 90,000 miles or 9 years.
→98MY: Replacement every 80,000 miles or
8 years (tensioner must also be replaced).
99MY→: Replacement every 80,000 miles or
8 years (tensioner must also be replaced).

Injection pump belt:
All years: Check and adjust every 20,000 miles or
2 years.
96/97MY additions: Check at 10,000 miles or 1 year,
50,000 miles or 5 years and 90,000 miles or
9 years.
98MY additions: Check at 10,000 miles or 1 year
and 90,000 miles or 9 years.
→98MY: Replacement at 40,000 miles or 4 years,
80,000 miles or 8 years and then every
80,000 miles or 8 years (tensioner must also be
replaced).
99MY→: Replacement every 80,000 miles or
8 years (tensioner must also be replaced).
*The previous use and service history of the vehicle
must always be taken into account.*
*Refer to Timing Belt Replacement Intervals at the
front of this manual.*

Check For Engine Damage

*CAUTION: This engine has been identified as an
INTERFERENCE engine in which the possibility of
valve-to-piston damage in the event of a timing belt
failure is MOST LIKELY to occur.*
*A compression check of all cylinders should be
performed before removing the cylinder head.*

Repair Times – hrs

Check and adjust:	
Timing belt	0,90
Injection pump belt (except V70 2000 MY→)	0,80
Injection pump belt (V70 2000 MY→)	0,80
Remove & install:	
Timing belt	2,10
Injection pump belt (except V70 2000 MY→)	1,10
Injection pump belt (V70 2000 MY→)	1,30

Special Tools

- Camshaft sprocket holding tool – Volvo No.999 5644.
- Crankshaft pulley locking tool – Volvo No.999 5645.
- Tensioner pulley spanner – Volvo No.999 5649.

Special Precautions

- Disconnect battery earth lead.
- DO NOT turn crankshaft or camshaft when timing belt removed.
- Remove glow plugs to ease turning engine.
- Turn engine in normal direction of rotation (unless otherwise stated).
- DO NOT turn engine via camshaft or other sprockets.
- Observe all tightening torques.
- Check diesel injection pump timing after belt replacement.

Removal

Timing Belt

1. Remove:
   - ❏ Engine upper cover.
   - ❏ Coolant expansion tank.
   - ❏ PAS reservoir (if required).
   - ❏ Turbocharger air hoses.
   - ❏ Auxiliary drive belt and tensioner.
   - ❏ Timing belt upper cover **1**.
   - ❏ Injection pump belt cover.
2. Raise and support front of vehicle.
3. Remove:
   - ❏ Splash guard.
   - ❏ RH front wheel.
   - ❏ RH wheel arch liner.
4. Turn crankshaft to TDC on No.1 cylinder. Ensure timing marks aligned **2** or **3**.
5. Ensure injection pump timing marks aligned **4**.
   NOTE: If not: Turn crankshaft one turn clockwise.
6. Slacken injection pump locking bolt **5**. Remove keeper plate **6**. Tighten locking bolt to 30 Nm **5**.
7. Hold camshaft sprocket. Use tool No.999 5644.
8. Slacken camshaft sprocket bolt one turn **7**.
9. Loosen camshaft sprocket from taper using a drift through hole in timing belt rear cover.
10. Lock crankshaft pulley. Use tool No.999 5645.
    NOTE: Ensure crankshaft does not move.
11. Slacken crankshaft pulley centre bolt **8**.
12. Slacken crankshaft pulley bolts **9**.
13. Remove:
    - ❏ Crankshaft pulley locking tool.
    - ❏ Crankshaft pulley bolts **9**.
    - ❏ Crankshaft pulley centre bolt **8**.
    - ❏ Crankshaft pulley **10**.
    - ❏ Timing belt lower cover **11**.
    - ❏ Tensioner.
    - ❏ Timing belt.

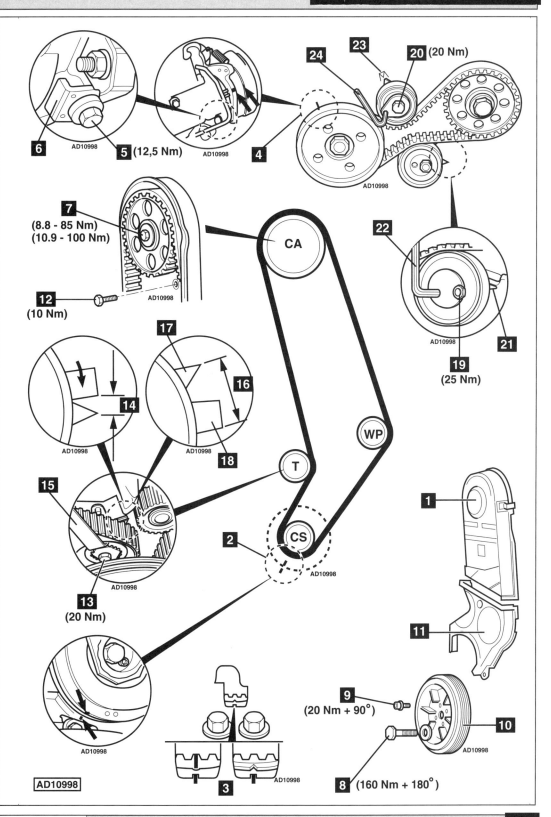

Installation

Timing Belt

1. Ensure rear timing belt cover screw is marked with white dot 🔢. If not:
   - ❏ Remove rear timing belt cover screw 🔢.
   - ❏ Clean threads. Apply locking compound.
   - ❏ Fit screw and tighten.
     Tightening torque: 10 Nm.
   - ❏ Mark screw with white dot.
2. Fit new tensioner.
3. Lightly tighten tensioner bolt 🔢.
4. Ensure timing marks aligned 🔢 & 🔢.
5. Ensure injection pump locked in position with bolt 🔢.
6. Fit timing belt in anti-clockwise direction, starting at crankshaft sprocket. Ensure belt is taut between sprockets.
   NOTE: Ensure engine temperature below 30°C before adjusting timing belt.
7. Slacken tensioner bolt 🔢. Adjust belt tension until pointer is 2 mm below fixed mark 🔢. Use tool No.999 5649 🔢.
   NOTE: If distance between pointer and fixed mark is exceeded: Release tensioner and repeat tensioning procedure.
8. Tighten tensioner bolt to 20 Nm 🔢.
9. Fit crankshaft pulley.
10. Fit crankshaft pulley bolts 🔢. Use new bolts.
11. Fit crankshaft pulley centre bolt. Use new bolt.
12. Lock crankshaft pulley. Use tool No.999 5645.
    NOTE: Ensure crankshaft does not move.
13. Tighten crankshaft pulley centre bolt.
    Tightening torque: 160 Nm + 180° 🔢.
14. Tighten crankshaft pulley bolts.
    Tightening torque: 20 Nm + 90° 🔢.
15. Remove crankshaft pulley locking tool.
    Tool No.999 5645.
16. Ensure timing marks aligned 🔢 or 🔢.
17. Hold camshaft sprocket. Use tool No.999 5644.
18. Tighten camshaft sprocket bolt according to identification number 🔢:
    - ❏ 8.8 – 85 Nm.
    - ❏ 10.9 – 100 Nm.
19. Slacken injection pump locking bolt 🔢.
20. Insert keeper plate 🔢. Tighten locking bolt to 12,5 Nm 🔢.
21. Turn crankshaft 2 turns clockwise.
22. Ensure timing marks aligned 🔢, 🔢 & 🔢.
23. Ensure tensioner pointer 2 mm below fixed mark 🔢.
24. If not: Repeat tensioning procedure.
25. Install components in reverse order of removal.

Removal

Injection Pump Belt

1. Access timing belt.
2. Remove injection pump belt cover.
3. Turn crankshaft to TDC on No.1 cylinder. Ensure timing marks aligned 🔢 or 🔢.
4. Ensure injection pump timing marks aligned 🔢.
   NOTE: If not: Turn crankshaft one turn clockwise.
5. If timing belt has not been replaced: Check distance between tensioner pointer and fixed mark 🔢:
   - ❏ If distance exceeds 14 mm: Renew timing belt and tensioner.
   - ❏ If distance less than 14 mm: Adjust tensioner until pointer and lower edge of fixed mark aligned 🔢 & 🔢.
   NOTE: Ensure camshaft sprocket can rotate freely during timing belt adjustment.
6. Slacken injection pump locking bolt 🔢. Remove keeper plate 🔢. Tighten locking bolt to 30 Nm.
7. Slacken guide pulley nut 🔢. Turn guide pulley away from belt.
8. Remove:
   - ❏ Tensioner.
   - ❏ Injection pump belt.

Installation

Injection Pump Belt

1. Fit new tensioner.
2. Lightly tighten tensioner bolt 🔢.
3. Ensure timing marks aligned 🔢 or 🔢.
4. Ensure injection pump timing marks aligned 🔢.
5. Ensure injection pump locked in position with bolt 🔢.
6. Fit injection pump belt. Ensure belt is taut between sprockets on guide pulley side.
7. Turn guide pulley until pointer aligned with cylinder head flange contour 🔢. Use Allen key 🔢.
8. Tighten guide pulley nut to 25 Nm 🔢.
9. Slacken injection pump locking bolt 🔢. Insert keeper plate 🔢. Tighten locking bolt to 12,5 Nm 🔢.
10. Turn tensioner pulley anti-clockwise until pointers aligned 🔢. Use Allen key 🔢.
    NOTE: If floating pointer passes fixed pointer: Release tensioner and repeat tensioning procedure.
11. Tighten tensioner bolt to 20 Nm 🔢.
12. Install components in reverse order of removal.

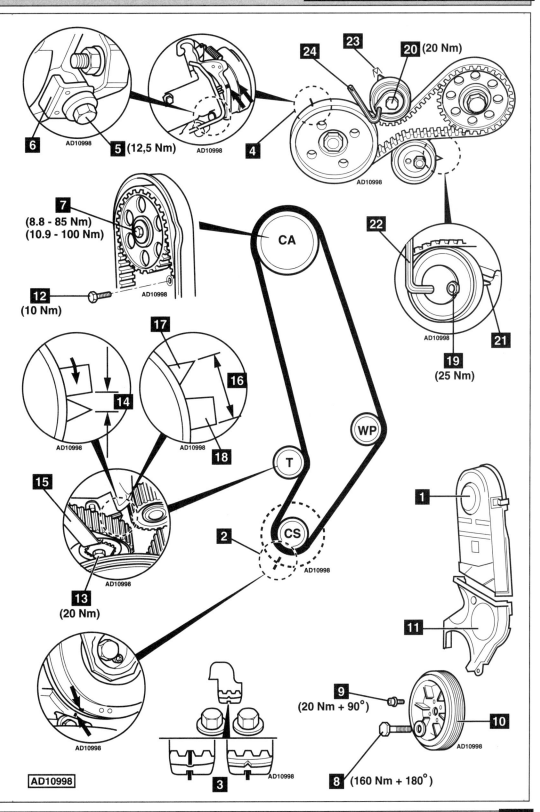

6

AD10998

5 (12,5 Nm)

AD10998

4

24

23

20 (20 Nm)

AD10998

22

7
(8.8 - 85 Nm)
(10.9 - 100 Nm)

AD10998

12
(10 Nm)

AD10998

21

19
(25 Nm)

17

14

16

18

AD10998

AD10998

15

T

WP

1

AD10998

CA

2

CS

AD10998

13
(20 Nm)

AD10998

11

AD10998

9
(20 Nm + 90°)

10

AD10998

3

AD10998

8 (160 Nm + 180°)

VOLVO

Model:	**940 D Turbo**
Year:	**1993-96**
Engine Code:	**D24TIC (EGR)**

Replacement Interval Guide

Volvo recommend replacement every
80,000 miles or 96 months.
*The previous use and service history of the vehicle
must always be taken into account.*
*Refer to Timing Belt Replacement Intervals at the front
of this manual.*

Check For Engine Damage

*CAUTION: This engine has been identified as an
INTERFERENCE engine in which the possibility of
valve-to-piston damage in the event of a timing belt
failure is MOST LIKELY to occur.*
*A compression check of all cylinders should be
performed before removing the cylinder head(s).*

Repair Times – hrs

Timing belt:	
Check & adjust	0,50
Remove & install	2,90
Injection pump belt:	
Check & adjust	0,40
Remove & install	1,50

Special Tools

- Tension gauge – Volvo No.5197.
- Camshaft extension spanner – Volvo No.5201.
- Camshaft setting bar – Volvo No.5190.
- Crankshaft extension spanner – Volvo No.5188.
- Crankshaft pulley locking tool – Volvo No.5187.
- Injection pump sprocket locking pin –
  Volvo No.5193.
- Sprocket holding wrench – Volvo No.5199.

Special Precautions

- Disconnect battery earth lead.
- DO NOT turn crankshaft or camshaft when timing belt removed.
- Remove glow plugs to ease turning engine.
- Turn engine in normal direction of rotation (unless otherwise stated).
- DO NOT turn engine via camshaft or other sprockets.
- Observe all tightening torques.
- Check diesel injection pump timing after belt replacement.

Removal

1. Remove:
   - ❏ Auxiliary drive belt(s).
   - ❏ Cooling fan cowling.
   - ❏ Cooling fan with spacer and pulley.
   - ❏ Intercooler outlet hose.
   - ❏ EGR pipe.
   - ❏ Inlet manifold extension pipe.
   - ❏ Crankcase ventilation hose.
   - ❏ Cylinder head cover.
   - ❏ Timing belt upper cover **1**.
   - ❏ Injection pump belt cover **2**.
2. Turn crankshaft to TDC on No.1 cylinder.
3. Ensure timing marks aligned **3** & **4**.
4. If injection pump timing marks not aligned **4**:
   Turn crankshaft one turn clockwise.
5. Lock crankshaft pulley. Use tool No.5187.
   NOTE: If necessary: Turn crankshaft slightly until tool locates on cooling fan mounting.
6. Slacken crankshaft pulley centre bolt **5**. Use tool No.5188.
7. Ensure timing marks aligned **3** & **4**.
8. Remove:
   - ❏ Crankshaft pulley centre bolt **5**.
   - ❏ Crankshaft pulley bolts (4 bolts) **6**.
   - ❏ Crankshaft pulley **7**.
   NOTE: Ensure crankshaft sprocket remains in position.
   - ❏ Timing belt lower cover **8**.
9. Hold camshaft rear sprocket **12**.
   Use tool No.5199 **9**.
10. Slacken bolt **10**. Use tool No.5201.
    NOTE: Ensure camshaft does not move.
11. Lock injection pump sprocket. Use tool No.5193 **11**.
12. Slacken injection pump mounting bolts.
13. Remove:
    - ❏ Injection pump belt.
    - ❏ Camshaft rear sprocket bolt **10**.
    - ❏ Camshaft rear sprocket **12**.
14. Fit setting bar No.5190 to rear of camshaft **13**.
15. Centralise camshaft using feeler gauges **16**.
16. Slacken tensioner bolt **14**. Move tensioner away from belt. Lightly tighten bolt.
17. Remove timing belt.

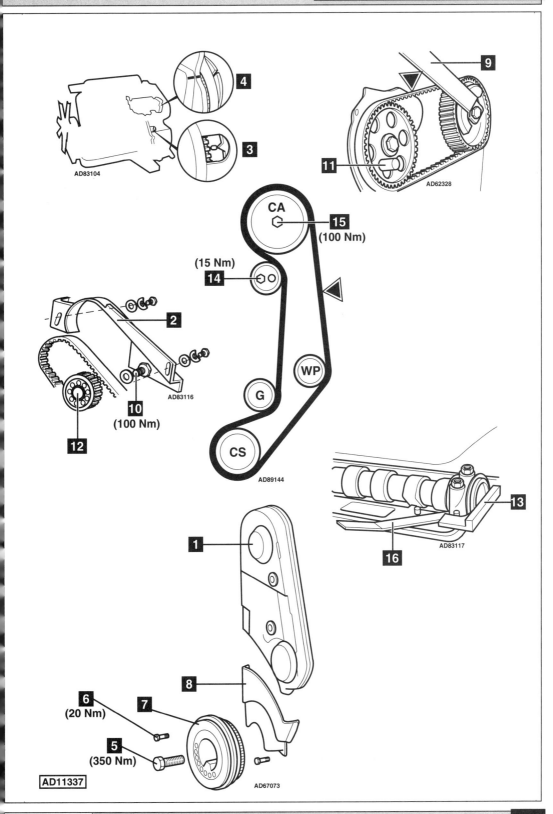

CA
(100 Nm)

(15 Nm)

(100 Nm)

(20 Nm)

(350 Nm)

Installation

1. Ensure flywheel timing marks aligned **3**.
2. Ensure locking pin located correctly in injection pump sprocket **11**.
3. Hold camshaft sprocket. Use tool No.5199.
4. Slacken bolt **15**. Loosen sprocket from taper using a drift.
5. Remove:
   - ❑ Camshaft sprocket bolt **15**.
   - ❑ Camshaft sprocket.
6. Fit timing belt.
   NOTE: Fit timing belt to camshaft sprocket and fit as an assembly to camshaft.
7. Fit camshaft sprocket bolt **15**. Lightly tighten bolt.
   NOTE: Sprocket should turn freely on taper but not tilt.
8. Slacken tensioner bolt **14**.
9. Install:
   - ❑ Timing belt lower cover **8**.
   - ❑ Crankshaft pulley **7**.
   - ❑ Crankshaft pulley bolts (4 bolts) **6**.
10. Tighten crankshaft pulley bolts to 20 Nm **6**.
11. Clean crankshaft pulley centre bolt.
12. Coat crankshaft pulley centre bolt with suitable thread locking compound.
13. Fit crankshaft pulley centre bolt **5**.
14. Lock crankshaft pulley. Use tool No.5187.
    NOTE: If necessary: Turn crankshaft slightly until tool locates on cooling fan mounting.
15. Tighten crankshaft pulley centre bolt **5**. Use tool No.5188. Tightening torque: 350 Nm.
    NOTE: Tightening torque correct when torque wrench and extension spanner in line.
16. Ensure flywheel timing marks aligned **3**.
17. Hold camshaft sprocket. Use tool No.5199.
18. Tighten camshaft sprocket bolt to 100 Nm **15**.
19. Remove camshaft setting bar **13**.
20. Turn crankshaft two turns clockwise to TDC on No.1 cylinder.
21. Tighten tensioner bolt just before TDC (while crankshaft is turning) **14**.
    Tightening torque: 15 Nm.
22. Attach tension gauge to belt at ▽.
    Tool No.5197.
23. Tension gauge should indicate 12-13 units.
24. If not: Adjust tensioner manually.
25. Ensure flywheel timing marks aligned **3**.
26. Install:
    - ❑ Camshaft rear sprocket **12**.
    - ❑ Camshaft rear sprocket bolt **10**.
    - ❑ Injection pump belt.
27. Lightly tighten camshaft rear sprocket bolt **10**.
    NOTE: Sprocket should turn freely on taper but not tilt.
28. Attach tension gauge to belt at ▽.
    Tool No.5197.
29. Move injection pump mounting until tension gauge indicates 12,5 units.
30. Tighten injection pump mounting bolts.
31. Apply firm thumb pressure to belt at ▽. Recheck belt tension.
32. If tension not as specified: Repeat tensioning procedure.
33. Hold camshaft rear sprocket **12**.
    Use tool No.5199 **9**.
34. Tighten camshaft rear sprocket bolt to 100 Nm **10**. Use tool No.5201.
    NOTE: Ensure camshaft does not move. Tightening torque correct when torque wrench at 90° to extension spanner.
35. Remove locking pin from injection pump sprocket **11**.
36. Install components in reverse order of removal.

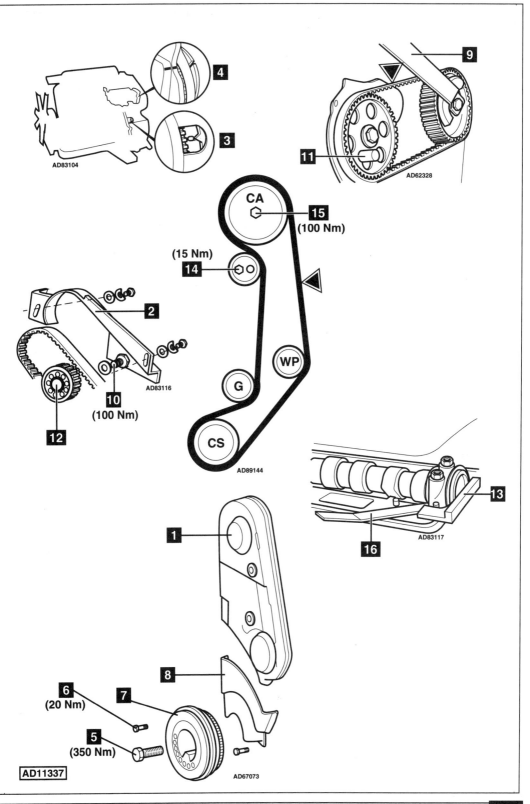

AD83104

4

3

9

11

AD62328

CA

15
(100 Nm)

(15 Nm)
14

2

AD83116

10
(100 Nm)

12

G

WP

CS

AD89144

13

AD83117

16

1

8

6
(20 Nm)

7

5
(350 Nm)

AD11337

AD67073

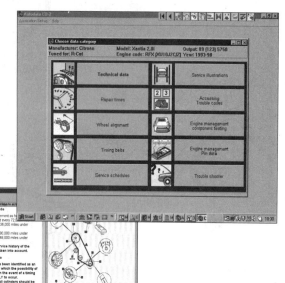
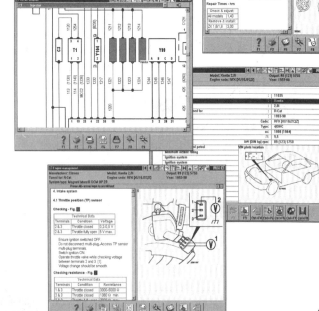